Sedimentary Provenance and Petrogenesis: Perspectives from Petrography and Geochemistry

edited by

José Arribas
Departamento de Petrologia y Geoquimica
Universidad Complutense de Madrid
C/Josè Antonio Novais s/n
28040 Madrid
Spain

Salvatore Critelli
Dipartimento di Scienze della Terra
Università degli Studi della Calabria
87036 Arcavacata di Rende (CS)
Italy

and

Mark J. Johnsson
California Coastal Commission
45 Fremont Street, Suite 2000
San Francisco, California 94105
U.S.A.

THE
GEOLOGICAL
SOCIETY
OF AMERICA®

Special Paper 420

3300 Penrose Place, P.O. Box 9140 ▪ Boulder, Colorado 80301-9140, USA

2007

Copyright © 2007, The Geological Society of America, Inc. (GSA). All rights reserved. GSA grants permission to individual scientists to make unlimited photocopies of one or more items from this volume for noncommercial purposes advancing science or education, including classroom use. For permission to make photocopies of any item in this volume for other noncommercial, nonprofit purposes, contact the Geological Society of America. Written permission is required from GSA for all other forms of capture or reproduction of any item in the volume including, but not limited to, all types of electronic or digital scanning or other digital or manual transformation of articles or any portion thereof, such as abstracts, into computer-readable and/or transmittable form for personal or corporate use, either noncommercial or commercial, for-profit or otherwise. Send permission requests to GSA Copyright Permissions, 3300 Penrose Place, P.O. Box 9140, Boulder, Colorado 80301-9140, USA.

Copyright is not claimed on any material prepared wholly by government employees within the scope of their employment.

Published by The Geological Society of America, Inc.
3300 Penrose Place, P.O. Box 9140, Boulder, Colorado 80301-9140, USA
www.geosociety.org

Printed in U.S.A.

GSA Books Science Editor: Marion E. Bickford and Abhijit Basu

Library of Congress Cataloging-in-Publication Data

Sedimentary provenance and petrogenesis : perspectives from petrography and geochemistry / edited by José Arribas, Salvatore Critelli, and Mark J. Johnsson.
    p. cm. (Special paper ; 420)
  Includes bibliographical references and index.
  ISBN-13 978-0-8137-2420-1 (pbk.)
  1. Petrogenesis. 2. Rocks, Sedimentary. 3. Geochemistry. 4. Sedimentology. I. Arribas, J. (José). II. Critelli, Salvatore. III. Johnsson, Mark J.

QE431.6.P4S43 2007
551'.03--dc22   2006052631

Cover: General view of the Cinca River near the medieval Ville of Ainsa (Huesca, Spain). The alluvial plain shows a braided pattern of gravel bars flowing to the South (ahead). Coarse-grained detritus is sedimentolithic and derived from the Pyrenees. Sources are constituted by sedimentary and metasedimentary rocks from Cenozoic (Palaeogene), Mesozoic (Triassic and Cretaceous), and minor Paleozoic terrains. The braided system cuts older thick alluvial deposits (left riverside) and Palaeogene deep marine deposits (right riverside) before flowing into the Mediano reservoir. At the background, a Paleogene (clastics and carbonates) monocline (deeping to the South) sedimentary sequence is observed.

10 9 8 7 6 5 4 3 2 1

Contents

Preface v

1. **Comparison of river and beach sand composition with source rocks, Dolomite Alps drainage basins, northeastern Italy** 1
   M. Dane Picard and Earle F. McBride

2. **Cyclic variations in sediment provenance from late Pleistocene deposits of the eastern Po Plain, Italy** 13
   Alessandro Amorosi, Maria Luisa Colalongo, Enrico Dinelli, Federico Lucchini, and Stefano Claudio Vaiani

3. **Geochemical and mineralogical proxies for grain size in mudstones and siltstones from the Pleistocene and Holocene of the Po River alluvial plain, Italy** 25
   Enrico Dinelli, Fabio Tateo, and Vito Summa

4. **Petrography of Paleogene turbiditic sedimentation in northeastern Italy** 37
   Cristina Stefani, Massimiliano Zattin, and Paolo Grandesso

5. **Alluvial sand composition as a tool to unravel late Quaternary sedimentation of the Modena Plain, northern Italy** 57
   Stefano Lugli, Simona Marchetti Dori, and Daniela Fontana

6. **Geochemistry and petrography of Western Tethys Cretaceous sedimentary covers (Corsica and Northern Apennines): From source areas to configuration of margins** 73
   Laura Bracciali, Michele Marroni, Luca Pandolfi, and Sergio Rocchi

7. **Petrographic analysis in regional geology interpretation: Case history of the Macigno (northern Apennines)** 95
   Piero Bruni, Enrico Pandeli, and Massimo Nebbiai

8. **Interpreting siliciclastic-carbonate detrital modes in foreland basin systems: An example from Upper Miocene arenites of the central Apennines, Italy** 107
   Salvatore Critelli, Emilia Le Pera, Fabrizio Galluzzo, Salvatore Milli, Massimiliano Moscatelli, Sonia Perrotta, and Massimo Santantonio

9. **Interpreting gypsarenites in the Rossano basin (Calabria, Italy): A contribution to the characterization of the Messinian salinity crisis in the Mediterranean** 135
   Mirko Barone, Rocco Dominici, and Stefano Lugli

10. **The onset of the sedimentary cycle in a mid-latitude upland environment: Weathering, pedogenesis, and geomorphic processes on plutonic rocks (Sila Massif, Calabria)** 149
    Fabio Scarciglia, Emilia Le Pera, and Salvatore Critelli

11. **Interpreting carbonate particles in modern continental sands: An example from fluvial sands (Iberian Range, Spain)** 167
    M.E. Arribas and J. Arribas

12. **Provenance discrimination of Lower Cretaceous synrift sandstones (eastern Iberian Chain, Spain): Constraints from detrital modes, heavy minerals, and geochemistry** 181
    M.A. Caja, R. Marfil, M. Lago, R. Salas, and K. Ramseyer

13. **Significance of geochemical signatures on provenance in intracratonic rift basins: Examples from the Iberian plate** 199
    M. Ochoa, M.E. Arribas, J. Arribas, and R. Mas

14. **Complex examination of the Upper Paleozoic siliciclastic rocks from southern Transdanubia, SW Hungary—Mineralogical, petrographic, and geochemical study** 221
    Andrea Varga, György Szakmány, Tibor Árgyelán, Sándor Józsa, Béla Raucsik, and Zoltán Máthé

15. **First-cycle sandstone composition and color of associated fine-grained rocks as an aid to resolve Gondwana stratigraphy in peninsular India** 241
    Prodip K. Dutta

16. **Sand and gravel provenance in the Waipaoa River system: Sedimentary recycling in an actively deforming forearc basin, North Island, New Zealand** 253
    Dawn E. James, Alissa M. DeVaughn, and Kathleen M. Marsaglia

17. **The petrology and provenance of sand in the Bounty submarine fan, New Zealand** 277
    Shawn A. Shapiro, Kathleen M. Marsaglia, and Lionel Carter

18. **Sediment sources of beach sand from the southern coast of the Baja California peninsula, Mexico—Fourier grain-shape analysis** 297
    J.M. Murillo-Jiménez, William Full, E.H. Nava-Sánchez, V. Camacho-Valdéz, and A. León-Manilla

19. **Detrital apatite geochemistry and its application in provenance studies** 319
    Andrew Morton and Greg Yaxley

20. **Predicting sand character with integrated genetic analysis** 345
    William A. Heins and Suzanne Kairo

Index 381

Preface

It has been two hundred twenty one years since James Hutton addressed the Royal Society of Edinburgh, giving birth to the concept of sedimentary provenance, as well as to that of deep time. Hutton read:

> …we are led to conclude, that, if this part of the earth which we now inhabit had been produced, in the course of time, from the materials of a former earth, we should, in the examination of our land, find the data from which to reason, with regard to the nature of that world, which had existed during the period of time in which the present earth was forming…

Since Hutton's address, thousands of geologists have followed in his footsteps, picking apart the detritus of such "former earths" in order to glean a viewpoint into what came before. In doing so, we have come to learn much not only about the configurations of ancient landscapes, now forever erased by erosion, but also of the processes of weathering, erosion, transport, deposition, and burial that have left imprints on this detritus.

Provenance is defined by the American Geological Institute as "a place of origin; specifically, the area from which the constituent materials of a sedimentary rock or facies are derived." It has come to take on a wider meaning, perhaps better characterized as "petrogenesis:" all of the conditions and processes that have acted on a sediment to produce its ultimate composition. Conversely, with skill and a little luck, many of these process and environments may be teased out of a sediment or sedimentary rock to shed light on many more characteristics of Hutton's "former earths" than merely their constituent materials.

The plate tectonic revolution in the Earth sciences impacted the study of sedimentary provenance and petrogenesis like it did most other areas of the science. A large group of researchers, with William R. Dickinson and his colleagues at their forefront, discovered a strong correlation between sandstone composition and plate tectonic setting. At the same time, other workers, including Paul Potter, Lee Suttner, Abhijit Basu, and Gian Gaspare Zuffa, started unraveling the climatic signal locked up in sands and sandstones. The roles of tectonic, chemical weathering, transportational, and diagenetic processes were explored in a NATO Advanced Studies Institute symposium held in 1984 in Calabria, Italy. This symposium lead to the publication in 1985 of a seminal volume edited by Gian Gaspare Zuffa. After a half dozen years, these processes had been further explored and the *system* controlling the composition was beginning to become clearer. A 1991 volume edited by A.C. Morton, S.P. Todd, and P.D.W. Haughton brought together many such studies. In 1991 a special session at the Geological Society of America Annual Meeting once again drew together an international body of researchers in clastic sedimentology, again leading to the publication of a summary volume, this one edited by Mark Johnsson and Abhijit Basu and published in 1993. In the following decade, the application of these principals has lead to dozens of provenance studies in which not only source rocks and tectonic setting, but also climate, geomorphic, and transport conditions have been revealed. The application of sophisticated analytical methods produced new approaches to unraveling provenance as exhibit papers compiled on a special issue in 1999 by Heinrich Bahlbury and Peter A. Floyd. Around the change of the millennium, the cutting edge of sedimentary petrology shifted to quantitative provenance analysis; the attempt to use compositional information to infer absolute scalar values—such as percentage of rock types in an ancestral catchment basin, precipitation or temperature regimes, dimensions of transport-depositional systems, or depths of burial. Two special issues, the first edited by Valloni and Basu and published in 2002; the other published in 2004 and edited by Gert Jan Weltje and Hilmar von Eynatten, bring together a large body of this work.

In the spirit of periodic international gatherings of researchers in the fields of sedimentary provenance and petrogenesis, we convened a symposium at the 32nd International Geologic Congress, held in Florence in August 2004. That symposium consisted of morning and afternoon oral sessions, and a lively poster session. It is from this symposium that the papers in this volume evolved. In organizing the papers in this volume, we at first thought we would follow the model of Johnsson and Basu (1993) and group them by which process affecting sediment composition was being explored. It soon became apparent, however, that most of these studies are so thoroughly multidimensional that such a grouping is impractical. Indeed, we see this as a sign of maturity in the field of sedimentary petrology—the majority of researchers not only recognize the complexity of the system controlling sediment composition, but actually are sorting out the components of that system—often in a quantitative manner—in application to specific field studies. Accordingly, we have chosen to group these papers regionally, culminating in the two papers that really do not take a regional focus.

Not surprisingly, given the venue of the symposium, the papers in this volume have an overwhelming European flavor, with ten of the twenty focusing on Italian study areas alone. Three papers focus on the Iberian peninsula, and our single eastern European paper examines rocks from the wonderfully names "Transdanubia," region of Hungary. The next four papers bring us from India, across the Indian Ocean to New Zealand, then across the Pacific Ocean to Mexico. The penultimate paper uses a variety of study areas to illustrate the utility of apatite geochemistry in provenance studies. Finally, a grand synthesis of the system controlling sand composition is modeled in the last paper.

The studies related in these papers are incredibly diverse; ranging from the use of "gypsarenites" as a proxy for the Messinian salinity crisis to provenance shifts accompanying Neolithic settlement. From carbonate particles to apatite geochemistry. From incipient weathering in uplands, to deep sea sedimentation in submarine fan systems. The common thread is the use of sediment character to infer something of Hutton's "former earths." We hope that you agree that this volume presents a multitude of ways of sorting through the refuse bin of the sedimentary record to uncover gems of understanding.

Mark Johnsson
José Arribas
Salvatore Critelli

SELECTED ANTECEDENT VOLUMES

Zuffa, G.G. (ed.), 1985, Provenance of Arenites, NATO Advanced Studies Institute: Dordrecht, The Netherlands, Reidel, 408 p.
Morton, A.C., Todd, S.P.H., and Haughton P.D.W. (eds.), 1991, Developments in Sedimentary Provenance Studies: Geological Society [London] Special Publication 57, 370 p.
Johnsson, M.J., and Basu, A. (eds.), 1993, Processes controlling the composition of clastic sediments: Geological Society of America Special Paper 284, 342 p.
Bahlburg, H., and Floyd, P.A. (eds), 1999, Advanced Techniques in Provenance Analysis of Sedimentary Rocks: Sedimentary Geology, v. 124, no. 1-4, 223 p.
Valloni, R., and Basu, A. (eds), 2002, Quantitative provenance studies in Italy: Rome, Instituto Poligrafico e Zecca della Stato, Presidenza del Consiglio dei Ministri, Dipartimento per i Servizi Tecnici Nazionali, Servicio Geologico Nazionale, Memorie Descrittive della Carta Geologica D'Italia, v. 61, 144 p.
Weltje, G.J., and von Eynatten, H. (eds), 2004, Quantitative Provenance Analysis of Sediments: Sedimentary Geology, v. 171, no. 1-4, 286 p.

ACKNOWLEDGMENTS

This volume grew out of a symposium dealing with "Sedimentary Provenance: Petrographic and Geochemical Perspectives" held in August 2004 at the 32nd International Geological Congress in Florence, Italy. The symposium was suggested by Gian Gaspare Zuffa, who invited us to organize what turned out to be a lively all-day interchange of ideas. We thank Gianni for his encouragement and help. We also are grateful to all of the organizers of the Congress, who made our role as conveners easy. Special thanks to those making presentations—some 56 in all—and their co-authors, all of whom made the symposium such a success. We acknowledge the hard work of the 63 authors who "went the extra mile" to translate their symposium presentations into formal papers for inclusion in this volume. We owe a special debt of gratitude to some three dozen reviewers who gave generously of their time to prepare thoughtful and constructive reviews. We thank the Complutense University of Madrid, University of Calabria, and the California Coastal Commission for allowing us the time and resources to devote to this volume. Finally, we thank the editorial staff of the Geological Society of America for their work in translating the manuscripts we provided them to the finished product that you are holding in your hands.

Comparison of river and beach sand composition with source rocks, Dolomite Alps drainage basins, northeastern Italy

M. Dane Picard[†]
Department of Geology and Geophysics, University of Utah, Salt Lake City, Utah 84112, USA

Earle F. McBride
University of Texas at Austin, Jackson School of Geosciences, Department of Geological Sciences,
1 University Station, C1100, Austin, Texas 78712, USA

ABSTRACT

We studied two short, high-gradient river systems draining the Dolomite Alps in northeastern Italy in order to determine which grain types survive transport and to what extent sand grain types reflect source rocks. Grains of all the labile rock types in the source areas survived to lower reaches of the rivers. In one drainage (Boite-Piave), they reached the Adriatic coast. Carbonate grains (largely dolomite) in the Gadera-Rienza Rivers decreased abruptly, largely by dilution, from >50% to trace amounts in 100 km of travel. Percentage of carbonate grains in the lower reaches of these rivers was generally less than one-half the areal percentage of limestone and dolostone exposure in the source areas. However, in the Boite-Piave Rivers (200 km long), enrichment of carbonate grains in beach sand at the expense of polycrystalline quartz and volcanic rock fragments results in dolostone sand at the beach reflecting 78% of its outcrop abundance and limestone (calcite) sand reflecting 68% of its outcrop abundance. Polycrystalline quartz and mafic volcanic rock fragments are less abundant in the beach because of dilution by longshore drift or the breakdown of these grains by wave abrasion. The relative resistance of carbonate textural grain types to abrasion is micrite > spar > mixed micrite/spar.

The results indicate that detritus from dominantly silicic and intermediate volcanic rocks can survive fluvial transport and at least moderate wave abrasion. Metamorphic rock fragments (mostly phyllite) in the Gadera-Rienza Rivers survived transport to the confluence with the Isarco River at Bressanone. In the Boite-Piave river system, metamorphic rock fragments survived fluvial transport to the beach plus some beach abrasion. They did so because the relatively rapid transport down the high-gradient, low-sinuosity streams did not permit extensive chemical weathering. Grains of calcite (micrite and spar), dolomite, and volcanic rock fragments increased in roundness by abrasion in the surf after undergoing only a few kilometers of transport along the coast.

Keywords: sand composition, detrital dolostone, roundness, provenance, abrasion.

[†]E-mail: picard@earth.utah.edu.

Picard, M.D., and McBride, E.F., 2007, Comparison of river and beach sand composition with source rocks, Dolomite Alps drainage basins, northeastern Italy, in Arribas, J., Critelli, S., and Johnsson, M.J., eds., Sedimentary Provenance and Petrogenesis: Perspectives from Petrography and Geochemistry: Geological Society of America Special Paper 420, p. 1–12, doi: 10.1130/2007.2420(01). For permission to copy, contact editing@geosociety.org. ©2007 Geological Society of America. All rights reserved.

INTRODUCTION

The composition of sand grains furnished to a stream is a function of bedrock composition and modification of the bedrock by weathering. The latter is strongly influenced by climate, relief, and slope (e.g., Krynine, 1950; Folk, 1974; Basu, 1976, 1985; Stallard, 1985; Valloni, 1985; Grantham, 1986; Grantham and Velbel, 1988; Johnsson, 1993; Ibbeken and Schleyer, 1991; Le Pera and Critelli, 1997; Arribas et al., 2000; Le Pera et al., 2001; Arribas and Tortosa, 2003).

However, as Palomares and Arribas said (1993, p. 313), "Little work has been done in quantifying the relative contributions of various source-rock types in a given source area to sand populations derived from the source area." To address this general problem, we compare specifically the areal extent of various bedrock types in the Dolomite Alps drainage basin with the abundance of sand grain types in the rivers that drain the basins. Thus, we examine the effects of weathering and stream abrasion on bedrock-derived detritus. We also examine the durability of different grain types and the modification of river sand by beach abrasion. Our study addresses two of the five facets of "quantitative provenance analysis" as defined by Basu (2002, p. 15): (1) identifying the types of sedimentary materials and their provenance, and (2) estimating the proportions of parent rocks represented in a sediment. We do not intend to assess the role of climate, slope, erosion rates, or human activity on the amount and types of sand reaching the rivers and beach.

The Dolomites of northeast Italy include a large amount of carbonate rock along with a variety of other bedrock. In the heart of the Dolomites stand the Marmolada Mountains, the loftiest of all the Dolomite Mountains, with their highest peaks rising to 3342 m. We chose this area, one of the world's most famous carbonate regions, to try to resolve the following questions for some high-gradient streams: (1) For rivers heading into the Dolomites, how far downstream can one recognize the existence of a dolostone and limestone source terrain? (2) To what extent does the light mineral fraction of sand reflect the relative abundance of bedrock types in the source areas? (3) What is the relative stability/durability of micritic versus sparry carbonate sand grains? (4) How rapidly does sand composition and roundness change downstream? (5) Does short-term shoreline abrasion leave an imprint on river sand delivered to the beach? Gazzi et al. (1973) showed that dolostone and limestone rock fragments are dominant grain types along the northern Adriatic coast of Italy south of our study area and of major rivers supplying detritus to the coast.

GEOLOGIC SETTING

The study area is composed of the drainage basins of the Boite River and the Piave River in the region of Veneto, and the drainage basins of the Rienza River and Gadera River, which are part of a major drainage to the north and west of the Dolomites, partly in Veneto and partly in the Trentino–Alto Adige region, northeastern Italy (Figs. 1 and 2). We sampled the Boite and

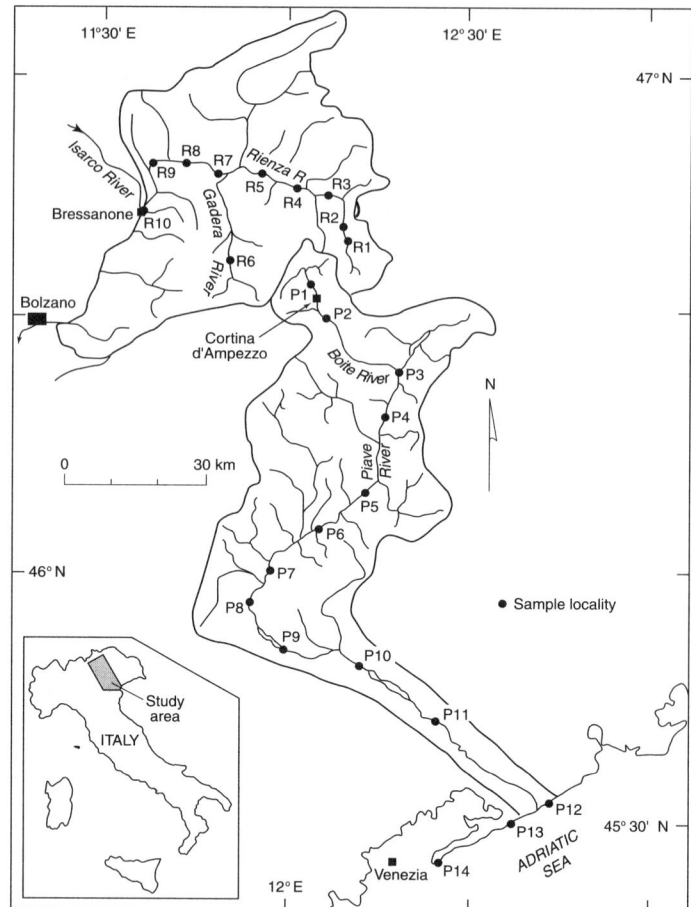

Figure 1. Location map of the study area. P series of numerals are sample sites for the Boite-Piave river system; R series of numerals are sample sites for the Gadera-Rienza river system.

Piave Rivers down to the Adriatic coast northeast of Venice. The Gadera-Rienza river system was sampled from its source in the Dolomites to where it turns southward approximately at Bressanone in Trentino–Alto Adige, at the confluence of the Rienza and Isarco Rivers. Longitudinal profiles of the rivers sampled are shown in Figure 3.

The geology of this part of the Dolomite Alps is particularly complex (Doglioni, 1987; Bertotti et al., 1993). Thick sequences of sedimentary rock, locally with pyroclastics, rest on or are overthrust by igneous and metasedimentary late Paleozoic rocks. The entire sequence is locally intruded by a variety of shallow Mesozoic and Tertiary intrusive rocks. The caps of high-standing peaks are principally dolostone, but limestone, marlstone, claystone, siltstone, sandstone, tuff, and conglomerate are exposed. Most of these rocks are Mesozoic in age. Basement rocks are variable gneisses and schists. There are also substantial outcrops of Quaternary alluvium, Alpine glacial deposits, and landslides. Figure 2 is a generalized bedrock map of the study area.

The large amount of carbonate rock in the source areas, diversity in all the exposed rocks, no dams, and the 200 km length

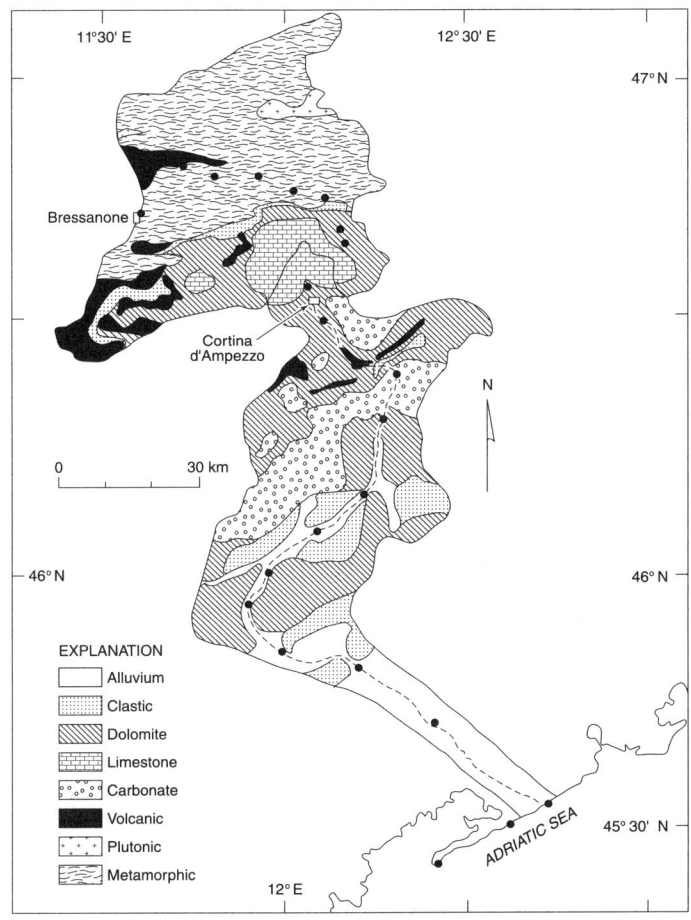

Figure 2. Generalized bedrock map of the study area. Dots are sample sites identified in Figure 1.

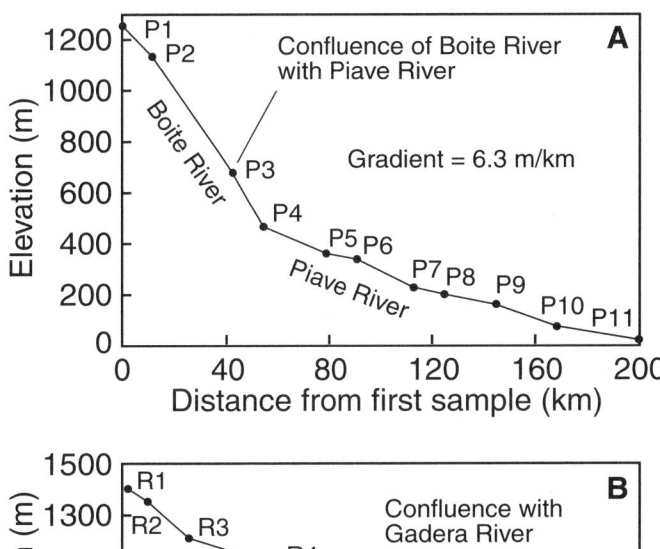

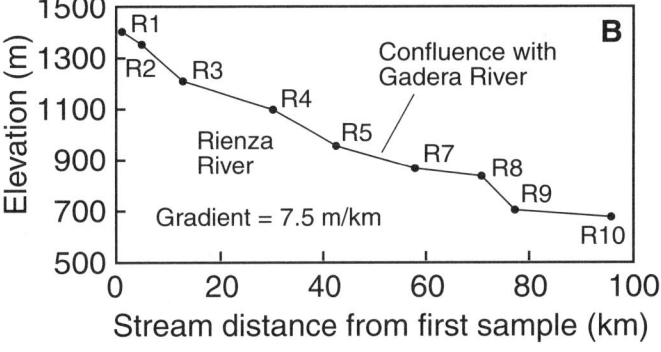

Figure 3. Profile of river systems studied: (A) Boite-Piave Rivers and (B) Gadera-Rienza Rivers.

of the Boite-Piave drainage from the drainage divide to the sea make the area attractive for a study of river and beach sands and source rocks. The Gadera-Rienza reach is shorter (100 km) and also free of dams. Given the long duration of human occupancy in the region, the rivers and their sediment have been affected by human modification. The effects of Alpine glaciation are an additional caveat.

In the region, the mean annual precipitation is between 100 and 200 cm/yr. Mean January temperature is 0 °C; the mean July temperature is 19 °C (Cantú, 1977). The main rainy period is during July and August with 12–17 thunderstorms recorded each year from 1880 to 1948.

PROCEDURES

We collected sand from longitudinal bars in the rivers and from the high-tide berm at beach localities along the Adriatic Sea near the mouth of the Piave River. From each sample, a split of the 1–2ϕ size fraction (medium sand), was made and it was impregnated in blue-dyed epoxy resin and thin sectioned (see Basu et al., 1975; Johnsson et al., 1988; Di Giulio et al., 2003;

Critelli and LePera, 2003). The sections were stained for K-feldspar with sodium cobaltinitrite and for calcite with alizarin red.

A total of 300 or 400 grains per sample was counted. Each grain was assigned to 1 of 21 categories shown in Table 1. Carbonate textural types were not assessed for the Gadera-Rienza samples. Unidentified rock fragments are defined as grains with nondiagnostic, fine-grained textures, or with alteration products that obscure the original grains.

We used the standard point-count method in which coarse-grained rock fragments are tabulated as lithic grains instead of their component minerals. There is little difference between results of such modal analyses and the Gazzi-Dickinson method (Gazzi, 1966; Dickinson, 1970; Ingersoll et al., 1984) in our study because coarse-grained siliciclastic rocks are nearly absent in the source area. We counted grains only if the microscope cross hairs fell on a part of a grain exposed at the upper surface of a thin section, thus avoiding the volume error that can result if the full width of grains embedded in transparent epoxy resin is included in the counting procedure (cf. Harrell, 1983).

We estimated the roundness of all grains of micrite and spar (to a maximum of 30 grains) in each of the samples we studied by comparing grain outlines with Powers' (1952) roundness images. We calculated the average roundness per grain type per sample using the logarithmic scale that Folk (1955) applied to

TABLE 1. SAND COMPOSITION OF THE BOITE-PIAVE AND GADERA-RIENZA RIVERS

Grain type/sample no.	1	2	3	4	5	6	7	8	9	10	11	12	13	14
Boite-Piave Rivers														
Quartz	0.0	0.6	2.0	3.0	2.0	3.6	5.6	3.6	6.6	8.6	8.6	4.3	6.6	8.6
Polyquartz	0.0	0.0	0.6	3.0	3.0	3.0	5.6	4.0	9.3	6.6	4.6	4.0	6.6	2.0
Feldspar	0.0	0.0	1.0	1.6	0.0	0.3	0.0	0.6	2.3	0.0	0.0	0.6	0.6	0.6
VRF	0.0	4.0	19.3	11.6	6.6	6.6	8.6	12.6	18.6	13.3	9.6	5.3	9.3	11.6
MRF	0.0	0.0	1.6	5.6	2.6	6.6	5.6	4.0	3.6	3.3	5.3	2.6	1.3	3.6
URF	0.0	0.0	0.6	0.6	0.0	0.0	0.3	0.0	2.3	0.0	0.6	0.0	0.0	1.0
Chert	0.0	1.0	5.6	5.0	2.0	3.6	3.6	1.6	7.6	3.6	24.6	2.0	6.3	9.3
Sandstone/siltstone	0.0	0.0	1.0	3.0	2.6	1.3	1.0	1.0	1.6	1.6	0.0	2.3	2.0	2.3
Mica	0.0	0.0	0.0	1.0	0.0	0.6	0.0	0.0	0.0	0.6	0.6	0.0	0.6	0.0
Heavies	0.0	0.3	0.6	1.0	4.0	0.0	0.6	1.0	0.0	1.0	1.3	1.3	1.0	0.0
Dolomite: total	79.2	69.3	42.9	45.2	48.2	49.0	49.5	47.3	28.1	30.8	23.5	40.8	35.4	25.8
Micrite	9.3	9.3	10.0	5.0	7.0	5.0	8.0	10.0	4.6	6.6	6.6	6.3	13.6	7.6
Spar	39.6	40.0	17.0	32.6	29.6	33.0	26.6	21.0	17.6	19.6	11.6	20.3	14.6	14.6
Mixed micrite/spar	30.0	20.0	15.9	7.6	11.6	11.0	13.3	15.3	5.3	4.6	5.3	13.6	6.6	3.6
Single crystal	0.3	0.0	0.0	0.0	0.0	0.0	1.6	1.0	0.6	0.0	0.0	0.6	0.6	0.0
Calcite: total	20.2	24.5	23.3	18.2	27.2	25.0	19.2	22.5	18.2	29.5	20.8	35.5	29.9	34.8
Micrite	11.3	7.0	10.0	3.6	13.0	13.0	8.6	12.6	9.3	16.6	12.6	6.6	14.3	17.6
Spar	5.3	10.6	9.0	8.6	2.6	6.0	6.6	3.3	4.3	7.6	1.6	13.6	5.3	6.6
Mixed micrite/spar	3.6	6.9	4.3	6.0	11.6	6.0	4.0	6.6	4.6	5.3	6.6	15.3	10.3	10.6
Total carbonate	99.4	93.8	66.2	63.4	75.4	74.0	68.7	69.8	46.3	60.3	44.3	76.3	65.3	60.6
Total	100.4	101.7	99.5	99.8	99.5	99.6	99.6	99.8	99.8	99.5	99.5	99.7	99.6	99.6
Modifed maturity index	0.0	1.6	8.9	12.4	7.5	11.4	17.4	10.1	30.7	23.2	60.8	11.5	24.2	24.8
Gadera-Rienza Rivers														
Monocrystalline quartz	0.0	0.0	0.7	13.5	13.5	0.5	29.2	22.5	28.2	18.7				
Polyquartz: <3 crystals	0.0	0.0	0.0	4.7	6.0	0.2	6.7	5.0	3.5	5.7				
Polyquartz: >3 crystals	0.0	0.0	1.7	10.5	15.2	0.0	17.5	13.5	8.7	13.7				
Plagioclase	0.0	0.0	0.2	11.2	2.5	0.7	2.5	5.2	7.7	3.0				
K-feldspar	0.0	0.0	0.0	8.7	5.7	0.0	4.0	4.2	17.5	4.0				
MRF	0.0	0.0	3.0	28.5	31.0	0.0	22.7	21.0	13.7	19.0				
VRF	0.0	0.2	0.0	0.5	3.5	4.0	2.0	4.2	3.0	3.5				
IRF	0.0	0.0	0.0	4.2	4.2	0.0	1.5	1.2	3.0	3.5				
Sandstone/siltstone	0.5	0.0	11.2	1.5	1.5	2.5	0.7	5.7	1.5	7.0				
Chert	0.0	0.0	0.0	0.0	0.7	0.0	0.0	0.0	0.0	0.0				
Mica	0.0	0.0	0.0	2.2	1.5	0.0	0.2	3.2	1.7	5.0				
Heavies	0.0	0.0	0.4	2.7	4.2	0.0	6.0	3.5	3.0	1.7				
Other	0.0	0.0	0.0	1.5	1.7	0.2	1.7	1.2	2.2	1.5				
Dolomite	99.0	99.4	79.6	8.0	2.0	60.4	0.2	3.0	1.2	1.0				
Calcite	0.4	0.0	2.8	2.0	6.6	31.2	4.7	6.2	4.7	12.5				
Total carbonate	99.4	99.4	82.4	10.0	8.6	91.6	4.9	9.2	5.9	13.5				
Total	99.9	99.6	99.6	99.7	99.6	99.8	99.7	99.6	99.6	99.8				
Modified maturity index	0.0	0.0	2.5	40.3	54.8	0.7	114.6	69.5	67.8	61.6				

Note: Values are % of total sample based on 400 counts per thin section for Rienza samples and 300 counts per section for Boite-Piave Rivers. VRF—volcanic rock fragments; MRF—metamorphic rock fragments; URF—unidentifiable rock fragments; IRF—igneous rock fragments.

Powers' (1952) roundness images (Table 2). Similarly, we examined roundness of monocrystalline quartz, polycrystalline quartz, chert, and metamorphic and volcanic rock fragments at several localities along the Piave River (Table 3). Previous testing of visual assessments of grain roundness with Fourier grain shape analyses indicated the reliability of the visual technique (Picard and McBride, 1993).

Drainage basins for the streams we sampled were determined from geologic maps that show topography. We used a planimeter to measure the map area of distinctive bedrock formations and Quaternary deposits exposed within each reach of the streams we sampled. Rock types are shown in Table 4. The planimeter data provide a measure of the map area of each map unit, but not the surface exposure area of the unit because we did not correct for topography. The error introduced by this technique may be significant on steep mountain faces. Sources of error in calculating the areal extent of specific rock types include: (1) failure to recognize some stratigraphic units of very small outcrop area on the geologic map, (2) errors in the planimetry procedure, (3) some formations contain more than a single rock type (i.e.,

TABLE 2. ROUNDESS OF MICRITE AND SPAR LIMESTONE GRAINS

Locality	Micrite			Spar		
	Number measured	Mean	SD	Number measured	Mean	SD
Boite-Piave System						
1	30	2.8	0.64	30	1.2	0.60
3	20	2.9	0.59	30	1.3	0.67
5	20	2.9	0.49	30	1.4	0.63
8	20	3.2	0.64	30	2.0	0.68
9	18	3.2	0.59	20	2.6	0.60
10	30	3.5	0.49	30	2.7	0.64
11	20	3.2	0.49	30	2.3	0.72
12 (beach)	20	3.8	0.47	30	3.1	0.72
13 (beach)	20	4.2	0.59	30	3.0	0.57
14 (beach)	20	3.9	0.49	30	3.1	0.68
Total	218			290		
Gadera-Rienza System						
1	30	3.6	0.50	30	2.7	0.59
2	20	3.6	0.50	30	2.7	0.66
3	25	3.6	0.49	25	2.9	0.70
5	8	3.9	0.52	23	2.8	0.56
6	20	3.6	0.60	30	2.2	0.59
7	10	3.6	0.32	29	2.8	0.63
9	11	3.9	0.67	28	3.3	0.63
10	19	4.1	0.50	30	3.6	0.64
Total	143			225		

TABLE 3. ROUNDNESS BY COMPOSITION AT VARIOUS LOCALITIES ALONG THE PIAVE RIVER

Grain type	Locality					
	3	8	10	11	13	14
Monocrystalline quartz			1.3		1.3	
Polycrystalline quartz			1.3		1.3	
Chert			1.5		1.4	
Metamorphic rock fragments			2.7		2.6	
Volcanic rock fragments	2.7	2.9	3.0	3.1	3.5	3.5
Spar	1.3	2.0	2.7	2.3	3.0	3.1
Micrite	2.9	3.2	3.5	3.2	4.2	3.9

TABLE 4. AREAL ABUNDANCE OF BEDROCK TYPES AT EACH SAMPLE SITE

Sample site	Catchment area	Dolomite	Carbonate	Phyllite	Volcanic	Noncarb.
Boite-Piave Rivers and Beach						
P 1	P 1	73	27	Tr	0	0
P 2	P 1 + P 2	67	26	Tr	7	0
P 3	P 1 to P 3	45	32	17	6	0
P 4	P 1 to P 4	53	33	10	5	0
P 5	P 1 to P 5	45	45	6	4	0
P 6	P 1 to P 6	37	54	5	3	1
P 7	P 1 to P 7	41	49	2	3	4
P 8	P 1 to P 8	38	53	3	2	4
P 9	P 1 to P 9	38	53	3	2	4
P 10	P 1 to P 10	38	53	3	2	4
P 11	P 1 to P 11	38	53	3	2	4
P 13; beach	P 13					
P 14; beach	P 14					
P 12; beach	P 12					
Gadera-Rienza Rivers						
R1	R1	39	61		0	0
R2	R1 + R2	77	23		0	0
R3	R1 to R3	67	13		0	20
R4	R1 to R4	32	12		Tr	56
R5	R1 to R5	27	9		2	62
R6	R6	50	36		Tr	14
R7	R1 to R7	23	8		4	65
R8	R1 to R8	22	8		<4	66
R9	R1 to R9	20	8		<2	70
R10	R1 to R10	19	8		<2	72

Note: Bedrock types are given in percent of total map exposure. Alluvium, landslide, and glacial deposits are assumed to have compositions proportional to bedrock exposures. Noncarb.—noncarbonate rocks. Most noncarbonate rocks in the Gadera-Rienza system are metamorphic.

sandstone and shale), (4) some formations of similar rock types (sandstone and siltstone; limestone and marlstone) were lumped for practicality, (5) rates of erosion of different bedrock types are not the same; and (6) we made no attempt to estimate what percent quartz and feldspar sand grains were generated from disintegration of coarse-grained igneous or metamorphic rocks. Phyllite is the only abundant metamorphic rock in the source area; most of its quartz grains are finer than medium-size sand, which was the grain size we analyzed.

We assessed the degree to which a particular grain type in river or beach sand represents its abundance in the source area by comparing its abundance in sand (point-count data) with the areal abundance of similar bedrock as planimetered (Figs. 4 and 5). We did not calculate the quantitative "sand generating index," which is a relative measure of the capacity of one bedrock type to generate sand with respect to another in a compound source area (Palomares and Arribas, 1993). This parameter is not intended for complex source areas such as the Dolomite Alps.

Quaternary deposits make up as much as 35% of some drainage reaches. We assume that the composition of the Quaternary deposits is proportional to the area of outcrop of the various bedrock types in each drainage reach. Thus, if limestone makes up 30% of the bedrock exposure in the drainage area, 30% of the alluvium, landslides, and moraines are considered to be limestone detritus. The possible error in this assumption because of selective loss of weathered minerals in terraces is unknown (cf. Johnsson and Meade, 1990; Robinson and Johnsson, 1997). Sites P10 and P11 were within the alluvial plain of the Piave, where there are no new contributions from bedrock.

The following geologic map sheets (scale 1:100,000) were used: Bressanone (sheet 4A), Dobbiaco (4B), Monte Marmolada (11), Cortina (12), Ampezzo (13), Feltre (22), Belluno (23), Bassano del Grappa (37), Conegliano (38), and Venezia (51).

SAND COMPOSITION TRENDS: BOITE-PIAVE SYSTEM

Bedrock for the Boite-Piave drainage system was subdivided into dolostone, limestone, phyllite, volcanic rock, and other rock. The map area of each rock type (as percent of total map drainage area) was compared with the abundance of sand grains of dolomite (dolostone), calcite (limestone), metamorphic rock fragments, volcanic rock fragments, and "other" grains at each sample site (Fig. 4). Exceptions to this are samples 11, 12, and 13, which were near where the Piave River debouches onto the beach (sample 11) and on the beach itself (samples 12 and 13).

Dolomite Grains

The general pattern of dolomite grains is one of decreasing percentages downstream: they decrease from an initial value of 79% at our highest locality in the Dolomite Mountains to 26% in the beach (Table 1). The dolomite content at the mouth of the river (sample P12) is anomalously high compared with upstream samples and nearby beach samples.

The distribution of dolomite spar closely resembles that of total dolomite (Table 1). The significant diminution of dolomite at locality P3, where the southeast-flowing Boite River enters the southerly flowing Piave River, occurs because the Boite has deposited a large amount of volcanic detritus there (~19%). Though less pronounced, the distribution of mixed micrite/spar dolomite (Table 1) is similar to that for total dolomite and sparry dolomite.

Of all the grain types, dolomite shows the best correlation of abundance in the source area with sand grain abundance at each sample site. A regression plot of the two parameters has a correlation coefficient of 0.84 and a 1:1 relationship (Fig. 6). The abundance of dolomite sand grains at the beach at site P14 is 78% of the areal fraction of dolostone exposure in the source area.

About 150 km south along the Adriatic coast from our sample P14 to where the Lamone River enters the sea north of Ravenna, dolomite in beach-ridge deposits is only 2.3% (Marchesini et al., 2000). In contrast, the mean amount of calcite-limestone at the same sample sites is 9.5%. During the Holocene transgression across the Romagna coastal plain, the shore migrated tens of kilometers to the west; the eastern Alpine sources furnished sediments to the

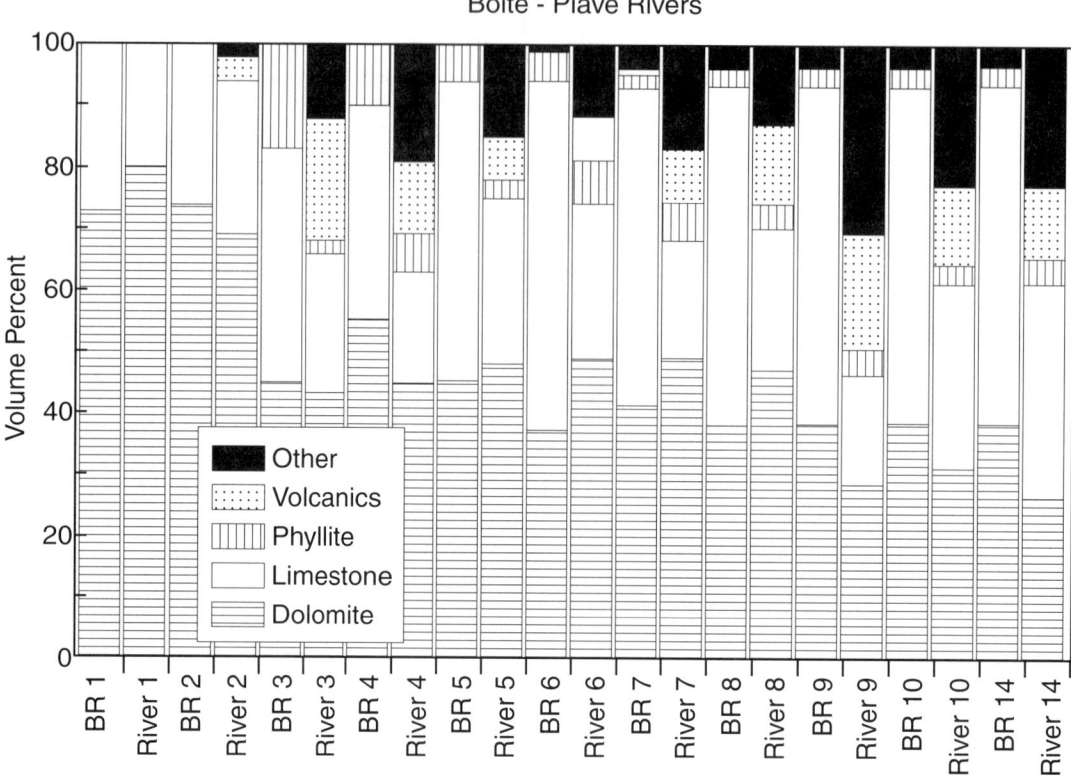

Figure 4. Sand composition versus bedrock type for upper part of the Boite-Piave system. BR—percent of bedrock; River—percent of river sand. Data for sites P10–P13 are omitted to simplify the figure.

intertidal zone. "This sediment supply," say Marchesini et al. (2000, p. 829), "continued during the early regressive phase and was cut off by a change in coastal morphology related to the development of the Po Delta." There followed a sediment supply completely related to the Po River basin.

Limestone Grains

The abundance of limestone is fairly uniform downstream (~20%), but increases to 30% to 35% in the beach. Of the textural types of limestone grains tabulated, micrite and, to a lesser extent, mixed micrite/spar are enriched in beach samples over river samples. Limestone is fairly well represented at the beach in that the abundance of limestone sand grains at the beach site is 64% of the value for the area of limestone exposure in the source area. There is no significant correlation of bedrock area and sand composition, however, for the sample sites collectively.

At the mouth of the Piave River and along the beach, the mean amounts of dolomite and limestone grains are almost equal (~34%, Table 1).

Quartz and Polycrystalline Quartz

There is no quartz at locality P1, the headwaters of the Piave River, but quartz increases toward the beach, where it reaches 9%. Polycrystalline quartz tracks total quartz, but does not increase in beach samples. At locality P9, polycrystalline quartz is ~9%. In this instance, polycrystalline quartz does not track the abundance of metamorphic rock fragments as it does in the Gadera-Rienza system.

Feldspar

K-feldspar is less than 2% at any locality, and plagioclase is absent. Feldspar is relatively high at locality P4 (1.6%) and somewhat higher at locality P9 (2.3%). K-feldspar has the same provenance as quartz. Its distribution is similar to that of unidentified rock fragments (Table 1), but the two have different sources.

Volcanic Rock Fragments

There is never less than 5.3% volcanic rock fragments downstream from locality P2, including the beach samples; the average is 10%. Volcanic rock fragments are grossly overrepresented in river sand at all but the first locality. For example, volcanic bedrock and tuff comprise only 2% of the mapped bedrock, but volcanic rock fragments at the beach make up 12% of the sand. Volcanic rock crops out to the northwest of the northern limit of the Boite drainage basin. We suggest that volcanic rock detritus was deposited in the Boite-Piave drainage basin by glaciers and

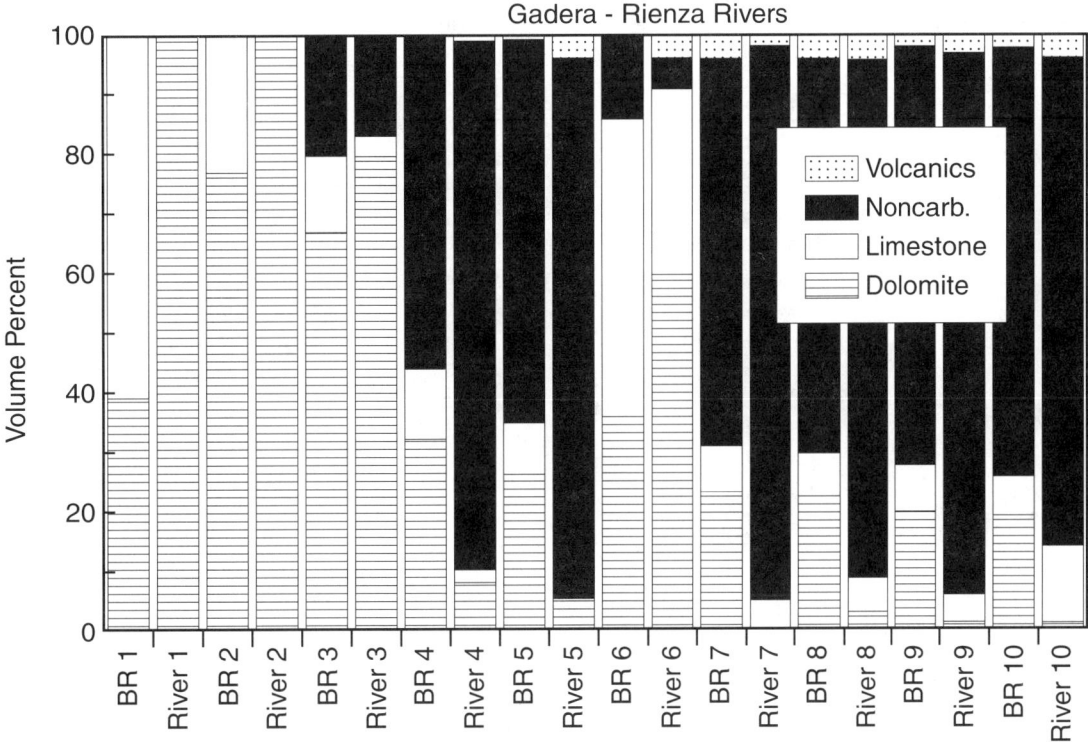

Figure 5. Sand composition versus bedrock type for Gadera-Rienza system. BR—percent of bedrock; River—percent of river sand.

resides in moraine and alluvium. Such detritus would not be represented by our data for bedrock.

Metamorphic Rock Fragments

Downstream, after these grains, largely phyllite, reach 5.6% at locality P4, they exceed 5.3% at each locality except for locality P5, where they amount to 2.6% metamorphic rock fragments. The correlation between metamorphic rock fragment abundance in the samples and metamorphic rock exposure in the basins is ~1:1 ($r = 0.73$: Fig. 7). Quartz and feldspar grains of metamorphic origin could not be uniquely identified to include them in the metamorphic rock fragment category. The abundance of metamorphic rock fragments at the beach compares very well with the exposure of metamorphic rocks in the map area, but the total amount of metamorphic grains is small compared with the major constituents.

Unidentified Rock Fragments

These are rock fragments, probably volcanic in origin, that are so heavily coated by iron oxides that they cannot readily be identified. They show no particular trend.

Chert

Chert exceeds 6% at localities P9, P13, and P14, and it is more than 24% at locality P11. Chert is derived from nodules in limestone and marlstone; its abundance in these source rocks is unknown.

Sandstone/Siltstone

Sandstone/siltstone grains are present in about three-fourths of the samples, averaging 1.8% where they occur. They are most abundant at localities P4, P5, P12, and P14.

Miscellaneous Grains

Trace amounts of opaque minerals are present in 10 of the 14 samples from the Piave River. Except for locality P5, where they amount to 4% of the sample, opaques do not exceed 1.3% at any locality. Mica is present in sufficient amounts to be noted in only four of the Piave River samples and in one of the beach samples (Table 1).

Other Components

This category of grains in Figure 4 includes quartz, feldspar, unidentified rock fragments, chert, sandstone/siltstone, and miscellaneous grains. These grains are overrepresented at the beach and in most river sands because most are more stable chemically and more durable than grains of carbonate and metamorphic rock fragments. It is predictable that these grains would be enriched during weathering and transport. In addition, some quartz and feldspar grains were derived from metamorphic rocks in the source area.

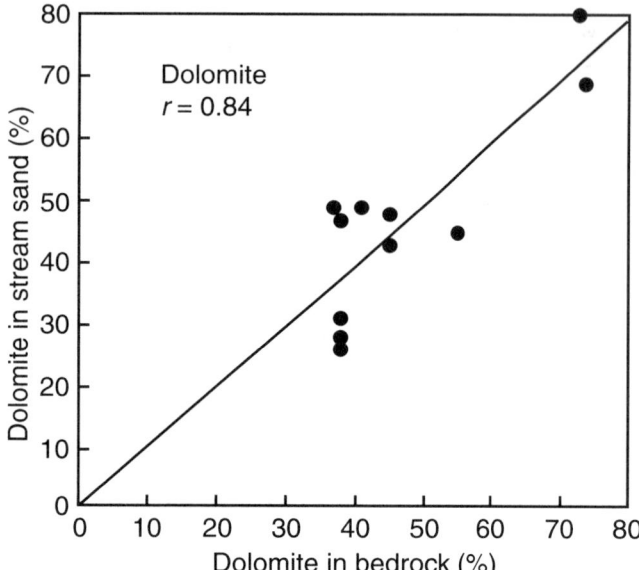

Figure 6. Plot of dolomite in river sand versus dolomite outcrop (map) area for Boite-Piave Rivers.

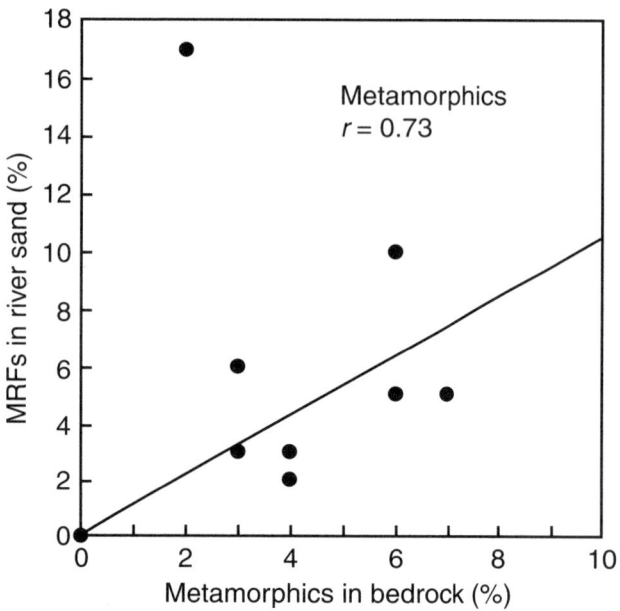

Figure 7. Plot of metamorphic rock fragments (MRFs) in river sand versus metamorphic rocks in outcrop area for Boite-Piave Rivers. Regression line ignores the outlier.

SAND COMPOSITION TRENDS: GADERA-RIENZA SYSTEM

The bedrock of the Gadera-Rienza drainage was divided into dolostone, limestone, volcanic rock, and "other" noncarbonate rock. Most of the latter is phyllite. Sand from site R10, at the confluence of the Rienza and the Isarco Rivers, includes detritus from both drainages. Bedrock data for this site reflect only the Gadera-Rienza drainage. Because of this, we use sand composition from locality R9 to assess how well sand composition reflects bedrock types in this drainage system. Sand composition versus bedrock type is shown in Figure 5.

Limestone and Dolostone

In upper reaches of the river system, dolomite sand-grain abundance is overrepresented relative to outcrop abundance, whereas calcite/limestone sand grains are greatly underrepresented. In lower reaches, dolomite grains are underrepresentative of bedrock, whereas limestone grains well represent bedrock. Limestone grains exceed 10% at localities R6 and R10, whereas dolomite exceeds 60% only upstream of locality R4 (except for R6). At locality R9, limestone reflects 63% of the abundance of bedrock in its drainage basin. The dolomite content of sand at locality R9 (only 1%) reflects only 5% of the dolostone outcrop in the source area. Carbonate sand in the Rienza River is diluted by sidestream contributions of noncarbonate detritus at locality R4.

The ratio of dolomite to calcite/limestone sand grains in this drainage basin does not correlate with the relative abundance of dolostone and limestone in outcrop. We restained thin sections with fresh alizarine red to verify our point count data and got only slightly different results. We cannot explain this anomaly.

Quartz and Polycrystalline Quartz

Only trace amounts appear until locality R4, where the total quartz content jumps to 28%. The large amount of polycrystalline quartz with >3 subgrains is commensurate with the abundance of metamorphic rock fragments in the sands.

Feldspar

Both K-feldspar and plagioclase first appear above trace amounts at locality R4. Both feldspar types are generally <5%; there are locally high values >15%.

Metamorphic Rock Fragments

These grains, phyllite and lesser schist, show a pattern similar to quartz, but exceed 28% at localities R4 and R5.

Igneous Rock Fragments

These are quartzose plutonic rock grains with distributions that track quartz. They are minor components (≤5%).

Volcanic Rock Fragments

These show no consistent downstream trend and no correlation with bedrock abundance. Volcanic rock fragments, however, comprise less than 5% of sand and bedrock. Sand at site R1 has

4% volcanic rock fragments, but no volcanic rocks are mapped in the drainage basin. If human activity is not responsible for this anomaly, and we found no evidence in the field to indicate this, we suggest that, like the upper reaches of the Piave, volcanic rock detritus was introduced into the drainage basin by glaciers during the Pleistocene.

Noncarbonate Grains

All noncarbonate grains except volcanic rock fragments were grouped to plot on Figure 5. These grains are mostly metamorphic rock fragments, quartz, and feldspar. Such grains are fairly well represented compared to outcrop extent in the upper reaches of the drainages, but they are minor components, and they are overrepresented compared to outcrop extent in the lower reaches. The regression for noncarbonate sand and bedrock has a correlation of 0.98. At locality R9, stream sand percentage of noncarbonate sand exceeds the percent of areal exposure of noncarbonate bedrock in the Gadera-Rienza drainage basin by 30%.

SAND MATURITY VALUES

A frequently used measure of downstream and beach enrichment of more durable grains is the mineralogical maturity of the samples. In 1957, Pettijohn (p. 509) introduced the mineralogical maturity index as:

$$MI = quartz + chert \times 100/feldspar + rock\ fragments,$$

where grain abundance is in volume percent. We (McBride and Picard, 1987; Picard and McBride, 1993) have introduced and used a modified maturity index (MMI) to include heavy minerals, micas, and skeletal grains in the labile category, because these grains also have implications for mineralogical maturity. The MMI is defined as:

$$MMI = quartz + chert \times 100/other\ grains.$$

Along both river systems and on the beach, all of the sands are much less mature (Table 1) than the average sandstone (~300). There is a relatively small variation in average maturity of the sands of the two river systems; the mean for the Boite-Piave is 18 ($\sigma = 15.4$) and the mean for the Gadera-Rienza is 41 ($\sigma = 40$; Table 1). The most immature samples are those at the headwaters of both river systems.

ROUNDNESS STUDY

We studied variations in roundness of volcanic rock fragments, micrite, and spar at several localities along the Gadera-Rienza and Boite-Piave drainages and along the Adriatic beach to determine whether abrasion was taking place during transport. No distinction was made between calcite and dolomite for assessing roundness of micrite versus spar.

Both carbonate grain types increase in roundness downstream. Roundness increases regularly in the Boite-Piave system for micrite from 2.8ρ to 3.2ρ and for spar from 1.2ρ to 2.7ρ (Table 2). In the Gadera-Rienza system, roundness changes little until close to the confluence with the Isarco River. Grains in the headwaters of the Gadera-Rienza Rivers have roundness values 1ρ higher than grains in the headwaters of the Boite-Piave Rivers.

Micrite grains are consistently better rounded than carbonate-spar grains, which is also the case for carbonate beach sands on Elba (Picard and McBride, 1993, p. 244). Spar grains are susceptible to cleaving. This generates angular corners during weathering and erosion, and it persists to some degree during transport. Micrite rock fragments lack this trait.

Volcanic rock fragments increase slightly in roundness downstream in the Boite-Piave system (Table 3).

RIVER VERSUS BEACH: COMPOSITION AND ROUNDNESS

The question arose whether beach abrasion leaves an imprint on the composition or roundness of the grains. To test this, we compared sand from river sample P10, which is 50 km upstream from the coast, with beach sample P14, which is 25 km from the mouth of the Piave River. The dominant longshore drift of beach sand along the coast east of Venice is to the southwest, as shown by the shape of the spit and by the work of Gazzi et al. (1973). Thus, sample site P14 is farther downdrift from the Piave River mouth than site P13. Site P12, however, is updrift from the river mouth. The composition of sample P12, which contains sand from rivers north of the Piave (Gazzi et al., 1973), is anomalous compared with P13 and P14.

Between localities P10 and P14 there are modest decreases in abundance of polycrystalline quartz, dolomite spar, and total dolostone (decreases of 4.6%, 5.0%, and 5.0%, respectively). There is no increase in the modified maturity index. The only grain type that shows a significant increase between the two sample sites is chert (5.7% increase). Between the same sample localities, there is also an increase in roundness of calcite micrite, calcite spar, dolomite, and volcanic rock fragments ($p < 0.05$).

DISCUSSION

This study differs from previous provenance studies of our own and of others in that Quaternary deposits, including glacial deposits, are common in the drainages. Glaciers apparently introduced volcanic rocks into both drainage basins; we found volcanic rock detritus to be consistently overrepresented compared to its outcrop abundance and found it in drainages where no volcanic bedrock is mapped. This suggests that one should consider the possible effect of Alpine glaciers when doing provenance studies of ancient deposits.

Detritus from all major bedrock types survived transport from mountainous source area to lower elevations and, for the

Boite-Piave river system, survived further transport a short distance along the coast. Dolomite from dolostone and calcite from limestone survived weathering and abrasion in spite of their softness and cleavage, as shown previously in this area by Gazzi et al. (1973) and for the Ligurian coast of Italy by Garzanti et al. (1998). The relative resistance of carbonate textural grain types to abrasion is micrite > spar > mixed micrite/spar. Between the coastal plain (sample P10) and the beach, there was some abrasion loss of dolomite spar and of polycrystalline quartz.

Dolomite grains, the most stable of carbonate grains (Arribas and Tortosa, 2003), survived weathering and abrasion better than calcite—pure dolomite is less soluble than calcite (Bathhurst, 1975). At the beach, limestone sand reflects 68% of its outcrop abundance, while dolomite reflects 78% of its outcrop abundance. In the Gadera-Rienza drainage, however, detritus from both carbonate rock types makes up less than one-half their outcrop abundance. The abrupt decrease in carbonate rock detritus at locality R4 is attributed to dilution by siliciclastic detritus from a tributary rather than to abrasion loss of carbonate grains. Dilution of specific grain types must be considered as a potentially significant perturbation process in studies such as ours.

The data further support the consensus of earlier studies that detritus from volcanic rocks, especially silicic and intermediate types, can survive fluvial transport (Webb and Potter, 1971; Cameron and Blatt, 1971; McBride and Picard, 1987) and at least moderate wave abrasion (McBride and Picard, 1987).

Metamorphic rock fragments have low survivability in low-gradient streams and humid climates where chemical weathering attacks the micas (Cameron and Blatt, 1971; Cleary and Connolly, 1971) and sands spend long periods in storage in bars (Johnsson, 1990). Our data for beach sands at Elba Island (Picard and McBride, 1993) are consistent with the earlier studies cited here, indicating that vigorous wave abrasion is highly destructive to these rock fragments. The latter conclusion has been documented for gravel-size gneiss clasts in Calabria, Italy, where gneiss fragments are reduced in abundance by 56% during only 7 km of beach transport (Ibbeken and Schleyer, 1991). In the present study area, which has an alpine Mediterranean climate, however, metamorphic rock fragments in the Boite-Piave river system survived fluvial transport to the beach plus some beach abrasion. Metamorphic rock fragments survived because the relatively rapid transport down the high-gradient streams (6–7 m/km) draining the Dolomites permitted neither extensive chemical weathering nor long-term storage in bars.

Spar and micrite increase in roundness downstream in both river systems. Abrasion is almost certainly the dominant cause, but some degree of selective roundness sorting cannot be ruled out.

We believe the decrease of polycrystalline quartz grains between river sample P10 and beach sample P14 is likely the result of dilution by sand introduced by longshore drift or by the disintegration of polycrystalline quartz through abrasion. The data indicate, however, that there is as much polycrystalline quartz at site P13 (6.6%) on the beach as there is at P10 upstream in the Piave River. Even though polycrystalline quartz is more durable than monocrystalline quartz based on abrasion experiments (Harrell and Blatt, 1978), dissolution of quartz at subgrain boundaries during chemical weathering and diagenesis leads to disintegration of weathered grains during subsequent transport (McBride and Abdel-Wahab, 1996). Though not large, the loss of sparry dolomite between P10 and P14 sample sites may be attributed to breakage of crystals along crystal boundaries and cleavage planes. The increase in roundness of calcite micrite, calcite spar, dolomite, and volcanic rock fragment grains between the river and beach is likely the result of abrasion in the surf zone rather than the result of selective roundness sorting.

In Liguria, Italy, we found that bedded chert readily breaks down to gravel-size clasts, the size and shape of which are controlled principally by the fracture network (McBride and Picard, 1987). Because of its toughness, resistance to chemical weathering, and absence of internal bedding planes, the bedded chert does not readily break down into sand-size particles, and thus is underrepresented in stream sand derived from it. We have no quantitative information on the abundance of chert nodules in limestone bedrock of the study area.

Our sample sites P10 (river) and P12, P13, and P14 (beaches) are essentially the same sites sampled by Gazzi et al. (1973). The carbonate fraction of the Piave River sample is identical in both studies (60%), but lower by 15%–20% for our beach samples compared with theirs. In their study, the dolomite/limestone ratio (determined by X-ray diffraction [XRD]) increased from the river to the beaches, whereas we found the reverse to exist from point-count data. Some of the carbonate grains at our common beach sample sites were introduced by rivers northeast of the Piave and transported southwestward by longshore processes (Gazzi et al., 1973). Extensive beach sand replenishment was in progress at our localities P13 and P14 when (1997) we collected our samples (R. Valloni, 2005, personal commun.). This complication and the different analytical techniques of the two studies (XRD versus point counts) may account for the discrepancy in data.

Beach sands of the Boite-Piave river system have about the same maturity as beach sands of Liguria, Italy (McBride and Picard, 1987; Garzanti et al., 1998) and are much less mature (mean of 20) than three-fourths of Elba Island beach sands (Picard and McBride, 1993, p. 239). Boite-Piave beach sands display greater maturity than the average beach sand of the northeast end of Papua New Guinea (Ruxton, 1970; see Basu, 1985, his Table 3). Sands of both river systems are comparable in maturity to some short-headed mountainous streams in Italy (McBride and Picard, 1987; Ibbeken and Schleyer, 1991), but they are more mature than short-headed streams from Papua New Guinea (Basu, 1985, his Table 3). The beach sands at the mouth of the Piave River on the Adriatic coast are considerably less mature than those beaches on broad passive continental margins (e.g., Hayes, 1965; Potter, 1986; Le Pera and Arribas, 2004), though passive margins also include carbonate beaches such as these.

Sand in beaches tend to be more quartz-rich and mineralogically mature in rivers that feed beaches (Basu, 1985; McBride and Picard, 1987; Ferree et al., 1988). We found no such difference here. Sand introduced to the beach from the northeast of our sampling sites by longshore drift obscures probable maturation effects taking place.

CONCLUSIONS

Grains of all labile rock types in source areas survived to lower reaches of the rivers, and in the Boite-Piave system, reached the Adriatic coast. Carbonate grains, largely dolomite, in the Gadera-Rienza rivers decrease abruptly from >50% to trace amounts in 100 km of fluvial transport, largely by dilution. Percentages of carbonate grains in the lower reaches of the rivers studied make up generally less than one-half of the areal fraction of limestone and dolostone exposure in the source area. In the Boite-Piave Rivers, however, dolomite sand at the beach reflects 78% of its outcrop abundance, and limestone sand reflects 68% of its outcrop abundance. Polycrystalline quartz and mafic volcanic rock fragments are less abundant in the beach than the river, although whether by selective abrasion in the beach or by dilution by longshore drift is uncertain.

The relative resistance of carbonate textural grain types to abrasion is micrite > spar > mixed micrite/spar.

The data support the interpretation that detritus from dominantly silicic and intermediate volcanic rocks can survive fluvial transport and at least moderate wave abrasion. Metamorphic rock fragments in the Gadera-Rienza Rivers survived transport to lower elevations, and in the Boite-Piave river system, they survived fluvial transport to the beach plus some beach abrasion. They did so because the relatively rapid transport down the high-gradient, low-sinuosity streams did not permit extensive chemical weathering. Calcite micrite, calcite spar, dolomite, and volcanic rock fragment grains increased in roundness by abrasion in the surf after undergoing only a few kilometers of transport along the coast.

ACKNOWLEDGMENTS

We dedicate this paper to Rodolfo Gelmini, our longtime Italian friend, field mate, and colleague on studies of Triassic rocks in Italy, now gone, but not forgotten. T.A. MacGillvary completed the point counts on the Rienza River samples; Dan McConnell did the planimetry. Sharon Christensen did the word processing on several drafts of the manuscript. We appreciate the helpful reviews of the manuscript by Abhijit Basu, William Cavazza, Patricia Cowley, R.L. Folk, Mark Johnsson, Emilia Le Pera, Craig Sanders, Elena Spadafora, and Renzo Valloni. The J. Nalle Gregory Chair in Sedimentary Geology of the University of Texas at Austin supported field work and manuscript preparation costs. We thank the Dipartimento di Scienze della Terra, Università degli Studi di Modena (especially Daniela Fontana) and the Dipartimento di Scienze della Terra, Università degli Studi di Bologna (especially Gianni Zuffa) for support of our sedimentologic studies in northern Italy. The hospitality of Gianni Zuffa and his family at Fontanelice was greatly appreciated.

REFERENCES CITED

Arribas, J., and Tortosa, A., 2003, Detrital modes in sedimenticlastic sands from low-order streams in the Iberian range, Spain: The potential for sand generation by different sedimentary rocks: Sedimentary Geology, v. 159, p. 275–303, doi: 10.1016/S0037-0738(02)00332-9.

Arribas, J., Critelli, S., Le Pera, E., and Tortosa, A., 2000, Composition of modern stream sand derived from a mixture of sedimentary and metamorphic source rocks (Henares River, central Spain): Sedimentary Geology, v. 133, p. 27–48, doi: 10.1016/S0037-0738(00)00026-9.

Basu, A., 1976, Petrology of Holocene fluvial sand derived from plutonic source rocks: Implications to paleoclimatic interpretation: Journal of Sedimentary Petrology, v. 46, p. 694–709.

Basu, A., 1985, Influence of climate and relief on compositions of sands released at source areas, in Zuffa, G.G., ed., Provenance of Arenites: Dordrecht, D. Reidel, p. 1–18.

Basu, A., 2002, A perspective on quantitative provenance analysis, in Valloni, R., and Basu, A., eds., Quantitative Provenance Studies in Italy: Memorie Descrittive Della Carta Geologica D'Italia, v. 51, p. 11–24.

Basu, A., Young, S.W., Suttner, L.S., James, W.C., and Mack, G.H., 1975, Re-evaluation of the use of undulatory extinction and polycrystallinity in detrital quartz for provenance interpretation: Journal of Sedimentary Petrology, v. 45, p. 873–882.

Bathhurst, R.G.C., 1975, Carbonate sediments and their diagenesis: Amsterdam, Elsevier, 658 p.

Bertotti, G., Picotti, V., Bernoulli, D., and Castellarin, A., 1993, From rifting to drifting: Tectonic evolution of the South-Alpine upper crust from the Triassic to the Early Cretaceous: Sedimentary Geology, v. 86, p. 53–76, doi: 10.1016/0037-0738(93)90133-P.

Cameron, K.L., and Blatt, H., 1971, Durabilities of sand-size schist and "volcanic" rock fragments during fluvial transport, Elk Creek, Black Hills, South Dakota: Journal of Sedimentary Petrology, v. 41, p. 565–576.

Cantú, V., 1977, The climate of Italy, in Wallen, C.C., ed., Climates of Central and Southern Europe: World Survey of Climatology, Volume 6: Amsterdam, Elsevier, 248 p.

Cleary, W.J., and Connolly, J.R., 1971, Distribution and genesis of quartz in a piedmont-coastal plain environment: Geological Society of America Bulletin, v. 82, p. 2755–2765.

Critelli, S., and Le Pera, E., 2003, Provenance relations and modern sand petrofacies in an uplifted thrust-belt, northern Calabria, Italy, in Valloni, R., and Basu, A., eds., Quantitative Provenance Studies in Italy: Memorie Descrittive Della Carta Geologica D'Italia, v. 51, p. 25–38.

Dickinson, W.R., 1970, Interpreting detrital modes of greywacke and arkose: Journal of Sedimentary Petrology, v. 40, p. 695–707.

Di Giulio, A., Ceriani, A., Ghia, E., and Zucca, F., 2003, Composition of modern stream sands derived from sedimentary source rocks in a temperate climate (northern Apennines, Italy): Sedimentary Geology, v. 158, p. 145–161, doi: 10.1016/S0037-0738(02)00264-6.

Doglioni, C., 1987, Tectonics of the Dolomites (southern Alps, northern Italy): Journal of Structural Geology, v. 9, p. 181–193, doi: 10.1016/0191-8141(87)90024-1.

Ferree, R.A., Jordan, D.W., Kertes, R.S., Savage, K.M., and Potter, P.E., 1988, Comparative petrographic maturity of river and beach sand and origin of quartz arenites: Journal of Geological Education, v. 36, p. 79–87.

Folk, R.L., 1955, Student operator error in determination of roundness, sphericity, and grain size: Journal of Sedimentary Petrology, v. 25, p. 297–301.

Folk, R.L., 1974, Petrology of Sedimentary Rocks: Austin, Texas, Hemphill Publishing Co., 182 p.

Garzanti, E., Scutello, M., and Vidimari, C., 1998, Provenance from ophiolites and oceanic allochthons: Modern beach and river sands from Liguria and the northern Apennines (Italy): Ofioliti, v. 23, p. 65–82.

Gazzi, P., 1966, Le arenarie del flysch sopracretacico dell'Appennino modenese: Correlazione con il flysch di Monghidoro: Mineralogia et Petrografica Acta, v. 12, p. 69–97.

Gazzi, P., Zuffa, G.G., Gandolfi, G., and Paganelli, L., 1973, Provenienza e dispersione litoranea delle sabbie delle spiagge Adriatiche fra le foci dell'Isonzo e del Foglia: Inquadramento regionale: Memorie della Società Geologica Italiana, v. 12, p. 1–37.

Grantham, J.H., 1986, The influence of climate and relief on lithic fragment abundance in modern fluvial sands of the southern Blue Ridge Mountains, North Carolina [abs.]: Society of Economic Paleontologists and Mineralogists, Annual Midyear Meeting, v. 3, p. 46–47.

Grantham, J.H., and Velbel, M.A., 1988, The influence of climate and topography on rock-fragment abundance in modern fluvial sands of the southern Blue Ridge Mountains, North Carolina: Journal of Sedimentary Petrology, v. 58, p. 219–227.

Harrell, J.A., 1983, Grain size and shape distributions, grain packing, and pore geometry within sand laminae: Characterization and methodologies [Ph.D. thesis]: Cincinnati, University of Cincinnati, 567 p.

Harrell, J.A., and Blatt, H., 1978, Polycrystallinity of quartz; effect on the durability of quartz: Journal of Sedimentary Petrology, v. 48, p. 25–30.

Hayes, M.O., 1965, Sedimentation in a semiarid, wave-dominated coast (south Texas), with emphasis on hurricane effects [Ph.D. thesis]: Austin, University of Texas, 350 p.

Ibbeken, H., and Schleyer, R., 1991, Source and Sediment: Berlin, Springer-Verlag, 286 p.

Ingersoll, R.V., Bullard, T.F., Ford, R.L., Grimm, J.P., Pickle, J.D., and Sares, S.W., 1984, The effect of grain size on detrital modes: A test of the Gazzi-Dickinson point-counting method: Journal of Sedimentary Petrology, v. 54, p. 103–116.

Johnsson, M.J., 1990, Tectonic versus chemical-weathering controls on the composition of fluvial sands in tropical environments: Sedimentology, v. 37, p. 713–726, doi: 10.1111/j.1365-3091.1990.tb00630.x.

Johnsson, M.J., 1993, The system controlling the composition of clastic sediments, in Johnsson, M.J., and Basu, A., eds., Processes Controlling the Composition of Clastic Sediments: Geological Society of America Special Paper 284, p. 1–19.

Johnsson, M.J., and Meade, R.H., 1990, Chemical weathering of fluvial sediments during alluvial storage: The Macuapanim Island point bar, Solimões River, Brazil: Journal of Sedimentary Petrology, v. 60, p. 827–842.

Johnsson, M.J., Stallard, R.F., and Meade, R.H., 1988, First-cycle quartz arenites in the Orinoco River basin, Venezuela and Colombia: The Journal of Geology, v. 96, p. 263–277.

Krynine, P.D., 1950, Petrology and stratigraphy of the Triassic sedimentary rocks of Connecticut: Connecticut State Geological and Natural History Survey Bulletin 73, 247 p.

Le Pera, E., and Arribas, J., 2004, Sand composition in an Iberian passive-margin fluvial course: The Tajo River: Sedimentary Geology, v. 171, p. 261–281, doi: 10.1016/j.sedgeo.2004.05.019.

Le Pera, E., and Critelli, S., 1997, Sourceland controls on the composition of beach and fluvial sand of the northern Tyrrhenian coast of Calabria, Italy: Implications for actualistic petrofacies: Sedimentary Geology, v. 110, p. 81–97, doi: 10.1016/S0037-0738(96)00078-4.

Le Pera, E., Arribas, J., Critelli, S., and Tortosa, A., 2001, The effects of source rocks and chemical weathering on the petrogenesis of siliciclastic sand from the Neto River (Calabria, Italy): Implications for provenance studies: Sedimentology, v. 48, p. 357–378, doi: 10.1046/j.1365-3091.2001.00368.x.

Marchesini, L., Amorosi, A., Cibin, U., Zuffa, G.G., Spadafora, E., and Preti, D., 2000, Sand composition and sedimentary evolution of a late Quaternary depositional sequence, northwestern Adriatic coast, Italy: Journal of Sedimentary Research, v. 70, p. 829–838.

McBride, E.F., and Abdel-Wahab, A., 1996, Evidence for destruction of detrital polycrystalline quartz grains by intrastratal dissolution: Geological Society of America Abstracts with Program, v. 28, no. 1, p. 53.

McBride, E.F., and Picard, M.D., 1987, Downstream changes in sand composition, roundness, and gravel size in a short-headed, high-gradient stream, northwestern Italy: Journal of Sedimentary Petrology, v. 57, p. 1018–1026.

Palomares, M., and Arribas, J., 1993, Modern stream sands from compound crystalline sources: Composition and sand generation index, in Johnsson, M.J., and Basu, A., eds., Processes Controlling the Composition of Clastic Sediments: Geological Society of America Special Paper 284, p. 313–322.

Pettijohn, F.J., 1957, Sedimentary Rocks (second edition): New York, Harper and Brothers, 718 p.

Picard, M.D., and McBride, E.F., 1993, Beach sands of Elba Island, Tuscany, Italy: Roundness study and evidence of provenance, in Johnsson, M.J., and Basu, A., eds., Processes Controlling the Composition of Clastic Sediments: Geological Society of America Special Paper 284, p. 235–245.

Potter, P.E., 1986, South America and a few grains of sand: Part 1. Beach sands: The Journal of Geology, v. 94, p. 301–319.

Powers, M.C., 1952, A new roundness scale for sedimentary particles: Journal of Sedimentary Petrology, v. 23, p. 117–119.

Robinson, R.S., and Johnsson, M.J., 1997, Chemical and physical weathering of fluvial sands in an Arctic environment: Sands of the Sagavanirktok River, North Slope, Alaska: Journal of Sedimentary Research, v. 67, p. 560–570.

Ruxton, B.P., 1970, Labile quartz-poor sediments from young mountain ranges in northeast Papua: Journal of Sedimentary Petrology, v. 40, p. 1262–1270.

Stallard, R.F., 1985, River chemistry, geology, geomorphology, and soils in the Amazon and Orinoco basins, in Drever, J.I., ed., The Chemistry of Weathering: Dordrecht, D. Reidel, p. 293–316.

Valloni, R., 1985, Reading provenance from modern marine sands, in Zuffa, G.G., ed., Provenance of Arenites: Dordrecht, D. Reidel, p. 309–332.

Webb, W.M., and Potter, P.E., 1971, Petrologia y geochemica de detritos derivados de un terreno riolitico de la region occidental de Chihuahua, Mexico: Boletin de la Sociedad Geologica Mexicana, v. 32, p. 45–61.

MANUSCRIPT ACCEPTED BY THE SOCIETY 9 AUGUST 2006

Cyclic variations in sediment provenance from late Pleistocene deposits of the eastern Po Plain, Italy

Alessandro Amorosi
Maria Luisa Colalongo
Enrico Dinelli
Federico Lucchini
Stefano Claudio Vaiani

Dipartimento di Scienze della Terra, University of Bologna, Via Zamboni 67, 40127 Bologna, Italy

ABSTRACT

A cyclic vertical succession of alluvial, littoral, and shallow-marine deposits is identified within two continuously cored boreholes (187-S1 and 204-S15) drilled to ~180 m beneath the present Po coastal plain, in northern Italy. Integrated sedimentologic, micropaleontologic (benthic foraminifers and ostracods), and geochemical studies allow the reconstruction of the paleogeographic evolution of the study area during the late Quaternary, with a special emphasis on major changes in provenance and sediment dispersal patterns.

Transgressive surfaces appear as the most readily identifiable stratigraphic features in the two cores, allowing identification of a series of transgressive-regressive sequences. The transgressive surfaces mark the onset of coastal to shallow-marine conditions, followed by delta and strand plain progradation and the reestablishment of continental environments. This cyclic pattern of facies is paralleled by distinctive cyclic variations in chemical composition of sediments, reflecting a systematic increase in Ni/Al within lower transgressive deposits, followed by a marked decrease in the overlying alluvial plain sediments. At relatively northern locations (core 187-S1), the maximum flooding surfaces identified within shallow-marine deposits on the basis of subtle, but consistent changes in microfaunal assemblages are characterized by anomalously high Mg/Al values.

The abrupt peaks in Ni/Al recorded at the transgressive surfaces are interpreted to reflect enrichments in mafic-ultramafic detritus, probably derived from the western Alps and the northwestern Apennines and supplied by the Po River to the coastal areas. These variations took place at the onset of brackish and littoral conditions, when direct connection with the sea favored sediment dispersal from the Po River mouth to lagoonal and coastal environments via the littoral drift. High Mg/Al values within open-marine deposits at the maximum flooding surfaces likely reflect an increasing contribution from eastern Alpine (dolomite-rich) sources at time of maximum shoreline migration.

Amorosi, A., Colalongo, M.L., Dinelli, E., Lucchini, F., and Vaiani, S.C., 2007, Cyclic variations in sediment provenance from late Pleistocene deposits of the eastern Po Plain, Italy, *in* Arribas, J., Critelli, S., and Johnsson, M.J., eds., Sedimentary Provenance and Petrogenesis: Perspectives from Petrography and Geochemistry: Geological Society of America Special Paper 420, p. 13–24, doi: 10.1130/2007.2420(02). For permission to copy, contact editing@geosociety.org. ©2007 Geological Society of America. All rights reserved.

The recurrent changes in geochemical composition recorded across the transgressive surfaces fully support the stratigraphic subdivision of late Quaternary deposits of the Po Basin into transgressive-regressive sequences, rather than depositional sequences. The sequence-bounding unconformities do not display distinctive geochemical signatures.

Keywords: sequence stratigraphy, micropaleontology, geochemistry, Po Plain, Quaternary.

INTRODUCTION

Provenance studies from clastic successions are traditionally based upon sandstone petrography (Zuffa, 1985; Ingersoll and Cavazza, 1991; Critelli and Le Pera, 1994; Busby and Ingersoll, 1995; Garzanti et al., 1996; Critelli et al., 1997, 2003; Le Pera and Arribas, 2004). In the last decades, with the development of sequence stratigraphic concepts, a growing number of papers has dealt with the relationships between cyclic sedimentation patterns and changes in petrofacies distribution (Zuffa et al., 1995; Marchesini et al., 2000; Arribas et al., 2003). Several recent studies have documented how whole-rock geochemistry may provide an invaluable tool to perform stratigraphic correlations and discern changes in provenance (Garver et al., 1996; Pearce et al., 1999; Cullers, 2000). In spite of this, little attention has been given to the use of variations in chemical composition of sediments to characterize the key surfaces in sequence-stratigraphic interpretation (Morad et al., 2000; Amorosi et al., 2002; Ketzer et al., 2003; Algeo et al., 2004).

Sediments emplaced during the late Quaternary sea-level cycles in the Po River basin (Fig. 1) constitute an excellent archive to this purpose, because of (1) the large amount of stratigraphic and sedimentological data available (Amorosi et al., 1999, 2003), (2) the very good chronologic control based upon radiocarbon dating, and (3) the existence of petrographic (Marchesini et al., 2000) and geochemical (Amorosi et al., 2002) data.

Sand petrography of late Quaternary deposits from the subsurface of Ravenna (Fig. 1) has shown that significant changes in sediment provenance took place at the onset of the Holocene transgression in response to the abrupt change from alluvial to shallow-marine conditions (Marchesini et al., 2000). A similar conclusion has been reported from the same area by Amorosi et al. (1999, 2002), based on integrated geochemical and mineralogical data from fine-grained (silt and clay) sediments. Particularly, Cr and Ni among the elements and serpentine among the minerals have been shown to represent unequivocal provenance indicators for the Po River basin (Fig. 1). The source material for Cr and Ni, delivered to the trunk river by a series of tributaries, comes from ultramafic (ophiolitic) complexes of the western Alps and northern (West Emilia) Apennines (Dinelli and Lucchini, 1999). By contrast, low Cr-Ni and serpentine values are invariably recorded in sediments derived from the northeastern (East Emilia and Romagna) Apennines (Fig. 1), which mostly drain Tertiary rocks of turbidite origin.

Although these studies have documented a close relationship between geochemical composition and facies distribution, changes in sediment provenance have been related to a comparatively short interval of time (the last 30 k.y.), coinciding with the upper part (~30 m) of the youngest glacial-interglacial cycle.

Detailed sedimentologic work in the Po Plain, integrated with micropaleontological (benthic foraminifers, ostracods, and pollen) studies, has revealed the depositional architecture of the uppermost 200 m, which consists of a cyclic alternation of alluvial and coastal–shallow-marine sediments that accumulated under a predominantly glacio-eustatic control (Amorosi et al., 2004). Sequence-stratigraphic analysis of this cyclic pattern of facies has led to identification of a series of transgressive-regres-

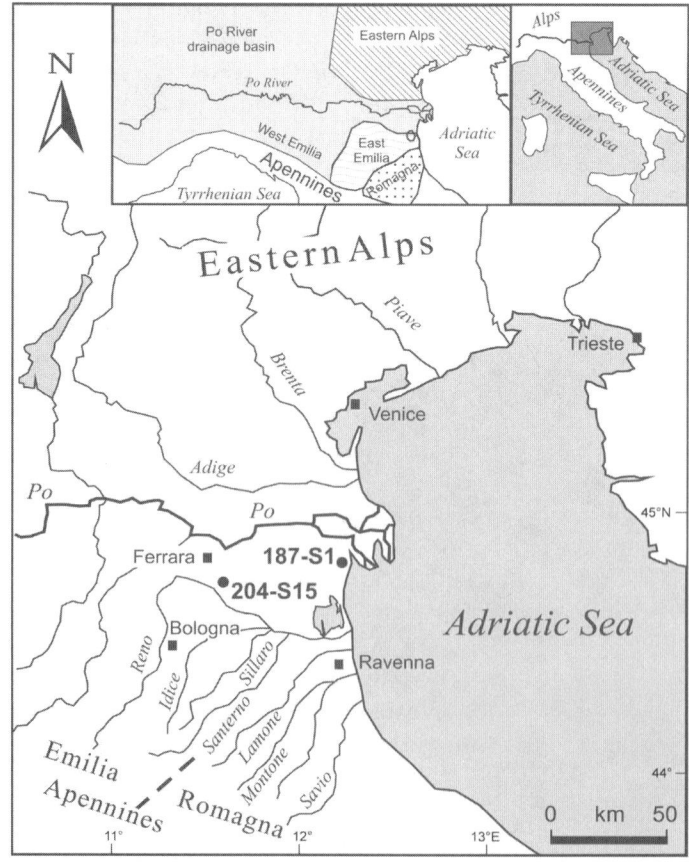

Figure 1. Location of the two study cores (187-S1 and 204-S15) and major provenance domains for the Po basin.

sive cycles, ~50–100 m thick, spanning intervals of time in the order of magnitude of 100 k.y. (Amorosi and Colalongo, 2005).

The major aim of this paper is the sequence-stratigraphic analysis of the relationships between cyclic sedimentation patterns and changes in sediment provenance, using detailed geochemical characterization of two continuously cored boreholes (187-S1 and 204-S15 in Fig. 1) that were drilled by the Geological Survey of Regione Emilia-Romagna to provide information for the construction of the new geological map of Italy. A specific objective of this study is to show how repetitive patterns in geochemical element distribution can provide fundamental information for unraveling cyclic changes in paleogeography through time.

METHODS

Facies analysis of cores includes description of lithology, grain size, and accessory components, such as mollusc shells, peat horizons, root traces, paleosols, etc. Micropaleontological data include a total of 398 samples (198 from core 187-S1, and 200 from core 204-S15). Approximately 200 g for each sample were dried for 8 h at 60 °C, soaked in water, and then washed through sieves of 63 μm (240 mesh). The >125 μm size fractions were analyzed for ostracods and foraminiferal assemblages; foraminiferal tests were separated by flotation using carbon tetrachloride (CCl_4).

Identification of benthic foraminifers and ostracods was based on the original microfossil descriptions and a series of key papers (von Daniels, 1970; Bonaduce et al., 1975; AGIP, 1982; Jorissen, 1988; Athersuch et al., 1989; Albani and Serandrei Barbero, 1990; Henderson, 1990; Cimerman and Langer, 1991; Sgarrella and Moncharmont Zei, 1993; Fiorini and Vaiani, 2001). Autecological information on species and paleoenvironmental significance of foraminiferal assemblages were provided by Blanc-Vernet (1969), Jorissen (1988), Albani and Serandrei Barbero (1990), Murray (1991), Barmawidjaja et al. (1992), Bellotti et al. (1994), Fiorini and Vaiani (2001), Donnici and Serandrei Barbero (2002), and Amorosi et al. (2003).

Interpretation of ostracod fauna was based upon papers by Colalongo (1969), Breman (1975), Bonaduce et al. (1975), Sokac (1978), Ciliberto and Pugliese (1980), Athersuch et al., (1989), Henderson (1990), Melis et al. (1995), and Montenegro and Pugliese (1995, 1996).

A total of 166 samples out of the larger micropaleontology set was collected for geochemical analyses (62 from core 187-S1, and 104 from core 204-S15). Samples were not regularly spaced, and their density reflected the different sedimentological attributes of the cores. All samples were oven dried at 40 °C and then homogenized in an agate mortar. Chemical determinations were obtained at the University of Bologna (Department of Earth Sciences) by X-ray fluorescence spectrometry (Philips PW 1480) on pressed powder pellets, following the matrix correction methods of Franzini et al. (1972, 1975), Leoni and Saitta (1976), and Leoni et al. (1982). The estimated precision and accuracy for trace-element determinations was better than 5%, except for those elements at 10 ppm and lower (10%–15%). LOI (loss on ignition) was evaluated after overnight heating at 950 °C. Geochemical data presented and discussed in the following sections refer uniquely to Mg and Ni—these elements were the most effective in the discrimination of the different facies associations. Since samples included a variety of textures (sands, silts, and clays), normalization of Ni and Mg to Al was used to ensure identification of the provenance signal, as discussed at length in Dinelli et al. (this volume).

STRATIGRAPHY, FACIES, AND GEOCHEMISTRY OF CORE 187-S1

Sedimentology

Core 187-S1 is located south of the modern Po River, close to the Adriatic Sea (Fig. 1). Facies interpretation for this core is provided in Bondesan et al. (2006). We show here a reinterpretation of the core based on sedimentological and stratigraphic similarities with adjacent areas (Amorosi et al., 2003), with a special emphasis on its sequence-stratigraphic interpretation (Fig. 2). This includes recognition of two stratigraphic intervals with marine deposits, bounded by two major transgressive surfaces at ~25 m and 130 m depth, respectively. Similar to what was observed in several deep cores from the southeastern Po Plain (Amorosi et al., 2004), the transgressive surfaces, which mark the abrupt boundaries between alluvial plain (fluvial-channel and floodplain) deposits and overlying back-barrier, coastal, and shallow-marine sediments, constitute the most readily identifiable surfaces in the study succession (Fig. 2). The two major transgressive surfaces coincide with the base of characteristic transgressive-regressive cycles (T-R sequences of Amorosi and Colalongo, 2005), which have been interpreted to reflect the last two glacial-interglacial transitions (base of oxygen isotope stages [OIS] 5e and 1, respectively; see Fig. 2).

Above the two major transgressive surfaces, the internal architecture of the transgressive-regressive cycles consists of transgressive deposits that show a deepening-upward tendency represented by vertical superposition of back-barrier, beach (transgressive barrier), and shallow-marine (offshore-transition) sediments. These show an upward transition to prodelta and delta front (beach-ridge) deposits, which display a shallowing-upward trend. This characteristic vertical stacking pattern of facies has been interpreted to reflect the landward migration of barrier-lagoon-estuary systems during sea level rise (transgressive systems tracts—TST), followed by delta progradation during the subsequent phases of sea-level highstand (highstand systems tracts—HST) (Amorosi et al., 1999, 2003). The upper part of the transgressive-regressive cycle of Tyrrhenian (OIS 5e) age is composed of coastal plain and alluvial plain deposits, representing the falling-stage and lowstand systems tracts (FST + LST).

The transgressive-regressive cycle attributed to OIS 7, which is marked by a minor transgressive surface at ~180 m depth, is truncated by an obvious erosional unconformity. This is overlain by fluvial-channel deposits related to the subsequent glacial period (OIS 6). Another minor transgressive surface, marking the onset of

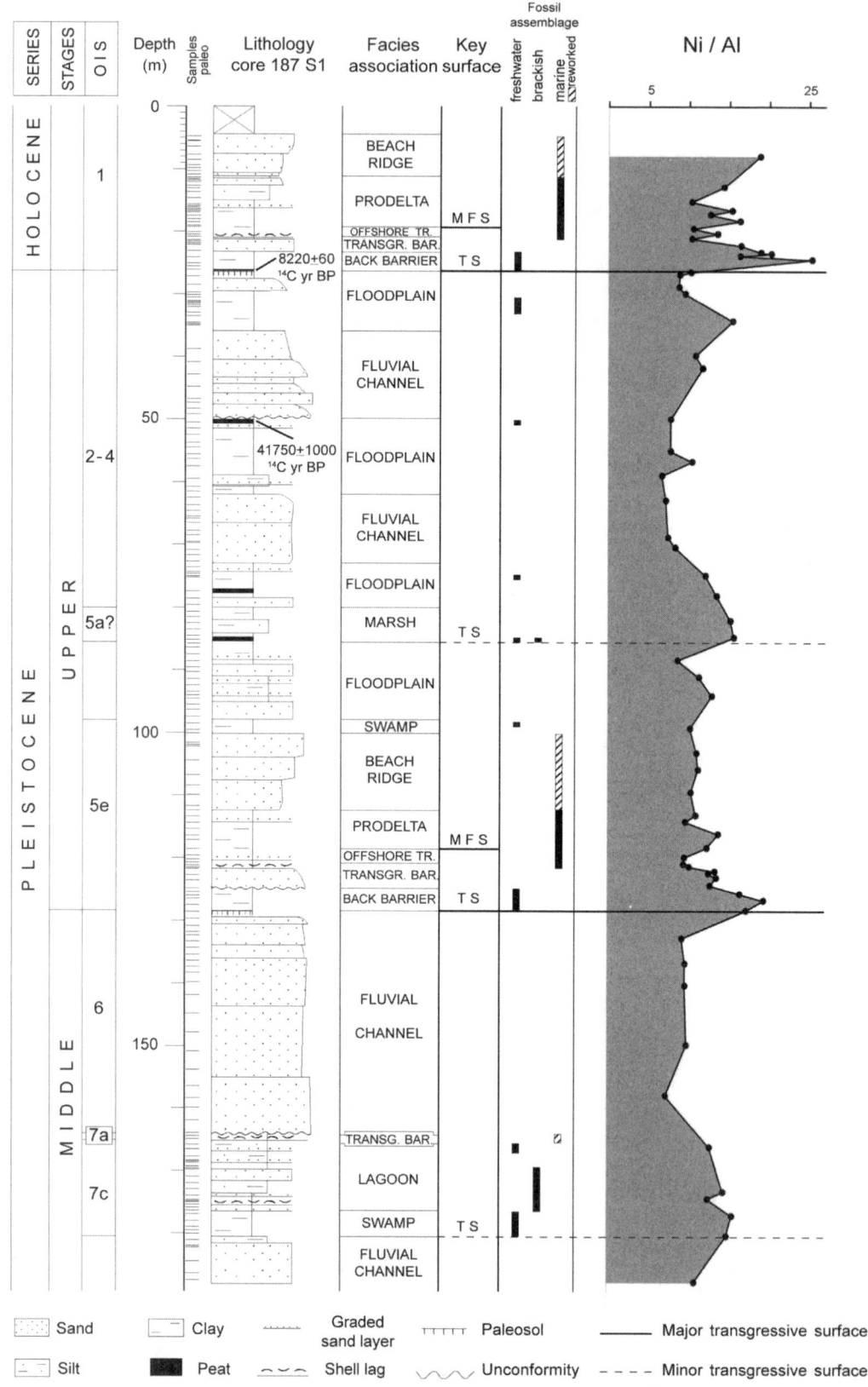

Figure 2. Sequence stratigraphy of core 187-S1 (see Fig. 1 for location), with indication of the transgressive surfaces (TS) and maximum flooding surfaces (MFS), microfossil characterization, and vertical profiles of Ni/Al. OIS—oxygen isotope stage.

brackish-water deposits on top of coastal plain sediments, is recorded at ~85 m. On the basis of stratigraphic correlations with adjacent southern areas (see Amorosi et al., 2004), this minor transgressive surface can be attributed tentatively to substage 5a transgression (Bondesan et al., 2006).

Micropaleontology

Previous paleontologic work on core 187-S1 includes data from ostracods and mollusks (Bondesan et al., 2006). In this paper, the quantitative analysis of benthic foraminifers from the same core provides useful information to refine facies attribution and the sequence-stratigraphic interpretation of this stratigraphic succession (Fig. 2).

It has been shown that vertical changes in microfaunal assemblages may provide a significant contribution to sequence stratigraphy, allowing the precise positioning of the maximum flooding surfaces within apparently homogeneous mud-dominated marine deposits (Amorosi and Colalongo, 2005). Application of these concepts to core 187-S1 (Fig. 3) shows that the two maximum flooding surfaces related to the OIS 1 (Holocene) and 5e (Tyrrhenian) transgressions can be traced in coincidence with diagnostic vertical changes in microfauna, at the turnaround from assemblages showing the highest contents and diversity of species characteristic of

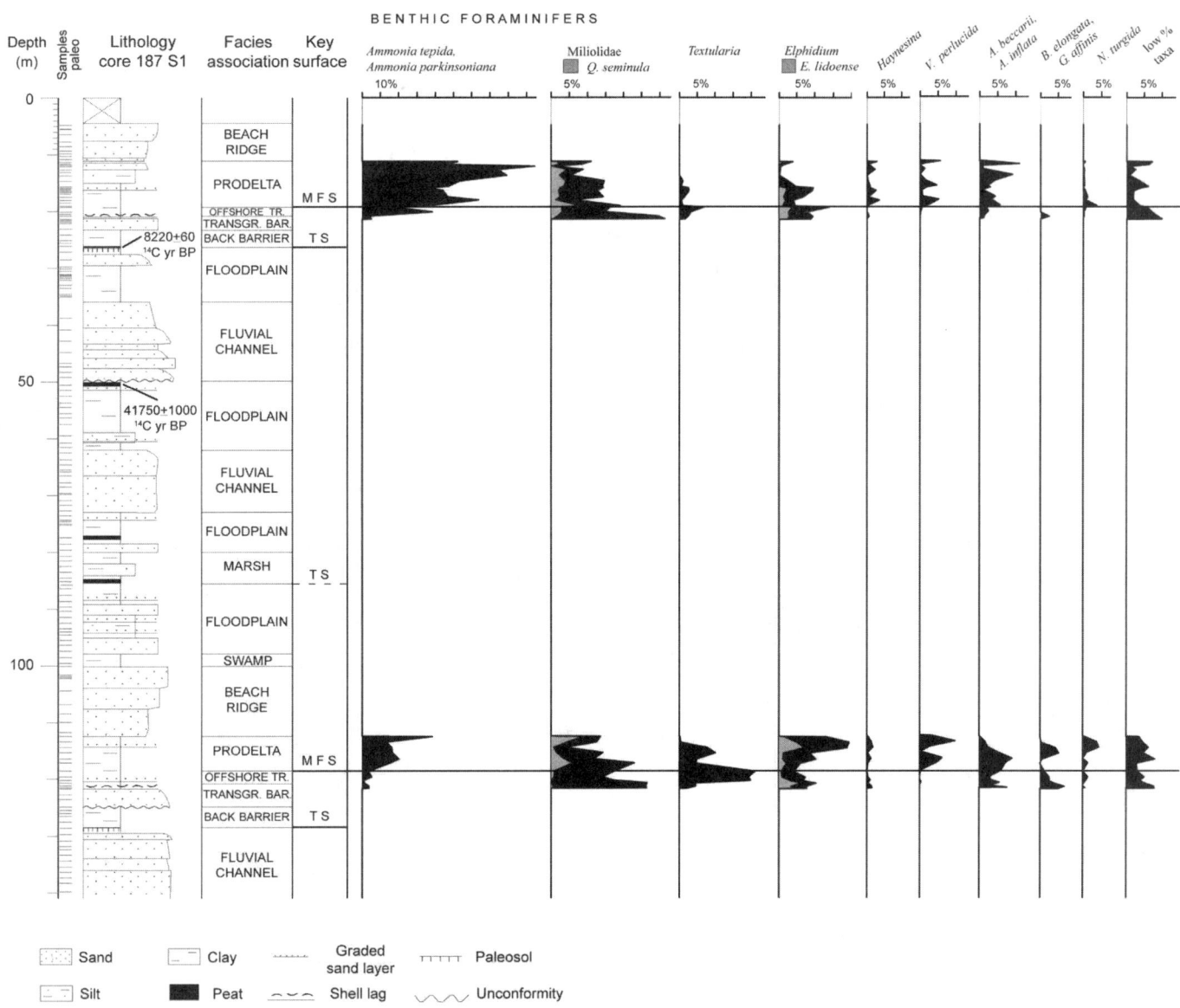

Figure 3. Detailed vertical distribution of benthic foraminifers, with indication of the transgressive surfaces (TS) and maximum flooding surfaces (MFS) related to oxygen isotope stages (OIS) 5e and 1. Genus abbreviations: A.—*Ammonia*, B.—*Bulimina*, E.—*Elphidium*, G.—*Globobulimina*, N.—*Nonionella*, Q.—*Quinqueloculina*, V.—*Valvulineria*.

relatively deep environments (transgressive systems tract) to assemblages showing a rapid upward increase in the amount of shallower and euryhaline species (highstand systems tract).

Below the maximum flooding surfaces, the transgressive barrier sands and the offshore-transition clay-sand alternations that form the two transgressiive systems tracts are characterized by high percentages of Miliolidae (including *Adelosina cliarensis, A. mediterranensis, Cycloforina costata, Pseudotriloculina rotunda, Quinqueloculina lata, Q. padana, Siphonaperta aspera, Triloculina gibba, T. schreiberiana,* and *T. trigonula*), *Textularia* spp. (including *Textularia agglutinans* and *T. bocky*), and *Elphidium* spp. (Fig. 4). This assemblage is considered to reflect a 20–40-m-deep marine environment. Benthic foraminifers from OIS 5e sediments reveal a marine environment slightly deeper than the one recorded by the Holocene deposits.

The sharp increase in species tolerant to low salinities, such as *Ammonia tepida, A. parkinsoniana, A. beccarrii, A. inflata,* and *Valvulineria perlucida,* is observed above the maximum flooding surfaces within clay-sand alternations interpreted as highstand (HST) prodelta deposits (Fig. 4). Increasing frequencies of *Ammonia tepida* and *A. parkinsoniana* are associated with significant amounts of other moderately euryhaline species, such as *Elphidium lidoense, Elphidium* spp., and *Quinqueloculina seminula*. This assemblage documents a regressive tendency within shallow-marine environments subjected to a progressively increasing freshwater influence. An important riverine influence is particularly recorded by the Holocene deposits, where *Ammonia tepida* and *A. parkinsoniana* reach 94% of benthic foraminiferal microfauna, thus indicating remarkably lower salinities. The overlying beach-ridge (delta front) sands (upper highstand systems tract) show the scattered occurrence of poorly preserved foraminiferal tests (mainly *Ammonia beccarii, Elphidium crispum,* and Miliolidae), probably transported from nearby environments or reworked from older units. This assemblage is typical of high-energy littoral environments.

OIS 7 deposits are characterized in their lower part by ostracods and mollusks indicative of freshwater environments (Fig. 2). This association is replaced upward by foraminiferal assemblages almost entirely composed (up to 95%) by *Ammonia tepida* and *A. parkinsoniana*. The common occurrence of brackish water ostracod *Cyprideis torosa* suggests deposition in a partially barred environment subjected to a high freshwater influence, such as a lagoon. At m 174.75 core depth, lower frequencies of *Ammonia tepida* and *A. parkinsoniana* (58%) combined with relatively high concentrations of *Elphidium* spp. (especially *E. granosum*) suggest a local increase in marine influence. A rapid return to freshwater conditions precedes the establishment of a high-energy, beach environment, including reworked marine fossils (Fig. 2). The maximum flooding surface of this transgressive-regressive cycle is lacking, owing to erosional truncation.

Finally, the minor positive sea-level shift observed at ~85 m is documented by the occurrence of freshwater and possibly transported brackish ostracods (Bondesan et al., 2006), which suggest the development of a marsh environment in partial connection with the sea.

Geochemistry (Ni/Al Profile)

Vertical profiles of Ni, normalized to Al in order to compensate for grain-size effects, show relatively low (< 10) values along the core, with the exception of the four transgressive surfaces, where a remarkable increase in the Ni-Al ratio (~15–25) is invariably recorded (Fig. 2). Maximum Ni/Al values for each transgressive-regressive cycle are recorded just above the transgressive surfaces and generally show a decreasing upward tendency throughout the cycle.

The two major transgressive pulses, attributed to OIS 5e and 1, display the highest Ni/Al values (Fig. 2). In terms of facies associations, these highest Ni/Al values are strictly related to deposits formed in back-barrier and coastal (transgressive barrier) environments. The role of Ni as an indicator of early transgressive deposits is also reflected by its high concentration within swamp and marsh deposits at ~180 and 65 m depth, corresponding to OIS 7 and 5a transgressive pulsations, respectively.

STRATIGRAPHY, FACIES, AND GEOCHEMISTRY OF CORE 204-S15

Sedimentology

Core 204-S15 is located ~50 km west of the present shoreline, in a comparatively landward position with respect to core 187-S1 (Fig. 1). No published data exist on this core. Its present location is very close to an abandoned distributary channel of the Po River that flowed through the town of Ferrara, coincident with the distal reaches of present Reno River (Fig. 1). This branch of Po River was active until the end of fourteenth century A.D. (Veggiani, 1974). Maximum landward migration of the shoreline in the Ferrara area during the Holocene was 25 km west of modern shoreline (Amorosi et al., 2003). This implies that core 204-S15 was not affected by marine transgression during the late Quaternary. Consistent with its geographic position, the uppermost 100 m of core 204-S15 do not show any marine influence (Fig. 4). Holocene sediments at this site consist uniquely of swamp and interdistributary area (delta plain) deposits, representing the landward equivalents of the beach-ridge (delta front) sediments recorded in core 187-S1. The stratigraphic link between Holocene fluvial and deltaic sediments in the Ferrara area is provided by the basal transgressive surface, which is recorded in core 204-S15 at ~12 m depth.

Due to the high degree of tectonic deformation affecting this part of the Po basin (see Pieri and Groppi, 1981) and lack of unequivocal stratigraphic markers, stratigraphic correlation of pre-Holocene units between core 204-S15 and core 187-S1 cannot be performed confidently on a physical basis only. For instance, it is not easy to establish whether or not the swamp intervals at ~50 m, 80 m, and 90 m depth may be laterally correlative with the transgressive back-barrier and beach deposits identified in core 187-S1. As a consequence, chronologic attribution of the late Quaternary deposits in this area is uncertain.

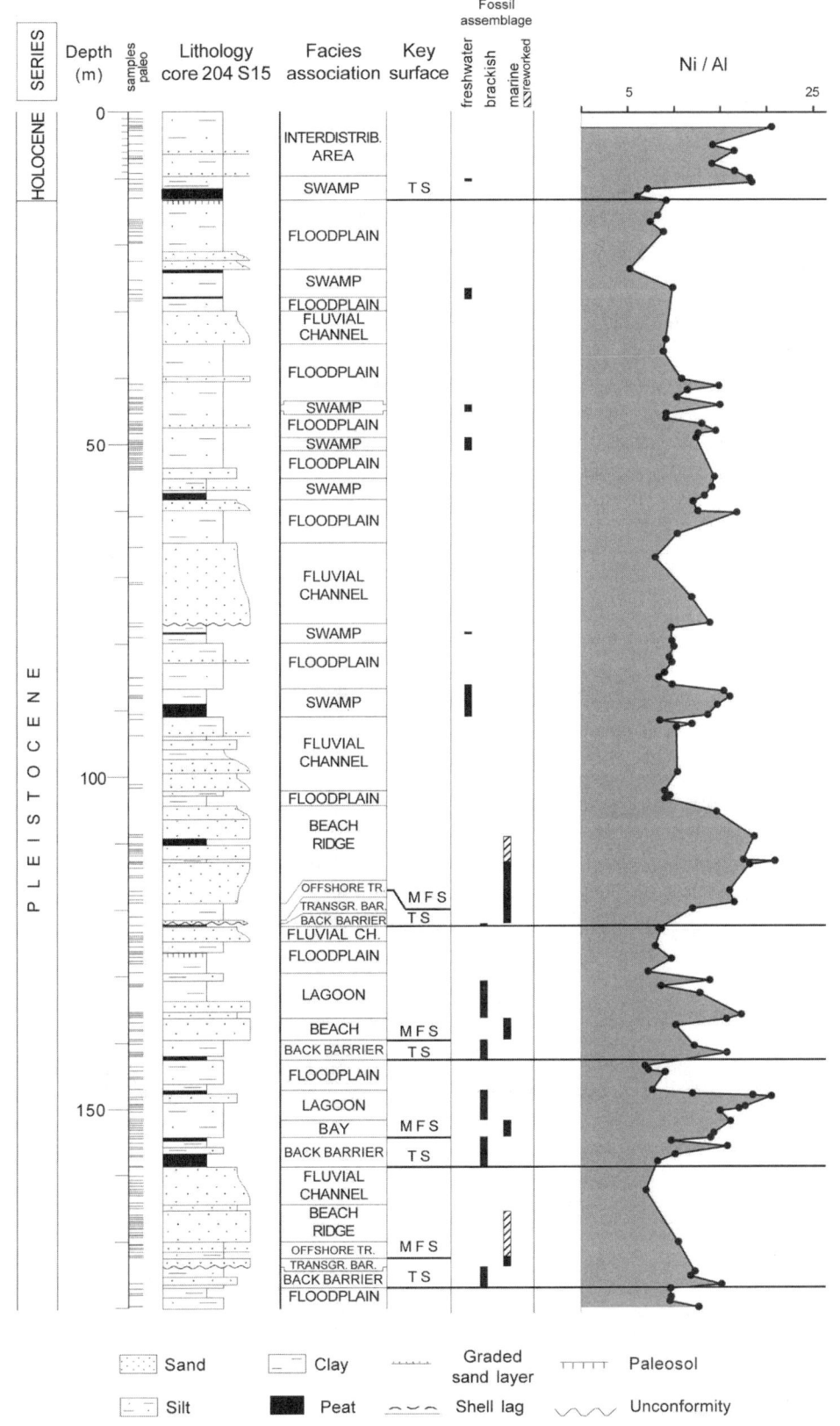

Figure 4. Sequence stratigraphy of core 204-S15 (see Fig. 1 for location), with indication of the transgressive surfaces (TS) and maximum flooding surfaces (MFS), microfossil characterization, and vertical profiles of Ni/Al.

Coastal to shallow-marine deposits are frequently encountered in the lower part of core 204-S15, between 100 and 180 m depth. Identification at this stratigraphic level of four transgressive surfaces allows subdivision of this part of the core into four transgressive-regressive cycles (Fig. 4). Back-barrier (lagoonal) deposits invariably form the lowermost portions of these cycles. In two instances, around 120 m and 170 m core depth, respectively, a deepening-upward trend is shown by vertical superposition of offshore-transition clay-sand alternations onto transgressive barrier sands. By contrast, at 140 m and 155 m core depths, maximum transgression led to the establishment of shallower (beach or bay) subenvironments. The upper part of transgressive-regressive cycles documents the reestablishment of continental conditions and consists of floodplain to delta plain deposits.

Micropaleontology

The uppermost 100 m of core 204-S15 lack any fossil evidence, with the exception of thin organic-rich clays (swamp facies associations), which are characterized by the presence of freshwater ostracod *Candona* spp.

Back-barrier deposits at the very base of the four transgressive-regressive cycles consist of brackish-water foraminiferal assemblages, including *Ammonia tepida, A. parkinsoniana, Miliolinella oblonga, Cribroelphidium pauciloculum*, along with ostracod *Cyprideis torosa*.

Maximum water depths are recorded between 121 and 119 m and 172–171 m core depths, where an abundant and varied microfauna can be identified. Foraminifers are characterized by high percentages of Miliolidae, including several species of *Adelosina* spp., *Quinqueloculina* spp., and *Pseudotriloculina* spp.; *Ammonia papillosa, A. beccarii, Elphidium crispum, E. macellum, E. granosum,* and *E. lidoense* are frequently encountered in this association. Small specimens of *Textularia* are also present. Concerning the ostracod fauna, the dominant species are *Ponthocythere turbida, Cytheridea neapolitana, Cytheretta subradiosa,* several species of *Semicytherura* (including *S. incongruens*), *Callistocythere,* and *Loxoconcha* (mostly *L. tumida*). This microfossil association, characteristic of the maximum flooding surface, suggests slightly shallower water depths with respect to core 187-S1.

Similar to what was observed in core 187-S1 (Fig. 2), regressive highstand deposits above maximum flooding surfaces record an increasing fluvial influence, documented by the highest concentration in *Ammonia*, with *A. papillosa* and *A. beccari* as dominant species, associated with abundant *Elphidium* (*E. crispum, E. macellum,* and *E. advenum*). Higher up in the stratigraphic column, this association is replaced by dominant *Ammonia tepida* and *A. parkinsoniana*, suggesting maximum riverine influx. The ostracod fauna is characterized by dominant *Ponthocythere turbida, Cytheretta adriatica,* and *Leptocythere bacescoi*. In the upper part of the marine successions, the regressive beach-ridge (delta front) sands are dominated by a transported and reworked microfaunal assemblage.

Transgressive-regressive cycles at 140 m and 155 m core depths display a different microfaunal assemblage. In particular, foraminifers include dominant *Ammonia tepida, A. parkinsoniana,* and subordinate *Cribroelphidium* spp., *Haynesina depressula, Miliolinella elongata, M. subrotunda, Quinqueloculina seminula, Q. stellifera, Triloculina affinis,* and *Adelosina cliarensis*. The ostracod fauna is commonly constituted by *Leptocythere* spp. (*L. levis, L.* cf. *multipunctata multipunctata*), *Loxoconcha elliptica, L. stellifera, L. tumida, Semicytherura costata,* with subordinate *Pontocythere turbida* and *Xestoleberis communis*. With respect to the previous transgressive-regressive cycles, this association is interpreted to reflect higher proximity to the coast and greater freshwater influence.

Geochemistry (Ni/Al Profile)

Similar to what was observed for core 187-S1, vertical profiles of Ni/Al in core 204-S15 show considerably higher values just above the five transgressive surfaces described in the previous sections (Fig. 4). Ni shows higher concentrations within lower transgressive deposits, and a rapid upward decrease in the upper parts of cycles.

In the lower part of core, between 100 and 180 m, high Ni/Al values are invariably recorded within sediments that accumulated under a direct marine influence (back-barrier clays, beach-ridge sands, and offshore-transition clay-sand alternations). In contrast, a marked decrease in Ni/Al is recorded at higher stratigraphic positions, within alluvial plain deposits (floodplain clays and fluvial-channel sands). In the upper 100 m, composed uniquely of continental deposits, Ni/Al values change significantly as a function of the different facies associations. Particularly, alluvial plain (floodplain and fluvial-channel) deposits generally display low Ni/Al values. In contrast, higher values are recorded coincident with swamp clays. Remarkably high Ni/Al values are observed in the uppermost 10 m, within interdistributary area clays attributed to the modern Po River.

PROVENANCE DOMAINS IN THE PO BASIN

Geochemical analyses of modern river sediment samples from the present fluvial network of the Po basin (Dinelli and Lucchini, 1999) show that selected geochemical ratios, such as Mg/Al and Ni/Al, are particularly effective in the distinction of three major sediment source areas (Fig. 5A), namely the Po River basin (including most Alps and West Emilia Apennines), East Emilia–Romagna Apennines, and the eastern Alps (cf. Fig. 1). Particularly, the combined use of these two parameters allows us to separate these three domains as follows:

1. The Po River is characterized by comparatively high (>18) Ni/Al values, and homogeneous (~0.36) Mg/Al.
2. Apenninic rivers are characterized by significantly lower (9–13) Ni/Al values. Despite the variability (0.22–0.39) of Mg/Al values, East Romagna rivers (Lamone, Montone, Savio) display comparatively higher values (>0.30) than rivers from the Emilia Apennines (Reno, Idice, Sillaro,

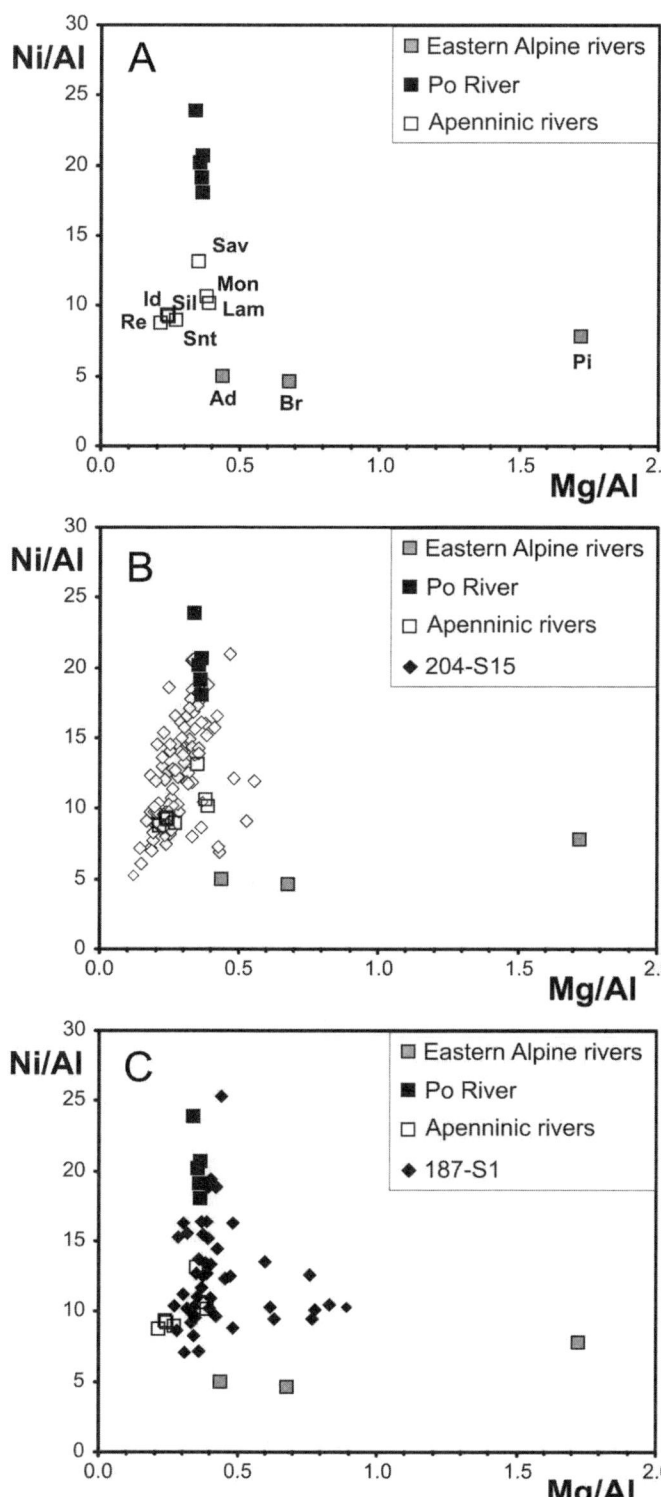

Figure 5. Mg/Al vs. Ni/Al diagrams, showing average composition of fluvial samples from the major rivers of the Po basin: (A) data from Dinelli and Lucchini (1999) and Amorosi et al. (2002) and their comparison with samples from core 204-S15 (B—this study) and core 187-S1 (C—this study). Apenninic rivers, from west to east: Re—Reno, Id—Idice, Sil—Sillaro, Snt—Santerno, Lam—Lamone, Mon—Montone, Sav—Savio. Eastern Alpine rivers: Ad—Adige, Br—Brenta, Pi—Piave (see Fig. 1 for location).

Santerno). This difference is consistent with different geology and rock composition in the drainage basins west and east of Sillaro River (Fig. 1), as pointed out by Amorosi et al. (2002).

3. Eastern Alpine rivers (Adige, Brenta, and Piave) are characterized by low (<8) Ni/Al values and variable, but generally high (>0.45) Mg/Al values. The lowest Mg/Al values are recorded in rivers (Adige and Brenta) draining mostly igneous and volcanic rocks, while very high (>1.7) values are recorded by Piave River sediments, which derive from a drainage basin rich in dolomitic rocks.

Samples from the two study cores (Figs. 5B and 5C) plot into similar fields of the Ni/Al-Mg/Al diagram, but with significant exceptions. Particularly, samples from core 204-S15 appear to plot along a mixing line that connects the Po River with an East Emilia Apenninic source area (Fig. 5B). On the other hand, samples from core 187-S1 display a higher dispersion, suggesting a mixed provenance from Po River drainage basin and Romagna Apennines, with a restricted number of samples pointing to anomalously high Mg/Al values (Fig. 5C).

CYCLIC CHANGES IN SEDIMENT PROVENANCE

Application of selected geochemical ratios from modern river deposits of northern Italy (Fig. 5A) to Ni/Al and Mg/Al profiles recorded in core 187-S1 (Fig. 6) allows us to establish the relationships between cyclic sedimentation patterns described in the previous sections and distinctive changes in provenance during the last 150 k.y. Normalization of Ni and Mg values to Al compensates for possible grain-size effects and ensures that the observed ratios are uniquely a function of sediment provenance, as discussed in Dinelli et al. (this volume).

The majority of samples from core 187-S1, especially those collected within fluvial-channel and floodplain deposits, display relatively low Ni/Al values. The characteristic Ni/Al peaks recorded just above the two major transgressive surfaces fall within the field of modern Po River composition (Ni/Al > 18) and invariably coincide with the onset of brackish (back-barrier, marsh) and then littoral conditions (Fig. 6). Similar cyclic fluctuations in Ni/Al values are recorded in the lower part of core 204-S15 (Fig. 4), where an obvious relationship between high (>18) Ni/Al values and the onset of marine sedimentation is observed. Interpretation of high Ni/Al values as an indication of sediment provenance from the Po River drainage basin is fully consistent with similar high Ni/Al values recorded in the topmost part of core 204-S15 (Fig. 4). This part of the core is representative of a recently abandoned Po delta plain, thus providing further confirmation of the possible use of high Ni/Al values as a geochemical indicator for a Po River origin.

If plotted against the Ni/Al profile, the Mg/Al profile reveals that the anomalously high values observed in Figure 5C are not randomly distributed but are concentrated at distinct stratigraphic levels in the lower part of both (OIS 5e and 1) transgressive-regressive cycles (Fig. 6). Unlike the Ni/Al profile,

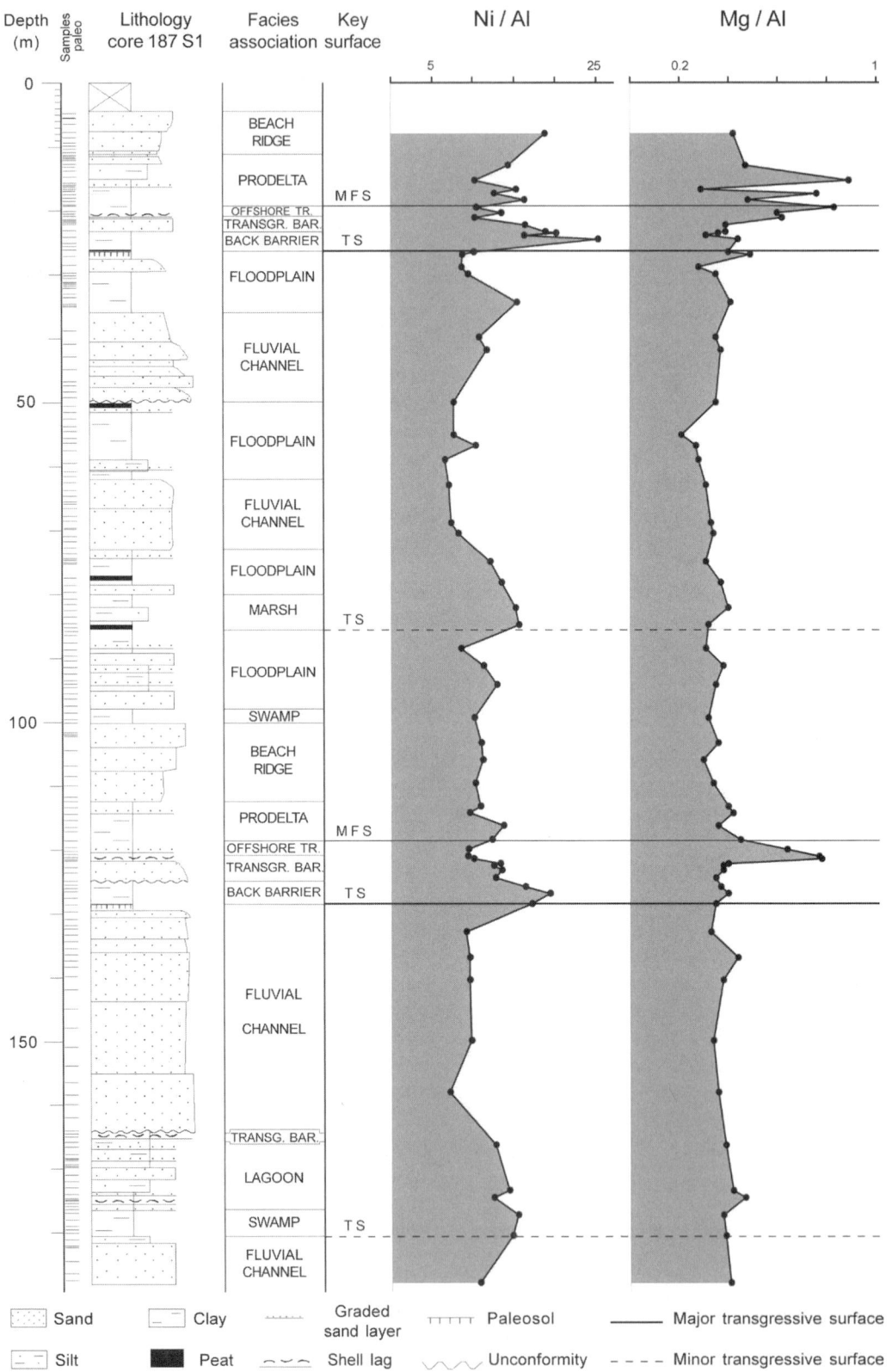

Figure 6. Geochemical characterization of the key surfaces for sequence-stratigraphic interpretation. TS—transgressive surface, MFS—maximum flooding surface.

however, peak concentrations of Mg/Al are not recorded at the transgressive surface, but are in a slightly higher stratigraphic position within shallow-marine deposits (Fig. 6). The two major peaks in Mg/Al, with values (>0.45) that are consistent with a possible sediment contribution from the eastern Alps, are associated with late transgressive deposits, i.e., they are very close to the maximum flooding surfaces identified on a micropaleontological basis.

In terms of paleogeographic evolution, it can be inferred that significant changes in sediment dispersal patterns occurred at the onset of major transgressive events. Particularly, at early stages of transgression (just above the transgressive surfaces in Fig. 6), formation of coastal environments in direct connection with the sea favored sediment dispersion from the Po River mouth to lagoonal and coastal environments via the littoral drift, resulting in high Ni/Al values and relatively low Mg/Al values (Fig. 6). With continuing transgression, shoreline migration led to a significant landward shift of Po River mouth. This was accompanied by the establishment of an open-marine environment in the study area. It is likely that an increasing sediment influx from the north, likely from the eastern Alpine rivers, occurred in this period, resulting in Mg-rich and Ni-poor marine sediments (Fig. 6). Progradation of Po River delta and adjacent strand plains during the subsequent highstand phase led to renewed increase in Ni concentration and progressive decrease in Mg content within prodelta deposits just above the maximum flooding surfaces (Fig. 6). Glacial periods were characterized by return to alluvial plain conditions, with relatively lower Ni/Al and Mg/Al. This can be interpreted as the result of the reestablishment of an alluvial plain drained by rivers of Apenninic provenance, although changes in basin physiography in response to variations in vegetation and ice cover cannot be ruled out.

CONCLUSIONS

1. Characteristic cyclic patterns of facies are identified through integrated sedimentological and micropaleontological analyses of two continuous cores, ~180 m deep, from the subsurface of the eastern Po Plain. Precise positioning of the transgressive surfaces allows us to subdivide late Quaternary deposits into transgressive-regressive sequences that formed under a predominantly glacio-eustatic control.
2. Geochemical characterization of the study units shows that major compositional changes occur across the bounding surfaces of transgressive-regressive sequences, with a systematic increase in Ni/Al just above the transgressive surfaces, and remarkably high Mg/Al values close to the maximum flooding surfaces. High Ni contents in early transgressive, coastal deposits are interpreted to reflect increasing sediment contribution from the Po River at the onset of transgressions, followed by significant sediment contribution from the eastern Alps at time of maximum flooding. Return to alluvial plain conditions during glacial periods is accompanied by an increasing sediment contribution from Apenninic sources.
3. The late Quaternary deposits of the Po basin provide a significant source of information in terms of sequence stratigraphy. Particularly, the transgressive surfaces carry unique geochemical fingerprints that support pragmatic subdivision of the late Quaternary succession into transgressive-regressive sequences, rather than depositional sequences.

ACKNOWLEDGMENTS

We are indebted to Prodip Dutta and Kathy Marsaglia for their careful review of manuscript.

REFERENCES CITED

AGIP, 1982, Foraminiferi Padani (Terziario e Quaternario): Milan, AGIP, Plate I–LII.

Albani, A., and Serandrei Barbero, R., 1990, I foraminiferi della laguna e del Golfo di Venezia: Memorie di Scienze Geologiche (Padova), v. 42, p. 271–341.

Algeo, T.J., Schwark, L., and Hower, J.C., 2004, High-resolution geochemistry and sequence stratigraphy of the Hushpuckney Shale (Swope Formation, eastern Kansas): Implications for climato-environmental dynamics of the late Pennsylvanian Midcontinent Seaway: Chemical Geology, v. 206, p. 259–288, doi: 10.1016/j.chemgeo.2003.12.028.

Amorosi, A., and Colalongo, M.L., 2005, The linkage between alluvial and marginal-marine successions: Evidence from the late Quaternary record of the Po River Plain, Italy, in Blum, M.D., Marriott, S.B., and Leclair, S.F., eds., Fluvial Sedimentology VII: International Association of Sedimentologists Special Publication 35, p. 257–275.

Amorosi, A., Centineo, M.C., Dinelli, E., Lucchini, F., and Tateo, F., 1999, Provenance changes across the Pleistocene/Holocene boundary in the south-eastern Po-Plain, in Ármannsson, H., ed., Geochemistry of the Earth Surface (GES-5, Reykiavik, p. 16–20 Agosto 1999): Balkema, Rotterdam, p. 23–26.

Amorosi, A., Centineo, M.C., Dinelli, E., Lucchini, F., and Tateo, F., 2002, Geochemical and mineralogical variations as indicators of provenance changes in late Quaternary deposits of SE Po Plain: Sedimentary Geology, v. 151, p. 273–292, doi: 10.1016/S0037-0738(01)00261-5.

Amorosi, A., Centineo, M.C., Colalongo, M.L., Pasini, G., Sarti, G., and Vaiani, S.C., 2003, Facies architecture and latest Pleistocene-Holocene depositional history of the Po delta (Comacchio area), Italy: The Journal of Geology, v. 111, p. 39–56, doi: 10.1086/344577.

Amorosi, A., Colalongo, M.L., Fiorini, F., Fusco, F., Pasini, G., Vaiani, S.C., and Sarti, G., 2004, Palaeogeographic and palaeoclimatic evolution of the Po Plain from 150-ky core records: Global and Planetary Change, v. 40, p. 55–78, doi: 10.1016/S0921-8181(03)00098-5.

Amorosi, A., Centineo, M.C., Colalongo, M.L., and Fiorini, F., 2005, Millennial-scale depositional cycles from the Holocene of the Po Plain, Italy: Marine Geology, v. 222, p. 7–18.

Arribas, J., Alonso, A., Mas, R., Tortosa, A., Rodas, M., Barrenechea, J.F., Alonso-Azcarate, J., and Artigas, R., 2003, Sandstone petrography of continental depositional sequences of an intra-plate rift basin: Western Cameros Basin (north Spain): Journal of Sedimentary Research, v. 73, p. 309–327.

Athersuch, J., Horne, D.J., and Whittaker, J.E., 1989, Marine and brackish water ostracods, in Kermack, D.M., and Barnes, R.S.K., eds., Synopses of the British Fauna (New Series) 43: New York, Brill E.J. Leiden, 343 p.

Barmawidjaja, D.M., Jorissen, F.J., Puskaric, S., and Van der Zwaan, G.J., 1992, Microhabitat selection by benthic foraminifera in the northern Adriatic Sea: Journal of Foraminiferal Research, v. 22, p. 297–317.

Bellotti, P., Carboni, M.G., Di Bella, L., Palagi, I., and Valeri, P., 1994, Benthic foraminiferal assemblages in the depositional sequence of the Tiber delta: Bollettino della Società Paleontologica Italiana, v. 2, Special, p. 29–40.

Blanc-Vernet, L., 1969, Contribution à l'étude des foraminifères de Méditerranée: Travaux de la Station Marine d'Endoume: Marseille, Station Marine d'Endoume, 281 p.

Bonaduce, G., Ciampo, and G., Masoli, M., 1975, Distribution of Ostracoda in the Adriatic Sea: Publicazione Stazione Zoologica di Napoli, v. 40, 304 p.

Bondesan, M., Cibin, U., Colalongo, M.L., Pugliese, N., Stefani, M., Tsakiridis, E., Vaiani, S.C., and Vincenzi, S., 2006, Benthic communities and sedimentary facies recording late Quaternary environmental fluctuations in a Po delta subsurface succession (northern Italy), in Coccioni, R., Lirer, F., and Marsili, A., eds., Proceedings of the Second Italian Meeting of Environmental Micropaleontology: The Grzybowski Foundation Special Publication.

Breman, E., 1975, The distribution of ostracodes in the bottom sediments of the Adriatic Sea [Ph.D. thesis]: Amsterdam, Free University of Amsterdam, 198 p.

Busby, C.J., and Ingersoll, R.V., editors, 1995, Tectonics of Sedimentary Basins: Oxford, Blackwell Sciences, 579 p.

Ciliberto, B.M., and Pugliese, N., 1980, Ostracodi bentonici del tratto di mare compreso tra Grado e Caorle (Adriarico Settentrionale): Gortania, Atti del Museo Friulano di Storia Naturale, v. 2, p. 65–80.

Cimerman, F., and Langer, M.R., 1991, Mediterranean foraminifera: Academia Scientiarum et Artium Slovenica (Ljubljana), Classis IV, v. 30, 118 p.

Colalongo, M.L., 1969, Ricerche sugli ostracodi nei fondali antistanti il delta del Po: Giornale di Geologia, v. 36, p. 335–362.

Critelli, S., and Le Pera, E., 1994, Detrital modes and provenance of Miocene sandstones and modern sands of the southern Apennines thrust-top basins (Italy): Journal of Sedimentary Research, v. 64, p. 824–835.

Critelli, S., Le Pera, E., and Ingersoll, R.V., 1997, The effects of source lithology, transport, deposition and sampling scale on the composition of southern California sands: Sedimentology, v. 44, p. 653–671, doi: 10.1046/j.1365-3091.1997.d01-42.x.

Critelli, S., Arribas, J., Le Pera, E., Tortosa, A., Marsaglia, K.M., and Latter, K.K., 2003, The recycled orogenic sand provenance from an uplifted thrust belt, Betic Cordillera, southern Spain: Journal of Sedimentary Research, v. 73, p. 72–81.

Cullers, R.L., 2000, Geochemistry of shales, siltstones and sandstones of Pennsylvanian-Permian age, Colorado, U.S.A.: Implication for provenance and metamorphic studies: Lithos, v. 51, p. 181–203, doi: 10.1016/S0024-4937(99)00063-8.

Dinelli, E., and Lucchini, F., 1999, Sediment supply to the Adriatic Sea basin from the Italian rivers; geochemical features and environmental constraints: Giornale di Geologia, v. 61, p. 121–132.

Dinelli, E., Tateo, F., and Summa, V., 2007, Is there any influence of grain-size in the provenance signal of "fine-grained" sediments? Results from Pleistocene to recent sediments of the Po Plain, northern Italy, in Arribas, J., Critelli, S., and Johnsson, M.J., eds., Sedimentary Provenance and Petrogenesis: Perspectives from Petrography and Geochemistry: Geological Society of America Special Paper 420, doi: 10.1130/2006.2420(03).

Donnici, S., and Serandrei Barbero, R., 2002, The benthic foraminiferal communities of the northern Adriatic continental shelf: Marine Micropaleontology, v. 44, p. 93–123, doi: 10.1016/S0377-8398(01)00043-3.

Fiorini, F., and Vaiani, S.C., 2001, Benthic foraminifers and transgressive-regressive cycles in the late Quaternary subsurface sediments of the Po Plain near Ravenna (northern Italy): Bollettino della Società Paleontologica Italiana, v. 40, p. 357–403.

Franzini, M., Leoni, L., and Saitta, M., 1972, A simple method to evaluate the matrix effects in X-Ray fluorescence analysis: X-Ray Spectrometry, v. 1, p. 151–154, doi: 10.1002/xrs.1300010406.

Franzini, M., Leoni, L., and Saitta, M., 1975, Revisione di una metodologia analitica per fluorescenza-X, basata sulla correzione completa degli effetti di matrice: Rendiconti della Società Italiana di Mineralogia e Petrologia, v. 31, p. 365–378.

Garver, J.I., Royce, P.R., and Smick, T.A., 1996, Chromium and nickel in shale of the Taconic foreland: A case study for the provenance of fine-grained sediments with an ultramafic source: Journal of Sedimentary Research, v. 66, p. 100–106.

Garzanti, E., Critelli, S., and Ingersoll, R.V., 1996, Paleogeographic and paleotectonic evolution of the Himalayan Range as reflected by detrital modes of Tertiary sandstones and modern sands (Indus transect, India and Pakistan): Geological Society of America Bulletin, v. 108, p. 631–642, doi: 10.1130/0016-7606(1996)108<0631:PAPEOT>2.3.CO;2.

Henderson, P.A., 1990, Freshwater ostracods, in Kermack, D.M., and Barnes, R.S.K., ed., Synopses of the British Fauna (New Series) 42: New York, Brill E.J. Leiden, 228 p.

Ingersoll, R.V., and Cavazza, W., 1991, Reconstruction of Oligo-Miocene volcaniclastic dispersal patterns in north-central New Mexico using sandstone petrofacies, in Fisher, R.V., and Smith, G.A., eds., Sedimentation in Volcanic Settings: Society for Sedimentary Geology (SEPM) Special Publication 45, p. 227–236.

Jorissen, F.J., 1988, Benthic foraminifera from the Adriatic Sea; principles of phenotypic variation: Utrecht Micropaleontological Bulletins, v. 37, 176 p.

Ketzer, J.M., Morad, S., and Amorosi, A., 2003, Predictive diagenetic clay-mineral distribution in siliciclastic rocks within a sequence stratigraphic framework, in Worden, R.H., and Morad, S., eds., Clay Mineral Cementation in Sandstones: International Association of Sedimentologists Special Publication 34, p. 42–59.

Leoni, L., and Saitta, M., 1976, X-ray fluorescence analysis of 29 trace elements in rock and mineral standard: Rendiconti della Società Italiana di Mineralogia e Petrologia, v. 32, p. 497–510.

Leoni, L., Menichini, M., and Saitta, M., 1982, Determination of S, Cl and F in silicate rocks by X-ray fluorescence analyses: X-Ray Spectrometry, v. 11, p. 156–158, doi: 10.1002/xrs.1300110404.

Le Pera, E., and Arribas, J., 2004, Sand composition in an Iberian passive-margin fluvial course: The Tajo River: Sedimentary Geology, v. 171, p. 261–281, doi: 10.1016/j.sedgeo.2004.05.019.

Marchesini, L., Amorosi, A., Cibin, U., Zuffa, G.G., Spadafora, E., and Preti, D., 2000, Sand composition and sedimentary evolution of a late Quaternary depositional sequence, northwestern Adriatic coast, Italy: Journal of Sedimentary Research, v. 70, p. 829–838.

Melis, R., Pugliese, N., and Degrassi, C., 1995, Ostracofauna del lago di mezzo e del lago inferiore (Mantova, Lombardia—Italia): Atti del Museo Geologico e Paleontologico di Monfalcone, Quaderno Speciale 3, p. 65–70.

Montenegro, M.E., and Pugliese, N., 1995, Ostracodi della laguna di Orbetello: Tolleranza ed opportunismo: Atti del Museo Geologico e Paleontologico di Monfalcone, Quaderno Speciale 3, p. 71–80.

Montenegro, M.E., and Pugliese, N., 1996, Autecological remarks on the ostracod distribution in the Marano and Grado Lagoons (northern Adriatic Sea Italy): Bollettino della Società Paleontologica Italiana, v. 3, Special, p. 123–132.

Morad, S., Ketzer, J.M., and De Ros, L.F., 2000, Spatial and temporal distribution of diagenetic alterations in siliciclastic rocks: Implications for mass transfer in sedimentary basins: Sedimentology, v. 47, Supplement 1, Millennium Reviews, p. 95–120.

Murray, J.W., 1991, Ecology and palaeoecology of benthic foraminifera: Harlow, UK, Longman Scientific and Technical, 397 p.

Pearce, T.J., Bealy, B.M., Wray, D.S., and Wright, D.K., 1999, Chemostratigraphy: A method to improve interwell correlation in barren sequences—Case study using onshore Duckmantian/Stephanian sequences (West Midlands, UK): Sedimentary Geology, v. 124, p. 197–220, doi: 10.1016/S0037-0738(98)00128-6.

Pieri, M., and Groppi, G., 1981, Subsurface geological structure of the Po Plain, Italy: Progetto Finalizzato Geodinamica Consiglio Nazionale delle Ricerche Pubblicazione 414, 23 p.

Sgarrella, F., and Moncharmont Zei, M., 1993, Benthic foraminifera of the Gulf of Naples (Italy): Systematics and autoecology: Bollettino della Società Paleontologica Italiana, v. 32, p. 145–264.

Sokac, A., 1978, Pleistocene ostracode fauna of the Pannonian Basin in Croatia: Palaeontologie Jugoslavia, v. 20, 51 p.

Veggiani, A., 1974, Le variazioni idrografice del basso corso del fiume Po negli ultimi 3000 anni: Padusa, v. 1–2, p. 39–60.

von Daniels, C.H., 1970, Quantitative ökologische analyse der zeitlichen und räumlichen verteilung rezenter foraminiferen im Limski Kanal bei Rovinj (nordliche Adria): Göttinger Arbeiten zur Geologie und Paläontologie, v. 8, 142 p.

Zuffa, G.G., ed., 1985, Provenance of Arenites: Dordrecht, Reidel Publishing Company, NATO-ASI Series, 408 p.

Zuffa, G.G., Cibin, U., and Di Giulio, A., 1995, Arenite petrography in sequence stratigraphy: The Journal of Geology, v. 103, p. 451–459.

MANUSCRIPT ACCEPTED BY THE SOCIETY 9 AUGUST 2006

Geochemical and mineralogical proxies for grain size in mudstones and siltstones from the Pleistocene and Holocene of the Po River alluvial plain, Italy

Enrico Dinelli[†]
Centro Interdipartimentale di Ricerca per le Scienze Ambientali, University of Bologna, Via Sant'Alberto 163, I-48100 Ravenna, Italy

Fabio Tateo
Istituto di Geoscienze e Georisorse, Consiglio Nazionale delle Ricerche, c/o Dipartimento di Geologia Paleontologia e Geofisica, University of Padua, Via Giotto 1, I-35122 Padova, Italy

Vito Summa
Istituto di Metodologie per l'Analisi Ambientale, Consiglio Nazionale delle Ricerche, Contrada S. Loja, Tito Scalo I-85050 Tito Scalo (Potenza), Italy

ABSTRACT

Recent studies carried out on fine-grained sediments recovered from boreholes in the eastern plain of the Po River demonstrate that significant mineralogical and geochemical changes in the provenance of sediments occurred in coincidence with the Pleistocene-Holocene transition. An increase in ultramafic-sourced sediment, related to more important inputs from the Po River, is evident at the beginning of the Holocene. The effects of grain-size distribution and provenance variation were investigated on recent unconsolidated sediments, mainly silts and clays. Sediments were collected from ten boreholes in the area, and the geochemical and mineralogical data were compared to the grain-size data. Among the chemical indexes, Zr/V, Y/Rb, Y/V, SiO_2/Al_2O_3, Fe_2O_3/SiO_2, Na/Al increase from pure clay to fine sand together with some mineralogical ratios, including quartz/interstratified illite-smectite and feldspar/interstratified illite-smectite. Some provenance indexes, both mineralogical and geochemical (Ni/Al, Cr/Al, serpentine/sheet silicates), were found to be independent from grain-size and are therefore valid for a wide textural range of sediments. Several geochemical and mineralogical proxies for grain size were identified. In the present case, all these indexes are independent from provenance influence and can be used as direct proxies for the grain size of the sediment, as confirmed by the multiple regression analysis performed to evaluate median and sorting. The equations included the most significant ratios and work well for median values <30 μm.

Keywords: geochemistry, provenance, grain-size, Po plain, Italy.

[†]E-mail: enrico.dinelli@unibo.it.

Dinelli, E., Tateo, F., and Summa, V., 2007, Geochemical and mineralogical proxies for grain size in mudstones and siltstones from the Pleistocene and Holocene of the Po River alluvial plain, Italy, *in* Arribas, J., Critelli, S., and Johnsson, M.J., eds., Sedimentary Provenance and Petrogenesis: Perspectives from Petrography and Geochemistry: Geological Society of America Special Paper 420, p. 25–36, doi: 10.1130/2007.2420(03). For permission to copy, contact editing@geosociety.org. ©2007 Geological Society of America. All rights reserved.

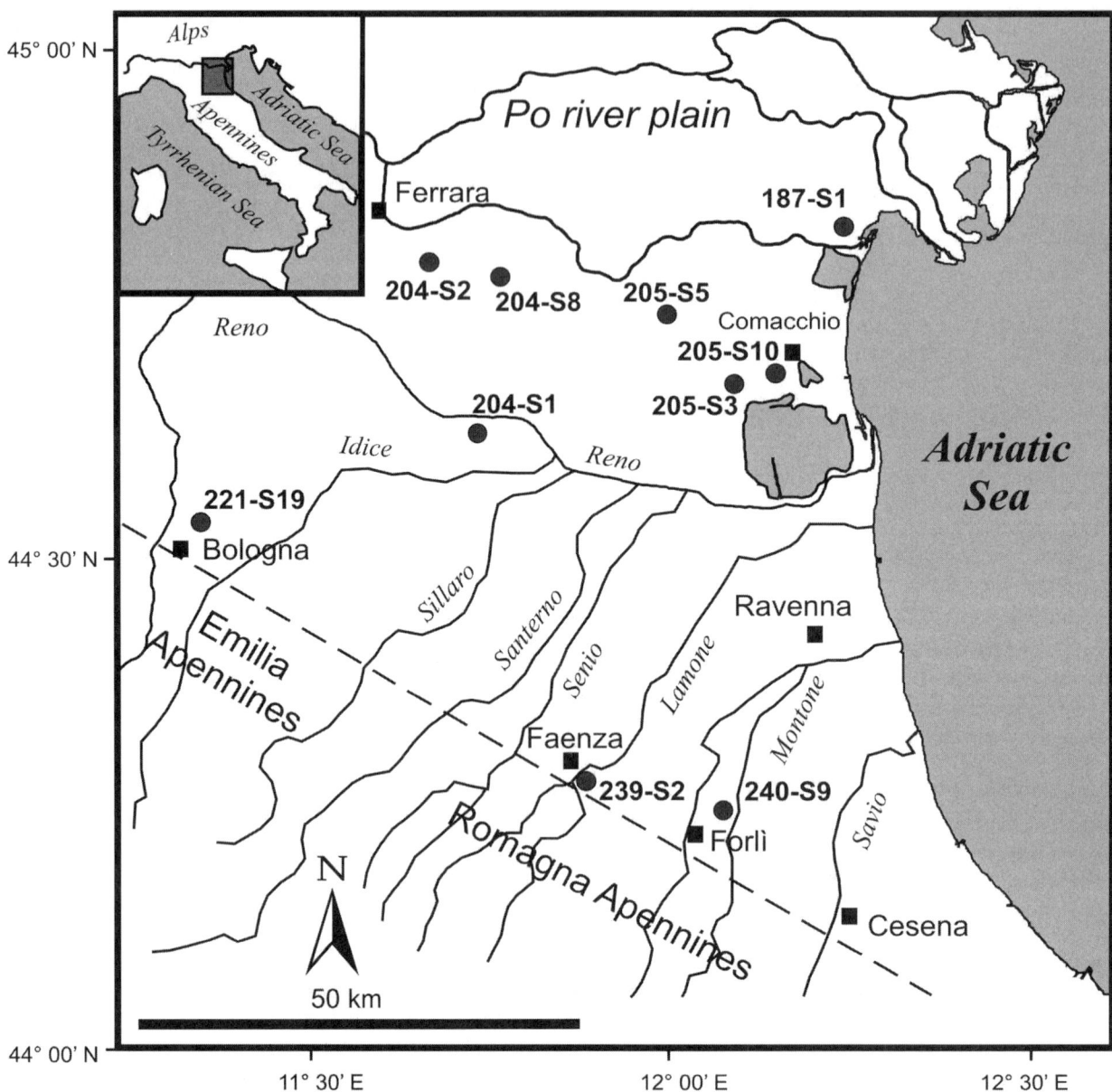

Figure 1. Location of the study area and sampled cores.

INTRODUCTION

The chemical and mineralogical composition of sediments and sedimentary rocks is influenced by many factors, including source rock composition, climatic conditions in the watershed, the length and energy of sediment transport, redox conditions in the depositional environment, and grain size (e.g., Bhatia and Cook, 1986; Jarvis and Higgs, 1987; Argast and Donnelly, 1987; McLennan et al., 1990, 1993; Pearce and Jarvis, 1992; Johnsson, 1993; Jones and Manning, 1994; Garver and Scott, 1995; Fralick and Kronberg, 1997; Thomson et al., 1993; Lopez-Buendıa et al., 1999; Dypvik and Harris, 2001; Ohta, 2004). Postdepositional processes such as weathering and diagenesis may alter the initial composition, so the study of recent unconsolidated sediments can provide an ideal situation to discriminate between grain-size and provenance signals using chemical and mineralogical parameters. In particular, changes in grain size in the composition of sediments and sedimentary rocks occur to the same extent as changes in provenance, weathering, and diagenesis (Bhatia, 1983; Johnsson, 1993), and their discrimination can represent a key step in the interpretation of provenance changes (see for example Ohta, 2004). Sedimentary particles are sorted according to dimension, shape, and specific gravity, creating variations in the chemical and mineralogical composition that can interfere with the signals derived from the source area, as documented, for example, by McLennan et al. (1993) for zircons in turbidites, which are in some cases affected by intense sedimentary sorting. Sorting effects are expected to be even greater in other depositional settings such as river channel deposits or beach sands, as reported for many elements (Zr, Ti, P, Cr, V, Y, Ce) by Garcia et al. (2004). For many of these elements, association with fine-grained sandstone and silt can be expected (e.g., McLennan, 1989; McLennan et al., 1993; Fralick and Kronberg., 1997; Dypvik and Harris, 2001), whereas other elements, such as V, Cr, Co, Ni, Cu, and Zn, are generally enriched in fine-grained sediments (Turekian and Wedephol, 1961; Horowitz, 1991).

The aim of this study was to investigate the relationships among compositional data (geochemistry and mineralogy) and grain size and sorting, with a focus on silt- and clay-sized sediments, even though some sands were included in our sampling. We discuss the significance of some geochemical ratios derived from bulk compositional data, as proxies for grain size, as well as possible combination of several parameters to quantitatively evaluate descriptive parameters of grain size (e.g., median and sorting). We selected for this study late Pleistocene to Holocene unconsolidated alluvial, transitional, and marine deposits in the eastern Po plain, northern Italy, the composition and evolution of which have been described in a previous paper (Amorosi et al., 2002). The study showed that changes in the composition of the sediment are related to different sediment sources, and a secondary aim of this work is to evaluate if the markers of provenance (e.g., Cr/Al_2O_3, Ni/Al, serpentine/sheet silicates) are influenced by grain-size variations.

STUDY AREA

The study area is located in northern Italy, in the eastern part of the Po River plain (Fig. 1), which is the largest alluvial plain in Italy and is drained by the east-flowing Po River. The Po drainage basin covers an area of ~75,000 km², and the river receives tributaries that drain both the Alps to the north and the Apennines to the south. Several minor rivers today flow directly into the Adriatic Sea from the Apennines, but during Pleistocene sea-level lowstand phases, they were tributaries of a major river system that flowed southward into the central Adriatic (Vai and Cantelli, 2004). Recent geological investigations associated with an extensive drilling campaign promoted by the Geological Survey of Regione Emilia-Romagna as part of the geological mapping project of Italy, have provided data for a better characterization of the late Quaternary deposits in the area. Amorosi et al. (1999b, 1999c, 2003, 2004) described in detail the architecture and evolution of the late Pleistocene–Holocene deposits (younger than 125 k.y.). They are characterized by a cyclic alternation of alluvial plain deposits of Pleistocene age overlain by a transgressive-regressive cycle that is made up of littoral to shallow-marine deposits of Holocene age. This cyclic evolution of sedimentary environments is fully exposed in the boreholes in the central Po River plain (cores labeled 187-S1, 204-S8, 205-S3, 205-S5, 205S-10), whereas in those cores closer to the Apennines chain (cores labeled 221-S19, 239-S2, and 240-S9), sedimentation is restricted to alluvial plain deposits.

MATERIALS AND METHODS

A total of 185 samples was collected from ten continuously cored boreholes in the eastern Po plain (Fig. 2), which have already been discussed in other works (Amorosi et al., 2002, 2007). Sampling was intended to characterize the fine-grained portions of the cores; it is irregularly spaced, but it describes the main facies recognized in the cores. The cores were sampled from the center in order to reduce possible contamination due to the coring operations. Slices of ~2 cm were collected and processed, except for some heterogeneous intervals that required restricted thickness. Chemical analyses on all the samples were obtained by X-ray fluorescence spectrometry (Philips PW 1480) on pressed powder pellets with methods described by Amorosi et al. (2002). The estimated precision and accuracy for trace-element determinations were better than 5%, except for those elements at 10 ppm and lower (10%–15%). Semiquantitative mineralogical data were obtained by X-ray diffraction (XRD); details of the method are in Amorosi et al. (2002).

For the grain-size analyses, 0.5 g of sample were dispersed in 50 mL of deionized water and 5 mL of H_2O_2 (30%) and shaken until effervescence stopped. Particle-size analyses were performed using a Malvern Master Sizer/E granulometer equipped with a monochromatic beam of helium-neon laser (λ = 633 nm) and an optical design that enabled a measurement

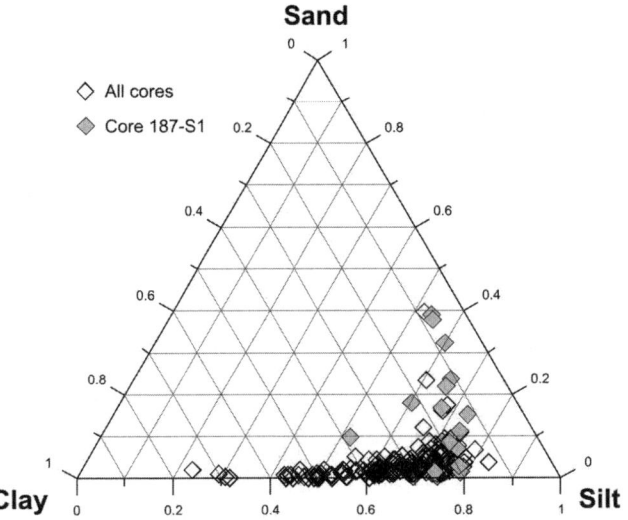

Figure 2. Ternary plot of the grain-size distribution. Samples from core 187-S1 are indicated by gray diamonds. Silt is dominant in the majority of the samples.

range between 0.1 and 600 µm. To evaluate the possible effect of calcite, 20 samples with variable carbonate content were reacted with diluted HCl (7%) until effervescence stopped, and then they were analyzed with the same analytical conditions. The use of cold diluted HCl removed mainly the calcite from the sample and did not affect dolomite.

In the statistical elaborations, samples with sand content >10% were not included, as well as all the samples from core 187-S1. The grain-size data from core 187-S1 will be presented in the following section and are used as a test for the proposed model.

Mineralogical, geochemical, and grain-size data for each sample are not reported here, but are available upon request from the authors.

RESULTS AND DISCUSSION

Grain-Size Data and Carbonate Effect

The investigated samples were for the most part mixtures of silt (63–4 µm) and clay (<4 µm) particles (Fig. 2). Only 4 of the samples had sand fraction (63 µm–2mm) > 10%, whereas 17 of the samples had a clay fraction >50%; this combination was present in every borehole but was more frequent in core 204-S1 (Fig. 3). Samples from core 187-S1 were mostly silts and also included many sands (7 samples out of 14).

The descriptive parameters of the grain-size distribution (median and sorting) are presented in Figure 3. Most of the

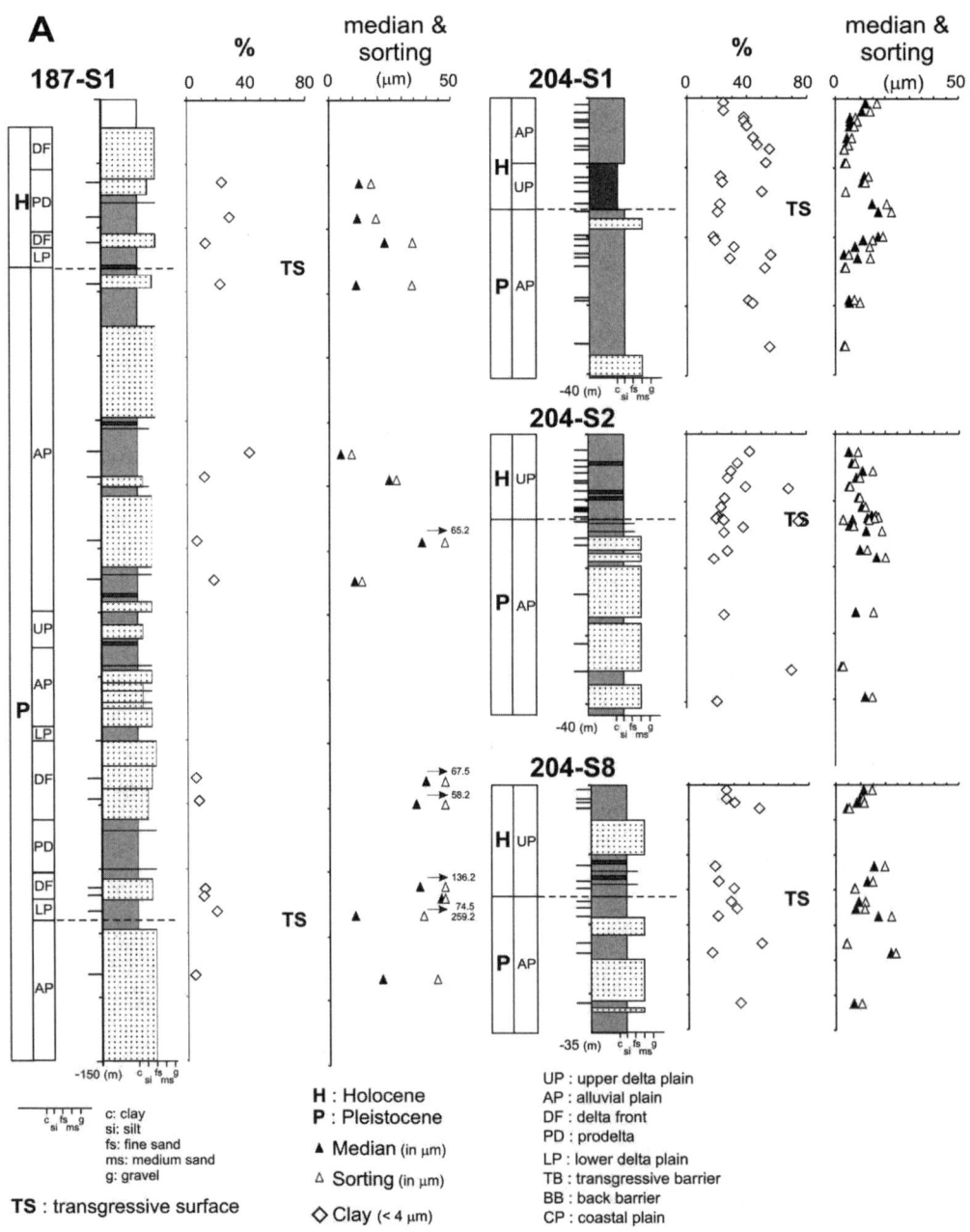

Figure 3 (A and B). Down-core variations of the percentage of clay and of the median and sorting in the studied cores. The field lithology is reported in pictorial form and deals with grain-size results from macroscopic observations.

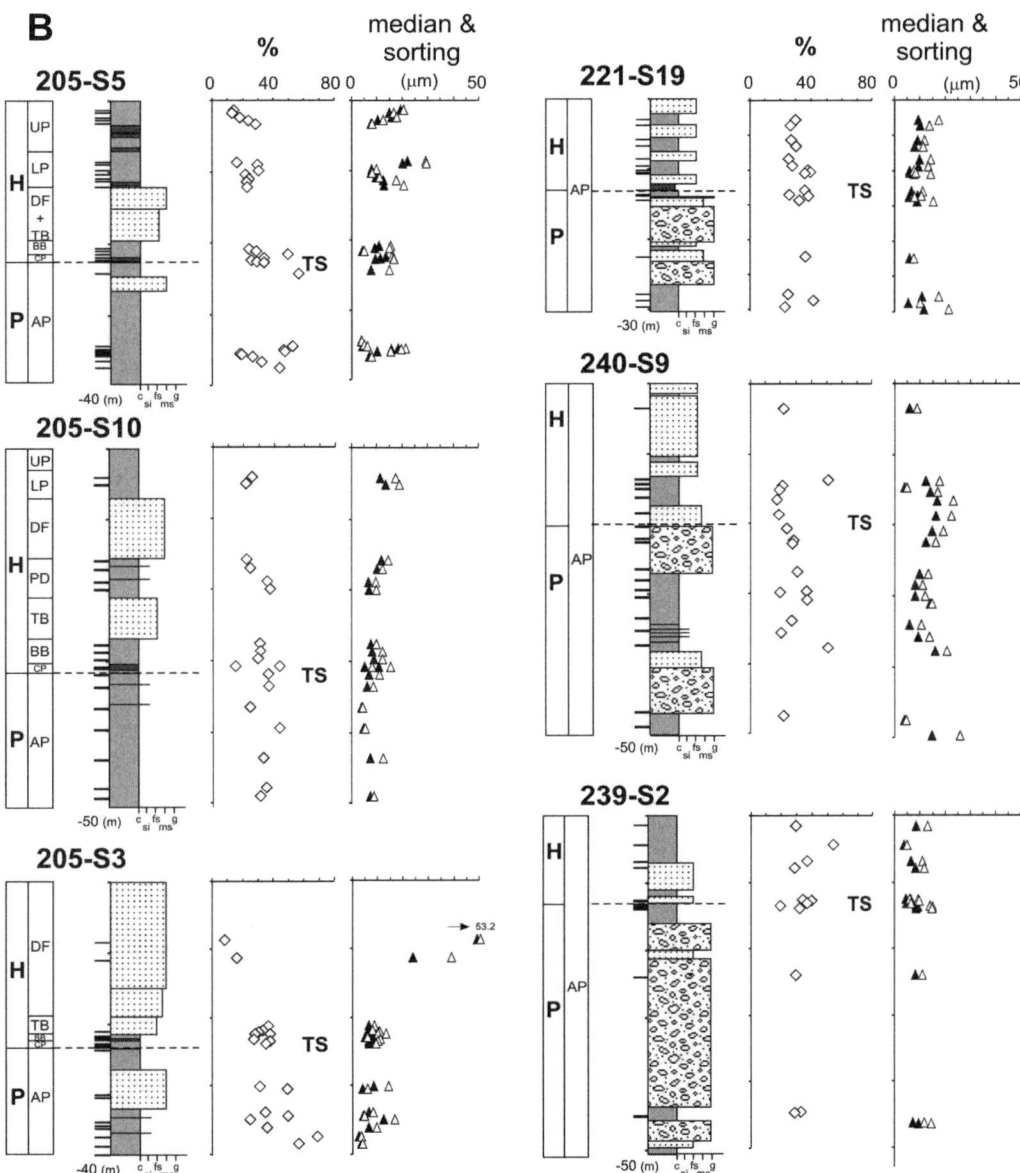

Figure 3B (*continued*).

median values range between 4 and 16 μm (83% of the data), and only 9% percent of the samples have finer median values, and 8% have median values >16 μm. Sorting, evaluated following Inman (1952) as (16 percentile − 84 percentile)/2, directly follows the median values due to the occurrence of variable amounts of coarse-grained sediments that create wider dispersion in the grain-size distribution. There is no systematic association of these parameters with facies changes, which depends on the sampling strategy. However coarse-grained intervals occur as channel sands in the alluvial plain deposits and gravels in the cores closer to the Apennines (e.g., cores 221-S19, 240-S9, 239-S2 in Fig. 3). Other coarse-grained deposits are the beach sands in the transgressive barrier facies and marine sands in the delta front deposits, which characterize the cores located near the coastline (e.g., cores 205-S3, 205-S5, 205-S10 in Fig. 3). Part of these coarse-grained intervals were considered in core 187-S1 and are characterized by high median values and lower sorting.

Calcite is common in all the analyzed samples (see data in Amorosi et al., 2002), and it is particularly abundant in the cores close to the Apennines (221-S19, 240-S9, and 239-S2). Its removal produced contrasting results: in samples with low calcite content (~20%), the acid attack did not produce macroscopic variation of the grain-size curve (Figs. 4A and 4B); other samples, with higher calcite abundance (up to 43%), showed more dramatic changes in size distribution (Figs. 4C–4F). All these samples, from core 239-S2 and characterized by high carbonate

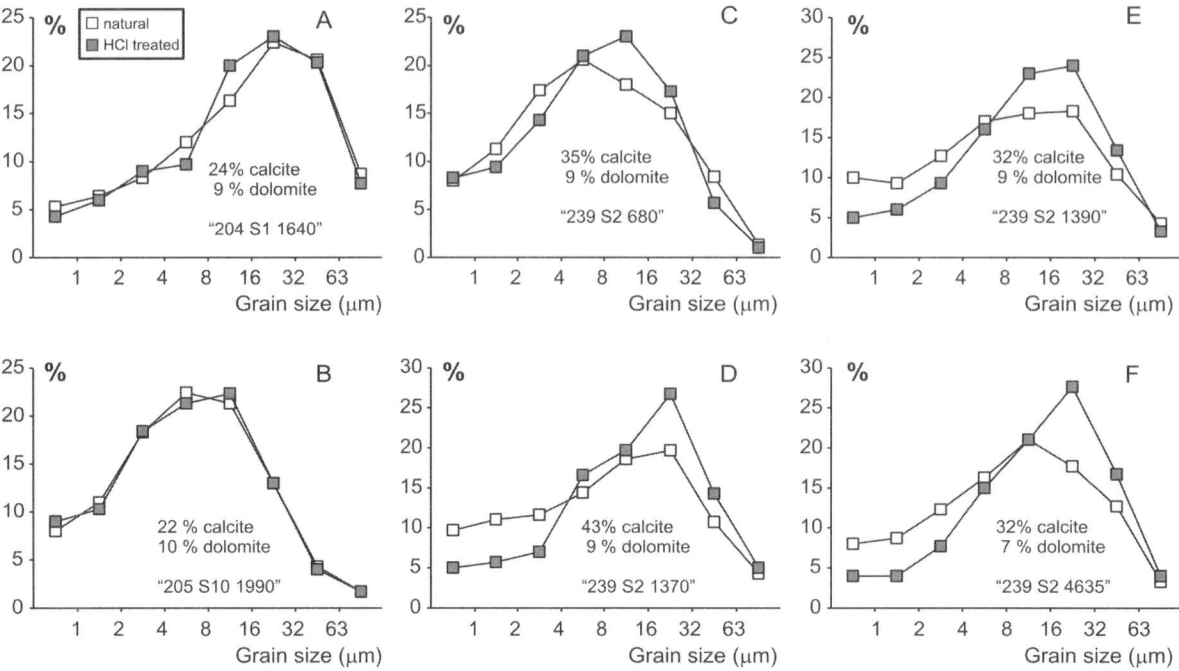

Figure 4. Effects of the HCl treatment on selected samples on the grain-size distribution. In cases C to F, a slight change in the shape of the distribution with increase in mode value and decrease in the fine-grain-size tail is apparent.

contents, showed a clear depletion of the finer grain sizes after carbonate removal.

Actually, the role of carbonate is twofold: it can be distributed randomly through all the grain sizes (leaving no effects after HCl digestion), and it can be concentrated in the clay fraction. The latter behavior, which could be surprising in cemented rock, is rather obvious for unconsolidated sediments. In our case, most of the calcite was detritic, originating from the sedimentary formations of the Apennines (Cavazza et al., 1993; Dinelli and Lucchini, 1998, 1999; Amorosi et al., 2002), and could be abundant in fine-grained fractions as a result of mechanical erosion and fragmentation during fluvial transport.

Grain-Size Relationships with Chemical and Mineralogical Data

Although better discrimination of the relationships between grain size and element distribution could be achieved through the analysis of separate grain-size fractions (Horowitz and Elrick, 1987; Weltje and Prins, 2003; Whitmore et al., 2004), the comparison of bulk compositional data with grain size has been applied. In the analysis of the trace-element distribution in sediments, both grain-size class abundance and statistical descriptive parameters of the grain-size distribution were considered (Horowitz, 1991; Zhao et al., 1999; Zhang et al. 2002; Viscosi-Shirley et al., 2003). We used these approaches to evaluate the possible effects of grain size on the distribution of minerals and chemical elements.

A correlation analysis was carried out on the whole data set, and correlation coefficients were calculated among element concentrations, mineral abundances, grain size classes (sand, coarse and fine silt, and clay), and descriptive parameters of the grain-size distribution (Table 1).

For the large number of samples ($n = 166$), correlations are significant at the 0.01 confidence level at absolute values of 0.22. Among the variables positively and significantly correlated with the clay fraction, although their correlation coefficients do not reach absolute high values, there are Fe_2O_3, V, Cu, Zn, Rb, La, Ce, interstratified illite-smectite and kaolinite. Only V among the chemical elements, and interstratified illite-smectite and kaolinite among the mineralogical parameters have lower scattering. Also positively correlated to the clay fraction are Al_2O_3, K_2O, TiO_2, Co, Nb, and Th. These associations are expected, relating the more abundant sheet silicates in the <4 μm fraction to the chemical elements that usually concentrate in shales.

Amphiboles K- and Ca-Na micas, and Zr, Na_2O, and P_2O_5 are positively correlated to the silt fraction, in particular to a coarse-silt fraction of the sediment (63–16 μm). The association of micas with silt agrees with the presence of coarse crystals within the sediment.

The association of Zr, P_2O_5, and Y is interpreted to be related to the occurrence of heavy minerals (zircon, garnet, monazite, apatite, hornblende) in fine-grained sediments, a feature which has been observed in other studies (Fralick and Kronberg, 1997; Dypvik and Harris, 2001; Garcia et al., 2004). These heavy

TABLE 1. PEARSON CORRELATION COEFFICIENTS BETWEEN COMPOSITION (CHEMISTRY AND MINERALOGY) AND GRAIN-SIZE CLASSES AND GRAIN-SIZE PARAMETERS

	Sand	C. Silt	F. Silt	Clay	Median	Sorting
SiO_2	**0.297**	**0.241**	–0.141	**–0.224**	**0.379**	**0.279**
TiO_2	0.009	–0.021	0.005	0.026	0.042	–0.052
Al_2O_3	–0.067	**–0.204**	–0.022	**0.245**	–0.069	–0.182
Fe_2O_3	**–0.317**	**–0.403**	0.186	**0.389**	**–0.395**	**–0.471**
MnO	–0.001	**0.179**	0.028	**–0.205**	0.032	0.100
MgO	–0.077	0.042	**0.175**	–0.119	0.054	–0.003
CaO	–0.075	0.018	0.021	–0.024	–0.135	–0.013
Na_2O	**0.252**	**0.396**	–0.091	**–0.414**	**0.467**	**0.393**
K_2O	–0.155	**–0.286**	0.019	**0.320**	**–0.207**	**–0.284**
P_2O_5	**0.367**	**0.375**	**–0.297**	**–0.301**	**0.359**	**0.342**
LOI	**–0.283**	**–0.248**	0.145	**0.226**	**–0.35**	**–0.259**
V	**–0.397**	**–0.527**	0.147	**0.551**	**–0.453**	**–0.547**
Cr	–0.163	–0.12	0.152	0.088	–0.061	–0.185
Co	**–0.26**	**–0.311**	0.188	**0.283**	**–0.227**	**–0.337**
Ni	**–0.200**	–0.095	**0.211**	0.036	–0.054	–0.179
Cu	**–0.292**	**–0.327**	0.067	**0.364**	**–0.39**	**–0.395**
Zn	**–0.343**	**–0.415**	0.141	**0.427**	**–0.402**	**–0.495**
Rb	**–0.286**	**–0.399**	0.098	**0.424**	**–0.348**	**–0.443**
Sr	–0.193	–0.124	0.124	0.093	**–0.252**	–0.172
Y	0.175	**0.236**	–0.167	–0.188	**0.331**	0.202
Zr	**0.502**	**0.529**	**–0.384**	**–0.442**	**0.628**	**0.573**
Nb	–0.080	–0.095	0.086	0.077	–0.102	–0.168
Ba	**0.201**	**0.208**	–0.195	–0.146	**0.216**	**0.222**
La	**–0.256**	**–0.305**	–0.053	**0.393**	**–0.296**	**–0.347**
Ce	–0.079	**–0.213**	–0.055	**0.273**	–0.158	**–0.235**
Pb	0.041	0.021	0.044	–0.046	0.128	0.043
Th	–0.094	–0.08	–0.006	0.107	–0.088	–0.117
S	0.054	–0.04	–0.05	0.057	–0.029	–0.015
Amp	**0.368**	**0.562**	**–0.3**	**–0.505**	**0.600**	**0.519**
Cal	**–0.271**	**–0.348**	0.143	0.178	**–0.344**	**–0.270**
Dol	0.093	0.131	0.072	–0.197	0.145	–0.187
Feld	**0.349**	**0.31**	**–0.304**	**–0.236**	**0.352**	**0.353**
Py	–0.082	–0.179	0.054	0.180	–0.127	–0.139
Qz	**0.360**	**0.233**	**–0.321**	–0.144	**0.315**	**0.303**
Chl	–0.201	–0.128	0.116	0.12	–0.081	–0.216
IS	**–0.323**	**–0.582**	–0.084	**0.72**	**–0.532**	**–0.556**
Kao	**–0.47**	**–0.703**	0.193	**0.734**	**–0.625**	**–0.680**
K-mica	**0.287**	**0.481**	–0.116	**–0.499**	**0.431**	**0.430**
Brittle mica	0.148	**0.346**	–0.076	**–0.348**	**0.363**	**0.308**
Serp	–0.200	–0.111	0.16	0.085	–0.069	–0.204
Talc	0.013	0.077	–0.026	–0.057	0.080	0.011

Note: Correlations significant at 0.01 are in bold. Abbreviations: Amp—amphiboles; Cal—calcite; Dol—dolomite; Feld—feldspars; Py—pyrite; Qz—quartz; Chl—chlorite; IS—interstratified illite-smectite; Kao—kaolinite; Serp—serpentine; LOI—loss on ignition. Sand: >63 μm; C. silt: Coarse silt 16–63 μm; F. silt: Fine silt 4–16 μm; Clay: <4 μm.

minerals were not revealed by bulk XRD but are common in the heavy mineral fraction of alluvial sediments of the area (Gazzi et al., 1973; Gandolfi et al., 1982; Marchesini et al., 2000).

Quartz, feldspar, amphibole, and K-mica, and SiO_2 and Na_2O positively correlate to the sand fraction of the samples.

The parameters descriptive of the grain size (median and sorting) positively correlate to the compositional variables associated with the coarse fractions of the sediment, with particular high correlation with Zr and Na_2O.

The results of these analyses open the possibility to propose and test geochemical and mineralogical proxies for bulk grain size of the sample that will be discussed in the following section. In particular:

- Na_2O, SiO_2, and amphiboles are representative of sands;
- amphiboles, Zr, K-mica, Na- and Ca-mica are representative of coarse silts;
- Zn, Fe_2O_3, Cu, Co, and V are representative of fine silts; and
- kaolinite, interstratified illite-smectite, V, Zn, Rb, and Fe_2O_3 are representative of clays.

Elements such as Cr and Ni did not correlate to any of the grain-size fractions or grain-size parameters (Table 1). It has been suggested that these elements were useful in the interpretation of sediment provenance in the area (Amorosi et al., 1999a, 2002) and clearly discriminate two populations in binary diagrams with Al_2O_3 (e.g., Figs. 5A and 5B). One population is influenced by sediments transported by the Po River (the black symbols in Fig. 5), and the other is mostly composed of sediment derived from the Apennines (open symbols in Fig. 5). These two populations cannot be clearly distinguished when other elements, such as V (Fig. 5C), are considered. Actually, if the data are subdivided according to the provenance attribution, the correlations became positive and significant (Figs. 5D and 5E), indicating that Cr and Ni are also controlled by the abundance of fine-grained material. Cr and Ni are controlled by the abundance of typical ophiolite-derived minerals, such as serpentine (Cr-serpentine correlation coefficient: 0.862; Ni-serpentine correlation coefficient: 0.856), or by the presence of ultramafic rock fragments.

Proxies for Grain Size

The results presented in the previous section show that there is strong partitioning of certain elements and minerals into different grain-size fractions, and so it is possible to derive several geochemical and mineralogical values that are useful proxies for grain size. We present some mineralogical proxies, but we are aware that published quantitative mineralogical data on fine-grained sediments are sporadic, even if the abundance of clay minerals by itself is a good indicator of the grain size of the sample. We concentrate the discussion mostly on geochemical proxies, based on both major and trace elements, which basically compare two or more elements that might have contrasting distribution compared to grain size. Some of the proxies have already been proposed and discussed in the literature (Argast and Donnelly, 1987; Herron, 1988; McLennan et al., 1993; Dypvik and Harris, 2001; Ohta, 2004); others have been derived from the results presented in Table 1, and their significance is evaluated here (e.g., Y/V, Zr/V, Y/Rb). Moreover, those ratios that were particularly effective in the discrimination of sediment provenance (Cr/Al, Ni/Al, Mg/Al) were tested to evaluate if their variation can be influenced by changes in grain size.

Some ratios significantly correlate with the sand content, such as SiO_2/Al_2O_3 and SiO_2/Fe_2O_3, and to a lower degree, Na_2O/Al_2O_3, $(Na_2O + K_2O)/Al_2O_3$, $(Zr + Ba)/Rb$, and Ba/Rb (Table 2). The ratios involving major elements reflect the presence of quartz (SiO_2) and feldspars (Na_2O and K_2O) compared to Al_2O_3, which is expected to be mostly controlled by the fine-grained sheet silicates. K_2O can be present also in different types of sheet silicates, so its significance cannot be precisely defined. Trace-elements ratios are less correlated to the sand content, although the ratios

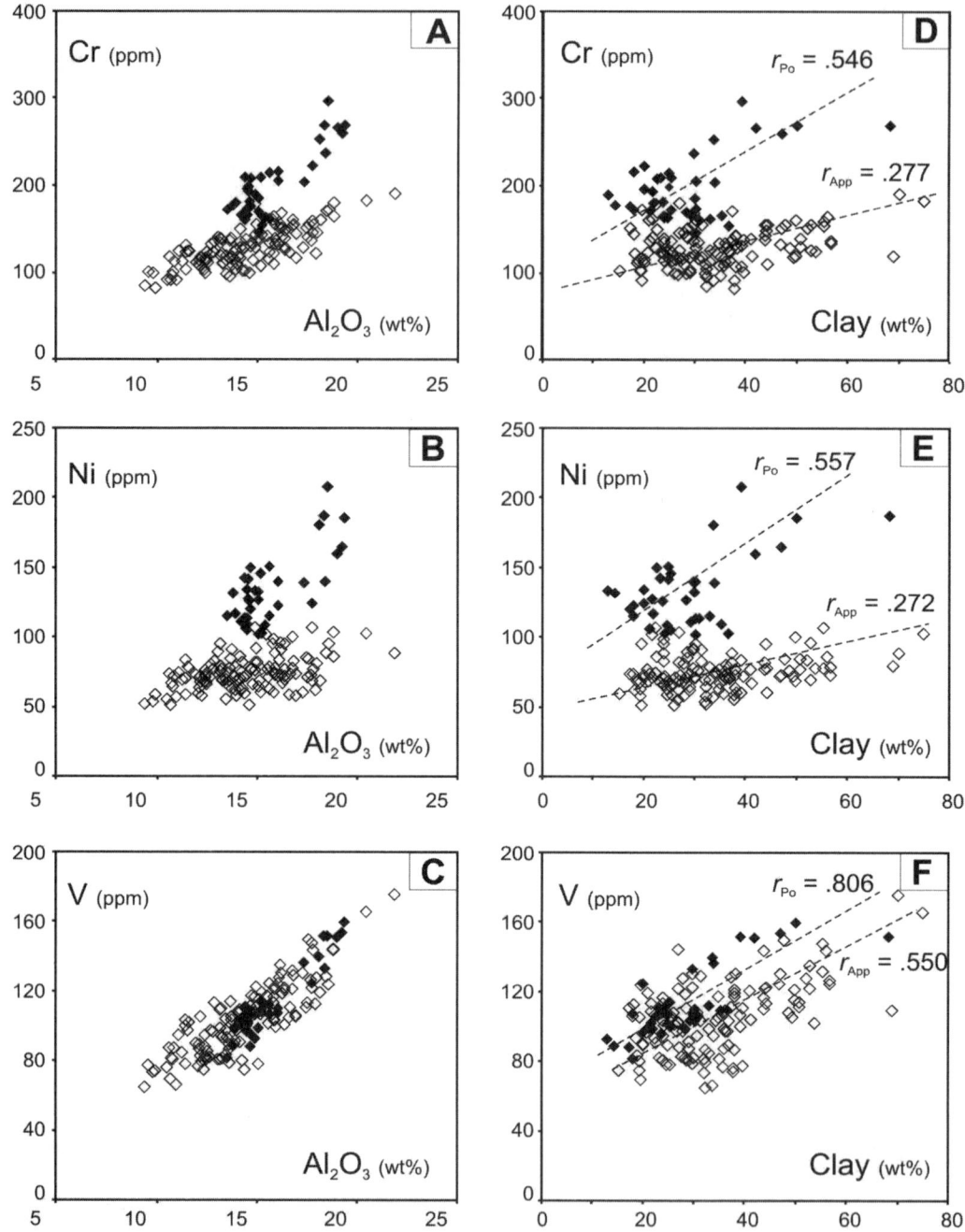

Figure 5. Plots of Cr, Ni, and V versus Al_2O_3 (A, B, C, respectively) and versus clay (D, E, F), Two populations with different provenance (open diamonds: Apennines provenance; filled diamond: Po River–influenced provenance) can be discriminated in A, B, D, and E, whereas the two populations are not clearly distinguished in C and F. It appears that Cr and Ni are better correlated to the clay content in the Po River–influenced sediments compared to the other source.

involving Ba, particularly Ba/Rb, seem to correlate to the sand content. Ba can substitute for K in feldspars and might be representative of a relatively coarse fraction, whereas Rb is mostly concentrated in the clay fraction, substituting for K in illites (Dypvik and Harris, 2001). This location of Rb is consistent with its covariance with interstratified illite-smectite ($r = 0.666$), which represents the smallest mineral phase. These proxies, particularly those involving Na and K, but also Ba and Rb, might be sensitive to the degree of feldspar weathering that might cause remobilization of the elements and possibly concentration in the residual product. In this case, feldspars are common in the sediment, suggesting that weathering is not so strong and the proxies might be

TABLE 2. PEARSON CORRELATION COEFFICIENTS BETWEEN GEOCHEMICAL AND MINERALOGICAL INDEXES AND GRAIN-SIZE CLASSES AND DESCRIPTIVE PARAMETERS

	Sand	C. silt	F. silt	Clay	Median	Sorting
SiO_2/Al_2O_3	**0.608**	**0.700**	−0.245	**−0.714**	**0.715**	**0.734**
SiO_2/Fe_2O_3	0.576	0.551	−0.314	−0.516	**0.688**	**0.654**
Ti/Al	0.109	0.281	0.010	−0.319	0.145	0.191
Na/Al	0.245	0.397	−0.109	−0.406	0.440	0.395
(Na + K)/Al	0.102	0.217	−0.044	−0.229	0.219	0.195
Zr/Rb	**0.673**	**0.714**	−0.473	**−0.623**	**0.791**	**0.806**
Zr/V	**0.690**	**0.723**	−0.478	**−0.633**	**0.794**	**0.798**
(Zr + Ba)/Rb	0.574	**0.633**	−0.418	−0.551	**0.628**	**0.710**
Ti/V	0.585	**0.683**	−0.290	**−0.670**	**0.662**	**0.698**
Y/Al	0.349	0.581	−0.254	−0.540	0.581	0.508
Y/Rb	0.542	**0.695**	−0.361	**−0.638**	**0.766**	**0.727**
Y/V	0.584	**0.726**	−0.376	**−0.671**	**0.780**	**0.731**
Ba/Al	0.246	0.348	−0.219	−0.298	0.222	0.342
Ba/Rb	0.422	0.488	−0.321	−0.422	0.435	0.543
Ce/Rb	0.278	0.243	−0.205	−0.198	0.236	0.280
Cr/V	0.179	0.356	0.012	−0.402	0.373	0.298
Cr/Ba	−0.23	−0.197	0.238	0.132	−0.147	−0.258
Cr/Al	−0.154	0.018	0.201	−0.086	−0.010	−0.093
Ni/Al	−0.171	0.030	0.229	−0.111	0.006	−0.088
Mg/Al	−0.039	0.122	0.118	−0.187	0.047	0.080
Qz/IS	**0.760**	**0.748**	0.595	**0.686**	−0.384	**−0.640**
(Qz + Fd)/IS	0.593	0.597	−0.431	−0.522	**0.632**	**0.664**
Fd/IS	0.569	**0.686**	−0.382	**−0.636**	**0.736**	**0.726**
Fd/Kao	0.428	0.445	−0.362	−0.365	0.424	0.495

Note: Correlations significant at 0.01 are underlined, in bold correlations with $r > 0.6$. Abbreviations: Qz—quartz; Fd—feldspar; IS—interstratified illite-smectite; Kao—kaolinite. Sand: >63 μm; C. silt: Coarse silt 16–63 μm; F. silt: Fine silt 4–16 μm; Clay: <4 μm.

valid. The limited number of sandy samples prevents confident extrapolation of these proxies to our data set; however, many of them (SiO_2/Al_2O_3, SiO_2/Fe_2O_3, and Na_2O/Al_2O) are reported in the literature to be high in sands.

SiO_2/Al_2O_3 and SiO_2/Fe_2O_3 are positively correlated with median and sorting, whereas the other ratios involving major elements have lower significance.

Those ratios involving Zr and Y are representative of the coarse fraction (coarse silt) of fine-grained sediments (Table 2), particularly when normalized to elements associated to a fine-grained fraction (e.g., Al, Rb, V). Ratios such as Zr/Rb, Zr/V, Zr/Al, Y/Al, Y/Rb, and Y/V correlate to high degree with the coarse silt fraction (Table 2) and reflect the occurrence of heavy minerals (zircon and garnet) in the sediments (Gazzi et al., 1973; Gandolfi et al., 1982; Marchesini et al., 2000). A similar explanation has been given for Zr/Rb by Dypvik and Harris (2001), but other ratios such as Zr/V, Y/Rb, and Y/V have the same significance and display slightly higher correlation coefficients with the coarse silt fraction. A slight difference is the higher coefficients of Zr/Rb and Zr/V to the sand fraction compared to the ratios involving Y. Ti is reported to be enriched in the coarse fraction of sediments in Aeolian sediments (e.g., Pye, 1987), in coarse sections of turbidites (e.g., Wehausen and Brumsack, 1999), or in extremely sorted clastic sediments (Garcia et al., 2004), so the ratio Ti/V is another good indicator of a coarse silt fraction of the sediment. Also some mineralogical indexes correlate to this fraction: amphibole/(interstratified illite-smectite), feldspar/(interstratified illite-smectite), and quartz/(interstratified illite-smectite). The first is controlled by the occurrence and abundance of a heavy mineral, whereas the others suggest that feldspars and quartz are preferentially concentrated into a coarse silt fraction.

Silt is the more important grain-size class in these samples, so all these ratios are important proxies. Zr/V, Zr/Rb, Y/V, and Y/Rb display large correlation coefficients ($r > 0.750$; Table 2) with median grain size, which is mostly in the silt range. The same ratio are also highly correlated with sorting of the sediment (Table 2).

Possible problems in the use of these ratios as grain-size proxies arise if the concentrations of the elements are altered by particular paleoenvironmental conditions, the occurrence of particular rock types, or if sorting of heavy minerals is particularly effective (Garcia et al., 2004). Under anoxic conditions and in sediments rich in organic matter (black shales and sapropels), V and Fe are significantly enriched (Vine and Tourtelot, 1970; Nijenhuis et al., 1999); furthermore, V has been used, in combination with Cr and Ni, as a paleoenvironmental indicator for redox state (Hatch and Leventhal, 1992; Jones and Manning, 1994). Fe concentrations could be influenced in reductive environments by precipitation of iron sulfides or carbonate in association with sediments rich in organic matter, but can be concentrated also in oxidizing conditions due to diagenetic remobilization (Froelich et al., 1979). Occurrence of alkaline volcanic rocks in the source area might alter the concentration on Zr, Y, and to a lesser degree K and Rb (Best and Christiansen, 2001), basaltic rocks might influence Fe, Ti, and V (Turekian and Wedephol, 1961), and ultramafic rocks can influence the distribution of Cr, Ni, and Mg, as already noted in the previous paragraph.

In the studied boreholes, Ni/Al, Cr/Al, and Mg/Al are the most significant provenance indicators, but as indicated in Table 2 and shown in detail in Figure 6, they do not show any correlation with the grain-size fraction or grain-size parameter. The negative correlation in the Mg/Al versus clay (Fig. 6C) shown by samples influenced by the Po River is related to the occurrence of detritic dolomite in samples poor in clay, which are more frequent in many of the cores close to the coastline that have a mixed provenance (Amorosi et al., 2002).

The identification of close relations between some geochemical indexes and grain-size classes and parameters describing the grain-size distribution, such as median and sorting, opens the possibility of using some geochemical indexes as direct proxy for grain size (Table 2). Cautions to the use a direct relation between a geochemical index and the median have been posed by Weltje and Prins (2003) because the grain-size distribution might be complex, and the median grain size might not fully represent the complete grain-size distribution observed. Sorting can help in the description of the sample heterogeneity and thus may account for complex grain-size distribution.

In order to further constrain the relation between median and sorting and geochemical indexes, we applied a stepwise multiple regression analysis to the data set, using the geochemical ratios as independent variables and median and sorting for dependent

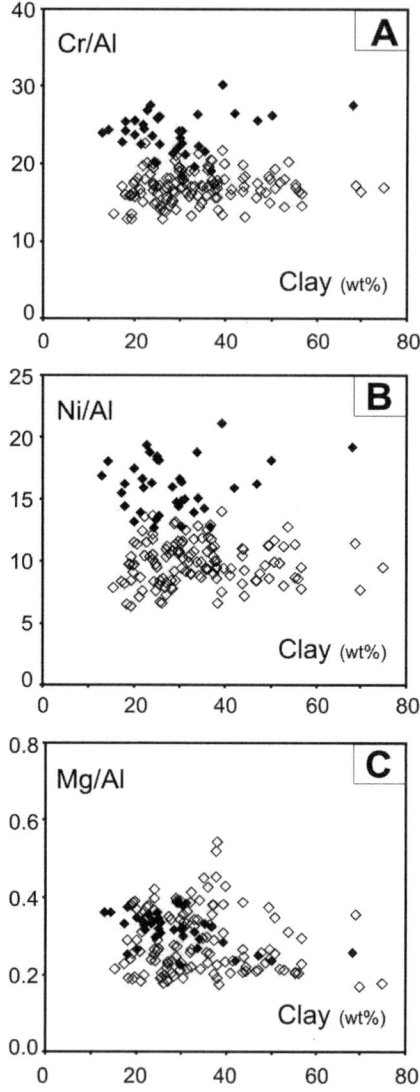

Figure 6. Plots of Cr/Al (A), Ni/Al (B) and Mg/Al (C) versus clay. These are the more effective geochemical indicators of provenance in the area (see Amorosi et al., 2002) that are not affected by changes in the clay content of the sediment.

variables. In the model, all the ratios presented in Table 2 were initially introduced, including those more sensitive to provenance changes, which, however, did not increase the variance. The two resulting equations for median and sorting are:

$$\text{Median} = -13.446 + \text{Zr/V} \times 1.555 + \text{Na/Al} \times 27.767 + \text{Ti/V} \times 0.221 + \text{SiO}_2/\text{Fe}_2\text{O}_3 \times 0.675 + \text{Y/Rb} \times 22.603 \quad (1)$$

and

$$\text{Sorting} = -14.85 + \text{Zr/V} \times 2.785 + \text{Na/Al} \times 37.271 + \text{Ba/Rb} \times 1.355 + \text{SiO}_2/\text{Fe}_2\text{O}_3 \times 0.638 + \text{Ti/V} \times 0.312, \quad (2)$$

and they account for 84 and 80% of the total variance, respectively, and provide good estimates of median and sorting as verified by the comparison of calculated and measured parameters for samples of core 187-S1. For fine-grained samples (median < 30 μm), Equation 1 works well (filled circles in Fig. 7A), whereas for a larger median value, the model fails because it is not constrained. The same applies also to sorting: Equation 2 is calculated using samples with sorting values (expressed in μm) <30 μm, whereas in many samples of 187-S1, far larger values are observed.

We are aware that this approach is at the moment qualitative, and more data, particularly on coarse-grained sediments, are needed to provide more reliable estimates, but the data presented suggest that at least for fine-grained sediments (silts and clays) some geochemical ratios might be an effective and consistent

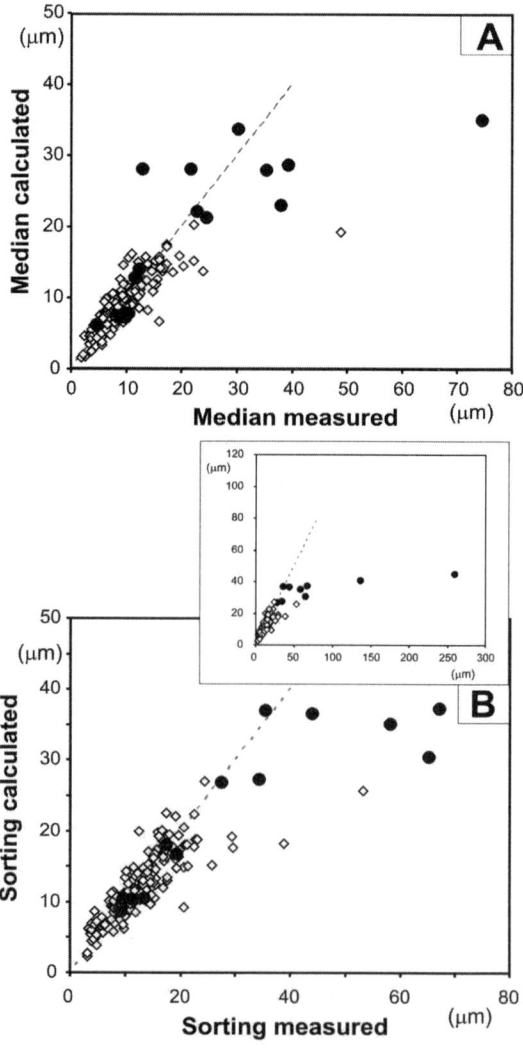

Figure 7. Comparison of measured versus calculated median (A) and sorting (B). Open diamonds are the samples used in the multiple regression analyses, including outliers; filled circles are samples from core 187-S1 used for comparison. The correspondence is good for median and sorting <30 μm, whereas for coarser and less-sorted samples, the model is inadequate.

proxy for the bulk grain size. In general, these results could provide useful insight into grain-size variations, particularly when working on siltstones and shales, for which lithification makes direct grain-size analyses difficult and petrographic tools can hardly help.

CONCLUSION

Effects of different grain-size distribution on mineralogical and geochemical parameters can be better analyzed in recent unconsolidated sediments that have not yet suffered postdepositional transformations related to diagenesis, compaction, and lithification. The comparison of bulk chemical and mineralogical composition with different grain-size classes in silts and clays has allowed the correlations between compositional and textural parameters to be outlined. The effectiveness of some mineralogical and geochemical ratios in describing the grain-size changes has been evaluated. Some indexes, such as quartz/interstratified illite-smectite; feldspar/interstratified illite-smectite; and SiO_2/Al_2O_3, Na/Al, Zr/Rb, Zr/V, Y/Rb, Y/V, and Ti/V are more strongly related to a coarse silt fraction and have higher values in more coarse-grained samples. Multiple regression analysis combining median and sorting with the geochemical ratios was applied and tested with a subset of samples from an auxiliary core and proved to be particularly effective in the quantification. The model is limited to silt and clays, and it does not work well for more coarse-grained samples, for which further work is necessary. The indications derived from the combined evaluation of these proxies can provide important clues in the evaluation of textural features of the sediment when direct analysis of grain size is not available and might be extremely useful in the analysis of siltstones and shales, when application of other techniques, such as direct grain-size analysis or optical evaluation, are difficult.

When sediment provenance is considered, it is important to assess the effect of grain-size sorting in the key parameters used in the reconstruction. In the study area, several chemical and mineralogical indexes of provenance of the sediment (Ni/Al, Cr/Al, Mg/Al; serpentine/silicate) are not obscured or affected by grain-size variations, and, for this reason, they can be used independently of the texture of the sample considered. Only Mg/Al value displays a correlation with the abundance of a coarse silt fraction, suggesting that dolomite is mostly concentrated in this grain size.

ACKNOWLEDGMENTS

We would like to acknowledge the editorial handling of Jose Arribas, and the useful comments of D. Garcia and M.R. Hounslow, which consistently improved the paper.

REFERENCES CITED

Amorosi, A., Centineo, M.C., Dinelli, E., Lucchini, F., and Tateo, F., 1999a, Provenance changes across the Pleistocene/Holocene boundary in the south-eastern Po-plain, in Ármannsson, H., ed., Geochemistry of the Earth Surface (GES-5, Reykiavik, 16–20 August 1999): Rotterdam, Balkema, p. 23–26.

Amorosi, A., Colalongo, M.L., Fusco, F., Pasini, G., and Fiorini, F., 1999b, Glacio-eustatic control of continental-shallow marine cyclicity from late Quaternary deposits of the south-eastern Po plain (northern Italy): Quaternary Research, v. 52, p. 1–13, doi: 10.1006/qres.1999.2049.

Amorosi, A., Colalongo, M.L., Pasini, G., and Preti, D., 1999c, Sedimentary response to late Quaternary sea-level changes in the Romagna coastal plain (northern Italy): Sedimentology, v. 46, p. 99–121, doi: 10.1046/j.1365-3091.1999.00205.x.

Amorosi, A., Centineo, M.C., Dinelli, E., Lucchini, F., and Tateo, F., 2002, Geochemical and mineralogical variations as indicators of provenance changes in late Quaternary deposits of SE Po plain: Sedimentary Geology, v. 151, p. 273–292, doi: 10.1016/S0037-0738(01)00261-5.

Amorosi, A., Centineo, M.C., Colalongo, M.L., Pasini, G., Sarti, G., and Vaiani, S.C., 2003, Facies architecture and latest Pleistocene–Holocene depositional history of the Po delta (Comacchio area), Italy: The Journal of Geology, v. 111, p. 39–56, doi: 10.1086/344577.

Amorosi, A., Colalongo, M.L., Fiorini, F., Fusco, F., Pasini, G., Vaiani, S.C., and Sarti, G., 2004, Palaeogeographic and palaeoclimatic evolution of the Po plain from 150-ky core records: Global and Planetary Change, v. 40, p. 55–78, doi: 10.1016/S0921-8181(03)00098-5.

Amorosi, A., Colalongo, M.L., Dinelli, E., Lucchini, F., and Vaiani, S.C., 2007, Cyclic variations in sediment provenance from late Pleistocene deposits of eastern Po plain, Italy, in Arribas, J., Critelli, S., and Johnsson, M.J., eds., Sedimentary Provenance and Petrogenesis: Perspectives from Petrography and Geochemistry: Geological Society of America Special Paper 420, doi:10.1130/2007.2420(02).

Argast, S., and Donnelly, T.W., 1987, The chemical discrimination of clastic sedimentary components: Journal of Sedimentary Petrology, v. 57, p. 813–823.

Best, M.G., and Christiansen, E.H., 2001, Igneous Petrology: Malden, Massachusetts, Blackwell Science, 458 p.

Bhatia, M.R., 1983, Plate tectonics and geochemical compositions of sandstones: The Journal of Geology, v. 91, p. 611–627.

Bhatia, M.R., and Cook, K.A.W., 1986, Trace element characteristics of graywackes and tectonic setting discrimination of sedimentary basins: Contributions to Mineralogy and Petrology, v. 92, p. 181–193, doi: 10.1007/BF00375292.

Cavazza, W., Zuffa, G.G., Camporesi, C., and Ferretti, C., 1993, Sedimentary recycling in a temperate drainage basin (Senio River, north-central Italy): Composition of source rock, soil profiles, and fluvial deposits, in Johnsson, M.J., and Basu, A., eds., Processes Controlling the Composition of Clastic Sediments: Geological Society of America, Special Paper 284, p. 247–261.

Dinelli, E., and Lucchini, F., 1998, Element dispersion patterns in the Reno River valley (northern Italy) evaluated by means of stream sediment geochemistry: Mineralogica Petrografica Acta, v. 41, p. 145–162.

Dinelli, E., and Lucchini, F., 1999, Sediment supply to the Adriatic Sea basin from the Italian rivers: Geochemical features and environmental constraints: Giornale di Geologia, v. 61, p. 121–132.

Dypvik, H., and Harris, N.B., 2001, Geochemical facies analysis of fine-grained siliciclastics using Th/U, Zr/Rb and (Zr+Rb)/Sr ratios: Chemical Geology, v. 181, p. 131–146, doi: 10.1016/S0009-2541(01)00278-9.

Fralick, P.W., and Kronberg, B.I., 1997, Geochemical discrimination of clastic sedimentary rock sources: Sedimentary Geology, v. 113, p. 111–124, doi: 10.1016/S0037-0738(97)00049-3.

Froelich, P.N., Klinkhammer, G.P., Bender, M.L., Luedtke, N.A., Heath, G.R., Cullen, D., Dauphin, P., Hammond, D., Hartman, B., and Maynard, V., 1979, Early oxidation of organic matter in pelagic sediments of the eastern equatorial Atlantic: Suboxic diagenesis: Geochimica et Cosmochimica Acta, v. 43, p. 1075–1090, doi: 10.1016/0016-7037(79)90095-4.

Gandolfi, G., Mordenti, A., and Paganelli, L., 1982, Composition and longshore dispersal of sands from the Po and Adige River since pre-Etruscan age: Journal of Sedimentary Petrology, v. 52, p. 797–805.

Garcia, D., Ravenne, C., Maréchal, B., and Moutte, J., 2004, Geochemical variability induced by entrainment sorting: Quantified signals for provenance analysis: Sedimentary Geology, v. 171, p. 113–128, doi: 10.1016/j.sedgeo.2004.05.013.

Garver, J.I., and Scott, T.J., 1995, Trace elements in shale as indicators of crustal provenance and terrane accretion in the southern Canadian

Cordillera: Geological Society of America Bulletin, v. 107, p. 440–453, doi: 10.1130/0016-7606(1995)107<0440:TEISAI>2.3.CO;2.
Gazzi, P., Zuffa, G.G., Gandolfi, G., and Paganelli, L., 1973, Provenienza e dispersione litoranea delle sabbie delle spiagge Adriatiche fra le foci dell'Isonzo e del Foglia: Inquadramento regionale: Memorie Società Geologica Italiana, v. 12, p. 1–37.
Hatch, J.R., and Leventhal, J.S., 1992, Relationship between inferred redox potential of the depositional environment and geochemistry of the Upper Pennsylvanian (Missourian) Stark Shale Member of the Dennis Limestone, Wabaunsee County, Kansas, USA: Chemical Geology, v. 99, p. 65–82, doi: 10.1016/0009-2541(92)90031-Y.
Herron, M.M., 1988, Geochemical classification of terrigenous sands and shales from core or log data: Journal of Sedimentary Petrology, v. 58, p. 820–829.
Horowitz, A.J., 1991, A primer on Sediment-Trace Element Chemistry (second edition): Chelsea, Michigan, Lewis Publishers Inc., 136 p.
Horowitz, A.J., and Elrick, K.A., 1987, The relation of stream sediment surface area, grain size and composition to trace element chemistry: Applied Geochemistry, v. 2, p. 437–451, doi: 10.1016/0883-2927(87)90027-8.
Inman, D.L., 1952, Measures for describing the size distribution of sediments: Journal of Sedimentary Petrology, v. 22, p. 125–145.
Jarvis, I., and Higgs, N., 1987, Trace-element mobility during early diagenesis in distal turbidites: Late Quaternary of the Madeira Abyssal Plain, N Atlantic, in Weaver, P.P.E., and Thomson J., eds., Geology and Geochemistry of Abyssal Plains: Geological Society [London] Special Publication 31, p. 179–209.
Johnsson, M.J., 1993, The system controlling the composition of clastic sediments, in Johnsson, M.J., and Basu, A., eds., Processes Controlling the Composition of Clastic Sediments: Geological Society of America Special Paper 284, p. 1–19.
Jones, B., and Manning, D.A.C., 1994, Comparison of geochemical indices used for the interpretation of paleoredox conditions in ancient mudstones: Chemical Geology, v. 111, p. 111–129, doi: 10.1016/0009-2541(94)90085-X.
Lopez-Buendía, A.M., Bastida, J., Querol, X., and Whateley, M.K.G., 1999, Geochemical data as indicators of palaeosalinity in coastal organic-rich sediments: Chemical Geology, v. 157, p. 235–254, doi: 10.1016/S0009-2541(98)00207-1.
Marchesini, L., Amorosi, A., Cibin, U., Zuffa, G.G., Spadafora, E., and Preti, D., 2000, Sand composition and sedimentary evolution of a late Quaternary depositional sequence, northwestern Adriatic coast, Italy: Journal of Sedimentary Research, v. 70, p. 829–838.
McLennan, S.M., 1989, Rare earth elements in sedimentary rocks: influence of provenance and sedimentary processes, in Lipin, B.R., and McKay, G.A., eds., Geochemistry and Mineralogy of Rare Earth Elements: Mineralogical Society of America, Review in Mineralogy, v. 21, p. 169–200.
McLennan, S.M., Taylor, S.R., McCulloch, M.T., and Maynard, J.B., 1990, Geochemical and Nd-Sr isotopic composition of deep-sea turbidites: Crustal evolution and plate tectonic associations: Geochimica et Cosmochimica Acta, v. 54, p. 2015–2050, doi: 10.1016/0016-7037(90)90269-Q.
McLennan, S.M., Hemming, S., McDaniel, D.K., and Hanson, G.N., 1993, Geochemical approaches to sedimentation, provenance, and tectonics, in Johnsson, M.J., and Basu, A., eds., Processes Controlling the Composition of Clastic Sediments: Geological Society of America Special Paper 284, p. 21–40.
Nijenhuis, I.A., Bosch, H.-J., Sinninghe Damsté, J.S., Brumsack, H.-J., and De Lange, G.J., 1999, Organic matter and trace element rich sapropels and black shales: A geochemical comparison: Earth and Planetary Science Letters, v. 169, p. 277–290, doi: 10.1016/S0012-821X(99)00083-7.
Ohta, T., 2004, Geochemistry of Jurassic to earliest Cretaceous deposits in the Nagato Basin, SW Japan: Implication of factor analysis to sorting effects and provenance signatures: Sedimentary Geology, v. 171, p. 159–180, doi: 10.1016/j.sedgeo.2004.05.014.
Pearce, T.J., and Jarvis, I., 1992, Applications of geochemical data to modelling sediment dispersal patterns in distal turbidites: Late Quaternary of the Madeira Abyssal Plain: Journal of Sedimentary Petrology, v. 62, p. 1112–1129.
Pye, K., 1987, Aeolian Dust and Dust Deposits: London, Academic Press, 334 p.
Thomson, J., Higgs, N.C., Croudace, I.W., Colley, S., and Hydes, D.J., 1993, Redox zonation of elements at an oxic/post-oxic boundary in deep-sea sediments: Geochimica et Cosmochimica Acta, v. 57, p. 579–595, doi: 10.1016/0016-7037(93)90369-8.
Turekian, K.K., and Wedephol, K.H., 1961, Distribution of the elements in some major units of the Earth's crust: Geological Society of America Bulletin, v. 72, p. 175–192.
Vai, G.B., and Cantelli, L., 2004, Litho-palaeoenvironmental maps of Italy during the last two climatic extremes: Bologna, Climex Maps Italy, Museo Geologico Cappellini, two maps 1:100,000, and 80 p, of explanatory notes.
Vine, J.D., and Tourtelot, E.B., 1970, Geochemistry of black shale deposits—A summary report: Economic Geology and the Bulletin of the Society of Economic Geologists, v. 65, p. 253–272.
Viscosi-Shirley, C., Mammone, K., Pisias, N., and Dymond, J., 2003, Clay mineralogy and multi element chemistry of surface sediments on the Siberian-Arctic shelf: Implications for sediment provenance and grain size sorting: Continental Shelf Research, v. 23, p. 1175–1200, doi: 10.1016/S0278-4343(03)00091-8.
Wehausen, R., and Brumsack, H.J., 1999, Cyclic variations in the chemical composition of eastern Mediterranean Pliocene sediments: A key for understanding sapropel formation: Marine Geology, v. 153, p. 161–176, doi: 10.1016/S0025-3227(98)00083-8.
Weltje, G.J., and Prins, M.A., 2003, Muddled or mixed? Inferring palaeoclimate from size distributions of deep-sea clastics: Sedimentary Geology, v. 162, p. 39–62, doi: 10.1016/S0037-0738(03)00235-5.
Whitmore, G.P., Crook, K.A.W., and Johnson, D.P., 2004, Grain size control of mineralogy and geochemistry in modern river sediment, New Guinea collision, Papua New Guinea: Sedimentary Geology, v. 171, p. 129–157, doi: 10.1016/j.sedgeo.2004.03.011.
Zhang, C., Wang, L., Li, G., Dong, S., Yang, J., and Wang, X., 2002, Grain size effect on multi-element concentrations in sediments from the intertidal flats of Bohai Bay, China: Applied Geochemistry, v. 17, p. 59–68, doi: 10.1016/S0883-2927(01)00079-8.
Zhao, Y., Marriot, S., Rogers, J., and Iwugo, K., 1999, A preliminary study of heavy metal distribution on the floodplain of the River Severn, UK by a single flood event: The Science of the Total Environment, v. 243–244, p. 219–231, doi: 10.1016/S0048-9697(99)00386-1.

Manuscript Accepted by the Society 9 August 2006

Geological Society of America
Special Paper 420
2007

Petrography of Paleogene turbiditic sedimentation in northeastern Italy

Cristina Stefani[†]

*Dipartimento di Geologia, Paleontologia e Geofisica, Università di Padova, Via Giotto 1, 35137 Padova, Italy
and Consiglio Nazionale delle Ricerche (CNR), Istituto di Geoscienze e Georisorse (Sezione di Padova),
C.so Garibaldi 37, 35137 Padova, Italy*

Massimiliano Zattin

*Dipartimento di Geologia, Paleontologia e Geofisica, Università di Padova, Via Giotto 1, 35137 Padova, Italy
and Dipartimento di Scienze della Terra e Geologico Ambientali,
Università di Bologna, Piazza di Porta San Donato 1, 40126, Bologna, Italy*

Paolo Grandesso

*Dipartimento di Geologia, Paleontologia e Geofisica, Università di Padova, Via Giotto 1, 35137 Padova, Italy
and Consiglio Nazionale delle Ricerche (CNR), Istituto di Geoscienze e Georisorse
(Sezione di Padova), C.so Garibaldi 37, 35137 Padova, Italy*

ABSTRACT

The Paleogene turbiditic sedimentation in the eastern Southern Alps represents the sedimentary response to tectonic activity related to the Mesoalpine phase, which involved the surrounding chains from Paleocene time onward. Field and petrographic analyses have allowed us to classify these turbiditic successions as multisource deposits, as demonstrated by the common presence of allochemical, mainly bioclastic detritus, associated with different types of terrigenous arenites. For all units, field data suggest more proximal sources for allochemical supply and distal sources for terrigenous material, characterized by the presence of chert, carbonate rocks, and metamorphic rock fragments. All the investigated successions display transparent heavy mineral associations, marked by the common presence of chrome spinel, alkaline amphibole, staurolite, epidote, and zoisite, which point to similar metamorphic sources. The location of the source of metamorphic rock fragments is uncertain, but inputs from the internal Dinaric belt are possible. The source of the allochemical detritus was located in the nearby reactivated Friuli Platform.

Keywords: turbidite deposits, stratigraphy, sandstone composition, heavy minerals, Southern Alps, Dinarides.

[†]E-mail: cristina.stefani@unipd.it.

Stefani, C., Zattin, M., and Grandesso, P., 2007, Petrography of Paleogene turbiditic sedimentation in northeastern Italy, *in* Arribas, J., Critelli, S., and Johnsson, M.J., eds., Sedimentary Provenance and Petrogenesis: Perspectives from Petrography and Geochemistry: Geological Society of America Special Paper 420, p. 37–55, doi: 10.1130/2007.2420(04). For permission to copy, contact editing@geosociety.org. ©2007 Geological Society of America. All rights reserved.

INTRODUCTION

This work discusses the petrographic composition of Paleogene turbiditic units that crop out in the eastern Southern Alps, between the Tagliamento and Piave Rivers (Figs. 1A and 1B). According to Doglioni and Bosellini (1987), these deposits represent the sedimentary response to the Mesoalpine tectonic phase. They indicated the Dinaric thrust belt as a possible sediment source without the support of petrographic analysis.

Over the last thirty years, many papers have been published on the stratigraphy of these deposits (Saint Marc, 1963; Gnaccolini, 1967, 1968; Di Napoli Alliata et al., 1970; Cousin, 1981; Doglioni and Bosellini, 1987; Stefani and Grandesso, 1991; Tunis and Venturini, 1992; Grandesso and Stefani, 1993), but only Piccoli and Proto Decima (1969), Richter (1970), Cousin (1981), Doglioni and Bosellini (1987), and Stefani and Grandesso (1991) linked the deposition of these units to the evolution of the Dinaric chain.

The aim of this work is to contribute to a better characterization of these deposits by means of petrographic data and attempt to reconstruct the paleogeographic and paleotectonic setting in the eastern Southern Alps during the early-middle Paleogene.

GEOLOGICAL SETTING

The investigated area is located in the northeastern corner of the Adria Plate (Channell and Horvath, 1976) and is bounded to the north by the Valsugana thrust system, to the east by the SW-verging Dinaric thrusts, to the south by the Apennine blind thrust belt, and to the west by the Lessini shelf (Fig. 1A).

The Mesozoic paleogeography of this area largely influenced the Paleogene sedimentation. According to Cati et al. (1989a, 1989b), at the end of the Cretaceous, the area was characterized by the presence of a Basin to the west (Belluno Basin) and a platform to the east (Friuli Platform), the southwesterly prosecution of which has recently been seismically delineated by Fantoni et al. (2002). During the Paleogene, the basinal area was affected by turbiditic sedimentation; the related deposits presently crop out in an E-W–striking area almost 150 km in length and 70 km in width.

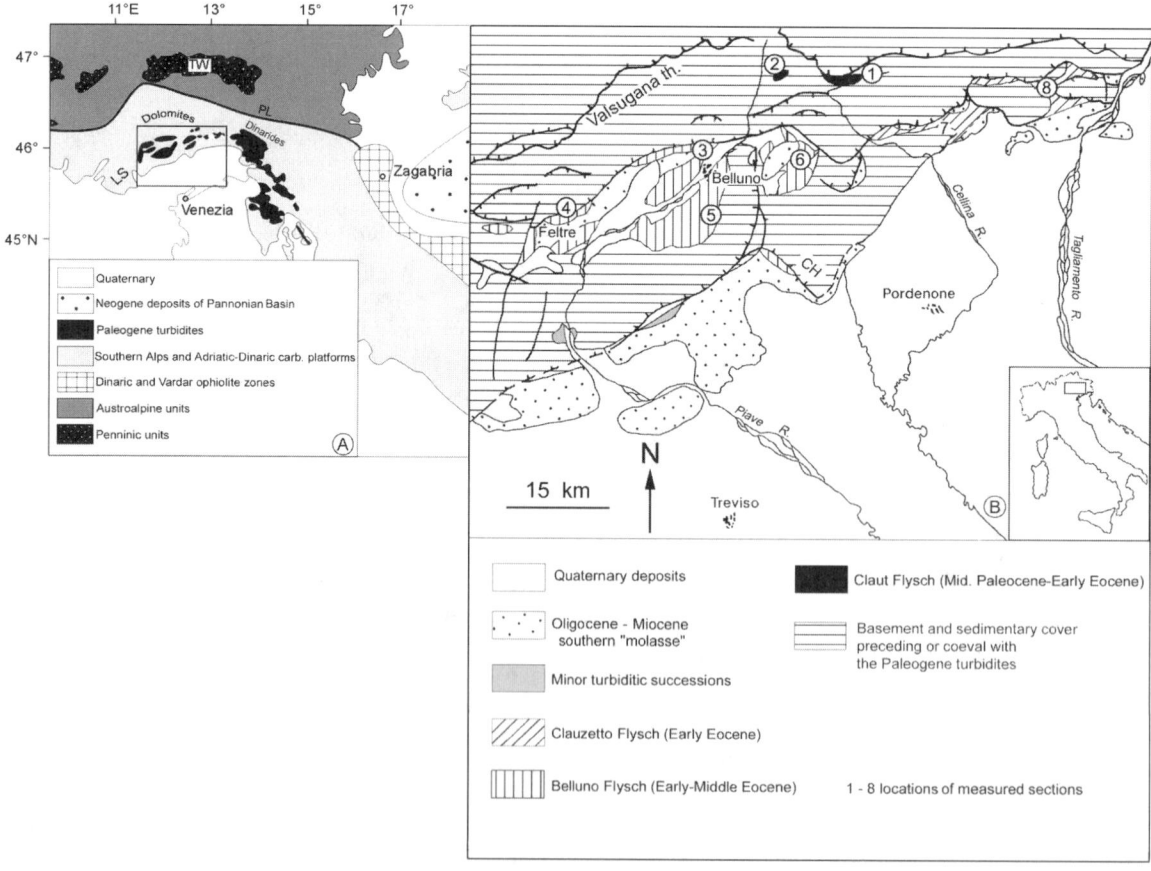

Figure 1. (A) Simplified geological sketch-map of the eastern Southern Alps and surrounding mountain chains. (B) Distribution map of the investigated Paleogene turbidite successions. Numbers refer to measured sections and localities cited in the text: (1) Claut, (2) Erto, (3) T. Medone, (4) T. Caorame, (5) T. Limana, (6) Alpago, (7) Fanna, (8) Travesio. LS—Lessini shelf; CH—Cansiglio High, TW—Tauern Window, PL—Periadriatic lineament.

Due to the postflysch tectonic activity, the preservation of Paleogene turbidites is poor in the eastern area (Fig. 1), particularly between Meduno and Maniago, while it is more extensive in the western area, where more sections have been considered (Fig. 1B). The northern margin of the basin is not preserved, as it was involved in the Southalpine thrust system, while the eastern side was largely affected by the southwestward migration of the Dinaric thrusts.

The turbidite units investigated in this work are the Claut flysch, Belluno flysch, and Clauzetto flysch (Fig. 1B).

STRATIGRAPHIC FRAMEWORK

All the investigated units were deposited above pelagic sediments like the Scaglia Rossa unit, which represents the pre-flysch facies (Aubouin, 1963); the succession is generally constituted by red marly shales, locally characterized by common carbonate intercalations and very scattered fine-grained sandy layers. The boundary between pre-flysch deposits and turbiditic sedimentation has been placed where the terrigenous debris becomes more abundant (Hsü, 1970).

In a few cases, e.g., Claut and Erto areas (Fig. 1B), the Paleogene turbiditic successions represent the most recent sediments cropping out; in other cases, the Oligocene-Miocene molasse sediments unconformably cover the turbiditic deposits.

Measurements and sampling were made in the most continuous sections, or in those less affected by tectonics; in most cases, the successions were reconstructed from a number of discontinuous outcrops. As a consequence, all the sections reported in Figures 2 and 3 are composite sections.

For each section, lithology, bed thickness, grain size, top and base geometry, and lateral continuity of beds were carefully examined. Samples were collected both for framework composition and heavy mineral and biostratigraphic analyses.

The units are made up of alternating dark gray arenitic divisions and subordinate gray marls. Bioclastic strata are present in all the successions and sometimes prevail on terrigenous beds.

Based on field observations, three types of turbiditic beds can be distinguished:

1. Arenitic terrigenous beds, made up of a centimeter-decimeter division, grading to a thicker argillaceous marly interval. The color varies from gray to ocher when altered. The sand/pelite ratio is commonly less than one. The arenitic division commonly displays cross- or convolute lamination (Tc; Bouma, 1962); planar-parallel lamination (Tb) is rarely present. The base of beds may be flat or erosive, with scarce paleoflow casts.
2. Bioclastic beds, mainly composed of carbonate allochemical grains, commonly thicker than the previous ones, with a ruditic-calcarenitic base grading upward to calcisiltites, capped by a thick carbonate-rich marly division. The sedimentary structures are represented by alternating planar-parallel and cross-laminations. The base is flat and commonly devoid of paleoflow indications.
3. "Mixed" beds; these are subordinate with respect to the other types, composed both of allochemical and terrigenous grains, characterized by an ruditic-arenitic basal interval passing upward into thicker gray marlstone. The more common internal structures of the ruditic/sandy interval are represented by the repetitive alternation of decimeter-scale packages of planar-parallel and cross-lamination. Convolute lamination due to water escape is common. Texture, composition, and structures of all the "mixed" beds are quite similar in all the examined units.

Thin, highly carbonate marls are commonly intercalated between the turbiditic beds; their recognition is easy due to lighter color and presence of a quite dense vertical bioturbation. Thin sections reveal that these beds are almost exclusively made up by planktonic foraminifers with a very scarce terrigenous silty fraction. These characters allow us to interpret them as hemipelagic layers and point out the deposition of all the units considered above the calcite compensation depth (CCD); these beds were preferentially sampled for the biostratigraphic determinations.

The Claut Flysch

The Claut flysch crops out in two sections along the Torrente Cellina near the village of Claut and southeast of Erto (Fig. 1B). The first section is located along a small tributary on the left side of the Cellina River, east of Claut village, and continues somewhat beyond the confluence. The second section was sampled S-SE of the Erto village along the main valley (right side) and in some small tributaries. Due to its inner position in the Southalpine chain, the Claut flysch was greatly involved in the tectonic deformation of the chain; the section exposure is not continuous and requires a correlation of separate segments. In both the considered sections, the boundary between the underlying Scaglia Rossa and the turbiditic facies is well exposed.

Near the village of Claut, the lower part of the unit, which consists of thin-bedded base-missing, moderately bioturbated Bouma sequences, grades upward into a highly pelitic middle portion, ~100 m thick, which displays sparse silty millimeter-size interbeds. The uppermost part of the section is characterized by an upward-increasing sandy content. Some scattered bioclastic beds are present and are interbedded with terrigenous layers very rich in chert grains. The top of the section is represented by a slump of pebbly fossiliferous sandstones well known in geological literature (Dainelli, 1910) (Fig. 2).

Paleocurrent data indicate transport mainly to the E-SE in the lower part and to the E-NE in the upper part (Fig. 2).

The total thickness of the unit has been evaluated at 600–700 m in both the examined sections.

The Belluno Flysch

The Belluno flysch crops out in the limbs of Belluno and Alpago synclines and near the town of Feltre (Fig. 1B). In the regional geological literature, different names are locally used for

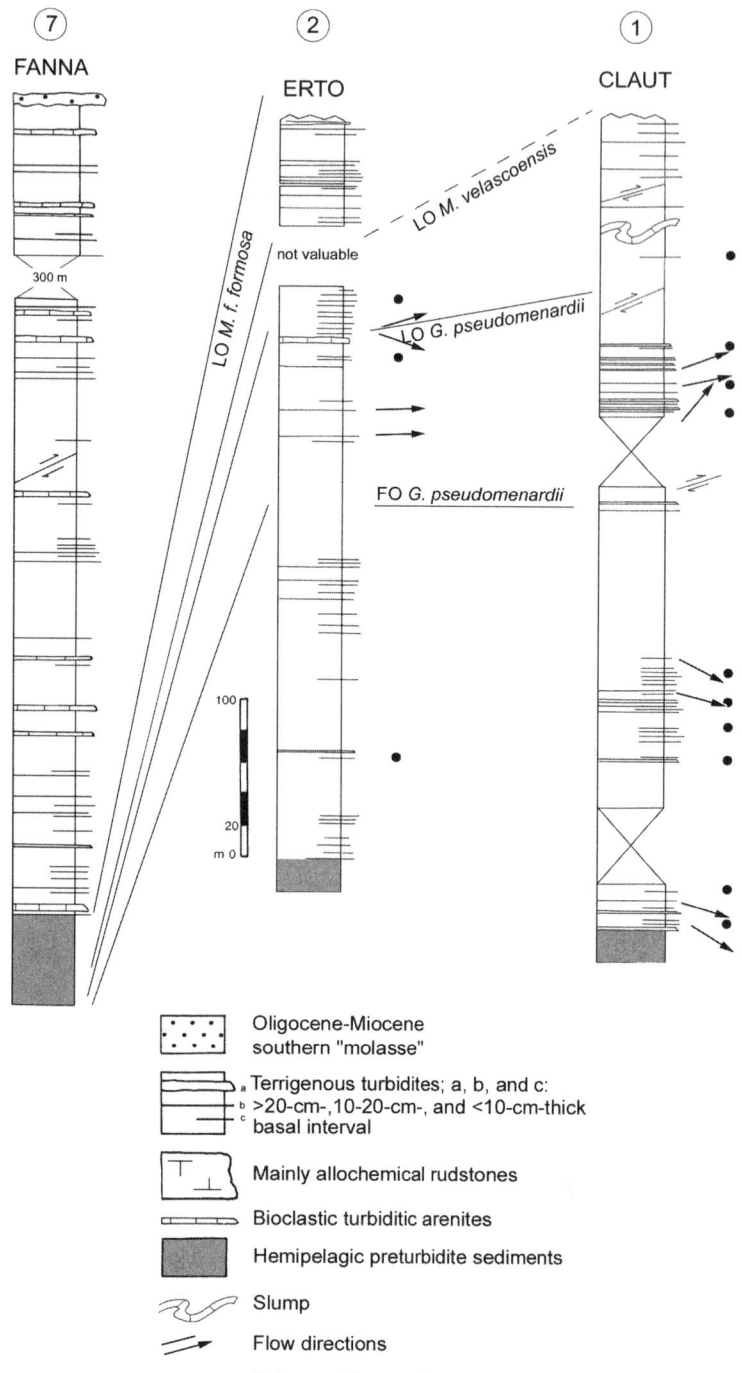

Figure 2. Schematic lithostratigraphic logs of the Claut flysch in the type locality and at Erto; for location see Figure 1. Arrows are flutes, and bars are grooves. The section of Fanna (Clauzetto flysch) is indicated only for age comparison with other sections of Figure 3. FO—first occurrence datum; LO—last occurrence datum. M.—*Morozovella*; G.—*Globanomalina*.

the flysch outcrops of Belluno-Feltre and Alpago areas, respectively (Gnaccolini, 1967, 1968), but this distinction is not supported by the biostratigraphy and lithostratigraphy.

The Belluno flysch is constituted by an almost regular alternation of turbiditic beds in which Ta-e Bouma sequences are very rare; base-missing sequences are common and are generally capped by few-millimeters- to few-centimeters-thick hemipelagic marls; the deposits are regarded as basin plain turbidites (Gnaccolini, 1967, 1968; Doglioni and Bosellini, 1987; Stefani and Grandesso, 1991). For this unit, four stratigraphic sections were considered (for location, see Fig. 1B). The first one (section 3, Figs. 1B and 3) is located immediately north of Belluno town, along the Medone and Ardo Rivers. The second section (number 4, Figs. 1B and 3) is located in the Feltre area, close to the western margin of the flysch basin. Due to the exposure conditions, the thicknesses of some segments of this section were evaluated only

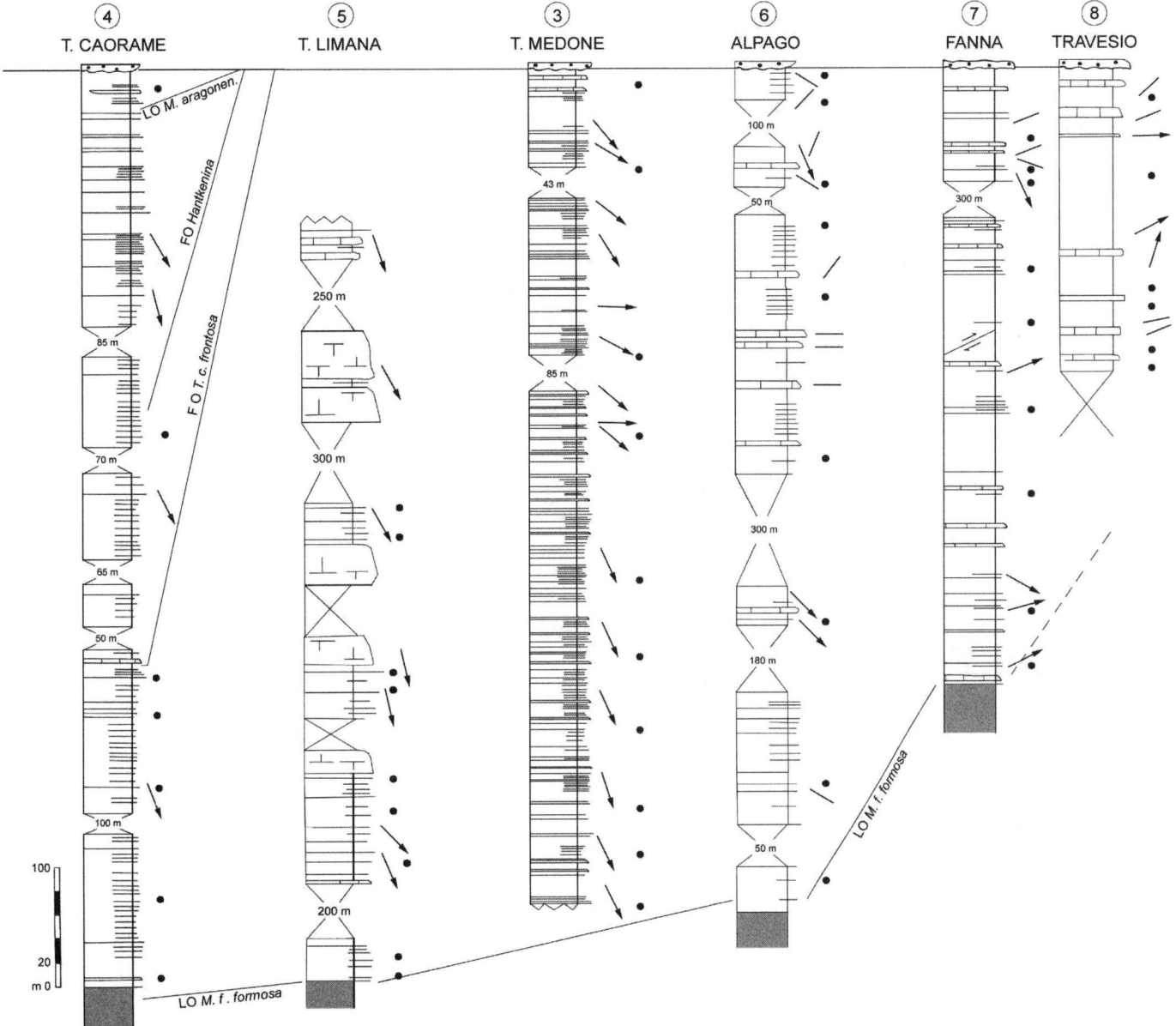

Figure 3. Schematic lithostratigraphic logs of the Belluno and Clauzetto flysch. See Figure 2 for symbol explanation and Figure 1B for location; FO—first occurrence datum; LO—last occurrence datum. M.—*Morozovella*; T.—*Turborotalia*.

approximately. The third section (T. Limana, number 5, Figs. 1B and 3) crops out in the southern flank of the Belluno valley and includes only the lower part of the flysch succession. The fourth section is located in the Alpago area (number 6, Figs. 1B and 3).

The grain size of the terrigenous beds shows a progressive fining westward, together with a significant increase in thickness of hemipelagic sediments interbedded with turbiditic layers. In the T. Caorame section (section 4), hemipelagic sediments are commonly thicker than 10 cm, with a maximum of 30–40 cm, testifying that this area was reached by very diluted turbidity currents.

Paleocurrents obtained from sole marks give NW-SE derivations, with only few exceptions. In the Feltre area, the paleocurrent indications become rare due to the high dilution of the turbidity currents. In the Alpago area, Gnaccolini (1967) reported a few paleoflow directions coming from the S-SE, i.e., from the Cansiglio High.

The total thickness of the Belluno flysch has been estimated to be at least one thousand meters in the well Sedico-1, where an early Eocene age has been determined (Costa et al., 1996); the succession becomes progressively thinner westward, where it shows onlapping relationships with the eastern slope of the

Lessini Shelf. The youngest strata (middle Eocene) have been recorded in the T. Caorame section (Grandesso, 1976; Stefani and Grandesso, 1991).

The Clauzetto Flysch

The Clauzetto flysch was sampled in two quarry roadcuts near Fanna and Travesio villages (Fig. 1B) where the unit is excellently exposed.

In the Fanna section, the Clauzetto flysch succession is complete from the base to the top, while in the Travesio quarry, only the upper part of the succession crops out (Fig. 3).

In both the analyzed sections, the unit is represented mostly by plane-parallel arenaceous intervals interbedded with prevalent gray marls and scattered millimeter-thick hemipelagic beds. The bioclastic beds represent ~50% of the total thickness and, in a few cases, show debris deposits with lenticular geometry.

Bioturbation is common, both in the hemipelagic layers and on the soles of arenitic intervals; according to Tunis and Uchman (1998), the predominant genera are *Planolites* and *Chondrites* in the pelites, while the arenitic soles are dominated by *Scolicia* and *Ophiomorpha*, indicating a moderately oligotrophic environment.

Paleocurrents show low variation and reflect currents flowing from W-SW. The total thickness of the unit is 500–600 m.

BIOSTRATIGRAPHY

Several samples were collected along the measured sections, mainly from hemipelagic layers, in order to obtain a biostratigraphical zonation according to planktonic foraminiferal scheme proposed by Berggren and Miller (1988) and Berggren et al. (1995). Results are synthetically exposed in Figures 2, 3, and 4. The following biostratigraphic events are considered: the total range of the *Globanomalina pseudomenardii*, the last occurrence (LO) of *Morozovella velascoensis*, *M. formosa formosa*, and *M. aragonensis*, and the first occurrence (FO) of *Turborotalia cerroazulensis frontosa* and *Hantkenina*.

In particular, the Claut flysch is referred, based on our data (Fig. 2 and 4), to the Selandian-Thanetian/Ypresian interval (P3-P5 zones of Berggren et al., 1995), in the Claut section and up to the P7 zone in the Erto section. The Claut section was previously referred by Marie and Cousin (1965) and Cousin (1981) to about the same stratigraphic interval (i.e., Selandian-Ypresian), but these authors considered a different formational boundary.

The Belluno flysch sedimentation took place in the Alpago and Belluno areas (sections Medone, Alpago, and Limana) during the Ypresian (uppermost P7–P9); our present data confirm substantially the age of Di Napoli Alliata et al. (1970). In the western sectors (T. Caorame section), the turbiditic sedimentation began later (P8 zone) and continued until the Lutetian (P12

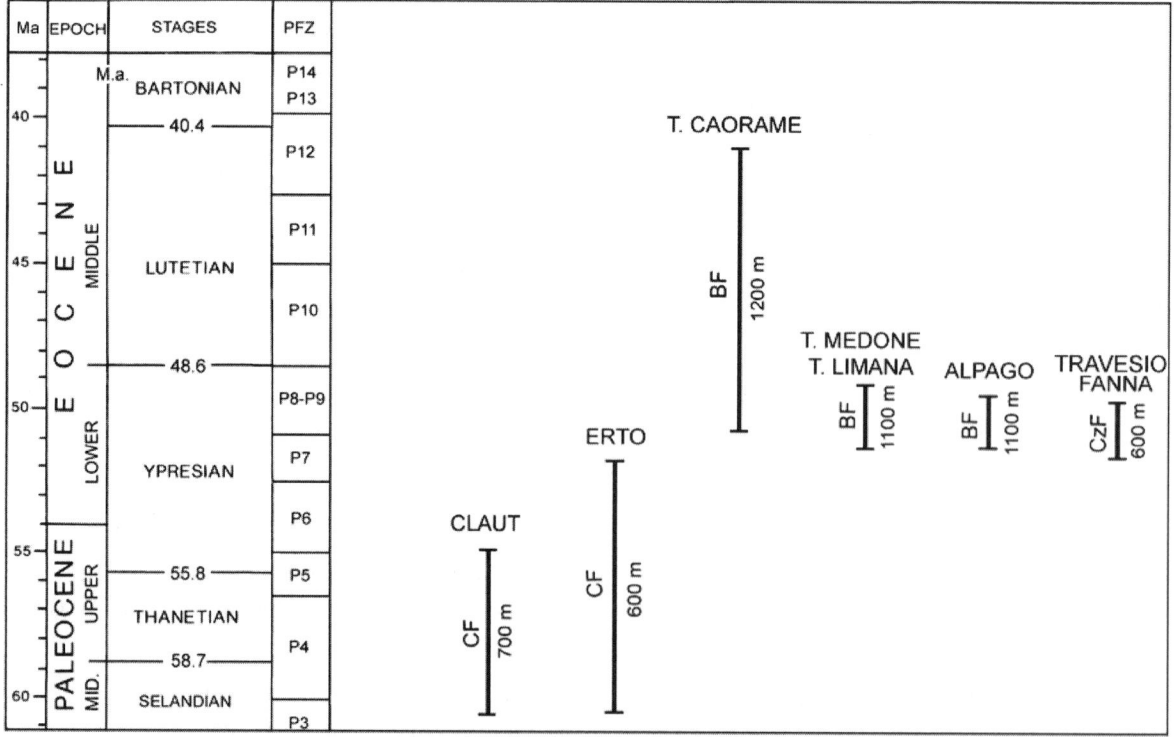

Figure 4. Age of the studied successions, according to zonation scheme of Berggren and Miller (1988) and Berggren et al. (1995). Time scale is from Gradstein et al. (2004), slightly modified. PFZ—planktonic foraminiferal zones; CF—Claut flysch; BF—Belluno flysch; CzF—Clauzetto flysch. See Figures 2 and 3 for location of the sections and lithostratigraphy.

zone) (Fig. 4) as recognized by Grandesso (1976) and Stefani and Grandesso (1991).

The Clauzetto flysch is referred to the Ypresian (from the top of P7 zone to P9). Previously the base of the unit, which crops out a few kilometers west of Fanna quarry, was referred to the Paleocene-Eocene boundary by Saint Marc (1963); Tunis and Uchman (1998) dubiously referred the Clauzetto flysch to the early Eocene (*Morozovella formosa* and *M. aragonensis* zones sensu Toumarkine and Luterbacher [1985], corresponding to intervals P7 to P9 of Berggren and Miller [1988]).

COMPOSITION OF THE TURBIDITE BEDS

Bioclastic Beds

These beds are generally the thickest in all the successions. The bases are commonly ruditic and grade upward into fine sand and silt; the finer portions of these beds are represented by highly carbonate marls. The carbonate beds may be classified as bioclastic rudstone-grainstone or packstone. The bioclastic content mainly consists of Nummulitidae, red algae, bryozoa, echinoderm debris, and, with few exceptions, intraclasts, peloids, green particles, all of which point to a provenance from carbonate shelf areas located in the photic zone.

The bioclastic supplies are particularly relevant for the Belluno flysch and Clauzetto flysch, both of which were deposited very close to the Friuli Platform, where, during the Paleogene, carbonate factories were active on the same sites of carbonate production during the Mesozoic (Cousin, 1981; Fantoni et al., 2002).

A peculiar situation has been recognized along the southern side of the Belluno valley (T. Limana section on Fig. 3) where the lower part of the Belluno flysch crops out. At least five megabeds, up to 40 m thick, are constituted mainly by coarse-grained breccias, entirely composed of almost coeval bioclastic material resedimented by gravity flows. According to Gnaccolini (1968), this material came from a southwestward prolongation of the Friuli Platform, as suggested by the rapid fining in the northerly direction. In a recent interpretation of the TRANSALP deep seismic profile, the prosecution of the Friuli carbonate platform has been clearly recognized in the subsurface (Fantoni et al., 2002; Bertelli et al., 2003).

Framework Composition of the Terrigenous Beds

Sixty-five samples of arenites representative of the total thickness of the three units were analyzed in order to obtain detailed information on composition. Five hundred points were counted for each thin section, stained for the distinction of K-feldspar and to discriminate the carbonate phases. Analyses were performed according to the Gazzi-Dickinson method (Gazzi, 1966; Dickinson, 1970), which tends to minimize the grain-size influence on arenite composition. Consequently, each grain composed of more than one crystal larger than the matrix limit (0.0625 mm) was classified during point-counting as mineral type and not as rock type. The point-counting results were tabulated (Table 1) and recalculated to obtain the diagrams of Figures 5 and 6.

Results of point counting are reported in Table 1; they are subdivided into four classes, and the recalculated proportions are represented in Figure 5A. According to Zuffa classification (1980, 1985), NCE represents the "noncarbonate extrabasinal grains," CE is the "extrabasinal carbonate grains," NCI is the "noncarbonate intrabasinal grains," and CI is the "carbonate intrabasinal grains." Figure 5B shows the composition of the terrigenous grain fraction, where Q comprises the total quartz, F, the feldspars, and L+CE, the fine-grained rock fragments and the extrabasinal carbonates. Chert was included in the L + CE pole due to the almost exclusive association with carbonate rocks in the sedimentary covers of the Southern Alps and Dinarides.

All investigated samples can be classified as carbonate or noncarbonate extra-arenites, according to the first-level classification of Zuffa (1980, 1985), while in the second-level classification, they can be described as "litharenites" (Fig. 5B). This term is not related to a particular classification, as for the use of the Gazzi-Dickinson method. However, most, if not all, classifications proposed in literature would provide this term for data placed on the bottom-right pole of the triangle. Consequently, we use the classifying term in inverted commas.

Quartz occurs both as monocrystalline and polycrystalline grains and as a phaneritic component in metamorphic and granitic/gneissic rock fragments. Feldspar grains are commonly present as single crystals and, in a few cases, within coarse-grained rock fragments derived from granitic/gneissic rocks. Low- to medium-grade metamorphic lithic grains include micaschists, epidote-schists, quartzites, phyllites, and a few metavolcanics. Chert grains consist of microquartz and scarce chalcedony; textural characters are frequently obliterated by Fe-oxides and dolomite rhombohedrons.

Volcanic rock fragments include rhyolites and rare intermediate grains with lathwork-like intersertal texture made up of euhedral plagioclase and rare glassy grains. Phyllosilicates are negligible. The other mineral grains include mainly garnet, zircon, epidotes, amphiboles, and spinel.

The most common lithotypes in the carbonate fraction are well-rounded grains of micritic limestone, followed by coarse- and fine-grained polycrystalline dolomitic clasts. In a few cases, some oolitic or fossiliferous limestone grains with planktonic foraminifers were recognized.

The Claut flysch displays the highest quartz content (mean value: Q 45.6, F 8.5, L + CE 45.9) and also the highest average amount of feldspar grains (Table 2). The rock fragment association displays high percentages of metamorphic, volcanic, and granitic and/or gneissic rock fragments ranging from 30% to 70% of the lithic population (Fig. 6). Other fractions of the rock fragments are represented by chert and subordinate dolostone and limestone.

TABLE 1. MODAL POINT COUNTS OF THE PALEOGENE TURBIDITIC SUCCESSIONS

Stratigraphic unit	Claut flysch														
Samples	1	2	3	4	5	6	7	8	9	10	11	12	13	14	15
NCE															
Q															
Quartz monocrystalline	20.8	16.5	14.0	13.0	16.0	20.8	19.3	15.0	22.0	22.8	1.5	9.8	13.0	16.3	9.5
Coarse-grained polycrystalline quartz	2.4	3.3	2.0	1.4	3.5	7.0	4.5	2.0	6.0	9.3	0.8	9.0	3.8	2.3	2.3
Fine-grained polycrystalline quartz	2.6	0.8	1.3	1.6	2.3	3.3	2.0	2.0	2.3	5.3	0.3	3.0	–	0.5	1.0
Quartz in acidic volcanic rock fragments	0.2	–	–	–	–	–	0.5	0.8	2.8	1.3	1.3	1.5	1.8	0.3	
Quartz in low-grade metamorphic rock fragments	0.8	0.8	0.5	0.6	–	–	–	–	–	–	–	–	–	–	–
Quartz in granitic and/or gneissic rock fragments	0.6	2.0	–	–	1.8	0.5	1.3	1.8	2.3	1.3	0.3	0.5	0.3	2.3	1.3
Calcite on monocrystalline quartz	1.4	2.8	3.3	1.6	6.0	4.0	2.5	7.0	8.3	1.8	0.8	4.3	4.0	6.5	6.5
Calcite on quartz in plutonic and/or gneissic rock fragments	–	0.5	0.5	0.4	1.5	0.8	0.5	0.5	0.5	–	–	0.3	0.3	0.8	–
F															
K-feldspar monocrystalline	0.4	0.3	1.3	0.8	0.8	3.5	2.0	–	0.3	0.3	–	–	1.0	–	0.5
K-feldspar in granitic and/or gneissic rock fragments	–	0.5	–	0.2	–	0.3	0.3	–	–	–	–	–	–	–	–
Calcite on K-feldspar (monocr.)	–	–	–	–	–	0.3	–	–	–	–	–	–	–	–	–
Calcite on K-feldspar in rock fragments	–	–	–	–	–	–	–	–	–	–	–	–	–	–	–
Plagioclase monocrystalline	0.4	3.8	2.3	3.0	1.3	1.8	4.5	4.3	2.3	6.5	–	5.5	0.5	3.0	1.5
Plagioclase in volcanic rock fragments	–	–	–	–	–	–	–	–	–	–	–	–	–	–	–
Plagioclase in low-grade metamorphic rock fragments	–	–	–	–	–	–	–	–	–	–	–	–	–	–	–
Plagioclase in granitic and/or gneissic rock fragments	0.8	–	0.3	0.6	1.8	0.3	1.3	0.5	2.0	1.8	0.3	–	0.5	0.3	1.8
Calcite on plagioclase (monocr.)	–	1.0	1.0	0.2	2.0	0.3	0.5	1.8	0.3	0.3	0.3	1.0	1.3	1.3	1.0
Calcite on plagioclase in rock fragments	–	–	–	–	–	–	–	–	–	–	–	–	–	–	–
L															
Acidic volcanic rock fragments	8.2	5.5	4.5	5.8	3.5	6.8	3.5	5.0	5.5	12.3	5.5	5.5	6.8	4.3	1.3
Low–grade metamorphic rock fragments	2.0	5.8	6.3	4.6	4.3	5.5	6.8	2.5	5.0	6.5	17.3	17.0	4.3	4.3	3.0
Cherts	3.6	2.8	2.0	1.4	1.0	2.8	1.8	0.8	1.8	6.0	9.0	–	2.0	2.3	1.8
Siltstone	0.2	–	–	0.4	–	–	–	–	–	–	–	2.3	–	–	–
Calcite on aphanitic rock fragments	0.6	1.8	1.5	0.2	1.3	1.5	0.8	1.0	2.5	1.0	7.3	–	2.5	5.3	3.3
Mica and chlorite	0.6	–	–	1.0	–	–	–	–	0.3	–	–	–	0.3	0.3	–
Mica and chlorite in granitic and/or gneissic rock fragments	0.2	–	–	–	–	–	–	–	–	–	–	–	–	–	–
Mica and chlorite in low-grade metamorphic rock fragments	–	–	–	–	–	–	–	–	–	–	–	–	–	–	–
Other minerals	0.2	–	–	–	–	–	–	–	–	0.3	–	–	0.3	–	–
Other minerals in rock fragments	–	–	0.3	–	–	–	–	–	–	–	–	–	0.3	–	–
CE															
Dolostone monocrystalline grain	0.6	0.3	0.3	2.0	0.3	–	–	–	–	0.5	–	1.5	2.0	–	
Coarse-grained polycrystalline dolostone	1.6	0.3	0.5	1.6	0.5	–	–	0.8	–	–	–	1.5	0.3	0.8	
Fine-grained polycrystalline dolostone	–	1.5	0.8	–	0.5	1.0	0.5	–	0.3	–	–	0.3	–	–	
Mudstone	3.6	5.8	8.5	1.4	6.0	4.0	4.5	4.3	1.5	2.8	9.5	7.0	6.8	7.8	10.5
Microspatitic limestone	2.8	1.0	4.8	1.8	2.0	0.3	2.8	5.8	1.0	1.0	9.0	1.8	4.3	3.8	5.3
Spatitic limestone	–	–	1.0	1.0	1.0	–	2.3	1.0	1.0	–	4.5	0.5	1.8	0.3	0.3
Cherty limestone/dolostone grain	–	–	–	–	–	–	–	–	–	–	–	–	–	–	–
Calcite on dolostone grain	–	0.3	–	–	–	–	–	–	–	–	–	–	–	–	–
NCI															
Green particles	0.6	–	–	0.2	0.3	–	–	–	–	–	–	–	–	–	–
Pelitic intraclasts	–	–	1.3	–	0.3	0.3	1.0	–	0.3	0.8	–	–	–	0.3	0.3
Silica sponge and radiolaria	0.8	0.8	0.3	0.8	0.5	–	–	0.3	0.3	–	0.5	0.3	0.8	0.3	0.5
CI															
Carbonate bioclasts	4.0	11.0	5.0	5.2	2.3	5.5	0.8	6.5	2.3	1.2	9.0	9.5	10.5	6.0	4.0
Limeclast	–	0.3	–	1.4	–	–	–	0.5	–	–	–	0.5	–	–	–
Siliciclastic matrix	4.2	0.3	3.3	6.0	0.8	0.8	2.0	0.5	1.3	1.0	–	0.5	–	0.3	0.5
Carbonate matrix	2.6	0.5	–	3.0	–	0.3	–	1.0	0.3	0.5	1.5	2.3	3.8	2.5	7.0
Calcite cement	19.2	2.5	7.8	13.4	11.0	3.8	10.8	7.5	4.0	3.8	8.8	2.0	3.8	5.8	7.5
Dolomite cement	–	–	–	–	–	–	–	–	–	–	–	–	–	–	–
Oxides	–	0.3	0.8	0.8	3.3	2.5	1.3	1.0	0.8	0.3	0.8	1.0	1.5	3.0	4.3
Undeterminate grain	1.2	–	–	0.6	1.3	–	0.5	0.8	0.3	–	0.3	–	1.5	0.5	1.0
Patchy calcite	2.8	24.0	21.3	3.8	20.0	20.3	18.3	22.3	24.3	9.5	9.8	12.8	14.5	13.8	18.0
Spar of calcite	5.0	1.5	3.3	7.8	3.3	2.5	3.5	3.0	0.5	1.3	1.5	1.3	5.5	0.8	4.0
Calcite on undeterminate grain	5.0	2.3	0.8	12.4	0.3	–	0.8	1.3	–	–	0.3	1.5	–	2.0	1.5
Total	100.0	100.0	100.0	100.0	100.0	100.0	100.0	100.0	100.0	100.0	100.0	100.0	100.0	100.0	100.0

(continued.)

TABLE 1. (Continued.)

Stratigraphic unit	Belluno flysch														
Samples	16	17	18	19	20	21	22	23	24	25	26	27	28	29	30
NCE															
Q															
Quartz monocrystalline	6.5	4.0	10.4	4.0	3.5	6.1	5.7	3.6	9.0	11.2	10.8	11.2	12.2	15.0	5.2
Coarse-grained polycrystalline quartz	2.8	0.4	0.3	1.3	0.7	4.8	6.2	2.8	3.6	2.8	4.8	4.6	4.0	6.4	2.8
Fine-grained polycrystalline quartz	0.4	0.9	1.4	2.3	1.1	0.3	4.7	0.6	0.6	0.8	2.2	0.8	1.6	0.2	1.0
Quartz in acidic volcanic rock fragments	1.2	0.2	–	–	0.7	–	–	–	–	0.2	0.2	0.4	–	–	0.2
Quartz in low-grade metamorphic rock fragments	1.6	–	0.3	–	0.7	2.0	1.0	0.4	1.0	1.6	3.0	3.0	2.0	5.6	0.4
Quartz in granitic and/or gneissic rock fragments	0.2	0.7	–	–	0.4	0.3	–	–	–	–	–	–	–	0.4	–
Calcite on monocrystalline quartz	0.8	0.7	4.5	1.0	1.0	2.4	0.5	0.4	2.8	3.4	2.8	1.8	9.6	5.2	3.0
Calcite on quartz in plutonic and/or gneissic rock fragments	–	–	–	–	–	–	–	0.4	–	0.8	1.4	0.2	0.8	1.0	0.6
F															
K-feldspar monocrystalline	0.2	0.4	–	1.0	1.4	1.0	–	–	0.4	0.4	0.8	–	–	0.6	–
K-feldspar in granitic and/or gneissic rock fragments	–	–	–	–	–	–	–	–	0.4	–	–	–	–	–	–
Calcite on K-feldspar (monocr.)	–	–	–	–	–	–	–	–	–	–	–	0.4	–	–	–
Calcite on K-feldspar in rock fragments	–	–	–	–	–	–	–	–	–	–	–	–	0.2	–	–
Plagioclase monocrystalline	3.4	2.5	–	2.7	1.4	6.5	4.3	–	–	0.4	1.2	0.4	0.2	1.0	–
Plagioclase in volcanic rock fragments	–	–	–	–	–	–	–	–	–	–	–	–	–	–	–
Plagioclase in low-grade metamorphic rock fragments	–	0.2	–	–	–	–	0.5	–	–	–	–	0.2	–	0.2	–
Plagioclase in granitic and/or gneissic rock fragments	–	–	–	–	–	–	–	–	–	–	–	–	–	–	–
Calcite on plagioclase (monocr.)	–	–	–	–	–	2.0	0.9	–	–	–	–	–	–	–	–
Calcite on plagioclase in rock fragments	–	–	–	–	–	–	–	–	–	–	–	–	–	–	–
L															
Acidic volcanic rock fragments	1.0	1.3	9.0	5.7	22.5	0.3	–	1.6	–	1.0	0.2	0.8	2.0	0.8	3.4
Low-grade metamorphic rock fragments	3.4	4.0	5.2	–	0.4	2.7	3.8	1.6	3.4	1.8	1.6	1.2	5.4	2.8	1.2
Cherts	1.2	2.0	3.8	5.0	12.6	–	2.4	1.0	0.8	0.4	–	0.2	–	–	0.4
Siltstone	–	–	–	–	–	–	4.4	0.4	2.6	0.6	0.6	–	1.6	2.2	
Calcite on aphanitic rock fragments	–	–	0.3	0.3	0.4	–	–	–	0.4	0.4	0.2	0.2	0.8	–	–
Mica and chlorite	0.2	0.2	–	–	–	–	0.5	–	0.6	0.4	1.6	0.2	0.8	0.4	0.6
Mica and chlorite in granitic and/or gneissic rock fragments	–	–	–	–	–	–	–	–	–	–	–	–	–	–	–
Mica and chlorite in low-grade metamorphic rock fragments	0.2	–	–	–	–	–	0.5	–	–	–	–	–	0.4	–	–
Other minerals	–	–	0.3	–	–	–	–	0.2	–	1.2	0.6	0.2	0.8	0.2	–
Other minerals in rock fragments	–	–	–	–	–	–	–	–	–	–	–	–	–	–	–
CE															
Dolostone monocrystalline grain	19.5	32.5	21.2	23.7	8.1	12.3	12.3	6.4	7.6	18.2	14.8	7.8	13.4	16.8	13.6
Coarse-grained polycrystalline dolostone	27.6	4.3	3.8	12.7	–	23.9	24.6	28.6	32.0	14.6	16.0	21.2	11.4	17.0	11.0
Fine-grained polycrystalline dolostone	2.2	1.8	10.1	1.3	0.7	3.1	3.3	7.6	7.2	3.8	5.2	5.2	2.0	3.4	2.8
Mudstone	2.6	1.3	2.1	3.0	7.4	–	0.9	–	–	–	–	–	–	–	–
Microspatitic limestone	0.3	0.9	–	2.0	–	–	0.5	–	–	–	–	–	–	–	–
Spatitic limestone	0.2	–	–	1.3	5.2	1.7	0.9	–	–	–	–	–	–	–	–
Cherty limestone/dolostone grain	–	3.8	–	1.3	1.0	–	–	–	–	–	–	–	–	–	–
Calcite on dolostone grain	–	–	–	–	–	–	–	0.4	–	1.0	0.2	0.8	1.4	1.0	1.6
NCI															
Green particles	–	–	–	–	–	–	–	–	–	–	–	–	–	–	–
Pelitic intraclasts	0.2	0.7	4.9	4.5	5.6	1.0	1.4	0.2	1.4	0.2	0.2	1.2	–	–	–
Silica sponge and radiolaria	0.2	–	1.0	–	0.3	0.7	–	–	–	–	–	–	–	–	–
CI															
Carbonate bioclasts	0.2	0.7	–	0.3	0.7	0.3	1.9	0.4	0.4	2.8	3.0	5.6	2.0	0.4	1.0
Limeclast	–	–	1.4	1.0	1.0	0.7	–	–	–	–	–	–	–	–	–
Siliciclastic matrix	1.0	1.3	–	0.7	0.7	0.7	0.5	–	0.2	1.8	1.2	0.6	3.0	2.2	2.4
Carbonate matrix	1.0	4.9	–	0.3	0.3	0.3	–	0.8	–	2.0	0.6	0.4	4.0	0.4	3.0
Calcite cement	19.1	20.9	0.9	1.7	2.4	4.5	3.8	22.8	21.6	20.2	18.6	15.6	13.8	15.4	30.2
Dolomite cement	0.2	–	–	2.4	0.3	6.8	1.4	–	–	–	–	0.2	–	–	–
Oxides	–	0.4	1.0	–	0.3	0.5	0.5	0.8	–	0.8	0.6	–	0.8	–	1.2
Undeterminate grain	–	–	–	2.1	0.7	–	–	1.0	0.8	0.8	–	1.6	1.0	0.6	1.8
Patchy calcite	–	0.7	–	–	–	–	–	–	–	–	–	–	–	–	–
Spar of calcite	–	–	–	–	–	–	–	–	–	–	–	–	–	–	–
Calcite on undeterminate grain	2.6	8.3	18.1	18.4	18.8	15.3	17.0	14.0	5.4	4.4	7.6	13.4	6.4	1.4	10.4
Total	100.0	100.0	100.0	100.0	100.0	100.0	100.0	100.0	100.0	100.0	100.0	100.0	100.0	100.0	100.0

(continued.)

TABLE 1. MODAL POINT COUNTS OF THE PALEOGENE TURBIDITIC SUCCESSIONS (Continued.)

Stratigraphic unit							Belluno flysch								
Samples	31	32	33	34	35	36	37	38	39	40	41	42	43	44	45
NCE															
Q															
Quartz monocrystalline	10.6	17.0	12.0	9.4	13.0	7.2	3.3	9.8	10.0	11.5	4.8	10.3	8.0	0.5	1.3
Coarse-grained polycrystalline quartz	1.2	6.0	3.4	1.8	4.2	2.0	11.3	4.8	7.0	8.0	1.3	2.3	2.3	3.0	2.3
Fine-grained polycrystalline quartz	3.0	1.0	1.4	1.0	2.8	0.8	6.3	4.5	3.8	3.3	3.8	1.8	1.8	1.0	1.5
Quartz in acidic volcanic rock fragments	–	0.4	–	0.2	0.6	–	0.8	–	–	–	–	–	–	–	–
Quartz in low-grade metamorphic rock fragments	1.2	1.4	1.4	1.0	4.4	1.0	–	–	0.3	0.5	–	–	–	–	–
Quartz in granitic and/or gneissic rock fragments	–	–	–	–	–	–	–	–	0.3	0.3	–	–	–	–	–
Calcite on monocrystalline quartz	6.4	6.6	5.6	6.2	3.4	–	–	–	–	–	–	–	–	–	–
Calcite on quartz in plutonic and/or gneissic rock fragments	0.6	0.2	2.6	–	2.0	0.8	0.3	–	–	–	–	–	–	–	–
F															
K-feldspar monocrystalline	1.0	0.2	0.8	–	0.4	–	–	1.8	1.0	1.8	0.3	0.5	3.8	–	–
K-feldspar in granitic and/or gneissic rock fragments	–	–	–	–	–	–	0.3	–	–	–	–	–	–	–	–
Calcite on K-feldspar (monocr.)	–	–	–	–	–	–	–	–	–	–	–	–	0.3	–	–
Calcite on K-feldspar in rock fragments	–	–	–	–	–	–	–	–	–	–	–	–	–	–	–
Plagioclase monocrystalline	0.2	0.4	0.8	–	0.2	–	–	–	0.3	–	–	0.3	–	–	–
Plagioclase in volcanic rock fragments	–	–	0.2	–	–	–	–	–	–	–	–	–	–	–	–
Plagioclase in low-grade metamorphic rock fragments	–	–	–	–	–	–	–	–	–	–	–	–	–	–	–
Plagioclase in granitic and/or gneissic rock fragments	–	–	–	–	–	–	–	–	–	–	–	–	–	–	–
Calcite on plagioclase (monocr.)	–	–	–	–	–	–	–	–	–	–	–	–	–	–	–
Calcite on plagioclase in rock fragments	–	–	–	–	–	0.6	–	–	–	–	–	–	–	–	–
L															
Acidic volcanic rock fragments	1.2	1.4	2.0	1.2	1.6	1.8	1.8	0.8	1.0	1.0	0.5	0.3	0.3	1.0	1.8
Low-grade metamorphic rock fragments	1.2	2.8	3.4	0.6	3.4	1.0	6.0	8.3	5.8	6.5	3.5	3.3	10.0	8.8	3.8
Cherts	–	–	–	0.2	–	1.2	5.8	1.3	1.0	0.5	1.3	1.3	–	6.0	5.0
Siltstone	1.8	0.2	1.2	0.8	1.0	2.8	0.3	0.3	–	–	0.3	–	–	0.5	0.3
Calcite on aphanitic rock fragments	–	0.8	–	1.2	0.6	0.2	0.5	–	–	–	–	–	–	–	–
Mica and chlorite	0.6	1.4	0.4	0.8	0.6	0.4	–	–	–	–	–	0.3	1.0	–	–
Mica and chlorite in granitic and/or gneissic rock fragments	–	0.2	–	–	–	–	–	–	–	–	–	–	–	–	–
Mica and chlorite in low-grade metamorphic rock fragments	–	0.2	–	–	0.4	–	–	–	–	–	–	–	–	–	–
Other minerals	0.2	–	0.2	0.4	–	0.6	–	–	0.3	0.3	–	–	0.3	–	0.3
Other minerals in rock fragments	–	–	–	–	–	–	–	–	–	–	–	–	–	–	–
CE															
Dolostone monocrystalline grain	16.2	18.8	13.6	20.6	18.0	16.6	3.5	10.8	16.0	12.3	17.0	21.0	14.0	1.3	4.0
Coarse-grained polycrystalline dolostone	10.0	16.8	15.4	22.0	15.8	21.0	30.8	18.0	22.5	18.0	30.3	28.0	21.0	31.0	34.8
Fine-grained polycrystalline dolostone	0.4	2.2	2.6	1.6	1.4	1.0	–	–	–	–	–	–	–	–	–
Mudstone	–	–	–	–	–	–	2.8	1.8	1.3	0.5	3.8	5.5	5.8	11.5	7.0
Microspatitic limestone	–	–	–	–	–	–	4.3	3.5	4.3	1.0	5.3	1.8	1.8	9.5	10.8
Spatitic limestone	–	–	–	–	–	–	0.3	3.3	1.0	2.3	2.5	0.3	0.3	1.5	1.8
Cherty limestone/dolostone grain	–	–	–	–	–	–	–	–	–	–	–	–	–	–	–
Calcite on dolostone grain	–	1.0	1.8	1.6	1.6	0.2	0.3	–	0.3	0.3	0.5	–	1.0	–	0.5
NCI															
Green particles	–	–	–	0.2	–	–	tr	tr	tr	–	–	–	–	tr	–
Pelitic intraclasts	–	–	–	–	–	–	–	0.3	–	0.3	0.5	1.3	–	0.3	0.3
Silica sponge and radiolarian	–	–	–	–	–	–	–	–	–	–	–	–	–	–	–
CI															
Carbonate bioclasts	6.2	0.2	0.4	3.4	0.6	2.2	6.0	12.0	6.8	4.8	4.0	4.3	4.3	5.0	3.8
Limeclast	–	–	–	–	–	–	–	–	–	–	–	0.3	–	–	–
Siliciclastic matrix	1.8	0.2	0.8	0.6	2.0	2.0	1.0	2.5	1.3	5.3	3.3	1.3	11.8	0.8	0.8
Carbonate matrix	0.2	0.4	0.4	0.2	0.6	1.6	–	–	–	0.3	–	–	0.5	–	0.5
Calcite cement	25.4	12.0	16.6	17.2	14.4	17.0	9.0	11.3	10.3	10.3	8.0	7.3	6.0	6.8	10.3
Dolomite cement	–	–	–	–	–	–	–	–	–	–	–	–	–	–	–
Oxides	0.8	–	0.6	0.6	0.8	1.2	–	–	0.3	1.0	0.5	0.3	0.8	–	0.5
Undeterminate grain	0.6	0.6	1.4	0.2	1.4	2.2	–	–	0.3	–	–	–	–	–	–
Patchy calcite	–	–	–	–	–	–	–	–	–	–	–	0.8	1.0	0.3	–
Spar of calcite	–	–	–	–	–	–	6.0	5.5	5.5	10.5	8.0	8.3	5.3	11.8	9.3
Calcite on undeterminate grain	9.2	7.6	11.0	7.0	4.8	14.6	–	–	–	–	–	–	–	–	–
Total	100.0	100.0	100.0	100.0	100.0	100.0	100.0	100.0	100.0	100.0	100.0	100.0	100.0	100.0	100.0

(continued.)

TABLE 1. (*Continued.*)

Stratigraphic unit	Belluno flysch					Clauzetto flysch								
Samples	46	47	48	49	50	51	52	53	54	55	56	57	58	59
NCE														
Q														
Quartz monocrystalline	2.3	9.5	4.5	5.0	13.2	5.8	4.8	2.3	12.5	8.8	8.3	5.5	2.3	7.5
Coarse-grained polycrystalline quartz	1.5	2.8	4.0	3.3	4.4	0.3	0.3	0.8	0.3	1.5	0.8	0.3	0.8	1.8
Fine-grained polycrystalline quartz	0.8	1.3	1.5	0.8	4.4	–	0.8	1.0	2.3	1.5	1.3	0.3	0.3	1.3
Quartz in acidic volcanic rock fragments	–	–	–	–	–	–	0.3	–	–	–	0.3	0.3	–	–
Quartz in low-grade metamorphic rock fragments	–	–	–	–	1.2	–	–	0.5	0.3	0.3	0.3	0.3	0.3	0.3
Quartz in granitic and/or gneissic rock fragments	–	–	–	–	–	–	–	0.5	–	0.3	–	0.3	–	0.3
Calcite on monocrystalline quartz	–	–	–	–	8.0	–	–	–	–	–	–	–	–	–
Calcite on quartz in plutonic and/or gneissic rock fragments	–	–	–	–	0.4	0.3	0.3	0.3	–	–	0.3	–	–	–
F														
K-feldspar monocrystalline	0.3	0.3	0.8	1.5	0.6	0.3	0.3	–	–	0.5	–	–	0.3	–
K-feldspar in granitic and/or gneissic rock fragments	–	–	–	–	–	–	–	–	–	–	–	–	–	–
Calcite on K-feldspar (monocr.)	–	–	–	–	–	–	–	–	–	–	–	–	–	–
Calcite on K-feldspar in rock fragments	–	–	–	–	0.2	–	–	–	–	–	–	–	–	–
Plagioclase monocrystalline	–	0.3	0.3	–	0.2	1.0	0.3	0.5	0.8	0.8	1.8	0.5	0.3	0.8
Plagioclase in volcanic rock fragments	–	–	–	–	–	–	–	–	–	–	–	–	–	–
Plagioclase in low-grade metamorphic rock fragments	–	–	–	–	–	–	–	–	–	0.3	–	–	–	–
Plagioclase in granitic and/or gneissic rock fragments	0.5	–	–	–	–	–	–	–	–	–	–	–	–	–
Calcite on plagioclase (monocr.)	–	–	–	–	–	–	–	–	–	–	–	–	–	–
Calcite on plagioclase in rock fragments	–	–	–	–	–	–	–	–	–	–	–	–	–	–
L														
Acidic volcanic rock fragments	0.8	1.8	1.3	0.3	4.8	1.6	2.3	2.3	4.1	1.3	2.8	4.3	2.8	4.3
Low-grade metamorphic rock fragments	5.3	11.3	17.8	7.0	4.4	3.0	2.5	1.0	4.3	4.3	1.8	2.8	0.3	6.8
Cherts	2.5	2.0	0.8	1.0	–	3.3	3.0	9.8	1.5	4.5	5.5	7.3	6.3	5.8
Siltstone	0.3	1.0	–	2.0	–	–	–	–	3.5	3.0	–	0.5	–	–
Calcite on aphanitic rock fragments	–	–	–	–	2.6	–	–	–	–	–	–	–	–	–
Mica and chlorite	–	–	–	0.3	0.2	0.3	0.3	–	0.5	0.3	0.5	–	0.3	0.3
Mica and chlorite in granitic and/or gneissic rock fragments	–	–	–	–	–	–	–	–	–	–	–	–	–	–
Mica and chlorite in low-grade metamorphic rock fragments	–	–	–	–	0.2	–	–	–	–	–	–	–	–	–
Other minerals	–	–	–	–	0.4	0.3	–	0.3	0.5	–	0.3	–	–	0.3
Other minerals in rock fragments	–	–	–	–	–	–	–	–	–	–	–	–	–	–
CE														
Dolostone monocrystalline grain	7.0	17.8	12.5	17.3	11.8	8.3	10.0	8.3	15.3	17.5	19.3	12.0	9.8	14.5
Coarse-grained polycrystalline dolostone	42.5	27.3	32.8	29.8	3.8	24.3	20.5	23.8	4.5	7.5	8.8	23.8	27.8	9.0
Fine-grained polycrystalline dolostone	–	–	–	–	0.6	9.0	6.3	6.3	–	3.5	1.8	4.0	7.0	1.8
Mudstone	8.3	2.5	3.5	1.3	–	13.5	21.3	13.8	16.0	10.5	10.3	10.8	11.3	11.1
Microspatitic limestone	11.0	3.8	4.5	1.3	–	4.5	1.3	9.5	1.3	3.3	2.0	4.5	6.3	4.0
Spatitic limestone	1.8	1.0	1.0	0.3	–	0.3	0.3	–	–	–	–	–	–	–
Cherty limestone/dolostone grain	–	–	–	–	–	1.0	–	0.3	0.5	0.3	–	0.3	0.5	0.3
Calcite on dolostone grain	0.3	0.3	0.5	–	–	–	–	–	–	–	–	–	–	–
NCI														
Green particles	–	–	–	–	–	–	–	–	–	–	–	–	–	–
Pelitic intraclasts	0.5	–	–	–	–	–	–	0.8	0.3	0.3	0.5	0.3	–	1.5
Silica sponge and radiolaria	–	–	–	–	–	–	–	–	–	–	–	–	–	–
CI														
Carbonate bioclasts	4.8	1.0	1.5	4.3	0.6	3.0	1.8	3.3	1.8	2.8	5.3	2.0	1.3	1.3
Limeclast	–	–	–	–	–	–	0.3	–	–	–	–	–	–	–
Siliciclastic matrix	1.0	3.8	1.0	7.0	5.2	0.5	1.8	0.5	8.1	4.2	0.8	–	1.3	2.5
Carbonate matrix	0.3	0.8	0.5	0.3	1.4	1.5	1.5	1.0	14.8	6.0	1.9	0.2	0.3	1.5
Calcite cement	4.0	4.3	5.3	8.0	17.8	4.8	7.3	7.1	1.4	4.0	9.3	4.3	6.8	9.8
Dolomite cement	–	–	–	–	–	2.0	1.3	0.3	–	1.3	1.0	1.5	1.3	2.5
Oxides	0.3	0.8	0.8	1.3	2.8	0.8	1.0	0.3	0.3	0.8	–	0.5	–	0.3
Undeterminate grain	–	–	–	–	3.4	0.5	0.8	–	0.3	–	–	–	–	–
Patchy calcite	–	–	–	–	–	7.3	7.0	4.3	4.0	7.5	13.0	12.3	11.3	8.8
Spar of calcite	4.5	7.0	5.5	8.5	–	1.5	0.3	0.3	–	0.3	1.0	0.3	0.5	0.8
Calcite on undeterminate grain	–	–	–	–	7.4	1.0	2.3	0.5	0.8	2.8	1.0	0.8	0.5	0.8
Total	100.0	100.0	100.0	100.0	100.0	100.0	100.0	100.0	100.0	100.0	100.0	100.0	100.0	100.0

(*continued.*)

TABLE 1. MODAL POINT COUNTS OF THE PALEOGENE TURBIDITIC SUCCESSIONS (*Continued.*)

Stratigraphic unit	Clauzetto Flysch					
Samples	60	61	62	63	64	65
NCE						
Q						
Quartz monocrystalline	6.8	5.3	3.8	6.0	8.0	7.8
Coarse-grained polycrystalline quartz	2.5	0.5	2.0	1.8	1.3	2.8
Fine-grained polycrystalline quartz	0.8	–	1.0	1.0	0.8	0.5
Quartz in acidic volcanic rock fragments	–	0.3	0.5	0.5	0.3	0.5
Quartz in low-grade metamorphic rock fragments	0.3	–	0.6	0.3	0.5	–
Quartz in granitic and/or gneissic rock fragments	–	–	–	0.5	–	–
Calcite on monocrystalline quartz	–	–	–	–	–	–
Calcite on quartz in plutonic and/or gneissic rock fragments	0.3	–	0.8	0.5	–	–
F						
K-feldspar monocrystalline	0.3	0.3	0.5	0.3	0.5	–
K-feldspar in granitic and/or gneissic rock fragments	–	–	–	–	–	–
Calcite on K-feldspar (monocr.)	–	–	–	–	–	–
Calcite on K-feldspar in rock fragments	–	–	–	–	–	–
Plagioclase monocrystalline	1.0	0.5	1.3	1.0	1.5	0.3
Plagioclase in volcanic rock fragments	–	–	0.5	0.3	0.3	–
Plagioclase in low-grade metamorphic rock fragments	–	–	0.3	–	–	–
Plagioclase in granitic and/or gneissic rock fragments	–	–	0.3	0.3	–	–
Calcite on plagioclase (monocr.)	–	–	–	–	–	–
Calcite on plagioclase in rock fragments	–	–	–	–	–	–
L						
Acidic volcanic rock fragments	4.5	5.6	4.3	4.5	4.8	2.3
Low-grade metamorphic rock fragments	5.5	4.8	5.3	4.3	4.8	5.3
Cherts	6.3	5.0	6.3	5.8	5.3	5.0
Siltstone	1.0	–	0.5	0.3	0.8	–
Calcite on aphanitic rock fragments	–	–	0.3	–	–	–
Mica and chlorite	0.5	0.3	0.3	0.3	0.8	0.5
Mica and chlorite in granitic and/or gneissic rock fragments	–	–	–	–	–	–
Mica and chlorite in low-grade metamorphic rock fragments	–	–	–	–	–	–
Other minerals	0.8	0.3	0.3	0.3	0.5	0.3
Other minerals in rock fragments	–	–	–	–	–	–
CE						
Dolostone monocrystalline grain	14.8	19.8	12.3	11.3	13.5	11.3
Coarse-grained polycrystalline dolostone	13.8	7.8	6.8	8.8	8.0	7.8
Fine-grained polycrystalline dolostone	1.5	1.0	1.5	1.8	2.3	1.0
Mudstone	11.8	13.5	11.5	7.3	11.3	10.0
Microspatitic limestone	10.8	12.0	5.3	7.8	6.8	3.5
Spatitic limestone	1.0	1.0	0.5	0.5	–	–
Cherty limestone/dolostone grain	1.0	–	0.3	0.3	0.5	0.8
Calcite on dolostone grain	–	–	–	–	–	–
NCI						
Green particles	–	–	0.3	0.3	–	–
Pelitic intraclasts	0.8	0.5	1.8	–	1.0	1.8
Silica sponge and radiolaria	–	–	–	–	–	–
CI						
Carbonate bioclasts	1.5	1.0	1.5	–	1.8	0.8
Limeclast	–	–	–	–	–	–
Siliciclastic matrix	1.5	2.0	3.0	6.3	3.0	3.8
Carbonate matrix	0.5	0.5	1.8	2.3	0.5	4.5
Calcite cement	3.8	3.3	9.7	8.2	12.7	9.5
Dolomite cement	3.3	1.0	3.0	2.5	0.5	–
Oxides	0.3	0.3	0.3	0.8	0.3	–
Undeterminate grain	–	–	–	–	–	2.5
Patchy calcite	2.5	12.3	10.0	13.3	6.8	16.3
Spar of calcite	–	0.8	0.5	0.5	–	0.8
Calcite on undeterminate grain	0.5	0.3	1.0	–	0.8	0.3
Total	100.0	100.0	100.0	100.0	100.0	100.0

Note: NCE—noncarbonated extrabasinal; CE—carbonate extrabasinal; NCI—noncarbonated intrabasinal; CI—carbonate intrabasinal.

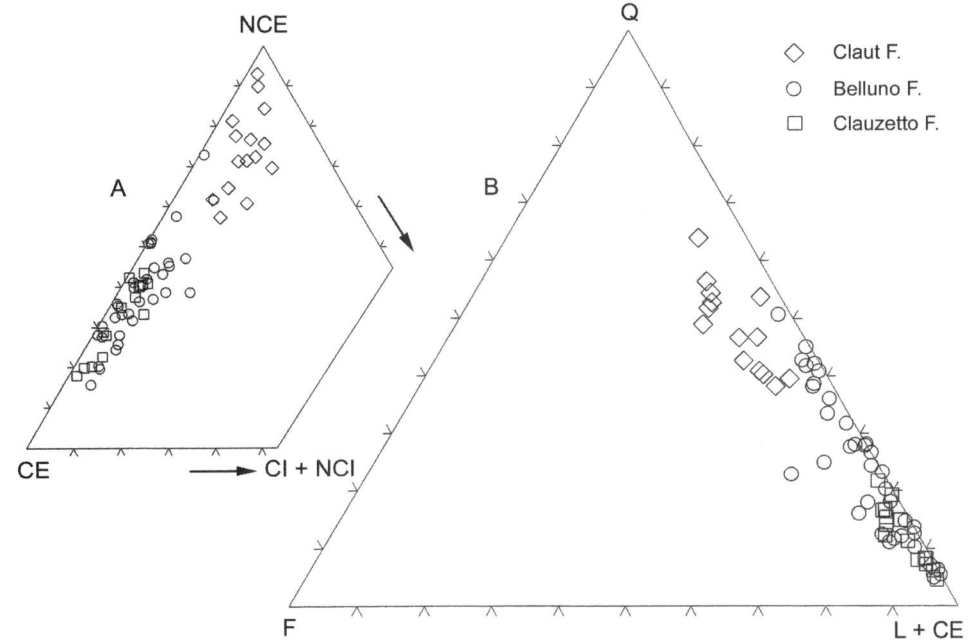

Figure 5. Petrography of the investigated successions showing: (A) the first-level classification with the whole intrabasinal and extrabasinal framework; (B) the extrabasinal framework; see text for the explanation of the vertices. NCE—noncarbonated extrabasinal; CE—carbonate extrabasinal; CI—carbonate intrabasinal; Q—quartz; F—feldspars; L + CE—fine-grained lithic fragments plus carbonate extrabasinal grains.

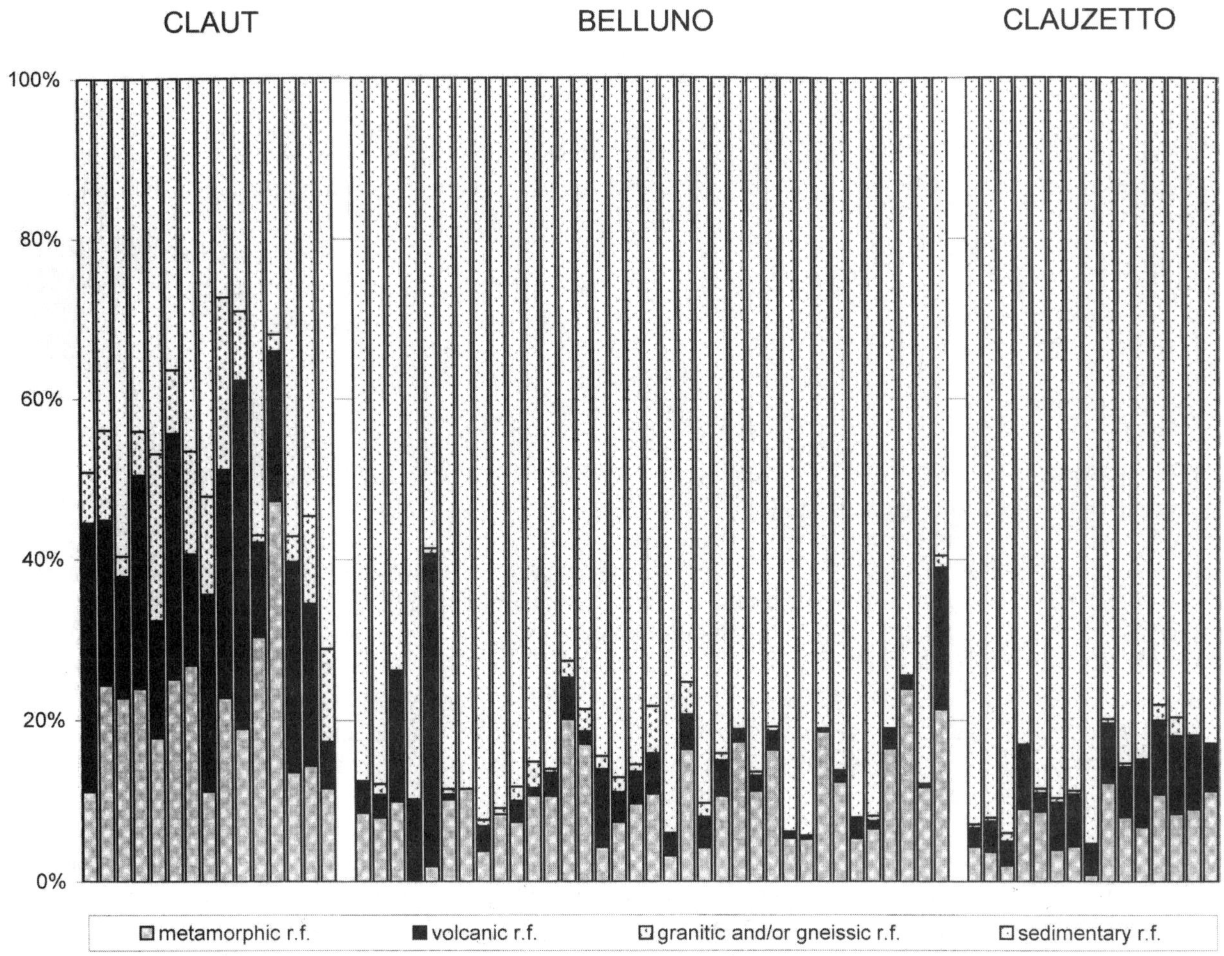

Figure 6. Total (coarse- + fine-grained) rock fragment (r.f.) populations in the Paleogene turbidite successions.

TABLE 2. FIELD AND LABORATORY DATA ON STUDIED SUCCESSIONS

Lithostratigraphic units	Age	Evaluated thickness (m)	Composition of the terrigenous arenite framework[†] (mean values)	Rock classification (number of samples)	Lithic fragments[‡]	Heavy minerals[‡]	Carbonate allochemical supplies	Paleoflows
Claut flysch (CF)	Middle Paleocene to early Eocene	Up to 700	QFL + CE 45.6–8.5–45.9	Quartz-rich "litharenite" (15)	CHERT, LIMESTONE GRANITIC AND /OR GNEISSIC Acidic volcanics Low-grade metamorphic	ZTR Cr-SPINEL Garnet	Scarse (<5%) and limited to the upper portion	From western sectors
Belluno flysch (BF)	Early Eocene to middle Eocene	Up to 1200	QFL + CE 25.0–2.2–72.8	"Litharenite" (35)	DOLOSTONE, LIMESTONE Chert Low-grade metamorphic Acidic volcanics	ZTR Cr-SPINEL EPIDOTE Garnet	Locally (T. Limana and Alpago sections) up to 30% on total sediment volume	Mainly from NW from S the main allochemical supplies
Clauzetto flysch (CzF)	Early Eocene	Up to 600	QFL + CE 13.1–1.7–85.2	"Litharenite" (15)	LIMESTONE, DOLOSTONE Chert Low-grade metamorphic Acidic volcanics	ZTR Cr-SPINEL EPIDOTE Garnet	Important (~50%) on total sediment volume	From western sectors

[†]Q—quartz; F—feldspars; L + CE—total fine-grained lithic fragments plus carbonate rock fragments.
[‡]Capital letters indicate abundant and characteristic components. ZTR index is the combined percentage of zircon, plus tourmaline, plus rutile (Hubert, 1962).

Both Belluno flysch and Clauzetto flysch show a significantly different composition and are "litharenitic" in composition (mean values, respectively, Q 25.0, F 2.2, L + CE 72.8, and Q 13.1, F 1.7, L + CE 85.2; Table 2) with a predominance of dolomite on the other rock fragment types. Low-grade metamorphic rock fragments include phyllite and quartzite; volcanic rock fragments of rhyolitic composition are also present.

Heavy Minerals

About 40 samples were selected for heavy mineral investigation and were prepared following the standard techniques described by Gazzi et al. (1973) and Mange and Maurer (1992). Repeated dilute HCl (10%) immersions and boiling with oxalic acid and aluminum were necessary to remove the high carbonate content and Fe-oxide grain coatings, respectively. The heavy minerals were separated from the 0.0625–0.250 mm sandy fraction using tetrabromoethan (density: 2.96 g/cm^3) and then optically analyzed in transmitted light after immersion in methylene iodide (n = 1.74); at least two hundred transparent grains were determined for each sample. Results are reported in Table 3 and plotted in Figure 7; the class "others" includes kyanite, chloritoid, olivine, anatase, and brookite.

The heavy mineral fractions are commonly <1.0% by weight of the grain-size fraction considered, and transparent heavy minerals average ~20% of the total heavy fraction.

All samples have a ZTR index (zircon-tourmaline-rutile; Hubert, 1962) (Table 3; Fig. 7) ranging from a few percent to over 40%, high amounts of reddish brown Cr-spinel (up to 70%), and variable amounts of garnets, ranging from 2% to 27%. Epidotezoisite grains are always present, and in two samples of Clauzetto flysch, they are very abundant (up to 60%). Moreover, we found minor or trace amounts of unstable heavy mineral species such as staurolite, kyanite, olivine, glaucophane s.s., and crystals of glaucophane-riebekite series. At a first glance, the occurrence of these mineral species is particularly indicative of various ranks of metamorphic rock sources. Moreover, the presence of small quantities of monazite and xenotime points to a granitic source.

The presence of abundant Cr-spinel suggests provenance from ophiolite-bearing units, even if the absence of serpentinite rock fragments points to possible recycling of the spinels. In fact, Zimmerle (1984) stressed the high stability of the brown spinel that "can also occur as solitary, without unstable serpentinite fragments." Also Kukharenko (1964) demonstrated the high stability and durability of this mineral that may be hydraulically concentrated in placers. In two recent papers, Lenaz et al. (2000, 2003) analyzed the geochemistry of spinel grains from the Claut flysch, pointing out high Cr content of the minerals. According to Lenaz et al., the analyzed minerals were supplied mainly by mantle peridotites, while other turbidite successions outcropping more to the east were mainly characterized by Cr-spinel derived from volcanic sources. This mineral is considered distinctive of ophiolites of the Vardar Ocean eroded from the Jurassic to Paleogene (Lenaz et al., 2003).

PALEOGENE PALEOGEOGRAPHY

On the basis of our field and laboratory analyses, the Paleogene turbiditic successions of eastern Southern Alps can be regarded as multisource units, made up of two different types of sources that controlled the basin fill without appreciable variation during sedimentation time. Two specific sources of sediments were contemporarily active and gave origin compositionally distinctive types of bed: carbonate beds made up of intrabasinal bioclastic detritus and terrigenous beds made up of siliciclastic-carbonate detritus. Mixed beds, derived by the mixing of sediments are also present. Bioclastic detritus was supplied by active carbonate shelf areas located in the photic zone. As depicted by Cati et al. (1989a, 1989b) on the basis of subsurface data, the region was characterized by large carbonate platforms in which small troughs elongated with a Dinaric trend developed (Fig. 8). The bioclastic detritus could have been supplied by various sectors

TABLE 3. HEAVY MINERALS OF CLAUT, BELLUNO, AND CLAUZETTO ARENITES

Stratigraphic unit	Claut flysch					Belluno flysch													
Sample	1	3	4	11	15	16	18	21	22	23	24	27	29	30	31	32	33	34	35
Heavy minerals/2–4 phi	0.6	1.1	1.5	0.9	0.7	1.4	1.4	0.9	0.2	0.2	0.1	1.4	0.5	1.1	1.0	1.3	0.8	0.6	1.1
Transparents	21.2	22.3	28.4	33.5	21.2	34.2	41.0	44.0	46.0	41.2	42.3	35.7	39.2	47.6	16.6	11.6	21.3	17.2	21.4
Opaques	45.8	63.7	56.8	48.9	65.3	49.6	44.0	39.6	37.9	45.4	38.2	36.7	41.0	38.1	65.3	58.9	57.4	57.3	60.2
Turbids	33.0	14.0	14.8	17.6	13.5	16.2	15.0	16.4	16.1	13.4	19.5	27.6	19.8	14.3	18.1	29.5	21.3	25.5	18.4
Zircon	12.0	17.0	7.5	10.0	7.5	13.0	2.0	9.0	8.5	8.0	5.5	7.0	12.5	14.0	4.0	9.5	12.5	11.0	11.0
Tourmaline	8.5	3.5	18.5	15.5	16.5	5.0	11.5	21.0	28.0	8.5	15.0	8.5	4.5	21.5	5.5	4.5	4.0	7.5	6.0
Rutile	11.5	7.5	11.0	10.5	5.0	10.5	16.5	3.5	4.0	4.5	11.5	8.5	9.0	3.5	4.5	8.0	3.0	9.5	7.0
Garnet	16.0	15.5	22.0	21.5	18.5	11.0	10.5	7.5	4.5	9.0	6.0	5.0	12.0	13.0	18.0	9.3	6.0	10.0	11.5
Epidote	1.0	1.5	–	2.5	2.0	–	–	0.5	–	–	0.5	tr	–	–	14.7	10.0	9.0	4.0	9.5
Alkaline amphibole	2.0	3.0	2.5	0.5	5.0	2.0	0.5	0.5	0.5	1.0	2.0	3.0	4.0	–	1.3	–	–	–	–
Olivine	–	–	–	–	–	–	–	–	–	–	–	–	–	–	0.5	–	–	–	–
Titanite	–	tr	–	–	–	–	–	1.5	–	–	–	–	–	–	–	–	–	–	–
Staurolite	1.0	–	1.5	0.5	1.0	–	1.0	1.5	2.0	0.5	3.0	1.0	2.0	1.5	0.5	–	0.5	–	–
Kyanite	–	–	1.0	–	–	tr	–	–	–	–	–	–	–	–	–	–	–	tr	–
Chloritoid	–	0.5	–	–	1.0	–	tr	–	–	–	–	tr	–	–	0.5	–	–	–	1.0
Cr-spinel	43.5	48.0	34.0	19.5	21.5	53.0	54.0	24.0	12.0	53.5	38.0	39.0	34.0	10.0	41.0	56.7	48.0	56.5	47.0
Anatase	0.5	–	–	3.5	7.5	1.0	3.0	19.5	28.5	7.0	10.0	13.0	19.0	24.5	4.5	1.0	6.0	–	3.0
Brookite	1.0	1.0	–	10.0	10.5	0.5	0.5	10.0	10.0	5.5	4.5	11.0	1.5	8.5	–	1.0	3.5	–	4.0
Monazite	2.0	1.0	1.0	3.0	1.0	–	0.5	1.0	–	0.5	1.0	1.0	1.0	0.5	2.0	–	1.5	0.5	–
Xenotime	1.0	0.5	1.0	1.5	2.0	4.0	–	1.0	0.5	2.0	1.0	1.0	0.5	2.5	–	–	–	–	–
Undetermined	–	1.0	–	1.5	1.0	–	–	1.0	–	–	2.0	2.0	–	0.5	3.0	–	6.0	1.0	–
Total	100.0	100.0	100.0	100.0	100.0	100.0	100.0	100.0	100.0	100.0	100.0	100.0	100.0	100.0	100.0	100.0	100.0	100.0	100.0

TABLE 3. (*Continued.*)

Stratigraphic unit	Belluno flysch			Clauzetto flysch														
Sample	36	37	38	51	52	53	54	55	56	57	58	59	60	61	62	63	64	65
Heavy minerals/2–4 phi	0.9	0.6	0.3	0.8	1.0	0.6	0.8	0.5	0.8	0.3	0.3	0.7	0.4	1.0	0.7	0.9	0.5	0.8
Transparents	19.7	11.6	12.9	12.7	15.1	12.8	24.4	25.8	3.6	37.7	69.0	21.2	30.3	23.5	21.6	9.5	5.4	28.6
Opaques	56.4	67.4	62.3	72.4	65.0	74.3	61.3	60.2	89.9	48.0	24.1	63.2	47.0	53.3	60.5	51.9	75.8	55.9
Turbids	23.9	21.0	24.8	14.9	19.9	12.9	14.3	14.0	6.5	14.3	6.9	15.6	22.7	23.2	17.9	38.6	18.8	15.5
Zircon	6.5	4.5	12.5	6.1	4.0	14.6	5.3	8.1	5.9	–	0.9	6.0	6.0	9.0	8.0	5.6	5.9	7.0
Tourmaline	1.5	3.0	5.5	20.2	8.9	8.8	8.7	10.7	15.3	7.6	0.9	18.1	11.5	13.0	21.5	11.1	15.7	16.5
Rutile	4.0	4.0	4.0	13.1	5.0	16.6	10.0	4.0	7.1	4.1	–	3.4	13.0	10.0	12.0	2.2	3.9	7.5
Garnet	3.0	3.0	5.0	4.0	3.0	7.8	9.3	9.4	27.1	2.5	2.6	2.0	2.0	2.5	2.0	–	9.8	6.0
Epidote	1.0	–	11.5	3.0	–	2.0	–	5.4	1.2	65.0	46.1	1.3	14.0	1.5	0.5	–	2.0	1.5
Alkaline amphibole	–	0.5	0.5	–	–	–	0.7	–	–	1.5	0.9	–	–	–	–	–	–	–
Olivine	–	–	–	–	–	–	–	–	–	–	–	–	–	–	–	–	–	–
Titanite	–	–	–	–	tr	–	–	–	–	–	–	–	–	–	–	–	–	–
Staurolite	–	–	–	–	tr	–	–	–	–	–	tr	–	–	–	–	–	–	–
Kyanite	–	0.5	–	–	–	–	–	–	–	–	–	–	–	–	–	–	–	–
Chloritoid	1.5	0.5	1.5	–	–	–	–	–	–	–	–	–	–	–	–	–	–	–
Cr-spinel	75.0	76.5	55.0	47.5	75.1	38.0	58.7	52.3	30.4	5.1	47.7	64.5	29.0	55.0	49.0	66.7	58.8	44.5
Anatase	2.0	3.5	2.0	–	–	–	–	–	2.4	0.5	–	–	–	–	–	–	–	0.5
Brookite	2.5	2.0	1.0	6.1	4.0	11.7	7.3	10.1	5.9	13.7	0.9	4.7	24.0	9.0	7.0	14.4	3.9	16.5
Monazite	1.0	–	0.5	–	–	–	–	–	1.2	–	–	–	–	–	–	–	–	–
Xenotime	–	–	–	–	–	–	–	–	3.5	–	–	–	–	–	–	–	–	–
Undetermined	2.0	2.0	1.0	–	–	0.5	–	–	–	–	–	–	0.5	–	–	–	–	–
Total	100.0	100.0	100.0	100.0	100.0	100.0	100.0	100.0	100.0	100.0	100.0	100.0	100.0	100.0	100.0	100.0	100.0	100.0

Note: tr indicates trace.

of the Friuli Platform. This platform was an active carbonate factory since the Late Jurassic and during the Cretaceous. According to some authors, it underwent repeated subaerial exposures during the Late Cretaceous and early Paleocene (Iaccarino and Roveri, 1964; Cousin, 1981; Cati et al., 1989a, 1989b). Many stratigraphic indications suggest a reestablishment of the carbonate production also during the Paleogene, as demonstrated by the presence of resedimented carbonate debris in the basinal or slope deposits, such in the Scaglia Rossa (Tunis and Uchman, 1998; Swinburne and Noacco, 1993) and the overlying Belluno flysch (Gnaccolini, 1967, 1968) and Clauzetto flysch. It cannot be excluded that other carbonate platforms, presently not preserved, were active in sectors proximal to the terrigenous supplies. The presence of "mixed" beds containing different kinds of grains (i.e., allochemical and terrigenous) indicates a parking area where these two types of sediments could have been mixed up, and suggests a location of a carbonate platform close to the entry points of the terrigenous detritus. On the contrary, the general distal facies associations that characterize all stratigraphic units point to very long-distance terrigenous supplies. Unfortunately, the compressional tectonics active during Neogene in the Southalpine domain probably destroyed the proximal portions of these turbiditic basins and their relation with the respect to the source areas.

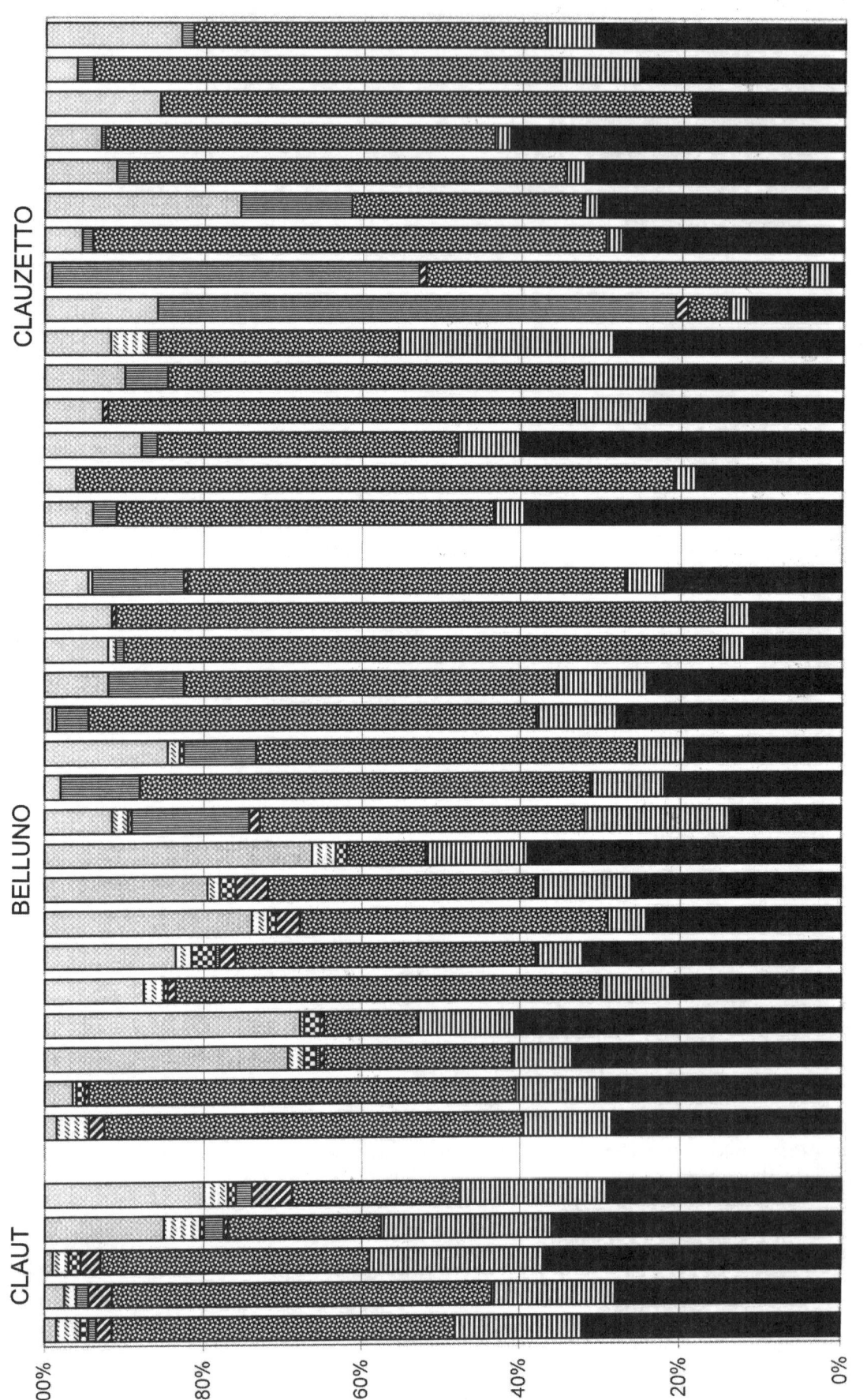

Figure 7. Transparent heavy mineral associations in the investigated successions. The class "others" includes kyanite, chloritoid, olivine, anatase, and brookite (see also Table 3). ZTR index is the combined percentage of zircon, plus tourmaline, plus rutile (Hubert, 1962).

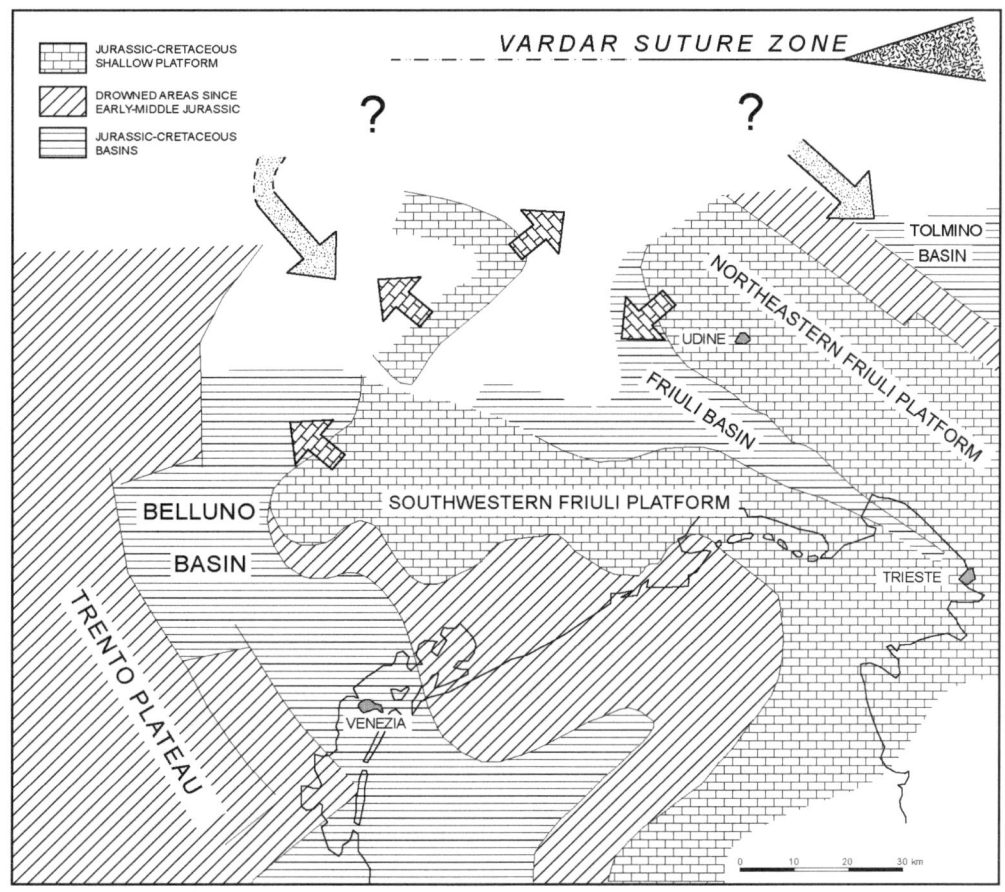

Figure 8. Paleogeographic reconstructions of the region encompassing the eastern Southern Alps and external Dinarides with the indication of the major dispersal patterns (modified by Cati et al., 1989b).

Westward of Feltre, the turbiditic facies gradually change into slope facies, mostly hemipelagic pelites, and the flysch succession onlaps the eastern slope of the Lessini Shelf (Bosellini, 1989; Doglioni, 1990; Grandesso and Stefani, 1993; Trevisani, 1994). Also, the westward-younging of the turbidites and the variation of sediment thickness (Fig. 3, T. Caorame section versus T. Limana; Grandesso, 1976; Stefani and Grandesso, 1991) testify to the depocenter migration in response to the Dinaric thrust propagation (Cousin, 1981; Doglioni and Bosellini, 1987).

In all the investigated units, the paleocurrent data indicate a predominant sediment transport parallel to the inferred NW-SE–trending Dinaric basin axis, with a provenance from the northern sector, where part of the terrigenous material was probably eroded.

The petrography of the terrigenous arenites clearly shows the presence of different sources for the detritus of the investigated successions. In fact, the Claut flysch, the oldest one, is more quartzolithic and displays a higher variety of basement-derived rock fragments with respect to the youngest successions (Belluno and Clauzetto F., Fig. 5B), where this types of terrigenous input is masked and diluted by high amounts of detritus derived from carbonate sedimentary cover (Fig. 6). Both the Dolomites and Dinaric areas are characterized by a carbonate sedimentary cover, but a thick carbonate cover also characterizes the Austroalpine units, presently cropping out north of the Periadriatic Lineament and bordering the Tauern Window (Fig. 1A).

The heavy mineral data indicate a relatively constant source area of the terrigenous material from middle Paleocene (sedimentation of Claut flysch) to early-middle Eocene (sedimentation of Belluno and Clauzetto flysch). This source was characterized by medium-grade metamorphic and granitic rocks, but a partial recycling from older arenitic successions cannot be ruled out.

An important source of sediment is represented also by metamorphic rocks of various rank that have furnished Cr-spinel, alkaline amphiboles, epidotes, and staurolite. The location of these metamorphic rocks is particularly intriguing.

For the provenance of alkaline amphiboles, we can hypothesize some inputs from the Variscan high-pressure successions that had been exposed since the middle Cretaceous, as evidenced by the presence of glaucophane-crossite associations in the Allgäu flysch deposits (Winkler and Bernoulli, 1986). A (re-)activation of the same source also during the Paleocene-Eocene cannot be excluded. However, a possible oceanic source can be supposed more to the south. In fact, the final closure of the Vardar/Meliata Ocean (Late Jurassic) led to the formation of an initial nappe pile at the southeastern margin of the Austroalpine microplate that included the Vardar/Meliata oceanic suture zone (Neubauer, 1994). Sediments started to be shed into the Austroalpine realm in the ?Berriasian/Valanginian from this nappe pile (Faupl and Tollmann, 1979).

Von Eynatten and Gaupp (1999) argued that a provenance from an oceanic crustal source is proved by the dominant presence of Cr-spinel in part of the Cretaceous sedimentary succession that crops out also in the Northern Calcareous Alps. According to their reconstruction, this source was no longer active since the Turonian/Coniacian.

Chromite is very abundant in the Aptian-Albian flysch of the Lienz Dolomites (Faupl, 1976) and in the Turonian to Coniacian flysch sequences of the Simme nappe (Flück, 1973). On the contrary, chromite is very rare or absent in the Upper Cretaceous Lombardian flysch (Bernoulli and Winkler, 1990) and, where present, it probably was derived from the ultramafic rocks cropping out westward.

The presence of Cr-spinel and alkaline amphiboles in our samples does not allow a clear distinction between these different sources, but the paleogeographic location of the investigated successions and the presence of NW-SE–trending troughs suggest that rocks derived from the Vardar Ocean could have been an important source (Fig. 8). It is important to stress that the turbiditic successions cropping out more eastward (Lenaz et al., 2003) show the significant presence of Cr-spinel grains, suggesting the Vardar zone as the most probable source of terrigenous detritus.

In conclusion, the slight similarity in composition of Claut flysch, quartz-rich "litharenites," with respect to the youngest Belluno flysch and Clauzetto flysch, carbonate-rich "litharenites," suggests that the terrigenous detritus supplied into the basins from late Paleocene to middle Eocene was similar, in particular in regard to heavy mineral associations, and, consequently, it may have been eroded from different sectors of the same orogenic belt, characterized by small lithological differences. The increasing content of carbonate rock fragments in the young successions provides evidence of an important involvement of the sedimentary cover with time. On the contrary, the basement supply became less significant with time, as underlined by the different proportions of "granitic and/or gneissic rock fragments" among the oldest Claut flysch and the youngest Belluno and Clauzetto flysch.

CONCLUSIONS

Biostratigraphic, lithostratigraphic, and petrographic data allow us to consider the Paleogene turbiditic successions cropping out in the eastern Southern Alps as the response to the tectonic phases that involved the Dinaric chain from Paleocene onward. Petrographic analyses demonstrated that these successions are prevalently composed by two types of sediments derived from well distinct sources. An important source was represented by the coeval allochemical detritus yielded by carbonate factories, most probably linked to the Cretaceous Friuli Platform, which were renewed during the Paleocene to middle Eocene.

The terrigenous detritus of Claut, Belluno, and Clauzetto flysch deposits derived from different sectors of the same orogenic belt, as demonstrated by similar arenite composition and heavy mineral associations. The terrigenous source areas were probably characterized by extensive carbonate covers above a metamorphic basement with remnants of oceanic crust as indicated by the common presence of Cr-spinels. The scattered but significant presence of alkaline amphibole in all the successions points to a high-pressure source related to the subduction and subsequent obduction of an oceanic crust.

ACKNOWLEDGMENTS

Stefani and Grandesso were supported by research grants of Padova University and by Consiglio Nazionale delle Ricerche (CNR), Istituto di Geoscienze e Georisorse, Padova. The manuscript strongly benefited from the critical, patient reviews of Daniela Fontana and Hilmar von Eynatten. Detailed sample locations will be provided on request.

REFERENCES CITED

Aubouin, J., 1963, Essai sur la paleogeographie post Triasique et l'evolution Secondaire et Tertiarie du versant sud des Alpes Orientales (Alpes Meridionales; Lombardie et Venetie, Italie; Slovenie Occidentale, Yougoslavie): Bolletin de la Societe Geologique de France, v. 5, p. 730–766.

Berggren, W.A., and Miller, K.G., 1988, Paleogene tropical planktonic foraminiferal biostratigraphy and magnetobiochronology: Micropaleontology, v. 34, p. 362–380, doi: 10.2307/1485604.

Berggren, W.A., Kent, D.V., Swisher, C.C., III, and Aubry, M.P., 1995, A revised Cenozoic geochronology and chronostratigraphy, in Berggren, W.A., Kent, D.V., Swisher, C.C., III, and Aubry, M.P., and Hardenbol, J., eds., Geochronology Time Scales and Global Stratigraphic Correlation: Society of Economic Paleontologists and Mineralogists Special Publication 54, p. 129–212.

Bernoulli, D., and Winkler, W., 1990, Heavy mineral assemblages from Upper Cretaceous South- and Autroalpine flysch sequences (northern Italy and southern Switzerland): Source terranes and palaeotectonic implications: Eclogae Geologicae Helvetiae, v. 83, p. 287–310.

Bertelli, L., Cantelli, L., Castellarin, A., Fantoni, R., Mosconi, A., Sella, M., and Selli, L., 2003, Upper crustal style, shortening and deformation age in the Alps along the southern sector of the TRANSALP profile: Memorie di Scienze Geologiche, v. 54, p. 123–126.

Bosellini, A., 1989, Dynamics of Tethyan carbonate platforms, in Crevello, P.D., et al., eds., Controls on Carbonate Platform and Basin Development: Society of Economic Paleontologists and Mineralogists Special Publication 44, p. 3–13.

Bouma, A.H., 1962, Sedimentology of Some Flysch Deposits: Amsterdam, Elsevier, 168 p.

Cati, A., Fichera, R., and Cappelli, V., 1989a, Northeastern Italy: Integrated processing of geophysical and geological data: Memorie della Società Geologica Italiana, v. 40, p. 273–288.

Cati, A., Sartorio, D., and Venturini, S., 1989b, Carbonate platforms in the subsurface of northern Adriatic area: Memorie della Società Geologica Italiana, v. 40, p. 295–308.

Channell, J.E.T., and Horvath, F., 1976, The African/Adriatic promontory as a palaeogeographic premise for Alpine orogeny and plate movements in the Carpatho-Balkan region: Tectonophysics, v. 35, p. 71–101, doi: 10.1016/0040-1951(76)90030-5.

Costa, V., Doglioni, C., Grandesso, P., Masetti, D., Pellegrini, G.B., and Tracanella, E., 1996, Note illustrative del Foglio 063 Belluno della Carta Geologica d'Italia alla scala 1:50.000: Roma, Istituto Poligrafico e Zecca dello Stato, p. 1–73.

Cousin, M., 1981, Les rapports Alpes-Dinarides—Les Confins de l'Italie et de la Yougoslavie: Societe Géologique du Nord, v. 5, I, p. 1–521, II, p. 1–521.

Dainelli, G., 1910, L'Eocene del Friuli occidentale: Bollettino della Società Geologica Italiana, v. 29, p. 1–54.

Dickinson, W.R., 1970, Interpreting detrital modes of graywacke and arkose: Journal of Sedimentary Petrology, v. 40, p. 695–707.

Di Napoli Alliata, E., Proto Decima, F., and Pellegrini, G.B., 1970, Studio geologico, stratigrafico e micropaleontologico dei dintorni di Belluno: Memorie della Società Geologica Italiana, v. 9, p. 1–28.

Doglioni, C., 1990, Thrust tectonics examples from the Venetian Alps: Studi Geologici Camerti, vol. spec. 1990, p. 117–129.

Doglioni, C., and Bosellini, A., 1987, Eoalpine and mesoalpine tectonics in the Southern Alps: Geologische Rundschau, v. 76, p. 735–754, doi: 10.1007/BF01821061.

Fantoni, R., Catellani, D., Merlini, S., Rogledi, S., and Venturini, S., 2002, La registrazione degli eventi deformativi Cenozoici nell'Avampaese Veneto-Friulano: Memorie della Società Geologica Italiana, v. 57, p. 301–313.

Faupl, P., 1976, Sedimentologische studien im Kreideflysch der Lienzer Dolomiten: Anzieger der Osterreichischen Akademie der Wissenschaften, Mathematisch-Naturwissenschaftliche Klasse, v. 113, p. 131–134.

Faupl, P., and Tollmann, A., 1979, Die Roßfeldschichten: Ein Beispiel für Sedimentation im Bereich einer tektonisch aktiven Tiefseerinne aus der kalkalpinen Unterkreide: Geologische Rundschau, v. 68, p. 93–120, doi: 10.1007/BF01821124.

Flück, W., 1973, Die Flysche der praealpinen decken im Simmental und Saanenland: Bern, Beiträge zur Geologischen Karte der Schweizer, Neuve Folge 146, p. 146.

Gazzi, P., 1966, Le arenarie del flysch sopracretaceo dell'Appennino modenese: Correlazioni con il flysch di Monghidoro: Mineralogica Petrografica Acta, v. 12, p. 69–97.

Gazzi, P., Zuffa, G.G., Gandolfi, G., and Paganelli, L., 1973, Provenienza e dispersione litoranea delle sabbie delle spiagge Adriatiche fra le foci dell'Isonzo e del Foglia: Inquadramento regionale: Memorie della Società Geologica Italiana, v. 12, p. 1–37.

Gnaccolini, M., 1967, Il Flysch dell'Alpago: Rivista Italiana di Paleontologia e Stratigrafia, v. 73, p. 889–906.

Gnaccolini, M., 1968, Caratteristiche sedimentologiche del Flysch del Vallone Bellunese: Rivista Italiana di Paleontologia e Stratigrafia, v. 74, p. 63–70.

Grandesso, P., 1976, Biostratigrafia delle formazioni Terziarie del Vallone Bellunese: Bollettino della Società Geologica Italiana, v. 94, p. 1323–1348.

Grandesso, P., and Stefani, C., 1993, La terminazione occidentale del flysch Bellunese: Osservazioni stratigrafiche e petrografiche: Pavia, Atti del Terzo Convegno Annuale del Gruppo Italiano di Sedimentologia, p. 11–12.

Gradstein, F.M., Ogg, J.O., and Smith, A.G., eds., 2004, A Geological Time Scale: Cambridge, UK, Cambridge University Press, 589 p.

Hsü, K.J., 1970, The meaning of the word Flysch—A short historical research, in Lajoie, J., ed., Flysch sedimentation in North America: Toronto, Geological Association of Canada Special Paper 7, p. 1–11.

Hubert, J.F., 1962, A zircon-tourmaline-rutile maturity index and interdependence of the composition of heavy mineral assemblages with the gross composition and texture of sandstones: Journal of Sedimentary Petrology, v. 32, p. 440–450.

Iaccarino, S., and Roveri, E., 1964, Sull'età della scaglia nella media valle dell'Arzino in destra Tagliamento (Udine): Bollettino della Società Geologia Italiana, v. 83, p. 3–20.

Kukharenko, A.A., 1964, Mineralogie des gisements alluvionnaires, Moscou 1961: Gosgoltekhizdat, v. 1, p. 318 (translation no. 4453, Paris, Bureau de recherches geologiques et minieres [BRGM], 1964).

Lenaz, D., Kamenetsky, V.S., Crawford, A.J., and Princivalle, F., 2000, Melt inclusions in detrital spinel from the SE Alps (Italy-Slovenia): A new approach to provenance studies of sedimentary basins: Contributions to Mineralogy and Petrology, v. 139, p. 748–758, doi: 10.1007/s004100000170.

Lenaz, D., Kamenetsky, V.S., and Princivalle, F., 2003, Cr-spinel supply in the Brkini, Istrian and Krk Island flysch basins (Slovenia, Italy and Croatia): Geological Magazine, v. 140, p. 335–342, doi: 10.1017/S0016756803007581.

Mange, M.A., and Maurer, H.F.W., 1992, Heavy Minerals in Colour: London, Chapman and Hall, 147 p.

Marie, P., and Cousin, M., 1965, Données micropaléontologiques sur le couches inférieures du flysch dans le Val Cellina (Alpes Méridionales, Udine, Italie): Bulletin de la Société Géologique de France, v. 6, p. 456–460.

Neubauer, F., 1994, Kontinentkollision in den Ostalpen: Geowissenschaften, v. 12, p. 136–140.

Piccoli, G., and Proto Decima, F., 1969, Ricerche biostratigrafiche sui depositi flyschoidi della regione Adriatica settentrionale e orientale: Memorie degli Istituti di Geologia e Mineralogia dell'Università di Padova, v. 27, p. 1–21.

Richter, D., 1970, Flysch und Molasse an der Sudalpen-Dinariden-Grenze zwischen Brenta und Isonzo: Geologische Mitteilungen, v. 9, p. 207–302.

Saint-Marc, P., 1963, Étude géologique de la région de Barcis (Alpes méridionales, province d'Udine, Italie): Bolletin de la Societe Géologique de France, s. 7, v. 5, p. 803–808.

Stefani, C., and Grandesso, P., 1991, Studio preliminare di due sezioni del Flysch Bellunese: Rendiconti della Società Geologica Italiana, v. 14, p. 157–162.

Swinburne, N., and Noacco, A., 1993, The platform carbonates of Monte Jouf, Maniago, and the Cretaceous stratigraphy of the Italian Carnian Prealps: Geologia Croatica, v. 46, p. 25–40.

Toumarkine, M., and Luterbacher, H., 1985, Paleocene and Eocene planktonic foraminifera, in Bolli, H.M., Saunders, J.B., and Perch-Nielsen, K., eds., Plankton Stratigraphy: Cambridge, Cambridge University Press, p. 87–154.

Trevisani, E., 1994, Evoluzione paleogeografica e stratigrafica sequenziale del margine orientale del Lessini Shelf durante l'Eocene inferiore-medio (Marosticano–Bassanese, Prealpi Venete): Memorie di Scienze Geologiche, v. 46, p. 1–15.

Tunis, G., and Uchman, A., 1998, Ichnology of Eocene flysch deposits in the Carnian pre-Alps (north-eastern Italy): Gortania, Atti Museo Friulano di Storia Naturale, v. 20, p. 41–58.

Tunis, G., and Venturini, S., 1992, Evolution of the southern margin of the Julian Basin with emphasis on the megabeds and turbidites sequences of the southern Julian Prealps (NE Italy): Geologia Croatica, v. 45, p. 127–150.

Winkler, W., and Bernoulli, D., 1986, Detrital high-pressure/low-temperature minerals in a late Turonian flysch sequence of the eastern Alps (western Austria): Implications for early Alpine tectonics: Geology, v. 14, p. 598–601, doi: 10.1130/0091-7613(1986)14<598:DHMIAL>2.0.CO;2.

von Eynatten, H., and Gaupp, R., 1999, Provenance of Cretaceous synorogenic sandstones in the eastern Alps: Constraints from framework petrography, heavy mineral analysis and mineral chemistry: Sedimentary Geology, v. 124, p. 81–111, doi: 10.1016/S0037-0738(98)00122-5.

Zimmerle, W.W., 1984, The geotectonic significance of detrital brown spinel in sediments: Mitteilungen Geologisch-Paläontologischen Institut Universität Hamburg, v. 56, p. 337–360.

Zuffa, G.G., 1980, Hybrid arenites: Their composition and classification: Journal of Sedimentary Petrology, v. 50, p. 21–29.

Zuffa, G.G., 1985, Optical analyses of arenites: Influence of methodology on compositional results, in Zuffa, G.G., ed., Provenance of Arenites: Dordrecht, Reidel, NATO ASI Series, p. 165–189.

MANUSCRIPT ACCEPTED BY THE SOCIETY 9 AUGUST 2006

Alluvial sand composition as a tool to unravel late Quaternary sedimentation of the Modena Plain, northern Italy

Stefano Lugli
Simona Marchetti Dori
Daniela Fontana
Dipartimento di Scienze della Terra, Università degli Studi di Modena e Reggio Emilia, Largo S. Eufemia n. 19, 41100 Modena, Italy

ABSTRACT

The Modena alluvial plain is located on the northern side of the northern Apennines fold-and-thrust belt, where streams draining the chain flow toward the northeast into the Po River. The alluvial plain is characterized by a spectacular abundance of archaeological sites of various ages and can be considered a natural laboratory for the reconstruction of the recent sedimentary evolution of the Po Plain. Detailed modal analyses of modern sands of the Modena Plain streams indicate that the provenance signal can be distinguished on the basis of key components, such as quartz, feldspar, carbonate, and lithic fragments. The compositional fields of the streams depend on the extent of the watershed, the recycling of older fluvial sediments, and the sediment input from tributary streams.

The modal analyses demonstrate that sand composition of the major rivers (Panaro and Secchia) has not changed during the Holocene, when sediment production, storage, and dispersal were probably dominated by colluvial aggradation in an environment characterized by dense vegetation cover.

In the late Pleistocene, fluvial sands were characterized by higher feldspar contents compared with modern and Holocene sands. This feldspar abundance could reflect a high-frequency signal in sediment supply rates linked to secular variations of weathering processes, and it reveals the strong denudation and sediment removal conditions of the last glacial stage (15–18 ka).

The implication of this study is that provenance of Holocene sediments now buried in the floodplain can be determined by a simple comparison with modern sand composition. Sand composition studies may represent a useful tool to reconstruct the Pleistocene-Holocene fluvial sediment supply and the evolution of human settlements as function of climate and drainage system changes.

Keywords: Modena, fluvial sand, provenance, Quaternary.

INTRODUCTION

Sand composition studies are useful for reconstructing the sedimentary history of a basin as a function of climatic-physiographic control of sediment production, supply, and dispersal (Basu, 1985; Critelli et al., 1997; Weltje et al., 1998). These studies have a particular significance in areas such as the late Pleistocene–Holocene Modena Plain, where fluvial sediments have buried a spectacular number of Neolithic, Iron Age, Bronze Age, Etruscan, Roman, and Longobardian archaeological sites. The evolution of these human settlements through time can be investigated in detail by reconstructing the paleogeography of the plain, which has been done mainly by stratigraphic (Cremaschi et al., 1989) and geomorphological studies (Cardarelli et al., 2004; Panizza et al., 2004). Paleochannel traces visible on the surface, for example, may help to reconstruct the ancient local drainage patterns, but the limitation of this method is that only the very late-stage evolution of the plain can be investigated in this manner. Sediment composition studies may extend our knowledge of the sedimentary supply further back in time through the stratigraphic column, provided that fluvial sediment provenance can be distinguished. Moreover, the good chronological control available for the stratigraphy of the area allows us to investigate in detail the sediment compositional variability through time as a function of late glacial-interglacial climate cycles.

This paper presents our research on the stream sediments in the Modena Plain, where we have compared modern and ancient sands to reconstruct their compositional evolution through time. The purpose of this study is to provide a contribution to the understanding of the fluvial sediment supply to test the possibility of investigating the evolution of human settlements as a function of climate and drainage system changes.

GEOLOGICAL SETTING

The Modena alluvial plain area is located on the northern side of the northern Apennines fold-and-thrust belt, where streams draining the chain flow toward the northeast into the Po River (Fig. 1). The northern Apennines formed mainly during the Tertiary as consequence of convergence between the European and the Adria plates. The convergence consumed the interposed Tethyan oceanic crust with the formation of an accretionary prism, which, during the subsequent collisional phase, produced an orogenic wedge consisting of the following tectono-stratigraphic units (Bettelli and De Nardo, 2001):
1. Tuscan-Umbria-Romagna units formed by deformation of the Adria passive margin; in the study area, these units consist mainly of late Oligocene–middle Miocene thick siliciclastic turbidite sequences and chaotic shaly assemblages (mélanges);
2. Sub-Ligurian units consisting of Paleocene–early Miocene siliciclastic and carbonate turbidite sandstones, conglomerates, and shales; and
3. Ligurian units generated by the subduction of the Tethys Ocean and represented mainly by Early Cretaceous–Early Tertiary chaotic deep-water shaly rocks (mélanges and olistostromes) and calcareous and arenaceous turbidite sandstones; these units contain also Late Jurassic–Early Cretaceous ophiolites, chert, and limestones derived from the former oceanic floor sequence.

The Ligurian thrust-nappe units overlie the Sub-Ligurian thrust-nappe units and both lie on the Tuscan-Umbria-Romagna fold-and-thrust belt units. On the northern side of the chain, the Ligurian units are unconformably overlain by the Epi-Ligurian Sequence and by Miocene-Pliocene and Quaternary terrigenous deposits of the Po Plain. The Epi-Ligurian sequence consists of a thick middle Eocene–early Messinian succession of matrix-supported breccias (olistostromes), deep-water varicolored shales and marls, muddy and sandstone turbidites, conglomerates, and shallow-water siliciclastic and bioclastic deposits. The Pliocene deposits consist mainly of marine deep-water mudstones.

The Po Plain is the syntectonic sedimentary wedge filling the Pliocene-Pleistocene Apennine foredeep. The total basin infill is up to 4 km thick, and the Quaternary deposits reach a thickness of 1.5 km.

CLIMATE AND GEOMORPHOLOGY

The study area has an approximate extent of 150 km^2 and is limited by two major rivers, the Secchia River to the west and the Panaro River to the east (see Table 1 for river data). The area is crossed by minor streams that mostly drain into the Panaro River (Fig. 2). The drainage basin climate is classified as "temperate subcontinental" (Mennella, 1972). Mean annual temperature varies from 8 to 12 °C, and the coldest and hottest months are January and July, respectively. Precipitation varies with elevation, ranging from 800 to 2400 mm/yr, and generally has two maxima during the year, at November and May, and a minimum in July.

As previously described, the streams drain areas that consist mainly of sedimentary rocks, such as arenites, siltstones, and shales (Fig. 1). Magmatic rocks crop out along the main valleys as isolated ophiolite blocks including diabase, serpentinite, and gabbro suites. No metamorphic rocks are present, with the exception of very limited areas at the extreme tip of the Secchia River watershed, where amphibolites and quartzites crop out. River drainage basin data and estimates of the relative proportion of erodible terranes in the catchment areas are reported in Table 1 (Cati, 1981). Although the Secchia River has a larger drainage basin than the Panaro River (2174 vs. 1784 km^2), the latter includes a larger proportion of erodible terranes and as a result is characterized by a higher suspended load (2030 vs. 1847 t/km^2) and a higher inferred soil erosion (0.769 vs. 0.684 mm/yr; Table 1).

The minor streams drain the northernmost side of the Apennine chain and cut mostly argillaceous and arenite sediments (Fig. 1). Their drainage system in the alluvial plain has been heavily modified since ancient time to limit the impact of floods on urban areas. The Cerca stream (Torrente Fossa), for example, was deviated into the Secchia River in the fifteenth or sixteenth century (Fig. 2).

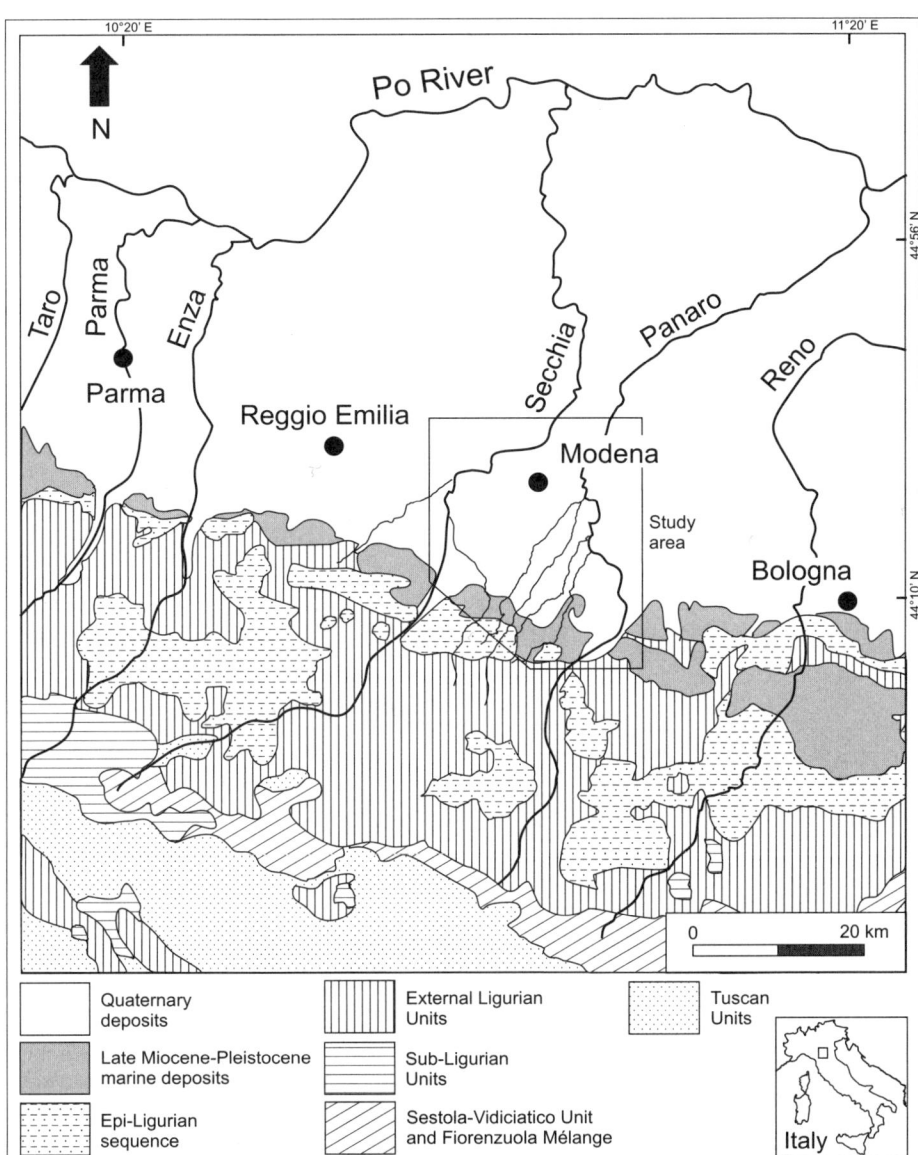

Figure 1. Geological sketch of the drainage basins of the Secchia and Panaro Rivers in the northern Apennines (simplified from Bettelli and De Nardo, 2001).

TABLE 1. PHYSIOGRAPHIC CHARACTERISTICS OF SECCHIA AND PANARO RIVER DRAINAGE BASINS

	Secchia River	Panaro River
Drainage basin		
Area (km^2)	2174	1784
Maximum elevation (m a.s.l.)	2121	2165
Minimum elevation (m a.s.l.)	14	10
Elevation range (m a.s.l.)	2107	2155
Mean elevation (m a.s.l.)	606	662
Highly erodible terranes (%)	30.7	45.3
Medium erodible terranes (%)	25.8	15
Low erodible terranes (%)	43.5	39.7
River channel		
Length (km)	170	160
Average discharge (m^3/s)	27.2	23.7
Suspended load (t/km^2)	1847	2030
Inferred soil erosion (mm/yr)	0.684	0.769

Note: m a.s.l.—meters above sea level.

Gasperi and Pizziolo (2007) report historical accounts that the Tiepido stream was deviated into the Panaro River; unfortunately the year when this deviation took place is unknown.

SAMPLES AND METHODS

Forty-two samples of modern sands were collected in active longitudinal bars of nine stream channels (Fig. 2). Sampling covered the channel segments between the point where the stream entered the floodplain and the last downstream occurrence of sand in the stream beds. The sampling network was designed to obtain sand specimens at approximately the same spacing distance for each stream and to include, where possible, upstream and downstream confluence points.

Nine ancient sand samples were collected from river cuts, archaeological sites, and cores located in the channel belt of the main rivers (Fig. 2). Unfortunately, no sections with archaeological

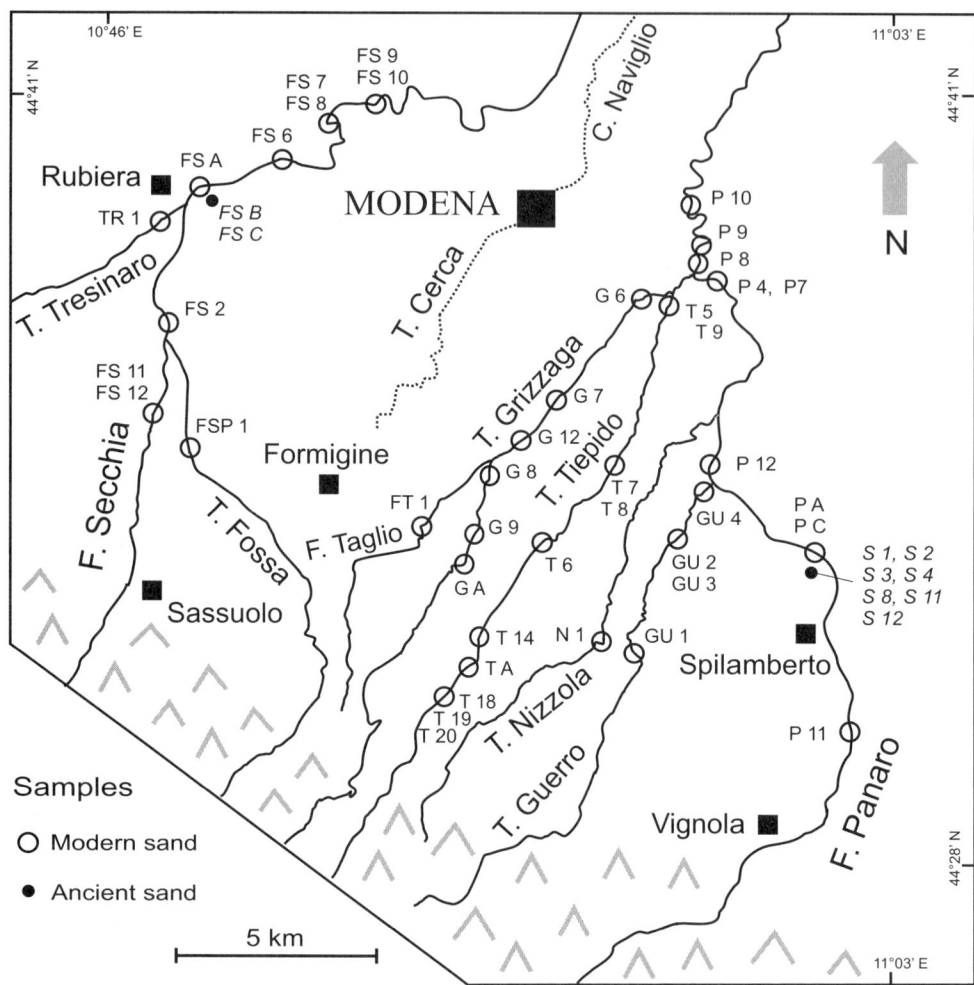

Figure 2. Simplified sketch of the fluvial drainage network in the Modena Plain with sampling sites of modern and ancient sands noted.

remains and ancient sand sediments were available for the minor streams. Hereafter, ancient sediments younger than late Pleistocene are called "Holocene" sands, whereas the term "modern" is restricted to present-day settings.

Sand samples were washed with dilute H_2O_2 to remove organic matter and were air dried and sieved to obtain the fine sand fraction (0.125–0.250 mm, 3–2 ϕ). The necessity to analyze the fine sand fraction was dictated by the lack of medium-coarse sand at some of the sampling sites. The fine sand fraction was impregnated in epoxy resin under vacuum, thin-sectioned, and stained for feldspar and carbonate identification.

Point counting under transmitted light microscopy was performed according to the Gazzi-Dickinson method (Gazzi, 1966; Dickinson, 1970; Zuffa, 1985; Ingersoll et al., 1984). At least 300 grains were point counted for each section to achieve modal composition, and 40 categories of grains were distinguished. Results of point counting are presented in Table 2. Components not related to the original sand composition, such as brick and pottery fragments, penecontemporaneous shell fragments, and authigenic carbonate nodules, were excluded from the final calculations.

Statistically rigorous confidence regions of the studied samples were calculated according to Weltje (2002). Error bars in the diagrams were calculated according to Howarth (1998).

RESULTS

Modern Sands

The modal analyses show that modern stream sands in the Modena Plain have similar overall compositions, but show significant variations in quartz, feldspar, carbonate, and lithic fragment contents (Table 2; Figs. 3 and 4). In a Q + F-total carbonates-lithic fragments ternary diagram (Q = total quartz + chert; F: plagioclase + k-feldspar, according to Dickinson, 1970) Q and F have different characters as explanation = and: respectively. The compositional fields reveal that sediments from different modern streams can be clearly distinguished (Fig. 4).

Streams draining the eastern sector of the study area (Panaro, Tiepido, Guerro) have a generally higher carbonate content than those draining the western part, which are enriched in quartz and feldspar (Secchia and Grizzaga).

TABLE 2. RESULTS OF PETROGRAPHIC MODAL ANALYSES

	Tresinaro	Secchia									Holocene		Fossa	Taglio	Grizzaga					
	Modern	Modern											Modern	Modern	Modern					
Sample	TR 1	FS 11	FS 12	FS 2	FS A	FS 6	FS 7	FS 8	FS 9	FS 10	FS B	FS C	FSP 1	FT 1	G A	G 9	G 8	G 12	G 7	G 6
NCE																				
Q																				
Quartz single crystal	11.8	9.6	11.2	18.9	15.2	17.7	13.5	21.2	19.2	20.2	13.1	17.0	12.4	14.4	22.3	20.8	9.8	19.5	15.2	20.8
Quartz polycrystalline coarse texture	2.2	3.9	2.1	1.6	3.8	4.0	2.7	1.9	2.7	3.4	2.2	1.6	2.1	1.6	4.1	1.7	0.9	2.9	3.6	3.9
Quartz polycrystalline fine texture	3.7	1.5	0.6	1.6	1.9	3.7	1.2	1.9	2.7	2.8	1.2	2.2	1.5	1.6	1.4	1.7	1.2	1.1	1.5	1.8
Chert	0.9	0.9	0.9	0.9	0.6	0.9	–	0.6	0.9	0.6	0.3	0.9	0.3	1.0	1.2	2.0	0.6	1.1	0.9	0.3
Quartz in plutonic-gneissic rock fragment	1.9	2.7	3.0	0.3	3.5	0.3	1.5	0.3	2.1	2.1	0.9	1.3	1.2	–	2.9	1.1	0.9	0.9	0.6	0.9
Quartz in metamorphic rock fragment	0.3	–	–	0.6	0.9	–	0.3	–	0.3	0.3	–	0.3	–	–	–	0.6	–	0.9	–	–
Quartz in volcanic rock fragment	–	–	–	–	–	–	–	–	–	–	–	–	–	–	–	–	–	–	–	–
Quartz in clastic rock fragment	0.3	1.2	0.9	1.9	0.9	1.8	2.4	0.3	0.9	0.9	0.9	1.6	1.8	1.6	1.4	2.5	2.4	1.4	1.2	0.3
K																				
K-feldspar single crystal	9.0	6.3	5.2	6.3	3.2	8.0	10.8	9.7	5.8	7.4	4.7	3.1	4.2	2.6	6.4	13.5	8.6	7.2	7.3	8.4
K-feldspar in plutonic-gneissic rock fragment	0.9	0.9	1.2	0.3	0.9	1.2	1.2	3.1	0.3	0.9	–	0.9	0.6	–	0.6	1.4	0.9	0.3	0.3	0.9
K-feldspar in metamorphic rock fragment	–	–	–	–	–	–	–	–	–	–	–	–	–	–	–	0.3	–	–	–	–
K-feldspar in volcanic rock fragment	–	–	–	–	–	–	–	–	–	–	–	–	–	–	–	–	–	–	–	–
K-feldspar in clastic rock fragment	–	–	–	–	–	0.3	0.3	–	0.6	0.6	–	–	0.3	–	0.3	–	–	0.3	–	–
P																				
Plagioclase single crystal	6.8	5.4	13.4	7.5	6.3	7.0	8.4	5.3	10.1	6.7	11.2	11.0	4.5	4.8	8.1	7.0	5.9	6.9	9.4	6.3
Plagioclase in plutonic-gneissic rock fragment	–	2.4	0.9	0.3	0.3	0.6	2.4	–	0.3	0.3	0.9	0.9	0.3	1.3	0.9	0.3	0.6	1.1	0.6	1.5
Plagioclase in metamorphic rock fragment	–	–	–	–	–	–	–	–	–	–	–	–	–	–	–	–	–	–	–	–
Plagioclase in volcanic rock fragment	–	–	–	–	–	–	–	–	–	–	–	–	–	–	–	–	–	–	–	–
Plagioclase in clastic rock fragment	–	0.9	0.3	–	0.3	–	0.3	0.3	0.3	0.6	–	–	0.3	0.3	0.6	0.3	1.2	0.6	0.3	–
L																				
Metamorphic rock fragment	–	–	–	0.3	0.6	1.2	0.3	0.6	1.2	1.5	–	0.3	–	0.3	0.6	–	–	–	–	–
Volcanic rock fragment	–	–	–	–	1.9	–	–	–	–	–	–	–	–	–	–	–	–	–	–	–
Hypabyssal lithic	0.3	–	–	–	0.3	0.9	1.5	1.9	1.8	3.4	–	–	–	–	–	–	–	0.3	–	–
Splite	5.6	1.5	1.5	0.6	2.5	3.4	1.5	2.8	1.5	1.8	1.9	0.6	0.6	1.9	0.6	0.6	0.6	1.4	0.6	0.6
Serpentinite	9.0	2.4	2.1	2.2	2.5	11.0	9.6	9.7	8.2	10.4	16.2	16.0	9.1	10.6	5.2	4.5	11.2	6.0	6.4	3.6
Clastic lithic: Shale	10.2	18.3	14.3	14.2	12.0	11.0	8.7	8.7	8.5	8.3	11.8	9.1	9.4	9.3	8.4	4.8	8.9	5.7	8.8	4.8
Siltstone		11.4	10.6	9.7	10.8	7.3														
M																				
Muscovite + chlorite single crystal	0.6	0.6	–	0.6	–	–	–	0.3	0.3	–	1.2	0.6	–	0.3	0.6	–	0.6	0.6	–	–
Muscovite + chlorite in rock fragment	–	–	–	–	–	–	–	–	0.3	–	–	–	–	–	–	–	–	–	–	–
Heavy mineral single crystal (unspecified)	–	0.3	–	–	–	0.3	0.6	–	–	–	0.3	–	–	0.3	0.3	–	–	–	–	–
Heavy mineral in rock fragment (unspecified)	–	–	–	–	–	–	–	–	–	–	–	–	–	0.3	–	–	–	–	–	–
Fe-oxide	0.6	–	0.3	–	–	0.3	0.9	0.6	0.3	–	0.6	–	–	–	–	–	0.6	–	–	–
CE																				
C																				
Calcite single crystal	12.7	11.7	9.4	11.6	12.7	14.7	15.0	13.1	16.5	15.0	13.1	11.6	15.8	16.3	10.4	12.1	14.5	16.4	10.6	14.2
Sparitic limestone	13.3	6.3	4.9	3.8	4.7	1.2	5.7	4.7	4.3	5.2	2.5	5.3	8.8	5.1	7.0	4.2	9.5	5.5	15.8	9.3
Silty arenitic limestone	–	–	–	2.5	2.2	–	–	–	–	–	4.0	1.9	–	4.8	1.4	1.4	–	1.1	–	–
Mudstone-wackestone	8.4	11.7	15.8	12.9	13.0	12.8	11.4	12.8	10.1	7.4	11.5	10.7	21.5	15.7	10.7	13.5	16.9	13.2	11.9	16.6
Bioclast (terrigenous)	0.6	–	0.6	0.6	0.9	0.9	–	0.3	0.3	0.3	0.6	0.3	2.4	3.8	4.6	3.1	2.1	3.2	3.6	4.5
NCI																				
Brick and pottery fragments	0.9	0.3	0.6	0.3	–	–	–	–	0.3	0.3	0.3	–	1.5	1.9	1.4	2.2	1.5	2.0	1.2	0.9
CI																				
Bioclast (penecontemporaneous)	–	–	–	–	–	–	–	–	–	–	–	0.6	–	–	–	–	–	–	–	0.3
Caliche	–	–	–	–	0.3	–	–	–	–	–	0.3	0.3	–	–	–	–	–	–	–	–
Undetermined	–	–	–	0.3	–	0.3	–	–	–	–	–	–	–	–	–	0.6	–	0.3	–	–
Total	100.0	100.0	100.0	100.0	100.0	100.0	100.0	100.0	100.0	100.0	100.0	100.0	100.0	100.0	100.0	100.0	100.0	100.0	100.0	100.0

(*continued.*)

TABLE 2. RESULTS OF PETROGRAPHIC MODAL ANALYSES (Continued.)

Sample	Tiepido Modern												Nizzola Modern	Guerro Modern						Panaro Modern		
	T 18	T 19	T 20	T A	T 14	T 6	T 5	T 7	T 8	T 9	N 1	GU 1	GU 2	GU 3	GU 4	P 11	P 12	P A	P C	P 4
NCE																				
Q																				
Quartz single crystal	7.7	10.6	10.4	11.6	11.3	10.3	10.4	8.8	11.9	7.3	22.7	8.4	11.7	8.4	10.7	9.6	12.9	13.2	11.7	12.3
Quartz polycrystalline coarse texture	1.7	1.4	1.7	1.2	1.5	3.1	1.5	2.5	2.7	1.5	4.0	0.9	1.8	3.1	2.2	2.8	3.4	2.5	1.6	2.4
Quartz polycrystalline fine texture	2.0	0.9	1.7	1.5	0.9	1.2	0.3	1.0	1.2	0.6	1.2	1.8	0.3	0.6	1.9	1.2	0.3	1.9	1.9	1.5
Chert	0.6	1.4	1.7	–	1.7	0.9	0.3	0.2	–	0.6	0.3	1.8	1.5	0.3	0.6	–	0.6	0.9	0.9	0.3
Quartz in plutonic-gneissic rock fragment	–	–	0.3	0.6	0.6	0.3	0.9	1.0	0.6	1.2	1.2	0.3	0.3	0.9	0.6	0.9	2.1	1.3	0.9	0.6
Quartz in metamorphic rock fragment	–	–	–	–	0.6	0.3	0.3	0.2	–	–	1.8	–	–	–	–	0.3	–	–	0.3	–
Quartz in volcanic rock fragment	–	–	–	–	–	–	–	–	–	–	–	–	–	–	–	–	–	–	–	–
Quartz in clastic rock fragment	3.7	2.3	1.4	1.5	2.6	0.9	1.2	1.5	1.2	4.1	0.3	3.9	1.5	1.3	2.2	2.5	2.1	0.9	1.9	3.0
K																				
K-feldspar single crystal	4.3	2.6	3.2	3.3	3.2	3.1	1.8	2.7	3.3	3.2	6.1	2.4	4.9	2.2	2.2	5.9	7.4	4.7	3.2	5.7
K-feldspar in plutonic-gneissic rock fragment	0.6	–	0.6	0.3	–	–	–	0.2	0.3	0.3	0.3	1.5	–	–	–	–	–	0.3	–	0.6
K-feldspar in metamorphic rock fragment	–	–	–	–	–	–	–	–	–	–	–	–	–	–	–	–	–	–	–	–
K-feldspar in volcanic rock fragment	–	–	–	–	–	–	–	–	–	–	–	–	–	–	–	–	–	–	–	–
K-feldspar in clastic rock fragment	–	0.6	0.3	0.3	0.6	–	–	–	0.3	–	–	0.6	0.3	–	–	0.3	–	–	0.3	0.6
P																				
Plagioclase single crystal	4.0	3.4	2.6	2.7	2.9	6.2	4.6	3.9	4.6	3.5	4.6	2.4	1.8	4.1	5.3	3.7	9.2	7.2	7.6	5.1
Plagioclase in plutonic-gneissic rock fragment	1.4	1.1	0.3	–	0.6	0.3	–	–	–	0.9	–	–	0.3	0.6	–	–	0.9	1.3	0.3	1.2
Plagioclase in metamorphic rock fragment	–	–	–	–	–	–	–	–	–	–	–	–	–	–	–	–	–	–	–	–
Plagioclase in volcanic rock fragment	–	–	–	–	–	–	–	–	–	–	–	–	–	–	–	–	–	–	–	–
Plagioclase in clastic rock fragment	–	0.9	0.6	0.3	0.6	–	0.3	0.2	–	0.3	0.3	–	–	–	0.9	0.3	0.6	1.6	0.6	0.3
L																				
Metamorphic rock fragment	–	–	–	–	–	–	–	–	–	–	–	–	–	–	–	–	–	–	–	–
Volcanic rock fragment	–	–	–	–	–	–	–	–	0.3	–	–	0.6	0.3	–	–	0.6	–	0.9	0.6	–
Hypabyssal lithic	–	–	–	–	–	–	–	–	–	–	–	–	–	–	–	–	–	0.3	–	–
Spilite	–	–	–	–	0.3	–	–	–	0.3	0.3	–	0.3	0.3	–	0.3	0.6	1.2	1.6	0.3	0.6
Serpentinite	0.6	0.3	–	0.3	0.3	0.6	0.9	1.0	0.6	–	–	–	–	–	–	0.6	–	–	0.6	–
Clastic lithic: shale	12.2	12.9	11.5	11.9	11.9	10.6	6.4	6.4	7.0	8.4	8.0	3.3	5.8	9.1	6.6	19.8	14.1	8.5	16.1	18.9
Siltstone	9.7	9.2	7.5	7.0	9.9	9.7	9.2	9.1	10.3	10.2	16.9	13.1	9.5	10.6	7.5	12.4	10.4	13.8	10.4	9.0
M																				
Muscovite + chlorite single crystal	1.4	0.3	0.6	–	0.3	0.3	0.3	–	–	0.6	–	0.6	0.3	0.9	–	0.3	0.6	–	0.9	1.2
Muscovite + chlorite in rock fragment	–	–	–	–	–	–	–	–	–	–	–	–	–	–	–	–	–	–	–	–
Heavy mineral single crystal (unspecified)	–	–	–	–	–	–	0.3	0.2	–	0.3	0.3	0.3	–	–	–	–	–	0.3	–	–
Heavy mineral in rock fragment (unspecified)	–	–	–	–	–	–	–	–	–	–	–	–	–	–	–	–	–	0.3	–	–
Fe-oxide	–	–	–	1.2	–	–	0.6	–	1.5	–	–	–	1.2	0.9	0.6	0.6	0.3	–	–	–
CE																				
C																				
Calcite single crystal	19.6	18.7	19.3	20.4	21.5	18.1	28.2	24.6	21.0	23.5	7.7	24.2	23.9	26.6	26.7	14.2	13.2	15.4	12.3	9.9
Sparitic limestone	6.0	5.5	8.1	7.3	8.1	12.8	9.8	13.8	13.1	9.9	4.0	20.6	15.6	14.1	14.8	7.7	6.4	3.8	5.4	7.2
Silty arenitic limestone	2.3	2.9	4.3	7.0	1.7	–	4.6	–	–	2.6	–	–	–	–	–	–	–	3.1	3.8	–
Mudstone-wackestone	17.6	20.7	20.5	17.0	15.4	17.8	14.1	19.7	15.2	18.9	16.9	12.5	16.3	15.0	16.0	14.2	13.5	13.8	16.1	17.4
Bioclast (terrigenous)	3.1	2.0	2.0	2.7	3.5	2.2	2.1	1.0	3.0	2.0	–	0.3	0.9	0.3	–	1.2	0.6	0.6	1.6	1.8
NCI																				
Brick and pottery fragments	1.7	1.7	1.4	1.2	0.6	1.2	0.9	2.0	0.9	–	2.8	1.2	1.5	0.6	0.6	0.3	–	0.9	–	0.6
CI																				
Bioclast (penecontemporaneous)	–	0.6	–	–	–	–	0.6	–	–	–	–	–	–	–	–	0.3	–	0.6	–	–
Caliche	–	–	–	–	–	–	–	–	–	–	–	–	–	–	–	–	–	–	–	–
Undetermined	–	–	–	–	–	–	–	–	0.6	–	0.6	–	–	0.3	–	–	–	0.3	0.3	–
Total	100.0	100.0	100.0	100.0	100.0	100.0	100.0	100.0	100.0	100.0	100.0	100.0	100.0	100.0	100.0	100.0	100.0	100.0	100.0	100.0

(continued.)

TABLE 2. (Continued.)

Sample	Panaro Modern				Holocene				Late Pleistocene		
	P 7	P 8	P 9	P 10	S 1	S 2	S 3	S 4	S 8	S 12	S 11
NCE											
<u>Q</u>											
Quartz single crystal	12.6	12.3	14.9	14.6	10.3	11.3	13.5	12.2	12.8	18.0	12.1
Quartz polycrystalline coarse texture	4.9	3.8	1.6	3.5	1.8	1.3	0.3	–	3.4	1.2	5.1
Quartz polycrystalline fine texture	0.6	0.9	0.6	1.3	0.3	1.3	1.5	3.3	1.1	1.5	0.6
Chert	0.6	0.9	0.6	0.6	0.6	0.3	0.3	0.7	0.6	0.3	0.6
Quartz in plutonic-gneissic rock fragment	1.5	0.9	0.6	0.9	0.6	0.3	1.5	1.3	2.5	0.6	0.6
Quartz in metamorphic rock fragment	0.3	–	–	0.3	0.3	–	0.6	–	0.3	–	–
Quartz in volcanic rock fragment	–	–	–	–	–	–	–	–	–	–	–
Quartz in clastic rock fragment	2.8	0.6	2.2	0.6	2.4	0.3	0.3	–	2.5	3.2	0.6
<u>K</u>											
K-feldspar single crystal	6.2	6.0	6.8	5.4	3.0	2.6	2.1	2.0	6.7	5.9	7.0
K-feldspar in plutonic-gneissic rock fragment	1.2	0.3	0.6	0.6	1.2	–	0.6	–	0.8	0.9	0.6
K-feldspar in metamorphic rock fragment	–	–	–	–	–	–	–	–	–	–	–
K-feldspar in volcanic rock fragment	–	–	–	–	–	–	–	–	–	–	–
K-feldspar in clastic rock fragment	–	0.9	0.6	0.6	0.6	0.3	–	–	–	0.3	0.3
<u>P</u>											
Plagioclase single crystal	5.8	4.7	4.7	5.1	7.0	6.8	5.2	7.6	8.1	8.3	8.3
Plagioclase in plutonic-gneissic rock fragment	0.3	0.6	–	–	0.6	0.3	1.8	1.3	1.1	1.2	1.6
Plagioclase in metamorphic rock fragment	–	–	–	–	–	–	–	–	–	–	–
Plagioclase in volcanic rock fragment	–	–	–	–	–	–	–	–	–	–	–
Plagioclase in clastic rock fragment	–	–	–	–	0.6	0.6	0.6	0.3	0.3	1.8	–
<u>L</u>											
Metamorphic rock fragment	–	–	–	0.6	0.3	–	–	0.3	–	0.3	–
Volcanic rock fragment	–	–	–	–	–	–	–	–	–	–	–
Hypabyssal lithic	–	–	–	–	–	–	–	–	–	–	–
Spilite	0.3	0.3	–	0.3	–	–	0.3	–	0.3	0.9	–
Serpentinite	–	0.9	0.9	0.3	0.6	–	0.3	–	0.8	1.2	1.0
Clastic lithic: shale	11.4	13.2	13.4	12.7	11.9	18.0	11.3	14.8	17.3	18.3	15.6
Siltstone	10.5	14.2	8.4	11.4	14.3	14.5	14.4	12.8	14.5	9.4	15.3
<u>M</u>											
Muscovite + chlorite single crystal	0.6	–	0.6	0.3	–	–	–	–	0.6	–	–
Muscovite + chlorite in rock fragment	–	–	–	–	–	–	–	–	–	–	–
Heavy mineral single crystal (unspecified)	–	–	–	–	–	–	–	–	–	–	–
Heavy mineral in rock fragment (unspecified)	–	–	–	–	–	–	–	–	–	–	–
Fe-oxide	–	–	–	–	–	0.3	–	0.3	–	–	–
CE											
<u>C</u>											
Calcite single crystal	16.3	17.3	18.0	15.5	13.7	17.4	14.4	16.8	13.1	9.7	14.6
Sparitic limestone	10.8	5.7	7.1	7.3	5.5	2.6	6.4	5.9	5.6	4.7	4.8
Silty arenitic limestone	–	–	–	–	3.6	1.9	4.6	4.9	–	2.4	–
Mudstone-wackestone	12.9	15.1	16.8	17.7	18.2	16.4	17.7	13.2	7.3	9.4	11.1
Bioclast (terrigenous)	0.3	0.6	1.2	0.3	–	1.3	0.6	0.3	0.3	0.6	–
NCI											
Brick and pottery fragments	–	0.6	0.3	–	0.6	–	0.6	2.0	–	–	–
CI											
Bioclast (penecontemporaneous)	–	–	–	–	–	–	–	–	–	–	–
Caliche	–	–	–	–	1.5	2.3	0.9	–	–	–	–
Undetermined	–	–	–	–	0.3	–	–	–	–	–	–
Total	100.0	100.0	100.0	100.0	100.0	100.0	100.0	100.0	100.0	100.0	100.0

Note: NCE—noncarbonated extrabasinal grains; CE—carbonate extrabasinal grains; NCI—noncarbonated intrabasinal grains; CI—carbonate intrabasinal grains (Zuffa, 1985).

The most distinctive compositions are those of Guerro and Tiepido, which have the highest carbonate content, mainly represented by single calcite crystals (Fig. 4; Table 2).

The Grizzaga stream shows the greatest variation in terms of the measured parameters, and therefore has the broadest compositional field (Fig. 4).

Holocene Sands (Younger than ca. 7 ka)

Ancient sands collected in the channel belt of the major rivers (ages given by Gasperi and Pizziolo, 2007) provide an exceptional and unique view of the compositional variations of the sediment through time. The Panaro River deposits at Spilamberto (Fig. 2) buried a spectacular series of Neolithic, Roman, and Longobardian sites, and we can observe a complete sedimentary section ranging in age from the late Pleistocene to present day (Fig. 5). The samples younger than ca. 7 ka plot very close to the compositional field of modern sands (Fig. 6). In particular, these sands are slightly enriched in carbonate and lithic fragments and are impoverished in quartz compared with the modern ones (Fig. 6; Table 2).

We analyzed two Holocene sands from the Secchia River: one has an inferred age between 3540 ± 50 and 4585 ± 95 yr B.P. (sample FS B) and the other is slightly older than 4585 ± 95 yr B.P.

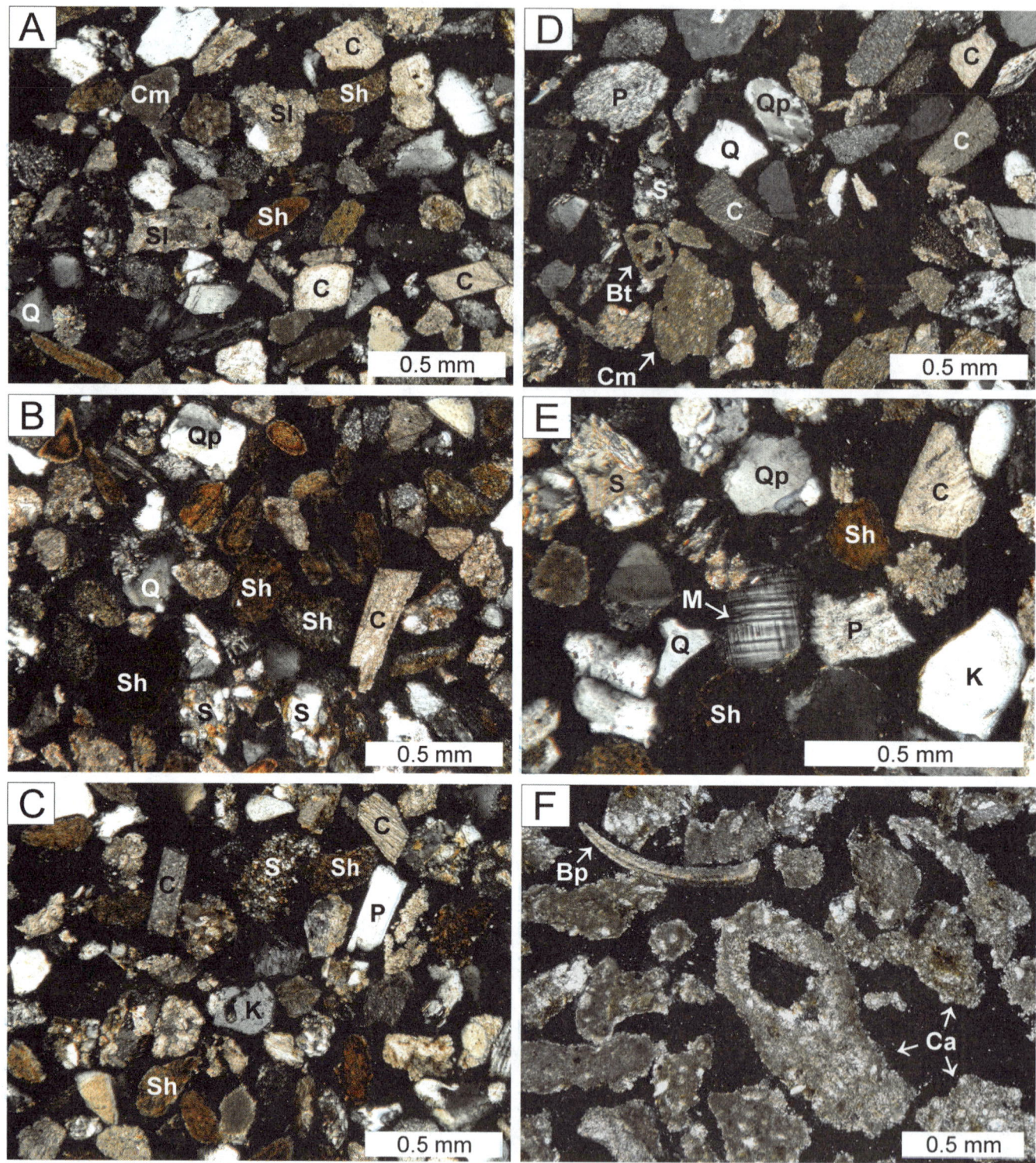

Figure 3. Photomicrographs of diagnostic grains in fluvial sands from the Modena Plain. (A) Modern sand from the Panaro River (sample P 9). Most single calcite grains (c) show distinct rhombohedral shape. (B) Holocene sand from the Panaro River (sample S 2); a large rhombohedral calcite grain is visible (c). (C) Late Pleistocene sand from the Panaro River (sample S 10); notice the rhombohedral calcite grains (c). (D) Modern sand from the Secchia River (sample FS 6); rhombohedral calcite grains (c) are common. (E) Altered plagioclase (p) and unaltered feldspar grains in late Pleistocene Panaro River sand (sample S 8). (F) Carbonate concretions (Ca) (sample from Modena urban area); notice the hollow concretion at center that probably developed around a root; a shell fragment from a thin-walled gastropod (Bp) is also visible. All photographs are in cross polarized light. C—single calcite crystal; Ca—caliche; Bp—bioclast (penecontemporaneous); Bt—bioclast (terrigenous); Cm—carbonate mudstone; K—K-feldspar; M—microcline; P—plagioclase; Q—quartz (monocrystalline); Qp—quartz (polycrystalline); S—siltstone; Sl—silty limestone; Sh—shale.

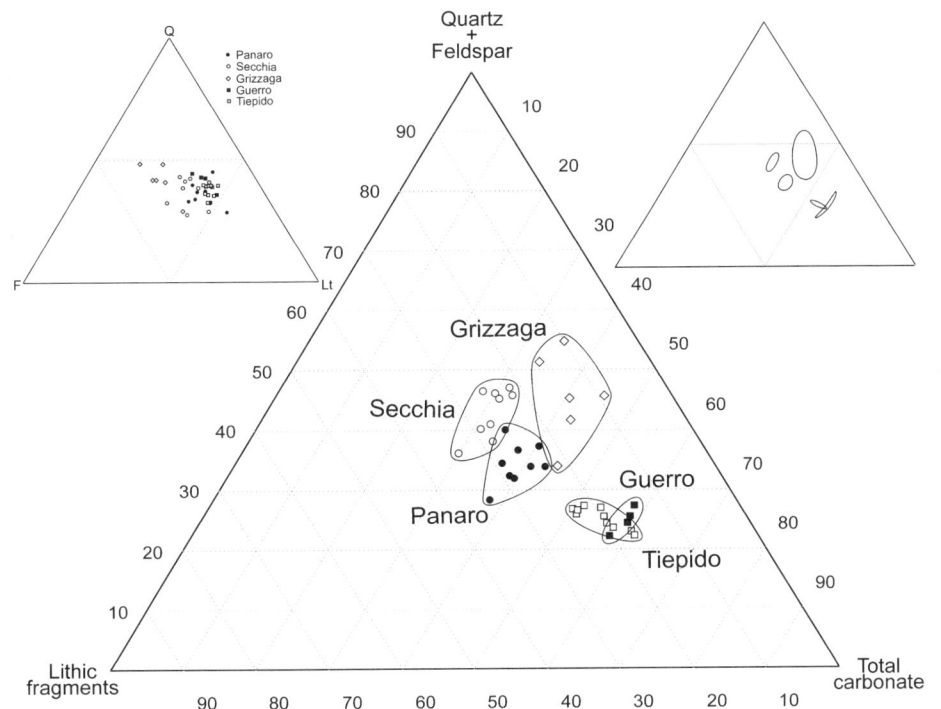

Figure 4. Ternary diagrams showing modern sand composition of the Modena Plain streams. Additional ternary diagram (above) shows 90% confidence regions of the mean for each stream sand (according to Weltje, 2002).

(sample FS C; Table 2; correlation with data from Cremaschi, 2000). These two samples plot directly into the compositional field of the modern sands (Fig. 6).

Late Pleistocene Sands (Older than 10–12 ka)

Ancient sands from fluvial deposits that are attributed to the last glacial event (Vignola Unit, late Pleistocene, age 15–18 ka; Gasperi and Pizziolo, 2007) have higher contents of quartz and feldspar compared to both modern and Holocene sands (Table 2). The feldspar content in late Pleistocene sands (age > 12 ka) from cores located a few kilometers north of Modena ranges from 13.5% to 30.8% (Lugli et al., 2004), whereas the range in the modern sands is from 6.9% to 21.3% (Table 2).

Higher feldspar abundances have also been detected in the late Pleistocene sands from the Panaro River channel belt at Spilamberto (Fig. 2). Here, as shown in Figure 7, feldspar content is highest in the Late Pleistocene (14.9%–18.4%; Table 2), decreases in the Holocene (10.5%–13.1%), and remains constant in the modern sediments (10.3%–15.3%), with the noticeable exception of sample P12 (18.1%; Table 2), which, as discussed later, is influenced by a direct supply of eroded Pleistocene sediments.

SEDIMENT DIAGENESIS

Correct interpretation of compositional data requires an understanding of the diagenetic processes that may have affected ancient sands. In the Modena Plain, postdepositional modifications that may have altered the provenance signal of the sands are: chemical weathering of feldspar grains to form clay minerals and Al-hydroxides, dissolution of carbonate clasts, and growth of carbonate concretions.

Survival of feldspar in the rock record is strongly influenced by climatic conditions during sediment production, storage, and dispersal (James et al., 1981; McBride et al., 1996; Critelli et al., 2003). Feldspar preservation during burial diagenesis may be roughly evaluated by the relative proportions of altered versus unaltered grains. Point counting of feldspar populations in the modern sands indicates that content of altered and unaltered feldspar grains is approximately the same (49%–66%). The proportion does not vary downstream in modern sands and remains approximately the same in the Holocene buried sediments (Panaro River). On the other hand, the most ancient sands that we analyzed (late Pleistocene, ages 15–18 ka) have more abundant feldspar than the modern and Holocene sands (ages younger than ca. 7 ka; Lugli et al., 2004), suggesting that postdepositional chemical weathering during burial has not caused any significant selective destruction of feldspar grains (Fig. 3E).

Carbonate sand grains can be dissolved by groundwater, but our observations suggest that dissolution phenomena have not significantly changed the composition of the analyzed sediments. Observations under the microscope show that most calcite clasts (single calcite crystals) are not corroded and show distinctive rhombohedral angular shapes (Figs. 3B and 3C), a feature that is incompatible with postdepositional dissolution. Removal of carbonate components also appears to have been negligible because buried ancient sands have generally higher carbonate content than the modern ones (Panaro River; Fig. 6). Limited partial dissolution preferentially affected microcrystalline carbonate grains, but

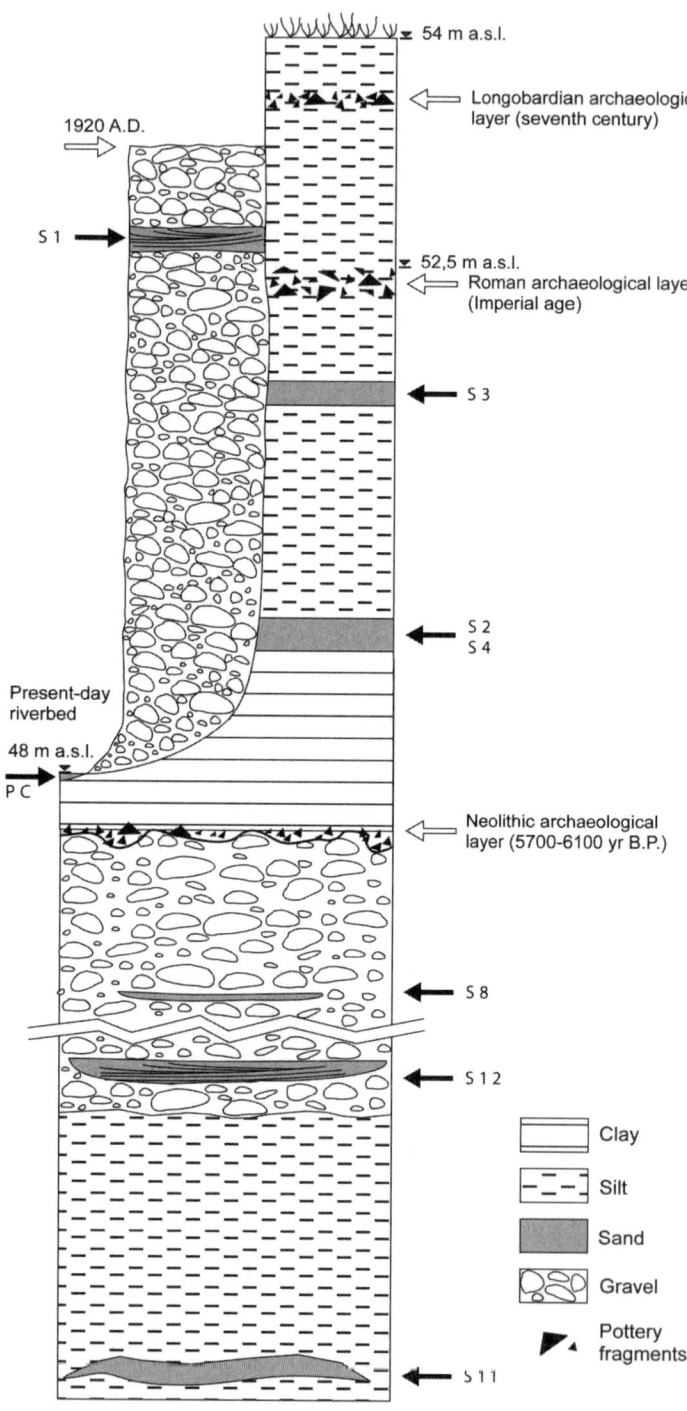

Figure 5. Simplified stratigraphic sketch of the Spilamberto quarry. Late Pleistocene braided channel gravels are overlain by Holocene overbank fines. The latter are in turn cut by a recent channel and by the present-day riverbed. Sampled sands are shown by black arrows. White arrows indicate archaeological layers. Vertical arrows show the general altimetry, not to scale.

total dissolution leading to grain removal was probably mainly concentrated in the very fine-grained fractions (<0.125 mm), and thus did not obliterate the original composition revealed by point counting of the 0.125–0.250 mm size fraction.

Caliche formation is the main diagenetic process in the ancient sediments, and it also affects relatively young sediments that buried Roman age artifacts (first to second centuries A.D.). These microcrystalline carbonate nodules range in size from fractions of millimeters to a few centimeters and are usually concentrated in thin layers. The caliches have grown displacively into the fine-grained sediments (clay and silt) and are commonly associated with roots (Fig. 3F). Very small nodules, up to a few fractions of millimeters in size, may be present also in the sand layers. Their formation is early and linked to groundwater table oscillations and to the development of soils (Maffei, 2001). As discussed earlier, their presence in the ancient sand does not compromise the definition of the original composition because they can be easily recognized under the microscope as secondary particles and excluded from the calculations. In particular cases, such as recycling of older sediments, some of the eroded caliche fragments may be difficult to distinguish from microcrystalline carbonate grains. As a consequence, point counting could overestimate the microcrystalline carbonate content of some modern sand, but, as discussed earlier, the carbonate content is slightly higher in Holocene than in modern sands (Panaro River; Table 2; Fig. 6).

DOWNSTREAM COMPOSITIONAL VARIATIONS: RECYCLING AND MIXING

The broad field composition of modern sand is due to both mixing of different sands at confluence points and recycling of older sediments along stream. Ancient sand deposits are cut and eroded by the main rivers (between sampling sites P 12 and P A for the Panaro River and between FS A and FS 9 for the Secchia River; Fig. 2). Because the compositions of the Holocene sands are similar to those of the modern sands, mixing of similar end members would not cause significant deviation of sediment composition downstream.

The only difference could be an increase of shale grains due to erosion of fine-grained sediments associated with the Holocene sands (Fig. 5). In contrast, late Pleistocene sands are not usually associated with fine-grained sediments and differ considerably in composition. For this reason, recycling of late Pleistocene sands produces a large compositional variation in modern sands.

The complex downstream compositional variations for the Secchia and Panaro Rivers are illustrated in Figures 8 and 9, respectively.

The Secchia River compositional diagram can be divided in two parts: upstream and downstream of sampling site FS A (Fig. 8). In the upstream part, the compositional pattern is characterized by a decrease in feldspar and lithic fragments and an increase in quartz content. This represents the expected trend of downstream labile grain destruction during transport.

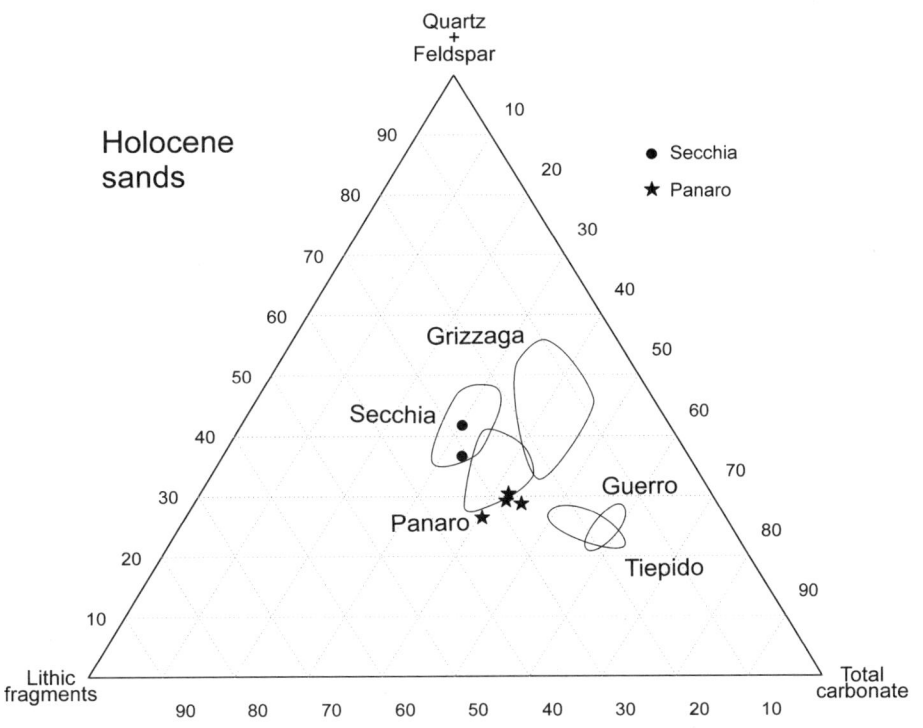

Figure 6. Ternary diagrams showing the composition of Holocene sand from the Modena Plain major rivers (Secchia and Panaro). Compositional fields of modern stream sands are also reported.

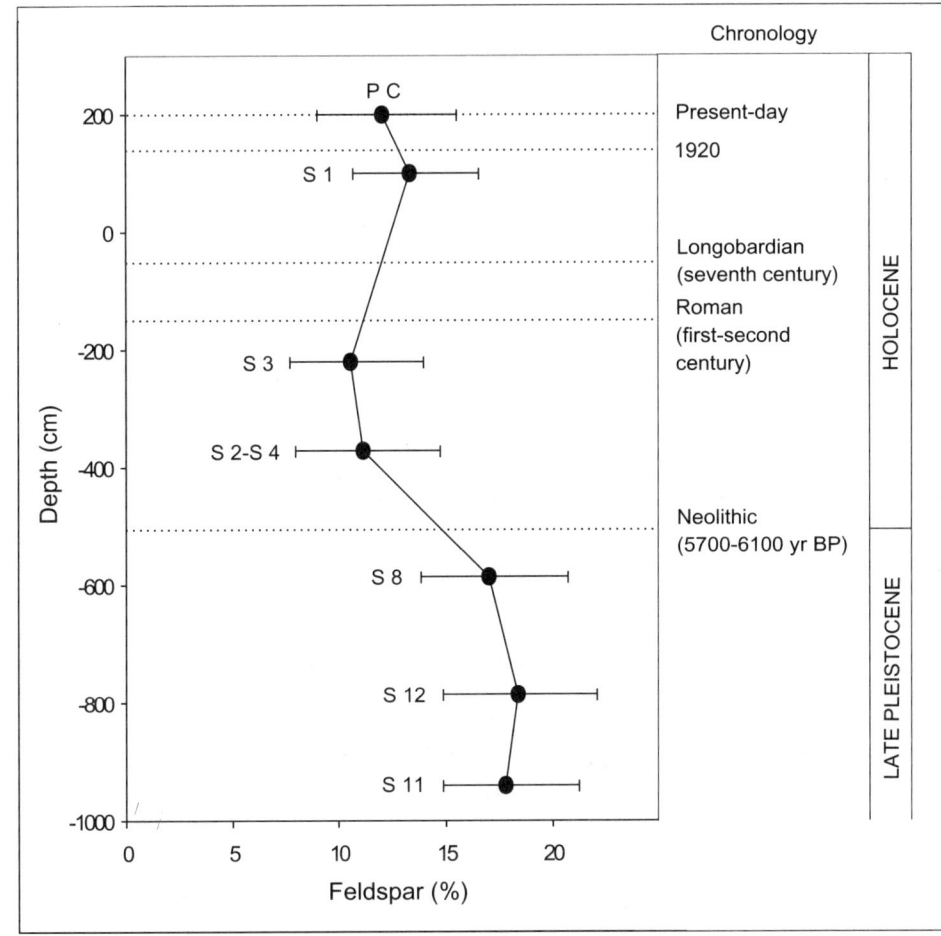

Figure 7. Plot of feldspar content in sands from the Spilamberto quarry as function of archaeological chronology. Samples younger than Longobardian age are plotted in an arbitrary vertical scale to show their relative age; true stratigraphic relationships are shown in Figure 5. Error bars were calculated according to Howarth (1998).

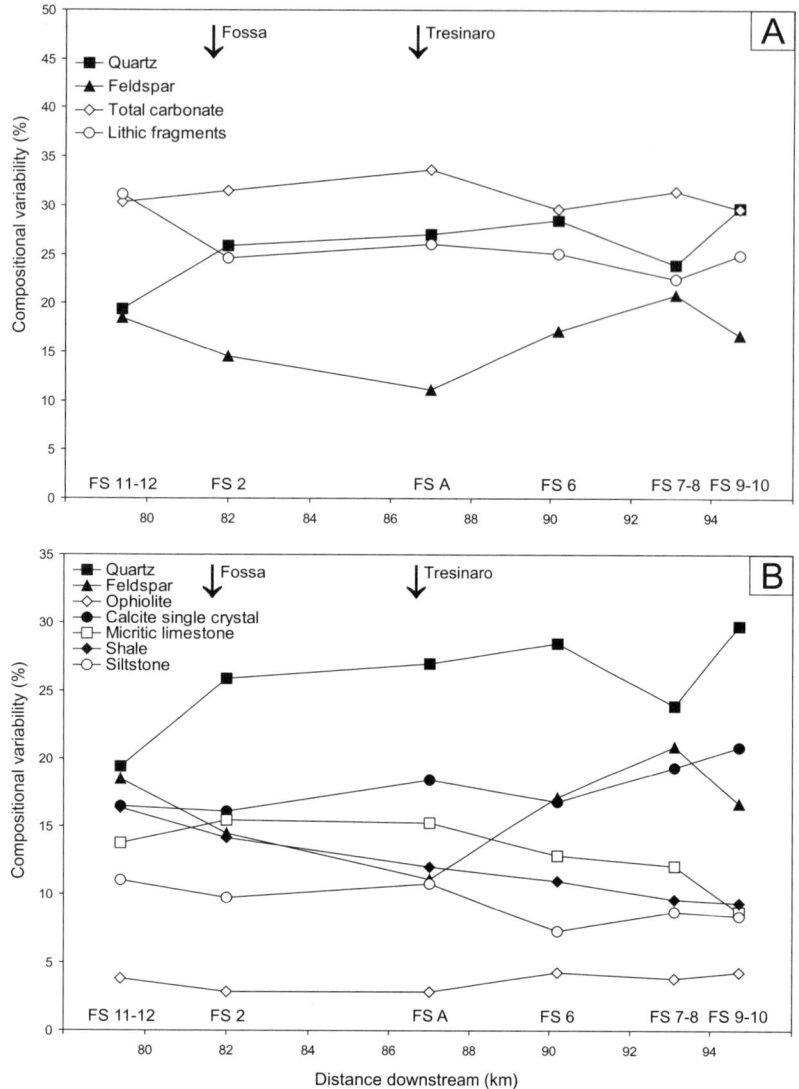

Figure 8. Plot of downstream compositional variability of Secchia River sands as a function of distance from source and confluence points of tributary streams (arrows). (A) Recalculated parameters. (B) All grain types. Errors bars are not reported for simplicity; error bar amplitudes are similar to those reported in Figure 9. Sites with more than one sample are reported as mean of measured parameters.

An exception is represented by the content of single calcite crystals, which increase downstream. This suggests that the main grain degradation mechanism is probably not abrasion, but mechanical breakage along cleavage planes. This process produces new smaller calcite rhombohedra, multiplying their content in the fine sand fraction. This hypothesis is supported by the observation that most single calcite grains show distinct rhombohedral shapes (Figs. 3A and 3D) in all sand fractions and that microcrystalline carbonate rock fragments decrease downstream, rather than increase. Mixing of sand input from the tributaries does not appreciably influence the composition of the Secchia River sediments, with the exception of the Fossa stream. The Fossa sand has a higher content of micritic carbonate grains, causing a reversal in the general decreasing trend shown by this grain type just after the confluence (between sampling sites FS 11 and FS 2; Fig. 8).

The downstream part of the Secchia River diagram shows a marked feldspar increase. This behavior is related to new detrital input resulting from erosion of late Pleistocene sediments, which have higher feldspar contents than modern and Holocene sands. This recycled sediment input, as we have verified in the field, is restricted to between sampling sites FS A and FS 7; at sample site FS 9 further downstream, the expected pattern of labile grain decline (Fig. 8) returns.

In the Panaro River, downstream compositional variations are more complex. Most labile grains, such as shale, siltstone, and micritic carbonate fragments, show such dramatic fluctuations that they cannot be simply related to mixing of different sands derived from the tributaries (Fig. 9). In this case, erosion of Holocene fine-grained river sediments results in multiple inputs of labile fragments that produce concurrent dilution effects exceeding by far the variations related to the tributary streams. The marked increase in feldspar content at sampling site P 12 is the result of recycling of feldspar-rich late Pleistocene sediments that are directly eroded in the riverbed. The feldspar signal is then attenuated after a few kilome-

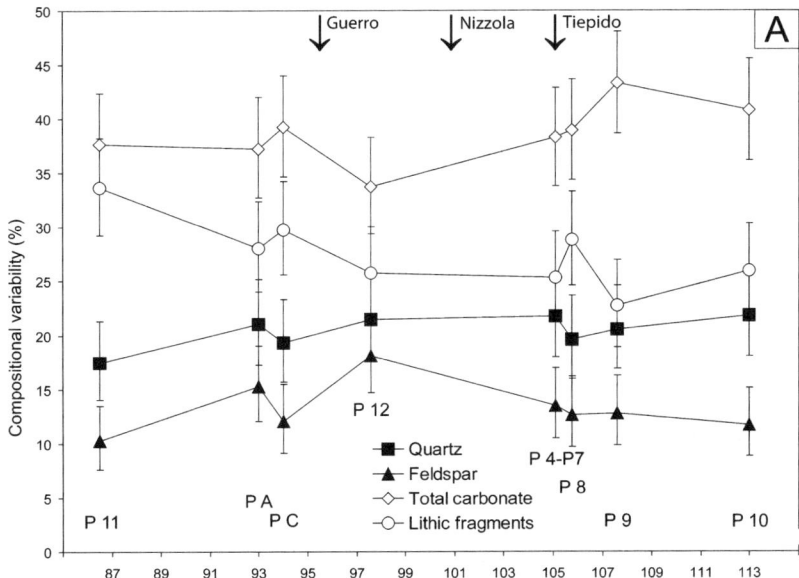

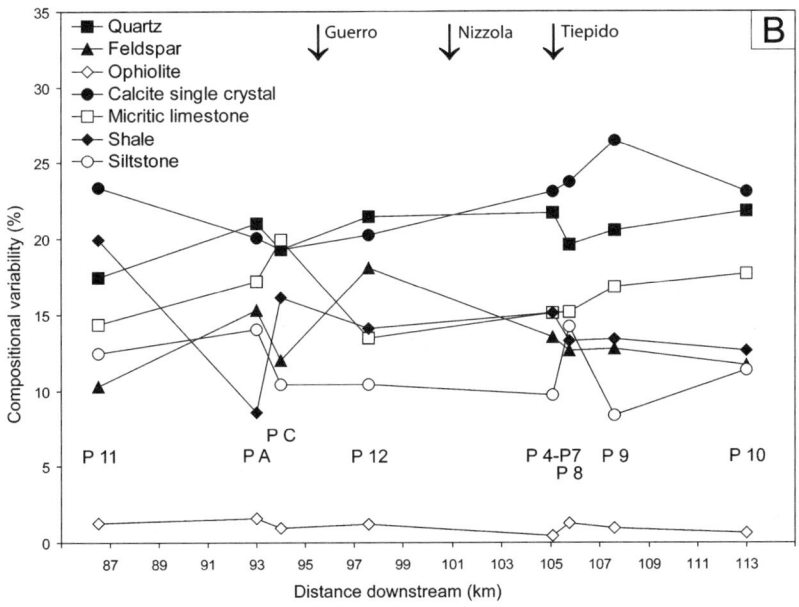

Figure 9. Plot of downstream compositional variability of Panaro River sands as a function of distance from source and confluence points of tributary streams (arrows). (A) Recalculated parameters. (B) All diagnostic grain types. Sites with more than one sample are reported as mean of measured parameters. Error bars were calculated according to Howarth (1998).

ters, possibly by grain abrasion/breakage, but most probably by downstream dilution, as evidenced by the increase of other labile grains, such as shale and siltstone fragments, derived through erosion of Holocene river sediments (Fig. 9).

Ophiolite fragments, which are by far the least abundant of the lithic fragments, appear to behave conservatively in both the Panaro and Secchia Rivers (Figs. 8 and 9). Their content does not significantly change downstream in all studied streams and appears to be transport invariant, as discussed by Weltje and Von Eynatten, (2004).

The complex downstream compositional variations of the sands are shown schematically in Figure 10. Although the interpretations proposed in this study should be confirmed by further analyses, it is important to note that the number of samples examined from this relatively small system is high compared with most similar studies reported in the literature.

DISCUSSION

Modern Sands Composition: Extent of Watershed and Recycling

On the basis of the available data, the compositional fields of the modern stream sands appear to be linked to: (1) the extent of the watershed, (2) recycling of ancient fluvial sediments, and (3) sediment input from tributary streams.

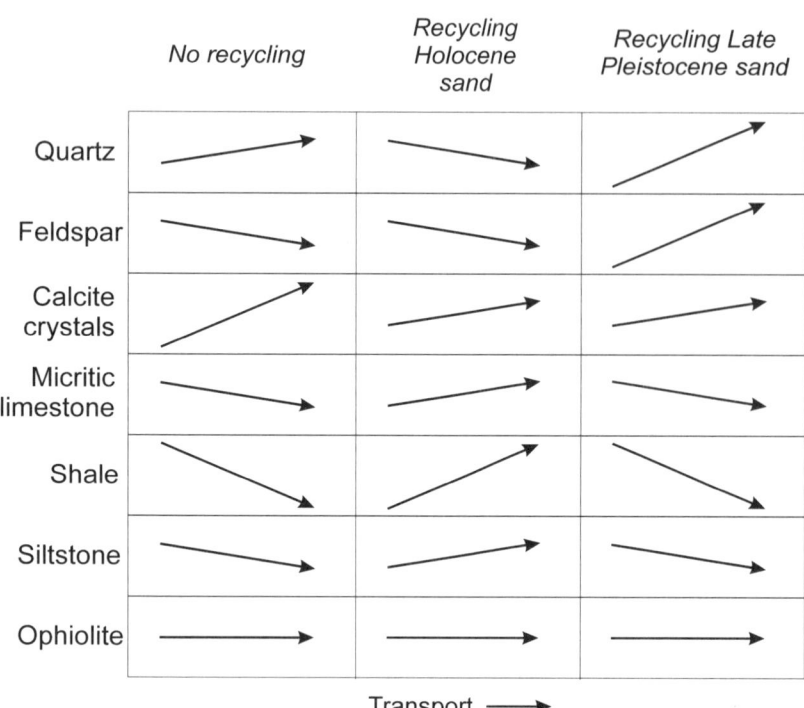

Figure 10. Sketch summarizing the downstream compositional variations of the modern stream sands as a function of transport and erosion of ancient sediments.

The extent of the watershed appears to be the dominant factor controlling sand composition. This effect is directly related to the setting of bedrock units that are mostly oriented parallel to the Apennine margin and normal to the drainage systems (Fig. 1). The most distinctive sand compositions are those of the streams characterized by smaller watersheds where sediment supply is derived from only one or a few geological formations. This is the case of the Guerro and Tiepido streams, which display the most extreme sand compositions, exemplified by their high carbonate content (Fig. 4).

Recycling of ancient sediments, in particular the late Pleistocene sands, which are particularly rich in feldspar, is the main factor that causes widening of the modern sediment composition fields. This effect is particularly evident for the Panaro River, which also receives a significant input of labile grains from erosion of ancient sediments (Fig. 9).

Sediment input from tributaries has a subordinate effect on sediment composition because minor streams generally provide sand that is not dramatically different in composition. However, the Nizzola stream, which is a tributary of the Panaro River (Table 2; Fig. 2), represents an exception to this rule: sands differ significantly because the Nizzola sand is higher in quartz and lower in carbonates, but sediment input at the confluence is very scarce and does not influence the Panaro River sand composition.

Holocene Sands: A Constant Composition

Our data demonstrate that sand composition of the main rivers has not varied significantly in the last 7 ka. Although constant sand composition for this time span can be demonstrated only for the major rivers (Panaro and Secchia), it seems reasonable that sediment composition of minor streams would also not have varied. On the other hand, paleochannel traces and historical records indicate that the minor streams changed flow patterns many times, both by natural avulsions and artificial diversions (Gasperi and Pizziolo, 2007). The result is that minor stream sediment supply patterns changed markedly through time, as a consequence of different confluence networks and different contributions of recycled older sand in their pathway. Thus, ancient sand compositions of minor streams are probably different to the modern streams. We are currently investigating this hypothesis by sampling paleochannel sediments.

Late Pleistocene Sand Composition: The Glacial Signal

The significant shift in sand composition for sediments older than ca. 12 ka, which have an overall higher content of quartz and feldspar components, could be explained by various factors, including (1) tectonics, (2) diagenesis, and (3) climatic effects.

A change in the bedrock lithology induced by neotectonics could modify the sediment supply to the basin. Tectonics and climate are considered to be the major controlling factors for the Quaternary alluvial architecture on the southern margin of the Po Basin in the Bologna area (Fig. 1; Amorosi et al., 1996). On the other hand, the latest documented tectonic uplift of the chain ended in the Reggio Emilia sector in the late Pleistocene, and the continental late Quaternary deposits onlapping the Apennine margin appear to be undeformed (Barbacini et al., 2002).

Moreover, as discussed earlier, postdepositional diagenetic changes, such as feldspar alteration and carbonate dissolution, do not seem to have varied the sand composition.

Climate changes related to glacial-interglacial phases appear to be the most important factor that modified sediment composition at ca. 12 ka by changing bedrock weathering rate and removal of weathering products. This shift in composition represents a clear supply signal in the basin-fill record. High-frequency variations in the rate of sediment supply (order of 10 ka) are recorded as secular variations in the extent of weathering (Weltje et al., 1998). The change in composition suggests that sediments are today subjected to more prolonged chemical weathering during storage in the source area, but that alteration was negligible during storage in the alluvial sequence, which is contrary to what is reported for tropical climate settings (Johnsson et al., 1991). Thus, the sediment supply shift could reflect changes from a stage of marked denudation, such as during the Last Glacial Maximum (15–18 ka; Orombelli and Ravazzi, 1996), to a stage of colluvial aggradation and soil formation with dense vegetation cover (since ca. 7 ka).

CONCLUSIONS

In the late Pleistocene–Holocene sedimentary record of the Modena floodplain, the composition of fluvial sands has provided a unique opportunity to recognize and isolate the signature of factors controlling sediment supply and the nature of their complex interactions.

Modal analyses of modern sands from the Modena Plain streams indicate that their provenance signal can be clearly distinguished. The compositional field of modern sands from each stream depends, in order of importance, on the extent of the watershed, the recycling of older fluvial sediments, and, subordinately, on the sediment input from tributary streams.

Our data demonstrate that sand composition of major rivers (Panaro and Secchia) has not varied during the last ca. 7 ka, a stage characterized by colluvial aggradation with dense vegetation cover. The direct implication is that provenance of older sediments buried in the floodplain can be determined by a simple comparison with modern sand composition, suggesting that we have a powerful tool to reconstruct the evolution of the main drainage system in the Holocene. The next step will be to extend this study to characterize the evolution of human settlements as a function of drainage system changes.

The higher feldspar content in late Pleistocene fluvial sands appears to reflect a high-frequency signal in sediment supply rate due to secular variations of weathering processes. This compositional shift reveals the strong denudation and sediment removal conditions of the last glacial stage (15–18 ka). For these reasons, sand composition studies represent a new, fundamental way to investigate the glacial-interglacial climatic influence on sediment supply and deposition in the northern Apennines floodplain.

ACKNOWLEDGMENTS

We thank N. Giordani (Soprintendenza per i Beni Archeologici dell'Emilia Romagna) for permission to sample archaeological sections. We are grateful to D. Vai and A. Ferrari for support in archaeological sites. The thorough and constructive reviews of A.C. Morton, M. Johnsson, and G.G. Zuffa greatly improved the manuscript.

REFERENCES CITED

Amorosi, A., Farina, M., Severi, P., Preti, D., Caporale, L., and Di Dio, G., 1996, Genetically related alluvial deposits across active fault zones: An example of alluvial fan-terrace correlation from the Upper Quaternary of the southern Po Basin, Italy: Sedimentary Geology, v. 102, p. 275–295, doi: 10.1016/0037-0738(95)00074-7.

Barbacini, G., Bernini, M., Papani, G., and Rogledi, S., 2002, Le strutture embricate del margine Appenninico Emiliano tra il T. Enza ed il F. Secchia- Prov. Reggio Emilia: Bologna, Cartografia Geologica, Atti III Seminario sulla Cartografia Geologica, 26–27 Febbraio 2002, p. 64–69.

Basu, A., 1985, Influence of climate and relief on compositions of sands released at source areas, in Zuffa, G.G., ed., Provenance of Arenites: Dordrecht/Boston/Lancaster, D. Reidel Publishing Company, NATO ASI series, v. 148, p. 1–18.

Bettelli, G., and De Nardo, M.T., 2001, Geological outlines of Emilia Apennines (Italy) and introduction to the rock units cropping out in the areas of landslides reactivated in the 1994–1999 period: Quaderni di Geologia Applicata, v. 8, no. 1, p. 7–26.

Cardarelli, A., Cattani, M., Labate, D., and Pellegrini, S., 2004, Archeologia e Geomorfologia. Un approccio integrato applicato al territorio di Modena: Per un Atlante Storico Ambientale Urbano, Comune di Modena, Ufficio Ricerche e Documentazione sulla: Storia Urbana, p. 65–77.

Cati, L., 1981, Idrografia e idrologia del Po: Ufficio Idrografico del Po, Istituto Poligrafico e Zecca dello Stato, Pubblicazione 19, 310 p.

Cremaschi, M., 2000, Manuale di Geoarcheologia: Roma, Editori Laterza, 386 p.

Cremaschi, M., and Gasperi, G., 1989, L'alluvione alto medioevale di Mutina (Modena) in rapporto alle variazioni ambientali oloceniche: Memorie delle Società Geologica Italiana, v. 42, p. 179–190.

Critelli, S., Le Pera, E., and Ingersoll, R.V., 1997, The effects of source lithology, transport, deposition and sampling scale on the composition of southern California sand: Sedimentology, v. 44, p. 653–671, doi: 10.1046/j.1365-3091.1997.d01-42.x.

Critelli, S., Arribas, J., Le Pera, E., Tortosa, A., Marmaglia, K.M., and Latter, K.K., 2003, The recycled orogenic sand provenance from an uplifted thrust belt, Betic Cordillera, southern Spain: Journal of Sedimentary Research, v. 73, p. 72–81.

Dickinson, W.R., 1970, Interpreting detrital modes of graywacke and arkose: Journal of Sedimentary Petrology, v. 40, p. 695–707.

Gasperi, G., and Pizziolo, M., 2007, Note illustrative della carta geologica d'Italia a scala 1:50.000, Foglio 201 "Modena": Regione Emilia-Romagna, Servizio Geologico d'Italia (in press).

Gasperi, G., Cremaschi, M., Mantovani Uguzzoni, M.P., Cardarelli, A., Cattani, M., and Labate, D., 1989, Evoluzione Plio-Quaternaria del margine Appenninico modenese e dell'antistante pianura: Note illustrative alla carta geologica: Memorie delle Società Geologica Italiana, v. 39, p. 375–431.

Gazzi, P., 1966, Le arenarie del flysh sopracretaceo dell'Appennino modenese; correlazioni con il flysh di Monghidoro: Acta Mineralogico-Petrographica, v. 12, p. 69–97.

Howarth, R.D., 1998, Improved estimators of uncertainty in proportions, point-counting, and pass-fail text results: American Journal of Science, v. 298, p. 594–607.

Ingersoll, R.V., Bullard, T.F., Ford, R.L., Grimm, J.P., Pickle, J.D., and Sares, S.W., 1984, The effect of grain size on detrital modes: A test of the Gazzi-Dickinson point-counting method: Journal of Sedimentary Petrology, v. 54, p. 103–116.

James, W.C., Mack, G.H., and Suttner, L.J., 1981, Relative alteration of microcline and sodic plagioclase in semi-arid and humid climates: Journal of Sedimentary Petrology, v. 51, p. 151–163.

Johnsson, M.J., Stallard, R.F., and Lundberg, N., 1991, Controls on the composition of fluvial sands from a tropical weathering environment; sands of the Orinoco River drainage basin, Venezuela and Colombia: Geological Society of America Bulletin, v. 103, p. 1622–1647, doi: 10.1130/0016-7606(1991)103<1622:COTCOF>2.3.CO;2.

Lugli, S., Marchetti Dori, S., Fontana, D., and Panini, F., 2004, Composizione dei sedimenti sabbiosi nelle perforazioni lungo il tracciato ferroviario ad

Alta Velocità: Indicazioni preliminary sull'evoluzione sedimentaria della media pianura modenese: Il Quaternario, v. 17, p. 379–390.

Maffei, M., 2001, Evoluzione recente della pianura modenese-reggiana, implicazioni paleoambientali, paleoclimatiche e geoarcheologiche [Ph.D. thesis]: Modena, University of Modena e Reggio Emilia.

McBride, E.F., Abel-Wahab, A., and McGilvery, T.A., 1996, Loss of sand-size feldspar and rock fragments along the south Texas barrier island, USA: Sedimentary Geology, v. 107, p. 37–44, doi: 10.1016/S0037-0738(96)00016-4.

Mennella, C., 1972, Il clima d'Italia: Volume secondo: Napoli, Italy, Fratelli Conte Editore, 803 p.

Orombelli, G., and Ravazzi, C., 1996, The late glacial and early Holocene chronology and paleoclimate: Il Quaternario, v. 9, p. 439–444.

Panizza, M., Castaldini, D., Pellegrini, M., Giusti, C., and Piacentini, D., 2004, Matrici geo-ambientali e sviluppo insediativo: Un'ipotesi di ricerca: Per un Atlante Storico Ambientale Urbano, Comune di Modena, Ufficio Ricerche e Documentazione sulla: Storia Urbana, p. 31–62.

Weltje, G.J., 2002, Quantitative analysis of detrital modes: Statistically rigorous confidence regions in ternary diagrams and their use in sedimentary petrology: Earth-Science Reviews, v. 57, p. 211–253, doi: 10.1016/S0012-8252(01)00076-9.

Weltje, G.J., and Von Eynatten, H., 2004, Quantitative provenance analysis of sediments: Review and outlook: Sedimentary Geology, v. 171, p. 1–11, doi: 10.1016/j.sedgeo.2004.05.007.

Weltje, G.J., Meijer, X.D., and De Boer, P.L., 1998, Stratigraphic inversion of siliciclastic basin fills: A note on the distinction between supply signals resulting from tectonic and climatic forcing: Basin Research, v. 10, p. 129–153, doi: 10.1046/j.1365-2117.1998.00057.x.

Zuffa, G.G., 1985, Optical analyses of arenites: Influence of methodology on compositional results, in Zuffa, G.G., ed., Provenance of Arenites: Dordrecht/Boston/Lancaster, D. Reidel Publishing Company, NATO ASI series, v. 148, p. 165–189.

MANUSCRIPT ACCEPTED BY THE SOCIETY 9 AUGUST 2006

Geochemistry and petrography of Western Tethys Cretaceous sedimentary covers (Corsica and Northern Apennines): From source areas to configuration of margins

Laura Bracciali[†]
Dipartimento di Scienze della Terra, Università di Pisa, 56126 Pisa, Italy

Michele Marroni
Luca Pandolfi
Sergio Rocchi
Dipartimento di Scienze della Terra, Università di Pisa, 56126 Pisa, Italy, and Istituto di Geoscienze e Georisorse, Consiglio Nazionale delle Ricerche (CNR), 56126 Pisa, Italy

ABSTRACT

Provenance studies most commonly apply the classical approach based on petrographic modal analysis of arenites. In this paper, a modal analysis of arenites is combined with both a petrographic study on conglomerate clasts and a geochemical investigation of major and trace elements of pelites. The Ligure-Piemontese oceanic basin, a branch of Western Tethys, and its continental margins were consumed during the Eocene collisional events that led to the formation of the Alpine-Apennine belt. Remnants of Cretaceous sedimentary successions supplied by the continental margins are today preserved as tectonic units in the Alpine-Apennine belt: Balagne Nappe in Alpine Corsica and Internal and External Ligurian units in the Northern Apennines. The petrography of pebbles from rudites and lithic fragments from arenites shows that Corsica and Internal Ligurian units contain debris from granitoids, low-grade metamorphic rocks, and carbonate platform rocks, while the External Ligurian units contain debris from low- to high-grade metamorphic rocks, a mantle-rock source, carbonate platform, and pelagic siliceous and carbonate rock sources. Geochemical data on pelites indicate a more mafic-ultramafic character for External Ligurian units (enrichment in Cr, Co, Ni, and Th/Sc/Cr/V/Ni relationships that show a systematic shift toward an ultramafic contribution). Petrographic and chemical data indicate that the source for sediments of Corsica and Internal Ligurian units was made up of the upper part of a continental basement and its carbonate sedimentary cover (the Corsica-Europe continental margin). On the other hand, the External Ligurian units were supplied by a source area where a complete lithospheric section

[†]E-mail: bracciali@dst.unipi.it.

was exposed, from the upper mantle up to the deep-sea sedimentary cover (the Adria continental margin).

These findings are useful in order to unravel the processes related to the opening mechanisms of the Ligure-Piemontese oceanic basin: among the different rifting models in existence, our data support an asymmetric mechanism dominated by a west-dipping detachment fault, with the Adria margin acting as the lower plate.

Keywords: Northern Apennines, Corsica, provenance, Western Tethys rifting, petrography, sediment geochemistry, Cretaceous.

INTRODUCTION

Provenance studies on clastic sedimentary rocks are generally performed following a classical approach based on petrographic modal analysis of arenites (Dickinson, 1970, 1985; Crook, 1974; Schwab, 1975; Dickinson and Suczek, 1979; Ingersoll and Suczek, 1979; Dickinson and Valloni, 1980; Valloni, 1985; Zuffa, 1980, 1985, 1987; Johnsson, 1993). However, the geochemical study of pelitic rocks, the most abundant rocks in sedimentary basins, would make it possible to harvest a wealth of additional data. It is well established that some relatively immobile elements that show very low concentrations in natural waters, (e.g., rare earth elements [REE], Th, Nb, Sc) are transferred nearly quantitatively throughout the sedimentary process from the parent rocks into the clastic sediments (Taylor and McLennan, 1985; Condie, 1991). Moreover, abundances and ratios of relatively immobile elements are generally unmodified during diagenesis or low-grade metamorphism, leading to the understanding that the geochemistry of pelitic sediments reflects the nature of their source rocks. Also, the efficiency of mixing during suspension transport causes a mud-derived provenance signal to be much more representative than a sand-based provenance signal. On this basis, the geochemical features of siliciclastic rocks have been fruitfully used in the last decade in provenance studies to constrain sediment provenance and source composition (e.g., van de Kamp and Leake, 1995; Plank and Langmuir, 1998; Owen et al., 1999; Cullers, 2002; Hofmann et al., 2003). As suggested and recommended in recent works dealing with provenance (Johnsson and Basu, 1993; Basu, 2003; Weltje and von Eynatten, 2004, and references therein), a multidisciplinary approach to siliciclastic provenance is highly desirable. Hence, in this work, we join a modal analysis of arenites and a mesoscopic-microscopic petrographic study on conglomerate clasts to a geochemical investigation of major-and trace-element distribution in pelites.

This approach has been applied in order to unravel a geological problem: the reconstruction of the configuration of continental margins of the Ligure-Piemontese oceanic basin (a branch of the Western Tethys that opened in Jurassic time between the Corsica-Europe and Adria plates), which would not be solvable using a provenance study based only on arenite petrography. We studied and compared sedimentary rocks supplied by the same tectonic setting, the passive margins of Western Tethys. The main goals of this work are: (1) to unravel the composition of the two margins, which in turn can provide further constraints for their configurations prior to their involvement in the collisional events; and (2) to apply our findings in order to discriminate among the different models existing for the opening history of the Ligure-Piemontese Ocean.

Remnants of the Ligure-Piemontese basin, today preserved as tectonic units within the Alpine-Apennine belt (which originated after the continental collision between the Corsica-Europe and Adria plates during Eocene), display well-developed Lower to Upper Cretaceous sedimentary successions. Paleogeographic reconstructions permit the pristine location of these units in the Ligure-Piemontese basin. In this paper, sedimentary successions derived from the following units are analyzed: (1) the Balagne Nappe (Corsica) and the Internal Ligurian units (Northern Apennines), both considered representative of an area of the oceanic basin located close to the Corsica-Europe continental margin; and (2) the External Ligurian units (Northern Apennines), which are interpreted to be derived from the domain that joined the oceanic area to the Adria continental margin.

GEOLOGICAL SETTING

Along the Northern Apennines–Corsica transect, the Alpine belt is characterized by the association of Jurassic ophiolites with Late Jurassic–middle Eocene sedimentary successions. They are regarded as fragments of the Ligure-Piemontese oceanic basin (a branch of Western Tethys) and its transition to the continental margins.

In the paleogeographic reconstructions of the Jurassic (e.g., Stampfli and Borel, 2004), the Ligure-Piemontese basin is generally depicted as a small, no more than 500-km-wide, oceanic area that opened between the Corsica-Europe and Adria continental margins. The opening of the Ligure-Piemontese oceanic basin occurred in Middle Jurassic time, as evidenced by both the age of the oldest radiolarites found at the top of the ophiolite sequence (e.g., Chiari et al., 2000) and isotopic dating of the intrusive sequences (e.g., Borsi et al., 1996). The spreading phase was predated by a long-lived rifting history, the inception of which is generally referred to as Early Triassic in age (e.g., Marroni et al., 1998). In the Late Cretaceous, probably during the late Campanian, a main change in Africa and Eurasia plate motion induced a large-scale convergence in the Ligure-Piemontese ocean, which caused the subduction and the related tectonic events. This continuous convergence led to a complete consumption of the oceanic crust involved in the subduction processes. From the Paleocene onward, the deformation affected the continental margins up to the final collision that occurred in the middle Eocene. During the continental collision, the remnants of the oceanic basin and the neighboring ocean-continent transition were enclosed in the Alpine-Apennine belt as deformed and/or metamorphosed slices. Among them, some units, even if strongly deformed, were affected only by very low-grade metamorphism, and their pristine stratigraphic, petrographic, and geochemical features can be detected. These units are located at the top of the nappe pile in both Alpine Corsica and the Northern Apennines (Fig. 1). In Alpine Corsica, these units are represented by the Balagne Nappe, which has been interpreted as a portion of the Ligure-Piemontese oceanic crust located close to the Corsica-Europe continental margin (Durand Delga et al., 1997, and references therein). In the Northern Apennines, the units derived from the oceanic basin and its transition to the Adria continental margin are represented by the Ligurian units. These units are recognized in two different lithostratigraphic and tectonic settings, corresponding to Internal Ligurian and External Ligurian units (Elter et al., 1966; Elter, 1972, 1975). The Internal Ligurian units are representative of an area of the Ligure-Piemontese oceanic basin located close to the Corsica-Europe continental margin (Abbate et al., 1980; Abbate and Sagri, 1982; Nilsen and Abbate, 1984, and references therein; Gardin et al., 1994), whereas the External Ligurian units are interpreted as having derived from the domain that joined the oceanic area to the Adria continental margin (Marroni et al., 2001a, and references therein).

In this work, only the deposits sedimented before the late Campanian were investigated. These deposits predated the main basin inversion, which probably affected both the continental margins starting from the late Campanian and created new source areas inside the basin, such as the subduction-related accretionary wedge. During this stage, the sediments dispersal pattern became more complicated and the sediment composition was influenced by the subduction and the continental collision-related processes. Therefore, study of the pre–late Campanian deposits can provide useful information for the reconstruction of the configuration of the continental margins, as inherited from the rifting process and not yet modified by the main collisional tectonic events.

Balagne Nappe from Alpine Corsica

In Alpine Corsica (Fig. 1), the best-preserved oceanic deposits crop out in the Balagne Nappe (Fig. 2). The Balagne Nappe is subdivided into two main tectonic units known as the Toccone and Navaccia units. Both units were affected by a polyphase deformation developed under very low-grade metamorphic conditions (Egal, 1992; Marroni and Pandolfi, 2003).

The succession of the Balagne Nappe (Fig. 2) can be reconstructed in the Navaccia unit, where a complete stratigraphic succession from Jurassic ophiolites up to the Upper Cretaceous sedimentary cover is recognized (Marroni and Pandolfi, 2003, and references therein). The ophiolite sequence is very similar to that of the Internal Ligurian units; it has a basement consisting of mantle lherzolites and gabbros covered by massive and pillow basalts interfingering with ophiolitic sedimentary breccias. In the Navaccia unit, the ophiolite sequence is overlain by radiolarian-bearing Bathonian to early Callovian Chert (Chiari et al., 2000; De Wever and Danelian, 1995) that grades upward to the San Colombano Limestone (Tithonian–early Berriasian). The San Colombano Limestone (cf. Calpionella Limestone) shows a transition to the San Martino Formation (early Berriasian–late

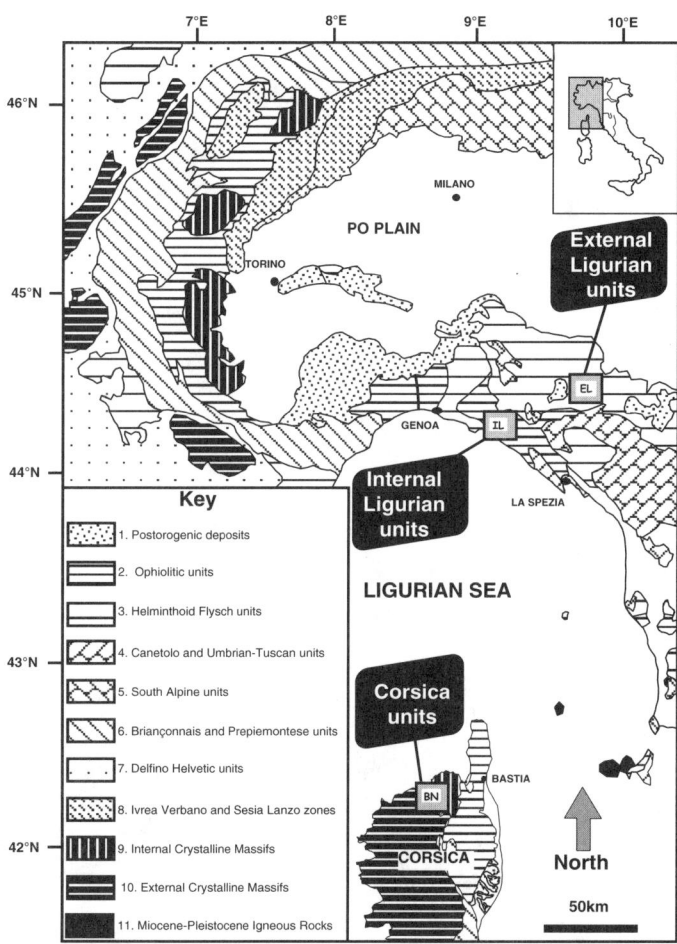

Figure 1. Tectonic sketch map of the Western Alps, Northern Apennines, and Corsica with location of the study sequences. (1) Postorogenic sedimentary sequences of the Tertiary Piemonte basin, Ranzano basin, and Miocene deposits of Corsica. (2) Alpine and Apennine ophiolitic units (IL—Internal Ligurian units, Sestri Voltaggio, Voltri Group, Piemontese units; BN—Balagne Nappe and Schistés Lustrées of Corsica). (3) Helminthoid Flysch and associated sedimentary complexes (EL—External Ligurian units, Mt. Antola unit, Autapie, and S. Remo–M. Saccarello units). (4) Canetolo and Umbrian-Tuscan units. (5) South Alpine units. (6) Briançonnais and Prepiemontes units, including sedimentary cover of Hercynian Corsica. (7) Delfino Helvetic units. (8) Ivrea Verbano unit and Sesia Lanzo zone. (9) Internal Crystalline massifs of the western Alps and Tenda Massif. (10) External Crystalline massifs of the Western Alps and Hercynian basement of Corsica. (11) Miocene-Pleistocene igneous and volcanic rocks.

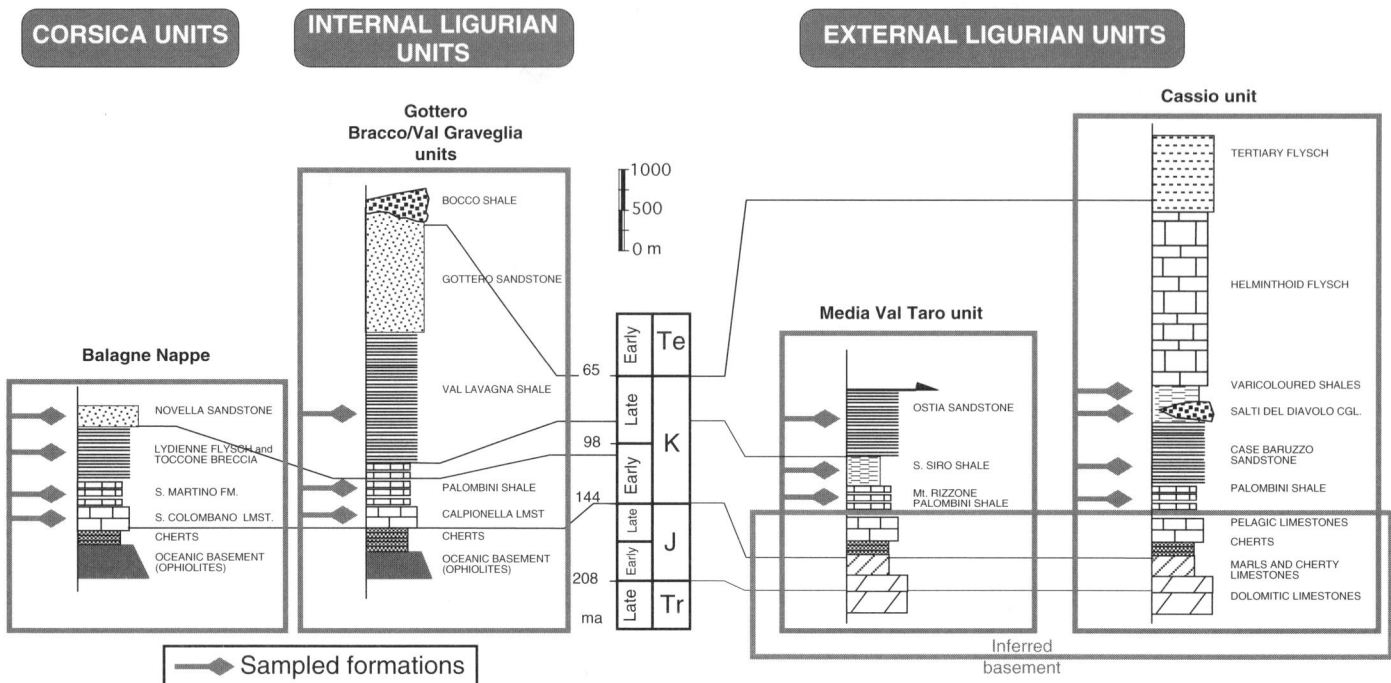

Figure 2. Stratigraphic log of the studied units with locations of samples. Age data from Corsica are from: Conti et al. (1985), De Wever and Danelian (1995), Marino et al. (1995), Marroni et al. (2000); Internal Ligurian units: Passerini and Pirini (1964), Decandia and Elter (1972), Monechi and Treves (1984), Marroni and Perilli (1990), Perilli and Nannini (1997), Chiari et al. (2000); External Ligurian units: Iaccarino and Rio (1972), Rio et al. (1983), Rio and Villa (1987), Marroni et al. (1992), Vescovi et al. (1999), Vescovi (2006). Te—Tertiary, K—Cretaceous, J—Jurassic, Tr—Triassic.

Hauterivian/early Barremian) that has been correlated with the Palombini Shale from Internal Ligurian units (Marroni et al., 2000). The San Martino Formation grades upward to the Lydienne Flysch, which consists of thin-bedded, mixed turbidites that show a vertical and lateral stratigraphic relationship with the Toccone Breccia and the Novella Sandstone (Nardi et al., 1978; cf. Gare de Novella Sandstone) of late Cenomanian age. The Novella Sandstone represents the youngest formation recognizable in the Balagne Nappe.

According to Gruppo Lavoro Ofioliti Mediterranee (1977) and Durand-Delga et al. (1997), the geochemical features of the basalts from the Balagne Nappe reveal a typical affinity of a crust developed in the first stage of the oceanic spreading. In addition, in the pillow basalt sequence, levels of terrigenous debris made up of quartz and minor feldspar fragments have been described. The typology of detrital zircons also found in the terrigenous debris clearly indicates a source area from Hercynian Corsica (Durand-Delga et al., 1997). These features indicate a paleotectonic position of the Navaccia unit succession close to the Corsica-Europe continental margin.

Internal Ligurian Units from Northern Apennines

In the Northern Apennines, the Internal Ligurian units occur as a complex stack of tectonic units that have been affected by a polyphase deformation developed under very low-grade metamorphic conditions (Marroni and Pandolfi, 1996; Marroni et al., 2004).

The Bracco–Val Graveglia and Gottero units are characterized by well-preserved successions that allow the reconstruction of the complete stratigraphic log of the Internal Ligurian units (Fig. 2). This succession includes an ophiolite sequence, represented by mantle ultramafic rocks and gabbros, covered by a thick volcano-sedimentary complex that consists of pillow lavas and massive basaltic flows (normal to transitional mid-ocean-ridge basalts [MORBs]; Venturelli et al., 1981, and references therein) that interfinger with ophiolite breccias and cherts (e.g., Decandia and Elter, 1972; Abbate et al., 1980). The overlying sedimentary cover consists of hemipelagic deposits represented by Cherts (Callovian-Tithonian), Calpionella Limestone (Berriasian-Valanginian), and the Palombini Shale (Valanginian-Santonian). The Palombini Shale grades upward to siliciclastic turbidites (Val Lavagna Shale and Gottero Sandstone; Campanian–early Paleocene) that are regarded as a deep-sea turbidite system (Nilsen and Abbate, 1984). The youngest deposit of the ophiolite sedimentary cover is represented by Lower Paleocene coarse-grained deposits known as the Bocco Shale, which are characterized by the occurrence of debris flows and slide deposits (Marroni and Pandolfi, 2001).

According to Abbate and Sagri (1982), Nilsen and Abbate (1984), and Pandolfi (1997), the turbidite sequences recognizable in the Internal Ligurian succession are characterized by turbiditic facies indicative of a connection between continental-margin and deep-sea deposits (cf. inner fan facies of Nilsen and Abbate, 1984). This feature indicates a location close to the source area represented by the Corsica-Europe continental margin.

External Ligurian Units from Northern Apennines

In the Alpine-Apennine framework, the External Ligurian units are regarded as representative of the domain that joined the Ligure-Piemontese oceanic basin to the Adria continental margin (e.g., Elter et al., 1966; Elter, 1975; Marroni et al., 2001a). The External Ligurian units successions are generally detached from their pristine substratum. The oldest parts of the External Ligurian successions mainly consist of Upper Cretaceous sedimentary complexes, reported in literature as "complessi di base" (hereafter referred to as basal complexes). The basal complexes consist of turbidite and hemipelagic deposits associated with sedimentary coarse-grained mélanges typically characterized by slide blocks of igneous, metamorphic, and sedimentary rocks. Some of these basal complexes preserve a gradual transition to Upper Cretaceous carbonate flysch (Helminthoid Flysch Auctt.). The External Ligurian units can be divided into two different groups according to the lithostratigraphic features of their basal complexes. The basal complexes of the first group include slide blocks and debris flow from mantle lherzolites, gabbros, and basalts. The basal complexes of the second group display a clear continental affinity, and the presence of mafic-ultramafic rocks has never been documented. According to their features, the first group (referred to as western) is considered to have been located at the transition to the Ligure-Piemontese oceanic domain, whereas the second group, (referred to as eastern) was located at the distal edge of the Adria continental margin (Marroni et al., 2001a). This work is focused on the basal complexes of the eastern External Ligurian units, particularly on the Cassio and Media Val Taro units, which show the best-preserved successions (Fig. 2).

In the Cassio unit (Fig. 2), a complete transition from the basal complex to Upper Cretaceous Helminthoid Flysch and Lower Tertiary, predominantly shaly deposits can be reconstructed. The studied basal complex includes the Palombini Shale (Hauterivian-Aptian), the Case Baruzzo Sandstone (Cenomanian arenites correlated to the Ostia Sandstone; Vescovi et al., 1999), and the Varicolored Shale (Cenomanian–late Campanian). The hemipelagic Varicolored Shale is characterized by intercalations of conglomerates and coarse-grained arenites, known as the Salti del Diavolo Conglomerate. The Varicolored Shale grades upward to the Upper Campanian–Maastrichtian Monte Cassio Flysch.

The Media Val Taro unit (Fig. 2; Vescovi et al., 1999) consists of a basal complex represented by the Mt. Rizzone Palombini Shale (Barremian-Aptian), the S. Siro Varicolored Shale (Aptian-Turonian), and the Ostia Sandstone (Coniacian-Santonian). The Ostia Sandstone consists of thin-bedded turbidites with modal analyses (Valloni and Zuffa, 1984) that point to a litharenitic-hybrid composition.

SAMPLING STRATEGY AND ANALYTICAL METHODS

Rudites, arenites, and pelites from each stratigraphic unit were investigated. Samples of the different grain-size classes were collected from the same formation and, whenever possible, from the same layer.

Petrographic analysis was performed on thin sections from conglomerates clasts, in order to identify representative rock types of the source areas (Fig. 2). The criteria for sampling included representativity of the lithological distribution and low degree of alteration of the sample (cf. Wandres et al., 2004).

A modal analysis was performed on the arenitic fraction in order to estimate the contribution to the sediment of the different rock types recognized in the ruditic fraction. Point counting (500 points) of arenites was performed using the Gazzi-Dickinson technique (Gazzi, 1966; Dickinson, 1970; Ingersoll et al., 1984; Zuffa, 1987) to minimize the dependence of arenite composition on grain size. All point-counted arenites were stained for plagioclase and K-feldspar using sodium cobaltinitrite (Houghton, 1980) and Alizarin-red-S plus potassium ferricyanide solutions for carbonate identification (Lindholm and Finkelman, 1972). Calculated grain parameters are reported in Table 1. Original data of the arenites from External Ligurian units have been integrated with other data from the literature (Daniele and Bianchi, 1995).

Postdepositional modifications that affected both the studied arenites and pelites are quite strong and pervasive. The collected samples are far from an unconsolidated sediment: they derive from several units that show different structural evolution, deformation patterns, and metamorphism ranging from the limits of diagenesis to very low-grade metamorphism (External Ligurian units) up to the limits between very low-grade and low-grade metamorphism (Internal Ligurian units) (Leoni et al., 1996, and references therein). The recrystallization processes affected mainly the fine-grained fraction (silt and mud) producing quartz + albite + chlorite + illite + paragonite + Fe-oxides. Due to these recrystallization processes and according to several authors (van de Kamp and Leake, 1995; Fedo et al., 1996; Cullers 2000; Totten et al., 2000), the chemical analyses were performed without separating the different grain-size fractions.

Chemical analyses were performed on powders from pelitic samples. Loss on ignition (LOI) was determined gravimetrically on preheated powders (110 °C) after 1 h ignition at 1000 °C in a microwave oven (MAS 300). Major elements were determined by X-ray fluorescence (XRF) (ARL 9400 XP®) on glass beads following the procedure of Tamponi et al. (2003). Fused beads were prepared after ignited powders in order to avoid sulfide minerals leading to the corrosion of the platinum crucible during the flux melting.

The sampling of pelites was carried out on carbonate-poor layers, to better investigate the composition of noncarbonate source areas by the analysis of clay minerals where REE and other key trace elements reside. The carbonate fraction, which results in a simple dilution effect because of its typically fairly

TABLE 1. RECALCULATED MODAL POINT COUNT DATA FOR THE STUDIED ARENITES

Sample	Formation–unit	Grain size	Sorting		NCE CI + NCI CE (%)			Q F L (%)			Q F L + C (%)			Lm Lv Ls (%)			Lm Lv Ls + C (%)		
					NCE	CI + NCI	CE	Q	F	L	Q	F	L + C	Lm	Lv	Ls	Lm	Lv	Ls + C
Corsica Units																			
PL20	NS-NAV	m/c	C		92.9	3.1	4.0	54.5	24.3	21.2	52.2	23.3	24.5	37.9	62.1	0	31.4	51.4	17.1
PL21	NS-NAV	m/c	C		73.4	2.0	24.6	55.0	27.3	17.7	41.2	20.5	38.4	54.7	45.3	0	18.8	15.6	65.6
PL22	NS-NAV	m	B		97.1	0.9	2.0	61.5	22.8	15.7	60.2	22.3	17.5	46.3	50.8	3.0	40.8	44.7	14.5
CO51	NS-NAV	m/c	C		82.5	1.8	15.7	34.3	24.3	41.4	28.8	20.4	50.8	32.0	68.0	0	21.9	46.4	31.7
CO52b	NS-NAV	m/c	C		95.0	0.9	4.1	55.5	22.1	22.4	54.1	21.2	24.7	52.3	47.7	0	43.4	39.6	17.0
CO53	NS-NAV	m/c	C		84.5	6.0	9.5	50.6	28.9	20.6	45.4	25.9	28.6	45.3	54.7	0	29.3	35.3	35.3
CO55	NS-NAV	m/c	B		87.5	4.6	7.9	54.6	30.5	15.0	51.0	27.9	21.1	50.0	50.0	0	30.2	30.2	39.5
PL7	TB-NAV	m/c	C		76.4	0	23.6	54.1	24.1	21.8	40.9	18.1	40.9	29.2	70.8	0	11.5	27.9	60.6
PL10	LF-NAV	m/c	B		75.2	0	24.8	24.0	32.0	44.0	17.9	23.9	58.2	13.6	86.4	0	7.7	27.9	43.6
PL11	LF-NAV	m/c	B		79.0	0.2	20.8	23.9	28.3	47.8	18.7	22.2	59.1	19.1	80.9	0	12.1	51.3	36.7
PL12	LF-NAV	m/c	B		62.1	0.8	37.1	24.4	29.0	46.6	15.1	17.9	67.0	9.0	91.0	0	3.9	39.2	57.0
PL16	LF-NAV	m/c	B		56.7	0.3	43.1	23.7	30.4	46.0	13.3	17.0	69.7	13.6	86.4	0	5.0	32.0	63.0
PL23	LF-NAV	m/c	B		74.6	3.5	21.9	63.8	20.1	16.0	49.1	15.5	35.4	34.0	66.0	0	11.9	23.0	65.2
CO49	LF-NAV	m/c	C		67.5	0	32.5	34.3	24.4	41.3	23.0	16.4	60.7	10.3	89.7	0	4.7	41.0	54.3
CO50	LF-NAV	m/c	B		74.8	0.2	24.9	34.9	32.2	32.9	25.9	23.9	50.3	25.8	74.2	0	12.5	36.0	51.5
CO61b	LF-NAV	m/c	B		78.1	0.2	21.6	23.3	25.2	51.5	18.1	19.6	62.4	15.1	84.9	0	9.6	54.2	36.1
CO62	LF-NAV	m/c	C		82.0	0	18.0	27.7	28.0	44.3	22.6	22.8	54.6	7.6	92.4	0	5.1	61.0	33.9
CO63	LF-NAV	m/c	B		66.7	0	33.3	32.1	29.7	38.2	21.2	19.6	59.2	22.2	77.8	0	9.5	33.2	57.3
CO64	LF-NAV	m/c	B		75.2	0	24.8	29.4	27.5	43.1	22.1	20.6	57.3	17.0	83.0	0	9.6	46.9	43.5
CO65	LF-NAV	m/c	B		68.7	0	31.3	44.2	25.3	30.5	30.1	17.2	52.7	15.9	84.2	0	6.3	33.2	60.6
				X	77.5	1.2	21.3	40.3	26.8	32.9	32.5	20.8	46.7	27.6	72.3	0.2	16.3	39.5	44.2
				SD	10.6	1.8	11.4	14.5	3.4	12.8	15.1	3.3	16.4	15.6	15.8	0.7	12.3	11.3	16.4
Internal Ligurian Units																			
LB14	VLS-GOT	m/c	C		100.0	0	0	61.2	35.9	2.9	63.0	35.0	2.0	55.1	41.4	3.4	54.5	41.0	4.5
LB15	VLS-GOT	m/c	C		100.0	0	0	35.6	57.6	6.8	37.9	61.3	0.8	100.0	0	0	100.0	0	0
LB103	VLS-GOT	m/c	B		98.0	1.0	1.0	46.2	33.3	20.5	50.4	36.3	13.3	42.4	24.2	33.3	37.8	21.6	40.5
LB111	VLS-GOT	m/c	C		98.9	0.8	0.3	43.1	44.8	12.2	44.8	46.6	8.6	42.1	47.4	10.5	40.0	45.0	15.0
LB190	VLS-GOT	m/c	B		96.3	3.7	0	38.5	39.7	21.8	39.3	40.5	20.3	59.5	40.5	0	59.5	40.5	0
LB193	VLS-GOT	m/c	B		96.5	3.0	0.5	40.2	30.4	29.4	41.9	31.7	26.3	75.0	25.0	0	72.7	24.2	3.0
LB194	VLS-GOT	m/c	B		97.3	2.8	0	39.5	32.9	27.5	42.9	35.7	21.4	64.2	35.9	0	64.2	35.9	0
LB196	VLS-GOT	m/c	C		98.8	1.2	0	45.4	29.5	25.1	47.7	31.0	21.3	65.1	33.3	1.6	65.1	33.3	1.6
LB198	VLS-GOT	m/c	B		99.0	0.5	0.5	40.5	32.7	26.8	42.8	34.5	22.8	57.1	41.1	1.8	55.2	39.7	5.2
LB201	VLS-GOT	m/c	B		97.6	2.4	0	43.0	31.3	25.7	44.6	32.5	22.9	66.0	34.0	0	66.0	34.0	0
LB206	VLS-GOT	m/c	C		99.5	0.5	0	53.4	25.2	21.5	55.9	26.4	17.7	26.0	70.0	4.0	26.0	70.0	4.0
LB207	VLS-GOT	m	B		98.6	1.4	0	42.2	34.9	22.9	43.8	36.3	20.0	72.6	27.5	0	72.6	27.5	0
LB208b	VLS-GOT	m/c	C		97.4	2.6	0	47.7	26.2	26.2	50.9	28.0	21.1	46.3	53.7	0	46.3	53.7	0
LB210	VLS-GOT	m/c	C		99.8	0.2	0	53.0	22.9	24.1	55.7	24.1	20.2	73.1	26.9	0	73.1	26.9	0
LB211	VLS-GOT	m/c	B		98.2	1.8	0	52.1	28.5	19.3	54.5	29.8	15.7	78.3	21.7	0	78.3	21.7	0
LB212	VLS-GOT	m/c	B		99.5	0	0.5	47.9	28.8	23.3	49.5	29.8	20.7	65.2	34.8	0	62.5	33.3	4.2
LB214	VLS-GOT	m	C		99.1	0.9	0	39.4	34.2	26.4	43.1	37.4	19.5	75.4	24.6	0	75.4	24.6	0
LB214b	VLS-GOT	m/c	C		98.8	1.2	0	42.8	31.6	25.6	46.5	34.3	19.2	54.7	45.3	0	54.7	45.3	0
LB215	VLS-GOT	m/c	B		97.8	2.2	0	45.2	36.9	17.9	46.7	38.1	15.2	68.6	31.4	0	68.6	31.4	0
LB216	VLS-GOT	m/c	C		99.8	0	0.2	43.5	34.7	21.9	44.7	35.7	19.6	65.5	34.5	0	64.4	33.9	1.7
				X	98.5	1.3	0.2	45.0	33.6	21.4	47.3	35.2	17.4	62.6	34.7	2.7	61.8	34.2	4.0
				SD	1.1	1.1	0.3	6.2	7.6	6.9	6.3	8.0	6.7	16.0	14.2	7.6	16.4	14.1	9.3
External Ligurian Units																			
PL43	SDC-CAS	m/c	B		71.2	0.2	28.6	41.1	29.4	29.4	28.9	20.7	50.4	9.6	78.3	12.1	4.0	32.5	63.5
LM202	SDC-CAS	m/c	B		79.1	0	20.9	35.5	40.4	24.1	27.7	31.5	40.9	5.4	66.2	28.4	2.5	30.4	67.1
DB33	CBS-CAS	m/c	–		83.6	8.2	8.2	62.1	11.4	26.5	54.6	11.0	34.5	16.4	26.2	57.4	12.1	19.3	68.7
DB34	CBS-CAS	m/c	–		79.8	3.5	16.7	57.8	12.1	30.2	47.5	9.9	42.6	23.6	13.6	62.7	13.0	7.5	79.5
DB55	CBS-CAS	m/c	–		68.8	4.5	26.6	62.5	2.9	34.6	45.0	2.1	52.9	28.3	2.2	69.6	13.1	1.0	85.9
DB57	CBS-CAS	m/c	–		67.9	3.0	29.2	48.1	7.7	44.3	33.6	5.3	61.1	25.7	34.3	40.0	12.1	16.1	71.8
DB58	CBS-CAS	m/c	–		66.0	8.0	26.1	58.0	5.7	36.3	41.6	4.1	54.3	29.8	8.8	61.4	14.3	4.2	81.5
DB59	CBS-CAS	m/c	–		67.4	9.7	22.9	51.7	13.4	34.9	38.5	10.0	51.5	28.3	26.4	45.3	13.4	12.5	74.1
DB5	OS-MVT	m/c	–		87.3	6.0	6.7	58.1	27.6	14.3	53.9	25.5	20.6	37.9	31.0	31.0	22.0	18.0	60.0
DB35	OS-MVT	m/c	–		77.8	3.9	18.3	52.9	17.5	29.7	42.9	14.2	42.9	42.9	24.7	32.5	23.7	13.7	62.6
DB36	OS-MVT	m/c	–		88.7	1.9	9.4	53.9	25.2	20.9	48.6	22.8	28.6	20.6	20.6	58.8	13.6	13.6	72.8
LM208	OS-MVT	m/c	C		52.2	0.2	47.6	62.9	17.1	20.0	32.4	8.8	58.8	47.6	40.5	11.9	9.2	7.8	83.0
LM220	OS-MVT	m/c	B		67.5	8.9	23.7	56.3	17.9	25.9	41.3	12.9	45.8	20.6	77.9	1.5	8.9	33.8	57.3
PL46	OS-MVT	m/c	B		69.3	8.2	22.5	55.6	24.3	20.1	44.1	16.3	39.5	39.7	37.9	22.5	13.6	14.3	72.1
PL47	OS-MVT	m/c	C		71.3	5.6	23.2	57.1	28.3	14.6	46.2	19.3	34.5	40.4	39.1	20.4	12.9	15.0	72.1
				X	73.2	4.8	22.0	54.2	18.7	27.0	41.8	14.3	43.9	27.8	35.2	37.0	12.6	16.0	71.5
				SD	9.6	3.3	10.1	7.7	10.4	8.4	8.3	8.4	11.3	12.3	23.1	21.4	5.5	9.8	8.5

Note: m—medium, m/c medium/coarse; X—mean; SD—standard deviation. Sorting parameters (C, B) are according to Longiaru (1987). Grain parameters: NCE—siliciclastic extrabasinal fragments; NCI—siliciclastic intrabasinal fragments; CI—carbonate intrabasinal fragments; CE—carbonate extrabasinal fragments; Q—total quartz; F—total feldspar; L—fine-grained lithic fragments; C—extrabasinal carbonate lithic fragments; Lm—fine-grained metamorphic lithic fragments; Lv—fine-grained volcanic lithic fragments; Ls—fine-grained siliciclastic sedimentary lithic fragments. LF—Lydienne Flysch; TB—Toccone Breccia; NS—Novella Sandstone; VLS—Val Lavagna Shale; CBS—Case Baruzzo Sandstone; SDC—Salti del Diavolo Conglomerate and Varicolored Shale; OS—Ostia Sandstone. NAV—Navaccia unit; GOT—Gottero unit; MVT—Media Val Taro unit; CAS—Cassio unit. Samples labeled DB derive from the work of Daniele and Bianchi (1995).

low trace-element contents, was avoided (Cullers et al., 1979; Taylor and McLennan, 1985; Cullers, 2002). In addition to a sampling concentrated on the carbonate-free layers, XRF major-element concentrations were corrected to eliminate the CaO that did not derive from noncarbonate sources. The correction for calcite content was based on the assumption that the composition of plagioclase in pelites is usually albitic and Ca is mainly hosted in the carbonate fraction of the sediment (the samples exhibited a positive correlation between CaO and LOI before the correction). The correction consisted of subtracting all the CaO of the sample from major-element concentrations, apart from that necessary to saturate phosphorus in apatite.

Selected pelitic samples were analyzed for 36 trace elements using inductively coupled plasma–mass spectrometry (ICP-MS, PQ2 Plus®). Samples were dissolved in screw-top PFA vessels on a hotplate at ~120 °C with HF-HNO$_3$ mixture. Analyses were performed by external calibration using basaltic geochemical reference samples as composition- and matrix-matching calibration solutions. Precision, evaluated by replicate dissolutions and analyses of in-house and international silicate rock reference samples, was generally between 2% and 5% RSD (Relative Standard Deviation) for most elements, except Cr, Ni, Cu, and Pb (6%–11% RSD). The elemental concentrations measured in the studied pelites were 10 to >1000 times the detection limit of the method used. Zirconium was determined via XRF (Philips PW 1480®) on pressed powder pellets, owing to possible incomplete dissolution of zircon during acid digestion of samples in ICP-MS analysis.

RESULTS

Petrography of Arenites and Rudites

Corsica Units

A modal analysis was performed on 20 thin sections from Lydienne Flysch, Toccone Breccia, and Novella Sandstone (Balagne Nappe; Fig. 2; Table 1). Moreover, pebbles from San Colombano Limestone, Toccone Breccia, and Novella Sandstone were analyzed by means of polarizing microscope (35 samples).

Pebbles were characterized by igneous (intrusive, volcanic, and subvolcanic rocks), low-grade metamorphic, and carbonate sedimentary rocks. Intrusive rocks were represented by fine-grained syeno- and monzogranites, while volcanic rocks were characterized by dacites and pyroclastic rhyolites with variable amounts of quartz. Pebbles of subvolcanic dacite and rhyolite porphyries were also observed. Clasts of aphyric to porphyritic basalts were observed over all the stratigraphic succession. These rocks were classified as transitional to tholeiitic within-plate basalts by Marroni et al. (2001b). Low-grade metamorphic pebbles are common; they are muscovite-bearing mica schists and muscovite and/or biotite-bearing gneisses. Limestone pebbles are represented by carbonate platform–derived rocks, mainly grainstones and mudstones of Triassic and Jurassic age. The allochems in the grainstone pebbles are peloids, ooids, minor benthonic foraminifera, and undeterminable macrofossil fragments.

No differences in framework composition were recognized among the arenites from Lydienne Flysch, Toccone Breccia, and Novella Sandstone. They are sublitharenites and subarkoses ($Q_{40}F_{27}L_{33}$) characterized by a mixed siliciclastic-carbonate framework composition ($NCE_{78}NCI + CI_1CE_{21}$; Fig. 3A) and by a volcaniclastic composition of the fine-grained lithic fragments ($Lm_{28}Lv_{72}Ls_{<1}$; Fig. 4).

The extrabasinal siliciclastic framework is characterized by a widespread presence of mono- and polycrystalline quartz (31 ± 14%), plagioclase (7 ± 3%), and K-feldspar (13 ± 3%). Lithic volcanic fragments are widespread (18 ± 9%) and include rhyolite and dacite fragments (Fig. 3A) with porphyritic texture. Intrusive-derived lithic fragments, such as granitoids, are widespread as coarse-grained rock fragments (7 ± 3%). Metamorphic rock fragments include coarse-grained gneisses and fine-grained low-grade schists, mica schists, and metaquartzites (fine-grained lithic fragments, 7 ± 2%). Lithic sedimentary fragments (fine-grained arenites and siltstones) are scarce or completely lacking. Ophiolite-derived fragments, such as serpentinites, gabbros, MORBs, or radiolarites are absent. Carbonate extrabasinal fragments represent an important part of the total framework (21 ± 11%). This group is represented by limestone fragments generally composed of grainstones (Fig. 3A) and mudstones.

Internal Ligurian Units

Modal analysis was performed on 20 samples from the lower part of the Val Lavagna Shale Group (Manganesiferous Shale, Monte Verzi Marls, and Zonati Shale formations). Due to the grain size of these deposits, no data are available from rudites.

The analyzed arenites are arkoses and subarkoses ($Q_{45}F_{34}L_{21}$) characterized by an almost complete siliciclastic framework ($NCE_{99}NCI + CI_1CE_{<1}$; Fig. 3B) and by a metamorphiclastic composition of the fine-grained lithic fragments ($Lm_{62}Lv_{35}Ls_3$; Fig. 4). The arenite framework is dominated by the presence of mono- and polycrystalline quartz (43 ± 6%), plagioclase (9 ± 3%), and K-feldspar (23 ± 6%). Lithic fragments of volcanic nature are common (4 ± 2%) and include porphyritic rhyolite and dacite fragments. Intrusive coarse-grained fragments such as granitoids are also common (16 ± 8%). Metamorphic rock fragments include low-grade schists and mica schists (12 ± 5%). Ophiolite-derived fragments are absent. Carbonate extrabasinal fragments are scarce (<1%), and they are represented by oolitic- and peloidic-grainstones and mudstones. It is worth noting that the overlying Gottero Sandstone (Valloni and Zuffa, 1984; van de Kamp and Leake, 1995; Pandolfi, 1997) is characterized by the same petrographic composition as the Val Lavagna Shale Group.

External Ligurian Units

Data were derived from the integration of literature data (Daniele and Bianchi, 1995) with new analyses (15 samples). The

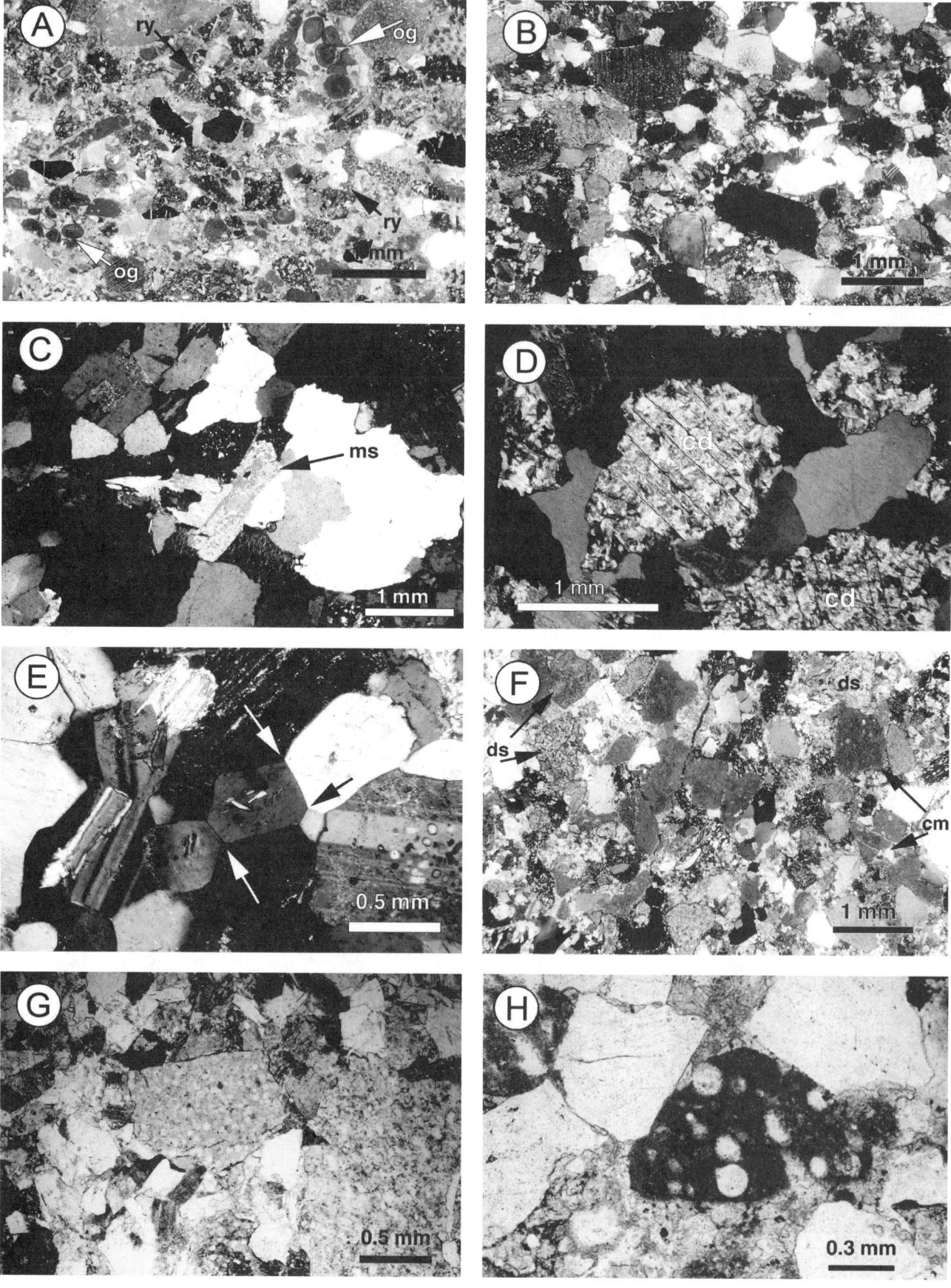

Figure 3. (A) Photomicrograph of arenite from Lydienne Flysch (Navaccia unit, Corsica). Note the mixed siliciclastic-carbonate framework composition and the widespread presence of carbonate platform–derived rock fragments (oolitic grainstones [og]) and acidic volcanic rock fragments (porphyritic rhyolite [ry]). Crossed polars. (B) Photomicrograph of arkose from Val Lavagna Shale (Internal Ligurian unit). Crossed polars. (C) Photomicrograph of peraluminous monzogranite where primary muscovite crystals (ms) are preserved. Crossed polars. Pebble is from Salti del Diavolo Conglomerate, Cassio unit, External Ligurian units. (D) Photomicrograph of cordierite-bearing gneiss. Crystals of cordierite (cd) altered to pinite are indicated. Crossed polars. Pebble is from Salti del Diavolo Conglomerate, Cassio unit, External Ligurian units. (E) Photomicrograph of cordierite-bearing gneiss characterized by widespread triple-point junctions (arrows). Crossed polars. Pebble is from Salti del Diavolo Conglomerate, Cassio unit, External Ligurian units. (F) Photomicrograph of arenite from Case Baruzzo Sandstone (Cassio unit, External Ligurian units). Crossed polars. Note the mixed siliciclastic-carbonate framework composition. Dolostones (ds) and *Calpionella*-bearing mudstones (cm) are indicated. (G) Photomicrograph of arenite from Case Baruzzo Sandstone (Cassio unit, External Ligurian units). Close-up of radiolarian-bearing siliceous packstone rock fragment. The presence of siliceous rock fragments is quite common in the External Ligurian units arenites. (H) Photomicrograph of arenite from Case Baruzzo Sandstone (Cassio unit, External Ligurian units). Close-up of radiolarian-bearing wackestone rock fragment.

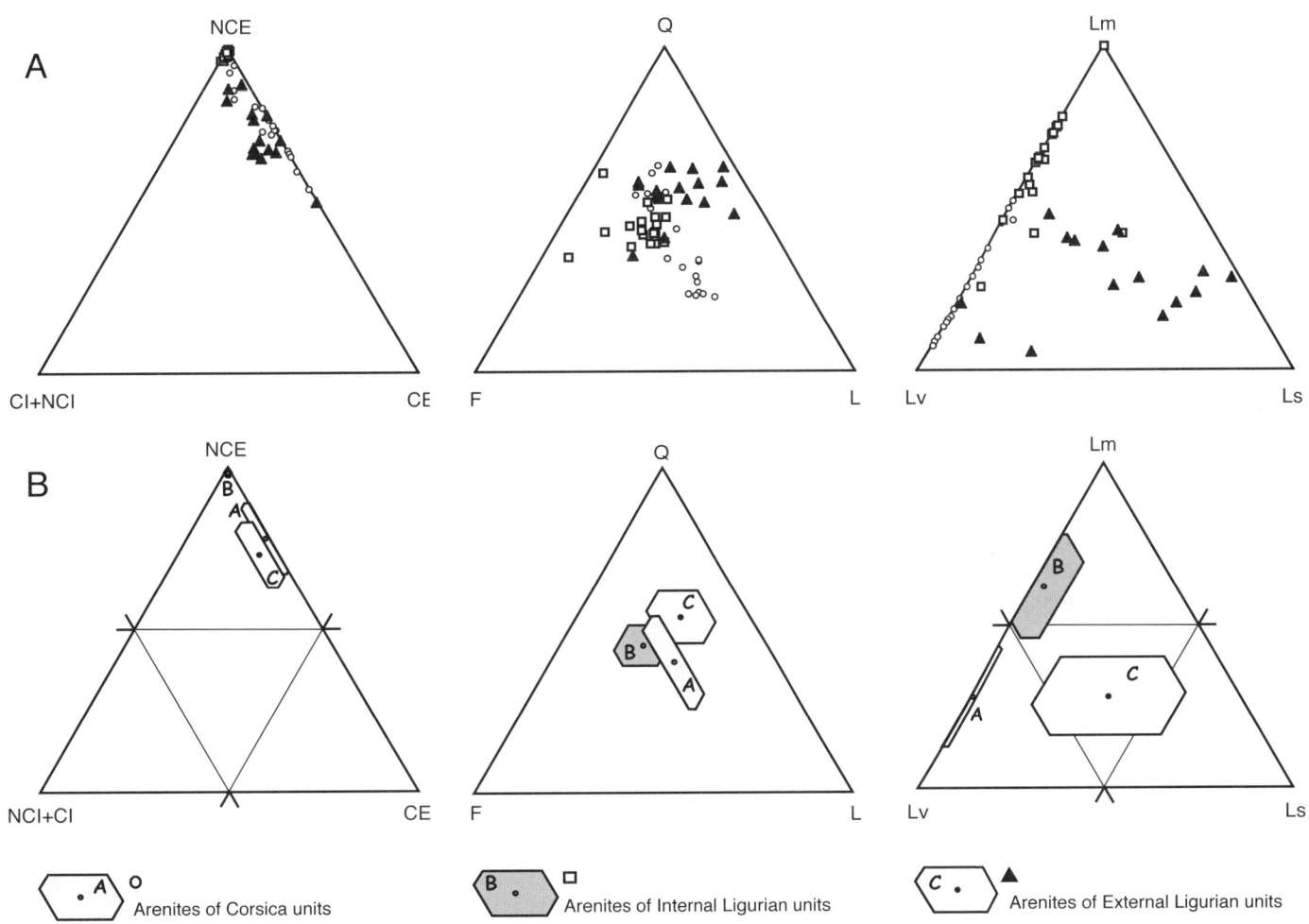

Figure 4. (A) Modal compositional data of arenites plotted on NCE–CI + NCI–CE (Noncarbonate Extrabasinal, Carbonate Intrabasinal + Noncarbonate Intrabasinal, Carbonate Extrabasinal grains after Zuffa, 1980), Q-F-L (Total quartzose grains, Feldspars grains, Lithic fragments after Dickinson, 1985), and Lm-Lv-Ls (fine-grained metamorphic, volcanic, siliclastic lithic fragments after Ingersoll and Suczek, 1979) ternary plots. Plots show framework modes of arenites from Corsica (20 samples), Internal (20 samples) and External (15 samples) Ligurian units. (B) Mean (circle) and standard deviation (polygon) are represented in triangular plots.

literature data concern arenites from pre–late Campanian formations of the Cassio unit (Case Baruzzo Sandstone, 6 samples) and the Media Val Taro unit (Ostia Sandstone, 3 samples). In addition, 6 thin sections from Ostia Sandstone (4 samples) and Salti del Diavolo Conglomerates–Varicolored Shale (2 samples) were analyzed. Moreover, pebbles from Salti del Diavolo Conglomerate (Cassio unit) and Ostia Sandstone (Media Val Taro unit) were analyzed by means of polarizing microscope (30 samples).

Pebbles are characterized by intrusive, volcanic, low- to medium-grade metamorphic, siliceous and carbonate sedimentary rocks (cf. Baldacci et al., 1972). Intrusive rocks are represented by medium- to coarse-grained syeno- and monzogranites. Peraluminous monzogranites, with primary muscovite crystals, were recognized (Fig. 3C). Volcanic rocks are characterized by dacites and pyroclastic rhyolites. Pebbles of metamorphic rocks are common. Low- to medium-grade metamorphic rocks, such as phyllites, schists, muscovite-biotite– and garnet-bearing mica schists, and gneisses, were recognized. Moreover cordierite-bearing gneisses are present (Fig. 3D) with common polygonal textures and widespread triple-point junctions (Fig. 3E). The peraluminous character of these rocks is evidenced by the presence of primary cordierite, muscovite, and garnet. Pebbles of carbonate rocks (mainly limestones and dolostones) and siliceous rocks are also present (cf. Baldacci et al., 1972). These carbonate pebbles are: mudstones, radiolarian-bearing wackestones and packstones, peloidic grainstones, and *Calpionella*-bearing mudstones. Recrystallized dolostones and arenites made up of carbonate platform fragments are also observed. Siliceous rocks are red and green radiolarites, silicified radiolarian-bearing mudstones and siltstones, radiolarian-bearing packstones, and cherty-limestones.

Arenites from Case Baruzzo Sandstone, Ostia Sandstone, Varicolored Shale, and Salti del Diavolo Conglomerate are sublitharenites ($Q_{54}F_{19}L_{27}$) characterized by a mixed siliciclastic-carbonate framework composition ($NCE_{73}NCI + CI_5CE_{22}$; Fig. 3F) and by a mixed composition of fine-grained lithic fragments ($Lm_{28}Lv_{35}Ls_{37}$; Fig. 4).The extrabasinal siliciclastic arenite framework is characterized by mono- and polycrystalline quartz (26 ± 6%), plagioclase (4 ± 2%), and K-feldspar (7 ± 6%) grains. Coarse-grained lithic fragments of granitoids are common (4 ± 3%). Minor porphyritic rhyolites are also present. Metamorphic rock fragments include low-grade schists, mica schists, and minor fragments of medium-grade, cordierite- and garnet-bearing gneisses (6 ± 2%).

Ophiolite-derived rock fragments were not observed, but the presence of millimeter sized crystals of picotite spinel was recognized in the Ostia and Case Baruzzo Sandstones (as previously indicated by the heavy-mineral analysis of Mezzadri, 1964).

Siliceous rock fragments are represented by radiolarian-bearing cherts (Fig. 3G), cherty-limestones, siliceous mudstones, and siliceous siltstones (4 ± 2%). Carbonate mudstones, radiolarian-bearing wackestones (Fig. 3H), and medium- to coarse-grained dolostones (Fig. 3F) are the most common extrabasinal carbonate rock fragments (16 ± 10%). *Calpionella*-bearing mudstones (Fig. 3F), oolitic grainstones, and radiolarian-bearing packstones are also present.

Geochemistry of Pelites

A positive correlation between Al_2O_3 and major (e.g., K_2O and TiO_2) and trace elements (e.g., Sc and Nb) exists (Table 2; Fig. 5), suggesting that detrital aluminous clay-minerals such as illite or smectite are the main host of these elements in pelites. In addition, the lack of correlation of major and trace elements with Zr and P_2O_5 indicates that accessory minerals such as zircon or apatite and monazite do not control REE distribution in pelites (Cullers et al., 1979; Taylor and McLennan, 1985, and references therein; Condie, 1991).

Major- and trace-element data of pelite fraction display a first-order chemical homogeneity, as shown by similar chondrite-normalized REE patterns for all the units (Fig. 6). All the samples are characterized by high concentrations, enrichment of light REE (LREE) to heavy REE (HREE) values, and a moderately negative Eu anomaly ($Eu/Eu^* = Eu_N/[Sm_N/Gd_N]^{1/2}$, regarded as evidence for a differentiated source indicating that processes of chemical fractionation occurred within the source igneous rocks from upper crust; Taylor and McLennan, 1985).

The REE patterns are typical of well-mixed average shales, such as ES (composite of Paleozoic European shales), PAAS (post-Archean average Australian shale), and NASC (North American shale composite), which in turn can be considered to reflect the composition of the upper continental crust exposed to weathering and erosion (Taylor and McLennan, 1985).

The overall trace-element composition of pelites is also shown in Figure 7, where concentrations of selected elements are normalized to PLUC (upper crustal composition of Plank and Langmuir, 1998). This array includes major and trace elements such as LILE (large ion lithophile elements, e.g., Cs, Rb, K, Ba), HFSE (high field strength elements, e.g., Hf, Zr, Nb, Ta, Ti, Th, U, Y), and REE. The abundance for all of these elements in the sediments is mainly controlled by their concentration in the detrital phases, i.e., the source region (Plank and Langmuir, 1998). The overall normalized values for all of the samples plot, as a general feature, around unity, with a more fractionated pattern shown by incompatible elements.

DISCUSSION

Nature of the Source Areas

The application of the classical approach in provenance studies based on modal analysis of arenites and a first-order chemical investigation of pelites results in a misleading petrographic and chemical homogeneity.

Modal analysis of arenites indicates that all the samples are characterized by a siliciclastic to mixed siliciclastic-carbonate composition ranging from arkose to litharenite. The ternary

TABLE 2. MAJOR- AND TRACE-ELEMENT ANALYSES OF PELITES

Unit	Corsica units Navaccia and Toccone units								Internal Ligurian units Gottero and Bracco–Val Graveglia units								
Sample	LM12	LM4	LM1	LM2	LM14	LM11	LM19	LM15	LM104	LM103	LM101	LM110	LM113	LM112	LM114	LM115	LM116
Formation	SMF	SMF	SMF	LF	LF	TB	LF	NS	CL	CL	PS	PS	VLS	VLS	VLS	VLS	VLS
Major elements (wt%)																	
SiO$_2$	61.75	56.37	60.57	70.64	82.73	63.04	75.33	49.81	73.08	56.07	54.24	55.74	56.81	60.26	59.13	60.10	59.49
TiO$_2$	0.67	0.89	0.79	0.83	0.42	0.93	0.78	0.78	0.67	0.88	1.33	0.91	1.11	1.03	0.98	0.98	1.01
Al$_2$O$_3$	15.94	20.55	18.92	15.32	9.24	18.85	12.69	18.22	16.61	18.37	25.89	21.16	22.77	20.92	20.72	21.34	20.93
Fe$_2$O$_3$^{T}	9.16	7.22	7.83	5.99	3.93	5.57	5.77	6.40	3.02	7.35	5.04	8.27	7.31	6.84	7.37	5.95	7.65
MnO	0.03	0.03	0.02	0.01	0.01	0.01	0.01	0.06	0.01	0.04	0.02	0.03	0.06	0.04	0.12	0.03	0.07
MgO	4.03	2.72	3.15	2.52	2.36	3.64	2.26	2.55	1.59	2.82	2.87	3.86	3.44	2.98	3.02	2.92	3.00
CaO	4.22	5.90	3.19	0.23	0.25	0.44	0.30	17.19	0.31	9.52	0.46	3.41	0.25	0.17	0.81	1.06	0.16
Na$_2$O	0.60	0.85	0.94	0.68	0.33	0.11	0.44	0.24	0.49	0.60	1.67	1.25	1.69	1.48	1.94	1.74	1.36
K$_2$O	2.86	4.34	3.70	3.46	2.14	7.08	2.51	5.48	4.22	3.82	6.07	4.15	5.09	4.72	4.31	4.80	5.13
P$_2$O$_5$	0.11	0.11	0.09	0.08	0.11	0.06	0.17	0.15	0.13	0.10	0.14	0.16	0.13	0.12	0.11	0.11	0.11
Σ	99.37	98.98	99.20	99.76	101.52	99.73	100.26	100.88	100.13	99.57	98.73	98.94	98.66	98.56	98.51	99.03	98.91
LOI	7.88	9.20	7.31	4.85	3.01	4.86	4.46	15.75	3.69	11.18	5.99	7.69	5.55	5.61	5.60	6.00	4.91
Trace elements (ppm)																	
Li	79	–	–	37	44	63	40	31.3	20.7	–	48	87	80	–	–	91	84
Be	2.60	–	–	2.34	1.55	5.2	2.10	2.21	2.78	–	3.7	3.13	5.5	–	–	4.0	4.4
Sc	16	–	–	14	9	16	13	17	20	–	29	21	18	–	–	20	18
V	117	–	–	109	86	160	132	132	102	–	183	149	164	–	–	152	152
Cr	92	–	–	80	64	120	93	95	88	–	137	121	116	–	–	154	106
Co	12	–	–	7	10	6	5	8	24	–	29	16	16	–	–	24	12
Ni	59	–	–	59	62	35	45	46	96	–	85	61	59	–	–	96	55
Ga	–	–	–	–	–	–	–	–	–	–	–	–	–	–	–	28.9	27.2
Rb	108	–	–	131	95	244	111	152	175	–	242	154	160	–	–	238	225
Sr	87	–	–	46	41	20.2	64	201	37	–	75	95	54	–	–	78	43
Y	19.3	–	–	24.4	20.5	19.9	34	20.6	16.7	–	19.4	25.0	13.1	–	–	31.2	32.8
Zr	112	–	–	186	116	271	240	141	129	–	235	138	128	–	–	137	128
Nb	11.6	–	–	14.1	7.6	38	14.7	14.6	12.0	–	21.4	15.1	19.4	–	–	20.0	18.9
Mo	1.26	–	–	0.39	–	0.215	1.07	0.70	0.110	–	1.66	0.310	0.235	–	–	–	–
Cs	5.2	–	–	11.4	5.6	17.8	4.9	6.7	10.5	–	19.3	9.9	15.2	–	–	15.7	9.0
Ba	273	–	–	336	239	462	429	386	295	–	667	454	216	–	–	531	656
La	30.3	–	–	32.0	22.6	40	42	26.3	32.8	–	56	43	26.7	–	–	47	41
Ce	55	–	–	60	42	87	83	49	78.	–	116	82	77	–	–	105	95
Pr	6.8	–	–	7.4	5.5	9.3	10.0	6.1	7.2	–	13.3	10.1	6.4	–	–	11.9	10.1
Nd	25.0	–	–	27.3	21.4	33.2	38	22.9	24.9	–	47	38	23.8	–	–	44	38
Sm	4.9	–	–	5.1	4.6	5.8	7.0	4.4	3.23	–	6.8	6.6	4.4	–	–	7.9	7.7
Eu	0.83	–	–	0.95	0.74	0.88	1.47	0.99	0.66	–	0.66	1.08	0.73	–	–	0.65	1.10
Gd	4.3	–	–	4.4	3.9	4.2	5.8	3.7	2.56	–	3.5	5.1	3.12	–	–	5.0	6.3
Tb	0.63	–	–	0.69	0.6	0.65	0.94	0.59	0.42	–	0.58	0.82	0.48	–	–	0.89	1.01
Dy	3.6	–	–	4.0	3.4	3.6	5.4	3.5	2.73	–	3.6	4.6	2.6	–	–	5.5	5.9
Ho	0.69	–	–	0.78	0.69	0.73	1.07	0.70	0.62	–	0.82	0.91	0.52	–	–	1.13	1.16
Er	1.90	–	–	2.22	1.84	2.07	2.91	1.92	1.94	–	2.49	2.50	1.35	–	–	3.07	3.04
Tm	0.266	–	–	0.34	0.267	0.315	0.44	0.268	0.296	–	0.42	0.37	0.215	–	–	0.48	0.46
Yb	1.76	–	–	2.09	1.62	2.01	2.78	1.78	1.85	–	2.85	2.43	1.34	–	–	2.97	2.73
Lu	0.257	–	–	0.298	0.241	0.298	0.41	0.257	0.277	–	0.42	0.35	0.180	–	–	0.40	0.40
Hf	3.3	–	–	5.5	3.5	8.1	7.1	4.2	3.8	–	7.0	4.1	3.8	–	–	4.1	3.8
Ta	0.85	–	–	1.21	0.60	3.07	1.11	0.96	0.89	–	1.60	1.20	1.84	–	–	1.94	1.66
Tl	0.45	–	–	0.76	0.59	2.50	0.50	1.42	1.49	–	1.36	0.76	1.33	–	–	1.89	1.32
Pb	–	–	–	–	11.8	–	–	–	–	–	–	–	–	–	–	38	33.0
Th	9.0	–	–	10.4	6.6	11.9	10.4	6.9	10.3	–	17.3	12.4	10.9	–	–	17.8	15.0
U	1.64	–	–	2.52	1.63	3.6	4.2	1.66	1.38	–	3.15	2.28	2.60	–	–	3.4	3.4

(continued.)

diagrams also indicate that the source areas of these sediments were characterized by a continental basement made up of metamorphic rocks, acidic volcanics, and granitoids, and the relative sedimentary cover represented by extrabasinal non-coeval carbonate rock successions (Fig. 4).

Nevertheless, a more accurate investigation based on the petrographic analysis of pebbles from rudites and lithic fragments from arenites indicates that prominent differences between Corsica–Internal Ligurian units and External Ligurian units exist. The results are summarized in Table 3, where the compositional features of pebbles and arenite rock fragments from the three sampled groups are shown. The main differences derive from the characteristics of the basement metamorphic rocks and the sedimentary rocks that formed the source areas of Corsica and Internal Ligurian units against those of the External Ligurian units. The first group is composed of a debris made up of low-grade metamorphic rocks, while the External Ligurian units are represented by low-, medium-, and high-grade metamorphic rock types. In the lower part of the Cassio and Media Val Taro units, millimeter-sized Cr-spinel, probably derived from a mantle-rock source, have been observed. Analogously, extrabasinal sedimentary rock fragments indicate two distinct source areas for Corsica–Internal Ligurian units and for External Ligurian units. Corsica and Internal Ligurian units contain carbonate platform rock fragments, while the External Ligurian units show both carbonate platform and pelagic siliceous/carbonate rock fragments. This suggests a different evolution of the External Ligurian units source area that was characterized during the Late Triassic–Early

TABLE 2. (Continued.)

	External Ligurian units											
Unit	Cassio unit							Media Val Taro unit				
Sample	LM205	LM204	LM216	LM209	LM207	LM206	LM203	LM215	LM219	LM213	LM218	LM212
Formation	PS	PS	PS	CBS	CBS	VS	SDC	PS	SSS	SSS	OS	OS
Major elements (wt%)												
SiO_2	60.25	61.59	57.31	51.14	52.30	62.29	60.81	73.21	46.91	69.94	52.25	57.27
TiO_2	1.05	1.05	0.68	0.59	0.45	0.91	0.82	0.60	0.49	0.69	0.62	0.71
Al_2O_3	21.59	20.89	14.53	13.37	9.57	19.77	18.58	14.31	13.12	16.75	13.16	15.95
$Fe_2O_3^T$	6.82	7.23	6.44	6.46	4.64	7.72	7.97	6.19	5.04	5.96	5.69	6.24
MnO	0.04	0.03	0.12	0.10	0.08	0.04	0.03	0.03	0.07	0.13	0.12	0.05
MgO	2.92	2.86	6.49	4.41	4.99	2.75	3.87	2.38	1.90	2.79	4.94	5.92
CaO	1.05	0.65	9.49	18.71	25.48	0.26	1.99	0.51	30.53	0.37	19.41	9.62
Na_2O	1.32	1.37	0.95	0.69	0.79	1.43	0.55	0.54	0.36	0.46	0.69	0.99
K_2O	3.98	3.78	3.08	2.66	1.83	3.90	4.75	2.62	1.23	3.05	2.82	3.48
P_2O_5	0.10	0.09	0.15	0.15	0.14	0.12	0.11	0.08	0.18	0.08	0.13	0.13
Σ	99.12	99.54	99.24	98.28	100.27	99.19	99.48	100.47	99.83	100.22	100.08	100.36
LOI	6.46	5.82	13.45	17.63	22.77	5.11	7.27	4.24	24.61	4.77	18.02	11.55
Trace elements (ppm)												
Li	–	80	–	71	–	83	49	59	–	61	–	67
Be	3.4	2.92	–	1.91	1.42	3.30	3.32	2.07	1.77	2.69	2.12	2.56
Sc	21	20	–	13	9	17	16	13	11	15	12	15
V	135	133	–	115	70	145	118	109	99	123	99	121
Cr	119	120	–	147	111	108	102	94	66	113	159	204
Co	23	21	–	18	12	24	13	26	11	18	13	13
Ni	55	56	–	136	82	67	55	92	36	82	105	141
Ga	24.2	24.1	–	15.3	9.8	24.9	21.7	17.9	14.2	20.5	13.8	18.0
Rb	160	159	–	107	68	182	193	119	47	140	106	153
Sr	206	177	–	273	315	237	97	63	371	98	183	253
Y	22.4	25.7	–	20.1	18.9	28.2	22.9	24.3	16.7	20.6	19.1	23.4
Zr	178	185	–	122	130	154	145	112	112	129	138	152
Nb	15.1	15.3	–	9.2	7.1	15.9	14.2	10.7	10.0	11.6	10.2	12.7
Mo	–	–	–	–	–	–	–	–	–	–	–	–
Cs	8.9	8.2	–	8.2	5.0	13.1	11.1	7.7	3.4	10.2	7.9	11.3
Ba	288	270	–	167	735	239	259	252	80	240	205	262
La	39	40	–	24.2	19.6	38	35	26.7	19.5	27.3	26.0	29.4
Ce	81	83	–	45	37	86	68	57	38	62	48	61
Pr	9.5	9.7	–	5.5	4.7	9.3	7.8	6.0	4.6	6.5	6.1	7.0
Nd	35	36	–	20.6	18.2	34	28.3	22.1	17.6	23.8	22.9	26.2
Sm	6.4	6.6	–	4.0	3.6	6.7	5.0	4.2	3.5	4.6	4.2	5.1
Eu	1.22	1.22	–	0.85	0.69	1.29	0.86	0.99	0.74	0.93	0.74	1.04
Gd	4.4	4.3	–	3.30	3.32	5.1	4.0	3.7	2.99	3.7	3.4	4.4
Tb	0.68	0.70	–	0.52	0.51	0.86	0.64	0.61	0.46	0.59	0.51	0.67
Dy	4.1	4.3	–	3.01	3.04	4.9	3.9	4.3	2.71	3.5	3.06	3.9
Ho	0.88	0.94	–	0.64	0.61	1.00	0.81	0.78	0.54	0.72	0.65	0.79
Er	2.56	2.78	–	1.76	1.69	2.66	2.32	2.15	1.48	2.02	1.84	2.24
Tm	0.41	0.43	–	0.277	0.261	0.41	0.36	0.320	0.226	0.314	0.289	0.34
Yb	2.56	2.72	–	1.79	1.61	2.58	2.22	1.97	1.33	1.9	1.79	2.08
Lu	0.38	0.40	–	0.259	0.238	0.37	0.328	0.296	0.194	0.287	0.268	0.305
Hf	5.3	5.5	–	3.6	3.9	4.6	4.3	3.3	3.3	3.8	4.1	4.5
Ta	1.16	1.15	–	0.66	0.56	1.28	1.11	0.82	0.71	0.90	0.79	1.06
Tl	1.07	1.06	–	0.74	0.46	1.50	1.15	0.92	0.43	1.11	0.7	0.88
Pb	17.8	15.1	–	18.2	12.6	25.0	19.3	5.2	9.9	9.5	10.2	14.8
Th	12.2	12.0	–	7.5	6.1	13.0	11.0	8.9	5.2	9.2	7.8	10.5
U	2.34	2.27	–	1.92	1.75	1.83	1.73	1.34	2.32	2.11	2.41	1.94

Note: Major element concentrations were measured on ignited powders; Loss on ignition (LOI) values are uncorrected for ferrous iron oxidation. SMF—San Martino Formation; LF—Lydienne Flysch; TB—Toccone Breccia; NS—Novella Sandstone; CL—Calpionella Limestone; PS—Palombini Shale; VLS—Val Lavagna Shale; CBS—Case Baruzzo Sandstone; VS—Varicoloured Shale; SDC—Salti del Diavolo Conglomerate; SSS—San Siro Shale; OS—Ostia Sandstone.

Cretaceous time span by a more widespread and significant subsidence.

On the other hand, the first-order chemical homogeneity of pelitic samples is overcome by the observation of distinctive features of the different groups of samples based on geochemical data.

A continental provenance for the whole of the samples can be assessed on the basis of the diagram La/Sm versus Yb/Sm (Fig. 8). The samples plot across the field of continental detritus, typically enriched in LREE and depleted in HREE. Contributions from a dissecting arc are excluded, because they would lead to higher values for the Yb/Sm ratio and lower values for the La/Sm ratio. In agreement with these observations, no dissecting arc-derived volcanic rock fragments have been recognized in the framework of arenites.

Overall, the main chemical types of sources that contribute to the sediment can be grouped as: felsic (upper-crustal rocks), mafic, and ultramafic (mantle rocks). Samples from External Ligurian units combine the less-conspicuous europium anomalies with the lowest values for the ratio La_N/Yb_N (8.3–10.5, i.e., a flatter REE pattern), which suggests a more mafic-ultramafic character if compared to samples from Corsica and the Internal

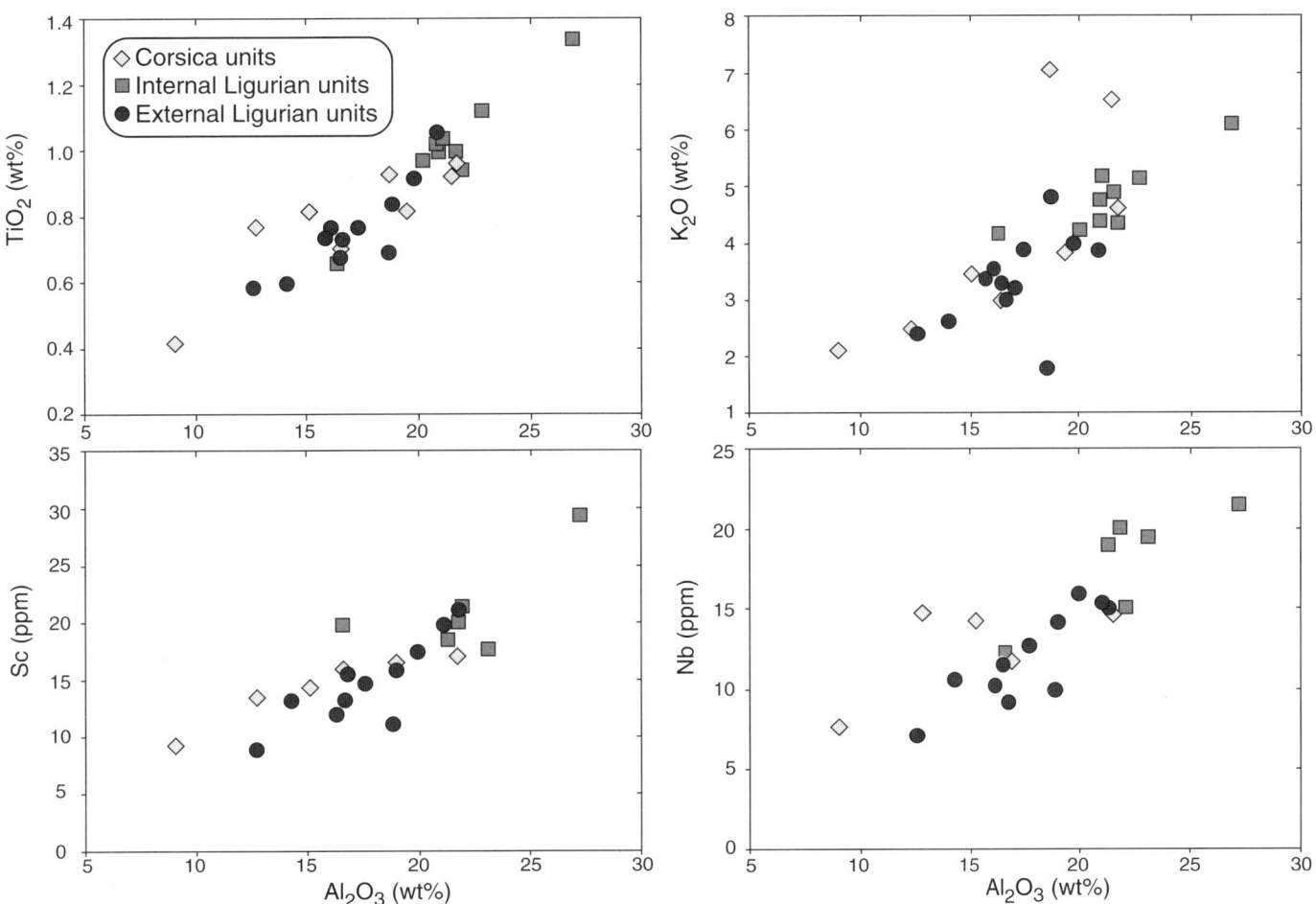

Figure 5. TiO_2, K_2O, Sc, and Nb versus Al_2O_3 in studied pelitic rocks. The positive correlations illustrate how major and trace elements (e.g., TiO_2, K_2O, Sc, and Nb) generally follow Al_2O_3 and thus the detrital phase (i.e., aluminous clay minerals of pelites).

Ligurian units (La_N/Yb_N = 9.5–13.6 for Corsica units, 10.3–13.5 for Internal Ligurian units). Similar differences can be highlighted by the low contents of Cr, Co, and Ni for samples from Corsica units, by the intermediate values for samples from Internal Ligurian units, and the highest for samples from External Ligurian units (Table 2).

The abundance of the normative mineral Mg-chlorite of the pelitic samples roughly increases with increasing Cr content (Fig. 9). All the samples plot between the felsic and mafic compositional types. Some samples from External Ligurian units plot very close to the mafic composition, which suggests an utmost contribution from a mafic source or, alternatively, a minor (even if significant) supply from an ultramafic source.

To better constrain the mafic or ultramafic versus felsic character of the detritus, elemental ratios such as Cr/Th and Th/Sc were considered (Fig. 10). High values of these ratios reflect an enrichment in mafic-ultramafic and felsic components, respectively (e.g., Hofmann et al., 2003). Samples from Corsica and the Internal Ligurian units fit a mixing hyperbolic curve between felsic and mafic end members, with a major contribution from the felsic end member. Samples from External Ligurian units are shifted toward a mixing hyperbolic curve between the same felsic and an ultramafic end member. These samples exhibit high Cr concentrations (up to 200 ppm). This is in agreement with the occurrence of millimeter-sized Cr-spinel fragments in the corresponding arenitic fraction (see also Mezzadri, 1964; Wildi, 1985), typically derived from mantle coarse-grained ultramafic rocks. The presence of ultramafic debris in samples from External Ligurian units can also be seen by the relative enrichment in Ni with respect to V and Th. In the ternary diagram V-Ni-Th*10 (Fig. 11), fields representative of ultramafic, mafic, and felsic rocks plot separately. As a general rule, all the samples plot between the mafic and the felsic compositions, but External Ligurian samples are shifted toward the ultramafic compositions (compare plot of Th/Sc versus Cr/Th, Fig. 10).

As a whole, petrographic and chemical data indicate that the sedimentary cover of the Corsica and Internal Ligurian units can be regarded as deriving from a similar source area that was made up of the upper part of a continental basement and its carbonate sedimentary cover. According to several authors (Gardin et al.,

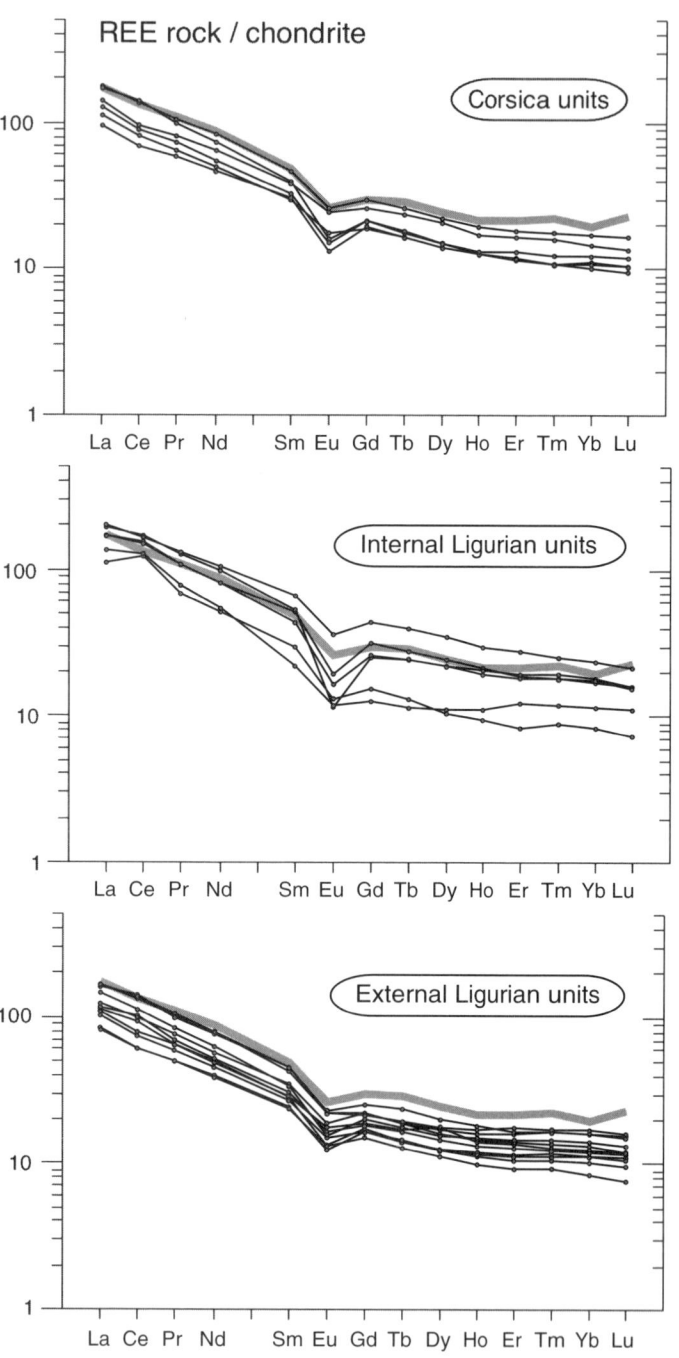
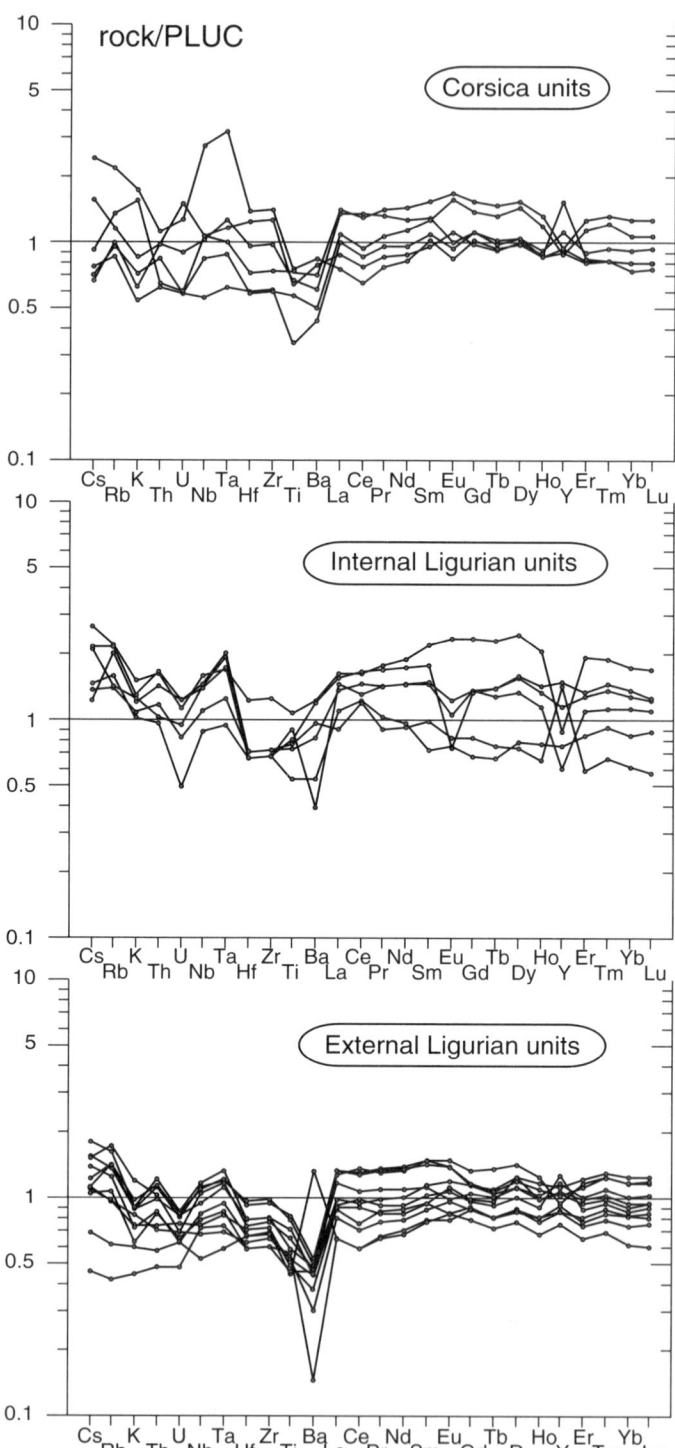

Figure 6. Rare earth element (REE) concentrations of pelitic samples normalized to chondritic values from McDonough and Sun (1995). ES, a composite of European shales of Paleozoic age, is plotted for comparison as a gray solid line [the lacking value for Dy has been interpolated as the expected value for a smooth chondrite-normalized REE pattern: $Dy_N = (Tb_N \times Ho_N)^{1/2}$]. Note that the overall shape of REE patterns is similar for all the samples, and only minor differences in concentration occur, except for a couple of samples from the Internal Ligurian units. Also, the patterns compare well to that of ES, which in turn reflects mean REE abundances in the upper continental crust (Taylor and McLennan, 1985).

Figure 7. Multielement diagram of pelitic samples normalized to PLUC (upper crust composition after Plank and Langmuir, 1998). The normalized values for the majority of the samples plot astride the line representing the mean value for PLUC, and the most incompatible elements have a more fractionated pattern. The significant Ba negative anomaly in all of the samples from the External Ligurian units might suggest high biological productivity or hydrothermal barite (Plank and Langmuir, 1998). Note also the positive Ta and Nb anomaly (around three times PLUC value) of the pelite from Toccone Breccia (Corsica units), probably related to a basaltic alkaline intraplate affinity provenance (see Discussion section for details).

TABLE 3. PETROGRAPHY OF PEBBLES AND LITHIC FRAGMENTS FROM ARENITES

Corsica units	Internal Ligurian units		External Ligurian units
Low-grade metamorphic rocks (mainly schists, mica schists, and gneisses)	Low-grade metamorphic rocks (mainly schists, mica schists, and gneisses)	Crystalline basement rocks	Low-grade metamorphic rocks (mainly schists, mica schists, and gneisses)
Granites (syeno-and monzogranites)	Granites (syeno- and monzogranites)		Medium- and high-grade metamorphic rocks (mainly cordierite-bearing gneisses)
			Granites ("pink" and "white" granites and peraluminous granites)
Carbonate platform rocks (ooids and peloidic grainstones, mudstones, and macrofossil fragments)	Carbonate platform rocks (ooids and peloidic grainstones, mudstones, and macrofossil fragments)	Sedimentary rocks	Carbonate platform rocks (ooids and peloidic grainstones, dolostones, mudstones, and macrofossil fragments)
			Pelagic rocks (*Calpionella*-bearing wackestones, radiolarian-bearing limestones, cherty-limestones, and radiolarites)
Acidic volcanic rocks (mainly rhyolites and dacites and minor intraplate basalts)	Acidic volcanic rocks (rhyolites and dacites)	Volcanic rocks	Acidic volcanic rocks (rhyolites and dacites)
Not detected	Not detected	Mantle ultramafic rocks	Millimeter-sized Cr-spinel

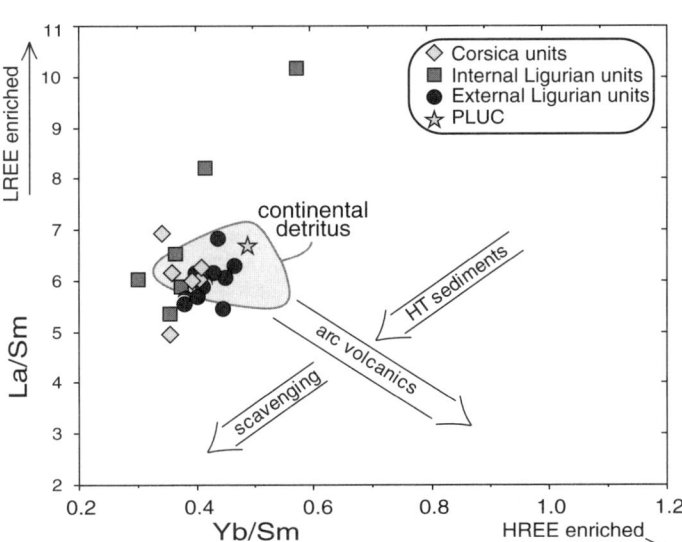

Figure 8. La/Sm versus Yb/Sm (Plank and Langmuir, 1998). Pelitic samples plot astride the area of continental detritus, without any evidence of contributions from volcaniclastic detritus or any relevant influence from processes such as hydrothermalism (HT) or scavenging by hydrogenous oxides and by apatite fish debris due to prolonged exposure to seawater. The composition of PLUC (upper crust composition after Plank and Langmuir, 1998) is plotted for comparison. LREE—light rare earth elements, HREE—heavy rare earth elements.

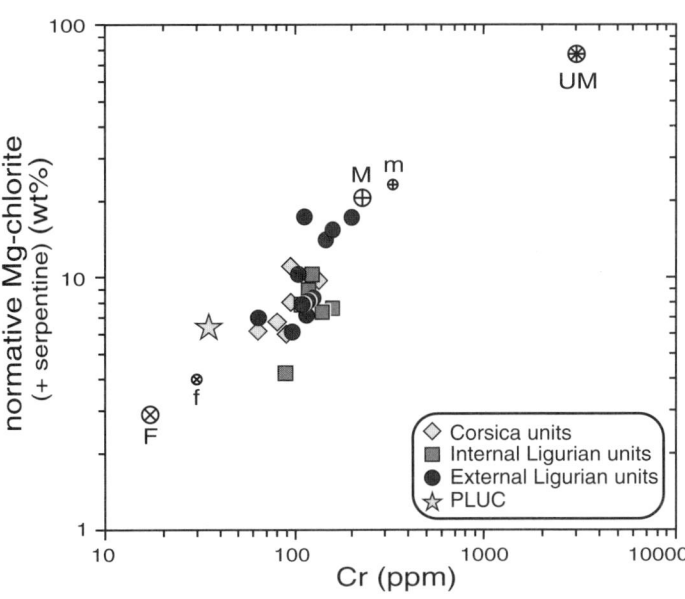

Figure 9. Plot of normative Mg-chlorite versus Cr in pelites. Normative Mg-chlorite is used here as a mafic-ultramafic index, in the same sense as van de Kamp and Leake (1995). For the ultramafic end member only, the normative mineral serpentine has been calculated. The normative mineralogy for pelitic samples was calculated following similar criteria to those used by Cohen and Ward (1991) for the program SEDNORM. Compositions of end members are averages of felsic (F), mafic (M), and ultramafic (UM) rock groups. Felsic samples are Hercynian granitoids from Sardinia and Corsica Islands, 22 samples; mafic and ultramafic rocks are basalts-gabbros and peridotite from Western Tethys, 64 and 43 samples, respectively. Data are from Cocherie et al. (1994), Tommasini et al. (1995), Di Vincenzo et al. (1996), Rampone et al. (1998), Poli and Tommasini (1999), Saccani et al. (2000), Melcher et al. (2002), Cauli and Mori (unpublished data from Elba Island). Felsic (f) and mafic (m) end-member compositions from Taylor and Mc Lennan (1985) are also plotted. The composition of PLUC (upper crust composition after Plank and Langmuir, 1998) is plotted for comparison.

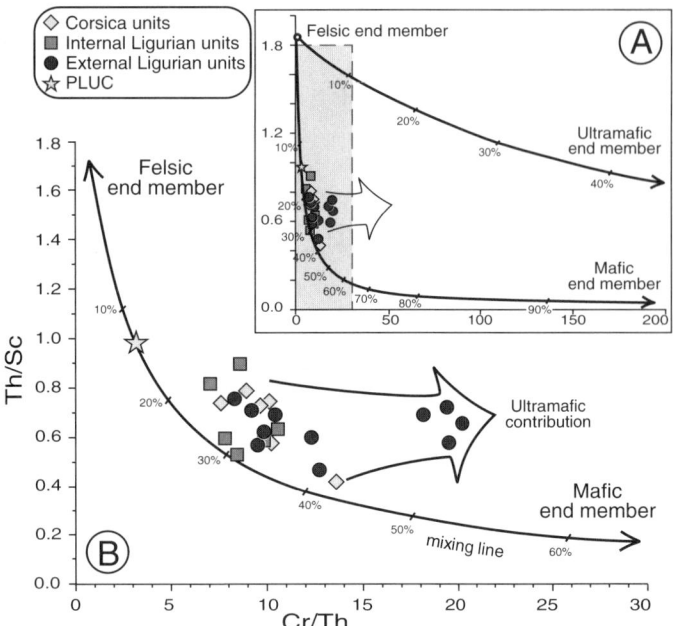
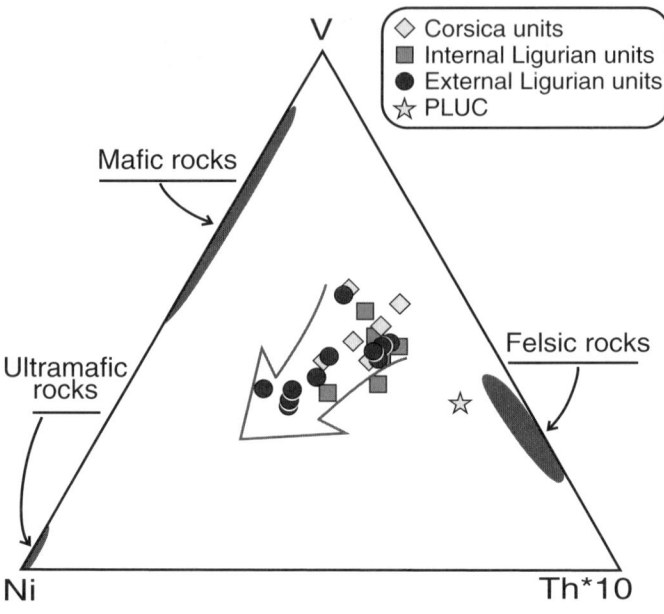

Figure 10. Plot of Th/Sc versus Cr/Th in pelitic samples (Condie and Wronkiewicz, 1990; Totten et al., 2000). B is the enlargement of shaded area in A. Two mixing curves have been calculated between a felsic and a mafic end member, and between a felsic and an ultramafic end member (the same as for Fig. 9). Percentages reported on the mixing curves represent the mafic end-member contribution to the mixing products. Corsica units and Internal Ligurian units compositions are consistent with a mixing between a mafic and a felsic end member, with a major contribution from a felsic source. The large arrow indicates the shifting of some External Ligurian units samples toward the ultramafic-felsic mixing line, suggesting a contribution from ultramafic detrital material. The composition of PLUC (upper crust composition after Plank and Langmuir, 1998) is plotted for comparison.

Figure 11. V-Ni-Th*10 plot of pelites. Concentrations of elements are in ppm. Shaded areas represent composition of the same felsic, mafic, and ultramafic rocks as for Figure 9. PLUC (upper crust composition after Plank and Langmuir, 1998) is plotted for comparison. The shifting of samples from External Ligurian units toward the Ni corner suggests an ultramafic component in the detritus from the External Ligurian units.

1994, and references therein), this source area is considered to be the Corsica-Europe continental margin. On the other hand, the sedimentary cover of the External Ligurian units was supplied by a source area where a complete lithospheric section, from the upper mantle up to the siliceous/carbonate sedimentary cover, was exposed. According to Marroni et al. (2001a), this source area is recognized as the Adria ocean-continent transition.

Implications for the Configuration of Margins of the Ligure-Piemontese Oceanic Basin

Our results outline a Jurassic picture of the Western Tethys area in which an oceanic basin (the Ligure-Piemontese Ocean) was bordered on one side by the Corsica-Europe plate, which was constituted by the shallowest level of the crust and its covers. On the other side, the complete lithospheric section of Adria plate was exposed.

Three contrasting models about the processes related to the opening of the Ligure-Piemontese basin have already been proposed (Decandia and Elter, 1969; Lemoine et al., 1987; Dal Piaz, 1993).

1. Decandia and Elter (1969), Piccardo (1977), and Lombardo and Pognante (1982) suggested that the opening occurred by a passive rifting, where the continental crust delamination was achieved by two opposite, landward-dipping detachment faults located at the crust-mantle boundary. A large area of subcontinental mantle was progressively unroofed and exposed on the seafloor after the rifting processes. The resulting oceanic area was bounded by a pair of symmetric continental margins, characterized by high-angle normal faults with opposite, seaward dipping angles.

2. Lemoine et al. (1987) proposed a second model where the opening of the Ligure-Piemontese oceanic basin was achieved by a passive, asymmetric extension of the lithosphere by simple shear (Wernicke, 1981, 1985; Lister et al., 1986, 1991). Continental crust was delaminated by a low-angle detachment fault cutting across the whole continental crust up to the lithospheric mantle. The low-angle detachment fault divided the continental

lithosphere into a lower and an upper plate that were characterized by an asymmetry in their structures and evolutionary paths. After the break-up, the lower-plate margin consisted of a large area where subcontinental mantle and the lowest levels of continental lithosphere were exposed at the seafloor. A striking feature of the lower-plate margin is therefore an ocean-continent transition generally characterized by large exposures of subcontinental mantle covered with remnants of the upper plate. The latter, referred to as extensional allohthons, were displaced along the detachment fault. By contrast, the upper plate consisted of upper continental crust affected by a major flexure with small, landward fault systems. Therefore, the oceanic basin derived by passive, asymmetric extension of the lithosphere by simple shear shows contrasting architecture and lithological features in its continental margins. The detachment fault was west-dipping, and the lower plate is represented by the Adria continental margin. A simple shear model with west-dipping detachment fault was also assumed by Bertotti (1991) and Hoogerduijn Strating et al. (1993). Subsequently Froitzheim and Manatschal (1996) and Marroni et al. (1998) proposed a rifting model subdivided into two distinct phases of Triassic to Jurassic age. The first phase, mainly developed in Triassic to Early Jurassic times, was characterized by high-angle normal faults (Eberli and Froitzheim, 1990; Bertotti, 1991; Bertotti et al., 1993), the geometry of which suggests a rifting process dominated by lithosphere stretching by pure shear extension. On the contrary, the subsequent phase (Middle to Late Jurassic age) produced large-scale exhumation of subcontinental mantle through asymmetric extension by west-dipping, low-angle detachment faults. However, this model also represents an asymmetric opening mechanism for the Ligure-Piemontese oceanic basin, with the lower plate represented by the Adria continental margin.

3. A third model was proposed by Dal Piaz (1993), Trommsdorff et al. (1993), and Rampone and Piccardo (2000) where the opening of the Ligure-Piemontese oceanic basin was achieved by a passive, asymmetric rifting. The rifting occurred without an earlier stage of pure shear extension, but only by a continuous extension achieved by simple shear mechanism. The main difference from the previously described models is the east-dipping detachment fault. Consequently, the lower continental crust and the subcontinental mantle were exposed in correspondence to the Corsica-Europe margin.

The three different models (symmetric, asymmetric with west-dipping detachment fault, and asymmetric with east-dipping detachment fault) suggest not only a different mechanism of oceanic opening but also a different configuration and lithological features of the continental margins. These margins were destroyed during the subduction and the subsequent continental collision. Their remnants, today preserved as tectonic units in the Alpine-Apennine belt, are insufficient to draw a complete picture of their paleogeographic domain. In this frame, useful information can be extracted from the deposits from the Ligure-Piemontese oceanic basin sedimented before the inception of major tectonic events in the continental margins.

The data collected in this paper indicate a different composition of the margins that supplied the Corsica and Internal Ligurian successions versus the External Ligurian deposits (Fig. 12). The first group was supplied by the uppermost part of the lithospheric section of Europe-Corsica plate, while the External Ligurian units were supplied by the complete lithospheric section of the Adria plate, including upper-mantle rocks. These data indicate an asymmetric mechanism for the Western Tethys opening characterized by a different configuration of the continental margins caused by the rifting processes (Fig. 13). Moreover the presence of debris derived from a complete lithospheric section in the sediment supplied by the Adria margin indicates that this margin was

| Main features of the two source areas ||
Corsica and Internal Ligurian units **EUROPE-CORSICA MARGIN**	External Ligurian units **ADRIA MARGIN**
CONTRIBUTION FROM CONTINENTAL DETRITUS (no arc)	CONTRIBUTION FROM CONTINENTAL DETRITUS (no arc)
CONTRIBUTION FROM THE UPPER CONTINENTAL CRUST (low-grade metamorphic rocks and granitoids)	CONTRIBUTION FROM A COMPLETE CONTINENTAL CRUST SECTION (low-, medium- and high-grade metamorphic rocks and granitoids)
CONTRIBUTION FROM A SHALLOW-WATER CARBONATE SEQUENCE (carbonate platform limestones)	CONTRIBUTION FROM SHALLOW-WATER AND PELAGIC CARBONATE SEQUENCES (carbonate platform dolostones/limestones and pelagic limestones, cherty limestones and siliceous rocks)
NO CONTRIBUTION FROM OCEANIC CRUST (i.e., no ophiolites)	CONTRIBUTION FROM MANTLE ROCKS (Cr-spinel from mantle ultramafics and geochemical evidences)

Figure 12. Main features of the two source areas inferred from petrographic and chemical data.

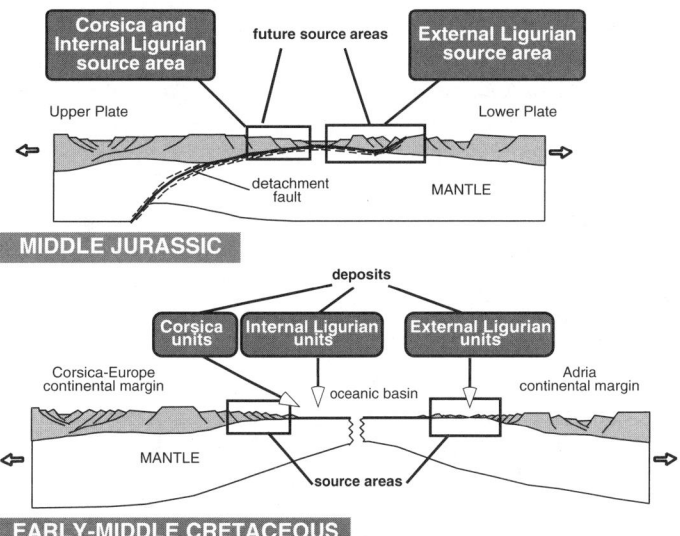

Figure 13. Reconstruction of Western Tethys margins along the Corsica–Northern Apennine traverse during Middle Jurassic and Early-Middle Cretaceous. The source areas and basin locations of the studied units are indicated.

the lower plate during the rifting processes, which were dominated by a west-dipping detachment fault (Fig. 13).

CONCLUSIONS

In this paper, we applied a multidisciplinary approach to investigate the pre–Late Campanian sedimentary covers of the Ligure-Piemontese oceanic basin in order to understand the nature of its source areas (the continental margins) and to provide further constraints for the reconstruction of its original configuration.

Petrographic data from rudites and arenites have been combined with geochemical data from pelites, and the main features of the source areas (continental margins) have been highlighted using the differences in composition inferred from these data. The source area of Corsica and Internal Ligurian units, regarded as the Corsica-Europe continental margin, was made up of the upper part of a continental basement and its carbonate sedimentary cover. The sedimentary record generated by its erosion is represented by turbidite deposits derived from low-grade metamorphic rocks (schists, mica schists, and gneisses), granites (syeno- and monzogranites), acidic volcanic rocks (rhyolites and dacites), and carbonate platform rocks (ooids and peloidic grainstones and mudstones). The geochemical data relative to the siliciclastic fraction of these deposits indicate a composition resulting from the mixing of felsic and mafic sources, where the felsic component is often prominent.

On the other hand, the source area of the External Ligurian units (the Adria margin) is represented by a complete lithospheric section, from the upper mantle to the siliceous/carbonate sedimentary cover. The erosion of this margin produced turbidite deposits derived from mantle rocks (millimeter-sized Cr-spinel), low- to high-grade metamorphic rocks (schists, mica schists, gneisses, and cordierite-bearing gneisses), granites (syenogranites, monzogranites, and peraluminous granites), acidic volcanic rocks (rhyolites and dacites), carbonate platform rocks (ooids and peloidic grainstones, mudstones, and dolostones), and pelagic sedimentary rocks (*Calpionella*-bearing wackestones, radiolarian-bearing limestones, cherty-limestones, and radiolarites). The geochemical data relative to the siliciclastic fraction of these deposits indicate an ultramafic source standing out from the mafic-felsic components that also contribute to the sediment.

On the whole, the differences in composition are thought to reflect different configurations of the two margins, which in turn provide further constraints to the understanding of the rifting process. Data collected in this work lend support to an asymmetric opening mechanism of Western Tethys, where the Adria margin acted as the lower plate during the rifting processes.

In conclusion, our multidisciplinary approach is useful when dealing with provenance. Moreover, the combined analysis of the different grain-size fractions of sedimentary rocks provides a comprehensive picture of the case study. The petrographic and geochemical investigation of sediments can be considered a valuable tool in paleotectonic reconstructions, especially in those cases in which the regional-scale features of the investigated geological framework have been lost due to major successive tectonic events.

ACKNOWLEDGMENTS

Massimo D'Orazio is kindly acknowledged for his help with inductively coupled plasma–mass spectrometry (ICP-MS) analyses carried out at the Dipartimento di Scienze della Terra, Pisa. An early version of the manuscript was improved following the comments and suggestions of Eduardo Garzanti. This research was supported by Consiglio Nazionale delle Ricerche, Istituto di Geoscienze e Georisorse, Pisa and Programmi di Ricerca di Interesse Nazionale grants.

REFERENCES CITED

Abbate, E., and Sagri, M., 1982, Le unità torbiditiche Cretaciche dell'Appennino settentrionale ed i margini continentali della Tetide: Memorie della Società Geologica Italiana, v. 24, p. 115–126.

Abbate, E., Bortolotti, V., and Principi, G., 1980, Apennine ophiolites: A peculiar oceanic crust, in Rocci, G., ed., Tethyan Ophiolites—Western Area: Ofioliti, Special Issue 5, p. 59–96.

Baldacci, F., Cerrina Feroni, A., Elter, P., Giglia, G., and Patacca, E., 1972, Il margine del paleocontinente nord-Appenninico dal Cretaceo all'Oligocene: Nuovi dati sulla ruga Insubrica: Memorie della Società Geologica Italiana, v. 11, p. 367–390.

Basu, A., 2003, A Perspective on Quantitative Provenance Analysis: Memorie Descrittive della Carta Geologica d'Italia, v. LXI, p. 11–22.

Bertotti, G., 1991, Early Mesozoic extension and Alpine shortening in the western Southern Alps: The geology of the area between Lugano and Menaggio (Lombardy, northern Italy): Memorie di Scienze Geologiche di Padova, v. 43, p. 17–123.

Bertotti, G., Picotti, V., Bernoulli, D., and Castellarin, A., 1993, From rifting to drifting: Tectonic evolution of the South-Alpine upper crust from the Triassic to the Early Cretaceous: Sedimentary Geology, v. 86, p. 53–76, doi: 10.1016/0037-0738(93)90133-P.

Borsi, L., Scharer, U., Gaggero, L., and Crispini, L., 1996, Age, origin and geodynamic significance of plagiogranites in lherzolites and gabbros of the Piedmont Ligurian Basin: Earth and Planetary Science Letters, v. 140, p. 227–242, doi: 10.1016/0012-821X(96)00034-9.

Chiari, M., Marcucci, M., and Principi, G., 2000, The age of the radiolarian cherts associated with the ophiolites in the Apennines (Italy) and Corsica (France): A revision: Ofioliti, v. 25, p. 141–146.

Cocherie, A., Rossi, P., Fouillac, A.M., and Vidal, P., 1994, Crust and mantle contributions to granite genesis; an example from the Variscan batholith of Corsica, France, studied by trace-element and Nd-Sr-O-isotope systematics: Chemical Geology, v. 115, p. 173–211, doi: 10.1016/0009-2541(94)90186-4.

Cohen, D., and Ward, C.R., 1991, SEDNORM—A program to calculate a normative mineralogy for sedimentary rocks based on chemical analyses: Computers & Geosciences, v. 17, p. 1235–1253, doi: 10.1016/0098-3004(91)90026-A.

Condie, K.C., 1991, Another look at rare earth elements in shales: Geochimica et Cosmochimica Acta, v. 55, p. 2527–2531, doi: 10.1016/0016-7037(91)90370-K.

Condie, K.C., and Wronkiewicz, D.J., 1990, The Cr/Th ratio in Precambrian pelites from the Kaapvaal craton as an index of craton evolution: Earth and Planetary Science Letters, v. 97, p. 256–267, doi: 10.1016/0012-821X(90)90046-Z.

Conti, M., Marcucci, M., and Passerini, P., 1985, Radiolarian cherts and ophiolites in the Northern Apennines and Corsica: Age correlations and tectonic frame of siliceous deposition: Ofioliti, v. 10, p. 201–225.

Crook, K.A.W., 1974, Lithogenesis and geotectonics: The significance of compositional variation in flysch arenites (graywackes), in Dott, R.H., and Shaver, R.H., eds., Modern and Ancient Geosynclinal Sedimentation: Society for Sedimentary Geology (SEPM) Special Publication 19, p. 304–310.

Cullers, R.L., 2000, The geochemistry of shales, siltstones and sandstones of

Cullers, R.L., Chaudhuri, S., Kilbane, N., and Koch, R., 1979, Rare-earths in size fractions and sedimentary rocks of Pennsylvanian-Permian age from the mid-continent of the U.S.A.: Geochimica et Cosmochimica Acta, v. 43, p. 1285–1302, doi: 10.1016/0016-7037(79)90119-4.

Cullers, R.L., 2002, Implications of elemental concentrations for provenance, redox conditions, and metamorphic studies of shales and limestones near Pueblo, CO, USA: Chemical Geology, v. 191, p. 305–327, doi: 10.1016/S0009-2541(02)00133-X.

Cullers, R.L., Chaudhuri, S., Kilbane, N., and Koch, R., 1979, Rare-earths in size fractions and sedimentary rocks of Pennsylvanian-Permian age from the mid-continent of the U.S.A.: Geochimica et Cosmochimica Acta, v. 43, p. 1285–1302, doi: 10.1016/0016-7037(79)90119-4.

Dal Piaz, G.V., 1993, Evolution of Austro-Alpine and Upper Penninic basement in the northwestern Alps from Variscan convergence to post-Variscan extension, in Von Raumer, J.F., and Neubauer, F., eds., Pre-Mesozoic Geology in the Alps: Berlin, Springer, p. 325–342.

Daniele, G., and Bianchi, L., 1995, Studio petrografico delle Arenarie di Ostia della media Val di Taro e loro confronto con arenarie di altre successioni: Memorie dell'Accademia Lunigianese delle Scienze "G Capellini", v. 64/65, p. 131–148.

Decandia, F.A., and Elter, P., 1969, Riflessioni sul problema delle ofioliti nell'Appennino settentrionale (nota preliminare): Atti della Società Toscana di Scienze Naturali, v. 76, p. 1–9.

Decandia, F.A., and Elter, P., 1972, La "zona" ofiolitifera del Bracco, nel settore compreso tra Levanto e la Val Graveglia: Memorie della Società Geologica Italiana, v. 11, p. 503–530.

De Wever, P., and Danelian, T., 1995, Supra-ophiolitic radiolarites from Alpine Corsica (France), in Baumgartner, P.O., et al., eds., Middle Jurassic to Lower Cretaceous Radiolaria of Tethys: Occurrences, Systematics, Biochronology: Lausanne, Mémoire Géologie, v. 23, p. 731–735.

Dickinson, W.R., 1970, Interpreting detrital modes of greywacke and arkose: Journal of Sedimentary Petrology, v. 40, p. 695–707.

Dickinson, W.R., 1985, Interpreting provenance relations from detrital modes of sandstones, in Zuffa, G.G., ed., Provenance of Arenites: NATO ASI (Advanced Study Institutes) series: Dordrecht, D. Reidel Publishing Company, p. 333–362.

Dickinson, W.R., and Suczek, C.A., 1979, Plate tectonics and sandstones composition: American Association of Petroleum Geologist Bulletin, v. 63, p. 2164–2182.

Dickinson, W.R., and Valloni, R., 1980, Plate settings and provenance of sands in modern ocean basins: Geology, v. 8, p. 82–86, doi: 10.1130/0091-7613(1980)8<82:PSAPOS>2.0.CO;2.

Di Vincenzo, G., Andriessen, P.A.M., and Ghezzo, C., 1996, Evidence of two different components in a Hercynian peraluminous cordierite-bearing granite: The San Basilio intrusion (central Sardinia, Italy): Journal of Petrology, v. 37, p. 1175–1206.

Durand-Delga, M., Peybernès, B., and Rossi, P., 1997, Arguments en faveur de la position, au Jurassique, des ophiolites de Balagne (Haute-Corse, France) au voisinage de la marge continentale Européenne: Comptes Rendus de l'Académie des Sciences, v. 325, p. 973–981.

Eberli, G., and Froitzheim, N., 1990, Extensional detachment faulting in the evolution of a Tethys passive continental margin, Eastern Alps, Switzerland: Geological Society of America Bulletin, v. 102, p. 1297–1308, doi: 10.1130/0016-7606(1990)102<1297:EDFITE>2.3.CO;2.

Egal, E., 1992, Structures and tectonic evolution of the external zone of Alpine Corsica: Journal of Structural Geology, v. 14, p. 1215–1228, doi: 10.1016/0191-8141(92)90071-4.

Elter, G., Elter, P., Sturani, P., and Weidmann, M., 1966, Sur la prolungation du domaine Ligure de l'Apennin dans le Monferrat e les Alpes et sur l'origine de la nappe de la Simme s.l. des Prealpes romandes et chaiblaisiennes: Archives des Sciences de la Société de Physique et d'Histoire Naturelle de Genève, v. 19, p. 1002–1012.

Elter, P., 1972, La zona ofiolitifera del Bracco nel quadro dell'Appennino settentrionale: Introduzione alla geologia delle Liguridi, in 66° Congresso della Società Geologica Italiana, Field Trip Guidebook, 35 p.

Elter, P., 1975, L'ensemble Ligure: Bulletin de la Société Géologique de France, v. 17, p. 984–997.

Fedo, C.M., Eriksson, K.A., and Krogstad, E.J., 1996, Geochemistry of shales from Archean (approximately 3.0 Ga) Buhwa greenstone belt, Zimbabwe; implications for provenance and source-area weathering: Geochimica et Cosmochimica Acta, v. 60, no. 10, p. 1751–1763, doi: 10.1016/0016-7037(96)00058-0.

Froitzheim, N., and Manatschal, G., 1996, Kinematics of Jurassic rifting, mantle exhumation, and passive-margin formation in the Austroalpine and Penninic nappes (eastern Switzerland): Geological Society of America Bulletin, v. 108, p. 1120–1133, doi: 10.1130/0016-7606(1996)108<1120:KOJRME>2.3.CO;2.

Gardin, S., Marino, M., Monechi, S., and Principi, G., 1994, Biostratigraphy and sedimentology of Cretaceous Ligurid Flysch: Paleogeographical implication: Memorie della Società Geologica Italiana, v. 48, p. 219–235.

Gazzi, P., 1966, Le arenarie del flysch sopracretaceo dell'Appennino modenese: Correlazioni con il flysch di Monghidoro: Mineralogica et Petrographica Acta, v. 12, p. 69–97.

Gruppo Lavoro Ofioliti Mediterranee, 1977, I complessi ofiolitici e le unità cristalline della Corsica Alpina: Ofioliti, v. 2, p. 265–324.

Hofmann, A., Bolhar, R., Dirks, P., and Jelsma, H., 2003, The geochemistry of Archaean shales derived from a mafic volcanic sequence, Belingwe greenstone belt, Zimbabwe; provenance, source area unroofing and submarine versus subaerial weathering: Geochimica et Cosmochimica Acta, v. 67, p. 421–440, doi: 10.1016/S0016-7037(02)01086-4.

Hoogerdujin Strating, E.H., Rampone, E., Piccardo, G.B., Drury, M.R., and Vissers, R.L.M., 1993, Subsolidus emplacement of mantle peridotites during incipient rifting and opening of the Mesozoic Tethys (Voltri Massif, NW Italy): Journal of Petrology, v. 34, p. 901–927.

Houghton, H.F., 1980, Refined technique for staining plagioclase and alkali feldspar in thin section: Journal of Sedimentary Petrology, v. 50, p. 629–631.

Iaccarino, S., and Rio, D., 1972, Nannoplancton calcareo e foraminiferi della serie di Viano (Val Tresinaro, Appennino settentrionale): Rivista Italiana di Paleontologia e Stratigrafia, v. 78, no. 4, p. 641–678.

Ingersoll, R.V., and Suczek, C.A., 1979, Petrology and provenance of Neogene sands from Nicobar and Bengala fans, DSDP Sites 211 and 218: Journal of Sedimentary Petrology, v. 49, p. 1217–1228.

Ingersoll, R.V., Bullard, T.F., Ford, R.L., Grimm, J.P., Pickle, J.D., and Sares, S.W., 1984, The effect of grain size on detrital modes: A test of the Gazzi-Dickinson point counting method: Journal of Sedimentary Petrology, v. 54, p. 212–220.

Johnsson, M.J., 1993, The system controlling the composition of clastic sediments, in Johnsson, M.J., and Basu, A., eds., Processes Controlling the Composition of Clastic Sediments: Geological Society of America Special Paper 284, p. 1–19.

Johnsson, M.J., and Basu, A., 1993, Processes Controlling the Composition of Clastic Sediments: Geological Society of America Special Paper 284, 342 p.

Lemoine, M., Tricart, P., and Boillot, G., 1987, Ultramafic and gabbroic ocean floor of the Ligurian Tethys (Alps, Corsica, Apennines): In search of a genetic model: Geology, v. 15, p. 622–625, doi: 10.1130/0091-7613(1987)15<622:UAGOFO>2.0.CO;2.

Leoni, L., Marroni, M., Sartori, F., and Tamponi, M., 1996, The grade of metamorphism in the metapelites of the Internal Liguride units (Northern Apennines, Italy): European Journal of Mineralogy, v. 8, p. 35–50.

Lindholm, R.C., and Finkelman, R.B., 1972, Calcite staining: Semiquantitative determination of ferrous iron: Journal of Sedimentary Petrology, v. 42, p. 239–242.

Lister, G.S., Etheridge, M.A., and Symonds, P.A., 1986, Detachment faulting and the evolution of passive margins: Geology, v. 14, p. 246–250, doi: 10.1130/0091-7613(1986)14<246:DFATEO>2.0.CO;2.

Lister, G.S., Etheridge, M.A., and Symonds, P.A., 1991, Detachment models for the formation of passive continental margins: Tectonics, v. 10, p. 1038–1064.

Lombardo, B., and Pognante, U., 1982, Tectonic implications in the evolution of the Western Alps ophiolite metagabbros: Ofioliti, v. 7, p. 371–394.

Longiaru, S., 1987, Visual comparators for estimating the degree of sorting from plane and thin section: Journal of Sedimentary Petrology, v. 57, p. 791–794.

Marino, M., Monechi, S., and Principi, G., 1995, New calcareous nannofossil data on the Cretaceous-Eocene age of Corsican turbidites: Rivista Italiana di Paleontologia e Stratigrafia, v. 101, p. 49–62.

Marroni, M., and Pandolfi, L., 1996, The deformation history of an accreted ophiolite sequence: The Internal Liguride units (Northern Apennines, Italy): Geodinamica Acta, v. 9, p. 13–29.

Marroni, M., and Pandolfi, L., 2001, Debris flow and slide deposits at the top of the Internal Liguride ophiolitic sequence, Northern Apennines, Italy: A record of frontal tectonic erosion in a fossil accretionary wedge: The Island Arc, v. 10, p. 9–21, doi: 10.1046/j.1440-1738.2001.00289.x.

Marroni, M., and Pandolfi, L., 2003, Deformation history of the ophiolite sequence from Balagne Nappe (northern Corsica): Insights in the tectonic evolution of the Alpine Corsica: Geological Journal, v. 38, p. 67–83, doi: 10.1002/gj.933.

Marroni, M., and Perilli, N., 1990, Nuovi dati sull'età del complesso di M. Penna/Casanova (unità Liguri esterne, Appennino settentrionale): Rendiconti della Società Geologica Italiana, v. 13, p. 139–142.

Marroni, M., Monechi, S., Perilli, N., Principi, G., and Treves, B., 1992, Late Cretaceous flysch deposits of the Northern Apennines, Italy: Age of inception of orogenesis-controlled sedimentation: Cretaceous Research, v. 13, p. 487–504, doi: 10.1016/0195-6671(92)90013-G.

Marroni, M., Molli, G., Montanini, A., and Tribuzio, R., 1998, The association of continental crust rocks with ophiolites (Northern Apennines, Italy): Implications for the continent-ocean transition: Tectonophysics, v. 292, p. 43–66, doi: 10.1016/S0040-1951(98)00060-2.

Marroni, M., Pandolfi, L., and Perilli, N., 2000, Calcareous nannofossil dating of the San Martino Formation from the Balagne ophiolite sequence (Alpine Corsica): Comparison with the Palombini Shale of the northern Apennine: Ofioliti, v. 25, p. 147–156.

Marroni, M., Molli, G., Ottria, G., and Pandolfi, L., 2001a, Tectono-sedimentary evolution of the External Liguride units (northern Apennine, Italy): From rifting to convergence history of a fossil ocean-continent transition zone: Geodinamica Acta, v. 14, p. 307–320, doi: 10.1016/S0985-3111(00)01050-0.

Marroni, M., Pandolfi, L., and Saccani, E., 2001b, Mafic rocks from the sedimentary breccias associated to the Balagne ophiolitic nappe (northern Corsica): Geochemical features and geological implications: Ofioliti, v. 26, p. 433–444.

Marroni, M., Meneghini, F., and Pandolfi, L., 2004, From accretion to exhumation in a fossil accretionary wedge: A case history from Gottero unit (Northern Apennines, Italy): Geodinamica Acta, v. 17, p. 41–53.

McDonough, W.F., and Sun, S.-s., 1995, The composition of the Earth: Chemical Geology, v. 120, p. 223–253, doi: 10.1016/0009-2541(94)00140-4.

Melcher, F., Meisel, T., Puhl, J., and Koller, F., 2002, Petrogenesis and geotectonic setting of ultramafic rocks in the Eastern Alps: Constraints from geochemistry: Lithos, v. 65, p. 69–112, doi: 10.1016/S0024-4937(02)00161-5.

Mezzadri, G., 1964, Petrografia delle « Arenarie di Ostia »: Rendiconti della Società Mineralogica Italiana, v. 17, p. 193–228.

Monechi, S., and Treves, B., 1984, Osservazioni sulle età delle arenarie del Gottero. Dati dal nannoplancton calcareo: Ofioliti, v. 9, no. 1, p. 93–96.

Nardi, R., Puccinelli, A., and Verani, M., 1978, Carta geologica della Balagne "sedimentaria" (Corsica) alla scala 1:25.000 e note illustrative: Bollettino della Società Geologica Italiana, v. 97, p. 3–22.

Nilsen, T.H., and Abbate, E., 1984, Submarine-fan facies associations of the Upper Cretaceous and Paleocene Gottero Sandstone, Ligurian Appennines, Italy: Geo-Marine Letters, v. 3, p. 193–197, doi: 10.1007/BF02462467.

Owen, A.W., Armstrong, H.A., and Floyd, J.D., 1999, Rare earth elements in chert clasts as provenance indicators in the Ordovician and Silurian of the Southern Uplands of Scotland: Sedimentary Geology, v. 124, p. 185–195, doi: 10.1016/S0037-0738(98)00127-4.

Pandolfi, L., 1997, Stratigrafia ed evoluzione strutturale delle successioni torbiditiche Cretacee della Liguria orientale (Appennino settentrionale) [Ph.D. thesis]: Pisa, Università di Pisa, 175 p.

Passerini, P., and Pirini, C., 1964, Microfaune Paleoceniche nella formazione dell'Arenaria del M. Ramaceto e degli Argilloscisti di Cichero: Bollettino della Società Geologica Italiana, v. 83, p. 211–218.

Perilli, N., and Nannini, D., 1997, Calcareous nannofossil biostratigraphy of the Calpionella Limestone and Palombini Shale (Bracco/Val Graveglia unit) in the eastern Ligurian Apennines (Italy): Ofioliti, v. 22, p. 213–225.

Piccardo, G.B., 1977, Le ofioliti dell'areale Ligure: Petrologia e ambiente geodinamico di formazione: Rendiconti della Società Italiana di Mineralogia e Petrologia, v. 33, p. 221–252.

Plank, T., and Langmuir, C.H., 1998, The chemical composition of subducting sediment and its consequences for the crust and mantle: Chemical Geology, v. 145, p. 325–394, doi: 10.1016/S0009-2541(97)00150-2.

Poli, G., and Tommasini, S., 1999, Geochemical modeling of acid-basic magma interaction in the Sardinia-Corsica Batholith; the case study of Sarrabus, southeastern Sardinia, Italy: Lithos, v. 46, p. 553–571, doi: 10.1016/S0024-4937(98)00082-6.

Rampone, E., and Piccardo, G.B., 2000, The ophiolite-oceanic lithosphere analogue: New insight from the Northern Apennines (Italy), in Dilek, Y., Moores, E.M., Elthon, D., and Nicolas, A., eds., Ophiolites and oceanic crust; new insights from field studies and the Ocean Drilling Program: Geological Society of America Special Paper 349, p. 21–34.

Rampone, E., Hofmann, A.W., and Raczek, I., 1998, Isotopic contrasts within the Internal Liguride ophiolite (N. Italy); the lack of a genetic mantle-crust link: Earth and Planetary Science Letters, v. 163, p. 175–189, doi: 10.1016/S0012-821X(98)00185-X.

Rio, D., and Villa, G., 1987, On the age of the "Salti del Diavolo" conglomerates and of the M. Cassio flysch "Basal complex" (Northern Apennines, Parma Province): Giornale di Geologia, v. 49, p. 63–69.

Rio, D., Villa, G., and Cantadori, M., 1983, Nannofossils dating of the Helminthoid Flysch units in the Northern Apennines: Giornale di Geologia, v. 45, p. 57–86.

Saccani, E., Padoa, E., and Tassinari, R., 2000, Preliminary data on the Pineto Gabbroic Massif and Nebbio basalts: Progress toward the geochemical characterization of Alpine Corsica ophiolites: Ofioliti, v. 25, no. 2, p. 75–85.

Schwab, F.L., 1975, Framework mineralogy and chemical composition of continental margin-type sandstones: Geology, v. 3, p. 487–490, doi: 10.1130/0091-7613(1975)3<487:FMACCO>2.0.CO;2.

Stampfli, G.M., and Borel, G.D., 2004, The TRANSMED transect in space and time: Constraints on the paleotectonic evolution of the Mediterranean domain, in Cavazza, W., Roure, F., Spakman, W., Stampfli, G.M., and Ziegler, P.A., eds., The TRANSMED Atlas—The Mediterranean Region from Crust to Mantle: Berlin, Springer, p. 53–80.

Tamponi, M., Bertoli, F., Innocenti, F., and Leoni, L., 2003, X-ray fluorescence analysis of major elements in silicate rocks using fused glass discs: Atti della Società Toscana di Scienze Naturali: Memorie Serie A, v. CVII, p. 73–80.

Taylor, S.R., and McLennan, S.M., 1985, The Continental Crust: Its Composition and Evolution: Oxford, Blackwell, 312 p.

Tommasini, S., Poli, G., and Halliday, A.N., 1995, The role of sediment subduction and crustal growth in Hercynian plutonism; isotopic and trace element evidence from the Sardinia-Corsica batholith: Journal of Petrology, v. 36, p. 1305–1332.

Totten, M.W., Hanan, M.A., and Weaver, B.L., 2000, Beyond whole-rock geochemistry of shales; the importance of assessing mineralogic controls for revealing tectonic discriminants of multiple sediment sources for the Ouachita Mountain flysch deposits: Geological Society of America Bulletin, v. 112, p. 1012–1022, doi: 10.1130/0016-7606(2000)112<1012:BWRGOS>2.3.CO;2.

Trommsdorff, V., Piccardo, G.B., and Montrasio, A., 1993, From magmatism through metamorphism to sea-floor emplacement of subcontinental Adria lithosphere during pre-Alpine rifting (Malenco, Italy): Schweizerische Mineralogische und Petrographische Mitteilungen, v. 73, p. 191–203.

Valloni, R., 1985, Reading provenance from modern marine sands, in Zuffa, G.G., ed., Provenance of Arenites: Dordrecht, Reidel Publishing, p. 309–332.

Valloni, R., and Zuffa, G.G., 1984, Provenance changes for arenaceous formations of the Northern Apennines (Italy): Geological Society of America Bulletin, v. 95, p. 1035–1039, doi: 10.1130/0016-7606(1984)95<1035:PCFAFO>2.0.CO;2.

van de Kamp, P.C., and Leake, B.E., 1995, Petrology and geochemistry of siliciclastic rocks of mixed feldspathic and ophiolitic provenance in the Northern Apennines, Italy: Chemical Geology, v. 122, p. 1–20, doi: 10.1016/0009-2541(94)00162-2.

Venturelli, G., Capedri, R.S., Thorpe, R.S., and Potts, P.J., 1981, Rare earth and trace elements characteristic of ophiolitic metabasalts from the Alpine-Apennine belt: Earth and Planetary Science Letters, v. 53, p. 109–123, doi: 10.1016/0012-821X(81)90032-7.

Vescovi, P., 2006, Carta Geologica d'Italia: Foglio 216, Borgo Val di Taro: Roma, Regione Emilia Romagna and Agenzia per la Protezione dell'Ambiente e per i Servizi Tecnici (APAT) scale 1:50,000.

Vescovi, P., Fornaciari, E., Rio, D., and Valloni, R., 1999, The basal complex stratigraphy of the Helminthoid Monte Cassio Flysch; a key to the Eoalpine tectonics of the Northern Apennines: Rivista Italiana di Paleontologia e Stratigrafia, v. 105, p. 101–128.

Wandres, A.M., Bradshaw, J.D., Weaver, S., Maas, R., Ireland, T., and Eby, N., 2004, Provenance analysis using conglomerate clast lithologies: A case study from the Pahau terrane of New Zealand: Sedimentary Geology, v. 167, p. 57–89, doi: 10.1016/j.sedgeo.2004.02.002.

Weltje, G.J., and von Eynatten, H., 2004, Quantitative provenance analysis of sediments: Review and outlook: Sedimentary Geology, v. 171, p. 1–11, doi: 10.1016/j.sedgeo.2004.05.007.

Wernicke, B., 1981, Low-angle normal faults in the Basin and Range Province: Nappe tectonics in an extending orogen: Nature, v. 291, p. 645–648, doi: 10.1038/291645a0.

Wernicke, B., 1985, Uniform-sense normal simple shear of the continental lithosphere: Canadian Journal of Earth Sciences, v. 22, p. 108–125.

Wildi, W., 1985, Heavy mineral distribution and dispersal pattern in Apenninic and Ligurian flysch basins (Alps, Northern Apennines): Giornale di Geologia, v. 47, p. 77–99.

Zuffa, G.G., 1980, Hybrid arenites: Their composition and classification: Journal of Sedimentary Petrology, v. 49, p. 21–29.

Zuffa, G.G., ed., 1985, Provenance of Arenites: Dordrecht, D. Reidel Publishing Company, NATO ASI (Advanced Study Institutes) series, 357 p.

Zuffa, G.G., 1987, Unravelling hinterland and offshore paleogeography from deepwater arenites, in Leggett, J.K., and Zuffa, G.G., eds., Marine Clastic Sedimentology: Concepts and Case Studies: London, Graham and Trotman, p. 39–61.

MANUSCRIPT ACCEPTED BY THE SOCIETY 9 AUGUST 2006

Petrographic analysis in regional geology interpretation: Case history of the Macigno (northern Apennines)

Piero Bruni[†]
Department of Earth Sciences, University of Florence, via G. La Pira 4, 50121 Florence, Italy

Enrico Pandeli[‡]
Department of Earth Sciences, University of Florence, via G. La Pira 4, 50121 Florence, Italy, and Institute of Earth Sciences and Earth Resources, Consiglio Nazionale delle Ricerche (CNR), Florence Section, via La Pira 4, 50121, Florence, Italy

Massimo Nebbiai
Department of Earth Sciences, University of Florence, via G. La Pira 4, 50121 Florence, Italy

ABSTRACT

The Macigno is a widely outcropping terrigenous turbidite succession, up to 3000 m thick, in the northern Apennines. It was deposited in an Upper Oligocene–Lower Miocene perisutural basin that flanked the Apennine chain during uplift. In the Abetone study area, the Macigno lithofacies are characterized downward by prevailing thick-bedded, frequently amalgamated coarse sandstones (lower Macigno) and upward by recurrent fine and thin-bedded turbidites and siltstones (upper Macigno). The paleocurrent indicators are generally oriented toward the southeast and east. In the upper Macigno, a 250-m-thick intercalation known as the "Monte Modino Olistostrome" is present, represented by Cretaceous to Oligocene, locally chaotic, varicolored shale, marl, and limestone. Some authors hold that the latter is really an olistostrome that briefly interrupted the turbidite sedimentation. Others interpret the intercalation as the stratigraphic base of a tectonic unit that was thrusted onto the lower Macigno. Detailed modal petrographic analyses (Gazzi-Dickinson) performed on medium- to coarse-grained sandstones of the turbidite succession, along with plots of the data following the stratigraphic order of the samples, help to solve the geologic debate on stratigraphic continuity or tectonic discontinuity of the succession. The new data show that: (1) the main (QFL + C) and secondary components (Lv, Lm, Ls + C) are substantially similar along the succession; (2) from the stratigraphic base to the top of the turbidite succession, the petrographic parameters are characterized by appreciable trends and variations; and (3) the shaly-marly-calcareous intercalation (i.e., the Monte Modino Olistostrome) does not interrupt these trends and fluctuations. Therefore, we suggest that: (1) there is an overall common source area for the turbidite beds, even if minor compositional variations occur; and (2) the Macigno is

---

[†]E-mail: bruni@geo.unifi.it.
[‡]E-mail: pandeli@geo.unifi.it.

Bruni, P., Pandeli, E., and Nebbiai, M., 2007, Petrographic analysis in regional geology interpretation: Case history of the Macigno (northern Apennines), *in* Arribas, J., Critelli, S., and Johnsson, M.J., eds., Sedimentary Provenance and Petrogenesis: Perspectives from Petrography and Geochemistry: Geological Society of America Special Paper 420, p. 95–105, doi: 10.1130/2006.2420(07). For permission to copy, contact editing@geosociety.org. ©2007 Geological Society of America. All rights reserved.

a thick, stratigraphically continuous succession, and the sedimentary emplacement of the Monte Modino Olistostrome briefly interrupted the turbidite sedimentation. The new results contribute to geological mapping, to local- and regional-scale correlations, and to a better definition of the paleogeographic and tectonic setting of the Oligocene-Miocene siliciclastic turbidite successions of the northern Apennines.

Keywords: turbidites, provenance, sedimentology, northern Apennines.

INTRODUCTION

The northern Apennine chain is the birthplace of studies on siliciclastic turbidite successions (e.g., Migliorini, 1943, 1950; Kuenen and Migliorini, 1950; Sestini, 1970; Mutti and Ricci Lucchi, 1972, 1975; see reviews in Sestini et al. [1986] and in Argnani and Ricci Lucchi [2001]), especially on the Oligocene-Miocene turbidite units, such as the Macigno and the Marnoso-Arenacea. These units represent the geological backbone of the chain and its thicker (over 3000 m) and more widely outcropping successions (Fig. 1). The Oligocene-Miocene turbidite units build up different tectonic-stratigraphic units (e.g., the Tuscan Nappe, the Cervarola-Falterona unit) of the structural pile of the chain that are related to the infill stages of the eastward-migrating foredeep in front of the advancing Apenninic orogenic stack of nappes (Merla, 1951; Bortolotti et al., 1970; Ricci Lucchi, 1975, 1986; Boccaletti et al., 1990; Conti and Gelmini, 1994; Argnani and Ricci Lucchi, 2001). Their stratigraphic architecture and paleogeographic setting are fundamental in the choice of a model for the chain's evolution and represent some of the main topics debated in the geological literature on the northern Apennines. Unfortunately, the intense involvement of the Oligocene-Miocene turbidite successions in the folds and thrusts of the chain makes their stratigraphic and tectonic reconstructions debatable. This is the case of the Macigno, the stratigraphic uppermost unit of the Tuscan Nappe, which has been interpreted as a thick, continuous turbidite succession, or as the tectonic pile of two stratigraphic successions (see reviews of Fazzuoli et al., 1985, 1994; Abbate and Bruni, 1987; Chicchi and Plesi, 1991). This paper aims to present the data obtained from lithological, sedimentary, and petrographic studies performed on the Macigno outcropping in the Abetone area (Fig. 1) and to outline

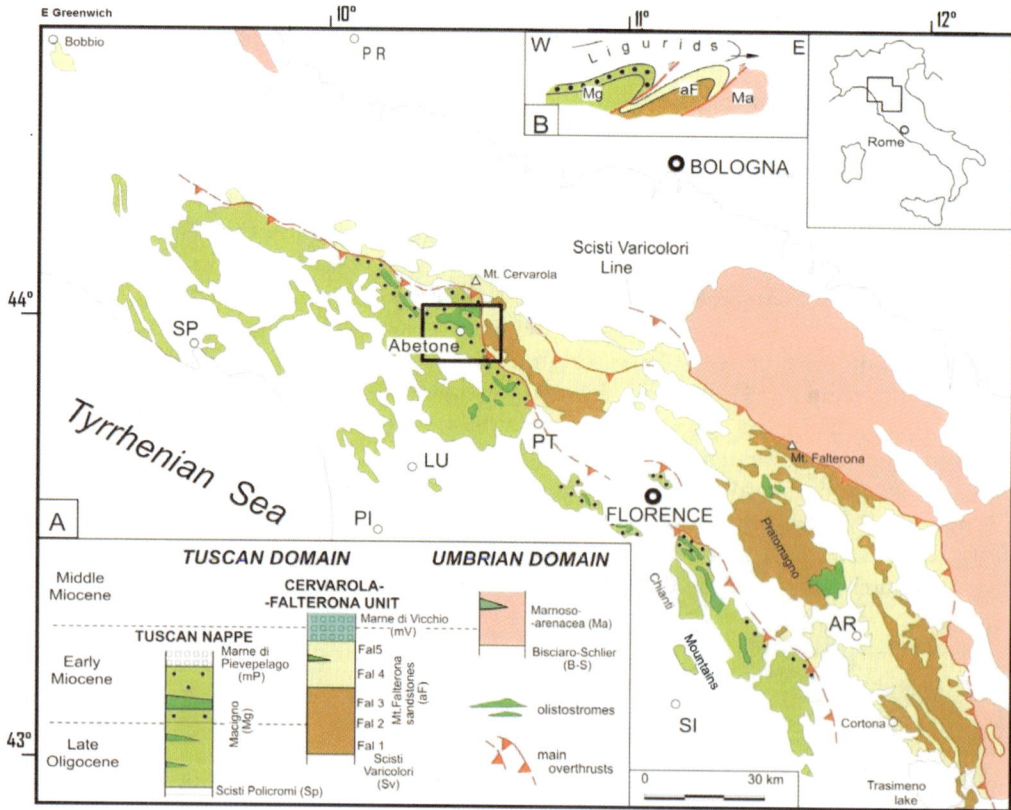

Figure 1. Areal distribution of the main Oligocene-Miocene turbidite units in the northern Apennines with Abetone study area marked with a box.

the importance of petrography in solving the geologic debate on stratigraphic-continuity versus tectonic-discontinuity of such thick turbidite successions.

GEOLOGICAL BACKGROUND

The Macigno lies above the pelagic, mainly shaly Tuscan Scaglia of Cretaceous–Lower Oligocene age and represents, together with the overlying Pievepelago Marl, the Middle–Upper Oligocene to Lower Miocene uppermost part of the Tuscan Nappe (Sestini, 1970; Dallan Nardi and Nardi, 1972; Catanzariti et al., 1991) (Figs. 2 and 3). This siliciclastic turbidite unit crops out on a regional scale in the northern Apennines (see Fig. 1) and is well exposed in the Abetone area, where it reaches a total thickness of ~3200 m. Its upper part includes an ~250-m-thick intercalation known as the "Monte Modino Olistostrome" (Fig. 3). This

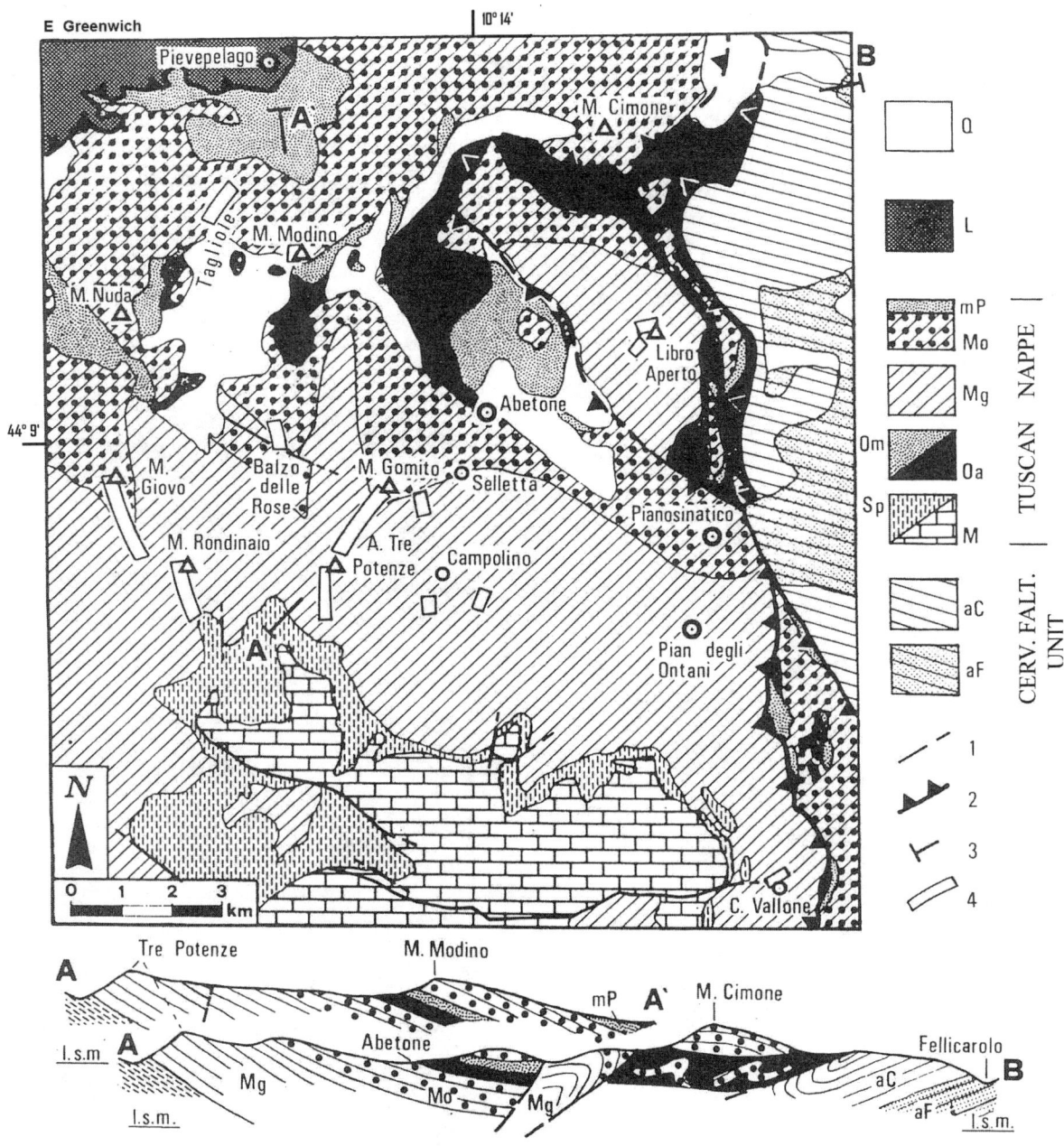

Figure 2. Geological sketch map of the Abetone area and location of the sections measured. Q—Quaternary sediments; L—Ligurian nappes; Tuscan Nappe: mP—Pievepelago Marl, Mo—upper Macigno, Mg lower Macigno, Om—Marmoreto Marl and Varicolored Fiumalbo Shale, Oa—Monte Modino Olistostrome, Sp—Polychromatic schists, M—Mesozoic formations; Cervarola-Falterona unit: aC—Monte Cervarola Sandstone, aF—Monte Falterona Sandstone. 1—faults, 2—main overthrusts, 3—geologic cross section, 4—location of the detailed measured sections, 4-location of the detailed measured sections; l.s.m.—sea level.

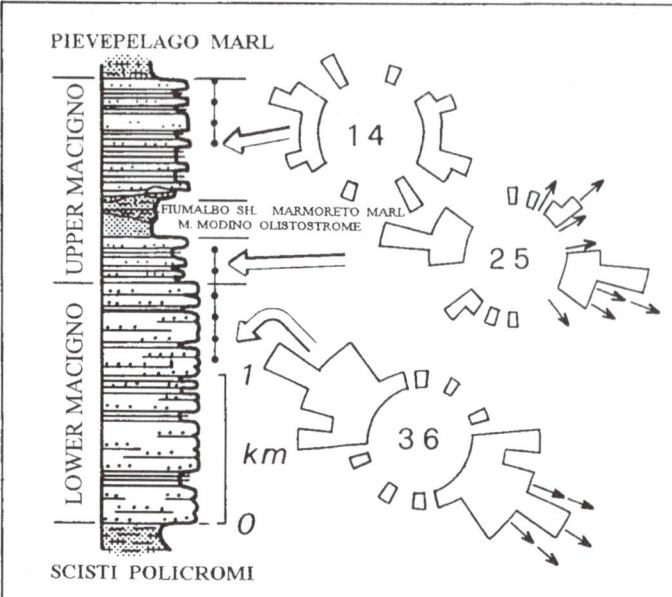

Figure 3. Simplified stratigraphic column of the Macigno and rose diagrams of the paleocurrents (numbers refer to readings).

shaly-marly-calcareous intercalation of Cretaceous-Oligocene lithologies with ophiolitic sandstones and breccias is capped by hemipelagic Upper Rupelian–Lower Aquitanian Marmoreto Marl and Fiumalbo Shale (Martini and Sagri, 1977; Chicchi and Plesi, 1991; Perilli, 1994; Mochi et al., 1995; De Libero, 1998). The relationship among these lithological units is a matter of debate. Some authors believe that the Macigno is a stratigraphically continuous unit and interpret the thick shaly-marly-carbonate intercalation as a synsedimentary slide (the Monte Modino Olistostrome) that came from the frontal part of the advancing orogenic wedge during its emplacement in the Tuscan turbidite foredeep (Abbate and Bortolotti, 1961; Giannini et al., 1962; Nardi and Tongiorgi, 1962; Nardi, 1964, 1965; Baldacci et al., 1967; Sestini, 1968; Dallan Nardi and Nardi, 1972, 1978; Sagri, 1975). Others, however, interpret the Monte Modino Olistostrome and overlying turbidite sedimentation as a tectonic-stratigraphic unit (the so-called Monte Modino unit) that thrusted over the Macigno from an innermost (Plesi, 1975; Reutter and Groscurth, 1978; Bettelli et al., 1987; Martini and Plesi, 1988; Chicchi and Plesi, 1991) or an outermost (Gunter and Rentz, 1968; Reutter, 1969; Zanzucchi, 1988) paleogeographic area.

The remarkable thickness, monotonous lithologies, wide range of bed thicknesses (Bruni and Pandeli, 1992), compositional similarities (Valloni et al., 1992; Gandolfi and Paganelli, 1993; Bruni et al., 1994a, 1994b; Pandeli et al., 1994), and the weak biostratigraphic signals of the turbidite sedimentation overlying and underlying the Monte Modino Olistostrome are not helpful in finding the solution to this problem. Moreover, the successions are frequently involved in complex tectonic structures. The geologic map and sections in Figure 2 show the main structural and stratigraphic features of the Abetone area, which are characterized by a monoclinal attitude in the westerly outcrops and by recumbent folds and reverse faults to the east, where the Tuscan Nappe thrusts onto the Cervarola-Falterona unit.

LITHOLOGIC AND SEDIMENTARY FEATURES OF THE MACIGNO

The description of the sedimentary features of the Abetone Macigno derives from detailed measurements and physical correlations of several stratigraphic sections (see their locations in Fig. 2). The reader is referred to Bruni and Pandeli (1992) for more information. The data show that the lower half of the Macigno is characterized by thick and coarse-grained arenaceous turbidites, whereas fine and thin-bedded turbidites become prevalent in the upper half. Detailed and laterally correlated sections at the transition point from the lower to upper Macigno are shown in Figures 4 and 5.

Figure 4. Panoramic view of the transition between the lower Macigno and upper Macigno along the Monte Giovo–Monte Rondinaio divide. For details and outcrop scale, see the Monte Giovo section in Figure 5.

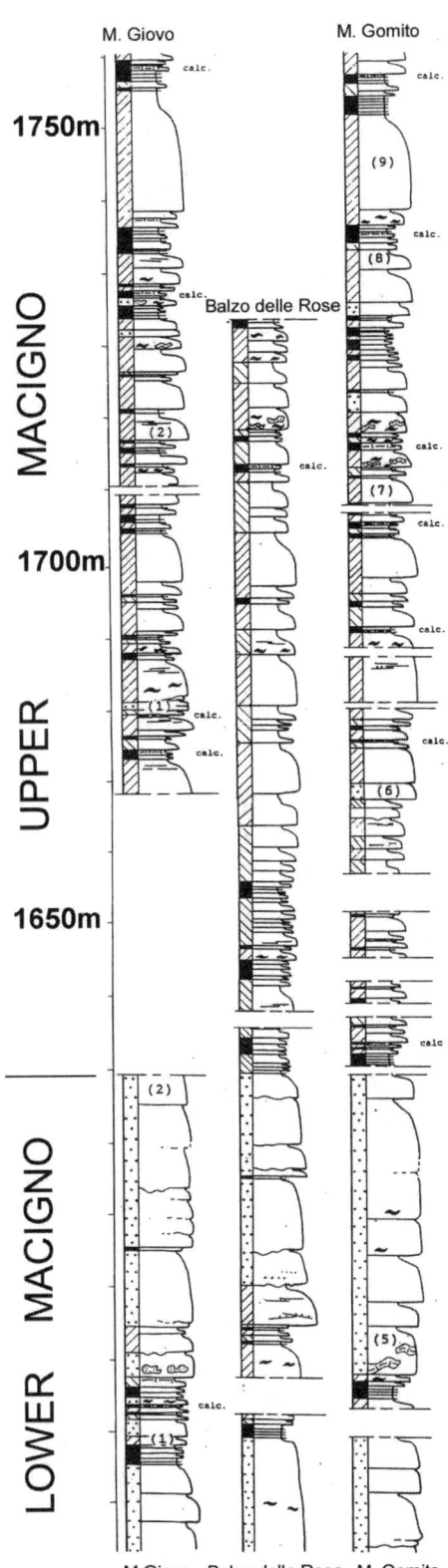

Figure 5. Detailed sedimentary logs from the transition between the lower Macigno and upper Macigno. For section locations, see Figure 2. The M Giovo section corresponds to the outcrop shown in Figure 4; the numbers identify the beds in the field.

Lower Macigno

This part consists of coarse-grained, frequently erosional or amalgamated arenaceous beds that are 1–4 m thick (Curcio et al. [1968] report a maximum bed thickness of ~50 m). The Bouma Ta lower division of the beds, frequently characterized by microconglomeratic lenses, coarse-tail grading, and fluid escape structures, represents at least 80% of the bed thickness. Centimeter-thick coarse-grained, plane-parallel or, sometimes, reverse-graded laminations also occur. Although the arenaceous beds are not evidently channelized, the lateral continuity of the beds is limited to a few tens or a few hundreds of meters (see the sections of the lower Macigno in Fig. 5). The thicker and coarse-grained turbidites often cluster in bed sequences up to a hundred meters thick, which can be followed laterally for several kilometers; they are separated by sequences that are a few meters thick and have prevailing thin-bedded fine-grained turbidite sandstones and siltstones. Paleocurrent data (Fig. 3) suggest that the turbidity currents flowed toward the southeast. In the lower Macigno, thin-bedded black shales and medium- to fine-grained terrigenous calcarenites (hybrid turbidites of Zuffa, 1980) from a few centimeters to one meter thick are sometimes present. These recur at the base of the Macigno and contain abundant macroforams (Azzaroli, 1955; Bortolotti and Pirini, 1965; Montanari and Rossi, 1983; Barelli and Mantovani Uguzzoni, 1985). Age determinations of these beds indicate Middle–Upper Oligocene ages (Bortolotti and Pirini 1965; Merla and Abbate, 1969), or more precisely, the Middle Chattian (Montanari and Rossi, 1983). At the top of the lower Macigno, the calcareous nannoplankton assemblages examined by Catanzariti et al. (1991) suggest an early Miocene age (NN1-NN2 biozone of Fornaciari and Rio, 1996).

Upper Macigno

Fine and thin-bedded (10–50 cm), sandstones and siltstones prevail, of which the base begins with turbidite Tb, Tc, or Td divisions (Bouma, 1962). Coarse-grained turbidite sandstones up to several meters thick are abundant or very common (see the sections of the upper Macigno of Fig. 5). The basal surfaces of the beds are generally flat, and the Bouma sequences are usually well developed and frequently complete. Meter-sized mud pebbles and clay chips are common in the middle and upper part of the Bouma Ta interval in the thicker beds that are laterally continuous for many kilometers. The paleocurrent data (Fig. 3) suggest average flow directions toward the east. In the upper Macigno, medium- to fine-grained calcareous turbidites and rare black shales are present. Their nannoplankton assemblages suggest that the upper Macigno, up to the overlying Pievepelago Marl, is referable to the Aquitanian–Lower Burdigalian (Catanzariti et al., 1991; Mochi et al., 1995). Instead, the Fiumalbo Shale and the Marmoreto Marl, which occur at the top of the Monte Modino Olistostrome, contain Paleocene to Rupelian calcareous nannoplankton assemblages (with condensed biostratigraphic events, but in good vertical order) that reach the Oligocene-Miocene boundary in the uppermost marly deposits of the

Marmoreto Marl (Catanzariti et al., 1991; Mochi et al., 1995). In other words, the shaly and marly sediments at the top of the olistostrome appear to be older than the turbidite sediments below and above them.

PETROGRAPHIC FEATURES OF THE MACIGNO

The medium- to coarse-grained sandstones of the Macigno were sampled in the lower Ta interval of the beds (48 samples: see Tables 1 and 2). The modal petrographic analysis was performed counting 500 points per section and following the Gazzi-Dickinson method (Dickinson, 1970; see also Chiocchini and Cipriani, 1992; Di Giulio and Valloni, 1992, and references therein). The huge amount of data obtained so far has been recorded in schedules (see Figure 7 of Aruta et al., 2004) by using the integrated software-hardware system of Cipriani et al. (2004). To detect and compare the clay mineral associations of the Tde intervals of the turbidite beds, the occasional black shale intercalations of the Macigno, and the shaly lithofacies of the Monte Modino Olistostrome, the samples were analyzed through X-ray diffraction (Moore and Reynolds, 1989) using the methodology of Cipriani (1958) and Cipriani and Malesani (1972).

Modal Analysis

The data (see Tables 1 and 2) are represented in the familiar QFL + C and LvLmLs + C diagrams (Figs. 6 and 7). None of the diagrams shows significant differences between lower and upper Macigno. Instead, many significant compositional differences are recognizable when the petrographic parameters are plotted following the stratigraphic order of the samples (see Fig. 8). This uncommon representation of petrographic data allows us to better appreciate and define the main components of the framework.

Quartz

Quartz is the most abundant mineral (from 30% to ~42%) in the modal composition and is present in monomineralic grains or in rock fragments (Tables 1 and 2). Monocrystalline quartz prevails over polycrystalline quartz, and the ratio (Qm/Q in Fig. 8) slightly increases from the lower to upper Macigno. More than 60% of the monocrystalline quartz displays undulose extinction, and the Qm nonundulose/Qm ratio (Fig. 8) is less variable in the lower Macigno than in the upper Macigno. The Q/Q + F ratio (Fig. 8) is stable (0.54–0.68) in the lower Macigno and is more variable (0.46–0.73) in the upper Macigno.

Feldspars

The feldspar content of the Macigno is on average ~24% of the modal composition (see Tables 1 and 2). However, the K-feldspar/total feldspar ratio (K/F in Fig. 8) is always >0.20 (max 0.60) in the lower Macigno and is more variable (from 0 to 0.55) in the upper Macigno (see Fig. 8). About 50% of the feldspars are altered: they are seritized in the lower Macigno and clay altered in the upper Macigno. In the feldspar grains, untwinned crystals

TABLE 1. DETAILED DATA OF THE MODAL PETROGRAPHIC COMPOSITION OF THE SAMPLES COLLECTED IN THE LOWER MACIGNO UNIT OF THE ABETONE AREA

Samples	1-4a	1-7	1-10	1-9-0-1	4-05	4-07	4-09	4-11	4-17	1-12	1-13	1-14	1-19	1-22	1-25	1-11	1-26a	1-26b	GO 1	GO 3	GO 4	GO 6	Ta-04	GO 8	GO 11	GO 12	Ta-02	GO 13
Qz xxm nonundulose	13.6	13.0	16.0	14.2	16.2	14.6	14.0	15.8	13.2	13.2	16.2	15.2	15.4	16.0	15.0	16.8	16.2	13.8	14.4	13.4	10.7	8.0	12.5	12.2	9.4	12.7	11.2	11.8	13.2
Qz xxm undulose	15.8	17.0	15.2	19.2	19.2	18.8	17.0	19.4	18.8	16.0	19.4	18.4	15.6	16.4	18.4	18.0	19.4	17.6	18.0	23.8	21.7	25.4	19.1	20.0	23.6	22.0	21.9	18.0	20.2
Qz monocrystalline	29.4	3C.0	31.2	33.4	35.4	33.4	31.0	35.2	32.0	29.2	35.6	33.6	31.0	32.4	33.4	34.8	35.6	31.4	32.4	37.2	32.4	35.6	31.6	32.2	33.0	33.2	34.6	29.8	33.4
Qz polycrystalline	8.6	1C.2	8.0	5.0	5.6	7.4	6.6	6.6	5.8	6.8	6.8	6.8	4.8	8.2	6.6	6.6	5.8	6.8	6.6	5.2	7.0	5.4	4.6	4.6	4.0	4.6	3.6	4.0	5.2
Total quartz	38.0	40.2	39.2	38.4	41.0	40.8	37.6	41.8	37.8	36.0	40.8	40.4	37.8	37.8	41.6	41.4	41.4	38.2	39.0	42.4	39.4	41.0	36.2	36.8	37.0	37.8	38.2	33.8	38.6
K-feldspars	7.4	5.8	6.5	7.3	7.9	7.6	8.0	8.5	10.3	9.6	10.5	6.2	7.4	6.8	8.2	6.6	5.4	8.6	11.3	7.0	10.8	9.0	5.2	11.8	12.8	9.8	8.0	17.0	11.0
Plagioclase	14.6	14.8	12.3	15.3	11.7	14.0	11.4	12.9	16.3	16.6	14.7	14.8	14.2	10.8	7.7	9.9	12.9	11.8	15.3	17.6	12.6	14.4	15.2	15.8	17.0	21.6	21.2	12.2	17.8
Total feldspars	22.0	20.6	18.8	22.6	19.6	21.6	19.4	21.4	26.6	26.2	25.2	21.0	21.6	17.6	15.9	16.5	18.3	20.4	26.6	24.6	23.4	23.4	20.4	27.6	29.8	31.4	29.2	29.2	28.8
Lit. intrusive	0.0	0.0	0.0	0.0	0.0	0.0	0.0	0.0	0.0	0.0	0.0	0.0	0.2	0.0	0.0	0.0	0.0	0.0	0.6	0.0	0.4	0.4	0.4	0.4	0.6	0.5	0.4	0.7	0.2
Lit. v. felsic	0.8	1.6	1.0	0.0	0.8	1.0	1.4	0.6	0.8	0.4	0.2	0.0	0.2	0.4	0.6	0.4	0.8	0.6	0.6	0.4	0.7	0.2	0.4	0.6	0.2	0.7	0.7	0.7	0.6
Lit. v. intermediate	0.0	0.4	0.8	0.8	0.4	0.8	0.6	2.0	1.2	0.8	0.4	0.0	1.6	0.6	0.4	0.8	0.4	0.6	0.6	0.6	1.1	0.9	0.2	0.2	1.3	1.1	1.1	0.7	1.1
Lit. volcanic	0.8	2.0	1.8	1.8	1.2	2.2	2.0	2.6	2.2	1.6	0.6	0.4	1.6	0.8	0.6	1.2	1.2	1.2	1.2	1.1	2.2	1.7	0.6	0.8	1.5	1.8	1.8	1.4	1.7
Lit. metamorphic	8.4	8.4	8.2	5.8	8.6	7.6	6.4	6.6	6.6	8.2	5.4	9.8	8.2	7.0	9.4	9.0	8.3	10.2	8.8	8.3	10.2	9.4	9.1	8.5	8.9	6.1	8.8	5.9	6.5
Serpentinite	0.6	1.4	0.6	1.4	1.6	1.8	3.4	3.0	3.2	2.6	3.2	1.8	3.2	5.4	4.8	4.0	0.6	1.8	1.4	0.4	0.6	0.0	0.0	0.6	0.0	0.2	0.2	0.0	0.3
Lit. siliciclastic	0.0	0.0	0.0	0.2	0.6	0.2	0.8	0.0	0.2	0.8	0.6	0.0	0.4	0.6	0.6	0.8	0.4	0.0	0.4	0.2	0.4	0.0	0.0	0.6	0.2	0.2	0.2	0.0	0.6
Lit. carbonate	0.6	0.4	0.4	0.6	0.4	0.0	0.0	0.0	0.6	0.2	0.0	0.2	0.0	1.4	0.2	0.2	0.4	0.0	0.4	0.2	0.0	0.2	0.0	0.0	0.2	0.0	0.0	0.9	0.6
Chert	0.0	0.0	0.0	0.0	0.0	0.0	0.0	0.0	0.0	0.0	0.6	0.2	0.6	0.8	0.6	0.6	0.2	2.0	0.0	0.0	0.4	0.0	0.0	0.2	0.0	0.0	0.0	0.0	0.6
Total lithic fragments	10.4	12.2	11.0	9.8	12.6	12.6	12.6	11.6	11.8	12.8	12.6	12.2	11.2	15.4	12.2	11.2	13.0	11.4	9.6	12.6	12.8	10.0	10.8	10.8	11.4	10.8	8.8	9.0	9.6
Muscovite	5.2	5.6	5.8	7.6	6.2	5.0	7.6	6.8	6.0	4.6	4.4	3.6	3.4	4.6	4.8	5.2	4.6	5.8	11.4	3.7	3.3	3.2	7.9	4.0	4.4	4.5	8.8	9.0	4.2
Biotite	3.6	2.6	3.0	3.0	3.6	2.2	3.2	1.6	2.0	1.6	1.4	2.2	1.4	3.2	1.4	1.4	2.8	2.2	2.2	2.2	2.1	2.1	5.4	1.9	2.2	2.9	4.5	4.3	2.6
Chlorite	3.2	1.4	1.8	3.4	3.0	2.6	2.6	2.4	4.4	2.8	2.6	2.8	2.0	4.4	2.0	3.6	2.2	3.0	2.8	2.7	4.4	4.5	3.4	3.5	3.6	3.9	2.9	2.6	2.4
Detrital matrix	1.6	2.2	2.6	1.8	2.6	2.8	2.6	2.4	4.4	3.4	3.2	1.4	2.8	2.6	3.2	2.0	2.4	3.0	2.2	5.6	3.6	3.6	4.0	2.8	4.2	2.8	2.0	2.4	3.8
Diagenetic matrix	4.6	5.0	5.0	4.6	3.8	5.0	5.6	3.4	3.2	4.4	2.6	2.8	6.6	3.4	4.0	4.6	5.2	4.6	6.2	6.4	6.6	6.4	4.0	6.4	6.6	9.2	6.2	5.0	6.2
Carbonate cement	4.8	2.8	5.6	3.0	2.2	1.6	0.0	0.0	2.4	2.4	3.4	3.2	1.6	4.0	7.4	5.2	4.0	4.0	7.2	8.0	7.4	4.2	10.6	4.2	6.6	0.0	1.0	6.2	1.2
Replacement calcite	4.4	5.2	3.8	3.0	3.2	4.0	2.4	4.0	2.6	1.8	0.6	0.2	0.8	0.8	0.4	1.2	1.4	0.6	1.4	1.4	0.4	3.4	0.0	3.4	4.2	0.0	0.2	3.6	1.8
Bioclast	0.0	0.0	0.4	0.0	0.0	0.0	0.0	0.0	0.4	0.0	0.0	1.4	1.0	2.2	1.4	2.2	2.6	2.6	2.6	0.0	1.0	0.0	0.0	4.2	0.0	0.0	0.4	0.0	0.0
Heavy minerals	2.2	2.2	3.0	2.8	2.2	2.0	3.6	2.6	2.8	4.0	3.8	4.2	3.6	2.8	4.0	4.0	2.8	1.6	2.6	1.2	1.4	1.2	1.6	1.6	0.8	0.2	1.4	1.0	0.8
Total	100	100	100	100	100	100	100	100	100	100	100	100	100	100	100	100	100	100	100	100	99.4	100	99.4	100	100	99.8	100	100	100

Note: See Figure 9 for the stratigraphic position of the samples. Abbreviations: Qz—quartz; xxm—monocrystalline; Lit.—aphanitic rock grains; v.—volcanic.

TABLE 2. DETAILED DATA OF THE MODAL PETROGRAPHIC COMPOSITION OF THE SAMPLES COLLECTED FROM THE UPPER MACIGNO UNIT IN THE ABETONE AREA

Samples	GO 16	GO 17	GO 20	GO 23	GO 24	Go-25	Go-26	191	1	2	3	4	Ta-6	Ta-8	Ta 10	Ta 19	Ta 28	Ta 33	Ta-39	Ta-41
Qz xxm.nonundulose	12.3	12.8	11.6	11.8	12.3	12.8	10.2	15.6	15.8	18.6	13.4	15.8	11.4	14.6	13.9	11.3	11.1	12.6	12.4	13.9
Qz xxm. undulose	20.9	19.8	17.0	18.8	19.3	16.2	13.8	15.0	16.6	14.2	14.2	16.2	15.8	15.6	18.9	15.5	11.9	20.6	18.2	13.8
Qz monocrystalline	33.2	32.6	28.6	30.6	31.6	29.0	24.0	30.6	32.4	32.8	27.6	32.0	27.2	30.2	32.8	26.8	23.0	33.2	30.6	26.8
Qz polycrystalline	4.4	4.8	4.2	4.4	4.0	3.6	3.8	4.0	4.4	5.0	4.2	3.2	4.2	2.4	3.8	2.4	1.6	4.4	4.6	3.8
Total quartz	37.6	37.4	32.8	35.0	35.6	32.6	27.8	34.6	36.8	37.8	31.8	35.2	31.4	32.6	36.6	29.2	24.6	37.6	35.2	30.6
K-feldspars	12.0	11.6	3.8	9.6	10.8	2.6	6.2	0.0	0.5	0.0	5.0	3.7	16.4	16.8	11.0	0.6	0.2	14.4	11.6	10.8
Plagioclase	20.8	17.8	12.4	19.0	24.8	13.8	11.2	16.4	19.7	18.2	16.2	13.1	20.0	14.0	17.8	10.4	10.2	19.2	14.8	19.4
Total feldspars	32.8	29.4	16.2	28.6	35.6	16.4	17.4	16.4	20.2	18.2	21.2	16.8	36.4	30.8	28.8	11.0	10.4	33.6	26.4	30.2
Lit. intrusive	0.0	0.3	0.0	0.0	0.0	0.0	0.0	0.2	0.0	0.0	0.2	1.8	0.3	0.8	0.6	0.2	0.1	0.0	0.2	0.3
Lit. v. felsic	0.0	0.3	0.3	0.0	0.6	0.4	1.0	0.6	0.0	1.0	0.0	0.4	0.5	1.3	0.6	0.2	0.2	0.7	0.7	0.0
Lit. v.intermediate	1.4	1.8	0.9	0.7	0.9	0.6	1.0	0.2	1.6	0.0	1.9	2.2	0.3	0.0	0.6	0.2	0.3	0.7	0.9	0.6
Lit. volcanic	1.4	2.1	1.3	0.7	1.5	1.0	1.0	1.2	1.6	1.0	1.9	5.1	1.1	1.3	1.2	0.2	0.3	1.4	1.6	0.6
Lit. metamorphic	5.3	5.1	5.9	7.5	4.9	5.0	9.4	6.3	8.9	6.8	5.4	1.9	6.3	5.3	7.0	6.9	3.7	7.6	5.6	4.5
Serpentinite	2.2	1.8	4.1	4.4	3.1	1.8	4.7	1.3	4.2	0.9	3.0	1.0	2.0	2.0	1.5	0.2	0.0	1.8	2.1	1.4
Lit. siliciclastic	0.3	0.9	0.6	0.0	1.8	0.2	0.0	0.4	4.9	0.5	0.2	0.0	1.0	0.5	0.0	0.0	0.1	0.7	0.2	0.6
Lit. carbonate	0.0	0.0	0.3	1.0	0.0	1.2	2.6	0.0	0.0	0.3	0.1	0.2	0.2	0.2	0.0	0.5	0.8	1.1	0.5	0.8
Chert	0.0	0.3	0.0	0.0	0.0	0.0	0.0	0.0	0.0	0.0	0.1	0.4	0.0	0.0	0.0	0.0	0.1	0.0	0.0	0.0
Total lithic fragments	9.2	10.6	12.2	13.6	10.4	9.2	17.8	9.4	19.6	9.2	10.8	10.4	8.8	9.8	9.8	7.8	5.0	12.6	10.2	8.2
Muscovite	4.0	4.2	5.8	4.2	1.5	8.8	6.6	6.4	5.0	6.6	7.4	7.8	2.2	4.0	4.6	9.9	3.3	2.2	2.8	4.0
Biotite	3.0	2.6	5.5	3.1	1.5	2.8	2.0	1.8	1.0	2.6	2.4	2.8	2.0	1.6	2.8	6.1	2.4	1.4	3.4	3.0
Chlorite	3.4	3.4	3.9	1.3	1.4	1.8	2.6	2.8	2.8	3.0	1.6	2.0	2.6	2.0	2.8	5.8	1.9	2.0	2.0	1.8
Detrital matrix	1.2	2.0	3.4	2.2	1.4	3.2	3.2	4.8	2.4	2.8	5.2	2.4	1.6	3.6	3.2	4.4	2.0	3.4	4.8	1.2
Diagenetic matrix	7.6	8.2	2.8	5.2	2.0	6.6	5.8	8.8	4.0	3.2	2.8	5.6	10.4	10.6	6.4	4.4	0.4	3.4	4.8	4.6
Carbonate cement	0.0	0.4	6.6	3.2	8.0	12.0	8.8	7.2	0.6	4.2	11.8	6.0	2.2	1.2	2.6	8.2	22.2	1.0	3.6	9.0
Replacement calcite	0.0	1.2	9.8	3.0	1.6	6.2	5.6	7.0	7.4	11.4	4.4	10.0	1.4	2.8	1.4	15.6	26.6	2.0	5.6	7.0
Bioclast	0.0	0.0	0.0	0.2	0.0	0.0	0.0	0.2	0.0	0.4	0.6	1.0	1.0	0.0	0.0	0.0	1.2	0.0	0.2	0.0
Heavy minerals	1.2	0.6	1.0	0.2	0.4	0.4	2.4	0.0	0.2	0.6	0.6	1.0	1.0	1.0	1.0	0.8	1.2	0.8	1.0	0.4
Total	100	100	100	100	100	100	100	100	100	100	100	100	100	100	100	100	100	100	100	100

Note: See Figure 9 for the stratigraphic position of the samples. Abbreviations: Qz—quartz; xxm—monocrystalline; Lit.—aphanitic rock grains; v.—volcanic.

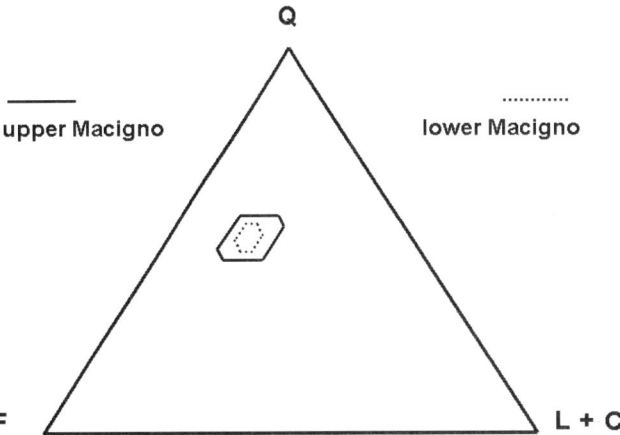

Figure 6. Petrographic composition of the Macigno turbidite sandstones: main detrital components according to Dickinson (1970) and Di Giulio and Valloni (1992). Q—quartz; F—feldspars; L + C—aphanitic and carbonate rock grains.

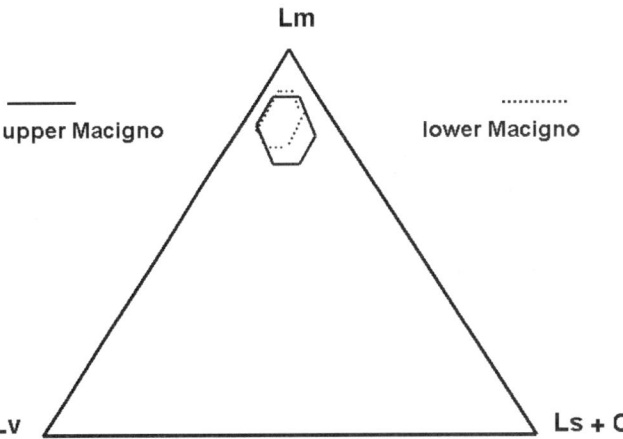

Figure 7. Petrographic composition of the Macigno turbidite sandstones: lithic grain composition. Lm—metamorphic grains; Lv—volcanic; Ls + c—sedimentary and carbonate grains.

predominate (~70%) over perthites (~25%) and microcline (5%). Plagioclase is mostly untwinned (65%), and 22% of plagioclase crystals display albite polysynthetic twinning.

Rock Fragments

Aphanitic rock grains are ~15% on average of the modal composition (see Tables 1 and 2) and are mainly represented by metamorphic lithic fragments (= Lm) (see LmLvLs + C diagram of Fig. 7). The latter are on average 70% (serpentinite grains excluded) in the lower Macigno and decrease to ~56% in the upper Macigno (Fig. 8). Serpentinite grain content is variable from 0% to 32%, and the log shape is specular to the log of the metamorphic lithic fragments. The percentage of sedimentary siliciclastic + carbonate lithic fragments (= Ls + C) is lower and more variable in the lower Macigno (up to 15%) than in the upper Macigno (up to 25%), and carbonate grains prevail over siliciclastic

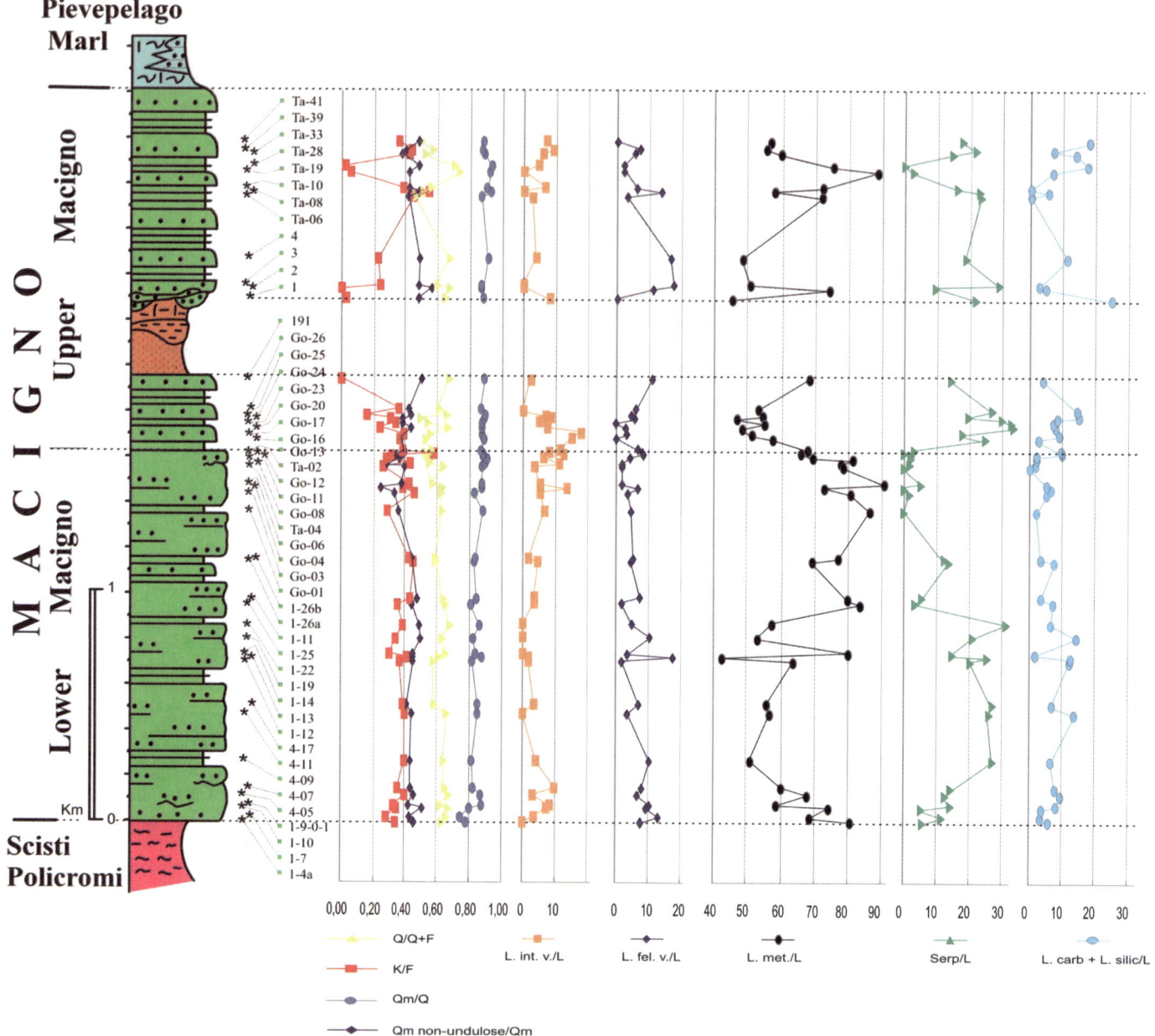

Figure 8. Thin-section petrography of the Macigno in the Abetone area (see Fig. 2). The figure indicates the stratigraphic position of the analyzed samples and synthesizes the vertical variations of some petrographic parameters of the Macigno. K/F—K-feldspar/total feldspars; Qm nonundulose/Qm stands for monocrystalline quartz characterized by nonundulose extinction/monocrystalline quartz; Qm/Q—monocrystalline quartz/total quartz and the percentage of the different types of lithics (L intermediate volcanics/L total, L acidic volcanics/L total, L metamorphic/L total, L serpentinite/L total, L carbonate + siliciclastic/L total).

grains (see Tables 1 and 2). Chert fragments are rare. The content of the intermediate and acidic volcanic lithics varies from 0% to 10%, with local increase up to 19% at the transition point from the lower to the upper Macigno.

The Lserp. and Lvolc. values do not fit the results of previous authors, e.g., the upward decrease of Lv in Valloni et al. (1992) and Pandeli et al. (1994). This discrepancy must be related to the different grouping method of L serp. values, which were included in Lv by these authors.

Micas and Chlorites

The average mica and chlorite content is ~10% and occurs prevalently as single crystals or, to a lesser degree, in metamorphic rock fragments. Most of the mafic mica is altered.

Matrix and Cement

The detrital matrix (Folk, 1968) is less than 6%, whereas the diagenetic matrix (cf. Dickinson, 1970) varies from 2% to 10%. This is mainly represented by epimatrix (Di Giulio and Valloni, 1992), which Valloni (1978) also indicates for the Macigno of the Val Gordana section. The cement is represented by calcite: 0%–7.2% in the lower Macigno and 0%–22% in the upper Macigno.

Clay Mineral Analysis

Fifty samples were collected from the Bouma T*de* interval at the top of the Macigno turbidite beds, 3 samples were collected from the black shale intercalations, and 2 samples were collected from the shale lithologies of the Monte Modino Olistostrome. These data, represented in the histograms of Figure 9, show different clay mineral associations. In the T*de* intervals of Macigno, kaolinite (K) and illite (I) predominate over chlorite (Cl), vermiculite (V), and mixed layers (chlorite-vermiculite). The clay-mineral association of the black shale of the Macigno and that of the olistostrome are instead characterized by illite with subordinate vermiculite and mixed layers (illite-smectite).

DISCUSSION AND CONCLUSIONS

The lithologic, sedimentary, and petrographic data presented here allow us to refine the stratigraphic setting of the Macigno in the Abetone area and restrict the structural setting of the Tuscan turbidite successions on a regional scale. The stratimetric data suggest that the lower Macigno is mainly characterized by thick, coarse-grained beds with frequently amalgamated or erosional basal contacts; T*c-e* and T*de* beds are subordinate. The paleocurrent indicators are slightly dispersed and oriented to the southeast (Fig. 3). The upper portion of the succession is characterized by thin-bedded T*c-e* and T*de* beds, while medium and thick beds are less frequent and display a complete Bouma sequence. The paleocurrent indicators are rather dispersed toward the east-southeast.

Despite the lithologic and sedimentary differences, the main and secondary petrographic components (Figs. 6 and 7) of the upper and lower Macigno are quite similar. These similarities, which have already been noted by other authors (Cipriani and Malesani, 1964, 1972; Sestini, 1970; Valloni, 1978; Valloni and Zuffa, 1984; Aruta et al., 1998), are also evidenced by the clay mineral associations (Fig. 9) of the T*de* intervals of thick and

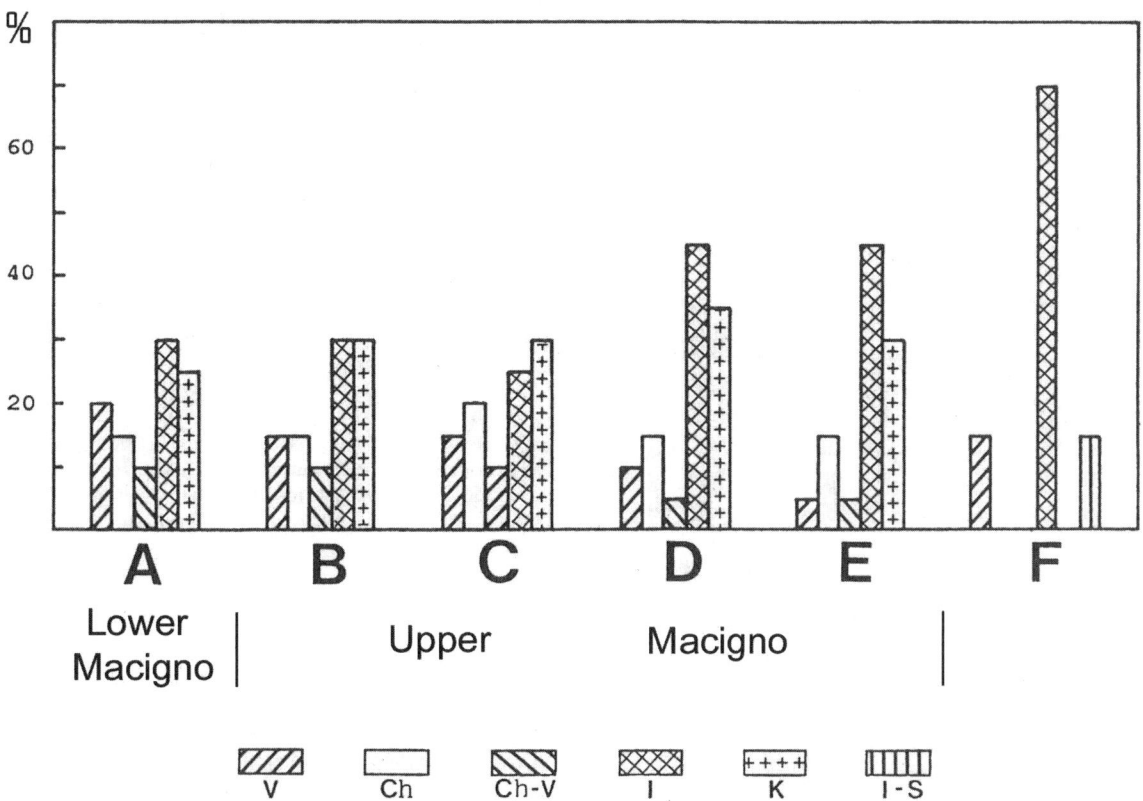

Figure 9. Clay minerals of the Bouma T*de* intervals of the Macigno: A—lower Macigno; B, C—coarse- to medium-grained and fine-grained, respectively, turbidite beds of the upper Macigno: lower part; D, E—coarse- to medium-grained and fine-grained, respectively, turbidite beds of the upper Macigno: upper part; F—clay minerals of the black shaly layers interbedded in the Macigno. V—vermiculite; Ch—chlorite; Ch-V—mixed layers chlorite/vermiculite; I—illite; K—kaolinite; I-S—mixed layers illite/smectite.

coarse sandstones and of thin and fine-grained sandstones and siltstones. Therefore, the data suggest a common source area for the lower and the upper Macigno, reasonably located in the central-western Alps (Bortolotti et al., 1970; Valloni, 1978; Di Giulio, 1999). The specular shape of the logs of Lmet./L ratio versus Lserp./L + Lcarb. + silic./L ratio (Fig. 8) as well as the different K-feldspar alteration suggest that the detrital input was likely conditioned by local lithologic differences and/or by climatic and tectonic variations of the same wide source area.

In spite of the similarity of the main modal composition, the lower and upper Macigno can be distinguished by plotting the petrographic parameters (K/F, Qm nonundulose/Qm, Qm/Q, serpentinite/L, L carbonate + siliciclastic/L) following the stratigraphic order of the samples (see Fig. 8). This representation of the petrographic data shows appreciable trends and variations that are not interrupted by the Monte Modino intercalation. Therefore, the Macigno represents a stratigraphically continuous turbidite succession, briefly interrupted by the olistostrome emplacement. The petrographic data and their elaborations, integrated with lithostratigraphic and biostratigraphic analyses, represent a useful tool for geological mapping, regional-scale correlations, and for defining the paleogeographic and tectonic setting of the siliciclastic turbidite successions.

ACKNOWLEDGMENTS

We wish to thank Geological Society of America reviewers A. Basu and R. Marfil for their careful critical comments and suggestions, which substantially improved the manuscript.

REFERENCES CITED

Abbate, E., and Bortolotti, V., 1961, Tentativo di interpretazione dei livelli di Argille scagliose intercalati nella parte alta del Macigno lungo l'allineamento M. Prado-Chianti (Appennino settentrionale) mediante colate sottomarine: Bollettino Società Geologica Italiana, v. 80, p. 335–342.

Abbate, E., and Bruni, P., 1987, Modino-Cervarola o Modino e Cervarola? Torbiditi Oligo-Mioceniche ed evoluzione del margine nord-Appenninico: Memorie Società Geologica Italiana, v. 39, p. 19–33.

Argnani, A., and Ricci Lucchi, F., 2001, Tertiary silicoclastic turbidite systems of the northern Apennines, in Vai, G.B., and Martini, I.P., eds., Anatomy of an Orogen: The Apennines and Adjacent Mediterranean Basins: Dordrecht NL, UK, Kluwer Academic Publishers, p. 327–350.

Aruta, G., Bruni, P., Cipriani, N., and Pandeli, E., 1998, The silicoclastic turbidite sequences of the Tuscan domain in the Val di Chiana–Val Tiberina area (eastern Tuscany and north-western Umbria): Memorie Società Geologica Italiana, v. 52, p. 579–593.

Aruta, G., Bruni, P., Buccianti, A., Cecchi, M., Cipriani, N., Monti, L., Nebbiai, M., Papini, M., Pandeli, E., and Reale, V., 2005, Integrated stratigraphic, petrographic and statistical data as a tool for mapping perisutural siliciclastic turbidite successions, in Pasquarè G., Venturelli M., and Groppelli G., eds., Geological Mapping in Italy: Rome, Agenzia per la Protezione dell'Ambiente e per I servizi tecnici (APAT), p. 208–214.

Azzaroli, A., 1955, L'Appennino Tosco-Emiliano dal Passo di Pradarena al Passo delle Forbici, e i nuclei Mesozoici di Corfino e di Soraggio: Bollettino Società Geologica Italiana, v. 74, p. 1–72.

Baldacci, F., Elter, P., Giannini, E., Giglia, G., Lazzarotto, A., Nardi, R., and Tongiorgi, M., 1967, Nuove osservazioni sul problema della Falda Toscana e sulla interpretazione dei flysch arenacei tipo "Macigno" dell'Appennino settentrionale: Memorie Società Geologica Italiana, v. 6, p. 213–244.

Barelli, G., and Mantovani Uguzzoni, M.P., 1985, Le microfacies dei ciottoli nelle brecce intercalate nelle formazioni della Scaglia Toscana e del Macigno nei pressi di Sassorosso (Prov. Lucca): Atti Società Natturali, Matematici di Modena, v. 115, p. 125–146.

Bettelli, G., Bonazzi, U., Fazzini, P., and Gelmini, R., 1987, Macigno, Arenarie di Modino e Arenarie di Monte Cervarola del Crinale Appenninico Emiliano: Memorie Società Geologica Italiana, v. 39, p. 1–17.

Boccaletti, M., Coli, M., Decandia, F., Giannini, E., and Lazzarotto, A., 1980, Evoluzione dell'Appennino settentrionale secondo un nuovo modello strutturale: Memorie Società Geologica Italiana, v. 21, p. 359–374.

Boccaletti, M., Calamita, F., Deiana, G., Gelati, R., Massari, F., Moratti, G., and Ricci Lucchi, F., 1990, Migrating foredeep-thrust belt system in the northern Apennines and southern Alps: Palaeogeography, Palaeoclimatology, Palaeoecology, v. 77, p. 3–14, doi: 10.1016/0031-0182(90)90095-O.

Bortolotti, V., and Pirini, C., 1965, Nota preliminare sull'età della base del Macigno: Bollettino Società Geologica It.aliana, v. 84, p. 29–36.

Bortolotti, V., Passerini, P., Sagri, M., and Sestini, G., 1970, The miogeosynclinal sequences, in Sestini, G., ed., Development of the Northern Apennines Geosyncline: Sedimentary Geology, v. 4, p. 341–444, doi: 10.1016/0037-0738(70)90019-9.

Bouma, A.H., 1962, Sedimentology of some Flysch Deposits☐A graphic approach to facies interpretation: Amsterdam, NL, Elsevier, 168 p.

Bruni, P., and Pandeli, E., 1992, Il Macigno e le Arenarie di M. Modino tra l'Abetone e il Lago Santo, in 76ª Riunione estiva S.G.I. "L'Appennino settentrionale" e Convegno S.I.M.P. "Minerogenesi Appenninica"–"Guida alla Traversata dell'Appennino settentrionale": Descrizione dell' Escursione del IV Giorno: Firenze, Centro Duplicazione Offset s.r.l., p. 187–196.

Bruni, P., Cipriani, N., and Pandeli, E., 1994a, Sedimentological and petrographical features of the Macigno and the Monte Modino Sandstone in the Abetone area (northern Apennines): Memorie Società Geologica Italiana, v. 48, p. 331–341.

Bruni, P., Cipriani, N., and Pandeli, E., 1994b, New sedimentological and petrographical data on the Oligocene-Miocene turbidite formations of the Tuscan domain: Memorie Società Geologica Italiana, v. 48, p. 251–260.

Catanzariti, R., Rio, D., Chicchi, S., and Plesi, G., 1991, Età e biostratigrafia a nannofossili calcarei nelle Arenarie di M. Modino e del Macigno nell'Alto Appennino Reggiano-Modenese: Memorie Descrittive Carta Geologica d'Italia, v. 46, p. 187.

Chicchi, S., and Plesi, G., 1991, Il Complesso di M. Modino–M. Cervarola nell'alto Appennino Emiliano (tra il Passo di Lagastrello e il M. Cimone) e i suoi rapporti con la Falda toscana, l'Unità di Canetolo e le Liguridi: Memorie Descrittive Carta Geologica d'Italia, v. 46, p. 139–163.

Chiocchini, U., and Cipriani, N., 1992, Provenance and evolution of Miocene turbidite sedimentation in the central Apennines, Italy: Sedimentary Geology, v. 77, p. 185–195, doi: 10.1016/0037-0738(92)90125-B.

Cipriani, C., 1958, Ricerche sui minerali costituenti le arenarie: Atti Società Toscana Scienze Naturali, Serie A, v. 65, p. 165–220.

Cipriani, C., and Malesani, P.G., 1964, Ricerche sulle arenarie 9. Caratterizzazione e distribuzione geografica delle arenarie Appenniniche Oligo-Mioceniche: Memorie Società Geologica Italiana, v. 4, p. 339–374.

Cipriani, C., and Malesani, P.G., 1972, Composizione mineralogica della frazione politica delle formazioni del Macigno e Marnoso Arenacea (Appennino settentrionale): Memorie Istituto di Geologia e Mineralogia dell'Università di Parma, v. 29, p. 1–24.

Cipriani, N., Nebbiai, M., and Nirta, G., 2004, An integrate software hardware system for counting as a useful modal analysis tool: 32nd International Geological Congress, Italy Abstract volume.

Conti, S., and Gelmini, R., 1994, Miocene-Pliocene tectonic phases and migration of foredeep-thrust belt system in the northern Apennines: Memorie Società Geologica Italiana, v. 48, p. 261–274.

Curcio, M., Pranzini, G., and Sestini, G., 1968, Stratimetria di una serie di Macigno (Oligocene, Appennino Tosco-Emiliano): Bollettino Società Geologica Italiana, v. 87, p. 623–641.

Dallan Nardi, L., and Nardi, R., 1972, Schema stratigrafico e strutturale dell'Appennino settentrionale: Memorie Accademia Lunigianese di Scienze "G Capellini", v. 42, p. 1–212.

Dallan Nardi, L., and Nardi, R., 1978, Il quadro paleotettonico dell'Appennino settentrionale: Un'ipotesi alternativa: Atti Società Toscana di Scienze Naturali, Memorie, Series A, v. 85, p. 289–297.

De Libero, C.M., 1998, Sedimentary vs. tectonic deformation in the "Argille Scagliose" of Mt. Modino (northern Apennines): Giornale Geologia, v. 60, p. 143–166.

Dickinson, W.R., 1970, Interpreting detrital modes of greywacke and arkose: Journal Sedimentary Petrography, v. 40, p. 695–707.

Di Giulio, A., 1999, Mass transfer from the Alps to the Apennines: Volumetric constraints in the provenance study of the Macigno-Modino source basin system, Chattian-Aquitanian, north-western Italy: Sedimentary Geology, v. 124, p. 69–80, doi: 10.1016/S0037-0738(98)00121-3.

Di Giulio, A., and Valloni, R., 1992, Sabbie e areniti, analisi ottica e classificazone: Analisi microscopica delle areniti terrigene: Parametri metrologici e composizionali modali: Acta Naturalia Ateneo Parmense, v. 28, p. 55–101.

Fazzuoli, M., Ferrini, G., Pandeli, E., and Sguazzoni, G., 1985, Le formazioni Giurassico-Mioceniche della Falda Toscana a Nord dell'Arno: Considerazioni sull'evoluzione sedimentaria: Memorie Società Geologica Italiana, v. 30, p. 159–201.

Fazzuoli, M., Pandeli, E., and Sani, F., 1994, Considerations on the sedimentary and structural evolution of the Tuscan domain since Early Liassic to Tortonian: Memorie Società Geologica Italiana, v. 48, p. 31–50.

Folk, R.L., 1968, Petrology of Sedimentary Rocks: Austin, Hemphill, 159 p.

Fornaciari, E., and Rio, D., 1996, Latest Oligocene to early-middle Miocene quantitative calcareous nannofossil biostratigraphy in the Mediterranean region: Micropaleontology, v. 42, p. 1–36, doi: 10.2307/1485981.

Gandolfi, G., and Paganelli, L., 1993, Le torbiditi arenacee Oligo-Mioceniche dell'Appennino settentrionale fra La Spezia ed Arezzo—Studio petrografico ed implicazioni paleogeografiche: Giornale di Geologia, v. 55, p. 93–102.

Giannini, E., Nardi, R., and Tongiorgi, M., 1962, Osservazioni sul problema della Falda Toscana: Bollettino Società Geologica Italiana, v. 81, p. 17–98.

Gunter, K., and Rentz, K., 1968, Contributo alla geologia della catena principale dell'Appennino Tosco-Emiliano tra Ligonchio, Civago e Corfino: L'Ateneo Parmense, Acta Naturalia, v. 4, p. 67–87.

Kuenen, Ph.H., and Migliorini, C.I., 1950, Turbidity currents as a course of graded bedding: The Journal of Geology, v. 58, p. 91–127.

Martini, G., and Plesi, G., 1988, Scaglie tettoniche divelte dal Complesso di Monte Modino e trascinate alla base delle Unità subligure e Ligure—Gli esempi del M. Ventasso e del M. Cisa (Appennino Reggiano): Bollettino Società Geologica Italiana, v. 107, p. 171–191.

Martini, P.I., and Sagri, M., 1977, Sedimentary filling of ancient deep-sea channel: Two examples from northern Apennines (Italy): Journal of Sedimentary Petrology, v. 47, p. 1542–1553.

Merla, G., 1951, Geologia dell'Appennino settentrionale: Bollettino Società Geologica Italiana, v. 70, p. 95–382.

Merla, G., and Abbate, E., 1969, Note illustrative della carta geologica d'Italia: Foglio 97, San Marcello: Roma, Servizio Geologico d'Italia, 54 p.

Migliorini, C.I., 1943, Sul modo di formazione di complessi tipo Macigno: Bollettino Società Geologica Italiana, v. 62, p. 48–49.

Migliorini, C.I., 1950, Dati a conferma della risedimentazione delle arenarie Macigno: Memorie Società Toscana Scienze Naturali, v. 57, p. 82–94.

Mochi, E., Plesi, G., and Villa, G., 1995, Biostratigrafia a nannofossili calcarei della parte basale della successione del M. Modino (nell' area dei Fogli 234 e235) ed evoluzione strutturale dell' unità omonima: Studi Geologici Camerti, v. 13, p. 39–73.

Montanari, L., and Rossi, M., 1983, Evoluzione delle unità stratigrafico-strutturali del Nord Appennino, 2—Macigno s.s. e Pseudomacigno: Nuovi dati cronostratigrafici e loro implicazioni: Memorie Società Geolocica Italiana, v. 25, p. 185–217.

Moore, D.M., and Reynolds, R.C., Jr., 1989, X-Ray Diffraction on the Identification and Analysis of Clay Minerals: Oxford, Oxford University Press, 332 p.

Mutti, E., and Ricci Lucchi, F., 1972, Le torbiditi dell'Appennino settentrionale: Introduzione all'analisi di facies: Memorie Società Geologica Italiana, v. 11, p. 161–199.

Mutti, E., and Ricci Lucchi, F., 1975, Turbidite facies and facies association: Examples of turbidite facies associations from selected formations of the northern Apennines: Nice, IX International Congress Sedimentology, Field Trip Guide A 11, p. 21–36.

Nardi, R., 1964, Contributo alla geologia dell'Appennino Tosco-Emiliano, 4. La geologia della valle dello Scoltenna tra Pievepelago e Montecreto (Appennino modenese): Bollettino Società Geologica Italiana, v. 83, p. 353–400.

Nardi, R., 1965, Schema geologico dell'Appennino Tosco-Emiliano tra il M. Cusna ed il M. Cimone e considerazioni sulle unità tettoniche dell'Appennino: Bollettino Società Geologica Italiana, v. 84, p. 35–91.

Nardi, R., and Tongiorgi, M., 1962, Contributo alla geologia dell'Appennino Tosco-Emiliano. 1. Stratigrafia e tettonica dei dintorni di Pievepalago (Appennino modenese): Bollettino Società Geologica Italiana, v. 81, p. 1–76.

Pandeli, E., Ferrini, G., and Lazzari, D., 1994, Lithofacies and petrography of the Macigno Formation from the Abetone to the Monti del Chianti areas (northern Apennines): Memorie Società Geologica Italiana, v. 48, p. 321–329.

Perilli, N., 1994, The Mt. Modino Olistostrome Auctorum (Appennino Modenese)—Stratigraphical and sedimentological analysis: Memorie Società Geologica Italiana, v. 48, p. 343–350.

Plesi, G., 1975, La giacitura del Complesso Bratica-Petrignacola nella Serie del Rio di Roccaferrara (Val Parma) e dei flysch arenacei tipo Cervarola dell'Appennino settentrionale: Bollettino Società Geologica Italiana, v. 94, p. 157–176.

Reutter, K.J., 1969, La geologia dell'alto Appennino modenese tra Civago e Fanano e considerazioni geotettoniche sulla Unità di M. Modino–M. Cervarola: Ateneo Permense: Acta Naturalia, v. 5, p. 3–86.

Reutter, K.J., and Groscurth, J., 1978, The pile of nappes in the northern Apennines, its unravelment and emplacement, in Closs H., Roeder D., and Schmidt K., eds., Alps, Apennines, Hellenides: Inter-Union Commission on Geodynamics, Scientific Report, v. 38, p. 234–243.

Ricci Lucchi, F., 1975, Miocene paleogeography and basin analysis in the Periadriatic Apennines, in Squyres, C.H., ed., Geology of Italy: Petroleum Exploration Society of Libya, Tripoli, v. 2, p. 129–236.

Ricci Lucchi, F., 1986, The Oligocene to Recent foreland basins of the northern Apennines, in Allen, P.A., and Homewood, P., eds., Foreland Basins: International Association of Sedimentologists Special Publication 8, p. 105–139.

Sagri, M., 1975, Ambienti di deposizione e meccanismi di sedimentazione nella successione Macigno-olistostroma-arenarie di M. Modino (Appennino modenese): Bollettino Società Geologica Italiana, v. 94, p. 771–788.

Sestini, G., 1968, Notes on the internal structure of the major Macigno Olistostrome (Oligocene, Modena and Tuscan Apennines): Bollettino Società Geologica Italiana, v. 87, p. 51–64.

Sestini, G., 1970, Flysch facies and turbidite sedimentology, in Sestini, G., ed., Development of the Northern Apennines Geosyncline: Sedimentary Geology, v. 4, p. 559–597, doi: 10.1016/0037-0738(70)90023-0.

Sestini, G., Bruni, P., and Sagri, M., 1986, The flysch basins of the northern Apennines: A review of facies and of Cretaceous-Neogene evolution: Memorie Società Geologica Italiana, v. 31, p. 87–106.

Valloni, R., 1978, Provenienza e storia post-deposizionale del Macigno di Pontremoli (Massa): Bollettino Società Geologica Italiana, v. 97, p. 317–326.

Valloni, R., and Zuffa, G.G., 1984, Provenance changes for arenaceous formations of the northern Apennines, Italy: Geological Society of America Bulletin, v. 95, p. 1035–1039, doi: 10.1130/0016-7606(1984)95<1035: PCFAFO>2.0.CO;2.

Valloni, R., Belfiore, A., Calzetti, L., Calzolari, M.A., Donagemma, V., Lazzari, D., and Pandeli, E., 1992, Evoluzione delle petrofacies arenacee nell'Oligo-Miocene d'avanfossa del Nord-Appennino, in 76ª Riunione estiva della Società Geologica Italiana: Firenze, 21–23 settembre 1992, Riassunti, p. 110–112.

Zanzucchi, G., 1988, Ipotesi sulla posizione paleogeografica delle "Liguridi esterne" Cretacico-Eoceniche, nell'Appennino settentrionale: Atti Ticinesi Scienze della Terra, v. 31, p. 327–339.

Zuffa, G.G., 1980, Hybrid arenites: Their composition and classification: Journal Sedimentary Petrology, v. 50, p. 21–29.

MANUSCRIPT ACCEPTED BY THE SOCIETY 9 AUGUST 2006

Interpreting siliciclastic-carbonate detrital modes in foreland basin systems: An example from Upper Miocene arenites of the central Apennines, Italy

Salvatore Critelli[†]
Emilia Le Pera[‡]
Dipartimento di Scienze della Terra, Università della Calabria, 87036 Arcavacata di Rende (CS), Italy

Fabrizio Galluzzo[§]
Agenzia per la Protezione dell'Ambiente e per i Servizi Tecnici (APAT), Via Vitaliano Brancati 48, 00144 Roma, Italy

Salvatore Milli[#]
Dipartimento di Scienze della Terra, Università di Roma "La Sapienza," Piazzale Aldo Moro, 5, 00185 Roma, Italy and *Istituto di Geologia Ambientale e Geoingegneria, Consiglio Nazionale delle Ricerche (CNR), Via Bolognola, 7, 00138 Roma, Italy*

Massimiliano Moscatelli[††]
Istituto di Geologia Ambientale e Geoingegneria, Consiglio Nazionale delle Ricerche (CNR), Via Bolognola, 7, 00138 Roma, Italy

Sonia Perrotta[‡‡]
Istituto di Geologia, Università di Urbino, località Crocicchia, 61029 Urbino, Italy, and Dipartimento di Scienze della Terra, Università della Calabria, 87036 Arcavacata di Rende (CS), Italy

Massimo Santantonio[§§]
Dipartimento di Scienze della Terra, Università di Roma "La Sapienza," Piazzale Aldo Moro, 5, 00185 Roma, Italy

ABSTRACT

In the central Apennines, interacting siliciclastic and carbonate marine clastic wedges filled the foreland basin system during the late Miocene. Conjunction of collisional thrust tectonics and prethrusting normal faults generated a complex foredeep with intrabasinal structural highs that represented additional source areas to the basin.

[†]E-mail: critelli@unical.it.
[‡]E-mail: emilia.lepera@unical.it.
[§]E-mail: fabrizio.galluzzo@apat.it.
[#]E-mail: salvatore.milli@uniroma1.it.
[††]E-mail: massimiliano.moscatelli@igag.cnr.it.
[‡‡]E-mail: soniaperrotta@yahoo.it.
[§§]E-mail: massimo.santantonio@uniroma1.it.

Critelli, S., Le Pera, E., Galluzzo, F., Milli, S., Moscatelli, M., Perrotta, S., and Santantonio, M., 2007, Interpreting siliciclastic-carbonate detrital modes in foreland basin systems: An example from Upper Miocene arenites of the central Apennines, Italy, *in* Arribas, J., Critelli, S., and Johnsson, M.J., eds., Sedimentary Provenance and Petrogenesis: Perspectives from Petrography and Geochemistry: Geological Society of America Special Paper 420, p. 107–133, doi: 10.1130/2006.2420(08). For permission to copy, contact editing@geosociety.org. ©2007 Geological Society of America. All rights reserved.

Detrital modes of the late Miocene central Apennines orogenic system range in composition from intrabasinal carbonate to quartzofeldspatholithic and calclithite arenites. The external zone of the foredeep is characterized by hemipelagic deposits, called the Orbulina Marl. Their arenite beds are composed by intrabasinal carbonate, with dominant bioclasts and minor intraclasts, and glauconite derived from an active shallow-marine carbonate source. These hemipelagic deposits are partly coeval with and partly overlain by siliciclastic turbidites of the Frosinone and the Argilloso-Arenacea Formations, and they represent deposition within local foredeep depocenters. Siliciclastic turbidite sandstones are quartzofeldspatholithic, which documents provenances from metamorphic, plutonic, ophiolitic, and sedimentary rocks. Carbonate intrabasinal structural highs were the main source for carbonate breccias, intrabasinal arenites, and calclithites of the Brecce della Renga Formation, the deposits of which are locally interbedded with the coeval siliciclastic turbidite sandstones.

Evolution of late Miocene sandstone detrital modes reflected the changing nature of the central Apennines thrust belt through time and the complex architecture of the foreland basin system; it records the history of accretion, deformation of the foredeep, and progressive areal reduction of carbonate-producing areas along with the sedimentary and structural evolution of local intrabasinal highs.

Keywords: arenites, provenance, carbonate-siliciclastic mixtures, foreland basin system, central Apennines, Italy.

INTRODUCTION

The composition of arenites is primarily controlled by source lithology, tectonics, climate, topography, weathering, transport, and depositional environment, and each of these factors are not always easily interpreted from optical analysis of sand particles (e.g., Johnsson, 1993). Quantitative analysis of arenites is mostly a descriptive approach, using mineralogical, textural, and paleontological characteristics, and only in a few cases are the genetic aspects of sand particles taken into account for detailed paleogeographic reconstructions of source or basin systems (e.g., Zuffa, 1985, 1987). More refined compositional studies for arenites have focused on detailed temporal and spatial deciphering of specific clastic populations, such as carbonate, volcanic, and other intrabasinal grains (e.g., Zuffa, 1980, 1985, 1987, 1991; Mount, 1985; Fontana et al., 1989; Garzanti, 1991; Amorosi, 1995; Critelli and Ingersoll, 1995). Mixing of carbonate and siliciclastic sediments is common in the stratigraphic record and has been documented as distinct interbedded strata or as mixtures in the same bed. Although the literature about mixtures of carbonate and terrigenous clastic sequences is vast (e.g., Doyle and Roberts, 1988; Dolan, 1989; Budd and Harris, 1990; Mount, 1984, 1985), detailed information on carbonate particles in arenites is found in only a few papers (e.g., Zuffa, 1985, 1987, 1991), in which carbonate particles are classified based on their age (coeval versus noncoeval) and location (extrabasinal versus intrabasinal) of source areas. This spatial/temporal approach in deciphering sand grains in arenites has been widely used to detail the basinal dispersal pathways in different geotectonic settings, wherever mixed silicate and carbonate terranes act as the major source rocks, from rifted-continental margins to collisional orogens (e.g., Zuffa, 1987, and references therein). The Alpine-Himalayan orogenic belt is a good case example in this respect (e.g., Gandolfi et al., 1983; Fontana et al., 1989; Fontana, 1991; Critelli and Le Pera, 1995, 1998; Garzanti et al., 1996; Critelli, 1999; Cibin et al., 2001). The terranes of the circum-Mediterranean and Alpine orogeny record a complete tectonostratigraphic history from early Mesozoic Neotethyan rifting to Tertiary collision, in which the precollisional submarine physiography of the Apulia-Adria and European-Iberian continental margins was very complex (e.g., Bernoulli and Jenkyns, 1974; Santantonio, 1993, 1994; Bosellini, 2004). Following continental collision between the European-Iberian and African-Adria plates, mixed siliciclastic and carbonate shallow- to deep-marine clastic wedges filled the foreland basin systems of the Apenninic orogenic belt. The interference of collisional thrust tectonics with prethrusting normal faults generated a complex physiography across some of these foreland basin systems, in which the foredeep basin floor was fragmented into several intrabasinal structural highs, which acted as additional sediment sources to the basin. This is the case with the late Miocene central Apennines foreland basin system (Bigi et al., 2003).

The focus of this paper is on interpretation of siliciclastic-carbonate arenite detrital modes within a portion of the late Miocene foreland basin system in the central Apennines. These upper Miocene relatively deep-marine strata illustrate the distinction between extrabasinal carbonate and silicate detritus from the fold-and-thrust belt, and intrabasinal coeval and noncoeval carbonate detritus from productive shallow-water shelves and isolated platforms and from submarine carbonate structural highs undergoing erosion. This study utilized petrological methods to discriminate extrabasinal carbonate and noncoeval intrabasinal carbonate detritus, and to attempt to decipher the sedimentary balance from diverse consanguineous sources in the specific geotectonic setting of the foreland basin system.

REGIONAL GEOLOGICAL SETTING

The central Apennines are a portion of the Apenninic orogenic belt (Fig. 1), which represents one segment of the circum-Mediterranean orogens. The central Apennines fold-and-thrust belt developed from the Miocene to early Pleistocene during convergence and subsequent collision between the European-Iberian plate and the westward-subducted Adria and Ionian oceanic plates (e.g., Parotto and Praturlon, 2004). The accretionary processes related to continental collision were responsible for deformation and tectonic transport of three main paleogeographic domains: a more internal domain with crustal rocks of the European-Iberian domain, characterized by Paleozoic plutonic-metamorphic rocks and a Mesozoic to Tertiary sedimentary cover; the oceanic domain (Ligure-Piedmont Ocean), characterized by a Middle-Upper Jurassic to

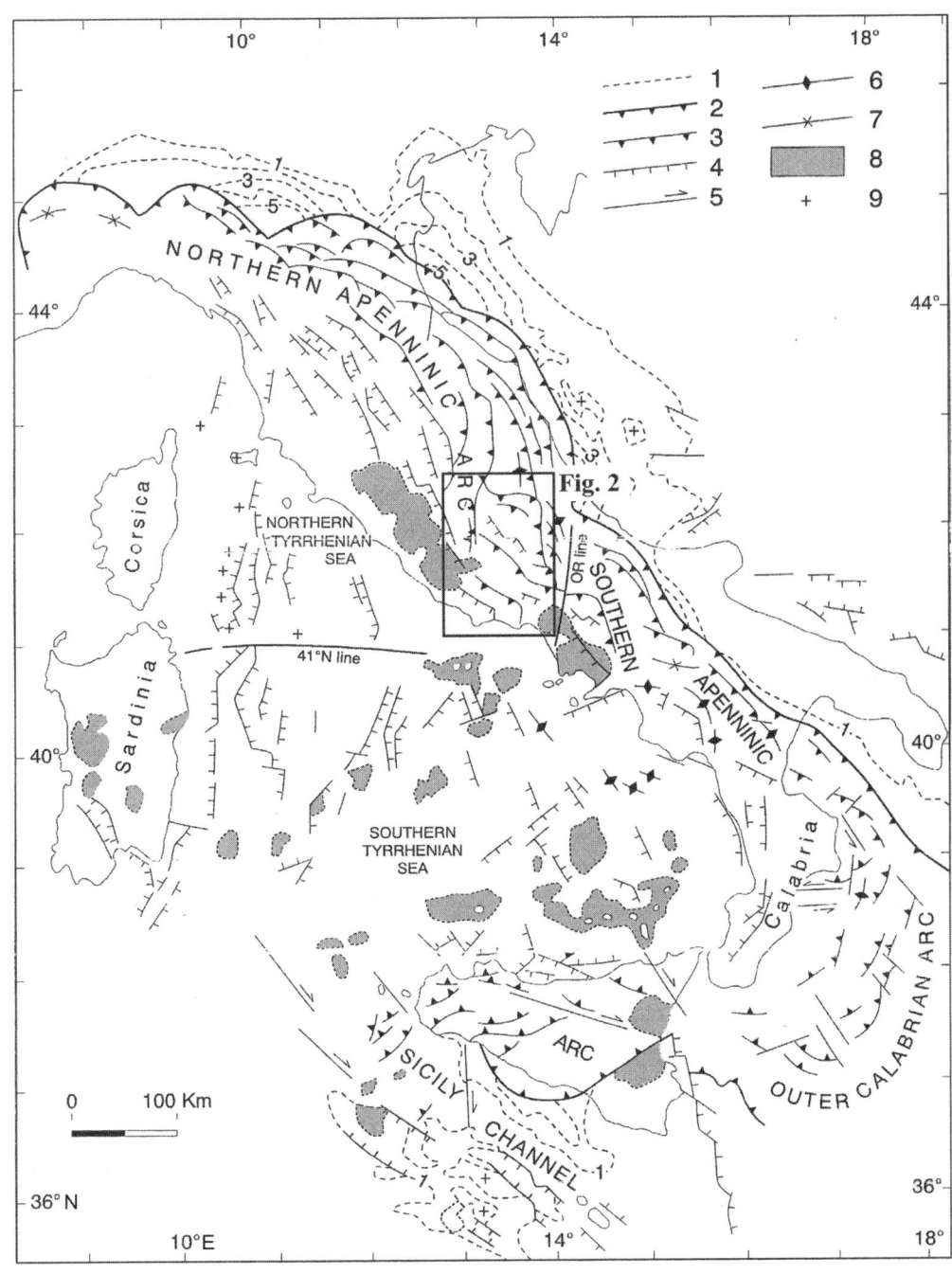

Figure 1. Structural map of Apennines and Tyrrhenian Sea. Legend: (1) base of Pliocene-Quaternary isobaths (in km); (2) front of thrust belt; (3) major post-Tortonian thrusts; (4) normal faults; (5) strike-slip faults; (6) antiforms; (7) synforms; (8) volcanoes; (9) intrusive bodies. OR line—Ortona-Roccamonfina line. Square indicates central Apennines area shown in Figure 2. (Modified after Patacca et al., 1993.)

Oligocene sedimentary succession resting on oceanic basement; and an external domain, characterized by a Triassic to Lower Miocene sedimentary succession deposited on the Adria-Apulia-Africa passive margin (Critelli, 1999; Elter et al., 2003; Parotto and Praturlon, 2004; and references therein). Several authors consider the Apennine chain as an accretionary prism that has a complex arcuate geometry that developed through the progressive eastward migration of the compressive front (toward the Adria-Apulia-Africa foreland) (e.g., Malinverno and Ryan, 1986; Patacca et al., 1990; Sartori, 2003; and references therein). Two coeval and related processes are considered to be responsible for development of the accretionary prism: (1) roll-back of the Adria plate and (2) opening of the Tyrrhenian backarc basin beginning in the late Miocene (e.g., Malinverno and Ryan, 1986; Patacca et al., 1990; Sartori, 2003; and references therein). The eastward migration of the Apennine chain resulted in the development of a complex foreland basin system since the early Miocene, essentially filled with thick successions of siliciclastic turbidites.

The central Apennines (Fig. 2) represent a key portion of the external domain, which was characterized by two main Triassic-Miocene paleogeographic sectors: (1) the Latium-Abruzzi carbonate platform to the south (stratal thickness of 5–6 km), and (2) the Umbria-Marche and Abruzzi-Sabina pelagic carbonate basin to the north and west (stratal thickness of 3–4 km). Both overlie a Permian-Triassic continental to shallow-marine succession that has a maximum thickness of 3 km (Parotto et al., 2004).

Two late Miocene foredeep clastic wedges, which onlap the southwestward tilted foreland ramp of the Adria plate, have been recognized: (1) the Upper Tortonian–Lower Messinian turbidite deposits on the Orbulina Marl; and (2) the Upper Messinian–Holocene turbidite deposits on the Gessoso-Solfifera Formation. Integration of structural, stratigraphic, and sedimentological data provides evidence for a complex foredeep geometry characterized by subbasins and intrabasin ridges. These elements controlled the structural evolution of the chain and the nature and stratigraphic architecture of the turbidite depositional systems (e.g., Milli and Moscatelli, 2000; Moscatelli 2003; Bigi et al., 2003, 2004; Milli et al., 2004). This interpretation is in contrast with that proposed by other authors (e.g., Cipollari and Cosentino, 1995; Cipollari et al., 1995), who consider that the Upper Miocene turbidite units were deposited in different foredeeps.

Thrust fronts involving the carbonate platform seem to be controlled by the prethrusting foreland deformation that occurred in an extensional regime during subduction-related lithospheric flexuring. Normal faults dissected the foreland during Tortonian-Messinian time, and they are presently best exposed along the margins of the deepest structural basins. The prethrusting normal faults, recently recognized in different sectors of the central Apennines, such as the Simbruini Mountains, Salto Valley, Gran Sasso, and Maiella (e.g., Compagnoni et al., 1991, 2005; Scisciani et al., 2001, 2002; Bigi and Costa Pisani, 2002), localize the positions of thrust propagation and depocenters at the front of the structures or are passively translated on the back side of the main thrust fronts.

STRATIGRAPHIC AND DEPOSITIONAL SETTING

In the studied area, the Upper Miocene Orbulina Marl (Serravallian?–Tortonian to Lower Messinian) is partly coeval with and partly overlain by siliciclastic turbidite deposits, the Frosinone (Upper Tortonian) and the Argilloso-Arenacea (Lower Messinian) Formations. Carbonate intrabasin ridges constituted an important morphological feature that supplied carbonate clastic coeval and noncoeval sediments to the basin (Brecce della Renga Formation) (Figs. 3, 4, 5, and 6).

Hemipelagic Deposits (Orbulina Marl)

The Orbulina Marl, which consists dominantly of marls and argillaceous marls with planktonic foraminifera, drapes the Miocene "temperate to subtropical-tropical–type" carbonate succession (Briozoi and Litotamni Limestones Formation) (e.g., Corda and Brandano, 2003, and references therein) of the Laziale-Abruzzese domain (Figs. 3 and 4). Three lithofacies of variable thickness have been recognized in these strata (Compagnoni et al., 1991, 2005). Lithofacies 1 consists of marls and marly limestone; it is separated from the underlying Briozoi and Litotamni Limestones Formation by a thin phosphatic hardground (e.g., Bergomi and Damiani 1976; Corda 1990; Brandano, 2001, 2002) that records rapid drowning of the Latium-Abruzzi Miocene carbonate platform. Lithofacies 2 consists of marls and clays that also bear intercalations of bioclastic calcarenites (Galluzzo and Santantonio, 1995) (Fig. 5A), which have a *Cylindrites* and *Zoophycos* ichnoassociation (Bellotti et al., 1984). Lithofacies 3 consists of alternating marls and clays with thin, very fine turbidite sandstone beds (Fig. 5B). Some authors (see Cosentino et al., 1997; Milli and Moscatelli, 2000) have moved lithofacies 3 into the basal Argilloso-Arenacea Formation. A genetic link of lithofacies 3 with a coeval siliciclastic turbidite system was suggested by Compagnoni et al. (1991), also based on the discovery of a stratigraphic hiatus at its base. Compagnoni et al. (1991) also mentioned subtle internal unconformities and wedging out of individual beds within this subunit. Their option to hold it as a part of the Orbulina Marl was essentially dictated by practical geological mapping purposes.

Pampaloni et al. (1994) indicated a Serravallian to early Messinian age for the Orbulina Marl, whereas Cipollari et al. (1993) and Cosentino et al. (1997) suggested a late Tortonian to early Messinian age, which is also supported by recent radiometric dating reported in Mariotti et al. (1999) and Brandano (2001, 2002). The uncertainties related to recognition of the Serravallian-Tortonian boundary through joint planktonic foraminifera and calcareous nannofossil stratigraphy are addressed in the paper by Compagnoni et al. (1991).

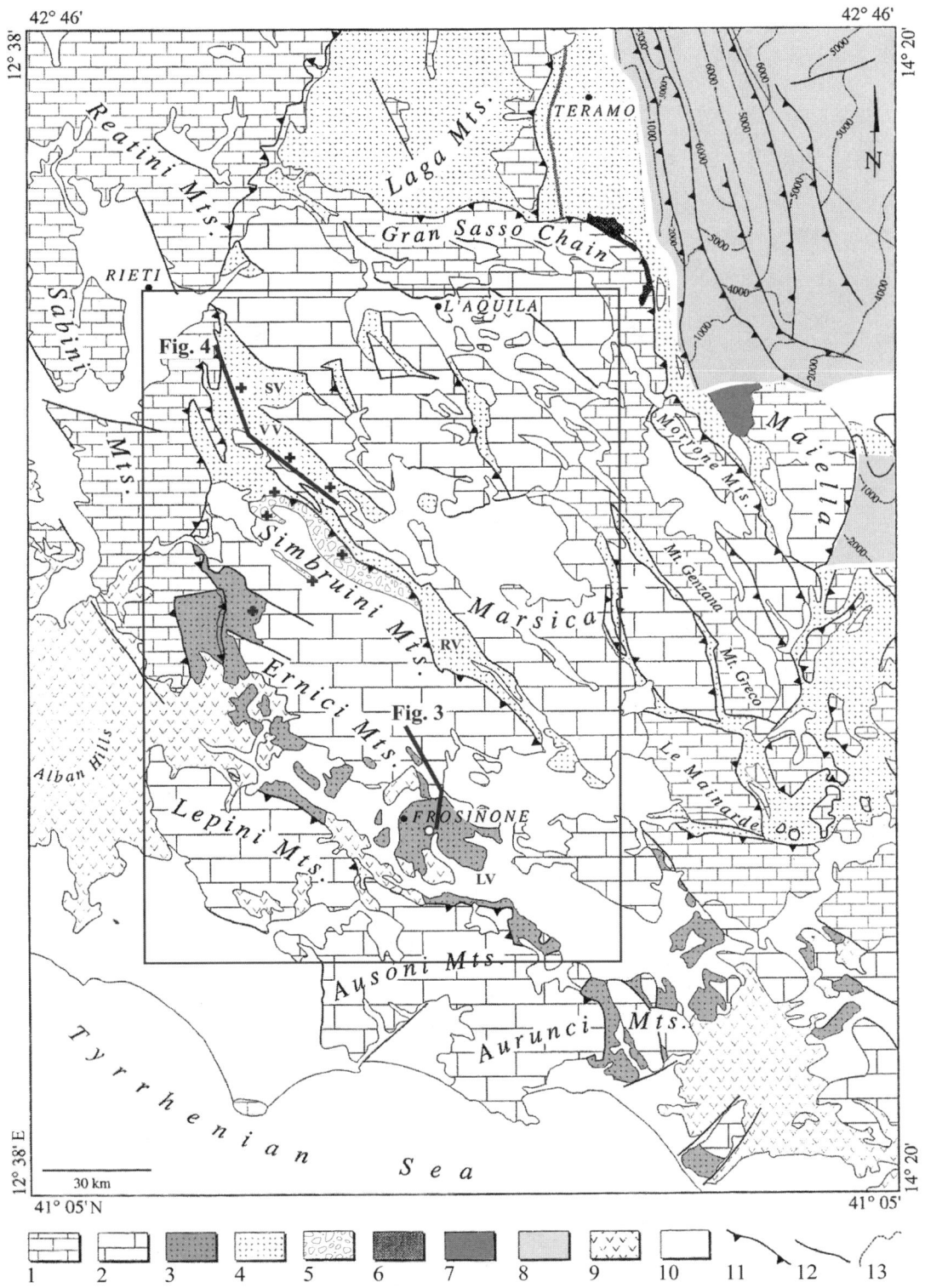

Figure 2. Structural map of central Apennines. Legend: (1) Mesozoic-Cenozoic slope and basin deposits; (2) Mesozoic-Cenozoic platform limestones; (3) Upper Tortonian siliciclastic turbidites (Frosinone Formation); (4) Lower Messinian–Lower Pliocene siliciclastic turbidites (Lower Messinian Argilloso-Arenacea Formation in studied area); (5) Tortonian–Lower Messinian carbonate clastic deposits of Brecce della Renga Formation; (6) clastic deposits of Messinian salinity crisis; (7) clastic deposits of Messinian–Lower Pliocene thrust-top basins; (8) buried Pliocene marine sediments; (9) Middle-Upper Pleistocene volcanics; (10) Pliocene-Pleistocene and Holocene marine and continental deposits; (11) thrusts; (12) undifferentiated faults; (13) isobath (in m) of the base of Pliocene strata. SV—Salto Valley; VV—Varri Valley; RV—Roveto Valley; LV—Latina Valley. Square indicates studied area. Crosses indicate sampling points and circle shows location of ENI-AGIP (Ente Nazionale Idrocarburi – Azienda-Generale Italiana Petroli) Frosinone 1 well utilized in correlation panel of Figure 3. (Modified after Cipollari et al., 1999.)

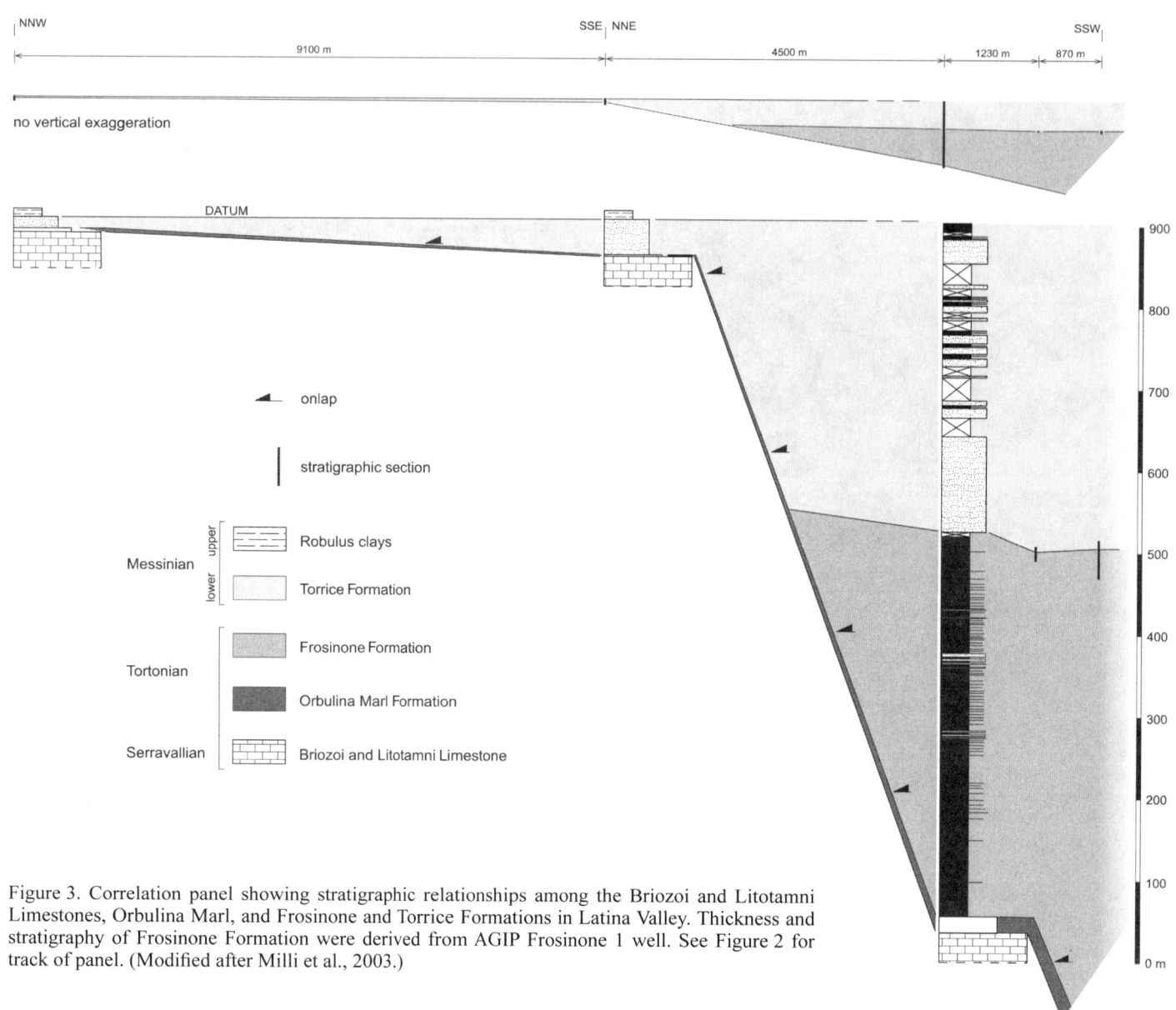

Figure 3. Correlation panel showing stratigraphic relationships among the Briozoi and Litotamni Limestones, Orbulina Marl, and Frosinone and Torrice Formations in Latina Valley. Thickness and stratigraphy of Frosinone Formation were derived from AGIP Frosinone 1 well. See Figure 2 for track of panel. (Modified after Milli et al., 2003.)

Carbonate Breccias and Calcarenites (Brecce della Renga Formation)

Carbonate breccias and calcarenites, ranging in age from the latest Serravallian–Tortonian to the early Messinian, crop out along the outer margin of the Simbruini Mountains (e.g., Compagnoni et al., 1990, 1991, 2005, and references therein) (Fig. 2) and cover an area in excess of 100 km². Recent studies (e.g., Compagnoni et al., 1990, 1991; Bigi et al., 2003) have distinguished lithofacies subdivisions within these deposits where breccias, calcarenites, and sandstones are present in variable proportions. These lithofacies generally constitute isolated outcrops, each containing rocks of different ages, so that no single stratigraphic section exists that covers the total time spanned by the whole unit. Facies and geometries indicate deposition through an array of gravity-driven mechanisms (Figs. 5C and 5D), including rockfall and grain flow for chaotic, locally megaclastic deposits resting unconformably on the carbonate bedrock, and turbidity currents for graded and laminated beds interbedded with hemipelagites or siliciclastic turbidites. Compagnoni et al. (1991) also reported the local occurrence of uppermost Serravallian–lowest Tortonian basin-margin breccias and calcarenites that onlaps the Cretaceous and Lower Miocene substrate and forms a submarine escarpment, most possibly the product of an underlying normal fault. In general, the architecture and composition of the various lithofacies of the Brecce della Renga document submarine erosion of an intrabasinal structural high, coupled with building of a carbonate apron (see Mullins and Cook, 1986) due to heterozoan-type carbonate production (e.g., James, 1997; Corda and Brandano, 2003), which must have persisted locally. Although a source area for such carbonate deposit must indeed have existed, no deposits

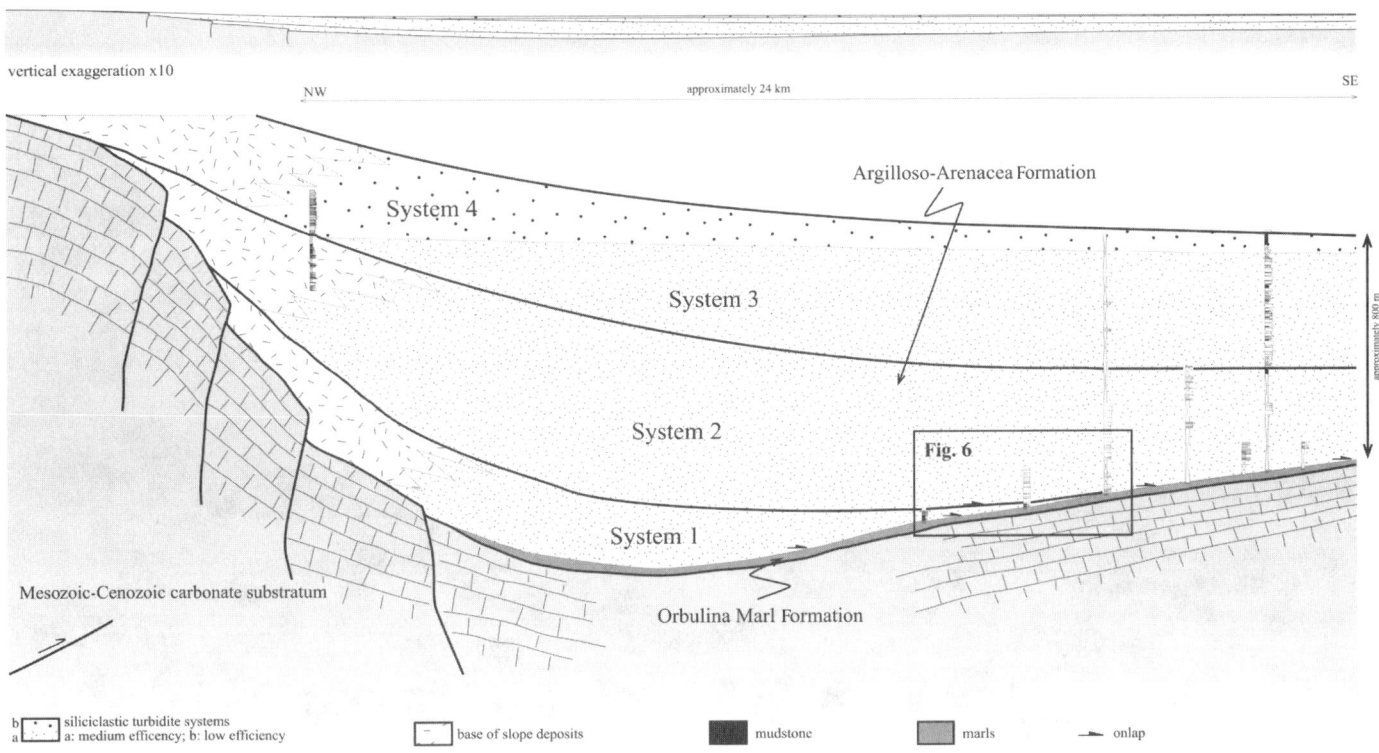

Figure 4. Correlation panel showing stratigraphic relationships among turbidite systems 1, 2, 3, and 4 developed inside the Argilloso-Arenacea Formation in the Salto Valley and Varri Valley. See Figure 2 for track of panel. (Modified after Moscatelli et al., 2004.)

documenting in situ benthic production in the late Miocene have so far been positively documented across the study area (Compagnoni et al., 2005). While this may be explained through postorogenic erosion and/or burial by thrust sheets, this is partly speculative, and room certainly exists for alternative interpretations.

It is conceivable that the youngest lithofacies, including the Messinian breccias and arenites interbedded with siliciclastic turbidites at the front of the Simbruini Mountains, record the switch to a contractional regime. The Tortonian to Messinian interval is a crucial one in the regional orogenic evolution, so it is probable that those clastic deposits, which we collectively call the Brecce della Renga Formation, were produced under changing tectonic conditions, and the footwall blocks of the extended foreland were later turned, along with their marginal clastic wedges, into hanging walls in the thrust system (Compagnoni, 2005).

Siliciclastic Turbidite Deposits (Frosinone and Argilloso-Arenacea Formations)

Siliciclastic deposits of the Frosinone (Upper Tortonian) and the Argilloso-Arenacea (Lower Messinian) formations represent the filling of local depocenters within the Upper Miocene Apenninic foreland basin system. The oldest unit is a turbidite succession, ~1200 m thick, that consists of thin beds of fine and very fine-grained sandstone alternating with muddy deposits (Fig. 5E). The apparent facies uniformity, coupled with the virtual absence of any discontinuity surfaces, does not allow subdivision of this formation, although the overall succession shows, from bottom to top, an initial thickening- and coarsening-upward trend that passes upward to a thinning- and fining-upward trend. The top of the Frosinone Formation is erosively truncated by the siliciclastic turbidite deposits of the Torrice Formation (Lower Messinian) (Fig. 3), which is a lithostratigraphic unit (not examined in this work) that shows compositional analogies with the Frosinone Formation.

Outcrops and well data record a clear, high-frequency cyclicity in the Frosinone Formation that is characterized by thinning- and fining-upward meter- to decameter-thick facies sequences. Generally, these sequences have a basal portion made of thick beds of fine seemingly massive sandstone (F8 facies in Mutti [1992] and Mutti et al. [1999]) or fine to very fine sandstone with traction-plus-fallout structures (F9 facies in Mutti [1992] and Mutti et al. [1999]), which pass upward to interbedded thin sandstone and mudstone beds (Fig. 5E). These strata are interpreted as having been deposited by a bipartite high- to low-density turbidity current, flowing from NW to SE, and supplied via flood-generated hyperpycnal flows. The Frosinone Formation onlaps the western margin of the Simbruini Mountains, a structural high of the Lazio-Abruzzi carbonate platform partially draped by the hemipelagic deposits of the Orbulina Marl (Fig. 3).

Recent studies on the Argilloso-Arenacea Formation (see Milli and Moscatelli, 2000, 2001; Bigi et al., 2003; Moscatelli, 2003; and references therein) permit subdivision of this unit

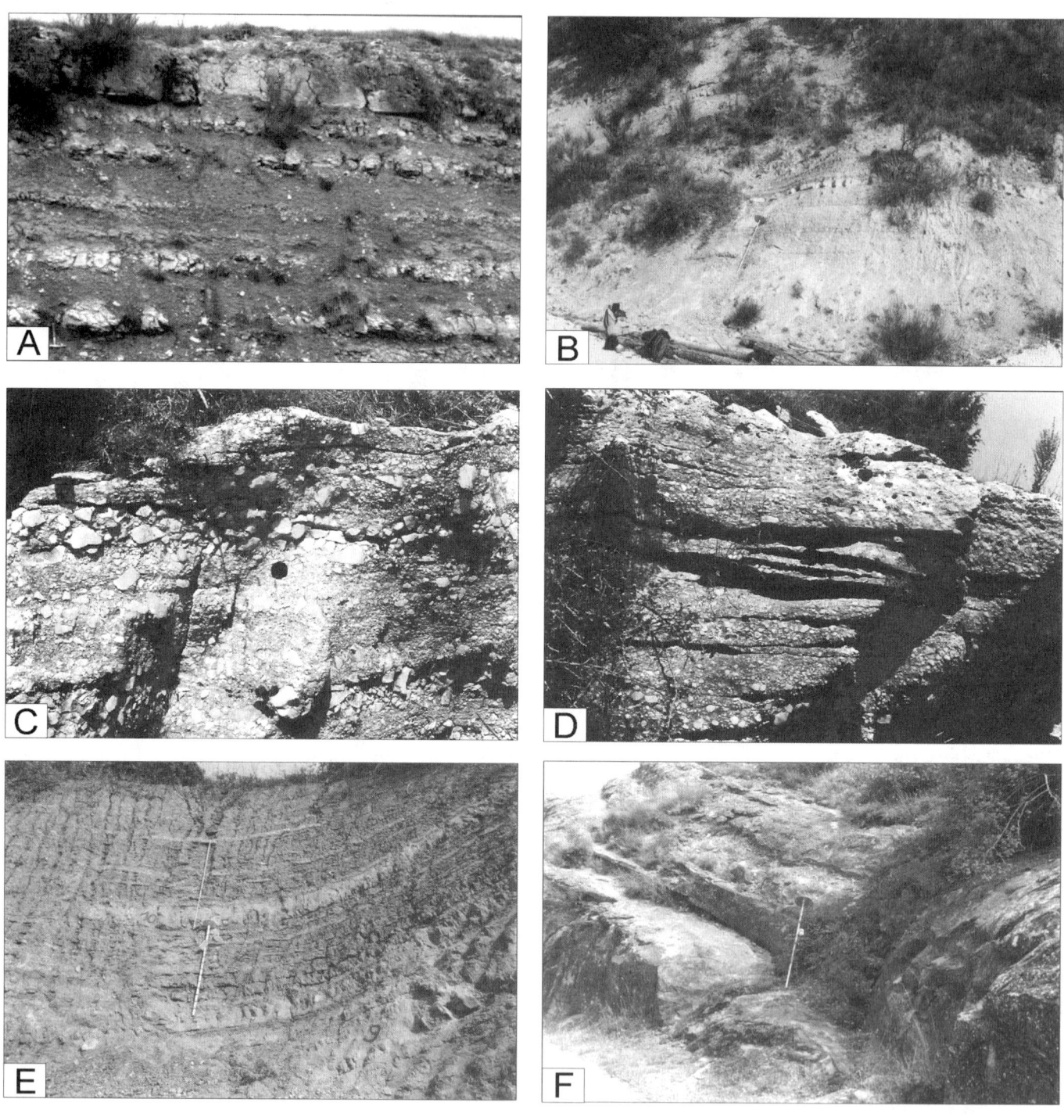

Figure 5. Lithologic and geometric characters of analyzed deposits. (A) Alternating clayey marls and marls related to Orbulina Marl in the Pescorocchiano area; prominent bed at section top is a fine-grained carbonate turbidite. See hammer at bottom left for scale. (B) Stratigraphic passage between Orbulina Marl (lithofacies 2) and base of Argilloso-Arenacea Formation (system 1) in the Salto Valley. (C–D) Conglomerate and coarse-grained facies of Brecce della Renga Formation placed through high-concentration gravity flows. (E) Interbedding of thin sandstone and mudstone beds (thin-bedded turbidites; F9 facies of Mutti, 1992) in Frosinone Formation. (F) Massive sands with frequent water-escape structures in Argilloso-Arenacea Formation (F5 and F8 facies of Mutti, 1992). The Jacob's ast for scale in E and F is 1.5 m long.

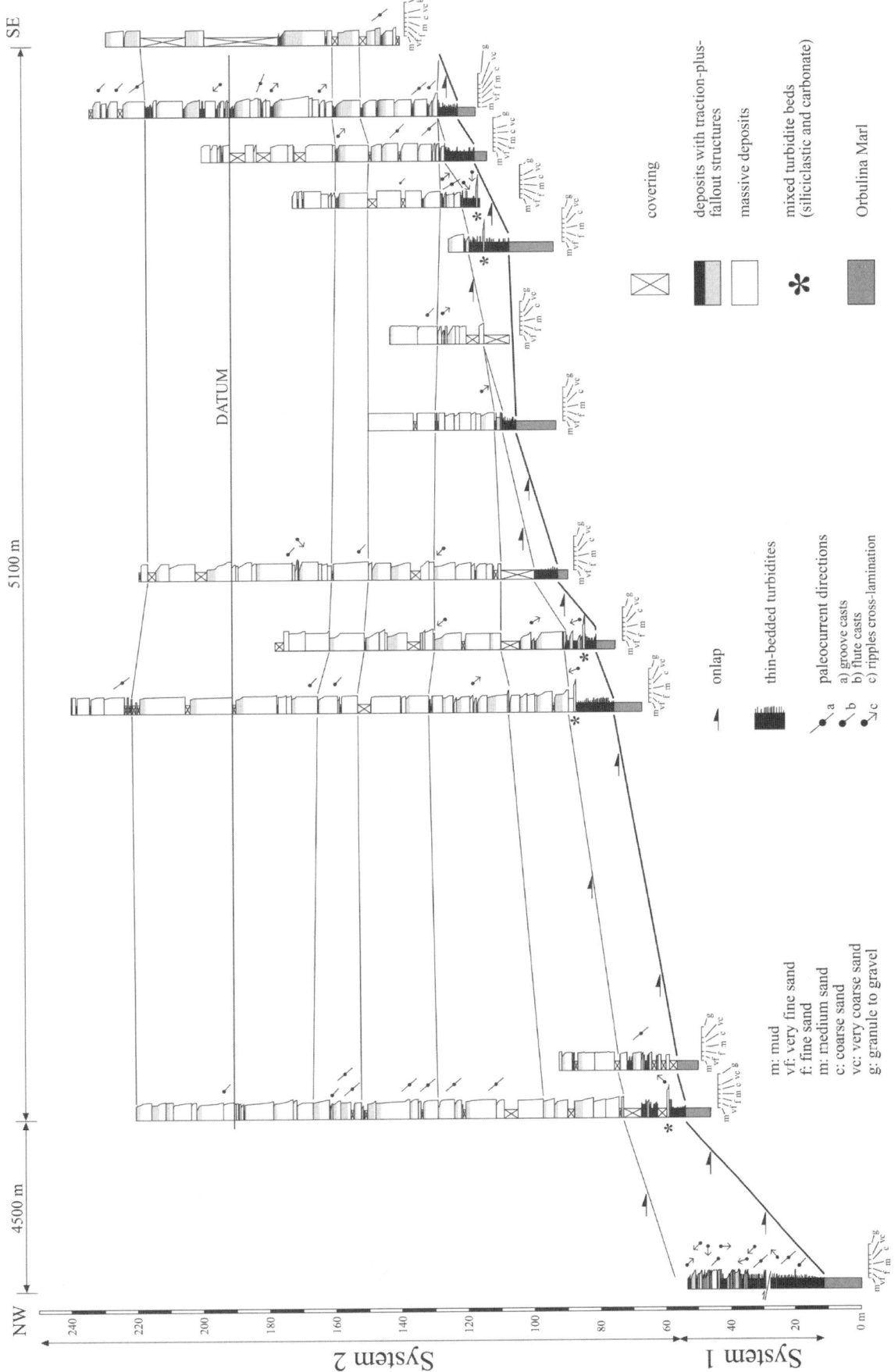

Figure 6. Detailed correlation panel showing stratigraphic relationships between Orbulina Marl, and systems 1 and 2 of Argilloso-Arenacea Formation. System 1 is here interpreted as the equivalent Lower Messinian portion of the Frosinone Formation in the Salto and Varri Valleys. It is worthy of note that mixed siliciclastic-carbonate beds occur just in system 1 and were supplied from the south (Simbruini high) and east sectors of the Lazio-Abruzzi carbonate platform (Marsica; see Fig. 1 for localities). (Modified after Bigi et al., 2003.)

into four turbidite depositional systems (systems 1, 2, 3, and 4) that define the Valle del Salto–Val di Varri Turbiditic Complex (Fig. 4). The first three systems crop out along the Val di Varri central sector and show paleocurrents directed toward the SE. The fourth system crops out in the Salto Valley and was supplied by a western source that had paleocurrent directions toward the E, NE, and SE. Each system shows a general fining-upward trend, although the formation is characterized, from bottom to top, by fining- to coarsening-upward trend. Cyclicity within these systems is defined by thinning- and fining-upward facies sequences, the correlations of which define simple and composite sedimentary bodies (lobes). Their deposition was likely connected to poorly efficient, variable-volume flows that gave rise to primarily massive sand with common water-escape structures and mud clasts (F5 and F8 facies of Mutti [1992] and Mutti et al. [1999]) (Fig. 5F), and subordinately to deposits with traction-plus-fallout structures (F9 facies of Mutti [1992] and Mutti et al. [1999]). Sedimentological and structural data seem to confirm turbidite deposition within confined basins where turbidite flow deflection and reflection occurred. Facies type and reconstructed facies tracts (sensu Mutti, 1992) indicate sedimentation within a low-efficiency, sand-rich turbidite system sourced by coeval deltaic systems (e.g., Milli and Moscatelli, 2000; Bigi et al., 2003; Moscatelli et al., 2004), which are entirely comparable to the "delta-fed systems" of Heller and Dickinson (1985) and to the "mixed systems" of Mutti et al. (2003). Tectonics controlled deposition of these sediments by triggering slope failures at delta fronts and prodelta slopes, and also modified submarine topography, which controlled turbidity-current flow direction and resulting facies. Due to the reduced size of basins, turbidity flows did not evolve downslope, giving rise to dominantly massive beds with common water-escape structures.

We accordingly interpret the oldest turbidite system of the Argilloso-Arenacea Formation (system 1) as the equivalent Lower Messinian portion of the Frosinone Formation in the Valle del Salto–Val di Varri sector (see also Moscatelli, 2003; Moscatelli et al., 2004) (Fig. 6). Such deposits occupied the area between the Sabini Mountains front, the northern sector of the Simbruini-Ernici structure, and Marsica, and thus constituted the first filling of the Valle del Salto depocenter. Following an early Messinian deformational phase, the basin depocenter migrated to the front of the Simbruini structure, giving rise to deposition of systems 2, 3, and 4. This interpretation is supported by structural and sedimentological data (i.e., Bigi et al., 2003), as well as by compositional data, which show that the petrographic character of system 1 was closely similar to that of the Frosinone Formation (see following).

METHODS

Seventy-six unaltered medium to coarse arenite samples that covered the sedimentary fill of the late Miocene central Apennines foreland basin system were selected for thin-section analysis: three were from the Orbulina Marl, five samples were collected from the Frosinone Formation, 43 from the Argilloso-Arenacea Formation, and 25 from the Brecce della Renga Formation. About 500 points were counted for each thin section according to the Gazzi-Dickinson method (Ingersoll et al., 1984; Zuffa, 1985, 1987). All thin sections were etched and stained for plagioclase and K-feldspar identification and to determine the mineralogy and Fe content of carbonate minerals according to the procedure of Lindholm and Finkelman (1972). Counted grains were assigned to the monomineralic and polymineralic compositional categories and were assigned to the spatial (extrabasinal versus intrabasinal) and temporal (coeval versus noncoeval) categories listed in Table 1. Point-count results are tabulated in Table A1 and have been recalculated as shown in Table A2. Grain types and recalculated parameters are those of Dickinson (1970, 1985), Zuffa (1985, 1987), and Critelli and Le Pera (1994).

Framework detritus of arenites was subdivided using criteria proposed by Zuffa (1980, 1985, 1987) into extrabasinal and intrabasinal components. Special attention was paid to the analysis of carbonate grains. In particular, carbonate grains were subdivided into coeval and noncoeval, both of which had provenance from intrabasinal (CI) or extrabasinal (CE) sources (Table 1; Table A1), using compositional, paleontological, and textural criteria (e.g., Zuffa, 1980, 1985, 1987, 1991; Fontana, 1991; Arribas and Tortosa, 2003).

Compositional characteristics of carbonate grains include various textural terms utilized in both Folk (1962) and Dunham (1962) classifications for carbonate rocks. Carbonate grains include allochemical particles, dominantly bioclasts or peloids; few include ooids and intraclasts. Interstitial components are micrite matrix and carbonate spar cement (dolomitic, Fe-calcite, Mg-calcite). Carbonate lithic fragments include mudstone, wackestone, and packstone fragments with abundant fossil tests; grainstone fragments are minor. Single unaltered allochemical particles are interpreted here as coeval intrabasinal grains.

Paleontological criteria were used to separate carbonate grains with different ages. In particular, lithic fragments include Jurassic to Lower Tertiary pelagic carbonate rocks, identified as mudstone, wackestone, and packstone with planktonic tests, and shallow-water fossiliferous carbonate rocks ranging from Late Triassic to early Miocene in age.

Textural criteria of carbonate grain types were used to distinguish the detrital supply of noncoeval carbonate lithic fragments with different provenances (i.e., extrabasinal and intrabasinal).

RESULTS

The analyzed deposits include siliciclastic turbidite, carbonate-clastic and mixed carbonate-siliciclastic arenites.

The carbonate-clastic strata are typical of the Brecce della Renga Formation, although a few of these strata are also interbedded within the turbidite deposits of the Argilloso-Arenacea Formation. The carbonate-clastic strata are conglomerate to coarse-grained arenites, mostly composed of carbonate detritus derived from a noncoeval intrabasinal source. The carbonate

TABLE 1. KEY TO COUNTED FRAMEWORK GRAINS AND RECALCULATED PARAMETERS

NCE CE CI NCI noncoeval vs.coeval	QmFLt; QmKP; LmLvLs; QpLvmLsm	QtFL	Rg Rs Rm Rv	Petrographic classes
				Noncarbonate extrabasinal (NCE)
				Q
NCE	Qm	Qt		Quartz (single crystals)
NCE	Qp	Qt		Polycrystalline quartz
NCE	Qm	Qt	Rv	Quartz in volcanic rock fragments
NCE	Qm	Qt	Rm	Quartz in metamorphic rock fragments
NCE	Qm	Qt	Rg	Quartz in plutonic rock fragments
NCE	Qm	Qt	Rg	Quartz in plutonic or gneissic rock fragments
NCE	Qm	Qt		Calcite replacement on quartz
				K
NCE	F-K	F		K-feldspar (single crystals)
NCE	F-K	F	Rv	K-feldspar in in volcanic rock fragments
NCE	F-K	F	Rm	K-feldspar in metamorphic rock fragments
NCE	F-K	F	Rg	K-feldspar in plutonic rock fragments
NCE	F-K	F	Rg	K-feldspar in plutonic or gneissic rock fragments
NCE	F-K	F		Calcite replacement on K-feldspar
				P
NCE	F-P	F		Plagioclase (single crystals)
NCE	F-P	F	Rv	Plagioclase in in volcanic rock fragments
NCE	F-P	F	Rm	Plagioclase in metamorphic rock fragments
NCE	F-P	F	Rg	Plagioclase in plutonic rock fragments
NCE	F-P	F	Rg	Plagioclase in plutonic or gneissic rock fragments
NCE	F-P	F		Calcite replacement on plagioclase
				M
NCE		M		Micas and chlorite (single crystals)
NCE		M	Rg	Micas and chlorite in plutonic rock fragments
NCE		M	Rm	Micas and chlorite in metamorphic rock fragments
				L
NCE	Lv,Lvm	L	Rv	Volcanic lithic with microlitic texture
NCE	Lv, Lvm	L	Rv	Volcanic lithic with felsitic granular texture
NCE	Lm, Lvm	L	Rm	Metavolcanic lithic
NCE	Lv, Lvm	L	Rm	Serpentinite
NCE	Lm, Lvm	L	Rm	Serpentine-schist
NCE	Lm, Lsm	L	Rm	Phyllite and slate
NCE	Lm, Lsm	L	Rm	Fine-grained schist
NCE	Ls, Qp	Qt	Rs	Impure chert
NCE	Ls, Lsm	L	Rs	Siltstone
NCE		DM		Dense and opaque minerals
NCE		DM	Rg	Dense minerals in plutonic rock fragments
NCE		DM	Rm	Dense minerals in metamorphic rock fragments
				Carbonate extrabasinal (CE)
CE	Ls, Lsm	L	Rs	Dolostone
CE	Ls, Lsm	L	Rs	Micritic limestone
CE	Ls, Lsm	L	Rs	Sparitic limestone
CE	Ls, Lsm	L	Rs	Microsparitic limestone
CE	Ls, Lsm	L	Rs	Foliated limestone
CE	Ls, Lsm	L	Rs	Biosparitic/grainstone limestone
CE	Ls, Lsm	L	Rs	Biomicritic/packstone limestone
CE	Ls, Lsm	L	Rs	Fossil in limestone-dolostone
CE	Ls, Lsm	L	Rs	Single spar (calcite)
				Noncoeval carbonate intrabasinal (CI noncoeval)
CI noncoeval	Ls, Lsm	L	Rs	Micritic limestone
CI noncoeval	Ls, Lsm	L	Rs	Sparitic limestone
CI noncoeval	Ls, Lsm	L	Rs	Microsparitic limestone
CI noncoeval	Ls, Lsm	L	Rs	Biomicritic/packstone limestone
CI noncoeval	Ls, Lsm	L	Rs	Biosparitic limestone
CI noncoeval	Ls, Lsm	L	Rs	Grainstone
CI noncoeval	Ls, Lsm	L	Rs	Fossil in limestone-dolostone
CI noncoeval	Ls, Lsm	L	Rs	Fossil (single skeleton)
CI noncoeval	Ls, Lsm	L	Rs	Silty-arenitic limestone
CI noncoeval	Ls, Lsm	L	Rs	Dolostone and dolomicrite
CI noncoeval	Ls, Lsm	L	Rs	Calcite and dolomite (single spar)
				Coeval carbonate intrabasinal (CI coeval)
CI coeval				Bioclast
CI coeval				Intraclast and peloid
				Noncarbonate intrabasinal (NCI)
NCI				Glauconite
NCI				Fe-oxide and phosphatic concretions
NCI				Rip-up clasts (argillaceous and siltitic)

Note: Qm (monocrystalline quartz)-F (total feldspar= K-feldspar+plagioclase)-Lt (aphanitic lithic fragments+polycrystalline quartz); Qt (total quartz = monocrystalline quartz+polycrystalline quartz)-F (total feldspar= K-feldspar+plagioclase)-L (aphanitic lithic fragments); Qm (monocrystalline quartz)-K (K-feldspar)- P (plagioclase); Rg (plutonic rock fragments)-Rs (sedimentary rock/lithic fragments)-Rm (metamorphic rock/lithic fragments)- Rv (volcanic rock/lithic fragments); Qp (polycrystalline quartz)-Lvm (volcanic and metavolcanic lithic fragments)-Lsm (sedimentary and metasedimentary lithic fragments); Lm (metamorphic lithic fragments)-Lv (volcanic lithic fragments)-Ls (sedimentary lithic fragments).

TABLE A1. RAW DATA

NCE Petrographic classes	Sample	SC 53	SC 54	SC 55	SC 56	SC 57	SC 58	SC 59	SC 60	SC 61	SC 62	SC 63	SC 64	SC 65	SC 66	SC 67	SC 68	SC 69	SC 70	SC 71	SC 38	SC 39	SC 40	SC 41	SC 43	SC 44
Q																										
Quartz (single crystals)		156	8	37	42	53	42	57	168	5	6	39	131	24	114	69	62	63	65	62	19	17	39	26	-	-
Polycrystalline quartz		13	-	-	-	3	3	3	4	2	-	4	7	-	3	10	8	13	8	10	2	2	2	1	-	-
Quartz in volcanic r.f.		-	-	-	-	-	-	-	-	-	-	-	-	-	-	-	-	-	-	-	-	-	-	-	-	-
Quartz in metamorphic r.f.		17	2	1	2	5	5	10	10	5	1	15	14	12	8	21	9	14	16	15	6	9	16	6	-	-
Quartz in plutonic r.f.		1	-	-	-	-	2	-	2	-	-	-	2	-	-	1	-	1	1	-	-	-	1	1	-	-
Quartz in plutonic or gneissic r.f.		-	-	-	-	-	-	-	-	-	-	-	-	-	-	4	2	-	4	5	-	-	1	-	-	-
Calcite replacement on quartz		-	-	-	-	-	-	-	-	-	-	-	-	-	-	-	-	-	-	-	-	-	-	-	-	-
K																										
K-feldspar (single crystals)		11	-	5	3	9	7	7	25	2	-	8	15	2	19	25	13	11	10	13	3	3	8	2	-	-
K-feldspar in in volcanic r.f.		-	-	-	-	-	-	-	-	-	-	-	-	-	-	-	-	-	-	-	-	-	-	-	-	-
K-feldspar in metamorphic r.f		1	-	-	-	-	-	-	-	-	-	-	-	-	-	2	-	1	1	-	1	-	-	-	-	-
K-feldspar in plutonic r.f.		7	-	-	-	2	1	3	3	1	1	2	5	1	2	16	2	12	8	7	-	-	2	-	-	-
K-feldspar in plutonic or gneissic r.f.		-	-	-	-	-	-	-	-	-	-	-	-	-	-	-	-	-	-	-	-	-	-	-	-	-
Calcite replacement on K-feldspar		-	-	-	-	-	1	-	-	-	-	-	-	-	1	-	-	1	-	-	-	-	-	-	-	-
P																										
Plagioclase (single crystals)		48	1	5	8	15	7	20	53	2	2	11	30	13	28	9	12	7	4	7	2	4	10	8	-	-
Plagioclase in in volcanic r.f.		-	-	-	-	-	-	-	-	-	-	-	-	-	-	-	-	-	-	-	-	-	-	-	-	-
Plagioclase in metamorphic r.f		7	-	3	-	-	1	6	-	1	3	3	-	3	2	1	5	-	4	1	1	7	1	-	-	-
Plagioclase in plutonic r.f.		4	-	1	-	1	-	2	1	2	-	2	6	-	11	5	2	5	1	2	2	-	2	2	-	-
Plagioclase in plutonic or gneissic r.f.		1	-	-	-	-	-	-	-	-	-	1	-	-	1	-	-	-	-	1	-	-	-	-	-	-
Calcite replacement on Plagioclase		-	-	-	-	-	2	-	-	-	-	-	1	-	-	-	-	-	-	-	-	-	-	-	-	-
M																										
Micas and Chlorite (single crystals)		33	1	10	11	15	6	15	73	2	2	11	26	11	29	5	2	3	4	3	1	3	3	4	-	-
Micas and Chlorite in plutonic r.f.		-	-	-	-	-	-	-	-	-	-	1	-	-	-	-	-	-	-	-	-	-	-	-	-	-
Micas and Chlorite in metamorphic r.f		6	-	-	-	-	-	-	3	-	-	-	1	-	-	-	-	-	-	-	1	-	-	-	-	-
L																										
Volcanic with microlitic texture		-	-	-	-	-	-	-	-	-	-	-	-	-	-	-	-	-	-	-	-	-	-	-	-	-
Volcanic with felsitic granular texture		2	-	-	-	-	-	-	1	-	-	-	1	-	2	3	2	3	-	3	-	-	1	-	-	-
Metavolcanic		-	-	-	-	-	-	-	-	-	-	-	-	-	2	-	-	-	-	-	-	-	-	-	-	-
Serpentinite		-	-	-	-	-	-	-	-	-	-	-	-	-	-	-	-	-	-	-	-	-	-	-	-	-
Serpentine-schist		3	-	-	-	-	-	-	2	-	1	3	-	3	-	-	-	-	-	2	-	-	-	-	-	-
Phyllite and Slate		16	2	6	6	7	8	11	18	7	4	12	16	7	31	20	13	15	14	13	9	5	15	11	-	2
Fine-grained Schist		7	-	-	-	-	-	-	2	-	-	-	5	-	3	1	-	1	-	3	2	5	1	-	-	-
Impura chert		-	-	-	-	-	-	-	-	-	-	-	-	-	1	4	2	4	3	1	-	-	-	-	-	-
Siltstone and Shale		-	-	-	-	-	1	-	-	-	-	-	-	-	1	9	4	1	3	2	-	-	-	-	-	-
Dense Minerals																										
Dense and opaque minerals		15	-	1	-	4	2	3	8	1	-	4	15	5	11	2	-	2	-	-	2	-	6	1	-	-
Dense minerals in plutonic r.f.		-	-	-	-	-	-	-	-	-	-	-	-	-	-	-	-	-	-	-	-	-	-	-	-	-
Dense minerals in metamorphic r.f.		1	-	-	-	-	-	-	-	-	-	-	2	-	-	-	-	-	-	-	-	-	-	-	-	-
CE																										
Dolostone		1	-	1	1	-	-	-	-	-	-	-	-	-	2	2	1	2	-	-	2	-	1	-	-	-
Micritic Limestone		8	2	5	5	10	6	5	11	2	3	6	14	7	6	16	8	21	11	11	4	4	4	6	5	5
Sparitic Limestone		-	1	4	2	6	3	-	2	2	1	-	-	2	3	1	1	5	-	2	3	1	4	3	1	-
Microsparitic Limestone		-	-	-	-	2	1	-	-	-	-	-	-	-	-	1	-	1	-	-	2	-	-	-	-	-
Foliated Limestone		-	-	-	-	-	-	-	-	1	-	-	-	-	-	-	-	-	3	-	-	-	-	-	-	-
Biosparitic Limestone and Grainstone		-	-	-	-	-	-	-	-	-	-	-	-	-	-	-	-	-	2	-	-	-	-	-	-	-
Biomicritic/packstone Limestone		2	1	3	2	6	3	2	1	1	2	2	4	1	3	8	7	14	3	10	5	3	3	3	1	-
Fossil in Limestone-Dolostone		-	-	-	-	-	-	-	-	-	-	-	-	-	-	-	-	-	-	-	-	-	-	-	-	-
Single spar (calcite)		-	-	-	-	-	-	-	-	-	-	-	-	-	-	-	-	-	-	-	-	-	-	-	-	-
non-coeval CI																										
Micritic Limestone		-	14	5	27	22	38	19	4	27	31	21	20	51	16	7	13	11	11	12	20	14	8	16	11	17
Sparitic Limestone		2	113	84	24	26	48	42	4	51	32	27	5	26	19	11	13	24	4	15	17	28	37	45	59	70
Microsparitic Limestone		4	36	25	20	17	12	12	1	9	32	15	8	6	6	15	11	19	12	17	18	12	15	19	21	22
Biomicritic/packstone Limestone		19	163	163	213	144	135	156	12	194	188	163	49	231	53	103	121	149	159	138	142	160	114	133	172	189
Biosparitic Limestone		1	18	13	10	4	10	8	2	27	18	12	-	3	2	5	34	10	11	10	15	49	46	61	19	21
Grainstone		1	4	8	3	-	4	2	-	2	7	1	1	-	1	5	3	-	8	2	4	6	5	2	5	10
Fossil in Limestone-Dolostone		5	25	11	9	21	30	42	2	50	54	50	6	39	17	6	33	29	31	32	32	53	35	81	38	48
Fossil (single skeleton)		1	-	-	-	2	1	1	-	2	1	-	-	1	-	2	2	1	2	1	3	3	3	6	2	4
Silty-arenitic Limestone		-	1	-	2	2	2	1	-	5	5	3	3	2	2	-	4	8	3	-	1	6	3	3	-	-
Dolostone and Dolmicrite		-	27	24	4	6	6	5	-	12	10	11	2	1	2	5	7	6	2	9	12	7	3	9	20	13
Calcite and Dolomite (Single spar)		-	3	5	6	8	11	-	-	1	1	3	-	1	-	7	7	6	4	4	2	-	-	5	8	-
coeval CI																										
Bioclast		4	28	22	17	27	20	24	3	40	31	28	7	19	14	56	67	59	48	61	57	63	27	54	90	66
Intraclast and Peloid		3	4	4	4	2	2	1	2	7	4	3	4	2	5	2	-	6	7	3	7	9	9	3	6	
NCI																										
Glauconite		-	-	-	1	-	-	-	-	-	-	-	-	-	1	-	-	-	-	-	1	1	-	-	-	-
Fe-Oxid and Phosphatic concretions		-	-	-	-	-	-	-	-	-	-	-	-	-	-	-	-	-	-	-	-	-	-	-	-	-
Rip-up clasts (argillaceous and siltitic)		-	-	-	-	-	-	-	-	-	1	-	-	-	-	-	-	-	-	2	-	1	-	-	-	-
Intersticies (Matrix)																										
Siliciclastic matrix		8	-	-	7	5	-	9	-	-	-	2	-	13	3	5	7	8	3	-	-	5	-	-	-	-
Carbonate matrix		-	20	20	37	24	9	6	-	8	7	17	-	19	6	7	4	14	14	14	-	10	17	20	3	6
Intersticies (Cement)																										
Carbonate cement (pore-filling)		-	26	24	22	25	41	22	3	31	53	17	5	9	2	8	14	22	3	12	99	25	27	8	46	21
Carbonate cement (patchy calcite)		-	-	-	-	-	-	-	-	-	-	-	-	-	-	-	-	-	-	-	-	-	-	-	-	-
Phyllosilicate and Oxid-Fe cements		1	-	-	-	-	1	-	2	-	-	-	-	-	-	-	-	-	-	-	3	-	-	-	-	-
Quartz overgrowth		-	-	-	-	-	-	-	-	-	-	-	-	-	-	-	-	-	-	-	-	-	-	-	-	-
Calcite replacement on undeter.grain		91	-	10	13	25	32	19	65	1	-	4	88	2	56	17	13	30	10	10	3	7	12	-	-	-
TOTAL		500	500	500	500	500	500	500	500	500	500	500	500	500	500	500	500	600	500	522	500	500	500	547	507	500

(continued.)

TABLE A1. (Continued.)

NCE Petrographic classes	Sample	Orbulina Marls Formation			Frosinone Formation (quartzofeldspathic subpetrofacies 1)					Argilloso-Arenacea Formation lower portion (quartzofeldspathic subpetrofacies 1)																
		SC 48	SC 32	SC 26	SC 27	SC 28	SC 29	SC 30	SC 31	SC 33	SC 34	SC 35	SC 36	SC 37	SC 42	SC 45	SC 46	SC 47	SC 49	SC 50	SC 51	SC 52	SC 72	SC 73	SC 74	SC 75
Q																										
Quartz (single crystals)		1	1	1	121	124	133	128	126	146	131	47	32	140	139	144	152	161	126	112	154	136	157	146	164	153
Polycrystalline quartz		-	1	-	8	8	13	8	13	9	8	5	2	15	7	6	3	7	12	15	13	8	5	6	11	6
Quartz in volcanic r.f.		-	-	-	-	-	-	-	-	-	-	1	-	-	-	-	-	-	-	-	-	-	-	-	-	-
Quartz in metamorphic r.f.		-	-	-	31	39	41	63	61	3	7	7	6	16	15	6	15	12	13	14	8	12	14	23	22	27
Quartz in plutonic r.f.		-	-	-	1	4	-	-	1	-	-	3	1	1	1	2	-	2	3	2	1	1	-	1	1	2
Quartz in plutonic or gneissic r.f.		-	-	-	2	1	1	2	2	-	-	-	-	-	-	-	1	3	1	-	-	-	1	-	-	-
Calcite replacement on quartz		-	-	-	-	-	-	-	-	-	-	-	-	-	-	-	-	-	-	-	-	-	-	-	-	-
K																										
K-feldspar (single crystals)		-	-	-	27	20	28	21	32	24	23	5	3	30	24	22	38	46	37	32	46	33	32	38	28	30
K-feldspar in in volcanic r.f.		-	-	-	-	-	-	-	-	-	-	-	-	-	-	-	-	-	-	-	-	-	-	-	-	-
K-feldspar in metamorphic r.f		-	-	-	2	2	-	2	3	-	2	-	-	1	-	2	1	-	-	-	-	2	1	-	-	2
K-feldspar in plutonic r.f.		-	-	-	2	4	3	2	2	3	5	1	-	4	3	5	5	6	13	11	7	5	6	6	4	5
K-feldspar in plutonic or gneissic r.f.		-	-	-	-	-	-	-	-	-	-	-	-	-	-	-	-	-	-	-	-	-	-	-	-	-
Calcite replacement on K-feldspar		-	-	-	-	-	-	-	-	1	-	-	-	1	1	1	2	1	3	2	-	-	1	-	1	-
P																										
Plagioclase (single crystals)		-	4	-	44	52	62	61	53	25	29	6	11	65	35	55	39	72	25	28	33	31	57	72	58	42
Plagioclase in in volcanic r.f.		-	-	-	-	-	-	-	-	-	-	-	-	-	-	-	-	-	-	-	-	-	-	-	-	-
Plagioclase in metamorphic r.f		-	-	-	11	9	14	15	9	1	1	3	-	1	3	2	5	3	1	2	5	2	4	7	6	3
Plagioclase in plutonic r.f.		-	-	-	5	5	6	6	1	2	-	6	3	4	3	7	7	7	5	1	5	2	7	5	7	9
Plagioclase in plutonic or gneissic r.f.		-	-	-	-	2	1	-	-	-	-	4	-	-	-	-	-	-	-	-	-	-	1	-	-	-
Calcite replacement on Plagioclase		-	-	-	-	-	-	-	1	3	-	-	1	-	-	-	-	1	-	1	1	-	-	3	2	-
M																										
Micas and Chlorite (single crystals)		1	-	-	33	31	26	36	32	36	68	10	2	32	42	33	43	34	9	9	11	18	46	56	58	54
Micas and Chlorite in plutonic r.f.		-	-	-	-	-	-	-	-	-	-	-	-	-	-	-	-	-	-	-	-	-	-	-	-	-
Micas and Chlorite in metamorphic r.f		-	-	-	3	2	3	7	4	-	5	1	-	3	3	6	-	3	-	2	-	-	4	3	4	1
L																										
Volcanic with microlitic texture		-	-	-	-	-	2	-	1	-	-	-	-	-	-	-	-	-	-	-	-	-	-	-	-	-
Volcanic with felsitic granular texture		-	-	-	3	1	-	3	1	-	-	-	-	3	1	2	1	-	-	6	2	1	-	1	-	-
Metavolcanic		-	-	-	1	-	-	-	2	-	-	-	-	-	-	-	-	-	-	2	-	-	-	-	1	-
Serpentinite		-	-	-	4	3	2	-	1	-	-	-	-	-	-	-	-	-	1	-	-	-	-	-	-	-
Serpentine-schist		-	-	-	5	5	8	9	6	1	1	-	1	4	6	3	5	1	-	3	-	-	2	4	2	2
Phyllite and Slate		-	-	-	27	26	28	34	27	17	17	14	15	24	23	18	25	28	7	10	10	10	17	25	12	12
Fine-grained Schist		-	-	-	6	6	6	10	8	2	4	1	6	2	7	7	8	14	-	3	1	1	8	1	5	6
Impura chert		-	-	-	1	1	-	3	-	-	-	-	-	-	-	-	-	-	-	1	3	-	-	-	-	-
Siltstone and Shale		-	-	-	1	2	-	-	-	-	1	-	-	3	1	1	2	1	9	6	5	8	-	-	-	2
Dense Minerals																										
Dense and opaque minerals		-	-	-	16	14	18	21	21	11	9	5	3	10	9	16	10	14	3	4	4	6	14	20	17	12
Dense minerals in plutonic r.f.		-	-	-	-	-	-	-	-	-	-	-	-	-	-	-	-	1	-	-	-	-	-	-	-	-
Dense minerals in metamorphic r.f.		-	-	-	-	3	-	-	1	1	-	-	1	-	-	1	-	1	-	-	-	-	1	1	-	1
CE																										
Dolostone		-	-	-	-	-	-	-	-	-	-	2	-	-	1	-	-	-	-	-	-	-	-	-	-	-
Micritic Limestone		3	12	11	19	10	20	12	22	35	15	24	21	18	16	16	22	5	35	32	37	30	9	13	16	12
Sparitic Limestone		-	2	7	15	20	15	9	18	7	1	28	35	-	14	7	11	12	21	11	22	16	2	3	2	11
Microsparitic Limestone		-	1	1	5	5	11	5	4	5	5	23	13	3	9	3	6	1	11	7	8	11	3	4	3	3
Foliated Limestone		-	-	-	-	-	-	-	-	-	-	-	-	-	-	-	-	-	-	-	-	-	-	-	-	-
Biosparitic Limestone and Grainstone		-	1	2	-	-	-	-	-	-	-	19	26	-	-	-	-	-	1	1	1	-	-	-	-	-
Biomicritic/packstone Limestone		17	3	25	6	3	-	1	3	3	-	49	31	-	2	-	-	-	6	20	1	10	-	1	-	2
Fossil in Limestone-Dolostone		1	-	1	-	-	-	-	-	-	1	58	83	-	-	-	-	-	-	3	-	1	-	-	-	-
Single spar (calcite)		-	12	1	5	-	-	-	-	4	1	9	19	-	13	6	8	-	3	10	3	9	3	1	8	7
non-coeval CI																										
Micritic Limestone		-	-	-	-	-	-	-	-	-	-	-	-	-	-	-	-	-	-	-	-	-	-	-	-	-
Sparitic Limestone		-	-	-	-	-	-	-	-	-	-	-	-	-	-	-	-	-	-	-	-	-	-	-	-	-
Microsparitic Limestone		-	-	-	-	-	-	-	-	-	-	-	-	-	-	-	-	-	-	-	-	-	-	-	-	-
Biomicritic/packstone Limestone		-	-	-	-	-	-	-	-	-	-	-	-	-	-	-	-	-	-	-	-	-	-	-	-	-
Biosparitic Limestone		-	-	-	-	-	-	-	-	-	-	-	-	-	-	-	-	-	-	-	-	-	-	-	-	-
Grainstone		-	-	-	-	-	-	-	-	-	-	-	-	-	-	-	-	-	-	-	-	-	-	-	-	-
Fossil in Limestone-Dolostone		-	-	-	-	-	-	-	-	-	-	-	-	-	-	-	-	-	-	-	-	-	-	-	-	-
Fossil (single skeleton)		-	-	-	-	-	-	-	-	-	-	-	-	-	-	-	-	-	-	-	-	-	-	-	-	-
Silty-arenitic Limestone		-	-	-	-	-	-	-	-	-	-	-	-	-	-	-	-	-	-	-	-	-	-	-	-	-
Dolostone and Dolmicrite		-	-	-	-	-	-	-	-	-	-	-	-	-	-	-	-	-	-	-	-	-	-	-	-	-
Calcite and Dolomite (Single spar)		-	-	-	-	-	-	-	-	-	-	-	-	-	-	-	-	-	-	-	-	-	-	-	-	-
coeval CI																										
Bioclast		324	296	305	5	1	2	-	4	7	4	114	134	7	4	-	-	3	14	10	11	10	1	1	-	2
Intraclast and Peloid		11	-	10	-	-	-	-	-	-	-	-	-	-	-	-	-	-	-	-	-	-	-	-	-	-
NCI																										
Glauconite		4	52	34	-	1	-	1	1	1	-	1	2	-	2	-	-	-	-	-	-	2	-	-	-	-
Fe-Oxid and Phosphatic concretions		-	3	2	-	-	-	-	-	-	-	-	-	-	-	-	-	-	-	-	-	-	-	-	-	-
Rip-up clasts (argillaceous and siltitic)		2	-	-	-	-	-	-	-	-	-	-	-	-	2	-	-	-	-	-	-	-	-	-	-	-
Intersticies (Matrix)																										
Siliciclastic matrix		-	2	2	1	6	2	7	-	6	16	3	-	25	10	4	5	4	3	4	2	9	1	6	5	11
Carbonate matrix		132	1	25	-	-	-	-	-	5	-	-	-	-	-	-	-	-	-	-	-	-	-	-	-	-
Intersticies (Cement)																										
Carbonate cement (pore-filling)		-	106	73	4	1	1	3	-	5	4	29	32	1	1	3	4	-	3	6	3	5	-	-	1	1
Carbonate cement (patchy calcite)		3	-	-	-	-	-	-	-	-	-	-	-	-	-	-	-	-	-	-	-	-	-	-	-	-
Phyllosilicate and Oxid-Fe cements		1	1	-	-	6	-	1	-	-	-	-	-	-	5	2	-	1	4	3	-	2	-	2	5	6	9
Quartz overgrowth		-	-	-	1	-	-	-	-	-	-	-	-	-	-	-	-	-	-	-	-	-	-	-	-	1
Calcite replacement on undeter.grain		-	2	-	82	83	54	26	40	135	143	20	16	83	102	124	78	54	130	125	103	121	99	50	56	72
TOTAL		500	500	500	497	500	500	497	500	498	500	500	500	500	500	500	500	500	500	497	500	500	500	500	500	500

(continued.)

TABLE A1. RAW DATA (Continued.)

NCE — Argilloso-Arenacea Formation middle-upper portions (quartzofeldspathic subpetrofacies 2)

Petrographic classes / Sample	SC 631	SC 632	SC 633	SC 634	SC 635	SC 636	SC 637	SC 638	SC 639	SC 641	SC 642	SC 643	SC 644	SC 645	SC 646	SC 647	SC 619	SC 620	SC 621	SC 622	SC 623	SC 624	SC 625	SC 626	SC 627	SC 628	SC 629	SC 630
Q																												
Quartz (single crystals)	78	85	73	79	79	58	86	25	96	90	52	80	68	79	62	82	63	30	42	31	51	90	38	81	44	43	55	79
Polycrystalline quartz	46	40	62	46	59	49	51	36	51	37	44	44	49	54	37	25	65	52	58	69	46	61	44	40	35	37	70	29
Quartz in volcanic r.f.	-	-	-	-	-	-	-	-	-	-	1	-	-	-	-	-	-	-	-	1	-	-	-	-	-	-	-	-
Quartz in metamorphic r.f.	22	27	31	27	48	41	23	36	25	10	26	17	31	22	11	7	37	35	57	30	23	20	18	16	22	20	38	10
Quartz in plutonic r.f.	25	27	25	18	29	29	8	5	22	10	19	10	26	20	17	5	34	43	26	45	22	17	14	15	27	22	25	6
Quartz in plutonic or gneissic r.f.	-	-	-	-	1	3	-	-	1	-	1	2	3	-	2	-	5	2	-	2	6	-	3	-	1	-	-	-
Calcite replacement on quartz	37	34	34	50	39	35	63	15	22	44	35	46	52	36	23	53	26	23	31	26	30	42	36	39	27	44	36	30
K																												
K-feldspar (single crystals)	10	8	3	9	11	20	17	8	23	14	17	28	3	5	17	6	1	7	8	4	14	11	6	5	15	7	6	7
K-feldspar in volcanic r.f.	-	-	-	-	1	-	-	-	-	-	-	-	-	-	-	-	-	-	-	-	-	-	-	-	-	-	-	-
K-feldspar in metamorphic r.f	2	1	2	6	17	12	2	2	5	2	5	7	5	2	4	-	-	5	7	3	4	2	4	2	4	6	4	1
K-feldspar in plutonic r.f.	5	5	5	11	15	9	6	5	11	5	15	9	-	4	6	-	-	10	6	7	3	2	9	4	13	7	5	9
K-feldspar in plutonic or gneissic r.f.	-	-	-	-	-	-	-	-	-	-	-	1	-	-	-	-	-	-	-	-	-	-	-	-	-	1	-	-
Calcite replacement on K-feldspar	1	1	1	2	5	2	-	-	5	4	2	6	-	1	2	1	-	1	-	-	2	-	-	-	4	3	1	-
P																												
Plagioclase (single crystals)	49	43	23	30	25	20	27	7	32	32	30	27	25	22	16	32	37	21	5	25	41	31	32	31	44	43	26	9
Plagioclase in volcanic r.f.	1	-	-	-	-	-	-	-	-	-	-	-	-	-	-	-	-	-	3	-	-	-	-	-	1	1	-	-
Plagioclase in metamorphic r.f	17	16	25	15	28	22	18	12	13	5	17	10	24	8	9	7	38	39	55	41	22	17	31	13	30	49	18	5
Plagioclase in plutonic r.f.	43	29	21	27	18	28	10	2	14	4	25	12	17	18	30	7	33	62	36	44	30	29	41	36	63	58	44	4
Plagioclase in plutonic or gneissic r.f.	-	-	-	-	-	1	-	-	-	-	-	-	3	2	-	-	2	5	1	-	4	-	1	-	-	2	-	-
Calcite replacement on Plagioclase	9	9	5	10	6	8	13	1	4	8	13	5	4	10	4	6	6	8	10	9	11	11	12	10	23	14	2	-
M																												
Micas and Chlorite (single crystals)	36	24	28	25	22	29	33	1	44	38	20	34	22	33	19	31	4	21	19	8	9	16	2	22	15	13	15	9
Micas and Chlorite in plutonic r.f.	-	2	1	-	2	1	3	-	2	1	3	3	4	3	5	-	4	5	1	-	3	1	-	-	2	1	1	-
Micas and Chlorite in metamorphic r.f	9	10	9	12	9	17	7	6	14	7	8	17	14	8	7	12	9	15	8	4	5	7	3	12	4	2	5	5
L																												
Volcanic with microlitic texture	-	-	-	-	-	-	-	-	-	-	-	-	-	2	1	-	-	-	-	-	-	-	-	-	-	-	-	-
Volcanic with felsitic granular texture	1	2	-	5	-	-	1	-	-	1	-	-	1	2	3	-	5	2	2	3	3	4	8	3	3	5	5	-
Metavolcanic	-	-	-	-	-	-	-	6	-	-	-	-	-	-	-	-	-	-	-	-	-	-	-	-	-	-	-	-
Serpentinite	-	1	-	-	-	-	-	-	1	-	-	1	-	-	-	2	-	2	-	2	2	-	-	-	-	1	-	-
Serpentine-schist	4	-	2	1	2	2	1	2	2	1	2	2	2	2	2	1	4	2	-	2	11	4	3	9	4	4	2	6
Phyllite and Slate	7	8	13	8	10	7	6	3	6	1	10	11	7	13	6	5	15	21	13	17	17	9	28	12	14	8	11	8
Fine-grained Schist	6	3	4	3	4	1	7	3	6	3	6	6	10	10	5	4	3	6	3	6	9	11	12	2	1	3	2	5
Impura chert	-	1	1	1	-	-	2	8	1	-	4	2	4	2	4	3	4	1	1	11	9	4	18	-	3	6	2	5
Siltstone and Shale	-	-	-	-	1	-	2	-	-	-	-	1	1	-	-	-	1	-	-	1	2	2	-	1	1	-	-	-
Dense Minerals																												
Dense and opaque minerals	8	8	7	1	8	7	14	1	11	11	9	12	19	15	14	10	6	3	7	8	10	20	3	13	2	1	10	4
Dense minerals in plutonic r.f.	2	-	-	-	-	-	1	-	-	1	1	-	-	1	1	-	-	3	-	1	-	1	-	-	2	-	-	-
Dense minerals in metamorphic r.f.	4	3	3	1	4	5	1	-	5	1	1	1	1	10	1	2	2	5	4	2	2	2	1	-	1	2	1
CE																												
Dolostone	-	-	-	-	-	-	-	5	-	-	-	-	-	-	-	-	-	-	-	-	-	-	-	-	-	-	-	5
Micritic Limestone	3	2	5	5	6	5	1	4	1	8	9	8	4	4	4	4	6	-	3	5	9	6	8	3	10	3	6	-
Sparitic Limestone	1	-	5	-	-	2	1	16	1	-	3	-	2	3	7	-	-	-	1	10	9	5	3	-	-	3	7	15
Microsparitic Limestone	-	4	1	2	5	1	5	10	3	1	4	5	1	2	5	2	1	-	6	11	11	7	6	2	8	4	13	2
Foliated Limestone	-	-	-	-	-	-	-	5	-	-	-	-	-	-	-	-	-	-	-	-	-	-	-	-	-	-	-	6
Biosparitic Limestone and Grainstone	1	-	1	1	1	2	-	36	1	6	4	2	-	-	-	-	-	-	-	-	-	3	-	1	3	-	1	-
Biomicritic/packstone Limestone	-	-	1	1	-	-	-	43	-	-	-	-	-	-	-	-	-	-	-	-	2	1	7	1	1	1	1	3
Fossil in Limestone-Dolostone	-	-	-	-	-	-	-	-	-	-	-	-	-	-	-	-	-	-	-	-	-	-	-	-	-	-	-	-
Single spar (calcite)	-	-	-	-	-	-	-	-	-	-	-	-	-	-	-	-	-	-	-	-	-	-	-	-	-	-	-	-
non-coeval CI																												
Micritic Limestone	-	-	-	-	-	-	-	-	-	-	-	-	-	-	-	-	-	-	-	-	-	-	-	-	-	-	-	-
Sparitic Limestone	-	-	-	-	-	-	-	-	-	-	-	-	-	-	-	-	-	-	-	-	-	-	-	-	-	-	-	-
Microsparitic Limestone	-	-	-	-	-	-	-	-	-	-	-	-	-	-	-	-	-	-	-	-	-	-	-	-	-	-	-	-
Biomicritic/packstone Limestone	-	-	-	-	-	-	-	-	-	-	-	-	-	-	-	-	-	-	-	-	-	-	-	-	-	-	-	-
Biosparitic Limestone	-	-	-	-	-	-	-	-	-	-	-	-	-	-	-	-	-	-	-	-	-	-	-	-	-	-	-	-
Grainstone	-	-	-	-	-	-	-	-	-	-	-	-	-	-	-	-	-	-	-	-	-	-	-	-	-	-	-	-
Fossil in Limestone-Dolostone	-	-	-	-	-	-	-	-	-	-	-	-	-	-	-	-	-	-	-	-	-	-	-	-	-	-	-	-
Fossil (single skeleton)	-	-	-	-	-	-	-	-	-	-	-	-	-	-	-	-	-	-	-	-	-	-	-	-	-	-	-	-
Silty-arenitic Limestone	-	-	-	-	-	-	-	-	-	-	-	-	-	-	-	-	-	-	-	-	-	-	-	-	-	-	-	-
Dolostone and Dolmicrite	-	-	-	-	-	-	-	-	-	-	-	-	-	-	-	-	-	-	-	-	-	-	-	-	-	-	-	-
Calcite and Dolomite (Single spar)	-	-	-	-	-	-	-	-	-	-	-	-	-	-	-	-	-	-	-	-	-	-	-	-	-	-	-	-
coeval CI																												
Bioclast	1	2	-	-	-	-	-	106	-	-	-	1	-	1	1	1	-	-	-	-	2	3	3	4	-	-	-	94
Intraclast and Peloid	-	-	-	-	-	-	-	1	-	-	-	-	-	-	-	-	-	-	-	-	-	-	-	-	-	-	-	3
NCI																												
Glauconite	-	1	-	-	-	-	-	-	-	-	-	-	1	-	1	-	-	-	-	-	-	-	-	-	-	-	-	-
Fe-Oxid and Phosphatic concretions	-	-	-	-	-	-	-	-	-	-	-	-	-	-	-	-	-	-	-	-	-	-	-	-	-	-	-	-
Rip-up clasts (argillaceous and siltitic)	-	-	-	-	-	-	-	-	-	-	-	-	-	-	-	-	-	-	-	-	-	-	-	-	-	-	-	-
Intersticies (Matrix)																												
Siliciclastic matrix	-	-	-	-	-	-	-	-	-	-	-	-	-	-	-	-	2	2	-	2	1	-	-	1	1	2	-	2
Carbonate matrix	-	1	-	-	-	-	1	-	7	-	-	-	-	-	-	-	2	2	-	1	-	-	-	2	1	1	-	-
Intersticies (Cement)																												
Carbonate cement (pore-filling)	11	9	15	14	14	8	10	5	7	14	8	11	4	6	-	2	4	1	7	3	11	13	12	18	8	7	10	5
Carbonate cement (patchy calcite)	36	19	30	33	20	13	21	91	46	62	61	75	57	72	87	110	28	19	27	10	14	37	38	43	23	18	18	97
Phyllosilicate and Oxid-Fe cements	16	14	18	19	11	11	23	2	12	8	5	5	10	7	15	4	29	18	14	16	11	18	7	21	11	16	14	8
Quartz overgrowth	-	-	-	-	-	-	-	-	-	-	-	-	-	-	-	-	-	-	-	-	-	-	-	-	-	-	-	2
Calcite replacement on undeter.grain	30	65	57	43	38	54	47	41	21	81	58	41	58	36	77	85	22	34	41	38	59	35	43	59	45	43	42	41
TOTAL	521	504	511	505	537	504	508	559	507	510	519	539	543	513	508	508	501	504	506	506	513	535	503	517	506	507	514	511

Note: NCE—extrabasinal non-carbonate; CE—extrabasinal carbonate; non-coeval CI—intrabasinal carbonate non-coeval; coeval CI—intrabasinal carbonate coeval; NCI—intrabasinal non-carbonate; Q—quartz; K—K-feldspar; P—plagioclase; M—micas and chlorite; L—aphanitic lithic fragments.

TABLE A2. RECALCULATED MODAL POINT-COUNT DATA FOR THE LATE MIOCENE ARENITES OF THE CENTRAL APENNINES FORELAND BASIN SYSTEM, ITALY. SEE TABLE 1 FOR DESCRIPTION OF THE RECALCULATED PARAMETERS

Sample no.	QmFLt (%)			QtFL (%)			QmKP (%)			QpLvmLsm (%)			LmLvLs (%)			RgRsRm (%)			NCE CE+Cln-co Clco (%)			P/F (%)
	Qm	F	Lt	Qt	F	L	Qm	K	P	Qp	Lvm	Lsm	Lm	Lv	Ls	Rg	Rs	Rm	NCE	CE+Cln-co	Clco	P/F
Brecce della Renga Formation (calclithite petrofacies)																						
SC 67	25	16	59	29	16	55	61	28	11	6	1	93	10	1	89	10	73	17	46	40	14	0.28
SC 68	19	8	73	21	8	71	70	15	15	3	1	96	5	1	94	2	90	8	29	56	15	0.5
SC 69	17	9	74	21	9	70	62	23	15	5	1	94	5	1	94	5	85	10	31	58	11	0.4
SC 70	21	6	73	24	6	70	78	17	5	4	0	96	5	0	95	4	86	10	31	57	12	0.21
SC 71	20	8	72	23	8	69	71	17	12	4	1	95	5	1	94	5	85	10	31	55	14	0.41
SC 43	0	0	100	0	0	100	0	0	0	0	0	100	0	0	100	0	100	0	0	80	20	0
SC 44	0	0	100	0	0	100	0	0	0	0	0	100	1	0	99	0	99	1	0	85	15	0
SC 38	8	3	89	8	3	89	73	12	15	1	1	98	5	0	95	1	92	7	13	72	15	0.56
SC 39	7	2	91	7	2	91	76	9	15	1	0	99	2	0	98	0	95	5	10	75	15	0.63
SC 40	15	7	78	15	7	78	66	12	22	1	0	99	7	0	93	2	86	12	27	65	8	0.66
SC 41	7	3	90	7	3	90	72	4	24	0	0	100	3	0	97	1	94	5	12	77	11	0.85
SC 54	2	0	98	2	0	98	91	0	9	0	0	100	0	0	100	0	99	1	3	90	7	1
SC 55	9	3	88	9	3	88	73	10	17	0	0	100	2	0	98	0	97	3	15	79	6	0.64
SC 56	11	3	86	11	3	86	80	5	15	0	0	100	2	0	98	0	98	2	17	78	5	0.73
SC 57	16	8	76	16	8	76	67	14	19	1	0	99	2	0	98	1	95	4	27	66	7	0.57
SC 58	13	4	83	13	4	83	74	12	14	1	0	99	3	0	97	1	95	4	20	75	5	0.53
SC 59	17	8	75	17	8	75	67	10	23	1	0	99	4	0	96	2	91	7	29	65	6	0.7
SC 61	2	2	96	3	2	95	58	18	24	1	1	98	2	0	98	1	95	4	7	83	10	0.57
SC 62	2	1	97	2	1	97	64	9	27	0	0	100	1	0	99	0	98	2	4	87	9	0.75
SC 63	13	7	90	14	7	79	67	12	21	1	0	99	4	0	96	1	90	9	25	68	7	0.63
SC 65	8	4	88	8	4	88	69	6	25	0	0	100	2	0	98	0	95	5	16	79	5	0.81
X	11	5	84	12	5	83	71	12	17	2	0	98	3	0	97	2	92	6	19	71	10	0.54
SD	±7	±4	±11	±8	±4	±12	±8	±7	±6	±2	0	±2	±2	±0	±3	±2	±7	±4	±12	±13	±4	±0.26
Brecce della Renga Formation: siliciclastic turbidites (quartzofeldspathic petrofacies)																						
SC 53	52	23	25	56	23	21	68	8	24	15	6	79	36	3	61	11	38	51	87	11	2	0.76
SC 60	54	27	19	55	27	18	68	10	22	6	2	92	33	2	65	7	47	46	90	9	1	0.68
SC 64	42	17	41	44	17	39	71	10	19	5	3	92	18	1	81	8	66	26	70	27	3	0.67
SC 66	33	18	49	35	18	47	65	12	23	2	4	94	22	1	77	7	67	26	65	31	4	0.66
X	45	21	34	48	21	31	68	10	22	7	4	89	27	2	71	8	55	37	78	20	2	0.69
SD	±10	±5	±14	±10	±5	±14	±2	±2	±2	±6	±2	±7	±9	±1	±9	±2	±14	±13	±12	±12	±1	±0.05
Orbulina Marls Formation (intrarenite petrofacies)																						
SC 48	4	0	96	4	0	96	100	0	0	0	0	100	0	0	100	0	100	0	1	6	93	0
SC 26	2	0	98	2	0	98	100	0	0	0	0	100	0	0	100	0	100	0	0	13	87	0
SC 32	3	11	86	5	11	84	20	0	80	3	0	97	0	0	100	0	100	0	2	9	89	0
X	3	4	93	4	4	92	73	0	27	1	0	99	0	0	100	0	100	0	1	9	90	0
SD	±1	±6	±6	±1	±6	±7	0	0	0	0	0	0	0	0	0	0	0	0	±1	±4	±3	0
Frosinone Formation (quartzofeldspathic subpetrofacies 1)																						
SC 27	44	26	30	46	26	28	63	13	24	8	12	80	44	3	53	6	34	60	87	12	1	0.66
SC 28	47	27	26	50	27	23	64	10	26	10	10	80	49	1	50	11	27	62	91	9	0	0.72
SC 29	44	29	27	48	29	23	60	11	29	12	11	77	48	2	50	7	29	64	89	10	1	0.73
SC 30	49	27	24	52	27	21	65	8	27	12	13	75	62	3	35	6	17	77	94	6	0	0.77
SC 31	48	25	27	52	25	23	65	13	22	12	10	78	47	2	51	3	27	70	89	10	1	0.63
X	46	27	27	50	27	23	63	11	26	11	11	78	50	2	48	6	27	67	90	9	1	0.7
SD	±2	±1	±2	±3	±1	±3	±2	±2	±3	±2	±1	±2	±7	±1	±7	±3	±6	±7	±3	±2	±1	±0.06
Argilloso-Arenacea Formation: lower portion (quartzofeldspathic subpetrofacies 1)																						
SC 33	52	20	28	55	20	25	72	13	15	11	1	88	27	0	73	6	64	30	82	16	2	0.53
SC 34	55	24	21	58	24	18	70	15	15	15	2	83	49	0	51	8	35	57	92	7	1	0.5
SC 37	47	31	22	52	31	17	60	13	27	21	10	69	53	5	42	11	29	60	93	5	2	0.67
SC 42	47	22	31	49	22	29	69	13	18	7	7	86	39	1	60	6	46	48	85	14	1	0.59
SC 45	49	29	22	51	29	20	63	11	26	9	7	84	44	3	53	16	37	47	91	9	0	0.7
SC 46	47	28	25	47	28	25	63	18	19	3	7	90	43	1	56	11	40	49	89	11	0	0.53
SC 47	46	36	18	48	36	16	57	17	26	10	1	89	69	0	31	18	19	63	95	4	1	0.6
SC 49	43	25	32	47	25	28	63	23	14	12	1	87	8	0	92	17	66	17	75	21	4	0.38
SC 50	38	23	39	43	23	34	62	22	16	14	8	78	15	5	80	10	65	25	74	23	3	0.42
SC 51	45	26	29	49	26	25	63	20	17	12	4	84	14	2	84	11	67	22	79	18	3	0.45
SC 52	45	23	32	48	23	29	66	18	16	8	1	91	11	1	88	7	70	23	76	21	3	0.47
SC 72	51	34	15	53	34	13	61	14	25	10	4	86	61	0	39	17	21	62	96	4	0	0.64
SC 73	48	36	16	49	36	15	57	15	28	10	8	82	56	2	42	13	22	65	95	5	0	0.66
SC 74	53	30	17	56	30	14	64	11	25	18	5	77	41	0	59	13	31	56	93	7	0	0.69
SC 75	54	27	19	56	27	17	66	14	20	10	3	87	35	0	65	15	35	50	90	9	1	0.59
X	48	28	24	51	27	22	64	16	20	11	5	84	38	1	61	12	43	45	87	12	1	0.56
SD	±4	±5	±7	±4	±5	±6	±4	±4	±5	±4	±3	±6	±19	±2	±19	±4	±18	±17	±8	±7	±1	±0.10
Quartzofeldspathic subpetrofacies 1 (Frosinone and lower portions of the Argilloso-Arenacea formations)																						
	48	27	25	50	27	23	64	15	21	11	6	83	41	2	57	11	39	50	88	11	1	0.6
	±4	±4	±6	±4	±4	±6	±4	±4	±5	±4	±4	±6	±18	±2	±18	±4	±18	±18	±7	±6	±1	±0.11

(continued.)

TABLE A2. (Continued.)

Sample no.	QmFLt (%)			QtFL (%)			QmKP (%)			QpLvmLsm (%)			LmLvLs (%)			RgRsRm (%)			NCE CE+Cln-co Clco (%)			P/F (%)
	Qm	F	Lt	Qt	F	L	Qm	K	P	Qp	Lvm	Lsm	Lm	Lv	Ls	Rg	Rs	Rm	NCE	CE+Cln-co	Clco	P/F
Argilloso-Arenacea Formation: middle-upper portions (quartzofeldspathic subpetrofacies 2)																						
02-SC 631	44	37	19	57	37	6	54	6	40	67	7	26	74	4	22	50	3	47	99	1	0	0.87
02-SC 632	50	32	18	62	32	6	61	5	34	67	5	28	57	10	33	45	5	50	98	2	0	0.87
02-SC 633	47	25	28	66	25	9	66	4	30	66	2	32	58	0	42	34	9	57	97	3	0	0.87
02-SC 634	49	31	20	62	31	7	61	10	29	65	8	27	44	19	37	40	7	53	98	2	0	0.75
02-SC 635	48	31	21	62	31	7	61	15	24	68	2	30	57	0	43	33	6	61	97	3	0	0.61
02-SC 636	46	34	20	58	15	27	58	15	27	70	3	27	48	0	52	37	6	57	98	2	0	0.65
02-SC 637	52	27	21	67	27	6	66	9	25	71	2	27	58	4	38	27	9	64	98	2	0	0.73
02-SC 638	27	12	61	42	12	46	69	13	18	24	5	71	10	0	90	6	61	33	45	29	26	0.59
02-SC 639	48	31	21	63	31	6	61	16	23	72	3	25	67	0	33	38	5	57	99	1	0	0.59
02-SC 641	54	26	20	67	11	22	68	11	21	64	3	33	24	5	71	32	22	46	96	4	0	0.66
02-SC 642	39	36	25	53	36	11	52	15	33	56	2	42	43	0	57	40	14	46	95	5	0	0.68
02-SC 643	46	30	24	59	31	10	60	19	21	57	4	39	54	0	46	29	13	58	96	4	0	0.52
02-SC 644	51	23	26	66	23	11	69	3	28	58	3	39	69	2	29	31	7	62	98	2	0	0.9
02-SC 645	49	22	29	66	22	12	69	5	26	60	4	36	64	5	31	35	9	56	98	2	0	0.83
02-SC 646	41	31	28	56	31	13	57	14	29	53	12	35	38	12	50	48	15	37	95	5	0	0.67
02-SC 647	59	23	18	70	23	7	72	3	25	62	5	33	50	5	45	20	15	65	98	2	0	0.88
02-SC 619	43	30	27	61	30	9	59	0	41	66	10	24	60	12	28	39	6	55	98	2	0	0.99
02-SC 620	35	42	23	5	42	8	46	8	46	62	5	33	88	6	6	50	1	49	100	0	0	0.85
02-SC 621	41	35	24	57	35	8	55	7	38	65	6	29	61	6	33	30	5	65	98	2	0	0.84
02-SC 622	33	32	35	52	32	16	51	5	44	55	11	34	47	4	49	39	15	46	94	6	0	0.89
02-SC 623	34	34	32	48	34	18	50	9	41	45	6	49	39	4	57	34	22	44	91	8	1	0.82
02-SC 624	44	27	29	61	27	12	62	6	32	58	6	36	44	8	48	34	17	49	96	4	0	0.85
02-SC 625	28	35	37	44	35	21	44	8	48	43	12	45	48	8	44	31	20	49	93	6	1	0.86
02-SC 626	47	32	21	59	32	9	60	4	36	59	10	31	64	11	25	44	6	50	97	2	1	0.89
02-SC 627	31	47	22	41	47	12	40	12	48	45	10	45	41	6	53	51	12	37	94	5	1	0.8
02-SC 628	32	50	18	43	50	7	39	7	54	59	10	31	36	14	50	45	10	45	97	3	0	0.88
02-SC 629	39	30	31	57	30	13	56	6	38	58	9	33	35	9	56	39	16	45	94	6	0	0.87
02-SC 630	52	15	33	66	16	18	77	11	12	43	0	57	43	0	57	21	41	38	64	9	27	0.54
X	44	30	26	59	28	13	59	9	32	60	5	35	51	5	44	36	13	51	94	4	2	0.78
SD	±8	±8	±9	±8	±9	±8	±8	±5	±9	±11	±3	±10	±16	±5	±16	±10	±12	±9	±12	±5	±7	±0.13

grains are coarse-grained, generally coarser than associated terrigenous clasts, mostly angular or subangular, and lack any alteration coating. These particles are dominantly fossiliferous limestone and dolostone of Mesozoic age. Textural characteristics of carbonate detritus derived from this intrabasinal structural high source contrast with the extrabasinal carbonate lithics in the siliciclastic sandstones of the Frosinone and Argilloso-Arenacea Formations. These latter extrabasinal carbonate fragments are dominantly altered, and rounded, and have the same grain size as noncarbonate detritus.

Extrabasinal Components

Sand grains that have an extrabasinal provenance are subdivided into noncarbonate and carbonate particles (Fig. 7); these grains are noncoeval with respect to the age of deposition.

Noncarbonate Extrabasinal Grains (NCE)

Noncarbonate extrabasinal grains include noncoeval siliciclastic minerals (quartz, feldspars, phyllosilicates, and dense minerals), and siliciclastic aphanitic lithic fragments and phaneritic rock fragments (Figs. 7A–7D). Quartz is abundant, particularly in turbidite sandstones of the Frosinone and Argilloso-Arenacea Formations; it is present as both monocrystalline and polycrystalline grains. Feldspars grains are also abundant in siliciclastic turbidites of the Frosinone and Argilloso-Arenacea Formations, including both plagioclase and K-feldspar (P/F ratio ranging from 0.6 to 0.78). Noncarbonate lithic fragments include abundant metasedimentary phyllite and fine-grained schist, and ophiolitic serpentinite with cellular texture, serpentine-schist, and glaucophane-bearing schist. Other lithic fragments include volcanic lithic fragments that have various textural attributes ranging in composition from rhyolitic to rhyodacitic and andesite; impure radiolarian chert and argillaceous chert, siltstone, and shale fragments are also present.

Phaneritic rock fragments are represented by felsic plutonic fragments, and medium- to high-grade metamorphic fragments, including sillimanite and garnet.

Siliciclastic detritus of the Brecce della Renga Formation is similar to turbidite strata of the Frosinone and Argilloso-Arenacea Formations.

Carbonate Extrabasinal Grains (CE)

Extrabasinal carbonate lithic fragments include diverse grain types, mostly carbonate mudstone or micritic limestone, wackestone, biomicritic limestone, sparite, and biosparite limestone and dolostone (Table A1; Figs. 7E and 7F). The extrabasinal carbonate lithic fragments are common in all the samples; they are noncoeval grains. Their source rocks include Mesozoic to Lower Tertiary pelagic successions, Cretaceous deep-water carbonate strata covering oceanic sequences, and Mesozoic to Lower Miocene shallow-water carbonate platform successions.

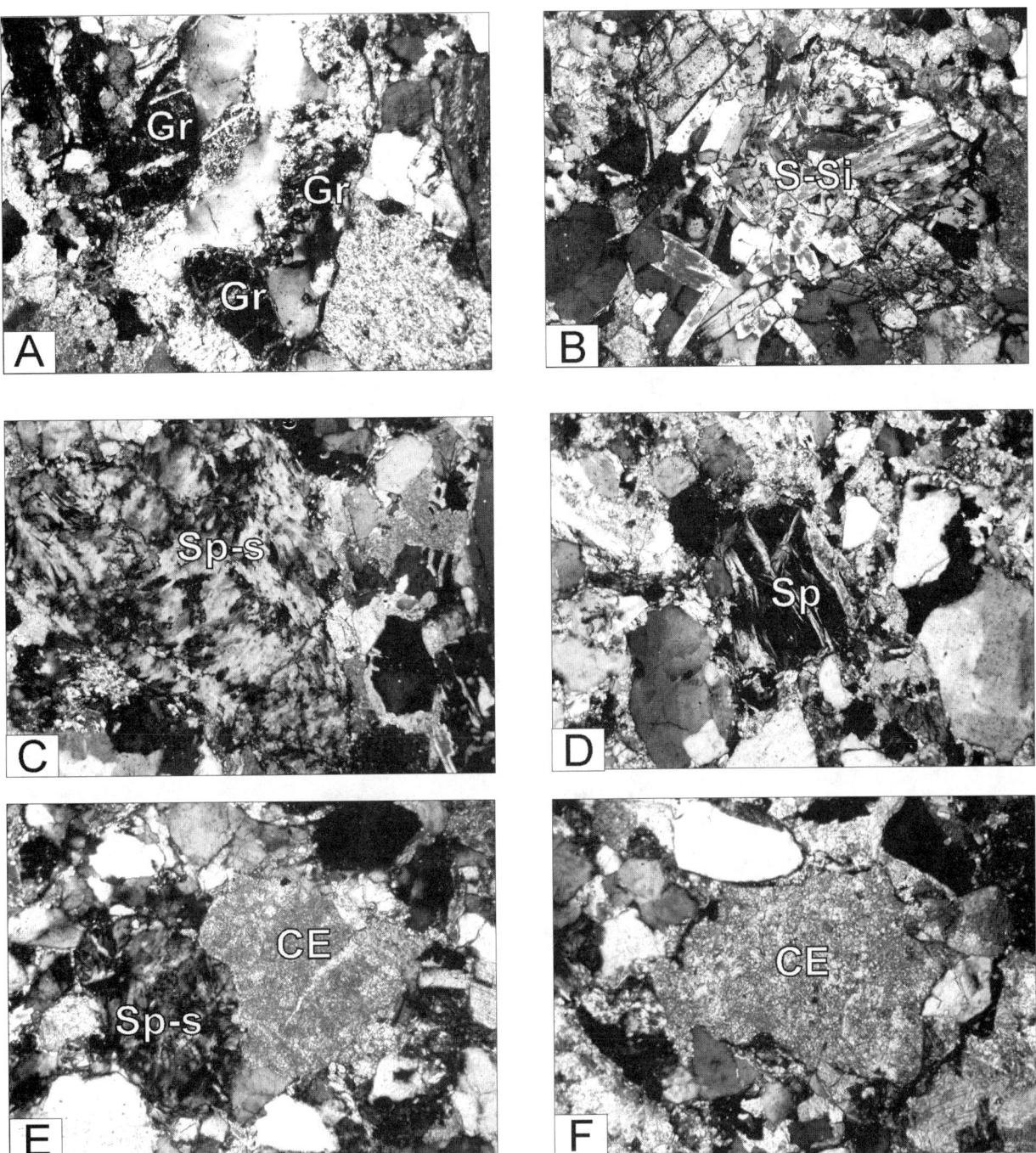

Figure 7. Photomicrographs of extrabasinal noncarbonate and carbonate sand grains in the quartzofeldspatholithic petrofacies of late Miocene arenites of central Apennines foreland basin system. (A–B) Coarse- to fine-grained metamorphic particles of (A) garnet-bearing (Gr) gneiss, and (B) sillimanite-bearing schist (S-Si). (C–D) Ophiolitic fragments of (C) serpentine schist (Sp-s) and (D) serpentinite (Sp). (E–F) Sedimentary carbonate lithic fragments (CE) of pelagic biomicritic (packstone) lithics. All photos are in cross-polarized light. Field size of photomicrograph is 1 × 0.7 mm.

Intrabasinal Components

Arenites of the studied successions contain different types of intrabasinal sand grains, both carbonate and noncarbonate in composition. Carbonate intrabasinal grains have been subdivided into coeval and noncoeval particles with respect to the age of the deposit (Fig. 8). The presence of noncoeval intrabasinal grains is peculiar to the sedimentary fill of the central Apennines foreland basin system.

Noncoeval Carbonate Intrabasinal Grains ($CI_{noncoeval}$)

Noncoeval carbonate intrabasinal grains are represented by carbonate coarse-grained sediments of the Brecce della Renga

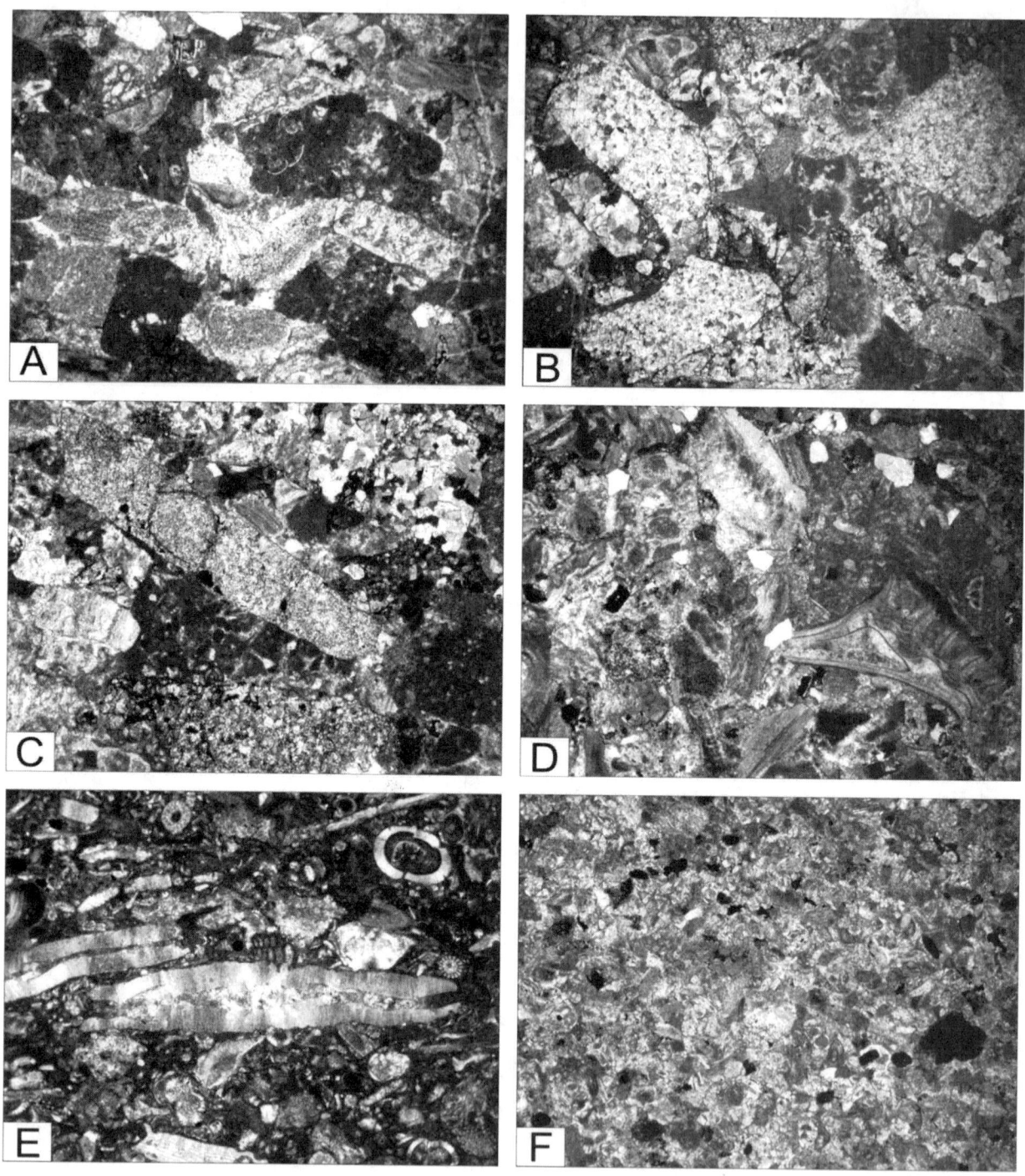

Figure 8. Photomicrographs of noncoeval and coeval intrabasinal carbonate and noncarbonate sand grains in calclithitic (A–D) and intra-arenitic (E–F) petrofacies of late Miocene arenites of central Apennines foreland basin system. Calclithite petrofacies of Brecce della Renga Formation consists of dominantly noncoeval intrabasinal carbonate particles that have generally larger grain size with respect to the extrabasinal grains. (A) Abundant biomicritic limestone particles that have angular to subangular shapes. (B–C) Mixing of dolostone, sparitic and biomicritic limestone particles. (D) Mixing of biomicritic limestone, coeval bioclasts, and noncarbonate extrabasinal particles. Intra-arenitic petrofacies of Orbulina Marl consists of dominantly coeval bioclasts (E), including glauconite (F) and minor extrabasinal grains. A and E are in plane-polarized light; B, C, D, and F are in cross-polarized light. Field size of photomicrograph is 1 × 0.7 mm.

Formation that represent a rather unique submarine deposit, interbedded with siliciclastic turbidite strata. The noncoeval carbonate grains, in deposits ranging from megabreccias to conglomerates and arenites, are represented by detritus that has a local intrabasinal source. Noncoeval carbonate intrabasinal particles are dominant in the Brecce della Renga Formation arenites, ranging from 99% to 40% of the whole rock. These grain types include several types of carbonate textures (Table A1; Figs. 8A–8D). Single noncoeval fossils are rare or minor. All these noncoeval carbonate intrabasinal particles are coarser-grained than the other associated terrigenous clasts, and they are mostly angular or subangular. These grains, which do not show alteration coating, mostly represent shallow-water carbonate-platform environments. Fossiliferous limestone/dolostone lithic fragments range from Jurassic to Lower Miocene, and typically include the following types of grain:

1. dolostone, mudstone, and oolitic limestones of probable Jurassic?–Early Cretaceous age;
2. mudstones and wackestones with *Orbitolina*, *Globotruncana appenninica*, and *Ovalveolina maccagnoi* of Aptian-Cenomanian age;
3. grainstones with rudist fragments, orbitoids, and echinoids of Maastrichtian age;
4. packstones and wackestones with nummulitids of Paleogene age; and
5. rare silty-arenitic limestones bearing echinoids, pectinids, and *Amphistegina* of early Miocene age.

Coeval Carbonate Intrabasinal Grains (CI_{coeval})

Turbidite beds of the Frosinone and Argilloso-Arenacea Formations, the Brecce della Renga Formation, and hemipelagic strata of the Orbulina Marl include variable proportions of coeval carbonate particles derived from active shallow-water shelf environments or from the pelagic (biological) factory (Fig. 8E). Shallow-water carbonate platforms at the time were of the temperate to subtropical-tropical type and were characterized by a heterozoan association (e.g., James, 1997; Corda and Brandano, 2003; and references therein), so coeval carbonate intrabasinal particles include planktonic foraminifera, coralline algae, bryozoa, worm tubes, undetermined shell fragments, peloids, and micrite intraclasts. Coeval carbonate intrabasinal detritus is dominant in arenites of the Orbulina Marl, representing 90% ($NCE_1 CE_9 CI_{90}$; Table A2) of the framework composition (more than 60% of the whole rock; Table A1). It is also an important component in arenites of the Brecce della Renga Formation, representing 10% of the framework composition, whereas it makes up only 2% of the siliciclastic turbidite sandstones of the Frosinone and Argilloso-Arenacea Formations.

Noncarbonate Intrabasinal Grains (NCI)

In the studied successions, coeval intrabasinal noncarbonate grains include glauconite, Fe-oxide, and phosphate concretions and argillaceous rip-up clasts. Glaucony (Fig. 8F) is particularly abundant in the arenitic strata of the Orbulina Marl, and it is associated with the presence of phosphatic and Fe-oxide concretions. Turbidite sandstones of the Frosinone and Argilloso-Arenacea Formations include intrabasinal mudclasts (rip-up clasts).

ARENITE DETRITAL MODES

The analyzed Upper Miocene arenites are subdivided into three key petrofacies, with further subdivisions into subpetrofacies based on clusters seen on the plots in Figure 9. These main petrofacies mostly coincide with lithostratigraphic subdivisions, and include: (1) intra-arenitic petrofacies, typical of arenite strata of the Orbulina Marls, (2) quartzofeldspatholithic petrofacies, which is mostly typical of siliciclastic turbidites of the Frosinone and the Argilloso-Arenacea Formations, with a few quartzofeldspatholithic strata within the Brecce della Renga Formation, and (3) calclithic petrofacies, which is typical of the submarine carbonate-apron deposits of the Brecce della Renga Formation. A few calclithic arenite strata are also interbedded within siliciclastic turbidites of the Argilloso-Arenacea Formation.

Intra-Arenitic Petrofacies

The intra-arenites of this petrofacies contain dominant coeval intrabasinal carbonate grains, represented by abundant bioclasts, and minor intraclasts and peloids; noncarbonate grains include glauconite and rare phosphatic grains. These intra-arenite strata are typical of the Orbulina Marl, even if isolated intra-arenite strata are also interbedded within the Brecce della Renga Formation and the Argilloso-Arenacea Formation.

Quartzofeldspatholithic Petrofacies

The quartzofeldspatholithic sandstone of this petrofacies ($Qm_{46}F_{27}Lt_{27}$) contains abundant quartz and equal proportions of feldspar and lithic fragments. Plagioclase grains are more abundant than K-feldspar (P/F = 0.7). Aphanitic lithic grains include abundant sedimentary and metasedimentary, and minor volcanic, metavolcanic, and ophiolitiferous lithic particles ($Lm_{42}Lv_2Ls_{56}$). Sedimentary lithic fragments consist of abundant extrabasinal carbonate fragments, mostly Jurassic to Lower Tertiary pelagic carbonate fragments, and minor radiolarian chert, siltstone, and shale fragments. Metasedimentary lithic fragments include abundant phyllite, slate, and fine-grained schist; ophiolitiferous detritus includes serpentine-schist, serpentinite, and metavolcanic (metabasalts) and volcanic (microlitic basalt) lithic fragments, whereas volcanic lithic fragments are minor and consist of lithic particles that have a felsitic granular texture. Phaneritic rock fragments (apportioned in Qm, P, K, micas, and dense minerals; e.g., Critelli and Le Pera, 1994; Critelli and Ingersoll, 1995; Critelli et al., 1995) include plutonic and metamorphic detritus; plutonic detritus consists of dominantly quartz-plagioclase-biotite particles of granodioritic to tonalitic compositions, and minor quartz-K-feldspar composite grains of granitic composition. Phaneritic metamorphic detritus consists of quartz-plagioclase-sillimanite,

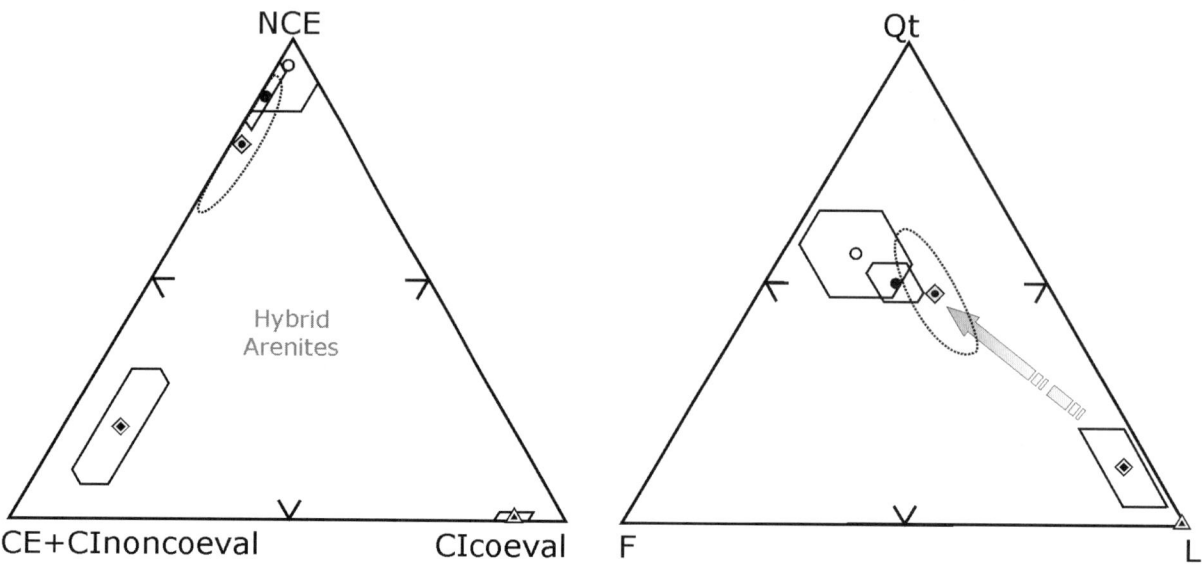
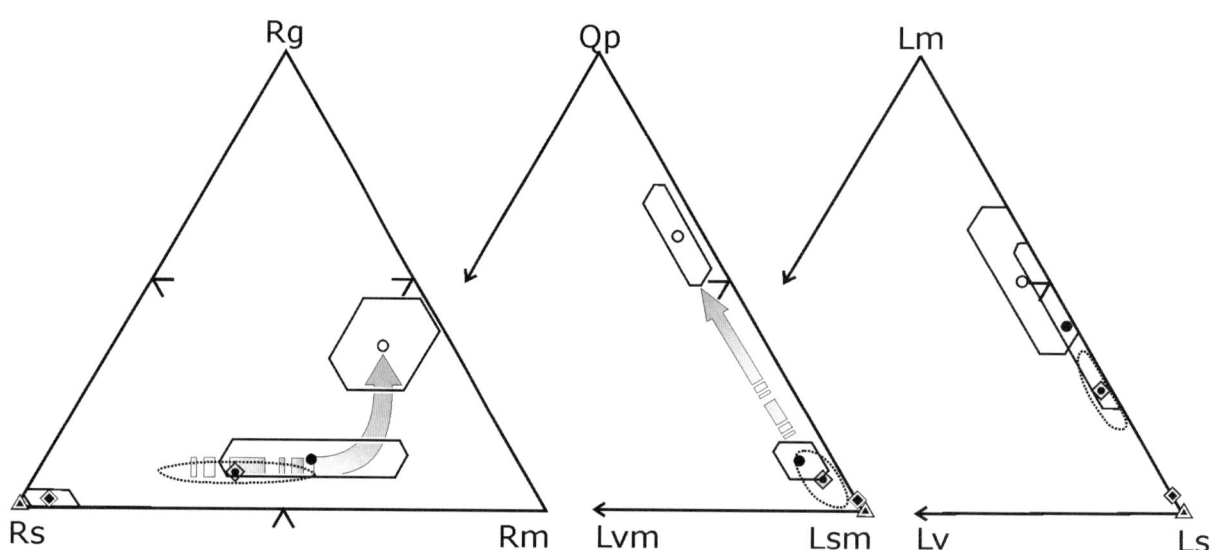

Quartzofeldspatholithic Petrofacies (Siliciclastic arenites)
○ Upper Argilloso-Arenacea Formation (early Messinian)
● Lower Argilloso-Arenacea (early Messinian) and Frosinone Formations (late Tortniann)
◆ Brecce della Renga Formation (Serravallian? to early Messinian)

 Evolutionary trends

Calclithite Petrofacies
◆ Brecce della Renga Formation (Serravallian? to early Messinian)

Intra-arenite Petrofacies
△ Orbulina Marl Formation (Serravallian? to early Messinian)

Figure 9. Ternary plots of NCE (noncarbonate extrabasinal)–CE (carbonate extrabasinal) + $CI_{noncoeval}$ (noncoeval carbonate intrabasinal)–CI_{coeval} (coeval carbonate intrabasinal); Qt (total quartz = monocrystalline quartz + polycrystalline quartz)–F (feldspars)–L (aphanitic lithic fragments); Rg (plutonic rock fragments)–Rs (sedimentary rock/lithic fragments)–Rm (metamorphic rock/lithic fragments); Qp (polycrystalline quartz)–Lvm (volcanic and metavolcanic lithic fragments)–Lsm (sedimentary and metasedimentary lithic fragments); and Lm (metamorphic lithic fragments)–Lv (volcanic lithic fragments)–Ls (sedimentary lithic fragments); data are for late Miocene arenites of central Apennines foreland basin system. Means and field of variations (polygons) are defined by 1 standard deviation on either side of mean.

and quartz-plagioclase-garnet composite grains of gneiss, mica schist, and epidote-rich schist.

The quartzofeldspatholithic sandstone petrofacies characterizes huge volumes of siliciclastic turbidite sediments filling the late Tortonian to early Messinian tectonic-controlled subbasin. The Frosinone Formation and the diverse turbidite depositional systems of the Argilloso-Arenacea Formation have dominantly quartzofeldspatholithic sandstone. Few strata of quartzofeldspatholithic sandstones of this petrofacies are interbedded within the Brecce della Renga Formation.

In spite of the rather homogeneous quartzofeldspatholithic compositions, some significant petrologic differences allow distinction of two main subpetrofacies, corresponding to the Frosinone Formation and the lower part of the Argilloso-Arenacea Formation (system 1), versus the middle-to-upper part of the Argilloso-Arenacea Formation (systems 2, 3 and 4). The boundary of the two quartzofeldspatholithic subpetrofacies corresponds to the system 1–system 2 boundary of the Argilloso-Arenacea Formation.

The lower quartzofeldspatholithic subpetrofacies (Frosinone Formation and system 1 of the Argilloso-Arenacea Formation) has a late Tortonian to early Messinian age. Sandstones include abundant quartz ($Qm_{47} F_{28} Lt_{25}$). The total detrital budget of both phaneritic and aphanitic particles suggests a mixing of provenance from sedimentary and metamorphic source terranes and minor plutonic sources ($Rg_9 Rs_{35} Rm_{56}$). Ophiolitic detritus represents serpentinized oceanic crust and its sedimentary cover (radiolarian chert, pelagic limestone, and shale fragments).

The upper quartzofeldspatholithic subpetrofacies ($Qm_{44} F_{30} Lt_{26}$; P/F = 0.8; systems 2, 3, and 4 of the Argilloso-Arenacea Formation) has an early Messinian age. The total detrital budget suggests an increase of plutonic detritus and an abrupt reduction of sedimentary detritus ($Rg_{36} Rs_{13} Rm_{51}$).

The quartzofeldspatholithic sandstones of the Brecce della Renga Formation are interbedded with the calclithite petrofacies. These quartzofeldspatholithic strata have an additional source with respect to the quartzofeldspatholithic sandstones of the Frosinone and Argilloso-Arenacea Formations, represented by a noncoeval intrabasinal carbonate area. The sedimentary budget of these Brecce della Renga quartzofeldspatholithic sandstones includes 82% extrabasinal detritus (78% noncarbonate and 4% carbonate) and 18% intrabasinal detritus (16% noncoeval, and 2% coeval).

Calclithite Petrofacies

The term "calclithite" (e.g., Folk, 1974) has traditionally been used for arenites that have abundant noncoeval extrabasinal carbonate lithic fragments (e.g., Zuffa, 1980, 1985, 1987; Fontana et al., 1989). In this study, we use the term "calclithite" as a compositional term, regardless of whether noncoeval carbonate detritus has an intrabasinal or an extrabasinal source. The calclithite petrofacies is exclusively in the Brecce della Renga Formation. Detrital composition of the calclithite petrofacies includes diverse carbonate and noncarbonate particles, both extrabasinal and intrabasinal, and coeval and noncoeval with respect to the age of the deposit ($NCE_{19\%}$–$CE + CI_{noncoeval\ 71\%}$–$CI_{coeval\ 10\%}$). The total sedimentary budget deduced from analysis of the framework composition for arenites of the Brecce della Renga Formation is 22% extrabasinal detritus (19% noncarbonate and 3% carbonate) and 78% intrabasinal detritus (67% noncoeval and 11% coeval).

Extrabasinal noncarbonate particles (NCE) are present in all samples except two. Extrabasinal carbonate (CE) detritus is similar to that of the quartzofeldspatholithic petrofacies.

Intrabasinal sources include two main types of carbonate particles, noncoeval and coeval. Noncoeval intrabasinal carbonate detritus is mostly carbonate lithic fragments, represented by biosparite and biomicrite limestone and dolostone, that have dominantly Jurassic, Cretaceous, and early Miocene ages. These noncoeval carbonate fragments reflect derivation from local sources, such as the shallow-marine successions of the Simbruini Mountain, a portion of the broader Mesozoic to Lower Tertiary Latium-Abruzzi carbonate platform. Coeval intrabasinal carbonate particles account for ~10% of the entire detrital sedimentary budget of the Brecce della Renga Formation and include single bioclasts, intraclasts, and peloids. Coeval intrabasinal noncarbonate particles represent <1%; they include glauconite and rip-up clasts.

SOURCE-BASIN PALEOGEOGRAPHY

Sedimentological and stratigraphic characteristics of the analyzed lithostratigraphic units, along with the arenite compositional data, provide support for a reconstruction of the late Miocene source-basin paleogeography of the more internal sector of the central Apennines foreland basin system. Several petrologic parameters indicate that the Orbulina Marl, Frosinone, Argilloso-Arenacea, and Brecce della Renga Formations are mixed silicic-carbonate-clastic detrital successions, in which carbonate, metamorphic, ophiolitic, and plutonic detritus is abundant.

Provenance of Arenites of the Central Apennines Foreland Basin System

The general detrital mode evolution of the late Miocene central Apennines foreland basin system reflects the interplay of two source areas: the orogenic system and the intrabasin ridges. The peculiar signal of the sedimentary infilling of the analyzed subbasins is the presence of an intrabasinal carbonate source, in the form of structural highs, where the Mesozoic to Lower Miocene carbonate succession was exposed at the seafloor and subjected to erosion.

Modal analyses of arenites for these formations indicate plutonic-metamorphic, ophiolitic, and sedimentary source terranes for the extrabasinal detritus, derived from the foreland fold-and-thrust belt. The intrabasinal detritus includes both older (noncoeval) and coeval carbonate detritus that was derived from erosion of the local limestone substrate and from areas with active benthic carbonate production, respectively.

The influences of intrabasinal and extrabasinal tectonics and the input of noncoeval carbonate detritus derived from both subaerial and submarine sources within the rapidly evolving late Miocene central Apennines foreland basin are revealed by the spatial- and time-dependent arenite petrofacies distribution of the Orbulina Marl, Frosinone, Argilloso-Arenacea, and Brecce della Renga Formations.

Early foreland sedimentation in the central Apennines was marked by deposition of the Orbulina Marl, a hemipelagic unit mantling the articulated substratum of the foredeep (Fig. 10). These hemipelagites consist of mudstone and wackestone in the lower portions, passing vertically into shale and turbiditic mudstone and arenite. Arenite strata have abundant coeval intrabasinal carbonate particles, including planktonic foraminifera and assorted bioclasts indicating shallow to deeper shelf environments. The abundance of coeval intrabasinal carbonate grains suggests an active carbonate platform source for the turbidite intra-arenites of the Orbulina Marl in the early Tortonian. The source areas for coeval carbonate detritus were most probably those more external sectors of the Latium-Abruzzi carbonate platform domain (Figs. 10 and 11), where flexure-related drowning of the Briozoi and Litotamni Limestone platform had not yet taken place (Galluzzo and Santantonio, 1995).

Growth of the fold-and-thrust belt is recorded by hemipelagic deposits of the Orbulina Marl at the base, passing laterally and vertically to turbidite sandstone of the Frosinone and Argilloso-Arenacea Formations. The onset of turbidite sedimentation marked the eastward propagation of the thrust belt (Bigi et al., 2003), which produced abrupt paleogeographic and paleotectonic changes and depocenter migration within the complex foredeep. In the internal sector of the chain, these major tectonic events were accomplished by deformation of the Umbria-Marche-Sabina pelagic successions, and by eastward displacement of large crustal slices of the Alpine internal crystalline and ophiolitic nappes following backarc extension of the Tyrrhenian Sea (e.g., Doglioni et al., 1998; Carminati et al., 2004). Quartzofeldspatholithic sandstones of the turbidite strata were derived from the accretionary fold-and-thrust belt, where ophiolitic and metamorphic detritus are well represented (see also compositional data reported in Chiocchini and Cipriani, 1989, 1992). In addition, extrabasinal carbonate grains had to be derived from both ophiolitic units and the pelagic rocks of the Umbria-Marche-Sabina units.

The conclusions that can be drawn from analysis of framework modes of the Frosinone and Argilloso-Arenacea Formations, by using traditional diagrams, are clearly in agreement with the regional geologic setting and reflect a recycled-orogenic source area (i.e., Dickinson, 1985). In order to highlight vertical trends, the RgRsRm plot is used to evaluate both aphanitic and phaneritic lithic/rock fragments (e.g., Critelli and Le Pera, 1994; Critelli et al., 1995). Figure 9 depicts clear evolutionary trends in the detrital modes of the Frosinone and Argilloso-Arenacea Formations that represent an upward increase of plutonic and metamorphic detritus during deposition of the siliciclastic turbidite systems. These vertical compositional trends evidently reflect progressive unroofing.

As evidenced by many structural and sedimentological data, the central Apennines foredeep physiography was complex (e.g., Ricci Lucchi, 1986; Argnani and Ricci Lucchi, 2001); subbasins were filled by siliciclastic turbidites, separated by intrabasinal structural highs bounded by prethrusting normal faults (Figs. 10 and 11).

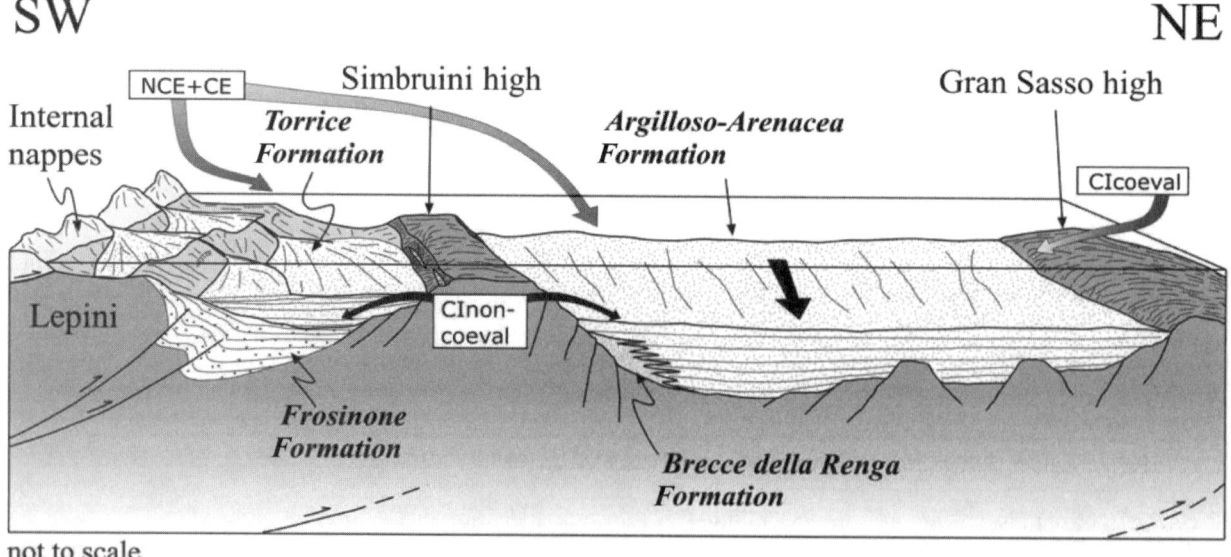

Figure 10. Block diagram showing central Apennines foreland basin system during the early Messinian. Arrows indicate main turbidity flow directions and diverse provenance terranes from extrabasinal and intrabasinal coeval and noncoeval sources. The Orbulina Marl is not represented in the figure, causing the reduced thickness of the mantle in the articulated substrate of the foredeep. (Modified after Milli et al., 2004.)

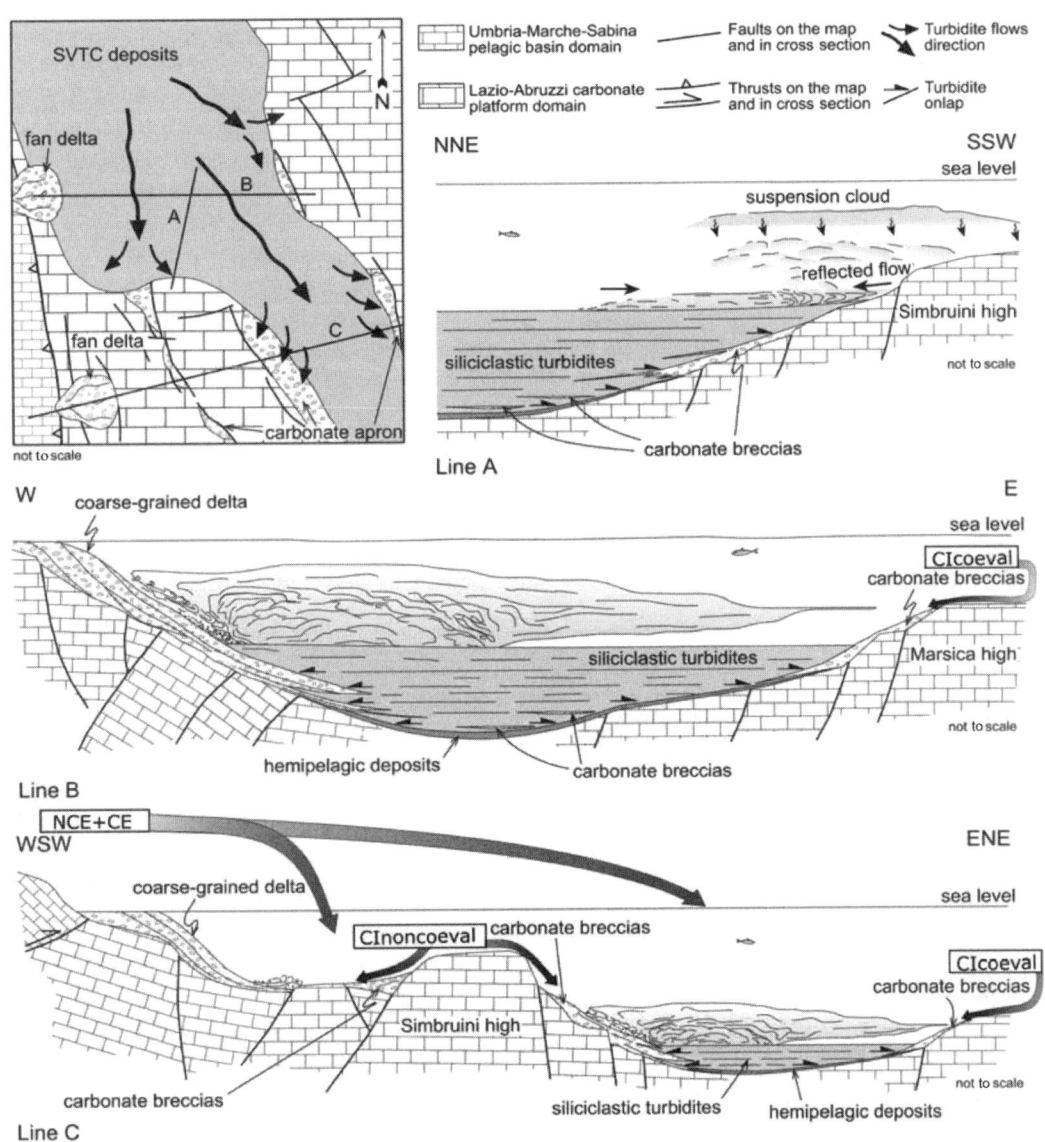

Figure 11. Depositional stratigraphic cross sections (lines A, B, C) showing paleogeographic relationships between Argilloso-Arenacea Formation and breccia and arenite deposits of the Brecce della Renga Formation that crop out along the northern and eastern sectors of the Simbruini Mountains. SVTC—Valle del Salto-Val di Varri Turbiditic Complex. (Modified after Milli and Moscatelli, 2001.)

One such structural high was the Simbruini structure, corresponding to a part of the Simbruini Mountains, which during the late Miocene was bordered by carbonate-clastic sediments of the Brecce della Renga Formation. Carbonate-clastic detritus is mainly conglomerate-to-breccia and coarse arenite, and it reflects local sourcing from the Mesozoic to Tertiary Latium-Abruzzi carbonate platform, which is documented in the calclithite fraction. The total sedimentary budget for clastics of the Brecce della Renga Formation includes a provenance of 78% from intrabasinal sources and 22% from extrabasinal sources (19% noncarbonate and 3% carbonate). Intrabasinal provenance is 67% from intrabasinal carbonate highs (noncoeval intrabasinal, $CI_{noncoeval}$) and 11% from coeval shelfal intrabasinal carbonates (CI_{coeval}). Extrabasinal provenance includes dominantly silicate provenance (extrabasinal noncarbonate, NCE), represented by metamorphic and ophiolitic, and minor plutonic detritus, and a subordinate carbonate provenance (CE), represented by dominantly pelagic carbonate fragments derived from both the Neotethyan ophiolitic units and from Adria margin pelagic sequences of the Umbrian-Marche-Sabina units.

Provenance of Carbonate Particles Based on Their Spatial and Temporal Attributes

Many attempts have been made to distinguish diverse carbonate particles in sand(stone) (e.g., Zuffa, 1980, 1985, 1987, 1991; Mount, 1984, 1985; Fontana et al., 1989; Fontana, 1991; Spence and Tucker, 1997; Arribas and Tortosa, 2003) for reconstructing source-basin relationships. Numerous studies in ancient and modern sediments have shown that carbonate particles are produced across a vast range of sedimentary systems, from continental to deep-marine environments, virtually in all geotectonic settings (e.g., Gazzi et al., 1973; Cavazza et al., 1993; Le Pera and Critelli, 1997; Arribas et al., 2000; Garzanti et al., 2002; Arribas and Tortosa, 2003; Critelli et al., 2003; Di Giulio et al., 2003; Fontana and Stefani, 2003; Fontana et al., 2003; Zuffa et al., 2003; Le Pera and Arribas, 2004).

Sand(stone) that has significant proportions of carbonate detritus may occur in rifted-continental margins (e.g., Zuffa et al., 1980; Arribas et al., 2003), continental blocks (e.g., Le Pera and Arribas, 2004), and during closure of remnant-ocean basins (e.g., Suczek and Ingersoll, 1985; Critelli et al., 1990; Fontana, 1991; Critelli, 1993; Fontana et al., 1994; Fontana and Stefani, 2003; Zuffa et al., 2003) and subsequent continental collision. During the evolution of foreland basin systems, carbonate particles can be a very significant detrital component both as extrabasinal or intrabasinal coeval and noncoeval grains (e.g., Gandolfi et al., 1983; Schwab, 1986; Ingersoll et al., 1987; Fontana et al., 1989; Critelli and Ingersoll, 1994; Critelli and Le Pera, 1994, 1998; Critelli, 1999; Cibin et al., 2001), and can be derived from both fold-and-thrust belts and foreland regions.

The case history presented here highlights the generation of different carbonate particles supplied to the basin as a result of structural control. The late Miocene physiography of the central Apennines foreland basin system (Figs. 10 and 11) was partially controlled by prethrusting normal faults and thrust tectonics, which produced:

1. a fold-and-thrust belt in which extrabasinal carbonate (CE) source rocks were dominantly pelagic, derived from sedimentary cover of ophiolitic units and from deformation of pelagic basins of the Adria margin (i.e., the Umbria-Marche-Sabina basin); and

2. a foredeep with a complex seafloor topography, in which noncoeval intrabasinal carbonate (breccia deposits) and coeval intrabasinal carbonate (bioclasts of shallow-water organisms, and minor intraclasts) particles were derived from structural highs (e.g., Simbruini high).

The extrabasinal carbonate particles (CE) are generally rounded to well-rounded, with oxide rims and weathering traces, and they have a grain size similar to that of other noncarbonate particles. These particles occur in the siliciclastic turbidites of the Frosinone and Argilloso-Arenacea Formations, and in the calcithite strata of the Brecce della Renga Formation, while they are minor or absent in arenites of the Orbulina Marl. The intrabasinal noncoeval carbonate particles ($CI_{noncoeval}$) are almost exclusively of the Brecce della Renga Formation, with only minor quantities derived from turbidite deposits. The $CI_{noncoeval}$ particles have generally larger grain size, and are more angular to subangular grains with respect to extrabasinal detritus (both carbonate and noncarbonate). These $CI_{noncoeval}$ particles do not have oxide rims or weathering traces, and they differ from extrabasinal carbonate particles (CE) in having been derived from an uplifted shallow-water carbonate succession.

CONCLUSIONS AND PALEOGEOGRAPHIC IMPLICATIONS

Arenites of the Orbulina Marl, Frosinone, Argilloso-Arenacea, and Brecce della Renga Formations provide detailed information on the distribution of siliciclastic and carbonate detritus, and record the tectonic-sedimentary evolution of the late Miocene central Apennines foreland basin system. Lithostratigraphic units preserve a complete record of accretionary processes and erosion of the fold-and-thrust belt along the Adria margin and sedimentation response in the foredeep. Since the middle-late Miocene boundary, the central Apennines have experienced deformation, which has produced a complex physiography of the foreland basin system with diverse extrabasinal and intrabasinal sources. Three key arenite petrofacies, reflecting the paleotectonic and paleogeographic evolution of the source-basin system, describe the dispersal pathways and provenance regions. The foreland basin was dominantly filled by quartzofeldspatholithic sandstone petrofacies of the Frosinone and Argilloso-Arenacea Formations, indicating provenance from the growing fold-and-thrust belt. This quartzofeldspatholithic petrofacies was preceded and accompanied by deposition of hemipelagic deposits with interbedded turbidite intra-arenite petrofacies of the Orbulina Marl, and it was accompanied by deposition of calclithite petrofacies of the Brecce della Renga Formation. The diachroneity of lithostratigraphic boundaries across the basin, one remarkable example being the base of the Orbulina Marl, which records the drowning of the heterozoan-type carbonate platform, is evidence for stepwise deformation of the foreland-foredeep system.

In spite of its homogeneous composition, the quartzofeldspatholithic sandstone petrofacies bears some relevant information about the structural evolution of the fold-and-thrust belt. The basal Argilloso-Arenacea Formation (system 1) has a composition similar to that of the Frosinone Formation. The passage from system 1 to system 2 in the low to middle portions of the Argilloso-Arenacea Formation is recorded in the field by a sharp discontinuity surface. The quartzofeldspatholithic sandstone petrofacies suggests uplift of the orogen. This is recorded by the change of crystalline detritus through time, which evolved from dominantly metamorphic provenance to plutonic provenance. This would indicate that the orogenic chain, representing the main source area, underwent deep erosion coupled with fast exhumation rates. The presence of ophiolitic detritus in the quartzofeldspatholithic sandstone petrofacies suggests the accretion and exhumation of the oceanic accretionary wedge (ophiolitic units); the presence of extrabasinal carbonate detritus is an important element in confirming that the Mesozoic and Tertiary pelagic successions had been accreted to the orogen by late Tortonian–early Messinian times.

The calclithite petrofacies of the Brecce della Renga Formation was clearly derived from the fragmented Mesozoic to Tertiary Latium-Abruzzi carbonate platform (Simbruini high). Carbonate clastic deposition initially occurred along the margins of intrabasinal structural highs, starting around the Serravallian–

Tortonian boundary, and continued parallel to inversion of pre-thrusting extensional structures in the Messinian.

Sedimentological and stratigraphic studies (e.g., Milli and Moscatelli, 2000; Moscatelli, 2003) have shown that the turbidite flows were deposited at an essentially constant depth of around a few hundred meters, in areas very proximal to the feeder systems and at the front of the Apennine thrust belt. The reduced size of subbasins forced the flows to impact the basin margins, giving rise to reflection and deflection processes that are recorded in turbidite strata (Fig. 11).

In conclusion, integration of sedimentology, stratigraphy, and petrostratigraphy for the Tortonian to early Messinian foreland basin system of the central Apennines documents the complex paleogeographical and paleotectonic relations of source-basin systems during foreland evolution. This study contributes to our understanding of dispersal pathways and provenance during orogenic evolution. The compositions of arenites from this foreland basin system provide an excellent example of the changing nature of the entire foreland region source area through space and time. Dispersal pathways of mixed siliciclastic and diverse carbonate sources resulted in significant changes in sandstone compositions that primarily reflect response to tectonism of a fold-and-thrust belt and foredeep. This study confirms that documentation of the compositional, spatial, temporal, and textural characteristics of carbonate grains in sand(stone) may be used to constrain paleogeographic and paleotectonic reconstructions of source-basin systems on a detailed scale.

ACKNOWLEDGMENTS

We are grateful to G.G. Zuffa, whose ideas were important to the methodological conception of this work. Many of the ideas presented in this work were stimulated or refined by discussions with J. Arribas, M.E. Arribas, M. Barone, S. Bigi, R. Dominici, D. Fontana, R.V. Ingersoll, F. Muto, C. Neri, M. Sonnino, and O. Stanzione. We are grateful to Geological Society of America reviewers Maria Eugenia Arribas and Ray Ingersoll for reviews, helpful discussions, and comments on an earlier version of the manuscript. This work was supported by funds of the Ministero dell'Università e della Ricerca Scientifica (to S. Critelli, E. Le Pera, S. Milli, and M. Santantonio, and PRIN2006 to S. Critelli) and the Italian National Council of Research (to S. Milli). This paper is dedicated to Peppe Cello, Lauro Morten, Tor Nilsen, and Piera Spadea, old friends and guides.

REFERENCES CITED

Amorosi, A., 1995, Glaucony and sequence stratigraphy: A conceptual framework of distribution in siliciclastic sequences: Journal of Sedimentary Research, v. B65, p. 419–425.

Argnani, A., and Ricci Lucchi, F., 2001, Tertiary siliciclastic turbidite systems of the northern Apennines, in Vai, G.B., and Martini, I.P., eds., Anatomy of an Orogen: The Apennines and Adjacent Mediterranean Basins: Amsterdam, Kluwer Academic Publishers, p. 327–350.

Arribas, J., and Tortosa, A., 2003, Detrital modes in sedimenticlastic sands from low-order streams in the Iberian Range, Spain: The potential for sand generation by different sedimentary rocks: Sedimentary Geology, v. 159, p. 275–303, doi: 10.1016/S0037-0738(02)00332-9.

Arribas, J., Critelli, S., Le Pera, E., and Tortosa, A., 2000, Composition of modern stream sand derived from a mixture of sedimentary and metamorphic source rocks (Henares River, central Spain): Sedimentary Geology, v. 133, p. 27–48, doi: 10.1016/S0037-0738(00)00026-9.

Arribas, J., Alonso, A., Mas, R., Tortosa, A., Rodas, M., Barrenechea, J.F., Alonso-Azcárate, J., and Artigas, R., 2003, Sandstone petrography of continental depositional sequences of an intraplate rift basin: Western Cameros basin (north Spain): Journal of Sedimentary Research, v. 73, p. 309–327.

Bellotti, P., Landini, B., and Valeri, P., 1984, Associazioni di facies e lineamenti evolutivi generali del "complesso torbiditico laziale-abruzzese": Bollettino della Società Geologica Italiana, v. 103, p. 311–326.

Bergomi, G., and Damiani, A.V., 1976, Diagenesi precoce nei depositi Serravalliano-Tortoniani del Lazio e considerazioni sulla evoluzione strutturale del bacino di sedimentazione Miocenico: Bollettino del Servizio Geologico d'Italia, v. 97, p. 35–66.

Bernoulli, D., and Jenkyns, H.C., 1974, Alpine, Mediterranean and central Atlantic Mesozoic facies in relation to the early evolution of the Tethys, in Dott, R.H., and Shaver, R.H., eds., Modern and Ancient Geosynclinal Sedimentation: Society of Economic Paleontologists and Mineralogists Special Publication 19, p. 129–160.

Bigi, S., and Costa Pisani, P., 2002, The "pre-thrusting" Fiamignano normal fault: Bollettino della Società Geologica Italiana, v. 122, p. 267–276.

Bigi, S., Costa Pisani, P., Milli, S., and Moscatelli, M., 2003, The control exerted by pre-thrusting normal faults on the early Messinian foredeep evolution, structural styles and shortening in the central Apennines (Lazio-Abruzzo, area, Italy): Studi Geologici Camerti, volume speciale 2003, p. 17–37.

Bigi, S., Costa Pisani, P., Milli, S., and Moscatelli, M., 2004, Active thrust front/foredeep depocenter migration vs flexure migration: The evolution of central Apennines: Florence, 32nd International Geological Congress, Abstracts, p. 695.

Bosellini, A., 2004, The western passive margin of Adria and its carbonate platforms, in Crescenti, U., D'Offizi, S., Merlino, S., and Sacchi, L., eds., Geology of Italy: Florence, Italian Geological Society, Special Volume for the 32nd International Geological Congress, p. 79–92.

Brandano, M., 2001, Risposta fisica delle aree di piattaforma carbonatica agli eventi più significativi del Miocene dell'Appennino centrale [Ph.D. thesis]: Roma, Università "La Sapienza," 180 p.

Brandano, M., 2002, La Formazione del "Calcari a Briozoi e Litotamni" nell'area di Tagliacozzo (Appennino centrale) e considerazioni paleoambientali sulle facies rodalgali: Bollettino della Società Geologica Italiana, v. 121, p. 179–186.

Budd, D.A., and Harris, P.M., eds., 1990, Carbonate-Siliciclastic Mixtures: Society of Economic, Paleontologists and Mineralogists Reprint Series 14, 272 p.

Carminati, E., Doglioni, C., and Scrocca, D., 2004, Alps vs. Apennines: Florence, Italian Geological Society, Special Volume for the 32nd International Geological Congress, p. 141–151.

Cavazza, W., Zuffa, G.G., Camporesi, C., and Ferretti, C., 1993, Sedimentary recycling in a temperate climate drainage basin (Senio River, north-central Italy): Composition of soil profiles and fluvial deposits, in Johnsson, M.J., and Basu, A., eds., Processes Controlling the Composition of Clastic Sediments: Geological Society of America Special Paper 284, p. 247–261.

Chiocchini, U., and Cipriani, N., 1989, The composition and provenance of the Tortonian and Messinian turbidites in the context of the structural evolution of the central Apennines along the "Ancona-Anzio" line: Sedimentary Geology, v. 63, p. 83–91, doi: 10.1016/0037-0738(89)90072-9.

Chiocchini, U., and Cipriani, N., 1992, Provenance and evolution of Miocene turbidite sedimentation in the central Apennines: Sedimentary Geology, v. 77, p. 185–195, doi: 10.1016/0037-0738(92)90125-B.

Cibin, U., Spadafora, E., Zuffa, G.G., and Castellarin, A., 2001, Continental collision history from arenites of episutural basin in the northern Apennines, Italy: Geological Society of America Bulletin, v. 113, p. 4–19, doi: 10.1130/0016-7606(2001)113<0004:CCHFAO>2.0.CO;2.

Cipollari, P., and Cosentino, D., 1995, Il sistema Tirreno-Appennino: Segmentazione litosferica e propagazione del fronte compressivo: Studi Geologici Camerti, volume speciale 1995/2, p. 125–134.

Cipollari, P., Cosentino, D., and Perilli, N., 1993, Analisi biostratigrafica dei depositi terrigeni a ridosso della linea Olevano-Antrodoco: Geologica Romana, v. 29, p. 495–513.

Cipollari, P., Cosentino, D., and Parotto, M., 1995, Modello cinematico-strutturale dell'Italia centrale: Studi Geologici Camerti, volume speciale 1995/2, p. 135–143.

Cipollari, P., Cosentino, D., Esu, D., Girotti, O., Gliozzi, E., and Praturlon, A., 1999, Thrust-top lacustrine-lagoonal basin development in accretionary wedges: Late Messinian (Lago-Mare) episode in the central Apennines (Italy): Palaeogeography, Palaeoclimatology, Palaeoecology, v. 151, p. 149–166, doi: 10.1016/S0031-0182(99)00026-7.

Compagnoni, B., Galluzzo, F., and Santantonio, M., 1990, Le "Brecce della Renga" (Monti Simbruini): Un esempio di sedimentazione controllata dalla tettonica: Memorie Descrittive della Carta Geologica d'Italia, v. 38, p. 59–76.

Compagnoni, B., Galluzzo, F., Pampaloni, M.L., Pichezzi, R.M., Raffi, I., Rossi, M., and Santantonio, M., 1991, Dati sulla lito-biostratigrafia delle successioni terrigene nell'area tra i Monti Simbruini e i Monti Carseolani (Appennino centrale): Studi Geologici Camerti, volume speciale 1991/2, p. 173–179.

Compagnoni, B., D'Andrea, M., Galluzzo, F., Giovagnoli, M.C., Lembo, P., Molinari, V., Chiocchini, U., Pampaloni, M.L., Pichezzi, R.M., Rossi, M., Salvati, L., Santantonio, M., and Raffi, I., 2005, Note Illustrative del F°367 Tagliacozzo, alla scala 1:50,000: Roma, Servizio Geologico d'Italia, 82 p.

Corda, L., 1990, L'hardground serravalliano di Tornimparte (L'Aquila): 1. Caratteri sedimentologici: Bollettino della Società Geologica Italiana, v. 109, p. 633–641.

Corda, L., and Brandano, M., 2003, Aphotic zone carbonate production on a Miocene ramp, central Apennines, Italy: Sedimentary Geology, v. 161, p. 55–70, doi: 10.1016/S0037-0738(02)00395-0.

Cosentino, D., Carboni, M.G., Cipollari, P., Di Bella, L., Florindo, F., Laurenzi, M.A., and Sagnotti, L., 1997, Integrated stratigraphy of the Tortonian/Messinian boundary: The Pietrasecca composite section (central Apennines, Italy): Eclogae Geologicae Helvetiae, v. 90, p. 229–244.

Critelli, S., 1993, Sandstone detrital modes in the Paleogene Liguride complex, accretionary wedge of the southern Apennines (Italy): Journal of Sedimentary Petrology, v. 63, p. 464–476.

Critelli, S., 1999, The interplay of lithospheric flexure and thrust accommodation in forming stratigraphic sequences in the southern Apennines foreland basin system, Italy: Accademia Nazionale dei Lincei Rendiconti di Scienze Fisiche e Naturali, v. 10, ser. 9, p. 257–326.

Critelli, S., and Ingersoll, R.V., 1994, Sandstone petrology and provenance of the Siwalik Group (northwestern Pakistan and western-southeastern Nepal): Journal of Sedimentary Research, v. A64, p. 815–823.

Critelli, S., and Ingersoll, R.V., 1995, Interpretation of neovolcanic versus palaeovolcanic sand grains: An example from Miocene deep-marine sandstone of the Topanga Group (southern California): Sedimentology, v. 42, p. 783–804, doi: 10.1111/j.1365-3091.1995.tb00409.x.

Critelli, S., and Le Pera, E., 1994, Detrital modes and provenance of Miocene sandstones and modern sands of the southern Apennines thrust-top basins (Italy): Journal of Sedimentary Research, v. A64, p. 824–835.

Critelli, S., and Le Pera, E., 1995, Tectonic evolution of the southern Apennines thrust-belt (Italy) as reflected in modal compositions of Cenozoic sandstone: The Journal of Geology, v. 103, p. 95–105.

Critelli, S., and Le Pera, E., 1998, Post-Oligocene sediment-dispersal systems and unroofing history of the Calabrian microplate, Italy: International Geology Review, v. 40, p. 609–637.

Critelli, S., De Rosa, R., and Platt, J.P., 1990, Sandstone detrital modes in the Makran accretionary wedge, southwest Pakistan: Implications for tectonic setting and long-distance turbidite transportation: Sedimentary Geology, v. 68, p. 241–260, doi: 10.1016/0037-0738(90)90013-J.

Critelli, S., Le Pera, E., Perrone, V., and Sonnino, M., 1995, Le successioni siliciclastiche nell'evoluzione tettonica Cenozoica dell'Appennino meridionale: Studi Geologici Camerti, volume speciale 1995/2, p. 155–165.

Critelli, S., Arribas, J., Le Pera, E., Tortosa, A., Marmaglia, K.M., Marsaglia, K.H., and Latter, K.K., 2003, The recycled orogenic sand provenance from an uplifted thrust belt, Betic Cordillera, southern Spain: Journal of Sedimentary Research, v. 73, p. 72–81.

Dickinson, W.R., 1970, Interpreting detrital modes of graywacke and arkose: Journal of Sedimentary Petrology, v. 40, p. 695–707.

Dickinson, W.R., 1985, Interpreting provenance relations from detrital modes of sandstones, in Zuffa, G.G., ed., Provenance of Arenites: Dordrecht, Netherlands, D. Reidel, NATO Advanced Study Institute Series, v. 148, p. 333–361.

Di Giulio, A., Cerini, A., Ghia, E., and Zucca, F., 2003, Composition of modern stream sands derived from sedimentary source rocks in a temperate climate (northern Apennines, Italy): Sedimentary Geology, v. 158, p. 145–161, doi: 10.1016/S0037-0738(02)00264-6.

Doglioni, C., Mongelli, F., and Pialli, G., 1998, Boudinage of the Alpine belt in the Apenninic back-arc: Memorie della Società Geologica Italiana, v. 52, p. 457–468.

Dolan, J.F., 1989, Eustatic and tectonic controls on deposition of hybrid siliciclastic/carbonate basinal cycles: Discussion with examples: American Association of Petroleum Geologists Bulletin, v. 73, p. 1233–1246.

Doyle, L.J., and Roberts, H.H., eds., 1988, Carbonate-Clastic Transition: Amsterdam, Elsevier, Developments in Sedimentology, v. 42, 304 p.

Dunham, R.J., 1962, Classification of carbonate rocks according to depositional texture, in Ham, W.E., ed., Classification of Carbonate Rocks—A Symposium: American Association of Petroleum Geologists Memoir 1, p. 108–121.

Elter, P., Grasso, M., Parotto, M., and Vezzani, L., 2003, Structural setting of the Apennine-Maghrebian thrust belt: Episodes, v. 26, p. 205–211.

Folk, R.L., 1962, Spectral subdivision of limestone types, in Ham, W.E., ed., Classification of Carbonate Rocks—A Symposium: American Association of Petroleum Geologists Memoir 1, p. 62–84.

Folk, R.L., 1974, Petrology of Sedimentary Rocks: Austin, Texas, Hemphill Publishing Company, 182 p.

Fontana, D., 1991, Detrital carbonate grains as provenance indicators in the Upper Cretaceous Pietraforte Formation (northern Apennines): Sedimentology, v. 38, p. 1085–1095, doi: 10.1111/j.1365-3091.1991.tb00373.x.

Fontana, D., and Stefani, C., 2003, Extrabasinal and intrabasinal sources in siliciclastic-carbonate turbidite systems of the northern Apennines (Italy), in Basu, A., and Valloni, R., eds., Quantitative Provenance Studies in Italy: Memorie Descrittive della Carta Geologica d'Italia, v. 61, p. 41–48.

Fontana, D., Zuffa, G.G., and Garzanti, E., 1989, The interaction of eustasy and tectonism from provenance studies of the Eocene Hecho Group turbidite complex (south-central Pyrenees, Spain): Basin Research, v. 2, p. 223–237.

Fontana, D., Spadafora, E., Stefani, C., Stocchi, S., Tateo, F., Villa, G., and Zuffa, G.G., 1994, The Upper Cretaceous Helminthoid Flysch of the northern Apennines: Provenance and sedimentation: Memorie della Società Geologica Italiana, v. 48, p. 237–250.

Fontana, D., Parea, G.C., Bertacchini, M., and Bessi, P., 2003, Sand production by chemical and mechanical weathering of well lithified siliciclastic turbidites of the northern Apennines (Italy), in Basu, A., and Valloni, R., eds., Quantitative Provenance Studies in Italy: Memorie Descrittive della Carta Geologica d'Italia, v. 61, p. 51–60.

Galluzzo, F., and Santantonio, M., 1995, Segnalazione di torbiditi carbonatiche nelle marne a Orbulina dei Monti Carseolani (Abruzzo): Bollettino del Servizio Geologico d'Italia, v. 114, p. 87–96.

Gandolfi, G., Paganelli, L., and Zuffa, G.G., 1983, Petrology and dispersal pattern in the Marnoso-Arenacea Formation (Miocene, northern Apennines): Journal of Sedimentary Petrology, v. 53, p. 493–507.

Garzanti, E., 1991, Non-carbonate intrabasinal grains in arenites: Their recognition, significance, and relationships to eustatic cycles and tectonic setting: Journal of Sedimentary Petrology, v. 61, p. 959–975.

Garzanti, E., Critelli, S., and Ingersoll, R.V., 1996, Paleogeographic and paleotectonic evolution of the Himalayan range as reflected by detrital modes of Tertiary sandstones and modern sands (Indus transect, India and Pakistan): Geological Society of America Bulletin, v. 108, p. 631–642, doi: 10.1130/0016-7606(1996)108<0631:PAPEOT>2.3.CO;2.

Garzanti, E., Canclini, S., Moretti Foggia, F., and Petrella, N., 2002, Unraveling magmatic and orogenic provenance in modern sand: The back-arc side of the Apennine thrust belt, Italy: Journal of Sedimentary Research, v. 72, p. 2–17.

Gazzi, P., Zuffa, G.G., Gandolfi, G., and Paganelli, L., 1973, Provenienza e dispersione litoranea delle sabbie delle spiagge Adriatiche fra le foci dell'Isonzo e del Foglia: Inquadramento regionale: Memorie della Società Geologica Italiana, v. 12, p. 1–37.

Heller, P.L., and Dickinson, W.R., 1985, Submarine ramp facies model for delta-fed, sand-rich turbidite systems: American Association of Petroleum Geologists Bulletin, v. 69, p. 960–976.

Ingersoll, R.V., Bullard, T.F., Ford, R.L., Grimm, J.P., Pickle, J.D., and Sares, S.W., 1984, The effect of grain size on detrital modes: A test of the Gazzi-Dickinson point-counting method: Journal of Sedimentary Petrology, v. 54, p. 103–116.

Ingersoll, R.V., Cavazza, W., Graham, S.A., and Indiana University Graduate Field Seminar Participants, 1987, Provenance of impure calclithites in the Laramide foreland of southwestern Montana: Journal of Sedimentary Petrology, v. 57, p. 995–1003.

James, N.P., 1997, The cool-water carbonate depositional realm, in James, N.P, and Clarke, J., eds., Cool-Water Carbonates: SEPM (Society for Sedimentary Geology) Special Publication 56 p. 1–20.

Johnsson, M.J., 1993, The system controlling the composition of clastic sediments, in Johnsson, M.J., and Basu, A., eds., Processes Controlling the Composition of Clastic Sediments: Geological Society of America Special Paper 284, p. 1–19.

Le Pera, E., and Arribas, J., 2004, Sand composition in an Iberian passive-margin fluvial course: The Tajo River: Sedimentary Geology, v. 171, p. 261–281, doi: 10.1016/j.sedgeo.2004.05.019.

Le Pera, E., and Critelli, S., 1997, Sourceland controls on the composition of beach and fluvial sand of the northern Tyrrhenian coast of Calabria, Italy: Implications for actualistic petrofacies: Sedimentary Geology, v. 110, p. 81–97, doi: 10.1016/S0037-0738(96)00078-4.

Lindholm, R.C., and Finkelman, R.B., 1972, Calcite staining: Semiquantitative determination of ferrous ion: Journal of Sedimentary Petrology, v. 42, p. 239–242.

Malinverno, A., and Ryan, W.B.F., 1986, Extension in the Tyrrhenian Sea and shortening in the Apennines as result of arc migration driven by sinking of the lithosphere: Tectonics, v. 5, p. 227–245.

Mariotti, G., Corda, L., Brandano, M., Castorina, F., and Civitelli, G., 1999, Sequence analysis of a Miocene carbonate platform: An example from the western sector of the central Apennines (Italy): Bellaria, 2nd Forum Italiano di Scienze della Terra (FIST), Abstracts, p. 86–88.

Milli, S., and Moscatelli, M., 2000, Facies analysis and physical stratigraphy of the Messinian ramp turbiditic complex in the Valle del Salto and Val di Varri (central Apennines): Giornale di Geologia, v. 62, p. 57–77.

Milli, S., and Moscatelli, M., 2001, The control of sea-floor topography on turbidite sedimentation—An example in the Lower Messinian turbidite deposits of the central Apennines: Nice, 9–11 September 2001, Research Meeting "Turbidite Sedimentation in Confined Setting," Abstracts, p. 43.

Milli, S., Moscatelli, M., and Falciani, F., 2003, Le arenarie di Torrice: Un esempio di depositi torbiditici di wedge top basin nell'ambito del sistema di avanfossa alto Miocenico dell'Appennino centro-meridionale: Alghero, Sardinia, 1st Congress of the Italian Association for Sedimentary Geology, Abstracts, p. 195–197.

Milli, S., Moscatelli, M., Stanzione, O., Gennari, G., and Marini, M., 2004, Sedimentology and physical stratigraphy of the pre-gypsum arenites deposits of the Laga Formation: Florence, 32nd International Geological Congress, Abstracts, p. 1451.

Moscatelli, M., 2003, La sedimentazione torbiditica e le sue relazioni con quella fluvio-deltizia nel sistema d'avanfossa alto-Miocenico dell'Appennino centro-meridionale [Ph.D. thesis]: Roma, Università "La Sapienza," 164 p.

Moscatelli, M., Milli, S., Stanzione, O., Marini, M., Gennari, G., and Vallone, R., con contributi di Artoni, A., Bigi, S., and Lugli, S., 2004, I depositi torbiditici del Messiniano inferiore dell'Appennino centrale: Bacini del Salto-Tagliacozzo e della Laga (Lazio, Abruzzo, Marche): Roma, 2nd Congress of the Italian Association for Sedimentary Geology, Field Trip Guidebook, 40 p.

Mount, J.F., 1984, The mixing of siliciclastic and carbonate sediments in shallow-shelf environments: Geology, v. 12, p. 432–435, doi: 10.1130/0091-7613(1984)12<432:MOSACS>2.0.CO;2.

Mount, J.F., 1985, Mixed siliciclastic and carbonate sediments: A proposed first-order textural and compositional classification: Sedimentology, v. 32, p. 435–442, doi: 10.1111/j.1365-3091.1985.tb00522.x.

Mullins, H.T., and Cook, H.E., 1986, Carbonate apron models: Alternatives to the submarine fan model for paleoenvironmental analysis and hydrocarbon exploration: Sedimentary Geology, v. 48, p. 37–79, doi: 10.1016/0037-0738(86)90080-1.

Mutti, E., 1992, Turbidite Sandstones: Parma, Agip-Istituto di Geologia Università di Parma, 275 p.

Mutti, E., Tinterri, R., Remacha, E., Mavilla, N., Angella, S., and Fava, L., 1999, An introduction to the analysis of ancient turbidite basins from an outcrop perspective: American Association of Petroleum Geologists Continuing Education Course Note, Ser. 39, 96 p.

Mutti, E., Tinterri, R., Benevelli, G., Di Biase, D., and Cavanna, G., 2003, Deltaic, mixed and turbidite sedimentation of ancient foreland basins: Marine and Petroleum Geology, v. 20, p. 733–755, doi: 10.1016/j.marpetgeo.2003.09.001.

Pampaloni, M.L., Pichezzi, R.M., Raffi, M., and Rossi, M., 1994, Calcareous planktonic biostratigraphy of the marne a Orbulina unit (Miocene, central Italy): Giornale di Geologia, v. 56, p. 139–153.

Parotto, M., and Praturlon, A., 2004, The southern Apennine arc, in Crescenti, V., D'Offizi, S., Merlino, S., and Sacchi, L., eds., Geology of Italy: Florence, Italian Geological Society, Special Volume for the 32nd International Geological Congress, p. 33–58.

Parotto, M., Cavinato, G.P., Di Luzio, E., Miccadei, E., Patacca, E., Scandone, P., Bigi, S., and Nicolich, R., 2004, Geological interpretation of the Crop 11 seismic line between the Adriatic coast and the Tiber valley: Florence, 32nd International Geological Congress, Abstracts, p. 228.

Patacca, E., Sartori, R., and Scandone, P., 1990, Tyrrhenian basin and Apenninic arcs: Kinematic relations since late Tortonian times: Memorie della Società Geologica Italiana, v. 45, p. 425–451.

Patacca, E., Sartori, R., and Scandone, P., 1993, Tyrrhenian basin and Apennines: Kinematic evolution and related dynamic constraints, in Boschi, E., Mantovani, E., and Morelli, A., eds., Recent Evolution and Seismicity of the Mediterranean Regions, Nato ASI Series C, v. 402: Doordrecht, Kluwer Academic Publishers, p. 161–171.

Ricci Lucchi, F., 1986, The Oligocene to recent foreland basin of the northern Apennines, in Allen, P.A., and Homewood, P., eds., Foreland Basins: International Association of Sedimentologists Special Publication 8, p. 105–139.

Santantonio, M., 1993, Facies associations and evolution of pelagic carbonate platform/basin systems: Examples from the Italian Jurassic: Sedimentology, v. 40, p. 1039–1067, doi: 10.1111/j.1365-3091.1993.tb01379.x.

Santantonio, M., 1994, Pelagic carbonate platforms in the geologic record: Their classification, and sedimentary and paleotectonic evolution: American Association of Petroleum Geologists Bulletin, v. 78, p. 122–141.

Sartori, R., 2003, The Tyrrhenian back-arc basin and subduction of the Ionian lithosphere: Episodes, v. 26, p. 217–221.

Schwab, F.L., 1986, Sedimentary signatures of foreland basin assemblages: Real or counterfeit?, in Allen, P.A., and Homewood, P., eds., Foreland Basins: International Association of Sedimentologists Special Publication 8, p. 395–410.

Scisciani, V., Calamita, F., Tavernelli, E., Rusciadelli, G., Ori, G.G., and Paltrinieri, W., 2001, Foreland-dipping normal faults in the inner edges of syn-orogenic basins: A case from the central Apennines, Italy: Tectonophysics, v. 330, p. 211–224, doi: 10.1016/S0040-1951(00)00229-8.

Scisciani, V., Tavernelli, E., Calamita, F., and Paltrinieri, W., 2002, Pre-thrusting normal faults within sin-orogenic basins of the outer central Apennines, Italy: Implication for Apennine tectonics: Bollettino della Società Geologica Italiana, volume speciale 1, p. 295–304.

Spence, G.H., and Tucker, M.E., 1997, Genesis of limestone megabreccias and their significance in carbonate sequence stratigraphic models: A review: Sedimentary Geology, v. 112, p. 163–193.

Suczek, C.A., and Ingersoll, R.V., 1985, Petrology and provenance of Cenozoic sand from the Indus Cone and the Arabian Basin, DSDP Sites 221, 222, and 224: Journal of Sedimentary Petrology, v. 55, p. 340–346.

Zuffa, G.G., 1980, Hybrid arenites: Their composition and classification: Journal of Sedimentary Petrology, v. 50, p. 21–29.

Zuffa, G.G., 1985, Optical analyses of arenites: Influence of methodology on compositional results, in Zuffa, G.G., ed., Provenance of Arenites: Dordrecht, Netherlands, D. Reidel, NATO Advanced Study Institute Series, v. 148, p. 165–189.

Zuffa, G.G., 1987, Unravelling hinterland and offshore paleogeography from deep-water arenites, in Leggett, J.K., and Zuffa, G.G., eds., Deep-Marine Clastic Sedimentology: Concepts and Case Studies: London, Graham and Trotman, p. 39–61.

Zuffa, G.G., 1991, On the use of turbidite arenites in provenance studies: Critical remarks, in Morton, A.C., Todd, S.P., and Haughton, P.D.W., eds., Developments in Sedimentary Provenance Studies: Geological Society [London] Special Publication 57, p. 23–29.

Zuffa, G.G., Gaudio, W., and Rovito, S., 1980, Detrital mode evolution of the rifted continental-margin Longobucco Sequence (Jurassic), Calabrian Arc, Italy: Journal of Sedimentary Petrology, v. 50, p. 51–61.

Zuffa, G.G., Fontana, D., Morlotti, E., Premoli Silva, I., Sighinolfi, G.P., Stefani, C., and Fontani, L., 2003, Anatomy of carbonate turbidite mega-beds (M. Cassio Formation, Upper Cretaceous, northern Apennines, Italy), in Basu, A., and Valloni, R., eds., Quantitative Provenance Studies in Italy: Memorie Descrittive della Carta Geologica d'Italia, v. 61, p. 129–144.

MANUSCRIPT ACCEPTED BY THE SOCIETY 9 AUGUST 2006

Interpreting gypsarenites in the Rossano basin (Calabria, Italy): A contribution to the characterization of the Messinian salinity crisis in the Mediterranean

Mirko Barone[†]
Rocco Dominici
Dipartimento di Scienze della Terra, Università degli Studi della Calabria, 87036 Arcavacata di Rende, Cosenza, Italy

Stefano Lugli
Dipartimento di Scienze della Terra, Università degli Studi di Modena e Reggio Emilia, Largo S. Eufemia 19, 41100 Modena, Italy

ABSTRACT

Detrital evaporites and mixed siliciclastic-gypsum arenites are present in the Gessi Formation from the Rossano Basin in Calabria, Italy. The detrital origin of the gypsum fragments in the quartzofeldspathic sandstones is revealed by crystal overgrowths that outline the shape of former gypsum clasts. The gypsum was subsequently transformed into anhydrite at burial conditions. During exhumation, anhydrite was hydrated back to gypsum, a gypsum overgrowth rich in F, Na, K, Cl, and Al formed on the original gypsum grains, and the pore spaces were filled with gypsum cement.

Detrital modes of Gessi Formation sandstones suggest complex source-basin relationships in this area during the Messinian salinity crisis. The clastic deposits are the result of deep unroofing of the crustal terranes of the Calabrian arc and the reworking of primary Messinian evaporite facies (selenite). This study indicates that detrital evaporites and mixed siliciclastic-gypsum arenites are more widespread in the Mediterranean area than generally described in the literature.

Keywords: provenance, detrital mode, gypsarenites, Messinian salinity crisis, southern Italy.

INTRODUCTION

Gypsum sediments can be eroded, transported, and redeposited in the same fashion as other clastic sediments (Hardie and Eugster, 1971; Parea and Ricci Lucchi, 1972; Ricci Lucchi, 1973; Schlager and Bolz, 1977; Schreiber, 1973; Kendall, 1992; Kendall and Harwood, 1996). In the case where the gypsum is reworked within turbidite systems, parent flow mainly composed of gypsum clasts will produce downcurrent transformations similar to siliciclastic turbidite facies (Manzi et al., 2005).

[†]E-mail: mbarone@unical.it.

The Messinian facies in some areas of the Mediterranean area show significant examples of clastic evaporites, and it has been recently suggested that evaporites present in Mediterranean submarine basins may consist of deep-water resedimented deposits, rather than shallow-water to supratidal primary evaporites (Manzi et al., 2005). This would suggest that the Mediterranean was not completely desiccated during the Messinian salinity crisis as first proposed by Hsü et al. (1973).

The presence of detrital gypsum grains in sandstones has not received much attention in the literature, although this component may play an important role in paleogeographical reconstructions of the Messinian in the Mediterranean. One limiting factor is the transformation that gypsum may experience during burial and exhumation. These transformations may completely obliterate the original clastic features, and this poses some fundamental questions in the classification of gypsarenites and mixed arenites: is the gypsum a late cement? Is it clastic? Is it completely recrystallized? And if so, why and how?

This paper illustrates the results of our studies on Messinian gypsiferous arenites from the Rossano Basin in Calabria. These arenites show features that may help to unravel the origin of mixed gypsum-siliciclastic systems and contribute to the understanding of the complexity of the sedimentary history of the Messinian salinity crisis in the Mediterranean.

GEOLOGIC SETTING OF THE ROSSANO BASIN

The Rossano Basin is located along the southwestern margin of Taranto Gulf (Fig. 1), which represents the southern segment of the Apennine foreland basin system according to Pescatore and Senatore (1986), Senatore (1987), and Critelli (1999). The structural evolution of the Rossano Basin is linked to the onset of the southeastward migration of the Calabrian arc as a result of the opening of the Tyrrhenian Basin during the Serravallian-Tortonian (Mattei et al., 2002), and the collision between the Calabrian fold-and-thrust belt with the Adria plate continental margin from the late Tortonian to the Pliocene. The geological setting of the Rossano Basin was controlled by a compressive phase during the late Burdigalian, and a rift phase in the Serravallian and Tortonian. From the late Tortonian to the late Messinian, the structural regime was controlled by activity on the Rossano–San Nicola transpressional and transtensional NW- to NE-trending left-lateral shear zone (Tortorici, 1981; Meulenkamp et al., 1986), which was related to the rift phase of the Magnaghi basin (Mascle and Rehault, 1990). The Rossano Basin represents a southern limb of the Apennine foreland basin system (Pescatore and Senatore, 1986; Senatore, 1987; Patacca et al., 1990; Critelli, 1999).

STRATIGRAPHY

The infill of the Rossano Basin is characterized by Tortonian to Pleistocene dominantly clastic sediments, including Messinian evaporite deposits. This sedimentary infill unconformably overlies a basement that consists of various terranes, including the Sila Unit, a complex assemblage of Paleozoic crystalline rocks and its Mesozoic sedimentary cover. The Paleozoic rocks consist of medium- to high-grade (Gariglione Complex), medium-grade (Mandatoriccio Complex), and low-grade (Bocchigliero Complex) metasedimentary and metavolcanic rocks, and late Paleozoic (Variscan) plutonites (Sila Batholith; Messina et al., 1994). The Mesozoic sedimentary cover of the Sila Unit is represented by the Longobucco Group, which consists of continental sediments and shelf, slope, and deep-sea turbidites. The Mesozoic sedimentary cover of the Sila Unit is represented by the Longobucco Group consisting of continental sediments and shelf, slope, and deep sea turbidites, which unconformably overlies the Sila Unit (Figs. 1 and 2).

The stratigraphy of the Rossano Basin was studied by Roda (1967) and Ogniben (1962). Ogniben recognized 10 lithostratigraphic units and 3 olistostromes. Recently, Dominici (2005) reassembled the lithostratigraphic units, based on stratigraphic studies and geological mapping, into seven unconformity-bounded stratigraphic units (Salvador, 1987), three olistostromes, and one chaotic complex (Fig. 2).

The lowermost unconformity-bounded stratigraphic unit 1 (Fig. 2) is composed of red conglomerate, sandstone, and fossiliferous sandstone that gradually pass up into cyclic alternation of marl and shale, in some places minor sandstone interbeds. These sediments are topped by the first olistostrome, only in the eastern sector of the study area. The unconformity-bounded stratigraphic unit 1 is a transgressive system with alluvial deposits overlain by nearshore sediments that gradually pass up-section into turbidite strata (Roveri et al., 1992; Dominici, 2005). The first olistostrome interval reflects basin instability related to a tectonic phase (Critelli, 1999). The unconformity-bounded stratigraphic unit 2 is composed of euxinic and biosiliceous shales (Tripoli Formation) overlain by limestone with gypsum and halite pseudomorphs and thin layers of varicolored marl-shale. The unconformity-bounded stratigraphic unit 2 records the transition from normal marine to hypersaline conditions related to fragmentation of a large marine basin into different subbasins with restricted circulation that favored the onset of an evaporitic event (Dominici, 2005). The unconformity-bounded stratigraphic unit 3 is known locally as Molassa di Castiglione and is present only in marginal areas of the Rossano Basin. The Molassa di Castiglione is a succession, up to 30 m thick, that consists of a basal conglomerate and breccia, and sandstones with clay chips and rip-up clasts with oblique and cross-bedding. The Molassa di Castiglione passes northeastward into a chaotic complex, which includes large blocks of carbonate, anhydrite, euxinic and biosiliceous shale rocks. It crops out along a narrow area, close the northern side of a morphostructure related to a compressive or transpressive phase during the middle Messinian. It consists of a syntectonic sedimentary wedge that overlies the Calcare di Base Formation along an erosional angular unconformity. Dominici (2005) interpreted the Molassa di Castiglione and the chaotic complex as the sedimentary products of

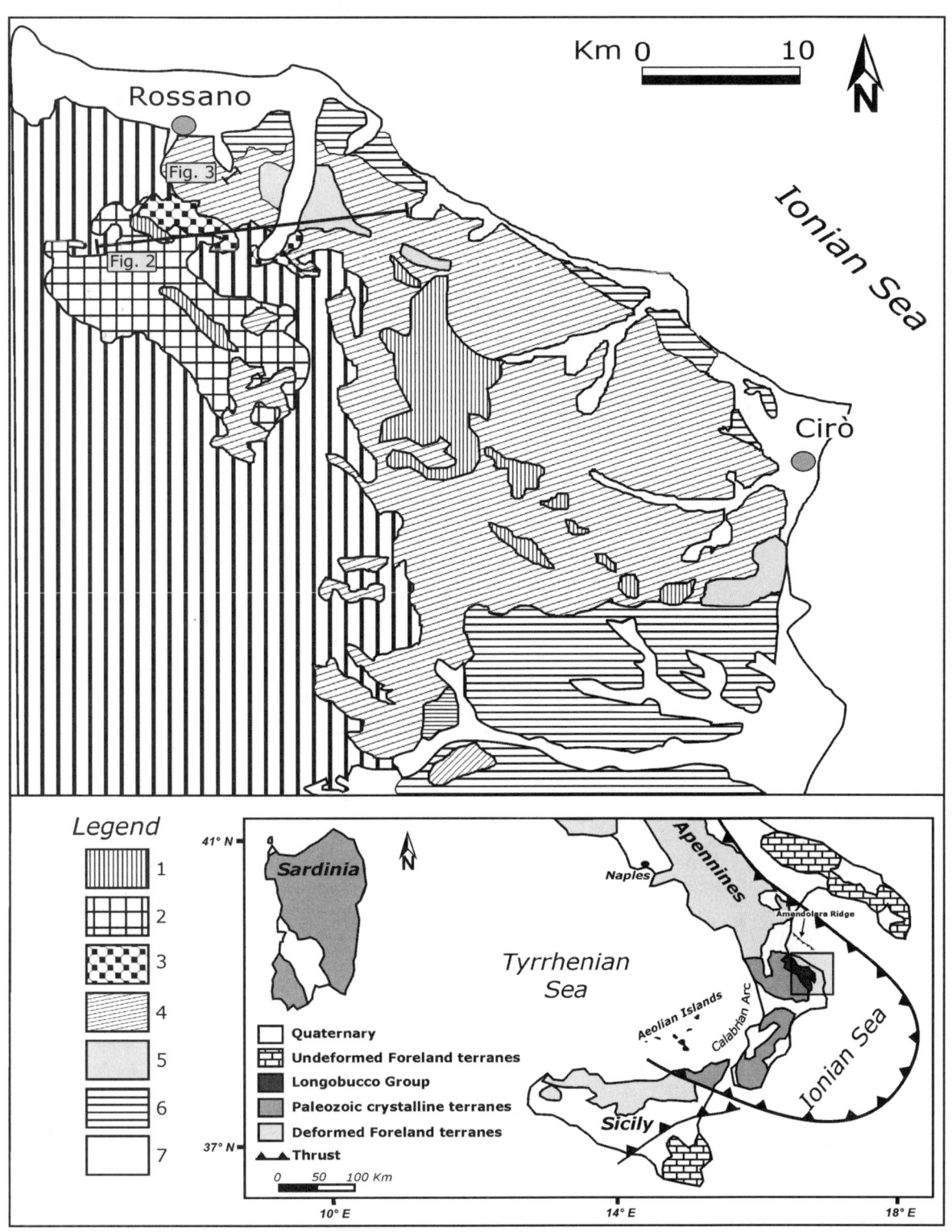

Figure 1. Simplified geological map of the southern Tyrrhenian basin and the Calabrian northeastern sector, showing (1) phyllite, schist, high-grade metamorphic rocks, and granite (Cambrian–Devonian); (2) conglomerate, arenite, and carbonate (Triassic–Cretaceous); (3) conglomerate and arenite (Upper Oligocene–Lower Miocene); (4) conglomerate, arenite, marl, shale, carbonate, gypsum, and halite (Tortonian–Lower Pliocene); (5) clay (Tortonian–Lower Messinian); (6) conglomerate, arenite, and shale (Middle Pliocene–Middle Pleistocene); and (7) conglomerate, arenite, and shale (Middle Pleistocene–Holocene).

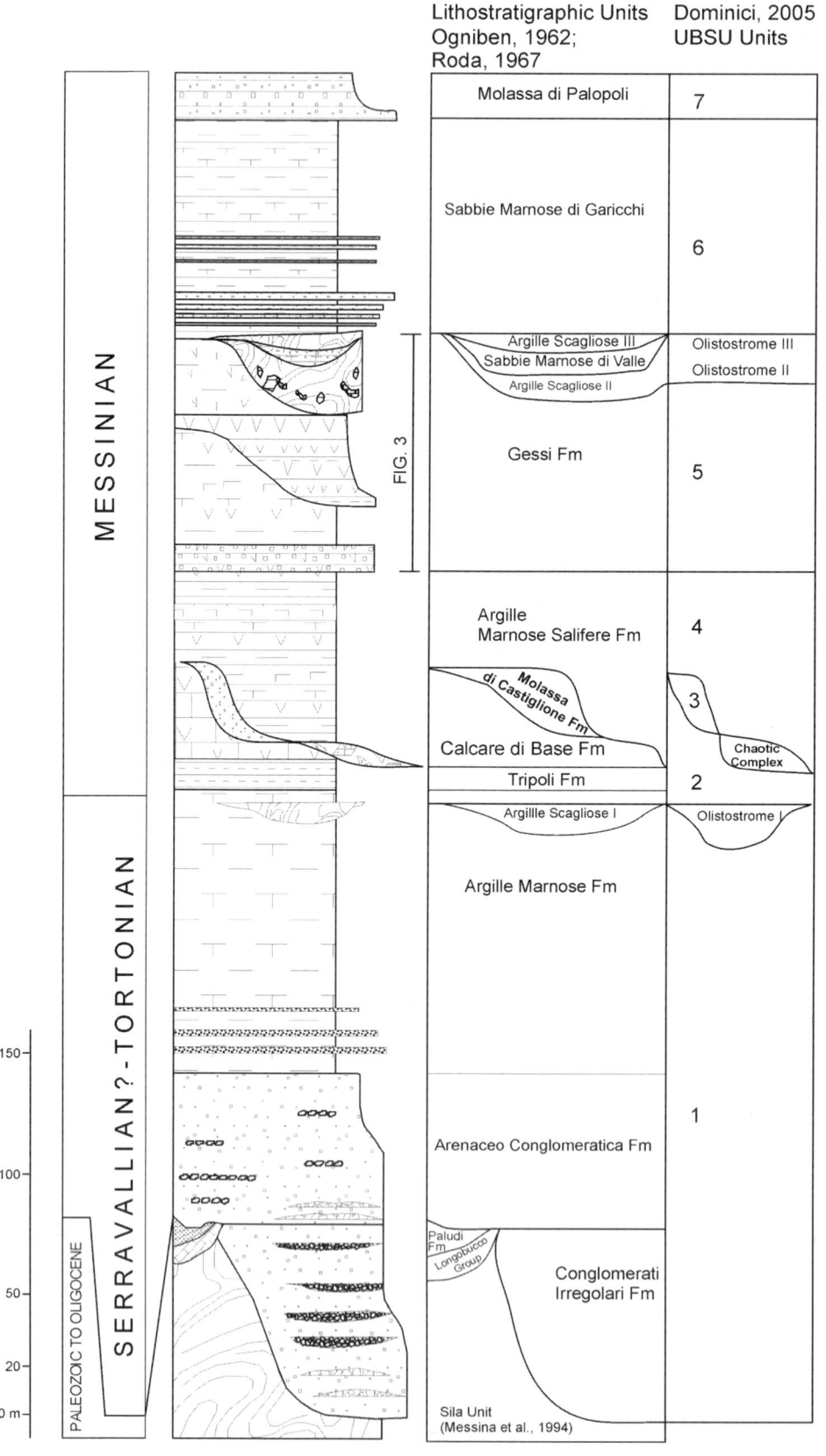

Figure 2. Stratigraphic column of the Rossano Basin located in Figure 1. UBSU—unconformity-bounded stratigraphic units.

intra-Messinian tectonics (Fig. 2). The unconformity-bounded stratigraphic unit 4, Argille Marnose Salifere Formation, onlaps unconformity-bounded stratigraphic units 2 and 3. It consists of alternating lenses of halite, shale-marl with gypsum, and anhydrite nodules and crystals, and includes a small lenticular olistostrome. This unconformity-bounded stratigraphic unit represents a transgressive depositional system that includes a wide range of evaporitic facies. Unconformity-bounded stratigraphic unit 5, the Gessi Formation, overlies the Argille Marnose Salifere Formation and consists of a succession of arenites, gypsarenites, carbonates, and halite. The evaporite sedimentation is closed by the second and third olistostromes. In the central sector of the Rossano Basin, gypsarenite and marl beds of the "di Valle Formation" are present between the olistostromes. Unconformity-bounded stratigraphic unit 5 shows a transgressive trend and a sudden influx of detrital material. The siliciclastic sedimentation was reinitiated with deposition of the Garicchi Formation (unconformity-bounded stratigraphic unit 6) and Palopoli Molassa (unconformity-bounded stratigraphic unit 7) during the late Messinian. The Garicchi Formation and Palopoli Molassa are widely distributed into a wedge-top basin confined between the Amendolara Ridge and the fold-and-thrust belt (Fig. 1). The Garicchi Formation consists of grayish fossiliferous marl and shale alternating with small sandy turbidite bodies that grade upward into, or in some places show an erosional contact with, conglomerate and arenites of the Palopoli Molassa.

EVAPORITE SEDIMENTATION IN THE ROSSANO BASIN

Messinian evaporites are found in two stratigraphic intervals in the Rossano Basin (unconformity-bounded stratigraphic units 4 and 5) that are separated at an angular unconformity covered by the syntectonic sedimentary wedge of the Molassa di Castiglione and a chaotic complex that include evaporite blocks (unconformity-bounded stratigraphic unit 3) (Fig. 2).

On the basis of physical characteristics and lithostratigraphic relationships, these intervals could be correlated to the first and second cycle of the Messinian evaporites in Sicily (Decima and Wezel, 1973). The basal shale in unconformity-bounded stratigraphic unit 2 records low clastic input, apart from the flanks of some morphostructural highs, where evaporite sedimentation occurred. The overlying clastic interval is characterized by resedimented gypsum and carbonates of the Calcare di Base Formation. In contrast, above the major unconformity, the second cycle (unconformity-bounded stratigraphic units 4–5) is characterized by evaporite and siliciclastic sedimentation. The sudden influx of detrital material probably reflects climatic (increased runoff) and/or paleogeographic changes related to a middle Messinian tectonic phase. In the marginal areas, unconformity-bounded stratigraphic units 4 and 5 pinch-out toward the southeast, onlapping the Molassa di Castiglione, the chaotic complex, and the Calcare di Base Formation. These record a rapid transgression into an evaporite wedge-top basin, probably during the late phase of the Messinian salinity crisis, as described for the marginal basins of the central-western Mediterranean (Butler et al., 1995).

The variety of facies suggests a complex depositional environment, from shallow to deep water. The lower siliciclastic subunit consists of a thin sequence of gypsiferous sandstone. The sandstone beds show normal-grading beds with erosional bases and clay chips, amalgamated facies, ripple- and cross-lamination deformed by water-escape structures and convolution. They are often characterized by a lateral contact with disorganized conglomerates (chaotic facies). The evaporite lithostratigraphic subunit consists of gypsiferous and saliferous marly-clay interbedded with lenticular halite deposits and thin-bedded, fine-grained gypsarenites and limestone. In these deposits, Dominici (2005) recognized several autoclastic breccias within thin-bedded, fine-grained gypsarenites and limestone possibly, related to subaerial exposure. The upper siliciclastic unit is composed of very coarse– to coarse-grained sandstone and conglomerates-breccias, including sandstone, limestone, gypsum nodules, and metamorphic pebbles-cobbles that show autoclastic structure. The breccias drape meter-deep depressions and pass vertically to coarse- to medium-grained sandstone characterized by cross-bedding. The lack of biostratigraphic markers makes it difficult to establish the depositional environment paleodepth. However, on the basis of sedimentary structures, we interpret the lower siliciclastic subunit as subaqueous deposits related to highly concentrated flows similar to those described in the northern Apennines (Manzi et al., 2005).

METHODS

A total of ten samples of arenites from the Gessi Formation were collected (Fig. 3), impregnated in epoxy resin under vacuum, thin-sectioned, and stained for K-feldspar identification. Quantitative petrographic analyses by point counting under transmitted-light microscopy were performed according to the Gazzi-Dickinson method (Gazzi, 1966; Dickinson, 1970; Zuffa, 1985; Ingersoll et al., 1984). About 500 points were counted for each thin section (Table 1). Results of point counting and recalculated grain parameters (Dickinson, 1970; Ingersoll et al., 1984; Critelli and Le Pera, 1994) are presented in Tables 1 and 2. Statistically rigorous confidence regions of the studied samples were calculated according to Weltje (2002). Further analyses were performed on polished thin sections with a STEREOSCAN 360 scanning electron microscope (SEM) by Cambridge Instruments, equipped with an energy dispersive X-ray microanalysis system (EDAX, Philips Electronics). Quantitative data were standardized with the correction ZAF system and are expressed in wt% of the correspondent oxides. The instrumental acquisition conditions were the following: EHT (Electrical High Tension) = 20 kV; tilt = 0.0°; take-off = 32.1°; WD (working distance) = 25 mm.

PETROGRAPHY AND GEOCHEMISTRY

Textural and Compositional Data

The sandstones of the Messinian Gessi Formation range from medium to coarse grained and from poorly to well sorted. They are constituted mainly by noncarbonate extrabasinal grains (NCE_{83}) with lesser carbonate extrabasinal grains (CE_8) and noncarbonate intrabasinal grains (NCI_9). They are quartzofeldspathic in composition ($Qm_{44}F_{32}Lt + CE + gy\text{-}clast_{24}$) (Fig. 4). The bulk composition of the Gessi Formation sandstone is little different from the composition at the time of deposition ($Qm_{37}F_{28}Lt + CE + gy\text{-}clast_{35}$) because most gypsum grains have been completely dissolved and precipitated as cement; for this reason, in order to compare the composition shift due to the gypsum grain dissolution, the gypsum cement component is included in the Lt class (Fig. 4). They have abundant quartz, both as single crystal and in polymineralic rock fragments. Plutonic fragments are represented by granodioritic-granitic types, whereas metamorphic fragments include schist and rare gneiss ($Rg_{17}Rs_{51}Rm_{32}$). The mean plagioclase/total feldspar ratio is 0.50. K-feldspar grains are present as microcline and orthoclase commonly altered along cleavage planes, whereas some plagioclase grains show typical albite twinning and sericite inclusions. Both quartz and feldspar detrital grains show brittle fractures, probably as a result of compaction. Aphanitic sedimentary lithic fragments are dominant, consisting of micritic to sparitic limestone fragment and gypsum grains ($Lm_{22}Lv_0Ls_{78}$). Metasedimentary lithics include fine-grained schist and phyllite. Siliciclastic volcanic and ophiolitic lithics fragments are rare to absent. The interstitial component in such sandstones is represented by micritic matrix, whereas the most common diagenetic phases include gypsum and carbonate cement.

The gypsum component in all samples is represented by gypsum detrital crystals except for two samples MB 04-44 and 04-78, in which these grains are totally absent. Gypsum is also present as pore-filling, syntaxial overgrowth (0%–19% of the whole rock) and commonly shows a poikilotopic texture. Calcite forms an early pore-filling and a patchy cement (7%–27%) and replaces detrital grains such as quartz, feldspar, and older gypsum cement. A matrix component is also present as micrite (0%–15%).

Gypsum Grains

Optical microscope and scanning electron microscopy and energy dispersive spectroscopy (SEM-EDS) analyses of the studied sandstones revealed a number of textural characters of the gypsum crystals that can be used to distinguish between detrital and cementing gypsum. The gypsum crystals are generally larger (up to 0.4 mm) than the siliciclastic grains and show some peculiar features (Fig. 5). The cores of the gypsum crystals are darker under the optical microscope under crossed nicols, whereas the outer parts are lighter. This feature is commonly related to the presence of opaque carbonaceous material in the inner core, which in some cases resembles disrupted

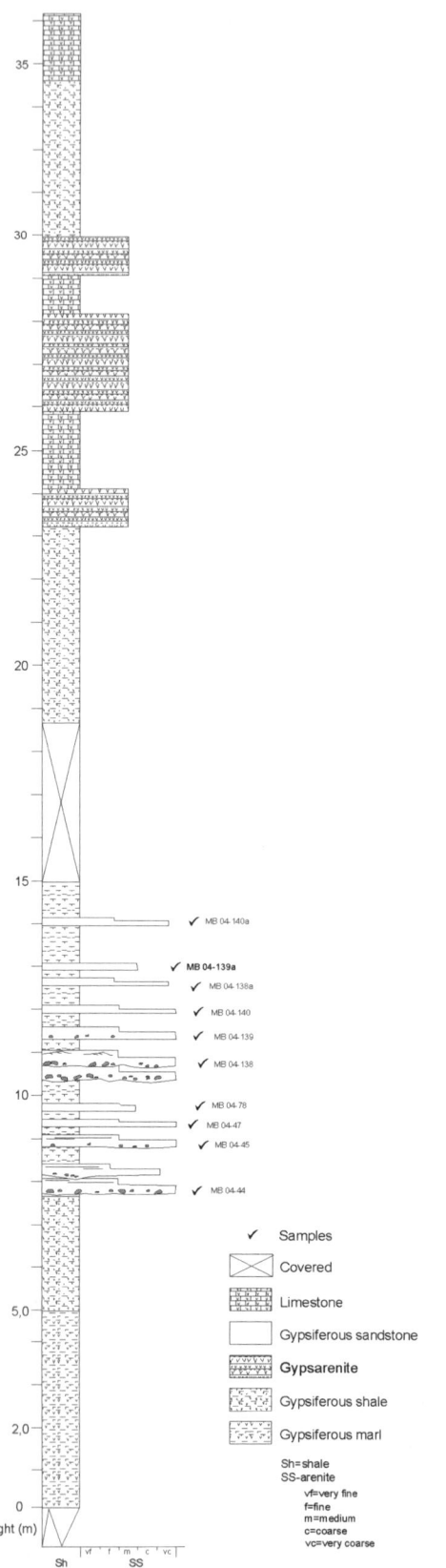

Figure 3. Detailed stratigraphic column of the Gessi Formation located in Figure 1.

TABLE 1. MODAL POINT COUNT DATA OF THE GESSI FORMATION SANDSTONE SAMPLES

Petrographic classes / Samples	MB 03–44	MB 03–45	MB 03–47	MB 04–138	MB 04–139	MB 04–140	MB 04–78	MB 04–138a	MB 04–139a	MB 04–140a
NCE										
Q										
Quartz (single crystals)	126	80	84	96	163	120	84	104	141	98
Polycrystalline quartz with tectonite fabric	8	26	2	2	2	–	–	8	6	6
Polycrystalline quartz without tectonite fabric	1	5	–	1	–	–	–	2	4	–
Quartz in metamorphic rock fragments	11	46	–	14	13	19	2	7	6	9
Quartz in plutonic rock fragments	7	17	–	6	4	8	13	11	3	6
Calcite replacement of quartz	7	–	6	4	6	1	5	6	2	4
K										
K-feldspar (single crystals)	44	28	21	33	46	38	110	32	41	42
K-feldspar in metamorphic rock fragments	4	17	–	1	4	–	3	1	–	–
K-feldspar in plutonic rock fragments	1	9	1	–	2	1	15	5	1	5
Calcite replacement of K-feldspar	1	–	–	–	–	2	2	–	1	–
P										
Plagioclase (single crystals)	46	58	43	34	41	25	80	28	34	29
Plagioclase in metamorphic rock fragments	3	24	–	2	1	3	2	1	2	1
Plagioclase in plutonic rock fragments	2	7	1	1	1	3	11	6	3	1
Calcite replacement of plagioclase	7	–	–	1	–	1	1	–	1	–
M										
Micas and chlorite (single crystals)	6	7	25	6	8	4	40	1	1	1
Micas and chlorite in plutonic rock fragments	–	1	–	–	–	–	–	2	–	–
Micas and chlorite in metamorphic rock fragments	1	4	–	1	1	1	1	–	–	–
L										
Volcanic lithic with felsitic seriate texture	–	–	–	–	1	–	–	–	–	–
Metavolcanic lithic	–	–	–	1	–	–	–	–	–	1
Serpentine schist	–	–	–	–	–	–	–	–	1	–
Phyllite	2	6	–	6	–	1	–	3	3	4
Fine-grained schist	1	2	1	7	23	21	1	2	15	5
Fine-grained arenite	–	–	–	–	–	–	–	–	–	–
Impure chert	–	–	1	–	–	–	–	–	1	–
Siltstone	–	1	–	–	–	–	–	–	–	–
Shale	–	–	3	–	1	–	–	–	–	–
Dense minerals (single crystals)	–	1	2	–	1	–	–	–	3	–
CE										
Micritic limestone	104	7	11	6	7	4	–	10	11	10
Sparitic limestone	–	2	4	3	10	3	–	8	7	10
Microsparitic limestone	5	4	8	1	9	2	–	3	5	4
Biosparitic limestone	–	–	–	–	–	–	–	1	1	2
Biomicritic limestone	–	–	1	–	–	–	–	1	–	–
Fossil fragment	1	–	–	4	2	–	1	–	–	–
Calcite single spar	–	–	–	–	1	–	–	–	–	–
CI										
Bioclast	–	1	–	2	–	–	–	2	2	1
Intraclast	–	5	–	7	–	3	–	–	–	–
NCI										
Gypsum clast	–	5	15	23	13	48	–	80	18	61
Rip-up clasts (argillaceous and silt)	–	–	–	3	–	–	–	–	–	–
Mx										
Siliciclastic matrix	10	6	1	–	–	–	–	1	2	–
Carbonate matrix	15	16	–	39	10	53	–	75	16	39
Cm										
Carbonate cement	5	–	–	–	–	–	–	–	–	–
Gypsum cement	–	55	169	94	30	64	–	49	52	65
Carbonate cement (patchy calcite)	67	45	99	85	120	117	136	50	103	95
Fe-oxides cement	12	4	1	6	–	–	–	4	15	2
Calcite replacement of undetermined grain	3	–	5	–	–	3	–	–	–	–
Alterite	–	1	1	3	–	1	–	–	–	–
Total	500	490	505	492	520	546	507	503	501	501

Note: NCE—noncarbonate extrabasinal grains; CE—carbonate extrabasinal grains; CI—carbonate intrabasinal grains; NCI—noncarbonate intrabasinal grains; Q—quartz; K—K-feldspar; P—plagioclase; M—mica; L—aphanitic lithic fragments; Mx— matrix; Cm—cement.

spaghetti-like cyanobacteria remains typical of selenite crystals (Vai and Ricci Lucchi, 1976; Rouchy and Monty, 1981).

Only the light portion of the gypsum crystals protrudes among the adjacent siliciclastic grains, filling the pore spaces. In some cases, many pore spaces are filled by the same light crystal, which assumes a poikilotopic appearance because it includes various siliciclastic grains (Fig. 5). The described dark-light zoning is visible also in backscattered-electron SEM images, and EDS analyses reveal that the light rim of the crystals is F-, Na-, K-, Cl-, and Al-rich compared to the dark core, which is composed of pure gypsum (Fig. 6). Only a few gypsum crystals that are completely surrounded by a micrite matrix contain anhydrite microrelicts. Late, thin (1 mm in thickness maximum) gypsum veins (satin spar) cut the rocks.

TABLE 2. RECALCULATED DETRITAL MODES OF THE GESSI FORMATION SANDSTONES

Samples	%				%			%			%			%			%			%			P/F
	NCE	CE	NCI	Qm	F	Lt + Gy-clast	Qm	F	*Lt + Gy-clast	Qm	K	P	Qp	Lvm	Lsm	Lm	Lv	Ls	Rg	Rs	Rm	
MB 04-44	72	28	0	40	28	32	40	28	32	58	19	23	7	0	93	3	0	97	7	73	20	0.54
MB 04-45	95	4	1	42	41	17	36	36	28	50	19	31	53	0	47	29	0	71	19	11	70	0.62
MB 04-47	83	10	7	44	33	23	24	18	58	58	14	28	7	0	93	2	0	98	4	90	6	0.67
MB 04-138	84	6	10	48	29	23	35	21	44	62	18	20	6	2	92	28	0	72	9	47	44	0.53
MB 04-139	88	8	4	53	27	20	49	25	26	66	19	15	3	1	96	34	1	65	7	46	47	0.45
MB 04-140	81	3	16	49	24	27	41	20	39	67	19	14	0	0	100	28	0	72	11	50	39	0.44
MB 04-78	100	0	0	31	68	1	31	68	1	32	39	29	0	0	100	50	0	50	80	2	18	0.42
MB 04-138a	68	7	25	40	23	37	35	20	45	64	19	17	9	0	91	5	0	95	16	69	15	0.48
MB 04-139a	86	8	6	49	27	24	42	23	35	65	18	17	16	0	84	28	0	72	8	52	40	0.48
MB 04-140a	71	9	20	39	26	35	32	22	46	60	24	16	6	1	93	10	0	90	10	69	21	0.4
Mean	83	8	9	44	32	24	37	28	35	58	21	21	11	0	89	22	0	78	17	51	32	0.50

Note: Grain parameters: NCE—noncarbonate extrabasinal grains; CE— carbonate extrabasinal grains; CI—carbonate intrabasinal grains; NCI—noncarbonate intrabasinal grains; Qm—monocrystalline quartz; Qp—polycrystalline quartz; Qt—Qm + Qp; K—K-feldspar; P—plagioclase; F—P + K; L—aphanitic lithic fragments; Lm—metamorphic lithic fragments; Lv—volcanic lithic fragments; Ls—sedimentary lithics + gypsum clast (Gy-clast); Lt—L + Qp + CE; Lvm—volcanic and metavolcanic lithic fragments; Lsm—sedimentary and metasedimentary lithic fragments + gypsum clast; Rg—plutonic rock fragments; Rs—aphanitic + phaneritic sedimentary rock fragments; Rm—aphanitic + phaneritic metamorphic rock fragments; * denotes recalculated parameters considering gypsum cement component as clast (Gy-clast).

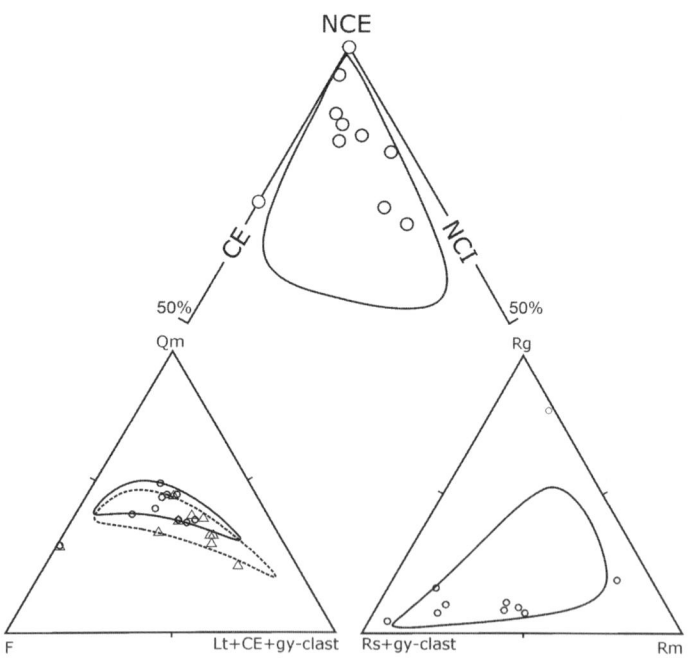

○ Samples where gypsum cement is omitted in the Lt component
△ Samples where gypsum cement is treated as gypsum clast (gy-clast)
--- 99% confidence regions of the mean

Figure 4. Ternary diagrams showing Gessi Formation sandstone compositions with 99% confidence regions of the mean after Weltje (2002): Qm (monocrystalline quartz), F (feldspar), and Lt (aphanitic lithic fragments and fine-grained polycrystalline quartz), K (K-feldspar), and P (plagioclase). Aphanitic lithic and phaneritic rock fragments: Rg (phaneritic plutonic fragments), Rs (aphanitic + phaneritic sedimentary fragments + gypsum clast), and Rm (aphanitic + phaneritic metamorphic fragments). NCE (extrabasinal noncarbonate grains), CE (extrabasinal carbonate grains), and NCI (noncarbonate intrabasinal grains).

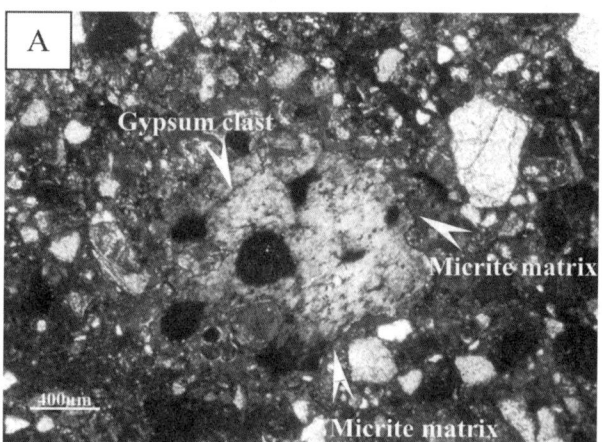

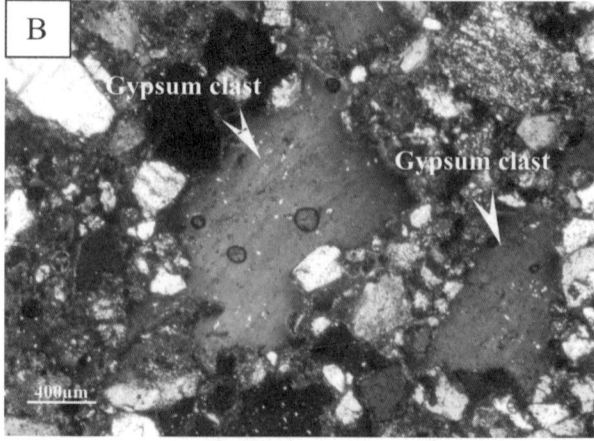

Figure 5. Photomicrographs of detrital and cementing gypsum: (A–D) oversized detrital gypsum grains surrounded by micrite matrix and with anhydrite inclusions. (*continued.*)

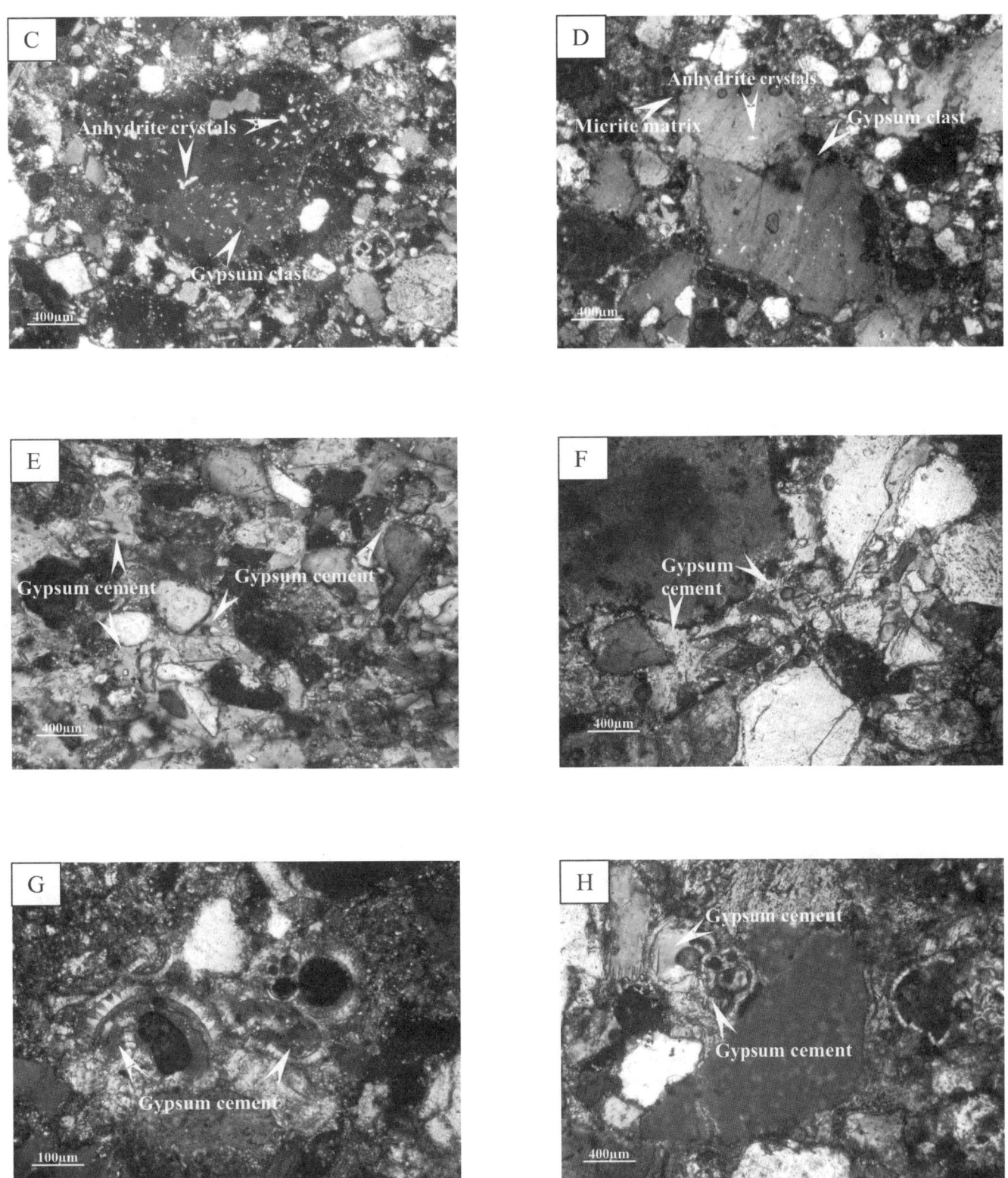

Figure 5. (continued) Photomicrographs of detrital and cementing gypsum: (A–D) oversized detrital gypsum grains surrounded by micrite matrix and with anhydrite inclusions; (E–F) cementing gypsum; and (G–H) cementing gypsum filling the bioclast and in some places replaced by later calcite (lc).

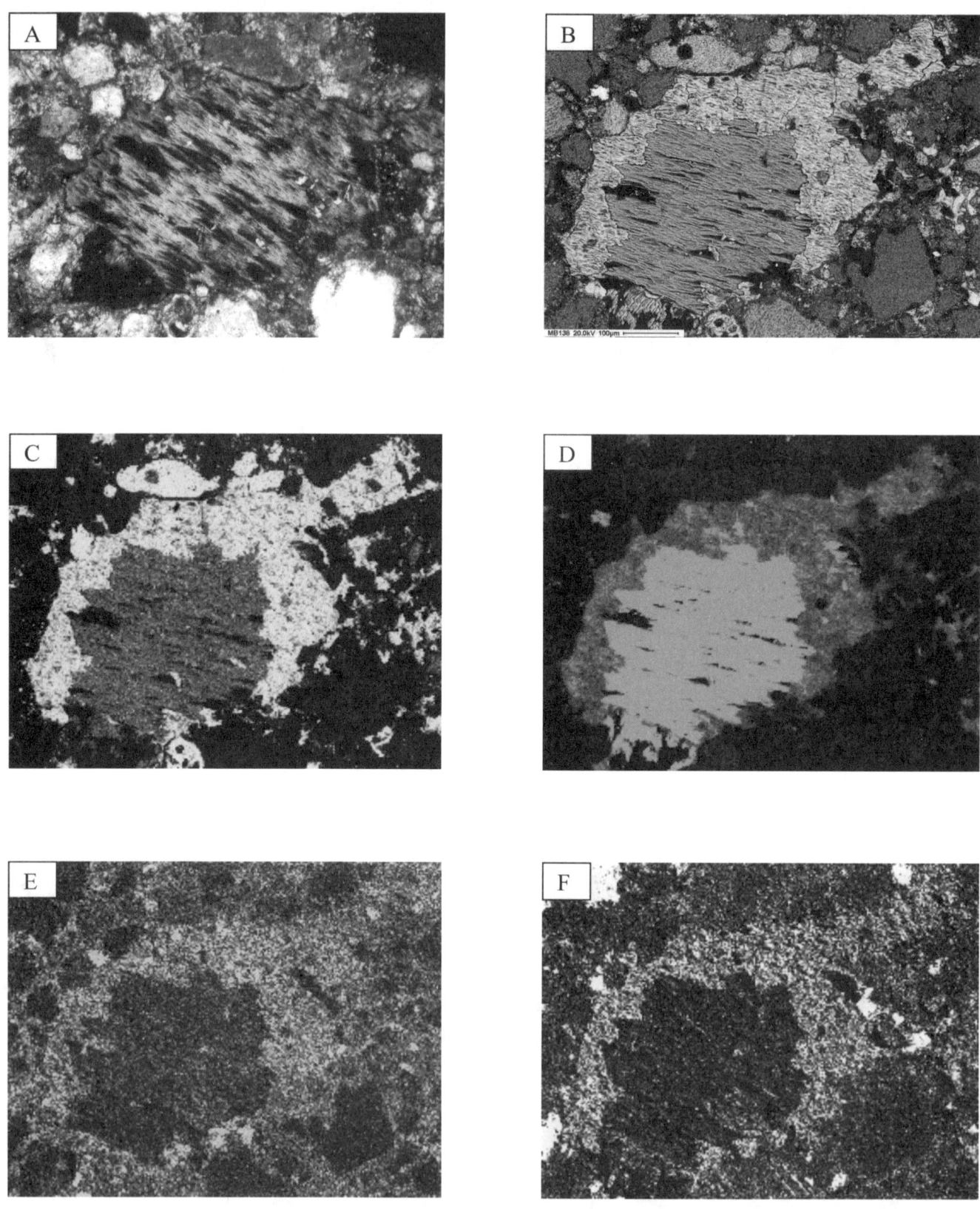

Figure 6. Gypsum grain photomicrograph (A), scanning electron microscope (SEM) backscattered electron image of the same crystal (B), and X-ray maps of various elements: calcium (C), sulfur (D), chlorine (E), sodium (F). (*continued.*)

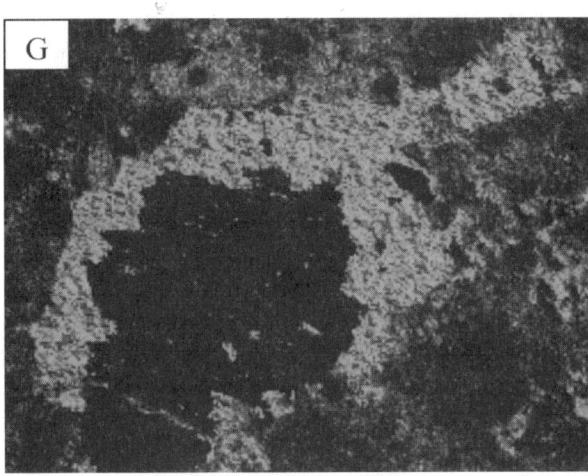
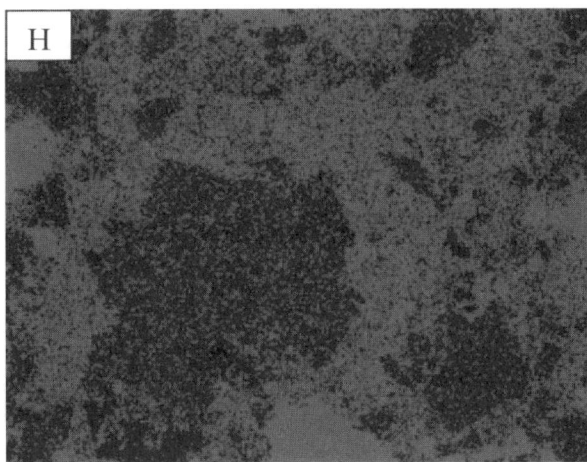

Figure 6. (*continued*) X-ray maps of various elements: fluorine (G), and potassium (H).

DISCUSSION

Gypsum in Arenites: General Considerations

Distinction between detrital and authigenic gypsum framework components in sandstone provides important information about the gypsum provenance and on the postdepositional processes that occur during diagenesis, such as sulfate dissolution and/or precipitation (Abdel, 1998).

Gypsum particles derived from an extrabasinal source area have a low-preservation potential due to their chemical and mechanical instability. For example, eolian gypsarenite dunes can be found only in particularly arid climates (Simpson and Loop, 1985). Studies on gypsum dissolution advocate transport by water as the most important limiting process, but also the transition from laminar to turbulent hydrodynamic regime causes an increase in dissolution rate (Raines and Dewers, 1997). Detrital gypsum preservation also depends on other factors, such as the water saturation grade, the original grain size, the transport duration, and the transport-deposition mechanism (Manzi et al., 2005).

The problems concerning recognition of detrital gypsum derived from erosion of primary evaporitic gypsum and/or gypsarenites have been addressed by many studies (Hardie and Eugster, 1971; Parea and Ricci Lucchi, 1972; Ricci Lucchi, 1973; Vai and Ricci Lucchi, 1976; Schlager and Bolz, 1977; Schreiber, 1973; Simpson and Loop, 1985; Kendall, 1992; Kendall and Harwood, 1996; Manzi et al., 2005). These studies focused on pure clastic sulfate sediments, but mixed sulfate-siliciclastic sediments have not received much attention in the literature, mostly because of the complex recognition of gypsum as true detrital particles in sandstones.

In the last two decades, only a few papers have focused on gypsum grains in arenites. In particular, the classification proposed by Zuffa (1980, 1985) offers a wide overview of the grain types grouped into classes using compositional (carbonate and noncarbonate: C, NC), temporal (coeval and noncoeval: C, NC), and spatial (intrabasinal and extrabasinal: I, E) criteria. Gypsum is included in the NCI class (noncarbonate intrabasinal), together with phosphates, glauconite, Fe-oxides, volcanics, rip-up clasts, which are grains that represent uncommon components with respect to the whole framework. These grains may be major framework constituents in condensed arenite intervals, suggesting specific paleoenvironmental conditions, climate, geodynamic setting, and sea-level fluctuations. Most of such grains have been thoroughly studied by numerous authors: Garzanti (1991) summarized general conditions for their formation; Critelli and Ingersoll (1995) emphasized textures and provenance of volcanic detritus; Hower (1961), Amorosi (1995), and other authors (Loutit et al., 1988; Baum and Vail, 1988; Vail et al., 1991; Mitchum and Van Wagoner, 1991) discussed the origin and stratigraphic significance of glauconite.

Detrital versus Cementing Gypsum in the Gessi Formation

Is gypsum a late precipitate filling pore spaces or was it present into the Gessi Formation arenites as detrital grains, or both? What are the criteria to distinguish among the different types of gypsum?

Because the light portion of the gypsum crystals protrudes among the adjacent grains, filling the pore spaces, it appears evident that this portion cannot represent a clastic feature but should be interpreted as a late overgrowth (Fig. 6). The shape of the dark core, on the contrary seems to mimic original gypsum detrital grain shape, and their sizes appear to be compatible with those of the other siliciclastic and carbonate grains (Fig. 6). Moreover, the dark core seems to contain inclusions of disrupted remnants of primary cyanobacterial filaments that are typically from vertically grown selenite crystals in a salina (Vai and Ricci Lucchi, 1976; Rouchy and Monty, 1999; Babel, 2004).

The chemical features revealed by SEM-EDS analyses are useful tool to interpret the origin of the gypsum. This is because trace elements in the gypsum lattice may reveal growth conditions

(Kushnir, 1980, 1982). As demonstrated by Kushnir (1982) and Lu et al. (1997), trace content of Cl in gypsum cannot be related to lattice substitution, but it can be related to fluid and/or solid inclusions. Na is significantly present in fluid inclusions, but can be also present in the gypsum lattice (Lu et al., 1997); Edinger (1973) demonstrated that Na$^+$ is preferentially adsorbed on the (111) face of gypsum and that increasing concentrations of Na$^+$ in the coexisting fluids result in a decrease of gypsum crystal size.

These considerations suggest two different growth mechanisms for the pure gypsum core and the F-, Na-, K-, Cl-, and Al-rich overgrowth. In particular, the lighter gypsum overgrowth may contain a considerably larger proportion of fluid inclusions, a feature suggesting a relatively higher growth rate with respect to the core zone. This is because fast crystal growth may produce more lattice defects that are filled with the coexisting fluids (negative crystals). Relatively fast growth is the possible result of anhydrite gypsification at subsurface conditions (Lugli, 2001). The presence of anhydrite relicts within some of the gypsum crystals suggests that the whole sulfate rock was transformed into anhydrite during burial by geothermal temperature increase. The anhydrite relicts are still preserved in the grains that have a micrite envelope, which possibly inhibited a complete grain hydration by reducing the local permeability conditions. The late light gypsum overgrowth is possibly the result of gypsification of anhydrite during exhumation and may represent the local volume increase connected with the transformation. This volume increase is apparent only in a high-porosity rock and is normally not visible in the pure sulfate rocks where the excess sulfate produced during the transition is removed by the hydrating fluids (Murray, 1964; Lugli, 2001).

In summary, we interpret the inner dark cores of the gypsum crystals as remnants of original detrital grains derived from the erosion of selenite that grew vertically in a Messinian salina. The gypsum was then transformed into anhydrite during burial, and the anhydrite was hydrated back to gypsum at subsurface conditions during exhumation. During or subsequent to this gypsification phase, the light gypsum overgrowth formed on the original detrital grains, filling and cementing the pore spaces (Fig. 7).

Detrital Gypsum Provenance

During the Messinian salinity crisis, large volumes of evaporite sediments were deposited in the Mediterranean basin due to tectonic and/or glacio-eustatic processes, which in some places progressively restricted and possibly isolated the Mediterranean Sea from the open ocean (Hsü et al., 1973; Vai, 1997; Krijgsman et al., 1999a, 1999b). The Rossano Basin experienced evaporite and detrital sedimentation, recording a transgressive/regressive cycle. The detrital evaporites have been interpreted as turbidites produced by mechanical reworking of earlier primary selenite and/or gypsarenite deposits (Dominici, 2005). At present, the location and the relative age of the primary evaporite deposit (lower or upper evaporites of the Mediterranean) are unclear.

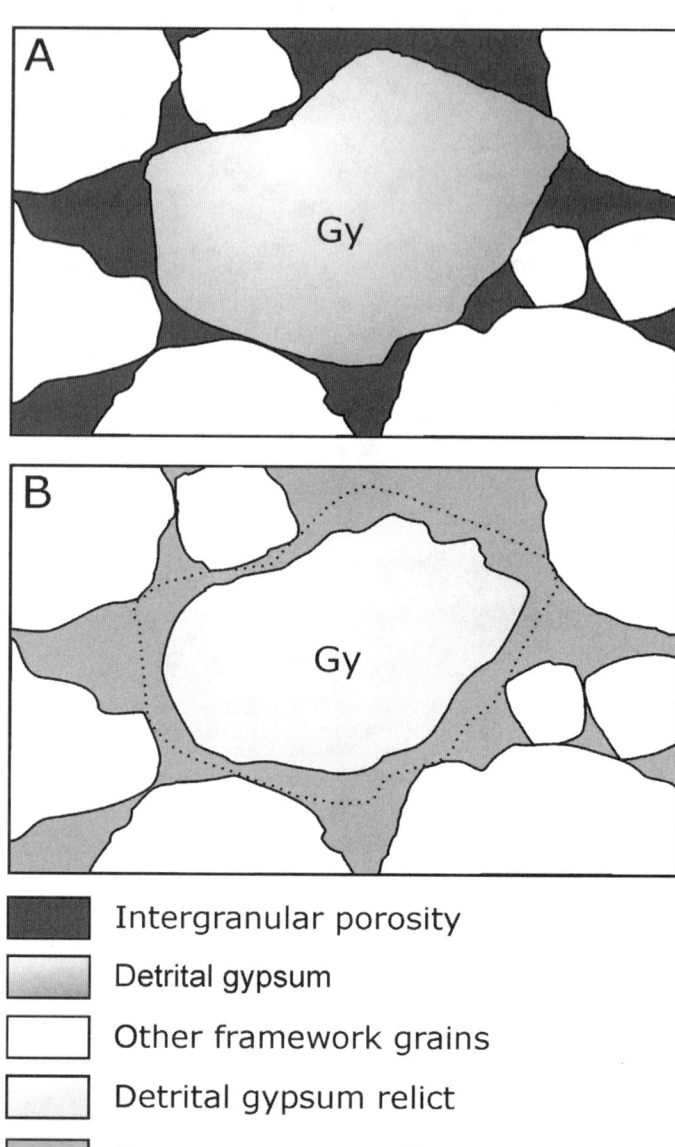

Figure 7. Schematic drawing showing: (A) detrital gypsum (Gy) in the Gessi Formation sandstones; and (B) the possible dissolution processes that affect gypsum crystals after deposition and obliterate the original detrital features.

Provenance of Siliciclastic Particles

The Messinian sandstones of the Gessi Formation from the Rossano Basin contain dominantly crystalline extrabasinal rock fragments, suggesting deep unroofing of the crustal terranes of the Calabrian arc (e.g., Critelli and Le Pera, 1995, 1998; Critelli, 1999) and minor input of an intrabasinal detrital gypsum component newly documented in this study. Compositional characteristics of such sandstones suggest that they were mainly derived from crystalline extrabasinal and subordinate intrabasinal source areas. The main petrographic parameters are consistent with such an origin, and the presence of aphanitic and phaneritic rock fragments indicates that detritus was mainly derived from the medium- to low-grade metamorphic rocks and from the plutonic sequence of the Sila batholith, which closely resemble the basement rocks of the Sila Unit allochthon. The abundance of schistose, phyllite, and medium-grade metamorphic detritus in the sandstones suggests a provenance from the Mandatoriccio Complex, with dominant metapelite and gneissic rocks, and the Bocchigliero Complex, with dominant phyllite (Messina et al., 1994), that crop out in the northeastern Calabrian sector.

The composition of the plutonic detritus is closely related to the granodiorite and tonalite rocks that constitute the Sila batholith (Messina et al., 1994). Progressive exhumation of the Calabrian crustal terranes during rapid uplift and erosion since middle to late Oligocene to ca. 10 Ma has been confirmed by fission-track results on apatite and zircon (Thomson, 1994, 1998). Reworking processes of older sedimentary successions, such as the Paludi Formation (Zuffa and De Rosa, 1978) and/or the Arenaceo Conglomeratica Formation (Ogniben, 1962), which are rich in crystalline particles, may have also supplied detritus.

CONCLUSIONS

We present here new evidence to asses the complexity of the Messinian salinity crisis in the Mediterranean. Detrital evaporites and mixed siliciclastic-gypsum arenites are more abundant than previously thought. They are present also in the Gessi Formation from the Rossano Basin in Calabria and are the result of deep unroofing of the crustal terranes of the Calabrian arc and the reworking of primary subaqueous evaporite facies (selenite).

The detrital gypsum origin in the quartzofeldspathic sandstones is revealed by gypsum crystal overgrowths that outline the shape of former gypsum clasts. The gypsum underwent a partial transformation into anhydrite during burial, and the anhydrite was hydrated back to gypsum during exhumation. During or subsequent to this gypsification phase, a gypsum overgrowth rich in F, Na, K, Cl, and Al formed on the original detrital grains and filled the pore spaces.

Detrital modes of the Gessi Formation sandstones in the Rossano Basin suggest complex source-basin relations in this area during the Messinian and record a break of evaporite sedimentation and the onset of a clastic sedimentation, as evidenced by the occurrence of both siliciclastic and gypsum detritus.

ACKNOWLEDGMENTS

We would like to thank S. Critelli, S. Perrotta, and E. Le Pera for their helpful suggestions and review of an early version of the manuscript, M. Davoli for support with scanning electron microscope–energy dispersive spectroscopy (SEM-EDS) analyses, and F. Colonnese for thin section preparation. We thank B.C. Schreiber, M. Babel, and K. Marsaglia for their reviews, which greatly improved the manuscript.

REFERENCES CITED

Abdel, W., 1998, Diagenetic history of Cambrian quartzarenites, Ras Dib-Zeit Bay area: Gulf of Suez eastern desert, Egypt: Sedimentary Geology, v. 121, p. 121–140, doi: 10.1016/S0037-0738(98)00076-1.

Amorosi, A., 1995, Glaucony and sequence stratigraphy: A conceptual framework distribution in siliciclastic sequences: Journal of Sedimentary Petrology, v. B65, p. 419–425.

Babel, M., 2004, Models for evaporite, selenite and gypsum microbialite deposition in ancient saline basins: Acta Geologica Polonica, v. 54, no. 2, p. 219–249.

Baum, G.R., and Vail, P.R., 1988, Sequence stratigraphy concept applied to Paleogene outcrops, Gulf and Atlantic basins, in Wilgus, C.K., Hastings, B.S., Kendall, C.G.St.C., Posamentier, H.W., Ross, C.A., and Van Wagoner, J.C., eds., Sea Level Changes: An Integrated Approach: Society for Sedimentary Geology (SEPM) Special Publication 42, p. 309–328.

Butler, R.W.G., Grasso, M., Peddley, H.M., and Rambert L., 1995, Tectonics and sequence stratigraphy in Messinian Basin Sicily: Constraints on the initiation and termination of the Mediterranean "salinity crisis": Geological Society of American Bulletin, v. 107, p. 425–439.

Critelli, S., 1999, The interplay of lithospheric flexure and thrust accommodation in forming stratigraphic sequences in the southern Apennines foreland basin system, Italy, in Rendiconti Lincei Scienze Fisiche e Naturali, serie IX, vol. X, fascicolo, v. 4, p. 1999.

Critelli, S., and Ingersoll, R.B., 1995, Interpretation of neovolcanic versus palaeovolcanic sand grains: An example from Miocene deep-marine sandstone of the Topanga Group (southern California): Sedimentology, v. 42, p. 783–804, doi: 10.1111/j.1365-3091.1995.tb00409.x.

Critelli, S., and Le Pera, E., 1994, Detrital modes and provenance of Miocene sandstones and modern sands of the southern Apennines thrust-top basins (Italy): Journal of Sedimentary Research, v. 64, no. A4, p. 824–835.

Critelli, S., and Le Pera, E., 1995, Tectonic evolution of the southern Apennines thrust-belt (Italy) as reflected in modal compositions of Cenozoic sandstone: The Journal of Geology, v. 103, no. 1, p. 95–105.

Critelli, S., and Le Pera, E., 1998, Post-Oligocene sediment-dispersal systems and unroofing history of the Calabrian microplate, Italy: International Geology Review, v. 40, no. 7, p. 609–637.

Decima, A., and Wezel, F.C., 1973, Late Miocene evaporites of the central Sicilian basin, Italy, in Ryan, W.B.F., Hsu, K.J., et al., eds., Initial Reports of Deep Sea Drilling Project Volume 13: Washington, D.C., U.S. Government Printing Office, p. 1234–1241.

Dickinson, W.R., 1970, Interpreting detrital modes of greywacke and arkose: Journal of Sedimentary Petrology, v. 40, p. 695–707.

Dominici, R., 2005, Relazioni tra sedimentazione clastica, evaporitica nel sistema di bacino di foreland in Calabria nord orientale [Ph.D. thesis]: Arcavacata di Rende, Cosenza, Università degli Studi della Calabria, 170 p.

Edinger, S.E., 1973, An investigation of the factors which affect the size and growth rate of gypsum: Journal of Crystal Growth, v. 18, p. 217–224, doi: 10.1016/0022-0248(73)90164-4.

Garzanti, E., 1991, Noncarbonate intrabasinal grain arenites: Their recognition, significance and relationship to eustatic cycles and tectonic setting: Journal of Sedimentary Petrology, v. 61, p. 959–975.

Gazzi, P., 1966, Le arenarie del flysch sopracretaceo dell'Appennino modenese; correlazioni con il flysch di Monghidoro: Mineralogica et Petrografica Acta, v. 12, p. 69–97.

Hardie, L.A., and Eugster, H.P., 1971, The depositional environment of marine evaporites: A case for shallow, clastic accumulation: Sedimentology, v. 16, p. 187–220, doi: 10.1111/j.1365-3091.1971.tb00228.x.

Hower, J., 1961, Some factors concerning the nature and origin of glauconite: The American Mineralogist, v. 46, p. 313–334.

Hsü, K.J., Cita, M.B., and Ryan, W.B.F., 1973, The origin of Mediterranean evaporates, in Ryan W.B.F., Hsü K., et al., Initial Reports of the Deep Sea Drilling Project, Leg 42A: Washington, D.C., U.S. Government Printing Office, p. 1053–1078.

Ingersoll, R.V., Bullard, T.F., Ford, R.L., Grimm, J.P., Pickle, J.D., and Sares, S.W., 1984, The effect of grain size on detrital modes: A test of the Gazzi-Dickinson point-counting method: Journal of Sedimentary Petrology, v. 54, p. 103–116.

Kendall, A.C., 1992, Evaporites, in Walker, R.G., and James, N.P., eds., Facies Models: Response to Sea-Level Changes: Geoscience Canada Reprint Series 1, p. 259–296.

Kendall, A.C., and Harwood, G.M., 1996, Marine evaporites: Arid shorelines and basins, in Reading, H.G., ed., Sedimentary Environments: Processes, Facies and Stratigraphy: Blackwell Science, p. 281–324.

Krijgsman, W., Hilgen, F.J., Marabini, S., and Vai, G.B., 1999a, New paleomagnetic and cycle stratigraphic age constraints on the Messinian of the northern Apennines (Vena del Gesso, Italy): Memorie della Società Geologica Italiana, v. 118, p. 25–33.

Krijgsman, W., Hilgen, F.J., Raffi, I., Sierro, F.J., and Wilson, D.S., 1999b, Chronology, causes and progression of the Messinian salinity crisis: Nature, v. 400, p. 652–655, doi: 10.1038/23231.

Kushnir, J., 1980, The coprecipitation of strontium, magnesium, sodium, potassium and chloride with gypsum, an experimental study: Geochimica et Cosmochimica Acta, v. 44, p. 1471–1482, doi: 10.1016/0016-7037(80)90112-X.

Kushnir, J., 1982, The composition and origin of brines during the Messinian desiccation event in the Mediterranean Basin as deduced from concentrations of ions coprecipitated with gypsum and anhydrite: Chemical Geology, v. 35, p. 333–350, doi: 10.1016/0009-2541(82)90010-9.

Loutit, T.S., Hardenbol, J., Vail, P.R., and Baum, G.R., 1988, Condensed section: The key to age determination and correlation of continental margin sequences, in Wingus, C.K., Hastings, B.S., Kendall, C.G.St.C., Posamentier, H.W., Ross, C.A., and Van Wagoner, J.C., eds., Sea Level Changes: An Integrated Approach: Society for Sedimentary Geology (SEPM) Special Publication 42, p. 183–213.

Lu, H., Meyers, W.J., and Schoonen, M.A., 1997, Trace and minor element analyses on gypsum: An experimental study: Chemical Geology, v. 142, p. 1–10, doi: 10.1016/S0009-2541(97)00070-3.

Lugli, S., 2001, Timing of post-depositional events in the Burano Formation of the Secchia Valley (Upper Triassic, northern Apennines); clues from gypsum-anhydrite transition and carbonate metasomatism: Sedimentary Geology, v. 140, p. 107–122, doi: 10.1016/S0037-0738(00)00174-3.

Manzi, V., Lugli, S., Ricci Lucchi, F., and Roveri, M., 2005, Deep-water clastic evaporite deposition in the Messinian Adriatic foredeep (northern Apennines, Italy): Did the Mediterranean ever dry out?: Sedimentology, v. 52, p. 875–902, doi: 10.1111/j.1365-3091.2005.00722.x.

Mascle, J., and Rehault, J.P., 1990, A revised seismic stratigraphy of the Tyrrhenian Sea: Implications for the basin evolution, in Kastens, K.A., Mascle, J., et al., Proceedings of the Ocean Drilling Program, Scientific Results, Volume 107: College Station, Texas, Ocean Drilling Program, p. 617–636.

Mattei, M., Cipollari, P., Cosentino, D., Argentieri, A., Rossetti, F., Speranza, F., and Di Bella, L., 2002, The Miocene tectono-sedimentary evolution of the southern Tyrrhenian Sea: Stratigraphy, structural and palaeomagnetic data from the on-shore Amantea basin (Calabrian arc, Italy): Basin Research, v. 14, p. 147–168, doi: 10.1046/j.1365-2117.2002.00173.x.

Messina, A., Russo, S., Borghi, A., Colonna, V., Compagnoni, R., Caggianelli, A., Fornelli, A., and Piccarreta, G., 1994, Il Massiccio della Sila. Settore settentrionale dell'arco Calabro-Peloritano: Bollettino della Società Geologica Italiana, v. 113, p. 539–586.

Meulenkamp, J.E., Hilgen, F., and Voogt, E., 1986, Late Cenozoic sedimentary-tectonic history of the Calabrian arc: Giornale di Geologia, serie III, v. 42, p. 345–359.

Mitchum, R.M., Jr., and Van Wagoner, J.C., 1991, High-frequency sequences and their stacking patterns: Sequence stratigraphic evidence of high-frequency eustatic cycles: Sedimentary Geology, v. 70, p. 131–160, doi: 10.1016/0037-0738(91)90139-5.

Murray, R.C., 1964, Origin and diagenesis of gypsum and anhydrite: Journal of Sedimentary Petrology, v. 34, p. 512–523.

Ogniben, L., 1962, Le Argille Scagliose ed i sedimenti Messiniani a sinistra del Trionto (Rossano, Cosenza): Geologica Romana, v. 1, p. 255–282.

Parea, G.C., and Ricci Lucchi, F., 1972, Resedimented evaporites in the Periadriatic trough (Upper Miocene, Italy): Israel Journal of Earth Sciences, v. 21, p. 125–141.

Patacca, E., Sartori, R., and Scandone, P., 1990, Tyrrhenian Basin and Apennine arcs: Kinematic relation since late Tortonian time: Memorie della Società Geologica Italiana, v. 45, p. 425–451.

Pescatore, T., and Senatore, M.R., 1986, A comparison between a present-day (Taranto Gulf) and a Miocene (Irpinian basin) foredeep of southern Apennines (Italy), in Allen, P.A., and Homewood, P., eds., Foreland Basins: International Association of Sedimentologists Special Publication 8, p. 169–182.

Raines, M.A., and Dewers, T.A., 1997, Mixed transport/reactions control of gypsum dissolution kinetics in aqueous solutions and initiation of gypsum karsts: Chemical Geology, v. 140, p. 29–48, doi: 10.1016/S0009-2541(97)00018-1.

Ricci Lucchi, F., 1973, Resedimented evaporites: Indicators of slope instability and deep-basins conditions in Periadriatic Messinian (Apennines foredeep, Italy): Messinian Events in the Mediterranean: Koninklijke Nederlandse Akademie Van Wetenshappen, Geodynamics Scientific Report 7, p. 142–149.

Roda, C., 1967, I sedimenti Neogenici ed alloctoni della zona di Cirò-Cariati (Catanzaro e Cosenza): Memorie della Società Geologica Italiana, v. VI, p. 137–149.

Rouchy, J.M., and Monty, C.L., 1981, Stromatolites and cryptalgal laminites associated with Messinian gypsum of Cyprus, in Monty, C.L., ed., Phanerozoic Stromatolites: Berlin, Springer-Verlag, p. 155180.

Rouchy, J.M., and Monty, C.L.V., 1999, Microbial gypsum sediments: Neogene and Modern examples, in Riding, R., and Awramik, S.M., ed., Microbial Sediments: Berlin, Springer-Verlag, p. 209–216.

Roveri, M., Bernasconi, A., Rossi, M.E., and Visentin, C., 1992, Sedimentary evolution of the Luna field area; Calabria, southern Italy, in Spencer, A.M., ed., Generation, Accumulation and Production of Europe's Hydrocarbons II: Special Publication of the European Association of Petroleum Geoscientist 2, p. 217–224.

Salvador, A., 1987, Unconformity-bounded stratigraphic units: Geological Society of America Bulletin, v. 98, p. 232–237, doi: 10.1130/0016-7606(1987)98<232:USU>2.0.CO;2.

Schlager, W., and Bolz, H., 1977, Clastic accumulation of sulfate evaporites in deep water: Journal of Sedimentary Petrology, v. 47, p. 600–609.

Schreiber, B.C., 1973, Survey of the physical features of the Messinian chemical sediments: Messinian Events in the Mediterranean: Koninklijke Nederlandse Akademie Van Wetenshappen, Geodynamics Scientific Report 7, p. 101–110.

Senatore, M.R., 1987, Caratteri sedimentari e tettonici di un bacino di avanfossa (Il Golfo di Taranto): Memorie della Società Geologica Italiana, v. 38, p. 177–204.

Simpson, E.L., and Loop, D.B., 1985, Amalgamated interdune deposits, White Sands, New Mexico: Journal of Sedimentary Petrology, v. 55, no. 3, p. 361–365.

Thomson, S.N., 1994, Fission track analysis of the crystalline basement rocks of the Calabrian arc, southern Italy: Evidence of Oligo-Miocene late orogenic extension and erosion: Tectonophysics, v. 238, p. 331–352, doi: 10.1016/0040-1951(94)90063-9.

Thomson, S.N., 1998, Assessing the nature of tectonic contacts using fission-track thermochronology: An example from the Calabrian arc southern Italy: Terra Nova, v. 10, p. 32–36, doi: 10.1046/j.1365-3121.1998.00165.x.

Tortorici, L., 1981, Analisi delle deformazioni fragili dei sedimenti postorogeni della Calabria settentrionale: Bollettino della Società Geologica Italiana, v. 100, p. 291–308.

Vai, G.B., 1997, Cyclostratigraphic estimate of the Messinian stage duration, in Montanari, A.G., Odin, S., and Coccioni, R., eds., Miocene Stratigraphy: An Integrated Approach: Amsterdam, Elsevier, p. 463–476.

Vai, G.B., and Ricci Lucchi, F., 1976, The Vena del Gesso Basin in northern Apennines, in Field Trip Guidebook: Gargano, Messinian Seminar no. 2, The International Geological Correlation Programme (IGCP) Pr. No 96 STEM-Mucchi Modena, p. 1–16.

Vail, P.R., Audemard, F., Bowman, S.A., Eisner, P.N., and Pérez-Cruz, C., 1991, The stratigraphic signature of tectonics, eustasy and sedimentology—An overview, in Einsele, G., Ricken, W., and Seilacher, A., eds., Cycles and Events in Stratigraphy: Berlin, Springer Verlag, p. 617–659.

Weltje, G.J., 2002, Quantitative analysis of detrital modes: Statistically rigorous confidence regions in ternary diagrams and their use in sedimentary petrology: Earth-Science Reviews, v. 57, p. 211–253, doi: 10.1016/S0012-8252(01)00076-9.

Zuffa, G.G., 1980, Hybrid arenites: Their composition and classification: Journal of Sedimentary Petrology, v. 50, no. 1, p. 21–29.

Zuffa, G.G., 1985, Optical analyses of arenites: Influence of methodology on compositional results, in Zuffa, G.G., ed., Provenance of Arenites: Dordrecht, Holland, D. Reidel Publishing Co., NATO ASI Series, p. 165–189.

Zuffa, G.G., and De Rosa, R., 1978, Petrologia delle successioni torbiditiche Eoceniche della Sila Nord Orientale (Calabria): Memorie della Società Geologica Italiana, v. 18, p. 31–35.

MANUSCRIPT ACCEPTED BY THE SOCIETY 9 AUGUST 2006

Geological Society of America
Special Paper 420
2007

The onset of the sedimentary cycle in a mid-latitude upland environment: Weathering, pedogenesis, and geomorphic processes on plutonic rocks (Sila Massif, Calabria)

Fabio Scarciglia[†]
Emilia Le Pera[‡]
Salvatore Critelli[§]

Dipartimento di Scienze della Terra, Università della Calabria, Via Pietro Bucci–Cubo 15B, 87036 Arcavacata di Rende, Cosenza, Italy

ABSTRACT

This work represents an integrated analysis of weathering landforms, including minor landform morphologies and soil profiles developed on granitoid terrains of the Sila Massif uplands (Calabria, southern Italy). The results of our analysis indicate that cryoclastic and thermoclastic processes, along with chemical weathering, are the main factors controlling rock degradation. Microscale features observed in primary minerals and parent rock fabrics, such as structural discontinuities, cleavage planes, fracturing patterns, and variations in chemical composition, play important roles in triggering weathering and, given sufficient time, progressively lead to grussification and soil development. Exfoliation, hydration, and splitting apart of biotite, as well as hydrolysis and etching of plagioclase and K-feldspar, appear to be prominent factors in the breakdown of bedrock.

Whereas time controls the degree of development of the main weathering features and climate influences type and intensity of the dominant processes, relief strongly influences the development and preservation/removal of the regolith/soil cover. Geomorphological evidence of severe surface erosion is quite good, especially along steep slopes where weathering products are quickly removed, although on the highest, dissected paleosurfaces (the oldest paleolandscape remnants in the Sila Massif), wide boulder fields represent relics of past, deep spheroidal weathering that have been exhumed by intense erosion. Erosive, depositional, or reworking phenomena, often enhanced by human activity, are well recorded by macro- and micromorphological features of soils, which show simple, poorly differentiated, rejuvenated profiles, buried or truncated horizons, abundant coarse-grained primary minerals or rock fragments, and pedorelicts. The soil clay mineralogy, characterized by illite,

---

[†]E-mail: scarciglia@unical.it.
[‡]E-mail: emilia.lepera@unical.it.
[§]E-mail: critelli@unical.it.

Scarciglia, F., Le Pera, E., and Critelli, S., 2007, The onset of the sedimentary cycle in a mid-latitude upland environment: Weathering, pedogenesis, and geomorphic processes on plutonic rocks (Sila Massif, Calabria), *in* Arribas, J., Critelli, S., and Johnsson, M.J., eds., Sedimentary Provenance and Petrogenesis: Perspectives from Petrography and Geochemistry: Geological Society of America Special Paper 420, p. 149–166, doi: 10.1130/2006.2420(10). For permission to copy, contact editing@geosociety.org. ©2007 Geological Society of America. All rights reserved.

chlorite, and vermiculite, and the dominance of coarse textures confirm a young pedogenetic stage of evolution, although highly weathered sand grains (quartz included) occur in rarely preserved mature paleosols. This interpretation is also consistent with the compositional immaturity of fluvial sands, which have undergone low to moderate transport.

Keywords: granite weathering, pedogenesis, geomorphic processes, sediment source, polygenesis, upland Mediterranean climate.

INTRODUCTION

Physical and chemical modifications of sediment prior to final deposition and burial reflect all aspects of the drainage area, including source rock texture and mineralogy, topography, climate, and transport processes (Basu, 1985; Johnsson, 1993; Le Pera et al., 2001; Arribas and Tortosa, 2003; Critelli et al., 2003). The contribution of various source-rock types to terrigenous sediments is largely dependent on the intensity of weathering, which may affect some rocks differently than others (e.g., Palomares and Arribas, 1993), and relief (Basu, 1985). For example, different types of bedrock may react differently to chemical weathering, resulting in variation in landscape and the development of weathering profiles (Ollier, 1971; Carson and Kirkby, 1972; Dixon and Young, 1981; Dejou et al., 1982; Pye, 1986; Twidale, 1990; Le Pera and Sorriso-Valvo, 2000b). Moreover, tectonics and climate, coupled with a combination of both chemical and physical weathering processes, may play a complementary role by acting as the main producers of sediment and regulating erosion rates (Le Pera and Sorriso-Valvo, 2000a; Riebe et al., 2000, 2001; Le Pera et al., 2001; Critelli and Le Pera, 2002; Scarciglia et al., 2005a). Furthermore, the genesis of clastic sediments and soils has been investigated in order to quantify processes occurring within source areas, including chemical and physical weathering, and textural and compositional modification of detritus during transition from bedrock to grus to soil to fluvial environment, in a well-controlled setting (e.g., Basu, 1985; Velbel, 1985; Grantham and Velbel, 1988; Cullers et al., 1988; Nesbitt and Young, 1989; Johnsson, 1993; Le Pera and Sorriso-Valvo, 2000b; Le Pera et al., 2001; Critelli and Le Pera, 2002; Girty et al., 2003). The Sila Massif (Fig. 1) is ideally suited to study relationships between landscape evolution and the genesis of clastic sediments and soils, and to calculate mass balances in a small, but tectonically active area characterized by very high sediment production (e.g., Ibbeken and Schleyer, 1991). In this work, we document modifications of source rocks through various stages of landform development within the Sila Massif and Neto River basin. Geologic (Critelli, 1999; Galli and Bosi, 2003) and morphologic (Critelli et al., 1991; Le Pera and Sorriso-Valvo, 2000a, 2000b; Le Pera et al., 2001; Molin et al., 2004; Scarciglia et al., 2005a, 2005b) evidence indicates that the study area is undergoing active landscape evolution.

GEOLOGICAL AND GEOMORPHOLOGICAL SETTING

The study area, located in the Sila Massif (Fig. 1), represents a section of the Hercynian orogenic belt of Western Europe, where allochthonous crystalline basement rocks during Miocene times were emplaced over Mesozoic to Cenozoic terrains of the southern Apennines. Rocks of the Sila Massif form the highest tectonic units (Calabrian arc) of the southern Italian fold-and-thrust belt (Amodio-Morelli et al., 1976) and consist of Paleozoic intrusive and metamorphic rocks, covered in places by unmetamorphosed Mesozoic sedimentary rocks (the Longobucco Group). Toward the Ionian Sea (to the east of the Sila Massif), a sedimentary succession of Miocene to Pleistocene age lies unconformably over the Paleozoic or Mesozoic rocks (Roda, 1964; Critelli, 1999). Paleozoic rocks consist of gneiss, amphibolite, schist, and phyllite, all of which have been affected by various Alpine metamorphic events and intruded by late Hercynian plutons (the Sila Batholith) (Messina et al., 1991). Gneiss consists of massive to migmatitic biotite-sillimanite-garnet–rich sills. A network of pegmatite dikes and an irregular thermal aureole of amphibolite facies rocks mark the contacts between Paleozoic country rocks and plutons (Caggianelli et al., 1994). The Sila Batholith consists of an array of intersecting intrusions, with different texture and fabric, that range in composition from granodiorite to gabbro and leucomonzogranite (Messina et al., 1991). Fission-track thermochronology of apatite and zircon from the basement rocks of the Sila Massif indicates a major period of exhumation from ca. 35 to ca. 15 Ma (Thomson, 1994).

The study area consists of flat to gently inclined paleosurfaces, i.e., uplifted and dissected landforms formed during Pliocene(?)–Lower Pleistocene time (Dramis et al., 1990; Sorriso-Valvo, 1993; Matano and Di Nocera, 1999). These paleosurfaces occur at elevations between 1700 and 1000 m above sea level and developed across Paleozoic plutonic rocks and their Miocene-Pliocene sedimentary cover. Quaternary lacustrine and fluvial deposits, often terraced, are morphologically entrenched within the paleosurfaces of older landscapes (e.g., Sorriso-Valvo, 1993; Scarciglia et al., 2005a, 2005b).

The upper reaches of the main drainage systems of the Sila uplands, i.e., the Mucone, Savuto, Trionto, Nicà, and Neto Rivers, mainly drain plutonic rocks of the Sila Batholith, where weathering and soil profiles were described and sampled for this study. Within the intermediate and lower coastal reaches, these rivers

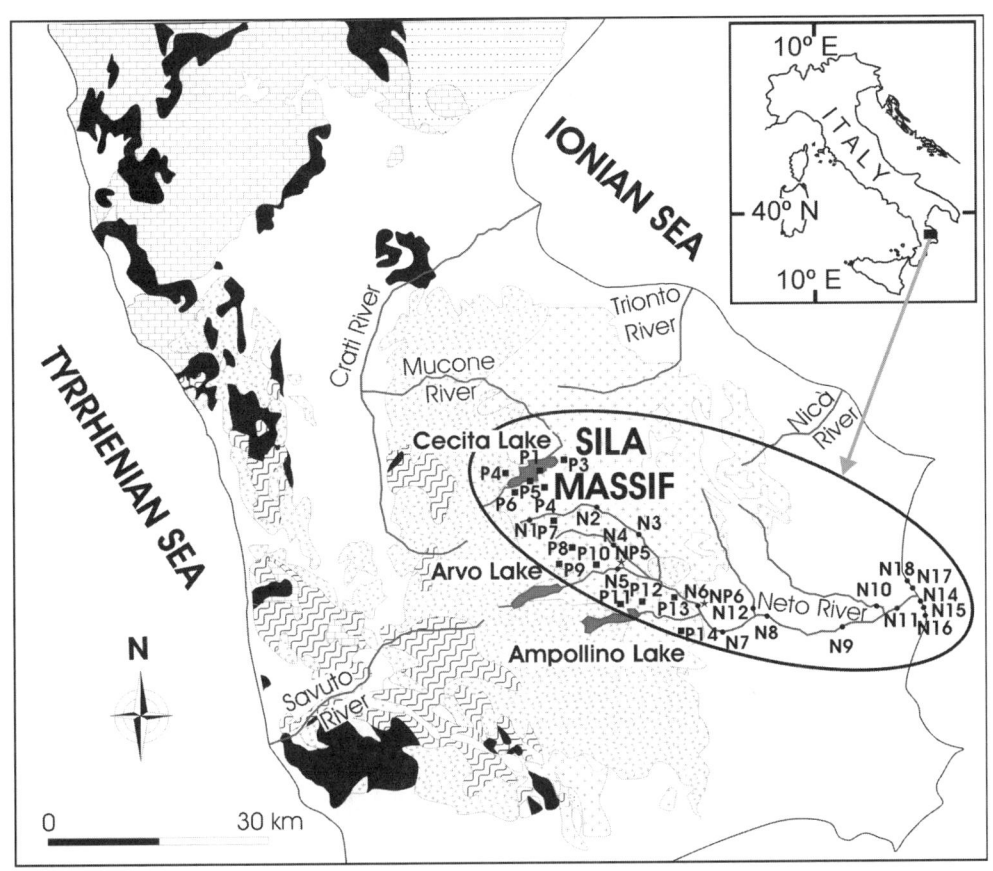

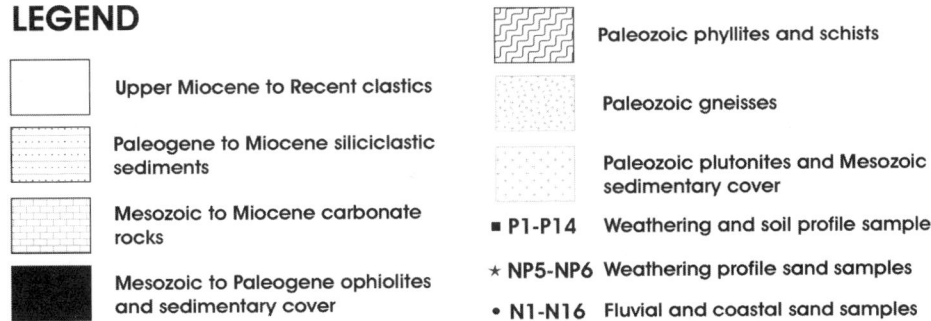

Figure 1. Geological sketch map of north-central Calabria with location of the study area (circled area) and locations of samples.

drain metamorphic and/or sedimentary terranes. The Neto River, selected for quantitative compositional analyses, drains areas underlain by essentially Paleozoic granites and minor gneiss in its upper reaches, and Miocene to Pleistocene sedimentary rocks (siliciclastic and minor evaporites) in the lower reaches, where it feeds a modern delta system (Le Pera et al., 2001).

CLIMATE AND VEGETATION

The study area has a moist temperate climate, typical of upland Mediterranean zones (Csb, sensu Köppen, 1936), with warm, humid, and short summers and relatively mild winters. Mean monthly temperatures of the coldest month (January) are close to −1 °C/1 °C, whereas in July or August, they reach 16–18 °C. From November to April, daily temperatures commonly fall a few degrees below zero, with absolute minima reaching −10 °C or lower values; during summer, absolute maxima may attain 30–32 °C. Rainfall is spread throughout the winter season, with mean annual rates between 1000 and 1400–1600 mm. During the summer, rainfall never exceeds 30 mm (Versace et al., 1989; Lulli et al., 1992; Colacino et al., 1997). Snowfall occurs mainly in areas above 1400–1600 m, where it persists for ~6 months (Lulli and Vecchio, 2000). According to the U.S. Department of Agriculture (2003), the pedoclimatic regime is mesic and udic (ARSSA, 2003).

Vegetation in the Sila uplands consists of grassland (often used for grazing), high mountain belt conifers (dominated by pine and fir) and/or beech forests, and cultivated fields. It is the result of a recent renaturalization and reforestation policy that was promoted during the last half-century after repeated earlier phases of severe deforestation, urban settlement, and poor agricultural practices: human activity had promoted progressively more extensive anthropogenic impact on the Sila Massif territories since the early prehistoric up to the classic (Greek and Roman) historical civilizations and modern times (e.g., Sorriso-Valvo, 1993; Scarciglia et al., 2005b).

METHODS

Macroscale field studies, as well as micromorphological investigations of weathering and soil profiles developed on granitoid rocks, were carried out in a broad upland plateau area, and along different flat hilltops of some river catchments (Fig. 1).

Undisturbed samples were collected from diagnostic soil horizons, grus, and bedrock, and were impregnated with a polyester crystic resin and hardened for thin section preparation. Weathering and pedogenetic features were described and classified according to the scheme of FitzPatrick (1984). Scanning electron microscopy and energy dispersive spectroscopy (SEM-EDS) were performed on sand-textured mineral grains and soil matrix clays. Samples were mounted on aluminum stubs, coated with gold, and examined with a Stereoscan 360 scanning electron microscope (Cambridge Instruments), equipped with an energy dispersive X-ray analyzer with a Si/Li-SUTW detector (EDAX, Philips Electronics). X-ray intensities were converted into weight percentages of oxides by applying the ZAF standard matrix correction procedure (Wilson, 1987); measured X-ray peaks were multiplied by factors dependent on atomic number (Z), absorption (A), and fluorescence (F) properties of specimens, to improve accuracy of chemical concentrations and mitigate errors. X-ray diffraction analysis (XRD) was performed on some selected soil horizons. Oriented specimens of the <0.2 μm clay fraction and random powders of the 2–0.2 mm sand fraction, which were separated by sieving and centrifugation following organic matter oxidation with H_2O_2, were scanned with a Philips 17/30 instrument (Cu-Kα radiation, 40 kW, 20 mA). X-ray diffractometer patterns were interpreted according to Berry (1974) and Brindley and Brown (1980). These approaches were utilized in order to recognize primary minerals and their pattern/degree of weathering, as well as pedogenetic clay minerals.

Finally, sand samples from the main channel of the Neto River were collected for point-count analysis (medium sand fraction, 0.50–0.25 mm).

WEATHERING FEATURES

The granitic rocks that crop out in the study area are characterized by extremely different degrees and patterns of physical and/or chemical weathering. The bedrock often shows a variety of physical discontinuities, such as shear zones, fault planes, pressure release and other types of jointing, pegmatite dikes, and minor lithologic or compositional changes. Such features sometimes intersect, creating blocks and wedges of different shapes and sizes. Overprinting this block and wedge pattern is an intense grussification, characterized by minute rock flawing and crumbling; this forms a gritty, loose material made of small polymineral aggregates and monomineralic particles (grain-by-grain disintegration, sensu Butzer [1976], or arenization, sensu Power and Smith [1994] and Teeuw et al. [1994]). Very often, especially where the soil or grus cover is extremely thin or completely absent, tree roots are observed to penetrate the bedrock or saprolite at a depth of 1–2 m, divaricating and deepening preexisting joints.

Saprolite is commonly some tens meters thick, but may reach 100 m in depth (Cascini et al., 1992; Matano and Di Nocera, 1999). Rounded core-stones and boulders (Butzer, 1976; Ollier, 1988; Le Pera and Sorriso-Valvo, 2000a), ranging from ~30 cm to 3–4 m in diameter (Fig. 2A, B), are often flaked, with an onion-like structure, and frequently occur within saprolite. They also may appear surrounded by a network of whitish to yellowish or light gray, bleached curved surfaces, intensely depleted in iron and fine particles (Fig. 2A). In contrast, other core-stones, and/or their grussified counterparts, often show reddish or blackish colors, related to Fe-Mn–oxide staining or clay illuviation. Similar features may coexist and form a mottled zone consisting of alternating subhorizontal or slightly wavy, whitish, eluvial surfaces, with yellowish brown, red, and/or black layers or elongated lenses. These features, even very close to the ground surface, may characterize the top of weathering profiles. In places, core-stones and boulders are partially (Fig. 2B) or completely exhumed (Figs. 2C and 2D) from the granular saprolite, and they frequently lie over the topographic surface, especially on the highest and oldest paleosurface remnants, and form wide boulder fields (Scarciglia et al., 2005a). Their unweathered inner core is usually surrounded by a weakly weathered, outer shell, with thin, millimeter- to centimeter-thick (Fig. 2D), fragmented exfoliation sheets (flaking; sensu Ollier, 1967, 1971) and encrusting lichens. The original translucent appearance of some primary feldspars in these external flakes has been replaced by a whitish to pale yellow alteration product with a dirty appearance, whereas biotite may exhibit cleaving and oxidation, coupled with local reddish Fe-staining as a possible biotite crystal derivative. On some boulders, massive exfoliation (large-scale spalling or sheeting; sensu Ollier, 1967, 1971) is present (Fig. 2C), with curved, edge-tapering sheets, 10–30 cm thick, as well as occasional splitting into large blocks along planar or curved cracks.

Saprolite and grus occasionally undergo coarse block rockfalls or trigger granular debris flows, mainly on the steepest slopes, where their stability thresholds are more easily overcome. Foot-slope scree taluses or detrital cones (Figs. 3A and 3B) are, in places, dominated by coarse rock fragments (gravels to boulders), while at other sites, they are mainly characterized by silt and sand to fine gravel and/or by a brownish to yellowish-brown or reddish (sometimes organic-rich) silty-clay matrix. Fallen tree trunks occur or lie over some scree talus accumulations (Fig. 3B).

Figure 2. (A) Rounded core-stones surrounded by a network of whitish, eluviated curved surfaces (hammer is 33 cm long). (B) Partially exhumed boulders, about 160 cm wide. (C) Completely exposed boulder, showing large outer spalling. (D) Thin, fragmented and weakly weathered exfoliation sheets around an exhumed core-stone (camera cap is 6 cm in diameter).

THE SOILS

The soil cover of the study area is not very homogeneous, and, in some zones, a complete lack of pedologic horizons allows the bare granite bedrock or saprolite to be exposed at the surface. Many soils are poorly differentiated and weakly structured and include a brown epipedon composed of an intimate mixture of humus and neoformed clay, which ranges from a few to some tens of centimeters in thickness (Fig. 4A). This organic-mineral horizon (A) is derived from the progressive degradation of fallen plant tissues and from the weak to moderate chemical weathering of primary minerals. Mainly, but not exclusively, on steep slopes, the tops of soils are often truncated or soils show abrupt lower boundaries with underlying bedrock. Soil profiles may reach a maximum depth of ~1.5–2 m, where a flat topography and/or a protective forest cover occur, and/or along colluvial footslope belts, where they are affected by strong reworking and include frequent rock fragments or are buried by detritus and younger soils. At these sites, they appear better differentiated into various soil horizons that consist of simple cambic (Bw) and organic (O and A) horizons (Fig. 4B). Micromorphological observations of soils in thin sections permitted the identification of the following relevant pedogenetic features: (1) occasional rounded pedorelicts, i.e., reworked fragments of soils exhibiting features different from those characterizing the soils in which they are now included (Fig. 4C); (2) rare clay coatings and infillings, with slightly grainy to intensely speckled extinction patterns in crossed polarized light; (3) isolated fragments of clay coatings or infillings (papules; sensu Brewer, 1976); and (4) scarce to frequent silt or silty-clay coatings and cappings.

Phyllosilicates, such as illite, chlorite, and/or vermiculite, as well as some mixed layer components, dominate clay mineral

Figure 3. (A) Fractured granite rock wall (1.5 m high) undergoing coarse block fragmentation and rockfalls, with some granular disintegration aggrading the basal scree talus. (B) Detrital cones at the base of a very steep slope, about 70 m in height: the left cone is dominated by coarse rock fragments, while the right cone is composed mainly of fine granular detritus dispersed in a pedogenized matrix. Fallen tree trunks are indicated by the white arrows.

assemblages. However, halloysite or possible short-range order aluminosilicate minerals are sometimes associated with these assemblages. More detailed results are available in Mirabella et al. (1996) and Scarciglia et al. (2005a, 2005b).

Bedrock and saprolite (R and Cr layers) are usually fragmented: very thin intragranular and transgranular microcracks may isolate small (millimetric to centimetric) mineral aggregates that are loosely bound together. Some feldspar grains in both bedrock and saprolite are quite weathered and have lost their original translucent aspect and acquired an almost powdery consistency, with dirty white to pale yellow colors. Micas, often affected by Fe-oxide staining, also exhibit distinct edge exfoliation or show an incipient disintegration by splitting along cleavage planes. Occasional to frequent Fe-depleted mottles and tongues, along with Fe-oxide segregations occur, as do rare to common clay coatings in intergranular voids.

Most of the soils are dominantly loamy sand to loam and sandy loam in texture. In addition, they display frequent skeletal grains (conversely having very low amounts of clay), umbric topsoils and cambic horizons, weakly to moderately acid pH, low cation exchange capacity, and small amounts of exchangeable bases, and an occasional lithic lower boundary toward the bedrock. In short, the main soil types are the orders Entisols and Inceptisols, sensu Soil Taxonomy (USDA, 2006); according to the WRB (World Reference Base for Soil Resources) classification (FAO et al., 1998), they are Leptosols, Cambisols, Umbrisols, Regosols, and Fluvisols (Dimase and Iovino, 1996; Lulli and Vecchio, 2000; ARSSA, 2003; Scarciglia et al., 2005a, 2005b).

More mature and older soils with other pedogenetic features, exceptionally reaching 5–6 m in depth, are exposed in extremely scattered and small sites (Dimase and Iovino, 1996; Lulli and Vecchio, 2000; ARSSA, 2003; Scarciglia et al., 2005a, 2005b). They mainly occur on terraced, granite-derived fluvial sediments, flat paleosurfaces, or colluvial slope deposits within protected landforms. Very often they may be truncated and/or buried by younger, less-developed soils, and they have a higher clay content and appear intensely rubified. Phyllosilicate clays include kaolinite (Fig. 5A) together with illite, and some chlorite and vermiculite minerals. Moreover, some are Alfisols (USDA, 2006) or Luvisols (FAO et al., 1998) with one or more argillic (Bt) horizon/s due to the abundant illuviation of fine material within pores and around soil aggregates. Microlaminated, crescentic clay coatings and infillings (Figs. 5B and 5C) were observed in thin sections. These pedofeatures frequently showed evidence of degeneration, with smooth-banded to speckled, grainy extinction patterns between crossed polars (Fig. 5B), evidence of fragmentation (Fig. 5C), and/or partial assimilation into the soil matrix (FitzPatrick, 1984).

CRYSTAL MICROTEXTURES

The primary minerals that characterize the granitoids of the Sila uplands were identified in thin sections as well as from bulk samples (sand fraction) by means of optical microscopy (OM) and scanning electron microscopy with X-ray spectral analysis. For some samples, X-ray diffraction was also performed (Scar-

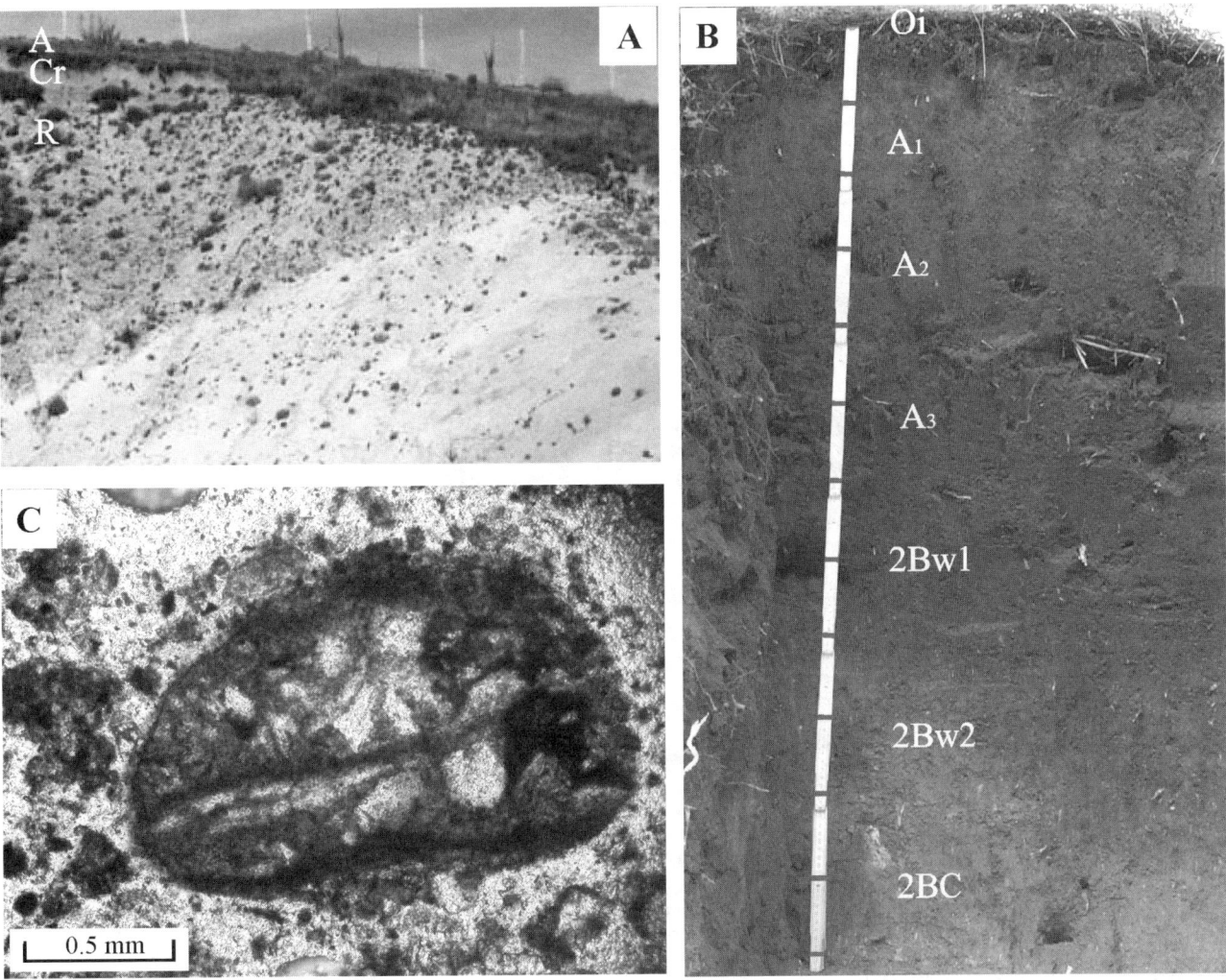

Figure 4. (A) A very thin soil profile (horizons A plus Cr reaching ~50 cm in depth) developed on granitic bedrock as a result of severe surface erosion. (B) A 120-cm-deep soil profile developed on a fluvial terrace under a pine forest. Two different sedimentary and pedogenetic cycles (horizons Oi to A_3 and $2Bw_1$ to 2BC) are evident. (C) Micrograph of a rounded, yellow to red, clayey and Fe-stained pedorelict included within a brown, silty-clay, organic-rich, microgranular soil matrix (plane polarized light).

ciglia et al., 2005a, 2005b), confirming the results from microscopy. The main components, both in bedrock and soil horizons, consist of quartz, K-feldspar, plagioclase, and mica (Fig. 6), and in soils, granite or granodiorite rock fragments. In particular, orthoclase and sometimes microcline occur, occasionally with Na-rich perthites, albite-twinned Na-plagioclase, biotite, and subordinate muscovite. In addition, accessory minerals, such as amphibole, apatite, epidote, Fe-oxide, and sphene, are present (Mirabella et al., 1996; Le Pera and Sorriso-Valvo, 2000a; Le Pera et al., 2001).

Many primary monocrystalline minerals and rock fragments, in the granite bedrock, grus, and soil horizons, show essentially mechanical weathering features, with irregular to linear or crossed fracture patterns. Regular fracture patterns follow main structural discontinuities, such as cleavage or twinning planes. Biotite crystals are frequently exfoliated and at least partly split into two or more fragments along cleavage planes (Figs. 6A and 6B). Small intragranular cracks often radiate from the cleaved biotite into surrounding feldspar and quartz crystals (Fig. 6A) and/or extend into intergranular cracks separating irregular polycrystalline rock fragments and monomineralic grains.

Weak to moderate chemical weathering affects crystals, more intensely in soil horizons than in grus layers and bedrock. Biotite may be oxidized, etched along its outer edges, or show evidence of early clay neogenesis (chloritization), as indicated by a local increase of Al and Mg in its chemical composition. K-feldspar and, above all, plagioclase frequently exhibit surface etch pitting, solution lines, Fe-oxide staining (Figs. 6C and 6D), and/or clay neogenesis. Na-perthite included within the K-feldspar component appears to be more highly weathered than the surrounding crystal (Fig. 6E). Occasionally, extremely weathered, unrecognizable pseudomorphs are found (Fig. 6F). More

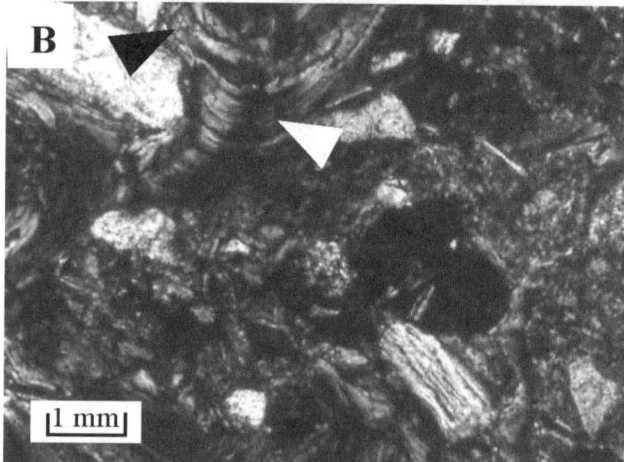
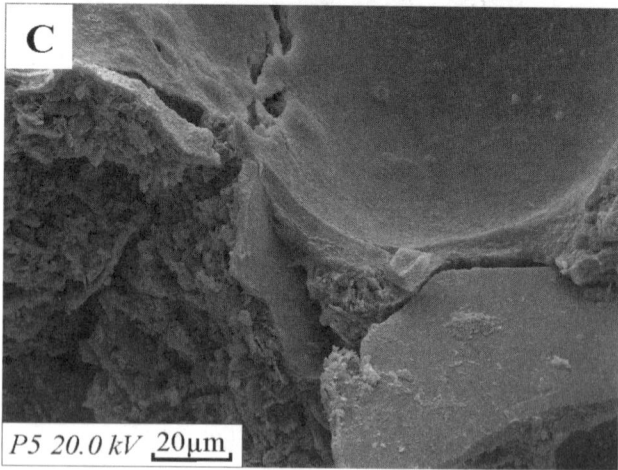

Figure 5. (A) Scanning electron micrograph (SEM) image of a kaolinite clay mineral with the typical pseudo-hexagonal habit; other irregularly shaped platelets of 2:1 phyllosilicate clay minerals are chaotically arranged over and around the kaolinite particle. (B) Photomicrograph of a microlaminated, crescentic clay infilling, with smooth-banded (white arrow) to grainy (black arrow) extinction patterns (crossed polarized light). (C) SEM image of fragmented clay coatings.

rarely, some quartz crystals in humic horizons show extremely superficial etch pits with subspherical, rounded shapes.

In the cambic or argillic horizons of the rare well-developed soils, e.g., those formed on the Middle to Upper Pleistocene terraced fluvial deposits near Cecita Lake (Scarciglia et al., 2005a, 2005b), quartz grains appear severely weathered. For example, some of them display physical fractures and signs of abrasion, while others exhibit microtextures that consist of irregular (Fig. 6G) or well-rounded to hexagonal, rhomboidal, and regularly oriented triangular etch pits, scarce linear, parallel-oriented grooves, and amorphous silica coatings or quartz overgrowths. Sometimes, solution features are particularly deep and may create an exceptionally vacuolar, cavernous structure in quartz grains (Fig. 6H).

FLUVIAL SAND, SOIL, GRUS, AND PARENT ROCK COMPOSITIONS

The detrital modes of the sand samples from the Neto River were compared to sands derived from grus and soil environments in the upper reaches of the same basin. Petrographic indices, such as the ratio of monocrystalline quartz (Qm) to total feldspar (F), and plagioclase (P) to total feldspar (F) ratio (P/F), indicate a nearly constant sand composition along the drainage basin of the Neto River.

Modal analysis of "grus-soil-fluvial sand" indicates that an increase in monocrystalline quartz grains (QmFLt) (Fig. 7) along with a decrease in plagioclase grains (QmKP) (Fig. 8) characterize the transition from bedrock to grus. Modal compositions of related grus and soil are characterized by an increase in monocrystalline quartz, coupled with a decrease in K-feldspar and plagioclase (QmFLt and QmKP diagrams, respectively) (Figs. 7 and 8). This trend has led to a mineralogic zonation of the soil profile (e.g., Nesbitt et al., 1997).

The Qm:F:Lt ratios in sands from the headwaters to the mouth of the Neto River are the same, although there are multiple source rocks throughout the basin (Le Pera et al., 2001). The

Figure 6. (A) Photomicrograph of cleaved biotite crystal showing intragranular microcracks radiating into the surrounding feldspar and quartz crystals. (B) Scanning electron micrograph (SEM) image displaying similar texture to that depicted in A. (C) Photomicrograph of feldspar grain with evidence of surface etch pitting and subparallel, linear (structure-controlled) solution features (white arrows). Black spots are Fe-oxide segregations. (D) SEM image of feldspar displaying solution pits and lines (white arrows). (E) Photomicrograph (crossed polarized light) of a K-feldspar (microcline) crystal, showing more intensely weathered Na-microperthites (white arrow). (F) Photomicrograph (crossed polarized light) of pseudomorphous grain due to extremely developed etch forms, clay neogenesis, and local Fe-oxide staining. (G–H) SEM images showing severely and irregularly etched quartz grains, sometimes with deep vacuolar, cavernous appearance, as in H.

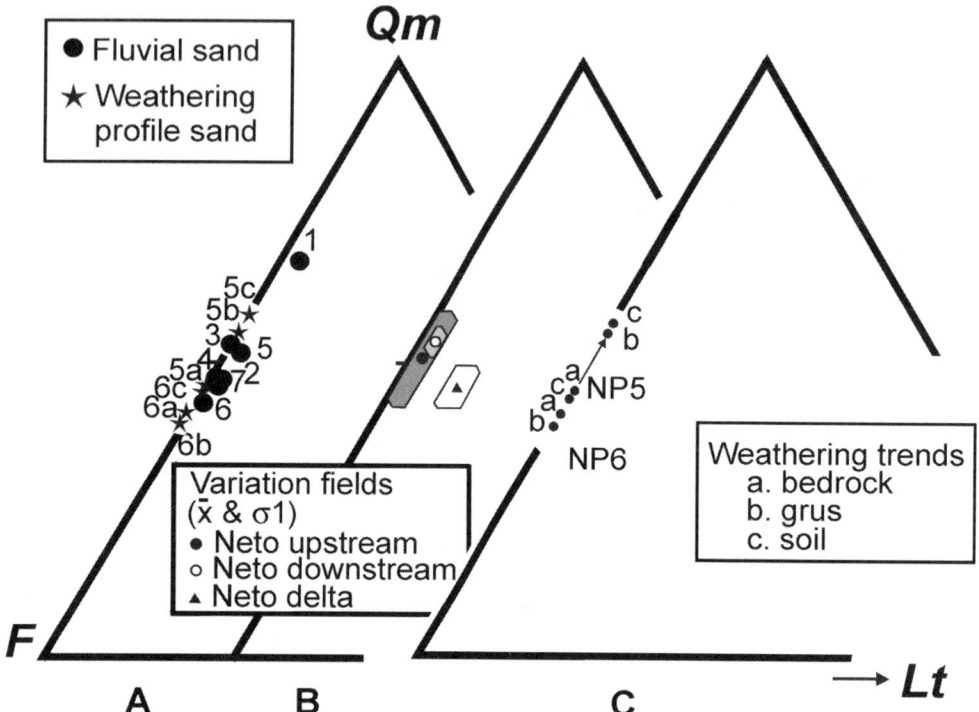

Figure 7. (A) Compositions of sand samples collected from the upper reaches of the Neto River plotted on a QmFLt (Qm—monocrystalline quartz; F—feldspars [plagioclase + K-feldspar]; Lt—aphanitic lithic fragments) diagram. (B) Variation fields (mean and one standard deviation) for upstream, downstream, and delta sand of the Neto River drainage basin, represented by polygons. (C) Weathering trends linking the composition of bedrock, grus, and soils. (Modified from Le Pera et al., 2001.) x—mean; σ—one standard deviation.

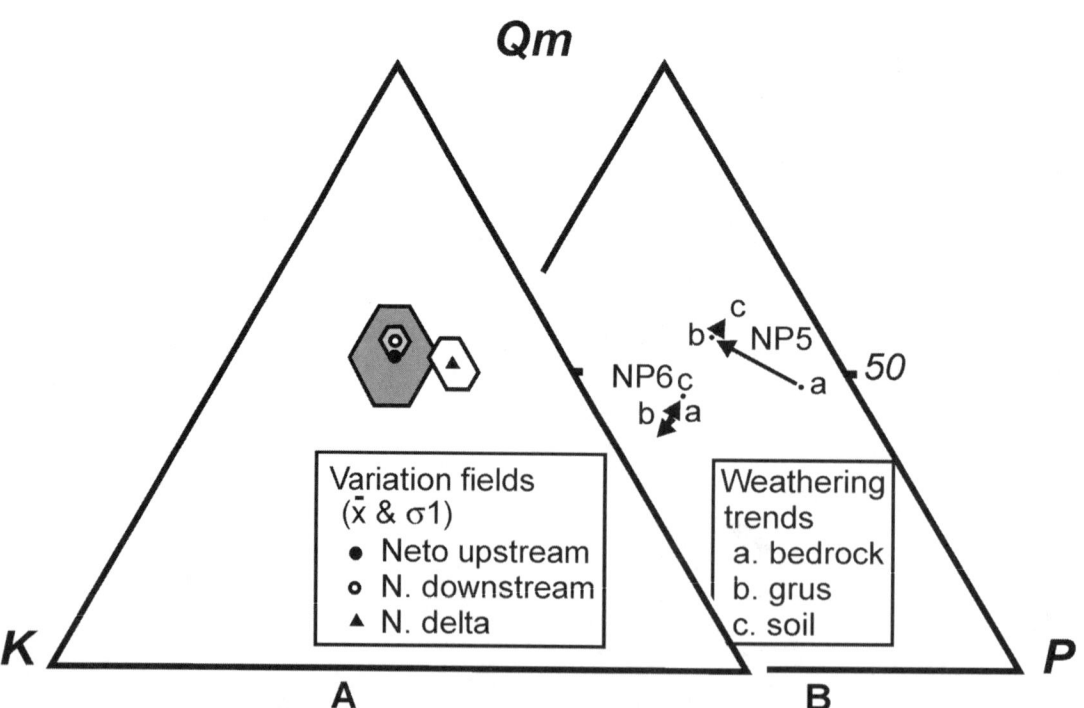

Figure 8. (A) Compositions of sand samples collected from the upper reaches of the Neto (N.) River plotted on a QmKP (Qm—monocrystalline quartz; P—plagioclase; K—K-feldspar) diagram. (B) Weathering trends linking the composition of bedrock, grus, and soils. (Modified from Le Pera et al., 2001.) x—mean; σ—one standard deviation.

composition of fluvial sand is nearly homogeneous and quartzofeldspathic (51%–48%–1% QmFLt in upstream sand; 53%–45%–2% QmFLt in downstream sand; 46%–45%–9% QmFLt in coastal sand); hence, there is no evidence that chemical weathering, sorting, and/or abrasion have affected feldspars relative to quartz in the fluvial system. Virtually all changes to the feldspar/quartz ratio must have taken place within weathering profiles mantling the bedrock of the source area and before incorporation into the Neto River. These results agree with findings from various other river systems (e.g., Nesbitt et al., 1996; Franzinelli and Potter, 1983), even if a downstream increase in compositional maturity cannot be ruled out (Johnsson et al., 1991; Robinson and Johnsson, 1997).

DISCUSSION

Physical Disintegration

The great variety of physical discontinuities affecting the granite bedrock, such as fault planes and shear zones, pressure release features, and lithologic changes, developed as a consequence of the complex tectonic history during the emplacement and uplift of the Sila Batholith. They all act as preferential weakness sites for the onset of both physical and chemical weathering processes.

Subspherical boulders and core-stones represent the inner, unweathered, or less-weathered portions of saprolite due to deep spheroidal weathering (e.g., Ollier, 1967; Twidale, 1986; Thomas, 1994; Migoń and Lidmar-Bergström, 2001), and they derive from progressive smoothing of the outer edges of jointed crystalline rocks to subspherical shapes, with complete isolation of rounded cores within a saprolitic to grussified groundmass. Spheroidal weathering appears to be the result of water percolation through rock fractures, a process that in turn enhances chemical reactions (hydration, solution, and mainly hydrolysis and oxidation) at the rock-water interface and thus widens and rounds off the sharp jagged edges of preexisting joints.

The discolored bleached zones surrounding core-stones and boulders within the saprolitic to grussified groundmass represent preferential pathways for downward-flowing water. Along such pathways, leaching of various components (fine particles or ions) and/or gleying are common processes. In contrast, the darker, reddish or blackish zones represent illuvial zones enriched in Fe- or Mn-oxides. The alternation of whitish and reddish/yellowish mottles in wavy, subhorizontal patterns indicates the contemporary occurrence of eluvial and illuvial zones, and/or it suggests alternating reducing and oxidizing conditions, possibly brought on by an oscillating water table. Where these features crop out near the ground surface, no evidence of a present-day water table is generally found. It can therefore be supposed that they represent signs of deep weathering, related to a paleohydrological (and presumably paleomorphological) feature, and that very intense erosion occurred, which removed the surface counterpart of such a feature (grus + soil cover). This interpretation is consistent with the occurrence of isolated spheroidal core-stones and boulders close to or at the ground surface of the oldest landforms in the Sila Massif, suggesting that they may have previously formed at considerable depths under constant volume conditions (Ollier, 1967, 1971) and were only later exhumed by erosion. At present, they denote an environment dominated by weathering-limited conditions, where transport rates of loose and mobile detritus strongly overcome those at which detritus is generated by weathering processes. The exfoliation of thick concentric shells (onionlike structure) observed on boulders is a typical feature of deep spheroidal weathering that occurs below the ground surface, where hydrolysis of silicate minerals dominates (Ollier, 1988). The efficacy of deep weathering in the study area can be related to high local relief, resulting from the strong tectonic uplift of the Sila Massif, and the concomitant high hydraulic gradient. The combination of these two features increases the deep penetration of groundwater for chemical reactions and the efficiency of frost weathering (see following). Once exposed at the surface, the weathering trend and modes presumably changed, involving volume increase, since boulders became free to expand. Boulder sheeting into thick slabs very likely represents an unloading feature (Ollier, 1967, 1971), whereas the large-scale splitting of some boulders along planar or curved cracks can be interpreted as a result of the induced tensile fracture mechanism proposed by Ollier (1978). Flaking, the surface type of exfoliation observed on these boulders, characterized by a small thickness of curved flakes, appears quite different from the spheroidal thick sheeting, and it is conceivably due to subaerial weathering (Ollier, 1967). It is very likely that weathering processes in the exposed environment preferentially act also on inherited (stress release and/or subsurface chemical weathering) features, possibly accelerating and amplifying their development.

Under the present-day climatic conditions, cryoclastic phenomena due to freeze-thaw cycles appear to be among the most effective physical factors for rock degradation during the coldest months, mainly widening and deepening joints and microcracks. Crack systems increase access of water to rock surfaces, accelerating the weathering processes and enhancing rock disintegration by frost weathering (Migoń and Thomas, 2002). Walder and Hallet (1985) demonstrated that crack growth is mostly effective if temperature reaches values between −4 and −15 °C, an uncommon temperature range in the study area. However, more recently Matsuoka (2001b) showed that at least partial filling of joints with water can minimize ice extrusion to open surfaces, thus concentrating volumetric expansion forces toward joint walls and enhancing the effectiveness of frost wedging even at not-so-low temperature values. This process is enhanced if unconfined water conditions are present (i.e., if water availability and migration through the pore-microfracture system occurs), even though oscillations around 0 °C do not take place (Matsuoka, 2001a). Although no data are available about amount and frequency of diurnal freeze-thaw cycles in the study area, oscillations a few degrees around zero can be presumed to occur rather frequently on the basis of the measured climatic parameters, along with the prolonged preservation of the snow cover during winter (especially on shadowed slopes or in protected patches). Such a conclusion is also supported by the occurrence of various cryonival morphologies and processes (Scarciglia et al., 2005b), for example, soil-grass hummocks (thufurs), shallow solifluction terraces, the growth of ice needles in topsoil horizons, the presence of illuvial silt pedofeatures, and the recurrent formation of night hoarfrost and its persistence for many hours after sunrise.

In addition, it can be assumed that cryoclastism was more efficient during the cold stadial phases of Quaternary glacial periods. The occurrence of poorly weathered but friable or disintegrated rock in the plateaus of the Sila Massif may potentially reflect cryogenic grussification (Butzer, 1976; Migoń and Thomas, 2002), which is at least partly inherited.

Similarly, thermoclastic processes likely play an important role for rock breakdown in the study area. According to Hall (1997) and Hall and André (2003), thermal stress fatigue and shock processes can cause rock shattering without the aid of freezing water, if very fast and frequent, low-wavelength, and low-amplitude thermal oscillations occur, possibly induced by wind refrigerant power. Also, diurnal to seasonal and yearly cycles of insolation can be assumed to be relevant. For example, different mineral species comprising the granitic bedrock have differential volumetric expansion/contraction responses to thermal variations, due to the different power of reflection-absorption-release of solar radiation by dark- and light-colored minerals (e.g., biotite and feldspar) (Ehlen, 2002). This results in different times and rates of grain microcracking. Moreover, insolation heating, which is thought to be particularly intense in upland areas, induces faster and larger temperature changes on the exposed rock surface than in its interior, thus producing wide temperature differences between the outer and inner parts. This implies that rapid temperature changes may lead to the continuous, differential expansion and contraction of rock minerals (Zhu et al., 2003), potentially creating thermal rock fatigue and breakdown. In addition, differential thermal expansion and displacement along different crystallographic axes typical of different mineral species may induce differential stress accumulation and release along preferential (structural) zones of weakness, also propagating into and destabilizing the surrounding rock mass.

These physical processes seem to be of prominent importance for the breakdown of granite, in particular, the small-scale exfoliation around boulders and the formation of grus. Furthermore, the effects of lichens (frequently observed on boulder surfaces) should not be neglected. It is well documented that they promote physical widening of fissures along their outer edges by adhering with their thallus and protruding their hyphae into the rock mass, and activate chemical dissolution by the production of acidic substances such as oxalic acids (Adamo and Violante, 2000).

Tree root penetration also plays a key role in rock wedging and fragmentation, as suggested by roots penetrating at depth within the cracked granite rock or saprolite, and occasional wind throw. The sporadic landslides triggered by rock degradation and breakage induce further fragmentation and comminution of rock particles. In particular, coarse debris accumulation on scree taluses and detrital cones is the result of sediment transport as rock falls and topples. Conversely, loose fine detritus (sometimes with higher clay and organic matter content) points to debris flow emplacement of intensely grussified rock and pedogenized material, probably enhanced by a prolonged period of weathering and pedogenesis, maybe under a stable forest cover (which is supported by the presence of fallen trees within the debris). All these processes produce sediment that enters the fluvial system.

Crystal microtextures are prominent in the weathering, fragmentation, and granular disintegration of the Sila Massif granitoids—they have an important upscaling effect—and in triggering or accelerating soil development. The main physical weathering features that affect mineral grains appear to be strictly controlled by major structural discontinuities (cleavage and twinning planes), or sometimes by irregular breakage patterns caused by sedimentary processes. A significant role in rock shattering seems to be played by biotite. Exfoliation and oxidation enhanced by hydration, coupled with progressive vermiculitization or chloritization due to hydrolysis, may lead to expansion and splitting apart of individual biotite flakes along cleavage planes (Taboada and García, 1999a, 1999b). This process creates a network of radiating intragranular and transgranular fractures that extend toward surrounding crystals (Isherwood and Street, 1976), which induce tension and consequent rock fatigue and breakdown. The observation that the amount of biotite crystals in soil sands derived from granites is higher than in correlated unweathered parent rock (Cullers, 1988; Le Pera et al., 2001) supports this hypothesis.

Chemical Decomposition

The principal chemical weathering features (etch pits or solution lines) and products of such processes (Fe-oxides and neoformed clay minerals) have been observed on feldspars (with plagioclase usually more weathered than K-feldspar) and micas. Etch pits and solution lines seem to be controlled by physical discontinuities (both natural and mechanically induced), coupled with possible differences in chemical composition (FitzPatrick, 1986), e.g., sodium microperthites within K-feldspar grains. Intragranular discontinuities become preferential sites for water migration and interaction with mineral surfaces, and, as a result, they enhance chemical attack. This attack results in the widening or deepening of etch pits and solution lines, with a progressive increase of available surface area, which in turn favors further chemical reactions and the weakening and disintegration of the whole rock mass. Etch forms grow and may coalesce eventually resulting in a complete pseudomorphic transformation. This general process is consistent with natural and experimental observations on alkali feldspars (Lee et al., 1998), which have shown a good correlation between dissolution rates and the density of microdiscontinuities, and they have shown that microtextures produce the greatest impact on mineral weathering rates during advanced stages of dissolution, when grains start to disintegrate to a microgranular material. The formation of secondary clay minerals along some microdiscontinuities could represent another source of fragmentation if they undergo expansion in the presence of water (cf. Frazier and Graham, 2000). The instability generated along intermineral contacts appears strongly favored in fine-grained rocks, which are characterized by a greater surface energy (Taboada and García, 1999a). Similarly, the occurrence of more intensely weathered mineral grains within soil horizons rather than in saprolite or parent rock suggests that, as soon as weathering processes develop, there is an increase in pedogenetic matrix produced by clay neogenesis and Fe oxidation. This increase in matrix, in turn, favors chemical reactions promoted by a prolonged interaction between mineral surfaces and the circulating soil solution. In addition, the activity of humic acids characteristic of organic-mineral horizons appears to enhance etching phenomena on primary minerals.

Some quartz grains from the most mature soils show cracks and signs of abrasion, probably related to crystal ruptures and impacts during sedimentary processes (Krinsley and Doornkamp, 1973; Al-Saleh and Khalaf, 1982) and in accord with their occurrence in soil profiles developed on fluvial deposits. The chemical microtextures on quartz grains display irregular to very regular outlines and distribution. Those grains that exhibit regular patterns clearly indicate a selective, crystallographic control over dissolution (Eswaran and Stoops, 1979; Al-Saleh and Khalaf, 1982; Howard et al., 1996). Some irregular solution features presumably represent the result of coalescing and deepening of previous smaller holes. The combination of secondary amorphous silica precipitates or quartz overgrowths (Krinsley and Doornkamp, 1973; Mazzullo and Magenheimer, 1987; Newsome and Ladd, 1999) with the quartz solution features suggests that they are conceivably the complementary facets of one main process.

Without excluding the well-assessed role of hydrothermal diagenetic processes to explain quartz dissolution (Dove and Nix, 1997; Dove, 1999), the severely etched quartz crystals can be interpreted to be the result of one or more of the following: (1) a highly acidic soil reaction initiated by humic compounds from organic-rich horizons (Krinsley and Doornkamp, 1973; Howard et al., 1996); (2) an aggressive pedoclimatic regime, under particularly warm and humid paleoclimates (e.g., tropical/subtropical environments), which enhanced chemical reactions and intense leaching phenomena and the ability to induce acidic pH values and silica solution (cf. Krinsley and Doornkamp, 1973; Eswaran and Stoops, 1979; Stoops, 1989; Summerfield, 1991; Pell and Chivas, 1995; Malengreau and Sposito, 1997; Moral Cardona et al., 1997); (3) a locally alkaline soil environment at the level of weathering microsites, with availability of bases (alkali and alkaline-earth cations) that promoted quartz solubility (Dove and Nix, 1997; Karlsson et al., 2001); and (4) a very long period of pedogenesis (Stoops, 1989; Al-Saleh and Khalaf, 1982; Howard et al., 1996; Schulz and White, 1999), which allowed polycyclic or polygenetic processes to occur.

Despite the fact that soils of the study area have subacid pH values (which favor only weak silica solution) and that only rarely do etched quartz grains occur in organic horizons, the role played by an acidic pedoenvironment can be of key importance for the formation of quartz chemical microtextures. For example, it is responsible for the hydrolysis of primary minerals, such as feldspars and micas, and thus increases the availability of bases (Scarciglia et al., 2005a); H^+ ions are incorporated into the original crystalline lattice as base elements are leached out. The leaching of bases in turn enhances an overall acidic soil environment, which produces additional hydrolysis. A concentration of base elements may induce a locally alkaline reaction at the soil solution–mineral surface, a result that would strongly increase silica solubility. Past climatic conditions warmer and more humid than today (presumably past Quaternary interglacials) could have been particularly favorable to the development of the quartz solution features at issue, possibly also increasing silica undersaturation of the soil solution by intense leaching, which would have in turn induced a thermodynamic disequilibrium of quartz (Schulz and White, 1999). The presence of etched quartz grains in the older, mature soils is consistent with other features of these soils, discussed in the following paragraphs. Alternatively, time appears to be a major cause of the severe etching of quartz. Even quite small, highly localized initial variations in weathering resistance (fractures, microtopographic features, biotic impacts) may operate as instability factors for the onset of etching, so that minor, apparently imperceptible solution features are cumulative and amplified through time (Howard et al., 1996; Turkington and Phillips, 2004). The extremely cavernous aspect of some quartz crystals could be an effect of such an amplification, possibly due to deepening, widening, and coalescence of initial microtextures. The size of etch features, weathering rates, and moisture concentration appear strictly correlated in a self-reinforcing mechanism. For example, the weathering of cavity inner walls tends to enlarge the hollows, in turn concentrating moisture in the interior, resulting in a further increase in weathering rates. Hence, a climatically controlled increase in moisture supply would have accelerated weathering rates and improved cavern growth.

Intensely etched quartz grains occur in mature or ancient soils, as some extremely weathered feldspar and plagioclase grains do. The latter minerals occur with fresh or weakly altered feldspars in some young soils. Significantly, both of these occurrences are found in soils that developed on fluvial deposits. This relationship suggests a cumulative effect of long-term processes that act only on selected mineral grains as a consequence of the multicycle origin of their parent material. Some degree of weathering of primary minerals conceivably was inherited from previous pedogenetic cycle(s), which would clearly enhance more advanced in situ alteration and pedogenesis.

The occurrence of pedorelicts supports the hypothesis of recycling (see following).

The coexistence of variable weathering patterns (macrocracking, granular disintegration, spheroidal boulder/corestone separation, mottling and bleaching, etc.), some of which occur in single features, sometimes as assemblages of more than one, and different degrees of weathering spatially very close to one another in the study area suggest a prominent microclimatic control. Even small differences in insolation and moisture availability, possibly influenced by small lithologic, relief (elevation, slope, and aspect), and vegetation changes, can be supposed to affect the dominant type(s), rate(s), and pattern(s) of weathering processes. In addition, it is also very likely that certain features related to present-day weathering processes are possibly superimposed or associated with others inherited from relict environments and processes.

Soil Development

The dominance of simple, poorly differentiated profiles, which consist of organic (O and A) and possibly cambic (Bw) horizons and overlie the parent rock or sediments (R or C), indicates a poor to moderate pedogenetic maturity. This weak development is consistent with coarse-grained textures, which still preserve abundant primary minerals or rock fragments, and therefore a strong influence by parent rock. Also, the low amount of clay (and associated low cation exchange capacity values) can be interpreted as resulting from poor pedogenetic development of clay minerals derived from an overall weak degree of weathering of primary components. Further support for this conclusion is provided by soil clay mineralogy. Illite, chlorite, vermiculite, and halloysite may represent weathering products of feldspars and micas (Barnhisel and Bertsch, 1989; Blum and Erel, 1997; Kretzschmar et al., 1997; Taboada and García, 1999b; Thomas et al., 1999; Sequeira Braga et al., 2002). Hydration and/or isomorphous substitution of cations in the structural units of primary micas commonly leads to the formation of illite, chlorite, and vermiculite in the early stages of weathering, and they are frequently found in granitic saprolites in temperate climates (e.g., Sequeira Braga et al., 2002). All these features, coupled with the occurrence of buried or truncated horizons, abrupt boundaries between different soil horizons or between soil horizons and the bedrock, rounded pedorelicts and papules, suggest a complex history of erosion, deposition, or reworking. These processes induce and enhance soil rejuvenation on steeper landforms and through human activity (forest clearance, tillage, and pasture) produce mobile material that in turn migrates and enters the sedimentary cycle within the drainage system, or stops and persists for a certain time span, and is recycled and undergoes further in situ weathering and/or pedogenesis.

As a whole, soils of the Sila plateaus formed in a highly leached pedoenvironment, as indicated by low base saturation, small amounts of exchangeable bases, and acidic soil reactions. These features are in accordance with the perhumid climate of the study area, where prolonged moisture availability and well-drained conditions result in an udic soil regime. This conclusion is consistent with the leaching factor (sensu Crowther, 1930) of the Sila uplands, which, based on primary climatic parameters, is the highest in Europe (Le Pera and Sorriso-Valvo, 2000a).

A higher degree of pedogenetic evolution is recorded by less widespread soils that have reddish colors and higher clay content as a consequence of an advanced weathering of primary minerals. As far as chemical reactions (hydration, solution, hydrolysis, and reduction/oxidation) occur, the release of iron from primary components and the subsequent crystallization of Fe-oxides and hydroxides in the soil matrix, as well as the genesis of secondary clay minerals, tend to increase. Sometimes the increase in clay fraction is due to illuviation, evidenced by abundant laminated clay coatings. In particular, these pedofeatures are relict, as they are frequently fragmented, degenerated, and partially assimilated into the soil matrix (FitzPatrick, 1984; Catt, 1989; Kemp, 1998; Scarciglia et al., 2003a, 2003b, 2005b). Both rubification and clay illuviation require warm and humid climates with xeric regimes (sensu USDA, 2006). Abundant rainfall and consequent soil moisture availability trigger chemical attack of primary minerals (and possible Fe release from Fe-bearing components), as well as downward migration of clays suspended in the soil solution. A marked seasonal contrast produces a water deficit under free drainage conditions in the dry season, leading to the oxidation of iron and the reddening of the matrix on one hand (cf. Diaz and Torrent, 1989; Schwertmann and Taylor, 1989), and the adhesion of clay coatings in pores on the other (Fedoroff, 1997). Such conditions appear in contrast with the present-day pedoenvironment, and in accordance with the relict significance, they are ascribed to the degenerated clay coatings; warmer and more humid climates than the present possibly occurred during past Quaternary interglacials (Scarciglia et al., 2003a, 2003b, 2005a, 2005b). Moreover, the presence of pedogenic kaolinite suggests a humid soil environment, where hydrolysis of primary minerals was relatively intense, and well-drained conditions, which promoted leaching of silica and soluble cations; this favored the formation of 1:1 clay minerals (Summerfield, 1991; Righi et al., 1999), possibly associated with a long period of pedogenesis (Bronger and Bruhn, 1989). Kaolinite could have been derived from the alteration of biotite (Kretzschmar et al., 1997; Sequeira Braga et al., 2002), possibly through an intermediate product such as vermiculite, as documented in highly leached soils and saprolites in tropical regions (Blum and Erel, 1997). These hypotheses are in good agreement with the quartz solution features described earlier in this paper. As a whole, the main features observed in the scattered remnants of mature and older soils imply a higher land surface stability, favored by relatively flat topography, where retention of movable grus and soil material was favored and rates of weathering and pedogenesis exceeded rates of erosion. Moreover, the longer the time span of geomorphologic stability (and pedogenesis), the higher was the possibility that different climatic conditions alternated during pedogenetic cycles, which therefore also recorded possibly polygenetic or polycyclic processes.

Fluvial Sand Composition as an Indication of Weathering Zone Erosion

The composition of sand from the Neto River is the result of chemical weathering and erosion of source rocks and of negligible transport within the riverine system (e.g., Le Pera et al., 2001).

Comparison of the detrital modes of fluvial and weathering profile sand emphasizes the petrogenesis of sediments in the absence of prolonged transport processes (e.g., Cullers et al., 1988; Critelli and Le Pera, 2002). Modal data from the Neto River sand samples, and especially the high F/Q ratio, suggest that the major source for most of the sand is grus developed on granodiorite and monzogranite of the Sila Batholith.

Sand composition is quartzofeldspathic and nearly homogeneous along the main channel of the Neto River, even where it cuts across a blanket of sedimentary cover. Thus, fluvial transport processes do not alter sand composition within the Neto drainage basin, and the nearly constant sand composition indicates minimal sediment maturation. These feldspar-rich sands indicate erosion mainly from grus rather than soil horizons of the weathering profile (e.g., Nesbitt et al., 1997). Moreover, the short and rapid sediment transport by the river, along with very short sediment storage, could have inhibited the effects of weathering, leading to remarkably immature fluvial sands. High-gradient slopes within the river drainage system, as assessed in another river draining the Sila uplands (Le Pera and Sorriso-Valvo, 2000b) and the severe mechanical erosion of source rocks are conditions that result in production of immature sediment (Nesbitt and Markovics, 1997; Nesbitt et al., 1997). An increase in monocrystalline quartz and mica grains and a decrease of plagioclase during the conversion of granite bedrock to regolith could suggest a rapid loss of plagioclase during grussification. The enrichment of biotite crystals from bedrock to grus to soil horizons is consistent with micromorphological observations, which show the expansion and splitting of weathered biotite, preferentially along cleavage planes, which lead to a greater amount of biotite flakes (cf. Cullers, 1988; Le Pera et al., 2001). The enrichment of quartz and the depletion in K-feldspar and plagioclase from grus to soil microenvironments indicate a further intensification of alteration processes acting on labile minerals. This result fits within the realm of modifications taking place in the pedogenetic environment, which include biological activity (e.g., Basu, 1981; Moulton and Berner, 1998), a prolonged interaction of the soil solution with sandy detritus, and strongly enhance chemical attack of primary minerals. In fact, abrasion during bedload transport of the Neto River appears to be insufficient to initiate the comminution of K-feldspar and plagioclase grains (Le Pera et al., 2001); hence, these minerals are preferentially destroyed in weathering and soil profiles (Nesbitt et al., 1996, 1997).

CONCLUSIONS

This work shows the great potential of combining complementary scientific disciplines to the study of weathering profiles and the derivative sand. Both physical and chemical weathering affect plutonic rocks of the Sila upland, where tectonics plays a key role as a predisposing source of (1) rock alteration or fragmentation through discontinuities and strain features; (2) local relief and slope produced by uplift, which act as driving forces of the main morphodynamic processes (erosion, transport, deposition), on one hand, reducing the time that detritus spends in the soil profile and concomitant rates of chemical weathering, and on the other, exposing deep unweathered bedrock to the surface, thus rejuvenating the weathering front and continuously producing newly weathered loose material; and (3) an enhanced, high hydraulic gradient that is able to promote a deeper percolation of water and consequent penetration of weathering at depth. Hydrolysis and oxidation were identified as important chemical processes operative in weathering profiles, whereas frost shattering and thermoclastism are the dominant processes of physical rock breakdown. These phenomena are enhanced by microtextures on mineral grains and macroscale discontinuities, which mutually interact to allow water penetration and access to rock surfaces and progressively lead to rock disintegration. Hydration, expansion, and cleaving of biotite and feldspar (sometimes even quartz) play important roles in grussification and soil formation. Although the present-day humid climate of the Sila Massif is very favorable to weathering processes, modal compositions of fluvial sand samples from the Neto River, and the weak soil development, indicate low pedogenetic and compositional maturity of sediments and soils in the studied drainage basins resulting from soil rejuvenation by erosion and rapid fluvial transport. In contrast, ancient and mature soils, rarely preserved on some flat paleosurfaces, fluvial terraces, or colluvial belts, clearly denote a higher degree of weathering, possibly due to a local multicycle origin of the soil parent material.

As a whole, a weathering-limited erosion regime characterizes the Sila uplands, where steep slopes and high local relief promote short residence times of weathered loose material, and potential sediment transport rates tend to exceed those at which weathering mantles are produced. In contrast, transport-limited conditions prevail on geomorphologically stable landforms (mainly terraced surfaces and depressions), where sediment storage and weathering and pedogenesis rates clearly overcome the efficiency of regolith and soil removal. Results discussed herein demonstrate that the composition of fluvial sand derived from weathering and soil profiles in the Neto basin is strongly dependent upon the extent of in situ chemical weathering of source rock. The composition of sands highlights that sediment of the Neto River is mainly derived from grus: the fluvial sands are composed dominantly of quartz and feldspar, which mimic that of sand-grade material studied from the grus. We conclude that first-cycle sands of the Neto River do not reflect the unweathered plutonic bedrock but the weathering profile environment, and specifically, the zone in which grus forms, widely mantling it. In this regard, our study provides confirmation (Grantham and Velbel, 1988; Suttner et al., 1981; Nesbitt et al., 1997; Le Pera et al., 2001; Girty et al., 2003) that chemical weathering and pedogenesis represent a fundamentally important control on the petrogenesis of siliciclastic sediments rather than provenance. Both climate and duration of weathering and pedogenesis appear to have a prominent influence on weathering rates and intensity.

ACKNOWLEDGMENTS

We are grateful to Mark J. Johnsson, Robert Cullers, and Gary H. Girty for their critical comments and suggestions, with special gratitude to the latter reviewer for his patience and editing care, which improved the quality of the manuscript. Many thanks are also due to Kathie Marsaglia for her fruitful revision of the final version of the manuscript.

REFERENCES CITED

Adamo, P., and Violante, P., 2000, Weathering of rocks and neogenesis of minerals associated with lichen activity: Applied Clay Science, v. 16, p. 229–256, doi: 10.1016/S0169-1317(99)00056-3.
Al-Saleh, S., and Khalaf, F.I., 1982, Surface textures of quartz grains from various recent sedimentary environments in Kuwait: Journal of Sedimentary Petrology, v. 52, no. 1, p. 215–225.
Amodio Morelli, L., Bonardi, G., Colonna, V., Dietrich, D., Giunta, G., Ippolito, F., Liguori, V., Lorenzoni, S., Paglionico, A., Perrone, V., Piccarreta, G., Russo, M., Scandone, P., Zanettin Lorenzoni, E., and Zuppetta, A., 1976, L'arco Calabro Peloritano nell'orogene Appenninico-Maghrebide: Memorie della Società Geologica Italiana, v. 17, p. 1–60.
Arribas, J., and Tortosa, A., 2003, Detrital modes in sedimenticlastic sand from low-order streams in the Iberian Range, Spain: The potential for sand generation by different sedimentary rocks: Sedimentary Geology, v. 159, p. 275–303, doi: 10.1016/S0037-0738(02)00332-9.
ARSSA, (Agenzia Regionale per lo Sviluppo e per i Servizi in Agricoltura), 2003, I suoli della Calabria: Carta dei suoli in scala 1:250,000 della regione Calabria: Soveria Mannelli (CZ), Rubbettino, ARSSA–Servizio Agropedologia, Monografia divulgativa, Programma Interregionale Agricoltura-Qualità—Misura 5, 387 p.
Barnhisel, R.I., and Bertsch, P.M., 1989, Chlorite and hydroxy-interlayered vermiculite and smectite, in Dixon, G.B., and Weed, S.B., eds., Minerals in Soil Environments (second ed.): Madison, Wisconsin, Soil Science Society of America, Book Series no. 1, p. 729–788.
Basu, A., 1981, Weathering before the advent of land plants: Evidence from unaltered detrital K-feldspar in Cambrian-Ordovician arenites: Geology, v. 9, p. 132–133, doi: 10.1130/0091-7613(1981)9<132:WBTAOL>2.0.CO;2.
Basu, A., 1985, Influence of climate and relief on compositions of sands released at source areas, in Zuffa G.G., ed., Provenance of Arenites: Dordrecht, D. Reidel, p. 1–18.
Berry, L.G., ed., 1974, Selected powder diffraction data for minerals: Philadelphia, Pennsylvania, JCPDS (Joint Committee on Powder Diffraction Standards) Publication DBM-1-23, 832 p.
Blum, J.D., and Erel, Y., 1997, Rb-Sr isotope systematics of a granitic soil chronosequence: The importance of biotite weathering: Geochimica et Cosmochimica Acta, v. 61, no. 15, p. 3193–3204, doi: 10.1016/S0016-7037(97)00148-8.
Brewer, R., 1976, Fabric and Mineral Analysis of Soils: Huntington, New York, Robert E. Krieger Publishing Company, 482 p.
Brindley, G.W., and Brown, G., eds., 1980, Crystal Structures of Clay Minerals and their X-Ray Identification: Mineralogical Society of London Monograph 5, 495 p.
Bronger, A., and Bruhn, N., 1989, Relict and recent features in tropical Alfisols from south India: CATENA, v. 16, supplement, p. 107–128.
Butzer, K.W., 1976, Geomorphology from the Earth: New York, Harper and Row Publishers, 463 p.
Caggianelli, A., Del Moro, A., and Piccarreta, G., 1994, Petrology of basic and intermediate orogenic granitoids from the Sila Massif (Calabria, southern Italy): Geological Journal, v. 29, p. 11–28.
Carson, M.A., and Kirkby, M.J., 1972, Hillslope Form and Process: Cambridge, Cambridge University Press, 475 p.
Cascini, L., Critelli, S., Di Nocera, S., Gullà, G., and Matano, F., 1992, Grado di alterazione e franosità negli gneiss del massiccio Silano: L'area di S. Pietro in Guarano (CS): Geologia Applicata e Idrogeologia, v. 27, p. 49–76.
Catt, J.A., 1989, Relict properties in soils of the central and north-west European temperate region: CATENA, v. 16, supplement, p. 41–58.
Colacino, M., Conte, M., and Piervitali, E., 1997, Elementi di climatologia della Calabria: Collana Progetto Strategico, in Guerrini A., ed., Clima, Ambiente e Territorio nel Mezzogiorno: Roma, CNR-IFA (Consiglio Nazionale delle Ricerche–Istituto di Fisica dell'Atmosfera), 218 p.
Critelli, S., 1999, The interplay of lithospheric flexure and thrust accommodation in forming stratigraphic sequences in the southern Apennines foreland basin system, Italy: Accademia Nazionale dei Lincei, Rendiconti Lincei Scienze Fisiche e Naturali, v. 10, p. 257–326.
Critelli, S., and Le Pera, E., 2002, Provenance relations and modern sand petrofacies in an uplifted thrust-belt, northern Calabria, Italy: Memorie Descrittive della Carta Geologica d'Italia, v. 51, p. 25–38.
Critelli, S., Di Nocera, S., and Le Pera, E., 1991, Approccio metodologico per la valutazione petrografica del grado di alterazione degli gneiss del massiccio Silano (Calabria settentrionale): Geologia Applicata ed Idrogeologia, v. 26, p. 41–70.
Critelli, S., Arribas, J., Le Pera, E., Tortosa, A., Marsaglia, K.M., and Latter, K.L., 2003, The recycled orogenic sand provenance from an uplifted thrust-belt, Betic Cordillera, southern Spain: Journal of Sedimentary Research, v. 73, no. 1, p. 72–81.
Crowther, E.M., 1930, The relationship of climate and geological factors to the composition of soil clay and the distribution of soil types: Proceedings of the Royal Society of London, Series B, v. 107, p. 1–30.
Cullers, R., 1988, Mineralogical and chemical changes of soil and stream sediment formed by intense weathering of the Danburg granite, Georgia, USA: Lithos, v. 21, p. 301–314, doi: 10.1016/0024-4937(88)90035-7.
Cullers, R., Basu, A., and Suttner, L.J., 1988, Geochemical signature of provenance in sand-size material in soils and stream sediments near the Tobacco Root Batholith, Montana, USA: Chemical Geology, v. 70, p. 335–348, doi: 10.1016/0009-2541(88)90123-4.
Dejou, J., Clement, P., and de Kimpe, C., 1982, Importance du site dans la genese des mineraux secondaires issus des alterations superficielle: Exemple des granites et gabbros du Mont Mégantic, Québec, Canada: CATENA, v. 9, p. 181–198, doi: 10.1016/S0341-8162(82)80014-3.
Diaz, M.C., and Torrent, J., 1989, Mineralogy of iron oxides in two soil chronosequences of central Spain: CATENA, v. 16, p. 291–299, doi: 10.1016/0341-8162(89)90015-5.
Dimase, A.C., and Iovino, F., 1996, I suoli dei bacini idrografici del Trionto, Nicà e torrenti limitrofi (Calabria), with 1:100,000 scale map: Florence, Accademia Italiana Scienze Forestali, 112 p.
Dixon, J.C., and Young, R.W., 1981, Character and origin of deep arenaceous weathering mantles on the Bega Batholith, southeastern Australia: CATENA, v. 8, p. 97–109, doi: 10.1016/S0341-8162(81)80007-0.
Dove, P.M., 1999, The kinetics of quartz in aqueous mixed cation solutions: Geochimica et Cosmochimica Acta, v. 63, no. 22, p. 3715–3727, doi: 10.1016/S0016-7037(99)00218-5.
Dove, P.M., and Nix, C.J., 1997, The influence of alkaline earth cations, magnesium, calcium and barium on the dissolution kinetics of quartz: Geochimica et Cosmochimica Acta, v. 61, no. 16, p. 3329–3340, doi: 10.1016/S0016-7037(97)00217-2.
Dramis, F., Gentili, B., and Pambianchi, G., 1990, Geomorphological scheme of the River Trionto Basin, in Sorriso-Valvo, M., ed., IGU-CoMTAG Symposium on Geomorphology of Active Tectonic Areas, CNR-IRPI, Rende (CS): Italy, Excursion Guide-Book: Geodata, v. 39, p. 71–75.
Ehlen, J., 2002, Some effects of weathering on joints in granitic rocks: CATENA, v. 49, p. 91–109, doi: 10.1016/S0341-8162(02)00019-X.
Eswaran, H., and Stoops, G., 1979, Surface textures of quartz in tropical soils: Soil Science Society of America Journal, v. 43, no. 2, p. 420–424.

FAO (Food and Agriculture Organization of the United Nations), ISRIC (International Soil Referene and Information Centre), and ISSS (International Society for Soil Science), 1998, World Reference Base for Soil Resources: Rome, FAO World Soil Resources Report, v. 84, 88 p.

Fedoroff, N., 1997, Clay illuviation in Red Mediterranean soils: CATENA, v. 28, p. 171–189, doi: 10.1016/S0341-8162(96)00036-7.

FitzPatrick, E.A., 1984, Micromorphology of Soils: London, Chapman and Hall, 433 p.

FitzPatrick, E.A., 1986, An Introduction to Soil Science (second ed.): Harlow, Longman Scientific and Technical, 256 p.

Franzinelli, E., and Potter, P., 1983, Petrology, chemistry, and texture of modern river sands, Amazon River system: The Journal of Geology, v. 91, p. 23–39.

Frazier, C.S., and Graham, R.C., 2000, Pedogenic transformation of fractured granitic bedrock, southern California: Soil Science Society of America Journal, v. 64, p. 2057–2069.

Galli, P., and Bosi, V., 2003, Catastrophic 1638 earthquakes in Calabria (southern Italy): New insights from paleoseismological investigation: Journal of Geophysical Research, v. 108, no. B1, doi: 10.1029/2001JB001713.

Girty, H.G., Marsh, J., Meltzner, A., McConnel, J.R., Nygren, D., Nygren, J., Prince, G.M., Randall, K., Johnson, D., Heitman, B., and Nielsen, J., 2003, Assessing changes in elemental mass as a result of chemical weathering of granodiorite in a Mediterranean (hot summer) climate: Journal of Sedimentary Research, v. 73, no. 3, p. 434–443.

Grantham, J.H., and Velbel, M.A., 1988, The influence of climate and topography on rock-fragment abundance in modern fluvial sands of the southern Blue Ridge Mountains, North Carolina: Journal of Sedimentary Petrology, v. 58, p. 219–227.

Hall, K., 1997, Rock temperatures and implications for cold region weathering. I: New data from Viking Valley, Alexander Island, Antarctica: Permafrost and Periglacial Processes, v. 8, p. 69–80, doi: 10.1002/(SICI)1099-1530(199701)8:1<69::AID-PPP236>3.0.CO;2-Q.

Hall, K., and André, M.-F., 2003, Rock thermal data at the grain scale: Applicability to granular disintegration in cold environments: Earth Surface Processes and Landforms, v. 28, p. 823–836, doi: 10.1002/esp.494.

Howard, J.L., Amos, D.F., and Daniels, W.L., 1996, Micromorphology and dissolution of quartz sand in some exceptionally ancient soils: Sedimentary Geology, v. 105, p. 51–62, doi: 10.1016/0037-0738(95)00133-6.

Ibbeken, H., and Schleyer, R., 1991, Source and Sediment: Berlin, Springer-Verlag, 283 p.

Isherwood, D., and Street, A., 1976, Biotite-induced grussification of Boulder Creek Granodiorite, Boulder County, Colorado: Geological Society of America Bulletin, v. 87, p. 366–370, doi: 10.1130/0016-7606(1976)87<366:BGOTBC>2.0.CO;2.

Johnsson, M.J., 1993, The system controlling the composition of clastic sediments, in Johnsson M.J., and Basu A., eds., Processes Controlling the Composition of Clastic Sediments: Geological Society of America Special Paper 284, p. 1–19.

Johnsson, M.J., Stallard, F.S., and Lundberg, N., 1991, Controls on the composition of fluvial sands from a tropical weathering environment: Sands of the Orinoco Rivera drainage basin, Venezuela and Colombia: Geological Society of America Bulletin, v. 103, p. 1622–1647, doi: 10.1130/0016-7606(1991)103<1622:COTCOF>2.3.CO;2.

Karlsson, M., Craven, C., Dove, P.M., and Casey, W.H., 2001, Surface charge concentrations on silica in different 1.0 M metal-chloride background electrolytes and implications for dissolution rates: Aquatic Geochemistry, v. 7, p. 13–32, doi: 10.1023/A:1011377400253.

Kemp, R.A., 1998, Role of micromorphology in paleopedological research: Quaternary International, v. 51–52, p. 133–141, doi: 10.1016/S1040-6182(97)00040-2.

Köppen, W., 1936, Das Geographische System der Klimate, in Köppen, W., and Geiger, R., eds., Handbuch der Klimatologie: Berlin, Gebrunder Borntraeger, v. 1, Part C, 46 p.

Kretzschmar, R., Robarge, W.P., Amoozegar, A., and Vepraskas, M.J., 1997, Biotite alteration to halloysite and kaolinite in soil-saprolite profiles developed from mica schist and granite gneiss: Geoderma, v. 75, p. 155–170, doi: 10.1016/S0016-7061(96)00089-4.

Krinsley, D.H., and Doornkamp, J.C., 1973, Atlas of Quartz Sand Surface Textures: London, Cambridge University Press, 91 p.

Lee, M.R., Hodson, M.E., and Parsons, I., 1998, The role of intragranular microtextures and microstructures in chemical and mechanical weathering: Direct comparisons of experimentally and naturally weathered alkali feldspars: Geochimica et Cosmochimica Acta, v. 62, no. 16, p. 2771–2788, doi: 10.1016/S0016-7037(98)00200-2.

Le Pera, E., and Sorriso-Valvo, M., 2000a, Weathering and morphogenesis in a Mediterranean climate, Calabria, Italy: Geomorphology, v. 34, p. 251–270, doi: 10.1016/S0169-555X(00)00012-X.

Le Pera, E., and Sorriso-Valvo, M., 2000b, Weathering, erosion and sediment composition in a high-gradient river, Calabria, Italy: Earth Surface Processes and Landforms, v. 25, p. 277–292, doi: 10.1002/(SICI)1096-9837(200003)25:3<277::AID-ESP79>3.0.CO;2-Z.

Le Pera, E., Arribas, J., Critelli, S., and Tortosa, A., 2001, The effects of source rocks and chemical weathering on the petrogenesis of siliciclastic sand from the Neto River (Calabria, Italy): Sedimentology, v. 48, p. 357–378, doi: 10.1046/j.1365-3091.2001.00368.x.

Lulli, L., and Vecchio, G., eds., 2000, I suoli della tavoletta "Lago Cecita" nella Sila Grande in Calabria: Catanzaro, Italy: Monografia Istituto Sperimentale per lo Studio e la Difesaa del Suolo, Progetto PANDA, Sottoprogetto 2, Serie 1, 78 p.

Lulli, L., Vecchio, G., Primavera, F., Gardin, L., Maletta, M., Napoli, R., and Calì, A., 1992, Contributo alla conoscenza dei suoli dell'altopiano Silano: Descrizione e commento della carta dei suoli centro sperimentale dimostrativo ESAC di Molarotta, Camigliatello Silano (CS): Calabria Verde, v. 11, p. 31–42.

Malengreau, N., and Sposito, G., 1997, Short-time dissolution mechanisms of kaolinitic tropical soils: Geochimica et Cosmochimica Acta, v. 61, no. 20, p. 4297–4307, doi: 10.1016/S0016-7037(97)00211-1.

Matano, F., and Di Nocera, S., 1999, Weathering patterns in the Sila Massif (northern Calabria, Italy): Il Quaternario—Italian Journal of Quaternary Sciences, v. 12(2), p. 141–148.

Matsuoka, N., 2001a, Direct observation of frost wedging in alpine bedrock: Earth Surface Processes and Landforms, v. 26, p. 601–614, doi: 10.1002/esp.208.

Matsuoka, N., 2001b, Microgelivation versus macrogelivation: Towards bridging the gaps between laboratory and field frost weathering: Permafrost and Periglacial Processes, v. 12, p. 299–313, doi: 10.1002/ppp.393.

Mazzullo, J., and Magenheimer, S., 1987, The original shapes of quartz sand grains: Journal of Sedimentary Petrology, v. 57, no. 3, p. 479–487.

Messina, A., Compagnoni, R., De Vivo, B., Perrone, V., Russo, S., Barbieri, M., and Scott, B., 1991, Geological and petrochemical study of the Sila Massif plutonic rocks (northern Calabria, Italy): Bollettino della Società Geologica Italiana, v. 110, p. 165–206.

Migoń, P., and Lidmar-Bergström, K., 2001, Weathering mantles and their significance for geomorphological evolution of central and northern Europe since the Mesozoic: Earth-Science Reviews, v. 56, p. 285–324, doi: 10.1016/S0012-8252(01)00068-X.

Migoń, P., and Thomas, M.F., 2002, Grus weathering mantles: Problems of interpretation: CATENA, v. 49, p. 5–24, doi: 10.1016/S0341-8162(02)00014-0.

Mirabella, A., Vecchio, G., and Risi, B., 1996, Caratterizzazione mineralogica dei suoli su granito e micascisto in Sila Grande: Calabria Verde, v. 2, p. 17–24.

Molin, P., Pazzaglia, F.J., and Dramis, F., 2004, Geomorphic expression of active tectonics in a rapidly-deforming forearc, Sila Massif, Calabria, southern Italy: American Journal of Science, v. 304, p. 559–589, doi: 10.2475/ajs.304.7.559.

Moral Cardona, J.P., Gutiérrez Mas, J.M., Sánchez Bellón, A., López-Aguayo, F., and Caballero, M.A., 1997, Provenance of multicycle quartz arenites of Pliocene age at Arcos, southwestern Spain: Sedimentary Geology, v. 112, no. 3–4, p. 251–261, doi: 10.1016/S0037-0738(97)00040-7.

Moulton, K.L., and Berner, R.A., 1998, Quantification of the effect of plants on weathering: Studies in Iceland: Geology, v. 26, p. 895–898, doi: 10.1130/0091-7613(1998)026<0895:QOTEOP>2.3.CO;2.

Nesbitt, H.W., and Markovics, G., 1997, Weathering of granodioritic crust, long-term storage of elements in weathering profiles, and petrogenesis of siliciclastic sediments: Geochimica et Cosmochimica Acta, v. 61, p. 1653–1670, doi: 10.1016/S0016-7037(97)00031-8.

Nesbitt, H.W., and Young, G.M., 1989, Formation and diagenesis of weathering profile: The Journal of Geology, v. 97, p. 129–147.

Nesbitt, H.W., Young, G.M., McLennan, S.M., and Keays, R.R., 1996, Effects of chemical weathering and sorting on the petrogenesis of siliciclastic sediments, with implications for provenance studies: The Journal of Geology, v. 104, p. 525–542.

Nesbitt, H.W., Fedo, C.M., and Grant, M.Y., 1997, Quartz and feldspar stability, steady and non-steady-state weathering, and petrogenesis of siliciclastic sands and muds: The Journal of Geology, v. 105, p. 173–191.

Newsome, D., and Ladd, P., 1999, The use of quartz grain microtextures in the study of the origin of sand terrains in Western Australia: CATENA, v. 35, p. 1–17, doi: 10.1016/S0341-8162(98)00122-2.

Ollier, C.D., 1967, Spheroidal weathering, exfoliation and constant volume alteration: Zeitschrift für Geomorphologie, v. 11, no. 1, p. 103–108.

Ollier, C.D., 1971, Causes of spheroidal weathering: Earth-Science Reviews, v. 7, p. 127–141, doi: 10.1016/0012-8252(71)90005-5.

Ollier, C.D., 1978, Induced fracture and granite landforms: Zeitschrift für Geomorphologie, v. 22, no. 3, p. 249–257.

Ollier, C.D., 1988, Deep weathering, groundwater and climate: Geografiska Annaler, v. 70A, no. 4, p. 285–290, doi: 10.2307/521260.

Palomares, M., and Arribas, J., 1993, Modern stream sands from compound crystalline sources: Composition and sand generation index, in Johnsson, M.J., and Basu, A., eds., Processes Controlling the Composition of Clastic Sediments: Geological Society of America Special Paper 284, p. 313–322.

Pell, S.D., and Chivas, A.R., 1995, Surface features of sand grains from the Australian Continental Dunefield: Palaeogeography, Palaeoclimatology, Palaeoecology, v. 113, p. 119–132, doi: 10.1016/0031-0182(95)00066-U.

Power, E.T., and Smith, B.J., 1994, A comparative study of deep weathering and weathering products: Case studies from Ireland, Corsica and southeast Brazil, in Robinson, D.A., and Williams, R.B.G., eds., Rock Weathering and Landform Evolution: New York, John Wiley and Sons, p. 21–40.

Pye, K., 1986, Mineralogical and textural controls on the weathering of granitoid rocks: CATENA, v. 13, p. 47–57, doi: 10.1016/S0341-8162(86)80004-2.

Riebe, C.S., Kirchner, J.W., Granger, D.E., and Finkel, R.C., 2000, Erosional equilibrium and disequilibrium in the Sierra Nevada, inferred from cosmogenic ^{26}Al and ^{10}Be in alluvial sediment: Geology, v. 28, no. 9, p. 803–806, doi: 10.1130/0091-7613(2000)028<0803:EEADIT>2.3.CO;2.

Riebe, C.S., Kirchner, J.W., Granger, D.E., and Finkel, R.C., 2001, Minimal climatic control on erosion rates in the Sierra Nevada, California: Geology, v. 29, no. 5, p. 447–450, doi: 10.1130/0091-7613(2001)029<0447:MCCOER>2.0.CO;2.

Righi, D., Terribile, F., and Petit, S., 1999, Pedogenic formation of kaolinite-smectite mixed layers in a soil toposequence developed from basaltic parent material in Sardinia (Italy): Clays and Clay Minerals, v. 47, no. 4, p. 505–514, doi: 10.1346/CCMN.1999.0470413.

Robinson, R.S., and Johnsson, M.J., 1997, Chemical and physical weathering of fluvial sands in an Artic environment: Sands of the Sagavanirktok River, north slope, Alaska: Journal of Sedimentary Research, v. 67, no. 3, p. 560–570.

Roda, C., 1964, Distribuzione e facies dei sedimenti Neogenici nel Bacino Crotonese: Geologica Romana, v. 3, p. 319–366.

Scarciglia, F., Terribile, F., and Colombo, C., 2003a, Micromorphological evidence of paleoenvironmental changes in northern Cilento (south Italy) during the late Quaternary: CATENA, v. 54, no. 3, p. 515–536, doi: 10.1016/S0341-8162(03)00124-3.

Scarciglia, F., Terribile, F., Colombo, C., and Cinque, A., 2003b, Late Quaternary climatic changes in northern Cilento (south Italy): An integrated geomorphological and paleopedological study: Quaternary International, v. 106–107, p. 141–158, doi: 10.1016/S1040-6182(02)00169-6.

Scarciglia, F., Le Pera, E., and Critelli, S., 2005a, Weathering and pedogenesis in the Sila Grande Massif (Calabria, south Italy): From field scale to micromorphology: CATENA, v. 61, no. 1, p. 1–29, doi: 10.1016/j.catena.2005.02.001.

Scarciglia, F., Le Pera, E., Vecchio, G., and Critelli, S., 2005b, The interplay of geomorphic processes and soil development in an upland environment, Calabria: South Italy: Geomorphology, v. 69, no. 1–4, p. 169–190, doi: 10.1016/j.geomorph.2005.01.003.

Schulz, M.S., and White, A.F., 1999, Chemical weathering in a tropical watershed, Luquillo Mountains, Puerto Rico III: Quartz dissolution rates: Geochimica et Cosmochimica Acta, v. 63, no. 3–4, p. 337–350, doi: 10.1016/S0016-7037(99)00056-3.

Schwertmann, U., and Taylor, R.M., 1989, Iron oxides, in Dixon, J.B., and Weed, S.B., eds., Minerals in Soil Environments (second ed.): Madison, Wisconsin, Soil Science Society of America, p. 379–438.

Sequeira Braga, M.A., Paquet, H., and Begonia, A., 2002, Weathering of granites in a temperate climate (NW Portugal): Granitic saprolites and arenization: CATENA, v. 49, p. 41–56, doi: 10.1016/S0341-8162(02)00017-6.

Sorriso-Valvo, M., 1993, The geomorphology of Calabria: A sketch: Geografia Fisica e Dinamica Quaternaria, v. 16, p. 75–80.

Stoops, G., 1989, Relict properties in soils of humid tropical regions with special reference to central Africa: CATENA, v. 16, supplement, p. 95–106.

Summerfield, M.A., 1991, Global Geomorphology: An Introduction to the Study of Landforms: London, Longman–Wiley, 537 p.

Suttner, L.J., Basu, A., and Mack, G.H., 1981, Climate and the origin of quartz arenites: Journal of Sedimentary Petrology, v. 51, p. 21–29.

Taboada, T., and García, C., 1999a, Pseudomorphic transformation of plagioclases during the weathering of granitic rocks in Galicia (NW Spain): CATENA, v. 35, p. 291–302, doi: 10.1016/S0341-8162(98)00108-8.

Taboada, T., and García, C., 1999b, Smectite formation produced by weathering in a coarse granite saprolite in Galicia (NW Spain): CATENA, v. 35, p. 281–290, doi: 10.1016/S0341-8162(98)00107-6.

Teeuw, R.M., Thomas, M.F., and Thorp, M.B., 1994, Regolith and landscape development in the Koidu basin of Sierra Leone, in Robinson, D.A., and Williams, R.B.G., eds., Rock Weathering and Landform Evolution: New York, Wiley, p. 303–320.

Thomas, M.F., 1994, Geomorphology in the tropics: A study of weathering and denudation in low latitudes: New York, John Wiley and Sons, 460 p.

Thomas, M.F., Thorp, M., and Mc Alister, J., 1999, Equatorial weathering, landform development and the formation of white sands in north western Kalimantan, Indonesia: CATENA, v. 36, p. 205–232, doi: 10.1016/S0341-8162(99)00014-4.

Thomson, S.N., 1994, Fission-track analysis of the crystalline basement rocks of the Calabrian arc, southern Italy: Evidence of Oligo-Miocene late-orogenic extension and erosion: Tectonophysics, v. 238, p. 331–352, doi: 10.1016/0040-1951(94)90063-9.

Turkington, A.V., and Phillips, J.D., 2004, Cavernous weathering, dynamical instability and self-organization: Earth Surface Processes and Landforms, v. 29, p. 665–675, doi: 10.1002/esp.1060.

Twidale, C.R., 1986, Granite landform evolution: Factors and implications: Geologische Rundschau, v. 75, no. 3, p. 769–779, doi: 10.1007/BF01820646.

Twidale, C.R., 1990, The origin and implications of some erosional landforms: The Journal of Geology, v. 98, p. 343–364.

USDA (U.S. Department of Agriculture), 2006, Keys to Soil Taxonomy (tenth ed.): Washington D.C., U.S. Department of Agriculture, Soil Survey Staff, Natural Resources Conservation Service, 333 p.

Velbel, M.A., 1985, Geochemical mass balances and weathering rates in forested watersheds of the southern Blue Ridge: American Journal of Science, v. 285, p. 904–930.

Versace, P., Ferrari, E., Gabriele, S., and Rossi, F., 1989, Valutazione delle piene in Calabria: Cosenza, CNR (Consiglio Nazionale delle Ricerche)-IRPI (Istituto di Ricerca per la Protezione Idrogeologica), Geodata, v. 30, 232 p.

Walder, J., and Hallet, B., 1985, A theoretical model of the fracture of rock during freezing: Geological Society of America Bulletin, v. 96, p. 336–346, doi: 10.1130/0016-7606(1985)96<336:ATMOTF>2.0.CO;2.

Wilson, M.J., ed., 1987, A Handbook of Determinative Methods in Clay Mineralogy: Glasgow, Blackie and Sons, 308 p.

Zhu, L., Wang, J., and Li, B., 2003, The impact of solar radiation upon rock weathering at low temperature: A laboratory study: Permafrost and Periglacial Processes, v. 14, p. 61–67, doi: 10.1002/ppp.440.

Manuscript Accepted by the Society 9 August 2006

Interpreting carbonate particles in modern continental sands: An example from fluvial sands (Iberian Range, Spain)

M.E. Arribas[†]
J. Arribas[‡]

Departamento de Petrología y Geoquímica, Facultad de Ciencias Geológicas, Universidad Complutense de Madrid, 28040, Madrid, Spain

ABSTRACT

We analyzed modern fluvial sands in the Iberian Range in order to obtain an accurate description of the different typologies of carbonate grains and to interpret their origin. Head streams of the Iberian Range mainly receive carbonate sediments as (1) fragments from ancient carbonate rocks, and (2) penecontemporaneous carbonate grains generated in the fluvial channels or in associated subenvironments. The erosion of proximal carbonate sources (Jurassic and Cretaceous in age) contributes to the generation of carbonate rock fragments. In addition, erosion of recent freshwater tufas, carbonate soils, and other recent carbonates produces an important volume of penecontemporaneous carbonate particles. Temperate to subhumid climate and short transport conditions promote good preservation of the composition and textures of carbonate grains in modern fluvial sands. Detailed petrographic analyses on penecontemporaneous carbonates provide diagnostic clues of their origin. Four main petrographic classes of penecontemporaneous grains have been established: (1) penecontemporaneous micritic grains, which are composed of microcrystalline calcite with a filamentous or laminated microfabric, are derived from erosion of recent freshwater carbonate tufas. Penecontemporaneous micritic grains with alveolar microfabric are derived from recent carbonate soils. (2) Penecontemporaneous sparitic grains, which are composed of single crystals or of mosaics with filamentous microfabric, are the result of erosion of carbonate tufas. Other penecontemporaneous sparitic grains include *Microcodium* and speleothems fragments. (3) Penecontemporaneous coated grains, which are composed of a nucleus plus a coating of penecontemporaneous carbonate, represent bioinduced carbonate particles (cyanoliths) that originate in streams. (4) Penecontemporaneous bioclasts, made from charophytes, ostracods, and mollusks, are rare. Identification of these grain categories in ancient deposits has implications for coeval carbonate supplies during fluvial sedimentation.

Keywords: provenance, coeval grains, carbonate grains, freshwater carbonate tufa, fluvial sands, Iberian Range, sand composition.

[†]E-mail: earribas@geo.ucm.es.
[‡]E-mail: arribas@geo.ucm.es.

Arribas, M.E., and Arribas, J., 2007, Interpreting carbonate particles in modern continental sands: An example from fluvial sands (Iberian Range, Spain), in Arribas, J., Critelli, S., and Johnsson, M.J., eds., Sedimentary Provenance and Petrogenesis: Perspectives from Petrography and Geochemistry: Geological Society of America Special Paper 420, p. 167–179, doi: 10.1130/2006.2420(11). For permission to copy, contact editing@geosociety.org. ©2007 Geological Society of America. All rights reserved.

INTRODUCTION

Modern provenance studies are mainly focused on the sandy products generated by erosion of plutonic, metamorphic, volcanic, and sedimentary rocks (Ibbeken and Schleyer, 1991; Marsaglia, 1993; Palomares and Arribas, 1993; Arribas and Tortosa, 2003, among others). These studies tend to acquire data from the effects that factors such as source rocks, weathering, and transport have on the composition of sands (Johnsson, 1993). Little effort has been made to undertake petrographic analysis of sandy particles from coeval formations or intrabasinal origin (Garzanti, 1991; Arribas and Tortosa, 1998; Arribas and Tortosa, 2003). Production of such particles is very common in marine environments, and their recognition is significant in provenance studies because of the relationship that exists between their occurrence and eustatic cycles and tectonic setting (Garzanti, 1991). Also, the erosion and degradation of carbonate sedimentary rocks in continental environments generate a high volume of carbonate grains, both extrabasinal (carbonate rock fragments) and intrabasinal (coeval carbonate grains) in origin. In these cases, the correct distinction between these particles is crucial during petrographic analysis for accurate provenance inferences. Characterization of intrabasinal and/or coeval grains permits deductions about the dynamics of sedimentary processes in the basin (Zuffa, 1985).

Fluvial systems that flow on carbonate terrain are capable of transporting Ca^{2+} and HCO_3^- in solution, and, consequently, carbonate may precipitate as fluvial tufas or other types of penecontemporaneous deposits, such as stromatolites and/or carbonate soils (Ordóñez and García del Cura, 1983; Pedley, 1990; Pedley et al., 1996; Fernández et al., 1998; Janssen et al., 1999; Arenas et al., 2000; Fernández et al., 2000; Valero-Garcés et al., 2001; Arenas et al., 2004). Moreover, rivers may erode these deposits and incorporate a high volume of carbonate fragments in their sediment. Also, new carbonate particles can be created from bio-induced precipitation (cyanoliths). Thus, degradation and dissolution of carbonate rocks at the source area and the subsequent bioinduced carbonate precipitation in associated fluvial subenvironments are significant processes for the origin of coeval carbonate grains in fluvial sands. Valuable information about physical and chemical processes in continental depositional environments can be obtained from penecontemporaneous carbonate particles.

Similarities in the general texture and composition among carbonate particles from several origins make difficult to deduce origins from a brief petrographic inspection. However, a detailed petrographic characterization of these grains may be decisive in provenance studies. In addition, these grains have not been carefully considered, and few examples about their recognition are known (Zuffa, 1985, 1987, 1991; Arribas and Arribas, 1991; Cavazza et al., 1993; Arribas and Tortosa, 1998, 2003).

Siliciclastic (sandstones and conglomerates) and carbonate rocks are the main bedrocks cropping out in the Iberian Range in central Spain (Fig. 1). The erosion of Jurassic and Cretaceous carbonate rocks in the catchment areas of the rivers generates carbonate rock fragment grains (limestone and dolostone rock fragments). In addition, these carbonate rocks show evidences of important degradation and dissolution processes with the development of karstification. At present, several rivers flow from these formations to the southwest, traversing a well-developed system of biogenic carbonates, which are similar to tufa deposits (Arribas and Tortosa, 1998; Fernández et al., 1998) such as occur in other Quaternary deposits in the Mediterranean region (Cipriani et al., 1977; Ordóñez and García del Cura, 1983; Freytet, 1992; Pedley et al., 1996).

These modern fluvial sands offer an excellent opportunity to analyze and characterize carbonate particles in order to determine their origin and thus use in the interpretation of provenance of poorly diagenized sandstones in the geological record.

TERMINOLOGY

The terms "intrabasinal" and "extrabasinal" refer to a sedimentary setting where source and basin are well defined (i.e., Zuffa, 1980). In studies of modern fluvial sands, "basin" is mainly used to refer to a hydrologic basin or subbasin, and so differs greatly from the sense of a sedimentary basin.

In addition, many authors use terms like "penecontemporaneous" and "coeval" to refer to sedimentary particles that have an origin related in time to the depositional processes (Zuffa, 1991; Cavazza et al., 1993; Arribas and Tortosa, 2003). Differences in the meaning of both terms are associated mainly with the time scale. "Coeval" is mainly used to describe sedimentary particles in the sedimentary record that originated during deposition of a sedimentary unit, usually a depositional sequence (Zuffa, 1991). Thus, this term has a geologic sense of time. On other hand, the term "penecontemporaneous" is used to describe sedimentary particles or even sedimentary deposits (i.e., tufas) that were generated recently with respect to sedimentary processes acting in association with modern environments (i.e., Cavazza et al., 1993; Arribas and Tortosa, 2003). In this work, we use the term "penecontemporaneous" to describe carbonate particles generated in a short interval of time, arbitrarily defined from the Quaternary period to the present time.

Terminology referring to different sedimentary calcium carbonate precipitates in continental realms (tufas, travertines, speleothems, and stromatolites) also varies in the literature. "Tufas" are often defined as a product of calcium carbonate precipitation under a cool-water regimen and typically contain the remains of micro- and macrophytes, invertebrates, and bacteria, while the term "travertine" is usually used to describe warm- and hot-water carbonates (Ford and Pedley, 1996; Cole et al., 2004). Also "travertine" is applied to well-lithified and older calcareous tufa deposits, but the term also is applied to atypical deposits devoid of macrophytes but dominated exclusively by heat-tolerant bacteria (Pedley, 1990; Valero-Garcés et al., 2001). In both cases, tufas and travertines are considered to be freshwater stromatolites (Freytet, 1992; Freytet and Verrecchia, 1998), and their origin is related to microbial mediation (Riding, 2000). On the other

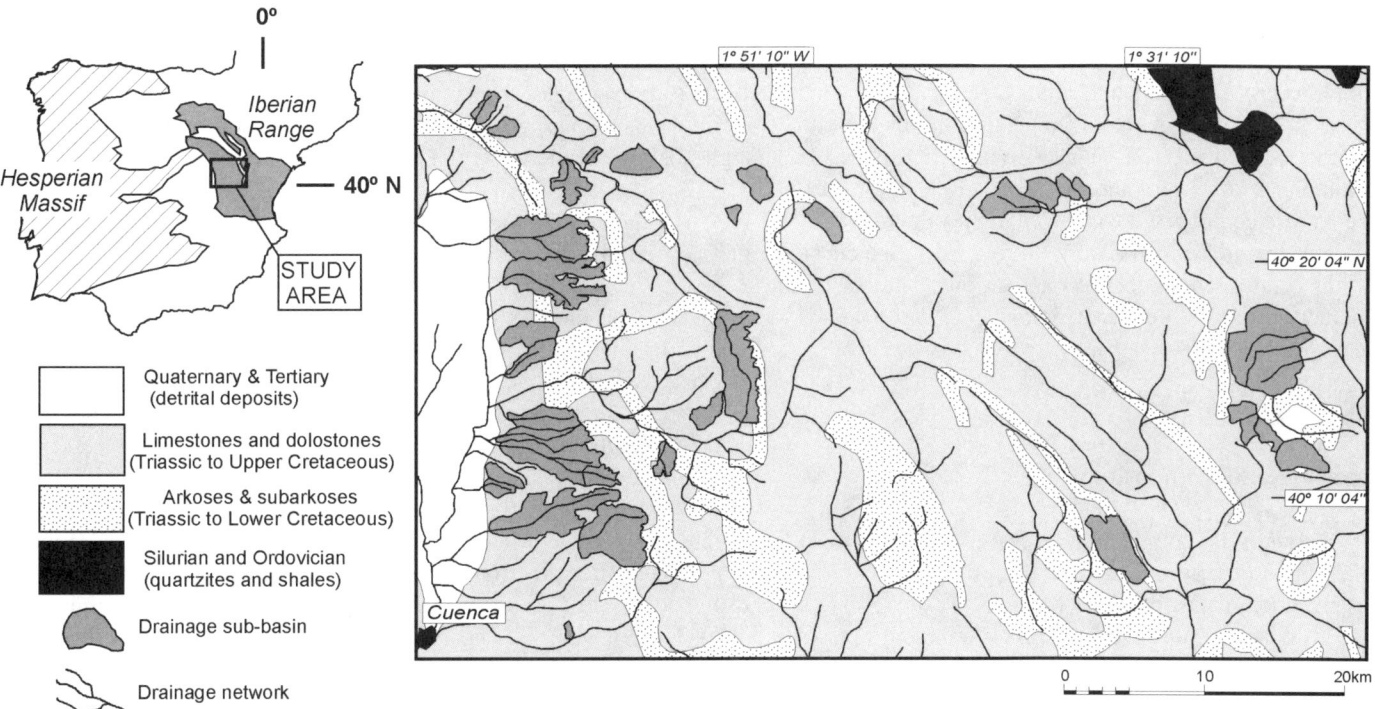

Figure 1. Simplified lithologic map with the location of the subbasins where sands were collected.

hand, speleothems are calcium carbonate precipitates that are formed in caves related to karst systems and differ from tufas in their inorganic origin (Pedley, 1990; Cole et al., 2004). Readers are referred to Pedley (1990), Freytet and Verrecchia (1998), and Cole et al. (2004) for more discussion about terminology.

In our case, recent biogenic carbonates overlie Mesozoic deposits (mainly Jurassic and Cretaceous) and form isolated outcrops mainly along the river banks as phytoherms, on the river bed as microbial mats, and as carbonate particles as coated grains (cyanoliths). These deposits have been considered as autochthonous tufa deposits (Pedley, 1990).

STUDY AREA

Mesozoic sedimentary rocks overlying Variscan basement are widely exposed in the Iberian Range in central Spain. The drainage areas considered in this paper are located in the northwestern sector of the Iberian Range (NE of the city of Cuenca), covering ~2500 km² (Fig. 1). In this area, the Mesozoic stratigraphic record consists of a sedimentary succession up to 2600 m thick. In the drainage basins, selected Jurassic units are represented by ~750 m of shallow-marine limestones, which have mud-supported and subordinately grain-supported textures, and dolostones. Lower Cretaceous rocks (300 m thick) consist of siliciclastic deposits (arkoses and subarkoses). Dolomitic lithologies (mainly dolosparites) prevail in the Upper Cretaceous strata (~550 m thick) (Vilas et al., 1982).

Recent carbonate deposits overlie Mesozoic deposits and constitute isolated tufas along main streams and are related to carbonate-rich springs. In this area of the Iberian Range, the Jurassic and Cretaceous carbonate bedrock (as well as Tertiary carbonates) provides the Ca^{2+} and HCO_3^- necessary for carbonate precipitation in the streams and springs as tufa deposits. Also, speleothems are present as precipitates in caves related to karstified carbonate source areas.

Water courses mainly have a seasonal regime, running during winter and spring, with occasional inputs from springs. The study area is characterized by a temperate to subhumid climate, with annual precipitation ranging from 500 to 800 mm and annual temperatures from 8 °C to 12 °C (IGN, 1991). Detailed physiographic information of the considered subbasins and their water courses is available in Arribas and Tortosa (2003).

METHODS

This work is based on the material collected by Arribas and Tortosa (2003). Thus, methods used in this collection and laboratory procedures (sieving, thin-sectioning, and selective staining for carbonates) are described there. From a total of 60 collected samples from 35 drainage subbasins, we selected samples with a significant penecontemporaneous carbonate clast content, generally higher than 25%.

The modal composition was determined by point counting following the criteria established by Gazzi (1966) and Dickinson (1970). The use of grain types corresponding to the so called "traditional" criteria (Ingersoll et al., 1984) can be considered by reevaluation of the point-count results. There were 200 to 400 grains counted per thin section that were classified into the 56 modal classes (Table 1 and Appendix A in Arribas and Tortosa, 2003). These authors defined four general categories of petrographic classes: ancient noncarbonate clast (AN), ancient carbonate clast (AC), penecontemporaneous carbonate clast (PC), and compound grains (CG). This last type is constituted by a single grain plus a contemporaneous carbonate coating. Counted points from these coatings were included in the "En" petrographic class of Arribas and Tortosa (2003). In this work, we considered CG grains as penecontemporaneous carbonate grains, because the characterization of penecontemporaneous carbonates is the main scope of the paper.

RESULTS

Stream Sand Composition

Modal sand composition in fine (0.062–0.25 mm), medium (0.25–0.5 mm), and coarse (0.5–1 mm) grain-size intervals varies considerably, from nearly pure siliciclastic sand (AN sands in Fig. 2) to nearly pure carbonate-rock-fragment sand (AC sands) to nearly pure penecontemporaneous carbonate sands (PC sands). Generally, concentration of penecontemporaneous carbonate grains does not exceed 50% of total grain population and occurs in sands with variable size fractions (Fig. 2). These variations in sand composition reflect a wide variety in sedimentary sources (AN and AC) in the different subbasins and variable productivity of penecontemporaneous carbonates.

Ancient clast population is quartzolithic in composition and plots near the QmLt line on Dickinson's QmFLt diagram (Fig. 5 in Arribas and Tortosa, 2003), which reflects, as expected, a "recycled orogen" signature provenance (Dickinson et al., 1983), but there is a clear dependence of sand composition on grain size. In many samples, penecontemporaneous carbonate grain (PC) content exceeds 25%, and sands can be considered to be hybrid arenites (Zuffa, 1980). Occasionally, this content exceeds 90% of the bulk sediment, causing considerable dilution of ancient population (AN + AC). Siliciclastic (ancient noncarbonate clast) grains consist mainly of quartz, feldspar, and mica. Ancient carbonate clast grains are limestone and dolostone fragments that show a wide variety of microfacies (Arribas and Tortosa, 2003).

Characterization of Penecontemporaneous Carbonate Grains

In the coarse sand grains (1–0.5 mm), a good preservation of internal microfabric in carbonate grains exists, which permits easier distinction between ancient (AC) and penecontemporaneous carbonate grains (PC). Ancient carbonate grains appear as well-rounded grains and show a large variation of textures (grainstones, packstones, wackestones, etc.), which correspond to those described by Vilas et al. (1982) in Mesozoic carbonates. Generally, these grains contain characteristic marine fossils. Fibrous cement inside grains as well as other evidences of early marine diagenesis are common. Moreover, ancient carbonate grains can be composed of a coarse crystalline dolomite and/or calcite mosaics reflecting different diagenetic processes. These grains appear in association with ancient noncarbonate clasts (AN) and usually possess coatings of penecontemporaneous carbonate.

On the other hand, penecontemporaneous carbonate grains are calcitic in composition and appear as micritic and/or sparitic porous grains with very irregular shapes. Usually, they show several microfabrics (filamentous, laminated), which permit their characterization as fragments of recent carbonate tufas. The fragile appearance of some crystalline carbonate grains can be indicative of a recent origin.

On the basis of textural and compositional criteria, four petrographic categories of penecontemporaneous carbonate grains are proposed (Table 1): (1) penecontemporaneous micritic grains; (2) penecontemporaneous sparitic grains; (3) penecontemporaneous coated grains; and (4) penecontemporaneous bioclasts.

TABLE 1. PETROGRAPHIC CHARACTERIZATION OF PENECONTEMPORANEOUS CARBONATE GRAINS AND THEIR ORIGIN FROM MODERN STREAM SANDS OF THE IBERIAN RANGE

Penecontemporaneous type	Structure	Microfabric	Morphology	Origin
Penecontemporaneous micritic grains (PMG) (Microcrystalline calcite)	Nonstructure	Clotted-spongy	Irregular Rounded Elongated	Erosion of carbonate tufa
	Structure	Alveolar	Irregular	Erosion of carbonate soil
		Filamentous (fan-like)	Irregular	Erosion of carbonate tufa
		Single filament	Elongated	Erosion of carbonate tufa
		Concentric laminae	Irregular	Microbial activity (cyanoliths)
Penecontemporaneous sparitic grains (PSG) (Mesocrystalline calcite)	Structure	Filamentous (fan-like)	Variable	Erosion of carbonate tufa
		Radial palisadic	Variable	Erosion of carbonate tufa
		Prismatic and radial (*Microcodium*)	Irregular	Erosion of carbonate soil
		Rossetes and longitudinal aggregates Euhedral calcite crystal	Variable	Erosion of speleothems?
	Nonstructure		Single crystal Cluster crystals	Uncertain Uncertain
Penecontemporaneous coated grains (PCG) (Cyanoliths)	Structure	Concentric laminae (Micritic and/or sparitic)	Rounded Elongated	Microbial activity
Penecontemporaneous bioclasts (PB) (Bioclasts)	Structure	Biogenic	Variable	Biogenic

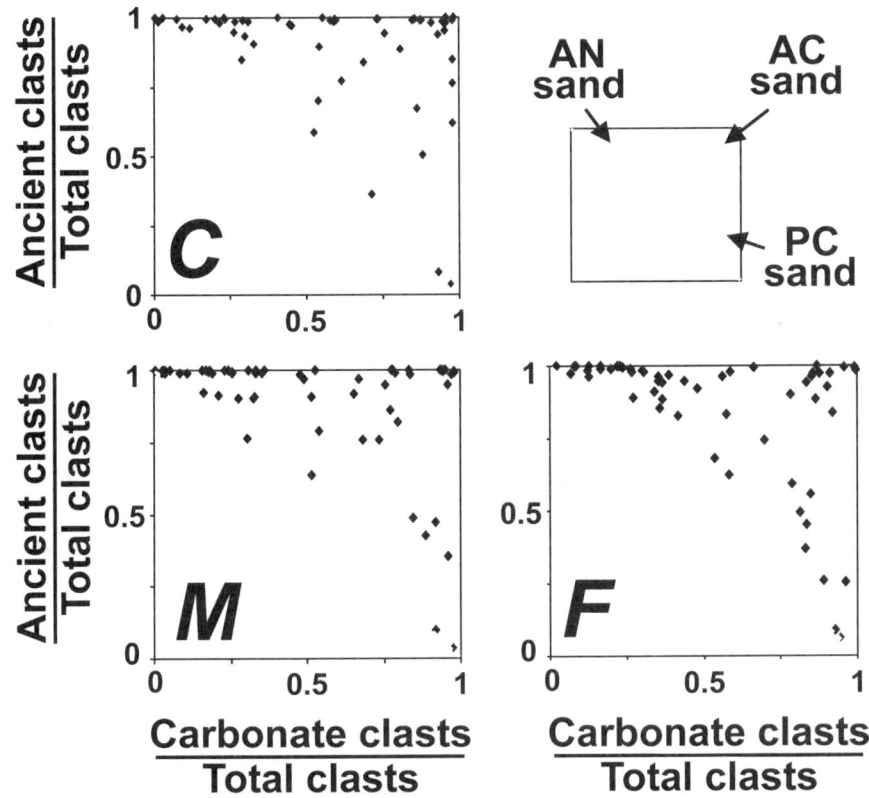

Figure 2. Compositional diagrams of analyzed sands using compositional (carbonate and noncarbonate) and provenance (ancient and penecontemporaneous) criteria (modified from Di Giulio and Valloni, 1992). C—coarse sand fraction (0.5–1 mm); M—medium sand fraction (0.5–0.25 mm); F—fine sand fraction (0.25–0.062 mm). Notice that sands are mainly "hybrid" (Zuffa, 1980), and their composition varies between three end members: AN sands (sands composed of ancient noncarbonate grains); AC sands (sands composed of ancient carbonate grains); and PC sands (sands composed of penecontemporaneous carbonate grains).

Penecontemporaneous Micritic Grains

Micritic grains are the most frequent penecontemporaneous grains and are composed of microcrystalline calcite (Fig. 3). Micrite production as seed crystals is abundant and very common in modern tufas (Freytet and Verrecchia, 1998). These grains are more abundant in the fine sand fraction, and they appear as homogeneous micrite grains (nonstructured penecontemporaneous micritic grains) or show an internal structure (structured penecontemporaneous micritic grains) (Table 1).

Nonstructured penecontemporaneous micritic grains appear as homogeneous clotted and spongy-porous grains that have irregular morphologies (Figs. 3A and 3B). Some elongated grains show a simple filament constituting a very delicate and porous framework (Fig. 3B). Clotted texture has been described as characteristic in some fluvial oncolite envelopments as well as in specific bands in microbial mats developed in the river beds (Ordóñez and García del Cura, 1983; Pedley et al., 1996). Clotted and spongy texture can be related to biomediated carbonate precipitation from cyanobacteria activity (Iron and Müller, 1968; Ordóñez and García del Cura, 1983). The precipitation of clotted or grumous calcite muds in recent tufa has been attributed to development of prokaryote–microphyte biofilms from static water or sluggish flow (Pedley et al., 1996). These biofilms are composed of variable numbers of coccoid and filamentous cyanobacteria, green algae, diatoms, and heterotrophic bacteria. Also, Freytet and Verrecchia (1998) pointed out that this texture may be related to the redistribution of organic matter in clots within crystallizations in filamentous algal structures.

In addition, penecontemporaneous micritic grains may show different microfabrics, such as filamentous, laminated, and alveolar fabrics. Filamentous penecontemporaneous micritic grains show an accurate filament structure that is very well preserved and made up of micrite microtubes (less than 30 μm in diameter) (Figs. 3C and 3D). This micrite is the bioinduced calcite crystallized around cyanobacterial filaments. In these cases, the grains can be composed by cyanobacterial colonies, which show a radiating pattern (fan-like, Figs. 3C and 3D) similar to those of *Rivularia* and *Phormidium* described by Schäfer and Stapf (1978). They appear as complete colonies or as fragments. Freytet (1992) described micrite calcite crystals (3–5 μm in size) growing over filaments ascribed to *Lyngbya, Phormidium,* or *Schizothrix* in bioinduced carbonates (tufas and travertines). Also, Freytet and Verrecchia (1998) showed that primary and micrite crystals coated filaments ascribed to *Phormidium incrustatum* and abundant *Schizothrix* spp. Sometimes these grains are partially replaced by microsparite, as an early diagenetic process.

Other penecontemporaneous micritic grains show an irregular concentric biolaminated structure around a micritic nucleus, and these are considered to be cyanoliths (Fig. 3E). Generally, morphology of these micritic grains is irregular and reproduces the morphology of their external lamina, but in some cases, these grains show their internal lamination, which is truncated by erosion.

Both filamentous and laminated micritic grains have been interpreted as a product of erosion of recent carbonate tufas (phytoherms and stromatolites) or by the organically induced precipitation of carbonate around a carbonate particle (cyanoliths), respectively.

Some grains show an alveolar microfabric made up of circumgranular porosity around micritic micronodules (Fig. 3F), but they are not common. The alveolar microfabric is a characteristic

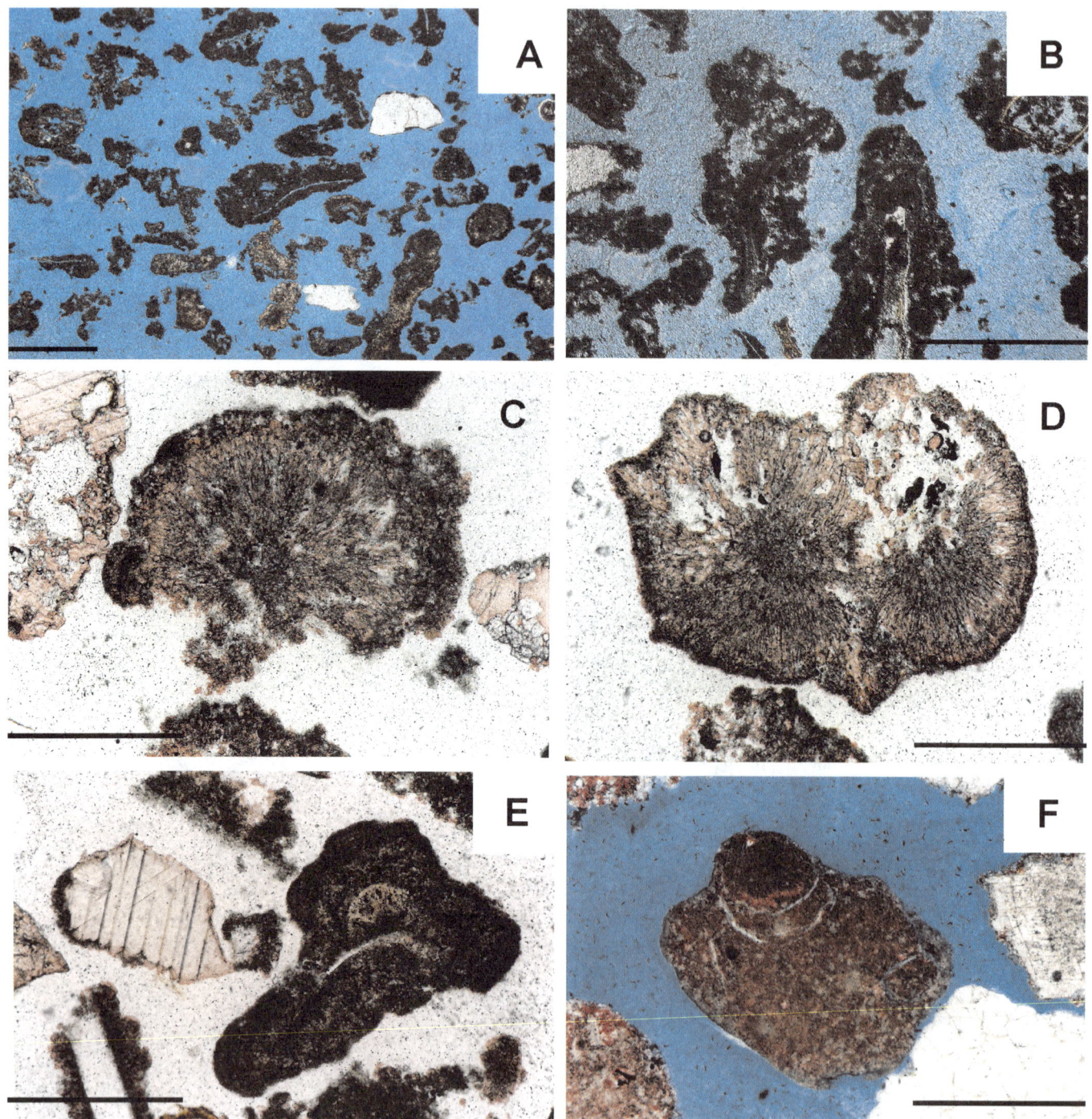

Figure 3. Penecontemporaneous micritic grains (PMG). (A) General view of sand composed principally of nonstructured penecontemporaneous micritic grains with a clotted and spongy-porous fabric. (B) Detailed view of grains with irregular morphology. (C–D) Structured grains showing a filamentous microfabric (fan-like) partially replaced by microsparite. (E) Grains showing an irregular concentric biolaminated structure around a nucleus (cyanolith). (F) Grains with circumgranular porosity around micritic micronodules. Scale bars = 0.5 mm.

pedogenetic feature (Freytet and Plaziat, 1982; Arribas, 1986). The occurrence of this microfabric permits the link of the origin of such clasts to the erosion of recent calcimorphic soils.

Scanning electron microscope (SEM) images of different micritic grains show that microcrystalline calcite mosaics growth together with cyanobacteria filaments and frustules of diatoms, constituting a complex network with a high microporosity (Fig. 4). This characteristic biotic community has been cited by several authors as appearing in recent bioinduced carbonates, who pointed out the important role of these organisms in carbonate fixation (Eggleston and Dean, 1976; Walter, 1976; Cameron et al., 1985; Chafetz, 1994; Gerdes et al., 1994; Braithwaite and Zedef, 1994; Riding, 1994; Pedley et al., 1996; Freytet and Verrecchia, 1998, among others). Riding (1994) indicated that diatoms, rather than algae, are the principal active trapping components in stromatolites.

Penecontemporaneous Sparitic Grains

These grains are abundant, and their composition is mainly mesocrystalline calcite (Fig. 5). They are composed of single crystals or as an association of them (>40 μm in size). Two types can be distinguished: structured and nonstructured sparitic grains (Table 1).

Structured penecontemporaneous sparitic grains that preserve a filamentous microfabric (Figs. 5A and 5B) are similar to those described in micritic grains, locally with a radial palisadic microfabric (Freytet and Verrecchia, 1999) (Fig. 5C). Freytet (1992) interpreted sparite calcite crystals (more than 100 μm) as the result of crystallization of several filaments of *Rivularea haematites*. Freytet and Verrecchia (1998) described many types of biomediated sparite and microsparite crystals in tufas from Western Europe, North Africa, and the Middle East. According with these authors, primary calcite crystals may occur early,

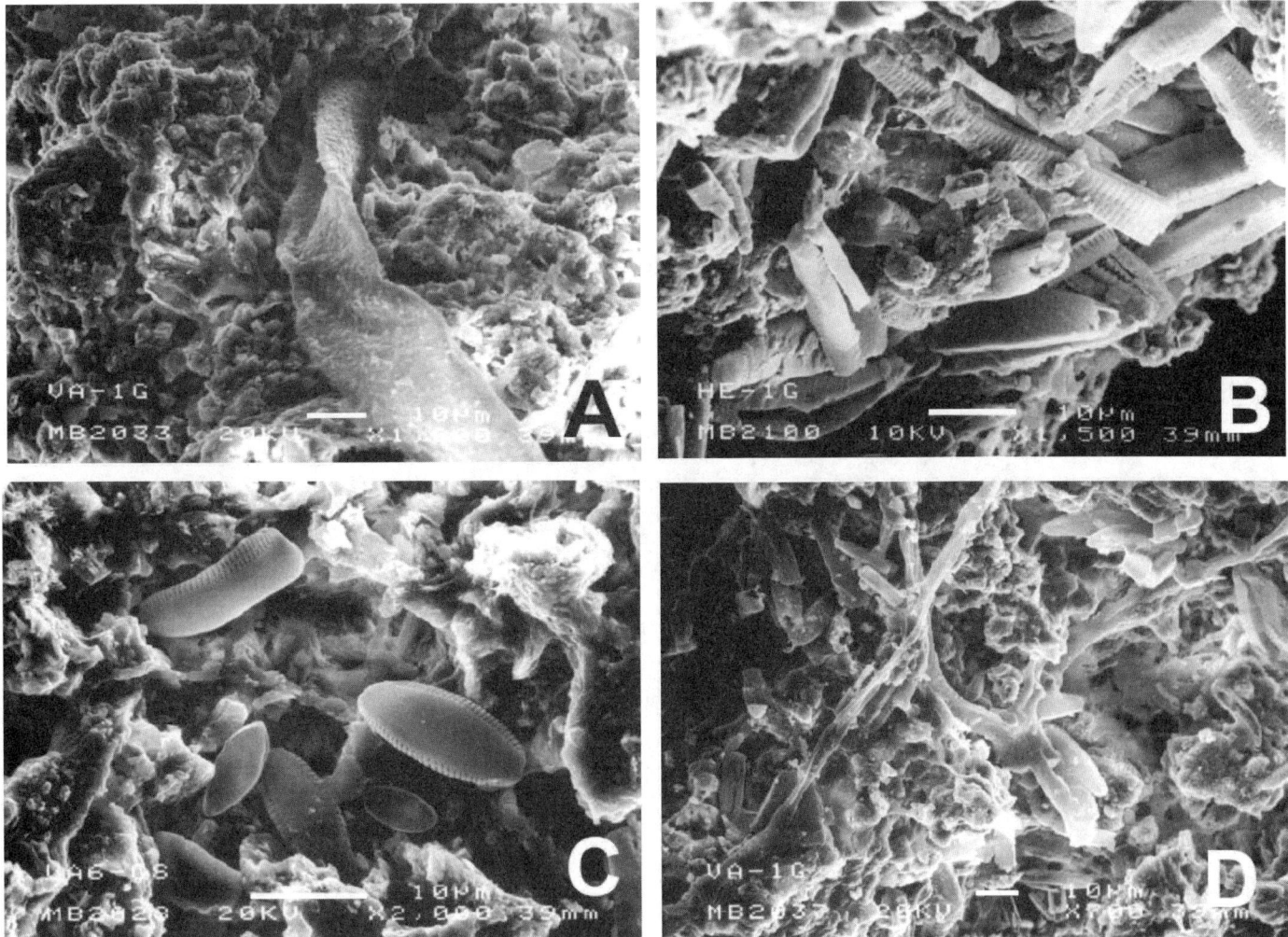

Figure 4. Scanning electron microscope (SEM) images of penecontemporaneous micritic grains. (A) Microcrystalline calcite mosaics with remains of cyanobacteria filaments. (B–C) Frustules of diatoms and calcite crystals. (D) Microcrystalline calcite mosaics, cyanobacteria filaments, and diatoms constitute a complex porous network.

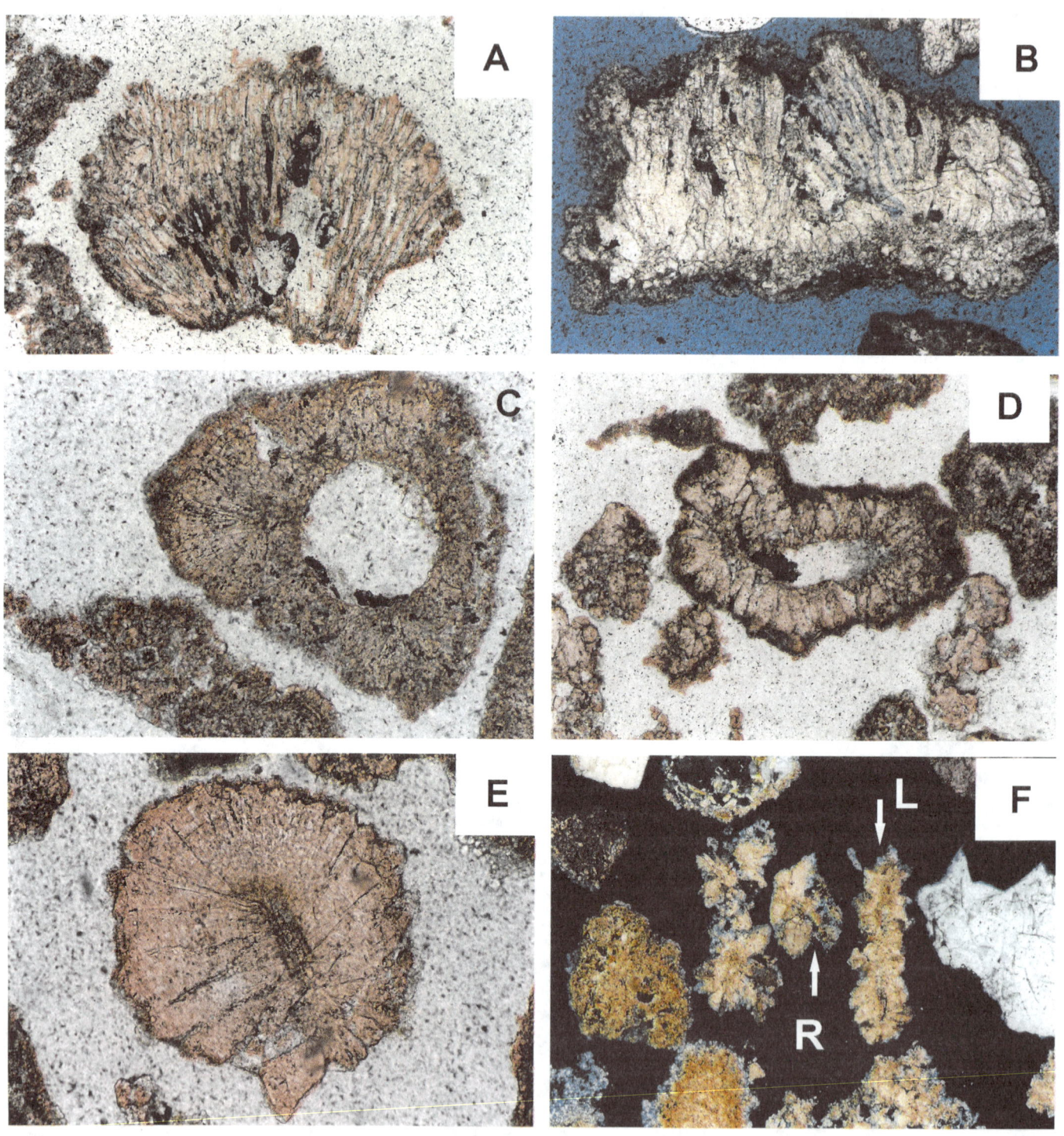

Figure 5. Penecontemporaneous sparitic grains (PSG). (A) Grain showing filamentous (fan-like) microfabric. (B) Partially recrystallized sparitic grains showing filamentous (fan-like) microfabric. (C) Grains exhibiting a radial palisadic microfabric developed on a rounded nucleus (now dissolved). (D) Grains composed by coarse sparitic crystals without a clear internal structure. (E) Grains showing prismatic and radial microstructure similar to *Microcodium* structures. (F) Grains constituted by euhedral calcite crystals as rosettes (R) or as longitudinal aggregates (L). Scale bar = 0.5 mm.

develop directly on the algae, and enclose bacteria, cells, and filaments, and seem to be related to algal secretions. Two species of cyanobacteria, *Rivularea haematites* and *Phormidium foveolarum*, are known to relate to organically induced precipitation of sparry calcite crystals (Janssen et al., 1999). Also calcite crystals can be the result of the sparitization of microbial micrite (Schäfer and Stapf, 1978; Freytet et al., 1996; Freytet and Verrecchia 1998, 1999). Diagenetic recrystallization of micrite and some primary sparite crystals produces crystalline habits similar to speleothems. Thus, stromatolite buildups of *Schizothrix* have been attributed to a purely physico-chemical origin (Freytet and Verrecchia, 1999). Other examples of filament remains composed of microsparite occur in many recent oncolites and other bioinduced continental carbonates (Freytet and Verrecchia, 1998; Janssen et al., 1999). These types of penecontemporaneous sparitic grains can be the result of early sparitization of original micrite filaments; this process partially affects micritic grains. However, it is possible that some types correspond to primary calcite crystal precipitating over filaments (Freytet and Verrecchia, 1998; Janssen et al., 1999). Sometimes these grains show round moldic pores inside them (Figs. 5C and 5D), possibly due to the decomposition of organic vegetable material (i.e., stems).

Some penecontemporaneous sparitic grains show a prismatic and radial microstructure like those ascribed to *Microcodium* structures (Fig. 5E). *Microcodium* structure is a pedogenetic feature diagnostic of subaerial exposure and has been interpreted as calcified roots (Esteban, 1974; Klappa, 1978; Jaillard, 1991; Arribas et al., 1996; Rossi, 1997; Alonso-Zarza et al., 1998; Kosir, 2004). Disintegration of *Microcodium* colonies may contribute to the formation of an important volume of penecontemporaneous carbonate grains in continental environments, such as occur in lacustrine Tertiary sediments in the Pyrenees (Arribas et al., 1996). In this case, penecontemporaneous sparitic grains are single calcite prisms or colony fragments (Fig. 5E). These grain types are not very common in the analyzed samples, but could be a very important type in the fossil record.

Sometimes penecontemporaneous sparitic grains are composed of euhedral calcite crystals in the form of rosettes or as longitudinal aggregates (Fig. 5F). In these cases, rosettes preserve a micritic nucleus or a micritic lamina in the longitudinal aggregates. The delicate framework of these grains established their penecontemporaneous origin. The origin of these fragile sparitic aggregates could be related to bioinduced precipitation of rhombohedral calcite crystals over filaments (Freytet and Verrecchia, 1998). Similar sparitic components have been described by Pomar et al. (1976) as the result of phreatic crystallizations. According to these authors, rhombohedrally crystallized floating calcite ("calcite rafts") precipitates in karst caves and grows rapidly around a nucleus (often of organic material). The presence of microorganisms favors the nucleation of these crystals. On the other hand, sparry crusts in tufas and stromatolites have been interpreted as primary deposits of purely physico-chemical origin (Iron and Müller, 1968; Braithwaite, 1979; Pedley, 1990).

Nonstructured penecontemporaneous sparitic grains were also observed as single euhedral crystals or mosaics without any clear structure. Crystal idiomorphism and the lack of abrasion features are the main features that lead us to ascribe these types to disintegration of other penecontemporaneous sparitic grains. However, identification of nonstructured penecontemporaneous sparitic grains may be hazardous, owing to the lack of conclusive microfabric features.

Penecontemporaneous Coated Grains

These grains correspond to the compound grains defined by Arribas and Tortosa (2003). They are very common and are composed of a nucleus (ancient or penecontemporaneous grain) plus a coating of penecontemporaneous carbonate (Fig. 6). Generally, coatings composed of a single micrite lamina are the most frequent texture (Figs. 6A, 6B, and 6C). Coatings may be composed of sparitic lamina alternating with micritic lamina, which develop a concentric microfabric (Figs. 6D, 6E, and 6F). Some coated grains are composed of incomplete laminae (Figs. 6A and 6D). Microfabrics in sparitic and micritic lamina are similar to those described in some penecontemporaneous grains derived from fragmentation of carbonate tufas, as well as those described in many oncolites and microbial mats (Freytet and Verrecchia, 1998; Janssen et al., 1999). Coatings develop where microbial activity grows over erratic grains (ancient or penecontemporaneous). Due to their origin, these grains can be considered as cyanoliths. Clastic deposits constituted by cyanoliths related to braided fluviatile sedimentation were named by Pedley (1990) as cyanolith "oncoidal" tufa.

Penecontemporaneous Bioclasts

They are not common, but some skeletal remains of gastropods, ostracods, and charophytes were also recognized. The low percentage in the sediment of this type of penecontemporaneous grain is indicative of the minor role that accumulation of skeletal grains has in the formation of carbonate grains in continental fluviatile environments.

SUMMARY AND CONCLUSIVE REMARKS

Fluvial systems that flow through carbonate sedimentary formations in the Iberian Range constitute important factories of penecontemporaneous carbonate sediments. In these environments, microbial activity is the principal factor that influences carbonate sediment production. This production includes (1) the development of carbonate buildups (tufa, stromatolites), which supply sandy carbonate fragments during fluviatile erosion; and (2) the production of isolated penecontemporaneous carbonate grains.

Modern karstification of the carbonate source areas provokes the dissolution of carbonate rocks and the increase in bicarbonate ions in the fluvial waters and the physico-chemical precipitation of calcite crystals as speleothems in the karstified carbonate sources. Meteoric diagenetic processes (dissolution + physico-chemical

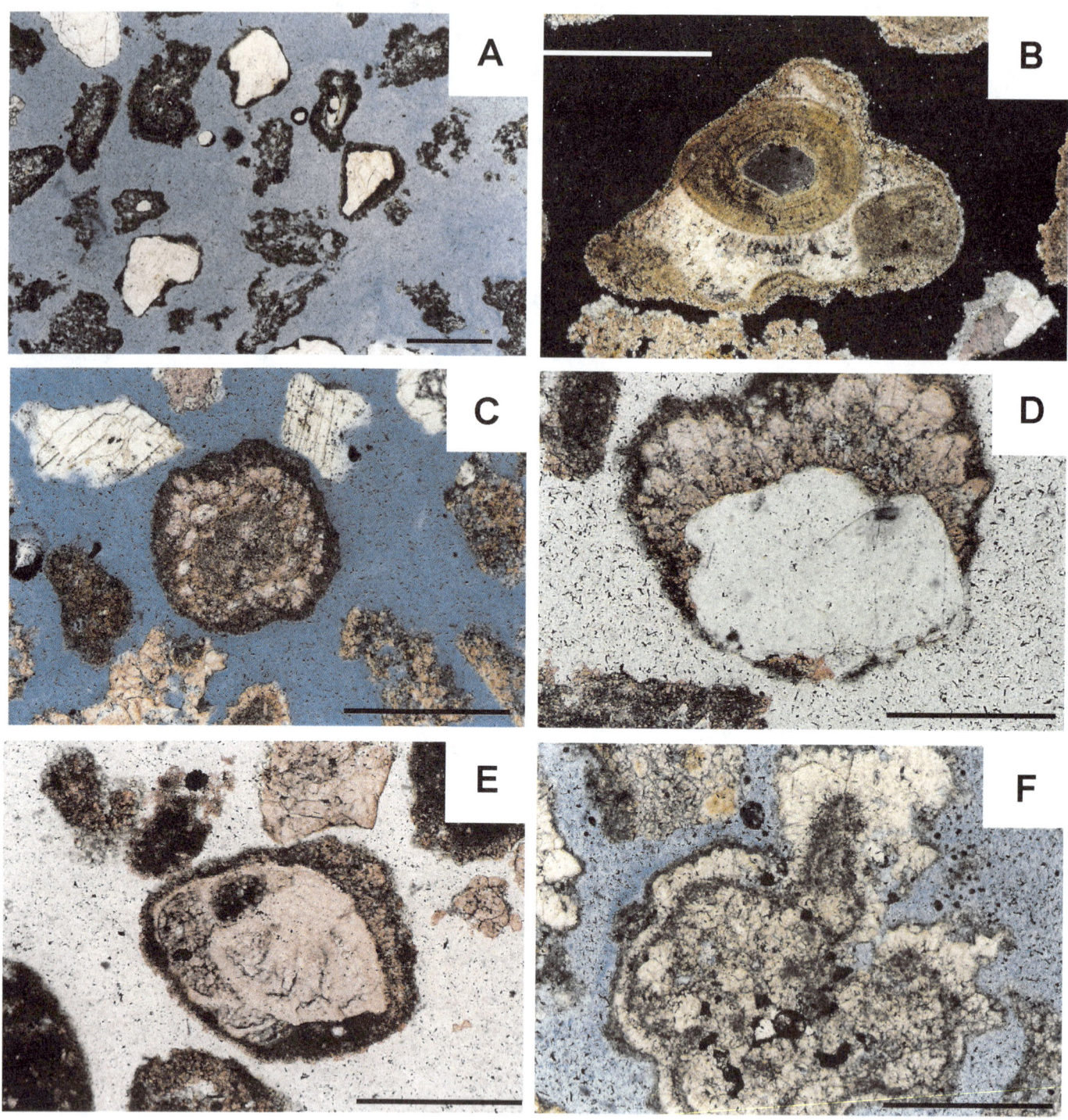

Figure 6. Penecontemporaneous coated grains (PCG). (A) General view of sand composed of nonstructured penecontemporaneous micritic grains (clotted and spongy microfabric) and penecontemporaneous coated grains with a single micritic lamina. Scale bar = 0.5 mm. (B) Grains composed of a nucleus (sands composed of ancient carbonate grains [AC]; oolitic grainstone with fibrous cement) plus a single micrite lamina. (C) Grains composed of a nucleus (penecontemporaneous bioclasts, charophyte) plus a single micrite lamina. (D) Grains showing incomplete laminae growing over a nucleus (sands composed of ancient noncarbonate grains [AN]; quartz grain). The first one corresponds to a sparitic lamina and the second one to a micritic lamina. (E) Grains composed of a nucleus (AC, foraminifera) plus micrite and sparite laminae. (F) Grains showing a coating composed of sparitic laminae alternating with micritic laminae developing a concentric microfabric. Nucleus is a fragment of carbonate tufa (penecontemporaneous sparitic grains). Scale bar = 0.5 mm.

precipitation) contribute to penecontemporaneous carbonate formation, such as bioinduced calcite from supersaturated waters or carbonate particles derived from erosion of karstified source areas.

The penecontemporaneous carbonate grains can be identified using textural and compositional criteria. Some of these features are: irregular morphology linked to delicate filamentous structure; fragile aggregates of calcite crystals; and high porosity of grains.

Four main types of penecontemporaneous carbonate grains have been distinguished: (1) penecontemporaneous micritic grains (PMG); (2) penecontemporaneous sparitic grains (PSG); (3) penecontemporaneous coated grains (PCG); and (4) penecontemporaneous bioclasts (PB).

Penecontemporaneous micritic grains are the most frequent type of penecontemporaneous grain. These grains are characterized by a clotted and spongy microfabric where filaments are often recognized. Some of these grains show a pristine filamentous structure outlining a biomediated carbonate precipitate linked to microbial activity. Their origin is related to the erosion of carbonate tufas developed along the river banks as well as microbial mats on the river beds. Eroded spring carbonate deposits may contribute to these types of grains. Some penecontemporaneous micritic grains with pedogenetic features (i.e., alveolar microfabric) can be related to the erosion of recent carbonate soils.

Penecontemporaneous sparitic grains are also abundant clastic components of the sands. These grains are composed of microsparite and/or sparite mosaics. Frequently, preservation of filamentous microfabric reflects organic participation of cyanobacteria. Coarse sparitic mosaics can form by early recrystallization of biomediated carbonate. This process produces a radial palisadic microfabric, and/or it can destroy the original biogenic microfabrics. The origin of these grains is related to the erosion of recent carbonate tufa (phytoherms and microbial mats). Other penecontemporaneous sparitic grains are prismatic and radial microstructure grains (*Microcodium* fragments). These grains can be related to the erosion of carbonate soils. Mosaics of euhedral calcite crystals (rosettes and longitudinal aggregates) may be released from karst systems associated with the carbonate sources.

Penecontemporaneous coated grains are chiefly abundant in those sands with abundant ancient grains (AC + AN). They are composed of a nucleus plus a coating of carbonate. Coatings are constituted by one or several micrite or sparite laminae. Penecontemporaneous coated grains are considered to be cyanoliths that formed as individual components in the fluvial channel.

Penecontemporaneous bioclasts consist of charophyte, ostracod, and gastropod remains. Their abundance in the fluvial sands is trivial.

Grains with different origins may show similar microfabrics. This convergence of microfabrics is emphasized in grains with laminated structure (microbial mats, cyanoliths, rhizoliths, and speleothems).

Early recrystallization and cementation are very frequent in continental environments and produce an increase of crystal size in the carbonate deposits. As a consequence, some penecontemporaneous grains show very similar textures to those in ancient carbonate rocks. This fact is emphasized in petrographic analysis on sandstones, when diagenesis masks the primary carbonate textures. Thus, it is easy to underestimate the presence of penecontemporaneous sparitic grains (formed as the result of early recrystallization, cementation, or bioinduced mediation) and to erroneously consider them to be ancient components.

ACKNOWLEDGMENTS

This work was funded by the Spanish DGICYT (Dirección General de Investigación Científica y Técnica) projects PB93-0178 and BTE2001-026 and CGL2005-07445-C03-02/BTE. We would like to thank Earle F. McBride and Daniela Fontana for their comments and suggestions, which substantially improved the manuscript.

REFERENCES CITED

Alonso-Zarza, A.M., Sanz, M.E., Calvo, J.P., and Estévez, P., 1998, Calcified root cells in Miocene pedogenic carbonates of the Madrid Basin: Evidence of the origin of *Microcodium* b: Sedimentary Geology, v. 116, p. 81–97.

Arenas, C., Gutiérrez, F., Osácar, C., and Sancho, C., 2000, Sedimentology and geochemistry of fluvio-lacustrine tufa deposits controlled by evaporite solution subsidence in the central Ebro Depression, NE Spain: Sedimentology, v. 47, p. 883–909, doi: 10.1046/j.1365-3091.2000.00329.x.

Arenas, C., Sancho, C., Osácar, M.C., Vázquez, M., and Auqué, L.F., 2004, La sedimentación tobácea actual en el Parque del Monasterio de Piedra (provincia de Zaragoza): Geotemas, v. 6, no. 2, p. 27–30.

Arribas, J., and Arribas, M.E., 1991, Petrographic evidence of different provenance in two alluvial fan systems (Paleogene of the northern Tajo Basin, Spain), *in* Morton, A.C., Todd, S.P., and Haughton, P.D.W., eds., Developments in Sedimentary Provenance Studies: Geological Society [London] Special Publication 57, p. 263–271.

Arribas, J., and Tortosa, A., 2003, Detrital modes in sedimentoclastic sands from low-order streams in the Iberian Range, Spain: The potential for sand generation by different sedimentary rocks: Sedimentary Geology, v. 159, p. 275–303, doi: 10.1016/S0037-0738(02)00332-9.

Arribas, M.E., 1986, Petrología y análisis secuencial de los carbonatos lacustres del Paleógeno del sector N de la cuenca Terciaria del Tajo: Cuadernos de Geología Ibérica, v. 10, p. 295–334.

Arribas, M.E., and Tortosa, A., 1998, Coeval carbonate grains in modern fluvial sands (Iberian Range, Spain): The activity of a continental carbonate factory, *in* 15th International Sedimentological Congress, Abstracts: Alicante, Spain, International Geological Congress, Abstracts book, Publicaciones de la Universidad de Alicante, 154 p.

Arribas, M.E., Estrada, R., Obrador, A., and Rampone, G., 1996, Distribución y ordenación de *Microcodium* en la Formación Tremp: Anticlinal de Camping (Pirineos Orientales, provincia de Barcelona): Revista de la Sociedad Geológica de España, v. 9, no. 1–2, p. 9–18.

Braithwaite, C.J.R., 1979, Crystal texture of recent fluvial pisolites and laminated crystalline crust in Dyfed, South Wales: Journal of Sedimentary Petrology, v. 49, p. 181–194.

Braithwaite, C.J.R., and Zedef, V., 1994, Living hydromagnesite stromatolites from Turkey: Sedimentary Geology, v. 92, p. 1–5, doi: 10.1016/0037-0738(94)90051-5.

Cameron, B., Cameron, D., and Jones, J.R., 1985, Modern algal mats in intertidal and supratidal quartz sands, northeastern Massachusetts, U.S.A., *in* Curran, H.A., ed., Biogenic Structures: Their Use in Interpreting Depositional Environments: Society of Economic Palaeontologists and Mineralogists Special Publication 35, p. 211–223.

Cavazza, W., Zuffa, G.G., Camporesi, C., and Ferretti, C., 1993, Sedimentary recycling in a temperate climate drainage basin (Senio River, north-central Italy): Composition of source rock, soil profiles, and fluvial deposits, in Johnsson, M.J., and Basu, A., eds., Processes Controlling the Composition of Clastic Sediments: Geological Society of America Special Paper 284, p. 247–261.

Chafetz, H.S., 1994, Bacterially induced precipitates of calcium carbonate and lithification of microbial mats, in Krumbein, W.E., Peterson, D.M., and Stal, L.J., eds., Biostabilization of Sediments: Oldenburg, Germany, Bibliotek und Informat der Carl von Ossietzky Universität Oldenburg, p. 149–163.

Cipriani, N., Malesani, P., and Vannucci, S., 1977, I travertini dell-Italia centrale: Bollettino del Servizio Geologico d'Italia, v. 98, p. 85–115.

Cole, J.M., Rasbury, E.T., Montañez, I.P., Pedone, V.A., Lanzirotti, A., and Hanson, G.N., 2004, Petrographic and trace element analysis of uranium-rich tufa calcite, middle Miocene Barstow Formation, California, USA: Sedimentology, v. 51, p. 433–453, doi: 10.1111/j.1365-3091.2004.00631.x.

Dickinson, W.R., 1970, Interpreting detrital modes of greywacke and arkose: Journal of Sedimentary Petrology, v. 40, p. 695–707.

Dickinson, W.R., Beard, L.S., Brakenridge, G.R., Erjavec, J.L., Ferguson, R.C., Inman, K.F., Knepp, R.A., Lindberg, F.A., and Ryberg, P.T., 1983, Provenance of North American Phanerozoic sandstones in relation to tectonic setting: Geological Society of America Bulletin, v. 94, p. 222–235, doi: 10.1130/0016-7606(1983)94<222:PONAPS>2.0.CO;2.

Di Giulio, A., and Valloni, R., 1992, Analisi microscópica delle areniti terrigene: Parametri petrologici e composición modali: Acta Naturalia de "L'Ateneo Parmese," v. 28, p. 55–101.

Eggleston, J.R., and Dean, W.E., 1976, Freshwater stromatolitic bioherms in Green Lake, New York, in Walter, M.R., ed., Stromatolites: Developments in Sedimentology, v. 20, p. 479–488.

Esteban, M., 1974, Caliche textures and "Microcodium": Bollettino Società Geologica Italiana, v. 92, p. 105–125.

Fernández, A., García del Cura, M.A., González Martín, J.A., and Ordóñez, S., 1998, Fluvial tufas in the Júcar River valley (Spain), in 15th International Sedimentological Congress, Abstracts: Alicante, Spain, International Geological Congress, Abstracts book, Publicaciones de la Universidad de Alicante, p. 324–325.

Fernández, A., García del Cura, M.A., González Martín, J.A., and Ordóñez, S., 2000, Morfogénesis y sedimentación carbonática Pleistocena en el Valle del Júcar (Albacete): Geotemas, v. 1, no. 3, p. 353–373.

Ford, T.D., and Pedley, H.M., 1996, A review of tufa and travertine deposits of the world: Earth-Science Reviews, v. 41, p. 117–175.

Freytet, P., 1992, Les cristallisations de calcite associées à des restes végétaux (algues, feuilles) en milieu fluviatile et lacustre, actuel et ansíen (tufs et travertins): Bulletin de la Société Botanique de France, v. 139, Actualités botaniques, 1, p. 69–74.

Freytet, P., and Plaziat, J., 1982, Continental carbonate sedimentation and pedogenesis: Contributions to Sedimentology, v. 11, 216 p.

Freytet, P., and Verrecchia, E.P., 1998, Freshwater organisms that build stromatolites: A synopsis of biocrystallization by prokaryotic and eukaryotic algae: Sedimentology, v. 45, p. 535–563, doi: 10.1046/j.1365-3091.1998.00155.x.

Freytet, P., and Verrecchia, E.P., 1999, Calcitic radial palisadic fabric in freshwater stromatolites: Diagenetic and recrystallized feature or physicochemical sinter crust?: Sedimentary Geology, v. 126, p. 97–102, doi: 10.1016/S0037-0738(99)00034-2.

Freytet, P., Kerp, H., and Broutin, J., 1996, Permian freshwater stromatolites associated with the conifer Cassinisia orobica Kerp et al.—A very peculiar type of fossilization: Reviews of Palaeobotany and Palynology, v. 91, p. 85–105.

Garzanti, E., 1991, Non-carbonate intrabasinal grains in arenites: Their recognition, significance, and relationship to eustatic cycles and tectonic setting: Journal of Sedimentary Petrology, v. 61, no. 6, p. 959–975.

Gazzi, P., 1966, Le arenarie del flysch sopracretaceo dell'Appennino modenese; correlazioni con il Flysch di Monghidoro: Mineralogica et Petrographica Acta, v. 12, p. 69–97.

Gerdes, G., Krumbein, W.E., and Reineck, H.-E., 1994, Microbial mats as architects of sedimentary surface structures, in Krumbein, W.E., Peterson, D.M., and Stal, L.J., eds., Biostabilization of Sediments: Oldenburg, Germany, Bibliotek und Informat. der Carl von Ossietzky Universität Oldenburg, p. 165–181.

Ibbeken, H., and Schleyer, R., 1991, Source and Sediment: A Case Study of Provenance and Mass Balance at an Active Plate Margin (Calabria, Southern Italy): Berlin, Springer-Verlag, 286 p.

IGN (Instituto Geográfico Nacional), 1991, Atlas Nacional de España; II.9, Climatología: Madrid, Instituto Geográfico Nacional, Ministerio de Obras Públicas y Transportes (MOPT), 24 p.

Ingersoll, R.V., Bullard, T.F., Ford, R.L., Grimm, J.P., Pickle, J.D., and Sares, S.W., 1984, The effect of grain size on detrital modes: A test of the Gazzi-Dickinson point-counting method: Journal of Sedimentary Petrology, v. 54, p. 103–116.

Iron, G., and Müller, G., 1968, Mineralogy, petrology, and chemical composition of some calcareous tufa from the Schwabische Alb, Germany, in Müller, G., and Friedman, G.M., eds., Recent Developments in Carbonate Sedimentology in Central Europe: New York, Springer-Verlag, p. 157–171.

Jaillard, R., 1991, Structure and composition of calcified roots and their identification in calcareous soils: Geoderma, v. 50, p. 197–210, doi: 10.1016/0016-7061(91)90034-Q.

Janssen, A., Swennen, R., Podoor, N., and Keppens, E., 1999, Biological and diagenetic influence in Recent and fossil tufa deposits from Belgium: Sedimentary Geology, v. 126, p. 75–95, doi: 10.1016/S0037-0738(99)00033-0.

Johnsson, M.J., 1993, The system controlling the composition of clastic sediments, in Johnsson, M.J., and Basu, A., eds., Processes Controlling the Composition of Clastic Sediments: Geological Society of America Special Paper 284, p. 1–19.

Klappa, C.F., 1978, Biolithogenesis of Microcodium: Elucidation: Sedimentology, v. 25, p. 489–522, doi: 10.1111/j.1365-3091.1978.tb02077.x.

Kosir, A., 2004, Microcodium revisited: Root calcification products of terrestrial plants on carbonate-rich substrates: Journal of Sedimentary Research, v. 74, no. 6, p. 845–857.

Marsaglia, K.M., 1993, Basaltic Island sand provenance, in Johnsson, M.J., and Basu, A., eds., Processes Controlling the Composition of Clastic Sediments: Geological Society of America Special Paper 284, p. 41–66.

Ordóñez, S., and García del Cura, M.A., 1983, Recent and Tertiary fluvial carbonates in central Spain, in Collinson, J.D., and Lewin, J., eds., Ancient and Modern Fluvial Systems: International Association of Sedimentologists Special Publication 6, p. 485–497.

Palomares, M., and Arribas, J., 1993, Modern stream sands from compound crystalline sources: Composition and sand generation index, in Johnsson, M.J., and Basu, A., eds., Processes Controlling the Composition of Clastic Sediments: Geological Society of America Special Paper 284, p. 313–322.

Pedley, H.M., 1990, Classification and environmental models of cool freshwater tufas: Sedimentary Geology, v. 68, p. 143–154, doi: 10.1016/0037-0738(90)90124-C.

Pedley, H.M., Andrews, J.E., Ordóñez, S., González Martín, J.A., García del Cura, M.A., and Taylor, D.M., 1996, Climatically controlled fabrics in freshwater carbonates: A comparative study of barrage tufas from Spain and Britain: Palaeogeography, Palaeoclimatology, Palaeoecology, v. 121, p. 239–257.

Pomar, L., Gines, A., and Fontarnau, R., 1976, Las cristalizaciones freáticas: Endines, v. 3, p. 3–25.

Riding, R., 1994, Stromatolite survival and change: Their significance of Shark Bay and Lee Stocking Island subtidal columns, in Krumbein, W.E., Peterson, D.M., and Stal, L.J., eds., Biostabilization of Sediments: Oldenburg, Germany, Bibliotek und Informat. der Carl von Ossietzky Universität Oldenburg, p. 183–202.

Riding, R., 2000, Microbial carbonates: Te geological record of calcified bacterial-algal mats and biofilms: Sedimentology, v. 47, p. 179–214, doi: 10.1046/j.1365-3091.2000.00003.x.

Rossi, C., 1997, Microcodium y trazas fósiles en invertebrados en facies continentales (Paleoceno de la Cuenca de Áger, Lérida): Revista de la Sociedad Geológica de España, v. 10, no. 3–4, p. 371–391.

Schäfer, A., and Stapf, K.R., 1978, Permian Saar-Nahe Basin and Recent Lake Constance (Germany): Two environments of lacustrine algal carbonates, in Matter, A., and Tucker, M.E., eds., Modern and Ancient Lake Sediments: International Association of Sedimentologists Special Publication 2, p. 83–107.

Valero-Garcés, B., Arenas, C., and Delgado-Huertas, A., 2001, Depositional environments of Quaternary lacustrine travertines and stromatolites from high-altitude Andean lakes, northwestern Argentina: Canadian Journal of Earth Sciences, v. 38, p. 1263–1283, doi: 10.1139/cjes-38-8-1263.

Vilas, L., Mas, R., García, A., Arias, C., Alonso, A., Meléndez, N., and Rincón, R., 1982, Ibérica suroccidental: El Cretácico de España:

Madrid, Universidad Complutense de Madrid, p. 457–514.
Walter, M.R., 1976, Hot-spring sediments in Yellowstone National Park, in Walter, M.R., ed., Stromatolites: Developments in Sedimentology, v. 20, p. 489–498.
Zuffa, G.G., 1980, Hybrid arenites: Their composition and classification: Journal of Sedimentary Petrology, v. 50, p. 21–29.
Zuffa, G.G., 1985, Optical analysis of arenites: Influence of methodology on compositional results, in Zuffa, G.G., ed., Provenance of Arenites: NATO ASI (Advanced Science Institutes) Series, v. 148, p. 165–189.
Zuffa, G.G., 1987, Unravelling hinterland and offshore paleogeography from deep-water arenites, in Leggett, J.K., and Zuffa, G.G., eds., Deep-Marine Clastic Sedimentology: London, Graham and Trotman, p. 39–61.
Zuffa, G.G., 1991, On the use of turbidite arenites in provenance studies: Critical remarks, in Morton, A.C., Todd, S.P., and Haughton, P.D.W., eds., Developments in Sedimentary Provenance Studies: Geological Society [London] Special Publication 57, p. 23–30.

MANUSCRIPT ACCEPTED BY THE SOCIETY 9 AUGUST 2006

Provenance discrimination of Lower Cretaceous synrift sandstones (eastern Iberian Chain, Spain): Constraints from detrital modes, heavy minerals, and geochemistry

M.A. Caja[†]
R. Marfil[‡]
Departamento de Petrología y Geoquímica, Facultad CC. Geológicas, Universidad Complutense de Madrid, 28040 Madrid, Spain

M. Lago[§]
Departamento de Ciencias de la Tierra, Área de Petrología y Geoquímica, Facultad CC. Geológicas, Universidad de Zaragoza, 50009 Zaragoza, Spain

R. Salas[#]
Departamento de Geoquímica, Petrologia i Prospecció Geológica, Facultat de Geología, Universitat de Barcelona, 08028 Barcelona, Spain

K. Ramseyer[††]
Institut für Geologie, Universität Bern, 3012 Bern, Switzerland

ABSTRACT

Petrography, geochemical whole-rock composition, and chemical analyses of tourmaline were performed in order to determine the source areas of Lower Cretaceous Mora, El Castellar, and uppermost Camarillas Formation sandstones from the Iberian Chain, Spain. Sandstones were deposited in intraplate subbasins, which are bound by plutonic and volcanic rocks of Permian, Triassic, and Jurassic age, Paleozoic metamorphic rocks, and Triassic sedimentary rocks.

Modal analyses together with petrographic and cathodoluminescence observations allowed us to define three quartz-feldspathic petrofacies and recognize diagenetic processes that modified the original framework composition. Results from average restored petrofacies are: Mora petrofacies = P/F >1 and $Q(r)_{70}\ F(r)_{22}\ R(r)_9$; El Castellar petrofacies = P/F >1 and $Q(r)_{57}\ F(r)_{25}\ R(r)_{18}$; and Camarillas petrofacies = P/F ~ zero and $Q(r)_{64}\ F(r)_{28}\ R(r)_7$ (P—plagioclase; F—feldspar; Q—quartz; R—rock fragments; r—restored composition).

[†]E-mail: miguelangel.caja@geo.ucm.es.
[‡]E-mail: marfil@geo.ucm.es.
[§]E-mail: mlago@unizar.es.
[#]E-mail: ramonsalas@ub.edu.
[††]E-mail: karl.ramseyer@geo.unibe.ch.

Caja, M.A., Marfil, R., Lago, M., Salas, R., and Ramseyer, K., 2007, Provenance discrimination of Lower Cretaceous synrift sandstones (eastern Iberian Chain, Spain): Constraints from detrital modes, heavy minerals, and geochemistry, *in* Arribas, J., Critelli, S., and Johnsson, M.J., eds., Sedimentary Provenance and Petrogenesis: Perspectives from Petrography and Geochemistry: Geological Society of America Special Paper 420, p. 181–197, doi: 10.1130/2007.2420(12). For permission to copy, contact editing@geosociety.org. ©2007 Geological Society of America. All rights reserved.

Trace-element and rare earth element abundances of whole-rock analyses discriminate well between the three petrofacies based on: (1) the Rb concentration, which is indicative of the K content and reflects the amount of K-feldspar modal abundance, and (2) the relative modal abundance of heavy minerals (tourmaline, zircon, titanite, and apatite), which is reproduced by the elements hosted in the observed heavy mineral assemblage (i.e., B and Li for tourmaline; Zr, Hf, and Ta for zircon; Ti, Ta, Nb, and their rare earth elements for titanite; and P, Y, and their rare earth elements for apatite). Tourmaline chemical composition for the three petrofacies ranges from Fe-tourmaline of granitic to Mg-tourmaline of metamorphic origin.

The three defined petrofacies suggest a mixed provenance from plutonic and metamorphic source rocks. However, a progressively major influence of granitic source rocks was detected from the lowermost Mora petrofacies toward the uppermost Camarillas petrofacies. This provenance trend is consistent with the uplift and erosion of the Iberian Massif, which coincided with the development of the latest Berriasian synrift regional unconformity and affected all of the Iberian intraplate basins. The uplifting stage of Iberian Massif pluton caused a significant dilution of Paleozoic metamorphic source areas, which were dominant during the sedimentation of the lowermost Mora and El Castellar petrofacies.

The association of petrographic data with whole-rock geochemical compositions and tourmaline chemical analysis has proved to be useful for determining source area characteristics, their predominance, and the evolution of source rock types during the deposition of quartz-feldspathic sandstones in intraplate basins. This approach ensures that provenance interpretation is consistent with the geological context.

Keywords: sandstone provenance, petrography, geochemistry, heavy minerals, Lower Cretaceous, intraplate basin, Iberian Chain, Spain.

INTRODUCTION

One of the most typical characteristics of intraplate basins that develop during rifting stages is the occurrence of several source areas, which can simultaneously provide detritus and therefore complicate provenance discrimination. The classical way to use modal analyses based on light-optical and cathodoluminescence (CL) microscopies defines petrofacies, which are based on individual characteristics, i.e., type and modal abundance of quartz (Q), feldspars (F), and rock fragments (R), or quartz luminescence, or heavy mineral assemblages. Unfortunately, the representation of the defined petrofacies in the classical tectonic-scale models of Dickinson et al. (1983) and Dickinson (1985) does not permit a clear differentiation for sandstones deposited in intraplate rift basins where multisource provenances are common. Moreover, weathering and diagenetic processes may affect the framework composition toward a more stable assemblage. Thus, it is essential to restore the original composition for any provenance interpretation (Valloni et al., 1991; Milliken, 1988). In the same way, tectonic-setting discrimination diagrams using major-element geochemistry do not work properly in rift settings (Armstrong-Altrin and Verma, 2005) because the major elements are prone to alteration and/or diagenetic processes (e.g., chloritization, kaolinitization, and calcitization; Zimmermann and Bahlburg, 2003). However, the use of minor, trace, and rare earth elements (REEs) eliminates these difficulties because they are transported without any important changes from the parent rock to the sedimentary basin (Taylor and McLennan, 1985; Bhatia and Crook, 1986; McLennan and Taylor, 1991; McLennan et al., 1993). In this way, whole-rock analyses of trace and REE elements, e.g., Rb as indicative of K-feldspar modal abundance or Zr, Hf, Ta, Nb, Ti, V, Y, Sr, P, Li, or B as indicative of type, origin, and modal abundance of heavy minerals (von Eynatten and Gaupp, 1999; Torres-Ruiz et al., 2003), allow us to overcome the shortcomings of modal analyses based only on petrofacies.

The main objective of the present paper is to determine the multisource provenances of quartz-feldspathic sandstones deposited in intraplate rift basins, which are bound by contrasting source areas (plutonic, volcanic, metamorphic, and sedimentary rocks), by linking petrographic data (mineral composition, detrital modes, and quartz CL characteristics) with whole-rock geochemical compositions, including chemical analysis of a specific heavy mineral (i.e., tourmaline).

GEOLOGICAL SETTING

The study area is located in the northeastern part of the Iberian Peninsula (Spain), in the eastern part of the Iberian Chain

(Fig. 1A), which corresponds to the linkage zone between the Catalan Coastal Chain and the Iberian Chain (Fig. 1B). The evolution of the eastern Iberian plate is related to the opening of the Tethys and the North Atlantic in the eastern part of the Iberian Peninsula. Two main rifting stages took place in the Iberian plate during Mesozoic times: (1) Late Permian–Triassic rifting, and (2) late Oxfordian–middle Albian rifting (Salas and Casas, 1993; Salas et al., 2001). The second rifting event, during the Late Jurassic–Early Cretaceous, was related to the opening of the North Atlantic and Bay of Biscay basins. This rifting stage controlled the development of the Upper Jurassic–Lower Cretaceous intraplate basins such as: Cameros, Maestrat, Columbrets, and South Iberian. The Maestrat Basin contains up to 5.8 km of Mesozoic sediments, dominantly carbonates, and it contains 4.3 km of the synrift Upper Jurassic–Lower Cretaceous sequence. In this basin, the rift structure consists of an extensional synsedimentary fault system, which has divided the basin into several subbasins (Fig. 1B), on top of tilted basement blocks (Salas and Guimerà, 1996).

This paper is focused on the westernmost subbasins (Aliaga, Galve, and Penyagolosa; Fig. 1B), where siliciclastic sediments were deposited in response to the second rifting event during Late Jurassic–Early Cretaceous times. Paleocurrents and paleogeographic reconstructions for this period (Salas et al., 2001) suggest that these subbasins acted during the Early Cretaceous as a detrital trap for sediments coming mainly from the N-NW and W (Fig. 2). These subbasins were filled by fluvial sediments with intercalations of lacustrine carbonates, which evolved toward the top to shallow-marine carbonates. This sedimentary record belongs to the so-called "Weald facies" (Hahne, 1930), which

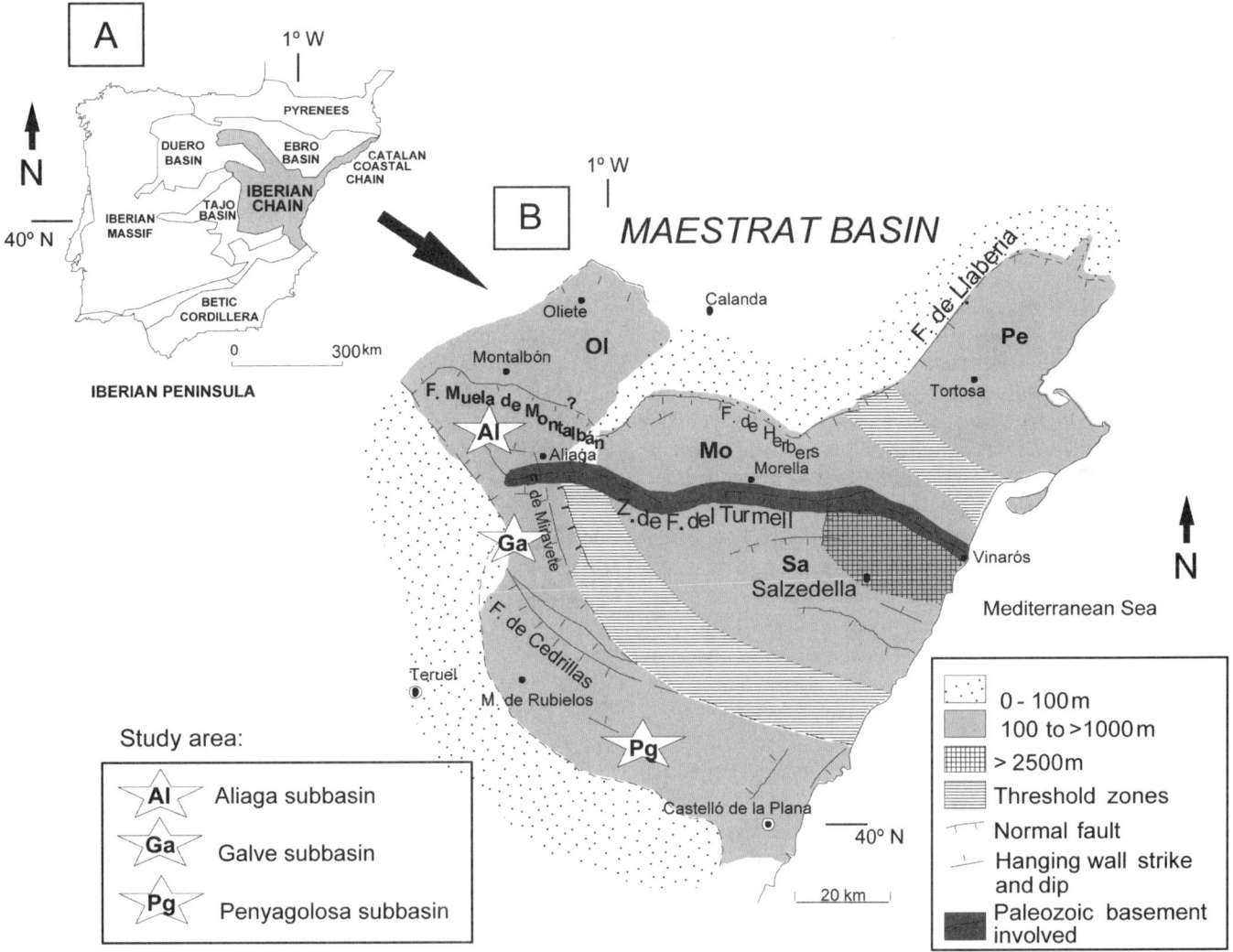

Figure 1. (A) Simplified map of the Iberian Peninsula showing the main structural units. (B) Detail corresponding to the Maestrat Basin located in the Iberian Chain (modified from Salas et al., 2001). The Maestrat Basin has been subdivided into the following seven subbasins: Oliete (Ol), Morella (Mo), Perelló (Pe), Salzedella (Sa), Penyagolosa (Pg), Galve (Ga), and Aliaga (Al). The last three subbasins are the study area. Thickness represented corresponds to Early Cretaceous sediments.

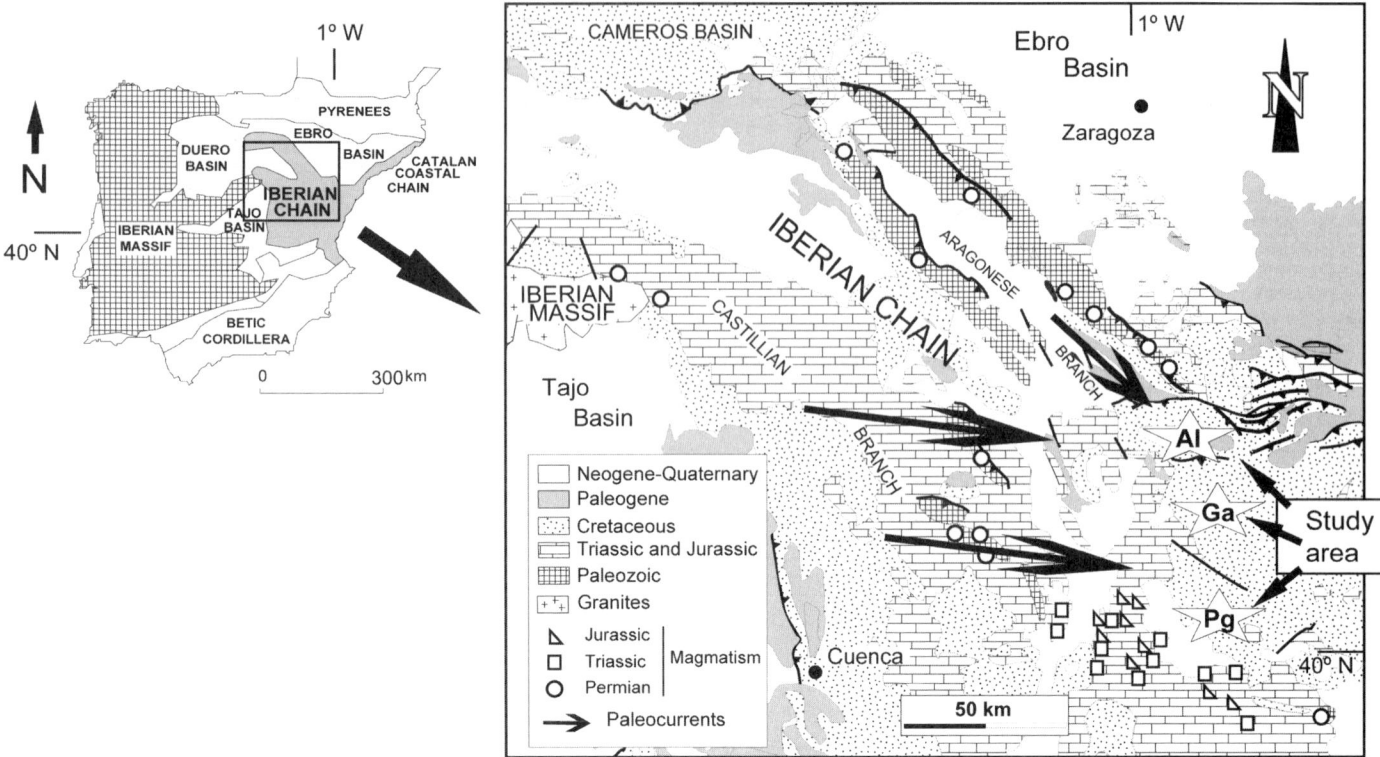

Figure 2. Simplified geologic map of the main possible source areas for the study area showing the main paleocurrent directions (Salas et al., 2001). Pg—Penyagolosa, Ga—Galve, and Al—Aliaga.

is composed of three lithostratigraphic units (Fig. 3): (1) Mora Formation sandstones, (2) El Castellar Formation limestones and sandstones, and (3) Camarillas Formation sandstones (Salas, 1987). The Mora Formation, which is Late Berriasian-Valanginian in age (up to 140 m of thickness; Fig. 3), is made up of white sandstones intercalated with red shales deposited in fluvial to transitional environments. These sediments were deposited during the first tilting movements of the Penyagolosa subbasin and are presently preserved in this subbasin only. The El Castellar Formation, which is late Valanginian–middle Hauterivian in age (80–140 m of thickness; Fig. 3), is made up of green shales and sandstones deposited in a muddy alluvial plain with channels, and these are overlain by lacustrine limestones. Sandstones are only well developed in the Penyagolosa subbasin. The Camarillas Formation, which is late Hauterivian–Barremian in age (125–250 m of thickness; Fig. 3), is characterized by yellow and white sandstones intercalated with red shales deposited in fluvial systems evolving toward the top to a deltaic system. These sandstones are very well developed in the western margin of the Maestrat Basin in the Aliaga, Galve, and Penyagolosa subbasins.

The boundaries between these three stratigraphic units are unconformities resulting in three depositional sequences (Fig. 3). D3 is a synrift major regional unconformity (latest Berriasian, ca. 140 Ma) recognized throughout the entire Iberian Basin. This unconformity is related to the Neocomian restructuring with uplift and compartmentalization of the Upper Jurassic rifted blocks (Salas et al., 2001). This scenario gave rise to major facies changes and the stacking pattern of parasequences, which can be subdivided into three second-order sequences (1–3, Fig. 3). The transgressive deposits of the upper sequence correspond to the lower part of the Artoles Formation, which consists of shallow-marine carbonates.

During Eocene-Oligocene times, the extensional faults were inverted due to the Alpine compressional tectonics, which gave rise to a reverse fault, fold-and-thrust system. This system is called the Iberian Chain; it is located in the eastern part of the Iberian Peninsula, and it shows a NW-SE trend (Salas et al., 2001).

METHODOLOGY

A total of 125 sandstone samples from Mora, El Castellar, and Camarillas Formations was collected in 10 stratigraphic sections. Thin sections were stained with alizarin red-S and sodium cobaltinitrite for carbonate (Lindholm and Finkelman, 1972) and feldspar (Chayes, 1952) distinction, respectively. Modal analyses were performed in 49 sandstones with fine to medium grain size (0.17–0.35 mm). Total points counted were up to 300–400, following the "Indiana method" (Basu, 1976; Suttner et al., 1981; Weltje, 2002).

A selection of seven representative samples from the three formations was studied with the high-sensitivity CL microscope at the Institute of Geological Sciences, University of Bern. Working conditions were 25 kV accelerating voltage and 0.4 $\mu A/mm^2$ beam current density (Ramseyer et al., 1989). Color

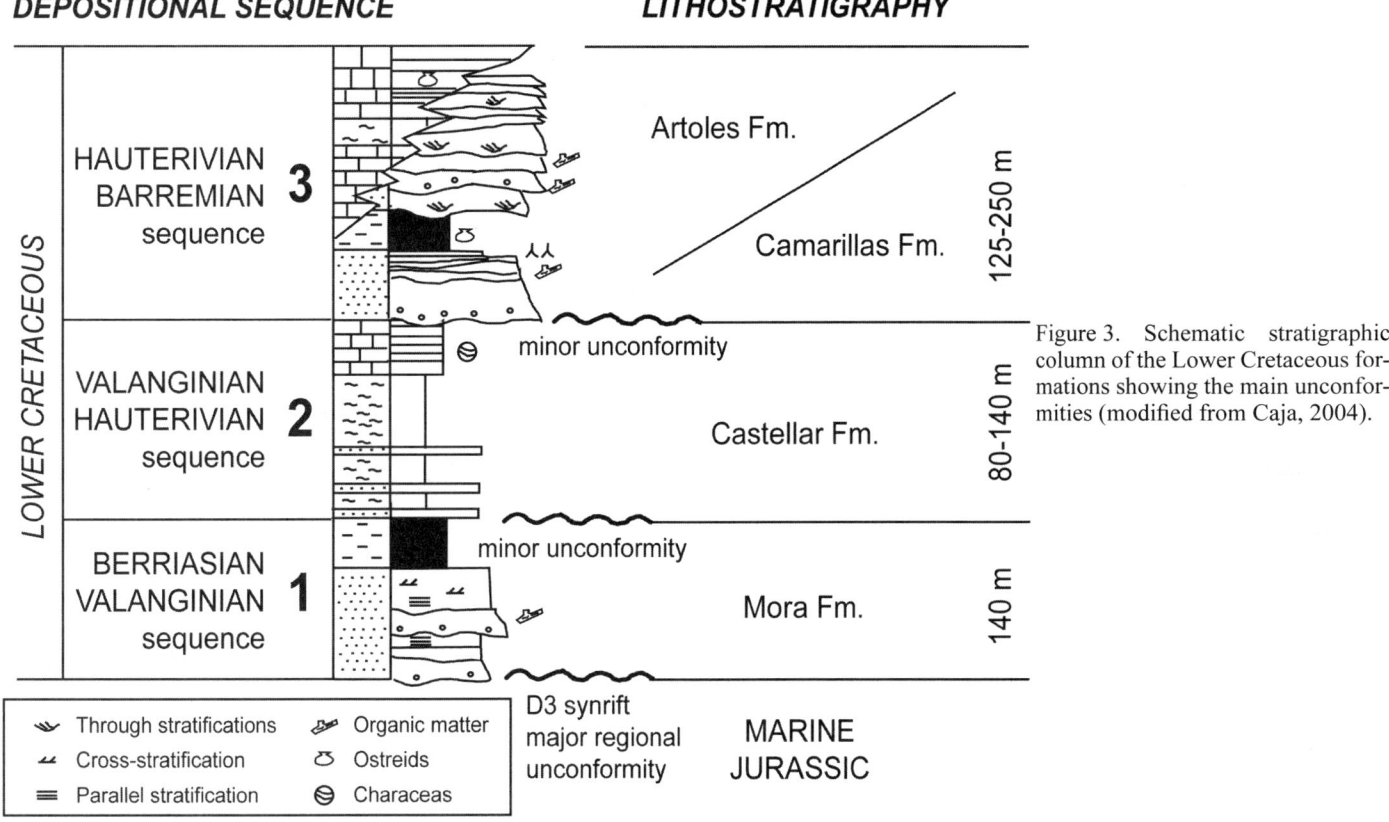

Figure 3. Schematic stratigraphic column of the Lower Cretaceous formations showing the main unconformities (modified from Caja, 2004).

microphotographs were taken with Ektachrome 400 ASA (American Standards Association) color transparency films, and they were developed at 800 ASA.

Whole-rock chemical composition analyses were performed on 18 sandstone samples with a high content of heavy minerals and no carbonate cements. Analyses were performed by ACTLABS (Canada) following the WRA+trace 4Lithoresearch routine with an inductively coupled plasma–atomic emission spectrometer (ICP-AES) for major elements and an inductively coupled plasma–mass spectrometer (ICP-MS) for trace and REE elements. B and Li were analyzed in 10 samples by PGNAA (pulsed gamma neutron activation analysis). Detection limits for major elements were <0.01% (except for TiO_2 and MnO, which had detection limits <0.001%). Detections limits for minor and trace elements were (in ppm): 30 for Zn; 5 for V, As, and Pb; 3 for Ba; 2 for Sr and Mo; 1 for Sc, Be, Co, Ga, Rb, Zr, and Sn; 0.5 for Y, Ge, and W; 0.2 for Nb and Sb; 0.1 for Cs, Hf, and Bi; 0.05 for La, Ce, Nd, Tl, and Th; 0.01 for Pr, Sm, Gd, Tb, Dy, Ho, Er, Ta, U, and Yb; 0.005 for Eu and Tm; and 0.002 for Lu.

The chemical composition of tourmaline was analyzed in 10 carbon-coated thin sections with a JEOL JXA-8900 M electron microprobe (working conditions: 15 kV accelerating voltage, 20 nA beam current, and 5µm beam diameter). Detections limits were: Ca 150 ppm, Mg 100 ppm, Fe 300 ppm, Mn 250 ppm, Ti 225 ppm, Na 150 ppm, K 175 ppm, Cr 250 ppm, P 150 ppm, F 500 ppm, Cl 200 ppm, Ni 150 ppm, Al 175 ppm, and Si 325 ppm.

B content was calculated from each chemical analyses based on tourmaline chemical formula.

RESULTS

Petrology

The framework composition of Mora Formation sandstones is classified as subarkose and sublitharenite (Pettijohn et al., 1973). Samples with matrix content higher than 15% are feldspathic graywackes and quartzwackes. Average detrital composition (Table 1) recalculated to 100% (Q—quartz; F—feldspars; R—rock fragments) is: Q_{87} F_8 R_5 (Fig. 4A). These sandstones are fine grained, moderate to poorly sorted, with subangular to subrounded grains. Clay matrix (up to 12.9%) is mainly derived from feldspar alteration to illite and kaolinite (epimatrix) and mechanical deformation of ductile framework grains (pseudomatrix). Thus, the restored composition is: $Q(r)_{70}$ $F(r)_{22}$ $R(r)_9$ (Fig. 4B; Table 1). The most representative intergranular cements are quartz overgrowth (<5.8%) and calcite cement (<13.9%). El Castellar Formation sandstones are classified as feldspathic graywackes and quartzwackes because of the presence of >15% matrix (average: Q_{86} F_{10} R_4; Fig. 4A). Sandstones with lower amounts of matrix are classified as subarkoses. Matrix is mostly characterized as pseudomatrix. Thus, the restored composition is: $Q(r)_{57}$ $F(r)_{25}$ $R(r)_{18}$ (Fig. 4B; Table 1). These sandstones are

TABLE 1. AVERAGE MODAL COMPOSITIONS

Formation:		Mora	Castellar	Camarillas			
Subbasin:		Penyagolosa	Penyagolosa	Penyagolosa	Galve	Aliaga	Penyagolosa, Galve, and Aliaga
Number of samples:		11	7	16	10	5	31
		mean	mean	mean	mean	mean	mean
Composition							
Quartz monocrystalline, undulosity <5°		8.1	6.2	8.6	7.1	7.4	7.7
Quartz monocrystalline, undulosity > 5°		32.1	28.5	29.0	33.4	37.8	33.4
Quartz polycrystalline 2–3 subgrains		14.0	12.2	10.8	12.2	6.3	9.7
Quartz polycrystalline >3 subcrystals		9.0	6.8	7.2	5.3	3.3	5.3
Chert		0.1	–	0.2	0.5	0.7	0.5
K-feldspar, single crystals		–	2.0	8.2	15.1	17.3	13.5
Plagioclase, single crystals		5.5	4.1	6.0	0.9	0.3	2.4
Biotite		0.1	1.3	1.4	–	–	0.5
Muscovite		0.7	1.1	0.7	0.6	0.8	0.7
Stable heavy minerals (zircon, titanite, apatite)		+	+	+	0.1	+	–
Tourmaline		0.6	0.2	0.1	0.2	0.2	0.2
Medium-rank metamorphic rock fragments		2.3	1.6	1.2	1.4	0.3	0.9
Low-rank metamorphic rock fragments		0.7	0.5	0.6	–	–	0.2
Plutonic rock fragments		0.4	0.5	1.9	0.1	0.2	0.7
Sedimentary rock fragments (limestones)		0.4	–	–	–	–	–
Argillaceous intraclasts		1.1	3.2	2.4	0.8	1.0	1.4
[Quartz overgrowths]		1.8	0.8	1.3	2.5	1.6	1.8
[Carbonate cement]		2.3	0.1	4.8	0.6	6.1	3.8
[Carbonate replacement on feldspars]	F(r)	5.4	10.7	6.0	–	–	6.0
[Barite cement]		0.1	–	–	–	–	–
[Kaolin cement]		0.1	–	0.9	4.1	2.4	7.4
[Clay coatings]		0.2	–	0.4	0.2	1.1	1.6
[Fe-oxides cement]		–	–	–	2.0	4.9	6.9
Infiltrated matrix		–	–	0.4	–	–	0.4
Clay minerals, replacement of feldspars (epimatrix)	F(r)	8.4	6.5	5.7	2.6	0.6	3.0
Pseudomatrix	R(r)	4.5	13.8	6.2	5.7	2.2	4.7
Primary porosity		0.9	–	0.5	1.1	1.7	1.1
Secondary porosity after feldspar	R(r)	–	–	0.9	–	–	0.3
Intraparticle after kaolin		0.5	–	–	3.0	2.9	2.0
Enlarged		0.8	–	1.6	0.3	0.4	0.8
Other types of porosity		–	–	0.2	0.3	0.5	0.4

Note: +—Observed. F(r) and R(r) were used for restoring modal composition.

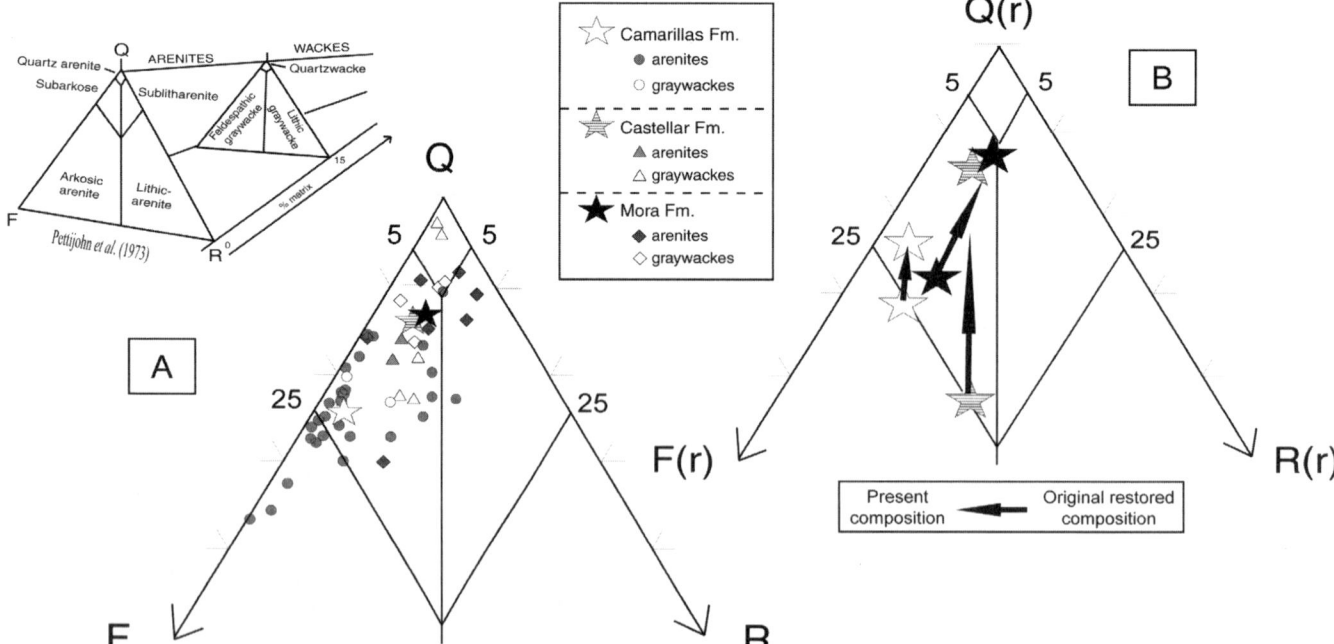

Figure 4. (A) QFR ternary diagram, after Pettijohn et al. (1973). Q—quartz; F—plagioclase + K-feldspar; R—rock fragments. Arenites are represented with a solid symbol, and graywackes are represented with open symbols. Stars correspond to average values of each formation. (B) Restored ternary diagram considering diagenetic processes that affected framework composition [F(r) = plagioclase + K-feldspar + epimatrix + replacements + moldic porosity of feldspar grains; R(r) = rock fragments + pseudomatrix].

fine grained, moderate to poorly sorted, with subangular to subrounded grains. Because of the high amount of matrix, the development of intergranular cements is scarce. Camarillas Formation sandstones are classified as subarkoses and arkoses. Framework average composition is $Q_{76} F_{21} R_{3}$ (Fig. 4A). These sandstones are medium to fine grained, moderate to poorly sorted, with subangular to subrounded grains. Average matrix (epimatrix and pseudomatrix) content ranges from 2.8% to 9.8%. The restored composition is: $Q(r)_{64} F(r)_{28} R(r)_{7}$ (Fig. 4B; Table 1). The most important cements are kaolin pore-filling (<9.8), quartz overgrowths (<5.5), and patchy calcite cement, which is inhomogeneously distributed.

In summary, Mora, El Castellar and Camarillas Formations underwent important modifications due to weathering and postdepositional diagenetic processes, which affected the modal abundance of feldspars and rock fragments (Fig. 4B).

In Mora Formation sandstones, monocrystalline quartz grains are more abundant than polycrystalline (average: Qm_{64} $Qp2-3_{22}$ $Qp+3_{14}$; where Qm is monocrystalline quartz, Qp2–3 is polycrystalline quartz with 2–3 crystals, and Qp + 3 is polycrystalline quartz with >3 crystals; Table 1). Monocrystalline quartz grains show embayments and a brown or occasionally violet luminescent color. In El Castellar Formation sandstones, the average distribution of quartz grains is: Qm_{66} $Qp2-3_{22}$ $Qp + 3_{12}$. Monocrystalline quartz grains have embayments and a dominantly brown luminescent color. Occasionally, some quartz grains have a yellow-brown luminescent color often found in quartz from hydrothermal veins. In Camarillas Formation sandstones, monocrystalline quartz grains are more abundant than polycrystalline (average: Qm_{74} $Qp2-3_{17}$ $Qp + 3_{9}$). However, quartz grains display dominantly a violet luminescence color and brown luminescence is rare. Relationships among types of quartz using the Basu et al. (1975) diagram modified by Tortosa et al. (1991) for the Iberian Massif (Fig. 5) indicate that the studied sandstones plot in the areas suggesting provenance from "granites" toward the "gneisses" field.

Feldspars in Mora Formation sandstones are exclusively untwinned plagioclase (0.9%–13%, average = 5.5%; Table 1), have a pure albite composition, are nonluminescent, and show a variable degree of alteration to illite and chlorite. Albite composition is in agreement with a provenance from albitized granitic rocks, or very low rank metamorphic rocks. Mora Formation sandstones had a higher plagioclase content than the present composition because of partial dissolution or replacement by chlorite and kaolin clay minerals (2.5%–18.2%; average = 8.4%) and by calcite (<14.2%, average = 5.4%). El Castellar Formation sandstones have moderate to low content of feldspars (1.2%–9.4%, average = 4.1%; Table 1), mainly idiomorphic plagioclase with or without twinning. Most of the plagioclase grains are nonluminescent albite. However, under CL, green luminescent plagioclase grains (~5%) and blue luminescent detrital K-feldspars (~1%–2%) are present in low abundances. El Castellar Forma-

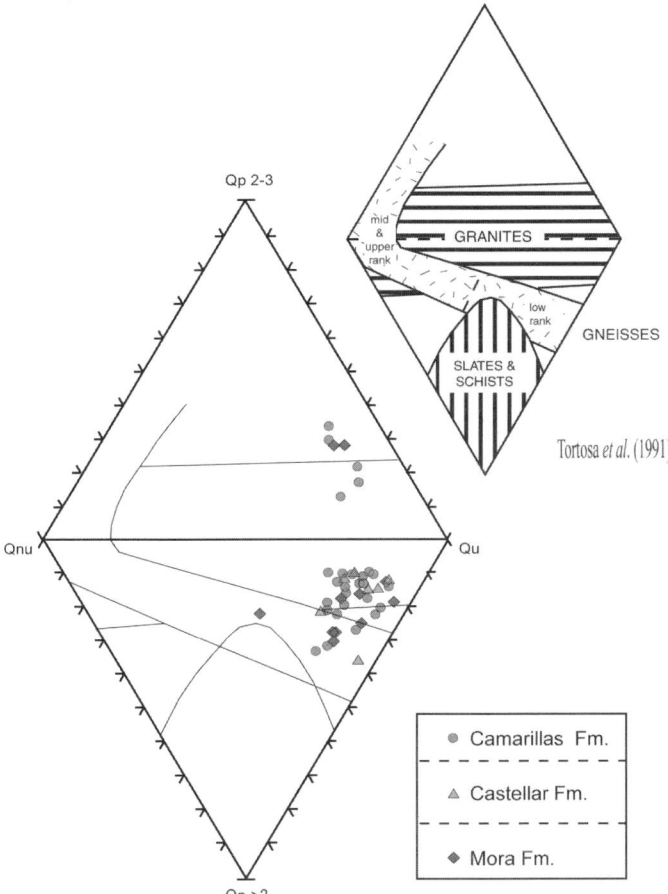

Figure 5. Quartz types plotted in the Basu et al. (1975) diagram modified by Tortosa et al. (1991) for the Iberian Massif. Qnu—nonundulatory monocrystalline quartz; Qu—undulatory monocrystalline quartz; Qp2–3—polycrystalline quartz with 2 or 3 subcrystals; Qp + 3—polycrystalline quartz with more than 3 subcrystals (including chert).

tion sandstones originally had more plagioclase, which is partially dissolved or replaced by kaolin (average = 6.5%) and by calcite (<22.4%; average = 10.7%). Camarillas Formation sandstones have abundant K-feldspar (<21.2%; average = 13.5%) and plagioclase with and without twinning (<10.9%, average = 2.4%; Table 1). K-feldspar grains occur from well-preserved to altered to illite and kaolin (epimatrix) and calcite replacement. Camarillas Formation also had more feldspar because of partial dissolution or kaolin replacement (<8.6%, average = 3%) and by calcite (<9.5%, average = 6%). Under CL, K-feldspar grains display blue luminescent colors with abundant nonluminescent albite patches. It is important to note that K-feldspar abundance in the studied sandstones increases toward the uppermost Camarillas Formation. In contrast, plagioclase is more abundant in the lower Mora Formation sandstones, and decreases toward the Camarillas Formation.

Low- and medium-grade metamorphic rock fragments are common in Mora Formation sandstones (<5.4%, average = 3%; Table 1). Ductile rock fragments deformed by mechanical compaction and transformed to pseudomatrix range from 1.3% to 7.4% (average = 4.5%). Occasionally, scarce micritic grains are present. In El Castellar Formation sandstones, clay chips are abundant and reach up to 9.9%. Medium-grade metamorphic rock fragments (up to 3%) and plutonic rock fragments (<1.9%) occur in low amounts (Table 1). Pseudomatrix derived from ductile rock fragments ranges from 1.8% to 27.4% (average = 13.8%). In Camarillas Formation sandstones, plutonic rock fragments are abundant (<8%, average = 0.7%) compared with low- and medium-grade metamorphic rock fragments (<4.2%; Table 1). Pseudomatrix ranges from 0.4% to 16.2% (average = 4.7%). In summary, low- and medium-grade metamorphic rock fragments are very common in Mora and El Castellar Formations, whereas plutonic rock fragments are dominant in the Camarillas Formation.

The assemblage of heavy minerals is made up of tourmaline, zircon, titanite, and apatite. Tourmaline is the most abundant heavy mineral up to 2.2% in the Mora Formation, <0.3% in the El Castellar Formation, and <0.6% in the Camarillas Formation. Tourmaline grains are characterized by coarse crystal size (<200 µm), high angularity, subidiomorphic shape, poor sorting, and brown to green pleochroism. These textural characteristics suggest a close proximity of the source area, fast transport, and scarce alteration. The other heavy minerals of the assemblage, i.e., zircon, titanite, and apatite, are present in amounts <1%, and they are poorly sorted and slightly rounded. Zircon and apatite have small crystal sizes (<30 µm), and titanite appears frequently disaggregated.

Whole-Rock Geochemistry

Results from whole-rock chemical analyses for major elements of the Lower Cretaceous sandstones are reported in Table 2. The chemical classification based on Fe_2O_3/K_2O versus SiO_2/Al_2O_3 ratios (Herron, 1988) corresponds to arkoses and subarkoses, which is similar to the obtained petrographic classification (Figs. 4A and 4B).

Weathering degree in the source area was estimated using geochemical indexes based on the relationships between nonmobile major components (Al^{3+}) and mobile cations (Ca^{2+}, Na^+, K^+), which are removed or mobilized during alteration and weathering processes (Nesbitt and Young, 1982). Average chemical index of alteration (CIA) values (Mora Formation = 58.2; El Castellar Formation = 51.4; Camarillas Formation = 58.3), as well as the chemical index of weathering (CIW) values (Mora Formation = 63.1; El Castellar Formation = 55.4; Camarillas Formation = 70) are comparable between the studied formations. In sandstones, the CIA and CIW suggest moderate weathering in the source area because average values are closer to fresh rocks (40) than to

TABLE 2. MAJOR-ELEMENT COMPOSITION OF REPRESENTATIVE SAMPLES OF LOWER CRETACEOUS SANDSTONES OF THE MAESTRAT BASIN (IBERIAN CHAIN)

Subbasin*	Fm.	Sample	SiO_2 (wt%)	Al_2O_3 (wt%)	Fe_2O_3 (wt%)	MnO (wt%)	MgO (wt%)	CaO (wt%)	Na_2O (wt%)	K_2O (wt%)	TiO_2 (wt%)	P_2O_5 (wt%)	LOI[†]	Total (%)
AL	Camarillas	AB.Cm.5	85.86	8.13	0.33	0.01	0.06	0.46	0.11	2.77	0.32	0.08	2.15	100.28
AL	Camarillas	AB.Cm.4	81.3	8.34	2.53	0.05	0.16	0.47	0.11	2.65	0.31	0.09	3.18	99.18
GA	Camarillas	M.Cm.11	82.39	8.44	1.68	0.07	0.22	0.33	0.61	2.29	0.37	0.1	2.68	99.19
GA	Camarillas	M.Cm.9	84.98	8.13	0.63	0.04	0.16	0.24	1.01	2.06	0.37	0.08	1.73	99.44
GA	Camarillas	M.Cm.8	81.9	8.06	0.95	0.06	0.24	1.09	1.02	2.14	0.40	0.11	3.02	98.98
PG	Camarillas	PÑ.Cm.4	54.94	5.22	2.02	0.16	0.98	17.81	0.59	1.59	0.16	0.07	16.55	100.09
PG	Camarillas	PÑ.Cm2	78.54	9.86	0.58	0.02	0.2	2.36	0.79	3.89	0.19	0.1	3.24	99.77
PG	Camarillas	MR.Cm.26	57.04	8.93	1.57	0.09	0.76	13.49	1.32	2.71	0.20	0.06	12.37	98.53
PG	Camarillas	MR.Cm.20	84.28	8.12	1.4	0.01	0.36	0.2	0.9	1.3	0.41	0.1	2.25	99.31
PG	Camarillas	MR.Cm.18	86.51	7.39	0.82	0.04	0.53	0.22	0.96	1.06	0.11	0.06	2.43	100.13
PG	Camarillas	MR.Cm.14	79.72	11.45	1.3	0.07	0.67	0.18	0.89	1.44	0.39	0.08	3.33	99.52
PG	Camarillas	MR.Cm.1	84.64	8.22	2.09	0.01	0.21	0.12	1.16	1.5	0.32	0.03	2.09	100.39
AL	Castellar	AB.Cs.0	93.62	2.85	1.02	0.04	0.08	0.43	0.01	0.21	0.43	0.02	1.49	100.2
PG	Castellar	MR.Cs.3	74.07	10.63	7.17	0.02	0.88	0.63	0.6	0.77	0.34	0.09	4.55	99.75
PG	Castellar	MR.Cs.1	91.79	3.57	1.7	0.01	0.36	0.24	0.07	0.35	0.11	0.05	1.56	99.79
PG	Mora	MR.Mo.9	87.00	6.15	0.45	0.07	0.34	1.11	0.33	0.54	0.1		3.08	100.15
PG	Mora	MR.Mo.7	85.99	6.5	0.71	0.01	0.79	0.27	0.31	0.75	0.91	0.15	2.29	98.66
PG	Mora	MR.Mo.1–2	72.15	6.62	1.53	0.13	0.84	7.71	0.19	1.18	0.90	0.06	8.12	99.42

*AL—Aliaga; GA—Galve; PG—Penyagolosa.
[†]LOI—loss on ignition.

residual sediments (100), which have undergone extreme weathering (Nesbitt and Young, 1982).

Mora sandstone petrofacies present the highest values in most of the trace elements (e.g., Y, Zr, Nb, Sn, La, Ce, Pr, Nd, Sm, Eu, Gd, Dy, Er, Tm, Yb, Lu, Hf, Ta, Th, and U) compared with El Castellar and Camarillas petrofacies (Table 3). If we compare bulk-rock trace-element and REE element abundances with the North American shale composite (NASC; Gromet et al., 1984), it can be observed that Mora petrofacies values are clearly different from Castellar and Camarillas petrofacies values (Fig. 6). However, no differences are observed between Castellar and Camarillas petrofacies values (Fig. 6).

In order to unravel provenance differences among the three defined petrofacies, the trace-element contents (including REE) have been plotted in ternary diagrams (Fig. 7). In all ternary diagrams, Rb has been selected as a pole because it is considered indicative of the K content and reflects the amount of K-feldspar abundance. The relationship between the Rb content and modal K-feldspar abundance is consistent with the high correlation index, up to 0.96, in Rb versus total feldspar modal percentage (Fig. 8). In the other two poles of the ternary diagrams, different trace and REE elements have been selected that are mainly hosted in the observed heavy mineral assemblage (Rollinson, 1993; Bea, 2001). In this way, the ternary diagram of Rb-Zr-(Y + Nb) emphasizes the abundance of zircon (Zr) over apatite (Y) and titanite (Nb) (Fig. 7A); the diagram for Rb-LREE-HREE (Fig. 7B) reflects apatite abundance (enriched in light [L] REEs) over zircon (high content in heavy [H] REEs); the Rb-ΣREE-(Ta + Nb) ternary diagram (Fig. 7C) considers the abundance of zircon (Ta) and titanite (Nb); the Rb-ΣREE-(Ti + P) ternary diagram (Fig. 7D) differentiates the three petrofacies based on the titanite (Ti) and apatite (P) abundance; and finally, the Rb-(B + Li)-(Y + Nb) diagram (Fig. 7E) allows the separation of the petrofacies due to the abundance in tourmaline (B + Li) and titanite (Nb). These ternary relationships between selected trace and REE elements agree with the three defined petrofacies, reflecting framework modal abundances (e.g., K-feldspar) and heavy mineral type and relative abundance. These element relationships (Fig. 7) allow the definition of a compositional and provenance trend.

Tourmaline Chemistry

Chemical composition of tourmalines ranges from dravite (Mg end member) to schorl (Fe end member) (Table 4). B_2O_3 versus Mg# (Mg# = Mg/Mg + Fe) cross plots (Fig. 9) show that Mg-tourmaline and Fe-tourmaline are well represented in the three defined petrofacies. B_2O_3 abundance varies slightly among analyzed tourmalines of the different formations. Tourmaline chemical compositions in Mora petrofacies have the maximum values in Mg content and the lowest Fe. This pattern is also observed in Camarillas petrofacies. Compositions of El Castellar petrofacies tourmalines are characterized by a group with high Mg content and another group with high Fe content. Tourmaline chemical compositions suggest an origin from "Li-poor granitic rocks, pegmatites and aplites" (Fig. 10; field 2 after Henry and Guidotti, 1985) and "metapelites and metapsammites with Al-saturation phase" (Fig. 10; field 4 after Henry and Guidotti, 1985). If analyzed tourmaline composition is compared with tourmaline from the possible source areas (Fig. 11), it is possible to rule out a provenance supply from calc-alkaline Lower Permian volcanic rocks of the Iberian Chain (Lago et al., 1993) because of their high Fe-tourmaline composition. However, tourmaline from granites of the Iberian Massif (Andonaegui, 1992) has Fe and Mg ranges comparable to the tourmaline analyzed in the Lower Cretaceous sandstones (Fig. 11).

DISCUSSION

During the Early Cretaceous, the marginal subbasins Aliaga, Galve, and Penyagolosa of the Maestrat Basin in the Iberian Chain (Fig. 1) acted as traps for detritus coming mainly from the N-NW and W (Fig. 2). This information has been obtained from paleocurrent data and paleogeographic reconstructions (Salas et al., 2001). Possible source rocks, outcropping in the N-NW and W of the study subbasins during the Early Cretaceous, are volcanic rocks from the Lower Permian (Lago et al., 2004a, 2004b), Upper Triassic (Lago et al., 2000), and Lower-Middle Jurassic (Martínez et al., 1997; Lago et al., 2004a), Triassic arkoses and subarkoses (Marfil et al., 1998), low-rank Precambrian metamorphic rocks and Paleozoic shales and sandstones (Bauluz et al., 2000), and several types of granitic and gneissic rocks (Villaseca and Herreros, 2001). Therefore, a typical characteristic of these intracratonic basins is a simultaneous basin infill from different sources (Weltje and von Eynatten, 2004).

Prior to any source-rock assignment, it was necessary to restore the original modal composition of the sandstones because diagenetic and weathering reactions altered the mineralogical and partly also the bulk chemical composition. This restoration of the original composition changed the modal compositions toward a more feldspathic and rock fragment–rich composition (Fig. 4). Based on these restored modal analyses and the heavy mineral assemblage, the following diagnostic compositional differences have been established among the three Lower Cretaceous formation sandstones (Fig. 12). There is (1) a dominant metamorphic quartz with brown luminescence (Ramseyer et al., 1988) in the Mora and El Castellar petrofacies but a dominance of plutonic quartz with violet luminescence (Ramseyer et al., 1988) in the Camarillas petrofacies, (2) a dominant occurrence of low- and medium-rank metamorphic rock fragments in the Mora and El Castellar petrofacies but plutonic rock fragments in the Camarillas petrofacies, (3) exclusive presence of albitized plagioclase in Mora petrofacies and dominant K-feldspar abundance in the Camarillas petrofacies, and (4) a heavy mineral assemblage with tourmaline, zircon, titanite, and apatite and dominantly tourmaline in the Mora petrofacies.

Therefore, petrological analyses point to an important low- and medium-rank metamorphic source during Mora and

TABLE 3. MINOR- AND TRACE-ELEMENT (RARE EARTH ELEMENT [REE]) COMPOSITION

Formation:					Camarillas								Castellar			Mora		
Subbasin:	AL	AL	GA	GA	GA	PG	PG	PG	PG	PG	PG	PG	AL	PG	PG	PG	PG	PG
Sample:	AB.Cm.5	AB.Cm.4	M.Cm.11	M.Cm.9	M.Cm.8	PÑ.Cm.7	PÑ.Cm.2	MR.Cm.26	MR.Cm.20	MR.Cm.18	MR.Cm.14	MR.Cm.1	AB.Cs.0	MR.Cs.3	MR.Cs.1	MR.Mo.9	MR.Mo.7	MR.Mo.1–2
Element (ppm)																		
B	66	n.a.	62	n.a.	83	n.a.	n.a.	n.a.	n.a.	n.a.	n.a.	37	86	36	20	232	148	84
Li	21	n.a.	18	n.a.	16	n.a.	n.a.	n.a.	n.a.	n.a.	n.a.	17	7	67	21	24	31	27
Sr	54	131	47	40	54	79	48	94	116	26	56	55	26	50	29	21	22	84
Y	13	24	13	11	12	7	9	19	13	5	14	8	18	12	6	26	26	30
Sc	2	4	3	2	2	1	3	2	2	2	4	2	<d.l.	5	<d.l.	1	2	3
Be	1	3	2	1	1	1	2	2	2	1	2	1	1	2	1	1	1	2
V	13	23	21	11	14	6	12	15	17	9	21	12	19	21	15	13	14	29
Co	<d.l.	58	28	3	3	<d.l.	1	2	4	4	5	<d.l.	1	6	1	4	2	3
Zn	7	203	<d.l.	<d.l.	<d.l.	4	<d.l.	<d.l.	<d.l.	6	47	48	<d.l.	60	<d.l.	<d.l.	<d.l.	<d.l.
Ga	1.0	8	7	7	7	0.7	10	8	8	6	12	9	3	12	4	6	6	7
Ge	<d.l.	1.6	1.1	1.3	1.2	<d.l.	1.5	0.9	1.5	1.0	1.4	1.3	1.2	2.2	2.4	1.5	1.5	1.2
As	101	57	53	<d.l.	<d.l.	<d.l.	<d.l.	<d.l.	<d.l.	<d.l.	<d.l.	14	<d.l.	<d.l.	8	10	<d.l.	<d.l.
Rb	193	106	90	84	89	60	169	115	72	47	83	66	18	59	19	25	32	58
Zr	5.0	119	158	123	168	81	60	145	159	45	112	104	287	86	32	390	344	404
Nb	<d.l.	4.7	5.9	5.1	6.3	2.8	4.3	3.9	5.9	2.5	6.7	5.3	5.8	7.0	2.9	18.1	12.5	12.1
Mo	5	<d.l.	<d.l.	<d.l.	<d.l.	<d.l.	<d.l.	<d.l.	<d.l.	147	<d.l.	4	<d.l.	<d.l.	<d.l.	<d.l.	<d.l.	5
Sn	0.4	4	5	3	3	2	5	3	4	4	6	5	3	7	1	13	8	4
Sb	4.1	2.3	1.3	0.4	0.3	0.3	0.3	<d.l.	0.5	0.3	0.3	0.3	0.6	0.5	0.4	0.6	0.6	0.8
Cs	420	4.7	4.2	4.5	4.2	2.6	6.3	5.4	4.5	1.8	5.5	2.1	2.6	4.1	1.4	1.5	2.0	4.2
Ba	29.1	271	249	226	232	98	279	1120	3750	229	311	123	322	225	90	163	541	2230
La	62.3	15.9	22.5	15.7	20.0	12.3	13.1	17.8	20.0	6.9	22.4	19.6	24.4	17.1	13.1	42.9	38.8	42.4
Ce	6.49	39.3	48.1	35.5	45.7	26.4	26.6	40.0	45.3	14.8	45.8	38.1	57.1	37.4	28.3	88.1	81.9	87.8
Pr	23.9	4.69	5.13	3.87	5.10	2.77	3.23	5.31	5.29	1.53	5.18	4.39	6.36	4.41	3.16	10.8	8.86	10.4
Nd	4.85	18.0	19.3	13.2	17.3	9.71	12.0	20.5	18.7	5.64	19.2	16.0	23.8	16.4	11.3	42.8	33.0	41.2
Sm	0.767	4.22	4.22	2.81	3.27	1.96	2.20	4.36	3.63	1.33	3.51	2.71	5.05	3.31	2.00	8.45	7.02	7.94
Eu	3.49	1.21	0.784	0.621	0.652	0.443	0.582	0.902	0.592	0.358	0.725	0.555	1.01	0.842	0.444	1.38	1.18	1.22
Gd	0.46	5.32	3.15	2.53	2.98	1.63	2.22	4.70	3.73	1.18	3.21	2.11	4.13	3.45	1.73	7.26	5.25	7.64
Tb	2.48	0.68	0.46	0.35	0.39	0.22	0.30	0.63	0.44	0.17	0.46	0.25	0.56	0.45	0.22	0.96	0.74	1.10
Dy	0.46	3.33	2.61	1.98	2.11	1.26	1.45	3.00	2.22	0.96	2.40	1.35	3.19	2.13	1.13	4.87	4.39	5.67
Ho	1.22	0.64	0.48	0.40	0.43	0.25	0.27	0.60	0.46	0.18	0.46	0.27	0.63	0.42	0.23	0.90	0.86	1.05
Er	0.190	1.71	1.37	1.10	1.20	0.72	0.82	1.72	1.35	0.47	1.45	0.87	1.67	1.19	0.69	2.75	2.41	3.16
Tm	1.34	0.230	0.212	0.157	0.184	0.103	0.118	0.247	0.201	0.072	0.223	0.134	0.252	0.174	0.105	0.432	0.394	0.511
Yb	0.206	1.36	1.38	1.04	1.16	0.67	0.73	1.49	1.23	0.48	1.34	0.78	1.72	1.04	0.63	2.61	2.61	2.98
Lu	4.7	0.210	0.219	0.179	0.199	0.109	0.111	0.221	0.194	0.079	0.212	0.138	0.283	0.154	0.095	0.396	0.413	0.458
Hf	1.05	2.9	3.6	2.9	3.9	1.9	1.6	3.9	3.9	1.1	2.9	2.9	6.6	2.2	0.9	9.9	8.2	10.3
Ta	1.5	0.56	0.89	0.73	0.75	0.40	0.62	0.47	0.72	0.36	0.95	0.69	0.80	0.90	0.21	4.73	2.24	1.58
W	0.67	1.5	2.2	1.7	1.6	1.2	1.6	1.1	1.8	0.8	2.4	1.4	1.9	2.3	0.5	4.3	3.1	3.0
Tl	11	0.92	0.55	0.69	0.61	0.40	1.32	0.65	0.29	0.33	0.56	0.35	0.12	0.27	0.14	0.29	0.22	0.32
Pb	0.1	24	24	18	8	9	17	9	<d.l.	<d.l.	5	31	<d.l.	7	<d.l.	112	5	9
Bi	10.8	0.2	0.1	0.1	<d.l.	<d.l.	<d.l.	0.1	0.2	<d.l.	<d.l.	0.2	0.2	0.2	<d.l.	42.9	0.3	0.2
Th	2.86	5.56	7.94	6.50	7.82	4.37	4.42	7.55	7.13	2.83	7.33	5.59	6.82	5.49	3.75	14.7	12.4	14.7
U	11,497	4.77	7.23	2.13	2.53	1.68	1.87	1.78	1.78	1.06	2.24	3.36	1.42	1.80	1.02	8.25	5.55	3.89
K*	1912	10,999	9505	8550	8883	6600	16,146	11,248	5396	4400	5977	6226	872	3196	1453	2241	3113	4898
Ti*	175	1834	2230	2242	2374	977	1121	1175	2446	647	2326	1912	2602	2044	671	5959	5449	5366
P*		196	218	175	240	153	218	131	218	131	175	65	44	196	109	218	327	131

Note: AL—Aliaga; GA—Galve; PG—Penyagolosa; <d.l.—below detection limit; n.a.—not available.
*Calculated from wt%.

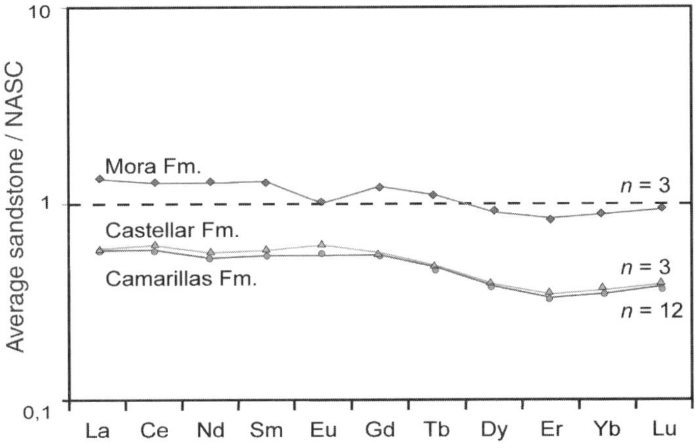

Figure 6. Trace-element and rare earth element (REE) average values of sandstones compared to the North American shale composition (NASC; Gromet et al., 1984). Note that Mora sandstones present the highest trace-element and REE abundances.

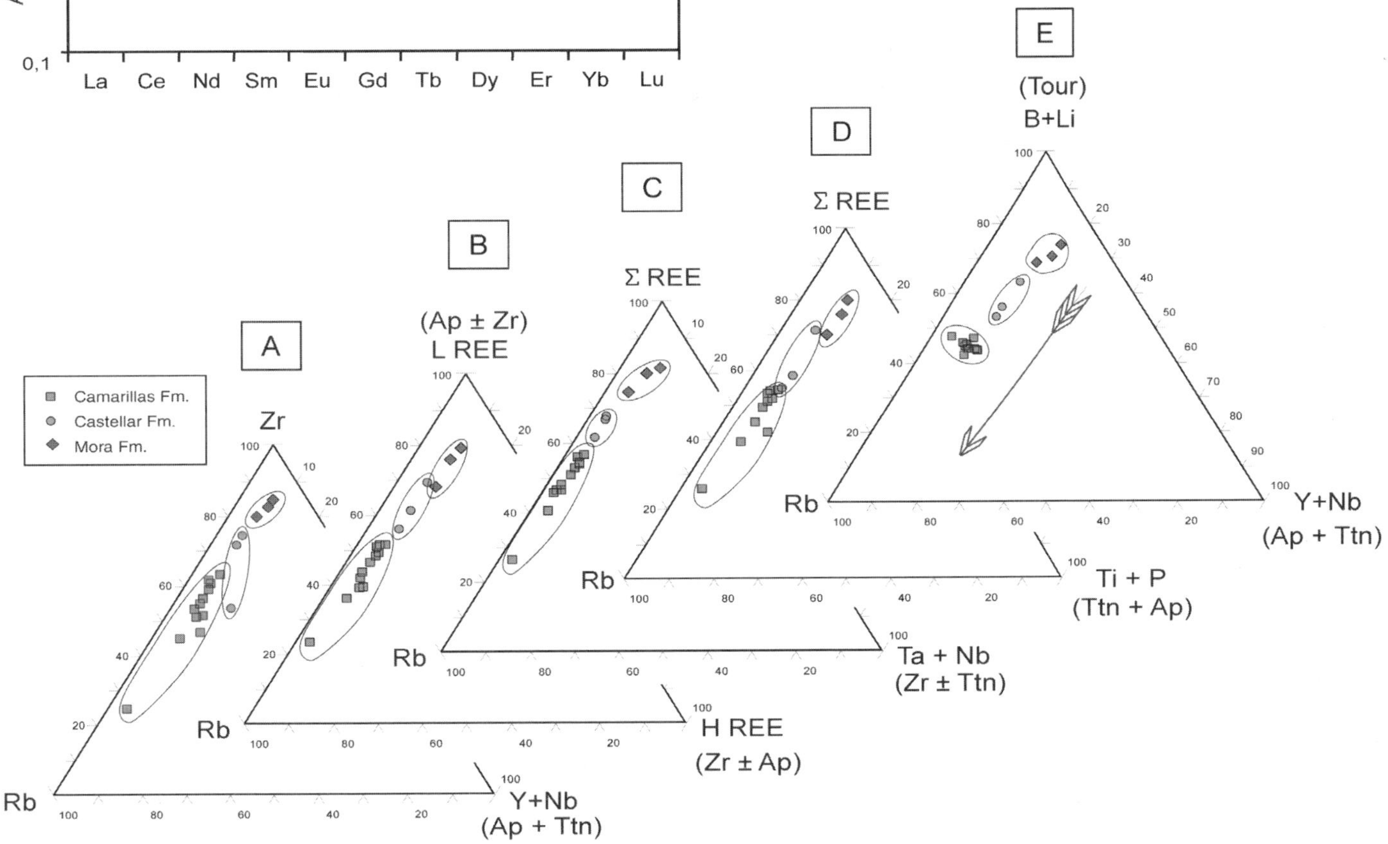

Figure 7. Relationships among selected trace elements (including rare earth elements [REEs]) based on their contribution to the heavy mineral assemblages and to the feldspar modal abundance: (A) zircon, apatite, and titanite; (B) zircon versus apatite; (C) titanite and zircon; (D) apatite and titanite; (E) tourmaline, apatite, and titanite. L—light, H—heavy.

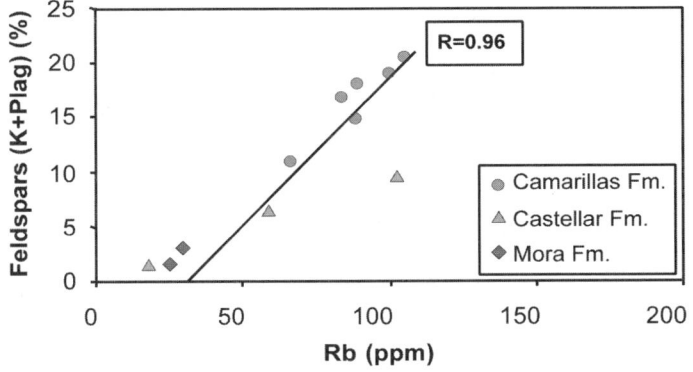

Figure 8. Rb versus modal feldspar abundance. Note the high correlation between both parameters.

TABLE 4. CHEMICAL COMPOSITION OF DETRITAL TOURMALINE

Formation:	Mora			Castellar			Camarillas		
Subbasin:*	PG	PG	PG	PG	PG	PG	AL	GA	PG
Sample:	MR.Mo.1–2	MR.Mo.6	MR.Mo.3	CA.Cs.4	CA.Cs.1	MR.Cs.1	AB.Cm.7	M.Cm.8	MR.Cm.33
Analyses (wt%)									
SiO_2	35.86	34.78	37.24	35.51	35.85	36.11	35.06	35.73	36.23
B_2O_3	10.39	10.07	10.79	10.28	10.38	10.46	10.16	10.35	10.49
Al_2O_3	34.40	32.97	33.76	33.68	35.65	33.17	33.88	34.41	33.54
Cr_2O_3	0.01	0.05	0.05	0	0.04	0	0	0.02	0.02
TiO_2	0.46	1.33	0.51	0.64	0.36	0.85	0.76	0.55	0.87
FeO	12.59	9.99	4.33	11.93	7.09	7.31	13.25	8.65	6.83
MnO	0.17	0.06	0.08	0.20	0.03	0.04	0.12	0.12	0.02
MgO	1.24	4.03	7.42	2.44	4.72	5.63	1.17	3.93	5.68
NiO	0	0.03	0	0.05	0	0	0.02	0	0.01
CaO	0.29	0.77	0.47	0.22	0.36	0.24	0.13	0.25	0.71
Na_2O	1.80	1.82	2.15	1.91	1.67	2.19	1.78	1.87	1.63
K_2O	0.02	0.05	0.02	0.02	0.01	0.05	0.04	0.03	0.03
P_2O_5	0.01	0.04	0	0.02	0	0	0	0	0.03
F	0.68	0.37	0.41	0.57	0.12	0.34	0.57	0.27	0.13
Cl	0	0	0.01	0.01	0	0	0.03	0.01	0
Total	97.63	96.19	97.06	97.25	96.24	96.24	96.71	96.08	96.17
(O = F)	0.29	0.15	0.17	0.24	0.05	0.14	0.24	0.11	0.06
Total(O = F)	97.34	96.04	96.89	97.01	96.19	96.10	96.47	95.97	96.11
Si	6.18	5.99	6.41	6.12	6.18	6.22	6.04	6.15	6.24
B	3.09	2.99	3.21	3.06	3.09	3.11	3.02	3.08	3.12
Al^{IV}	5.73	6.01	5.38	5.83	5.74	5.67	5.94	5.77	5.64
Sum T	15	15	15	15	15	15	15	15	15
Al^{VI}	1.25	0.68	1.47	1.01	1.50	1.06	0.94	1.22	1.17
Ti	0.06	0.17	0.07	0.08	0.05	0.11	0.10	0.07	0.11
Cr	0	0.01	0.01	0	0.01	0	0	0	0
Mg	0.32	1.03	1.90	0.63	1.21	1.45	0.30	1.01	1.46
Fe^{2+}	1.81	1.44	0.62	1.72	1.02	1.05	1.91	1.25	0.98
Mn	0.03	0.01	0.01	0.03	0	0.01	0.02	0.02	0
Ni	0	0	0	0.01	0	0	0	0	0
Ca	0.05	0.14	0.09	0.04	0.07	0.04	0.02	0.05	0.13
Na	0.60	0.61	0.72	0.64	0.56	0.73	0.59	0.63	0.54
K	0.01	0.01	0	0.01	0	0.01	0.01	0.01	0.01
P	0	0.01	0	0	0	0	0	0	0
Sum M	4.12	4.10	4.90	4.16	4.42	4.46	3.89	4.24	4.41
F	0.37	0.20	0.22	0.31	0.07	0.18	0.31	0.15	0.07
Cl	0	0	0	0	0	0	0.01	0	0
Total	19.50	19.30	20.12	19.47	19.48	19.65	19.21	19.39	19.49
Mg#[†]	0.15	0.42	0.75	0.26	0.54	0.58	0.13	0.44	0.60

*AL—Aliaga; GA—Galve; PG—Penyagolosa.
[†]Mg# = (Mg/Mg + Fe).

El Castellar Formation sedimentation but suggest a dominant supply of plutonic components during Camarillas Formation sedimentation. Contributions of volcanic and sedimentary rocks are below or at trace quantity, respectively (Table 1). Source areas provided metamorphic and plutonic rocks at the same time during Mora and El Castellar Formation deposition. Thus, the petrological approach allows a very good discrimination of major rock types present in the source areas (Grantham and Velbel, 1988). However, it must be considered that modal analyses are characterized by methodological errors in the quantification of the different categories, and petrographic observations, by the nonobjective criteria among different petrographers (Weltje, 2002).

The observed heavy mineral assemblage of tourmaline, zircon, titanite, and apatite may be associated not only to acid-intermediate granitic rocks, but also to metamorphic rocks and even to mature siliciclastic sediments (von Eynatten and Gaupp, 1999; Torres-Ruiz et al., 2003). Textural characteristics of zircon, titanite, and apatite (e.g., roundness) suggest a high transport distance and maybe recycling processes. In contrast, textural features of tourmaline (e.g., high angularity) are consistent with a relatively proximal source area and/or a first-cycle origin from crystalline rocks. Thus, petrological heavy mineral analyses are consistent with metamorphic and granitic source areas. Moreover, the chemical composition of tourmaline agrees with a provenance from granitic rocks (cf. Iberian Massif; Fig. 11). Still, there are no data available for metamorphic tourmalines of the Iberian Massif, but the chemical compositions shown in Figure 10 indicate, after Henry and Guidotti (1985), for some tourmalines, a metamorphic origin. In addition, bulk-rock and single-grain chemical analyses (Fig. 11) rule out the delivery of accountable detritus from the nearby outcropping Permian to Middle Jurassic volcanic rocks.

Selected trace-element and REE element relationships as shown on Figures 7A–7E confirm the formerly defined petrofacies and the observed compositional trend, whereby Mora petrofacies displays the lowest Rb contents due to the absence of K-feldspar. In the studied sandstones, the Rb content not only shows a good positive correlation with K-feldspar, but also with K_2O (Tables 1 and 2). This suggests that Rb is closely related to feldspars and not to shales because of the low correlation index (0.50) with Al_2O_3 (Bauluz et al., 2000). The exclusive occurrence

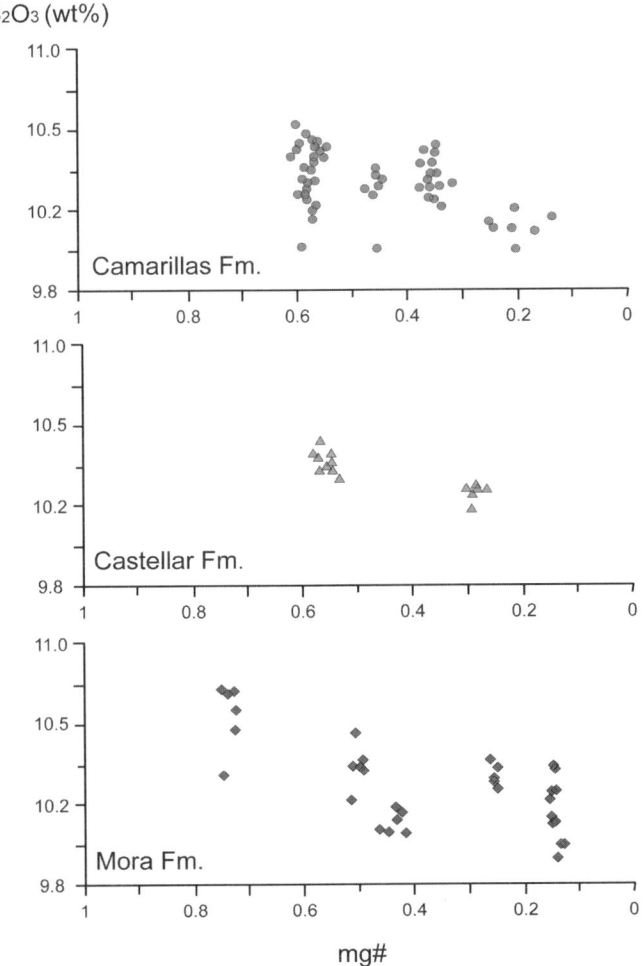

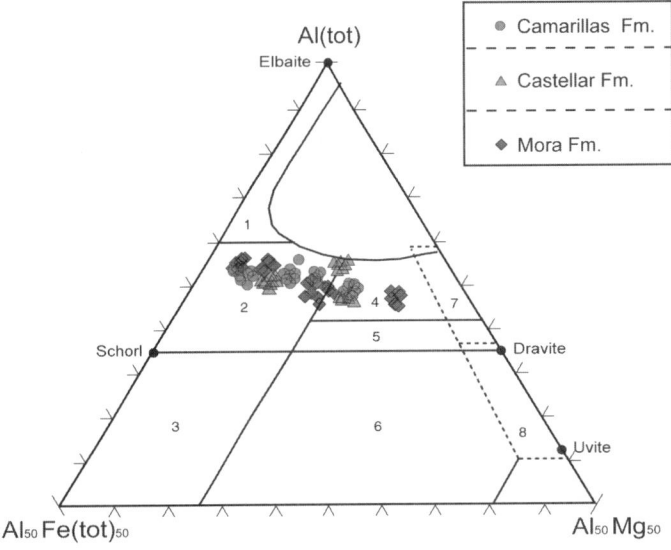

Figure 10. Chemical composition of the analyzed tourmaline. Diagram and fields after Henry and Guidotti (1985): (1) Li-rich granitoid pegmatites and aplites; (2) Li-poor granitoids rocks, pegmatites, and aplites; (3) Fe^{3+}-rich quartz-tourmaline rocks; (4) metapelites and metapsammites with an Al-saturation phase; (5) metapelites and metapsammites; (6) Fe^{3+}-rich quartz-tourmaline rocks, calc-silicate rocks, and metapelites; (7) low-Ca meta-ultramafics and Cr- and V-rich metasediments; (8) metacarbonates and metapyroxenites.

Figure 9. B_2O_3 versus mg# (mg# = Mg/Mg + Fe) cross plot of the analyzed tourmaline.

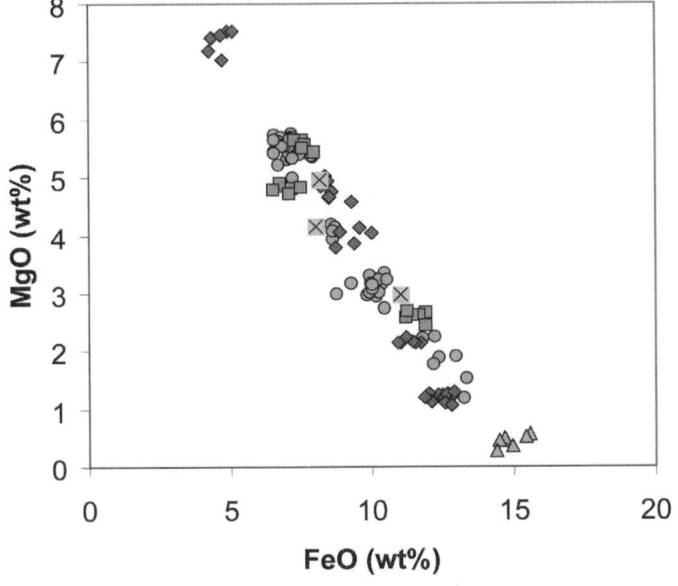

Figure 11. MgO versus FeO cross plot of the analyzed tourmaline, with tourmaline composition of two possible source areas, the calc-alkaline Lower Permian volcanic rocks of the Iberian Chain (Lago et al., 1993) and the granites of the Iberian Massif (Andonaegui, 1992).

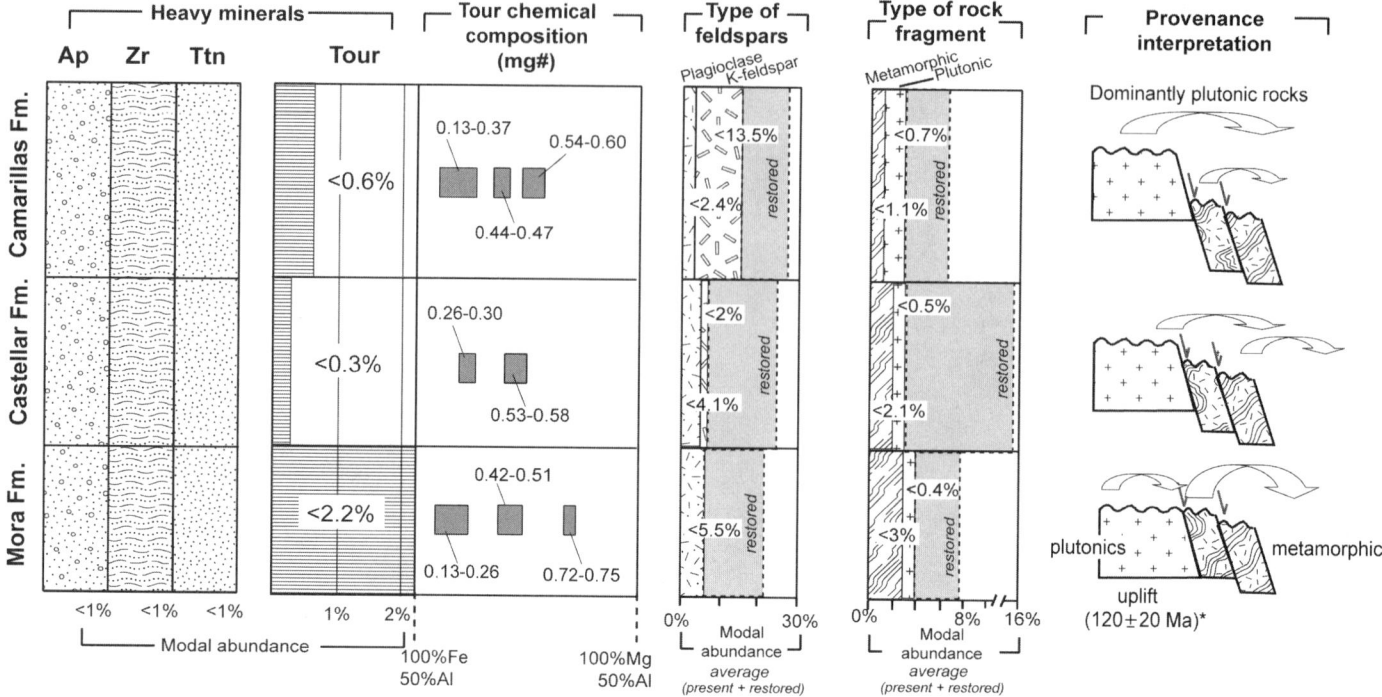

Figure 12. Synthesis of the type and heavy minerals modal abundance, tourmaline chemistry, and type and modal abundance of feldspars and rock fragments (considering diagenetic effects). Provenance interpretation involves the uplift and erosion event of the Iberian Massif during synrift Lower Cretaceous sandstones sedimentation in the Maestrat Basin (*Bruijne and Andriessen, 2000; Barbero et al., 2005). Ap—apatite, Zr—zircon, Ttn—titanite, Tour—tourmaline. mg# (mg# = Mg/Mg + Fe).

of albite in Mora petrofacies can be explained by two processes: (1) either a provenance from rocks with exclusive Na-plagioclase, and/or (2) a diagenetic alteration effect of feldspars during burial of the sandstones. The first possibility is not in agreement with the potential source rocks because pure albite is only found in very few granitic rocks from the Iberian Massif (Pérez-Soba, 1992). Therefore, the diagenetic alteration of feldspars is more plausible (Caja et al., 2003). However, the precursor of albite is not clear. It could be K-feldspar, which is common in granitic rocks and is diagenetically albitized after plagioclase (Saigal et al., 1988; Morad et al., 1990), or it could be Ca-plagioclase, which is common in low-rank metamorphic rocks (Trevena and Nash, 1981) and is prone to albitization during burial above 100 °C (Ramseyer et al., 1992; Milliken, 2003).

Moreover, the REE and trace elements reflect the heavy mineral type and abundance (Chang et al., 1996; Deer et al., 1982, 1997). The proposed relationships shown in Figure 7 have proved to be useful for distinguishing the Mora Formation in comparison with El Castellar and Camarillas Formations. The fact that the three petrofacies have comparable low chemical alteration (CIA) and weathering (CIW) indexes suggests that the compositional differences observed between the three petrofacies may be caused by diagenesis and/or by differences in the nature of the source area rocks. Thus, these element relationships are appropriate for provenance studies of sandstones with original heavy mineral assemblages.

Mora sandstones correspond to the first tilting movements of the Penyagolosa subbasin and were deposited only in this subbasin. Therefore, Mora petrofacies record the reactivation of the Late Jurassic–Early Cretaceous rifting stage that slowed during the Neocomian age. Thus, Mora petrofacies reflect the erosion of the Paleozoic medium- and low-grade metamorphic rocks from the Iberian Massif (Figs. 2 and 12). An alternative source of low-rank metamorphic rocks is the Precambrian and Paleozoic shales and sandstones exposed in the Aragonese branch of the Iberian Chain (Bauluz et al., 2000), which is located at the NW of the studied subbasins. Light and heavy REE abundances in the source rocks are comparable to studied sandstones. However, the petrofacies analyses do not reflect a provenance from slates and schists. Castellar petrofacies, which is only well developed in the Penyagolosa subbasin, reflects the progressive change from isolated subbasins (sedimentation of Mora Formation) to regional tectonic subsidence affecting the whole Maestrat Basin (Salas et al., 2001). During Camarillas Formation sedimentation, the tectonic subsidence was acting in the whole Maestrat Basin, and Camarillas sandstones are well developed with a significant thickness in all the western margin subbasins (i.e., Aliaga, Galve, and Penyagolosa). Thus, Camarillas petrofacies recorded a generalized erosion stage of plutonic rocks from the Iberian Massif that crop out in the west side of the studied subbasins (Figs. 2 and 12).

The Iberian Massif uplift and erosion event has been characterized by apatite fission-track analyses, which have dated an important cooling stage for the Mesozoic around 120 ± 20 Ma (Bruijne and Andriessen, 2000; Barbero et al., 2005). The most reasonable explanation is that the Iberian Massif uplift occurred when the major D3 regional synrift unconformity was developed during latest Berriasian times and affected all of the Iberian intraplate basins (Salas et al., 2001). In the Maestrat Basin, this uplift would be represented by a progressively major influence of granitic source rocks toward the uppermost Camarillas Formation (Fig. 12). A comparable provenance source trend and petrofacies have been reported for the Cameros intraplate rift basin (Iberian Peninsula). It developed during the same rift stage as the Maestrat Basin (Late Jurassic–Early Cretaceous). Arribas et al. (2003) characterized one quartz-feldspathic petrofacies with P/F >1 at the base of the Tithonian-Berriasian. These rocks derived from the erosion of low- to medium-grade metamorphic terranes of the West Asturian–Leonese Zone of the Iberian Massif. They characterized another quartz-feldspathic petrofacies (Barremian to early Albian) that had P/F near zero and a very low concentration of metamorphic rock fragments. The latter derived from the erosion of coarse crystalline plutonic rocks located in the Central Iberian Zone of the Iberian Massif. These similarities in petrofacies support the availability of the Iberian Massif as source area in both Cameros and Maestrat rift basins and the progressive contribution of granitic rocks. Differences between Cameros and Maestrat petrofacies and provenance interpretation could be due to the fact that the rifting stage started earlier in the Maestrat Basin (Salas et al., 2001). Moreover, the Cameros Basin experienced more subsidence than any other basin in the Iberian Rift system (Arribas et al., 2003).

CONCLUSIONS

Modal analyses together with petrographic and geochemical data allow us to define a specific petrofacies for each of the three Lower Cretaceous synrift sandstones (i.e., Mora, El Castellar, and Camarillas Formations) in the eastern Iberian Chain, Spain.

Trace-element and REE abundances of whole-rock analyses discriminate well among the three petrofacies based on: (1) the Rb concentration, which is indicative of the K content and reflects the amount of K-feldspar modal abundance, and (2) the relative modal abundance of heavy minerals (tourmaline, zircon, titanite, and apatite), which is reproduced by the elements hosted in the observed heavy mineral assemblage (i.e., B and Li for tourmaline; Zr, Hf, and Ta for zircon; Ti, Ta, Nb, and their REEs for titanite; and P, Y, and their REEs for apatite).

The three petrofacies defined are consistent with a mixed provenance from plutonic and metamorphic source rocks. No major contributions of volcanic and sedimentary rocks were detected. The uppermost Camarillas petrofacies reflects a dominant stage of plutonic source rocks, which is consistent with the uplift and erosion of granites from the Iberian Massif. This implies a significant dilution of the Paleozoic metamorphic sources, which was dominant during the sedimentation of the Mora and El Castellar petrofacies.

Linking petrographic data (modal analyses, heavy mineral assemblages, CL of quartz) with whole-rock geochemical compositions (major, minor, trace, and REE), including tourmaline chemical analysis has proved to be useful for determining source areas characteristics, their predominance, and the evolution of source rock types during deposition of quartz-feldspathic sandstones in intraplate basins. This approach provides a high reliability for source area characterization and evolution, and it is consistent with the geological and geodynamic context.

ACKNOWLEDGMENTS

Funding was provided by research project DGICYT BTE2000-0574-C03-02 to R. Marfil, and program Juan de la Cierva from the Spanish Ministry of Education and Science to M.A. Caja. Comments and suggestions by S. Critelli, G.J. Weltje, and S.A. Marenssi were of great help to improve the manuscript. Special thanks are due to the thin section preparation laboratory of the Departamento de Petrología y Geoquímica (Universidad Complutense de Madrid). A. Fernández-Larios is thanked for his assistance with the electron microprobe.

REFERENCES CITED

Andonaegui, P., 1992, Geoquímica y geocronología de los granitoides del sur de Toledo: [Ph.D. thesis]: Madrid, Universidad Complutense de Madrid, 366 p.

Armstrong-Altrin, J.S., and Verma, S.P., 2005, Critical evaluation of six tectonic setting discrimination diagrams using geochemical data of Neogene sediments from known tectonic settings: Sedimentary Geology, v. 177, p. 115–129, doi: 10.1016/j.sedgeo.2005.02.004.

Arribas, J., Alonso, A., Mas, R., Tortosa, A., Rodas, M., Barrenechea, J.F., Alonso-Azcárate, J., and Artigas, R., 2003, Sandstone petrography of continental depositional sequences of an intraplate rift basin: Western Cameros basin (north Spain): Journal of Sedimentary Research, v. 73, p. 309–327.

Barbero, L., Glasmacher, U.A., Villaseca, C., López, J.A., and Martín-Romera, C., 2005, Long-term thermo-tectonic evolution of the Montes de Toledo area (Central Hercynian belt, Spain): Constraints from apatite fission-track analysis: International Journal of Earth Sciences, v. 94, p. 193–203, doi: 10.1007/s00531-004-0455-y.

Basu, A., 1976, Petrology of Holocene fluvial sand derived from plutonic source rocks: Implications to paleoclimatic interpretation: Journal of Sedimentary Petrology, v. 46, p. 694–709.

Basu, A., Young, S.W., Suttner, L.J., James, W.C., and Mack, G.H., 1975, Re-evaluation of the use of undulatory extinction and polycrystallinity in detrital quartz for provenance interpretation: Journal of Sedimentary Petrology, v. 45, p. 873–882.

Bauluz, B., Mayayo, M.J., Fernández-Nieto, C., and Gonzalez, J.M., 2000, Geochemistry of Precambrian and Paleozoic siliciclastic rocks from the Iberian Range (NE Spain): Implications for source-area weathering, sorting, provenance, and tectonic setting: Chemical Geology, v. 168, p. 135–150, doi: 10.1016/S0009-2541(00)00192-3.

Bea, F., 2001, The influence of accessory minerals on the geochemistry of granite rocks, in Lago, M., Arranz, E., and Galé, C., eds., Abstract volume (Volumen de Actas del Congreso): Zaragoza, Spain, III Congreso Ibérico de Geoquímica y VIII° Congreso de Geoquímica de España, p. 17–33.

Bhatia, M.R., and Crook, K.A.W., 1986, Trace element characteristics of graywackes and tectonic setting discrimination of sedimentary basins: Contributions to Mineralogy and Petrology, v. 92, p. 181–193, doi: 10.1007/BF00375292.

Bruijne, C.H., and Andriessen, P.A., 2000, Interplay of intraplate tectonics and surface processes in the Sierra de Guadarrama (central Spain) assessed by apatite fission track analysis: Physics and Chemistry of the Earth, v. 25, p. 555–563, doi: 10.1016/S1464-1895(00)00085-5.

Caja, M.A., 2004, Procedencia y diagénesis de los sedimentos del Jurásico superior–Cretácico inferior (facies Weald) en las subcuencas occidentales de la Cuenca del Maestrazgo, Cordillera Ibérica Oriental [Ph.D. thesis]: Madrid, Universidad Complutense de Madrid, 293 p.

Caja, M.A., Marfil, R., and Salas, R., 2003, Procesos de albitización en feldespatos detríticos de las areniscas del Cretácico inferior (subcuencas occidentales de la Cuenca del Maestrazgo), in Neiva, A.M.R., Neves, L.J.P.F., Silva, M.M.V.G., and Gomes, E.M.C., eds., Abstract volume: Coimbra, Portugal, IV Congresso Ibérico de Geoquímica y XIII Semana de Geoquímica, p. 171–173.

Chang, L.L.Y., Howie, R.A., and Zussman, J., 1996, Non-silicates: Sulphates, carbonates, phosphates, halides: Rock-Forming Minerals, v. 5B, p. 297–335.

Chayes, F., 1952, Notes on the staining of potash feldspar with sodium cobaltinitrite in thin section: The American Mineralogist, v. 37, p. 337–340.

Deer, W.A., Howie, R.A., and Zussman, J., 1982, Orthosilicates: Rock-Forming Minerals, v. 1A, p. 418–466.

Deer, W.A., Howie, R.A., and Zussman, J., 1997, Disilicates and ring silicates: Rock-Forming Minerals, v. 1B, p. 559–603.

Dickinson, W.R., 1985, Interpreting provenance relations from detrital modes of sandstones, in Zuffa, G.G., ed., Provenance of Arenites: NATO Advanced Study Institutes Series, v. 148, p. 333–361.

Dickinson, W.R., Beard, L.S., Brakenridge, G.R., Erjavec, J.L., Ferguson, R.C., Inman, K.F., Knepp, R.A., Lindberg, F.A., and Ryberg, P.T., 1983, Provenance of North American Phanerozoic sandstones in relation to tectonic setting: Geological Society of America Bulletin, v. 94, p. 222–235, doi: 10.1130/0016-7606(1983)94<222:PONAPS>2.0.CO;2.

Grantham, J.H., and Velbel, M.A., 1988, The influence of climate and topography on rock-fragments abundance in modern fluvial sands of the southern Blue Ridge Mountains, North Carolina: Journal of Sedimentary Petrology, v. 58, p. 219–227.

Gromet, L.P., Dymek, R.F., Haskin, L.A., and Korotev, R.L., 1984, The 'North American shale composite': Its compilation, major and trace element characteristics: Geochimica et Cosmochimica Acta, v. 48, p. 2469–2482, doi: 10.1016/0016-7037(84)90298-9.

Hahne, C., 1930, La Cadena Celtibérica al este de la línea Cuenca-Teruel-Alfambra (Translation for San Miguel de la Cámara): Publicaciones Alemanas sobre Geología de España, v. II, p. 7–50.

Henry, D.J., and Guidotti, C.V., 1985, Tourmaline as a petrogenetic indicator mineral: An example from the staurolite-grade metapelites of NW Maine: The American Mineralogist, v. 70, p. 1–15.

Herron, M.M., 1988, Geochemical classification of terrigenous sands from core or log data: Journal of Sedimentary Petrology, v. 58, p. 820–829.

Lago, M., Auqué, L., Arranz, E., Gil, A., and Pocovi, A., 1993, Caracteres de la fosa de Bronchales (Stephaniense-Pérmico) y de la turmalinización asociada a las riolitas calco-alcalinas (Provincia de Teruel): Cuadernos del Laboratorio Xeolóxico de Laxe, v. 18, p. 65–79.

Lago, M., Galé, C., Arranz, E., Gil, A., Pocovi, A., and Vaquer, R., 2000, The Triassic alkaline dolerites of the Valacloche-Camarena (SE Iberian Chain, Teruel): Geodynamic implications: Estudios Geológicos, v. 56, p. 211–228.

Lago, M., Arranz, E., Gil, A., and Pocovi, A., 2004a, Magmatismo asociado (Cordilleras Ibérica y Costero-Catalana), in Vera, J.A., ed., Apartado 5.4: Madrid, Geología de España (SGE)/Instituto Geológico y Minero de España (IGME), p. 522–525.

Lago, M., Arranz, E., Pocovi, A., Galé, C., and Gil-Imaz, A., 2004b, Lower Permian magmatism of the Iberian Chain, central Spain, and its relationship to extensional tectonics, in Wilson, M., Neumann, E.-R., Davies, G.R., Timmerman, M.J., Heeremans, M., and Larsen, B.T., eds., Permo-Carboniferous Magmatism and Rifting in Europe: Geological Society [London] Special Publication 223, p. 465–490.

Lindholm, R.C., and Finkelman, R.B., 1972, Calcite staining: Semiquantitative determination of ferrous iron: Journal of Sedimentary Petrology, v. 44, p. 428–440.

Marfil, R., Hall, A., García-Gil, S., and Stamatakis, M.G., 1998, Petrology and geochemistry of diagenetically altered tuffaceous rocks from the middle Triassic of central Spain: Journal of Sedimentary Research, v. 68, p. 391–403.

Martínez, R.M., Lago, M., Valenzuela, J.I., Vaquer, R., Salas, R., and Dumitrescu, R., 1997, El volcanismo Triásico y Jurásico del sector SE de la Cadena Ibérica y su relación con los estadios de rift Mesozoicos: Boletín Geológico y Minero, v. 108, p. 367–376.

McLennan, S.M., and Taylor, S.R., 1991, Sedimentary rocks and crustal evolution revisited: Tectonic setting and secular trends: The Journal of Geology, v. 99, p. 1–22.

McLennan, S.M., Hemming, S., McDaniel, D.K., and Hanson, G.N., 1993, Geochemical approaches to sedimentation, provenance and tectonics, in Johnsson, M.J., and Basu, A., eds., Processes Controlling the Composition of Clastic Sediments: Geological Society of America Special Paper 284, p. 21–40.

Milliken, K.L., 1988, Loss of provenance information through subsurface diagenesis in Plio-Pleistocene, northern Gulf of Mexico: Journal of Sedimentary Petrology, v. 58, p. 992–1002.

Milliken, K.L., 2003, Late diagenesis and mass transfer in sandstone-shale sequences, in Mackenzie, F.T., ed., Treatise on Geochemistry: Sediments, Diagenesis and Sedimentary Rocks, v. 7, p. 159–190.

Morad, S., Bergan, M., Knarud, R., and Nystuen, J.P., 1990, Albitization of detrital plagioclase in Triassic reservoir sandstones from the Snorre Field, Norwegian North Sea: Journal of Sedimentary Petrology, v. 60, p. 411–425.

Nesbitt, H.W., and Young, G.M., 1982, Early Proterozoic climates and plate motions inferred from major element chemistry of lutites: Nature, v. 299, p. 715–717, doi: 10.1038/299715a0.

Pérez-Soba, C., 1992, Petrología y geoquímica del macizo granítico de La Pedriza, Sistema central español [Ph.D. thesis]: Madrid, Universidad Complutense de Madrid, 225 p.

Pettijohn, F.J., Potter, P.E., and Siever, R., 1973, Sand and Sandstones: Berlin, Springer-Verlag, 617 p.

Ramseyer, K., Baumann, J., Matter, A., and Mullis, J., 1988, Cathodoluminescence color of alpha quartz: Mineralogical Magazine, v. 52, p. 669–677.

Ramseyer, K., Fischer, J., Matter, A., Eberhardt, P., and Geiss, J., 1989, A cathodoluminescence microscope for low intensity luminescence: Journal of Sedimentary Petrology, v. 59, p. 619–622.

Ramseyer, K., Boles, J.R., and Lichtner, P.C., 1992, Mechanism of plagioclase albitization: Journal of Sedimentary Petrology, v. 62, p. 349–356.

Rollinson, H.R., 1993, Using Geochemical Data: Evaluation, Presentation, Interpretation: New York, Longman Scientific and Technical, 352 p.

Saigal, G.C., Morad, S., Bjørlykke, K., Egeberg, P.K., and Aagaard, P., 1988, Diagenetic albitization of detrital K-feldspars in Jurassic, Lower Cretaceous and Tertiary clastic reservoir rocks from offshore Norway, I. Textures and origin: Journal of Sedimentary Petrology, v. 58, p. 1003–1013.

Salas, R., 1987, El Malm y el Cretaci inferior entre el Massis de Garraf y la Serra d´Espadà [Ph.D. thesis]: Barcelona, Universidad de Barcelona, 345 p.

Salas, R., and Casas, A., 1993, Mesozoic extensional tectonics, stratigraphy, and crustal evolution during the Alpine cycle of the eastern Iberian basin: Tectonophysics, v. 228, p. 33–55, doi: 10.1016/0040-1951(93)90213-4.

Salas, R., and Guimerà, J., 1996, Main structural features of the Lower Cretaceous Maestrat Basin (eastern Iberian Range): Geogaceta, v. 20, no. 7, p. 1704–1706.

Salas, R., Guimerà, J., Mas, R., Martín-Closas, C., Meléndez, A., and Alonso, A., 2001, Evolution of the Mesozoic central Iberian Rift system and its Cenozoic inversion (Iberian Chain), in Ziegler, P.A., Cavazza, W., Robertson, A.H.F., and Crasquin-Soleau, S., eds., Peri-Tethys Memoir 6: Peri-Tethyan Rift/Wrench Basins and Passive Margins: Paris, Mémoires du Mus. Muséum National d'Histoire Naturelle, v. 186, p. 145–185.

Suttner, L.J., Basu, A., and Mack, G.H., 1981, Climate and the origin of quartz arenites: Journal of Sedimentary Petrology, v. 51, p. 1235–1246.

Taylor, S.R., and McLennan, S.M., 1985, The Continental Crust: Its Composition and Evolution: Oxford, Blackwell, 312 p.

Torres-Ruiz, J., Pesquera, A., Gil-Crespo, P.P., and Velilla, N., 2003, Origin and petrogenetic implications of tourmaline-rich rocks in the Sierra Nevada (Betic Cordillera, southeastern Spain): Chemical Geology, v. 197, p. 55–86, doi: 10.1016/S0009-2541(02)00357-1.

Tortosa, A., Palomares, M., and Arribas, J., 1991, Quartz grain types in Holocene deposits from the Spanish Central System: Some problems in provenance analysis, in Morton, A.C., Todd, S.P., and Haughton, P.D.W., eds., Developments in Sedimentary Provenance Studies: Geological Society of London Special Publication 57, p. 47–54.

Trevena, A.S., and Nash, W.P., 1981, An electron microprobe study of detrital feldspar: Journal of Sedimentary Petrology, v. 51, p. 137–150.

Valloni, R., Lazzari, D., and Calzolari, M.A., 1991, Selective alteration of arkose framework in Oligo-Miocene turbidites of the northern Apennines foreland: Impact on sedimentary provenance analysis, in Morton, A.C., Todd, S.P., and Haughton, P.D.W., eds., Developments in Sedimentary

Provenance Studies: Geological Society of London Special Publication 57, p. 125–136.

Villaseca, C., and Herreros, V., 2001, A sustained felsic magmatic system; the Hercynian granitic batholith of the Spanish Central System: Transactions of the Royal Society of Edinburg, Earth Science, v. 91, p. 207–219.

von Eynatten, H., and Gaupp, R., 1999, Provenance of Cretaceous synorogenic sandstones in the eastern Alps: Constraints from framework petrography, heavy mineral analysis and mineral chemistry: Sedimentary Geology, v. 124, p. 81–111, doi: 10.1016/S0037-0738(98)00122-5.

Weltje, G.J., 2002, Quantitative analysis of detrital modes: Statistically rigorous confidence regions in ternary diagrams and their use in sedimentary petrology: Earth-Science Reviews, v. 57, p. 211–253, doi: 10.1016/S0012-8252(01)00076-9.

Weltje, G.J., and von Eynatten, H., 2004, Quantitative provenance analysis of sediments: Review and outlook: Sedimentary Geology, v. 171, p. 1–11, doi: 10.1016/j.sedgeo.2004.05.007.

Zimmermann, U., and Bahlburg, H., 2003, Provenance analysis and tectonic setting of the Ordovician clastic deposits in the southern Puna Basin, NW Argentina: Sedimentology, v. 50, p. 1079–1104, doi: 10.1046/j.1365-3091.2003.00595.x.

MANUSCRIPT ACCEPTED BY THE SOCIETY 9 AUGUST 2006

Significance of geochemical signatures on provenance in intracratonic rift basins: Examples from the Iberian plate

M. Ochoa[†]
M.E. Arribas
J. Arribas
Departamento de Petrología y Geoquímica, Universidad Complutense de Madrid, 28040 Madrid, Spain

R. Mas
Departamento de Estratigrafía, Universidad Complutense de Madrid, 28040 Madrid, Spain

ABSTRACT

Following the Variscan orogeny, the Iberian plate was affected by an extensional tectonic regime from Late Permian to Late Cretaceous time. In the central part of the plate, NW-SE–trending rift basins were created. Two rifting cycles can be identified during the extensional stage: (1) a Late Permian to Hettangian cycle, and (2) a latest Jurassic to Early Cretaceous cycle. During these cycles, thick clastic continental sequences were deposited in grabens and half grabens. In both cycles, sandstone petrofacies from periods of high tectonic activity reveal a main plutoniclastic (quartzofeldspathic) character due to the erosion of coarse-grained crystalline rocks from the Hesperian Massif, during Buntsandstein (mean $Qm_{72}F_{25}Lt_3$) sedimentation and during Barremian–early Albian times (mean $Qm_{81}F_{18}Lt_1$). Geochemical data show that weathering was more intense during the second rifting phase (mean chemical index of alteration [CIA]: 80) due to more severe climate conditions (humid) than during the first rifting phase (mean CIA: 68) (arid climate).

Ratios between major and trace elements agree with a main provenance from passive-margins settings in terms of the felsic nature of the crust. However, anomalies in trace elements have been detected in some Lower Cretaceous samples, suggesting additional basic supplies from the north area of the basin. These anomalies consist of (1) low contents in Hf, Th, and U; (2) high contents in Sc, Co, and Zr; and (3) anomalous ratios in Th/Y, La/Tb, Ta/Y, and Ni/V. Basic supplies could be related to the alkaline volcanism during Norian-Hettangian and Aalenian-Bajocian times. Geochemical composition of rift deposits has been shown to be a useful and complementary tool to petrographic deduction in provenance, especially in intensely weathered sediments. However, diagenetic processes and hydrothermalism may affect the original detrital deposits, producing changes in geochemical composition that mislead provenance and weathering deductions.

Keywords: provenance, sandstones, geochemical composition, fluvial deposits, rift basins, Permian-Triassic, Jurassic, Early Cretaceous, Iberian Range.

[†]E-mail: mochoa@geo.ucm.es.

Ochoa, M., Arribas, M.E., Arribas, J., and Mas, R., 2007, Significance of geochemical signatures on provenance in intracratonic rift basins: Examples from the Iberian plate, *in* Arribas, J., Critelli, S., and Johnsson, M.J., eds., Sedimentary Provenance and Petrogenesis: Perspectives from Petrography and Geochemistry: Geological Society of America Special Paper 420, p. 199–219, doi: 10.1130/2006.2420(13). For permission to copy, contact editing@geosociety.org. ©2007 Geological Society of America. All rights reserved.

INTRODUCTION

Studies in provenance have been mainly performed according to classic sandstone petrography. At present, several models are in common usage to deduce provenance parameters (source rock lithology, climate, weathering, transport, geotectonic setting) from petrographic analysis on sandstone framework (e.g., Basu et al., 1975; Dickinson and Suczek, 1979; Dickinson, 1985).

During the last two decades, the use of geochemical data for provenance inferences has experienced an important development (see McLennan et al., 1993), and models have been elaborated to decipher aspects concerning source lithology (Floyd and Leveridge, 1987; Gu et al., 2002), weathering (e.g., Nesbitt and Young, 1982; Taylor and McLennan, 1985), maturation during transport (Bhatia, 1983; Gu et al., 2002; Whitmore et al., 2004), and geotectonic setting (Bhatia and Taylor, 1981; Maynard et al.,

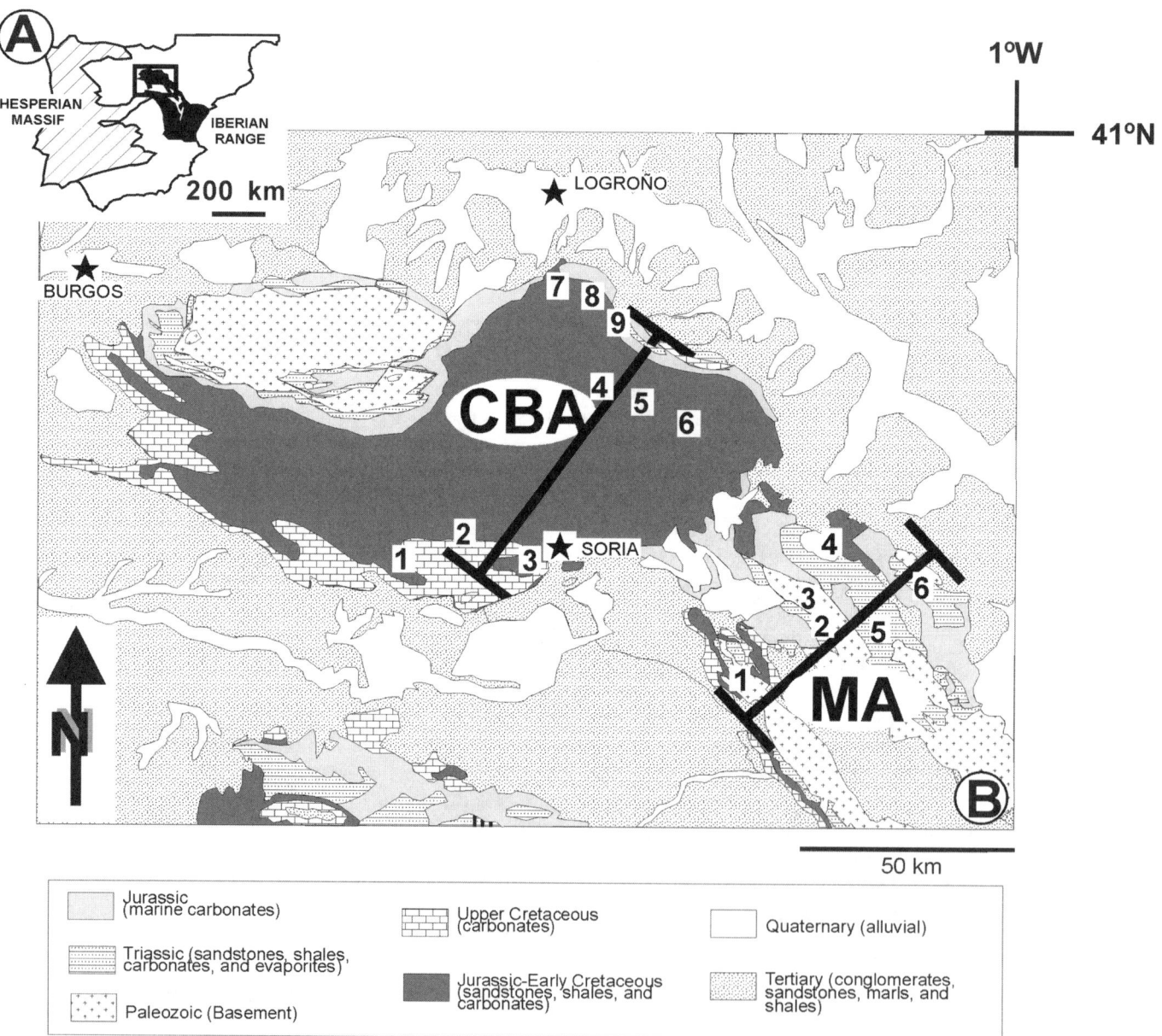

Figure 1. (A) Study area in the Iberian Peninsula. (B) Simplified geological map showing the location of sections in Permian-Triassic deposits on Moncayo area (MA): 1—Alameda section, 2—Aranda del Moncayo section, 3—Beratón section, 4—Moncayo section, 5—Tierga section, and 6—Tabuenca section; and the location of sections in Lower Cretaceous deposits on Cameros Basin area (CBA): 1—Muriel section, 2—Cidones-Abejar section, 3—Trinchera del Ferrocarril section, 4—Yanguas section, 5—San Pedro Manrique section, 6—Valdemadera section, 7—Trevijano section, 8—Jubera section, and 9—Arnedillo section.

1982; Bhatia, 1983, 1984; Roser and Korsch, 1985, 1986, 1988; Bhatia and Crook, 1986; McLennan and Taylor, 1991; Gu et al., 2002). Geochemical procedures generate quick and objective data, and they can be used on whole rock of clastic deposits from a wide range of grain sizes (Eynatten et al., 2003). In spite of these advantages, geochemical analysis may indicate a consistent loss of information about textures. Provenance origin of clasts (intrabasinal or extrabasinal, Zuffa, 1980; coeval or noncoeval, Zuffa, 1991) is indecipherable by whole-rock chemical analysis. In addition, diagenetic products are mixed with detrital material, and this fact may produce biased inferences on the provenance of original clastic material (García et al., 2004).

The Iberian Range is a linear structure trending NW-SE in the northeast edge of the Iberian microplate (Fig. 1); it is an intracratonic, folded segment of the Alpine Chain that developed as a rift basin (Iberian Basin) in two phases (e.g., Salas et al., 2001): the first phase was generated from the Early Permian to the Late Triassic, and the second phase of rifting occurred from Late Jurassic to early Albian time (Fig. 2). During these cycles, thick clastic sequences, from alluvial to lacustrine at the top, were deposited in grabens and half grabens. Both cycles evolved to periods of postrift thermal subsidence, where predominant shallow-marine carbonate sedimentation took place. During the Paleogene and Lower–Middle Miocene, compressive events caused structural inversion, folding, and thrusting.

Stratigraphy, sedimentology, and petrography of sediments generated during the two active rifting phases (Arribas, 1984; Arribas et al., 2003; Benito et al., 2001; Martín-Closas and Alonso-Millán, 1998; Mas et al., 2003; Ochoa et al., 2004) and the tectonic evolution of the basin (Guimerà et al., 1995; Salas

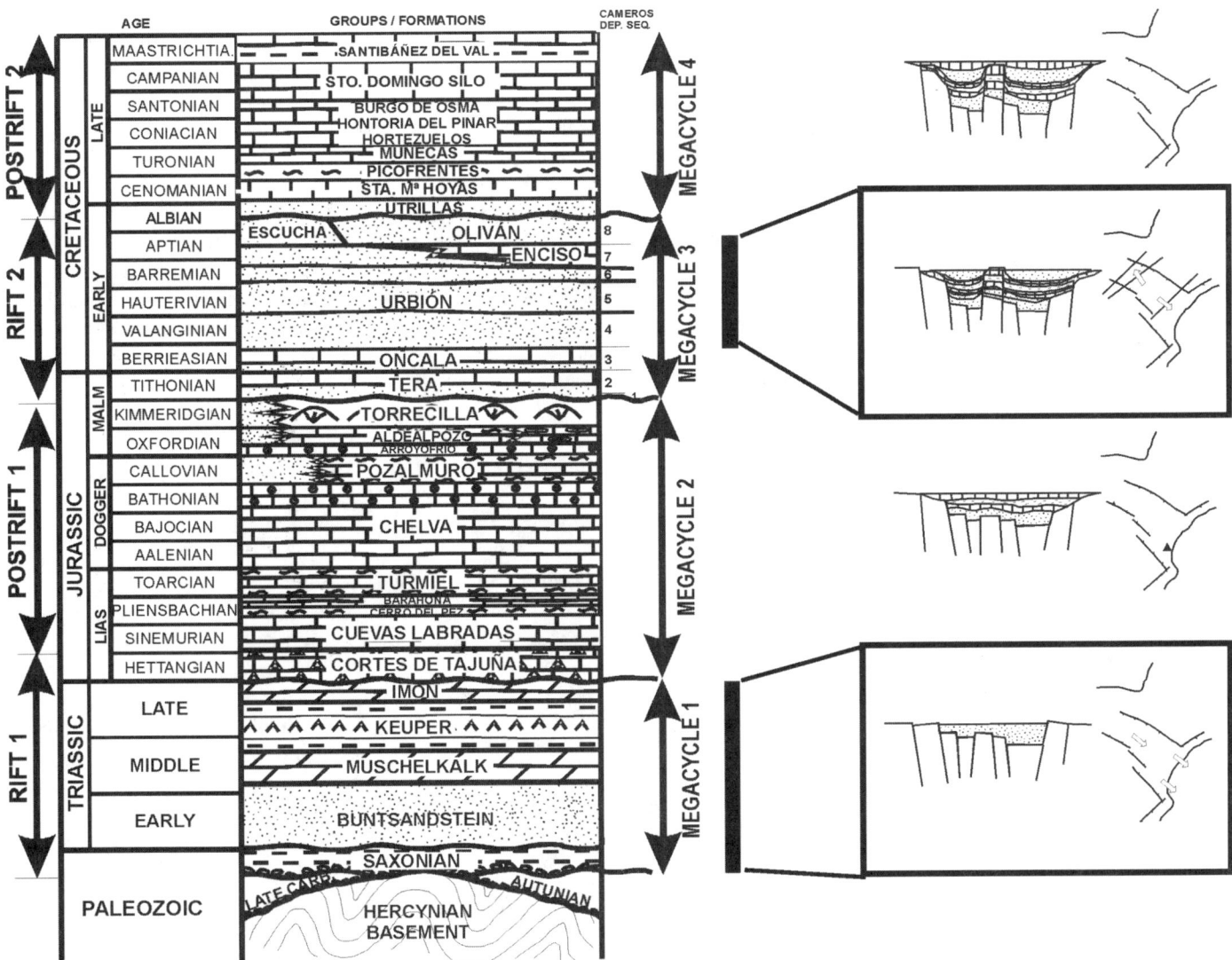

Figure 2. Synthetic sketches and stratigraphic section showing the stratigraphic record in the studied area and rift cycles on the evolution of the Iberian Basin (modified from Salas et al. [2001] and Mas et al. [2003]).

et al., 2001; Guimerà et al., 2004) have been consistently analyzed. Thus, these deposits represent an excellent opportunity to contrast geochemical data analysis with valuable background information.

The principal aim of this paper is to evaluate the informative power of geochemical data from petrographically well-known examples of clastic sediments generated at different times, but in a similar geotectonic scenario: the intracratonic Iberian Rift. Furthermore, the contrast of geochemical signatures between sediments generated during the two active stages of rifting may contribute to a better understanding of the evolution of the Iberian Rift. Finally, the data obtained in this paper will increase general knowledge of intracratonic rift basins and will be applicable to future models of such basin types.

GEOLOGICAL SETTING

The study area is located in the northwest sector of the Iberian Range in central Spain, and it includes clastic deposits from the first (Permian to Triassic) and second (Late Jurassic to Early Cretaceous) rifting stages. These deposits outcrop in two different areas: the Moncayo area, which is composed of Permian to Triassic deposits, and Cameros Basin, where Upper Jurassic to Lower Cretaceous deposits appear (Fig. 1).

During the Late Permian–Triassic extensional stage (rift 1, Fig. 2), the reactivation of the wrench faults as normal faults induced the propagation of rift systems in the Iberian plate as the Iberian Trough (Sopeña and Sánchez-Moya, 1997; Mas et al., 2003). This extensional stage corresponds to the beginning of the Alpine sedimentary cycle and is represented by the clastic Saxonian and Buntsandstein facies. These facies are continental clastic sediments that infilled asymmetrical half grabens (Fig. 3A). Buntsandstein deposits are mainly continental red beds that form the base of a sequence that evolve into siliciclastic and carbonate tidal sediments (Muschelkalk facies). The clastic infill in the Moncayo area consists mainly of arkosic deposits arranged into five main lithostratigraphic units (Arribas, 1984, 1985). The thickness of these deposits varies from 100 m to more than 900 m due to differential subsidence of troughs.

The second stage of rifting (Late Jurassic to Early Cretaceous) was related to the opening of the Central Atlantic (rift 2, Fig. 2). As a consequence, the Cameros Basin was formed as an extensional rift basin above a south-dipping ramp on a blind, low-angle normal fault several kilometers deep in the basement (Alonso and Mas, 1993; Guimerà et al., 1995; Salas et al., 2001). According to Alonso and Mas (1993) and Mas et al. (2003), the sedimentary record (Tithonian–early Albian) constitutes a large megasequence bounded by two main unconformities at the base and at the top, and it can be further subdivided into eight depositional sequences separated by minor unconformities (DS-1 to DS-8 in Fig. 2). This study is focused on the maximum synrift filling stage (DS-4 to DS-7 [late Berriasian to early Aptian]; Fig. 3B) related to maximum tectonic activity of this rifting phase. The infill of the basin varies drastically in thickness, from nearly 100 m in the marginal areas of the basin (toward NE and SW) to 2200 m in the depocentral areas (central sector) (Fig. 3B). This record is constituted by fluvial sequences that consist of coarse deposits of conglomerates and channelized fluvial (mainly braided) sandstone bodies in proximal areas evolving to meandering and lacustrine facies in distal areas (Mas et al., 2003; Ochoa et al., 2004). During the Middle to Late Cretaceous, a low-grade metamorphic event (hydrothermalism) took place in depocentral areas and affected the sedimentary record (Casquet et al., 1992; Barrenechea et al., 1995; Alonso-Azcárate et al., 1999).

METHODS

A total of 53 sandstone and shale samples was collected for geochemical analysis from several stratigraphic sections from both Permian-Triassic deposits in the Moncayo area (24 samples) and from Lower Cretaceous deposits (DS-4 to DS-7) in the Cameros Basin (29 samples). Sample locations are shown in Figure 3. Major and trace elements were determined for all samples. Analyses were performed at the Actlabs Laboratories (Canada) by Code 4Lithoresearch. All geochemical data are reported in Tables 1, 2, and 3.

For petrographic analysis of Permian-Triassic sandstones, one of the authors (J. Arribas) examined several databases from previous works (Arribas, 1984, 1987). This author provided thin sections of sandstones for analysis. For this paper, a new point counting method was performed on these samples following Gazzi-Dickinson criteria to obtain geotectonic inferences (Dickinson, 1985). In addition, Lower Cretaceous sandstones were analyzed petrographically following the same procedures and point-counting methods as those used for Permian-Triassic sandstones (Arribas et al., 2003; Ochoa et al., 2004).

RESULTS

Petrography

Permian and Triassic Sandstones

Framework composition of sandstones varies from quartzose (mean $Qm_{97}F_0Lt_3$; Qm—monocrystalline quartz; F—feldspars; Lt—total lithics) at the base of the succession (Saxonian facies; PS in Figs. 3A and 4A) to quartzofeldspathic (mean $Qm_{72}F_{25}Lt_3$) petrofacies at the top of the sedimentary sequence (Buntsandstein facies; B-1 and B-2 in Figs. 3A and 4A).

Quartzose Saxonian petrofacies (Fig. 5A) are very mature texturally; they show evidence of maturation during transport (very well-sorted sediments and high values of quartz grain roundness) and recycling of metasediments in the Variscan basement (e.g., presence of inherited quartz overgrowth). Lithic rock fragments are scarce and consist mainly of low-grade metamorphic fragments (shales and chert); some quartzose sandstone fragments also occur. Syntaxial quartz overgrowth is the main interstitial cement in the sandstones. Framework composition of Saxonian sandstones is very homogeneous in all the Moncayo

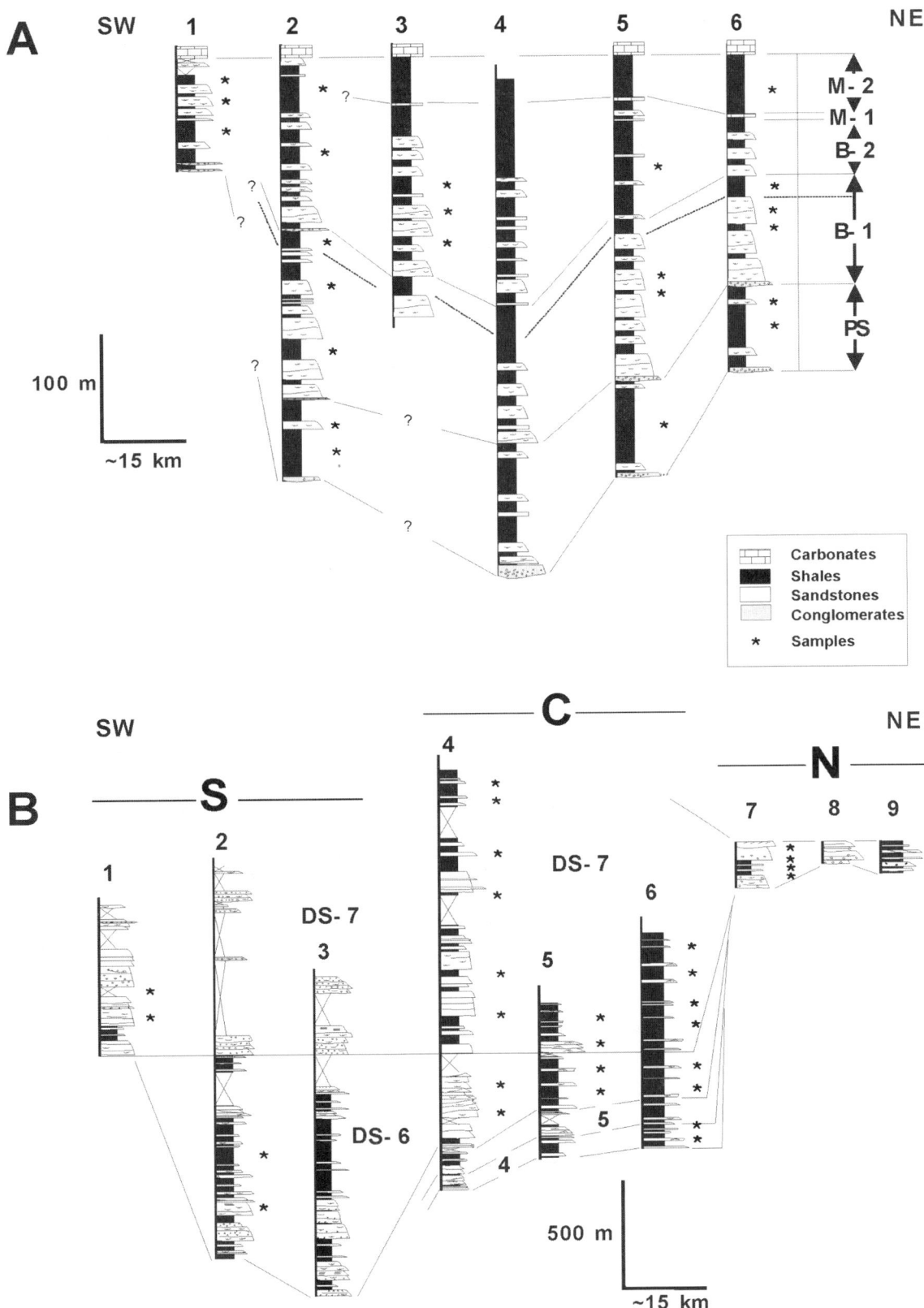

Figure 3. (A) SW-NE–trending stratigraphic correlation of analyzed sections in Permian-Triassic deposits in Moncayo area (MA, Fig. 1). Depositional sequences: PS—Saxonian facies; B-1 and B-2—Buntsandstein facies; M-1 and M-2—Muschelkalk facies. (B) Stratigraphic correlation of depositional sequences (from DS-4 to DS-7) in Lower Cretaceous sediments in Cameros Basin area (CBA, Fig. 1); S—southern area; C—central area; N—northern area. Note different vertical scales in A and B.

TABLE 1. MAJOR-ELEMENT COMPOSITION AND RELATED PARAMETERS FROM PERMIAN-TRIASSIC AND CRETACEOUS CLASTIC DEPOSITS IN THE NORTHWESTERN IBERIAN RANGE

Sample	Lithology[†]	SiO_2	Al_2O_3	Fe_2O_3	MnO	MgO	CaO	Na_2O	K_2O	TiO_2	P_2O_5	Fe_2O_3 + MgO	Al_2O_3/SiO_2	Na_2O/K_2O	SiO_2/Al_2O_3	Fe_2O_3/K_2O	CIA
Permian-Triassic (rift 1)																	
Tierga																	
BTL-1	sh	60.94	18.14	6.05	0.06	1.63	1.06	0.23	5.38	0.86	0.17	7.68	0.30	0.04	3.36	1.12	73.12
BTL-2	sh	58.43	19.93	7.64	0.03	2.25	0.39	0.28	5.84	0.81	0.17	9.89	0.34	0.05	2.93	1.31	75.38
BTL-3	sh	59.43	17.97	7.34	0.03	2.53	0.45	0.25	6.66	0.77	0.17	9.87	0.30	0.04	3.31	1.10	70.94
BTL-4	sh	51.98	20.19	8.24	0.03	4.42	0.46	0.25	6.96	0.71	0.19	12.66	0.39	0.04	2.57	1.18	72.47
BTA-1	ss	80.14	9.60	2.33	0.02	0.75	0.28	0.16	4.08	0.82	0.21	3.08	0.12	0.04	8.35	0.57	67.99
BTA-2	ss	76.49	10.44	2.94	0.02	1.14	0.59	0.21	4.82	0.92	0.21	4.08	0.14	0.04	7.33	0.61	65.01
Beraton																	
BBL-1	sh	58.17	20.25	7.54	0.02	1.88	0.29	0.29	6.57	0.94	0.19	9.42	0.35	0.04	2.87	1.15	73.91
BBL-2	sh	76.63	10.90	3.66	0.14	0.93	0.17	0.16	4.42	0.84	0.12	4.59	0.14	0.04	7.03	0.83	69.65
BBA-1	ss	78.38	10.88	1.51	0.01	0.64	0.15	0.17	5.81	0.59	0.14	2.15	0.14	0.03	7.20	0.26	63.96
Alameda																	
ALL-5	sh	57.86	16.96	6.35	0.07	3.00	1.67	0.26	5.18	0.86	0.22	9.35	0.29	0.05	3.41	1.23	70.46
ALL-11	sh	56.62	18.84	7.39	0.03	2.59	0.17	0.25	7.42	0.83	0.10	9.98	0.33	0.03	3.01	1.00	70.61
ALA-10	ss	57.85	6.19	1.41	0.13	0.28	16.05	0.17	4.64	0.06	0.06	1.69	0.11	0.04	9.35	0.30	22.88
Aranda																	
BRL-1	sh	74.95	10.96	4.94	0.06	0.81	0.70	0.14	3.32	0.85	0.14	5.75	0.15	0.04	6.84	1.49	72.49
BRL-2	sh	56.29	21.57	7.10	0.02	1.28	0.52	0.32	6.60	0.80	0.22	8.38	0.38	0.05	2.61	1.08	74.35
BRL-3	sh	64.26	16.06	5.62	0.04	1.34	1.12	0.21	4.99	0.88	0.17	6.96	0.25	0.04	4.00	1.13	71.76
BRL-3/R	sh	64.31	16.08	5.62	0.04	1.33	1.17	0.21	5.04	0.88	0.17	6.95	0.25	0.04	4.00	1.12	71.47
BRL-4	sh	73.51	8.88	2.89	0.09	1.15	3.38	0.16	3.80	0.87	0.17	4.04	0.12	0.04	8.28	0.76	54.75
BRL-5	sh	60.40	16.18	6.90	0.04	2.10	1.33	0.23	5.66	0.82	0.31	9.00	0.27	0.04	3.73	1.22	69.15
BRA-1	ss	89.47	4.62	1.93	0.04	0.32	0.14	0.09	1.48	0.46	0.09	2.25	0.05	0.06	19.37	1.30	72.99
BRA-2	ss	89.07	3.11	1.61	0.04	0.16	0.27	0.08	0.99	0.27	0.13	1.77	0.03	0.08	28.64	1.63	69.89
Tabuenca																	
BAL-1	sh	62.38	17.57	6.86	0.04	1.68	0.26	0.29	4.63	0.99	0.15	8.54	0.28	0.06	3.55	1.48	77.23
BAL-2	sh	64.30	16.44	6.35	0.04	2.18	0.47	0.23	4.82	0.90	0.16	8.53	0.26	0.05	3.91	1.32	74.86
BAL-3	sh	65.53	15.84	6.35	0.02	1.78	0.40	0.26	4.53	0.85	0.18	8.13	0.24	0.06	4.14	1.40	75.32
BAA-11	ss	88.82	4.49	1.55	0.11	0.54	0.82	0.11	1.50	0.24	0.10	2.09	0.05	0.07	19.78	1.03	64.88
Early Cretaceous (rift 2)																	
San Andrés																	
SAN-4	ss	92.97	2.46	2.04	0.13	0.20	0.40	0.10	0.21	0.38	0.04	2.59	0.03	0.03	0.48	9.71	77.60
Yanguas																	
YNG-L6	sh	57.43	23.91	6.44	0.02	0.86	0.30	1.30	3.80	1.04	0.08	23.93	0.42	0.42	0.34	1.69	81.58
YNG-L8	sh	56.14	23.56	8.19	0.04	0.93	0.26	1.24	3.14	1.04	0.10	23.60	0.42	0.42	0.39	2.61	83.55
YNG-L12	sh	78.35	10.86	3.45	0.04	0.54	0.50	0.68	1.44	0.93	0.06	10.90	0.14	0.14	0.47	2.40	80.56
YNG-L49	sh	55.76	25.75	6.76	0.07	0.49	0.12	0.75	4.16	0.85	0.06	25.82	0.46	0.46	0.18	1.63	83.66
YNG-6/R	ss	91.75	3.68	1.78	0.01	0.21	0.05	0.24	0.33	0.41	0.03	3.69	0.04	0.04	0.73	5.39	85.58
YNG-8	ss	96.27	1.86	0.93	0.01	0.09	0.04	0.14	0.12	0.20	0.03	1.87	0.02	0.02	1.17	7.75	86.11
YNG-12	ss	95.54	1.46	2.30	0.04	0.20	0.03	0.03	0.09	0.06	0.05	1.50	0.02	0.02	0.33	25.56	90.68
YNG-49	ss	88.82	5.83	1.98	0.01	0.26	0.10	0.45	0.85	0.36	0.06	5.84	0.07	0.07	0.53	2.33	80.64
San Pedro Manrique																	
SPM-3	ss	91.33	1.56	1.29	0.09	0.13	2.75	0.15	0.17	0.13	0.03	1.65	0.02	0.02	0.88	7.59	33.69
SPM-9	ss	89.68	5.52	1.13	0.02	0.14	0.26	0.44	0.79	0.78	0.05	5.54	0.06	0.06	0.56	1.43	78.74
SPM-15	ss	68.18	18.02	6.39	0.03	0.46	0.17	0.62	2.12	1.06	0.11	18.05	0.26	0.26	0.29	3.01	86.10
SPM-30	ss	89.96	3.97	3.68	0.02	0.65	0.03	0.10	0.42	0.52	0.04	3.99	0.04	0.04	0.24	8.76	87.83
Valdemadera																	
VLM-L2	sh	58.17	21.79	8.11	0.02	1.25	0.26	1.09	2.99	0.96	0.10	21.81	0.37	0.37	0.36	2.71	83.39
VLM-L7	sh	53.11	26.47	6.87	0.05	0.57	0.83	1.26	3.63	1.14	0.12	26.52	0.50	0.50	0.35	1.89	82.23
VLM-L11R	sh	69.22	17.91	4.16	0.09	0.31	0.08	0.78	2.56	1.08	0.08	18.00	0.26	0.26	0.30	1.63	83.97
VLM-L12	sh	61.13	20.74	6.57	0.05	0.73	0.19	0.71	3.61	0.89	0.11	20.79	0.34	0.34	0.20	1.82	82.14
VLM-2	ss	91.98	4.68	0.35	0.00	0.06	0.03	0.41	0.74	0.41	0.02	4.68	0.05	0.05	0.55	0.47	79.86
VLM-7	ss	91.94	4.20	1.55	0.01	0.21	0.08	0.34	0.43	0.45	0.05	4.21	0.05	0.05	0.79	3.60	83.17
VLM-11	ss	96.11	1.76	1.05	0.01	0.14	0.12	0.16	0.25	0.18	0.05	1.77	0.02	0.02	0.64	4.20	76.86
VLM-12	ss	61.13	20.72	6.53	0.05	0.73	0.20	0.75	3.62	0.90	0.11	20.77	0.34	0.34	0.21	1.80	81.93
Trevijano																	
TRE-L6	sh	29.74	9.45	3.45	0.07	0.79	27.25	0.07	2.15	0.45	0.06	4.37	0.32	0.32	3.15	1.60	24.28
TRE-L7	sh	33.54	8.38	3.16	0.04	0.64	26.84	0.05	1.89	0.43	0.06	4.94	0.25	0.25	4.00	1.67	22.55
TRE-5	ss	76.08	5.35	0.71	0.05	0.35	7.93	1.20	0.49	0.47	0.04	2.03	0.07	0.07	14.22	1.45	35.74
TRE-8	ss	62.76	0.95	0.77	0.02	0.22	19.31	0.16	0.11	0.21	0.02	3.50	0.02	0.02	66.06	7.00	4.63
Cidones																	
CID-33	sh	58.01	19.96	8.66	0.01	1.05	0.24	0.24	4.54	0.85	0.08	8.25	0.34	0.34	2.91	1.91	79.90
CID-31	ss	95.16	2.46	0.52	0.00	0.04	0.02	0.01	0.75	0.04	0.03	13.00	0.03	0.03	38.68	0.69	75.93
Gan																	
GAN-15	sh	59.56	23.62	2.49	0.01	0.81	0.04	0.30	4.60	1.09	0.05	3.07	0.40	0.40	2.52	0.54	82.70
GAN-13	ss	92.22	3.91	0.37	0.00	0.03	-0.01	0.03	1.55	0.22	0.04	12.33	0.04	0.04	23.59	0.24	71.35

Note: CIA—chemical index of alteration.
[†]ss—sandstones; sh—shales.

TABLE 2. TRACE-ELEMENT CONCENTRATIONS IN PPM AND RELATED PARAMETERS FROM PERMIAN-TRIASSIC AND CRETACEOUS CLASTIC DEPOSITS IN THE NORTHWESTERN IBERIAN RANGE

Sample	Lithology[†]	Th	U	Sc	Zr	Co[‡]	Hf	Rb	Cs	Ba	Sr	Y	Ta	Ni[‡]	V	K	La/Th	Co/Th	La/Sc	Th/Co	La/Co	Th/U
Permian-Triassic (rift 1)																						
Tierga																						
BTL-1	sh	16.56	3.66	17.00	213.81	6.06	6.46	180.00	18.98	597.47	148.81	31.40	1.46	21.00	94.00	24.03	2.64	0.37	2.57	2.73	7.21	4.52
BTL-2	sh	17.19	2.82	17.00	197.63	11.45	5.91	197.00	20.88	747.39	265.87	29.80	1.47	29.00	100.00	26.08	2.59	0.67	2.62	1.50	3.89	6.09
BTL-3	sh	16.67	4.09	15.00	206.16	13.65	6.27	203.00	19.65	707.42	86.37	30.30	1.57	29.00	92.00	29.74	2.45	0.82	2.72	1.22	2.99	4.08
BTL-4	sh	16.55	4.15	17.00	122.35	14.54	3.65	243.00	25.68	664.58	92.77	26.30	1.56	31.00	107.00	31.08	2.53	0.88	2.47	1.14	2.88	3.99
BTA-1	ss	13.49	2.66	4.00	273.33	4.12	7.66	130.00	8.31	297.90	55.72	26.00	1.34	–20.00	34.00	18.22	2.73	0.31	9.22	3.27	8.95	5.07
BTA-2	ss	14.27	2.57	6.00	275.41	5.25	7.51	129.00	8.30	341.19	80.45	27.50	1.40	–20.00	40.00	21.53	2.60	0.37	6.19	2.72	7.08	5.55
Beratón																						
BBL-1	sh	19.51	3.67	18.00	255.25	12.05	7.47	221.00	22.78	719.59	98.95	35.80	1.71	34.00	131.00	29.34	3.16	0.62	3.43	1.62	5.12	5.32
BBL-2	sh	15.73	3.67	8.00	434.94	9.67	12.41	125.00	10.96	457.96	38.57	33.30	1.46	–20.00	53.00	19.74	2.32	0.61	4.57	1.63	3.78	4.29
BBA-1	ss	9.93	1.91	3.00	234.43	2.45	6.49	146.00	6.62	386.45	41.37	19.00	0.99	–20.00	18.00	25.95	3.69	0.25	12.20	4.06	14.96	5.20
Alameda																						
ALL-5	sh	16.23	3.88	15.00	250.99	11.92	7.49	165.00	20.58	558.22	181.51	33.70	1.38	–20.00	76.00	23.13	3.32	0.73	3.59	1.36	4.52	4.18
ALL-11	sh	15.76	2.97	16.00	207.40	6.88	6.03	223.00	25.69	598.44	71.05	29.50	1.65	–20.00	85.00	33.14	3.31	0.44	3.26	2.29	7.58	5.30
ALA-10	ss	2.40	3.43	2.00	36.35	1.36	1.04	117.00	5.23	469.85	36.78	6.90	0.15	–20.00	9.00	20.72	3.48	0.57	4.18	1.76	6.15	0.70
Aranda																						
BRL-1	sh	14.64	3.64	10.00	365.25	4.68	10.52	126.00	13.17	446.09	91.72	35.80	1.38	–20.00	55.00	14.83	3.27	0.32	4.79	3.13	10.24	4.02
BRL-2	sh	16.12	3.14	19.00	160.29	5.19	4.79	238.00	31.02	746.91	385.47	27.90	1.49	–20.00	106.00	29.48	3.55	0.32	3.01	3.11	11.02	5.13
BRL-3	sh	14.42	3.17	14.00	247.14	7.86	7.27	172.00	18.89	558.38	200.36	32.40	1.39	–20.00	81.00	22.29	3.38	0.55	3.49	1.83	6.21	4.55
BRL-3R	sh	14.34	3.16	14.00	242.10	7.91	7.14	170.00	18.39	568.77	201.86	32.70	1.33	–20.00	82.00	22.51	3.37	0.55	3.45	1.81	6.11	4.54
BRL-4	sh	15.98	3.54	6.00	525.61	6.68	14.69	111.00	8.11	864.14	90.70	35.30	1.55	–20.00	32.00	16.97	3.06	0.42	8.14	2.39	7.31	4.52
BRL-5	sh	16.11	3.68	13.00	273.60	8.77	7.87	198.00	22.50	544.90	184.54	35.50	1.55	–20.00	74.00	25.28	3.26	0.54	4.04	1.84	5.99	4.38
BRA-1	ss	9.12	2.31	3.00	438.26	1.44	12.07	50.00	3.23	1970.00	93.32	21.10	0.82	–20.00	27.00	6.61	3.81	0.16	11.58	6.32	24.04	3.95
BRA-2	ss	5.57	1.06	3.00	160.93	2.70	4.56	30.00	2.49	13,400.00	202.95	14.40	1.00	–20.00	16.00	4.42	3.45	0.49	6.41	2.06	7.11	5.26
Tabuenca																						
BAL-1	sh	15.87	3.81	16.00	282.62	10.68	8.02	169.00	13.85	519.94	160.98	36.00	1.53	21.00	110.00	20.68	3.43	0.67	3.40	1.49	5.10	4.17
BAL-2	sh	15.56	3.83	13.00	295.94	10.78	8.50	181.00	17.27	486.39	127.97	34.40	1.50	–20.00	89.00	21.53	3.41	0.69	4.08	1.44	4.92	4.06
BAL-3	sh	16.63	3.07	13.00	308.30	9.59	8.70	169.00	15.51	580.54	155.31	33.50	1.54	–20.00	98.00	20.23	2.87	0.58	3.67	1.73	4.98	5.42
BAA-11	ss	3.85	1.05	3.00	90.34	4.21	2.49	49.00	2.61	225.55	29.46	11.30	0.40	–20.00	20.00	6.70	3.86	1.09	4.95	0.92	3.53	3.67

(continued.)

TABLE 2. TRACE-ELEMENT CONCENTRATIONS IN PPM AND RELATED PARAMETERS FROM PERMIAN-TRIASSIC AND CRETACEOUS CLASTIC DEPOSITS IN THE NORTHWESTERN IBERIAN RANGE *(Continued.)*

Sample	Lithology†	Th	U	Sc	Zr	Co‡	Hf	Rb	Cs	Ba	Sr	Y	Ta	Ni‡	V	K	La/Th	Co/Th	La/Sc	Th/Co	La/Co	Th/U
Early Cretaceous (rift 2)																						
San Andrés																						
SAN-4	ss	3.61	1.19	1.00	153.09	6.00	4.21	10.20	0.73	27.19	14.07	8.76	0.62	−20.00	9.37	0.94	3.99	1.66	14.39	0.60	2.40	3.04
Yanguas																						
YNG-L6	sh	18.65	3.70	21.00	200.19	16.00	5.91	259.05	28.78	982.91	145.84	36.76	1.97	45.47	140.17	16.97	3.84	0.86	3.41	1.17	4.48	5.04
YNG-L8	sh	18.99	0.64	21.00	200.66	15.00	5.98	224.41	27.04	880.69	130.64	41.97	1.93	42.55	135.77	14.02	5.42	0.79	4.91	1.27	6.87	29.68
YNG-L12	sh	13.93	0.55	9.00	330.56	9.00	9.64	85.72	7.72	345.25	73.41	25.25	1.47	23.26	51.95	6.43	3.63	0.65	5.62	1.55	5.62	25.33
YNG-L49	sh	18.88	3.58	20.00	165.23	10.00	5.04	262.58	21.31	1450.00	87.71	31.03	1.78	21.29	118.47	18.58	3.07	0.53	2.90	1.89	5.79	5.27
YNG-6/R	sh	3.55	1.03	2.00	140.23	5.00	3.76	20.04	1.18	75.90	29.52	5.02	0.68	−20.00	20.42	1.47	4.43	1.41	7.86	0.71	3.14	3.44
YNG-8	ss	1.83	0.64	1.00	53.79	3.00	1.48	12.22	1.06	39.53	13.04	4.37	0.33	−20.00	9.37	0.54	4.75	1.64	0.00	0.61	2.89	2.85
YNG-12	ss	1.80	0.55	0.00	23.84	3.00	0.69	3.57	0.84	19.00	4.20	3.41	0.12	−20.00	8.51	0.40	2.87	1.66	0.00	0.60	1.73	3.28
YNG-49	ss	4.25	1.70	3.00	106.18	5.00	2.87	42.25	2.45	114.53	31.27	12.01	0.72	20.13	19.88	3.80	3.92	1.18	5.56	0.85	3.34	2.50
San Pedro Manrique																						
SPM-3	ss	1.53	0.53	0.00	35.97	3.00	1.03	5.56	0.65	30.19	15.94	4.55	0.26	−20.00	5.98	0.76	3.35	1.97	0.00	0.51	1.71	2.88
SPM-9	ss	8.07	1.90	4.00	269.79	2.00	7.43	45.86	2.28	187.37	41.88	17.40	1.10	−20.00	29.70	3.53	5.53	0.25	11.16	4.04	22.33	4.25
SPM-15	ss	16.65	3.97	17.00	268.26	2.00	8.14	142.78	10.76	506.12	67.51	47.90	1.84	30.03	98.33	9.47	3.05	0.12	2.99	8.33	25.40	4.20
SPM-30	ss	4.96	1.38	2.00	157.84	9.00	4.34	22.10	1.37	97.63	14.32	12.08	0.85	30.76	17.20	1.88	3.30	1.81	8.18	0.55	1.82	3.59
Valdemadera																						
VLM-L2	sh	16.65	2.49	19.00	189.59	15.00	5.81	192.89	23.00	595.00	202.00	36.30	1.73	57.84	117.17	13.35	1.66	0.90	1.46	1.11	1.85	6.69
VLM-L7	sh	22.03	3.25	23.00	183.70	10.00	5.73	225.35	26.00	684.00	224.00	34.65	2.34	27.49	136.67	16.21	2.28	0.45	2.18	2.20	5.01	6.78
VLM-L11R	sh	18.76	3.62	15.00	334.01	7.00	9.83	156.18	16.00	412.00	66.00	38.77	2.21	−20.00	85.25	11.43	2.13	0.37	2.66	2.68	5.70	5.18
VLM-L12	sh	20.01	6.90	17.00	246.62	16.00	7.49	211.95	16.00	788.00	109.00	41.05	1.83	44.86	105.56	16.12	2.44	0.80	2.87	1.25	3.05	2.90
VLM-2	ss	2.12	0.65	1.00	70.22	−1.00	1.88	43.95	2.70	118.38	60.52	4.39	0.69	−20.00	13.74	3.30	5.99	−0.47	12.71	-	3.05	3.26
VLM-7	ss	5.18	1.07	3.00	143.87	3.00	4.06	29.16	2.18	73.20	60.04	12.07	0.70	21.25	19.81	1.92	4.43	0.58	7.65	1.73	7.65	4.84
VLM-11	ss	1.59	0.54	1.00	34.84	2.00	0.98	10.49	0.75	24.79	17.05	3.00	0.34	−20.00	7.00	1.12	4.03	1.26	6.40	0.79	3.20	2.94
VLM-12	ss	19.99	6.88	17.00	245.54	15.00	7.43	209.03	15.47	858.01	107.30	40.73	1.87	45.17	102.65	16.17	2.42	0.75	2.85	1.33	3.23	2.91
Treviano																						
TRE-L6	sh	7.52	2.06	9.00	98.13	4.79	3.07	135.00	12.50	133.20	245.77	25.28	0.89	30.29	121.79	9.60	5.28	0.64	4.42	1.57	8.30	3.65
TRE-L7	sh	7.52	2.11	8.00	122.52	2.50	3.68	121.00	13.24	170.54	170.11	27.35	0.97	23.74	107.35	8.44	5.00	0.33	4.70	3.01	15.04	3.56
TRE-5	ss	5.05	3.20	3.00	131.54	3.31	3.86	28.00	2.80	89.41	77.35	10.75	0.80	−20.00	26.88	2.19	3.75	0.66	6.32	1.53	5.73	1.58
TRE-8	ss	2.90	0.91	0.00	158.92	0.00	4.54	7.00	0.56	296.76	100.32	11.33	1.14	−20.00	6.68	0.49	5.14	0.00	0.00	-	6.73	3.19
Cidones																						
CID-33	sh	14.75	3.00	18.00	209.42	10.58	6.38	248.00	26.25	683.56	101.07	32.82	1.80	50.79	58.19	20.28	3.78	0.72	3.10	1.39	5.27	4.92
CID-31	ss	1.44	2.05	1.00	21.32	2.42	0.75	36.00	2.16	76.69	21.81	4.12	0.12	28.11	7.58	3.35	5.88	1.68	8.48	0.60	3.50	0.70
Gan																						
GAN-15	sh	16.40	3.72	20.00	258.64	11.10	7.73	242.00	29.01	770.45	98.46	38.66	2.46	29.65	62.03	20.54	3.31	0.68	2.72	1.48	4.90	4.41
GAN-13	ss	2.37	0.86	2.00	50.23	1.04	1.48	66.00	3.12	150.95	20.59	5.19	0.48	−20.00	6.85	6.92	4.98	0.44	5.89	2.28	11.34	2.75

†ss—sandstones; sh—shales.
‡Negative values indicate less than the reporting limit.

TABLE 3. RARE EARTH ELEMENT COMPOSITIONS AND RELATED PARAMETERS FROM PERMIAN-TRIASSIC AND CRETACEOUS CLASTIC DEPOSITS IN THE NORTHWESTERN IBERIAN RANGE

Sample	Lithology[†]	La	Ce	Pr	Nd	Sm	Eu	Gd	Tb	Dy	Ho	Er	Tm	Yb	Lu	REE[‡]	LREE	HREE	LREE/HREE	La/Yb	(La/Yb)$_n$	(La/Sm)$_n$	(Gd/Yb)$_n$	Eu/Eu*
Permian-Triassic (rift 1)																								
Tierga																								
BTL-1	sh	43.70	85.02	9.38	35.38	7.24	1.49	6.20	1.01	5.68	1.13	3.17	0.51	3.31	0.49	203.71	180.72	21.50	8.40	13.18	8.91	3.80	1.52	0.68
BTL-2	sh	44.49	85.21	9.66	37.10	7.64	1.50	6.21	0.96	5.55	1.06	3.01	0.48	3.02	0.45	206.34	184.10	20.74	8.88	14.74	9.96	3.67	1.67	0.67
BTL-3	sh	40.86	78.53	8.72	32.77	6.81	1.33	5.66	0.97	5.50	1.07	2.98	0.47	2.99	0.45	189.13	167.69	20.11	8.34	13.65	9.22	3.78	1.53	0.66
BTL-4	sh	41.94	81.69	9.19	34.94	7.05	1.35	5.50	0.90	4.85	0.92	2.53	0.40	2.57	0.36	194.20	174.82	18.03	9.69	16.35	11.05	3.74	1.74	0.66
BTA-1	ss	36.87	72.85	7.91	30.07	6.20	0.97	5.11	0.86	4.84	0.92	2.52	0.40	2.59	0.37	172.47	153.90	17.60	8.74	14.25	9.63	3.74	1.60	0.53
BTA-2	ss	37.15	74.55	8.26	31.80	6.60	1.05	5.49	0.91	4.98	0.96	2.63	0.43	2.63	0.39	177.83	158.35	18.42	8.60	14.13	9.55	3.55	1.69	0.54
Beratón																								
BBL-1	sh	61.68	118.92	12.97	49.52	10.00	1.91	7.73	1.15	6.48	1.27	3.59	0.58	3.70	0.54	280.05	253.09	25.04	10.11	16.68	11.27	3.88	1.69	0.66
BBL-2	sh	36.56	71.60	7.83	30.32	6.54	1.15	6.10	1.04	6.02	1.17	3.35	0.54	3.51	0.55	176.27	152.84	22.28	6.86	10.41	7.03	3.52	1.41	0.55
BBA-1	ss	36.59	74.72	8.35	31.50	5.47	0.81	3.84	0.58	3.57	0.72	2.00	0.31	2.00	0.30	170.76	156.63	13.32	11.76	18.28	12.35	4.21	1.55	0.54
Alameda																								
ALL-5	sh	53.81	104.46	11.52	43.69	8.77	1.77	7.47	1.16	6.75	1.30	3.55	0.55	3.63	0.53	248.96	222.25	24.94	8.91	14.84	10.03	3.86	1.67	0.67
ALL-11	sh	52.18	104.76	12.12	45.89	8.96	1.74	7.08	1.01	5.78	1.10	3.04	0.47	3.01	0.46	247.58	223.90	21.94	10.20	17.35	11.72	3.66	1.91	0.67
ALA-10	ss	8.36	16.52	1.93	7.15	1.61	0.38	1.44	0.22	1.22	0.23	0.62	0.10	0.62	0.08	40.48	35.57	4.53	7.86	13.55	9.16	3.27	1.89	0.77
Aranda																								
BRL-1	sh	47.92	95.01	10.45	39.91	7.95	1.53	7.17	1.13	6.66	1.33	3.66	0.58	3.75	0.57	227.61	201.24	24.84	8.10	12.78	8.63	3.80	1.55	0.62
BRL-2	sh	57.16	113.28	12.24	45.70	9.11	1.92	7.38	1.06	5.69	1.08	2.94	0.45	2.92	0.44	261.37	237.49	21.96	10.82	19.56	13.22	3.95	2.05	0.72
BRL-3	sh	48.80	94.93	10.51	39.05	7.89	1.59	6.89	1.06	6.12	1.23	3.38	0.54	3.44	0.53	225.97	201.18	23.20	8.67	14.20	9.60	3.89	1.63	0.66
BRL-3R	sh	48.35	93.39	10.28	38.69	7.79	1.59	6.90	1.05	6.22	1.23	3.37	0.52	3.43	0.52	223.35	198.51	23.25	8.54	14.09	9.52	3.91	1.63	0.66
BRL-4	sh	48.84	96.79	10.82	40.82	8.28	1.33	7.47	1.12	6.66	1.30	3.65	0.59	3.75	0.57	231.99	205.54	25.11	8.19	13.01	8.79	3.71	1.61	0.52
BRL-5	sh	52.49	101.20	11.24	42.64	8.80	1.83	7.82	1.18	6.90	1.31	3.49	0.54	3.46	0.52	243.42	216.37	25.22	8.58	15.17	10.25	3.75	1.83	0.67
BRA-1	ss	34.73	69.39	7.43	27.59	5.31	0.95	4.31	0.64	3.89	0.78	2.17	0.35	2.26	0.36	160.17	144.45	14.76	9.79	15.39	10.40	4.12	1.55	0.61
BRA-2	ss	19.22	39.27	4.23	16.04	3.40	0.56	3.19	0.49	2.75	0.52	1.41	0.21	1.31	0.20	92.81	82.16	10.08	8.15	14.62	9.88	3.55	1.97	0.52
Tabuenca																								
BAL-1	sh	54.46	106.77	11.82	44.37	8.98	1.79	7.78	1.15	6.60	1.34	3.65	0.57	3.72	0.56	253.56	226.39	25.38	8.92	14.62	9.88	3.82	1.69	0.66
BAL-2	sh	53.02	107.51	11.56	43.44	8.48	1.64	6.93	1.07	6.27	1.27	3.49	0.57	3.60	0.55	249.40	224.02	23.75	9.43	14.72	9.95	3.93	1.56	0.65
BAL-3	sh	47.75	92.65	10.10	37.32	7.29	1.44	6.39	1.02	6.01	1.17	3.26	0.53	3.46	0.52	218.93	195.11	22.37	8.72	13.80	9.33	4.12	1.50	0.65
BAA-11	ss	14.84	30.34	3.39	13.18	2.74	0.60	2.64	0.40	2.18	0.42	1.10	0.17	1.10	0.16	73.24	64.48	8.17	7.90	13.44	9.08	3.42	1.94	0.68

(continued.)

TABLE 3. (Continued.)

Sample	Lithology†	La	Ce	Pr	Nd	Sm	Eu	Gd	Tb	Dy	Ho	Er	Tm	Yb	Lu	REE‡	LREE	HREE	LREE/HREE	La/Yb	(La/Yb)n	(La/Sm)n	(Gd/Yb)n	Eu/Eu*
Early Cretaceous (rift 2)																								
San Andrés																								
SAN-4	ss	14.39	29.75	3.24	12.54	2.57	0.53	2.13	0.31	1.66	0.31	0.90	0.15	0.95	0.14	69.59	62.50	6.57	9.52	15.15	10.15	3.47	1.82	0.70
Yanguas																								
YNG-L6	sh	71.69	140.95	15.69	60.43	12.60	2.47	9.76	1.35	7.01	1.32	3.60	0.59	3.64	0.52	331.62	301.36	27.79	10.84	19.68	13.18	3.53	2.17	0.69
YNG-L8	sh	103.01	200.81	21.50	82.55	16.20	2.72	11.33	1.58	8.16	1.47	4.03	0.64	3.90	0.57	458.45	424.07	31.66	13.40	26.42	17.70	3.94	2.35	0.62
YNG-L12	sh	50.61	95.12	10.25	38.08	7.38	1.18	5.58	0.84	4.76	0.94	2.69	0.44	2.84	0.43	221.14	201.44	18.52	10.88	17.81	11.93	4.25	1.59	0.57
YNG-L49	sh	57.92	109.94	12.08	44.56	8.71	1.84	7.03	1.08	5.96	1.13	3.16	0.51	3.21	0.47	257.59	233.21	22.54	10.35	18.06	12.10	4.12	1.77	0.73
YNG-6/R	sh	15.71	28.10	3.14	11.75	2.17	0.32	1.48	0.16	0.89	0.18	0.55	0.09	0.62	0.10	65.26	60.87	4.07	14.97	25.23	16.90	4.50	1.92	0.55
YNG-8	ss	8.68	16.82	1.83	6.98	1.35	0.23	1.03	0.15	0.77	0.15	0.43	0.07	0.41	0.06	38.95	35.65	3.07	11.59	21.05	14.10	3.99	2.03	0.60
YNG-12	ss	5.18	10.89	1.20	4.45	1.02	0.26	0.73	0.11	0.62	0.12	0.32	0.05	0.30	0.04	25.28	22.73	2.29	9.93	17.24	11.55	3.16	1.97	0.93
YNG-49	ss	16.69	31.67	3.34	12.51	2.55	0.49	2.52	0.42	2.31	0.45	1.24	0.20	1.19	0.17	75.77	66.77	8.51	7.85	14.01	9.39	4.05	1.72	0.60
San Pedro Manrique																								
SPM-3	ss	5.12	10.83	1.25	5.04	1.19	0.22	1.10	0.17	0.89	0.16	0.42	0.06	0.40	0.05	26.91	23.43	3.26	7.19	12.82	8.59	2.66	2.23	0.59
SPM-9	ss	44.66	86.42	9.38	35.29	6.50	1.08	4.26	0.59	3.18	0.62	1.74	0.28	1.79	0.27	196.04	182.24	12.72	14.33	24.94	16.71	4.26	1.93	0.64
SPM-15	ss	50.81	97.01	10.88	41.86	8.91	2.23	8.35	1.47	8.39	1.65	4.57	0.73	4.37	0.65	241.88	209.47	30.18	6.94	11.63	7.79	3.54	1.55	0.80
SPM-30	ss	16.37	32.25	3.40	12.82	2.67	0.45	2.27	0.39	2.15	0.44	1.20	0.20	1.23	0.19	76.05	67.52	8.07	8.36	13.26	8.88	3.80	1.49	0.57
Valdemadera																								
VLM-L2	sh	27.71	51.67	5.50	20.89	4.69	1.13	5.38	1.08	6.77	1.36	3.75	0.61	3.76	0.54	134.84	110.45	23.26	4.75	7.37	4.94	3.66	1.16	0.70
VLM-L7	sh	50.13	95.90	10.43	39.51	7.71	1.25	6.49	1.11	6.55	1.30	3.59	0.58	3.68	0.54	228.78	203.68	23.85	8.54	13.62	9.13	4.03	1.43	0.55
VLM-L11R	sh	39.87	79.42	8.66	32.76	6.44	1.08	5.98	1.10	6.58	1.34	3.86	0.64	4.08	0.62	192.42	167.15	24.20	6.91	9.76	6.54	3.84	1.19	0.54
VLM-12	sh	48.75	94.45	10.35	38.63	8.14	1.69	7.45	1.28	7.38	1.41	3.93	0.63	3.99	0.58	228.67	200.33	26.65	7.52	12.21	8.18	3.71	1.51	0.67
VLM-2	sh	12.71	25.00	2.82	10.52	1.80	0.33	1.19	0.16	0.84	0.16	0.45	0.07	0.48	0.07	56.59	52.84	3.41	15.48	26.36	17.66	4.38	2.00	0.70
VLM-7	ss	22.96	46.47	5.18	19.56	3.61	0.47	2.79	0.41	2.31	0.44	1.23	0.19	1.18	0.17	106.97	97.78	8.73	11.20	19.38	12.99	3.94	1.91	0.45
VLM-11	ss	6.40	13.01	1.43	5.65	1.21	0.22	0.84	0.13	0.64	0.12	0.31	0.05	0.31	0.04	30.37	27.70	2.45	11.33	20.53	13.75	3.28	2.19	0.67
VLM-12	ss	48.46	93.62	10.10	38.69	8.14	1.66	7.33	1.28	7.31	1.45	3.97	0.62	3.98	0.58	227.19	199.01	26.52	7.50	12.18	8.16	3.69	1.49	0.67
Treviiano																								
TRE-L6	sh	39.75	53.72	7.51	31.42	5.98	1.16	5.08	0.82	4.05	0.81	2.45	0.33	2.10	0.28	155.46	138.38	15.91	8.70	18.97	12.82	4.18	1.96	0.09
TRE-L7	sh	37.60	52.25	7.17	30.10	5.82	1.15	5.07	0.81	4.09	0.84	2.56	0.35	2.25	0.30	150.38	132.95	16.28	8.17	16.72	11.30	4.06	1.83	0.09
TRE-5	ss	18.96	33.78	4.03	17.01	3.32	0.60	2.62	0.42	2.08	0.40	1.22	0.18	1.08	0.16	85.86	77.10	8.16	9.45	17.62	11.91	3.59	1.98	0.09
TRE-8	ss	14.93	21.88	2.99	12.30	2.21	0.36	2.00	0.33	1.73	0.35	1.11	0.16	1.00	0.14	61.51	54.31	6.83	7.95	14.95	10.11	4.26	1.63	0.09
Cidones																								
CID-33	sh	55.77	100.04	11.78	47.86	9.00	1.74	7.36	1.25	6.25	1.25	4.00	0.56	3.73	0.50	251.08	224.45	24.89	9.02	14.97	10.11	3.90	1.60	0.09
CID-31	ss	8.48	14.06	1.84	7.78	1.62	0.33	1.25	0.17	0.87	0.17	0.51	0.07	0.50	0.07	37.70	33.77	3.60	9.38	17.12	11.57	3.29	2.04	0.09
Gan																								
GAN-15	sh	54.36	95.91	11.20	45.41	8.70	1.74	7.45	1.36	7.33	1.47	4.59	0.66	4.37	0.57	245.13	215.59	27.81	7.75	12.45	8.41	3.93	1.38	0.09
GAN-13	ss	11.77	23.84	2.94	12.48	2.50	0.50	1.62	0.24	1.11	0.21	0.66	0.09	0.65	0.09	58.70	53.52	4.68	11.44	18.16	12.27	2.97	2.03	0.09

†ss—sandstones, sh—shales.
‡Rare earth element (REE) chondrite-normalizing factors are from Taylor and McLennan (1985); REE = La-Lu; light (L) REE = La-Sm; heavy (H) REE = Gd-Lu.

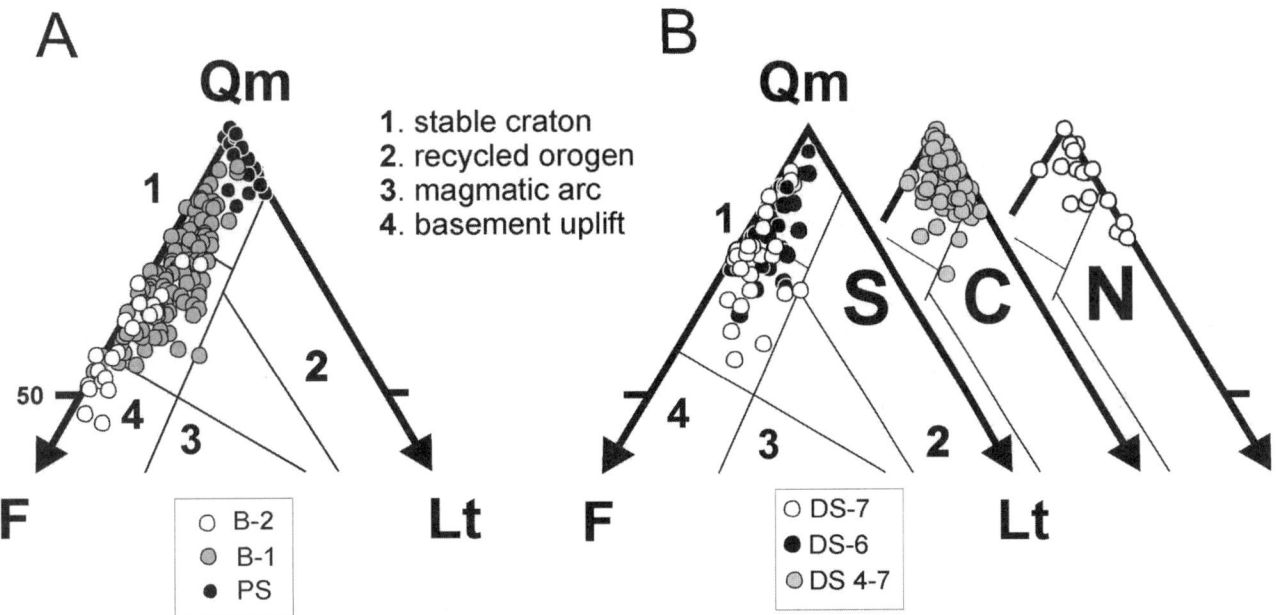

Figure 4. (A) QmFLt ternary plot (Dickinson, 1985) showing the evolution of petrofacies in Permian-Triassic sandstones. (B) QmFLt ternary plots (Dickinson, 1985) showing the variations of Berriasian to Lower Aptian petrofacies throughout the basin. PS—Saxonian facies; B-1 and B-2—Buntsandstein facies; DS-4 to DS-7—depositional sequences in Lower Cretaceous sediments.

area. These sandstone petrofacies represent the initial stage of rift 1, when metasedimentary cover was eroded.

Sandstone composition of Buntsandstein facies represents a drastic change from the provenance of underlying sediments. Quartzofeldspathic petrofacies suggest that the contribution of coarse crystalline rocks from the Hesperian Massif (Fig. 1A) diluted the supplies from metasedimentary rocks (Arribas et al., 1985). The content of feldspar varies, showing a consistent increase of K-feldspar to the top (Figs. 4A and 5B). The presence of these K-feldspar–rich petrofacies may suggest both arid conditions of Buntsandstein sedimentation (Arribas, 1984) and nearness to source area. Syntaxial K-feldspar and quartz overgrowths and minor carbonate are the main interstitial cements. Matrix is constituted by illite and minor kaolinite minerals and is mainly of diagenetic origin: epimatrix (alteration of K-feldspars), pseudomatrix (lithic rock fragment disaggregration), kaolinite pore-filling and illite pore-lining (Arribas, 1987).

Cretaceous Sandstones

The sandstone framework composition of depositional sequences 4, 5, 6, and 7 is very quartzose, with variable amounts of K-feldspar and lithics (Fig. 4B). In proximal areas (sections 1, 2, and 3 from CBA in Figs. 1 and 3), sandstone composition is quartzofeldspathic (mean $Qm_{81}F_{18}Lt_1$; S in Figs. 4B and 5C), and it evolves to more mature quartzose sandstones in depocentral areas (mean $Qm_{96}F_3Lt_1$; C in Figs. 4B and 5D). This fact suggests an important maturation during transport (~50 km) in humid climate (Rat, 1982). Quartzofeldspathic petrofacies are indicative of a plutoniclastic origin from coarse crystalline sources from the Hesperian Massif (Arribas et al., 2003). In addition, in the northeast edge of the basin, local supplies from Triassic and Jurassic sedimentary rocks (carbonate and clastics) produce quartzolithic sandstones petrofacies (mean $Qm_{93}F_1Lt_6$; N in Fig. 4B). Framework replacements by carbonate and kaolinite on K-feldspars have been identified. In addition, few metamorphic lithic grains were crushed producing pseudomatrix. As mentioned earlier, a low-grade metamorphic event (hydrothermalism) took place in depocentral areas of the basin, provoking some mineralogical changes in original sands. Some of these changes are silicification and chloritization of feldspars, metamorphic lithic grains, and intrabasinal argillaceous grains, as well as the growth of chloritoid crystals on these deposits (Barrenechea et al., 1995; Alonso-Azcárate et al., 1999; Mantilla-Figueroa, 1999).

Geochemical Composition

Major Elements

Major-element compositions of the shales and sandstones with derived geochemical parameters and indices are given in Table 1. Absolute concentrations in the different major elements (expressed in oxides) are higher in shales than in sandstones, except for SiO_2 (Table 1; Fig. 6).

Sandstones show an intermediate to high content in SiO_2, generally between 76.49% and 89.47% in Permian-Triassic (rift 1) sandstones and between 61.13% and 96.27% in Lower Cretaceous (rift 2) sandstones, and both can be considered as mature sandstones (Pettijohn et al., 1973). In some cases the SiO_2 content is extraordinarily low (57.85%, sample ALA–10, Table 1) due to pervasive carbonate cementation. In comparison, shales show a typical intermediate content in SiO_2 (between 51.98%

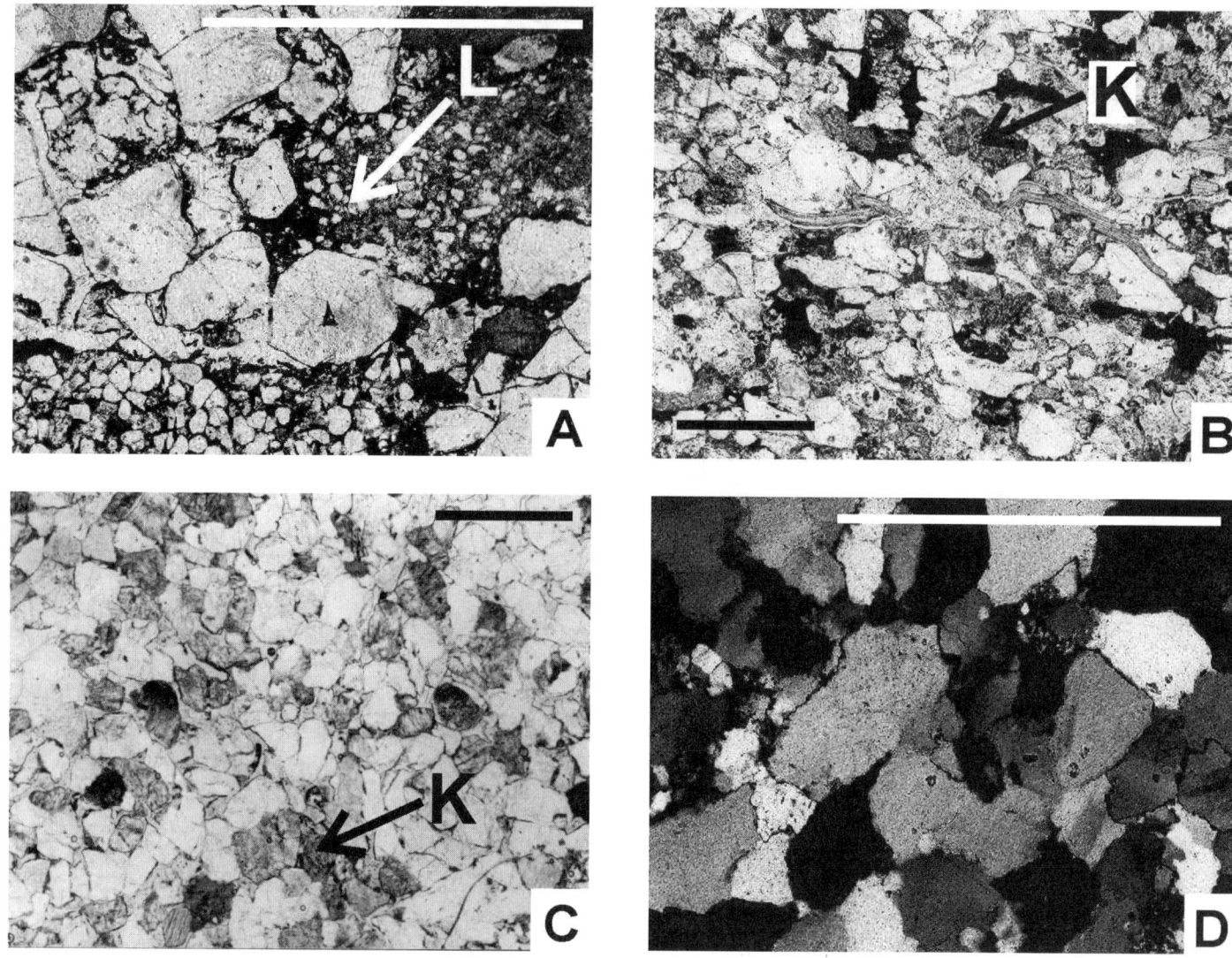

Figure 5. Microphotographs showing general aspects of petrofacies from Permian-Triassic and Lower Cretaceous sandstones. (A) Quartzolithic petrofacies of Saxonian sandstones. L—sedimentary lithic fragment. Parallel nichols. (B) Quartzofeldspathic petrofacies of Buntsandstein sandstones. K—K-feldspar. Parallel nichols. (C) Quartzofeldspathic petrofacies of Lower Cretaceous sandstones from southern zone of the basin. K—K-feldspar. Parallel nichols. (D) Quartzose petrofacies of Lower Cretaceous sandstones from center zone of the basin. Crossed nichols. Scale bars = 1 mm.

and 76.63% in rift 1 shales and between 29.74% and 78.35% in rift 2 shales).

Both shales and sandstones have an intermediate Al_2O_3/SiO_2 ratio (0.03–0.39 in rift 1 deposits and 0.02–0.50 in rift 2 deposits). This ratio in sandstones (0.09 in rift 1 sandstones and 0.06 in rift 2 sandstones) shows values lower than shales (0.27 and 0.35 in rift 1 and rift 2, respectively). The high clay minerals content in shales is reflected by a high percentage in Al_2O_3 (8.88%–21.57% in rift 1 shales and between 8.38% and 26.47% in rift 2 shales; Table 1). Lower values in $Fe_2O_3 + MgO$ are found in sandstones (1.69–4.08, mean 2.44 in rift 1; 1.50–20.77, mean 6.53 in rift 2) while shale values are higher (4.04–12.66, mean 8.22 in rift 1; 3.07–26.52, mean 16 in rift 2).

Harker diagrams for SiO_2 versus Al_2O_3 show a negative correlation (Fig. 6). Other negative correlations can be observed between SiO_2 versus Fe_2O_3, K_2O, MgO, Na_2O, and TiO_2. Nevertheless, marked positive correlations exist for Al_2O_3 versus K_2O (correlation coefficient $r = 0.86$ and 0.95 in rift 1 and rift 2 deposits), Al_2O_3 versus MgO (correlation coefficient $r = 0.77$ and 0.76 in rift 1 and rift 2 deposits), and Al_2O_3 versus TiO_2 (correlation coefficient $r = 0.69$ and 0.90 in rift 1 and rift 2 deposits) (Fig. 6).

CaO and MnO versus SiO_2 do not present a clear correlation. A CaO-SiO_2 scatter could result from carbonate cementation during diagenesis (Gu et al., 2002). Most samples present low CaO percentages (<1 wt%), though in some cases (ALA-10 sample, Table 1), high content is observed (near 16 wt%) due to the presence of carbonate cement. On the other hand, Al_2O_3

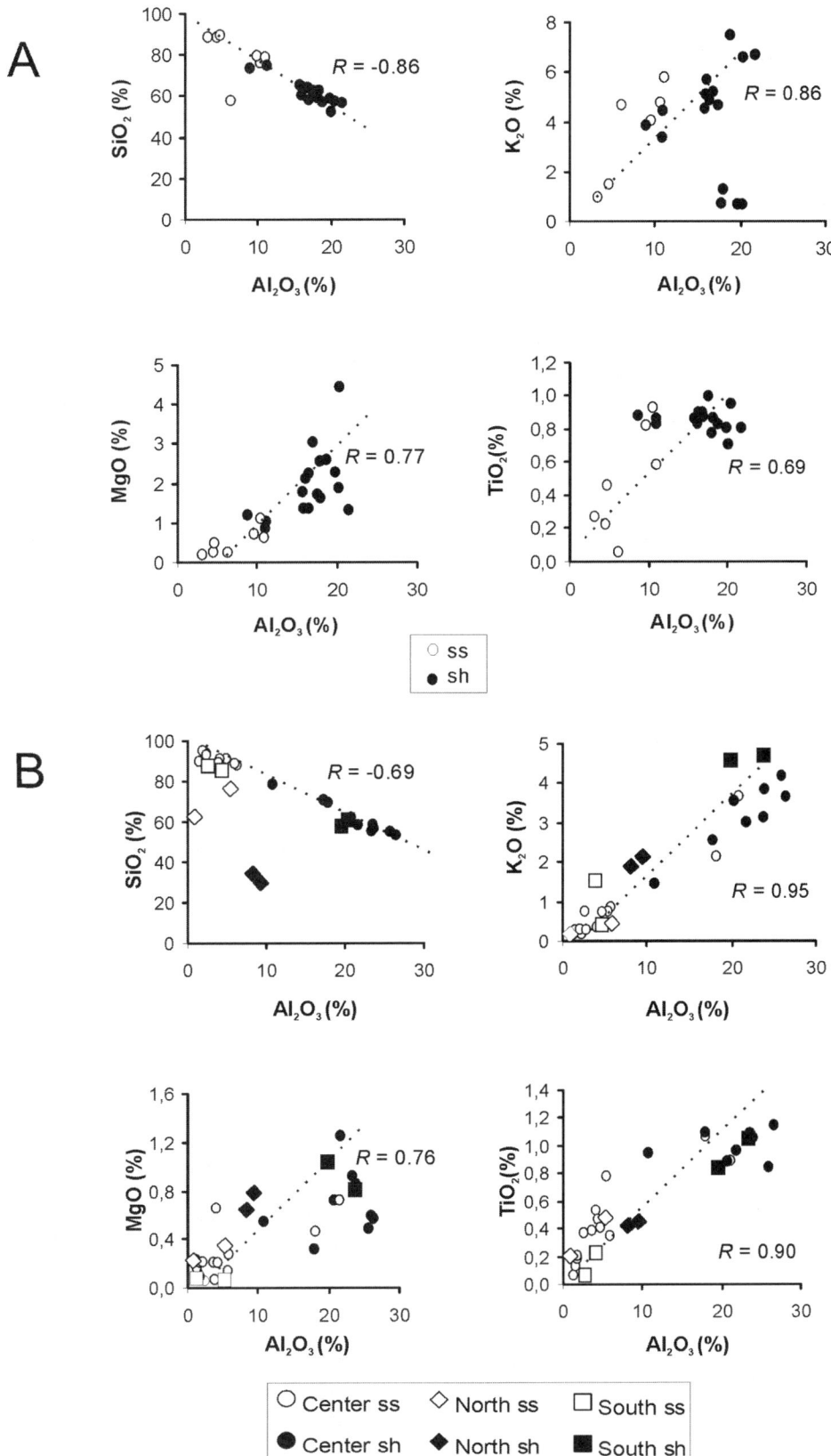

Figure 6. Harker diagrams showing major-element variation for both zones. (A) Permian-Triassic deposits, and (B) Lower Cretaceous deposits. Data are from Table 1; ss—sandstones, sh—shales.

versus CaO presents a negative correlation (correlation coefficient $r = -0.30$ and -0.18 in rift 1 and rift 2 deposits) in shales because Ca is leached during chemical weathering (Nesbitt et al., 1980).

Large Ion Lithophile Elements (LILEs)

The contents of Rb (30–242, mean 156 ppm in rift 1 deposits; 3–262, mean 106 ppm in rift 2 deposits), Cs (30–119, mean 15 ppm in rift 1 deposits; 0.60–29, mean 10 ppm in rift 2 deposits), Ba (25–13, mean 1144 ppm in rift 1 deposits; 19–1450, mean 368 ppm in rift 2 deposits), and Sr (29–385, mean 130 ppm in rift 1 deposits; 4–246, mean 81 ppm in rift 2 deposits) show a considerable scatter (Table 2), but their average values are comparable with the North American shale composite (NASC) (Gromet et al., 1984) or Average Upper Crust (AUC) (Taylor and McLennan, 1981). On the whole, shales display higher Rb, Cs, Ba, and Sr contents than sandstones.

Positive correlations are observed in K versus Rb (correlation coefficient $r = 0.31$ and 0.97 in rift 1 deposits and rift 2 deposits, respectively), K versus Sr (correlation coefficient $r = 0.31$ and 0.38 in rift 1 deposits and rift 2 deposits, respectively), K versus Cs (correlation coefficient $r = 0.25$ and 0.90 in rift 1 deposits and rift 2 deposits, respectively), and K versus Ba (correlation coefficient $r = 0.31$ and 0.78 in rift 1 deposits and rift 2 deposits, respectively). This suggests that K-rich clay minerals (such as illite) control the presence of these trace elements (McLennan et al., 1983; Feng and Kerrich, 1990). Sr content is low in rift 1 and rift 2 deposits, which would imply that the source rocks were poor in plagioclase (Feng and Kerrich, 1990).

Rare Earth Elements (REEs)

In Table 3, REE compositions are shown for shales and sandstones in each stratigraphic section. La to Lu elements were considered to determine absolute concentrations and several characteristic parameters (ΣREE, ΣLREE, ΣHREE, ΣLREE/ΣHREE, Eu/Eu*, La/Yb, La/Sm, Gd/Yb, where L is light and H is heavy.).

Samples generally show uniform and similar values to the NASC (Gromet et al., 1984) and the AUC (Taylor and McLennan, 1981). The most significant signatures are: (1) high LREE (La to Sm), (2) moderate HREE (Gd to Lu), and (3) a negative Eu anomaly. Values for shales are slightly higher than for sandstones, as seen in the chondrite-normalized REE diagrams (Fig. 7). The general tendency between sandstones and shales and among all the stratigraphic sections is very similar in all diagrams.

The enrichment in LREE in both rift 1 and rift 2 deposits is reflected by a high ratio of $(La/Yb)_n$ (7.03–13.22, mean 9.93 and between 4.94 and 17.70, mean 11.34, respectively), $(La/Sm)_n$ (3.27–4.21, mean 3.78 and 2.66–4.50, mean 3.79, respectively), and ΣLREE/HREE (6.86–11.76, mean 8.92 and 6.94–15.48, mean 9.70, respectively). A significant negative Eu anomaly (Eu/Eu*) is marked in the diagrams with values between 0.52 and 0.77, mean 0.63, in rift 1 deposits and 0.09–0.93, mean 0.49, in rift 2 deposits.

DISCUSSION

Weathering of the Source Area

The positive correlations between Al_2O_3 versus K_2O, Al_2O_3 versus MgO, and Al_2O_3 versus TiO_2 (Fig. 6) indicate that weathering was an important control in the source area in rift 1 and rift 2 deposits (Feng and Kerrich, 1990). Al and Ti are stable or residual elements during chemical weathering, while K and Mg are fixed in the clay minerals and Ca is leached (Nesbitt et al., 1980). Sandstones and shales in rift 1 and rift 2 deposits show variable degrees of negative correlations for SiO_2 versus Al_2O_3 related to the increase of mineralogical maturity (Bhatia, 1983;

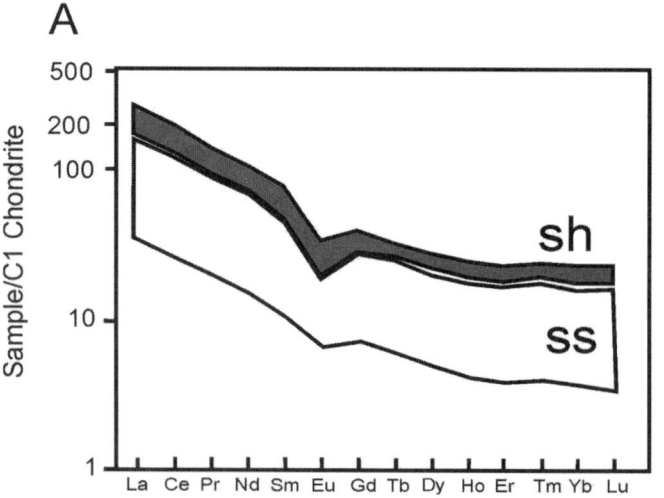 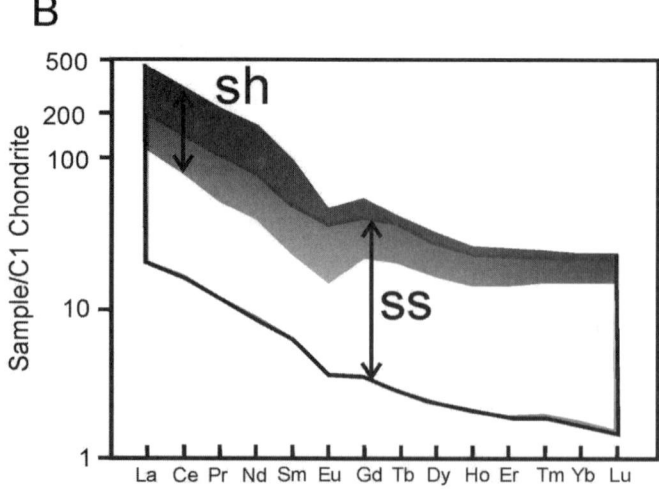

Figure 7. Chondrite-normalized rare earth element (REE) diagram showing interval values from shales and sandstones from (A) Permian-Triassic deposits and (B) Lower Cretaceous deposits. Data are from Table 2. Chondrite-normalizing values are from Taylor and McLennan (1985); ss—sandstones, sh—shales.

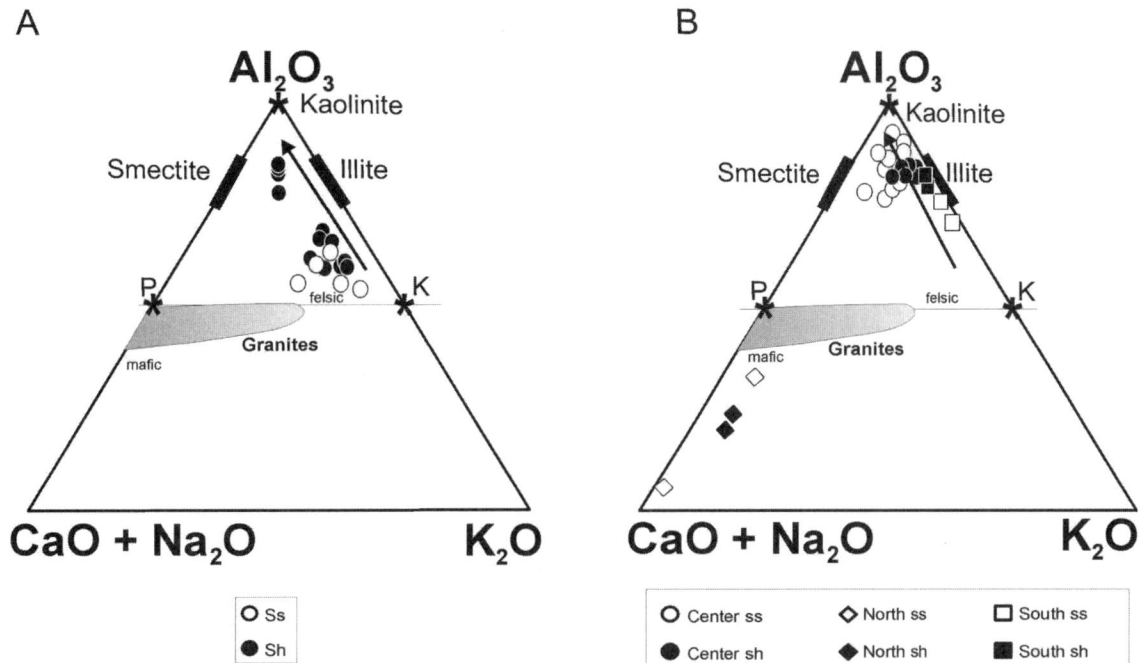

Figure 8. Chemical index of alteration (CIA) ternary plots of molecular proportions Al_2O_3-$(Na_2O + CaO)$-K_2O showing the weathering trend in (A) Permian-Triassic deposits and (B) Lower Cretaceous deposits. P—plagioclase; K—K-feldspar (after Nesbitt and Young, 1982); ss—sandstones; sh—shales.

Gu et al., 2002). These data are supported by the petrographic observations (Fig. 5).

The positive correlations between K versus Ba, K versus Rb, and K versus Cs indicate a clear relationship with alteration of minerals enriched in K (McLennan et al., 1983; Feng and Kerrich, 1990). Illite is enriched in K and Al, and as a consequence, an increase in illite corresponds to a greater abundance of Al, K, and their positively correlated elements.

The chemical index of alteration (CIA) of Nesbitt and Young (1982) was calculated to estimate the degree of weathering of source rocks (Table 1; Fig. 8). Values for sandstones vary between 63 and 73 in rift 1. Values in rift 2 sandstones are generally higher (71–91). These CIA values are according to petrographic compositional data from both rift sediments (Fig. 5). A better preservation of K-feldspar during the Triassic is observed. This fact may reflect a more intense weathering during Early Cretaceous (rift 2) than in Triassic (rift 1) times, as a consequence of climate conditions (Rat, 1982).

Rift 1 shales present higher CIA values than sandstones (70–78), suggesting that the main weathering products concentrate in shales as clay minerals. Graphic expression of the CIA suggests that weathering produces the alteration of K-feldspar to clay minerals as illite (Fig. 8A). Lower values in rift 1 sandstones can be explained as an intermediate value between those estimated for the unweathered source area (idealized value for granites, see Fig. 8A) and the shales values, outlining the weathering sequence.

CIA values of rift 2 sandstones are extremely high and plot near rift 2 shales (Fig. 8B), probably due to humid climate during sedimentation of rift 2 deposits (Rat, 1982).

In spite of these observations, the presence of hydrothermal chlorite minerals in rift 2 deposits could have increased the content of Al_2O_3 and biased interpretations about weathering (Grigsby, 2001). In addition, diagenetic alteration of K-feldspars (epimatrix) in rift 1 and rift 2 sandstones (Arribas, 1987; Mata, 1997) would increase the illite content in sandstones and thus increases in Al_2O_3 and CIA values can be expected.

Anomalous values consisting of high $CaO + Na_2O$ content are detectable in some rift 2 sandstones in the northern area of Cameros Basin (TRE-8, TRE-5, and SPM-3; Table 1; Fig. 8B). This can be explained by the nature of source rock (Triassic and Jurassic sedimentary carbonates) and interstitial carbonate cements (Ochoa et al., 2004). Furthermore, sedimentary provenance for samples in this area produces supplies with low contents in siliciclastic minerals.

Several authors (Taylor and McLennan, 1985; McLennan et al., 1995; Gu et al., 2002) have used the Th/U ratio to decipher the weathering history due to the oxidation and loss of uranium during the weathering process. Sediments of rift 1 deposits show a cluster slightly above the upper crust value, with a short weathering trend in shales (arrows in Fig. 9A). This could mean that weathering conditions were constant during Permian-Triassic sedimentation. On the other hand, rift 2 sandstones display anomalous low ratios of Th/U with low Th and U contents, and

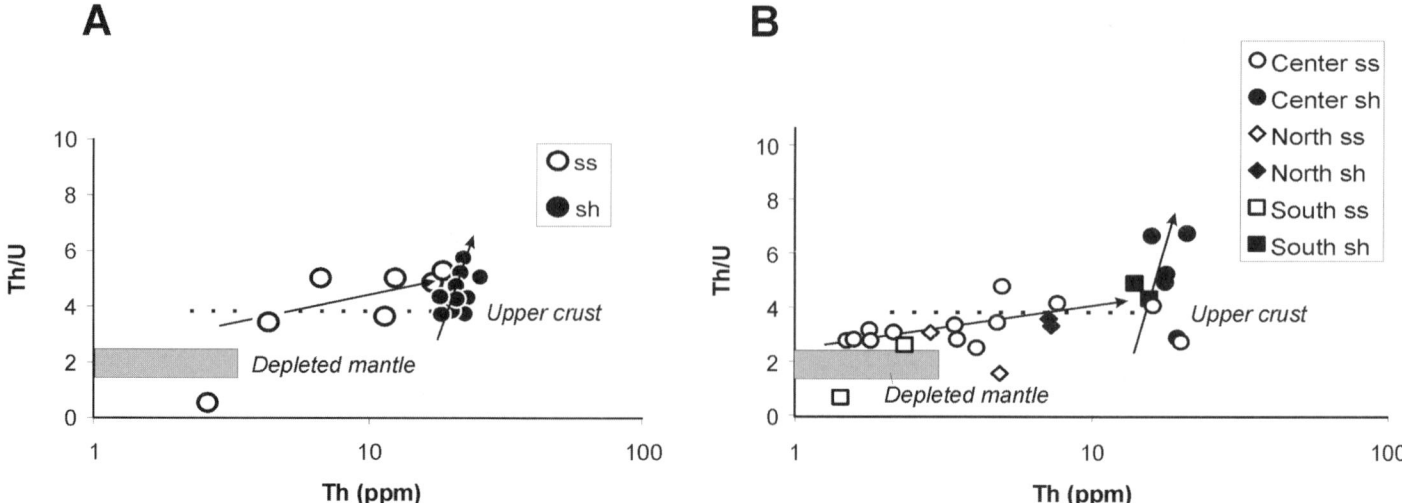

Figure 9. Elemental ratio Th/U and Th abundances (Taylor and McLennan, 1985) in (A) Permian-Triassic deposits and (B) Lower Cretaceous deposits. Arrows show weathering trends in sandstones (ss) and shales (sh).

they plot below the upper crust mean value (Fig. 9B). These anomalies could be related to provenance imprints, and probably to a coarser-grained texture in some samples that produces an important decrease of Th-rich dense minerals, as discussed next. Additionally, weathering trends in rift 2 sandstones and shales are visible and correspond to more intense weathering (higher values of Th/U ratios) than in Permian-Triassic deposits.

Tectonic Setting and Nature of Source Rocks

Using major elements, Bhatia (1983) and Roser and Korsch (1986) determined broad tectonic settings: oceanic and continental island arcs, and active and passive continental margins. According to these authors, both rift 1 and rift 2 deposits plot in or near the passive continental margin field (Fig. 10). Intracratonic grabens (aulacogens, e.g., Iberian Basin) are similar to passive-margin settings in terms of the nature of the crust (Bhatia and Crook, 1986).

Trace elements such as La, Th, Sc, Co, and Zr are transferred into clastic sediments during primary weathering, due to their low mobility. Thus, they are useful tools for provenance and tectonic discrimination (Bhatia, 1985; Taylor and McLennan, 1985; McLennan and Taylor, 1991; Bhatia and Crook, 1986; Gu et al., 2002).

We used La-Th-Sc, Th-Co-Zr/10, and Th-Sc-Zr/10 discrimination plots of Bhatia and Crook (1986) to characterize the tectonic setting (Fig. 11). In the La-Th-Sc ternary diagram, rift 1 and rift 2 samples plot in the field defined by Bhatia and Crook (1986) for graywackes from continental margins. However, no discrimination between passive and active continental margins can be observed in this diagram. Shales show a higher content in Sc than sandstones due to the increase in clay minerals where this element is fixed (Mata et al., 2000). In Th-Sc-Zr/10 and Th-Co-Zr/10 ternary plots, rift 1 and 2 sandstones plot mainly in the continental passive-margin field (Bhatia and Crook, 1986), but shales and some rift 2 sandstones plot in the continental island-arc field. This fact is again possibly due to the higher content of Sc in clay minerals. In addition, rift 2 sandstones from the central and northern area of the Cameros Basin show high contents in Zr and Co. The high contents in Zr could be associated with sorting and maturation processes during transport (McLennan et al., 1993; García et al., 2004; Whitmore et al., 2004). High content in Co can be related to provenance, as will be discussed in the following sections.

In rift 1 and 2 deposits, enrichment in LREEs, the characteristic negative Eu anomalies, and the flat HREE patterns suggest derivation from an old upper continental crust composed chiefly of felsic components (Gu et al., 2002).

Floyd and Leveridge (1987) proposed the La/Th versus Hf diagram to discriminate between different source compositions. Most rift 1 and 2 samples plot in the felsic source field (Fig. 12). However, low contents in Hf in some rift 2 samples force them to be plotted in the andesitic field (Fig. 12B).

La/Co average ratios are 10.25 and 6.63 for rift 1 and 2 sandstones, respectively (Table 2). On the other hand, Th/Co average ratios are 3.02 and 1.67 for rift 1 and 2 samples, respectively (Table 2; Fig. 13). Sands derived from granitoid sources show higher La/Co and Th/Co values than those sands derived from basaltic sources (Cullers and Berendsen, 1998). Rift 1 and rift 2 sandstones plot in an average range corresponding to sediments derived from upper continental crust.

All these general inferences about tectonic setting and nature of source rocks using minor elements in rift 1 and rift 2 deposits agree with an upper continental crust main provenance. However, some important anomalies are observed specially in some Cretaceous sandstone from the central and northern area of the basin. These anomalies are: (1) high content in Sc, Co, and Zr; (2) low content in Hf, Th, and U, and (3) anomalies in ratios like

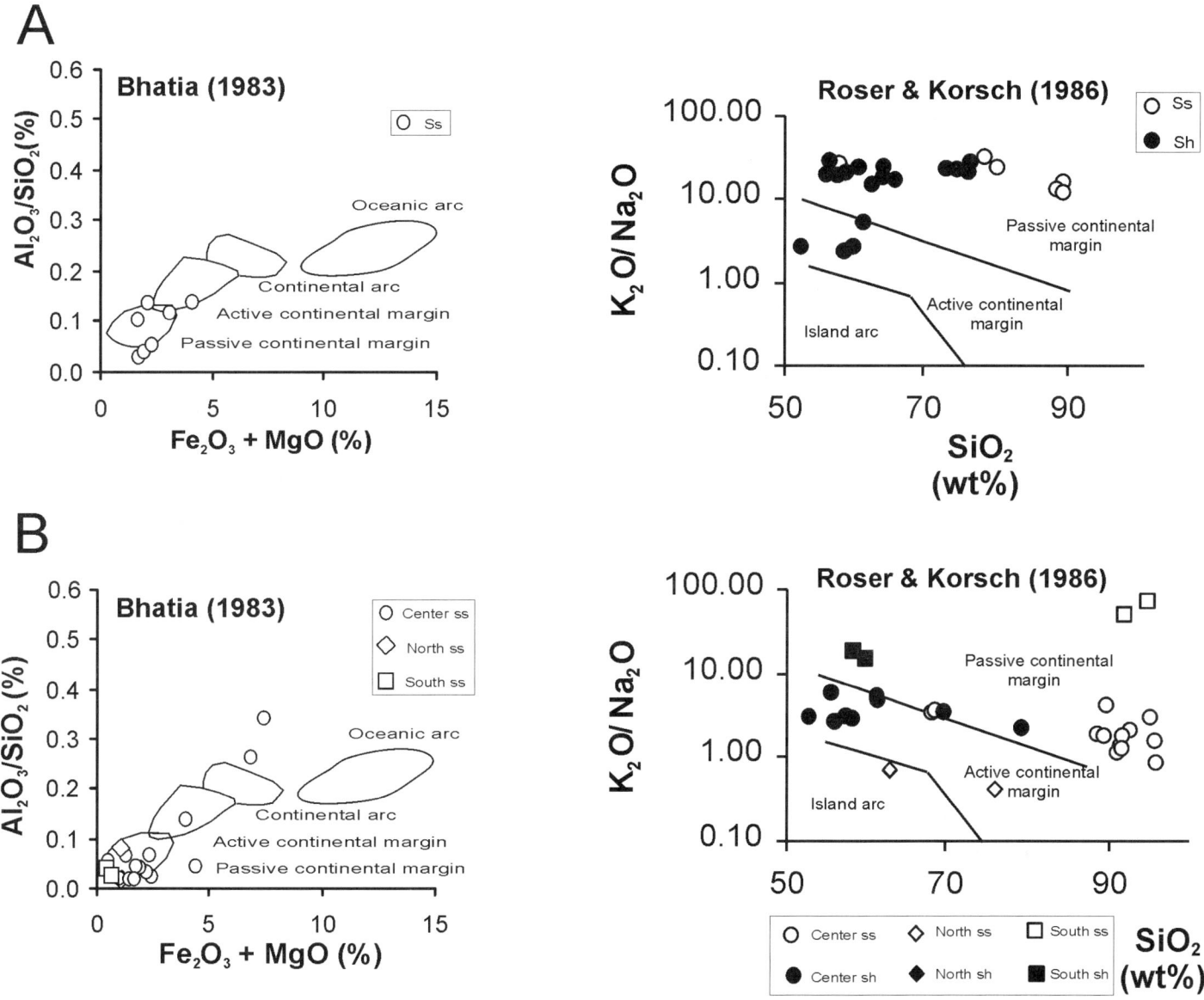

Figure 10. Tectonic-setting discrimination diagrams (ss—sandstones, sh—shales) for Permian-Triassic (A) and Lower Cretaceous samples (B).

Th/Y (mean 2.9), La/Tb (mean 35.53), Ta/Y (mean 1.11), and Ni/V (mean 0.72). These data produce important shifts in diagnostic provenance diagrams to intermediate-basic source fields (e.g., continental island arc, Bhatia and Crook, 1986; andesitic source, Floyd and Leveridge, 1987; basalt, Cullers and Berendsen, 1998). However, a clear compositional elemental spectrum that characterizes basic sources is observed, suggesting a mixture of a main felsic source with minor contribution from a mafic source (e.g., alkaline intermediate magmatism). The mafic sources must be considered in relation with a post-Buntsandstein magmatic activity, because these anomalies are not observed in the Triassic deposits (rift 1). Along the northern and southern margins of the Iberian Rift system, and during Norian to Bajocian times, several episodes of basaltic volcanic activity occurred (Salas and Casas, 1993; Martínez-González et al., 1996; Salas et al., 2001; González Menéndez and Suárez, 2005).

The presence of anomalous samples in the northern and central area of the Cameros Basin suggests that basic sources could be associated with Triassic and Jurassic sedimentary rocks located to the north of this basin that acted as sources during Early Cretaceous times.

In summary, petrography and geochemical data suggest that during the first stage of rifting (Permian-Triassic) sources were related to felsic coarse-grained rocks, associated with upper continental crust provenance (Hesperian Massif). In the second stage of rifting, during the most active phase (Berriasian–early Aptian), felsic upper crust provenance was maintained and located in the SW part of the basin (Arribas et al., 2003). In addition, in the NE

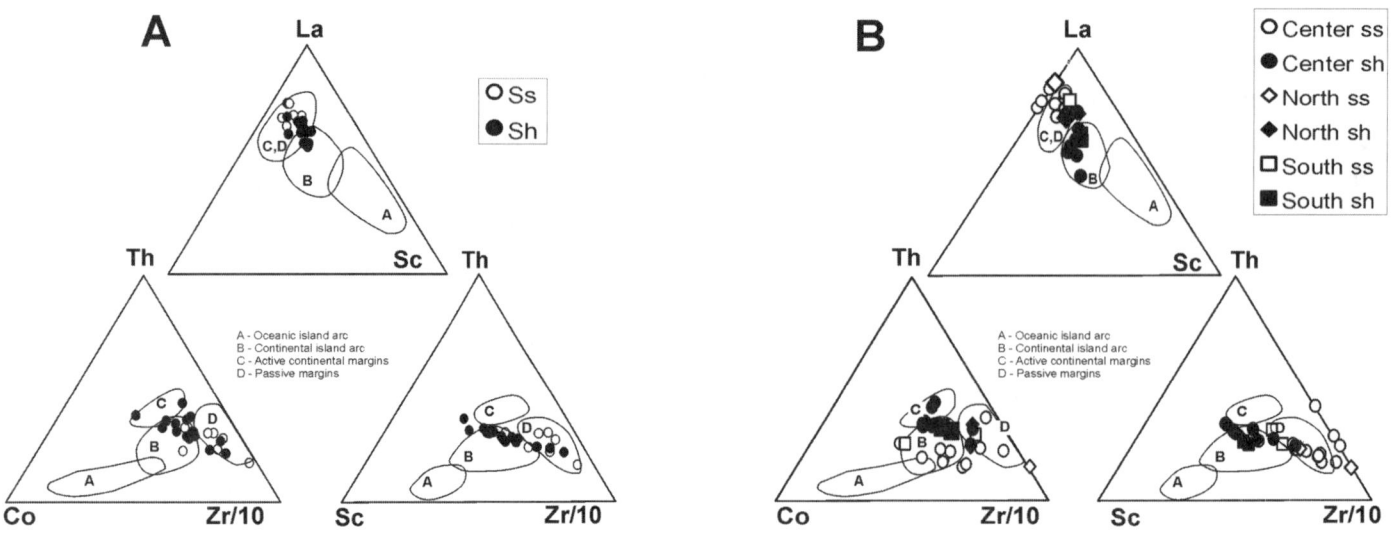

Figure 11. Tectonic-setting discrimination diagrams (ss—sandstones, sh—shales) (Bhatia and Crook, 1986) for Permian-Triassic (A) and Lower Cretaceous samples (B).

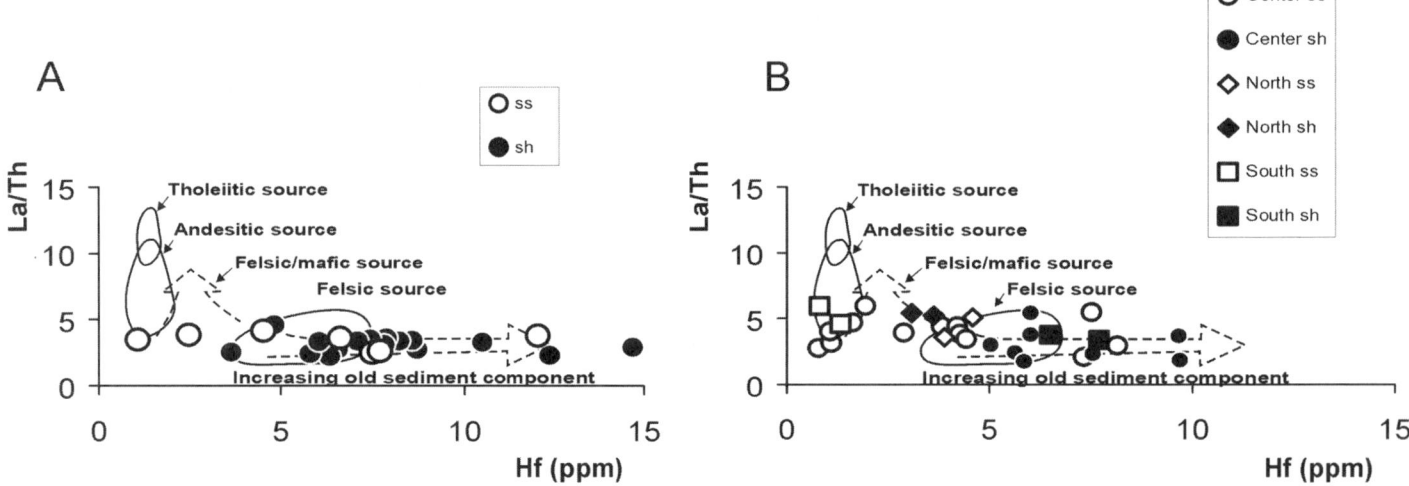

Figure 12. Source rock discrimination diagram (ss—sandstones, sh—shales) (Floyd and Leveridge, 1987) for Permian-Triassic (A) and Lower Cretaceous samples (B).

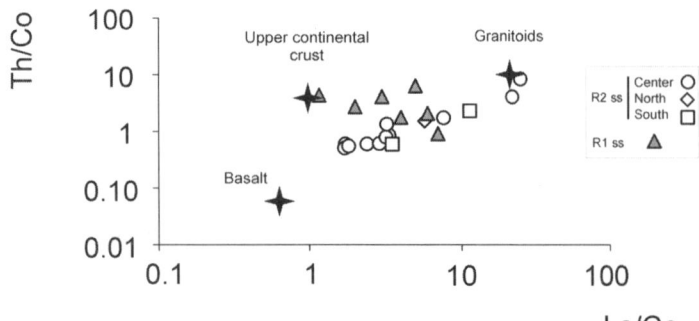

Figure 13. Source-rock discrimination binary diagram (Cullers and Berendsen, 1998), for Permian-Triassic sandstones (rift 1 ss) and Lower Cretaceous sandstones (rift 2 ss). Data about average of granitoids, upper continental crust, and basalts are also plotted.

part of the basin, contemporaneous supplies from a sedimentary cover (Permian-Triassic and Jurassic) were found in association with a volcanic imprint.

CONCLUSIONS

During the most active stages of rifting in the intracratonic Iberian Basin (rift 1: Permian-Triassic; rift 2: Late Jurassic to early Albian), quartzofeldspathic plutoniclastic petrofacies were generated in fluvial-lacustrine environments. Composition of Permian-Triassic sandstones varies in time from very quartzose quartzolithic/quartzofeldspathic at the base of the succession (Saxonian facies) to K-feldspar–rich quartzofeldspathic petrofacies at the top (Buntsandstein facies). This suggests sedimentation in arid conditions and poor maturation during transport. Composition of Berriasian–early Albian sandstones shows variations from proximal areas (quartzofeldspathic petrofacies) to depocentral zones of the basin, as a result of maturation during transport in a humid climate. In addition, sedimentoclastic petrofacies are found in the northern part of the basin. Both Permian-Triassic and Lower Cretaceous deposits are related with a provenance from the Hesperian Massif and its sedimentary cover.

Weathering inferences from geochemical data agree with petrographic deductions. Thus, CIA values in Permian-Triassic sandstones vary between 63 and 73, while in Lower Cretaceous sandstones, these values vary between 71 and 90, reflecting differences in weathering by climate conditions. However, values of CIA can be modified by (1) diagenetic processes in sandstones (illite epimatrix), which increase the Al_2O_3 content; (2) sedimentary supplies from the source area (increasing CaO + MgO content); and (3) allochemical hydrothermalism, which produces an increase in Al_2O_3 by the growth of chlorite minerals.

The use of Th/U ratio could describe a short weathering trend in Permian-Triassic deposits due to the persistent arid conditions during sedimentation. In Lower Cretaceous sandstones, weathering trends are more evident, but with very low ratios and low contents of Th and U.

Geochemical data (major, trace, and REE elements) from both rift 1 and rift 2 deposits fit well in most diagnostic diagrams used for tectonic setting and nature of source rocks.

Ratios between major (Al_2O_3, SiO_2, MgO, K_2O, Na_2O, and Fe_2O_3) and trace elements (La, Th, Sc, Co, and Zr) are in agreement with data from passive-margin settings, in terms of the nature of the crust (Bhatia, 1983; Bhatia and Crook, 1986; Roser and Korsch, 1986). REE values show an enrichment in LREE, a flat HREE pattern, and the characteristic negative Eu anomaly in both rift 1 and rift 2 deposits, suggesting a derivation from an old upper continental crust of felsic nature.

Anomalies such as (1) high content in Sc, Co, and Zr; (2) low content in Hf, Th, and U; and (3) anomalies in ratios like Th/Y, La/Tb, Ta/Y, and Ni/V in some Lower Cretaceous sediments suggest an additional basic source related to alkaline volcanism during Norian-Hettangian and Aalenian-Bajocian times. These volcanic sources could be related to the sedimentary cover (Permian-Triassic and Jurassic) located to the north of the Cameros Basin.

Finally, geochemical composition of rift deposits has manifested to be a useful and complementary tool to petrographic deductions, especially in throwing light on provenance from highly weathered sediments under different climate conditions and maturation during transport. However, many processes affecting the original detrital deposits (e.g., diagenetic processes, hydrothermalism) may produce changes in composition that could bias the provenance and weathering deductions.

ACKNOWLEDGMENTS

The authors thank E. Garzzanti and K. Sircombe for their comments and suggestions in a detailed review of the manuscript. We are grateful to M. Muñoz, M.J. Huertas, C. Villaseca, and J. Escuder for helpful suggestions and their remarks about geochemical data, which consistently improved an early version of the manuscript. Funding for this research was provided by the Spanish Government research projects BTE2001-026 and CGL2005-07445-C03-02.

REFERENCES CITED

Alonso-Millán, A., and Mas, R., 1993, Control tectónico e influencia del eustatismo en la sedimentación del Cretácico inferior de la Cuenca de Los Cameros: Cuadernos de Geología Ibérica, v. 17, p. 285–310.

Alonso-Azcárate, J., Rodas, M., Botrell, S.H., Raiswell, R., Velasco, F., and Mas, J.R., 1999, Pathways and distances of fluid flow during low-grade metamorphism: Evidence from pyrite deposits of the Cameros Basin, Spain: Journal of Metamorphic Geology, v. 17, p. 339–348, doi: 10.1046/j.1525-1314.1999.00202.x.

Arribas, J., 1984, Sedimentología y diagénesis del Buntsandstein y del Muschelkalk de la rama aragonesa de la Cordillera Ibérica (provincias de Soria y Zaragoza) [Ph.D. thesis]: Madrid, Universidad Complutense, 354 p.

Arribas, J., 1985, Base litoestratigráfica de las facies Buntsandstein y Muschelkalk en la rama aragonesa de la Cordillera Ibérica, zona norte: Estudios Geológicos, v. 41, no. 1–2, p. 47–57.

Arribas, J., 1987, Origen y significado de los cementos en las areniscas de las facies Buntsandstein (Rama Aragonesa de la Cordillera Ibérica): Cuadernos de Geología Ibérica, v. 11, p. 535–556.

Arribas, J., Marfil, R., and de la Peña, J.A., 1985, Provenance of Triassic feldspathic sandstones in the Iberian Range (Spain); significance of quartz types: Journal of Sedimentary Petrology, v. 55, no. 6, p. 864–868.

Arribas, J., Alonso-Millán, A., Mas, R., Tortosa, A., Rodas, M., Barrenechea, J.F., Alonso-Azcárate, J., and Artigas, R., 2003, Sandstone petrography of continental depositional sequences of an intraplate rift basin: Western Cameros Basin (north Spain): Journal of Sedimentary Research, v. 73, no. 2, p. 309–327.

Barrenechea, J.F., Rodas, M., and Mas, J.R., 1995, Clay mineral variations associated with diagenesis and low-grade metamorphism of Lower Cretaceous sediments in the Cameros Basin, Spain: Clay Minerals, v. 30, p. 119–133.

Basu, A., Young, S.W., Suttner, L.J., James, W.C., and Mack, G.H., 1975, Re-evaluation of the use of undulatory extinction and polycrystallinity in detrital quartz for provenance interpretation: Journal of Sedimentary Petrology, v. 45, no. 4, p. 873–882.

Benito, M.I., Lohmann, K.C., and Mas, R., 2001, Discrimination of multiple episodes of meteoric diagenesis in a Kimmeridgian reefal complex, north Iberian Range, Spain: Journal of Sedimentary Research, v. 71, no. 3, p. 380–393.

Bhatia, M.R., 1983, Plate tectonics and geochemical compositions of sandstones: The Journal of Geology, v. 91, p. 611–627.

Bhatia, M.R., 1984, Composition and classification of Paleozoic flysch mudrocks of the Eastern Australia: Implications in the provenance and tectonic setting interpretation: Sedimentary Geology, v. 41, p. 249–268, doi: 10.1016/0037-0738(84)90065-4.

Bhatia, M.R., 1985, Rare earth element geochemistry of Australian Palaeozoic greywackes and mudrocks: Provenance and tectonic control: Sedimentary Geology, v. 45, p. 97–113, doi: 10.1016/0037-0738(85)90025-9.

Bhatia, M.R., and Crook, K.A.W., 1986, Trace elements characteristics of greywackes and tectonic setting discrimination of sedimentary basins: Contributions to Mineralogy and Petrology, v. 92, p. 181–193, doi: 10.1007/BF00375292.

Bhatia, M.R., and Taylor, S.R., 1981, Trace-element geochemistry and sedimentary provinces: A study from the Tasman Geosyncline Australia: Chemical Geology, v. 33, p. 115–125, doi: 10.1016/0009-2541(81)90089-9.

Casquet, C., Galindo, C., González-Casado, J.M., Alonso-Millán, A., Mas, J.R., Rodas, M., García, E., and Barrenechea, J.F., 1992, El metamorfismo en la cuenca de los Cameros: Geocronología e implicaciones tectónicas: Geogaceta, v. 11, p. 22–25.

Cullers, R.L., and Berendsen, P., 1998, The provenance and chemical variation of sandstones associated with the Mid-Continent Rift System, U.S.A.: European Journal of Mineralogy, v. 10, no. 5, p. 987–1002.

Dickinson, W.R., 1985, Provenance relations from detrital modes of sandstones, *in* Zuffa, G.G., ed., Provenance of Arenites: NATO Advanced Science Institutes Series, C-148, p. 333–362.

Dickinson, W.R., and Suczek, C.A., 1979, Plate tectonics and sandstone compositions: American Association of Petroleum Geologists Bulletin, v. 63, p. 2164–2182.

Eynatten, H., Barceló-Vidal, C., and Pawlowsky-Glahn, V., 2003, Composition and discrimination of sandstones; a statistical evaluation of different analytical methods: Journal of Sedimentary Research, v. 73, no. 1, p. 47–57.

Feng, R., and Kerrich, R., 1990, Geochemistry of fine-grained clastic sediments in the Archean Abitibi greenstone belt, Canada: Implications for provenance and tectonic setting: Geochimica et Cosmochimica Acta, v. 54, p. 1061–1081, doi: 10.1016/0016-7037(90)90439-R.

Floyd, P.A., and Leveridge, B.E., 1987, Tectonic environment of the Devonian Gramscatho Basin, South Cornwall; framework mode and geochemical evidence from turbiditic sandstones: Journal of the Geological Society of London, v. 144, no. 4, p. 531–542.

García, D., Ravenne, C., Maréchal, B., and Moutte, J., 2004, Geochemical variability induced by entrainment sorting: Quantified signals for provenance analysis: Sedimentary Geology, v. 171, p. 113–128, doi: 10.1016/j.sedgeo.2004.05.013.

González Menéndez, L., and Suárez, O., 2005, Caracterización petrológica y geoquímica de los diques basálticos de Cadavedo (Valdés, Asturias): Trabajos de Geología, Universidad de Oviedo, v. 24, p. 81–89.

Grigsby, J.D., 2001, Origin and growth mechanism of authigenic chlorite in sandstones of the lower Vicksburg Formation, South Texas: Journal of Sedimentary Research, v. 71, no. 1, p. 27–36.

Gromet, L.P., Dymek, R.F., Haskin, L.A., and Korotev, R.L., 1984, The North American Shale Composite: Its composition, major and trace element characteristics: Geochimica et Cosmochimica Acta, v. 48, p. 2469–2482, doi: 10.1016/0016-7037(84)90298-9.

Gu, X.X., Liu, J.M., Zheng, M.H., Tang, J.X., and Qi, L., 2002, Provenance and tectonic setting of the Proterozoic turbidites in Hunan, south China: Geochemical evidence: Journal of Sedimentary Research, v. 72, p. 393–407.

Guimerà, J., Alonso-Millán, A., and Mas, R., 1995, Inversion of an extensional-ramp basin by a newly formed thrust: The Cameros Basin (N Spain), *in* Buchanan, J.G., and Buchanan, P.G., eds., Basin Inversion: Geological Society [London] Special Publication 88, p. 433–453.

Guimerà, J., Mas, R., and Alonso-Millán, A., 2004, Intraplate deformation in the NW Iberian Chain; Mesozoic extension and Tertiary contractional inversion: Journal of the Geological Society of London, v. 161, no. 2, p. 291–303.

Mantilla-Figueroa, L.C., 1999, El metamorfismo hidrotermal de la Sierra de Cameros (La Rioja, España): Petrología, geoquímica, geocronología y contexto estructural de los procesos de interacción fluido-roca [Doctoral thesis]: Madrid, Universidad Complutense, 361 p.

Martín-Closas, C., and Alonso-Millán, A., 1998, Estratigrafía y bioestratigrafía (Charophyta) del Cretácico inferior en el sector occidental de la Cuenca de Cameros (Cordillera Ibérica): Revista de la Sociedad Geológica de España, v. 11, no. 3–4, p. 253–269.

Martínez-González, R., Vaquer, R., and Lago, M., 1996, El volcanismo Jurásico de la Sierra de Javalambre (Cadena Ibérica, Teruel): Teruel, v. 86, no. 1, p. 43–61.

Mas, R., Benito, M.I., Arribas, J., Serrano, A., Guimerà, J., Alonso-Millán, A., and Alonso-Azcárate, J., 2003, The Cameros Basin: From Late Jurassic–Early Cretaceous Extensión to Tertiary Contractional Inversion—Implications of Hydrocarbon Exploration: Northwest Iberian Chain, North Spain: Barcelona, Geological Field Trip, 11, American Association of Petroleum Geologists International Conference and Exhibition 56 p.

Mata, M.P., 1997, Caracterización y evolución mineralógica de la Cuenca Mesozoica de Cameros (Soria-La Rioja) [Ph.D. thesis]: Universidad de Zaragoza, 349 p.

Mata, M.P., López-Aguayo, F., and Osácar, M.C., 2000, Una aproximación al área fuente del Weald de Cameros: Datos geoquímicas: Geotemas, v. 1, no. 3, p. 263–265.

Maynard, J.B., Valloni, R., and Yu, H.S., 1982, Composition of modern deep sea sands from arc-related basins, *in* Leggett, J.K., ed., Sedimentation and Tectonics on Modern and Ancient Active Plate Margins: Geological Society [London] Special Publication 10, p. 551–561.

McLennan, S.M., and Taylor, S.R., 1991, Sedimentary rocks and crustal evolution: Tectonic setting and secular trends: The Journal of Geology, v. 99, p. 1–21.

McLennan, S.M., Taylor, S.R., and Erriksson, K.A., 1983, Geochemistry of Archean shales from the Pilbara Supergroup, Western Australia: Geochimica et Cosmochimica Acta, v. 47, p. 1211–1222, doi: 10.1016/0016-7037(83)90063-7.

McLennan, S.M., Hemming, S., McDaniel, D.K., and Hanson, G.N., 1993, Geochemical approaches to sedimentation, provenance and tectonics, *in* Johnsson, M.J., and Basu, A., eds., Processes Controlling the composition of clastic sediments: Geological Society of America Special Paper 284, p. 21–40.

McLennan, S.M., Hemming, S.R., Taylor, S.R., and Eriksson, K.A., 1995, Early Proterozoic crustal evolution: Geochemical and Nd-Pb isotopic evidence from metasedimentary rocks, southwestern North America: Geochimica et Cosmochimica Acta, v. 59, p. 1153–1177, doi: 10.1016/0016-7037(95)00032-U.

Nesbitt, H.W., and Young, G.W., 1982, Early Proterozoic climates and plate motions inferred from major element chemistry of lutites: Nature, v. 299, p. 715–717, doi: 10.1038/299715a0.

Nesbitt, H.W., Markovics, G., and Price, R.C., 1980, Chemical processes affecting alkalis and alkaline earths during continental weathering: Geochimica et Cosmochimica Acta, v. 44, p. 1659–1666, doi: 10.1016/0016-7037(80)90218-5.

Ochoa, M., Arribas, J., and Mas, R., 2004, Changes in sandstone composition during Lower Cretaceous syn-rift fluvial sedimentation (Cameros Basin, Spain): Florence, Italy, 32nd International Geological Congress, Abstract CD, Session 242-34.

Pettijohn, F.J., Potter, P.E., and Siever, R., 1973, Sand and Sandstones: New York, Springer-Verlag, 618 p.

Rat, P., 1982, Factores condicionantes en el Cretácico de España: Cuadernos de Geología Ibérica, v. 8, p. 1059–1076.

Roser, B.P., and Korsch, R.J., 1985, Plate tectonics and geochemical composition of sandstones: A discussion: The Journal of Geology, v. 93, p. 81–84.

Roser, B.P., and Korsch, R.J., 1986, Determination of tectonic setting of sandstone-mudstone suites using SiO_2 content and K_2O/Na_2O ratio: The Journal of Geology, v. 94, p. 635–650.

Roser, B.P., and Korsch, R.J., 1988, Provenance signatures of sandstone-mudstone suites determined using discriminant function analysis of major-element data: Chemical Geology, v. 67, p. 119–139, doi: 10.1016/0009-2541(88)90010-1.

Salas, R., and Casas, A., 1993, Mesozoic extensional tectonics, stratigraphy, and crustal evolution during the Alpine cycle of the eastern Iberian basin: Tectonophysics, v. 228, p. 33–55, doi: 10.1016/0040-1951(93)90213-4.

Salas, R., Guimerá, J., Mas, J.R., Martín-Closas, C., Meléndez, A., and Alonso-Millán, A., 2001, Evolution of the Mesozoic Central Iberian Rift System and its Cenozoic inversion (Iberian Chain), *in* Cavazza, W., Robertson, A.H.F.R., and Ziegler, P., eds., Peri-Tethyan Rift/Wrench Basins and Passive Margins: Mémoires du Muséum National d'Histoire Naturelle, v. 186, p. 145–185.

Sopeña, A., and Sánchez-Moya, Y., 1997, Tectonic systems tract and depositional architecture of the western border of the Triassic Iberian Trough (central Spain): Sedimentary Geology, v. 113, p. 245–267, doi: 10.1016/S0037-0738(97)00069-9.

Taylor, S.R., and McLennan, S.M., 1981, The composition and evolution of the continental crust: Rare earth element evidence from sedimentary

rocks: Philosophical Transactions of the Royal Society of London, v. A3, p. 381–399.
Taylor, S.R., and McLennan, S.M., 1985, The Continental Crust: Its Composition and Evolution: Oxford, Blackwell Scientific Publication, 312 p.
Whitmore, G., Crook, K., and Johnson, D., 2004, Grain size control of mineralogy and geochemistry in modern river sediment, New Guinea collision, Papua New Guinea: Sedimentary Geology, v. 171, p. 129–157, doi: 10.1016/j.sedgeo.2004.03.011.

Zuffa, G.G., 1980, Hybrid arenites: Their composition and classification: Journal of Sedimentary Petrology, v. 50, no. 1, p. 21–29.
Zuffa, G.G., 1991, On the use of turbidite arenites in provenance studies: Critical remarks, *in* Morton, A.C., Todd, S.P., and Haughton, P.D.W., eds., Developments in Sedimentary Provenance Studies: Geological Society [London] Special Publication 57, p. 23–29.

MANUSCRIPT ACCEPTED BY THE SOCIETY 9 AUGUST 2006

Complex examination of the Upper Paleozoic siliciclastic rocks from southern Transdanubia, SW Hungary—Mineralogical, petrographic, and geochemical study

Andrea Varga[†]
György Szakmány[‡]
Tibor Árgyelán
Sándor Józsa
Department of Petrology and Geochemistry, Eötvös University, Pázmány Péter sétány 1/C, H–1117 Budapest, Hungary

Béla Raucsik[§]
Department of Earth and Environmental Sciences, University of Veszprém, P.O. Box 158, H–8201 Veszprém, Hungary

Zoltán Máthé[#]
Mecsek Ore Environment, P.O. Box 121, H–7614 Pécs, Hungary

ABSTRACT

A vertical section of Upper Paleozoic sandstones from southern Transdanubia (Mecsek-Villány area, Tisza mega-unit, Hungary) has been analyzed for major and trace elements, including rare earth elements (REEs). In addition, the clay mineralogy of the sandstone samples and the petrography and geochemistry of gneiss and granitoid clasts extracted from the associated conglomerates have been determined.

Geochemistry of the sandstone samples analyzed in this study shows that these rocks were predominantly derived from a felsic continental source; nevertheless, compositions vary systematically up-section. The Pennsylvanian (Upper Carboniferous) Téseny Formation has higher SiO_2 and lower Na_2O, CaO, Sr, high field strength element (HFSE), and ΣREE contents relative to the Permian strata. Its high K_2O and Rb contents together with the presence of abundant illite-sericite suggest a potassium metasomatism in this formation. Clay mineralogy and large ion lithophile element (LILE) contents of the Lower Permian Korpád Formation vary spatially and are interpreted as local variations in composition of the source region and postdepositional conditions. Zr and Hf abundances and REE patterns, however, show that this formation was derived from mature upper continental crust. The Upper Permian Cserdi

---

[†]E-mail: raucsikvarga@freemail.hu.
[‡]E-mail: gyorgy.szakmany@geology.elte.hu.
[§]E-mail: raucsik@almos.vein.hu.
[#]E-mail: mathezoltan@mecsekerc.hu.

Formation has higher TiO_2, Th, U, Y, Cr, and heavy (H) REE contents, and higher Cr/Th and Cr/Zr ratios relative to the underlying formations. These trends can be explained by a sedimentary system dominated by highly weathered detritus derived from combined recycled-orogen, basement-uplift, and volcanic-arc provenance in the Téseny Formation, with an increased proportion of less weathered detritus derived from combined volcanic and basement-uplift provenances in the Permian formations. Characteristics of the Cserdi unit may reflect relatively proximal derivation from a felsic volcanic source.

Keywords: late Paleozoic, clay mineralogy, geochemistry, provenance, European plate, Hungary.

INTRODUCTION

Sandstone geochemistry is widely considered to be a powerful tool for determining the sediment source areas and tectonic settings of ancient terrigenous deposits (Bhatia, 1985; Bhatia and Crook, 1986; Bauluz et al., 1995; Garver and Scott, 1995; Cullers and Berendsen, 1998; Zimmermann and Bahlburg, 2003). Whole-rock chemistry can also detect variations in elements that are not picked up in modal analysis, for example, rare earths elements (REE), Th, Zr, Sc, and Cr (Willan, 2003). In addition, variations in clay-mineral assemblages also may be useful for detecting changes in the source areas and for recognizing the diagenetic conditions of sandstone units (Weaver, 1989; Bauluz et al., 1995; Arribas et al., 2003). On the other hand, relatively few papers deal specifically with clast composition of coexisting conglomerates in the same locality (Floyd et al., 1991; Brügel et al., 2003; Varga et al., 2003; Noda et al., 2004); however, the clast types can provide a wealth of information concerning the provenance and geological evolution of the sediment source areas. Therefore, integrated examination of siliciclastic rocks, including clay mineralogy, whole-rock geochemistry, and clast petrography and geochemistry, is particularly fruitful in provenance analysis and paleogeographic reconstruction.

In this paper, results of clay mineralogical, petrographic, and geochemical studies of the late Paleozoic siliciclastic successions from southern Transdanubia (Mecsek-Villány area, Tisza mega-unit, SW Hungary) are presented.

Previous research on Paleozoic siliciclastic rocks in southern Transdanubia has been restricted to conventional sedimentological and petrographic analyses (Jámbor, 1969; Balogh and Barabás, 1972; Hetényi and Ravasz-Baranyai, 1976; Fazekas, 1987; Barabás and Barabás-Stuhl, 1998). Recently, Varga et al. (2001, 2003, 2004) and Varga and Szakmány (2004) reported the provenance and chemical composition of the Pennsylvanian sedimentary rocks. They showed that these sediments are predominantly composed of material from a recycled orogenic area with small amounts of volcanic-derived detritus. On the other hand, no major and trace element data for sandstones of the Permian formations are available at present. For this reason, this study was aimed at providing a contribution to a database on the elemental concentrations of the Upper Paleozoic sandstones in this region.

We concentrated this study on clay mineralogical, petrological, and geochemical analyses of sandstone and conglomerate samples. Examination of detritus in Pennsylvanian to Permian sediments in the Mecsek-Villány area is useful for gathering information about the provenance and reconstructing the paleogeographic setting of the late Paleozoic southern margin of the European plate.

GEOLOGIC AND PALEOGEOGRAPHIC SETTING

Hungary is located in the central Carpathian-Pannonian area (Fig. 1). The Mid-Hungarian line, a key element in the tectonics of the intra-Carpathian area, subdivides its pre-Tertiary basement in two parts: the Alcapa (Alpine–west Carpathian–Pannonian) block on the north and the Tisza mega-unit (Tisza-Dacia unit) on the south (Csontos and Nagymarosy, 1998; Csontos et al., 1992, 2002). Paleogeographically, at the end of the Variscan cycle, the polymetamorphic complexes of the Tisza mega-unit belonged to the southern part of the Moldanubian zone (Variscan orogenic belt), which formed the European margin of Paleotethys (Haas et al., 1999; Buda et al., 2000). During the late Paleozoic interval, the position of the Tisza mega-unit has been determined to be at the southern margin of the European plate, east of the Bohemian Massif and the western Carpathians (Fig. 2). Within it, the Mecsek-Villány area was probably located in the most external position (Haas et al., 1999).

Following the Variscan orogeny, the Tisza mega-unit was affected by an extensional tectonic regime. Variscan postorogenic sedimentation began earlier in the Villány area than in the Mecsek area and produced a Pennsylvanian nonmetamorphic (locally anchimetamorphic) molasse-type overstep sequence (Téseny Sandstone Formation), which was draped over the eroded surface of the crystalline basement (Hetényi and Ravasz-Baranyai, 1976; Fülöp, 1994; Barabás and Barabás-Stuhl, 1998; Szederkényi, 2001). This formation was deposited in a foreland basin (Jámbor, 1969; Hetényi and Ravasz-Baranyai, 1976; Varga et al., 2003). In the Early Permian, continental sedimentation continued in the previously formed molasse basins and initiated in the newly formed extensional rift troughs (Turony Formation,

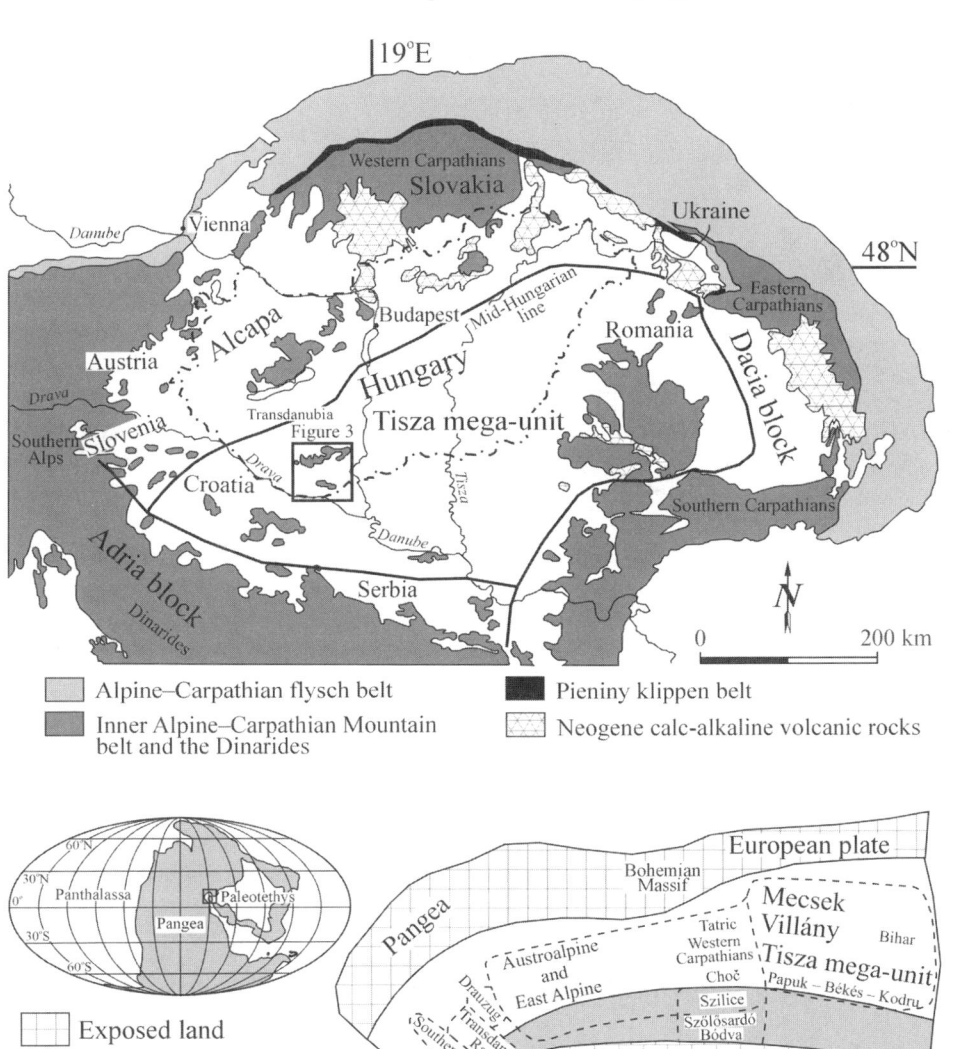

Figure 1. Geologic framework and major tectonic units of the Carpathian–Pannonian area, after Csontos et al. (1992, 2002).

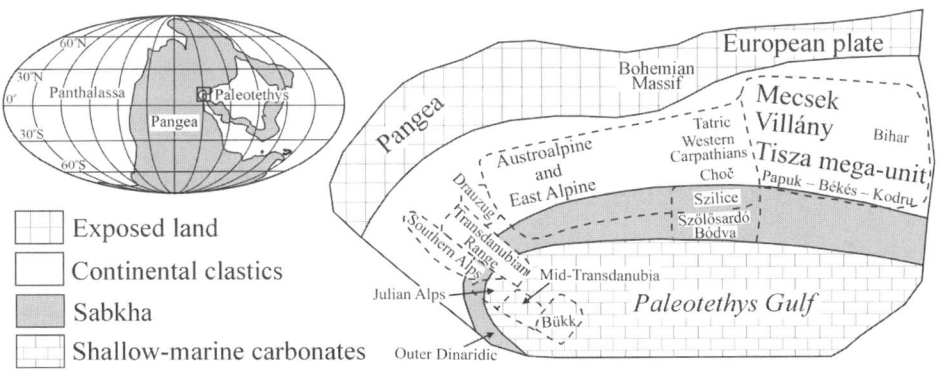

Figure 2. Paleogeographic reconstruction for the Late Permian, modified from Haas et al. (1999). Index map shows the plate tectonic configuration for Pangea (ca. 255 Ma) with the location of the Carpathian–Pannonian area, after Golonka and Ford (2000).

Korpád Sandstone Formation). At the end of the Early Permian, deposition of continental siliciclastic sediments was punctuated by an intense acidic magmatic activity (Gyűrűfű Rhyolite Formation). In the Late Permian, continental red beds (Cserdi Formation, Boda Siltstone Formation, and Kővágószőlős Formation) characterized the development of the Mecsek-Villány area (Fülöp, 1994; Barabás and Barabás-Stuhl, 1998; Haas et al., 1999). These formations tend to be barren in terms of their fauna, and accurate and reliable biostratigraphic correlations are often impossible. Therefore, correlation of the continental Paleozoic sequences of the Tisza mega-unit in the subsurface is very difficult (Barabás and Barabás-Stuhl, 1998).

Within the crystalline basement of the Tisza mega-unit, three terranes have been distinguished: (1) Slavonia-Drava unit, which can be subdivided into the Babócsa (or Görgeteg; after Fülöp, 1994) and Baksa subunits, (2) Kunság unit, including Variscan granitoids of the Mórágy Complex, and (3) Békés unit (Szederkényi, 2001). In southern Transdanubia, the study area includes the western Mecsek Mountains, which are part of the Kunság unit, and the western flank of the Villány Mountains, which are part of the Slavonia-Drava unit. They are composed of Variscan crystalline basement and/or Carboniferous granitoids, late Paleozoic siliciclastic cover, and a Mesozoic sequence (Figs. 3 and 4).

The Paleozoic formations studied in this paper are the Téseny, Korpád, and Cserdi Formations (Figs. 4 and 5). The Téseny Formation (Pennsylvanian), which is interpreted as fluvial system deposits, unconformably overlies the crystalline basement (Babócsa and Baksa metamorphic complexes) and has a maximum thickness of ~1500 m. This formation is found partly in the western flank of the Villány Mountains and also in the Drava Basin. It is composed of conglomerate, sandstone, and

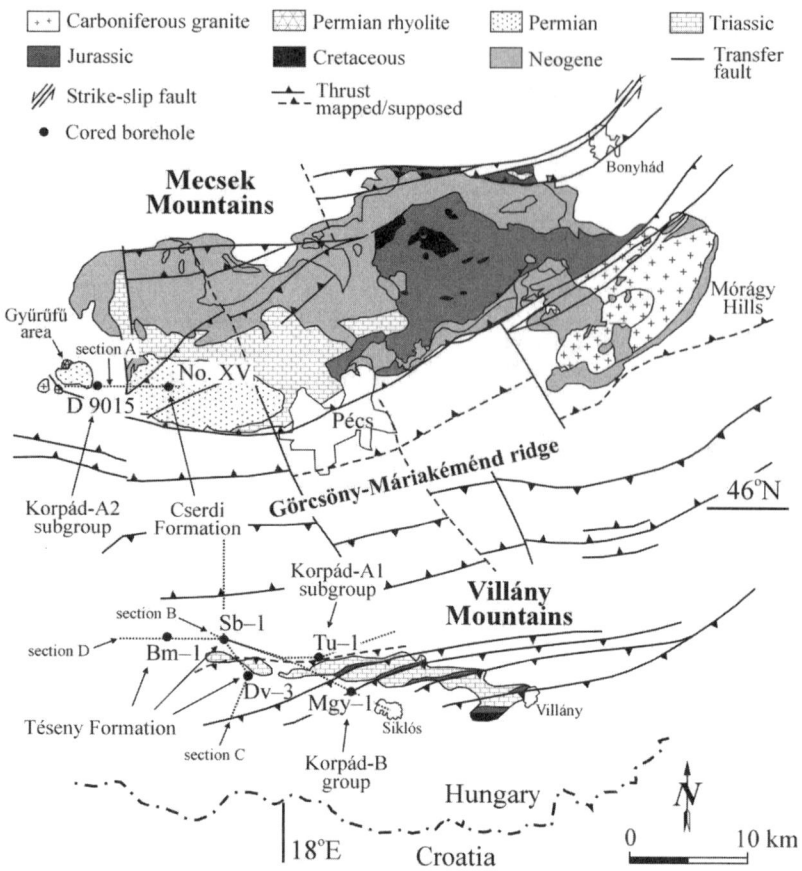

Figure 3. Generalized geologic map of the Mecsek-Villány area (southern Transdanubia, Hungary), showing the localities where analyzed samples were collected in this study; cross sections A–D are shown in Figure 4 (modified after Nagy [1968] with structural geological data of Csontos et al. [2002]).

siltstone; shale and coal seams also occur (Jámbor, 1969; Hetényi and Ravasz-Baranyai, 1976; Varga et al., 2003). These rocks contain a Namurian–Westphalian flora composed of *Pecopteris, Sphenopteris, Neuropteris, Alethopteris, Sphenophyllum, Annularia, Calamites* assemblage and Westphalian palynomorphs (Hetényi and Ravasz-Baranyai, 1976). The alluvial Korpád Formation (Lower Permian) ranges up to 700 m in thickness and consists of polymictic basal conglomerate, breccia, sandstone, siltstone, and claystone. However, extreme variation in thickness has been documented (Balogh and Barabás, 1972; Fazekas, 1987; Barabás and Barabás-Stuhl, 1998). Rocks of this formation contain a sparse Lower Permian plant flora (e.g., *Pecopteris, Voltzites*) and a lowermost Permian microflora composed of *Potonieisporites* and *Vittatina* assemblage (Barabás and Barabás-Stuhl, 1998). The Cserdi Formation (Upper Permian) consists of up to 1000 m of polymictic conglomerate, sandstone, and siltstone beds representing debris flow–dominated alluvial fan deposits (Balogh and Barabás, 1972; Fazekas, 1987; Barabás and Barabás-Stuhl, 1998). Due to the lack of palynomorphs within these continental sediments, precise dating and biostratigraphic correlation among natural outcrops and boreholes is poorly constrained (Barabás and Barabás-Stuhl, 1998).

SAMPLING AND ANALYTICAL METHODS

The Pennsylvanian and Lower Permian formations occur in the subsurface in southern Transdanubia. Téseny rocks are known from numerous boreholes in the western flank of the Villány Mountains (Figs. 3–5), for example, in boreholes Bogádmindszent-1 (Bm-1), Diósviszló-1 (Dv-3), and Siklósbodony-1 (Sb-1). In the study area, the Lower Permian Korpád Sandstone was penetrated by boreholes Turony-1 (Tu-1), Máriagyűd-1 (Mgy-1), and Dinnyeberki 9015 (D 9015). The Upper Permian sediments crop out in the western part of the Mecsek Mountains, where a cored exploration borehole (No. XV) that penetrated Cserdi clastics was chosen for detailed observations (Figs. 3–5). A total of 72 sandstone and conglomerate-breccia core samples from these boreholes and 5 representative samples of pebble-sized gneiss clasts from borehole Mgy-1 were newly collected for this study (Table DR1).[1] In addition, the results from previous petrological and geochemical studies of the Téseny Formation in this area (Varga and Szakmány, 2004; Varga et al., 2001, 2003, 2004) have been integrated into this paper (65 sandstone and conglomerate core samples; Tables DR1–DR3 [see footnote 1]).

[1]GSA Data Repository item 2007168, data tables DR1 to DR6, is available on the Web at http://www.geosociety.org/pubs/ft2007.htm. Requests may also be sent to editing@geosociety.org.

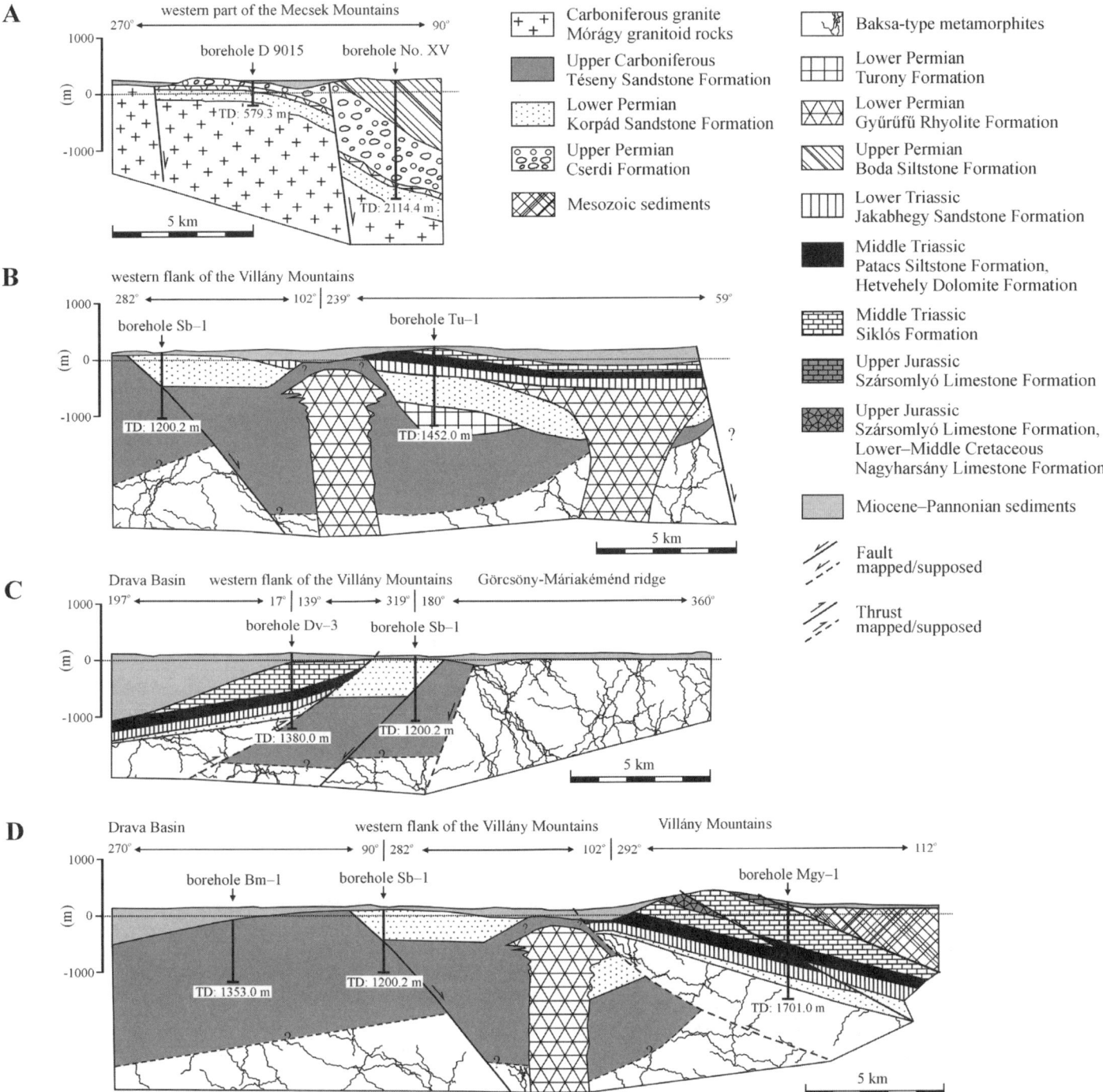

Figure 4. Schematic cross sections of the Mecsek-Villány area, showing the stratigraphic and structural relationships in the field area (modified after Barabás and Barabás-Stuhl [1998] and Barabás-Stuhl [2000], personal commun.). TD—total depth below the surface.

A total of 42 sandstone samples, including Téseny rocks, were selected for mineralogy (Table DR2). The mineralogical analyses of the whole rocks and their clay-size fraction were performed at the Department of Earth and Environmental Sciences of University of Veszprém (Hungary) by X-ray diffraction (XRD) using a Philips PW 1710 diffractometer.

Twenty-four representative rock samples and five gneiss clasts were used for geochemistry (Tables DR4–DR6, see footnote 1). Major and trace element (Rb, Sr, Ba, Pb, Y, V, Ni, Zn, and Co) abundances were established by X-ray fluorescence (XRF) analysis using a Bruker AXS S4 Pioneer instrument in the laboratory of the University of Tübingen (Department of

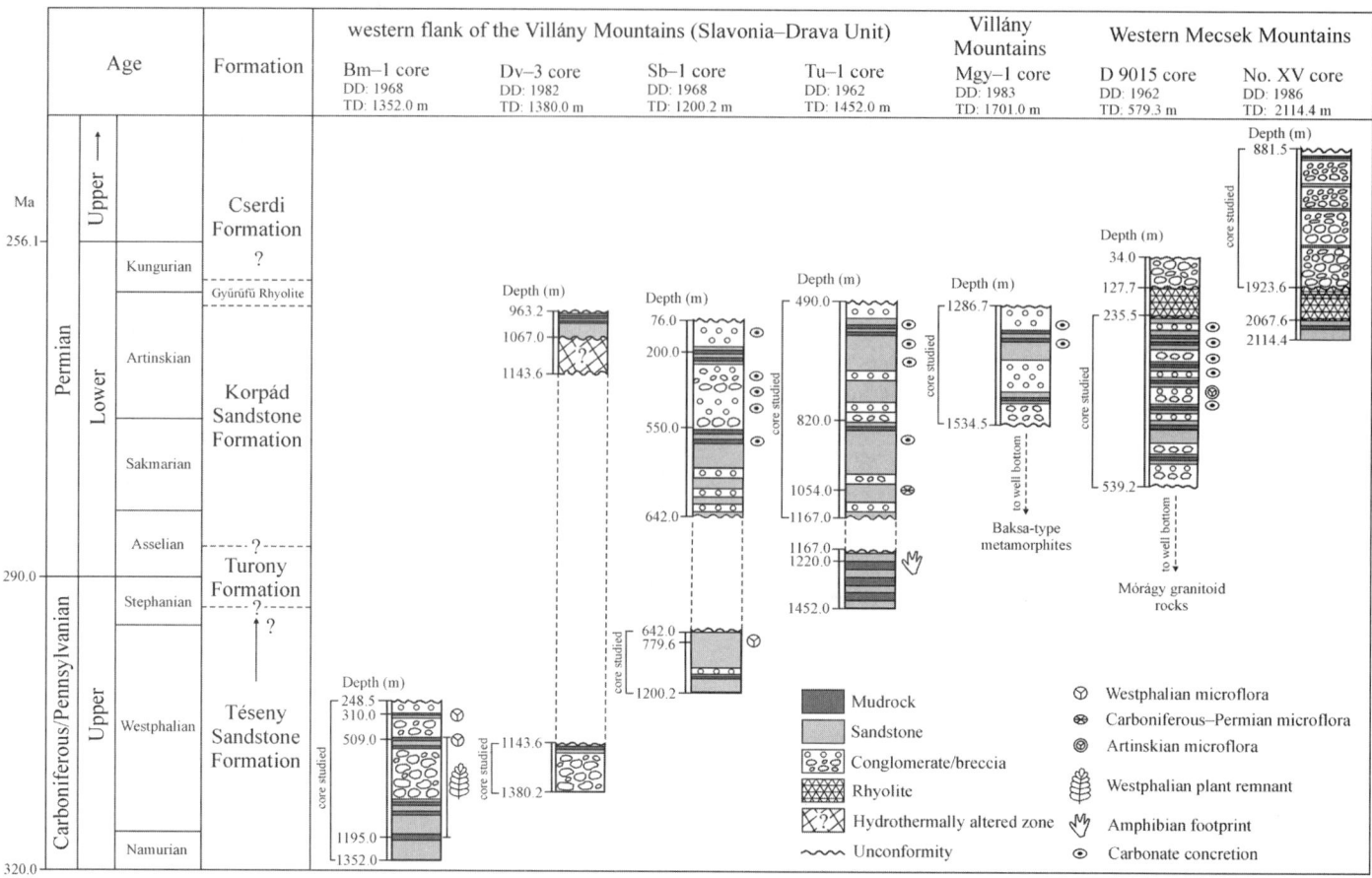

Figure 5. Schematic lithologic logs of the boreholes, showing the stratigraphic relationships of the cores studied and the position of the investigated section within the hole (modified after Fülöp [1994], Barabás and Barabás-Stuhl [1998], and Varga et al. [2004]). DD—date of drilling; TD—total depth below the surface.

Geochemistry, Germany). Chemical analyses of some minor elements were performed at the ACME Analytical Laboratories (Vancouver, Canada) using the following techniques: REE, Zr, and Nb were quantified by inductively coupled plasma–mass spectrometry (ICP-MS); Sc, Th, U, Ta, Hf, and Cr were quantified by neutron activation analysis (NAA). In additional, we have included the geochemical results of five gneiss and granitoid clasts collected from Téseny conglomerate to supplement our data set (Table DR6, see footnote 1). Analytical procedures are reported in Varga et al. (2003).

The results of previous petrological studies (Jámbor, 1969; Hetényi and Ravasz-Baranyai, 1976; Fazekas, 1987; Barabás and Barabás-Stuhl, 1998; Varga et al., 2001, 2003; Árgyelán, 2004, 2005) indicated that rocks of the Pennsylvanian and Lower Permian formations studied in this paper are not composed of detrital carbonate rock fragments. However, within these sedimentary rocks, fracture-controlled vein systems dominated by quartz and carbonates are present. In three samples (samples 15/17, MGY/7, and XV/55) with more than 4.5 wt% of CaO, increased CaO content is accompanied by increased LOI (loss on ignition) content (up to 11.41 wt%), suggestive of the presence of postdepositional carbonate minerals. Chemical composition of these samples does not reflect the primary composition of detritus entering the depositional basin and cannot be used to interpret the provenance area and tectonic setting; therefore, these samples were avoided.

A complete description of the methods used for XRF and XRD analyses is available in the Appendix.

RESULTS

Mineralogy and Petrography

Téseny Sandstone Formation

The samples are poorly to moderately sorted lithic-feldspathic wacke to arenite and arkose (Folk, 1968). They are composed of variable amounts of mono- and polycrystalline quartz grains, plagioclase, K-feldspar, micas, chlorite, clays, Fe-oxides, and lithic grains, such as quartz-rich metamorphic, acidic–intermediate volcanic, granitoid, and siliciclastic rock fragments (Fig. 6A). Moreover, there are some accessory minerals such as zircon, tourmaline, apatite, rutile, and opaque grains. Illite-sericite

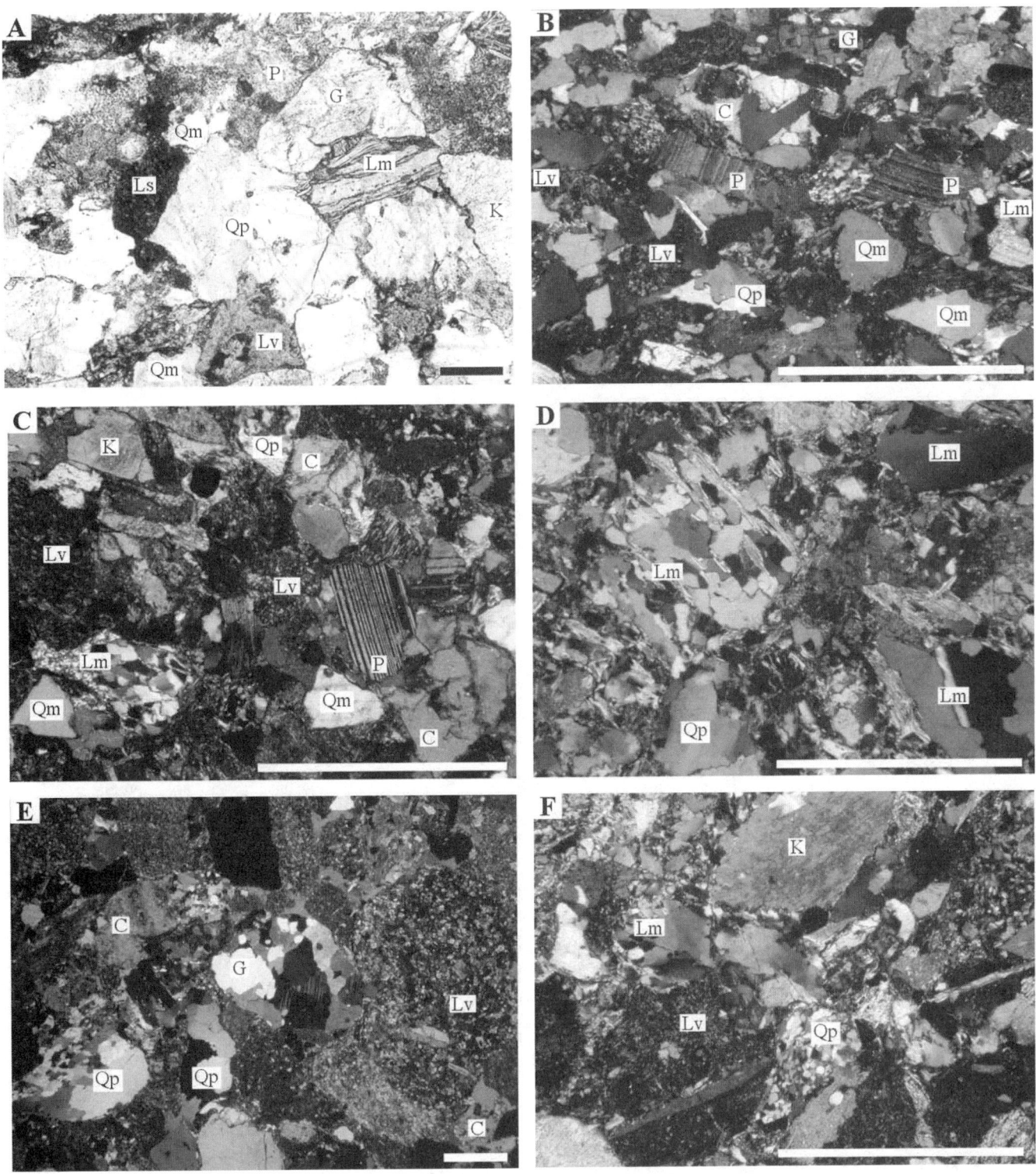

Figure 6. Thin-section photomicrographs of detrital components and textures in southern Transdanubia siliciclastic rocks. Scale bar in all photographs is 1 mm. (A) Very coarse-grained Téseny sandstone (Bm-1 core). The framework exhibits dense packing generated by compaction. Plane-polarized light. (B) Medium-grained Korpád-A sandstone (Tu-1 core). Cross-polarized light. (C) Medium-grained Korpád-A sandstone (D 9015 core). Cross-polarized light. (D) Coarse-grained Korpád-B sandstone (Mgy-1 core) exhibiting a poorly sorted quartzose framework with metamorphic lithic fragments. Cross-polarized light. (E) Fine-pebble Korpád-A conglomerate (D 9015 core) exhibiting a great variety of clasts. Cross-polarized light. (F) Coarse-grained Cserdi sandstone (No. XV core). Cross-polarized light. Abbreviations: Qm—monocrystalline quartz; Qp—polycrystalline quartz; P—plagioclase; K—K-feldspar; M—muscovite; Ls—siliciclastic rock fragment; Lm—metamorphic rock fragment; Lv—volcanic fragment; G—granitoid fragment; C—carbonate.

TABLE 1. PETROGRAPHIC DATA OF THE SANDSTONE SAMPLES

Formation	Core	Lithology (Folk, 1968)	Sorting	Q types	F types	R types	Accessory minerals	Diagenetic replacement	Cements
Téseny	Bm-1	Arkose, subarkose, sublitharenite	Poor to moderate	Qp > Qm	K > P, acidic K: microcline, orthoclase	Lv: acidic to intermediate, (G) Lm: Q-Ab-mu schist, paragneiss Ls: mudrocks, chert	mu, bio, chl, zir, tou, ap, rut, (gar), opaque minerals	P → ill/ser, ka (K → ill/ser, ka) bio → chl	ill/ser, silica, (chl, cc, dol, sid)
	Dv-3	Subarkose, litharenite, sublitharenite	Moderate	Qp > Qm	K > P, acidic K: microcline, orthoclase	Lv: acidic to intermediate Lm: Q-Ab-mu schist, paragneiss Ls: mudrocks, chert	mu, bio, chl, tou, rut, zir, ap, pyr, opaque minerals	F → ill/ser, ka bio → chl hydrothermally pyritized	ill/ser, silica, (Fe-oxide, chl)
	Sb-1	Litharenite, sublitharenite	Poor	Qp >> Qm	Ab >> K K: orthoclase	Lv: (acidic to intermediate) Lm: Q-Ab-mu schist, paragneiss, quartzite Ls: (mudrocks, chert)	mu, bio, chl, zir, rut, ap, opaque minerals	P → ill/ser bio → chl	ill/ser, silica, Fe-oxide, chl
Korpád	Tu-1 (A1)	Feldspathic-lithic arenite, subarkose, arkose	Poor to well	Qm > Qp	K ≥ P, acidic K: orthoclase	Lv: acidic to intermediate, (G) Lm: low-grade metamorphites Ls: mudrocks	mu, bio, zir, rut, ap, tou, mon, (gar), opaque minerals	F → ill/ser, ka bio → chl	cc, dol, Fe-oxide, ill/ser, chl
	D 9015 (A2)	Arkose, subarkose, lithic arenite	Poor to well	Qm > Qp	K ≥ P, acidic K: orthoclase, (microcline)	Lv: acidic to intermediate, (G) Lm: (Q-mu schist) Ls: mudrocks	mu, bio, chl zir, tou, rut, (gar), opaque minerals	F → ill/ser bio → chl	Fe-oxide, ill/ser, cc, dol, (chl)
	Mgy-1 (B)	Feldspathic-lithic wacke, subarkose, arkose	Poor	Qm > Qp	K ≥ P K: orthoclase	Lm: gneiss, Q-mu schist, Q-F-mu schist Ls: chert	mu, (bio), rut, zir, tou, opaque minerals	F → ill/ser, ka, cc recrystallized K bio → chl	Fe-oxide, ill/ser, cc
Cserdi	No. XV	Litharenite, lithic-feldspathic wacke	Poor to moderate	Qm > Qp (resorbed Qm)	P > K K: orthoclase, microcline, (sanidine, perthite)	Lv: acidic to intermediate, mafic(?) glass, (G) Lm: Q-mu schist, Q-F-mu schist, sericite schist (orthogneiss)	(mu), bio, tou, rut, zir, ap, mon, (gar), opaque minerals	Lm → chl Lv glass → chl bio → chl	Fe-oxide, cc, dol, ill/ser, chl

Notes: Abbreviations: Q—quartz; F—feldspar; R—rock fragment; Qm—monocrystalline quartz; Qp—polycrystalline quartz; P—plagioclase; K—K-feldspar; Ab—albite; Lv—volcanic fragment; Lm—metamorphic rock fragment; Ls—sedimentary rock fragment; G—granitoid fragment; mu—muscovite; bio—biotite; chl—chlorite; zir—zircon; tou—tourmaline; ap—apatite; gar—garnet; rut—rutile; pyr—pyrite; mon—monazite; ill—illite; ser—sericite; cc—calcite; dol—dolomite; sid—siderite. Component identified in minor amounts is given in parentheses.

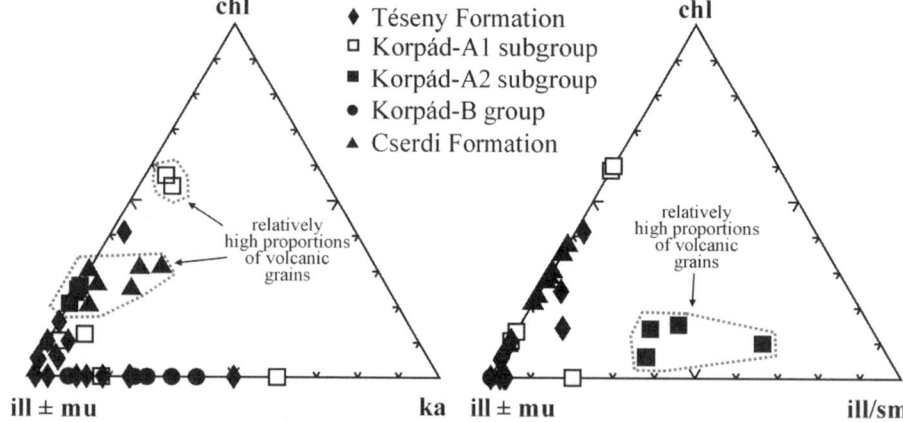

Figure 7. Triangular plots of sandstone samples based on relative abundance (%) in the clay fraction (<2 μm) of clay minerals. Abbreviations: ill ± mu—illite ± muscovite; chl—chlorite; ka—kaolinite; ill/sm—mixed-layer illite-smectite. Clay-size X-ray diffraction (XRD) data are available in Table DR2 (see text footnote 1).

replacement of feldspar grains (mainly plagioclase) is typical. Illite-sericite, silica, and chlorite cements are common. Additionally, hematite, siderite, and rare calcite cements also occur (Table 1). The clay-mineral assemblage consists predominantly of illite ± muscovite with minor proportions of chlorite with an Fe(II)-rich interlayer, kaolinite, and mixed-layer illite-smectite (Fig. 7).

Greater amounts of K-feldspar and volcanic rock fragments are noted in boreholes Bm-1 and Dv-3. In borehole Sb-1, plagioclase-rich metamorphic lithic fragments are the main components, and K-feldspar and volcanic grains appear in low proportions.

The poorly to moderately sorted polymictic Téseny conglomerate samples are characterized by metamorphic (gneiss, quartz-muscovite-albite schist, phyllite, mylonite, metagranitoid, metaquartzite), sedimentary (mudrock, sandstone, chert), and acidic–intermediate volcanic (rhyolite, dacite, trachyandesite, andesite) rock clasts (Table 2).

Korpád Sandstone Formation

Two groups of Korpád clastics are recognized on the basis of particle composition. One is volcanic-rich (Korpád-A) sediment from boreholes Tu-1 and D 9015; the other is metamorphic lithic-rich (Korpád-B) rock from borehole Mgy-1.

TABLE 2. PETROGRAPHIC DATA OF THE CONGLOMERATE AND BRECCIA SAMPLES

Formation	Core	Clast types	Roundness	Sorting	Fabric	Grain size (mm) Range	Grain size (mm) Average
Téseny	Bm-1	Rv—rhyolite, dacite, trachyandesite, andesite; Rm—gneiss, Q-mu schist, Q-mu-Ab schist, phyllite, mylonite, metagranitoid, metaquartzite/quartzite; Rs—mudrock, sandstone, chert; Others—silicified caustobiolithic rocks, feldspar megacryst	(Subangular), Subrounded, (rounded)	Poor to well	Grain supported	4–50	14
	Dv-3	Rv—rhyolite, dacite, trachyandesite, andesite; Rm—gneiss, Q-mu schist, Q-mu-Ab schist, phyllite, mylonite, metagranitoid, metaquartzite/quartzite; Rs—mudrock, sandstone, chert; Others—silicified caustobiolithic rocks, feldspar megacryst	Rounded, well rounded	Poor to moderate	Grain supported	6–80	19
	Sb-1	Rv—(rhyolite, dacite, trachyandesite, andesite); Rm—gneiss, Q-mu schist, Q-mu-Ab schist, phyllite, mylonite, metagranitoid, metaquartzite/quartzite; Rs—(mudrock, sandstone, chert)	Angular, subangular, (subrounded)	Poor (to moderate)	Grain supported, slightly oriented texture	4–45	12
Korpád	Tu-1 (A1)	Rv—recrystallized acidic–intermediate volcanites; Rm—P gneiss, Q-mu schist, metaquartzite, metasandstone, (granitoid/metagranitoid); Rs—(resedimented claystone, wacke, chert)	Subrounded	Moderate to well	Grain supported	3–18	5
	D 9015 (A2)	Rv—recrystallized acidic–intermediate volcanites, (andesite); Rm—P gneiss, Q-mu schist, metaquartzite, metasandstone, (granitoid/metagranitoid); Rs—(resedimented claystone, wacke, chert)	Subrounded	Poor	Grain supported	3–30	12
	Mgy-1 (B)	Rm—gneiss, quartzite, (garnet-bearing mica schist), (metagranitoid)	Angular, (subrounded, rounded)	Very poor	Matrix supported slightly oriented texture	4–70	15
Cserdi	No. XV	Rv—rhyolite, dacite, (andesite); Rm—(gneiss, Q-mu schist, Q-mu-F schist, mylonite, metaquartzite/quartzite), (granitoid/metagranitoid)	Subangular, subrounded	Poor to moderate	Grain supported (matrix supported) (grading)	4–110	25

Notes: Abbreviations: Rv—volcanic rock clast; Rm—metamorphic rock clast; Ls—sedimentary rock clast; Q—quartz; F—feldspar; P—plagioclase; Ab—albite; mu—muscovite. Component identified in minor amounts is given in parentheses.

Korpád-A sandstones are poorly to well-sorted feldspathic-lithic arenite to subarkose and arkose (Folk, 1968). Most of the studied rocks are rich in Fe-oxides. They are composed of mono- and polycrystalline quartz, plagioclase, K-feldspar, acidic–intermediate volcanic grains, and lesser amounts of granitoid and metamorphic rock grains and recycled mudrock fragments (Figs. 6B and 6C). Intermediate hypabyssal lithic fragments are occasionally present. Detrital muscovite and a small proportion of biotite also occur in most of the samples. Accessories are zircon, monazite, rutile, apatite, tourmaline, opaque grains, and rare garnet. Calcite and dolomite cements are common. Additionally, illite-sericite and chlorite cements also occur (Table 1). In the middle part of borehole Tu-1 and in the lower part of borehole D 9015, greater amounts of granitic and metamorphic rock fragments and K-feldspar grains are present.

Korpád-A samples can be subdivided into two subgroups on the basis of clay-mineral assemblage. In the western flank of the Villány Mountains, samples from borehole Tu-1 (Korpád-A1) consist dominantly of illite ± muscovite and chlorite with a Mg-rich interlayer. Minor proportions of kaolinite and mixed-layer illite-smectite also occur (Fig. 7). In the western Mecsek Mountains, samples from borehole D 9015 (Korpád-A2) are composed of variable amounts of illite ± muscovite, ISII (i.e. illite/smectite/illite/illite using Wantanabe [1981]'s nomenclature) ordered variety of mixed-layer illite-smectite and chlorite, with an Fe(II)-rich interlayer (Fig. 7). As discussed in detail in the following, compositional differences between the subgroups can be explained as a product of diagenesis.

Korpád-B samples are poorly sorted feldspathic-lithic wacke to subarkose and arkose (Folk, 1968) and contain angular detrital grains. They consist mostly of mono- and polycrystalline quartz, plagioclase, K-feldspar, muscovite, and lithic fragments with minor biotite, opaque grains, Fe-oxides, clayey matrix, and calcite cement. Tourmaline, zircon, and rutile occur as accessories. Feldspar grains are partially or totally altered and replaced by clay minerals or calcite (Table 1). The lithic fragments are dominated by unstable detritus derived from quartz-rich metamorphic rocks (Fig. 6D). Illite ± muscovite and kaolinite are the only phyllosilicates present in the clay-mineral assemblage (Fig. 7).

In Korpád-A conglomerate samples, acidic–intermediate volcanic and metamorphic (plagioclase gneiss, quartz-muscovite schist, metaquartzite, metasandstone) rock clasts are the main components; felsic plutonic (granitoid/metagranitoid), and intraformational sedimentary (claystone, wacke) lithic clasts are subordinate. A small number of intermediate hypabyssal rock clasts are also present (Table 2).

The matrix-supported Korpád-B breccia is characterized by a predominance of metamorphic clasts. Gneiss and quartzite clasts are the main components; garnet-bearing mica schist and metagranitoid clasts are subordinate (Table 2). Volcanic rock clasts were not identified in this unit.

Cserdi Formation

The samples are poorly to moderately sorted lithic-feldspathic wacke to litharenite (Folk, 1968) and contain angular to subangular detrital grains. Acidic volcanic lithic fragments and feldspar (more plagioclase than K-feldspar) grains are prominent; quartz grains are subordinate in most of the samples (Fig. 6F). In the lower part of this unit, resorbed monocrystalline quartz grains are locally abundant. Polycrystalline quartz grains, intermediate volcanic, metamorphic, and granitoid lithic fragments appear in low proportions. Typical accessory minerals are tourmaline, rutile, zircon, apatite, monazite, opaque grains, and rare garnet. Illite-sericite, chlorite, calcite, and dolomite occur as cement, in some cases, replaced by other minerals. Locally, metamorphic grains and volcanic glass fragments are strongly chloritized (Table 1). Based on XRD, anhydrite is present only in sample XV/55. The clay-mineral assemblage includes illite ± muscovite, swelling chlorite, and occasionally minor amounts of kaolinite (Fig. 7).

The clast-supported conglomerates with volcanic pebbles are the main lithofacies of this formation, which has infiltrated silty and sandy matrix that has grains with the same immature detrital compositions as interbedded sandstones. Volcanic rock (rhyolite, dacite, and lesser amounts of andesite) clasts are the main components; metamorphic (quartz-muscovite schist, quartz-muscovite-feldspar schist, mylonite, metagranitoid, gneiss) and granitoid clasts are subordinate (Table 2).

Gneiss and Granitoid Clasts

Téseny Conglomerate. Nonfoliated and foliated gneiss clasts have typical gneissic structure and coarse-grained granoblastic texture. Some gneiss samples are slightly mylonitized. Rock-forming minerals are quartz, plagioclase, K-feldspar, muscovite, and biotite. Strong alteration of feldspars (mainly sericitization) is typical. In certain cases, K-feldspar shows some relic magmatic features such as perthitic intergrowths and tabular shape. Newly grown minerals, including well-developed muscovite flakes, are common (Fig. 8A). Biotite is generally highly degraded to chlorite, sericite, white mica, and opaque minerals. Accessories are apatite, zircon, rutile, opaque grains, and rare strongly altered garnet.

Granitoid clasts, including aplite and quartz diorite, are present as a small proportion of total clasts but are persistent in boreholes Bm-1 and Dv-3. Aplite clasts have fine-grained panallotriomorphic-granular (i.e. all the observed minerals are xenomorphic) that is occasionally obliterated by mylonitization. Rock-forming minerals are quartz, K-feldspar, and plagioclase. Few grains of muscovite also occur. Accessory minerals are zircon, apatite, and rare rutile. K-feldspars are relatively fresh or weakly sericitized; the plagioclases are frequently altered to sericite. Quartz diorite clasts have hypidiomorphic-granular texture and contain quartz, plagioclase, and muscovite as major constituents. Biotite occurs as a subordinate mineral with accessory apatite, zircon, and titanite. Plagioclase is partially altered and replaced by sericite and clay minerals. Biotite is generally highly degraded to chlorite. Siderite and pyrite were also identified as accessories in this clast type. Their presence is due to hydrothermal mineralization.

Korpád-B Breccia. Gneiss clasts from Korpád-B breccia show great similarity concerning mineralogical composition; however, their two types are recognized on the basis of proportion of certain minerals and rock texture. One is a foliated, mica-rich biotite gneiss (type I; Fig. 8B); the other is a mica-poor gneiss/metagranite (type II), which generally is nonfoliated (Fig. 8C). Type I gneiss clasts have lepidoblastic, lepido-granoblastic, occasionally blasto-poikilitic texture. Type II gneiss clasts have granoblastic, lepido-granoblastic, occasionally blasto-poikilitic texture. Original igneous textures are commonly recognizable in this clast type. Stages of increasing metamorphic recrystallization from none (granite protolith) to complete (granoblastic gneiss with newly grown crystals) were observed.

Their rock-forming minerals are quartz, plagioclase, K-feldspar, and two micas. Plagioclase crystals are generally subhedral and myrmekitic. In certain cases, their lamellar twinning is recognizable. Strong alteration of plagioclases is typical. K-feldspar crystals are mostly euhedral or subhedral. These grains are predominantly untwinned; however, twinned grains are also present in inconsiderable amounts. Occasionally perthitic intergrowths were observed. They are partially altered and replaced by sericite; in some cases, they are strongly sericitized along the grain boundaries. Muscovite and biotite coexist in most samples, although biotite is generally highly degraded to chlorite, white mica, and opaque minerals. In a few cases, relic magmatic textures such as lath-shaped or tabular habit of biotite were preserved. Kink-band structure and undulose extinction are common on white mica bunches. Zircon, apatite, rutile, titanite, ilmenite, and rare tourmaline occur as accessories. They appear as inclusions or as a matrix mineral. Calcite and hematite are widespread as secondary minerals in some specimens.

Geochemistry

Chemical data of sandstone and clast samples studied from southern Transdanubia are reported in Tables DR3–DR6 (see footnote 1). In this study, the median is used as the summary statistical parameter (Table 3) because it is independent of outliers, and this value provides a robust estimate of central tendency for data sets drawn from a population with an unknown distribution pattern (Cox et al., 1995; Lee, 2002). Values of some elemental ratios and associated parameters are listed in Table 4 for sandstones and in Table 5 for clasts.

 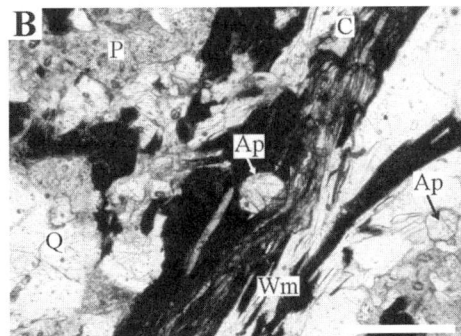 

Figure 8. Thin-section photomicrographs of gneiss clasts. Scale bar in all photographs is 1 mm. (A) Foliated Téseny gneiss clast (Sb-1 core). Cross-polarized light. (B) Type-I Korpád-B gneiss clast (Mgy-1 core). Plane-polarized light. (C) Type-II Korpád-B gneiss clast (Mgy-1 core). Cross-polarized light. Abbreviations: Q—quartz; P—plagioclase; K—K-feldspar; Mu—muscovite; Ap—apatite; Wm—white mica; C—carbonate.

TABLE 3. MEDIAN VALUES FOR SANDSTONE SAMPLES

Unit No.	Téseny (18)	Korpád-A1 (6)	Korpád-A2 (3)	Korpád-B (4)	Cserdi (8)	UCC
SiO_2	78.89	73.77	72.51	73.06	72.12	66
TiO_2	0.35	0.38	0.66	0.54	0.59	0.68
Al_2O_3	11.96	13.06	14.02	15.36	13.56	15.2
Fe_2O_3	2.49	2.19	4.01	2.97	4.15	5.03
MnO	0.04	0.07	0.07	0.02	0.04	0.08
MgO	1.08	1.46	1.95	1.12	1.71	2.2
CaO	0.29	2.69	2.26	0.76	0.61	4.2
Na_2O	1.24	4.27	3.49	1.73	3.49	3.9
K_2O	3.31	1.56	2.12	3.37	3.09	3.4
P_2O_5	0.06	0.09	0.09	0.14	0.15	0.15
Rb	133	63	87	123	148	112
Sr	69	210	196	65	134	350
Ba	546	388	428	536	552	550
Th	8.1	9.0	6.8	7.9	12.0	10.7
U	2.5	2.6	1.9	1.9	3.3	2.8
Zr	94	180	157	174	179	190
Hf	3	6	5	6	6	6
Nb	8	8	9.3	7	11	12
Y	18	24	21	28	34	22
Sc	6.0	7.8	9.3	7.8	10.2	13.6
V	N.D.	40	47	37	64	107
Cr	42	34	45	30	95	83
Co	6	5	9	3	10	17
Ni	16	L.D.	L.D.	L.D.	41	44
La	20	31	23	30	30	30
Ce	36	46	52	58	64	64
Pr	N.D.	5.5	6.1	6.7	7.4	7.1
Nd	N.D.	25	24	25	30	26
Sm	2.9	4.5	4.5	5.1	6.3	4.5
Eu	0.50	0.94	0.94	1.01	1.11	0.88
Gd	N.D.	3.7	3.3	3.9	5.1	3.8
Tb	N.D.	0.62	0.52	0.64	0.86	0.64
Dy	N.D.	3.2	3.0	3.8	4.9	3.5
Ho	N.D.	0.7	0.6	0.8	1.0	0.8
Er	N.D.	2.1	1.8	2.0	2.7	2.3
Tm	N.D.	0.30	0.28	0.33	0.43	0.33
Yb	1.2	1.9	1.6	2.0	2.6	2.2
Lu	0.13	0.30	0.25	0.32	0.39	0.32

Notes: Major oxides are in wt%, trace elements, in ppm. Total iron is listed as Fe_2O_3. Major element data are recalculated volatile-free. No.—number of samples; N.D.—no data; L.D.—lower detection limit; UCC—upper continental crust (McLennan, 2001).

TABLE 4. ELEMENTAL RATIOS AND PARAMETERS FOR SANDSTONES

Unit No.	Téseny (18)	Korpád-A1 (6)	Korpád-A2 (3)	Korpád-B (4)	Cserdi (8)	UCC
Th/Sc	1.34	1.15	0.73	1.01	1.18	0.79
La/Sc	3.25	4.01	2.43	3.80	2.91	2.21
Co/Th	0.75	0.52	1.32	0.39	0.86	1.59
Cr/Th	5.16	3.81	6.62	3.80	7.92	7.76
Cr/Zr	0.44	0.19	0.29	0.17	0.53	0.44
La/Co	3.25	6.65	2.51	9.56	2.85	1.76
Cr/Ni	2.63	N.D.	N.D.	N.D.	2.32	1.89
La_N/Sm_N	4.23	4.37	3.16	3.70	2.97	4.20
La_N/Yb_N	10.98	10.86	9.60	10.14	7.59	9.21
Gd_N/Yb_N	N.D.	1.55	1.67	1.60	1.57	1.40
ΣREE	60	126	121	139	156	146
Eu/Eu*	N.D.	0.70	0.75	0.69	0.60	0.65

Notes: La_N/Sm_N, La_N/Yb_N, and Gd_N/Yb_N ratios were calculated from chondrite-normalized values. ΣREE were calculated from concentrations of each element analyzed in ppm. Eu/Eu* represents the Eu anomaly according to the method of McLennan (1989). No.—number of samples; N.D.—no data; REE—rare earth element; UCC—upper continental crust (McLennan, 2001).

Geochemical Characteristics of the Sandstone Samples

Using the chemical classification scheme of Pettijohn et al. (1972), the Téseny and Korpád-B samples are classified as arkose and litharenite (Fig. 9). Korpád-A and Cserdi samples are classified mainly as graywacke and litharenite. Moreover, if we consider this diagram in the light of petrographic observations, it would emerge that the high Na_2O/K_2O values, corresponding to the graywacke field, reflect the abundance of volcanic rock fragments and plagioclase grains. The low Na_2O/K_2O values of some arkosic samples suggest that their behavior is controlled by the high mobility of Na_2O, especially during chemical weathering, diagenesis, and secondary alteration processes (Gaillardet et al., 1999; McLennan, 2001; Varga and Szakmány, 2004).

The sandstone samples collected for this study show moderate variations for SiO_2 and Al_2O_3. Medians of SiO_2 range from 72.12 wt% for Cserdi samples to 78.89 wt% for Téseny rocks. On average, Al_2O_3 content of the sandstone units is widely uniform, and medians lie in a narrow range from 11.96 wt% for Téseny Formation to 15.36 wt% for Korpád-B group (Table 3). In comparison with the upper continental crust (McLennan, 2001), which provides a consistent normalizing scheme for sandstone geochemistry (Floyd et al., 1991; Gaillardet et al., 1999; Zimmermann and Bahlburg, 2003), all the samples studied are slightly enriched in SiO_2 (Fig. 10A). Apart from the Korpád-A group, sandstone units analyzed have similar median values of K_2O abundance to upper continental crust. In addition, they are depleted in TiO_2, Al_2O_3, Fe_2O_3, MnO, MgO, CaO, Na_2O, and

TABLE 5. ASSOCIATED PARAMETERS OF GNEISS AND GRANITOID CLASTS

Sample number	Gn1	Gn2	Gn3	Apl	Qdi	Gn4 Type II	Gn5 Type II	Gn6 Type I	Gn7 Type I	Gn8 Type I–II
A/CNK	1.55	1.31	1.42	1.10	1.65	2.01	1.84	2.09	2.28	2.29
A/NK	1.67	1.45	1.58	1.21	1.81	2.13	1.97	2.36	2.48	2.48
Na_2O/K_2O	1.77	2.15	1.92	0.58	1.34	1.05	1.45	0.65	0.43	0.42
La_N/Sm_N	1.51	1.57	2.10	4.96	2.10	2.50	2.82	2.87	2.83	2.70
La_N/Yb_N	8.11	3.15	8.11	50.01	5.41	2.92	2.95	5.51	5.28	2.38
Gd_N/Yb_N	N.D.	N.D.	N.D.	N.D.	N.D.	1.05	1.06	1.43	1.46	N.D.
ΣREE	54.35	36.16	49.42	108.23	36.05	84.00	51.62	150.65	104.73	56.66
Eu/Eu*	N.D.	N.D.	N.D.	N.D.	N.D.	0.36	0.60	0.63	0.51	N.D.

Notes: N.D.—no data. Gn1–Gn3—gneiss (Téseny conglomerate, borehole Sb-1); Apl—aplite (Téseny conglomerate, borehole Dv-3); Qdi—quartz diorite (Téseny conglomerate, borehole Dv-3); Gn4–Gn8—gneiss (Korpád-B breccia, borehole Mgy-1). A/CNK—$Al_2O_3/[CaO + Na_2O + K_2O]$ molar; A/NK, $Al_2O_3/[Na_2O + K_2O]$ molar. See Table 4 for an explanation of rare earth element (REE) parameters.

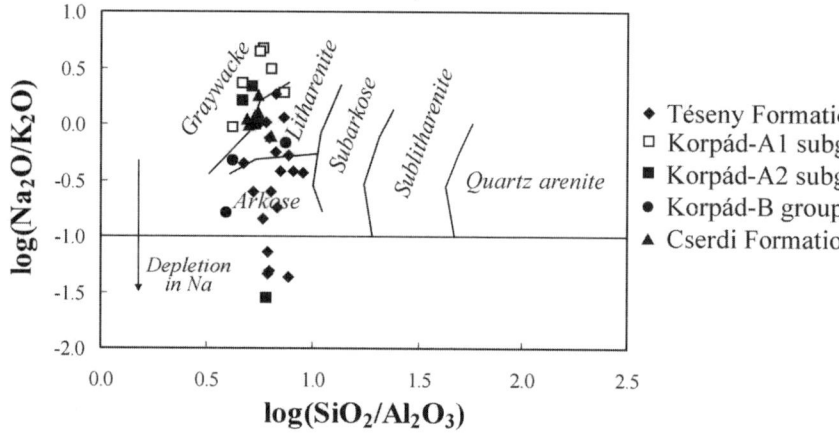

Figure 9. Chemical classification scheme of clastic sediments based on major elements: $\log(Na_2O/K_2O)$ versus $\log(SiO_2/Al_2O_3)$ diagram after Pettijohn et al. (1972). The symbols are the same as those in Figure 7.

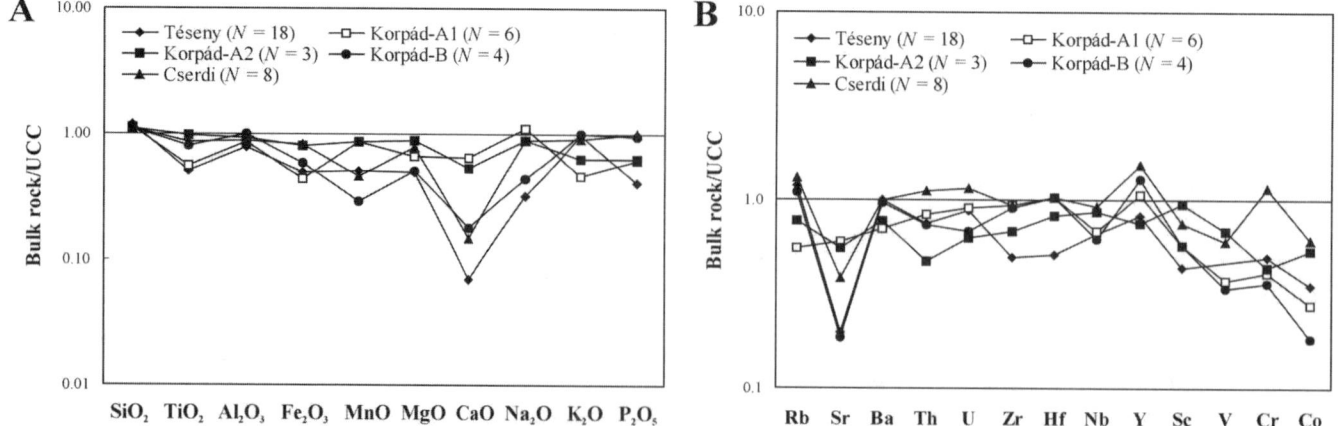

Figure 10. Upper continental crust (UCC)–normalized (McLennan, 2001) distribution of major (A) and trace (B) elements of southern Transdanubia sandstones. Major element data are recalculated volatile-free.

P_2O_5 relative to the upper continental crust. These differences are probably due to the quartz dilution effect. Low values of CaO and Na_2O would also reflect the differential effects of source-area weathering. The Téseny Formation, which shows extremely low Ca and Na contents (0.29 and 1.24 wt%, respectively) relative to all other rocks, was deposited under humid climatic conditions indicated by the presence of coal seams (Hetényi and Ravasz-Baranyai, 1976). In addition, it has lower TiO_2 and P_2O_5 contents relative to the other sedimentation intervals (Fig. 10A), probably because of its lower contents in heavy minerals. However, as has been discussed previously, the Téseny Formation has elevated SiO_2 content relative to the other units studied, reflecting the presence of abundant detrital quartz. High median of SiO_2 and low TiO_2 and P_2O_5 values are, therefore, attributed to quartz dilution. On the other hand, the Korpád-A samples (Korpád-A1 and Korpád-A2 subgroups) have higher medians of MnO, CaO, and Na_2O, and lower median of K_2O (Table 3), probably because of their higher contents in carbonate cement and sodic plagioclase, and lower contents in K-bearing mineral phases relative to the other sandstone units.

With respect to the large ion lithophile elements (LILE), the samples are slightly enriched in Rb (except for Korpád-A group), and they are slightly depleted in Ba relative to the upper continental crust (Fig. 10B). In addition, all the units studied show extremely low Sr abundances (65–210 ppm) relative to the upper continental crust (350 ppm). However, the Korpád-A samples are distinctive in terms of lower Rb and Ba and higher Sr abundances compared to the other units (Fig. 10B). This pattern, as with the major elements such as K_2O and CaO, confirms that the LILE distribution of Korpád-A group reflects its lower contents in K-bearing minerals and higher contents in diagenetic carbonates relative to the other sequences.

In general, the high field strength elements (HFSE) do not display significant lithostratigraphically dependent variations; however, the Téseny Formation has the lowest HFSE contents, and the medians of Y increase from Téseny to Cserdi samples (Table 3). This pattern is similar to the TiO_2 data distribution (Fig. 8A), suggesting that their behavior is mainly controlled by the detrital heavy mineral fraction. In the Korpád-A1, Korpád-B, and Cserdi sandstone samples, Zr and Hf have normalized values similar to the upper continental crust, whereas Th and U display considerable scatter. These units show minor negative Nb and positive Y anomalies (Fig. 8B).

The samples studied are strongly depleted in transition metals, including Sc, V, Cr (except for Cserdi Formation), Ni, and Co with respect to the upper continental crust (Table 3). On the other hand, median of Cr in Cserdi rocks exhibits a minor but significant positive anomaly, indicating a change in the source material (Fig. 10B).

The REE generally show systematic enrichment throughout the sequence, from Pennsylvanian to Upper Permian (Table 4). In a chondrite-normalized (McLennan, 1989) REE diagram (Fig. 11A), the units studied display light REE (LREE) enrichment trends, variable degrees of development of a negative Eu anomaly (Eu/Eu*), and near-flat heavy REE (HREE) patterns. They show strong similarity to the upper continental crust and thus reflect derivation from typical fractionated upper continental crust. The SiO_2-rich Téseny Formation has very low total REE (ΣREE of 60 ppm), presumably due to quartz dilution (Table 4). Its REE fractionation, expressed by the La_N/Yb_N ratio, is higher (10.98) compared with that of the upper continental crust (9.21). The median of La_N/Sm_N ratio is 4.23, which suggests that the LREE fractionation is similar to that of the upper continental crust (4.20). The Korpád samples show quite uniform compositions, with significant LREE enrichment (La_N/Sm_N ratios of 3.16–4.37) and high ΣREE (~121 ppm and higher; Table 4). Eu/Eu* stays in a narrow range (0.69–0.75). The Cserdi Formation also has fractionated REE with La_N/Yb_N of 7.59 and high ΣREE (156 ppm), but it is slightly enriched in HREE and has a larger

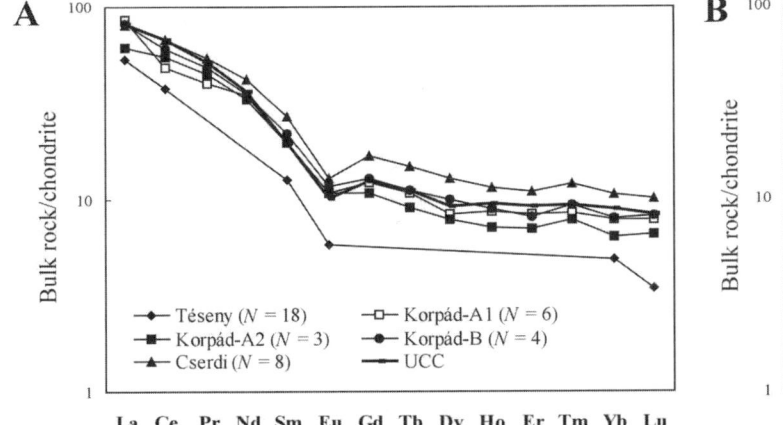
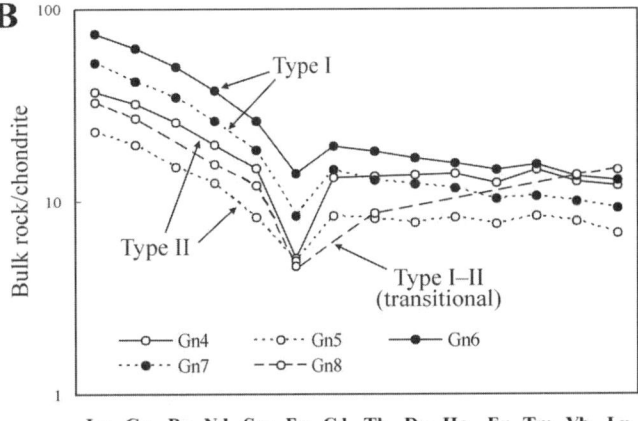

Figure 11. Chondrite-normalized (McLennan, 1989) rare earth element (REE) patterns of sandstone samples (A) and gneiss clasts from Korpád-B breccia (B). The upper continental crust (McLennan, 2001) pattern is given as a reference. Note fractionated light (L)REE, negative Eu anomaly, and flat heavy (H)REE pattern.

Eu anomaly (Eu/Eu* = 0.6) in comparison with the other units and the upper continental crust (Table 4; Fig. 11A).

Geochemistry of Extracted Clasts

The relic magmatic textures of the Téseny and Korpád-B gneiss clasts discussed already suggest that they can be classified as orthogneiss; therefore, discrimination diagrams for discerning the character of magmatic rocks (Shand, 1943) and the tectonic setting of granites (Pearce et al., 1984) are relevant for determining the geochemical character of their protolith.

The clasts studied have peraluminous character (Fig. 12A). The Téseny and Korpád-B gneiss and Téseny quartz diorite clasts are dissimilar to the main granitoids of the Mórágy Complex (Buda et al., 2000) or Görgeteg-type metamorphites recovered from numerous boreholes in southwestern Transdanubia (Fülöp, 1994). On the other hand, the plotted area of Baksa-type basement rocks from borehole Mgy-1 (Török, 1986; Árgyelán, 2004) is very similar to the area of the Téseny and Korpád-B clasts. The Téseny aplite clast plots in the Mórágy microgranite field (Buda et al., 2000).

It is noteworthy, however, that compositions of the Baksa-type metamorphic rocks published by Török (1986) and the samples studied show a linear trend (Fig. 12A), reflecting A/CNK = A/NK values (where A/CNK = $Al_2O_3/[CaO + Na_2O + K_2O]$ molar; A/NK = $Al_2O_3/[Na_2O + K_2O]$ molar). Petrographic features of basement rocks indicate extensive hydrothermal alteration (Török, 1986). In the clasts extracted, strong alteration of feldspars and presence of secondary minerals were observed. Additionally, postdepositional albitization of feldspars cannot be excluded. Therefore, the chemical composition of clasts studied might have a postdepositional origin as well. For these reasons, the mobile Ca might have been partially removed from these rocks (Fig. 12B), resulting in negligible differences between the A/CNK and A/NK ratios.

With respect to the alkali elements, both types of clasts ($Na_2O/K_2O > 1$ and < 1) were recognized (Fig. 12B). Apart from the aplite sample, the Téseny clasts are Na enriched. In addition, type II gneiss clasts from Korpád-B sediments are slightly enriched in Na_2O relative to K_2O. In contrast, type I gneiss clasts and sample Gn8 with transitional (type I–II) composition are depleted in Na_2O, reflecting their lower sodic plagioclase and higher white mica contents. The Téseny aplite clast also shows K-enrichment; however, it is mineralogically distinct from all other clasts in terms of higher K-feldspar content. Relatively high Si_2O and K_2O contents (77.30 and 4.65 wt%, respectively) are characteristic features of this sample, suggesting a high-K character.

Apart from the aplite sample, the Téseny clasts have very low ΣREE and low La_N/Sm_N ratios (Table 5). The aplite clast is enriched in LREEs and has relatively high values of ΣREE, indicating a more differentiated character compare with the gneiss and quartz diorite clasts. The chondrite-normalized REE patterns of Korpád-B gneiss clasts are very similar for both rock types, which have a characteristic negative Eu anomaly (Fig. 11B). However, type I clasts have higher ΣREE values and La_N/Yb_N ratios (Table 5), probably because of their higher contents in micas and accessories relative to the type II clasts.

Trace-element analysis after Pearce et al. (1984) indicates that the gneiss and quartz diorite clasts are mainly volcanic arc origin (Fig. 10C); however, they fall very close to the field of within-plate granites. The plotted area of the Korpád-B gneiss clasts is very similar to the area of the Baksa-type metamorphites (Árgyelán, 2004). The composition of the aplite clast plots in the field of syncollisional granites. Its value of Rb is

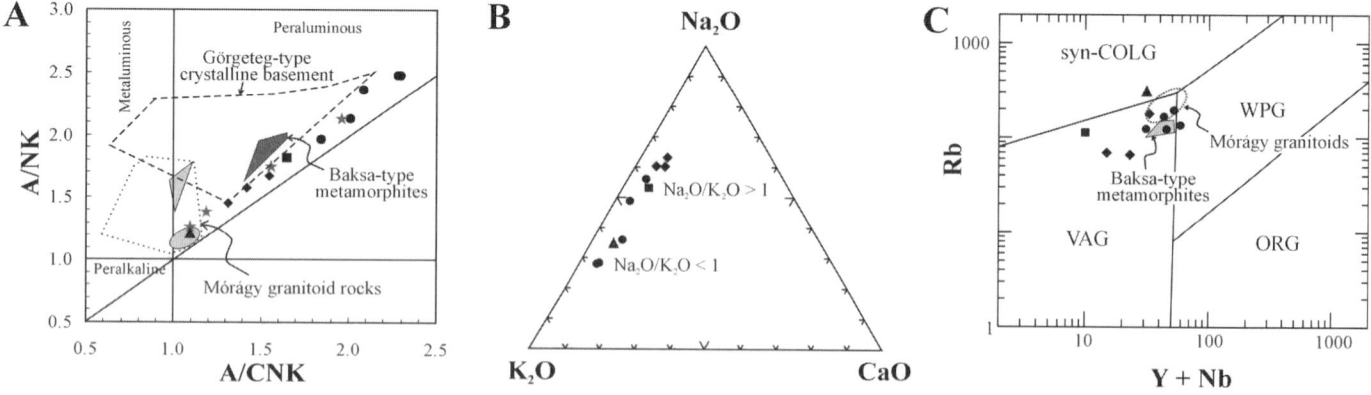

Figure 12. Chemical features of gneiss and granitoid clasts from Téseny and Korpád-B clastics. (A) Shand's indices (A/CNK, $Al_2O_3/[CaO + Na_2O + K_2O]$ molar; A/NK, $Al_2O_3/[Na_2O + K_2O]$ molar) of clasts. (B) $Na_2O–K_2O–CaO$ ternary diagram. (C) Rb versus (Y + Nb) diagram for discerning the tectonic setting of granites (after Pearce et al., 1984), showing the fields of syncollisional granites (syn-COLG), within-plate granites (WPG), volcanic arc granites (VAG), and ocean-ridge granites (ORG). Diamond—Gn1-3; triangle—Apl; square—Qdi; circle—Gn4-8; star—basement rocks from borehole Mgy-1 after Török (1986); shaded triangle—Mórágy microcline megacryst-bearing granitoids; shaded ellipse—Mórágy microgranite. Fields for Mórágy granitoid data come from Buda et al. (2000). Field for Görgeteg-type crystalline rocks comes from Fülöp (1994). Field for Baksa-type metamorphites from borehole Mgy-1 comes from Árgyelán (2004).

different from the other clasts studied, and it is similar to those of the Mórágy granitoid rocks (Buda et al., 2000).

DISCUSSION

Provenance and Tectonic Setting Based on Sandstone Geochemistry

Selected immobile trace element (e.g., Th, Zr, Hf, Sc, Co, Ni, Cr, and REEs) parameters and discriminatory diagrams form a useful tool to characterize the provenance of clastic sedimentary rocks and to decipher the tectonic setting of sandstones (Bhatia, 1985; Bhatia and Crook, 1986; Bauluz et al., 1995; Garver and Scott, 1995; Cullers and Berendsen, 1998). Previous provenance studies (Cullers, 1995; McLennan, 1989; Lee, 2002; Willan, 2003) have shown that Th, Zr, Hf, and REEs are enriched in felsic rather than in mafic rocks because they are highly incompatible during most igneous melting and fractionation processes. On the other hand, input from mafic and ultramafic source areas would result in an enrichment of Sc, Co, Cr, and Ni (Garver and Scott, 1995; Zimmermann and Bahlburg, 2003). These elements are typically much more compatible than REEs, Th, and Zr. Correspondingly, the Th/Sc, La/Sc, La/Co, Co/Th, Cr/Th, and Cr/Zr ratios of siliciclastic rocks are very sensitive provenance indicators (Bhatia and Crook, 1986; Bauluz et al., 1995; Cullers and Berendsen, 1998). In addition, anomalously high Cr and Ni absolute concentrations with Cr/Ni ratios of ~1.2–1.6 suggest that these elements were derived from a source with ultramafic rocks. Higher Cr/Ni ratios are probably indicative of derivation of these elements from mafic volcanic rocks (Garver and Scott, 1995, and references therein).

Additionally, the REE patterns may differ in different sources (Bhatia, 1985; McLennan, 1989; Cullers 1995; Garver and Scott, 1995; Zimmermann and Bahlburg, 2003; Willan, 2003). In general, these studies show that siliciclastic sediment derived from mature continental crust is characterized by LREE enrichment and high ΣREE values. On the other hand, sediment derived from young, undifferentiated oceanic arcs has lower La_N/Sm_N ratios than either continental arc or old continental crust, has lower ΣREE values, and can lack an Eu anomaly.

As discussed in detail in the previous sections, the HFSE and REE contents of the Paleozoic formations analyzed in this study are similar to average upper crustal values. None of these units shows anomalous concentrations of Sc, V, Cr, Co, or Ni, so it is unlikely that much, if any, of the source region was composed of ultramafic rocks. The ubiquitous depletion of compatible elements (Table 3; Fig. 10B) indicates a relatively felsic source area. In addition, the sandstone samples show REE patterns typical of continental-derived sediments, with compositions comparable to the upper continental crust (Fig. 11A).

The key trace element ratios also support a silicic source area (Table 4). All the units studied have high Th/Sc, La/Sc, and La/Co ratios, and have higher values than those of the upper continental crust. Moreover, medians of the Co/Th ratio are significantly lower in the samples than in the upper continental crust. These parameters are in the range of intermediate to silicic source rocks (Cullers, 1995). The low values of Cr/Th and Cr/Zr ratios in Téseny and Korpád formations also provide no support for significant amounts of mafic or ultramafic rocks in the source area. The Cr/Th and Cr/Zr ratios in the Cserdi Formation, however, may signal a change in sediment provenance. Notably, Cr and Ni anomalies do not exist in this unit (Table 3), and the median of Cr/Ni ratio is 2.32, reflecting a volcanic provenance. In the sandstone samples, totally chloritized volcanic glass fragments appear in low proportions, which could be related to a mafic source area. It is not possible, however, that significant amounts of the Cserdi detritus were derived from mafic volcanic rocks. We suspect that these features reflect the higher amounts of mafic minerals, especially biotite and rare pyroxene, derived from felsic volcanites, an interpretation that is supported by high ΣREE values, high La_N/Sm_N ratio, and sediment composition.

Ternary discrimination diagrams after Bhatia and Crook (1986) have been widely applied to the provenance of sandstones but were developed using small, geographically restricted data sets (Willan, 2003). According to the La-Th-Sc plot (Bhatia and Crook, 1986), the samples studied plot in the fields of continental island arcs and continental margins (Fig. 13). In the Th-Sc-Zr/10 diagram, almost all of the samples plot in the continental island arc and active continental margin fields, whereas in the Th-Co-Zr/10 plot, the majority of the Cserdi samples conform to the continental island arc range, with the Téseny sediments distributed toward the Co apex; the Korpád samples have the strongest passive margin signature (Fig. 13). Consequently, tectonic discrimination diagrams based on these elements and those developed for graywackes of Paleozoic turbidite sequences of eastern Australia (Bhatia and Crook, 1986) are probably not generally applicable to continental sedimentary formations in the Mecsek-Villány area.

The diagrams do, however, display the compositional trends among the samples in this study and confirm that these sediments were dominantly derived from a mixture of various proportions of felsic volcanic rocks, corresponding to the continental island arc field, and granite-gneisses and siliceous volcanics of the uplifted basement, corresponding to the active continental margin field. Additionally, rift-bounded grabens are included in the passive margin–type tectonic setting (Bhatia and Crook, 1986); therefore, the Korpád samples, which were deposited in continental environments of intraplate rift basins (Barabás and Barabás-Stuhl, 1998; Haas et al., 1999), show passive margin affinity.

Source-Area Interpretation

Téseny Sandstone Formation

The coal-bearing Pennsylvanian Téseny sediments are interpreted as having formed in a molasse foreland basin proximal to the Variscan fold belt (Hetényi and Ravasz-Baranyai, 1976; Varga et al., 2003). The rock and mineral fragments in the

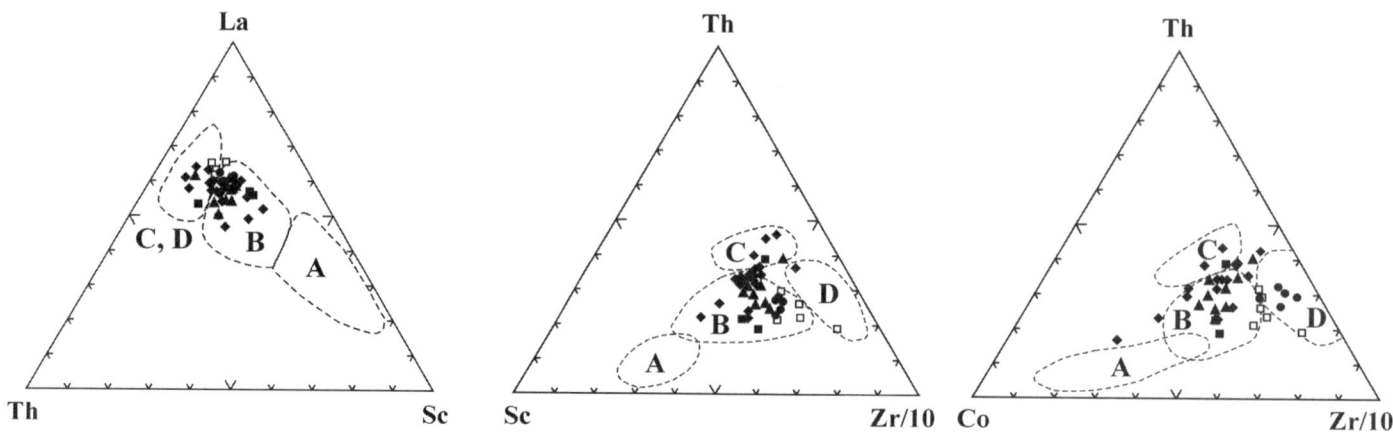

Figure 13. Tectonic setting of the samples studied in La–Th–Sc, Th–Sc–Zr/10, and Th–Co–Zr/10 diagrams after Bhatia and Crook (1986). A—oceanic island arc; B—continental island arc; C—active continental margin; D—passive margin. Symbols are as in Figure 9.

sands and conglomerates are identifiable as coming from three sources: (1) a recycled Variscan orogenic area (collision suture and fold-thrust belt; Varga et al., 2003), indicated by the presence of metamorphic and sedimentary lithic fragments, (2) an uplifted plutonic (granite-gneiss) basement, and (3) an old (probably Variscan) magmatic arc, indicated by the lesser amounts of siliceous volcanic rocks. The local nature of the source areas is manifested by local variations in feldspar, lithic fragment, and clay-mineral contents (Tables 1 and 2). The presence of abundant detrital micas and high proportion of illite ± muscovite in the clay-mineral assemblage (Fig. 7) confirm the predominance of metamorphic sources for this formation. Chlorite with an Fe(II)-rich interlayer, which was detected in the clay-size fraction, is present in minor quantity due to its coarser grain size; essentially, it was formed by alteration of biotite, and less commonly was formed as matrix material.

Our geochemical results also support a silicic source area of the Téseny Formation (Fig. 13). The relatively high median SiO_2 value and low HFSE and REE contents in this formation (Figs. 10 and 11A) reflect source rocks more mature in composition than those that supplied detritus to the Permian strata. It is important to note that the relatively quartzose nature of Téseny samples, with extremely low medians of Na_2O, CaO, and Sr, and high medians of K_2O and Rb, indicates an intense weathering in the source region, where Na, Ca, and Sr are preferentially leached, and K and Rb are fixed in clays (Nesbitt et al., 1980). Alternatively, the secondary clay minerals may have been more aluminous, such as kaolinite, and were later enriched in K to form illite (Fedo et al., 1995). The latter scenario is supported by the presence of abundant illite-sericite (Fig. 7), both as matrix material between grains and as alteration of weathered feldspar grains, which strongly suggests a potassium metasomatism in the Téseny clastics.

Korpád Sandstone Formation

As discussed in detail in the previous sections, characteristics of the Lower Permian Korpád Formation vary spatially and are interpreted as local variations in composition of the source areas and diagenetic environments. Relatively immobile trace element relations, however, clearly show that the Korpád Formation, including Korpád-A and Korpád-B groups, was derived from mature upper continental crust (Figs. 11A and 13).

During the Early Permian rifting of the southern margin of the European plate, the Korpád-A group represents the erosion of acidic–intermediate volcanic rocks with variable amounts of metamorphic and coarse crystalline plutonic rocks from the uplifted Variscan basement (Tables 1 and 2; Fig. 13). With regard to the phyllosilicates, the phases present in both Korpád-A1 and Korpád-A2 subgroups are illite ± muscovite and, occasionally, mixed-layer illite-smectite. Kaolinite and chlorite with a Mg-rich interlayer appear only in the Korpád-A1 subgroup (Fig. 7). On the other hand, chlorite with an Fe(II)-rich interlayer occurs in the Korpád-A2 subgroup. This might be explained by the possibility that the postdepositional environments of these subgroups were different.

The relationship between the relative abundance of volcanic fragments and chlorite in the <2 μm fraction (Fig. 7) suggests that fine-grained chlorite might have been formed by alteration of volcanic glass. Mg-rich chlorite is a stable phase under evaporitic conditions (Weaver, 1989), which is consistent with the presence of dolomite cement in the Korpád-A1 subgroup. As has been pointed out in the literature, kaolinite disappears from the clay-mineral parageneses during the burial of the sedimentary series due to such factors as (1) combination with Mg from destabilized dolomite to produce chlorite; combination with other phases to produce (2) illite and chlorite or (3) mixed-layered minerals (Weaver, 1989; Huang, 1993; Bauluz et al., 1995). In the Korpád-A2 subgroup, this third option is supported by the presence of abundant ISII ordered variety of mixed-layer illite-smectite, corresponding to 150–200 °C burial heating (Weaver, 1989), and Fe(II)-rich chlorite together with dolomite cement.

The provenance of the Korpád-B group represents sources exclusively from uplifted basement rocks (metagranite-gneiss, mica schist). The phyllosilicates present in this group are illite ± muscovite and kaolinite (Fig. 7) as feldspar alteration products of crystalline sources.

Cserdi Formation

The petrographic features and the clay-mineral association (Fig. 7) of the Upper Permian sandstone and conglomerate samples suggest that the Cserdi Formation was derived mainly from acidic volcanic rocks. The presence of minor low-grade to medium-grade metamorphic and plutonic rock fragments indicates additional input from the uplifted Variscan basement. In the clay fraction of the Cserdi Formation, swelling chlorite is a common phase, apparently because of an abundance of volcanic material (Weaver, 1989).

Geochemistry of the Cserdi Formation, especially its high TiO_2, P_2O_5, Th, U, Y, and Cr contents and evolved HREE pattern relative to the other sedimentation intervals (Figs. 10 and 11A), is compatible with textural immaturity and mineralogy of its terrigenous detritus noted previously and may reflect relatively proximal derivation from a felsic volcanic source (Fig. 13). Tourmaline, rutile, zircon, apatite, and monazite occur as dominant accessory phases in this formation. Titanium and P are network-forming cations in rutile and phosphorus minerals (e.g., apatite and monazite), respectively. The major trace element substitutions occurring in zircon include Hf^{4+}, U^{4+}, and Th^{4+} substituting directly for Zr^{4+}. Yttrium and the HREEs can enter the zircon lattice through a coupled substitution scheme involving P and Si (McLennan, 1989; Floyd et al., 1991; Preston et al., 1998). Yttrium and REEs are also commonly present in apatite, substituting directly for Ca. Additionally, the LREEs, Y, and Th are major network-forming cations in monazite (McLennan, 1989; Preston et al., 1998). It is possible, that the whole-rock TiO_2, P_2O_5, Th, U, Y, and HREE budgets are controlled by the abundances and compositions of the heavy minerals. On the other hand, elevated Cr content of the Cserdi samples could be related to the totally chloritized (probably mafic) volcanic glass fragments. Another source for Cr content in these rocks could be the higher amounts of mafic minerals (e.g., biotite and pyroxene) derived from felsic volcanites.

Source-Area Location

There are relatively well-known metamorphic terrains in the Tisza mega-unit, but their detailed description, including whole-rock chemistry, is still missing. Furthermore, the metamorphic and coarse crystalline basement rocks were subsequently strained (mylonitization, cataclasis, etc.) and affected by hydrothermal alteration (Török, 1986; Buda et al., 2000; Szederkényi, 2001), so deciphering original relationships among the basement terrains and overlying siliciclastic units has been difficult.

Based on clast composition, low-grade to medium-grade metamorphic source components might be derived from local sources in the Slavonia-Drava terrane, which consists mostly of gneiss, mica schist, and migmatite (Szederkényi, 2001). Within it, the Baksa subunit is the likely main source area for gneiss/metagranitoid and mica schist clasts from Téseny and Korpád-B sediments (Fig. 12). Unfortunately, however, with our present data set, there is no way to determine the source of Korpád-A and Cserdi metamorphic clasts.

Crystalline plutonic rocks of Mississippian age (340–350 Ma) are well represented in the Mórágy Complex (Kunság unit) and show great similarities to granitoids of the European Variscides, especially to the Moldanubian zone of the Bohemian Massif (Buda et al., 2000). Variscan quartz monzonite, monzogranite, microgranite, and pegmatite are the main bedrock lithologies in the Mórágy Complex. Outcrops are located in the eastern part of the Mecsek Mountains (Mórágy Hills) and in the westernmost part of the area studied (Fig. 2). This latter area, near the boreholes D 9015 and No. XV, represents the principal coarse crystalline plutonic source that fed the Permian rift basins during deposition of Korpád-A and Cserdi units. Additionally, the inferred sources of fine-grained plutonic rocks (Téseny aplite clasts) are microgranite dikes of the Mórágy Complex (Fig. 12). However, the location of parent rocks of Téseny quartz diorite clasts is unknown. Nonetheless, important information about the provenance of Korpád-A and Cserdi clasts will result from future geochemical and correlation studies aimed at determining the similarities or differences among these units and deduced parent rocks.

Unfortunately, there is no evidence for volcanites older than Early Permian in the Mecsek-Villány area. Therefore, the origin of the acidic–intermediate volcanic source component of the Pennsylvanian Téseny sediments and the lowermost Permian Korpád-A unit is obscure. However, Varga et al. (2003) showed that the characteristics of rhyolite, dacite, trachyandesite, and andesite clasts collected from the Téseny conglomerate reflect convergent, active continental margin affinity. Their features are very similar to those of the Lower to Upper Mississippian to Pennsylvanian calc-alkaline volcanites from the intra-Sudetic Basin, SW Poland (Awdankiewicz, 1999). On the other hand, the petrographic nature of felsic volcanic fragments from the Cserdi unit suggests a provenance from Early Permian rhyolite (Gyűrűfű Rhyolite Formation) located in the western part of the Mecsek Mountains (Fig. 3) and associated felsic dikes present within some basements (Fülöp, 1994; Barabás and Barabás-Stuhl, 1998). Future geochemical work on the Cserdi volcanic clasts will provide an important test of this hypothesis.

CONCLUSIONS

There are distinct contrasts in provenance and source among the late Paleozoic siliciclastic formations analyzed in this study.

The Pennsylvanian Téseny Formation consists of relatively metamorphic lithic-rich sandstones with a combined recycled-orogen, basement-uplift, and volcanic-arc provenance. Low-grade to medium-grade metamorphic source components might be derived from local sources in the Baksa subunit. The inferred sources of fine-grained plutonic rocks are microgranite dikes of the Mórágy Complex (Kunság unit). The high median SiO_2 values and low Na_2O, CaO, Sr, HFSE, and REE contents in this formation reflect that source rocks were more mature than those that supplied detritus to the Permian strata. Our data also indicate a more intense weathering in the source region. Additionally, high medians values of K$_2$O and Rb, together with the presence of abundant illite-sericite, suggests a potassium metasomatism in the Téseny clastics.

Characteristics of the Lower Permian Korpád Formation vary spatially and are interpreted as local variations in composition of the source areas and diagenetic conditions. Relatively immobile trace element relations, however, clearly show that this formation was mainly derived from mature upper continental crust. The Korpád-A group represents combined acidic–intermediate volcanic and basement-uplift provenances. A possible source for this group is the Mississippian granitoid basement in the western Mecsek Mountains. On the other hand, the provenance of the Korpád-B group represents sources exclusively from the uplifted Baksa-type metamorphic rocks.

The petrographic, clay-mineralogical, and geochemical features of the Upper Permian Cserdi Formation, especially the presence of abundant acidic volcanic fragment and chlorite, together with the high TiO$_2$, Th, U, Y, and Cr contents, and evolved HREE pattern relative to the underlying formations, may reflect relatively proximal derivation from a felsic volcanic source. It may be possible to link huge amounts of felsic detritus in Cserdi sediments to the Gyűrűfű Rhyolite Formation.

These upward variations to less mature deposits might be related to such factors as increasing aridity and favorable conditions for weathering-limited erosion in the source regions and the increase of tectonic activity, including intense acidic volcanism during the Early Permian rifting in the southern margin of the European plate.

APPENDIX

Description of XRD Methods

The mineralogical analyses of the whole-rock samples and their clay-size fraction were performed at the Department of Earth and Environmental Sciences of University of Veszprém (Hungary) by X-ray diffraction (XRD), using a Philips PW 1710 diffractometer, Cu-K$_\alpha$ radiation, and diffracted-beam graphite single-crystal monochromator.

Rock samples were disaggregated under standard conditions using a jaw crusher. Random powder of the bulk sample was used for characterization of the whole-rock mineralogy. Prior to clay fraction (<2 µm) separation, the samples were treated to remove the easily soluble carbonates by using 10% acetic acid and properly dispersed by using ultrasonic deflocculation. The clay fraction was separated from the stabilized aqueous suspension, and oriented specimens were prepared by smearing a paste of <2 µm fraction onto a glass slide to minimize size fractionation of the clay particles. The semiquantitative mineralogical analysis of the oriented clay-size fraction was made after air drying at room temperature and analyzed again after keeping the samples under ethylene-glycol-solvated conditions for 4 h at 80 °C. Analytical procedure was based on methods developed by Kübler (1968) and Árkai (1991). The clay minerals were identified from their characteristic basal reflections (Weaver, 1989). Smectite content of mixed-layer illite-smectite was estimated by the method of Watanabe (1981). Samples containing nonexpanding 14 Å phases were heated at 350, 450, 550, and 640 °C under atmospheric pressure to determine the dominant interlayer cations (Bailey, 1988). After the heat-treatment, specimens were stored in an exsiccator. Semiquantitative estimations of the relative concentrations of the clay minerals were based on the peak area method (Biscaye, 1965) using the intensity factors of Rischák and Viczián (1974) and Árkai (1991). The relative abundance of phyllosilicates was determined by the peak area ratio of the 001/001 reflection of mixed-layer illite-smectite and the 001 reflection of illite, chlorite, and kaolinite after glycolation. Mixed-layer phases close to pure illite were corrected by multiplying by a factor 2, while those close to smectite were multiplied by a factor 0.5. Peak area of discrete illite was corrected in a similar manner by a factor 2. Kaolinite and chlorite had factor 1.

Description of XRF Methods

Major and selected trace element analyses were performed at the Department of Geochemistry of the University of Tübingen (Germany).

SiO$_2$ (240 ppm), TiO$_2$ (12 ppm), Al$_2$O$_3$ (244 ppm), Fe$_2$O$_3$ (180 ppm), MnO (5 ppm), MgO (88 ppm), CaO (48 ppm), Na$_2$O (75 ppm), K$_2$O (24 ppm), P$_2$O$_5$ (14 ppm), Rb (2.9 ppm), Sr (3.0 ppm), Ba (11.1 ppm), Pb (10.3 ppm), Y (1.8 ppm), V (2.6 ppm), Ni (3.3 ppm), Zn (3.0 ppm), and Co (1.6 ppm) abundances were established by X-ray fluorescence (XRF) analysis of glass fusion disks and pressed powder pellets using a Bruker AXS S4 Pioneer instrument with a rhodium X-ray source (detection limits are given in parentheses). Samples (1.5000 g) were mixed with 7.5000 g of Spectromelt Fluxing agent (Merck A12, dilithium tetraborate/lithium metaborate [66:34]) and melted, using the OxiFlux-System of the Firm CBR Analyze Service at 1200 °C to obtain homogeneous tablets (H. Taubald and F. Pintér, 2004, personal commun.). Results were evaluated with the computer program Traces, which uses 32 standards for calibration. Total Fe content is reported as Fe$_2$O$_3$. Loss on ignition (LOI) was determined after heating powdered samples at 1000 °C for 1 h.

ACKNOWLEDGMENTS

The Mecsek Ore Environment Company (Pécs, Hungary) made the core samples from different boreholes available for study. The authors are grateful to G. Hámos, L. Merényi, H. Taubald, and F. Pintér for their constructive help in field and laboratory work. Funding for this work came from several sources, including the Hungarian Scientific Research Fund (OTKA) projects T 022938 and T 034924 to György Szakmány, the International Association of Sedimentologists grant 2003 to Andrea Varga, and the Pro Renovanda Cultrura Hungariae (PRCH) Student Science Foundation. This study also forms part of Andrea Varga's Ph.D.

research at Eötvös University (Budapest, Hungary). Thoughtful reviews by W.A. Heins and G. Yaxley contributed significant improvement to the manuscript.

REFERENCES CITED

Árgyelán, T., 2004, A Korpádi Homokkő Formáció kavicsanyagának kőzettani és geokémiai vizsgálata a Máriagyűd–1 számú fúrásban [Petrology and geochemistry of metamorphic clasts of the Lower Permian conglomerate unit from borehole Máriagyűd–1] [National Scientific Conference of Students thesis]: Budapest, Eötvös University, 73 p.

Árgyelán, T., 2005, A XV. szerkezeti fúrás által feltárt Cserdi Konglomerátum formáció gneisz és granitoid kavicsanyagának kőzettani és geokémiai vizsgálata [Petrology and geochemistry of gneiss and granitoid clasts of Cserdi Conglomerate Formation from borehole No. XV] [M.S. thesis]: Budapest, Eötvös University, 117 p.

Árkai, P., 1991, Chlorite crystallinity: An empirical approach with illite crystallinity, coal rank and mineral facies as exemplified by Palaeozoic and Mesozoic rocks of northeast Hungary: Journal of Metamorphic Geology, v. 9, p. 723–734.

Arribas, J., Alonso, Á., Mas, R., Tortosa, A., Rodas, M., Barrenechea, J.F., Alonso-Azcárate, J., and Artigas, R., 2003, Sandstone petrography of continental depositional sequences of an intraplate rift basin: Western Cameros Basin (north Spain): Journal of Sedimentary Research, v. 73, p. 309–327.

Awdankiewicz, M., 1999, Volcanism in a late Variscan intramontane trough: The petrology and geochemistry of the Carboniferous and Permian volcanic rocks of the intra-Sudetic Basin: SW Poland: Geologica Sudetica, v. 32, p. 83–111.

Bailey, S.W., 1988, Chlorites: Structures and crystal chemistry, in Bailey, S.W., ed., Hydrous Phyllosilicates (Exclusive of Micas): Reviews in Mineralogy, v. 19, p. 347–403.

Balogh, K., and Barabás, A., 1972, The Carboniferous and Permian of Hungary: Acta Mineralogica-Petrographica, Szeged, v. 20, no. 2, p. 191–207.

Barabás, A., and Barabás-Stuhl, Á., 1998, A Mecsek és környezete perm képződményeinek rétegtana [Stratigraphy of the Permian formations in the Mecsek Mountains and its surroundings], in Bérczi, I., et al., eds., Magyarország Geológiai Képződményeinek Rétegtana [Stratigraphy of Geological Formations of Hungary]: Budapest, MOL Hungarian Oil and Gas Company and Geological Institute of Hungary, p. 187–215.

Bauluz, B., Mayayo Burillo, M.J., Fernandez-Nieto, C., and Gonzalez Lopez, J.M., 1995, Mineralogy and geochemistry of Devonian detrital rocks from the Iberian Range (Spain): Clay Minerals, v. 30, p. 381–394.

Bhatia, M.R., 1985, Rare earth element geochemistry of Australian Paleozoic graywackes and mudrocks: Provenance and tectonic control: Sedimentary Geology, v. 45, p. 97–113, doi: 10.1016/0037-0738(85)90025-9.

Bhatia, M.R., and Crook, K.A.W., 1986, Trace element characteristics of graywackes and tectonic setting discrimination of sedimentary basins: Contributions to Mineralogy and Petrology, v. 92, p. 181–193, doi: 10.1007/BF00375292.

Biscaye, P.E., 1965, Mineralogy and sedimentation of recent deep-sea clay in the Atlantic Ocean and adjacent seas and oceans: Geological Society of America Bulletin, v. 76, p. 803–832.

Brügel, A., Dunkl, I., Frisch, W., Kuhlemann, J., and Balogh, K., 2003, Geochemistry and geochronology of gneiss pebbles from foreland molasse conglomerates; geodynamic and paleogeographic implications for the Oligo-Miocene evolution of the eastern Alps: The Journal of Geology, v. 111, no. 5, p. 543–563, doi: 10.1086/376765.

Buda, G., Puskás, Z., Gál-Sólymos, K., Klötzli, U., and Cousens, B.L., 2000, Mineralogical, petrological and geochemical characteristics of crystalline rocks of the Üveghuta boreholes (Mórágy Hills, south Hungary): Budapest, MÁFI Évi Jelentése az 1999 évről [Annual Report of the Geological Institute of Hungary, 1999], p. 231–252.

Cox, R., Lowe, D.R., and Cullers, R.L., 1995, The influence of sediment recycling and basement composition on evolution of mudrock chemistry in the southwestern United States: Geochimica et Cosmochimica Acta, v. 59, no. 14, p. 2919–2940, doi: 10.1016/0016-7037(95)00185-9.

Csontos, L., and Nagymarosy, A., 1998, The Mid-Hungarian line: A zone of repeated tectonic inversions: Tectonophysics, v. 297, p. 51–71, doi: 10.1016/S0040-1951(98)00163-2.

Csontos, L., Nagymarosy, A., Horváth, F., and Kovác, M., 1992, Tertiary evolution of the intra-Carpathian area: A model: Tectonophysics, v. 208, p. 221–241, doi: 10.1016/0040-1951(92)90346-8.

Csontos, L., Benkovics, L., Bergerat, F., Mansy, J., and Wórum, G., 2002, Tertiary deformation history from seismic section study and fault analysis in a former European Tethyan margin (the Mecsek-Villány area, SW Hungary): Tectonophysics, v. 357, p. 81–102, doi: 10.1016/S0040-1951(02)00363-3.

Cullers, R.L., 1995, The controls on the major- and trace-element evolution of shales, siltstones and sandstones of Ordovician to Tertiary age in the Wet Mountains region, Colorado, U.S.A.: Chemical Geology, v. 123, p. 107–131, doi: 10.1016/0009-2541(95)00050-V.

Cullers, R.L., and Berendsen, P., 1998, The provenance and chemical variation of sandstones associated with the Mid-Continent Rift system, USA: European Journal of Mineralogy, v. 10, p. 987–1002.

Fazekas, V., 1987, A mecseki perm és alsótriász korú törmelékes formációk ásványos összetétele (Mineralogical composition of Permian and Lower Triassic clastics from the Mecsek Mts.): Földtani Közlöny (Bulletin of the Hungarian Geological Society), v. 117, no. 1, p. 11–30.

Fedo, C.M., Nesbitt, H.W., and Young, G.M., 1995, Unraveling the effects of potassium metasomatism in sedimentary rocks and paleosols, with implications for paleoweathering conditions and provenance: Geology, v. 23, p. 921–924, doi: 10.1130/0091-7613(1995)023<0921: UTEOPM>2.3.CO;2.

Floyd, P.A., Shail, R., Leveridge, B.E., and Franke, W., 1991, Geochemistry and provenance of Rhenohercynian synorogenic sandstones: Implications for tectonic environment discrimination, in Morton, A.C., et al., eds., Developments in Sedimentary Provenance Studies: Geological Society [London] Special Publication 57, p. 173–188.

Folk, R.L., 1968, Petrology of Sedimentary Rocks: Austin, Texas, Hemphill's, 170 p.

Fülöp, J., 1994, Magyarország geológiája: Paleozoikum II [Geology of Hungary: Paleozoic, II]: Budapest, Akadémiai Kiadó, 447 p.

Gaillardet, J., Dupré, B., and Allègre, C.J., 1999, Geochemistry of large river suspended sediments: Silicate weathering or recycling tracer: Geochimica et Cosmochimica Acta, v. 63, no. 23–24, p. 4037–4051, doi: 10.1016/S0016-7037(99)00307-5.

Garver, J.I., and Scott, T.J., 1995, Trace elements in shale as indicators of crustal provenance and terrane accretion in the southern Canadian Cordillera: Geological Society of America Bulletin, v. 107, p. 440–453, doi: 10.1130/0016-7606(1995)107<0440:TEISAI>2.3.CO;2.

Golonka, J., and Ford, D., 2000, Pangean (Late Carboniferous–Middle Jurassic) paleoenvironment and lithofacies: Palaeogeography, Palaeoclimatology, Palaeoecology, v. 161, p. 1–34, doi: 10.1016/S0031-0182(00)00115-2.

Haas, J., Hámor, G., and Korpás, L., 1999, Geological setting and tectonic evolution of Hungary: Geologica Hungarica Series Geologica, v. 24, p. 179–196.

Hetényi, R., and Ravasz-Baranyai, L., 1976, A baranyai antracittelepes felsőkarbon összlet a Siklósbodony 1. és a Bogádmindszent 1. sz. fúrás tükrében [The anthracitiferous Upper Carboniferous sequence of Baranya, South Hungary, in the light of boreholes Siklósbodony-1 and Bogádmindszent-1]: Budapest, MÁFI Évi Jelentése az 1973 évről [Annual Report of the Geological Institute of Hungary, 1973], p. 323–361.

Huang, W.L., 1993, The formation of illitic clays from kaolinite in KOH solution from 225°C to 350°C: Clays and Clay Minerals, v. 41, no. 6, p. 645–654, doi: 10.1346/CCMN.1993.0410602.

Jámbor, Á., 1969, Karbon képződmények a Mecsek és a Villányi-hegység közötti területen [Carboniferous deposits in the area between the Mecsek and Villány Mountains]: Budapest, MÁFI Évi Jelentése az 1967 évről [Annual Report of the Geological Institute of Hungary, 1967], p. 215–221.

Kübler, B., 1968, Evaluation quantitative du métamorphisme par la cristallinité de l'illite: Bulletin de Centre Recherche Pau-SNPA, v. 2, p. 385–397.

Lee, Y.I., 2002, Provenance derived from the geochemistry of late Paleozoic–early Mesozoic mudrocks of the Pyeongan Supergroup, Korea: Sedimentary Geology, v. 149, p. 219–235, doi: 10.1016/S0037-0738(01)00174-9.

McLennan, S.M., 1989, Rare earth elements in sedimentary rocks: Influence of provenance and sedimentary processes, in Lipin, B.R., et al., eds., Geochemistry and Mineralogy of Rare Earth Elements: Reviews in Mineralogy, v. 21, p. 169–200.

McLennan, S.M., 2001, Relationships between the trace element composition of sedimentary rocks and upper continental crust: Geochemistry, Geophysics, Geosystems, v. 2, doi: 10.1029/2000GC000109.

Nagy, E., 1968, A Mecsek hegység triász időszaki képződményei [Triassic formation of the Mecsek Mountains]: MÁFI Évkönyv [Annales of the Geological Institute of Hungary], v. 51, 198 p.

Nesbitt, H.W., Markovics, G., and Price, R.C., 1980, Chemical processes affecting alkalines and alkaline earths during continental weathering: Geochimica et Cosmochimica Acta, v. 44, p. 1659–1666, doi: 10.1016/0016-7037(80)90218-5.

Noda, A., Takeuchi, M., and Adachi, M., 2004, Provenance of the Murihiku terrane, New Zealand: Evidence from the Jurassic conglomerates and sandstones in Southland: Sedimentary Geology, v. 164, p. 203–222, doi: 10.1016/j.sedgeo.2003.10.003.

Pearce, J., Harris, N.B.W., and Tindle, A.G., 1984, Trace element discrimination diagrams for the interpretation of granitic rocks: Journal of Petrology, v. 25, p. 956–983.

Pettijohn, F.J., Potter, P.E., and Siever, R., 1972, Sand and Sandstone: New York, Springer-Verlag, 618 p.

Preston, J., Hartley, A., Hole, M., Buck, S., Bond, J., Mange, M., and Still, J., 1998, Integrated whole-rock trace element geochemistry and heavy mineral chemistry studies: Aids to the correlation of continental red-bed reservoirs in the Beryl Field: UK North Sea: Petroleum Geoscience, v. 4, p. 7–16.

Rischák, G., and Viczián, I., 1974, Agyagásványok bázisreflexióinak intenzitását meghatározó ásványtani tényezők [Mineralogical factors determining the intensity of basal reflections of clay minerals]: Budapest, MÁFI Évi Jelentése az 1972 évről [Annual Report of the Geological Institute of Hungary, 1972], p. 229–256.

Shand, S.J., 1943, Eruptive Rocks: Their Genesis, Composition, Classification and their Relations to Ore-Deposits (second ed.): New York, Wiley, 444 p.

Szederkényi, T., 2001, Tisza mega-unit, in Haas, J., ed., Geology of Hungary: Budapest, Eötvös University Press, p. 148–169.

Török, K., 1986, Adatok a Dél-Dunántúl kristályos aljzatának felépítéséhez [Composition of the crystalline basement rocks from southern Transdanubia, Hungary] [M.S. thesis]: Budapest, Eötvös University, 75 p.

Varga, A.R., and Szakmány, Gy., 2004, Geochemistry and provenance of the Upper Carboniferous sandstones from borehole Diósviszló-3 (Téseny Sandstone Formation, SW Hungary): Acta Mineralogica-Petrographica, Szeged, v. 45, no. 2, p. 7–14.

Varga, A., Szakmány, Gy., Józsa, S., and Máthé, Z., 2001, A nyugat-mecseki alsó-miocén konglomerátum karbon homokkő kavicsainak és a Tésenyi Homokkő Formáció képződményeinek petrográfiai és geokémiai összehasonlítása [Petrographic and geochemical comparison between the Carboniferous sandstone pebbles of the Lower Miocene conglomerate from western Mecsek Mountains and Téseny Sandstone Formation]: Földtani Közlöny [Bulletin of the Hungarian Geological Society], v. 131, p. 11–36.

Varga, A.R., Szakmány, Gy., Józsa, S., and Máthé, Z., 2003, Petrology and geochemistry of Upper Carboniferous siliciclastic rocks (Téseny Sandstone Formation) from the Slavonian-Drava unit (Tisza megaunit, S Hungary)—Summarized results: Acta Geologica Hungarica, v. 46, no. 1, p. 95–113.

Varga, A.R., Raucsik, B., and Szakmány, Gy., 2004, A Siklósbodony Sb–1 mélyfúrás feltételezett karbon-perm határképződményeinek ásványtani, kőzettani és geokémiai jellemzői [Mineralogical, petrographic and geochemical characteristics of siliciclastic rocks from the supposed Carboniferous–Permian boundary in borehole Siklósbodony Sb-1, southwestern Hungary]: Földtani Közlöny [Bulletin of the Hungarian Geological Society], v. 134, no. 3, p. 321–343.

Watanabe, T., 1981, Identification of illite/montmorillonite interstratifications by X-ray powder diffraction: Journal of Mineralogical Society of Japan, v. 15, Special Issue, p. 32–41.

Weaver, C.E., 1989, Clays, Muds, and Shales: Amsterdam, Elsevier, 819 p.

Willan, R.C.R., 2003, Provenance of Triassic–Cretaceous sandstones in the Antarctic Peninsula: Implications for terrane models during Gondwana breakup: Journal of Sedimentary Research, v. 73, no. 6, p. 1062–1077.

Zimmermann, U., and Bahlburg, H., 2003, Provenance analysis and tectonic setting of the Ordovician clastic deposits in the southern Puna Basin, NW Argentina: Sedimentology, v. 50, no. 6, p. 1079–1104, doi: 10.1046/j.1365-3091.2003.00595.x.

MANUSCRIPT ACCEPTED BY THE SOCIETY 9 AUGUST 2006

Geological Society of America
Special Paper 420
2007

First-cycle sandstone composition and color of associated fine-grained rocks as an aid to resolve Gondwana stratigraphy in peninsular India

Prodip K. Dutta[†]
Department of Geography and Geology, Indiana State University, Terre Haute, Indiana 47809, USA

ABSTRACT

A stratigraphic problem in most Gondwana basins in peninsular India has persisted for more than a century. Originally, in the type area of Damodar valley, the stratigraphy was established based on the order of superposition without the aid of fossils. However, fossils were used to classify the Gondwana succession into Lower Gondwana (Early Permian to Late Triassic), characterized by *Glossopteris* flora (considered to have become extinct by the end of Triassic), and Upper Gondwana, which has *Ptilophyllum* flora (Jurassic). The problem started as the work extended further outside the type area, where *Glossopteris* flora was discovered in Upper Gondwana rocks of Early Jurassic age. At this point, fossils took precedence in determining the relative age of strata under the assumption that *Glossopteris* flora could not extend beyond the Triassic. Therefore, *Glossopteris*-bearing Upper Gondwana–looking rocks were assumed to be of pre-Jurassic age. Thus, Upper Gondwana rocks were relegated to Lower Gondwana status. The confidence on fossils to determine the relative age of strata was so strong that the necessity to confirm their relative age based on the law of superposition was thought unnecessary. This happened in spite of the fact that ages based on a fossil are conceptual entities, whereas establishing the order of superposition is the ultimate proof of relative age of strata.

Gondwana sedimentation coincided with the global climatic change from an icehouse state in the late Paleozoic to a greenhouse state in the Mesozoic. Climate played the most important role in controlling sandstone composition and other lithological attributes. A strong correlation between sand/sandstone composition and climate is well established. Any perceptible change in climate brings a perceptible change in sand/sandstone composition. Such compositional transition would be expected in the stratigraphic column as the climate changed between the two extremes. Sandstone composition of the Gondwana succession in the Raniganj basin, India, supports such a hypothesis. In addition, color of fine-grained rocks in the Gondwana succession, which to some extent is also influenced by climate, uniquely identifies each lithostratigraphic unit.

[†]E-mail: gedutta@isugw.indstate.edu.

Dutta, P.K., 2007, First-cycle sandstone composition and color of associated fine-grained rocks as an aid to resolve Gondwana stratigraphy in peninsular India, *in* Arribas, J., Critelli, S., and Johnsson, M.J., eds., Sedimentary Provenance and Petrogenesis: Perspectives from Petrography and Geochemistry: Geological Society of America Special Paper 420, p. 241–252, doi: 10.1130/2006.2420(15). For permission to copy, contact editing@geosociety.org. ©2007 Geological Society of America. All rights reserved.

Stratigraphy must be based on the law of superposition. But sandstone composition along with the color of fine-grained rocks could serve as additional criteria to stratigraphic interpretation. Such an approach may be useful in structurally overturned sequences and in reconstructing the subsurface stratigraphy from borehole cores.

Keywords: sandstone composition, physical stratigraphy, Gondwana.

INTRODUCTION

The hierarchical approach, beginning with the physical criteria of defining lithostratigraphic units and establishing their order of superposition, followed by biostratigraphy and chronostratigraphy, is a fundamental stratigraphic procedure (Grabau, 1960). By using such an approach, many stratigraphic pitfalls can be avoided. Surprisingly, such standard procedure was not followed in some Gondwana basins in peninsular India; instead, physical criteria were bypassed and fossils were used to determine the relative age of strata. The relative age of strata must be based on physical criteria whenever possible because relative age dating of strata based on fossils is a conceptual entity and therefore may lead to erroneous interpretation, as has been the case of Gondwana stratigraphy in India. Any solution to this problem will need some understanding as to why physical evidence was overlooked in the first place and how paleontological criteria played the decisive role in stratigraphic reconstruction!

The objective of this paper is to develop a new physical stratigraphic method based on sandstone composition and the color of fine-grained rocks as an aid to the method based on the law of superposition. The rationale for using these parameters as fingerprints to place each facies in their respective stratigraphic position is also offered. This new tool can be used not only to resolve the stratigraphic problems usually encountered in many continental sediments, such as in peninsular Gondwana basins of India, but it can also be used in structurally overturned sequences and in reconstructing the subsurface stratigraphy from borehole cores.

Geological Setting of the Gondwana Succession in Peninsular India

The Gondwana succession, discussed in this paper, ranges in age from Early Permian to Early Jurassic and is represented by continental sediment that is nearly 3000 m thick (Pascoe, 1975). Based on lithological associations, the entire succession has been differentiated into four easily identifiable lithostratigraphic units. Such physically diverse units within a single continental succession have led to speculation about the importance of various controlling factors such as the source rock composition, the environment of deposition, rigor of transport, and climate in molding the nature of sediment, particularly the sandstone composition ranging from arkose to quartz arenite.

Following Dickinson and Suczek (1979), peninsular India during Gondwana sedimentation may be classified as a continental block provenance. The basin formation is related to reactivation of preexisting structural grains of the Precambrian shield, which generated block-faulted grabens and half-grabens (Acharyya, 2000; Datta et al., 1983; Casshyap and Tewari, 1988; Chaterji and Ghosh, 1970). These depressions, aligned along rectilinear belts, became the repository of first-cycle sediment derived from adjacent uplands of mostly granitic composition (Fig. 1). During deposition of the entire succession, the source rock composition remained constant (Suttner and Dutta, 1986).

Sedimentation in embryonic basins was initiated with ice-contact deposits of tillite and other coarser sediments. Subsequent to the melting of glaciers, a braided stream and glacial lake system developed in the outwash plain during the Early Permian (Sakmarian) and deposited conglomerate, sandstone, shale, and glacial varve (Ghosh and Mitra, 1975). These glaciogenic sediments are designated as Facies A.[1] As the glacial climate changed to temperate humid conditions late in the Early Permian (Kungurian), the periglacial outwash plain gave rise to a meandering fluvial system with well-defined channels and floodplains. Coal, carbonaceous shale, and gray shale/siltstone were deposited within the vast floodplain, with its extensive coal swamps, where the sands were deposited within the meandering channels. The meandering system continued, though the coal-forming environment dwindled, possibly as a result of an arid trend. The coal-forming conditions were terminated by the end of Permian (Mitra, 1991). The coal-bearing sedimentary package is designated as Facies B. The same meandering system continued, and a distinctly different sedimentary package, mostly characterized by channel sandstone and floodplain deposits of greenish/red shale. This assemblage is designated as Facies C. An abrupt termination of Facies C is observed almost at the end of Triassic. This episode is related to a tectonic event that uplifted the source area and changed the

---

[1]Gondwana succession in peninsular India has been divided into various lithostratigraphic units with formal names. Names of many of the units differ from basin to basin. In this paper, for better communication, brevity, and to avoid confusion with too many unfamiliar local names, the sedimentary packages that have similar lithological attributes will be referred to as facies. However, the names of standard lithostratigraphic units used in Indian stratigraphy corresponding to each facies are mentioned in parenthesis: Facies A (Talchir Formation); Facies B (Damuda Group); Facies C (Panchet/Maleri/Tiki Formation); Facies D (Mahadeva/Supra-Panchet, Kota, Parsora Formation).

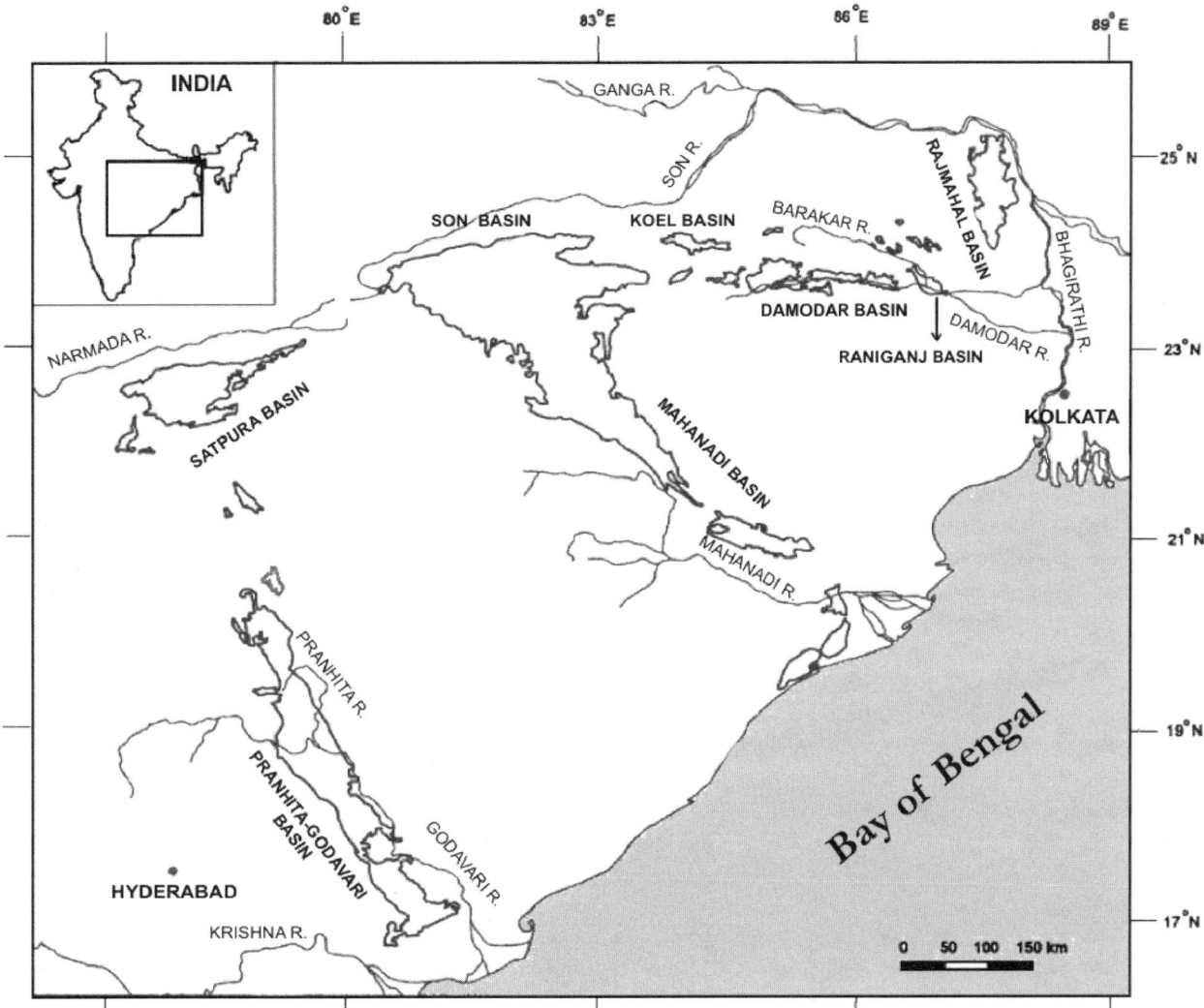

Figure 1. Geological map of the Gondwana basins in peninsular India. The basins are aligned parallel to the preexisting structural grains of the basement rock.

landscape. The meandering system was replaced by an extensive braided stream system in warm humid climate. Braided rivers deposited mostly conglomerate and sandstone with very little fine-grained sediment. This arenaceous unit is designated as Facies D. The transition from meandering to braided system resulted in a widespread erosional surface where Facies D overlapped on to the older rocks (Dutta, 2002). Except the ice-contact deposits at the base of the succession, the entire Gondwana sediments were deposited within a fluvial system. In such depositional setting, compositional modification of sand does not take place (Mack, 1977).

Paleogeographic reconstruction of individual Gondwana basins or basin belts suggests that these repositories had a limited drainage area, and, consequently, the maximum distance of transport of detritus was restricted to a few tens of kilometers to a few hundreds of kilometers (Dutta, 1983). Transport over such short distances, mostly along a braided and meandering system, had little modifying effect on sand composition (Hayes, 1962; Pollock, 1961; Russel, 1937).

Floral, miofloral, and faunal evidence indicate a climatic change from a cold-glacial phase during the Early Permian to warm humid conditions in Late Triassic and Early Jurassic time, with an intervening arid trend during the Late Permian and Early Triassic (Lele, 1969; Kar, 1976; Shah, 1976; Surange, 1966). Mainly based on paleontological evidence and the presence of climate-sensitive rocks, such as tillite, coal, and evaporites, Frakes (1979) postulated a similar climatic change on a global scale from a glacial climate in late Paleozoic time to a warm and humid climate in the Mesozoic. These observations on climate fit into Fischer's (1982) model of global climate change from an icehouse state during the late Paleozoic to a greenhouse state in the Mesozoic.

This analysis shows that during Gondwana sedimentation, the source rock remained constant while the environment of deposition and the rigor of transport played no part in modifying the composition of sand. The only variable that changed was the climate, between two extremes. Thus, it seems reasonable to infer that climate was responsible for generating different lithological associations and imparting color to fine-grained rocks and, most importantly, sand composition.

What Caused Misinterpretations of Gondwana Stratigraphy in India?

In the Damodar basins, superposition of the four facies, viz. A, B, C, and D can be easily established almost by default because the moderately dipping beds display a saucer-shaped layered-cake succession (Fig. 2). However, in the Son, Pranhita-Godavari, and in some other basins, confusion about the stratigraphic relationship between D- and C-like facies indicated by X and Y, respectively, in Figure 3 continued because:

1. The order of superposition between C- and D-like facies in these basins was not as obvious as in the Damodar basins. Gentle dip in a rolling topography camouflaged the order of superposition. Only a suspecting geologist would dig a trench to confirm the relative age of the strata (Fig. 3).
2. Throughout peninsular India, a widespread unconformity exists between Facies D and the underlying rocks. The

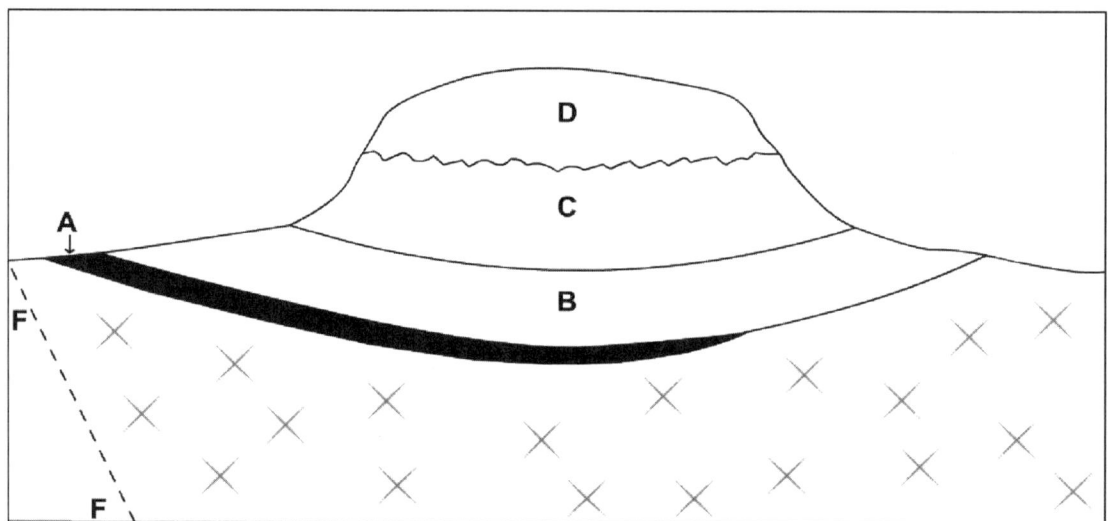

Figure 2. A schematic diagram showing the stratigraphy in the Damodar basins shown in Figure 1. Topography and the moderately dipping strata of Facies A, B, and C helped to establish the order of superposition; Facies D overlies the rest with an unconformity shown by the wiggly line. F—fault.

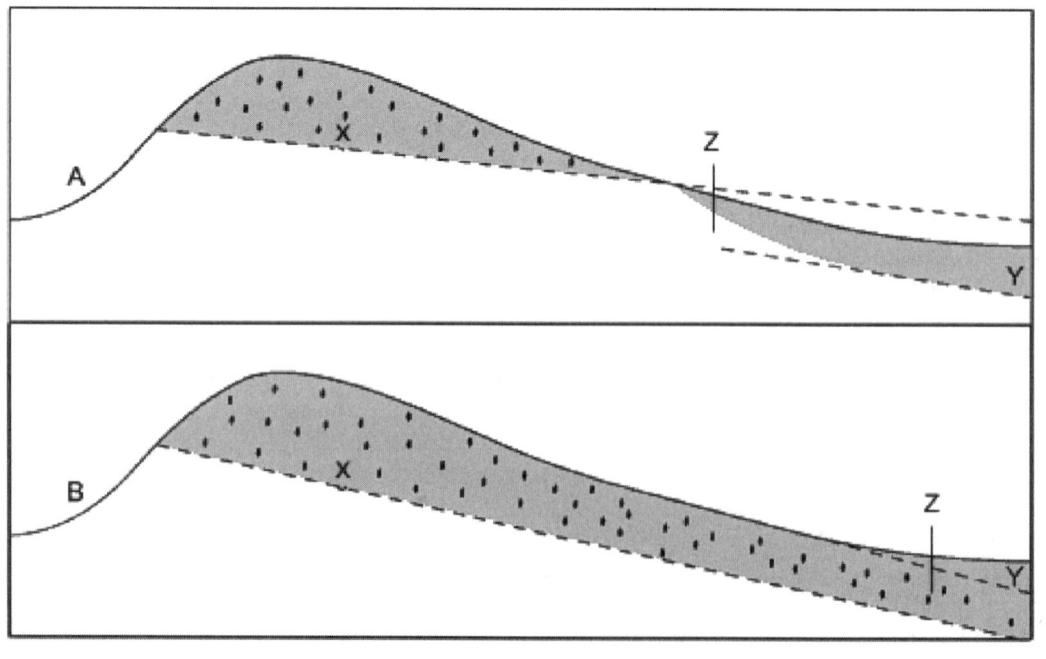

Figure 3. Sketch showing the difficulty in establishing the order of superposition of a gently dipping bed on rolling topography. To an unsuspecting observer, bed X could go either way, above or below Y. A shallow trench at Z would have resolved the problem long ago.

presence of this unconformity was not recognized in the Son and Pranhita-Godavari valley basins (Dutta, 2002).

3. The most obvious problem resulted from fossil occurrences: in 1881, Hughes and King, working in the Son and Pranhita-Godavari basins, respectively, observed a *Glossopteris* floral assemblage in D-like facies (outcrop c in Fig. 4) overlying the upper part of Facies B and Facies C (outcrops b and d, respectively, in Fig. 4). Hughes (1881) referred to such relationship in the Son basin where he observed Lower Gondwana fossil in Upper Gondwana rocks. It was their joint decision to relegate the isolated outcrops of D-like facies and merge them with the upper part of Facies B; they named the unit the "Kamthi Formation." They did this without verifying the order of superposition, even when it was obvious that the isolated outcrops of D-like facies were at a higher elevation along hills and hummocks, while the Facies B and C were forming the valleys at lower elevation. The presence of D-like facies in the midst of Facies C (outcrop e in Fig. 4) was explained by faults. Without the knowledge of the unconformity, the outcrops shown by lower case letters in Figure 4 were treated either as different lithostratigraphic units (Kutty et al., 1987—also see Table 1; Raiverman et al., 1985), or

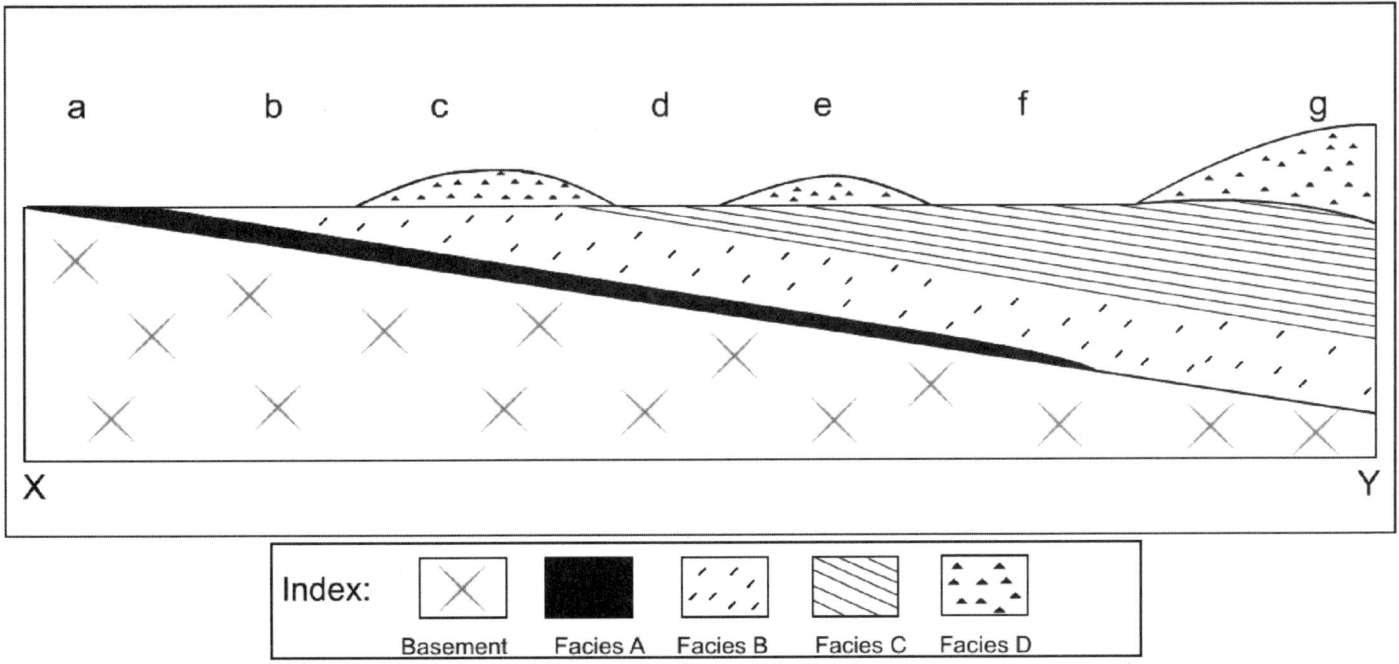

Figure 4. A schematic diagram showing the overlap of Facies D on older sediments. Without suspecting this unconformity, seven lithostratigraphic units, a, b, c, d, e, f, and g, would be identified from the outcrop pattern, instead of four Facies A, B, and C, and D. A *Glossopteris* floral assemblage was observed in an outcrop similar to outcrop c in the diagram. Without the unconformity, the outcrop becomes the upper part of Facies B.

TABLE 1. GONDWANA STRATIGRAPHY IN THE PRANHITA-GODAVARI BASIN

Formation name	Lithological description	Corresponding facies name (after Dutta, 2002)
Kota	*Coarse pebbly sandstone, fine-grained white sandstone, red clay, puplish siltstone, and a limestone band	Facies D
Dharmaram	Coarse sandstone and red clays	Facies C
Maleri	Red clay, fine- to medium-grained sandstone, and lime pellet rock	Facies C
Bhimaram sandstone	Medium-, coarse-, and fine-grained sandstone; some red clay	Facies C
Yerrapalli	Mainly red and violet clay with sandstone and lime pellet rock	Facies C
Kamthi	*Ferruginous nonfeldspathic or slightly feldspathic sandstone and purplish siltstone	Facies D
Infra Kamthi		
Lithozone 4	Feldspathic sandstone, red mudstone, limonitic shale	Facies C
	*Purplish siltstone	Facies D
Lithozone 3	Feldspathic sandstone, red, green, and yellow mudstone and shale	Facies C
Lithozone 2	Feldspathic sandstone, carbonaceous or coaly shale, and thin coal	Facies B
Lithozone 1	Feldspathic sandstone, shale with little carbonaceous matter	Facies B
Barakar	Feldspathic sandstone, carbonaceous shale, and coal	Facies B

Note: After Kutty et al. (1987). In this interpretation, repetition of C- and D-like facies is common. Facies D is indicated by asterisks and can easily be recognized by its nonfeldspathic and purple siltstone. In addition, Facies D tends to form mounds, hills, and hill ranges.

Facies D was shown as an uplifted block surrounded by Facies C (Sen Gupta, 1970; King, 1881).

In the Son basin, the stratigraphic relationship between Facies C and D flip-flopped (Fig. 5), and in the Pranhita-Godavari basin, every individual or group of researchers had their own version of stratigraphy (Fig. 6). It is apparent that there is a fundamental misunderstanding in our approach to resolve the problem and yet rarely does one find this issue discussed in the literature.

THE RATIONALE FOR USING SANDSTONE COMPOSITION AS A STRATIGRAPHIC TOOL

Weathering disintegrates the lithosphere into finer particles, the source of all clastic sedimentary rocks. The process is mainly accomplished through chemical weathering driven by climate. It is in weathering profiles that mineralogical modification of the parent rock takes place (Girty et al., 2003; Nesbitt and Markovics, 1997; Loughnan, 1969; Strakhov, 1967). Documentation in chemical weathering from varied climatic belts and in different source rock terrain under different relief conditions support this inference (Girty et al., 2003; Ugolini, 1986; Velbel, 1985; Ruxton, 1970; Feth et al., 1964; Garner, 1959). Once the precursors of clastic sediments are made in weathering/soil profiles, subsequent erosion and transport bring the detritus to the site of deposition for eventual burial and lithification. During this journey, little compositional modification takes place except in high-energy transport along mountainous streams (Ethridge, 1977; Pitman, 1969; Plumley, 1948). Mineralogical modification also takes place in high-energy environments as in eolian (Dutta et al., 1993; McKee, 1983; Ahlbrandt, 1979) and littoral environments (Mack, 1977; Flores, 1972; Sweet et al., 1971; Folk, 1960).

Evidence of climatic control on both first-cycle and multicycle fluvial sand is extensive (Johnsson and Meade, 1990; Johnsson et al., 1988; Velbel, 1985; Franzinelli and Potter, 1983; Suttner et al., 1981; Potter, 1978; Basu, 1976; Young, 1975). Climatic control on fluvial sand is best understood in first-cycle sediment since the source rock composition is better controlled compared to a mixed bag of source rocks. Significant compositional variations have been observed in sediments derived from granitic source rocks in different climatic belts (Girty et al., 2003; Johnsson et al., 1988; Basu, 1976). Even moderate differences in climate produce distinguishable imprints in first-cycle sand (Suttner et al., 1981).

Subsequent to deposition, mineralogical modification of original sand can take place during diagenesis. The degree of diagenetic destruction of original sand grains is mainly a function of burial depth. However, except in very deep burial diagenesis, the climatic control on nonmarine sediments is still preserved (Dutta, 1997; Suttner and Dutta, 1986). Moreover, it is also possible to estimate the diagenetic destruction of framework grains through petrographic studies (McBride, 1985) and reconstruct the original mineral grains and corresponding climatic signature.

It seems climate not only leaves an indelible imprint on sand through its composition, but its signature is also preserved in sandstone. During Gondwana sedimentation, the climate varied between two extremes, from glacial to warm and humid, encompassing all other climatic zones in between. Sandstones generated in this manner, as a response to a specific climatic milieu, may therefore serve as a stratigraphic datum.

COMPOSITION OF GONDWANA SANDSTONE IN THE RANIGANJ BASIN AND ITS STRATIGRAPHIC SIGNIFICANCE

In a study of the Gondwana sandstone composition in the Raniganj basin, the easternmost area of the Damodar basin belt

HUGHES (1881)	COTTER (1917)	FOX (1923)	RAO and SUKLA (1954)	LELE (1968)	DUTTA and GHOSH (1993)	MITRA (1993)	KUNDU et al. (1993)	SHAH (1994)
B	PS	B	PS	B	PS	B	PS	B
TIKI (MALERI)	PARSORA	TIKI	PARSORA	TIKI	PARSORA	TIKI	PARSORA	TIKI
PARSORA (MAHADEVA)	TIKI	PARSORA	TIKI	PARSORA	PALI-TIKI	PARSORA	TIKI	PARSORA

B = Biostratigraphy
PS = Physical Stratigraphy

Figure 5. Two approaches of biostratigraphy and physical stratigraphy led to opposing stratigraphic interpretations in Son basin.

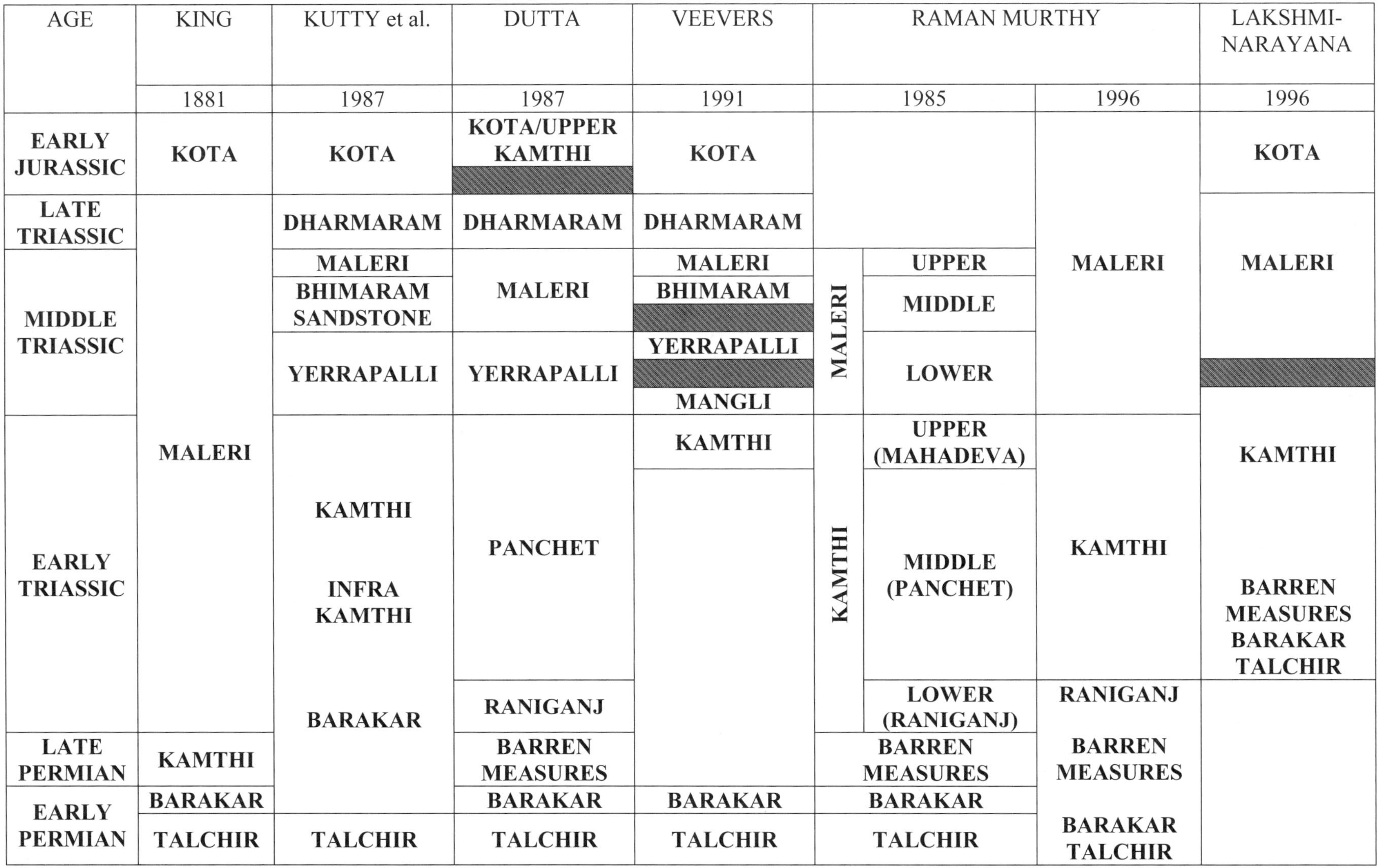

AGE	KING 1881	KUTTY et al. 1987	DUTTA 1987	VEEVERS 1991	RAMAN MURTHY 1985	RAMAN MURTHY 1996	LAKSHMI-NARAYANA 1996
EARLY JURASSIC	KOTA	KOTA	KOTA/UPPER KAMTHI	KOTA			KOTA
LATE TRIASSIC		DHARMARAM	DHARMARAM	DHARMARAM	MALERI — UPPER	MALERI	MALERI
MIDDLE TRIASSIC		MALERI	MALERI	MALERI	MALERI — MIDDLE		
		BHIMARAM SANDSTONE		BHIMARAM			
		YERRAPALLI	YERRAPALLI	YERRAPALLI	MALERI — LOWER		
	MALERI			MANGLI			
EARLY TRIASSIC		KAMTHI	PANCHET	KAMTHI	KAMTHI — UPPER (MAHADEVA)	KAMTHI	KAMTHI
		INFRA KAMTHI			KAMTHI — MIDDLE (PANCHET)		
			RANIGANJ		KAMTHI — LOWER (RANIGANJ)	RANIGANJ	BARREN MEASURES
LATE PERMIAN	KAMTHI	BARAKAR	BARREN MEASURES	BARAKAR	BARREN MEASURES	BARREN MEASURES	BARAKAR
EARLY PERMIAN	BARAKAR		BARAKAR		BARAKAR	BARAKAR	TALCHIR
	TALCHIR	TALCHIR	TALCHIR	TALCHIR	TALCHIR	TALCHIR	

Figure 6. Stratigraphic columns showing the lack of unanimity in interpreting the stratigraphy of the Pranhita-Godavari basin. Shaded areas represent unconformity.

shows a systematic change in framework mineralogy (Suttner and Dutta, 1986). Significant, easily recognizable compositional changes can be observed at certain stratigraphic levels throughout the succession, and each change may be considered as a stratigraphic datum (Table A1; Table 2). Since the formation of each sedimentary package is influenced by global climate, the bounding surface of each facies would represent an isochronous datum in a continental setting, as each facies is a product of a specific climate. For example, Facies A is a product of glacial climate, and the upper bounding surface of this facies represents the end of a glacial phase and the beginning of the humid temperate climate that generated the coal-bearing facies. Thus, the boundary between Facies A and B represents an isochronous surface.

Most immature sandstone with Q (Quartz): F (Feldspars): R (Rock Fragments) values of 54:42:4 are observed in Facies A. This immaturity is expected, since the intensity of chemical weathering was minimal due to a cold glacial climate. A sharp compositional change is observed having Q:F:R values of 88:10:2, in the lower part of Facies B,[2] where abundant coal seams are present designated as subfacies B′. This compositional change indicates a rapid climatic change from glacial to temperate humid conditions. It is noteworthy that even in a relatively cold temperate climate, the intensity of chemical weathering eliminated nearly 32% of feldspar and increased the quartz population by 34% compared to Facies A.

Sandstone composition again shows a sharp change in subfacies B″ to Q:F:R values of 61:34:5; a loss of 27% quartz, and an increase of 24% feldspar. This is due to the onset of a seasonal climate with a strong arid component punctuated by humid spells that generated enough vegetation for coal formation but in a restricted way, as indicated by fewer and thinner coal seams compared to subfacies B′. The arid trend continued, possibly with marginally more moist conditions, as indicated by a slight increase in sandstone maturity with Q:F:R values of 64:32:4, in subfacies C′. A compositional change is observed again in subfacies C″, which shows Q:F:R values of 83:16:1, indicating humid conditions that increased sediment maturity by eliminating 16% feldspar with a consequent increase of 19% quartz.

The most dramatic change in mineralogy can be observed between subfacies C″ and the overlying Facies D, the latter has Q:F:R values of 99:1:0. Intense chemical weathering in warm humid conditions eliminated almost all the unstable minerals, making the sandstone a first-cycle super-mature quartz arenite.

In addition to mineralogy, all four facies can also be differentiated by the color of the fine-grained rocks they are associated with: khaki green (Facies A), black/dark gray (Facies B), bluish green/red (Facies C), and purple (Facies D). Again, climate, to some extent, either directly or indirectly influenced the color of shale/siltstone/clay bed in each facies. The khaki-green color

TABLE A1. GONDWANA SANDSTONE COMPOSITION IN THE RANIGANJ BASIN, INDIA

Sample number	Quartz (Q) (%)	Total feldspar (F) (%)	Rock fragment (R) (%)
A			
01	50	46	4
02	52	46	2
03	55	40	5
04	53	41	6
05	48	49	3
06	53	44	3
07	59	35	6
08	57	42	1
09	56	39	5
10	57	37	6
B′			
11	92	6	2
12	97	3	0
13	91	6	3
14	95	5	0
15	93	6	1
16	96	3	1
17	92	7	1
18	85	13	2
19	81	16	3
20	82	14	4
21	85	11	4
22	93	5	2
23	85	14	1
24	85	15	0
25	83	15	2
26	83	14	3
27	79	14	7
B″			
28	61	35	4
29	72	25	3
30	54	38	8
31	61	35	4
32	62	33	5
33	55	37	5
34	60	35	5
35	59	38	3
36	66	32	2
37	55	41	4
38	64	31	5
39	64	32	4
40	66	31	3
41	58	38	4
42	59	34	7
43	65	30	5
44	58	40	2
45	65	30	5
C′			
46	63	34	3
47	62	32	6
48	68	30	2
49	61	35	4
50	59	38	3
51	70	29	1
C″			
52	88	11	1
53	79	20	1
54	82	16	2
55	79	19	2
56	78	20	2
57	82	16	2
58	91	8	1
D			
59	100	0	0
60	99	1	0
61	99	1	0
62	100	0	0
63	100	0	0
64	99	1	0
65	99	1	0

---

[2]Facies B has been defined as a coal-bearing unit, but within this unit, there is a wide compositional difference between the lower part, which has mature sandstone and abundant coal seams, designated as subfacies B′, and the upper part, as immature sandstone and restricted occurrence of coal and is designated as subfacies B″. On similar grounds, Facies C is also subdivided into two parts. The lower part is made up of immature sandstone with green/red shale (subfacies C′), while the upper part consists of mature sandstone with abundant red shale (subfacies C″).

TABLE 2. SANDSTONE COMPOSITION AND CLIMATIC INTERPRETATION FROM RANIGANJ BASIN, INDIA

Facies/ subfacies	Average Q:F:R	Range of Q:F:R	Standard deviation			Color of fine-grained rock	Prevailing climate
			Q	F	R		
D	99:1:0 n = 7	100:0:0 99:1:0	0.2	0.5	0	Purple	Warm humid
C″	83:16:1 n = 7	91:8:1 78:20:2	4.9	4.6	0.6	Red	Semihumid
C′	64:32:4 n = 6	70:29:1 59:38:3	3.1	3.3	1.7	Red/light green	Semiarid (oxidizing)
B″	61:34:5 n = 18	72:25:3 54:38:8	7.1	4.1	1.5	Black/gray	Semiarid (reducing)
B′	88:10:2 n = 17	97:3:0 79:14:7	6.7	4.7	1.7	Black/gray	Temperate humid
A	54:42:4 n = 10	48:49:3 57:42:1	3.4	4.4	1.6	Khaki-green	Glacial

Note: Sharp compositional variations are observed at the transition of each facies or subfacies. No perceptible mineralogical change is observed between subfacies B″ and C′. The reason for keeping them separate is their distinct lithological association of coal in subfacies B″ and green and red shale in subfacies C′. Q—quartz; F—feldspars; R—rock fragments.

(Warming trend ↑ indicated alongside climate column)

in Facies A is due to the strong presence of chlorite common in cold climates (Tardy, 1971; Griffin et al., 1968). Black and gray colors in Facies B is due to the presence of coal, carbonaceous shale/siltstone, and their formation in reducing environments in bogs and swamps. In Facies C, the light bluish green color at the base represents the transition from a reducing environment to an oxidizing environment in the Early Triassic, possibly as a result of sea-level retreat that increased relief (Haq et al., 1988). Formation of red shale may be due to increased moisture that is necessary for the formation of red beds (Van Houten, 1973). The reason for the purple hue in fine-grained rocks in Facies D is still not clearly understood. But whatever may be the process involved in imparting color to fine-grained rocks, each color identifies each facies in a unique way and can be used as a stratigraphic marker.

Petrographic data presented herein identify four sharp mineralogical changes as a function of climate; their respective average Q:F:R values are 54:42:4 (glacial; Facies A) → 88:10:2 (temperate humid; Facies B′) → 62:35:3 (cool arid; subfacies B″ and C′) → 83:16:1 (semihumid; subfacies C″) → 99:1:0 (warm humid; Facies D). Thus, knowing the sandstone composition makes it possible to identify the stratigraphic datum of a facies (lithostratigraphic unit). For example, if the average quartz content of sandstone is nearly 100%, the unit would be placed at the top of the succession. If the quartz content is nearly 50% with an equally high percentage of feldspar, the unit would be placed at the bottom of the succession. However, it may not be possible to differentiate between subfacies B″ and C′ based on sandstone composition alone, where their average Q:F:R values are in the sixties. But by using sandstone composition in combination with the color of fine-grained rocks, a clear distinction can be made. Sandstones with an average quartz content in the sixties associated with black/gray fine-grained rocks would be identified as subfacies B″. Sandstone with similar composition but with green/red–colored fine-grained rocks would be designated as subfacies C′. Even if there are doubts about sandstone composition with quartz content in eighties, as in subfacies B′ and C″, the color could be used to differentiate subfacies B′, which has black/gray fine-grained shale/siltstone, from subfacies C″, which has red shale (Table 2).

GONDWANA STRATIGRAPHY IN SON AND PRANHITA-GODAVARI BASINS

The stratigraphic controversy in the Son and Pranhita-Godavari basins is centered on the relationship between Facies C and Facies D. Facies C, as mentioned already, can be further divided into two subfacies: subfacies C′ is an arkose with green and red shale, while subfacies C″ is subarkose with red shale. Facies D stands out in a unique way as a first-cycle quartz arenite with purple siltstone.

It has been argued that for the generation of first-cycle quartz arenite, a warm humid climate is necessary, as exemplified in the present-day tropics, where first-cycle quartz arenite has been reported (Johnsson et al., 1988). During Gondwana sedimentation, a warm humid climate prevailed for the first time during the deposition of Facies D in the Early Jurassic. This was preceded by a relatively arid and subhumid climate that produced arkose and subarkose, respectively. There has been no evidence of repetition of climate, and, consequently, no repetition of sandstone with similar compositions along with similar colors of fine-grained rocks that has been observed in the stratigraphic column.

Petrographic data of the sandstone composition of Facies C and D in the Pranhita-Godavari and Son basins from various workers (Dutta and Ghosh, 1993; Raman Murthy, 1985; Sen Gupta, 1970) show a similar trend to that observed in the Raniganj basin. The published data available on Facies C in the Pranhita-Godavari basin are meager and cannot be differentiated into subfacies C′ and C″ and are therefore lumped together as Facies C, which has compositions ranging from arkose to subarkose (Table 3). The data in Table 3 are consistent across all three basins and clearly differentiate two facies: Facies C and Facies D. This interpretation was independently arrived at on the basis of the order of superposition by field geologists (Rao and Sukla, 1954; Dutta, 1987; Dutta and Ghosh, 1993; Kundu et al., 1993). Therefore, stratigraphy established on physical criteria fundamentally differs from the interpretation that emanates from fossil-based relative age dating (Hughes, 1881; King, 1881; Fox, 1923; Lele, 1969, Mitra, 1993).

TABLE 3. STRATIGRAPHY OF THE GONDWANA SUCCESSION IN PENINSULAR INDIA BASED ON SANDSTONE COMPOSITION AND COLOR OF FINE-GRAINED SEDIMENTS

Raniganj basin		Son basin		Pranhita-Godavari basin		Color of fine-grained rocks
Facies	Sandstone composition Q:F:R	Facies	Sandstone composition Q:F:R	Facies	Sandstone composition Q:F:R	
D	99:1:0 $n = 7$	D	98:2:0 $n = 17$	D	99:1:0 $n = 11$	Purple
C″	83:16:1 $n = 7$	C″	86:13:1 $n = 10$	C	Arkose-subarkose $n = 4$	Red
C′	64:32:4 $n = 6$	C′	65:32:2 $n = 6$	C	Arkose-subarkose $n = 4$	Red and green

Note: Abbreviations: Q—quartz; F—feldspars; R—rock fragments.

DISCUSSION

Using fossils to constrain the relative age of strata has led to this long, drawn-out stratigraphic controversy. This approach contradicts the fundamental principles of stratigraphy because it negates the hierarchical approach, which begins with the order of superposition followed by biostratigraphy and then by chronostratigraphy (Grabau, 1960).

Biostratigraphy is an important tool in delineating biostratigraphic units, tracing the evolutionary pattern of biota, and for correlation. However, biostratigraphic analysis is possible only when the range of a taxon or range of an assemblage is well established and the fossil-based stratigraphy is well nested within the physical stratigraphic framework.

Usually, fossil assemblages are used to determine the relative age in marine as well as in nonmarine sequences where the order of superposition is difficult, if not impossible, to establish. Age dating based on fossils is a conceptual entity. It is dependent on available data at a certain point of time, which may change when new data are available. For example, originally, *Glossopteris* was considered to be a late Paleozoic flora, and, subsequently, it was extended to the Early Triassic (Brongniart, 1882; Srivastava, 1971, quoted by Delevoryas and Person, 1975). *Glossopteris* or *Glossopteris*-like leaves, *Mexiglossa*, have been found in sediments of Middle Jurassic age (Delevoryas and Person, 1975). *Glossopteris* has also been reported from the Parsora Formation (Facies D) of Early Jurassic age in the Son basin (Dutta and Ghosh, 1993). It was this *Glossopteris* flora contained in D-like facies that led to this stratigraphic misunderstanding, because in the late nineteenth century, the idea of the last appearance datum (LAD) of *Glossopteris* by the end of the Triassic was sacrosanct.

In the late nineteenth century, fossils played the most dominant role in relative age dating. Most workers accepted the paradigm and therefore did not see the necessity of establishing the order of superposition. Only a handful of field geologists did establish the order of superposition, but their efforts went unheeded. It is indeed intriguing why the controversy continued for so long when a single joint field trip by all concerned might have resolved the problem long ago? Or why even a short stratigraphic borehole could not be drilled to establish the order of superposition between these two facies in question when an extensive exploratory drilling program for coal resources in these basins has continued for nearly half a century?

Communication is the key to scientific advancement, and yet there never has been any attempt to address the problem collectively to resolve the issue. The new stratigraphic tool, based on sandstone composition and the color of fine-grained rocks proposed here, requires a critical analysis from all concerned. To break the stalemate, a joint field trip can be arranged where all can examine the order of superposition, color of fine-grained rocks, and determine the sandstone composition at the outcrop.

CONCLUSIONS

An attempt is made here to point out that there is a problem in our understanding of the geological history of the Gondwana succession in parts of peninsular India. The problem and the basic differences in approaches are exemplified by Figures 5 and 6. Geological interpretations are not always accepted by all. But the most logical interpretation typically prevails. However, in this case, the disagreements are of fundamental nature and have a bearing on all geological investigations. This disagreement is unusual in that the reasons behind it are primarily historical and date back to the late nineteenth century.

The relative age of strata must be based on physical criteria whenever possible, because fossil-based age determination may lead to erroneous interpretation. However, this may not be possible in structurally complicated sequences where relative age–based on fossils are the only alternative. Once the physical stratigraphic framework is established, index fossils can be used to fine tune the stratigraphy and trace the evolutionary path of the biota.

Sandstone composition has been used to understand various aspects of sedimentary geology, including plate tectonic setting and climate. For the first time, sandstone composition along with color of fine-grained rock have been used as stratigraphic indicators. This methodology can be used to reconstruct stratigraphy in many geological settings, such as in other Gondwana basins (Dutta and Wheat, 1993; Dutta and Santos, 1991) as well as in overturned sequences and in subsurface borehole cores.

ACKNOWLEDGMENTS

My thanks are due to Tony Rathburn, Rónadh Cox, and an unknown reviewer for reviewing the manuscript, which has helped to improve the quality of the paper. However, for any misconceptions and misinterpretations, the author is fully responsible. Bharat Ganesh-Babu helped in drawing the diagrams. The work was partly supported by a grant from Indiana State University.

REFERENCES CITED

Acharyya, S.K., 2000, Coal and Lignite Resources of India: An Overview: Bangalore, Geological Society of India, 50 p.
Ahlbrandt, T.S., 1979, Textural parameters of eolian deposits, in McKee, E.D., ed., A Study of Global Sand Seas: U.S. Geological Survey Paper 1052, Chapter B, p. 21–51.
Basu, A., 1976, Petrology of Holocene fluvial sand derived from plutonic source rocks: Implications to paleoclimatic interpretation: Journal of Sedimentary Petrology, v. 46, p. 694–709.
Brongniart, A., 1828, Prodrome d'une histoire des vegetaux fossils: Paris 223 p.
Casshyap, S.M., and Tiwari, R.C., 1988, Depositional models and tectonic evolution of Gondwana basins: Palaeobotanist, v. 36, p. 59–66.
Chaterji, G.C., and Ghosh, P.K., 1970, Tectonic framework of peninsular Gondwanas of India: Geological Survey of India Record, v. 98, p. 1–5.
Cotter, G., de P., 1917, A revised classification of the Gondwana System: Record Geological Survey of India, v. 48, pt. 1, p. 23–33.
Datta, N.R., Mitra, N.D., and Bandyopadhyay, S.K., 1983, Recent trends in the study of Gondwana basins of peninsular and extra peninsular India: Petroleum Asia Journal, v. 1, p. 159–170.
Delevoryas, T., and Person, C.P., 1975, *Mexiglossa* Varia Gen. Nov., a new genus of Glossopteroid leaves from the Jurassic of Oaxaca, Mexico: Paleontographica Abt. B, v. 154, p. 114–120.
Dickinson, W.R., and Suczek, C.A., 1979, Plate tectonics and sandstone composition: American Association of Petroleum Geologists Bulletin, v. 63, p. 2164–2182.
Dutta, P.K., 1983, The role of climate in the evolution of detrital and authigenic mineralogy in sandstone from the Gondwana Supergroup, India [Ph.D. thesis]: Bloomington, Indiana University, 169 p.
Dutta, P.K., 1987, Upper Kamthi: A riddle in the Gondwana stratigraphy of peninsular India, in McKenzie, G.D., ed., Gondwana Six: Stratigraphy, Sedimentology, and Paleontology: American Geophysical Union Geophysical Monograph 41, p. 299–238.
Dutta, P.K., 1997, Interpreting glacial climate from detrital minerals in sediments, in Martini, I.P., ed., Late Glacial and Post-Glacial Environmental Changes: New York, Oxford University Press, p. 271–275.
Dutta, P.K., 2002, Gondwana lithostratigraphy of peninsular India: Gondwana Research, v. 5, p. 540–553, doi: 10.1016/S1342-937X(05)70742-5.
Dutta, P.K., and Ghosh, S.K., 1993, The century-old problem of the Pali-Parsora-Tiki stratigraphy and its bearing on the Gondwana classification of peninsular India: Journal of the Geological Society of India, v. 42, p. 17–31.
Dutta, P.K., and Santos, P.R., 1991, Provenance of Gondwana sandstones of the Parana basin, Brazil: Dallas, American Association of Petroleum Geologists, Annual Convention Abstracts, p. 566.
Dutta, P.K., and Wheat, R.W., 1993, Climatic and tectonic control on sandstone composition in the Permo-Triassic Sydney basin, eastern Australia, in Johnsson, M.J., and Basu, A., eds., Processes Controlling the Composition of Clastic Sediments: Geological Society of America Special Paper 284, p. 187–202.
Dutta, P.K., Zhou, Z., and Santos, P.R., 1993, A theoretical study of mineralogical maturation of eolian sand, in Johnsson, M.J., and Basu, A., eds., Processes Controlling the Composition of Clastic Sediments: Geological Society of America Special Paper 284, p. 203–209.
Ethridge, F.G., 1977, Petrology, transport, and environment in isochronous Upper Devonian sandstone and siltstone units, New York: Journal of Sedimentary Petrology, v. 47, p. 53–65.
Feth, J.R., Roberson, C.E., and Polzer, W.L., 1964, Source of mineral constituents in water from granitic rocks, Sierra Nevada, California and Nevada: U.S. Geological Survey Water Supply Paper I1535, 70 p.

Fischer, A.G., 1982, Long Term Climatic Oscillation Recorded in Stratigraphy: Climate in Earth History: Washington, D.C., National Academy Press, p. 97–104.
Flores, R.M., 1972, Delta front-delta plain facies at the Pennsylvanian Haymond Formation, northeast Marathon Basin, Texas: Geological Society of America Bulletin, v. 83, p. 3415–3424.
Folk, R.L., 1960, Petrography and origin of Tuscarora, Rose Hill, and Keefer Formations, Lower and Middle Silurian of western Virginia: Journal of Sedimentary Petrology, v. 30, p. 1–58.
Fox, C.S., 1923, The Gondwana system and the related formations: Calcutta, Geological Survey of India Memoir 21, 103 p.
Frakes, L.A., 1979, Climates through Geologic Time: Amsterdam, Elsevier, 310 p.
Franzinelli, E., and Potter, P.E., 1983, Petrology, chemistry, and texture of modern river sands, Amazon River system: The Journal of Geology, v. 91, p. 23–39.
Garner, H.F., 1959, Stratigraphic-sedimentary significance of contemporary climate and relief in four regions of the Andes Mountains: Geological Society of America Bulletin, v. 70, p. 1327–1368.
Ghosh, P.K., and Mitra, N.D., 1975, History of Talchir Sedimentation in Damodar Valley Basins: Geological Survey of India Memoir 105, 117 p.
Girty, G.H., Marsh, J., Meltzner, A., McConnell, J.R., Nygren, D., Nygren, J., Prince, G., Randall, K., Johnson, D., Heitman, B., and Nielsen, J., 2003, Assessing changes in elemental mass as a result of chemical weathering of granodiorite in a Mediterranean (hot summer) climate: Journal of Sedimentary Research, v. 73, p. 434–443.
Grabau, A.W., 1960, Principles of Stratigraphy: New York, A.G. Seiler and Co. Dover Publication, 1185 p.
Griffin, J.J., Windom, H., and Goldberg, E.D., 1968, The distribution of clay minerals in the world ocean: Deep Sea Research and Oceanographic Abstracts, v. 15, p. 433–459, doi: 10.1016/0011-7471(68)90051-X.
Haq, B.U., Hardenbol, J, and Vail, P.R., 1988, Mesozoic and Cenozoic chronostratigraphy and cycles of sea-level change, in Wilgus,C.K., Hastings, B.S., Kendall, C.G., Posamentier, H.W., Ross, C.A., and Van Wagoner, J.C., eds., Sea-level changes: an integrated approach: Society for sedimentary Geology (SEPM) Special Publication 42, p. 71-108.
Hayes, J.R., 1962, Quartz and feldspar content in South Platte, Platte and Missouri River sands: Journal of Sedimentary Petrology, v. 32, p. 793–800.
Hughes, T.W.H., 1881, Notes on the South-Rewa Gondwana basin: Geological Survey of India Record, v. 14, p. 126–138.
Johnsson, M.J., and Meade, R.H., 1990, Chemical weathering of fluvial sediments during alluvial storage: The Macuapanim Island point bar, Solimoes River, Brazil: Journal of Sedimentary Petrology, v. 60, p. 827–842.
Johnsson, M.J., Stallard, R.F., and Meade, R.H., 1988, First-cycle quartz arenite in the Orinoco River basin, Venezuela and Colombia: The Journal of Geology, v. 96, p. 263–277.
Kar, R.K., 1976, Miofloristic evidences for climatic vicissitudes in India during Gondwana: Geophytology, v. 6, no. 2, p. 230–245.
King, W., 1881, The geology of the Pranhita-Godavari valley: Geological Survey of India Memoir 19, p. 153–311.
Kundu, A., Pillai, K.R., and Thanvelu, C., 1993, Elucidation of stratigraphic inter-relationship of Pali-Tiki-Parsora Formations of South-Rewa Gondwana basins, Madhya Pradesh: Gondwana Geological Magazine, Special Volume, p. 49–59.
Kutty, T.S., Jain, S.L., and Roy Chowdhury, T., 1987, Gondwana sequence of the northern Pranhita-Godavari valley: Its stratigraphy and vertebrate faunas: The Paleobotanists, v. 36, p. 214–229.
Lele, K.M., 1969, The problem of Middle Gondwana in India, in Sundaram, R.K., ed., Proceedings of the 22nd International Geology Congress: New Delhi, International Geology Congress, v. 9, p. 181–202.
Loughnan, F.C., 1969, Chemical Weathering of the Silicate Minerals: New York, Elsevier, 154 p.
Mack, G.H., 1977, The effects of depositional environment on detrital mineralogy: The Cutler-Cedar Mesa facies transition near Moab, Utah [Ph.D. thesis]: Bloomington, Indiana University, 152 p.
McBride, E.F., 1985, Diagenetic processes that affect provenance determination in sandstone, in Zuffa, G.G., ed., Provenance of Arenites: D. Reidel, Dordrecht, NATO Advanced Study Institute Ser. C., v. 148, p. 95–113.
McKee, E.D., 1983, Eolian sand bodies of the world, in Brookfield, M.E., and Ahlbrandt, T.S., eds., Eolian Sediments and Processes: Development in Sedimentology, v. 38, p. 1–25.
Mitra, N.D., 1991, The sedimentary history of Lower Gondwana coal basins of peninsular India, in Ulbrich, H., and Rocha-Campos, A.C., eds., Proceed-

ings Gondwana Seven: Sao Paulo, Instituto de Geociências da Universidade de Sao Paulo, p. 273–288.
Mitra, N.D., 1993, Stratigraphy of Pali-Parsora-Tiki Formations of South-Rewa Gondwana basin and Permo-Triassic boundary problem, in Dutta, K.K., and Sen S., eds., Scientific Papers of Birbal Sahni Centenary National Symposium on Gondwana of India: Gondwana Geology Magazine, Special Volume, p. 41–48.
Nesbitt, H.W., and Markovics, G., 1997, Weathering of granodioritic crust, long-term storage of elements and petrogenesis of siliciclastic sediments: Geochimica et Cosmochimica Acta, v. 61, p. 1635–1670, doi: 10.1016/S0016-7037(97)00029-X.
Pascoe, E.W., 1975, A Manual of the Geology of India and Burma, Volume II: New Delhi, Manager of Publications, Government of India, p. 485–1343.
Pitman, E.D., 1969, Destruction of plagioclase twins by stream transport: Journal of Sedimentary Petrology, v. 39, p. 1432–1437.
Plumley, W.J., 1948, Black Hills terrace gravels: A study in sediment transport: The Journal of Geology, v. 56, p. 526–577.
Pollock, J.M., 1961, Significance of compositional and textural properties of South Canadian River channel sands, New Mexico, Texas and Oklahoma: Journal of Sedimentary Petrology, v. 31, p. 15–37.
Potter, P.E., 1978, Petrology and chemistry of modern big river sands: Journal of Geology, v. 86, p. 423–449.
Raiverman, V., Rao, M.R., and Pal, D., 1985, Stratigraphy and structure of the Pranhita- Godavari graben: Petroleum Asia Journal, v. 8, p. 174–190.
Raman Murthy, B.V., 1985, Gondwana sedimentation in Ramagundam-Mantheni area, Godavari Valley basin: Journal of the Geological Society of India, v. 26, p. 43–55.
Raman Murthy, B.V., and Rao, C.M., 1996, A new lithostratigraphy classification of Permian (Lower Gondwana) succession of Pranhita-Godavari basin with special reference to Ramagundam coalbelt, Andhra Pradesh, India, in Guha, P.K.S., Sengupta, S., Ayyasami, K., and Ghosh, R.N., eds., Ninth International Gondwana Symposium: New Delhi–Calcutta, Geological Survey of India, and Oxford, IBH Publishing Co., p. 67–78.
Rao, C.N., and Sukla, R.N., 1954, Examinations of the Supra-Barakars and Parsora-Tiki tract in Shadol District, Madhya Pradesh: Progress Report for the Field Season 1953–54: Calcutta, Geological Survey of India, p. 1–54.
Russel, R.D., 1937, Mineral composition of Mississippi River sands: Geological Society of America Bulletin, v. 48, p. 1307–1348.
Ruxton, P.B., 1970, Labile quartz-poor sediments from young mountain ranges in northwest Papua: Journal of Sedimentary Petrology, v. 40, p. 1262–1270.
Sen Gupta, S., 1970, Gondwana sedimentation around Bheemaram (Bhimaram), Pranhita- Godavari valley, India: Journal of Sedimentary Petrology, v. 40, p. 140–170.
Shah, B.A., 2004, Gondwana lithostratigraphy of peninsular India: Comment: Gondwana Research, v. 7, p. 600–607, doi: 10.1016/S1342-937X(05)70810-8.
Shah, S.C., 1976, Climates during Gondwana Era in peninsular India: Geophytology, v. 6, no. 2, p. 186–206.
Srivastava, P.N., 1971, Some gymnospermic remains from the Triassic Nidpur, Sidhi District, Madhya Pradesh: Palaeobotanist, v. 18, p. 280–296.
Strakhov, N.M., 1967, Principles of Lithogenesis, Volume I: New York, Consultants Bureau, 245 p.
Surange, R.K., 1966, Distribution of *Glossopteris* flora in the Lower Gondwana formations in India: Lucknow, Birbal Sahni Institute of Palaeobotany; Stratigraphy; Gondwanaland, p. 55–68.
Suttner, L.J., and Dutta, P.K., 1986, Alluvial sandstone composition and paleoclimate: I: Framework mineralogy: Journal of Sedimentary Petrology, v. 56, p. 329–345.
Suttner, L.J., Basu, A., and Mack, G.H., 1981, Climate and the origin of quartz arenite: Journal of Sedimentary Petrology, v. 51, p. 1235–1246.
Sweet, K., Klein, G.D.V., and Smit, D.E., 1971, A Cambrian tidal sand body—The Eriboll sandstone of northeast Scotland: An ancient-recent analog: The Journal of Geology, v. 79, p. 400–415.
Tardy, E., 1971, Characterization of the principal weathering types by the geochemistry of waters from some European and African crystalline massifs: Chemical Geology, v. 7, p. 253–271, doi: 10.1016/0009-2541(71)90011-8.
Ugolini, F.C., 1986, Processes and rates of weathering in cold and polar desert environments, in Colman, S.M., and Dethier, D.P., eds., Rates of Chemical Weathering of Rocks and Minerals: London, Academic Press, p. 193–235.
Van Houten, F.B., 1973, Origin of red beds: A review 1961–1972: Annual Review of Earth and Planetary Sciences, v. 1, p. 39–61, doi: 10.1146/annurev.ea.01.050173.000351.
Veevers, J.J., 1991, Mid-Triassic lacuna on the Gondwanaland platform during the final coalescence and incipient dispersal of Pangaea, in Ulbrich, H., and Rocha Campos, C., eds., Gondwana Seven Proceedings: Sao Paulo, Universidade de Sao Paulo, p. 603–613.
Velbel, M.A., 1985, Geochemical mass balances and weathering rates in forested watersheds in the southern Blue Ridge: American Journal of Science, v. 285, p. 904–930.
Young, S.W., 1975, Petrography of Holocene fluvial sand derived from regionally metamorphosed source rocks [Ph.D. thesis]: Bloomington, Indiana University, 144 p.

Manuscript Accepted by the Society 9 August 2006

Sand and gravel provenance in the Waipaoa River system: Sedimentary recycling in an actively deforming forearc basin, North Island, New Zealand

Dawn E. James[†]
Alissa M. DeVaughn[‡]
Kathleen M. Marsaglia[§]
Department of Geological Sciences, California State University–Northridge, 18111 Nordhoff Street, Northridge, California 91330-8266, USA

ABSTRACT

The sand and gravel within the Waipaoa River system of North Island, New Zealand, elucidate the provenance and sedimentary delivery system of an actively deforming continental forearc. In this region, Mesozoic to Cenozoic forearc basin sedimentary successions are being uplifted, eroded, and recycled into younger deposits. The identification of specific characteristics of the drainage basin bedrock can be linked to sediment composition, which in turn can document the processes of provenance mixing and dilution.

Despite the proximity of the Waipaoa River to an active volcanic arc (~250 km), its sediment load includes a relatively small proportion of volcanic debris. Sand detrital modes and clast count data indicate that Waipaoa River sediments are dominated by the crushed and sheared sedimentary rocks of the East Coast allochthon, which is located in the upper portion of the catchment. Higher volcanic percentages in the Waimata and Te Arai Rivers suggest that the volcanic contribution in the Waipaoa River has likely been diluted by the great volume of sediment brought into the system from the large gullies in the catchment headwaters.

Petrographic analysis of sand subfractions (very fine to very coarse) of Waipaoa River system sediment samples indicates that there is a distinct relationship between the size of the grains and their respective composition. Quartz and feldspar grains are concentrated in the finer fractions, reflecting the grain size of detrital sand liberated from Mesozoic to Cenozoic sandy source rocks. Thus, their distribution is at least in part an inherited feature related to sediment recycling within the system. Most of the sediment (gravel and sand) consists of mudstone and claystone lithic fragments.

[†]Present address: MWH Americas, Inc., 2503 Del Prado Boulevard, Suite 430, Cape Coral, Florida 33904, USA; e-mail: dawnjames@gmail.com.
[‡]Present address: Chesapeake Energy, Inc., 6100 North Western Avenue, Oklahoma City, Oklahoma 73118, USA; e-mail: adevaughn@chkenergy.com.
[§]Corresponding author present address: Department of Geological Sciences, California State University Northridge, 18111 Nordhoff Street, Northridge, California 91330-8266, USA; e-mail: kathie.marsaglia@csun.edu.

James, D.E., DeVaughn, A.M., and Marsaglia, K.M., 2007, Sand and gravel provenance in the Waipaoa River system: Sedimentary recycling in an actively deforming forearc basin, North Island, New Zealand, *in* Arribas, J., Critelli, S., and Johnsson, M.J., eds., Sedimentary Provenance and Petrogenesis: Perspectives from Petrography and Geochemistry: Geological Society of America Special Paper 420, p. 253–276, doi: 10.1130/2006.2420(16). For permission to copy, contact editing@geosociety.org. ©2007 Geological Society of America. All rights reserved.

In the Waipaoa River, the sand is dominated by noncalcareous mudstone lithic fragments, some of which are smectitic. In the Waimata River, the sand grains are mainly calcareous mudstone lithic fragments. This difference in rock-fragment type reflects differences in mud rock source lithologies. Similar trends are observed in the gravel fractions of the Waipaoa River, with noted increases in calcareous gravel components in the river where it (or tributaries) traverse calcareous lithologies.

These compositional differences suggest that in future studies of older alluvium and offshore sedimentary sections, it may be possible to distinguish coarse sediment derived from a purely Cenozoic source from that derived from Mesozoic rocks in the upper catchment of the Waipaoa River system, allowing us to interpret drainage evolution and denudation patterns.

Keywords: New Zealand, forearc basin, Waipaoa River, sand, gravel, provenance.

INTRODUCTION

Source rocks exert a first-order control on clastic sediment composition (Johnsson, 1993). In order to understand how the source rock signature is manifested in the composition of sand derived from it, it is important to look at relatively simple modern and ancient settings. Modern and ancient sand provenance studies (e.g., Ibbeken and Schleyer, 1991; Ridgeway and DeCelles, 1993; Critelli et al., 2000; Arribas et al., 2000) often involve source terrains with mixed rock types, but several have addressed the nature of sediment derived from specific plutonic (e.g., Girty, 1991; Heins, 1993; Palomares and Arribas, 1993), volcanic (e.g., Marsaglia, 1992, 1993), metamorphic (e.g., Palomares and Arribas, 1993; Garzanti et al., 2004; Shapiro et al., this volume), and sedimentary systems (e.g., Cavazza et al., 1993; Di Giulio et al., 2003). The Waipaoa River system is located on uplifted sedimentary units in an actively deforming forearc basin along the eastern margin of North Island, New Zealand (Figs. 1 and 2). It is an excellent location to study the compositional variation of sediment derived from mud-rich forearc lithologies.

The results of this study will have more far-reaching implications in that the Waipaoa River system has been designated as a primary focus site for the MARGINS Source-to-Sink Initiative sponsored by National Science Foundation. The goal of the Source-to Sink initiative is to use selected margins like the Waipaoa River system as templates for creating universal models for continental margin sedimentation (Driscoll and Nittrouer, 2000; Gomez et al., 2001a). Data collected from onshore river systems like the Waipaoa River system will ultimately be used to interpret the provenance of coarse sediments delivered to offshore shelf, slope, and trench basins. Our study provides fundamental information about the possible fingerprints (tracers) of sediment flux events (e.g., cyclones and associated flooding, gullying, and landsliding events), which may be translated to the offshore shelf component of the system. Data reported herein will help constrain planned theoretical models based on the synthesis of geomorphological and oceanographic databases collected in the Waipaoa River system. Furthermore, results from our study can be applied to other active margins where uplifted muddy sedimentary units are the main sediment sources.

TECTONIC, GEOLOGIC, AND PHYSIOGRAPHIC SETTING OF THE STUDY AREA

New Zealand consists of partially submerged continental blocks that straddle the Indian-Australian and Pacific plate boundary, which has undergone oblique convergence during the last 3–5 m.y. (Walcott, 1987; Bran, 1995; King, 2000; Fig. 2). Active subduction north of the transform Alpine fault, beneath the North Island, is marked by intermediate depth seismicity, arc-related volcanism, and active extension within the Taupo volcanic zone (Cole, 1990; Black et al., 1996; Fig. 2). Thrusting, folding, and uplift of forearc Cenozoic sedimentary rocks onto North Island results from the subduction of the Pacific plate and associated Hikurangi Plateau westward below the Australian plate at the Hikurangi Trough (Pettinga, 1982; Lewis and Pettinga, 1993).

Basement in the study area is composed of the late Paleozoic to Mesozoic Torlesse Supergroup (Mazengarb and Speden, 2000). This unit consists of relatively hard mudstone and quartzofeldspathic lithic sandstone or "greywacke," which is mainly derived from a calc-alkaline plutonic source (Suggate, 1978; Grapes, et al., 2001). It is unconformably overlain by conglomerate, sandstone, and olistostrome breccia of Early to Late Cretaceous age rocks of the Matawai Group (Isaac et al., 1994; Speden, 1972, 1975). Matawai Group rocks are in turn overlain by Tertiary forearc sequences, which include massive and thinly bedded mudstones and shelly sandstone deposited in bathyal to shelf environments (Mazengarb and Speden, 2000; Fig. 3).

Moderately dipping thrust structures within the study area are associated with erosional remnants of the allochthonous units (East Coast allochthon) present in the headwaters of the Waipaoa River system (Mazengarb and Speden, 2000). The siliciclastic sedimentary rocks within the allochthon range from Cretaceous to Oligocene in age (Fig. 3) and are thrust over in situ Cretaceous to earliest Miocene sequences (Mazengarb and Speden, 2000; Fig. 3). The allochthonous rocks are deeper-water facies

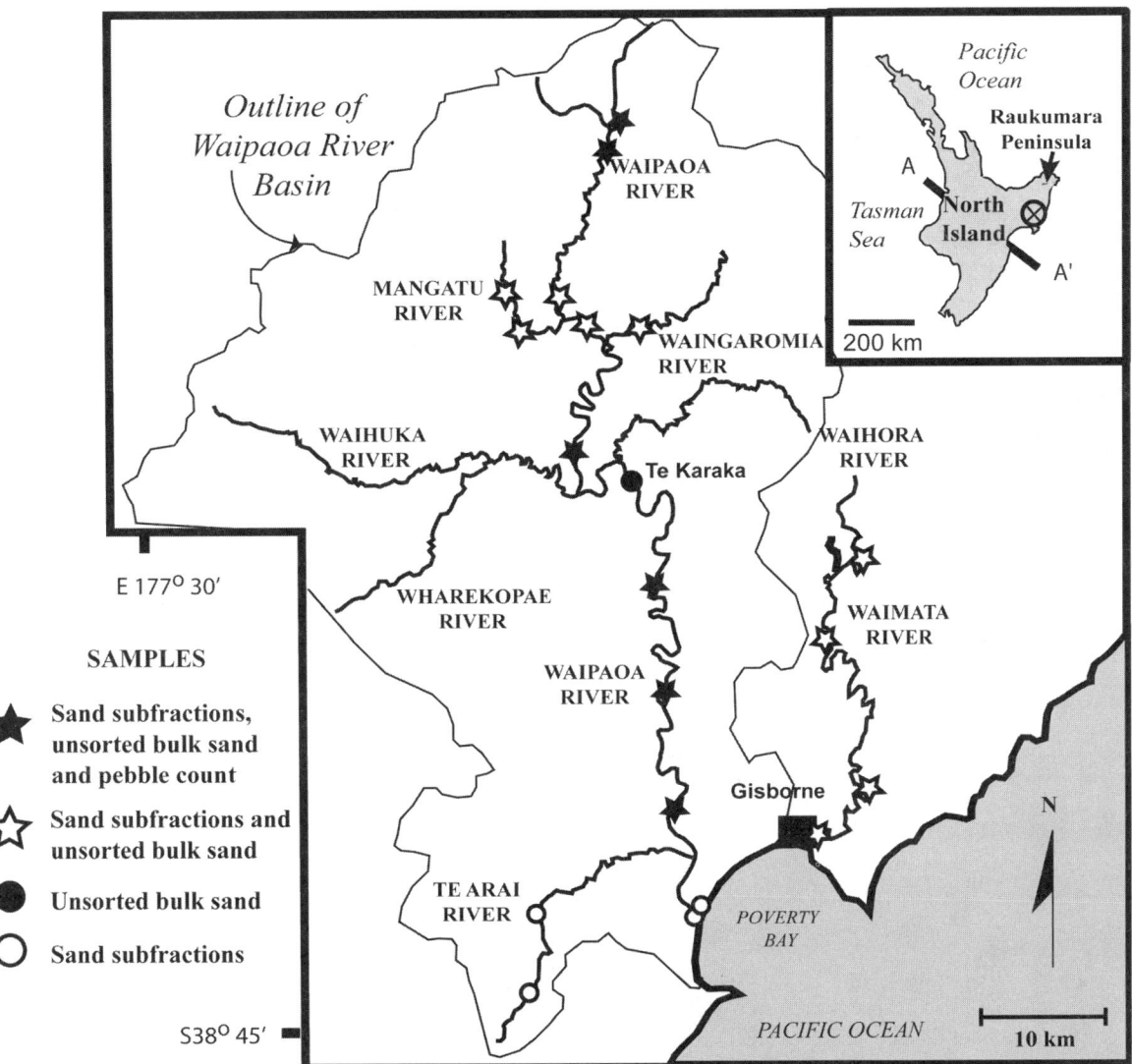

Figure 1. Location map showing the Waipaoa River drainage, East Cape, New Zealand. The circled "x" marks the approximate location of the Waipaoa drainage basin on the larger-scale map. Sample locations are keyed to the types of analyses performed. This figure was adapted from Mazengarb and Speden (2000).

equivalents of underlying autochthonous units thought to have been emplaced toward the south-southwest in the early Miocene, synchronous with the onset of andesitic volcanism in the Northland Coromandel volcanic arc and ophiolite obduction (Field et al., 1997; Mazengarb and Speden, 2000).

During the early Pliocene, the convergent plate margin rotated to the south-southeast, resulting in subduction at the Hikurangi Trench and the development of the Taupo volcanic zone (Field et al., 1997; King, 2000). Active intermediate to felsic volcanic centers associated with the Taupo volcanic zone are located to the west of the study area, separated from it by a topographic divide (Mazengarb and Speden, 2000; Fig. 2). Ash deposits sourced from these volcanic centers have repeatedly covered much of the northern half of North Island, including the study area, over the past million years (Healy et al., 1964; Scott, 1997).

The Waipaoa River system consists of the Waipaoa River and a series of tributary streams, as well as the Waimata River (Fig. 1), which likely merged with the Waipaoa River during the last sea-level lowstand. The total length of the Waipaoa River from the northern headwaters in the Mangatu Forest to the present-day coastline in Poverty Bay is 110 km (Berryman et al., 2000). The fluvial morphology of the Waipaoa River can be divided into three parts (Rosser, 1997; Gomez et al., 2001b): the upper reaches (110 km to 92 km from the sea), where the river exhibits a multithread braided morphology; the middle reaches, where the river flows through a narrow gorge for ~15 km then downstream through Pleistocene terraces and low-lying rolling hills; and the lower reaches (45 km to the sea), where the river forms a single channel that meanders across the Poverty Bay flats. At high flow within the middle reaches, the river exhibits a

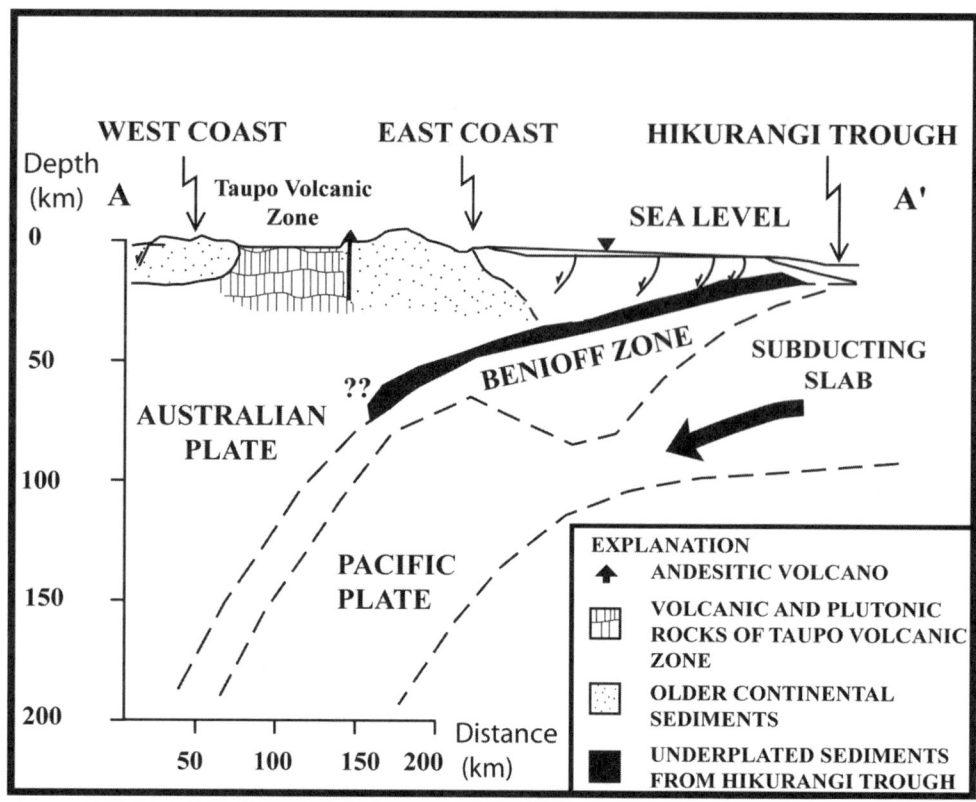

Figure 2. Plan-view tectonic schematic illustrating the Hikurangi subduction margin. This figure was adapted from Ferriere and Chanier (1993).

single-thread meandering morphology; however, it braids during periods of low flow.

The total catchment area for the Waipaoa system is 2380 km^2 (Gomez et al., 2001a, 2001b; Mazengarb and Speden, 2000). The rocks and sediments of the catchment area are Cretaceous to Holocene in age (Berryman et al., 2000; Mazengarb and Speden, 2000; Fig. 3). The loose to unconsolidated, fractured, and pervasively jointed nature of the Cretaceous argillites and sandstone in the headwaters of the Waipaoa River (Matawai Group) gives rise to landsliding, slumps, and gullies (Gomez et al., 1999). Outcrops within the lower catchment area are characterized by Miocene-Pliocene mudstone and sandstone. These rocks are relatively more competent and tend to support steeper slopes that also undergo mass wasting. The Quaternary units include poorly consolidated lacustrine, fluvial, and lagoonal deposits (Mazengarb and Speden, 2000).

Landslides and gully erosion are important mechanisms for sediment generation in the Waipaoa River system (Hicks et al., 2000). Foster and Carter (1997) related the high sediment load of the Waipaoa River (~15 million tons of suspended load annually; Hicks et al., 2000) and subsequent mud-dominated sedimentation on the continental shelf to highly erodible Tertiary sediments, active tectonism, meteorological fluxes, and changes in land use. They found that offshore sediment accumulation rates in the late Holocene were five times lower than that of the modern mud supply rate and equated this modern increase to terrestrial erosion following European deforestation in the late nineteenth century.

The effects of such events on the sand composition and supply are unknown.

PREVIOUS PETROLOGICAL WORK AND GOALS OF STUDY

It is important to look at the entire range of sediment sizes, gravel to mud (Ibbeken and Schleyer, 1991), because provenance signals may be variably preserved in different size fractions. There have been no prior studies of sand provenance in the Waipaoa River drainage area; however, there have been some detailed studies of the modern mud and gravel composition within the system. Gomez et al. (2001b) and Rosser (1997) examined gravel size variation and downstream fining in the Waipaoa River. Their general compositional results indicated little overall variation and no systematic downstream variation in the proportion of sandstone, siltstone, and argillite clasts in the subsurface bedload (Gomez et al., 2001b). D'Ath (2002) characterized the clay mineralogy within the Waipaoa River system. She sought to determine if the clay mineralogies differed between sediment derived from landslide erosion and that of gully erosion and if erosion style was influenced by the clay mineralogy of the bedrock. D'Ath (2002) concluded that there was no consistent mineralogical difference between areas where landsliding is prevalent and areas that experience gullying.

We focused on the composition of the gravel and sand fractions of the Waipaoa River system. Our goals were to document sediment composition within the Waipaoa River system, establish

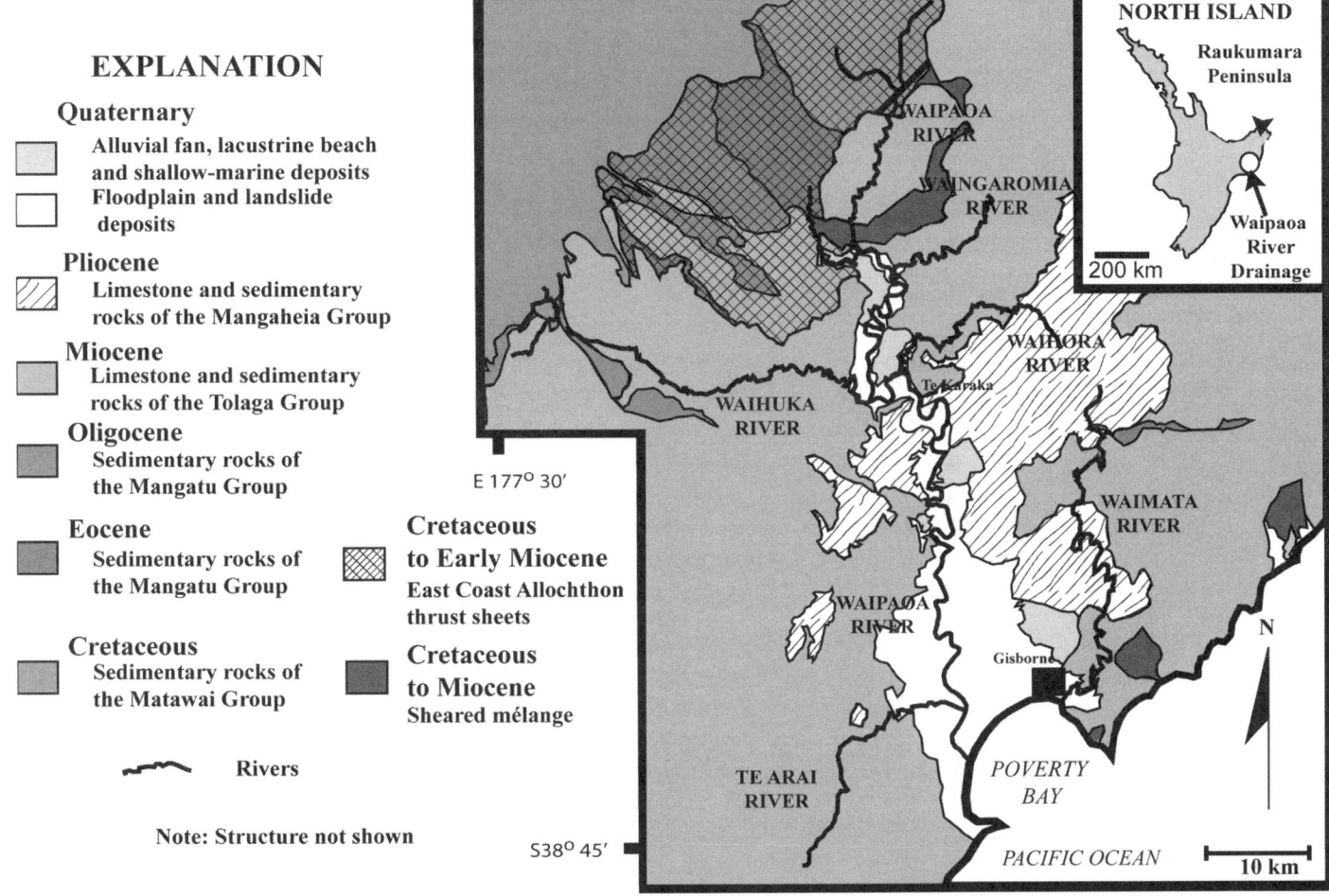

Figure 3. Simplified geologic map of the Waipaoa River drainage basin. This figure was adapted from Mazengarb and Speden (2000) and does not contain structural information on folds and faults.

whether patterns in sediment composition exist, and to relate those patterns to the geologic units that crop out in the drainage basin and the stream morphology. Owing to their dominance in the system, we emphasized mudstone clast lithologic variations in the sand and gravel fractions. We hoped that such a thorough characterization would ultimately help us recognize tectonic (e.g., volcanic eruptions), climatic (e.g., major storm events), and anthropogenic (e.g., deforestation) signals in the offshore record of the Waipaoa River system. On a broader level, we sought to characterize the compositional variations of sand that has a mostly sedimentary mudstone provenance, a characteristic of tectonically disrupted forearc regions (e.g., Marsaglia et al., 1992, 1995, 1999).

METHODS

Field Sampling

Sediment samples were collected from the Waipaoa River and tributary streams during periods of low to moderate flow in January and June of 2002 and 2003 (Fig. 1). Waipaoa River samples were obtained at ~1–5 km intervals from 1 km north of the confluence of the Te Weraroa stream and the Waipaoa River to the beach at Poverty Bay, a total distance of ~100 km (Fig. 1). Samples were collected mostly at bridges where public access to rivers was easiest. Owing to limited accessibility and sandy bar exposure, only four tributaries were sampled: the Mangatu, the Te Arai, the Waingaromia, and the Waihuka Rivers. Additional samples were collected from the Waimata River. One beach sand was collected near the mouth of the Waipaoa River.

The size and lithology of bedload clasts were recorded at six sites between 5 and 96 km downstream (Fig. 1). Clast counts were located in areas upstream and downstream of confluences with major sediment-contributing tributary streams. Grid sampling methods were used to ensure random clast selection (Smale, 1980; Wolman, 1954). The clasts were collected using a sampling grid net with fixed 12 cm spacing, several times larger than the mean particle size. An exception to this technique was made at 96 km in the Waipaoa River, where very sparsely distributed clasts were randomly gathered from the surface of an exposed bar. One hundred clasts between 20 mm and 250 mm in length were randomly selected and analyzed at each site. Clast dimensions (*x*, *y*, and *z*) were measured and recorded in the field along with grain size, lithology, and the wet and dry colors of each clast (see Table 4.2 in James [2004] for additional information) using a GSA Rock Color Chart (Rock-Color Chart Committee, 1991) and a 10× hand lens. Representative clasts from each group at each location were collected for thin section preparation and petrographic analysis.

To characterize the composition of sand-sized fluvial sediment across the Waipaoa River system, representative samples were collected from exposed sandy bars. Eighty-four sandy sediment samples (along with representative granule- to cobble-size clasts) were collected from the banks and sand bars of tributary streams as much as 25 km away from their confluence with the Waipaoa River, as well as from the Waimata River. Some of these were locations where gravel clast counts were also performed.

The overall strategy was to characterize downstream variability and influence of tributary streams on sediment composition in the Waipaoa River system. To help link sediment with source rocks, bedrock samples were also obtained from local outcrops within the drainage basin (James, 2004).

Laboratory Analysis

Eighty-four unconsolidated sandy samples were dry-sieved in order to separate the sand (0.0625 mm to 2 mm) from the gravel and mud fractions. Preliminary petrographic analysis of thin sections (point counts of 15 samples) from these bulk sand samples suggested significant relationships between grain size and composition. To study this relationship in greater detail, 19 of the bulk sand samples (Waipaoa River [8], Waimata River [4], Mangatu River [2], Te Arai River [2], Waingaromia River [2], and the beach at the mouth of the Waipaoa River [1]) were further sieved to separate five subfractions (very coarse [1–2 mm], coarse [0.5–1 mm], medium [0.25–0.5 mm], fine [0.25–0.125 mm], and very fine [0.0625–0.125 mm]); thin sections were produced from these subfractions. Note that all five size classes were not present in all of the 19 samples. Thin sections were also prepared from representative gravel clasts and outcrop samples. All thin sections were then stained to differentiate potassium- and calcium-bearing feldspar grains according to the method outlined in Marsaglia and Tazaki (1992).

Twenty-four groups of gravel clast lithologies, first determined by color, grain size, and lithology in the field, were then refined or confirmed on the basis of texture and mineralogical composition of representative clasts in thin section (James, 2004). The mineralogy and classification according to Blair and McPherson (1999) were determined for thin sections of 103 gravel clasts (James, 2004).

The light mineralogy of 105 sand thin sections (15 bulk, 90 subfractions) was determined petrographically with a Swift automated-stage point-counting system. The Gazzi-Dickinson point-counting method was used to minimize grain-size dependence on composition (Dickinson, 1970; Ingersoll and Suczek, 1979; Ingersoll et al., 1984). Four hundred points were counted for each thin section. Point-count categories and recalculated parameters are defined in Tables 1 and 2 (sedimentary lithic categories) and presented in Table A1.

RESULTS

Gravel Counts

Gravel clasts in the Waipaoa River reflect the spectrum of lithologies in the Cretaceous to Pliocene sedimentary units that crop out in the drainage basin. They are variably calcareous, owing to the presence of carbonate cement, matrix, and bioclastic debris, and range from relatively pure claystone to sandstone end members. Details of these categories can be found in James (2004) and DeVaughn (2005), as well as data and statistical analysis for clast sizes, estimated volumes, and recalculated percentages of the claystone, mudstone, and sandstone clasts. Herein, we present general data for the relative proportions/volumes of claystone, mudstone, and sandstone clasts (Fig. 4). In general, the clast populations are dominated by mudstone and exhibit variable ratios of sandstone and claystone (Fig. 4). Figure 5 shows the downstream occurrence of specific clast types distinguished by their color and texture in hand specimen (e.g., glauconitic sandstone, mafic volcanic tuff, and calcareous lithologies, including micritic claystone, calcareous siltstone, and calcareous sandstone); with one exception, their proportions are relatively constant.

TABLE 1. POINT-COUNT CATEGORIES AND DEFINITIONS OF RECALCULATED PARAMETERS

Counted categories	
Qm	Monocrystalline quartz
Qp	Polycrystalline quartz
P(A/D)	Plagioclase feldspar
K	Potassium feldspar
Lv	Volcanic glass
Lmt	Quartz-mica tectonite lithic
Ls	Sedimentary lithic[†]
NonOpD	Nonopaque dense mineral
Mc	Micrite
Foram	Foraminifera
Glau	Glauconite
Org	Organics
Pyrite	Pyrite
Metased	Unknown metasedimentary lithic fragment
Mus	Muscovite
Unk	Unidentified grain

Recalculated parameters
QFL%Q = 100* Q/(Q + F + L)
QFL%F = 100* F/(Q + F + L)
QFL%L = 100*L/(Q + F + L)
QmKP%Qm = 100*Qm/(Qm + P + K)
QmKP%K = 100*K/(Qm + P + K)
QmKP%F = 100*P/(Qm + P + K)
L%Ls = 100*Ls/L
L%Lv = 100*Lv/L
L%Lm = 100*Lm/L
P/F = P/(K + P)

Note: Q—total quartzose grains (Qm + Qp); L—total lithic grains (Lm + Lv + Ls); Lm—metamorphic lithic grains (Lma + Lmt); F—P + K.
[†]Refer to Table 2 for detailed definitions of lithic types.

Category	Name	Description
Claystone		
IA	Claystone	Pure end-member claystone/argillite; <10% silt, no birefringence, noncalcareous; no pyrite, no mica, no feldspar, ~3% quartz
IA1	Claystone w/ pyrite	Pure end-member claystone/argillite; <10% silt, no birefringence, noncalcareous; >10% pyrite, no mica, no feldspar, ~3% quartz
1A2	Claystone w/ organic matter	Pure end-member claystone/argillite; <10% silt, no birefringence, noncalcareous; >10% organics, no mica, no feldspar, ~3% quartz
IB	Calcareous claystone	Calcareous clay, >3% pyrite/organic matter, <10% silt including mica, high birefringence, mass extinction
Siltstone		
IIA	Quartz-rich siltstone	Quartz-rich siltstone; ~3–5% feldspar, >50% quartz with ~3% pyrite
IIB	Mica-rich siltstone	Mica-rich siltstone; ~3–5% feldspar and quartz, <50% mica, with >3% pyrite
IIC	Calcareous siltstone	Carbonate-rich (>10%) siltstone; >3% pyrite/organics, <10% clay
IID	Plain siltstone	Siltstone; >10% silt, <10% clay, >5% mica, trace pyrite/organics, low birefringence
IIE	Feldspathic siltstone	>10% silt, < 0% clay, has plagioclase/calcium feldspar and pyrite
Mudstone		
IIIA	Mudstone/siltstone with >10% pyrite	Mixed mudstone/siltstone; low birefringence, >5% mica, >10% pyrite, >10% silt, no carbonate
IIIA1	Mudstone/siltstone with carbonate and pyrite	Mixed mudstone/siltstone; low birefringence, >5% mica, >10% pyrite, >10% silt, with >10% carbonate
IIIB	Mudstone/siltstone with >5% organics	Mixed mudstone/siltstone; low birefringence, >5% mica, >5% pyrite, >10% silt, >5% organics
IIIB1	Mudstone/siltstone with >5% organics and carbonate	Mixed mudstone/siltstone; low birefringence, >5% mica, >5% pyrite, >10% silt, >5% organics, with >10% carbonate
IIIC	Plain mudstone	Mudstone; low birefringence, noncalcareous, no feldspar, no mica, no pyrite, <10% silt
IIIC1	Plain mudstone with carbonate	Mudstone; calcareous, no birefringence, no feldspar, no mica, no pyrite, <10% silt
IIID	Mudstone/siltstone with high birefringence, mass extinction	Mixed mudstone/siltstone; high birefringence, >5% mica, >10% silt, >3% pyrite, mass extinction, no calcareous
III D1	Calcareous mudstone/siltstone with high birefringence, mass extinction	Mixed mudstone/siltstone; high birefringence, >5% mica, >10% silt, >3% pyrite, mass extinction, with carbonate
IIIE	Mudstone/siltstone with little to no mica	Mixed mudstone/siltstone; low birefringence, >5% pyrite/organics, >10% silt, <3% mica
IIIE1	Mudstone/siltstone with little to no pyrite	Mixed mudstone/siltstone; low birefringence, <5% pyrite/organics, >10% silt, >5% mica
IIIF	Siliceous mudstone	>10% clay and silt, no pyrite
IIIF1	Siliceous mudstone with pyrite	>10% clay and silt, >5% pyrite
IIIF2	Siliceous mudstone with organics	>10% clay and silt, >5% organics
Sandstone		
VIA	Sandstone with clay, carbonate-rich	>10% sand and clay, >5% carbonate, <10% silt
VIB	Sandstone with clay, without carbonate	>10% sand and clay, <5% carbonate, <10% silt
VID	Sandstone with clay, silt, and pyrite	>10% sand, clay, and silt, <5% pyrite
VII	Mixed calcareous siltstone/ silty microsparite	
Carbonate		
VIII	Microcrystalline carbonate of unknown origin	
IX	Microcrystalline vein carbonate	
X	Coarse carbonate of unknown origin	

TABLE 2. DETAILED DESCRIPTION OF SEDIMENTARY AND CARBONATE LITHIC FRAGMENTS

Although clast dimensions vary substantially, mean clast size (and standard deviations) ultimately decrease downstream. This is illustrated in Figure 6, which shows the downstream distribution of mean maximum clast lengths. Sandstone clasts vary only slightly in their mean maximum clast length. In contrast, the mean maximum clast length for the mudstones decreases steadily from 25 km to 80 km downstream. The mean maximum claystone clast lengths are much more variable, perhaps a statistical artifact associated with the small percentage of these clasts at 25 km and 80 km downstream.

Clast volumes, estimated by multiplying the three measured axes of each gravel clast, were averaged for the general claystone, mudstone, and sandstone categories for each clast count location (James, 2004; DeVaughn, 2005). Figure 7 shows that the mean clast volumes for the sandstone and mudstone categories decrease downstream; however, the mean claystone clast volumes are variable.

Sand Petrography

In general, sand samples collected from the Waipaoa River system are composed predominantly of sedimentary lithic fragments, and lesser quartz, feldspar, and nonsedimentary lithic fragments are concentrated in the finer sand fractions (Fig. 8; Table A1). The sedimentary lithic fragments are overwhelmingly claystone/argillite, shale/mudstone, and siltstone clasts with variable carbonate, pyrite, and organic matter content. Glauconite, micrite/microsparite, and coarse carbonate are common in some

TABLE A1. POINT-COUNT DATA AND RECALCULATED PARAMETERS

Categories	Grain size	Downstream (km)	Qm	P	P (A/D)	K	Qp	IA	IA1	IA2	IB	IIA	IIB	IIC	IID	IIE	IIIA	IIIA1	IIIB	IIIB1	IIIC	IIIC1	IIID	IIID1	IIIE	IIIE1	IIIF	IIIF1	IIIF2	Mc	Foram	VIA	VIB
Sample no.																																	
NZ 2-64	vf-vc	5	25	1	0	0	0	13	0	0	0	0	164	3	19	0	0	12	0	29	4	0	114	6	0	0	0	0	0	0	0	1
NZ 2-64.V	vf	5	107	5	0	1	5	19	0	2	0	0	80	1	51	0	0	7	0	36	5	3	54	0	0	0	0	0	0	0	0	0
NZ 2-64.W	f	5	23	2	0	1	2	10	0	3	0	0	133	6	58	1	0	4	0	12	11	1	102	4	0	0	0	0	4	0	7	0
NZ 2-64.X	m	5	14	0	0	1	0	5	0	3	0	0	128	11	48	0	0	10	0	27	12	0	127	0	0	0	0	0	2	0	3	1
NZ 2-64.Y	c	5	13	0	0	0	0	2	0	6	0	0	185	15	40	3	0	16	0	16	30	0	41	0	0	0	0	0	16	0	5	2
NZ 2-64.Z	vc	5	0	0	1	0	0	0	0	0	0	0	47	7	3	0	0	16	0	13	0	0	19	0	0	0	0	0	2	0	5	0
NZ 2-63	vf-vc	7	101	8	0	0	1	0	0	0	0	0	91	7	27	0	0	8	0	46	11	0	14	11	0	0	0	0	0	0	8	8
NZ 2-63.V	vf	7	184	15	0	8	2	15	0	0	1	0	36	8	9	0	0	13	0	9	0	1	14	10	0	0	0	0	7	0	8	8
NZ 2-63.W	f	7	97	17	0	2	2	35	0	0	2	0	52	4	26	0	0	0	2	19	17	2	22	18	0	0	0	0	3	0	8	12
NZ 2-63.X	m	7	3	22	0	6	1	16	0	0	2	0	165	2	28	1	0	23	0	11	46	3	29	18	0	0	0	0	20	1	8	12
NZ 2-63.Y	c	7	0	1	0	3	0	8	0	0	6	0	151	50	8	1	5	13	0	6	44	3	25	9	0	0	0	0	1	0	7	3
NZ 2-63.Z	vc	7	3	0	0	0	0	4	0	0	0	0	30	5	5	0	0	20	2	2	7	1	17	8	0	0	0	0	8	0	17	1
NZ 1-12	vf-vc	25	40	4	0	1	2	12	0	2	0	0	123	9	48	0	0	10	0	31	36	1	16	6	0	0	0	0	20	0	13	6
NZ 1-12.V	vf	25	183	17	5	10	13	24	0	0	1	0	15	1	19	0	0	3	0	0	1	0	20	47	0	0	0	0	2	0	0	0
NZ 1-12.W(2)	f	25	78	12	0	3	2	4	0	3	3	0	117	21	10	2	0	2	0	11	1	0	58	18	0	0	0	0	1	0	0	0
NZ 1-12.X	m	25	65	5	0	4	5	6	0	2	2	0	110	24	45	1	0	12	0	9	12	0	37	6	0	0	0	0	17	2	9	6
NZ 1-12.Y	c	25	7	3	0	0	0	2	0	0	4	0	180	4	17	0	0	29	2	63	5	27	2	6	0	0	0	0	6	0	4	13
NZ 1-12.Z	vc	25	3	0	0	0	0	0	0	0	1	0	42	8	0	2	0	19	0	3	0	0	0	0	0	0	0	0	2	0	1	2
NZ 1-10	vf-vc	27	32	2	0	1	0	1	2	0	0	0	117	52	65	13	0	0	1	15	0	0	8	12	0	0	0	0	9	0	2	4
NZ 1-10.V	vf	27	133	3	2	2	0	9	0	3	0	0	49	25	31	8	0	0	0	5	1	0	17	5	0	0	0	0	5	1	1	0
NZ 1-10.W	f	27	55	4	2	6	1	9	5	1	24	0	62	31	55	1	0	0	1	3	1	0	37	2	0	0	0	0	10	2	1	0
NZ 1-10.X	m	27	14	3	0	4	2	0	15	12	8	0	137	34	53	2	0	10	0	5	3	1	25	3	0	0	0	0	16	0	3	2
NZ 1-10.Y	c	27	4	4	0	0	0	78	61	6	11	0	61	13	37	3	0	12	0	0	1	2	21	0	0	0	0	0	42	1	1	3
NZ 1-10.Z	vc	27	0	1	0	0	0	3	20	6	6	0	29	19	5	9	0	0	0	0	0	0	0	0	0	0	0	0	2	0	3	0
NZ 2-26	vf-vc	27	4	0	0	0	0	0	0	0	0	0	167	30	42	1	0	12	0	17	0	0	81	0	0	0	0	0	2	0	3	7
NZ 2-26.V	vf	42	151	6	0	0	6	2	0	0	0	0	45	31	22	26	0	0	0	0	0	0	13	13	0	0	0	0	2	1	2	0
NZ 2-26.W	f	42	38	3	0	1	2	1	0	0	2	0	108	82	26	19	0	7	0	6	3	0	14	4	0	0	0	0	0	0	2	1
NZ 2-26.X	m	42	15	2	0	1	1	1	0	0	4	0	129	70	43	15	0	12	0	19	0	0	14	4	0	0	0	0	11	0	11	8
NZ 2-26.Y	c	42	7	4	0	0	0	0	0	0	3	0	210	32	39	5	0	11	0	15	0	0	30	5	0	0	0	0	2	0	3	0
NZ 2-26.Z	vc	42	0	0	0	0	1	4	4	0	5	0	39	14	10	0	0	6	0	10	0	0	4	0	0	0	0	0	2	0	3	7
NZ 2-23	vf-vc	52	25	7	0	0	1	4	0	0	0	0	118	25	53	3	0	13	0	21	1	0	61	5	0	0	0	0	10	0	2	6
NZ 2-20	vf-vc	65	35	10	0	2	0	25	0	0	0	0	97	0	42	0	0	13	0	7	0	63	55	12	0	0	0	0	1	0	6	0
NZ 2-20.V	vf	65	136	14	0	6	4	25	0	3	0	0	29	0	43	1	0	6	0	4	0	29	18	31	0	0	0	0	4	0	3	3
NZ 2-20.W	f	65	62	8	0	8	2	21	14	7	3	0	81	4	34	0	1	11	0	17	2	2	45	18	0	0	0	0	11	1	5	4
NZ 2-20.X	m	65	3	29	0	0	0	1	0	0	10	5	183	0	3	0	0	16	0	23	14	4	0	0	0	0	0	0	0	0	4	7
NZ 2-20.Y	c	65	2	3	0	0	1	6	0	3	0	0	177	0	29	0	0	15	0	41	0	7	76	2	0	0	0	0	4	1	0	3
NZ 2-20.Z	vc	65	3	2	0	0	0	12	0	0	2	0	21	4	30	3	0	10	0	0	1	0	0	6	0	0	0	0	3	0	3	2
NZ 2-72	vf-vc	80	32	5	0	8	0	1	0	0	0	0	112	8	55	1	0	5	0	26	13	0	45	6	0	0	0	0	16	1	2	6
NZ 2-72.V(2)	vf	80	138	15	0	3	0	15	0	0	1	0	73	17	13	0	0	8	0	3	0	0	27	12	0	0	0	0	7	0	1	1
NZ 2-72.W	f	80	41	13	0	0	3	14	0	1	1	0	80	7	48	3	0	8	0	8	19	3	40	15	0	1	0	0	24	1	7	4
NZ 2-72.X	m	80	9	23	0	4	0	15	0	2	1	0	100	39	23	5	0	3	5	18	47	3	13	2	0	0	0	0	18	0	4	4
NZ 2-72.Y	c	80	0	2	0	1	2	6	0	0	2	0	105	29	26	2	0	10	0	10	52	2	58	1	0	0	0	0	13	1	1	9
NZ 2-72.Z	vc	80	0	0	0	0	2	5	0	0	0	0	32	4	17	6	0	6	0	1	1	0	11	0	0	0	0	0	4	0	11	3
NZ 3-34A.Y	vf	96	173	2	0	0	0	7	0	4	7	0	0	0	2	0	0	6	0	10	1	0	8	3	43	3	0	0	0	0	0	5	3
NZ 3-34A.X	f	96	113	3	0	0	8	4	4	0	3	5	11	2	16	1	0	4	0	23	0	0	79	12	52	12	0	0	0	0	0	4	7
NZ 3-34A.W	m	96	16	7	0	0	0	0	3	0	4	28	104	13	74	5	0	4	0	25	3	0	83	15	9	6	0	0	0	0	4	3
NZ 3-34A.V	c	96	5	4	0	0	10	2	0	3	1	63	31	14	91	16	0	7	0	62	7	7	15	15	6	7	0	0	3	0	3	2
NZ 3-34A.U	vc	96	4	0	0	0	0	2	10	0	7	0	20	3	22	1	0	10	0	13	2	17	2	8	4	7	0	0	16	0	2	6
NZ 1-6-1.Y	f	110	55	18	4	2	6	15	0	0	3	0	15	3	82	9	0	5	0	5	6	1	72	6	0	1	0	0	7	1	2	1
NZ 1-6-1.X	m	110	35	26	0	0	6	2	0	0	2	0	31	8	106	11	0	2	0	8	0	1	86	10	8	1	0	0	6	1	0	0
NZ 1-6-1.W	c	110	3	2	0	0	1	2	0	2	2	4	14	6	81	7	0	4	0	4	5	6	14	2	0	2	0	0	4	0	0	4
NZ 1-7.Z	vf	110b	44	20	1	11	1	2	0	2	0	0	1	0	4	0	0	0	0	4	1	0	32	14	1	1	0	0	5	0	0	1
NZ 1-7.Y	f	110b	39	26	2	7	0	7	0	7	0	1	10	0	47	3	1	7	0	4	0	6	50	13	1	1	0	0	2	0	0	0
NZ 1-7.X	m	110b	11	24	0	5	0	1	0	0	0	7	8	0	103	3	0	7	0	18	0	0	157	9	1	3	0	0	0	1	0	0
NZ 1-7.W	c	110b	4	3	1	0	0	0	0	0	0	29	75	7	145	8	0	19	0	38	3	2	36	8	0	0	0	0	0	0	0	0

(continued.)

TABLE A1. (Continued.)

Categories Sample no.	Grain size	Downstream (km)	VID	VII	VIII	IX	X	Lv	Glau	Org	Pyrite	metased	Mus	Non Op D	Lmt	unk	Q	F	K	P	L	Lv	Qm	Ls	Lm	P/F	%Lm	%Lv	%Ls	%Q	%F	%L	QFL	%Qm	%K	%P	QmKP
NZ 2-64	vf-vc	5	0	1	0	1	1	4	0	0	0	0	0	0	0	0	25	1	0	1	371	4	25	367	0	1.00	0.0	1.1	98.9	6.3	0.3	93.5	96.2	0.0	3.8
NZ 2-64.V	vf	5	0	0	1	0	1	9	0	0	0	0	2	4	4	0	112	6	1	5	274	9	107	261	4	0.83	1.5	3.3	95.3	28.6	1.5	69.9	94.7	0.9	4.4
NZ 2-64.W	f	5	0	0	4	0	1	7	3	0	0	0	0	1	1	0	23	3	1	2	358	3	23	354	1	0.67	0.3	0.8	98.9	6.0	0.8	93.2	88.5	3.8	7.7
NZ 2-64.X	m	5	0	0	6	0	0	1	0	0	0	0	0	0	0	0	14	0	0	0	378	1	14	377	0	0.00	0.0	0.3	99.7	3.6	0.0	96.4	100.0	0.0	0.0
NZ 2-64.Y	c	5	3	0	0	0	0	4	0	0	0	0	0	0	0	0	13	0	0	0	363	0	13	363	0	0.00	0.0	0.0	100.0	3.5	0.0	96.5	100.0	0.0	0.0
NZ 2-64.Z	vc	5	0	1	0	0	0	0	0	0	0	0	0	0	0	0	0	0	0	1	98	0	0	98	0	1.00	0.0	0.0	100.0	0.0	1.0	99.0	0.0	0.0	100.0
NZ 2-63	vf-vc	7	0	0	0	2	11	4	0	0	0	0	0	3	1	1	102	16	8	8	261	4	101	256	1	0.50	0.4	1.5	98.1	26.9	4.2	68.9	86.3	6.8	6.8
NZ 2-63.V	vf	7	0	0	1	0	11	25	0	0	0	0	0	1	0	1	186	17	2	15	177	25	184	138	14	0.88	7.9	14.1	78.0	48.9	4.5	46.6	91.5	1.0	7.5
NZ 2-63.W	f	7	0	0	5	0	13	6	0	0	0	0	0	0	14	0	98	23	6	17	236	6	97	229	1	0.74	0.4	2.5	97.0	27.5	6.4	66.1	80.8	5.0	14.2
NZ 2-63.X	m	7	0	0	0	3	8	11	0	0	1	0	0	1	0	0	3	25	3	22	360	11	3	349	0	0.88	0.0	3.1	96.9	0.8	6.4	92.8	10.7	10.7	78.6
NZ 2-63.Y	c	7	13	17	3	0	11	0	0	0	0	0	0	1	0	0	0	0	0	1	349	0	0	349	0	1.00	0.0	0.0	100.0	0.0	0.3	99.7	0.0	0.0	100.0
NZ 2-63.Z	vc	7	0	0	0	0	3	0	0	0	0	0	0	0	0	0	3	0	0	0	86	0	3	86	0	0.00	0.0	0.0	100.0	3.4	0.0	96.6	100.0	0.0	0.0
NZ 1-12	vf-vc	25	0	2	0	4	10	1	0	0	0	0	1	0	1	0	42	5	1	4	314	2	40	312	1	0.80	0.3	0.3	99.4	11.6	1.4	87.0	88.9	2.2	8.9
NZ 1-12.V	vf	25	0	0	2	5	14	2	1	0	0	0	3	0	2	1	196	32	10	22	140	2	183	135	3	0.69	2.1	1.4	96.4	53.3	8.7	38.0	85.1	4.7	10.2
NZ 1-12.W(2)	f	25	0	0	6	0	10	22	1	1	0	0	0	0	1	1	80	15	3	12	281	22	78	256	3	0.80	1.1	7.8	91.1	21.3	4.0	74.7	83.9	3.2	12.9
NZ 1-12.X	m	25	0	0	1	8	2	6	0	0	0	0	2	0	0	0	70	9	4	5	289	6	65	281	2	0.56	0.7	2.1	97.2	19.0	2.4	78.5	87.8	5.4	6.8
NZ 1-12.Y	c	25	0	0	3	8	10	0	0	0	0	0	0	0	0	0	7	3	0	3	358	0	7	358	0	1.00	0.0	0.0	100.0	1.9	0.8	97.3	70.0	0.0	30.0
NZ 1-12.Z	vc	25	0	0	5	3	0	0	0	0	0	0	0	0	0	0	3	0	0	0	92	0	3	91	1	0.00	1.1	0.0	98.9	3.2	0.0	96.8	100.0	0.0	0.0
NZ 1-10	vf-vc	27	0	2	2	5	22	10	6	0	4	0	0	4	0	0	32	3	1	2	315	10	32	297	8	0.67	2.5	3.2	94.3	9.1	0.9	90.0	91.4	2.9	5.7
NZ 1-10.V	vf	27	0	0	2	5	36	29	5	0	0	0	0	1	12	0	134	5	2	3	200	29	133	159	12	0.60	6.0	14.5	79.5	39.5	1.5	59.0	96.4	1.4	2.2
NZ 1-10.W	f	27	0	0	14	5	27	20	15	0	0	1	0	0	1	0	56	6	2	3	256	20	55	235	1	1.00	0.4	7.8	91.8	17.6	1.9	80.5	90.2	0.0	9.8
NZ 1-10.X	m	27	0	0	4	1	26	5	18	0	0	0	2	0	1	0	14	3	0	3	309	5	14	304	0	1.00	0.0	1.6	98.4	4.3	0.9	94.8	82.4	0.0	17.6
NZ 1-10.Y	c	27	0	10	8	0	10	3	3	0	0	0	0	0	0	0	4	0	0	0	331	0	4	328	3	0.00	0.9	0.0	99.1	1.2	0.0	98.8	100.0	0.0	0.0
NZ 1-10.Z	vc	27	0	3	0	3	1	0	1	0	0	0	0	0	1	0	0	0	0	0	94	0	0	94	0	0.00	0.0	0.0	100.0	0.0	0.0	100.0	0.0	0.0	0.0
NZ 2-26	vf-vc	42	0	2	0	2	13	8	0	0	0	0	0	0	0	0	4	0	0	1	370	8	4	361	1	1.00	0.3	2.2	97.6	1.1	0.3	98.7	80.0	0.0	20.0
NZ 2-26.V	vf	42	0	0	9	4	16	34	6	0	1	0	0	0	14	0	157	8	1	7	203	34	151	155	14	0.88	6.9	16.7	76.4	42.7	2.2	55.2	95.0	0.6	4.4
NZ 2-26.W	f	42	0	2	11	1	34	12	4	0	1	0	0	1	6	1	40	4	1	3	296	12	38	284	0	0.75	0.0	4.1	95.9	11.8	1.2	87.1	90.5	2.4	7.1
NZ 2-26.X	m	42	0	0	4	8	19	7	4	1	0	0	0	1	0	0	15	2	0	2	343	7	15	336	0	1.00	0.0	2.0	98.0	4.2	0.6	95.3	88.2	0.0	11.8
NZ 2-26.Y	c	42	0	12	0	9	26	7	4	0	0	0	0	0	0	0	7	3	2	1	367	4	7	363	0	1.00	0.0	1.1	98.9	1.9	0.8	97.3	87.5	0.0	12.5
NZ 2-26.Z	vc	42	0	1	0	4	2	0	0	0	0	0	0	1	0	0	0	0	0	0	96	0	0	96	2	0.00	2.0	0.0	98.0	0.0	0.0	100.0	0.0	0.0	0.0
NZ 2-23	vf-vc	52	0	0	4	4	14	13	4	0	2	0	0	0	1	0	26	8	1	7	331	13	25	317	1	0.88	0.3	3.9	95.8	7.1	2.2	90.7	75.8	3.0	21.2
NZ 2-20	vf-vc	65	0	1	0	1	6	7	3	0	0	0	0	0	1	0	39	12	2	10	337	7	35	329	1	0.83	0.3	2.1	97.6	10.1	3.1	86.9	74.5	4.3	21.3
NZ 2-20.V	vf	65	0	0	0	3	15	1	6	0	0	5	0	0	6	1	138	20	6	14	213	1	136	206	2	0.70	2.8	0.5	96.7	37.5	5.4	57.4	87.2	3.8	9.0
NZ 2-20.W	f	65	0	0	2	0	8	15	6	0	0	0	1	2	0	0	65	16	1	15	271	15	62	256	0	0.50	0.0	5.5	94.5	18.5	4.5	77.0	79.5	10.3	10.3
NZ 2-20.X	m	65	0	0	14	0	26	7	8	0	0	0	2	0	0	0	4	31	2	29	329	7	3	322	0	0.94	0.0	2.1	97.9	1.1	8.5	90.4	8.8	5.9	85.3
NZ 2-20.Y	c	65	0	0	0	9	4	6	0	0	0	0	1	0	0	0	2	3	0	3	375	5	2	370	0	1.00	0.0	1.3	98.7	0.5	0.8	98.7	40.0	0.0	60.0
NZ 2-20.Z	vc	65	0	0	0	4	0	2	0	0	0	0	0	0	0	0	0	0	0	0	90	0	0	90	0	0.00	0.0	0.0	100.0	0.0	0.0	100.0	0.0	0.0	0.0
NZ 2-72	vf-vc	80	0	0	0	4	16	12	4	0	0	2	0	0	1	2	33	13	8	5	310	12	32	296	2	0.38	0.6	3.9	95.5	9.3	3.7	87.1	71.1	17.8	11.1
NZ 2-72.V(2)	vf	80	0	0	8	0	17	20	7	0	0	0	0	1	8	0	141	15	0	15	198	20	138	168	10	1.00	5.1	10.1	84.8	39.8	4.2	55.9	90.2	0.0	9.8
NZ 2-72.W	f	80	0	1	0	1	28	9	9	0	2	0	0	1	0	0	41	17	4	13	279	9	41	268	2	0.76	0.7	3.2	96.1	12.2	5.0	82.8	70.7	6.9	22.4
NZ 2-72.X	m	80	0	0	0	3	23	11	0	0	0	0	0	0	6	0	11	24	1	23	321	11	9	310	0	0.96	0.0	3.4	96.6	3.1	6.7	90.2	27.3	3.0	69.7
NZ 2-72.Y	c	80	0	0	0	9	4	6	0	0	0	0	0	0	0	0	5	3	0	2	361	6	5	355	0	1.00	0.0	1.7	98.3	1.3	1.0	97.7	100.0	0.0	0.0
NZ 2-72.Z	vc	80	0	0	2	0	0	1	0	0	0	0	0	0	0	0	0	0	0	0	91	0	0	90	1	0.00	1.1	0.0	98.9	0.0	0.0	100.0	0.0	0.0	0.0
NZ 3-34A.Y	vf	96	0	0	0	4	27	10	18	0	0	2	1	1	0	2	181	2	0	2	164	10	173	164	5	0.79	2.1	4.1	93.8	52.2	0.6	47.3	98.9	0.0	1.1
NZ 3-34A.X	f	96	0	1	0	6	19	7	1	0	2	0	1	0	0	0	123	3	0	3	227	2	113	225	3	0.81	1.1	2.5	96.5	34.8	0.9	64.3	97.4	0.0	2.6
NZ 3-34A.W	m	96	0	0	2	2	0	45	2	0	2	0	0	0	4	4	17	3	0	3	341	45	16	295	1	1.00	0.3	13.2	86.5	4.7	0.8	94.4	69.6	0.0	30.4
NZ 3-34A.V	c	96	0	0	0	1	0	12	1	0	0	0	0	0	0	0	5	4	0	4	375	12	5	363	0	1.00	0.0	3.2	96.8	1.3	1.0	97.7	55.6	0.0	44.4
NZ 3-34A.U	vc	96	0	0	0	0	5	1	0	0	0	0	0	0	0	0	4	0	0	0	155	1	4	154	0	0.00	0.0	0.6	99.4	2.5	0.0	97.5	100.0	0.0	0.0
NZ 1-6-1.Y	f	110	0	3	0	0	27	10	18	0	0	2	0	0	4	2	58	28	6	22	243	10	55	228	5	0.79	2.1	4.1	93.8	17.6	8.5	73.9	66.3	7.2	26.5
NZ 1-6-1.X	m	110	0	4	1	0	19	7	4	0	0	0	0	3	0	0	37	32	6	26	283	7	35	273	3	0.81	1.1	2.5	96.5	10.5	9.1	80.4	52.2	9.0	38.8
NZ 1-6-1.W	c	110	0	0	2	0	1	1	0	0	0	0	0	1	0	1	3	2	0	2	157	1	3	155	1	1.00	0.6	0.6	98.7	1.9	1.2	96.9	60.0	0.0	40.0
NZ 1-7.Z	vf	110b	0	2	2	2	15	175	6	14	7	0	0	35	0	0	46	32	11	21	233	175	44	58	0	0.66	0.0	75.1	24.9	14.8	10.3	74.9	57.9	14.5	27.6
NZ 1-7.Y	f	110b	0	3	6	0	55	22	50	4	1	2	0	40	3	0	39	35	7	28	162	22	39	137	3	0.80	1.9	13.6	84.6	16.5	14.8	68.6	52.7	9.5	37.8
NZ 1-7.X	m	110b	0	1	2	0	19	2	1	2	0	0	0	2	0	0	12	29	5	24	328	2	11	324	2	0.83	0.6	0.6	98.8	3.3	7.9	88.9	27.5	12.5	60.0
NZ 1-7.W	c	110b	0	0	2	0	4	3	0	0	0	0	0	0	0	0	5	4	0	4	378	3	4	375	0	1.00	0.0	0.8	99.2	1.3	1.0	97.7	50.0	0.0	50.0

(continued.)

TABLE A1. POINT-COUNT DATA AND RECALCULATED PARAMETERS (Continued.)

Categories	Grain size	Downstream (km)	Qm	P	P (A/D)	K	Qp	IA	IA1	IA2	IB	IIA	IIB	IIC	IID	IIE	IIIA	IIIA1	IIIB	IIIB1	IIIC	IIIC1	IIID	IIID1	IIIE	IIIE1	IIIF	IIIF1	IIIF2	Mc	Foram	VIA	VIB
NZ 2-27	vf-vc	N. Mangatu R.	14	2	0	0	0	3	0	1	1	1	0	0	2	0	162	29	36	0	0	0	28	0	35	2	0	0	19	1	0	0	3
NZ 2-27.V	vf	N. Mangatu R.	116	1	0	0	4	6	1	0	2	1	0	0	0	0	41	52	28	14	8	0	0	0	15	12	1	0	0	0	0	0	0
NZ 2-27.W	vf	N. Mangatu R.	20	4	0	0	0	0	0	0	3	1	0	2	1	0	129	47	39	15	1	0	7	0	15	4	0	0	9	0	1	1	4
NZ 2-27.X	vf	N. Mangatu R.	7	2	0	0	1	0	0	0	1	0	0	2	4	0	205	42	34	4	0	0	1	0	33	2	0	0	9	0	0	0	0
NZ 2-27.Y	vf	N. Mangatu R.	3	0	0	0	0	0	21	5	15	0	0	0	2	0	154	43	60	4	0	0	24	5	8	1	0	0	16	1	1	1	5
NZ 2-27.Z	c	N. Mangatu R.	0	0	0	0	0	12	0	1	5	0	0	4	3	0	22	29	12	4	1	0	0	0	3	0	0	0	4	0	0	0	1
NZ 2-28	vf-vc	S. Mangatu	17	1	0	0	1	5	0	0	6	0	0	4	0	0	125	56	42	7	6	0	17	0	6	18	0	1	11	1	6	0	6
NZ 2-28.V	vf	S. Mangatu	144	8	0	0	0	2	0	2	2	6	0	1	12	0	33	53	14	20	2	0	0	0	9	4	1	0	1	1	0	0	6
NZ 2-28.W	f	S. Mangatu	36	2	0	0	1	5	4	0	19	0	0	1	0	0	98	7	52	0	9	0	3	0	58	4	0	0	2	0	0	1	1
NZ 2-28.X	m	S. Mangatu	12	2	0	0	0	14	35	6	10	0	0	4	2	0	83	67	18	25	4	0	2	0	13	4	0	0	23	0	1	5	3
NZ 2-28.Y	c	S. Mangatu	6	0	0	0	0	34	29	9	15	3	0	3	5	0	108	66	31	10	1	0	2	0	8	4	0	0	14	0	5	5	3
NZ 2-28.Z	vc	S. Mangatu	0	0	0	0	0	52	4	0	2	0	0	0	3	0	50	13	4	0	0	0	0	0	0	0	0	0	14	0	1	1	0
NZ 2-43	vf-vc	Waingaromia R.	29	10	0	0	9	9	0	4	4	0	0	2	18	0	14	24	30	57	1	0	20	9	0	9	2	0	21	6	7	3	10
NZ 2-43.V	vf	Waingaromia R.	150	13	2	0	0	15	3	3	7	1	0	0	0	0	14	30	39	42	1	0	0	0	11	3	0	0	0	0	3	0	0
NZ 2-43.W	f	Waingaromia R.	39	19	3	0	2	2	1	3	3	1	0	0	0	0	7	60	10	160	1	0	0	0	11	3	0	0	0	0	3	0	1
NZ 2-43.X	m	Waingaromia R.	35	65	0	0	6	13	4	7	14	3	0	3	4	0	16	13	59	45	1	0	0	0	10	7	0	0	7	1	9	3	1
NZ 2-43.Y	c	Waingaromia R.	23	7	0	0	8	10	9	5	14	2	0	1	4	0	25	44	43	45	3	0	1	0	9	20	1	0	23	6	4	4	3
NZ 2-43.Z	vc	Waingaromia R.	9	4	1	0	2	0	0	0	0	2	0	1	0	0	4	20	15	10	1	0	0	0	0	3	0	0	4	0	1	1	1
NZ 2-36	vf-vc	Waimata km upstream	105	24	0	1	17	11	0	0	2	2	1	1	0	0	12	21	9	24	8	0	12	0	31	11	0	1	0	8	3	3	1
NZ 2-36.V	vf	37	192	13	0	6	2	13	1	3	12	1	0	1	0	0	4	11	12	46	1	1	0	0	4	7	0	0	0	2	0	0	0
NZ 2-36.W	f	"	151	33	0	4	0	13	0	3	2	0	0	8	0	0	6	14	8	43	3	2	0	0	12	6	1	0	0	1	0	0	0
NZ 2-36.X	m	"	92	12	0	0	0	9	0	0	1	0	0	0	0	0	23	79	5	23	5	0	0	0	4	6	1	0	0	4	0	1	0
NZ 2-36.Y	c	"	74	8	1	1	0	2	0	0	0	0	0	0	0	0	16	44	31	50	14	5	0	0	4	2	1	0	0	3	0	0	0
NZ 2-36.Z	vc	"	9	2	0	1	0	1	0	0	3	0	0	0	0	0	0	13	0	46	6	0	0	0	1	1	0	0	1	2	0	0	0
NZ 2-31	vf-vc	28	55	21	0	0	5	14	0	0	3	0	0	5	23	0	3	56	6	66	3	0	8	3	21	5	0	0	3	5	16	0	4
NZ 2-31.V	vf	"	142	7	2	5	1	11	0	2	4	0	0	0	0	0	2	13	7	62	1	0	0	0	0	12	0	0	3	5	0	0	0
NZ 2-31.W	f	"	112	10	0	1	0	5	0	1	0	6	0	1	0	0	15	109	9	59	3	1	0	0	2	10	2	0	0	5	0	0	1
NZ 2-31.X	m	"	30	9	0	0	2	1	0	0	2	0	0	2	0	0	2	75	3	139	1	0	0	0	10	6	2	0	0	14	1	1	1
NZ 2-31.Y	c	"	23	12	0	0	0	3	3	5	2	0	0	1	2	0	50	85	41	75	4	0	0	0	6	5	2	0	3	3	2	2	3
NZ 2-31.Z	vc	"	2	1	0	0	0	0	1	0	4	0	0	0	4	0	2	4	0	38	1	0	0	0	10	1	0	0	3	1	2	0	1
NZ 2-42	vf-vc	17	37	41	0	0	2	4	0	2	4	0	0	2	6	0	10	92	10	24	1	0	0	0	10	3	0	0	5	3	1	0	5
NZ 2-42.V	vf	"	158	24	1	3	2	2	0	0	3	0	0	0	0	0	9	30	3	61	1	1	0	0	7	6	0	0	2	5	0	0	0
NZ 2-42.W	f	"	33	9	0	0	1	3	0	0	0	0	0	1	0	0	42	127	31	88	5	1	0	0	1	4	0	0	5	1	0	0	1
NZ 2-42.Y	c	"	26	29	0	1	0	1	8	4	19	6	0	1	1	0	5	86	11	77	1	2	0	0	4	2	0	1	5	1	3	0	0
NZ 2-42.Z	vc	"	15	7	0	0	0	71	0	6	11	2	0	5	0	0	27	52	23	21	2	0	3	0	1	8	2	0	8	5	0	3	2
NZ 2-37	vf-vc	7	0	1	0	0	0	28	6	0	13	0	0	3	0	0	2	7	2	3	0	0	2	1	1	0	0	0	3	0	0	0	2
NZ 2-37.V	vf	"	25	12	1	0	1	4	1	0	3	1	0	1	10	0	28	118	13	49	4	5	2	1	8	0	0	0	1	1	7	4	3
NZ 2-37.W	f	"	135	12	0	2	2	7	1	0	6	0	0	0	0	0	9	37	4	88	1	1	0	0	22	6	0	0	0	7	14	0	3
NZ 2-37.X	m	"	40	6	0	1	1	0	0	2	0	1	0	0	0	0	14	164	16	74	19	0	0	0	3	3	0	0	1	8	0	0	0
NZ 2-37.Y	m	"	18	20	0	0	0	2	5	5	33	0	0	0	0	0	8	95	1	153	0	0	0	0	2	0	0	1	0	1	1	0	0
NZ 2-37.Z	vc	"	15	13	0	0	0	5	1	0	0	3	0	0	1	0	50	204	23	27	2	0	0	0	3	4	0	1	1	1	0	5	2
NZ-3-37.U	vc	Te Arai River	8	2	0	0	9	0	0	3	1	0	0	1	0	0	2	22	0	33	2	0	0	0	3	4	0	0	9	1	0	0	0
NZ-3-37.V	vc	Te Arai River	20	3	0	4	0	18	0	4	2	4	0	1	1	0	18	20	36	18	0	0	0	0	15	0	0	0	0	0	2	0	0
NZ-3-37.W	m	Te Arai River	65	8	4	4	4	6	0	0	3	3	0	0	0	0	44	89	53	9	0	0	0	0	33	7	3	0	1	0	0	0	0
NZ-3-37.X	f	Te Arai River	89	19	16	0	8	1	0	0	0	0	0	0	1	0	5	21	38	10	0	0	0	0	32	31	0	9	1	0	0	1	0
NZ-3-37.Y	vf	Te Arai River	137	33	26	4	0	28	0	4	0	0	0	0	0	0	6	11	33	19	0	0	0	0	16	30	0	14	0	0	0	0	0
NZ-3-38B.U	vc	Te Arai River	0	0	2	0	0	4	0	0	0	0	0	1	0	0	1	0	5	8	1	0	0	0	4	1	0	0	0	0	0	0	1
NZ-3-38B.V	c	Te Arai River	14	12	0	4	1	4	0	0	6	3	0	2	1	0	27	19	21	17	2	0	0	1	10	3	0	0	1	2	8	8	1
NZ-3-38B.W	m	Te Arai River	17	36	0	0	1	1	0	0	0	0	0	1	1	0	5	33	68	94	0	0	0	0	10	3	0	1	0	2	3	0	0
NZ-3-38B.Y	vf	Te Arai River	115	37	10	9	0	0	0	0	0	0	0	0	0	0	15	7	31	39	0	0	0	0	5	15	4	0	0	2	0	0	0

(continued.)

TABLE A1. (Continued.)

Categories	Grain size	Downstream (km)	VID	VII	VIII	IX	X	Lv	Glau	Org	Pyrite	metased	Mus	Non Op D	Lmt	unk	Q	F	K	P	L	Lv	Qm	Ls	Lm	P/F	%Lm	LmLvLs %Lv	LmLvLs %Ls	QFL %Q	QFL %F	QFL %L	QmKP %Qm	QmKP %K	QmKP %P
NZ 2-27	vf-vc	N. Mangatu R.	1	8	0	25	0	4	15	4	0	1	0	0	2	0	15	2	0	2	308	4	14	304	1	1.00	7.8	1.3	98.4	4.6	0.6	94.8	87.5	0.0	12.5
NZ 2-27.V	vf	N. Mangatu R.	1	30	0	14	0.19	13	3	0	0	0	0	0	1	17	120	1	0	1	217	19	116	182	17	1.00	8.8	8.8	83.4	35.5	0.3	64.2	99.1	0.0	0.9
NZ 2-27.W	vf	N. Mangatu R.	5	19	0	42	0.17	15	1	0	0	0	0	1	0	0	20	4	0	4	284	17	20	272	0	1.00	0.0	6.0	94.0	6.5	1.3	92.2	83.3	0.0	16.7
NZ 2-27.X	vf	N. Mangatu R.	0	22	0	20	0.14	4	1	0	0	0	0	0	1	1	8	2	0	2	343	14	7	328	1	1.00	0.3	4.1	95.6	2.3	0.6	97.2	77.8	0.0	22.2
NZ 2-27.Y	vf	N. Mangatu R.	7	0	2	8	0.5	1	0	0	0	0	0	0	0	0	3	0	0	0	361	5	3	363	0	0.00	0.0	1.4	98.6	0.8	0.0	99.2	100.0	0.0	0.0
NZ 2-27.Z	c	N. Mangatu R.	0	0	0	3	0	0	0	2	0	0	0	0	1	0	0	0	0	0	93	0	0	93	0	0.00	0.0	0.0	100.0	0.0	0.0	100.0	100.0	0.0	0.0
NZ 2-28	vf-vc	S. Mangatu	1	17	0	17	0	6	5	2	0	0	0	0	1	0	18	1	0	1	326	6	17	321	1	1.00	1.8	1.8	98.2	5.2	0.3	94.5	94.4	0.0	5.6
NZ 2-28.V	vf	S. Mangatu	0	15	3	21	0.33	11	3	0	0	0	0	0	0	12	146	8	0	8	191	33	144	146	12	1.00	6.3	17.3	76.4	42.3	2.3	55.4	94.7	0.0	5.3
NZ 2-28.W	f	S. Mangatu	2	5	1	24	0.11	13	0	0	0	0	0	0	0	3	37	2	0	2	292	11	36	280	3	1.00	1.0	3.8	95.2	11.2	0.6	88.2	94.7	0.0	5.3
NZ 2-28.X	m	S. Mangatu	2	6	0	10	0.28	5	1	0	0	1	0	0	0	0	12	2	0	2	343	28	12	317	2	1.00	0.6	8.2	91.8	3.4	0.6	96.1	85.7	0.0	14.3
NZ 2-28.Y	c	S. Mangatu	4	6	0	15	0.1	3	1	0	0	0	0	0	0	0	6	0	0	0	351	1	6	354	0	0.00	0.0	0.3	99.7	1.7	0.0	98.3	100.0	0.0	0.0
NZ 2-28.Z	vc	S. Mangatu	4	0	0	2	0	0	0	0	0	0	0	0	0	0	0	0	0	0	86	0	0	90	0	0.00	0.0	0.0	100.0	0.0	0.0	100.0	100.0	0.0	0.0
NZ 2-43	vf-vc	Waingaromia R.	31	33	0	35	0.4	3	4	2	0	1	0	0	0	1	38	10	0	10	217	4	29	243	1	1.00	0.5	1.8	97.7	14.3	3.8	81.9	74.4	0.0	25.6
NZ 2-43.V	vf	Waingaromia R.	1	0	0	7	0.24	5	5	0	0	0	0	0	1	11	150	15	0	15	206	24	150	171	11	1.00	5.3	11.7	83.0	40.4	4.0	55.5	90.9	0.0	9.1
NZ 2-43.W	f	Waingaromia R.	0	11	1	15	0.42	3	4	0	0	0	0	0	0	0	41	22	3	19	302	42	39	260	2	0.86	0.0	13.9	86.1	11.2	6.0	82.7	63.9	4.9	31.1
NZ 2-43.X	m	Waingaromia R.	10	4	1	15	0.35	4	1	0	0	0	0	0	2	2	41	65	0	65	248	35	35	221	2	1.00	0.8	14.1	85.1	11.6	18.4	70.1	35.0	0.0	65.0
NZ 2-43.Y	c	Waingaromia R.	9	4	3	58	0.11	2	6	0	0	0	0	0	0	0	31	7	0	7	250	11	23	248	0	1.00	0.0	4.4	95.6	10.8	2.4	86.8	76.7	0.0	23.3
NZ 2-43.Z	vc	Waingaromia R.	9	6	0	5	0	0	1	1	0	0	0	0	0	0	9	5	1	4	59	0	9	68	0	0.80	0.0	0.0	100.0	12.3	6.8	80.8	64.3	7.1	28.6
NZ 2-36	vf-vc	Waimata km upstream	1	5	0	12	0.66	2	1	0	0	3	0	0	3	11	112	25	1	24	229	66	105	150	14	0.96	6.1	28.8	65.1	30.6	6.8	62.6	80.8	0.8	18.5
NZ 2-36V	vf	37	0	5	0	5	0.29	3	4	0	2	0	0	2	0	19	194	19	6	13	166	29	192	116	21	0.68	12.7	17.5	69.9	51.2	5.0	43.8	91.0	2.8	6.2
NZ 2-36.W	f	"	0	25	0	23	0.28	6	5	0	0	0	0	4	3	7	151	37	4	33	145	28	151	110	7	0.89	4.8	19.3	75.9	45.3	11.1	43.5	80.3	2.1	17.6
NZ 2-36.X	m	"	0	9	0	4	0.122	3	5	0	2	0	0	2	1	0	92	12	0	12	269	122	92	147	0	1.00	0.0	45.4	54.6	24.7	3.2	72.1	88.5	0.0	11.5
NZ 2-36.Y	c	"	0	20	0	8	1.95	2	16	0	2	0	0	0	1	0	74	10	1	9	264	95	74	168	1	0.90	0.4	36.0	63.6	21.3	3.2	75.9	88.1	1.2	10.7
NZ 2-36.Z	vc	"	0	0	0	2	0.12	1	5	0	0	0	0	0	0	0	9	2	0	2	77	12	9	65	1	1.00	1.6	15.6	84.4	10.2	2.3	87.5	81.8	0.0	18.2
NZ 2-31	vf-vc	28	11	15	0	11	0.23	2	5	0	0	0	0	0	1	0	60	21	0	21	267	23	55	255	8	1.00	0.0	8.6	91.4	17.2	6.0	76.7	72.4	0.0	27.6
NZ 2-31.V	vf	"	0	36	0	17	0.34	8	6	0	3	0	0	2	0	7	143	14	5	9	166	34	142	124	8	0.64	4.8	20.5	74.7	44.3	4.3	51.4	91.0	3.2	5.8
NZ 2-31.W	f	"	3	14	0	7	1.28	0	4	3	0	0	0	0	0	0	112	11	1	10	244	28	112	219	0	0.91	0.0	11.5	88.5	30.5	3.0	66.5	91.1	0.8	8.1
NZ 2-31.X	m	"	5	47	0	12	0.27	2	5	5	3	0	0	3	0	0	32	9	0	9	268	27	30	246	0	1.00	0.0	10.1	89.9	10.4	2.9	86.7	76.9	0.0	23.1
NZ 2-31.Y	c	"	7	13	0	13	5.22	2	5	5	2	0	0	0	0	0	23	12	0	12	313	22	23	298	0	1.00	0.0	7.0	93.0	6.6	3.4	89.9	65.7	0.0	34.3
NZ 2-31.Z	vc	"	11	5	0	5	1.5	0	4	7	2	0	0	5	0	1	2	1	0	1	69	12	2	75	2	1.00	0.8	7.2	92.8	2.8	1.4	95.8	66.7	0.0	33.3
NZ 2-42	vf-vc	17	3	16	5	35	0.19	3	3	0	3	0	0	2	1	1	39	41	0	41	245	19	37	227	2	1.00	0.8	7.8	91.4	12.0	12.6	75.4	47.4	0.0	52.6
NZ 2-42.V	vf	"	0	12	0	9	0.35	0	2	0	0	0	0	2	0	13	160	28	3	25	173	35	158	124	14	0.89	8.1	20.2	71.7	44.3	7.8	47.9	84.9	1.6	13.4
NZ 2-42.W	f	"	0	7	0	13	0.23	3	3	0	0	0	0	0	0	0	33	9	0	9	323	23	33	300	1	1.00	0.0	7.1	92.9	9.0	2.5	88.5	78.6	0.0	21.4
NZ 2-42.X	m	"	2	14	3	60	0.20	4	6	7	3	0	0	7	0	1	26	30	1	29	238	20	26	219	0	0.97	0.4	8.4	91.2	8.8	10.2	81.0	46.4	1.8	51.8
NZ 2-42.Y	c	"	2	8	0	81	1.19	0	9	0	6	0	0	9	0	2	15	7	0	7	261	19	15	244	2	1.00	0.8	7.3	91.9	5.3	2.5	92.2	68.2	0.0	31.8
NZ 2-42.Z	vc	"	0	0	0	16	6.0	0	7	0	0	0	0	0	0	0	0	1	0	1	73	0	0	73	0	1.00	0.0	0.0	100.0	0.0	1.4	98.6	0.0	0.0	100.0
NZ 2-37	vf-vc	7	6	7	1	31	0.23	1	8	0	0	2	0	3	1	1	25	13	1	12	295	23	25	276	2	0.92	0.7	7.8	91.5	7.5	3.9	88.6	65.8	2.6	31.6
NZ 2-37.V	vf	"	0	13	0	10	0.17	6	4	0	1	0	0	1	4	6	137	14	2	12	205	17	135	182	6	0.86	2.9	8.3	88.8	38.5	3.9	57.6	90.6	1.3	8.1
NZ 2-37.W	f	"	0	7	0	10	2.15	4	6	0	6	0	0	6	5	1	40	8	1	7	316	15	40	301	0	0.88	0.0	4.7	95.3	11.0	2.2	86.8	83.3	2.1	14.6
NZ 2-37.X	m	"	0	5	0	24	0.7	3	6	0	5	0	0	5	0	1	18	20	0	20	312	7	18	305	0	1.00	0.0	2.2	97.8	5.1	5.7	89.1	47.4	0.0	52.6
NZ 2-37.Y	f	"	0	4	0	18	4.7	0	7	0	7	1	0	7	1	0	15	13	0	13	334	7	15	329	2	1.00	0.6	2.1	97.3	4.1	3.6	92.3	53.6	0.0	46.4
NZ 2-37.U	vc	"	0	4	0	6	2.3	0	4	0	0	0	0	0	0	2	8	2	0	2	69	2	8	64	2	1.00	2.9	4.3	92.8	10.1	2.5	87.3	80.0	0.0	20.0
NZ 3-37.W	m	Te Arai River	0	9	4	0	4.42	0	1	0	1	0	0	1	0	2	20	4	1	3	165	42	20	122	2	0.75	0.6	25.5	73.9	10.6	2.1	87.3	83.3	4.2	12.5
NZ 3-37.X	f	Te Arai River	0	2	6	0	16.32	2	5	0	0	0	0	0	0	2	69	12	4	8	283	32	65	249	2	0.67	0.7	11.3	88.0	19.0	3.3	77.7	84.4	5.2	10.4
NZ 3-37.Y	f	Te Arai River	0	46	12	26	15	6	4	0	5	0	0	0	0	2	97	35	16	19	169	15	89	151	7	0.54	1.8	8.9	89.3	32.2	11.6	56.1	71.8	12.9	15.3
NZ 3-38B.U	vc	Te Arai River	0	13	5	0	11.6	0	6	0	5	0	0	0	0	2	141	59	26	33	142	6	137	129	3	0.56	4.9	4.2	90.8	41.2	17.3	41.5	69.9	13.3	16.8
NZ 3-38B.V	vc	Te Arai River	0	4	0	0	0	1	0	0	0	0	0	0	0	6	0	2	2	0	22	0	0	15	0	0.00	0.0	31.8	68.2	0.0	8.3	91.7	0.0	100.0	0.0
NZ 3-38B.W	m	Te Arai River	0	8	0	0	15.71	0	4	0	4	0	0	0	0	9	15	16	4	12	180	71	14	100	9	0.75	5.0	39.4	55.6	7.1	7.6	85.3	46.7	13.3	40.0
NZ 3-38B.Y	vf	Te Arai River	0	23	2	0	15.72	0	4	0	4	0	0	0	0	7	18	36	0	36	294	72	17	215	7	1.00	2.4	24.5	73.1	5.2	10.3	84.5	32.1	0.0	67.9
NZ 3-38B.Y	vf	Te Arai River	0	42	0	28	4	1	5	0	0	0	0	0	0	0	124	47	10	37	120	4	115	116	0	0.79	0.0	3.3	96.7	42.6	16.2	41.2	71.0	6.2	22.8

Note: vc—very c; c—c; m—mium; f—fine; vf—very fine.

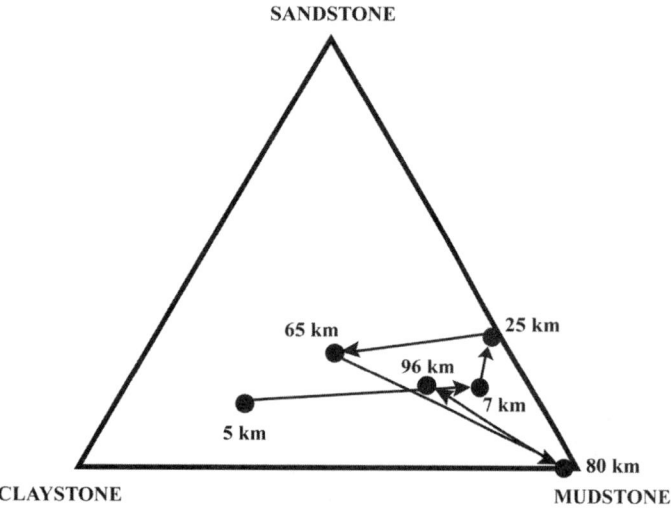

Figure 4. Ternary plot of Waipaoa River clast count proportions. Arrows indicate downstream direction. The data are presented in Table 4.3 in James (2004) and in DeVaughn (2005).

samples. Fragments of altered microlitic, vitric, felsitic, and tuffaceous debris are rare. Fresh volcanic glass (vitric) is also rare, except in Waimata and Te Arai River samples. The few metamorphic lithic fragments present are mostly quartz-mica tectonite.

The monomineralic sand components are dominated by quartz, with lesser feldspar and traces of nonopaque dense minerals (Fig. 8; Table A1). The quartz grains exhibit undulose to straight extinction and generally occur as individual monocrystalline grains and sand-sized components of sedimentary lithic fragments. Polycrystalline quartz grains, including chert, make up <5% of the framework grains. Both potassium and plagioclase feldspar are present in minor amounts. The feldspar grains mostly occur as individual monocrystalline grains, but some are sand-size components in sedimentary and plutonic (rare) lithic fragments; most are fresh, but some feldspar grains are altered to sericite and kaolinite. Some samples have a minor percentage of opaque dense minerals and a moderate to minor percentage of nonopaque dense minerals, including hornblende, zircon, and sphene. Other minor components include plant fragments and muscovite.

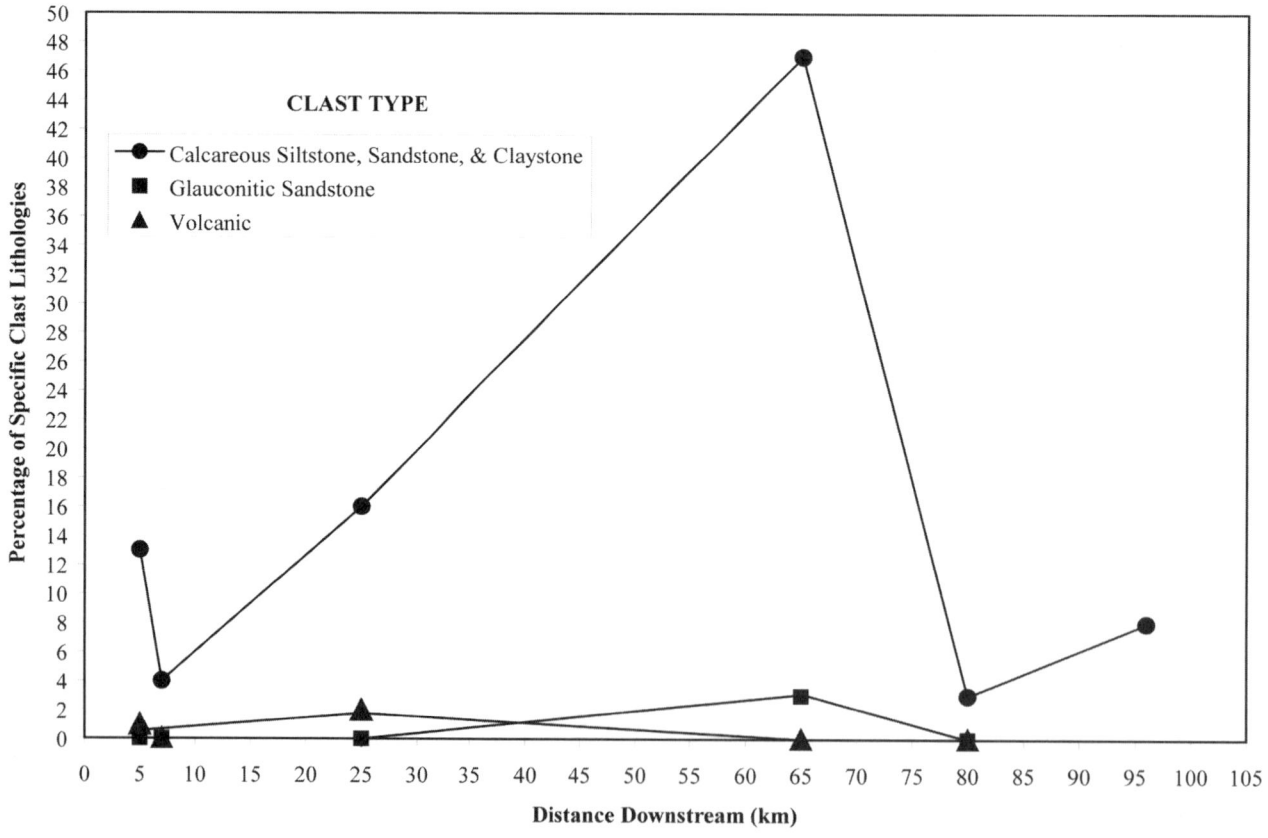

Figure 5. Distribution of specific clast lithologies, including calcareous siltstone, sandstone, and claystone, as well as glauconitic sandstone and volcanic rocks. Data are presented in Table 4.3 of James (2004) and in DeVaughn (2005).

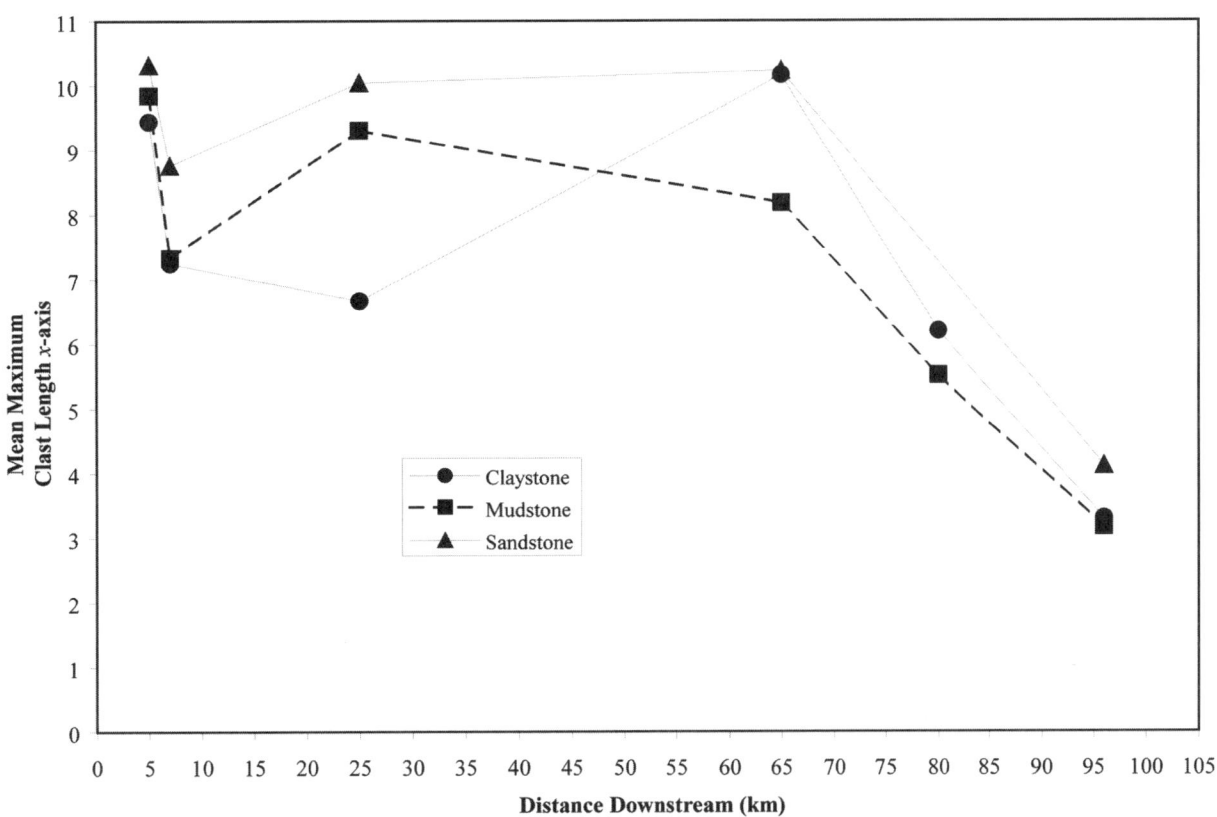

Figure 6. Distribution of mean clast lengths, including claystone, mudstone, and sandstone data. The data are presented in Table 4.4 of James (2004) and DeVaughn (2005).

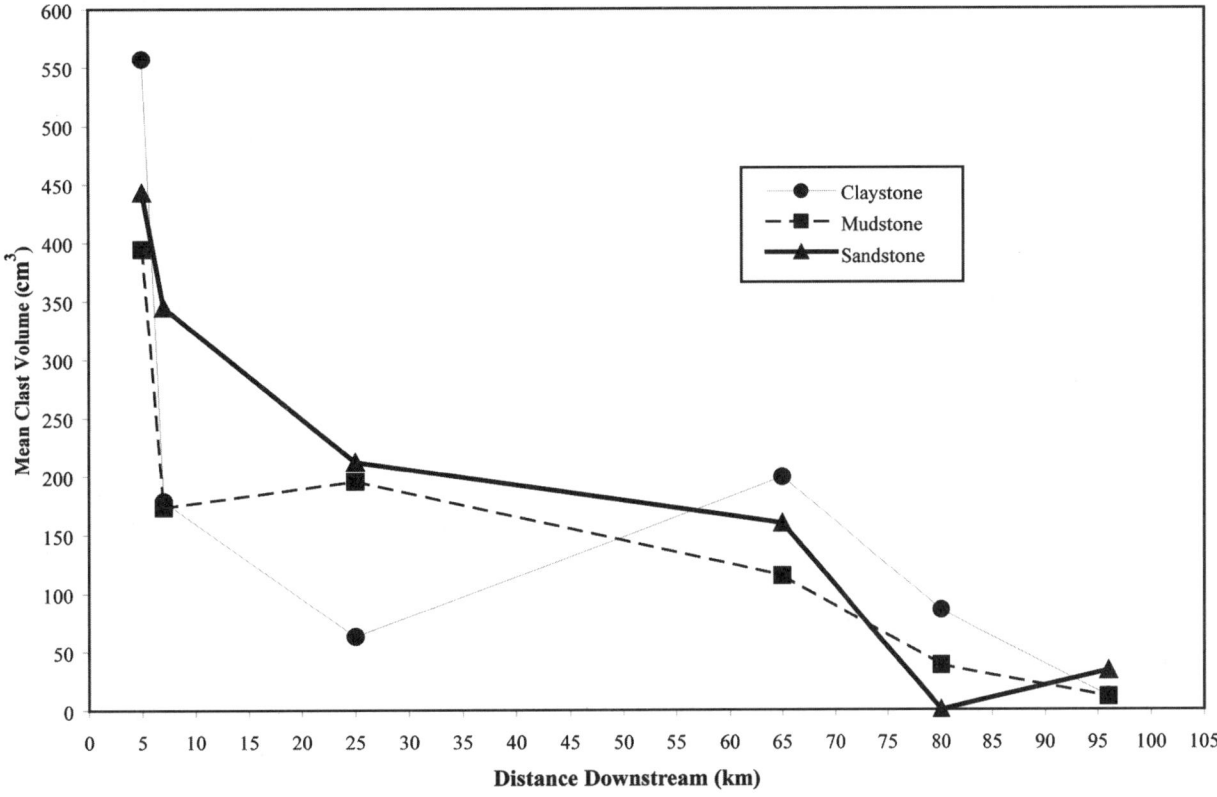

Figure 7. Distribution of mean clast volumes, including claystone, mudstone, and sandstone data. The data for this figure are presented in Table 4.4 of James (2004) and DeVaughn (2005).

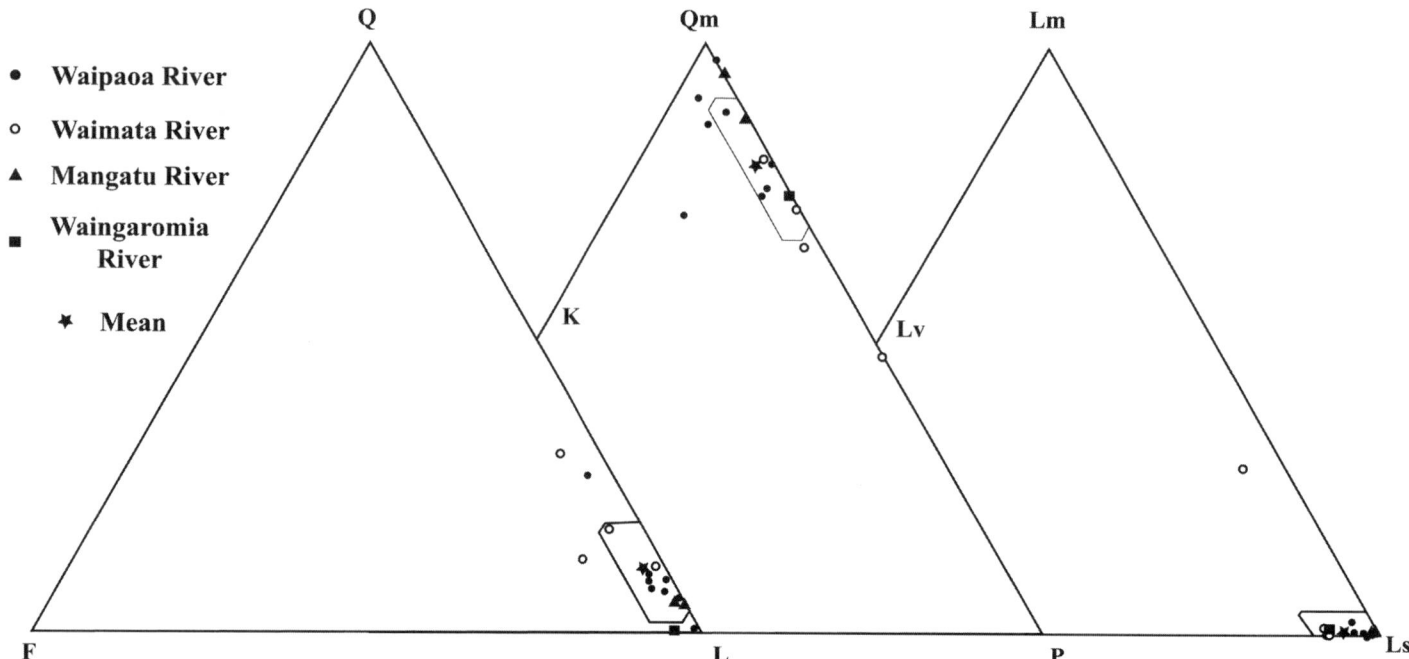

Figure 8. QFL, QmKP, and LmLvLs (see Table 1) plots of unsorted bulk sand data for Waipaoa River, Waimata River, Mangatu River, and the Waingaromia River (data from Appendix 8.6 of James [2004]). The mean (star) and fields of variation (polygons) are indicated for each river. The fields of variation (polygons) are formed by one standard deviation on each side of the mean.

Sand Detrital Modes

The 15 bulk sand samples from the Waipaoa River, Waimata River, and tributary streams are composed mainly of argillaceous sedimentary lithic fragments, quartz, and lesser plagioclase and volcanic components (Fig. 8). QFL (Table 1) detrital modes for the Waipaoa, Te Arai, and Waimata Rivers all show a linear distribution of mean values, with increasing quartz and to a lesser degree feldspar content with decreasing grain size (Figs. 9, 10, and 11). The QFL detrital modes for the Mangatu and Waingaromia tributary streams are not pictured, but are similar to those of Waipaoa and Waimata Rivers (Table A1). The feldspars are dominantly plagioclase, and proportions (P/F) are more variable in the Waipaoa River than in the Waimata (Fig. 12). Downstream distributions of the volcanic lithic proportions (L%Lv) for the sand subfractions from the Waipaoa, Waimata, and Te Arai Rivers (Fig. 13) indicate that the Waimata and Te Arai River sands are much more enriched in volcanic lithic components, including colorless volcanic glass (vitric) and other altered volcanic grains.

Several of the more common mudstone lithic types further illustrate compositional differences between the Waipaoa and Waimata Rivers (see Table 2 and James [2004] for lithic descriptions). These are: (1) sedimentary lithic type IIIA, a mixed mudstone/siltstone, (2) sedimentary lithic type IIIA1, a mixed mudstone/siltstone similar to IIIA, except that it contains >10% calcareous material, and (3) sedimentary lithic category IIIF(total), the sum of siliceous mudstones subtypes IIIF, 11F1, and IIIF2. Mean values for each sand subfraction are plotted for the Waipaoa and Waimata Rivers on the ternary diagram in Figure 14. Note how the samples from each river form a distinct cluster; the Waimata River samples are more enriched in calcareous mudstone. This distinction is also apparent when the overall calcareous lithic fraction is considered. Figure 15 shows the distribution of the total percentages of calcareous lithics in the lithic fraction for the Waipaoa, Waimata, Mangatu, Te Arai, and Waingaromia Rivers. The Waipaoa and Waimata River data generally plot at opposite ends of the spectrum of values, with the Waipaoa River containing a much lower fraction of calcareous lithics than the Waimata River. The three tributary rivers generally show intermediate or transitional compositions between the Waipaoa and Waimata Rivers.

DISCUSSION

Evidence for Contribution of Gravel Clasts to the Waipaoa River from Tributary Streams

According to previous studies (Rosser, 1997; Gomez et al., 1999; Page et al., 1999), the primary source of sediment for the modern Waipaoa River is from the upper catchment, where gully erosion and shallow landsliding are pervasive. In this area, the Mangatu and Tarndale gullies contribute significant amounts of coarse sediment derived from the crushed and sheared sedimentary rocks of the East Coast allochthon (Mazengarb and Speden, 2000). Rosser (1997) and Gomez et al. (2001b) found no significant

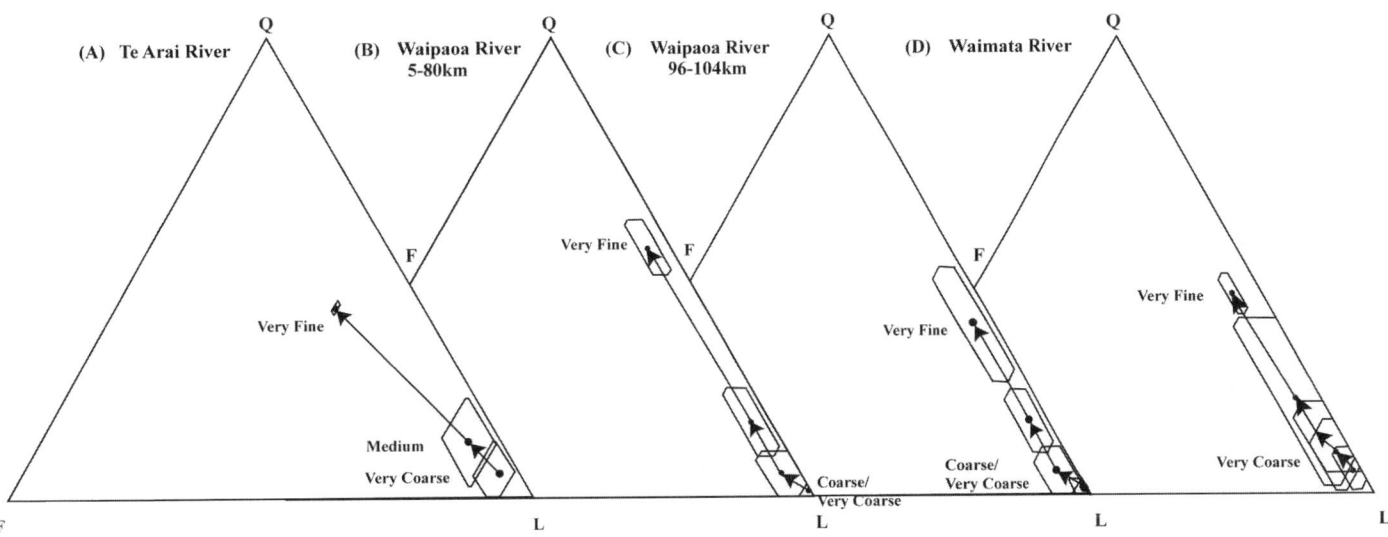

Figure 9. QFL (see Table 1) plots of mean sand subfraction data for the Waipaoa, Te Arai, and Waimata Rivers. Arrows indicate decreasing grain size.

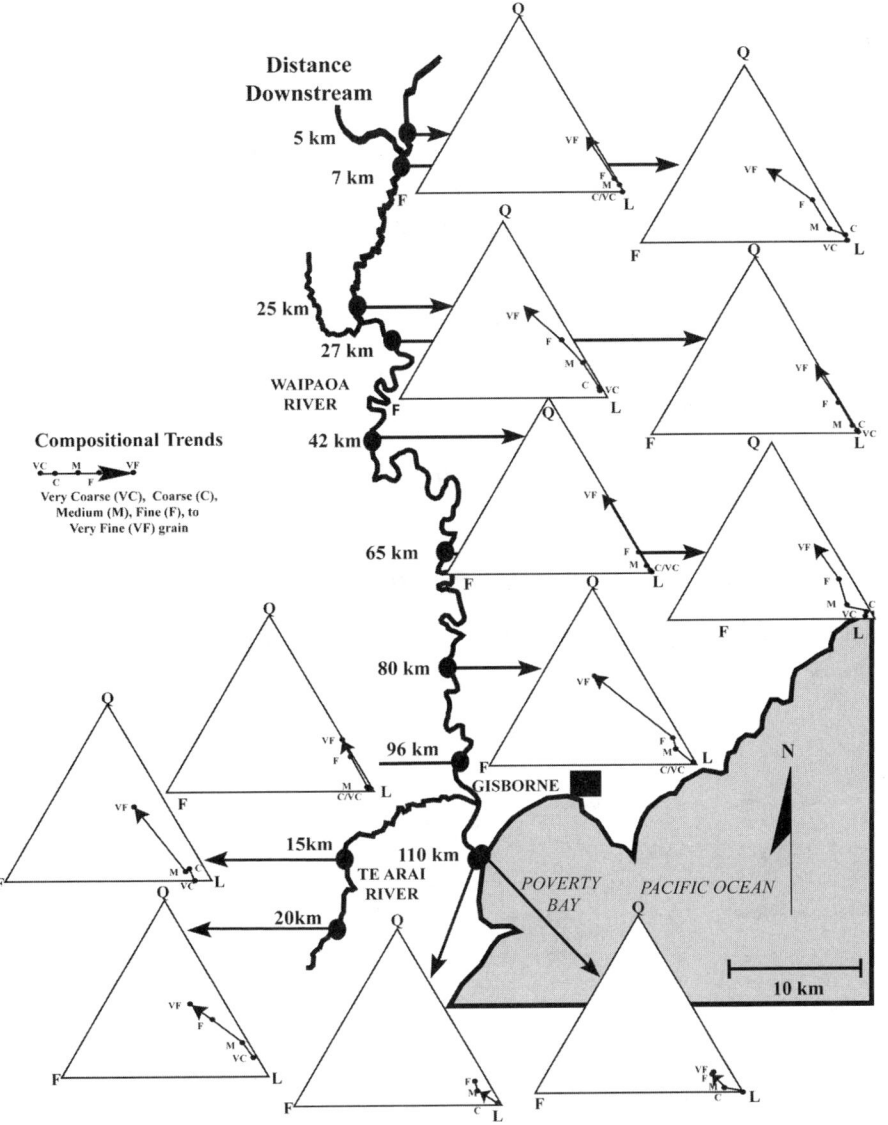

Figure 10. Distribution and location of Waipaoa and Te Arai River sand subfraction data. Arrows indicate decreasing grain size.

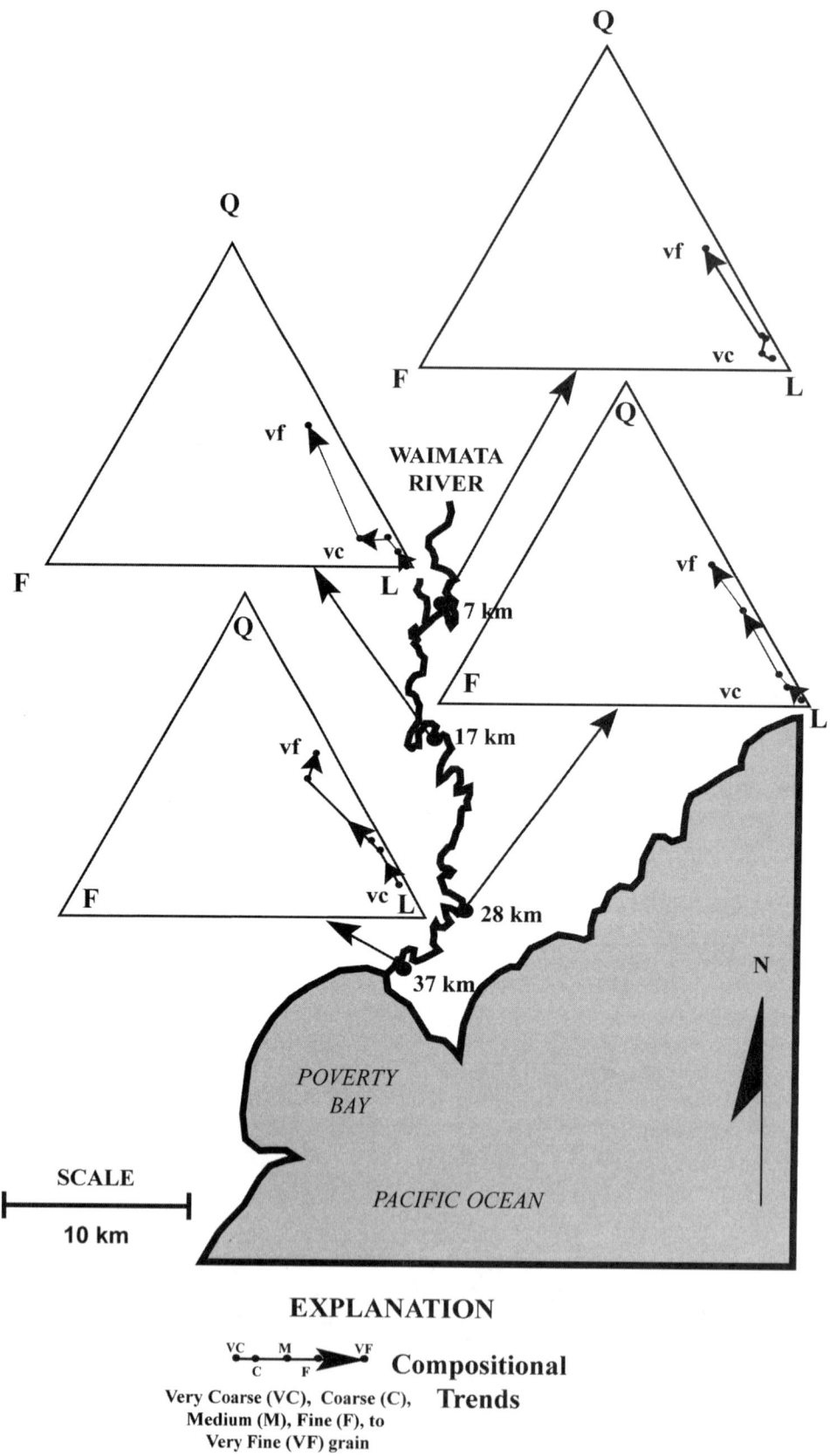

Figure 11. Distribution and location of Waimata River sand subfraction data. Arrows indicate decreasing grain size.

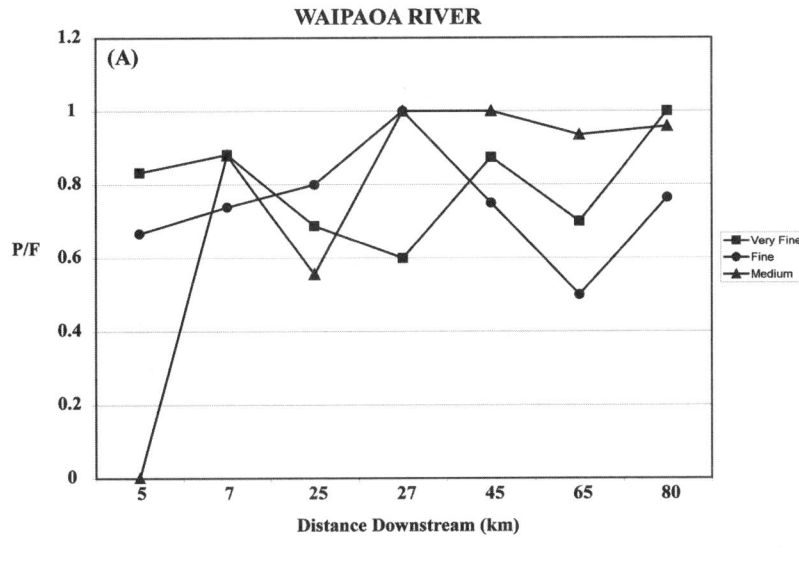

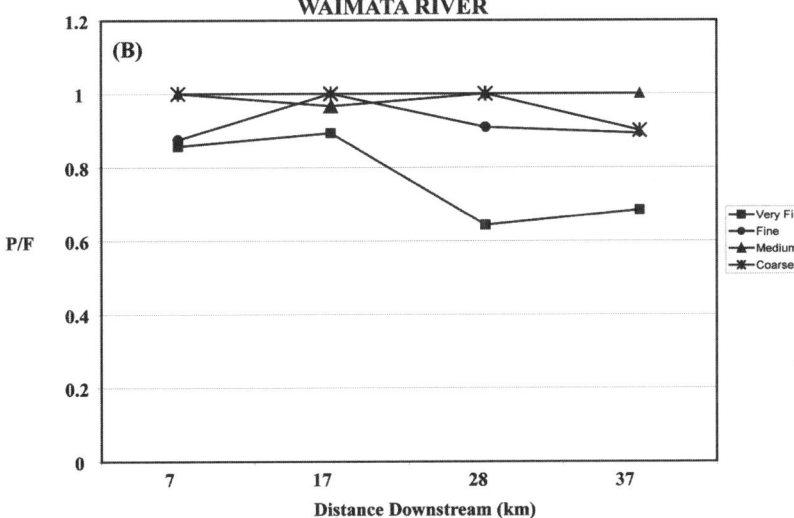

Figure 12. Graph representing the plagioclase to total feldspar ratio (P/F) versus distance downstream for the Waipaoa River (A) and the Waimata River (B).

downstream trends in the proportions of sandstone, argillite, and siltstone gravel clasts within the river, and they attributed downstream fining over the length of the Waipaoa River to sorting rather that abrasion.

We have found evidence that the patterns of downstream fining and sediment composition are locally disrupted by sediment contributions from tributary streams. In general, at the six sites examined in this study, the sandstone gravel clasts are slightly longer (Fig. 6) and, with one exception (80 km) where an anomalously high percent of mudstone (97%) and no sandstone clasts were found, the percentages of sandstone gravel clasts remains relatively constant from the headwaters to the lower reaches of the river (Fig. 4). This may be a function of the relatively greater durability of sandstone with respect to softer mudstone and claystone clasts, and local input of mudstone clasts at 80 km.

In addition, at 65 km downstream, the claystone clasts are larger than the sandstone clasts, possibly owing to influxes of claystone gravel from the tributary streams near this location.

This interpretation is supported by the significant increase in calcareous clasts at this site (Fig. 5). We were able to document this increase through petrographic examination of the clast lithologies, which in turn allowed us to key mudstone and claystone color to texture and composition and to further subdivide clast types according to carbonate content (see James [2004] for details). The population at this location contains 25% micritic claystone and 22% calcareous sandstone clasts, suggesting a local input of calcareous debris. This location is downstream of the confluence of both the Waihora and the Waihuka Rivers, which are tributary rivers underlain by Miocene and Pliocene units that are primarily massive calcareous mudstone with interbedded fine-grained sandstone, mudstone, and conglomerate with minor tuff and limestone. Thus, the significant increase in calcareous lithologies in the clast population recorded at 65 km this study is likely due to sediment being derived from tributaries that drain the more calcareous lithologies. Perhaps previous workers (Rosser, 1997; Gomez et al., 2001b) did not record a

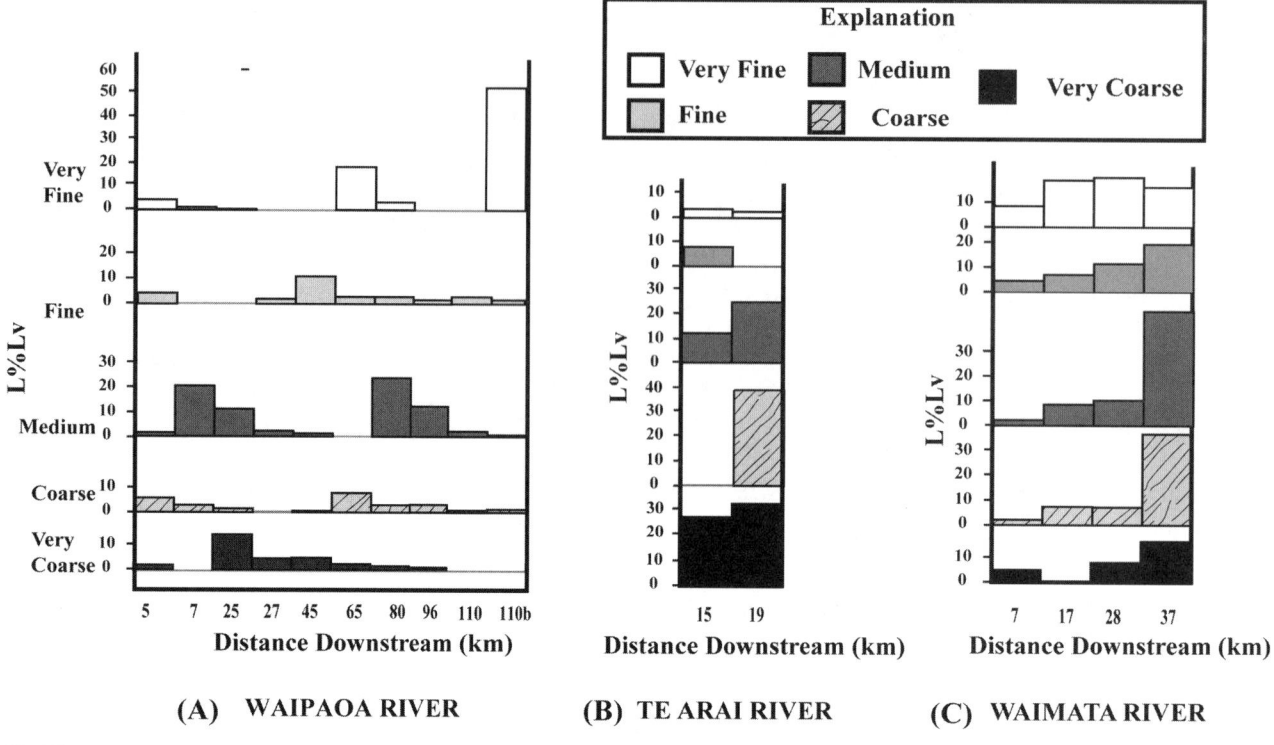

Figure 13. Histograms representing downstream distribution of L%Lv values for sand subfractions (very fine, fine, medium, coarse, and very coarse) from Waipaoa, Te Arai, and Waimata River data, where L%Lv is the percentage of volcanic lithics out of the total lithic population.

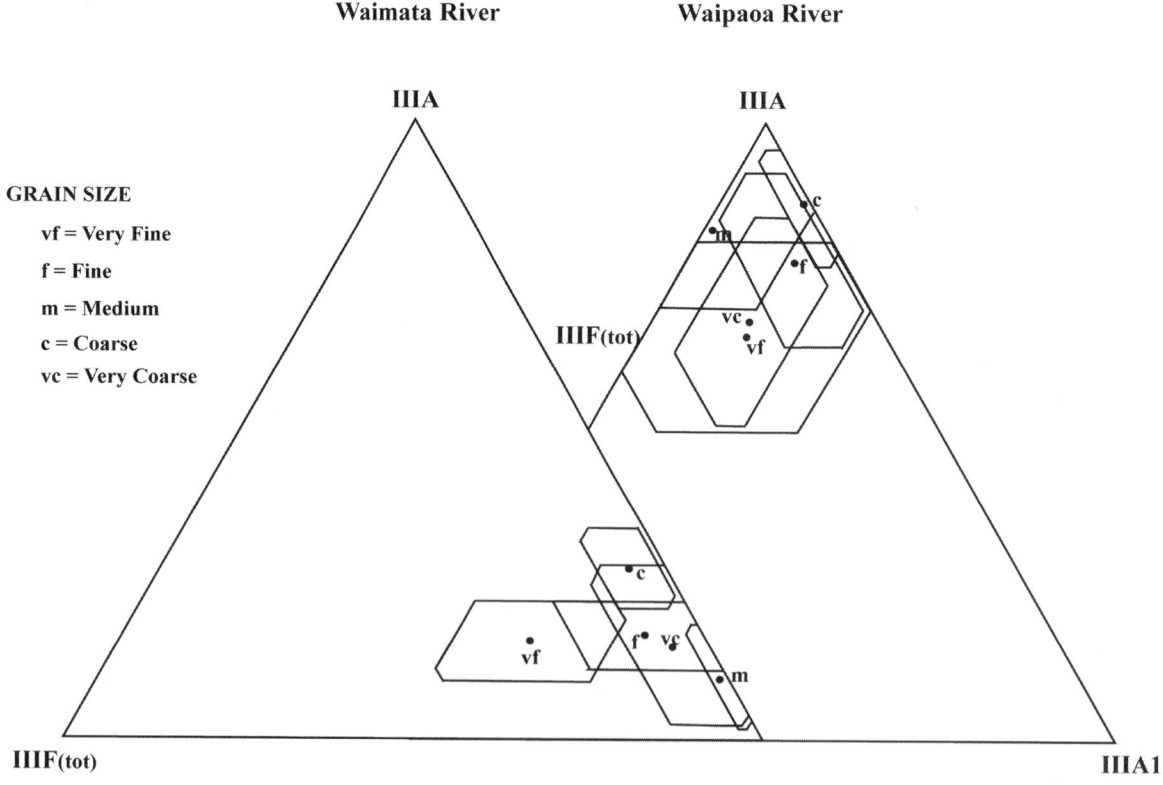

Figure 14. Mudstone lithic fragment ternary diagram for the Waimata and Waipaoa Rivers (see Table 4.7 in James [2004] for data and lithic type definitions and descriptions). Type IIIA is a mudstone/siltstone with pyrite, type IIIA1 is a mudstone/siltstone with carbonate and pyrite, and type IIIF (total) is siliceous mudstone.

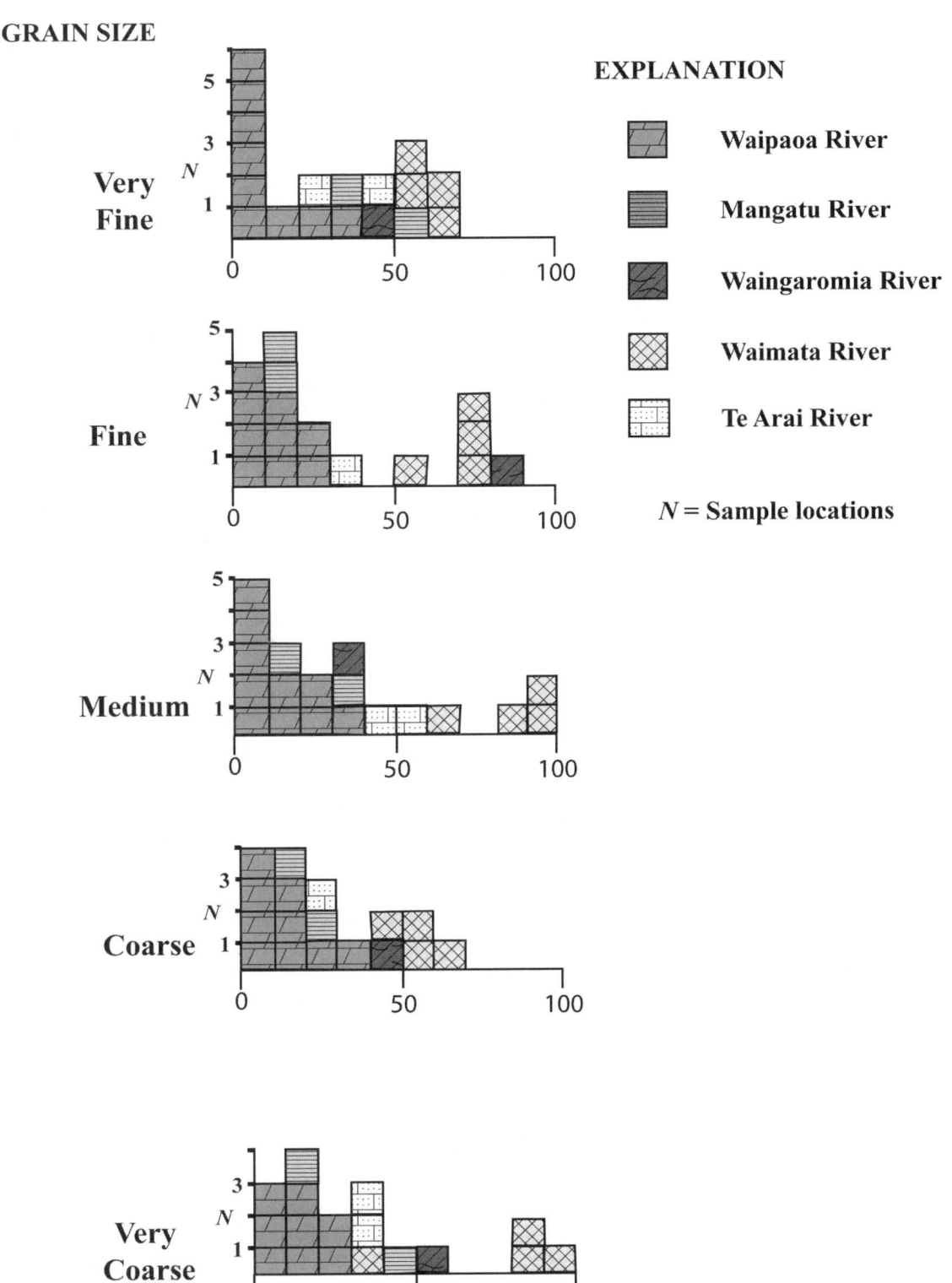

Figure 15. Histograms showing distribution of the proportion of calcareous lithic fragments (L% calcareous lithics) for the sand subset data from all rivers. These calcareous lithics include carbonate cemented siltstone, calcareous mudstone, limestone, chalk, and coarse carbonate; they are concentrated in the Waimata River, which drains Pliocene to Miocene calcareous lithologies. N is the number of sample locations. Data used to construct this figure are given in Table 4.7 and Appendix 8.6 of James (2004).

similar change in composition at this location in the river because their lithologic categories were too general.

Sand Composition in the Waipaoa River Catchment

The Waipaoa River and its tributary streams (Mangatu, Te Arai, and Waingaromia Rivers) exhibit similar sand detrital modes; the major components are sedimentary lithic fragments and quartz (Fig. 8). As shown in the following, however, some of the variations in the bulk sand sample compositions can be explained by differences in average grain size among samples. For example, the compositions of the coarser bulk samples are on average skewed toward more lithic end members, in spite of our use of the Gazzi-Dickinson method of point counting to minimize the dependence of composition on grain size (see Ingersoll et al., 1984). This led us to examine and count sand subfractions to demonstrate the pronounced difference in composition among the Udden-Wentworth sand-size classes (Figs. 9, 10, and 11). Quartz (QFL%Q from 1% to 41%) and to a lesser extent feldspar (QFL%F from 0.1% to 4%) proportions, on average, increase linearly with decreasing grain size. This compositional dependence on grain size occurs in both the Waipaoa and Waimata Rivers, as well as within all the tributary streams analyzed.

Within the Waipaoa bulk sand and sand subfraction samples, there are no significant linear downstream trends in general (QFL) or monomineralic (QmKP) or lithic (LmLvLs) proportions (Figs. 10, 12, and 13). The concentration of the monomineralic quartz and feldspar in the very fine, fine, and sometimes medium fractions is likely a function of the size of these components in the original sedimentary source rocks. Petrographic observations of sandy source rocks collected from the Waipaoa River system support this interpretation (James, 2004). The trace amounts of nonundulose quartz and fresh feldspar in the coarse and very coarse fractions likely reflect minor volcanic input from Taupo ashes (see further discussion next). Lastly, the very fine to fine sand fractions comprise a portion of the suspended load of the river, and their compositional variability from site to site (Figs. 10 and 11) may in part be due to periodic storm-induced flux of fine material through the system from various parts of the drainage basin.

Comparison of Waipaoa, Waimata, and Te Arai River Sediments

Although there are similarities (Figs. 8 and 9), the greatest differences in sediment composition within the streams of the Waipaoa River system were noted between the Waipaoa River and the Waimata River samples (Figs. 12–15). These differences can be linked to the lithology of the sedimentary sources (e.g., sheared Cretaceous rocks in the Waipaoa versus significant Pliocene calcareous rocks in the Waimata), as well as to the nature of sediment delivery to the stream (e.g., gullying as a major point source in the Waipaoa versus shallow landsliding providing sediment from across the catchment in the Waimata). The Te Arai River compositionally overlaps these two end members (Figs. 8, 9, 13, and 15). It drains only Miocene sedimentary rocks characterized by shallow landsliding.

In general, the unsorted bulk sand samples collected and analyzed along the Waipaoa and Waimata Rivers are dominated by sedimentary lithic fragments, and the Waimata River has a slightly higher quartz content (Fig. 8; note no bulk samples were analyzed for Te Arai River). Size fractions from the Waimata, Waipaoa, and Te Arai Rivers show similar trends (Fig. 9). There are significant differences in the lithic populations, however. The Waipaoa River only has a maximum of only a few percent volcanic lithics, whereas the percentage is much higher in the Waimata and Te Arai Rivers. In the Waimata River, the proportion of volcanic lithic fragments increases downstream in all size fractions (Fig. 13). The volcanic debris in the samples is dominated by vitric (glass) volcanic fragments and to a lesser extent by altered microcrystalline volcanic lithic fragments. The relatively fresh vitric components indicate recent volcanic sources, likely the erosional remnants of arc-derived ash deposits that cap the hills in this region. The altered to slightly altered volcanic lithic fragments may have been recycled from the East Coast allochthon in the upper reaches of the system, or from volcaniclastic intervals in the Miocene and Pliocene as described by Mazengarb and Speden (2000).

Gravel- to sand-sized mudstone lithic fragments from the Waimata and Waipaoa Rivers also exhibit compositional differences. Detailed gravel clast counts were only performed in the Waipaoa River sample sites. No gravel was observed at the Te Arai sample sites, but representative clasts were collected from the Waimata sites. The clast colors in the Waimata River are similar to, but less variable than those seen in the Waipaoa River. The Waimata River pebbles are limited to lighter colors, indicating that they are more calcareous. This difference is also apparent in specific mudstone lithic populations, as shown in the ternary plot of calcareous, siliceous, and pyrite-rich mudstone proportions in Figure 14; within the sand fractions, the mudstone lithic population within the Waimata River is dominated by calcareous mudstone clasts, while the Waipaoa River is dominated by noncalcareous mudstone clasts. Note that all size fractions for the Waimata and Waipaoa group distinctly on this plot. These compositional differences between the two rivers seem to reflect the differences in source rock lithologies in their respective drainage basins. The Waimata River only drains Pliocene and Miocene calcareous mudstone, sandstone, and to a lesser degree limestone. However, the majority of sediment for the Waipaoa River is derived from the upper reaches of the system, which is dominated by the relatively less calcareous Cretaceous lithologies, including pyrite-rich mudstones as described by Pearce and Black (1981).

Finally, the Waipaoa River is also characterized by the presence of birefringent mudstone lithic fragments (types IIID and IIID1 of James, 2004) that are smectite-rich, as determined by X-ray diffraction analysis of a pebble (A. Palmer, 2003, personal commun.), and exhibit similar petrographic characteristics in thin section. A high percentage of these birefringent smectitic sand fragments (Ls% ~ 15) is found near the headwaters of the Waipaoa River, just south of the Tarndale slip (samples NZ 2-64, NZ 2-63), where sheared rocks of the East Coast allochthon are reportedly rich in smectite (Pearce and Black, 1981; Mazengarb and Speden, 2000; D'Ath, 2002). Outcrops of smectitic mélange units may also contribute some fragments of this type. Only trace amounts of these birefringent smectitic fragments are present in the middle to lower reaches of the Waipaoa River. The survivability of these fragments is likely low because they would easily disintegrate during fluvial transport and alternate wetting and drying on exposed river banks. During extreme sediment-producing events, such as Cyclone Bola (e.g., Gomez et al., 2001a), however, sediment residence times would have been significantly diminished, perhaps resulting in more of these lithic types being carried to offshore depocenters.

Sediment Recycling

The Waipaoa River system produces sediment with a largely mudrock provenance because that is the lithology on which it is largely developed. In other published examples, fine-grained sedimentary units underlie only a portion of the drainage basin and are not proportionately represented in the sand fraction. For example, Cavazza et al. (1993) analyzed sedimentary recycling and the composition of sands shed from Miocene sedimentary source rock terranes in the Senio River drainage basin, Italy. In this region, the composition of the sands derived from a tectonically active source area reflected the various sedimentary source rocks, except for poorly indurated mudstones. McBride and Picard (1987) analyzed downstream changes in sand composition in a high-gradient stream in northwestern Italy. They concluded that although the Cretaceous shale and limestone bedrock yielded large amounts of sand, the proportion of sand grains from these rock types decreased rapidly as a result of dilution rather than abrasion. In addition, they found a larger proportion of sand grains from more resistant lithologies than from more friable lithologies (e.g., siltstones and shales).

The high proportion of sedimentary rock fragments in the Waipaoa River system as compared to the Italian rivers is a function of the limited fine-grained sedimentary source terrains in the latter. The concentration of monomineralic grains in the fine and very fine sand subfractions indicates that clast abrasion or, alternatively, disintegration during sediment transport, plays an important role in shaping the composition of the Waipaoa River sand fraction. However, this study has shown that in the Waipaoa River there is likely some downstream dilution through input of sediment derived from calcareous Tertiary outcrops. In the case of the purely Tertiary sedimentary sources (Waimata River), their semilithification and lack of diluting elements means that abrasion or disintegration during weathering is more important. Thus, sediment recycling is the theme today in the Waipaoa River system and may also have been in the past.

It is likely that the folded and uplifted Cretaceous to Neogene sedimentary sequences within the Waipaoa catchment are composed of detritus recycled, at least in part, from Torlesse rocks (Field et al., 1997). There are no published modal data for the Cretaceous to Pliocene units to compare with the modern Waipaoa River system sand data presented here, but there are data for Torlesse terrane sandstones just north and west of the Waipaoa River system in the Raukumara Ranges of North Island (Mortimer, 1994) (Fig. 16). Torlesse sandstones are feldspatholithic, and they plot within the dissected arc and transitional arc fields of Dickinson et al. (1983). In contrast, owing to their high mudstone lithic population, the Waipaoa River system samples plot within a provenance field ascribed to undissected arc settings (Fig. 16), whereas in fact, the sediments are dominated by sedimentary lithic rather than volcaniclastic debris. Higher sedimentary lithic proportions are associated with forearc deformation during aseismic and seismic ridge subduction elsewhere in the Pacific (e.g., Marsaglia et al., 1995, 1999), and in the Waipaoa River system, these proportions accurately reflect forearc deformation associated with subduction of the Hikurangi Plateau and seamounts. The Torlesse sandstones are much more feldspathic than the Waipaoa River system sediments, but increasing sediment maturity (quartz at expense of feldspar) cannot be simply attributed to sediment recycling because of the likely addition of volcanic quartz and feldspar debris to the Waipaoa River system from Holocene and Cenozoic volcaniclastic units.

CONCLUSIONS

Perhaps the most significant and unexpected result of this study was the close relationship between sand composition and grain size within the Waipaoa River system. The higher quartz and to a lesser degree feldspar contents of the fine to very fine sand fractions of the stream samples are likely inherited from their Cretaceous to Cenozoic source rocks. The main components are fragments of mud rocks and, to a lesser degree, volcanic lithic fragments. Within the Waipaoa River system, subtle variations in mud rock lithology are key to fingerprinting their provenance. This is true of both the gravel and sand fractions.

For example, like previous workers (Rosser, 1997; Gomez et al., 1999; Page et al., 1999), we found that the primary source of sediment for the modern Waipaoa River is from gully erosion in the upper catchment. However, by looking in more detail at the clast types, specifically the proportion of calcareous gravel clasts, we documented evidence of sediment input from tributary rivers that drain predominately Pliocene and Miocene calcareous mudstone and local limestone. Calcareous input from these units can also be seen in the sand fractions of the streams. The Waimata River, with a catchment largely underlain by these more calcareous lithologies, produces sand dominated by calcareous mudstone fragments, whereas the Waipaoa River, largely fed by gully erosion of Cretaceous units in the headwaters, produces distinctly different sand population dominated by noncalcareous mudstone fragments. Another compositional difference between the two rivers is the percentage of volcanic lithic fragments.

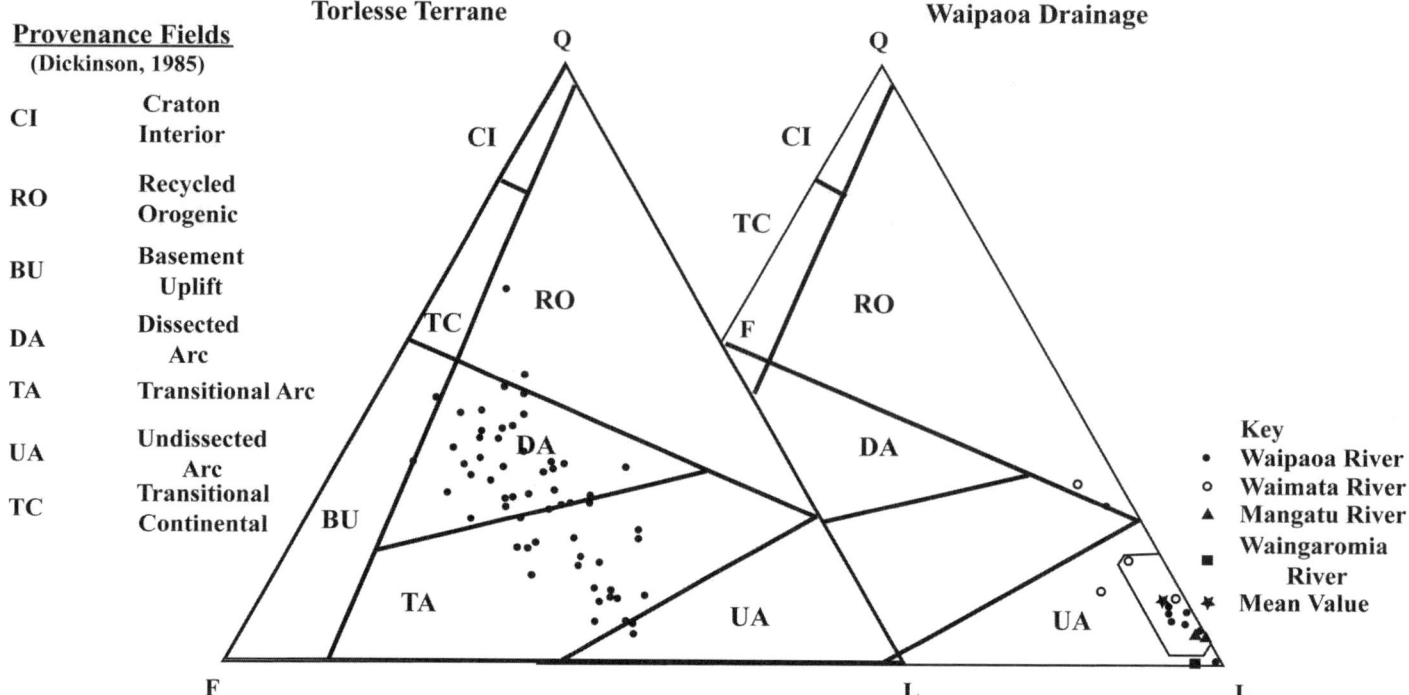

Figure 16. QFL (see Table 1) plot comparing the Torlesse terrane data (Mortimer, 1994) and Waipaoa River drainage sand data. Provenance fields of Dickinson (1985) are superimposed.

The sand and clast count data show that despite the proximity of the Waipaoa River to an active volcanic arc (~250 km), its sediment load contains a relatively small proportion of volcanic debris. Instead, the Waipaoa River sediments are dominated by the crushed and sheared sedimentary rocks of the East Coast allochthon, which is located in the upper portion of the catchment. The high sediment load caused by gully erosion in the upper reaches of the Waipaoa River may be responsible for diluting the volcanic signal. In contrast, the higher percentage of volcanic debris in the Waimata and Te Arai Rivers may reflect either more input of recycled volcanic debris from Quaternary tephras or from the Miocene-Pliocene section; alternatively, there may be less dilution of the ash signal in the Waimata and Te Arai Rivers by sedimentary lithic debris because of slower rates of denudation as compared to the rapidly eroding Cretaceous rocks in the headwaters of the Waipaoa River system. These results suggest that in future studies of offshore core samples and older alluvium, it may be possible to distinguish sand derived from a purely Miocene-Pliocene source from that derived from the East Coast allochthon.

The Waipaoa River system is an excellent location to examine the effects of sediment recycling on the rock record of active margins. The Waipaoa River system example indicates that when units are likely derived mainly from mud rocks, one must consider inherited compositional variations that may unequally affect detrital modes of sand fractions. Furthermore, detailed petrographic characterization of mud rock lithic types may help fingerprint sediment sources.

ACKNOWLEDGMENTS

Funding for this project was provided by a National Science Foundation grant (award 0119936) in support of the Catalyst Program at California State University–Northridge. The manuscript benefited from insightful reviews by Susanne Kairo and Miguel Ángel Caja. We extend our thanks to the following people, without whom this project would not have been possible: Basil Gomez provided us with encouragement, some preliminary samples, information on the field area, and constructive comments on our preliminary interpretations. Michael Marden provided information on regional and local geology, assistance in the field, and use of facilities at Landcare Research. Alan Palmer introduced us to the ash stratigraphy of the region, as well as provided some X-ray diffraction data for a selected gravel clast. Brenda Rosser and Michelle D'Ath graciously shared their Waipaoa River sediment data sets with us. Terry Dunn facilitated the project with her cheery clerical and logistical assistance.

REFERENCES CITED

Arribas, J., Critelli, S., Le Pera, E., and Tortosa, A., 2000, Composition of modern stream sand derived from a mixture of sedimentary and metamorphic source rocks: Sedimentary Geology, v. 133, p. 27–48, doi: 10.1016/S0037-0738(00)00026-9.

Berryman, K., Marden, M., Eden, D., Mazengarb, C., Ota, Y., and Moriya, I., 2000, Tectonic and paleoclimatic significance of Quaternary river terraces of the Waipaoa River, east coast, North Island, New Zealand: New

Zealand Journal of Geology and Geophysics, v. 43, p. 229–245.

Black, T., Shane, P., Westgate, J., and Froggart, P., 1996, Chronological and palaeomagnetic constraints on widespread welded ignimbrites of the Taupo volcanic zone, New Zealand: Bulletin of Volcanology, v. 58, no. 2–3, p. 226–238, doi: 10.1007/s004450050137.

Blair, T., and McPherson, J., 1999, Grain-size and textural classification of coarse sedimentary particles: Journal of Sedimentary Research, v. 69, p. 6–19.

Cavazza, E., Zuffa, G., Camporesi, C., and Ferretti, C., 1993, Sedimentary recycling in a temperate climate drainage basin (Senio River, north-central Italy): Composition of source rock, soil profiles, and fluvial deposits, in Johnsson, M.J., and Basu, A., eds., Processes Controlling the Composition of Clastic Sediments: Geological Society of America Special Paper 284, p. 247–261.

Cole, J.W., 1990, Structural control and origin of volcanism in the Taupo volcanic zone, New Zealand: Bulletin Volcanologique, v. 52, p. 445–459, doi: 10.1007/BF00268925.

Critelli, S., le Pera, E., and Tortosa, A., 2000, Composition of modern stream sand derived from a mixture of sedimentary and metamorphic source rocks (Henares River, central Spain): Sedimentary Geology, v. 133, p. 27–48, doi: 10.1016/S0037-0738(00)00026-9.

D'Ath, M., 2002, The clay mineralogy and erosion of the Waipaoa River catchment, Gisborne, New Zealand [M.S. thesis]: Palmerston North, New Zealand, Massey University, 134 p.

DeVaughn, A.M., 2005, Sand and gravel provenance of Te Arai River, Waipaoa River, and associated Quaternary alluvial terrace deposits, North Island, New Zealand [M.S. thesis]: Northridge, California, California State University Northridge, 81 p.

Dickinson, W., 1970, Interpreting detrital modes of greywacke and arkose: Journal of Sedimentary Petrology, v. 40, p. 695–707.

Dickinson, W.R., 1985, Interpreting provenance relations from detrital modes of sandstones, in Zuffa, G.G., ed., Provenance of arenites: Dordrecht, Reidel Publishing, Advanced Study Institute Series, v. 148, p. 333–361.

Dickinson, W., Beard, L., Brakenridge, G., Erjavec, J., Ferguson, R., Inman, K., and Ryberg, P., 1983, Provenance of Phanerozoic sandstones in relation to tectonic setting: Geological Society of America Bulletin, v. 94, p. 222–235, doi: 10.1130/0016-7606(1983)94<222:PONAPS>2.0.CO;2.

Di Giulio, A., Ceriani, A., Ghia, E., and Zucca, F., 2003, Composition of modern stream sands derived from sedimentary source rocks in a temperate climate (northern Apennines, Italy): Sedimentary Geology, v. 158, p. 145–161, doi: 10.1016/S0037-0738(02)00264-6.

Driscoll, N., and Nittrouer, C., 2000, Source to sink studies: Margins Newsletter, no. 5, University of Hawaii, p. 1–3.

Ferriere, J., and Chanier, F., 1993, La tectonique des plaques à l'epreuve de la realité; SW Pacifique et Nouvelle-Zelande: Geochronique, v. 45, p. 14–20.

Field, B., Uruski, C., Beu, A.G., Browne, G.H., Crampton, J.S., Funnell, R., Killops, S., Laird, M.G., Mazengarb, C., Morgans, H.E.G., Rait, G.J., Smale, D., and Strong, C.P., 1997, Cretaceous-Cenozoic geology and petroleum systems of the East Coast region, New Zealand: Institute of Geological and Nuclear Sciences Monograph 19, 7 enclosures, 301 p.

Foster, G., and Carter, L., 1997, Mud sedimentation on the continental shelf at an accretionary margin—Poverty Bay, New Zealand: New Zealand Journal of Geology and Geophysics, v. 40, p. 157–173.

Garzanti, E., Vezzoli, G., Lombardo, B., Ando, S., Mauri, E., Monguzzi, S., and Russo, P., 2004, Collision-orogen provenance (western Alps): Detrital signature and unroofing trends: The Journal of Geology, v. 112, p. 145–164, doi: 10.1086/381655.

Girty, G.H., 1991, A note on the composition of plutoniclastic sand produced in different climatic belts: Journal of Sedimentary Petrology, v. 61, p. 428–433.

Gomez, B., Eden, D., Hicks, D., Trustrum, N., Peacock, D., and Wilmshurst, J., 1999, Contribution of floodplain sequestration to the sediment budget of the Waipaoa River, New Zealand, in Marriot, S., and Alexander, J., eds., Floodplains: Interdisciplinary Approaches: Geological Society [London] Special Publication 163, p. 69–88.

Gomez, B., Fulthorpe, C., Carter, L., Berryman, K., Browne, G., Green, M., Hicks, M., and Trustrum, N., 2001a, Continental margin sedimentation to be studied in New Zealand: Eos (Transactions, American Geophysical Union), v. 82, p. 161, 166–167.

Gomez, B., Rosser, B., Peacock, D., Hicks, D., and Palmer, J., 2001b, Downstream fining in a rapidly aggrading gravel bed river: Water Resources Research, v. 37, p. 1813–1823, doi: 10.1029/2001WR900007.

Grapes, R., Roser, B., and Kashai, K., 2001, Composition of monocrystalline detrital and authigenic minerals, metamorphic grade, and provenance of Torlesse and Waipaoa graywacke, central North Island, New Zealand: International Geology Review, v. 43, p. 139–175.

Healy, J., Vucetich, C., and Pullar, W., 1964, Stratigraphy and chronology of late Quaternary volcanic ash in Taupo, Rotorua and Gisborne Districts: New Zealand Geological Survey Bulletin, v. 73, p. 145–163.

Heins, W.A., 1993, Source rock texture versus climate and topography as controls on the composition of modern, plutoniclastic sand, in Johnsson, M.A., and Basu, A., eds., Processes Controlling the Composition of Clastic Sediments: Geological Society of America Special Paper 284, p. 135–146.

Hicks, D., Gomez, B., and Trustrum, N., 2000, Erosion thresholds and suspended sediment yields, Waipaoa River Basin, New Zealand: Water Resources Research, v. 36, p. 1129–1142, doi: 10.1029/1999WR900340.

Ibbeken, H., and Schleyer, R., 1991, Source and sediment: Berlin, Springer-Verlag, 286 p.

Ingersoll, R., and Suczek, C., 1979, Petrology and provenance of Neogene sand from Nicobar and Bengal Fans, DSDP Sites 211 and 218: Journal of Sedimentary Petrology, v. 49, p. 1217–1228.

Ingersoll, R., Fullard, T., Ford, R., Grimm, J., Pickle, J., and Sares, S., 1984, The effect of grain size on detrital modes: A test of the Gazzi-Dickinson point-counting method: Journal of Sedimentary Petrology, v. 54, p. 103–116.

Isaac, M.J., Herzer, R.H., Brook, F.J., and Hayward, B.W., 1994, Cretaceous and Cenozoic Geology of Northland, New Zealand: Institute of Geological and Nuclear Sciences Monograph 8, 203 p.

James, D.E., 2004, Sand provenance in the Waipaoa River system, North Island, New Zealand [M.S. thesis]: Northridge, California, California State University–Northridge, 105 p.

Johnsson, M.J., 1993, The system controlling the composition of clastic sediments, in Johnsson, M.A., and Basu, A., eds., Processes Controlling the Composition of Clastic Sediments: Geological Society of America Special Paper 284, p. 1–19.

King, P.R., 2000, Tectonic reconstructions of New Zealand: 40 Ma to the Present: New Zealand Journal of Geology and Geophysics, v. 43, p. 611–638.

Lewis, K., and Pettinga, J., 1993, The emerging, imbricate frontal wedge of the Hikurangi Margin, in Balance, P.F., ed., South Pacific Sedimentary Basins: Sedimentary Basins of the World: Amsterdam, Elsevier, v. 2, p. 225–250.

Marsaglia, K.M., 1992, Petrography and provenance of volcaniclastic sands recovered from the Izu-Bonin arc, Leg 126, in Taylor, B., Fujioka, K., et al., Proceedings of the Ocean Drilling Program, Scientific Results, Volume 126: College Station, Texas, Ocean Drilling Program, p. 139–154.

Marsaglia, K.M., 1993, Basaltic island sand provenance, in Johnsson, M.A., and Basu, A., eds., Processes Controlling the Composition of Clastic Sediments: Geological Society of America Special Paper 284, p. 41–66.

Marsaglia, K.M., and Tazaki, K., 1992, Diagenetic trends in Leg 126 sandstones, in Taylor, B., Fujioka, K., et al., Proceedings of the Ocean Drilling Program, Scientific Results, Volume 126: College Station, Texas, Ocean Drilling Program, p. 125–138.

Marsaglia, K.M., Ingersoll, R.V., and Packer, B.M., 1992, Tectonic evolution of the Japanese islands as reflected in modal compositions of Cenozoic forearc and backarc sand and sandstone: Tectonics, v. 11, p. 1028–1044.

Marsaglia, K.M., Torrez, X., Padilla, I., and Rimkus, K., 1995, Provenance of Pleistocene and Pliocene sand and sandstone, ODP Leg 141, Chile margin: Proceedings of the Ocean Drilling Program, Scientific Results, Volume 141: College Station, Texas, Ocean Drilling Program, p. 133–151.

Marsaglia, K.M., Mann, P., Hyatt, R., and Olson, H., 1999, Evaluating the influence of aseismic ridge subduction and accretion(?) on the detrital modes of forearc sandstone: An example from the Kronotsky Peninsula, Kamchatka forearc: Lithos, v. 46, p. 17–42, doi: 10.1016/S0024-4937(98)00054-1.

Mazengarb, C., and Speden, I.G. (compilers), 2000, Geology of the Raukumara area: Institute of Geological and Nuclear Sciences Geological Map 6, scale 1:250,000, 1 sheet, 60 p.

McBride, E., and Picard, M., 1987, Downstream changes in sand composition, roundness, and gravel size in a short-headed, high-gradient stream, northwestern Italy: Journal of Sedimentary Petrology, v. 57, p. 1018–1026.

Mortimer, N., 1994, Origin of the Torlesse terrane and coeval rocks, North Island, New Zealand: International Geology Review, v. 36, p. 891–910.

Page, M., Reid, L., and Lynn, I., 1999, Sediment production from Cyclone Bola landslides, Waipaoa catchment: New Zealand Journal of Hydrology, v. 38, p. 289–308.

Palomares, M., and Arribas, J., 1993, Modern stream sands from compound crystalline sources; composition and sand generation index, in Johnsson, M.J., and Basu, A., eds., Processes Controlling the Composition of

Clastic Sediments: Geological Society of America Special Paper 284, p. 313–322.

Pearce, A., Black, R., and Nelson, C.S., 1981, Lithologic and weathering influences on slope form and process, eastern Raukumara Range, New Zealand, *in* Davies, T.R.H., and Pearce, A.J., eds., Erosion and Sediment Transport in Pacific Rim Steeplands: International Association of Hydrological Sciences–Association Internationale des Sciences Hydrologiques (IAHS-AISH) Publication 132, p. 95–122.

Pettinga, J.R., 1982, Upper Cenozoic structural history, coastal southern Hawke's Bay, New Zealand: New Zealand Journal of Geology and Geophysics, v. 25, p. 149–191.

Ridgeway, K.D., and DeCelles, P.G., 1993, Petrology of Mid-Cenozoic strike-slip basins in an accretionary orogen, St. Elias Mountains, Yukon Territory, Canada, *in* Johnsson, M.A., and Basu, A., eds., Processes Controlling the Composition of Clastic Sediments: Geological Society of America Special Paper 284, p. 67–89.

Rock-Color Chart Committee, 1991, Geological Society of America Rock Color Chart: Boulder, Geological Society of America, 10 p.

Rosser, B., 1997, Downstream Fining in the Waipaoa River: An Aggrading, Gravel-Bed River, East Coast, New Zealand [M.S. thesis]: Palmerston North, New Zealand, Massey University, 160 p.

Scott, B., 1997, Volcanic impacts—Gisborne District Council: GNS client report 44692D.17: Lower Hutt, Institute of Geological and Nuclear Sciences Ltd. (unpublished), 2 p.

Smale, D., 1980, Akatarawa conglomerate (Permian), Lake Aviemore, South Canterbury: New Zealand Journal of Geology and Geophysics, v. 23, no. 3, p. 279–292.

Speden, I., 1972, New fossil localities in the Torlesse Supergroup, western Raukumara Peninsula, New Zealand: New Zealand Journal of Geology and Geophysics, v. 15, p. 433–445.

Speden, I., 1975, Cretaceous stratigraphy of Raukumara Peninsula: Wellington, New Zealand Geological Survey Bulletin, Department of Scientific and Industrial Research, v. 91, 70 p.

Suggate, R., (ed), 1978, The geology of New Zealand, Volume 1: Wellington, New Zealand Geological Survey, Department of Scientific and Industrial Research, p. 1–342.

Walcott, R.I., 1987, Geodetic strain and the deformational history of the North Island of New Zealand during the late Cainozoic: Transactions of the Royal Philosophical Society of London, v. A321, p. 163–181.

Wolman, M., 1954, A method of sampling coarse riverbed material: American Geophysical Union Transactions, v. 35, p. 951–956.

Manuscript Accepted by the Society 9 August 2006

The petrology and provenance of sand in the Bounty submarine fan, New Zealand

Shawn A. Shapiro[†]
Kathleen M. Marsaglia
Department of Geological Sciences, California State University–Northridge, 18111 Nordhoff Street, Northridge, California 91330-8266, USA

Lionel Carter
National Institute of Water and Atmosphere, P.O. Box 14 901, Wellington, New Zealand

ABSTRACT

Off the east coast of South Island, New Zealand, the Bounty Fan lies in the most seaward axial deep of the Bounty Trough, a remnant continental rift. The north levee of the Bounty Channel was cored at Ocean Drilling Program (ODP) Site 1122 on Leg 181. The 617 m section, which is divided into three units and nine subunits, consists of a Quaternary fan turbidite sequence that transitions downward into a Pleistocene to Pliocene mixed turbidite to contourite facies, which unconformably overlies a Miocene contourite-pelagic succession. Petrographic analysis (point counts) of 55 fine to very fine sand samples across the cored interval shows them to be quartzofeldspathic with moderate mica and minor metamorphic lithic components (mean values = $Q_{44}F_{44}L_{12}$, $Qm_{50}K_3P_{47}$, $Lm_{70}Lv_4Ls_{26}$, total%mica$_{14}$). Mean recalculated parameters for Site 1122 units and subunits cluster on QFL, QmKP, and LmLvLs ternary plots with little compositional variation. Proportions of biotite, muscovite, chlorite, and various metamorphic lithic types also show little compositional variation among the units and subunits of Site 1122. Furthermore, there are no significant trends among thickness, grain size, composition, and depth of Site 1122 sand samples, except that thicker beds tend to contain slightly more metamorphic rock fragments. The generally homogeneous composition of Site 1122 sand indicates that it may have had a relatively uniform source back into the early Miocene. Thus, the up-section change from sandy contourite to turbidite deposits at Site 1122 is not reflected in sand composition. This suggests that the sand provenance remained constant while the depositional processes of sand at Site 1122 changed. Sand detrital modes at Site 1122 most closely match those of the Clutha River, especially in terms of QFL, QmKP, mica, and lithic proportions, suggesting that it was a major source of sand at Site 1122. However, admixing of sand from the Waitaki River and other sources cannot be completely ruled out.

Keywords: Bounty Fan, New Zealand, petrology, sedimentology.

[†]Present Address: Schlumberger Information Solutions, 5599 San Felipe, Suite 1700, Houston, Texas 77056, USA.

Shapiro, S.A., Marsaglia, K.M., and Carter, L., 2007, The petrology and provenance of sand in the Bounty submarine fan, New Zealand, *in* Arribas, J., Critelli, S., and Johnsson, M.J., eds., Sedimentary Provenance and Petrogenesis: Perspectives from Petrography and Geochemistry: Geological Society of America Special Paper 420, p. 277–296, doi: 10.1130/2006.2420(17). For permission to copy, contact editing@geosociety.org. ©2007 Geological Society of America. All rights reserved.

INTRODUCTION

One of the major features of the southwest Pacific is the submarine Bounty Fan system (>525,000 km²) (Fig. 1). The head of the Bounty Fan lies ~830 km east of South Island, New Zealand, at a depth of 4430 m, whereas the lower fan merges with the 5000+-m-deep floor of the southwestern Pacific Basin at least 1110 km from shore (Carter and Carter, 1987, 1993, 1996). Approximately half of the Bounty Fan lies within the most seaward axial deep of the Bounty Trough, a remnant continental rift thought to have formed in the late Cretaceous (Carter and Carter, 1996). This rift is flanked by the largely submerged Chatham Rise and Campbell Plateau (Fig. 1).

In 1998, Ocean Drilling Program (ODP) Leg 181 drilled Site 1122 on the north bank levee of the Bounty Fan channel (Fig. 2). The cores recovered from Site 1122 range in age from early Miocene to late Pleistocene (Shipboard Scientific Party, 1999b). The stratigraphy at Site 1122 (holes A–C) was described and divided by the Shipboard Scientific Party (1999b) into three major units (I–III) (Fig. 3). Unit I (ca. 0–1.40 Ma) consists mainly of silty clay and fine sand, with sharp basal contacts and normal grading interpreted by the Shipboard Scientific Party (1999b) as turbidite sequences. Unit II (ca. 1.40–12.34 Ma) consists mainly of silty-bioturbated clay and some fine sand with sharp, scoured contacts and laminations interpreted by the Shipboard Scientific Party (1999b) as possible contourite deposits. Unit III (ca. 12.34–17.4 Ma) consists mainly of silt to silty clay with interbedded fine sand, and some nannofossil-bearing foraminifer sand beds interpreted by the Shipboard Scientific Party (1999b) as contourites.

The Bounty Fan is a major sediment "sink" along South Island, New Zealand. The purpose of this study is to characterize sand detrital modes of the Bounty Fan in order to constrain sandy sediment source(s) for the fan and to relate fan development to the geological evolution of the South Island, New Zealand. There

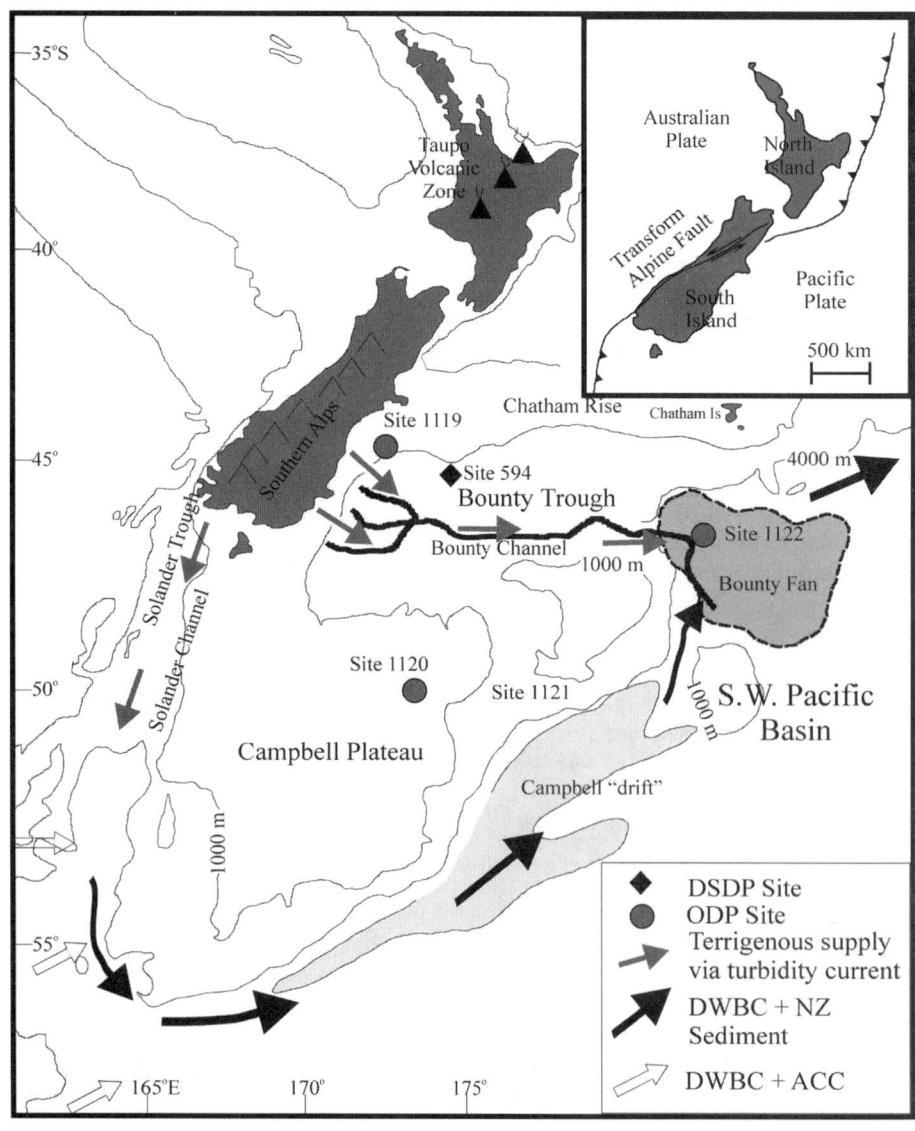

Figure 1. Map showing location of Ocean Drilling Program (ODP) Leg 181 Site 1122 and the Bounty Fan in relation to North and South Island, New Zealand (modified from Carter et al., 1999). DWBC—Deep Western Boundary Current; ACC—Antarctic Circumpolar Current; NZ sediment—New Zealand sediment from the southern portion of South Island; DSDP—Deep Sea Drilling Project. In insert, the Australian and Pacific plate boundary is barbed where convergent and arrowed where a transform.

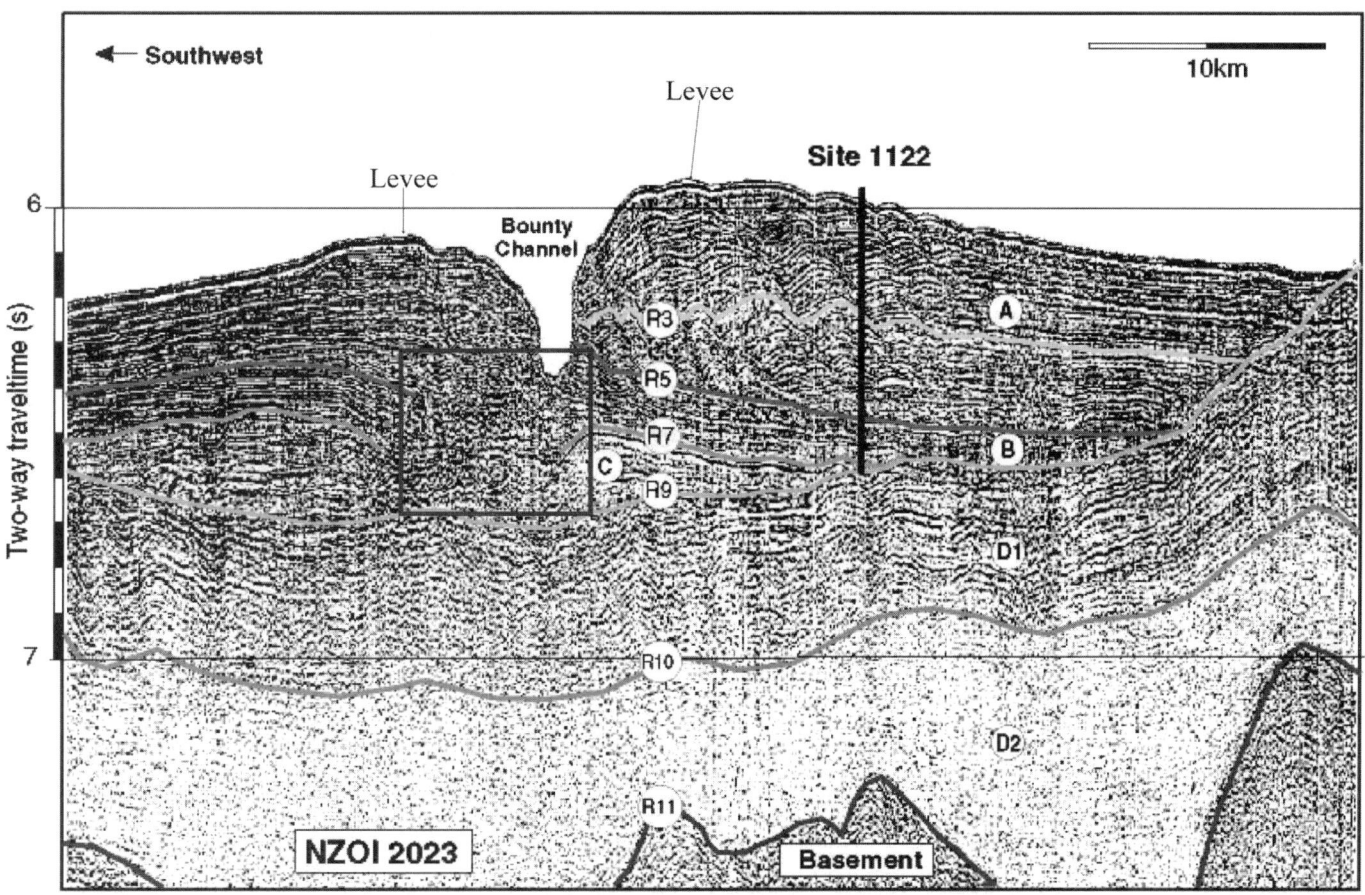

Figure 2. Seismic cross section NZOI 2023 showing Ocean Drilling Program (ODP) Site 1122 in relation to the Bounty Channel (modified from Shipboard Scientific Party, 1999b). R3, R5, R7, R9, R10, and R11 are reflectors, and A, B, C, D, and D1 are seismic units identified prior to ODP Leg 181. See text for discussion of probable channel facies.

are three potential sources for sediment in the Bounty Fan: near-fan sources, north-flowing deep ocean currents, and/or onshore river(s) (Fig. 1).

Near-fan sources are the wholly (Campbell Plateau) to partly (Chatham Rise) submerged continental fragments that flank the Bounty Trough (Fig. 1). They are not likely prominent sources because the terrigenous input to the shelves is minimal, there are no rivers of consequence, the shelf sediments are mainly biogenic carbonate, and both shelves have no obvious conduit to the fan. Simple wave suspension of shelf material, even at low sea level, would not be likely because the shelves are ~250 km from Site 1122.

The Campbell Plateau also influences local deep ocean currents. The Deep Western Boundary Current (DWBC) flows north from Antarctica at depths below 2000 m and encounters the continental margins of the Campbell Plateau and Chatham Rise (Fig. 1) off the South Island (Carter et al., 1996; Carter and Carter, 1996; Carter and McCave, 2002). It then follows along the slope east of the island across the Bounty Fan, eventually making its way to the subduction zone north of New Zealand (Fig. 1). This current is known to have reworked Clutha-derived sand, documented by Carter and Mitchell (1987), and to have transported material from the deep south (McCave and Carter, 1997). Thus, it may have carried in extrabasinal detritus (Shipboard Scientific Party, 1999b). This current also may have reworked Bounty Fan turbidite sand.

Small submarine canyons extending eastward from the shelf edge near the mouths of the Clutha (Molyneux Canyon) and Waitaki (Waitaki Canyon) Rivers feed into larger channels (North, Central, and South; Fig. 4) that merge to form the main Bounty Channel along the continental slope (Carter and Carter, 1993). Thus, probable onland sources of Bounty Fan sediment are the Clutha and Waitaki Rivers (Fig. 4), which are known to supply sediment to the continental shelf of South Island, New Zealand (Carter et al., 1985). These rivers drain significantly different geological terranes as outlined in the following sections.

TECTONIC EVOLUTION AND GEOLOGY OF SOUTH ISLAND, NEW ZEALAND

New Zealand sits on the boundary between the Pacific and Indian-Australian plates (Fig. 1; Kamp, 1987). This plate boundary is expressed as the strike-slip Alpine fault on South Island. Apatite and zircon fission-track analyses in the vicinity of the Alpine

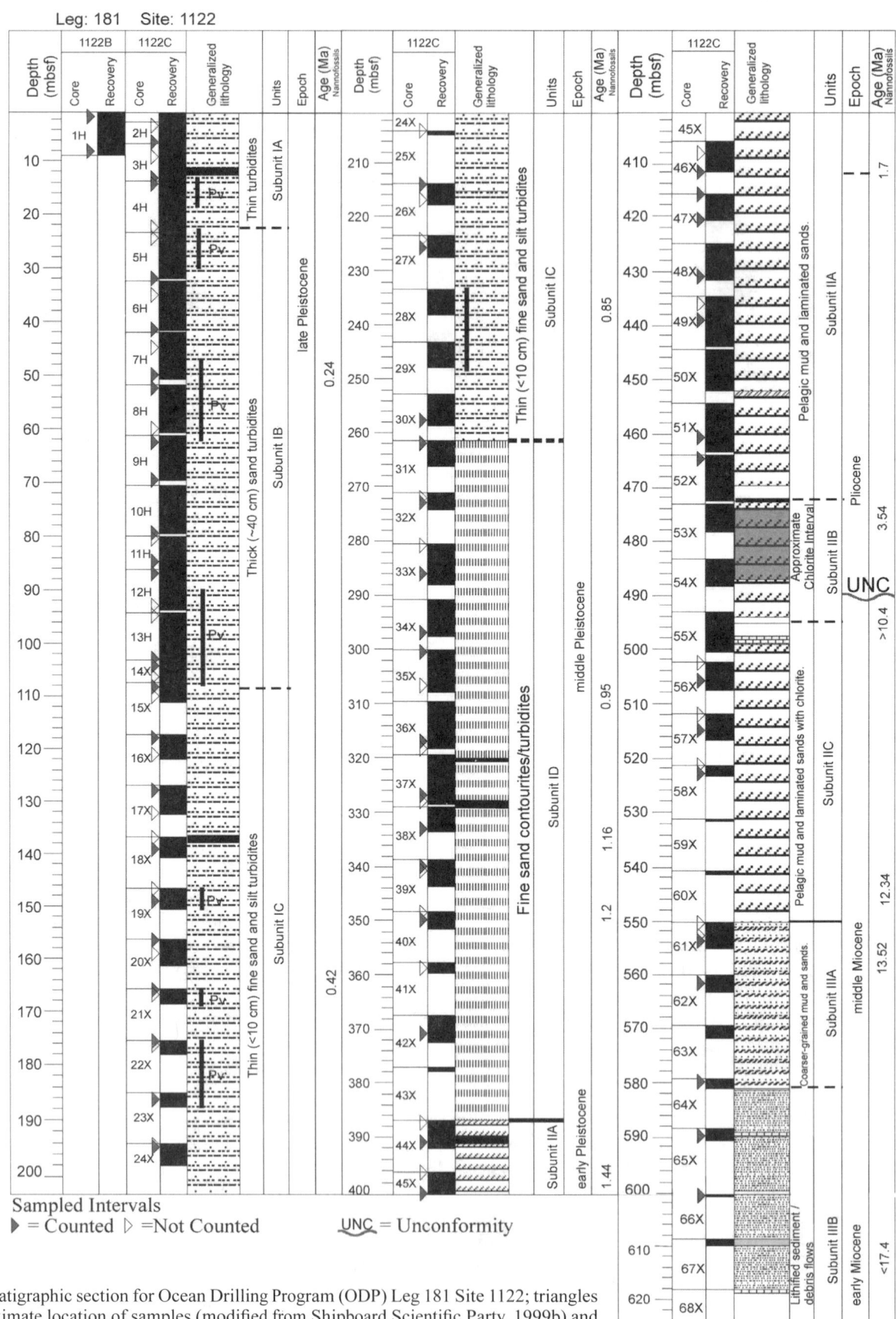

Figure 3. Stratigraphic section for Ocean Drilling Program (ODP) Leg 181 Site 1122; triangles show approximate location of samples (modified from Shipboard Scientific Party, 1999b) and mbsf is meters below seafloor.

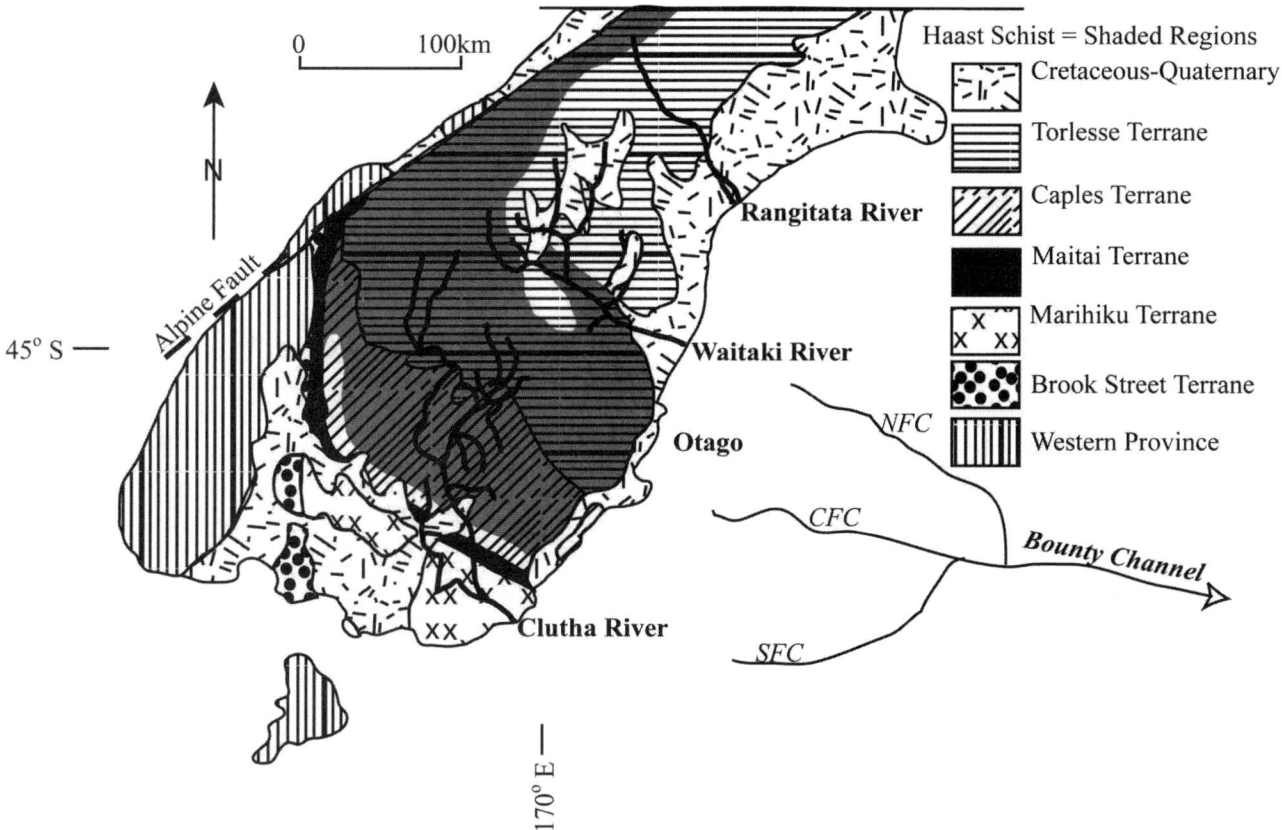

Figure 4. Map showing the terranes that comprise the geology of South Island, New Zealand (modified from Mortimer, 1993), and the channels that feed the Bounty Channel (SFC—south feeder channel; NFC—north feeder channel; CFC—central feeder channel) (Carter and Carter, 1993, 1996). Site 1122 is located off the map.

fault show two cooling trends, one ca. 100 Ma and the other at ca. 20 Ma (Kao, 2001). These may represent uplift and erosion events of the Southern Alps (Kamp, 2001). The uplift accelerated ca. 6.5 Ma when the pole of rotation shifted, increasing compression along the Alpine fault (Walcott, 1998).

South Island geology includes several terranes composed of sedimentary, igneous, and metamorphic rocks, the most extensive of which are the Caples and Torlesse terranes (Fig. 4). The Permian to Jurassic(?) Caples terrane is composed of metasedimentary rocks of volcaniclastic (mafic to continental arc) provenance with some sandstones thought to be submarine fan deposits (Mortimer, 1993). The Permian to Triassic Torlesse terrane is composed of quartzofeldspathic sandstones that contain some argillite and conglomerate (Mortimer, 1993). The Torlesse consists of graywacke-dominated turbidite sequences (Adams et al., 1998), which are part of a submarine accretionary prism complex (Mortimer, 1994); they show an active continental margin, volcanic-plutonic provenance (Mortimer, 1993).

The Otago-Haast Schist encompasses rocks of the Caples and Torlesse terranes, which were metamorphosed during the Early Jurassic (Bishop et al., 1985; Mortimer, 1993). The schist was uplifted during the Cretaceous (Kamp, 2001; Forster and Lister, 2003). The Otago-Haast Schist is composed of psammitic and pelitic gray schist and some greenschist (Mortimer, 1993). Quartzite, marble, and ultramafic schist are present within some greenschist bands (Mortimer, 1993). The metamorphic grade ranges from prehnite-pumpellyite facies in graywackes, through pumpellyite-actinolite facies to greenschist facies in the center of the schist outcrop (Mortimer, 1993).

The oldest cover rocks of the schist are Upper Cretaceous terrestrial fanglomerate deposits, and deltaic to shallow-marine sandstones (Mortimer, 1993). Upper Cretaceous to Miocene marine strata rest on the schist and graywacke in eastern Otago and on Miocene lacustrine strata in central Otago (Mortimer, 1993). Intraplate mafic volcanic and volcaniclastic rocks of Oligocene to Miocene age are interbedded with the marine strata near the coast (Mortimer, 1993). Pliocene to lower Quaternary piedmont gravels locally overlie the Cretaceous to Miocene cover strata at basin margins (Bishop, 1974).

Glaciation of the South Island began 2.5 Ma (Chinn, 1996) with 12 major cycles of glaciation within the last 750 k.y. (Nelson et al., 1985).

METHODS

Carter, a shipboard scientist on ODP Leg 181, collected the 110 sandy core samples from ODP Holes 1122B and 1122C, which we have analyzed in this study (Fig. 3). The 5 cm^3 samples were air-dried and subdivided using a sample splitter, and then the sample splits were gently sieved to separate the sand-sized fraction (0.0625–2 mm). Of the 110 samples, 98 contained sufficient sand for thin section preparation. Thin sections from the sand-sized fractions were stained for potassium and calcium feldspar. From the 98 thin sections, 55 samples evenly distributed across the recovered section were selected for point-count analysis (Fig. 3). Point-count data (Appendix A) were collected using a Nikon Eclipse E600 Pol microscope and a Prior Model G data collector. Four hundred points were counted on each slide. The Gazzi-Dickinson method of point counting (Ingersoll et al., 1984) was used with various monocrystalline and polycrystalline grain categories, such as quartz, potassium feldspar, plagioclase (including albite), and metamorphic lithic fragments (Table 1). Using this method, sand-sized mineral grains in a fragment were counted as monocrystalline components. Major lithic categories were then further subdivided; for example, micaceous metamorphic fragments were subdivided into quartz-mica tectonite, quartz-mica-feldspar aggregate, and polycrystalline mica. If possible, these grains were then categorized (Table 1) using the low- to medium-grade metamorphic classification scheme devised by Garzanti and Vezzoli (2003). This classification scheme first categorizes fragments as meta-siltstone, meta-claystone, meta-sandstone, meta-carbonate, and meta-basalt. Mineral grain size of the fragment is then used to determine the metamorphic grade (see Table 1 for description of categories). In matrix-rich samples, sand-sized fragments of matrix material created during dry-sieving were identified and petrographically characterized by their color, carbonate content, birefringence, and grain size. These characteristics were then applied to all other potential sedimentary lithic fragments to differentiate fragments of matrix from true detrital sedimentary lithic fragments. The former were not included in the point counts. The point-count data were entered into a spreadsheet to recalculate the percentages to allow further analysis on ternary and depth plots (Tables 1–4).

Sandy samples from the two major possible source rivers (Clutha and Waitaki) were collected by Marsaglia in January 2002. The Clutha River drains mainly Otago-Haast Schist,

TABLE 1. COUNTED AND RECALCULATED PARAMETERS

Qm	Monocrystalline quartz
Qp	Polycrystalline quartz
P	Plagioclase feldspar
Altered plag	Altered plagioclase feldspar
K	Potassium feldspar
Fu	Unstained feldspar
Lma	Quartz-feldspar-mica aggregate lithic (unclassified)
Lmf2*	Fine- to medium-grained meta-siltstone/sandstone with quartz, feldspar, and mica
Lmf3*	Medium-grained meta-siltstone/sandstone with quartz, feldspar, and mica
Lmt	Quartz-mica-tectonite lithic (unclassified)
Lmf2*	Fine- to medium-grained meta-siltstone/sandstone with quartz and mica
Lmf3*	Medium-grained meta-siltstone/sandstone with quartz and mica
Lmf4*	Coarse-grained meta-siltstone/sandstone with quartz and mica
Lmp2*	Fine- to medium-grained meta-claystone with quartz and mica
Lmp4*	Medium-grained meta-claystone with quartz and mica
Lmm	Polycrystalline mica lithic (unclassified)
Lmp2*	Fine- to medium-grained polycrystalline mica
Lmp3*	Medium-grained polycrystalline mica
Lmp4*	Medium- to coarse-grained polycrystalline mica
Rmp5*	Coarse-grained polycrystalline mica
Lm other	Other metamorphic lithic
Lma Q + P + other	Quartz-plagioclase-other aggregate lithic
Lma Q + dense	Quartz-dense aggregate lithic
Lma P + dense	Plagioclase-dense aggregate lithic
Lma K + dense	Potassium feldspar-dense aggregate lithic
Lma Q + F	Quartz-feldspar aggregate lithic
Lsi	Siltstone lithic
Lsa	Argillite-shale lithic
Lv	Volcanic lithic fragment (glass only)
M	Muscovite
B	Biotite
Chl	Chlorite
Malt	Altered mica
D	Dense minerals (unspecified)
Glau	Glauconite
Epi	Epidote
Carb	Carbonate
Foram	Foraminifers
Dia	Diatoms
Spong	Sponge spicules (siliceous)
Echino	Echinoderm spines
Shell frag	Mollusc (?) fragments
Other bio	Other bioclasts of unknown origin
Other	Other unknown mineral or lithic fragment

Note: Q = Qp + Qm; Lma = Lma + Lmf$_2$ + Lmf$_3$ + (Lma Q + P other) + (Lma Q + dense) + (Lma P + dense) + (Lma K + dense) + (Lma Q + F); Lmm = Lmm + Lmp$_2$ + Lmp$_3$ + Lmp$_4$ + Rmp$_5$; Lmt = Lmt + Lmf$_2$ + Lmf$_3$ + Lmf$_4$ + Lmp$_2$ + Lmp$_4$; Lm = Lma + Lmm + Lmt + (Lm other); Ls = Lsi + Lsa; F = P + K + Fu; L = Lm + Ls + Lv.

TABLE 2. RECALCULATED PARAMETERS

Sample Hole-core-section-top of interval (cm)	QFL Q (%)	QFL F (%)	QFL L (%)	QmKP Qm (%)	QmKP P (%)	QmKP K (%)	LmLvLs Lm (%)	LmLvLs Lv (%)	LmLvLs Ls (%)	Total M (%)	Total B (%)	Mica Chl (%)	Total Chl (%)
1122B-1H-2–19	38.4	52.5	9.1	41.7	50.0	8.3	92.9	3.6	3.6	3.8	5.0	20.0	0.8
1122B-1H-6–102	51.6	38.4	10.0	57.3	39.5	3.1	87.5	0.0	12.5	5.8	3.5	26.1	1.5
1122C-2H-6–135	42.0	49.1	8.9	46.0	52.3	1.7	65.6	6.3	28.1	4.8	3.3	36.8	1.8
1122C-3H-5–130	36.7	50.2	13.1	41.7	53.5	4.8	72.2	2.8	25.0	10.3	11.0	17.1	1.8
1122C-4H-7–39	39.1	47.6	13.3	45.6	48.7	5.7	57.4	0.0	42.6	3.3	0.8	46.2	1.5
1122C-5H-6–96	40.7	52.0	7.3	43.0	52.7	4.3	95.8	0.0	4.2	9.3	2.0	40.5	3.8
1122C-6H-6–58	42.2	43.9	13.9	48.1	44.2	7.8	75.5	2.0	22.4	2.8	1.3	54.5	1.5
1122C-7H-6–18	41.3	51.9	6.9	44.1	52.5	3.4	76.9	7.7	15.4	3.5	1.5	28.6	1.0
1122C-8H-1–51	43.4	48.0	8.6	46.6	51.1	2.3	80.0	0.0	20.0	3.5	2.5	14.3	0.5
1122C-9H-1–13	49.5	41.8	8.6	53.9	43.1	3.1	85.7	0.0	14.3	11.3	0.5	24.4	2.8
1122C-9H-6–70	43.8	46.5	9.6	48.0	44.9	7.1	97.0	3.0	0.0	3.0	2.5	25.0	0.8
1122C-10H-6–84	49.5	39.8	10.7	54.4	41.7	3.9	94.3	0.0	5.7	6.3	3.8	24.0	1.5
1122C-11H-4–7	50.5	43.9	5.6	53.5	43.7	2.8	70.6	0.0	29.4	11.5	4.8	28.3	3.3
1122C-12H-1–10	48.0	43.0	9.0	51.0	44.1	4.9	44.4	30.6	25.0	16.3	3.3	26.2	4.3
1122C-13H-6–84	36.8	53.2	10.0	39.7	52.7	7.6	88.9	0.0	11.1	15.0	6.0	21.7	3.3
1122C-14X-1–106	48.7	44.7	6.6	52.0	35.8	12.2	52.4	0.0	47.6	5.8	2.3	8.7	0.5
1122C-15X-1–10	32.1	50.8	17.1	38.3	59.1	2.6	58.8	5.9	35.3	44.0	4.0	19.9	8.8
1122C-16X-1–74	37.4	49.6	13.0	43.4	53.1	3.5	50.0	5.6	44.4	53.5	6.0	13.6	7.3
1122C-17X-1–86	45.8	45.1	9.1	49.6	44.4	6.0	100.0	0.0	0.0	22.8	6.3	18.7	4.3
1122C-18X-3–33	43.6	45.2	11.2	49.1	47.4	3.5	60.5	2.6	36.8	6.8	3.0	37.0	2.5
1122C-19X-3–86	33.3	39.9	26.8	45.5	49.7	4.8	83.6	0.0	16.4	30.3	6.3	9.9	3.0
1122C-20X-1–6	37.6	50.2	12.2	42.6	50.4	7.0	55.6	0.0	44.4	7.5	2.3	23.3	1.8
1122C-21X-1–18	37.5	42.4	20.1	46.2	48.9	5.0	69.5	3.4	27.1	15.0	2.5	28.3	4.3
1122C-22X-1–128	43.0	46.8	10.2	47.5	48.7	3.8	65.6	6.3	28.1	14.0	2.3	10.7	1.5
1122C-23X-2–56	43.7	47.8	8.5	48.1	47.0	4.9	77.8	0.0	22.2	8.3	1.8	27.3	2.3
1122C-24X-1–65	48.7	40.4	10.9	53.6	41.1	5.3	39.5	0.0	60.5	4.0	1.0	56.3	2.3
1122C-26X-1–28	34.0	59.0	7.0	36.4	56.5	7.1	66.7	6.7	26.7	28.5	10.5	21.9	6.3
1122C-27X-2–133	41.8	48.2	9.9	45.4	51.8	2.8	75.0	0.0	25.0	11.8	2.8	38.3	4.5
1122C-30X-5–8	49.8	40.5	9.6	54.7	43.7	1.6	78.6	0.0	21.4	11.3	2.0	24.4	2.8
1122C-31X-1–40	45.3	43.2	11.5	51.7	44.5	3.8	69.2	2.6	28.2	9.5	1.3	31.6	3.0
1122C-32X-1–25	44.6	43.0	12.4	50.0	46.3	3.7	89.7	0.0	10.3	11.3	1.3	22.2	2.5
1122C-33X-4–107	35.9	53.0	11.0	42.4	50.0	7.6	50.0	8.8	41.2	5.3	5.5	38.1	2.0
1122C-34X-5–120	45.5	41.8	12.7	51.1	44.4	4.4	28.6	0.0	71.4	28.5	2.8	18.4	5.3
1122C-35X-1–92	49.1	45.3	5.7	52.0	48.0	0.0	33.3	0.0	66.7	29.8	46.0	10.1	3.0
1122C-36X-6–88	37.4	46.6	16.0	45.4	50.2	4.4	64.3	0.0	35.7	3.0	0.5	33.3	1.0
1122C-37X-6–92	42.1	40.5	17.4	50.5	45.5	4.0	40.7	22.2	37.0	58.8	4.0	8.5	5.0
1122C-39X-2–47	44.5	38.7	16.8	53.3	41.4	5.2	100.0	0.0	0.0	25.5	0.8	7.8	2.0
1122C-40X-2–128	47.9	33.1	19.0	57.8	38.4	3.8	65.0	1.7	33.3	9.8	1.0	25.6	2.5
1122C-42X-1–130	47.8	41.9	10.4	53.6	42.3	4.2	82.1	0.0	17.9	12.8	1.5	13.7	1.8
1122C-44X-4–56	47.1	36.6	16.3	55.3	42.4	2.3	68.5	0.0	31.5	6.0	0.8	16.7	1.0
1122C-45X-4–91	52.8	38.4	8.8	57.5	40.3	2.2	64.7	5.9	29.4	4.0	0.3	31.3	1.3
1122C-46X-4–134	42.0	54.0	4.0	43.7	52.1	4.2	50.0	0.0	50.0	4.3	0.8	23.5	1.0
1122C-47X-1–56	40.0	44.2	15.8	46.5	48.5	5.1	89.5	0.0	10.5	28.5	19.0	17.5	5.0
1122C-47X-4–77	50.1	43.7	6.2	53.2	45.3	1.5	100.0	0.0	0.0	1.0	2.3	0.0	0.0
1122C-48X-1–97	43.2	37.9	18.9	53.8	43.1	3.1	59.1	3.0	37.9	5.5	0.3	31.8	1.8
1122C-49X-4–15	52.5	37.4	10.2	56.6	41.2	2.2	89.7	6.9	3.4	18.3	6.0	20.5	3.8
1122C-51X-5–111	56.9	36.9	6.3	60.5	36.1	3.4	80.0	0.0	20.0	14.5	1.5	20.7	3.0
1122C-52X-1–93	44.0	39.0	17.0	51.4	46.9	1.7	81.6	2.6	15.8	27.3	11.3	10.1	2.8
1122C-56X-4–6	44.9	41.6	13.4	51.0	48.2	0.8	65.3	16.3	18.4	13.8	1.3	12.7	1.8
1122C-57X-3–19	41.4	31.5	27.0	55.7	43.7	0.6	45.2	3.2	51.6	36.5	1.3	6.8	2.5
1122C-58X-2–37	53.6	37.0	9.3	57.9	40.5	1.7	90.9	3.0	6.1	8.0	0.5	21.9	1.8
1122C-61X-3–43	49.8	45.9	4.2	51.4	46.9	1.6	69.2	15.4	15.4	5.5	18.5	22.7	1.3
1122C-62X-1–12	37.3	39.5	23.2	47.4	51.3	1.3	72.3	10.8	16.9	8.3	3.3	6.1	0.5
1122C-64X-1–130	46.5	35.8	17.7	55.6	43.7	0.8	69.0	3.4	27.6	6.3	5.0	24.0	1.5
1122C-65X-1–21	44.4	43.8	11.7	50.4	46.5	3.2	85.0	5.0	10.0	9.0	2.5	5.6	0.5
1122C-66X-1–71	39.5	34.3	26.2	54.0	44.6	1.5	30.7	0.0	69.3	0.8	4.0	33.3	0.3

Note: QFL%Q = 100 × Q/(Q + F + L); QFL%F = 100 × F/(Q + F + L); QFL%L = 100 × L/(Q + F + L); QmKP%Qm = 100 × Qm/(Qm + K + P); QmKP%K = 100 × K/(Qm + K + P); QmKP%P = 100 × P/(Qm + K + P); LmLvLs%Lm = 100 × Lm/(Lm + Lv + Ls); LmLvLs%Lv = 100 × Lv/(Lm + Lv + Ls); LmLvLs%Ls = 100 × Ls/(Lm + Lv + Ls); LsLmmLmt%Ls = 100 × Ls/(Ls + Lmm + Lmt); LsLmmLmt%Lmm = 100 × Lmm/(Ls + Lmm + Lmt); LsLmmLmt%Lmt = 100 × Lmt/(Ls + Lmm + Lmt); MuscoviteBiotiteChlorite%Muscovite = 100 × Muscovite/(M + B + Chl); MuscoviteBiotiteChlorite%Biotite = 100 × Biotite/(M + B + Chl); MuscoviteBiotiteChlorite%Chlorite = 100 × Chlorite/(M + B + Chl); MicaLmmLmt%Mica = 100 × Mica/(M + B + Chl + Malt + Lmm + Lmt); MicaLmmLmt%Lmm = 100 × Lmm/(M + B + Chl + Malt + Lmm + Lmt); MicaLmmLmt%Lmt = 100 × Lmt/(M + B + Chl + Malt + Lmm + Lmt); Total%M = 100 × (B + M + Chl)/400; Total%B = 100 × (Foram + Diat + Spong + Echino + Shell Frag + Other Bio)/400; Mica%Chl = 100 × Chl/(M + B + Chl + Malt); Total%Chl = 100 × Chl/400. Classifications are according to Garzanti and Vezzoli (2003).

TABLE 3. COMPARISION OF ONSHORE RIVER DATA WITH OCEAN DRILLING PROJECT SITE 1122

	QFL			QmKP			LmLvLs		
	Q (%)	F (%)	L (%)	Qm (%)	K (%)	P (%)	Lm (%)	Lv (%)	Ls (%)
Rangitata River[†]	41	6	53	84	5	11	86	0	14
Waitaki River[†]	33	16	51	64	8	28	82	0	18
Clutha River[†]	58	32	9	64	0	36	77	3	20
Site 1122 average[‡]	44	44	12	50	4	47	71	4	26

[†]Very fine sand fraction.
[‡]Very fine to fine sand.

whereas the Waitaki River drains mostly Torlesse terrane with minor schist. A sample from the Rangitata River to the north was included even though it has no distinct connection to the Bounty feeder channel system, because it represents a pure Torlesse provenance. The samples were taken from exposed sandy bars, where Highway 1 intersects each of the three rivers ~5–25 km upstream from the coast to minimize possible tidal influences. These samples were sieved, separating the coarse to very fine sand fractions. Thin sections were made, stained, point counted, and analyzed using the same methods as described already for the ODP samples. However, point-count data were collected only for the very fine sand fraction of each river sample in order to use the same grain size as the majority of Site 1122 samples.

RESULTS

Ocean Drilling Program Site 1122 Sand Samples

Sand Description

In general, the counted sand samples from Site 1122 are very similar texturally and compositionally. Individual grains range from very fine-grained to a maximum of medium-grained sand, with an average grain size of very fine-grained sand. The medium end member of the grain-size range occurs more commonly in the top one-third of the section, whereas very fine-grained sand dominates the lower two-thirds of the section. Sand grains range from angular to subangular. Sand components are dominantly monocrystalline quartz and feldspar, with moderate mica and minor metamorphic lithic fragments (Fig. 5; Table 2).

Quartz grains generally exhibit straight extinction, but some grains exhibit undulose extinction; they occur as monocrystalline grains and sand-sized grains in lithic fragments (Fig. 6). A small percentage of the quartz grains contains small inclusions of non-opaque dense minerals. Polycrystalline quartz (Fig. 5) makes up <7% of the quartz fraction total and is absent in some samples. Plagioclase/albite feldspar (Fig. 5) is present in significant amounts and occurs mostly as single monocrystalline grains, although some are present as sand-sized grains in metamorphic lithic fragments. Feldspar grains rarely exhibit albite twinning. Some plagioclase/albite grains are variably altered to clay minerals. Potassium feldspar is present (Fig. 5) but makes up generally <10% of the feldspar total, except in a few samples, where it reaches 25% of the feldspar total. Muscovite and biotite grains (Fig. 5) on average make up a moderate proportion (14% and 4%, respectively) of the grains in most samples. There is some alteration of biotite grains to opaque minerals and chlorite. Single monocrystalline grains of pale green chlorite (Fig. 5) are generally rare to moderately common, averaging 2% of the total grains.

The lithic fraction is dominated by low-grade to medium-grade metamorphic lithic fragments (Fig. 5). Most of the metamorphic lithic fragments are quartz-mica-tectonite, with lesser amounts of quartz-mica-feldspar aggregate, and even lesser amounts of polycrystalline mica. Finely crystalline mica grains within these lithic fragments are predominately muscovite. Less than 5% of the metamorphic lithic fraction is classifiable into one of the subcategories of Garzanti and Vezzoli (2003). This may be due to the very fine-grained nature of the Bounty Fan samples as compared to the coarser-grained examples used to define the Garzanti and Vezzoli classification scheme. Of the metamorphic lithic grain types, quartz-mica-tectonite fragments were most easily classified (e.g., fine- to medium-grained meta-siltstone/sandstone with quartz and mica [Lmf2]) using the Garzanti and Vezzoli scheme. Sedimentary lithic fragments such as shale and siltstone are present, but do not make up a significant amount of the total lithic fraction (<30% on average). Volcanic lithic fragments are very rare; these are limited to volcanic glass and microlitic volcanic lithics.

Bioclastic material (Fig. 5) is present in each slide in varying amounts. Calcareous bioclasts are most commonly foraminifers,

TABLE 4. OTHER RECALCULATED PARAMETERS

		LsLmmLmt			MicaLmmLmt			MuscoviteBiotiteChlorite		
		Ls (%)	Lmm (%)	Lmt (%)	Mica (%)	Lmm (%)	Lmt (%)	Muscovite (%)	Biotite (%)	Chlorite (%)
Site 1122 unit means										
Unit I[†]	Avg.	39.3	16.4	44.2	78.9	2.2	18.9	48.6	24.2	27.3
	St dev.	26.5	11.1	22.5	17.4	3.7	17.1	14.5	12.6	12.6
Unit II[†]	Avg.	44.6	17.7	37.7	78.1	1.9	20.0	53.1	26.7	20.2
	St dev.	24.8	10.3	27.5	17.9	2.8	18.1	19.6	24.3	12.5
Unit III[†]	Avg.	29.8	17.9	52.3	73.7	2.8	23.5	55.3	23.5	21.2
	St dev.	22.7	17.4	21.2	24.9	4.8	20.3	11.4	17.0	13.9
River Samples										
Rangitata[‡]		17	3	80	1	4	95	0	100	0
Waitaki[‡]		27	11	62	2	14	84	0	100	0
Clutha[‡]		41	24	35	55	18	27	60	20	20

[†]Sample sets.
[‡]Single sample.

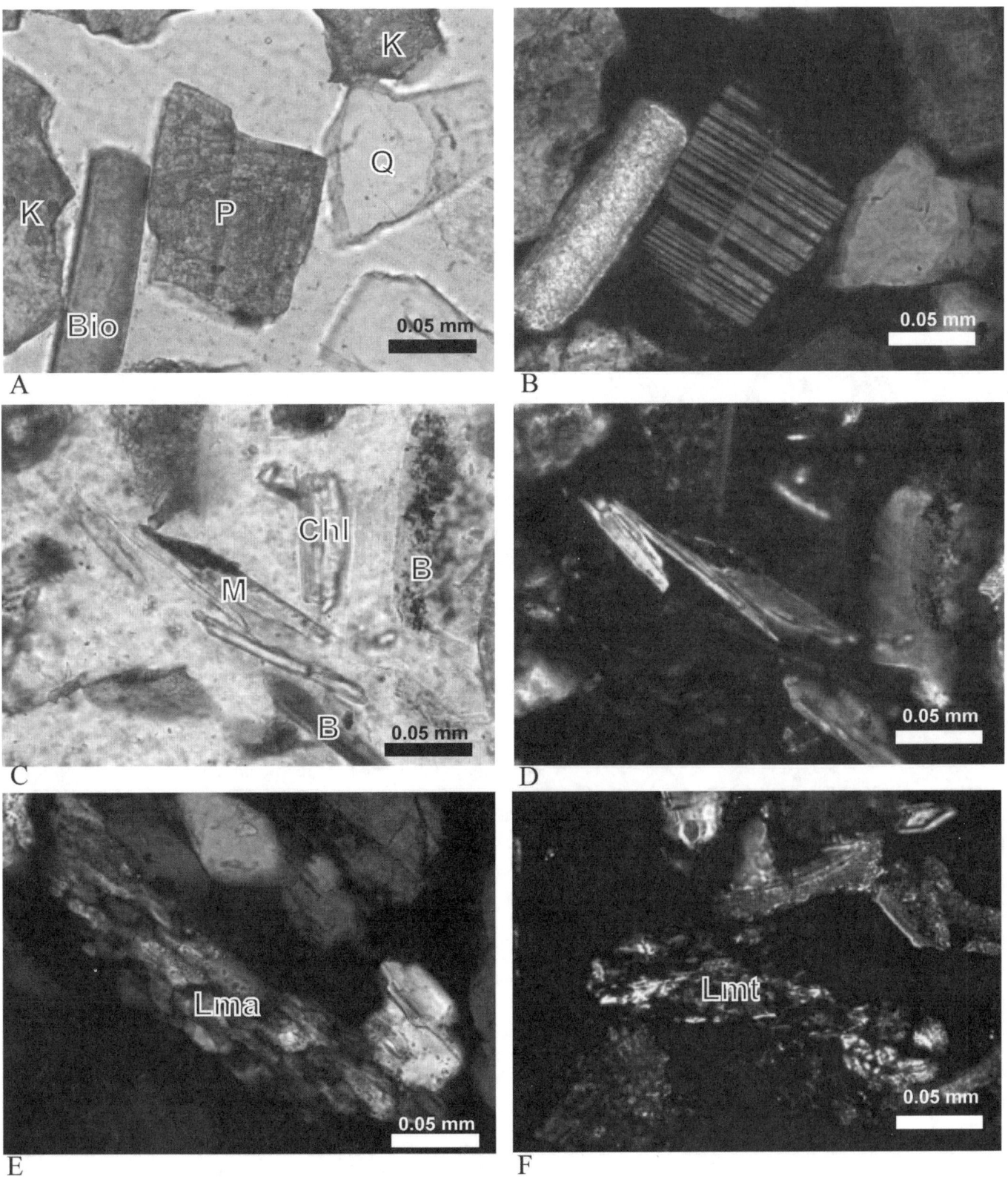

Figure 5. Representative photomicrographs of Ocean Drilling Program (ODP) Site 1122 samples. (A) View showing potassium feldspar (K), plagioclase/albite feldspar (P), quartz (Q) grains, and bioclast (Bio) under plane-polarized light. (B) Same view as in A but with nicols crossed. (C) View showing muscovite (M), biotite (B), and chlorite (Chl) grains under plane-polarized light. (D) Same view as in C but with nicols crossed. (E) View showing quartz-feldspar-mica aggregate (Lma) fragment under crossed nicols. (F) View showing quartz-mica-tectonite (Lmt) fragment with nicols crossed. Note: potassium feldspar takes both stains in these samples (artifact of staining technique?).

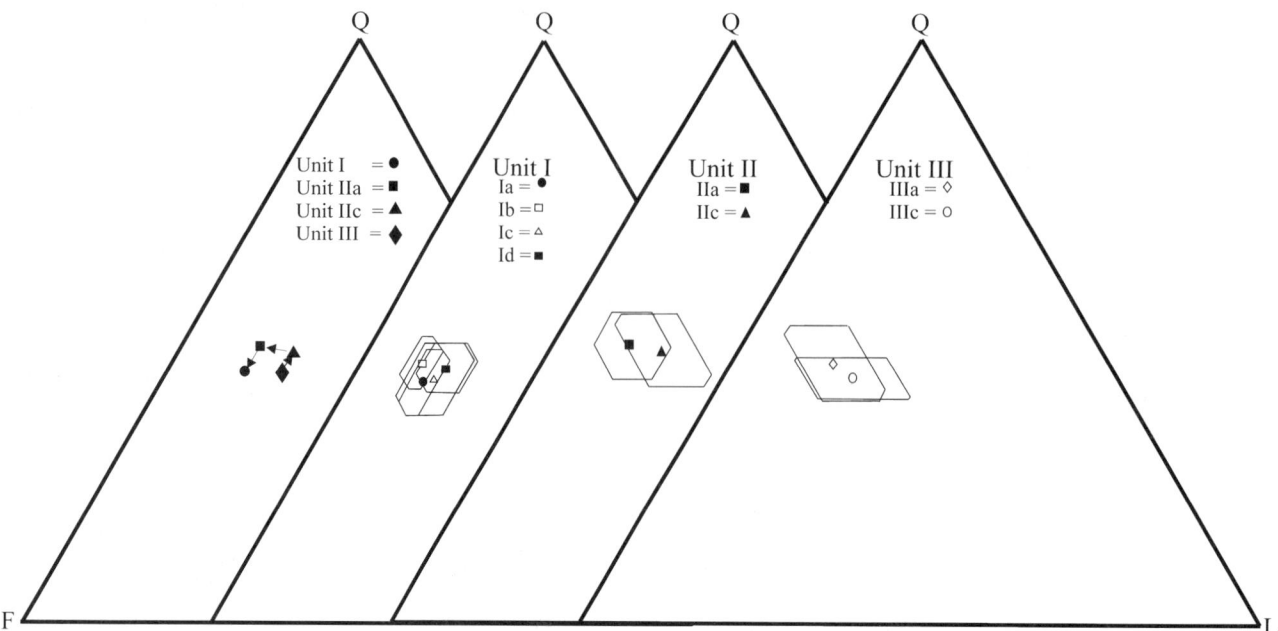

Figure 6. Diagram of the QFL (quartz-feldspar-lithics) ternary plots for Ocean Drilling Program (ODP) Site 1122. The three plots to the right show the average values for the subunits within each unit. The plot to the left shows the average for each unit. Subunit IIA lies above and IIC lies below an unconformity. Q, F, and L percentages are defined in Table 1 and presented in Table 2. Polygons are constructed using standard deviations.

and less frequently mollusca and echinoderm fragments. Siliceous bioclasts include sponge spicules, diatoms, and radiolarians. Carbonate of unknown affinity is present in moderate proportions. These grains may be bioclastic in origin, or perhaps carbonate cement or vein minerals. Also present in rare amounts is glauconite.

Dense minerals, both opaque and nonopaque, are rare, except in one sample (1122C-34X-5–120), where the opaque minerals were concentrated. The most common nonopaque dense mineral is epidote, with lesser amphibole, pyroxene, and zoisite.

Sand Detrital Modes

The range of sand compositions (Table 2) is shown in Figures 6, 7, and 8, where the data have been averaged by subunit, and Figures 9 through 12, where recalculated parameters are plotted versus sample depth (meters below seafloor [mbsf]). There is very little variation of the subunit averages on the QFL (quartz-feldspar-lithics) ternary plot (Fig. 6), except that there are, on average, more lithic fragments present in unit III and subunit IIC. Most, if not all, of the samples are quartzofeldspathic or lithic arkoses in the terminology of Folk (1968). The QFL%Q percentages vary greatly within subunits, ranging from 32.1% to 56.9% (Fig. 9). The downhole distribution of QFL%F roughly mirrors that of QFL%Q. The QmKP (monocrystalline quartz-potassium feldspar-plagioclase feldspar) ternary plot (Fig. 7) shows how poor the samples are in potassium feldspar. The samples exhibit subequal amounts of monocrystalline quartz and plagioclase/albite, and again subunit means are fairly similar. QmKP%K values are variable, ranging from 0% in subunit IC to a high of 12.2% in subunit IB (Fig. 10). Figure 8 shows the dominance of metamorphic fragments in the lithic population, with varying amounts of sedimentary fragments and only a minor volcanic component.

There are no distinct downhole trends in the lithic proportions (Table 2). Metamorphic lithic fragments are present in all the samples, and they are the dominant lithic type in most samples. LmLvLs%Lm (metamorphic lithics-volcanic lithics-sedimentary lithics) values range from a low of 28.6% in subunit ID to a maximum of 100% in subunits IC, ID, and IIA. The largest amount of volcanic fragments occurs in subunit IB, which has a total of 30.6%. Numerous samples scattered throughout the section contain no volcanic lithic fragments. In general, there are higher percentages of volcanic fragments in the lower part of the section, below the unconformity (Fig. 12). Sedimentary lithic fragments are slightly more abundant in the middle of the section and have a downhole distribution that mirrors that of the metamorphic fraction (Shapiro, 2004).

Bioclasts, although present in significant amounts, show no major overall trends in percentage with depth. There are several samples with high bioclast percentages, the largest with a value of 46% at ~320 m, and another five between 10% and 20% at ~560 m, 420 m, 465 m, 210 m, and 15 m. For the most part, these high values appear to occur in semiregular depth intervals (~100–150 m), with one missing at about the 110 mbsf where there is a spike in mica content.

In general, mica (muscovite, biotite, and chlorite) is less than 20% of the overall total grains, but in a few sections, it represents

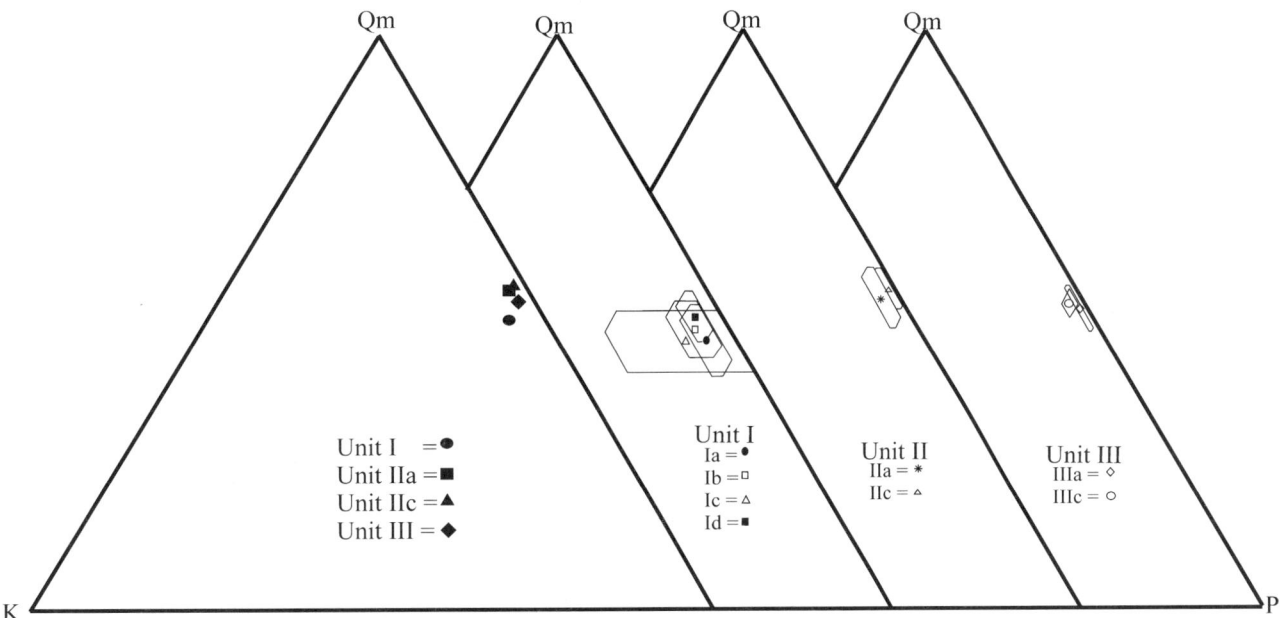

Figure 7. Diagram of the QmKP (monocrystalline quartz-potassium feldspar-plagioclase feldspar) ternary plots for Ocean Drilling Program (ODP) Site 1122. The three plots to the right show the average values for the subunits within each unit. The plot to the left shows the average for each unit. Subunit IIA lies above and IIC lies below the unconformity. Qm, K, and P percentages are defined in Table 1 and presented in Table 2. Polygons are constructed using standard deviations.

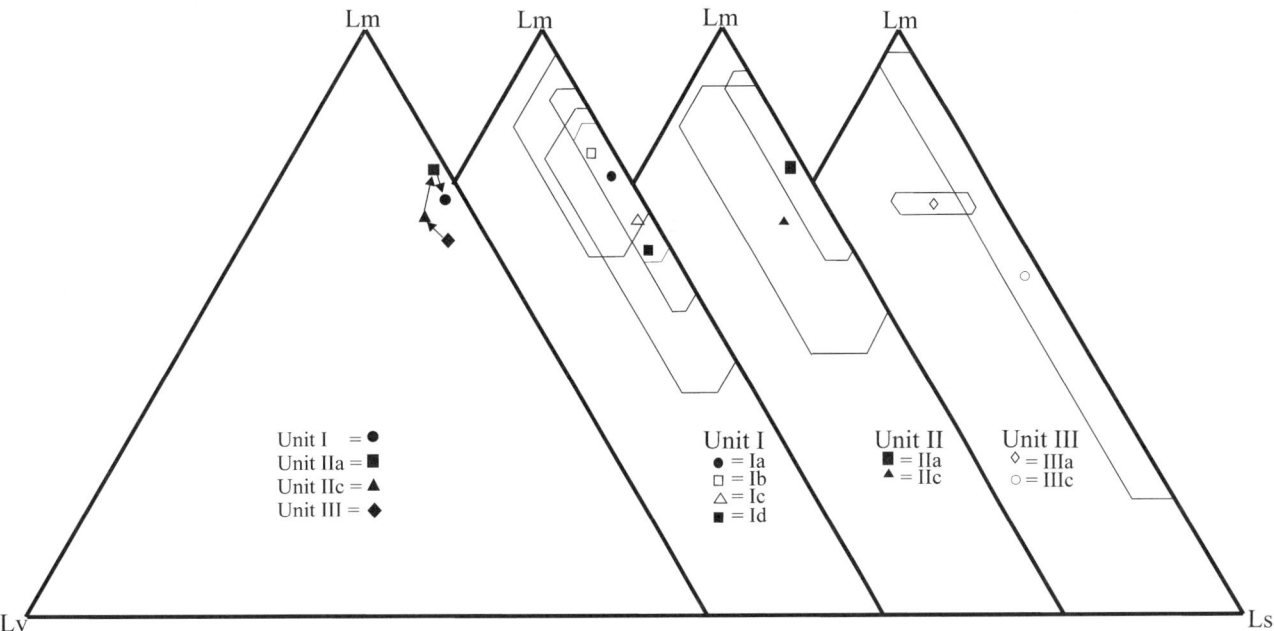

Figure 8. Diagram of the LmLvLs (metamorphic lithics-volcanic lithics-sedimentary lithics) ternary plots for Ocean Drilling Program (ODP) Site 1122. The three plots to the right show the average values for the subunits within each unit. The plot to the left shows the average for each unit. Subunit IIA lies above and IIC lies below the unconformity. Lm, Lv, and Ls percentages are defined in Table 1 and presented in Table 2. Polygons are constructed using standard deviations.

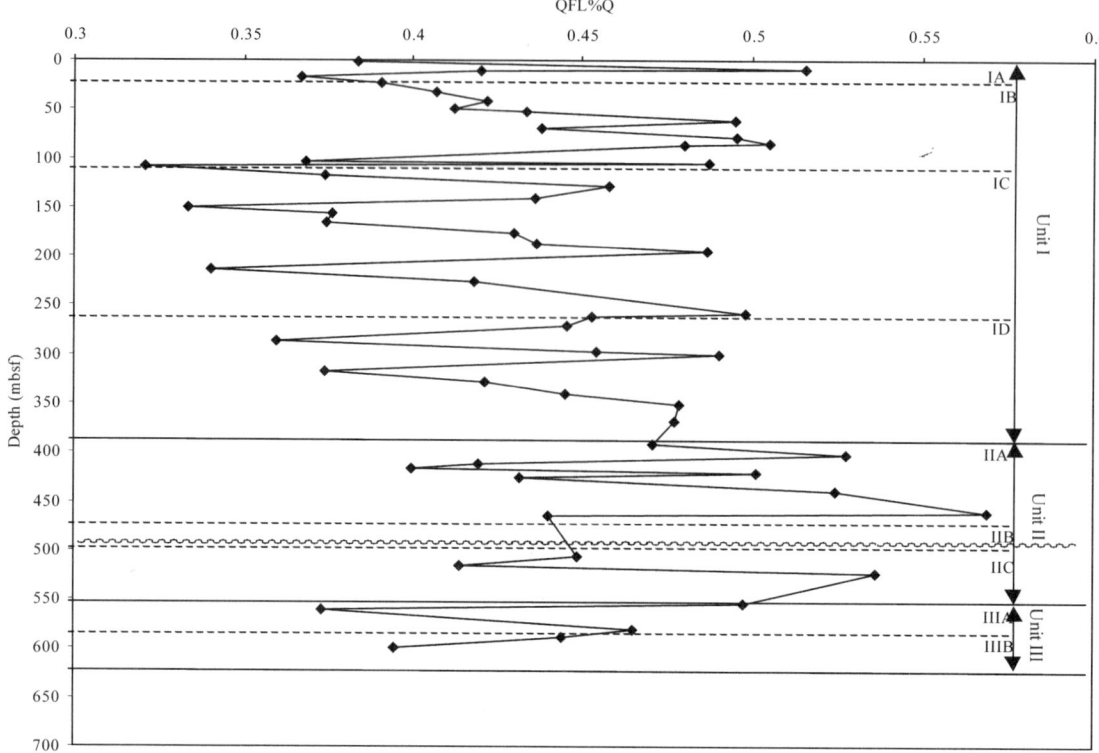

Figure 9. Downhole plot of QFL%Q. Arrow in upper 100 m outlines trend in unit I. The major unconformity is represented by the wavy line just above 500 meters below seafloor (mbsf). Dark solid lines indicate unit boundaries, whereas subunit boundaries are indicated by dashed lines. Data are presented in Table 2.

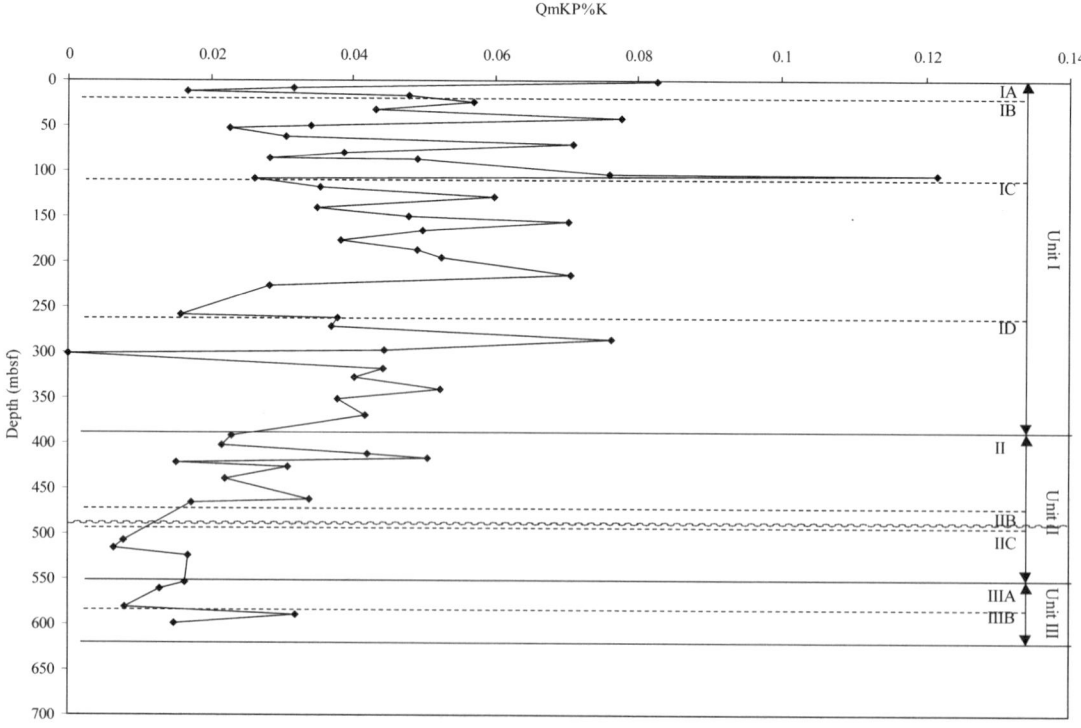

Figure 10. Downhole plot of QmKP%K. The major unconformity is represented by the wavy line just above 500 meters below seafloor (mbsf). Dark solid lines indicate unit boundaries, whereas subunit boundaries are indicated by dashed lines. Data are presented in Table 2.

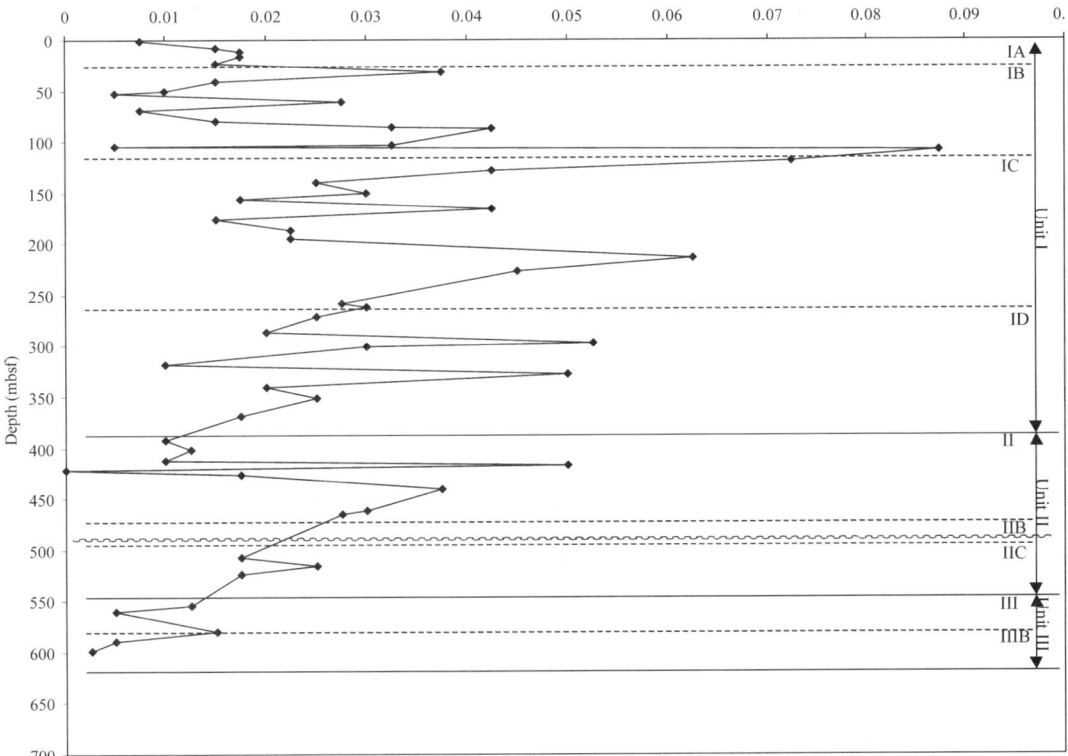

Figure 11. Downhole plot of total%chlorite. The major unconformity is represented by the wavy line just above 500 meters below seafloor (mbsf). Dark solid lines indicate unit boundaries, whereas subunit boundaries are indicated by dashed lines. Data are presented in Table 2.

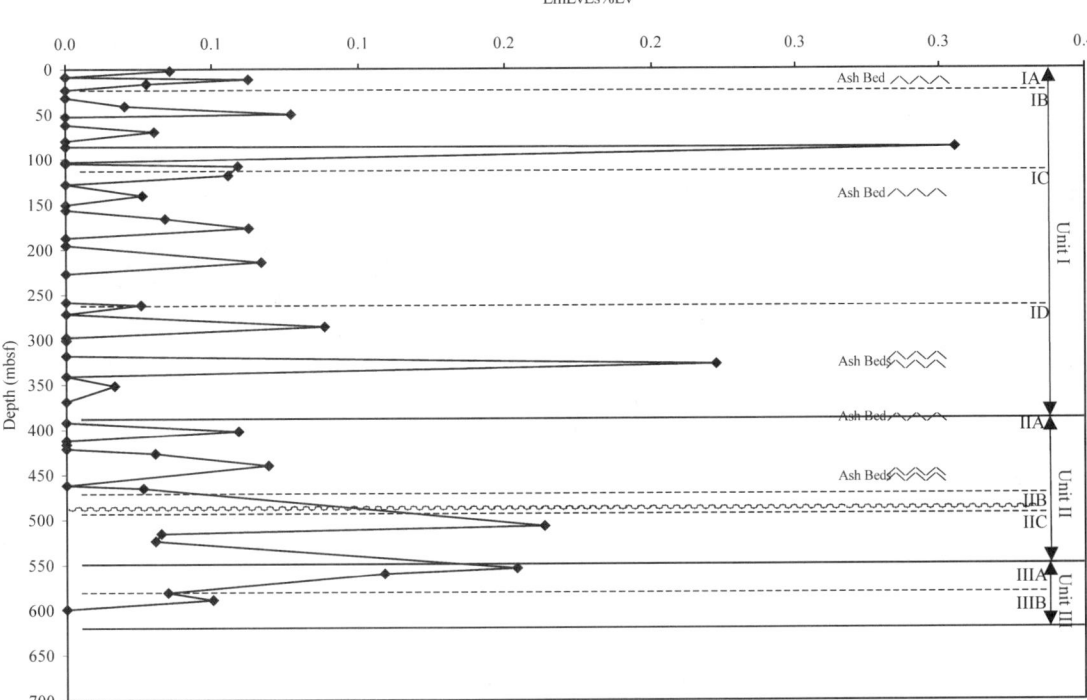

Figure 12. Downhole plot of LmLvLs%Lv. Ash (tuff) bed intervals are from Shipboard Scientific Party (1999b). The major unconformity is represented by the wavy line just above 500 meters below seafloor (mbsf). Dark solid lines indicate unit boundaries, whereas subunit boundaries are indicated by dashed lines. Data are presented in Table 2. Note that only one ash bed (327.92 mbsf; sample 1122C-37X-6 cm) corresponds to a volcanic-rich sand interval.

a significant (>50%) amount of the sand fraction of the sample. A few other samples are in the 30% to 40% mica range. There is no overall trend in mica percentage with depth. However, Figure 11 shows a gradual increase in chlorite (total%chl) from the bottom of the section up through subunit IIA.

Relationships among Grain Size, Bed Thickness, and Composition

Given the possibility of multiple sand sources, compositional trends in the samples were evaluated by grain size and bed thickness, with the idea that coarser and/or thicker beds may have had a unique (more proximal?) source that would provide a distinct compositional fingerprint. As stated earlier, visual estimates of average grain size indicate that the Site 1122 samples are very fine-grained sand and fine-grained sand, and the dominant grain size is very fine-grained sand. Sand bed thicknesses, as determined from shipboard measurements (Table 3 in Shipboard Scientific Party, 1999b) or estimated from core photographs, are more variable, ranging from a few centimeters to decimeters in scale. Samples were organized by bed thickness, and averages were calculated in ascending order of bed thickness; bed thickness was also compared to composition and grain size.

River Sand Samples

Very fine sand fractions of samples from the Clutha, Waitaki, and Rangitata Rivers were petrographically examined in this study (Table 3). The Clutha River sample is quartzofeldspathic, dominated by quartz grains, with minor amounts of feldspar and lesser amounts of lithic grains (Fig. 13). The Waitaki and Rangitata River samples are quartzolithic, dominated by metamorphic lithic fragments with moderate amounts of quartz grains and minor amounts of feldspar grains (Fig. 13). Quartz and feldspar grains in the three samples are subangular to angular, whereas lithic fragments are subangular to subrounded.

Most of the quartz grains in the river samples exhibit straight extinction, but some grains exhibit undulose extinction. Polycrystalline quartz is rare. Individual quartz grains are rare in the Rangitata River sample. Plagioclase/albite feldspar is more common than potassium feldspar in all river samples. The Rangitata and Waitaki Rivers contain higher proportions of quartz and potassium feldspar, whereas the Clutha River has a lower proportion of potassium feldspar (Fig. 13). Muscovite as individual grains is nearly absent in the Rangitata River sample, but it is more common in the Clutha River. Dense minerals are rare, except for epidote, which is common in the Clutha River sample.

The lithic fractions are dominated by low- to medium-grade metamorphic fragments (Fig. 13). These metamorphic fragments are predominantly quartz-mica tectonite, with moderate content of quartz-feldspar-mica aggregate and minor polycrystalline mica. Owing to their very fine-grained size, <5% of the metamorphic grains were classifiable into one of the subcategories of Garzanti and Vezzoli (2003). The Clutha River has the lowest metamorphic lithic proportions.

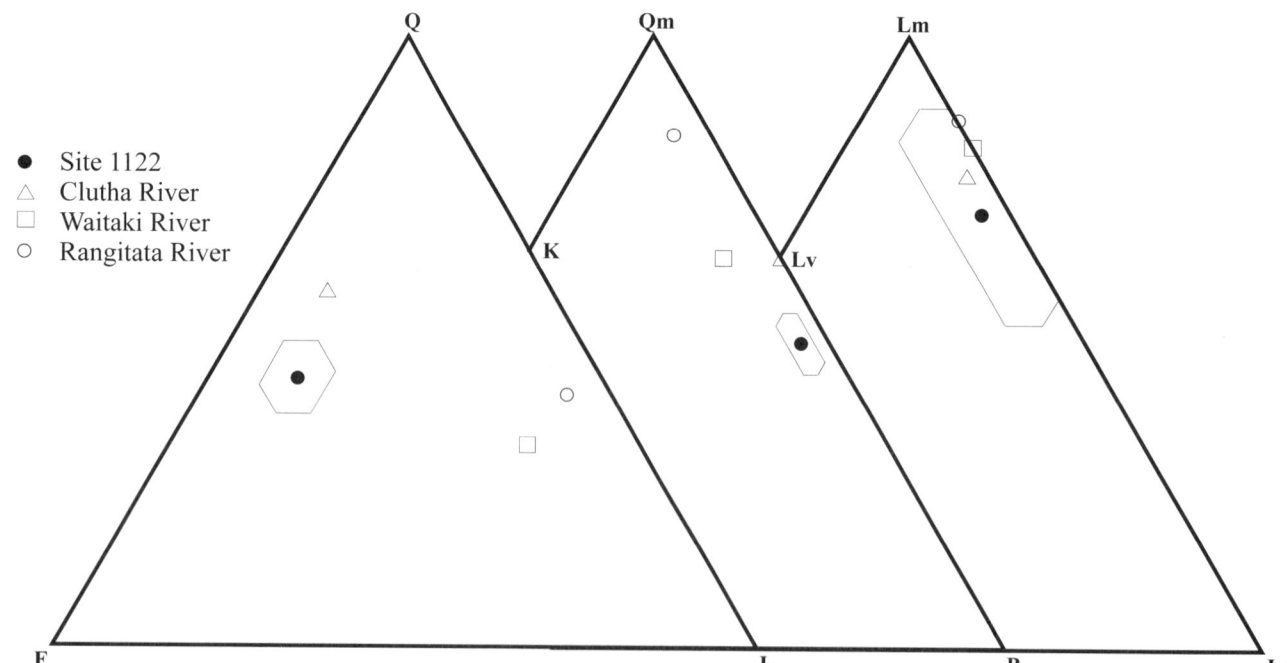

Figure 13. Ternary plots comparing onshore river sand data with Ocean Drilling Program (ODP) Site 1122C data. See Table 1 for parameter definitions and Tables 2 and 3 for data. Polygon for Site 1122 data was constructed using one standard deviation on either side of mean for all three parameters.

Metamorphic Lithic and Mica Proportions

Several ternary plots were constructed from data in Tables 3 and 4 to highlight trends relating to differences in metamorphic grade (mineralogy and texture) of the detritus comprising the Site 1122 and river sand samples (Fig. 14). With respect to the proportion of sedimentary, polycrystalline mica, and quartz-mica-tectonite fragments, Site 1122 unit means all plot fairly close to each other, with no major trends or differences in composition. Unit III contains the highest proportion of sedimentary lithics. The Clutha River sample plots within the field of variation for Site 1122 units I and II. The Waitaki and Rangitata Rivers have higher concentrations of quartz-mica-tectonite lithics.

Site 1122 samples contain high proportions of mica, moderate amounts of quartz-mica-tectonite, and minor amounts of polycrystalline mica fragments (Fig. 14). All mean values plot within the fields of variation for each unit. The Rangitata and Waitaki Rivers contain a very high proportion of quartz-mica-tectonite, moderate proportions of polycrystalline mica, and very low proportions of mica. None of the river samples plots with the fields of variation for Site 1122 units; however, the Clutha River plots closest to the Site 1122 samples. The concentration of mica within the sand is likely related to the hydrodynamic properties of mica flakes, rather than a provenance signal. However, the proportion of micaceous minerals (including chlorite) may retain more provenance information. The muscovite, biotite, and chlorite proportions are plotted in Figure 14. Site 1122 units contain mostly muscovite with less, but nearly equal proportions of biotite and chlorite. The Rangitata and Waitaki River samples are dominated by biotite, whereas the Clutha River plots within the fields of variation for the Site 1122 samples, but with a slightly higher proportion of muscovite.

DISCUSSION

Compositional Trends

The ternary plots in Figures 6, 7, 8, and 14 demonstrate that there are no major compositional differences among sand samples from the units and subunits at Site 1122. The subunit and unit mean values are generally clustered and all fall within the fields of variation for adjacent subunits. This suggests a relatively uniform source of detritus (sand) for the Bounty Fan at Site 1122.

Relationships among sand bed thickness, grain size, and composition (e.g., Marsaglia et al., 1996) can indicate multiple sources of sediment. However, at Site 1122, there are no significant differences in distribution among the fine-grained sand and very fine-grained sand samples with respect to bed thickness. Compositional distributions for the very fine-grained versus fine-grained samples are similar except that the very fine-grained sand shows a wider distribution with lower LmLvLs%Lm values. Additionally, sand composition does not show a distinct relationship to bed thickness except for a tendency for the thicker beds

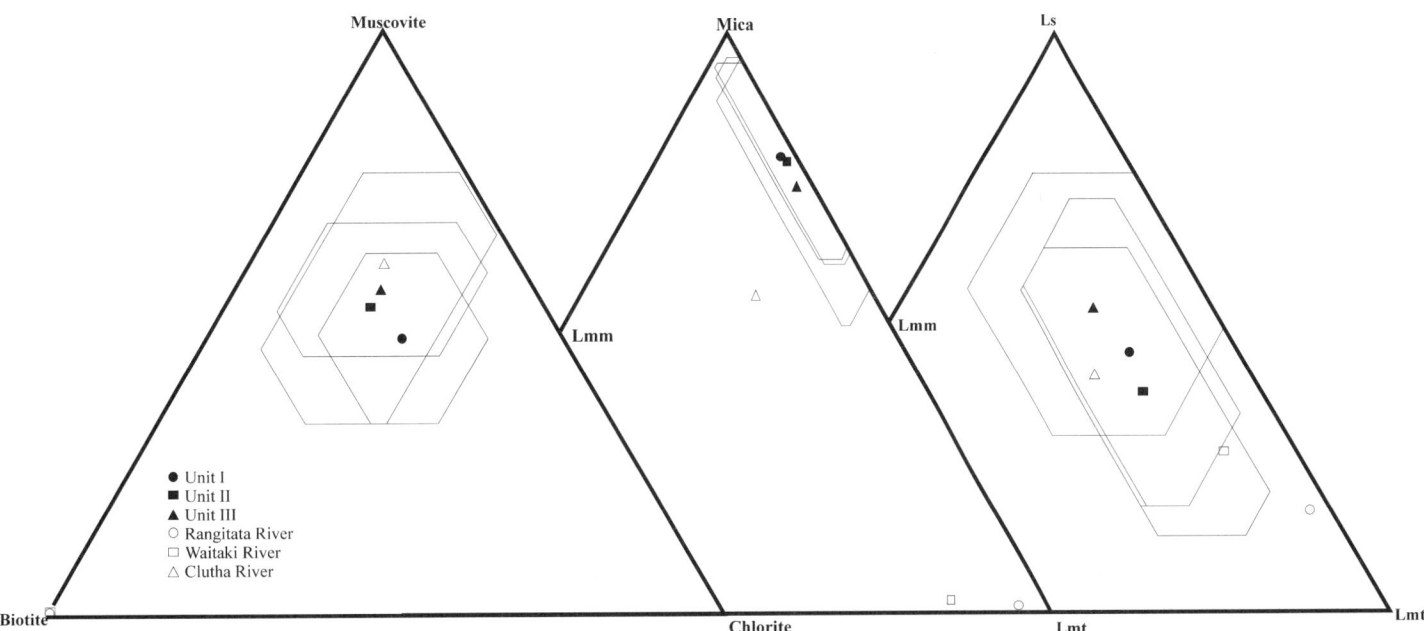

Figure 14. Ternary plots of metamorphic (polycrystalline mica and quartz-mica-tectonite) and sedimentary lithic proportions, metamorphic components (mica, polycrystalline mica, and quartz-mica-tectonite), and muscovite, biotite, and chlorite for sand samples from onshore rivers and Ocean Drilling Program (ODP) Site 1122 units (means). See Table 1 for definitions. Polygons for Site 1122 data were constructed using one standard deviation on either side of mean for all three parameters.

to contain more metamorphic rock fragments (LmLvLs%Lm > 50%). These data also support a uniform, relatively homogeneous source of the Bounty Fan sediment.

Sediment Sources

As discussed previously, there are three possible sediment sources for the Bounty Fan: the onshore rivers, near fan sources, and deep ocean currents. Additional input may have come from volcanic eruptions. Each of these source options is discussed here.

River and Shelf Input

Three major river systems, the Clutha, Waitaki, and Rangitata, drain the eastern side of the southern half of South Island (Fig. 4). Note that the Clutha and Waitaki Rivers are known sources for shelf sediment near the feeder canyons for the Bounty Channel (Carter et al., 1985; Carter and Carter, 1990), so they can be considered as proxies to the shelf sands, which were not analyzed in this study. Although the Rangitata River does not feed sediment into the Bounty system, it provides an important provenance end member for comparison to the Clutha and Waitaki Rivers because each of these three rivers has a drainage basin characterized by different types and proportions of rock units (Figs. 4): the Rangitata River has a drainage basin composed of unmetamorphosed Torlesse rocks (100% sedimentary); a mix of Torlesse and metamorphosed Torlesse (Otago-Haast Schist) comprise the drainage basin for the Waitaki River (25% metamorphosed, 75% sedimentary); and the Clutha River drains an area that is predominantly composed of metamorphosed Torlesse (Otago-Haast Schist) rocks (95% metamorphosed, 5% sedimentary).

Sand transported by these rivers reflects the geology of the respective drainage basins. Torlesse sandstones are feldspatholithic and contain metamorphic, volcanic, and sedimentary lithic fragments (MacKinnon, 1983; Mortimer, 1993; Adams et al., 1998). Sand derived from the Torlesse rocks (Rangitata and Waitaki River samples) is quartzolithic, with grains of quartz and plagioclase/albite feldspar, and fragments of low-grade metamorphic lithics (Figs. 13 and 14). The latter may be recycled lithic components from the Torlesse meta-sandstones, or metasedimentary fragments derived from Torlesse meta-argillite interbeds. In contrast, the Clutha River composition reflects the mineralogy of the Otago-Haast Schist, which includes quartz, albite, muscovite, biotite, chlorite, and epidote (Il et al., 2000; McMillian and Wilson, 1997; Mortimer and Roser, 1992), as well as vein-filling minerals such as quartz, chlorite, and carbonate (Smith and Yardley, 1999).

The similarity of the monocrystalline and lithic proportions of the Site 1122 and Clutha River samples (Figs. 13 and 14) suggests that the Clutha River may be the major river source of Bounty Fan sand. The Waitaki River, like the Clutha River, has a high sand content (~33%) and a prominent submarine channel leading to the Bounty Fan (Carter and Carter, 1993), so there is a likelihood of some sediment mixing from the two sources within the Bounty Channel. However, the distinct differences in QFL (Fig. 13) and mica and lithic (Fig. 14) proportions suggest that, on average, the Waitaki River has less of an influence on Bounty Fan sand composition than the Clutha River. Recent studies of Clutha River sediment geochemistry (Martin, 2005, personal commun.) are consistent with the compositional analysis reported herein (mineralogy dominated by quartz and albite and mica with some chlorite) and indicate that the medium to very fine sand fractions are relatively similar in composition along the lower reaches (100 km) of the Clutha River. Martin found distinct compositional differences among sediment size fractions, so our choice of the very fine size fraction from the Clutha River sample to compare to the Site 1122 samples was prudent. Possible reasons for the slight differences between our modern Clutha sample and the Site 1122 samples are: inherent compositional variability within the fluvial system, differences in size and/or geology between modern and ancient river drainage basins, effects of modern dams on the river sediment flux and composition, compositional sorting from the fluvial down through the submarine system, and mixing of sediment from different sources including those discussed next.

Deep Ocean Current and Near-Fan Input

Another potential source of sand at Site 1122 is material transported northward by the DWBC from the "Campbell drift" (Fig. 1). The Shipboard Scientific Party (1999b, 1999c) linked green-hued sediments rich in bright-green chlorite in subunit IIB at Site 1122 (Fig. 3) to a scoured sediment source of the "Campbell drift" drilled at the base of the Campbell Plateau (Site 1121; Fig. 1). They suggested unroofing of chlorite-bearing metamorphic units on the South Island as an alternate source for this chlorite, which, given the chlorite proportions in the Clutha River (Fig. 14), are a likely source. Data collected herein suggest a progressive increase in percentage of sand-sized chlorite grains from unit III and subunit IIC across the 6 m.y. unconformity into subunit IIA (Fig. 11). On average, unit I sand has a higher chlorite content expressed as a proportion of the total mica than sand in units II or III (Fig. 14; Table 4).

Lastly, sand samples in units II and III are slightly more bioclastic, which the Shipboard Scientific Party (1999a) attributed to reworking and winnowing by the deep ocean currents or by submarine mass wasting off the Campbell Plateau and Chatham Rise. The sand intervals rich in bioclasts could be shallower sediments enriched in bioclasts that were carried downslope by turbidity currents, or debris flows associated with some catastrophic event(s). Alternatively, they could represent periods of increased biologic productivity, but there is a mix of deep and shallower water (mid-shelf and nearshore) species of microfossils in these beds, which supports redeposition by turbidites or debris flows (Shipboard Scientific Party, 1999b). A detailed micropaleontological study of the types of bioclasts and their ages within the sand beds might help explain their occurrence.

Input of Volcanic Ash

There is a greater amount of volcanic lithics, mainly glass, in units II and III, although unit I does have a few significant spikes in percentage of volcanic components (Fig. 12). This volcanic debris may represent ash fall and settling in a submarine environment, or recycled vitric ash components from onshore drainage basins. Carter et al. (2003) showed that the volcanic ash in units I and II is from rhyolitic eruptions in the Coromandel and Taupo volcanic zones on North Island, New Zealand, that began ca. 12 Ma and 1.7 Ma, respectively. Miocene mafic volcanic centers on the Banks and Otago Peninsula (New Zealand Geological Survey, 1972) are alternative sources for the volcanic debris in unit III. Carter et al. (2003) listed ash beds recovered during Leg 181; only one ash bed correlates with a nearby sand bed (20 cm below the ash bed) rich in volcanic glass (see Shapiro, 2004), but given that the sand beds were randomly sampled for this study, this is not surprising.

Age of the Bounty Fan

According to Carter et al. (2004), the sedimentary record at Site 1122 suggests that: (1) at ca. 16 Ma, the Bounty Fan had not yet advanced to the present location, (2) sedimentation from 16 Ma to 10.4 Ma was dominated by deposition of contourites under a strong abyssal flow, which may have imported sediment from the south or reworked previously deposited sediment, (3) an unconformity from 10.4 Ma to 3.5 Ma is related to an increase in deep ocean currents that scoured the bottom, erasing the sedimentary record, and (4) at ca. 3.5 Ma, sediment made its way from the west down the Bounty Channel via turbidites, ultimately reaching the Deep Western Boundary Current (DWBC) and forming the Bounty Fan.

Compositional data presented herein suggest that Site 1122 had a relatively uniform source of sandy sediment throughout most of its history. This supports the idea of a long-lived Bounty "Channel" (Carter and Carter, 1987), but perhaps with a limited Bounty "Fan" (Shipboard Scientific Party, 1999b; Carter et al., 2004) at Site 1122 prior to the unconformity (ca. 3.5 Ma). Arguments for a limited western supply of sediment to the Bounty Fan prior to 6 Ma come from Deep Sea Drilling Project (DSDP) Site 594, located west of ODP Site 1122, on the southern flank of the Chatham Rise (Fig. 1; Shipboard Scientific Party, 1986). At Site 594, the first terrigenous influx (silty and clay) into the pelagic section occurred at ca. 6 Ma (Shipboard Scientific Party, 1986) at the inception of uplift of the Southern Alps. It is conceivable that sandy sediment from the Clutha River could have been funneled into the southerly feeder channels of the Bounty Channel prior to 6 Ma while pelagic sediment accumulated to the north at Site 594. The influx of muddy sediment at Site 594 might instead reflect input from the Waitaki River. Unfortunately, the clay mineral assemblages (Robert and Acquivia, 1986) within Site 594 muddy sediment, which include stilpnomelane, a low-grade metamorphic clay mineral, are consistent with derivation from rocks exposed (New Zealand Geological Survey, 1972) in either the Clutha or the Waitaki River basins.

In reviewing the seismic-reflection line across Site 1122 (Fig. 2), there is further evidence for a Bounty "Channel" during deposition of units II and III. Reflectors R7 and R5 in the seismic section suggest a series of stacked channels, however, filled down to at least reflector R9 (box, Fig. 2). This would then indicate that turbidity currents reached the area for most, if not all, of the time represented by the recovered section at Site 1122. This suggests that contour currents were more of a reworking mechanism rather than a primary depositional input mechanism during deposition of units II and III. However, this does not rule out short-term changes in deep-sea conditions, such as those suggested for subunit IIB by the Shipboard Scientific Party (1999b). Perhaps the pre–3.5 Ma section might be best described as a Bounty Channel–fed contourite succession, which evolved into the Bounty submarine fan system. We suggest that the mechanism for this change from contourite to turbidite facies at Site 1122 may have been linked to the rising Southern Alps, which resulted in a higher influx of sediment (40 cm/k.y.) at ca. 0.9 Ma that continues to the present (Carter et al., 1999). The relative importance of glacio-eustatic versus tectonic influences on sediment input is less clear. For example, turbidite deposition continued on the Bounty Fan during both glacial and interglacial periods, but turbidite frequency increased during glacial periods (Carter et al., 1999).

CONCLUSIONS

Quartzofeldspathic Miocene to Pleistocene sand recovered at Site 1122 is relatively uniform in composition with no major compositional differences among mean unit and subunit detrital modes. Also, there are no significant correlations among grain size, bed thickness, and composition except that there is a tendency for thicker beds to incorporate slightly more metamorphic rock fragments. This similarity in composition and lack of correlation among grain size, composition, and bed thickness suggest a relatively homogeneous source for Site 1122 sand. Similarity of the monocrystalline and lithic proportions between Site 1122 samples and the Clutha River sample suggests that it is/was likely a major source of sand at Site 1122. Although the Waitaki and Clutha River samples are both similar to Site 1122 sand in terms of QmKP and LmLvLs proportions, significant differences in QFL, mica, and lithic proportions between the Waitaki and the Site 1122 sands suggest that the average Site 1122 sand composition could not be easily explained by subequal mixing of sand from these two rivers. Shipboard analyses suggest some southerly input of sediment by deep ocean currents, but we found that the upsection change from sandy contourite to turbidite deposits at Site 1122 is not reflected in sand composition. This suggests that with respect to the sand fraction, the deep ocean currents appear to have been mostly a reworking mechanism or means of sediment removal (unconformity), rather than direct suppliers of sand to Site 1122.

ACKNOWLEDGMENTS

The manuscript was improved in response to reviews by Mark Johnsson, María Ochoa Rodríguez, and Richard Smosna.

REFERENCES CITED

Adams, C.J., Campbell, H.J., Graham, I.J., and Mortimer, N., 1998, Torlesse, Waipapa and Caples suspect terranes of New Zealand: Integrated studies of their geological history in relation to neighboring terranes: Episodes, v. 21, no. 4, p. 235–240.

Bishop, D.G., 1974, Stratigraphic, structural and metamorphic relations in the Dansey Pass area, Otago: New Zealand Journal of Geology and Geophysics, v. 17, p. 301–355.

Bishop, D.G., Bradshaw, J.D., and Landis, C.A., 1985, Provisional terrane map of South Island, New Zealand, in Howell, D.G., ed., Tectonostratigraphic Terranes of the Circum-Pacific Region: American Association of Petroleum Geologists Circum-Pacific Council for Energy and Mineral Resources Earth Science Series 1, p. 515–521.

Carter, L., and Carter, R.M., 1990, Lacustrine sediment traps and their effect on continental shelf sedimentation—South Island, New Zealand: Geo-Marine Letters, v. 10, p. 93–100, doi: 10.1007/BF02431026.

Carter, L., and Carter, R.M., 1993, Sedimentary evolution of the Bounty Trough: A Cretaceous rift basin, southwestern Pacific Ocean, in Balance, P.F., ed., South Pacific Sedimentary Basins: Sedimentary Basins of the World 2: Amsterdam, Elsevier Science Publishers, p. 51–67.

Carter, L., and McCave, I.N., 2002, Eastern New Zealand drifts, Miocene-Recent, in Stow, D.A.V., et al., eds., Deep-Water Contourite Systems: Modern Drifts and Ancient Series, Seismic and Sedimentary Characteristics: Geological Society [London] Memoir 22, p. 385–407.

Carter, L., and Mitchell, J.S., 1987, Late Quaternary sediment pathways through the deep ocean, east of New Zealand: Paleoceanography, v. 2, p. 409–422.

Carter, L., Shane, P.A., Alloway, B.A., Hall, I.R., Harris, S., and Westgate, J., 2003, A demise of one volcanic zone and birth of another—A 12 Ma marine record of major rhyolitic eruptions from New Zealand: Geology, v. 31, p. 493–496, doi: 10.1130/0091-7613(2003)031<0493:DOOVZA>2.0.CO;2.

Carter, L., Carter, R.M., and McCave, I.N., 2004, Evolution of the sedimentary system beneath the deep Pacific inflow off eastern New Zealand: Marine Geology, v. 205, p. 9–27, doi: 10.1016/S0025-3227(04)00016-7.

Carter, R.M., and Carter, L., 1987, The Bounty Channel system: 55-million-year-old sediment conduit to the deep sea, southwest Pacific Ocean: Geo-Marine Letters, v. 7, p. 183–190, doi: 10.1007/BF02242770.

Carter, R.M., and Carter, L., 1996, The abyssal Bounty Fan and lower Bounty Channel: Evolution of a rifted-margin sedimentary system: Marine Geology, v. 130, p. 181–202, doi: 10.1016/0025-3227(95)00139-5.

Carter, R.M., Carter, L., Williams, J.J., and Landis, C.A., 1985, Modern and relict sedimentation on the South Otago continental shelf, New Zealand: New Zealand Oceanographic Institute Memoir 93, p. 43.

Carter, R.M., Carter, L., and McCave, I.N., 1996, Current controlled sediment deposition from the shelf to the deep ocean: The Cenozoic evolution of circulation through the SW Pacific gateway: Geologische Rundschau, v. 85, p. 438–451.

Carter, R.M., McCave, I.N., Richter, C., Carter, L., et al., 1999, Proceedings Ocean Drilling Program, Initial Reports, Leg 181: College Station, Texas, Ocean Drilling Program [CD-ROM].

Chinn, T.J., 1996, The Southern Hemisphere glacial record; Antarctica and New Zealand, Southern Hemisphere late Neogene climates: Papers and Proceedings of the Royal Society of Tasmania, v. 130, p. 17–24.

Folk, R.L., 1968, Petrology of Sedimentary Rocks: Austin, Texas, Hemphill's Book Store, 170 p.

Forster, M.A., and Lister, G.S., 2003, Cretaceous metamorphic core complexes in the Otago Schist, New Zealand: Australian Journal of Earth Sciences, v. 50, p. 181–198, doi: 10.1046/j.1440-0952.2003.00986.x.

Garzanti, E., and Vezzoli, G., 2003, A classification of metamorphic grains in sands based on their composition and grade: Journal of Sedimentary Research, v. 73, p. 830–837.

Il, G., Essene, E.J., Peacor, D.R., and Coombs, D.S., 2000, Reactions leading to the formation and breakdown of stilpnomelane in the Otago Schist: New Zealand Journal of Metamorphic Geology, v. 18, p. 393–407.

Ingersoll, R.V., Bullard, T.F., Ford, R.L., Grimm, J.P., Pickle, J.D., and Sares, S.W., 1984, The effect of grain size on detrital modes: A test of the Gazzi-Dickinson point-counting method: Journal of Sedimentary Petrology, v. 54, p. 103–116.

Kamp, P.J., 1987, Age and origin of the New Zealand orocline in relation to Alpine fault movement: Journal of the Geological Society of London, v. 144, p. 641–652.

Kamp, P.J., 2001, Possible Jurassic age for part of Rakaia terrane; implications for tectonic development of the Torlesse accretionary prism: New Zealand Journal of Geology and Geophysics, v. 44, no. 2, p. 185–203.

Kao, M., 2001, Thermo-tectonic history of the Marlborough region, South Island, New Zealand: Terrestrial, Atmospheric and Oceanic Sciences, v. 12, no. 3, p. 485–502.

MacKinnon, T.C., 1983, Origin of the Torlesse terrane and coeval rocks, South Island, New Zealand: Geological Society of America Bulletin, v. 94, p. 967–985, doi: 10.1130/0016-7606(1983)94<967:OOTTTA>2.0.CO;2.

Marsaglia, K.M., Garcia y Barragan, J.C., Padilla, I., and Milliken, K.L., 1996, Evolution of the Iberian passive margin as reflected in sand provenance, in Whitmarsh, R.B., Sawyer, D.S., Klaus, A., and Masson, D.G., et al., Proceedings of the Ocean Drilling Program, Scientific Results, Volume 149: College Station, Texas, Ocean Drilling Program, p. 269–280.

McCave, I.N., and Carter, L., 1997, Recent sedimentation beneath the Deep Western Boundary Current off northern New Zealand: Deep-Sea Research, v. 44, p. 1203–1237, doi: 10.1016/S0967-0637(97)00011-3.

McMillian, S.G., and Wilson, G.J., 1997, Allostratigraphy of coastal south and east Otago: A stratigraphic framework for interpretation of the Great South Basin, New Zealand: New Zealand Journal of Geology and Geophysics, v. 40, p. 91–107.

Mortimer, N., 1993, Geology of the Otago Schist and Adjacent Rocks: Institute of Geological and Nuclear Sciences Geological Map 7, 1 sheet, scale 1:500,000.

Mortimer, N., 1994, Origin of the Torlesse terrane and coeval rocks, North Island, New Zealand: International Geology Review, v. 36, p. 891–910.

Mortimer, N., and Roser, B.P., 1992, Geochemical evidence for the position of the Caples-Torlesse boundary in the Otago Schist, New Zealand: Journal of the Geological Society of London, v. 149, p. 967–977.

Nelson, C.S., Hendy, C.H., Jarrett, G.R., and Cuthbertson, A.M., 1985, Near-synchroneity of New Zealand Alpine glaciations and Northern Hemisphere continental glaciations during the past 750 kyr: Nature, v. 318, p. 361–363, doi: 10.1038/318361a0.

New Zealand Geological Survey, 1972, Geological Map of New Zealand, South Island (1st ed.): Wellington, New Zealand, Department of Scientific and Industrial Research, scale 1:1,000,000.

Robert, C., and Acquaviva, M., 1986, Cenozoic evolution and significance of clay associations in the New Zealand region of the South Pacific, Deep Sea Drilling Project, Leg 90, in Kennett, J.P., von der Borch, C.C., et al., Initial Reports Deep Sea Drilling Project, Volume 90, part 2: Washington, D.C., Government Printing Office, p. 1225–1238.

Shapiro, S.A., 2004, Sand provenance of the Bounty submarine fan off the eastern coast of South Island, New Zealand [M.S. thesis]: Northridge, California State University, 90 p.

Shipboard Scientific Party, 1986, Site 594: Chatham Rise, in Kennett, J.P., von der Borch, C.C., et al., Initial Reports Deep Sea Drilling Project, Volume 90: Washington, D.C., Government Printing Office, p. 653–744.

Shipboard Scientific Party, 1999a, Leg 181 summary: Southwest Pacific paleoceanography, in Carter, R.M., McCave, I.N., Richter, C., Carter, L., et al., Proceedings of the Ocean Drilling Program, Initial Reports, Leg 181: College Station, Texas, Ocean Drilling Program, p. 1–80.

Shipboard Scientific Party, 1999b, Site 1122: Turbidites with a contourite foundation, in Carter, R.M., McCave, I.N., Richter, C., Carter, L., et al., Proceedings of the Ocean Drilling Program, Initial Reports, Leg 181: College Station, Texas, Ocean Drilling Program, [CD-ROM].

Shipboard Scientific Party, 1999c, Site 1121: The "Campbell drift," in Carter, R.M., McCave, I.N., Richter, C., Carter, L., et al., Proceedings of the Ocean Drilling Program, Initial Reports, Leg 181: College Station, Texas, Ocean Drilling Program, p. 1–62 [CD-ROM].

Smith, M.P., and Yardley, W.D., 1999, Fluid evolution during metamorphism of the Otago Schist, New Zealand: (I) Evidence from fluid inclusions: Journal of Metamorphic Geology, v. 17, p. 173–186, doi: 10.1046/j.1525-1314.1999.00189.x.

Walcott, R.I., 1998, Modes of oblique compression: Late Cenozoic tectonics of the South Island of New Zealand: Reviews of Geophysics, v. 36, p. 1–26, doi: 10.1029/97RG03084.

Manuscript Accepted by the Society 9 August 2006

TABLE A1. RAW POINT COUNT DATA

Hole-Core-Section-top of interval (cm)	Qm	Qp	P	Altered Plag	K	Fu	Lma	Lma Lmf2	Lma Lmf3	Lmt	Lmt Lmf2	Lmt Lmf3	Lmt Lmf4	Lmt Lmp2	Lmt Lmp4	Lmm	Lmm Lmp2	Lmm Lmp3	Lmm Lmp4	Lmm Rmp5	Lm other	Lma Q+P Other
1122B-1H-2-19	111	3	133	1	22	0	0	0	3	5	3	3	0	0	1	6	0	0	0	0	6	0
1122B-1H-6-102	164	1	113	0	9	1	4	2	0	6	4	0	0	0	0	0	0	0	0	0	8	0
1122C-2H-6-135	139	3	158	2	5	1	3	0	1	0	4	7	0	0	0	0	0	0	3	0	3	0
1122C-3H-5-130	96	2	123	0	11	0	3	0	0	8	0	4	0	1	0	2	0	1	1	0	3	0
1122C-4H-7-39	136	2	145	3	17	3	6	0	0	10	0	0	0	0	0	0	0	0	0	0	11	0
1122C-5H-6-96	129	5	158	0	13	0	9	0	0	2	0	0	4	0	0	5	0	0	0	0	1	1
1122C-6H-6-58	136	10	125	3	22	2	4	1	0	20	1	0	0	0	0	5	0	0	0	0	4	2
1122C-7H-6-18	143	1	170	0	11	2	1	1	0	1	0	0	0	1	0	2	0	0	0	0	9	1
1122C-8H-1-51	145	6	159	1	7	0	4	0	2	5	4	0	0	0	0	0	0	0	0	0	2	0
1122C-9H-1-13	159	2	127	0	9	0	0	1	0	4	2	1	0	0	1	6	0	2	0	0	8	0
1122C-9H-6-70	142	4	133	1	21	0	3	0	0	14	6	0	2	1	0	2	0	1	0	0	1	0
1122C-10H-6-84	154	8	118	1	11	0	0	0	0	5	9	3	4	0	2	4	0	0	0	0	2	0
1122C-11H-4-7	152	1	124	0	8	1	1	0	0	0	0	0	0	2	0	3	0	0	0	0	3	0
1122C-12H-1-10	125	9	108	0	12	0	0	0	0	4	2	4	0	0	0	1	0	0	2	0	4	0
1122C-13H-6-84	94	5	125	0	18	0	2	3	0	0	1	0	2	2	2	3	1	0	0	0	6	0
1122C-14X-1-106	154	1	106	0	36	0	2	0	0	4	0	0	0	0	0	0	0	0	0	0	6	0
1122C-15X-1-10	59	1	91	0	4	0	0	0	0	0	0	1	0	0	0	1	0	0	0	0	14	1
1122C-16X-1-74	49	0	60	0	4	1	1	0	0	1	0	0	0	0	0	2	0	0	0	0	4	0
1122C-17X-1-86	116	5	104	1	14	0	2	5	0	10	2	1	2	0	1	2	0	2	0	0	2	1
1122C-18X-3-33	141	3	136	3	10	0	3	6	0	1	2	3	0	0	0	2	0	1	5	0	6	0
1122C-19X-3-86	76	0	83	0	8	0	6	0	2	5	4	1	0	2	0	3	0	0	0	4	17	1
1122C-20X-1-6	109	2	129	0	18	1	0	0	0	0	0	0	0	0	0	0	0	0	0	0	10	1
1122C-21X-1-18	102	4	108	0	11	1	6	0	0	2	6	0	1	0	0	2	0	1	7	0	8	2
1122C-22X-1-128	124	2	127	1	10	0	2	3	0	5	1	1	1	0	0	2	0	2	0	0	5	0
1122C-23X-2-56	137	2	134	0	14	4	0	0	0	3	0	0	0	0	2	3	0	0	0	0	4	2
1122C-24X-1-65	163	7	125	0	16	4	0	0	0	12	5	1	0	0	0	6	0	1	0	0	5	0
1122C-26X-1-28	67	1	104	0	13	1	6	1	0	4	2	1	1	0	0	2	0	0	0	0	3	1
1122C-27X-2-133	113	5	129	0	7	0	0	0	0	3	0	1	1	0	0	1	1	0	0	0	1	0
1122C-30X-5-8	139	6	111	1	4	2	5	0	0	2	5	1	0	1	0	2	4	0	2	1	2	3
1122C-31X-1-40	150	0	129	0	11	3	2	0	0	2	0	1	1	1	0	0	0	4	0	0	10	0
1122C-32X-1-25	135	5	125	0	10	0	4	2	0	6	5	3	2	0	0	3	0	1	0	2	6	0
1122C-33X-4-107	100	1	118	10	18	3	3	0	0	0	7	3	0	1	0	9	3	1	0	0	1	0
1122C-34X-5-120	23	2	20	1	2	0	1	0	1	0	1	0	0	0	0	4	1	0	0	0	0	0
1122C-35X-1-92	26	0	24	0	12	0	0	0	0	0	0	1	0	0	0	2	0	0	0	0	6	0
1122C-36X-6-88	123	8	136	13	12	2	6	0	0	12	0	0	0	0	0	6	0	0	3	0	6	1
1122C-37X-6-92	50	1	45	0	4	0	0	0	1	4	5	1	1	0	0	2	0	1	0	0	4	0
1122C-39X-2-47	112	2	87	1	11	2	6	0	0	14	4	1	0	0	0	2	4	0	1	0	3	0
1122C-40X-2-128	137	12	91	3	9	0	4	0	0	6	6	3	0	2	0	3	2	1	2	0	4	3
1122C-42X-1-130	128	1	101	0	10	2	0	2	0	7	3	1	0	0	0	9	0	0	0	2	3	0
1122C-44X-4-56	145	11	111	4	6	0	8	0	0	5	5	5	2	1	0	3	3	0	0	0	7	0
1122C-45X-4-91	187	4	131	1	7	0	5	0	1	6	7	0	0	0	0	4	1	0	0	0	1	0
1122C-46X-4-134	145	1	173	0	14	1	6	0	0	0	1	0	0	0	0	4	0	0	0	0	0	0
1122C-47X-1-56	46	2	48	0	5	0	2	0	0	6	0	0	0	1	0	2	0	0	0	0	6	2
1122C-47X-4-77	176	3	150	0	5	1	1	2	0	3	5	1	0	0	4	0	0	3	3	0	5	0
1122C-48X-1-97	140	6	112	1	8	7	6	3	0	19	6	1	0	0	0	4	0	0	0	0	6	0
1122C-49X-4-15	129	10	94	0	5	0	4	2	0	3	5	0	1	0	0	2	2	1	1	0	6	0
1122C-51X-5-111	179	3	107	0	10	1	3	0	0	4	0	1	0	4	0	3	0	0	0	0	6	0
1122C-52X-1-93	90	6	82	0	3	0	4	3	1	10	4	0	4	1	0	2	1	1	2	0	9	0
1122C-56X-4-6	131	6	124	1	2	1	8	0	0	11	0	1	0	0	0	3	1	0	0	0	2	3
1122C-57X-3-19	88	4	69	0	1	0	5	2	1	6	6	3	4	1	0	2	1	0	2	0	4	0
1122C-58X-2-37	173	11	121	0	5	1	3	0	0	6	1	2	0	0	0	3	0	1	0	0	3	3
1122C-61X-3-43	126	3	115	0	4	0	1	2	1	1	0	0	0	1	0	0	0	0	0	2	2	0
1122C-62X-1-12	111	8	120	0	3	3	34	0	0	2	0	2	0	0	0	12	0	0	0	0	11	0
1122C-64X-1-130	140	7	110	0	2	1	2	1	0	3	2	1	1	4	0	5	0	0	1	0	14	0
1122C-65X-1-21	142	2	131	0	9	1	8	3	0	3	3	0	2	0	0	4	0	1	0	0	2	0
1122C-66X-1-71	109	4	90	4	3	1	3	7	1	1	1	3	0	0	0	1	0	0	1	0	0	0
River samples																						
Rangitata	111	46	15	0	6	2	15	0	0	89	15	2	3	19	0	5	0	0	0	0	22	0
Waitaki	108	19	48	0	13	2	35	0	0	75	5	0	1	2	0	13	1	0	0	0	22	0
Clutha	207	4	115	0	1	0	2	0	0	6	0	0	0	0	0	3	1	0	0	0	11	0

(continued.)

TABLE A1. (Continued.)

Hole-Core-Section-top of interval (cm)	Lma Q+ Dense	Lma P+ Dense	Lma K+ Dense	Lma Q+F	Lsi	Lsa	Lv	M	B	Chl	Malt	D	Glau	Epi	Carb	Foram	Dia	Spong	Echino	Shell frag	Other bio	Other	Total points
1122B-1H-2-19	1	1	0	0	1	0	1	4	8	3	0	3	1	6	56	14	0	1	0	0	5	1	400
1122B-1H-6-102	0	0	0	3	3	1	0	5	10	6	2	2	0	5	35	9	0	0	0	0	5	1	400
1122C-2H-6-135	0	2	0	2	8	1	2	10	1	7	1	1	2	5	21	6	0	0	0	0	7	1	400
1122C-3H-5-130	1	1	0	0	7	2	1	15	16	7	3	0	1	6	41	30	0	0	0	0	5	0	400
1122C-4H-7-39	0	0	0	0	14	6	0	4	2	6	1	12	0	1	15	2	0	0	0	0	14	2	400
1122C-5H-6-96	1	0	0	0	1	0	0	15	5	15	0	2	0	9	14	6	0	1	0	0	0	1	400
1122C-6H-6-58	1	0	0	0	9	2	1	1	7	6	4	5	0	8	23	3	0	0	0	1	2	4	400
1122C-7H-6-18	0	3	0	1	3	1	2	6	3	4	2	0	0	9	11	2	0	0	0	0	4	3	400
1122C-8H-1-51	0	2	0	0	6	0	0	7	7	2	2	5	2	4	18	7	0	0	0	0	3	0	400
1122C-9H-1-13	0	1	0	1	3	0	1	24	7	11	3	3	2	10	14	2	0	0	0	0	0	3	400
1122C-9H-6-70	0	1	0	1	0	1	0	3	4	3	2	5	0	11	25	5	0	0	0	0	3	3	400
1122C-10H-6-84	4	0	0	1	4	1	1	14	4	6	1	1	0	3	28	6	0	1	0	0	8	1	400
1122C-11H-4-7	0	3	0	3	4	1	0	26	4	13	3	10	0	6	16	16	0	0	0	0	3	0	400
1122C-12H-1-10	0	0	0	1	5	4	11	33	14	17	1	16	0	2	14	8	0	0	0	0	5	0	400
1122C-13H-6-84	1	0	0	3	2	0	0	14	29	13	4	8	2	6	35	15	0	0	0	0	8	1	400
1122C-14X-1-106	2	0	0	1	8	2	2	9	7	2	5	5	0	6	38	6	2	0	0	0	0	1	400
1122C-15X-1-10	0	0	0	1	11	1	2	98	40	35	3	5	1	1	13	11	0	0	0	0	5	1	400
1122C-16X-1-74	0	0	0	0	7	1	1	113	59	29	13	3	0	2	17	14	5	0	0	0	5	0	400
1122C-17X-1-86	0	0	0	0	0	0	0	42	26	17	6	10	0	5	15	20	5	1	0	0	4	1	400
1122C-18X-3-33	1	0	0	0	10	4	1	9	9	7	1	4	1	0	15	11	1	0	0	0	0	0	400
1122C-19X-3-86	0	0	0	2	9	1	0	71	7	10	1	9	0	5	12	17	0	0	0	0	8	3	400
1122C-20X-1-6	3	3	0	2	16	0	2	14	33	12	5	6	1	4	37	8	0	0	0	0	1	3	400
1122C-21X-1-18	0	2	0	1	16	0	0	33	9	17	6	12	0	16	29	5	0	0	0	0	5	4	400
1122C-22X-1-128	2	1	0	1	7	2	0	26	21	6	1	7	1	5	33	8	0	0	0	0	1	1	400
1122C-23X-2-56	2	0	0	3	5	1	2	13	8	9	3	0	0	6	21	5	0	0	0	0	2	0	400
1122C-24X-1-65	1	2	0	0	17	6	0	4	1	4	0	5	1	12	15	3	0	0	0	0	1	3	400
1122C-26X-1-28	2	1	0	0	2	2	6	57	44	20	0	3	0	13	41	32	3	0	0	1	6	0	400
1122C-27X-2-133	2	2	0	4	6	2	0	18	29	8	7	2	2	0	47	8	0	0	0	0	3	4	400
1122C-30X-5-8	3	3	0	1	5	1	1	20	9	11	5	3	3	4	38	6	1	0	0	0	2	1	400
1122C-31X-1-40	0	2	0	4	10	1	2	17	3	12	6	7	0	7	14	3	0	0	0	0	2	1	400
1122C-32X-1-25	2	2	0	0	4	0	0	27	5	10	3	1	0	9	20	3	3	1	0	0	2	1	400
1122C-33X-4-107	1	5	0	1	4	10	3	6	7	8	0	6	0	6	52	17	0	0	1	0	0	3	400
1122C-34X-5-120	0	0	0	1	5	0	0	60	17	21	16	11	2	1	2	8	21	0	3	4	2	1	400
1122C-35X-1-92	0	0	0	0	0	2	2	67	29	12	11	215	2	3	38	146	1	0	1	0	16	0	400
1122C-36X-6-88	4	0	0	4	7	13	1	6	1	4	1	3	0	1	18	1	0	0	0	1	2	6	400
1122C-37X-6-92	0	0	0	1	7	3	6	151	44	20	20	9	2	3	15	12	1	0	1	0	3	0	400
1122C-39X-2-47	0	1	0	4	0	0	0	56	31	8	7	4	0	4	32	2	2	0	0	0	0	1	400
1122C-40X-2-128	4	2	0	5	16	4	1	22	5	10	2	7	0	3	24	2	1	2	0	0	2	0	400
1122C-42X-1-130	2	1	0	1	5	0	2	31	6	7	0	9	3	4	65	2	0	1	0	0	2	1	400
1122C-44X-4-56	1	0	0	1	17	0	0	12	8	4	2	5	0	12	24	5	1	0	1	0	0	1	400
1122C-45X-4-91	1	1	0	4	9	1	1	7	2	5	1	2	1	7	10	3	0	0	0	0	1	0	400
1122C-46X-4-134	0	1	0	1	6	1	0	9	3	4	2	2	0	6	23	3	0	2	3	0	12	1	400
1122C-47X-1-56	2	1	0	2	1	0	3	55	38	20	1	6	3	5	74	54	7	0	3	4	0	2	400
1122C-47X-4-77	0	2	0	4	0	0	0	0	3	0	0	1	0	4	23	8	0	0	1	0	0	3	400
1122C-48X-1-97	0	0	0	4	12	13	2	6	2	7	7	4	0	0	31	1	0	0	3	0	0	1	400
1122C-49X-4-15	2	1	0	4	1	0	2	46	8	15	4	3	1	4	28	16	2	0	6	0	6	2	400
1122C-51X-5-111	1	3	0	3	5	3	9	27	13	12	6	4	1	0	4	2	0	0	0	2	1	3	400
1122C-52X-1-93	2	3	0	0	13	3	1	68	16	11	8	0	3	11	15	42	3	0	1	0	3	1	400
1122C-56X-4-6	0	0	0	1	3	1	2	38	8	7	2	3	0	9	21	2	0	0	0	0	1	1	400
1122C-57X-3-19	0	0	0	2	32	0	0	97	35	4	4	1	1	2	21	0	0	2	3	0	2	1	400
1122C-58X-2-37	1	1	0	3	2	1	1	18	4	7	3	1	0	10	11	3	0	0	1	0	1	0	400
1122C-61X-3-43	0	0	0	2	2	2	2	8	8	5	1	3	0	9	34	62	0	0	6	2	6	3	400
1122C-62X-1-12	1	0	0	3	1	9	9	14	10	2	7	1	1	5	15	10	0	0	1	0	1	1	400
1122C-64X-1-130	1	3	0	0	5	3	1	46	8	6	8	3	3	9	24	11	2	0	7	0	2	0	400
1122C-65X-1-21	0	0	0	9	13	1	2	27	13	12	6	4	0	13	4	2	0	0	7	0	1	2	400
1122C-66X-1-71	1	1	0	4	44	8	0	2	0	1	0	1	0	8	84	13	0	3	0	0	0	2	400
River samples																							
Rangitata	3	0	0	0	0	2	0	0	2	2	0	5	1	10	0	0	0	0	0	0	0	2	400
Waitaki	5	2	0	0	23	13	0	0	0	0	0	1	0	7	1	1	0	0	0	0	0	2	400
Clutha	4	0	0	0	6	1	1	6	2	2	2	1	0	21	0	0	0	0	0	0	0	4	400

Sediment sources of beach sand from the southern coast of the Baja California peninsula, Mexico—Fourier grain-shape analysis

J.M. Murillo-Jiménez[†]
Departamento de Oceanología, Centro Interdisciplinario de Ciencias Marinas, Instituto Politécnico Nacional (CICIMAR-IPN), Laboratorio de Geología Marina, Apartado Postal 592, La Paz, Baja California Sur, 23000, México

William Full
Tramontane, Inc., P.O. Box 58082, Salt Lake City, Utah 84158-0082, USA

E.H. Nava-Sánchez
V. Camacho-Valdéz
A. León-Manilla
Departamento de Oceanología, Centro Interdisciplinario de Ciencias Marinas, Instituto Politécnico Nacional (CICIMAR-IPN), Apartado Postal 592, La Paz, Baja California Sur, 23000, México

ABSTRACT

The purpose of the study is to (1) identify the sources of sediment in various environments, (2) define the history and transport processes of the sediments, and (3) better understand the erosion and potential replenishment of the local beaches along the southern coast of the Baja California peninsula. For the purpose of this study, six naturally defined areas were studied separately: El Cardonal, El Arco, San Lucas, El Tiburón, El Tule, and San José.

Two main sedimentary provinces were identified via Fourier grain-shape analysis, El Médano and Los Cabos. El Médano sedimentary province includes the El Cardonal and El Arco areas, which are influenced by the dynamics associated with the Pacific Ocean dominated by northwesterly winds, waves, and longshore transport. Beaches from this province have a source mostly from marine material from the shallow shelf, and they are dominantly affected by longshore transport. Secondarily, they are dominated by old and recent aeolian material dissected by intermittent arroyos and local arroyo material from intrusive rocks. The Los Cabos sedimentary province includes the other four areas, and it is influenced by the dynamics of the Gulf of California. In this province, dominant southerly waves are present. Sediment transport occurs along the coast from southwest to northeast; although, some beaches contain material from northern areas, probably related to the direction of waves and sediment transport direction during meteoric events such as hurricanes. Beaches from this province have a source mostly from local arroyo material from intrusive rocks.

[†]E-mail: jmurillo@ipn.mx.

Murillo-Jiménez, J.M., Full, W., Nava-Sánchez, E.H., Camacho-Valdéz, V., and León-Manilla, A., 2007, Sediment sources of beach sand from the southern coast of the Baja California peninsula, Mexico—Fourier grain-shape analysis, *in* Arribas, J., Critelli, S., and Johnsson, M.J., eds., Sedimentary Provenance and Petrogenesis: Perspectives from Petrography and Geochemistry: Geological Society of America Special Paper 420, p. 297–318, doi: 10.1130/2006.2420(18). For permission to copy, contact editing@geosociety.org. ©2007 Geological Society of America. All rights reserved.

Other beach material results from longshore transport and some material comes from the El Médano sedimentary province in the El Arco boundary area.

Grain-shape data and the information associated with elongation (harmonic 2) show that marine samples (beach, shallow, and deep inner continental shelf) from Los Médanos sedimentary province contain high frequencies of grains with low elongation, opposite of the arroyo samples. This suggests that the low elongation grain source may be farther north of this province. In the Los Cabos sedimentary province, the local arroyos and the longshore transport have been identified as the major factors that nourish and distribute the beach material along the coast. The results of this study parallel those found in similar geographic regions where storms rather than steady currents dominant.

Keywords: quartz, elongation, beaches, sources, sediment, transport, Mexico.

INTRODUCTION

The purpose of the present study is to identify the sources of beach sand along the southern coast of the Baja California peninsula, Mexico. The importance of this study is both geologic and economic; there is special interest in this area to understand the sediment dynamics because anthropogenic settlements and tourist infrastructure are present along the coast. This coastal area is characterized by high-energy waves, especially during the hurricane and summer storm season, when even higher-energy waves are present. These waves are responsible for rapid changes observed along the beach profile. The present lack of knowledge of the provenance of sand that nourishes local beaches is accentuated by the loss of sediment and beach erosion. It has been assumed that beach replenishment has been altered because the natural paths of beach sand from local arroyos are being continuously changed by civil engineering construction projects, which modify the flow of sediment to the related beach (San Lorenzo Dam, growing of urbanization, concrete walls over small creeks, among other projects).

For the present work, we used Fourier grain-shape analysis to identify the potential sources of sediment to local beaches as well as sediment paths along the coastal zone. This approach has been successfully applied to northern coastal areas of Baja California (Murillo et al., 1994) and other sedimentary environments in the United States, Italy, and Australia (Tortora et al., 1986; Evangelista et al., 1988, 1994, 1996, 2003; Civitelli et al., 1992; Osborne et al., 1994). The advantage of using Fourier grain-shape analysis over other techniques such as petrographic analysis in this area is that the overall composition of the sedimentary material does not vary greatly over the study area, whereas changes in quartz shapes have been found to vary in this area (Camacho-Valdéz, 2003).

STUDY AREA

The study area is located along the southern coast of the Baja California peninsula (Fig. 1) and is part of the El Cabo geological subprovince, which includes mostly Cretaceous and Tertiary intrusive rocks and Quaternary sandstones and conglomerates. Sediments from different environments, such as beaches and arroyos, are also present, as well as old and recent aeolian material, which extends inland and along the coast (Fig. 2; geological chart: INEGI, 1999). Along the coastline, wide and narrow sandy beaches are present, some of which are intercepted by rocky promontories. During the rainy season (August–September), and eventually during tropical storms and hurricanes, intermittent arroyos drain into the Pacific Ocean and Gulf of California and carry sediment derived from different types of rocks (Fig. 3; Table 1). In the study area, the larger basins are: San José (1194 km^2), Migriño (186 km^2), El Salto (181 km^2), and El Tule (111 km^2); (Fig. 3; topographic chart: INEGI, 1985; Defense mapping Agency, 1984). The beach sediment along the Pacific Ocean coastal area is coarse- to medium-size sand, white to yellowish in color, and is composed mostly of quartz, feldspars, and amphiboles from granitic rocks (Table 2; Camacho Valdéz, 2003).

El Cardonal area (Fig. 3) is located along the western coast of the southern portion of the Baja peninsula, along the Pacific Ocean. It presents wide and long beaches, with extended inland dune fields of parabolic and frontal dunes (Camacho-Valdéz, 2003). The shape of the coast is convex, oriented northwest-southeast, and is directly influenced by high-energy waves induced by dominant northwestern winds and less frequent southern winds. In the continental shelf, the 100 m isobath is ~2.5 km off the coastline. The largest drainage basin is Migriño, which has an area of 186 km^2 (Fig. 3). During heavy rains, arroyo flows dissect the beach in several places. In this area, there are old and recent, large aeolian deposits distributed inland and along the coast, respectively. Old dunes were dated (samples 274d and 276d) by luminescence technique and gave ages of 2.6 ± 0.7 ka and 3.2 ± 0.4 ka (C.D. Peterson, 2004, personal commun.).

El Arco area (Fig. 3) is located along the southernmost coast of the peninsula, along the Pacific Ocean, and is limited to the east by San Lucas Cape, which marks the boundary between the Pacific Ocean and the Gulf of California. It exhibits wide and long sandy beaches of medium-size sand, it is delimited by rocky promontories, and frontal dunes are observed along the coast. The shape of

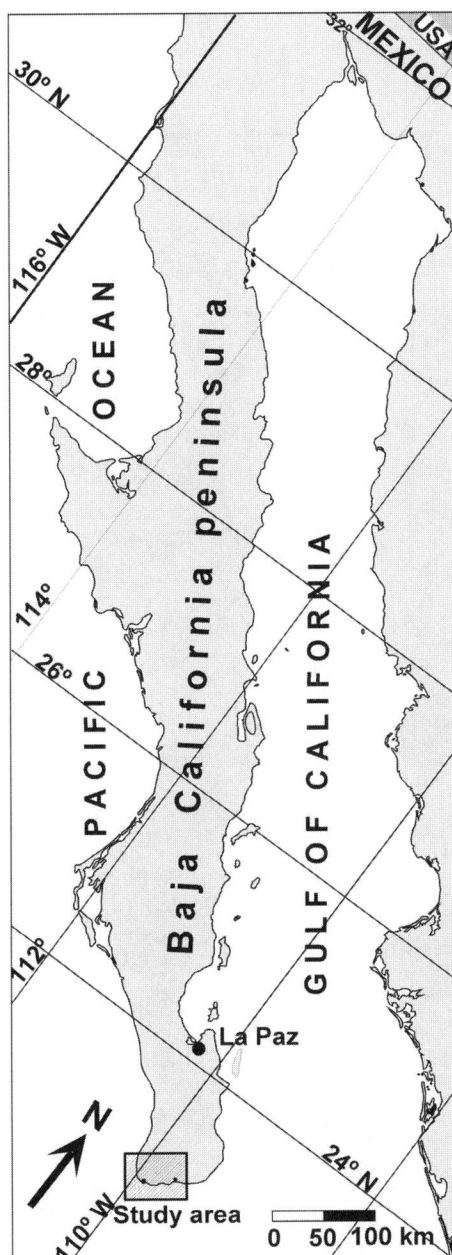

Figure 1. Location map of the study area in the southern portion of the Baja California peninsula, Mexico. Latitude is north of the equator and longitude is west of Greenwich.

the coast is linear, oriented west-east, and is influenced by high-energy waves induced by northwestern and southern winds. On the continental shelf, the 100 m isobath is at ~3 km off the coastline. The largest drainage basin is Palma Seda, which has an area of 7.9 km^2 (Fig. 3).

The San Lucas area (Fig. 3) is located on the southernmost portion of the peninsular margin of the Gulf of California and, thus, is partially protected from the Pacific Ocean. The largest drainage basin is El Salto, which has an area of 181 km^2 (Fig. 3). The west portion presents a 7 km beach, which is concave in shape and forms a bay oriented 67°Az (Cabo San Lucas Bay) delimited by rocky promontories. Frontal dunes are present along the coast. At the western side of the beach there is an artificial marina—Cabo San Lucas Marina. The seafloor morphology is highlighted by a feature called Cabo San Lucas Canyon. The axis of the canyon is in line with the El Salto arroyo, oriented 320°Az. The inner continental shelf shows that the 100 m isobath is at ~3.5 km from the coastline on the eastern portion, and gets closer to the San Lucas Cape (0.5 km offshore at this point), where the upper canyon area is present. Dominant wave directions are from the south and southeast (Troyo-Dieguez, 2003). An aerial photograph obtained in March 1993 (INEGI, 1993) shows that northwesterly and dominant local waves are refracted at the Cabo San Lucas Cape, and they enter the western portion of the bay as southeasterly-easterly waves. The eastern side of this area presents 3 km of a lightly convex coast with large rocky promontories forming small embayments with narrow beaches oriented 83°Az. The 100 m isobath is at ~2.5 km offshore at this point. Beach sediments are coarse and medium sand. The stabilized dunes in this area were dated by the luminescence technique (at the University of Wollongong in Australia by David Price) and they gave an age of 0.8 ± 0.3 ka (sample no. 560d, lab. no. W2837; Murillo-Jiménez et al., 2001).

The highlight of the El Tiburón area (Fig. 3) is a straight, 6-km-long coastline oriented 50°Az; this feature is likely structurally controlled. Narrow beaches intercepted by rocky promontories are common. At the southern portion of the area, a large rocky promontory delimits the area and forms a bay. The largest drainage basin is El Tiburón, which has an area of 9 km^2 (Fig. 3). The 100 m isobath is at ~2.5 km offshore from coast. Beach sediment is mostly coarse sand.

The El Tule area (Fig. 3) presents an embayment that includes three bays. The south bay, located south of the El Tule arroyo, is mostly rocky and has a sandy beach to the northeast, up to a rocky promontory. The second bay, from El Tule arroyo to Cerro Blanco Cape, has a long, wide, irregular sandy beach with some rocks outcropping at the beach and shallower waters. The east limits of this bay are rocks outcropping on a wide sandy beach (Cerro Blanco Point). The third bay, on the eastern portion of this area from Cerro Blanco to Punta Palmilla Sur, contains a long, wide sandy beach. From Punta Palmilla Sur to Punta Palmilla, a series of small bays is present with narrow sandy beaches dissected by rocky promontories. The general orientation of the three bays is 50°Az. The largest drainage basin within this area is the El Tule arroyo, which has an area of 111 km^2 (Fig. 3). Several smaller drainage basins are also present.

The San José area (Fig. 3) represents the easternmost study area. The largest embayment, San José del Cabo Bay, and drainage basin, San José, are located within this area. The bay is 15 km long, oriented 56° Az, with long, wide, medium-sized sandy beaches. The drainage basin supplying this bay has an area of 1194 km^2 (Fig. 3). Within this embayment, the San José arroyo displays a deltaic sediment wedge, and to the east of this delta, there is a rocky area with a wide sandy beach that delimits another embayment. Waves generally arrive from the southeast

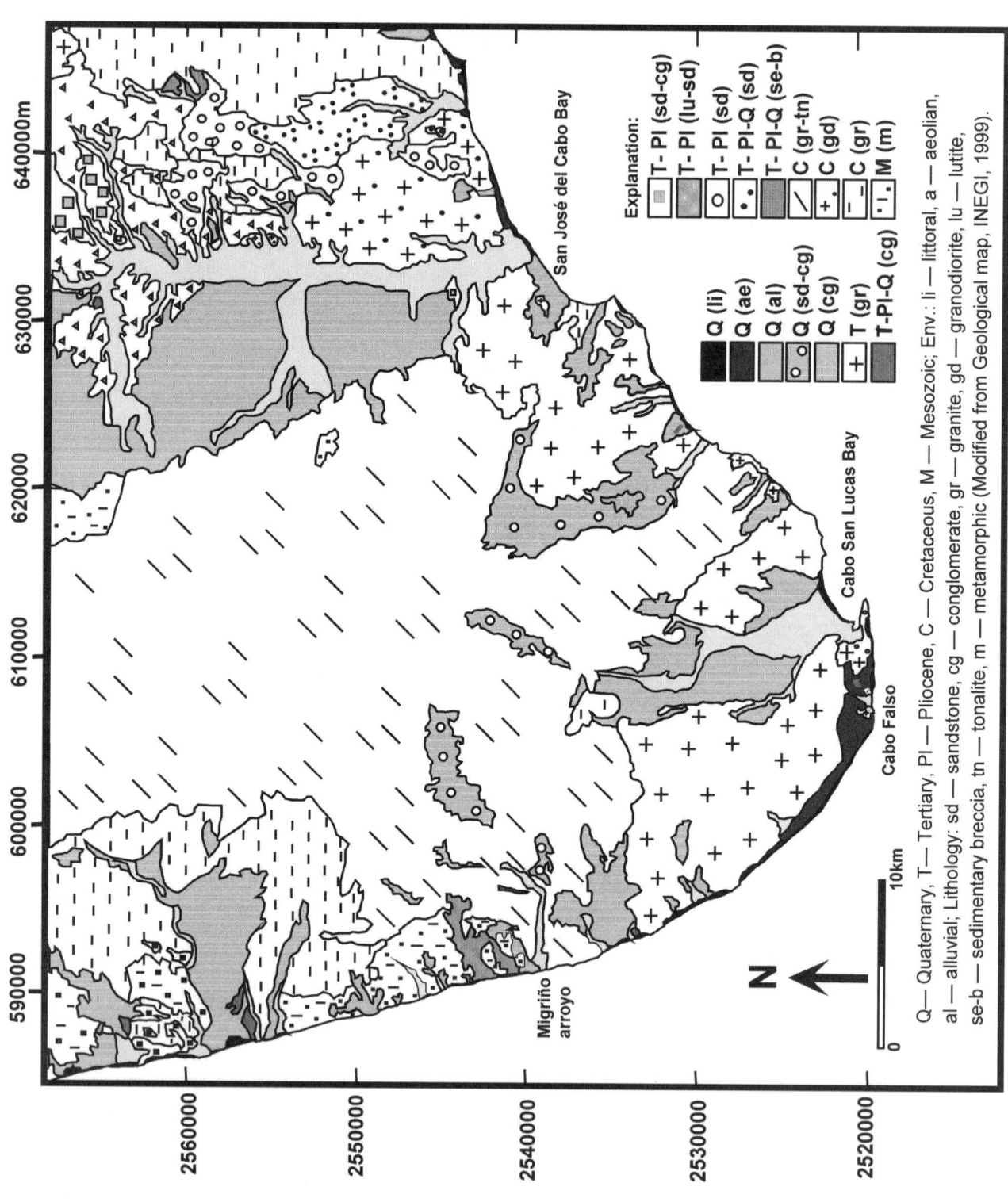

Figure 2. Geological map of the study area. Southern portion of the Baja California peninsula. México.

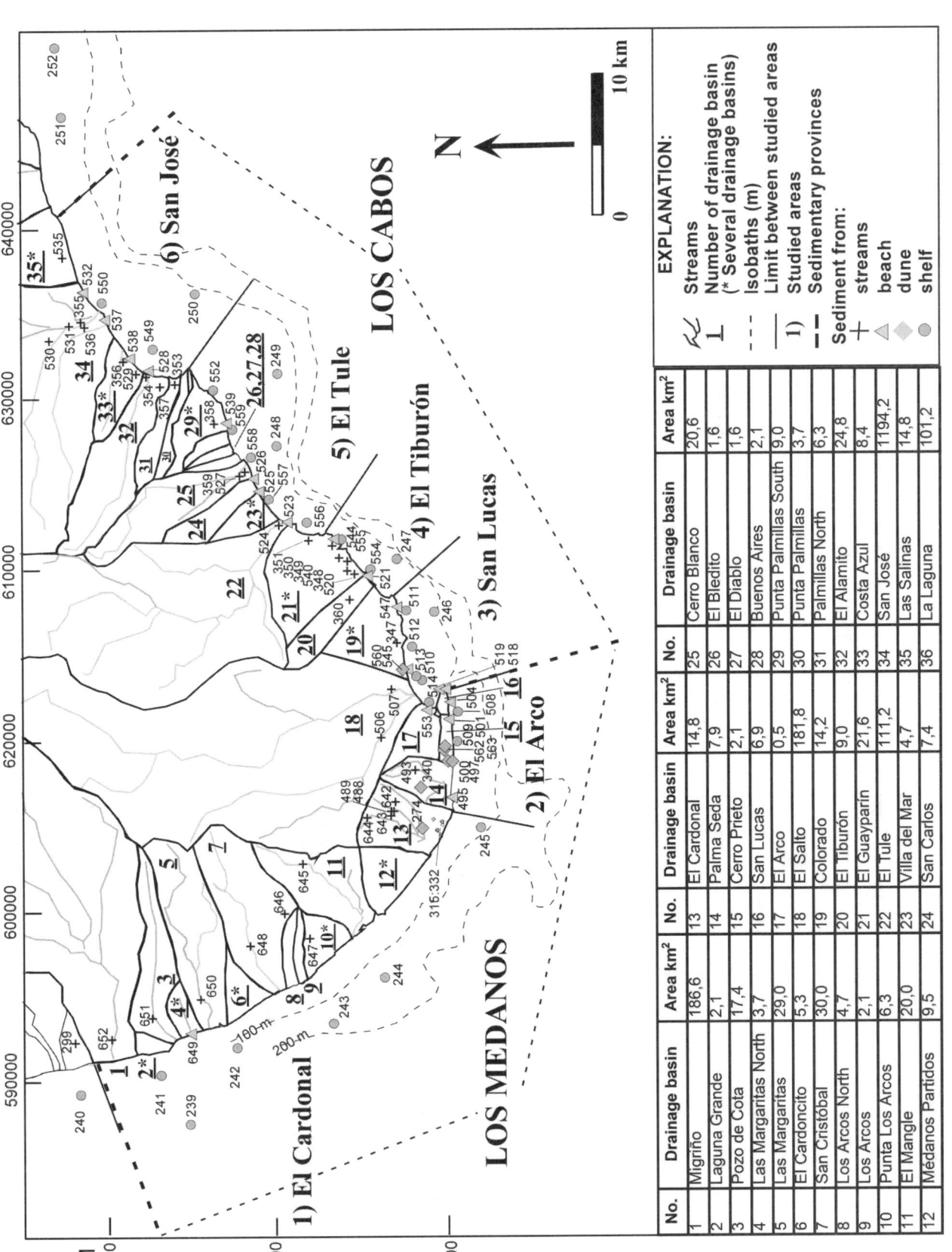

Figure 3. Drainage basin location and limits between geomorphological areas for the most southern portion of the Baja California peninsula, México. Sediment samples from arroyos, beaches, dunes and continental shelf for this study are also located.

Drainage basins from Topographic chart, INEGI, 1985; Isobaths from Defense Mapping Agency, 1984.

TABLE 1. LITHOLOGY WITHIN DRAINAGE BASINS WHERE ARROYO SEDIMENT SAMPLES WERE COLLECTED

Drainage basin	Area	Sample number	K(gd, tn)	K(gr)	K(gd)	T(gr)	TPL (lu-s)	TPL (s-cg)	Q(s-cg)	Q(cg)	Q(al)
El Cardonal	1	488				•			•		
El Cardonal	1	489				•			•		
El Cardonal	1	642				•			•		
El Cardonal	1	643				•					
El Cardonal	1	644				•					
El Mangle	1	645				•			•		
San Cristobal	1	648				•					
Las Margaritas	1	650	•	•		•					
Pozo de Cota	1	651	•	•					•		
Migriño	1	652	•	•					•		
Palma Seda	2	493				•					
Colorado	3	347				•			•	•	
El Salto	3	506				•			•	•	
El Salto	3	507				•			•	•	
El Guayparín	4	348	•							•	
El Guayparín	4	349	•								
El Guayparín	4	350	•			•			•		
El Guayparín	4	351	•								
El Tiburón	4	360				•			•	•	
El Guayparín	4	520							•		
El Guayparín	4	540	•			•			•		
Cerro Blanco	5	359			•	•			•		
El Tule	5	524	•			•			•		
Cerro Blanco	5	527				•			•		
Palmillas Norte	6	353							•		
El Alamito	6	354				•					
San José	6	355	•	•			•	•	•		•
Costa Azul	6	356							•		
El Alamito	6	357	•	•					•		
Punta Palmillas	6	358	•						•		
Costa Azul	6	529				•			•		
San José	6	530	•	•				•	•		•
San José	6	531	•	•					•		•
Las Salinas	6	535				•			•		
San José	6	536	•	•			•	•	•	•	•

Note: Abbreviations: Cretaceous (K), Tertiary (T), Tertiary–Pleistocene (TPL), Quaternary (Q), granodiorite (gd), tonalite (tn), granite (gr), sandstone (s), conglomerate (cg), lutite (lu), alluvial (al). Data are from geological map, La Paz, scale 1:250,000 (INEGI, 1999). •—present.

TABLE 2. MINERAL COMPOSITION OF BEACH MATERIAL WITHIN EL ARCO AREA

Sampling date		Quartz (%)	Feldspar (%)	Mica (%)	Amphibole (%)
March	2001	73	9.34	1.66	16
June	2001	76.66	9.34	5	9
November	2001	88	1.34	1.33	9.33
February	2002	89	4	2	5

Note: Modified from Table 7, profile 1, of Camacho-Valdéz (2003).

(135°Az) (Troyo-Dieguez, 2003), and in the western portion of the bay, an aerial photograph obtained in March 2003 (INEGI, 1993) shows that the rocky promontory (Palmilla Cape) refracts the waves toward the northwest, and they generally enter the bay at 290°Az.

METHODOLOGY

Sediment Sampling and Sample Preparation

Representative samples from 115 locations were obtained from different sedimentary environments (Fig. 3; Table 3) as follows. A Tyler splitter was used at the sites where fluvial samples were collected. Beaches and dunes were sampled with an aluminum cube (10 cm^3 capacity), and the inner continental shelf samples were taken from a Coast Guard vessel at ~10–115 m water depth with a VanVeen dredge (70 cm^3 capacity).

In order to track the mother source of the minerals and sediment paths within different sedimentary environments, two mineral-shape characteristics have to be acquired: mineral elongation and mineral roughness. The elongation of the minerals is engraved during the mineral formation, and the resultant shape is related to the physical conditions at the moment of the crystal formation. When the minerals are released from the source rock, they have rough surfaces that are progressively smoothed with weathering and transport processes. Quartz minerals, because of the high hardness, preserve the elongation shape characteristic, and, thus, the history of abrasion and weathering during transport registers on the mineral surface. Because of this, the present analysis looked solely at quartz material. For grain-shape analysis, the surface of the quartz grains must to be cleaned; therefore, the sand samples were bathed in a chemical solution to dissolve carbonate and/or Fe-oxide coatings that may be present at the surface of the grains. HCl (10%) was used to remove the carbonates, and a SnCl (5%) mixture was used to remove the Fe-oxide coatings. From previous work (Sundborg, 1956), it has been established

TABLE 3. LOCATION OF SEDIMENT SAMPLES FROM DIFFERENT SEDIMENTARY ENVIRONMENTS—SOUTHERN PORTION OF THE BAJA CALIFORNIA PENINSULA, MEXICO

Sample number	Env.	Latitude (N)	Longitude (W)	Water depth (m)	Sample number	Env.	Latitude (N)	Longitude (W)	Water depth (m)
239	s	23°0.06'	110°8.2'	84	501	b	22°52.44'	109°54.75'	
240	s	24°3.6'	111°7.4'	51	504	b	22°52.5'	109°54.1'	
241	s	23°1'	110°6.8'	49	506	a	22°54.48'	109°55.38'	
242	s	22°58.6'	110°5.7'	64	507	a	22°54.36'	109°53.67'	
243	s	22°55.5'	110°4.8'	105	508	s	22°52.14'	109°54.48'	20
244	s	22°54'	110°3.1'	103	509	s	22°52.14'	109°55.43'	15
245	s	22°50.99'	109°58.18'	115	510	s	22°53.34'	109°53.4'	20
246	s	22°52.9'	109°51.2'	63	511	s	22°53.82'	109°51.12'	10
247	s	22°54'	109°49.4'	78	512	s	22°53.7'	109°52.30'	12
248	s	22°57.6'	109°45.5'	73	513	s	22°53.52'	109°53.41'	13
249	s	22°57.5'	109°43.2'	49	514	s	22°53.1'	109°54.28'	12
250	s	23°0'	109°40.4'	87	518	b	22°52.56'	109°53.81'	
251	s	23°3.9'	109°34.4'	58	519	b	22°52.62'	109°53.79'	
252	s	23°4'	109°32.1'	87	520	a	22°55.2'	109°49.96'	
274	da	22°53.39'	109°58.17'		521	b	22°54.96'	109°49.96'	
299	a	22°3.90'	110°5.83'		523	b	22°57.42'	109°48.01'	
300	a	22°4.50'	110°6.12'		524	a	22°57.54'	109°48.11'	
315	b	22°52.64'	109°58.38'		525	b	22°58.26'	109°47.01'	
316	b	22°52.51'	109°58.54'		526	b	22°58.5'	109°46.56'	
317	b	22°52.50'	109°58.1'		527	a	22°58.5'	109°46.51'	
318	b	22°52.64'	109°58.38'		528	b	23°1.74'	109°42.86'	
319	b	22°52.51'	109°58.54'		529	a	23°1.8'	109°42.86'	
320	b	22°52.50'	109°58.1'		530	a	23°4.68'	109°41.68'	
321	b	22°52.64'	109°58.38'		531	a	23°4.02'	109°41.24'	
322	b	22°52.51'	109°58.54'		532	b	23°3.54'	109°40.19'	
323	b	22°52.50'	109°58.1'		535	a	23°4.02'	109°39'	
324	b	22°52.64'	109°58.38'		536	a	23°3.54'	109°41.29'	
325	b	22°52.51'	109°58.54'		537	b	23°2.81'	109°41.15'	
326	b	22°52.50'	109°58.1'		538	b	23°2.22'	109°42.46'	
327	d	22°52.64'	109°58.36'		539	b	22°59.22'	109°44.76'	
328	d	22°52.50'	109°58.1'		540	a	22°55.5'	109°49.28'	
329	d	22°52.64'	109°58.36'		544	b	22°55.86'	109°48.86'	
330	d	22°52.51'	109°58.52'		545	b	22°53.82'	109°53.07'	
331	d	22°52.77'	109°58.03'		547	b	22°54.06'	109°51.04'	
332	d	22°52.77'	109°58.03'		549	s	23°1.44'	109°42.15'	12
340	da	22°53.3'	109°56.8'		550	s	23°2.94'	109°40.55'	5
347	a	22°53.96'	109°52.22'		552	s	22°59.52'	109°43.56'	12
348	a	22°55.55'	109°49.73'		553	b	22°53.04'	109°54.37'	
349	a	22°55.78'	109°49.27'		554	s	22°54.84'	109°49.76'	10
350	a	22°55.86'	109°48.9'		555	s	22°55.74'	109°48.89'	8
351	a	22°56.71'	109°48.56'		556	s	22°56.76'	109°48.07'	20
353	a	23°0.69'	109°43.34'		557	s	22°57.9'	109°47.20'	11
354	a	23°1.76'	109°43.02'		558	s	22°58.44'	109°45.83'	10
355	a	23°3.62'	109°41.14'		559	s	22°59.04'	109°44.85'	13
356	a	23°2.34'	109°42.34'		560	d	22°53.88'	109°53.06'	
357	a	23°1.10'	109°43.43'		562	d	22°52.44'	109°55.78'	
358	a	22°59.54'	109°44.68'		563	da	22°52.44'	109°55.78'	
359	a	22°58.70'	109°46.57'		642	a	22°53.938'	109°57.46'	
360	a	22°55.41'	109°50.6'		643	a	22°54.07'	109°57.90'	
488	a	22°54.06'	109°57.78'		644	a	22°54.824'	109°58.02'	
489	a	22°54.06'	109°57.78'		645	a	22°56.819'	109°59.76'	
493	a	22°53.34'	109°56.52'		646	a	22°57.3'	110°1.2'	
495	b	22°52.44'	109°57.21'		647	a	22°56.52'	110°1.8'	
497	b	22°52.38'	109°55.95'		648	a	22°58.72'	110°2.43'	
500	d	22°52.44'	109°56.02'		650	a	23°0.167'	110°4.32'	

Note: Environment (Env.) = arroyo (a), beach (b), dune (d), old dune (da), and shelf (s).

that the 0.25–0.50 mm sand-size fraction is the most mobile sand grain fraction in the different sedimentary environments. Therefore, this size fraction should be the most representative to work with because it is the fraction that has acquired most of the roughness information. The 0.25–0.50 mm sand size fraction was separated from the sample by using Tyler wet sieves.

Once the samples were cleaned, and the required sand fraction was obtained, 320 quartz grains were picked with the help of a microscope and air mounted in a slide. A sample size between 300 and 400 quartz grains was found to capture most, if not all, of the information for a given sample in the coastal environment (Evangelista et al., 1996). The quartz grains were digitized using a video camera adapted to a petrographic microscope, a Matrox II video board, and the Forma program (Nelson et al., 1996). The Forma program obtains the grain image, calculates a grain center, and obtains polar coordinates of the grain boundary. Data for the

320 grains per sample are stored in a shape file (i.e., sample1.shp). The shape file is used as an input file for the Runfour program (Ehrlich and Full, 1987) using the techniques of Full and Ehrlich (1982); this preserves homology and the ability to compare one grain to the next. Additionally, this technique calculates the amplitude and phase values for the first 24 harmonics for each of the grains. Fourier data are stored in an independent file (i.e., sample1.f). The data are oversampled to avoid aliasing problems (Evangelista et al., 1996).

Optimization Approaches

The large collection of Fourier coefficients needs to be cast in optimal form. The approach used is based on the maximum entropy method (MEM) defined in Full et al. (1984, 1985). Each collection of sample harmonics is cast into frequency distributions. The result of this step is a collection of 23 (the first harmonic represents an error term associated with center finding) harmonic amplitude distributions for each sample. The question then arises where to look for statistically significant information and how many intervals to choose for the distributions. The statistical properties of the MEM provide answers to these questions (Full et al., 1984), and we used a refinement of this approach using the chi-square nonparametric statistic as discussed in Evangelista et al. (1996). The results of the optimization produced a constant sum data set that was used in the subsequent linear analysis. The programs used were unpublished C programs written by Nelson and Full and largely based on the FORTRAN programs in Full (1982).

From the grain-shape data, harmonic 2 is related to grain elongation, and for this work, these data were distributed along 10 optimal ranges or intervals, where interval 1 represented the frequency of grains with low elongation (low harmonic values), and interval 10 represented the frequency of grains with high elongation (higher-energy harmonic values). Frequency distributions were graphed for every studied area (Table 4); we created two graphs for each of the areas, where both graphs included the same beach samples, but one of them included the arroyo samples and the other one included the dune and inner continental shelf samples (Figs. 4–9). Each of the studied areas also includes some samples from nearby areas to identify the sediment path direction along the coast (Table 4).

Unmixing Analysis

For this analysis, three types of frequency distributions were used. The first consisted of the lower-frequency event (harmonic 2, larger features on the grain shape), the second consisted of the total amount of harmonic energy (sum of harmonics 2 through 24), while the third class consisted of the sum of the higher-frequency events (smaller features on the grain shape) created by summing the harmonic energies represented by the 16th through 24th harmonic. This approach follows that outlined in Civitelli et al. (1992) and Evangelista et al. (1996). These combinations of harmonics were submitted to the subsequent unmixing analysis, for each of the six sets of samples, which represent each studied area (Table 5).

Programs used in our present analysis have been called by different names for different applications. Three of these names are VECTOR, PVA, and SAWVEC, and all three names have appeared in literature. All of these programs are essentially the same pair of programs: EXTENDED CABFAC (Full et al., 1981) and EXTENDED/FUZZY QMODEL (Full et al., 1981, 1982) were presented in Full (1982). For the sake of this report, two parts of the unmixing programs are called VECTOR1 and VECTOR2, which correspond to the previously mentioned programs.

The program VECTOR1 was used to determine the number of subpopulations or end members. This program generates output that is subsequently used in the VECTOR2 program. The VECTOR2 program defines each end member and determines the relative frequency of each end member in each sample. Each end member is "sample-like" in that it is represented by a frequency distribution and may actually represent a real sample (Healy-Williams et al., 1997). The statistics determined by both of these programs are analyzed to determine an optimal number of end members and the proper definition of the frequencies and composition of each end member.

The results of the unmixing analysis strongly suggested that the optimal solution was four end members. The VECTOR2 program defined the frequencies of four end members for the set of samples within each studied area (Table 6). With the obtained frequencies of the beach samples, frequency distributions were created giving three plots per area, one for each harmonic combination (Figs. 10–15).

RESULTS

Frequency Distribution of the Quartz Elongation (QE´FD)

El Cardonal Area

Plots that include the Quartz Elongation Frequency Distribution (QE´FD) (harmonic 2, Figs. 4A and 4B; data in Table 4) of sediment samples within this area show that arroyos contain higher frequencies of high intervals (high elongation), while coastal dunes, inner continental shelf, and beaches areas have higher frequencies of low intervals (low elongation). Samples from the shelf show two types of quartz shape distributions: samples 243s (−103 m deep), 244s (−105 m), and 245s (−115 m) with high frequencies in the low intervals, and samples 239s, 240s, 241s, and 242s without dominant elongation frequency peaks.

El Arco Area

Plots that include the QE´FD (harmonic 2, Figs. 5A and 5B; data in Table 4) of sediment samples within this area, show that a local arroyo (sample 493a) is similar to the QE´FD of a local

TABLE 4. FREQUENCY DISTRIBUTION OF QUARTZ ELONGATION FOR 10 INTERVALS

Area	Sample	Env.	Area*	\multicolumn{10}{c}{Elongation interval[†]}									
				1	2	3	4	5	6	7	8	9	10
Arroyo samples													
1	488	a		1.27	3.16	7.59	9.18	9.18	13.61	12.97	14.56	14.87	13.61
1	489	a		0.67	3.67	3.67	6.00	8.67	12.67	10.33	14.00	20.00	20.33
1	642	a		0.31	3.77	3.77	5.97	11.95	10.38	13.84	15.41	19.18	15.41
1	643	a		0.32	3.19	6.07	7.67	12.46	10.86	8.95	12.78	17.89	19.81
1	644	a		0.32	3.53	3.21	5.77	9.29	12.50	8.65	14.74	20.51	21.47
1	645	a		0.63	0.95	5.40	9.21	8.25	10.79	11.11	15.24	17.14	21.27
1	648	a		0.64	5.45	5.13	7.37	10.26	11.54	13.14	16.99	16.35	13.14
1	650	a		0.97	2.59	8.09	6.15	10.68	13.27	13.27	13.92	15.53	15.53
1	651	a		1.59	4.13	7.94	9.84	10.16	10.79	11.43	14.60	17.14	12.38
1	652	a		1.60	4.79	4.15	5.43	8.63	11.50	16.61	16.29	14.70	16.29
2	493	a		1.33	3.33	6.67	8.33	7.00	13.33	14.33	17.67	18.00	10.00
3	347	a		1.51	4.82	6.33	7.83	9.94	12.35	11.45	14.46	16.27	15.06
3	506	a		2.27	3.25	9.74	9.74	12.34	10.39	10.39	12.01	13.96	15.91
3	507	a		1.66	2.33	6.64	7.64	8.31	7.31	12.62	15.28	17.94	20.27
4	348	a		0.60	2.38	8.93	6.85	11.90	12.20	12.50	17.26	16.37	11.01
4	349	a		1.22	2.43	5.17	8.21	10.64	11.85	13.07	14.59	13.07	19.76
4	350	a		1.52	4.56	6.99	6.99	12.16	8.21	13.98	13.37	14.59	17.63
4	360	a		0.89	5.04	5.04	10.68	8.31	11.28	12.17	13.95	16.62	16.02
4	520	a		3.03	4.24	4.55	10.00	8.18	12.42	11.82	14.55	16.97	14.24
4	540	a		2.30	4.34	7.65	10.71	7.65	9.69	13.27	13.78	13.78	16.84
5	351	a	4	1.80	3.89	7.78	8.68	11.08	12.87	14.07	11.08	13.47	15.27
5	358	a		3.61	3.92	9.94	13.55	12.95	12.65	11.45	11.14	9.94	10.84
5	359	a		0.91	3.02	6.95	9.37	12.69	10.27	16.62	10.57	11.78	17.82
5	524	a		27.04	14.78	14.47	9.12	9.12	6.29	6.60	6.29	4.40	1.89
5	527	a		5.19	7.79	10.71	10.71	12.99	10.06	10.06	9.42	11.36	11.69
6	353	a		0.90	4.19	6.29	10.18	12.87	10.78	14.67	14.07	16.77	9.28
6	354	a		1.55	4.33	8.98	9.29	12.38	13.62	10.22	11.46	13.93	14.24
6	355	a		0.59	2.67	7.72	7.72	11.57	13.95	14.54	12.46	13.65	15.13
6	356	a		0.90	2.41	7.83	9.04	11.45	12.65	13.55	14.76	13.25	14.16
6	357	a		0.90	1.50	5.11	6.91	12.01	13.21	13.51	15.92	16.22	14.71
6	529	a		1.66	5.32	4.65	9.30	10.30	10.63	13.62	13.62	15.61	15.28
6	530	a		2.00	4.00	10.33	7.00	10.00	11.00	11.67	13.00	13.33	17.67
6	531	a		2.27	4.21	5.83	9.06	7.44	11.33	12.95	12.62	17.15	17.15
6	535	a		8.97	8.65	11.86	8.65	9.62	15.38	13.78	9.94	9.29	3.85
6	536	a		1.27	4.11	6.33	8.23	9.49	13.92	13.61	16.46	12.03	14.56
Beach samples													
1	315	b		14.94	18.51	11.69	12.01	11.36	9.74	6.17	7.14	4.22	4.22
1	316	b		14.80	14.20	12.39	10.27	12.69	10.27	7.55	5.74	6.34	5.74
1	317	b		15.48	22.58	13.87	10.97	7.42	8.39	8.06	4.84	5.16	3.23
2	495	b	1	16.98	17.30	15.72	12.58	12.58	9.12	5.66	4.40	3.46	2.20
2	497	b		3.86	6.11	8.36	9.65	10.29	14.47	9.97	11.25	12.54	13.50
2	501	b	3	26.90	14.87	9.81	11.71	12.03	7.59	6.01	4.11	2.85	4.11
2	504	b	3	20.82	17.35	12.93	15.46	8.20	7.57	6.31	4.42	3.47	3.47
3	518	b	2	13.48	11.91	11.60	12.85	9.72	10.66	12.23	6.58	6.90	4.08
3	519	b	2	11.11	12.06	13.02	14.92	11.75	7.94	10.16	6.35	5.71	6.98
3	545	b		16.50	15.51	13.53	10.56	11.22	6.60	6.60	9.90	3.96	5.61
3	547	b	4	2.68	4.01	5.02	6.35	9.03	12.71	9.70	14.05	15.05	21.40
3	553	b	2	11.00	9.33	9.33	9.00	13.33	11.00	7.33	8.67	11.33	9.67
4	521	b		15.28	11.96	11.30	7.31	6.98	8.64	12.29	12.62	5.65	7.97
4	544	b		3.32	4.65	8.31	8.97	9.63	15.28	13.29	10.30	14.29	11.96
5	523	b		5.08	9.84	7.62	9.84	11.75	10.48	11.75	9.21	13.02	11.43
5	525	b		24.29	18.30	14.51	13.25	8.52	6.94	4.42	3.79	4.42	1.58
5	526	b		16.46	10.13	13.29	14.87	9.49	9.18	8.86	5.70	6.33	5.70
5	539	b		32.33	14.00	12.33	7.33	9.00	6.33	5.00	6.67	4.00	3.00
6	528	b		16.33	12.67	9.67	8.67	9.67	5.67	11.67	8.67	9.33	7.67
6	532	b		4.95	5.94	5.28	6.27	8.91	10.89	15.51	14.52	12.87	14.85
6	537	b		5.43	10.22	12.46	13.10	14.70	9.90	12.14	8.95	7.99	5.11
6	538	b		11.67	10.00	12.67	10.33	10.00	5.67	9.67	9.67	9.33	11.00

(Continued.)

beach (sample 497b), which is located to the east of the mouth of the arroyo. The QE´FD of the beach (sample 495b) located to the west of the arroyo mouth is different than the QE´FD of the arroyo. In the area of Cerro Prieto, where no arroyo samples were collected, beach (samples 501b and 504b), dune (samples 500d, 562d, and 563d) and inner continental shelf samples (sample 508s from 20 m deep and 509s from 15 m deep), show similar QE´FD. These samples are characterized by high frequencies in the low intervals (low elongation). The QE´FD values for beaches at the Cabo San Lucas Cape (samples 518b and 519b) are similar to inner continental shelf sample 246s, located to the south of the Cabo San Lucas Bay, and also to a beach sample (sample 553b) from the western side of the bay.

San Lucas Area

Plots that include the QE´FD (harmonic 2, Figs. 6A and 6B; data in Table 4) of sediment samples within this area show that beach samples within San Lucas Bay are different than the local

TABLE 4. (Continued.)

Area	Sample	Env.	Area*	Elongation interval†									
				1	2	3	4	5	6	7	8	9	10
Shelf samples													
1	239	s		9.27	14.57	10.60	11.26	8.94	13.91	11.92	6.95	7.95	4.64
1	241	s		5.11	8.31	13.10	10.54	10.54	10.54	10.54	9.58	12.46	9.27
1	242	s		10.70	10.70	13.38	9.70	11.71	8.36	9.03	7.69	9.70	9.03
1	243	s		17.28	14.62	10.63	10.96	8.31	9.30	8.64	7.64	6.31	6.31
1	244	s		17.88	15.89	9.93	10.93	7.95	10.26	6.62	6.29	6.95	7.28
1	245	s	2	23.92	20.93	15.28	10.96	7.64	5.32	5.32	5.32	3.65	1.66
2	508	s	3	21.93	12.62	12.96	11.96	7.97	5.32	5.98	6.98	5.32	8.97
3	246	s		14.62	11.96	11.30	10.30	10.63	11.63	7.31	7.64	7.31	7.31
3	509	s	2	30.69	12.21	12.21	12.21	8.25	7.92	4.95	4.62	2.97	3.96
3	510	s		2.33	4.32	7.31	9.30	11.96	11.63	9.97	13.95	11.30	17.94
3	511	s	4	6.62	8.28	6.95	9.93	5.96	10.26	12.25	12.25	11.59	15.89
3	512	s		7.95	8.94	8.61	6.95	8.28	11.92	10.26	13.25	15.89	
3	513	s		5.41	7.01	7.32	11.47	8.60	10.51	11.78	11.78	14.33	11.78
3	514	s		6.31	5.32	7.97	9.30	7.97	12.62	14.62	10.63	11.63	13.62
4	247	s		3.33	8.67	11.33	7.67	11.33	11.33	11.67	11.67	9.00	14.00
4	554	s		8.97	9.97	9.63	8.64	9.97	7.97	10.30	11.96	11.96	10.63
4	555	s		3.57	3.57	7.79	9.74	11.04	11.36	12.34	11.69	12.99	15.91
4	556	s	5	2.68	5.69	4.35	10.03	12.71	10.03	14.72	12.37	13.38	14.05
5	557	s		4.65	9.97	9.97	6.31	9.30	10.30	10.96	19.27	7.64	11.63
5	249	s		6.33	8.67	12.00	14.00	12.00	8.67	7.00	8.33	11.67	11.33
5	248	s		10.54	10.22	7.99	7.99	11.18	9.58	10.54	10.86	10.86	10.22
5	558	s		10.96	11.63	11.96	10.30	10.30	8.31	9.30	9.30	5.65	11.96
5	559	s		21.93	14.62	16.28	10.96	10.30	5.98	6.31	6.64	4.32	2.66
	240	s		5.30	5.96	6.62	9.60	9.93	9.60	11.59	13.91	12.58	14.90
6	250	s		9.40	9.72	10.66	12.23	10.97	11.29	10.34	9.72	8.46	7.21
6	549	s		4.65	8.31	8.31	10.63	6.98	13.95	11.63	11.63	11.63	12.29
6	550	s		0.66	5.32	6.31	5.32	10.96	8.97	13.62	17.94	16.28	14.62
6	552	s	5	20.00	10.67	12.67	11.33	8.00	10.33	7.67	7.00	6.67	5.67
Dune samples													
1	274	da		25.16	20.44	15.72	10.06	10.38	6.60	4.40	4.09	1.89	1.26
1	327	d		28.75	20.31	11.88	11.56	8.75	4.69	7.19	2.50	3.44	0.94
1	328	d		24.13	21.59	18.10	12.06	7.94	6.03	4.76	1.59	2.86	0.95
1	330	d		21.56	21.56	15.31	12.81	9.38	6.88	6.88	2.50	1.88	1.25
1	331	d		19.81	19.50	13.84	14.47	9.75	7.55	5.03	4.09	4.09	1.89
2	340	da		24.78	22.12	17.11	9.44	8.55	6.19	7.08	1.77	2.36	0.59
2	500	d		23.90	17.92	14.78	12.26	9.75	6.60	4.40	3.46	4.72	2.20
2	562	d		24.84	20.75	15.09	14.47	7.86	5.03	4.09	3.46	2.20	2.20
2	563	da		22.47	18.04	18.04	11.71	9.81	8.23	3.16	4.75	1.58	2.22
3	560	da		20.13	16.98	12.58	12.26	11.95	6.60	8.49	5.35	2.83	2.83

Note: Environment (Env.) = arroyo (a), beach (b), dune (d), old dune (da), and shelf (s).
*Area where sample was also included for analysis.
†Interval: 1—low elongation, 10—high elongation.

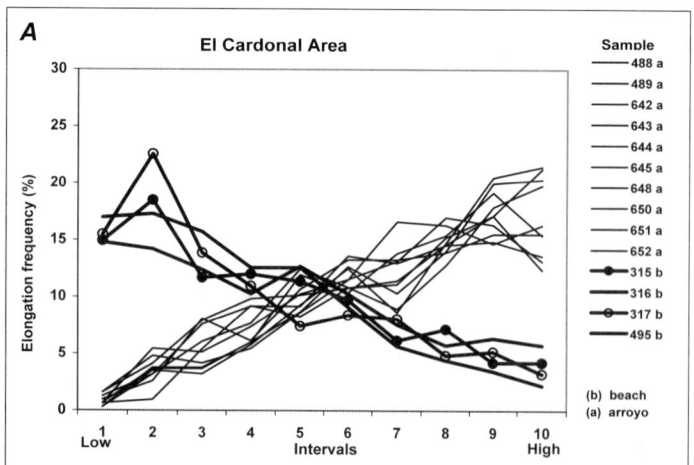
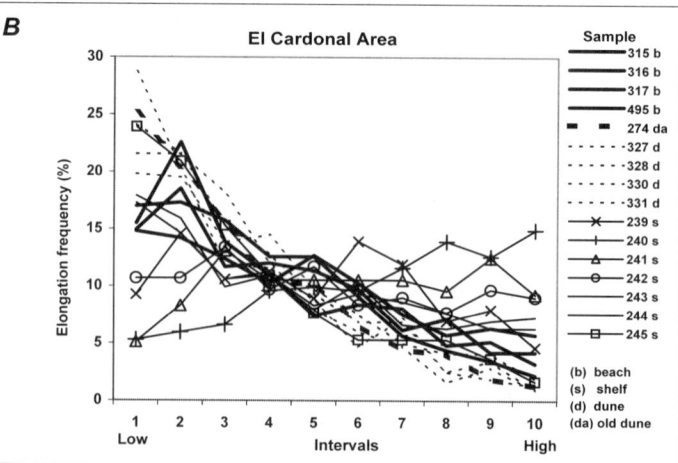

Figure 4. Area 1, El Cardonal. Quartz elongation frequency distributions along 10 intervals for harmonic 2: (A) arroyo and beach samples, and (B) beaches, dunes, old stabilized dune, and shelf samples.

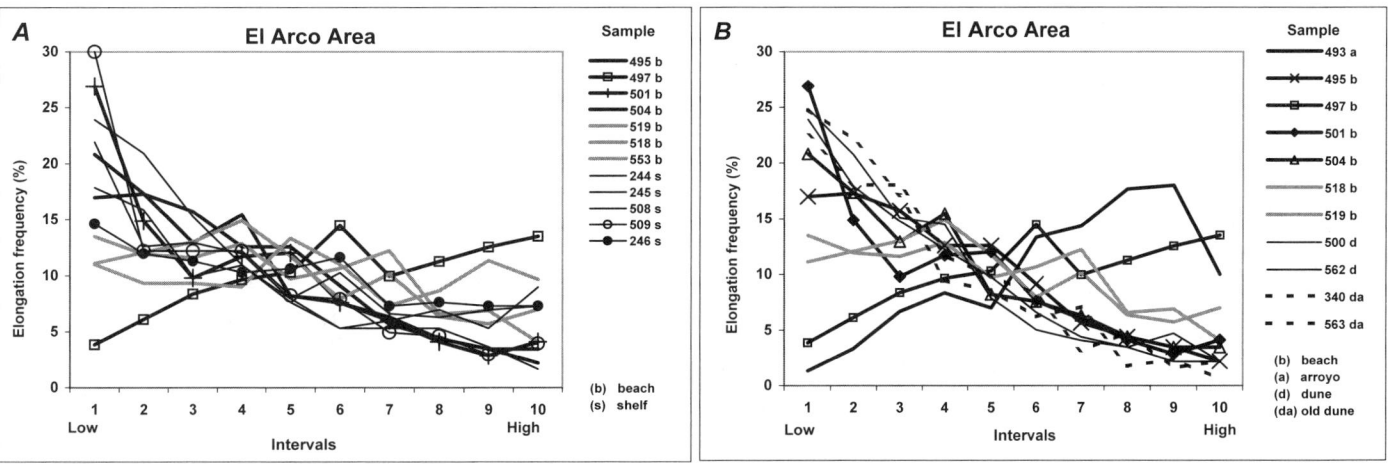

Figure 5. Area 2, El Arco. Quartz elongation frequency distributions along 10 intervals for harmonic 2: (A) beach and shelf samples, and (B) arroyo, beach, dunes, and old stabilized dune samples.

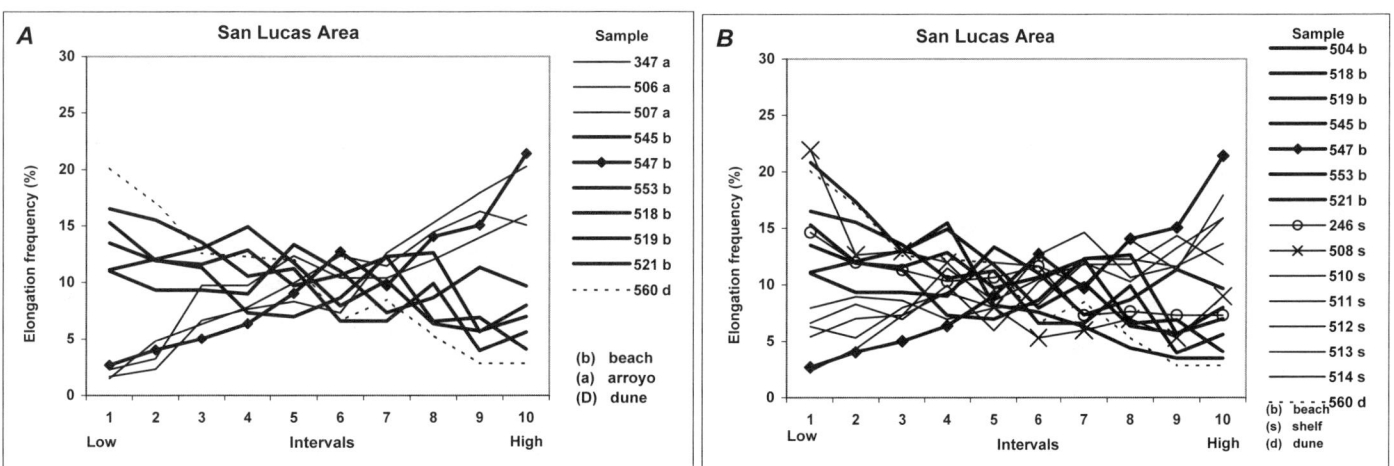

Figure 6. Area 3, San Lucas. Quartz elongation frequency distributions along 10 intervals for harmonic 2: (A) arroyo, beach, and dune samples, and (B) beach, shelf, and dune samples.

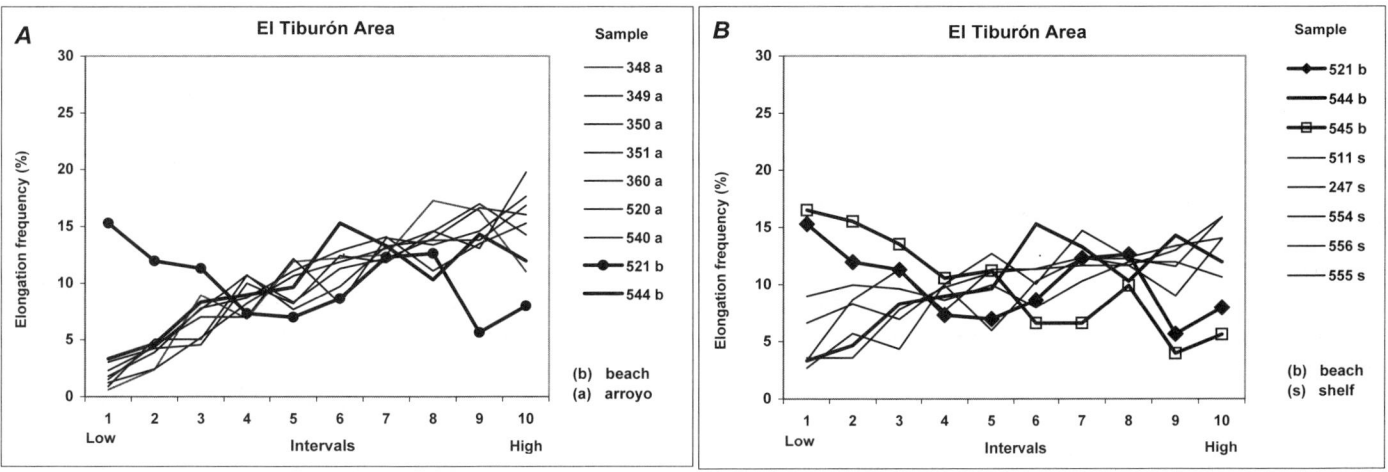

Figure 7. Area 5, El Tiburón. Quartz elongation frequency distributions along 10 intervals for harmonic 2: (A) arroyo and beach samples, and (B) beach and shelf samples.

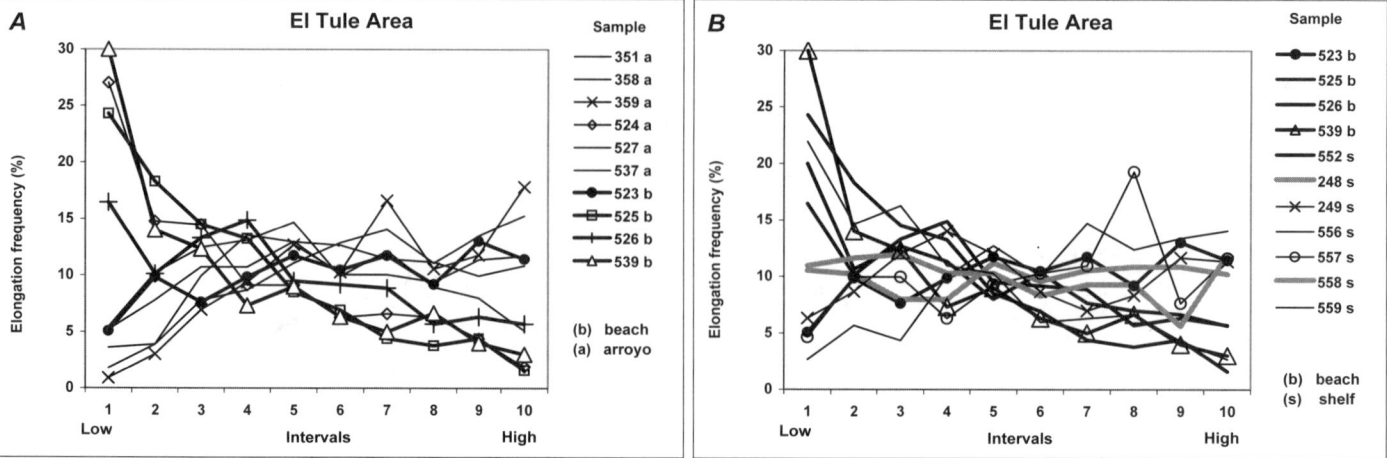

Figure 8. Area 5, El Tule. Quartz elongation frequency distributions along 10 intervals for harmonic 2: (A) arroyo and beach samples, and (B) beach and shelf samples.

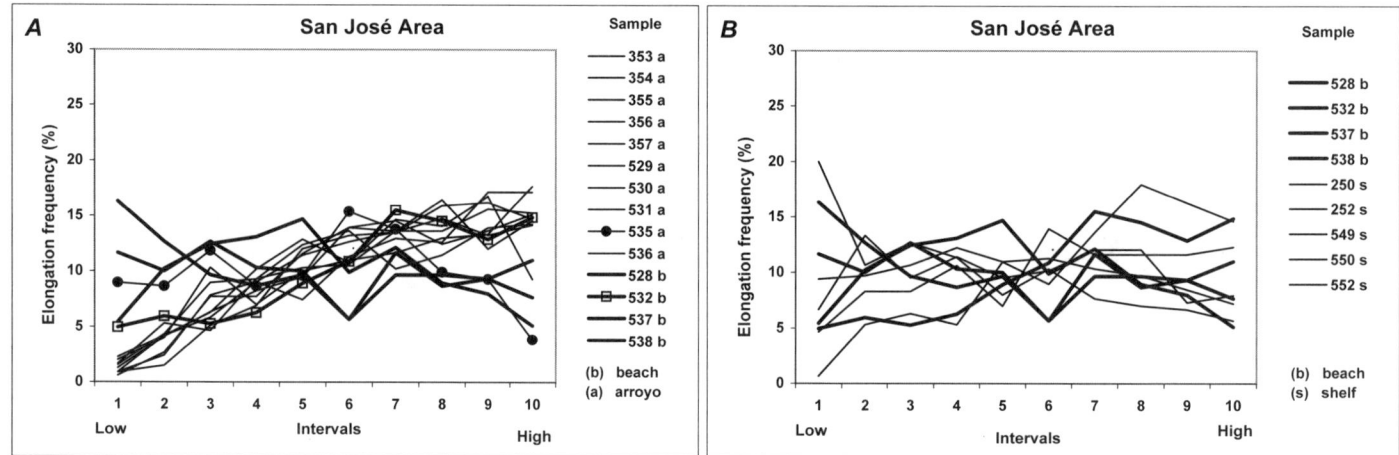

Figure 9. Area 6, San José. Quartz elongation frequency distributions along 10 intervals for harmonic 2: (A) arroyo and beach samples, and (B) beach and shelf samples.

arroyos and shallow shelf samples. Samples from the beaches, dune (samples 553b, 545b, and 560d), and shelf (sample 246s, 63 m deep) are more similar to the QE´FD of El Arco Beach samples (from El Arco area, to the west, samples 501b and 504b) than to local arroyo sediments. This is also observed for beach samples at the outer and inner coast of the San Lucas Cape. The QE´FD value of deep shelf material (sample 246s) within the Cabo San Lucas Bay is similar to the QE´FD value of western open-ocean coast material. The QE´FD values of fluvial sediments from the El Salto drainage basin are similar to the ones from shallow, inner continental shelf (samples 510s, 511s, 512s, 513s, and 514s).

El Tiburón Area

Plots that include the QE´FD (harmonic 2, Figs. 7A and 7B; data in Table 4) of sediment samples within this area show that a beach sample at the mouth of the El Tiburón drainage basin (sample 521b) is not similar to the QE´FD of adjacent arroyo samples (520a and 360a). This sample is more similar to a beach sample from the Cabo San Lucas Bay (sample 545b). The QE´FD value of the beach sample 544b, is similar to the QE´FD values of local arroyos (348a and 349a, among others).

El Tule Area

Plots that include the QE´FD (harmonic 2, Figs. 8A and 8B; data in Table 4) of sediment samples within this area show that beach samples located to the east of the El Tule arroyo (samples 525b, 526b, and 539b) are similar to the QE´FD of the arroyo sample from El Tule (sample 524a). For beach sample 523b, which is located to the southwest of the El Tule arroyo mouth, the observed QE´FD is not similar to the El Tule arroyo sample. The QE´FD of the shelf sample 248s is similar to the QE´FD of the shelf sample 558s; both samples contain around 10% of frequency for most of the intervals, which shows that there is not

TABLE 5. SAMPLES USED IN STUDY FOR ANALYSIS

El Cardonal	El Arco	San Lucas	El Tiburón	El Tule	San José
239 s	245 s	246 s	247 s	248 s	250 s
241 s	340 d	347 a	348 a	249 s	353 a
242 s	493 a	501 b	349 a	351 a	354 a
243 s	495 b	504 b	350 a	358 a	355 a
244 s	497 a	506 a	351 a	359 a	356 a
245 s	500 d	507 a	360 a	523 b	357 a
274 d	501 b	508 s	511 s	524 a	528 b
315 b	504 b	509 s	520 a	525 b	529 a
316 b	508 s	510 s	521 b	526 b	530 a
317 b	509 s	511 s	540 a	527 a	531 a
327 d	518 b	512 s	544 b	539 b	532 b
328 d	519 b	513 s	547 b	552 s	535 a
330 b	553 b	514 s	554 s	556 s	536 a
331 d	562 d	518 b	555 s	557 s	537 b
488 a	563 d	519 b	556 s	558 s	538 b
489 a		545 b		559 s	549 s
495 b		547 b			550 s
642 a		553 b			552 s
643 a		560 d			
644 a					
645 a					
648 a					
650 a					
651 a					
652 a					

Note: Samples are from shelf (s), dune (d), beach (b), arroyo (a).

a dominant quartz elongation frequency. Most of the shelf and arroyo samples within this area show similar QE´FD values.

San José Area

Plots that include the QE´FD (harmonic 2, Figs. 9A and 9B; data in Table 4) of sediment samples within this area show that a beach sample (sample 532b), at the northeast of the intermittent San Jose arroyo mouth, is similar to the QE´FD of samples from this arroyo. The fluvial samples from San José have similar QE´FD values, with the sole exception of sample 535a, which is located on the eastern portion of this area and displays a slightly higher frequency of grains in the lower intervals. The QE´FD values of the beach samples 537b, 528b, and 538b are more similar to the QE´FD of shallow shelf samples than to local arroyo samples. The QE´FD value of deeper shelf sample 250s (80 m) is similar to the QE´FD values of deep shelf samples within El Tule area, and it has similar frequencies between the low and high intervals. Beach samples from this area, along all the intervals, have the most irregular histograms.

TABLE 6. END-MEMBER RELATIVE FREQUENCIES OF BEACH SAMPLES FOR HARMONIC 2, HARMONICS 2–24, AND HARMONICS 16–24

Beach samples	Harmonic 2 end members				Harmonic 2–24 end members				Harmonic 16–24 end members			
Area 1	274 d	643 a	239 s	650 a	317 b	489 a	488 a	315 b	330 d	489 a	239 s	243 s
315 b	43.41	36.44	20.15	0.00	0.00	12.61	14.97	72.42	32.94	6.75	23.20	37.11
316 b	51.42	39.63	8.96	0.00	40.97	32.87	0.00	26.16	31.37	12.25	21.82	34.56
317 b	29.37	16.80	27.25	26.58	66.54	12.61	14.97	5.88	33.67	5.87	15.91	44.54
495 b	46.78	27.03	26.19	0.00	30.08	1.88	22.67	45.38	39.00	2.67	15.81	42.52
Area 2	553 b	508 s	518 b	495 b	562 d	493 a	518 b	504 b	508 s	493 a	509 s	495 b
495 b	8.76	8.13	8.30	74.81	33.86	0.00	66.14	0.00	27.86	7.98	17.32	46.84
501 b	14.56	17.71	9.16	58.57	33.90	17.21	0.87	48.01	40.67	15.89	31.79	11.65
504 b	0.00	12.31	27.52	60.17	0.00	14.79	2.70	82.51	38.01	13.47	24.82	23.69
518 b	8.76	8.13	73.18	9.93	0.00	14.79	73.78	11.43	0.00	35.26	31.16	33.58
519 b	54.98	0.00	45.02	0.00	25.83	14.50	59.67	0.00	24.25	34.26	14.89	26.59
553 b	14.29	24.11	48.69	12.90	13.31	44.25	0.00	42.44	36.57	51.48	6.21	5.74
Area 3	560 d	507 a	506 a	504 b	509 s	511 s	518 b	547 b	507 a	509 s	519 b	508 s
501 b	38.41	19.82	2.83	38.93	53.19	22.60	17.30	6.91	3.67	70.80	9.69	15.84
504 b	4.93	4.71	7.39	82.97	28.00	23.17	48.74	0.09	0.00	52.10	18.06	29.84
518 b	50.18	10.62	0.00	39.20	6.14	3.56	67.35	22.95	10.89	41.84	42.98	4.29
519 b	69.42	0.00	0.00	30.58	22.86	0.00	44.67	32.47	8.47	24.65	43.60	23.28
545 b	43.39	25.43	10.15	21.02	29.57	23.50	37.21	9.72	6.12	37.68	29.87	26.33
547 b	55.61	44.39	0.00	0.00	8.54	14.35	6.29	70.82	60.42	6.35	27.95	5.28
553 b	17.76	23.46	0.71	58.07	21.55	32.72	26.49	19.25	21.08	22.97	42.34	13.61
Area 4	350 a	540 b	547 s	349 a	511 s	521 b	360 a	547 b	360 a	521 b	351 a	547 b
521 b	7.99	0.00	40.76	51.26	16.22	79.92	0.00	3.86	8.74	91.26	0.00	0.00
540 b	10.71	83.62	5.67	0.00	0.63	43.46	23.79	32.11	41.72	10.58	18.03	29.68
544 b	32.85	0.00	30.20	36.95	12.45	28.04	2.52	56.99	43.68	17.95	38.37	0.00
547 b	10.71	16.07	73.22	0.00	16.22	0.00	0.00	83.78	8.74	7.46	0.00	83.80
Area 5	525 b	558 s	559 s	359 b	556 s	525 b	523 b	249 s	359 a	539 b	249 s	559 s
523 b	22.11	6.54	22.15	49.21	3.38	13.26	62.05	21.31	41.55	5.88	36.00	16.57
525 b	72.83	14.61	12.56	0.00	0.00	54.72	40.62	4.65	0.00	38.41	27.92	33.68
526 b	51.44	0.00	0.49	48.07	0.00	43.02	13.64	43.35	14.57	23.55	34.33	27.55
539 b	40.82	0.00	3.13	56.05	0.00	43.29	15.39	41.33	23.63	56.24	3.93	16.20
Area 6	537 b	357 a	535 a	550 s	532 b	535 a	528 b	549 s	357 a	552 s	537 b	532 b
528 b	0.00	31.15	27.36	41.49	0.84	10.77	72.54	15.85	1.15	69.37	9.09	20.39
532 b	7.20	60.13	23.20	9.47	59.23	13.27	17.71	9.79	6.63	22.47	18.36	52.54
537 b	65.32	12.39	13.33	8.96	11.87	47.86	33.92	6.35	6.72	33.83	58.82	0.63
538 b	33.09	42.48	0.00	24.43	2.79	7.88	60.30	29.03	0.00	51.01	21.01	27.97

Note: Samples are from arroyo (a), beach (b), dune (d), shelf (s). Data are for Figures 10–15.

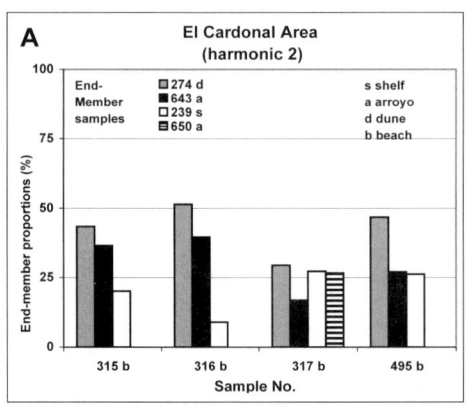 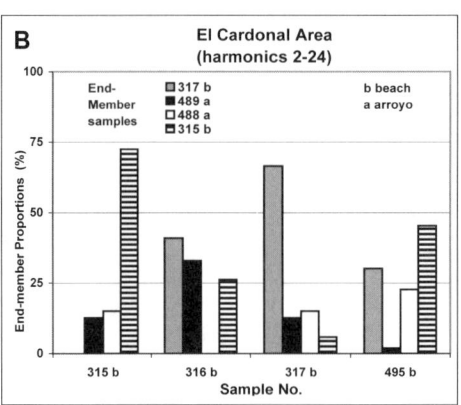 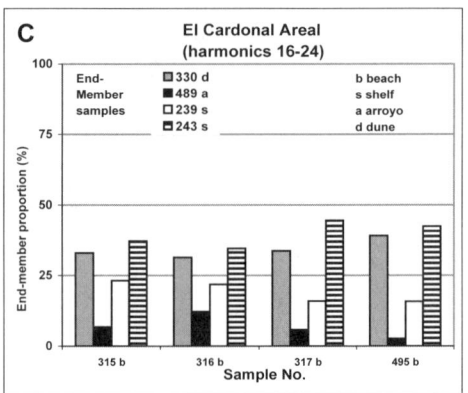

Figure 10. Area 1, El Cardonal. End-member relative frequency (%) of beach samples, for (A) harmonic 2, (B) harmonics 2 through 24 (2–24), and (C) harmonics 16 through 24 (16–24).

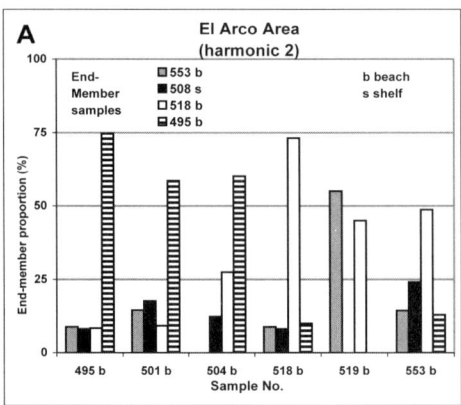 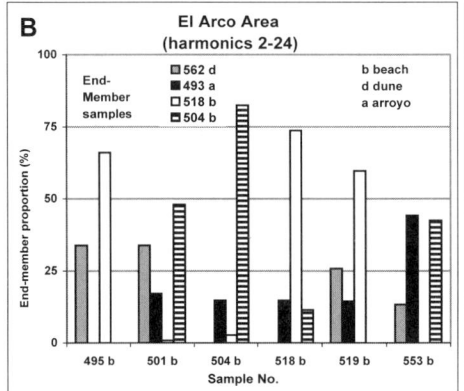 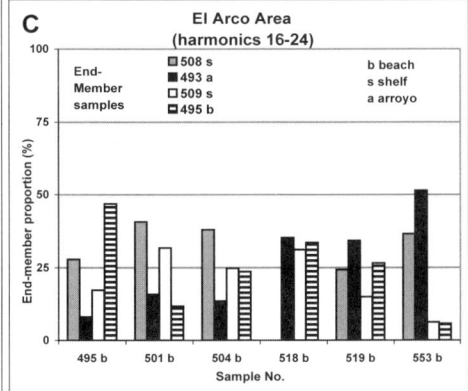

Figure 11. Area 2, El Arco. End-member relative frequency (%) of beach samples for (A) harmonic 2, (B) harmonics 2 through 24 (2–24), and (C) harmonics 16 through 24 (16–24).

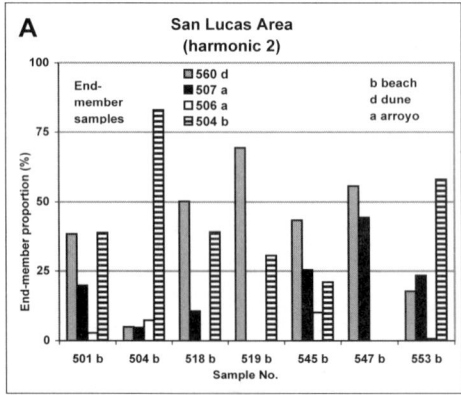 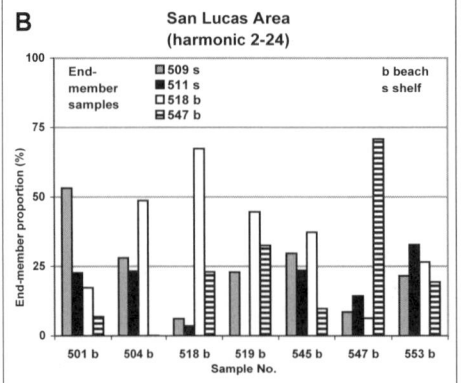 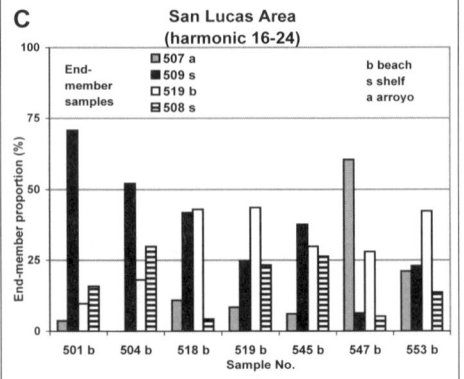

Figure 12. Area 3, San Lucas. End-member relative frequency (%) of beach samples for (A) harmonic 2, (B) harmonics 2 through 24 (2–24), and (C) harmonics 16 through 24 (16–24).

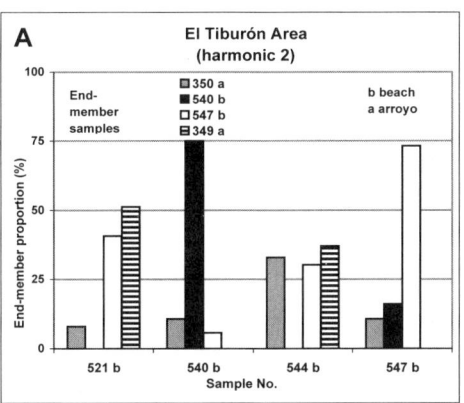
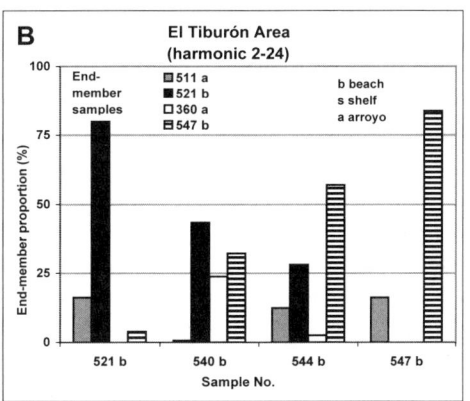
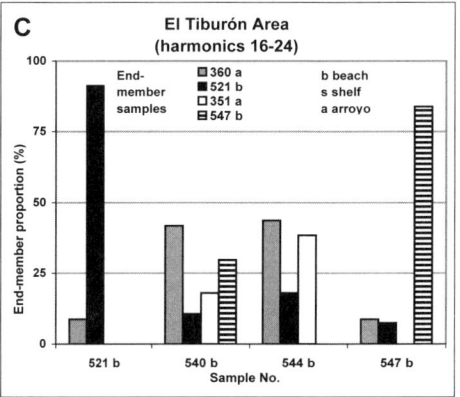

Figure 13. Area 4, El Tiburón. End-member relative frequency (%) of beach samples for (A) harmonic 2, (B) harmonics 2 through 24 (2–24), and (C) harmonics 16 through 24 (16–24).

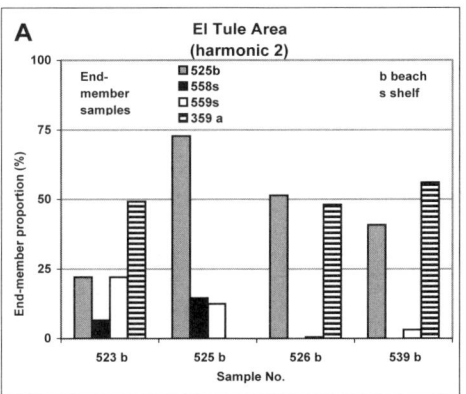
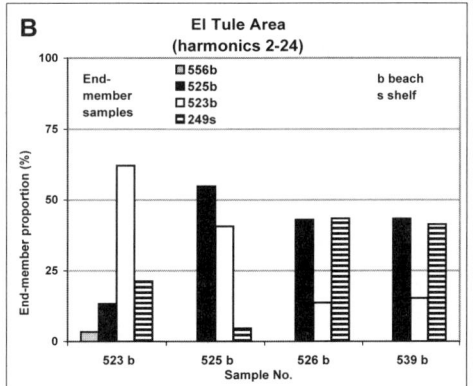
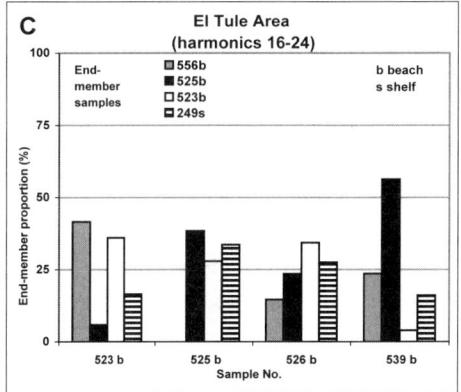

Figure 14. Area 5, El Tule. End-member relative frequency (%) of beach samples for (A) harmonic 2, (B) harmonics 2 through 24 (2–24), and (C) harmonics 16 through 24 (16–24).

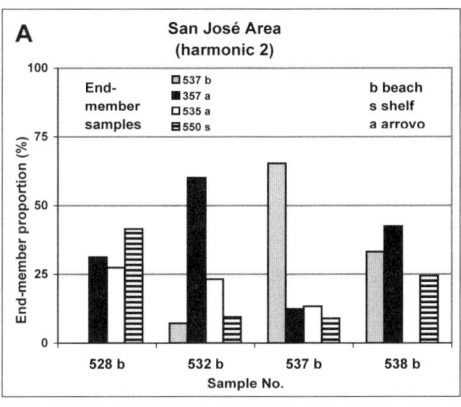
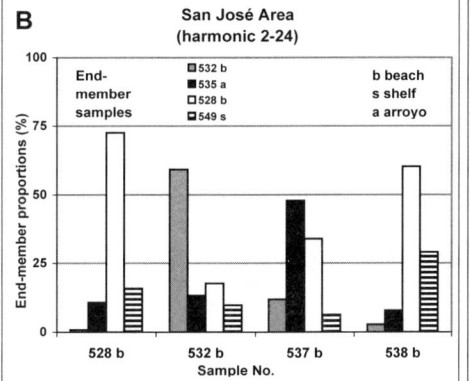
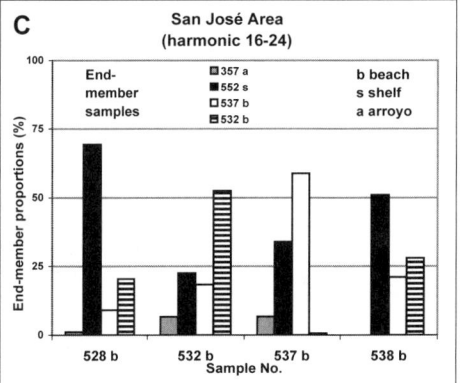

Figure 15. Area 6, San José. End-member relative frequency (%) of beach samples for (A) harmonic 2, (B) harmonics 2 through 24 (2–24), and (C) harmonics 16 through 24 (16–24).

End-Member Samples

The end-member samples are the samples that contain the most different shape information among all the samples. For a better understanding of the grain-shape composition data, the end-member samples were obtained from single harmonic information, harmonic 2, and group harmonic information, harmonics 2–24, and harmonics 16–24 (Table 6). Low harmonics give information about the grain elongation, which is acquired during crystal formation, and high harmonics give information about the grain-surface roughness, which is acquired during transport.

El Cardonal Area

For harmonic 2, the end-member samples are: 274d, 643a, 239s, and 650a. For harmonics 2–24, the end-member samples are: 317b, 489a, 488a, and 315b. For harmonics 16–24, the end-member samples are: 330d, 489a, 239s, and 243s.

El Arco Area

For harmonic 2, the end-member samples are: 553b, 508s, 518b, and 650a. For harmonics 2–24, the end-member samples are: 562d, 493a, 518b, and 504b. For harmonics 16–24, the end-member samples are: 508s, 493a, 509s, and 495b.

San Lucas Area

For harmonic 2, the end-member samples are: 560d, 507a, 506a, and 504b. For harmonics 2–24, the end-member samples are: 509s, 511s, 518b, and 547b. For harmonics 16–24, the end-member samples are: 507a, 509s, 519b, and 508s.

El Tiburón Area

For harmonic 2, the end-member samples are: 350a, 540b, 547s, and 349a. For harmonics 2–24, the end-member samples are: 511s, 521b, 360a, and 547b. For harmonics 16–24, the end-member samples are: 360a, 521b, 351a, and 547b.

El Tule Area

For harmonic 2, the end-member samples are: 525b, 558s, 559s, and 359b. For harmonics 2–24, the end-member samples are: 556s, 525b, 523b, and 249s. For harmonics 16–24, the end-member samples are: 359a, 539b, 249s, and 559s.

San José Area

For harmonic 2, the end-member samples are: 537b, 357a, 535a, and 550s. For harmonics 2–24, the end-member samples are: 532b, 535a, 528b, and 549s. For harmonics 16–24, the end-member samples are: 357a, 552s, 537b, and 532b.

DISCUSSION

El Cardonal Area

Beaches within this area have different sedimentary environment sources. The QE´FD values show that the shelf is the main source for the beach where samples 243s, 244s, and 245s are located, at the north and west shelf of El Cardonal drainage basin area. The end-member samples for harmonic 2 (Fig. 10A) indicate that old aeolian material (sample 274d) is the main source of the beaches in this area (samples 315b, 316b, 317b, and 495b), followed by local arroyos (643a, 650a) and the inner continental shelf (sample 239s). The end-member samples for harmonics 16–24 (Fig. 10C) show that the beach source is mostly recent aeolian material (40%), followed by the inner continental shelf (samples 243s and 239s), and a minor contribution from local arroyos (sample 489a).

For this area, it can be concluded that the dunes and inner continental shelf are important sources for the beach material, with a minor contribution from local or surrounding arroyos. The explanation of how material from the inner continental shelf and dunes could be the main source of sediment may be related to sea-level rise, which has moved the old beach inland, or possibly offshore sand movement and coastal dune erosion during high-energy storms, as identified by Camacho-Valdéz (2003) within this area. Relict sediment is probably part of the material of recent beach along this area. The contribution of the old dune material to the beach probably occurs through the erosion of the dunes by intermittent streams, which eventually dissect the beach, discharging material into the coast.

El Arco Area

The QE´FD values show that a local arroyo (sample 493a) is the source of the beach located to the east of the mouth of the arroyo. The beach QE´FD (sample 495b) located to the west of the arroyo mouth is different than the QE´FD of the arroyo, which suggests a different source. In the area of Cerro Prieto, where no arroyo samples were collected, beach (samples 501b and 504b), dune (samples 500d, 562d, and 563d), and inner continental shelf samples (sample 508s from 20 m deep and 509s from 15 m deep) show similar QE´FD values, which suggest that the nearby inner continental shelf is the major source of the beach and dune samples. These samples are characterized by high frequencies in the low intervals (low elongation). The QE´FD values for beaches at the Cabo San Lucas Cape (samples 518b and 519b) are similar to a inner continental shelf sample, 246s, located to the south of the Cabo San Lucas Bay, and also to a beach sample (sample 553b) from the western side of the bay. This similarity suggests that material from the cape travels around the San Lucas Cape toward the western side of this bay.

The end members from harmonic 2 data (Fig. 11A) show that the beach sample 495b is an end member (75%). Material from this beach location is supplying material to the eastern beaches (samples 501b and 504b). The beach sample 518b is another end member (73%), which shows that the beach where this sample was obtained has some contribution from a local arroyo (sample 493a), and also from the beach to the west (sample 504b). The beach (sample 518b) closer to the point of the Cabo San Lucas Cape is the source for the beach immediately to the north (sample 519b) and is also the source for the western beach of the Cabo San Lucas Bay (sample 553b). This suggests longshore transport from west to east that influences of the outer west coast into the Cabo San Lucas Bay.

The end-member samples from harmonics 2–24 data (Fig. 11B) show that the source of the western beach (sample 495b), which was collected to the west of the Palma Seda arroyo mouth, is not from the Palma Seda arroyo (sample 493a), but this beach has some contribution of material from the east portion of El Arco area. This suggests that there is probably also littoral transport from east to west in this area. This may happen during discharge periods of the arroyo, when the west beach does get some material from this arroyo, but it may depend on the littoral transport direction at that time. Alternatively, littoral transport may bypass the beach area entirely during high arroyo discharge. A more detailed sampling of the immediate region offshore of this portion would elucidate the correct mechanism. The source for the beach where sample 501b was collected is a close beach to the east (sample 504b), followed by material from a nearby old dune (sample 562d, thermoluminescence [TL] age 3.6 ± 0.6 ka; C.D. Peterson, 2004, personal commun.) and from the Palma Seda arroyo (sample 493a). Intermittent arroyos are probably eroding the old dune and bringing material to the coast. The beach on the east portion (sample 504b) is an end-member sample (80%) and is interpreted to have some contribution from the Palma Seda arroyo (sample 493a) and from the east beach (sample 518b). The beach where sample 518b was collected is another end member (74%) and has some contribution from Palma Seda arroyo (sample 493a) and the west beach (sample 504b). The source for the beach where sample 519b was collected is mainly the beach immediately to the south (sample 518b, 60%), followed by the old dune (sample 562d) and the Palma Seda arroyo (sample 493a).

The end-member samples for harmonics 16–24 (Fig. 11C) suggest that the western beach (sample 495b) is a relative end member (46%), with some contribution from the shelf sediment (samples 508s and 509s) and a smaller contribution from Palma Seda (sample 493a). The sources for the beaches where samples 501b and 504b were collected are mostly the shelf (sample 508s), followed by the western beach (sample 495b) and Palma Seda arroyo (sample 493a). Sources for beaches where samples 518b and 519b were collected are mainly beach, arroyo, and shelf material from the outer coast (samples 493a, 508s, and 509s), which suggest sediment transport from west to east.

From this information, it can be conclude that source material for the Los Arcos area is provided mainly by longshore transport from the northwest, followed by local arroyo material, along with influence of longshore transport from east to west.

San Lucas Area

Plots that include the frequency distribution of the quartz elongation (harmonic 2, Figs. 6A and 6B; data in Table 4) of sediment samples within this area show that beach samples within San Lucas Bay are different from the local arroyos and shallow shelf samples. Samples from the beaches, dune (samples 553b, 545b, and 560d), and shelf (sample 246s, 63 m deep) are more similar to the QE´FD of El Arco beach samples (from El Arco area, to the west, samples 501b and 504b) than to local arroyo sediments. This is also observed for beach samples at the outer and inner coast of the San Lucas Cape. Thus, it is assumed that littoral transport occurs from the El Arco area (San Lucas Cape) into the San Lucas Bay (from the western outer coast into the bay; Fig. 16), although a common sediment source from a much lower sea level or onshore transport during a very large storm cannot be ruled out. The wave diffraction at San Lucas Cape helps to distribute the sand along the western beaches of the bay. The QE´FD value of deep shelf material (sample 246s) within the Cabo San Lucas Bay is similar to the QE´FD value of western open ocean coast material and may have had relict beach material when the sea level was lower, perhaps related particularly to the time of the Flandrian transgression (18 ka to the present). If the ocean dynamic has been similar to the one from modern time since the Flandrian transgression, which is a reasonable supposition, the El Arco area may have been supplying sediment to the bay. The QE´FD values of fluvial sediments from the El Salto drainage basin are similar to the ones from the shallow inner continental shelf (samples 510s, 511s, 512s, 513s, and 514s). From this observation, it can be concluded that within this area (San Lucas area), the littoral transport from west to east is the dominant factor for nourishing the beaches, and that local intermittent arroyos are the dominant source for the shallow shelf sediments within the bay via shore bypass during seasonal heavy rains.

The end-member sample proportions for the different harmonic combinations (2, 2–24, 16–24; Figs. 12A, 12B, and 12C) show that the western beach (sample 553b) has high proportions of sediment from the El Arco area, followed by the El Salto Seco arroyo, stabilized dunes (sample 560d), located at the center of the bay, and beach and shallow shelf (samples 547b and 511s) from the eastern side of the bay. End-member samples for the beach from the central portion of the bay (sample 545b) show that the stabilized dunes (sample 560d) there are the primary source of material, followed by the El Salto Seco arroyo, the outer coast from the El Arco area, and the eastern shallow shelf and beach. The source for the beach at the eastern side of the bay (sample 547b) for harmonic 2, is mainly a stabilized old dune (samples 560d, 55%) and the El Salto Seco arroyo (sample 507a, 45%). For harmonics 2–24, this beach is considered an end-member sample (70%), with some contribution from the shallow shelf (samples 509s and 511s) and from El Arco outer beach (sample 518b). For harmonics 16–24, the main source is the El Salto Seco arroyo (60%), followed by the shallow shelf and beach from the El Arco area. All of these observations suggest that the source material for this beach (sample 547b) mostly is provided by littoral transport from the west of the bay, with some contribution of material from the El Arco area. Thus, the potential sources show that El Arco area supplies in part,

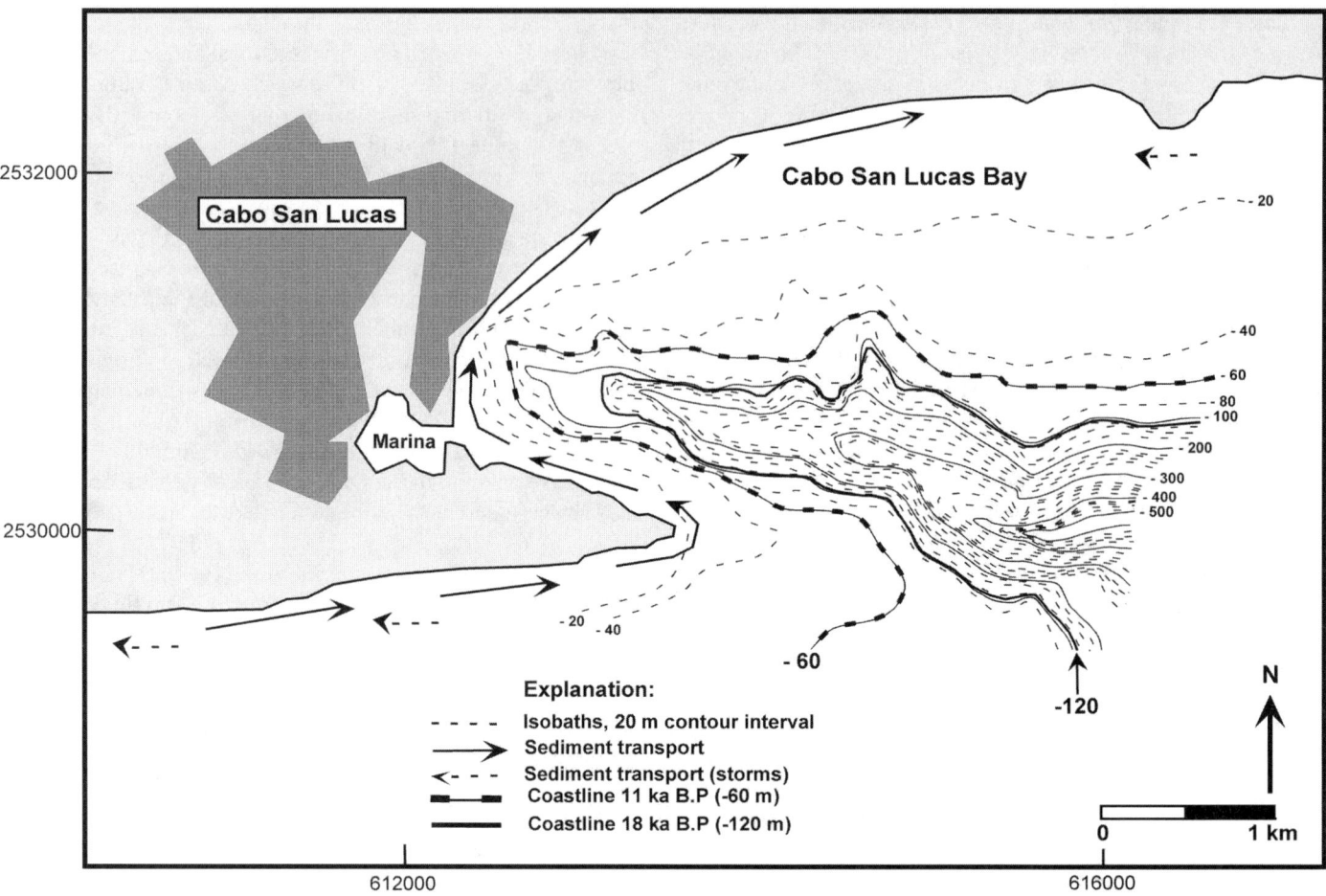

Figure 16. Dominant sediment littoral paths along El Arco and Cabo San Lucas areas, inferred from quartz elongation (harmonic 2). Water extension according to sea level data given by Shackleton (1987). Bathymetric map of Cabo San Lucas bay from Nava-Sánchez et al. (2000). Marine Geology Laboratory, CICIMAR-IPN, La Paz, B.C.S., Mexico.

by littoral transport, sand to the beaches within this area (San Lucas area, Fig. 16).

El Tiburón Area

Plots that include the frequency distribution of the quartz elongation (harmonic 2, Figs. 13A and 13B; data in Table 4) of sediment samples within this area show that a beach sample QE´FD value at the mouth of the El Tiburón drainage basin (sample 521b) is not similar to the QE´FD values of adjacent arroyo samples (520a and 360a). This sample is more similar to a beach sample from the Cabo San Lucas Bay (sample 545b), which suggests sediment transport from the western part of to this area. The QE´FD value of the beach sample 544b is similar to the QE´FD values of local arroyos (348a and 349a, among others); this suggests that the main sources of this beach are the adjacent local arroyos.

The end-member proportions obtained for El Tiburón beach (sample 521b), harmonic 2 (Fig. 13A), shows that the dominant sources for this beach are the arroyos to the northeast where samples 349a (51%) and 350a (8%) were collected and a beach that is close to the southern limit of this area (sample 547b, 41%). There appears to be little contribution from an arroyo (end-member sample 540a) that is close to the other arroyo sources. Harmonics 2–24 (Fig. 13B) show that this beach area represents an end member (80%) and has some contribution from the southern area of the beach and shallow shelf (samples 511s and 547b). There appears to be no significant contribution from an arroyo at the northern limit of this area (end-member sample 351a). Harmonics 16–24 (Fig. 13C) show that El Tiburón beach (sample 521b) is end-member (90%) material and has some contribution from El Tiburón drainage basin (sample 360a). Harmonic 2 data suggest that the source of El Tiburón beach material is a beach located toward the west, within the San Lucas area (end-member sample 547b). The different harmonic combinations show that source material for El Tiburón beach (sample 521b) is provided by littoral transport from the southwest (samples 547b and 511s) and northeast (sample 349a), but the presence of the high elongation frequencies in a sample from this beach (end-member, sample 521b) may suggest a local source that was not sampled at the western area of this site.

The end-member proportions for harmonic 2 at the El Guayparín beach (sample 544b, Fig. 13A) suggest that the source material for this beach is from a local arroyo (samples 349a and 350a) and a beach to the east of Cabo San Lucas Bay, southwest of this beach (sample 547b). For harmonics 2–24 (Fig. 13B), the dominant source is also the beach from the southwest (sample 547b), followed by a southwest beach within the area (sample 521b), the shallow shelf (sample 511s), and some contribution from El Tiburón drainage basin (sample 360a). For harmonics 16–24 (Fig. 13C), the beach sources are also the El Tiburón arroyo and beach (samples 360a and 521b) and an arroyo within the north area (El Tule area, sample 351a). From these data, it is inferred that the source material for El Guayparín beach is mostly provided by local arroyos and material from littoral transport from southwest beaches and shallow shelf, with some contribution of local arroyos from the northeast area.

El Tule Area

Plots that include the frequency distribution of the quartz elongation (harmonic 2, Figs. 8A and 8B; data in Table 4) of sediment samples within this area show that beach sample QE′FD values located to the east of the El Tule arroyo (samples 525b, 526b, and 539b) are similar to the QE′FD value of the arroyo sample from El Tule (sample 524a). This implies that this arroyo is the main source for material to the beaches in the northeast of this area. For beach sample 523b, which is located to the southwest of the El Tule mouth, the observed QE′FD value is not similar to the El Tule arroyo sample. The QE′FD value of shelf sample 248s is similar to the QE′FD value of the shelf sample 558s; both samples show no dominant quartz elongation frequency (~10% of frequency for most of the intervals). Most of the shelf and arroyo samples within this area show similar QE′FD values, which suggest that the shallow shelf is mostly influenced by local arroyo discharges.

The end-member proportions of beach samples for harmonic 2 (Fig. 14A) show that the source of the beach at the mouth of El Tule arroyo (sample 523b) is sediment from the northeast, which includes the local arroyos (sample 359a), followed by beach material (sample 525b), and shallow shelf (samples 559s and 558s). For harmonics 2–24 (Fig. 14B), this beach (sample 523b) is a relative end member (54%), which suggests that local material is probably the main source for the beach, followed by contribution of beach and shelf material from the northeast of this beach (samples 525b and 249s), and also contribution of the shelf from the south (sample 556s). For harmonics 16–24 (Fig. 14C), this beach (sample 523b) is closest to a dune end member (36%), with contribution from the south shallow shelf (sample 556s) and minor contributions of deep shelf (sample 249s) and northeast beach (sample 525b) sediment. The previous observations suggest that the source material for this beach comes dominantly from southwest and northeast littoral transport and has a contribution of local arroyo material and adjacent arroyo material from the El Guayparín drainage basin.

The beach located at the Villa del Mar drainage basin (sample 525b) is an end member for harmonic 2 (73%, Fig. 14A), harmonics 2–24 (55%, Fig. 14B), and harmonics 16–24 (38%, Fig. 14C). For harmonic 2, the main source of material is the beach itself, which implies sediment from nearby features and probably a major contribution from El Tule drainage basin (sample 524a), the closer and larger drainage basin located to the south of this beach. The beach also has some minor contributions from the northeast shallow shelf (samples 558s and 559s). For harmonics 2–24 and harmonics 16–24, the main source must be local, because this beach is a dominant end member. This beach has contributions from the beach located to the southwest (sample 523b), and there is also a contribution of deep shelf material (sample 249s), which is minor for harmonics 2–24 (5%) and larger for harmonics 16–24 (34%). From this information, it is inferred that the source material for this beach is mostly local material, probably from the Villa del Mar, San Carlos, and El Tule drainage basins, with contributions from the shallow and deep shelf.

For the beach (sample 526b) located at the San Carlos drainage basin, harmonic 2 distribution (Fig. 14A) suggests that the main sources of this material are the beach to the southwest (sample 525b, 51%) and Cerro Blanco arroyo (sample 359a, 48%). For harmonics 2–24 (Fig. 14B), the main sources are the beach to the southwest (sample 525b, 43%) and the deep shelf to the east (sample 249s, 43%). The main source for harmonics 16–24 (Fig. 14C) is the deep shelf to the east (sample 249s, 34%), followed by beach material from the northeast (samples 559s, 28%, and 539b, 24%) and a nearby arroyo (sample 359a, 14%). From this combination of harmonic observations, the sources for this beach are the local Cerro Blanco arroyo (sample 359b) and coastal material from the southwest and northeast of this beach.

For the beach (sample 539b) at the south of the Punta Palmilla Sur drainage basin, harmonic 2 distributions (Fig. 14A) suggest that the main sources are the Cerro Blanco arroyo (sample 359a, 56%) and a beach from the southwest (sample 525b, 40%). For harmonics 2–24 (Fig. 14B), the main sources are beaches from the southwest (samples 525b, 43%, and 523b, 15%) and the deep shelf (sample 249s, 41%). Harmonics 16–24 distributions (Fig. 14C) suggest that the main source is also the beach from the southwest (sample 525b, 56%), with some contribution from the shallow shelf (sample 556s, 23%) and a lesser contribution from the deep shelf (sample 249s, 16%). From these observations of the different harmonic combinations, it is inferred that the source material for the beach mostly comes from the southwest (sample 525b), followed by the shallow (sample 556s) and deep shelf (sample 249s). This suggests dominant littoral transport from the southwest.

San José Area

Plots that include the frequency distribution of the quartz elongation (harmonic 2, Figs. 9A and 9B; data in Table 4) of sediment samples within this area show that QE′FD values for a beach sample (sample 532b) at the northeast of the intermittent

San Jose arroyo mouth are similar to the QE´FD values of samples from this arroyo. This suggests that this portion of beach is being supplied by the discharge of this arroyo. The fluvial samples from San José have similar QE´FD values, with the sole exception of sample 535a, which is located on the eastern portion of this area and displays a slightly higher frequency of grains in the lower intervals. The QE´FD values of the beach samples 537b, 528b, and 538b are more similar to the QE´FD values of shallow shelf samples than to local arroyo samples, which suggest that the shallow shelf is the dominate source for beaches in this area. The QE´FD value of the deeper shelf sample 250s (80 m) is similar to the QE´FD values of deep shelf samples within El Tule area and has similar frequencies between the low and high intervals. Beach samples from this area, along all the intervals, have the most irregular histograms, which suggest that beaches are composed of a mix of sources, in contrast to the El Cardonal area.

For the beach (sample 528b) located between the El Alamito and Costa Azul drainage basins, the end-member proportions defined for harmonic 2 (Fig. 15A) imply that the sources of sediment are the shallow shelf (sample 550s), El Alamito arroyo (sample 357a), and a northeast arroyo (sample 535a) within the Las Salinas drainage basin. For harmonics 2–24 (Fig. 15B), this beach (sample 528b) represents an end-member source (75%), dominant local source, with contribution from the shallow shelf (sample 549s) and a beach from the northeast area (sample 532b). For harmonics 16–24 (Fig. 15C), the main source is the shallow shelf within El Tule area (sample 552s, 73%), followed by some contribution from the beach at the mouth of the San José arroyo (samples 537b and 532b). The different harmonic combinations suggest that the source of this beach is mostly shallow shelf sediments from the southwest (El Tule area, sample 552s and 549s) with contributions of material from the northeast, the El Alamito arroyo (sample 357a), the beach (samples 537b and 532b), and the shallow shelf within the San José drainage basin (sample 550s).

For the beach located at the Costa Azul drainage basin (sample 538b), the harmonic 2 analysis (Fig. 15A) defines the following sources: El Alamito arroyo (to the south; sample 357a) with 42%, the beach to the east (sample 537b) with 33%, and the northeast shallow shelf (sample 550s) with 24%. For the harmonics 2–24 (Fig. 15B), the main source is a beach from the southwest (sample 528b, 72%), with a significant contribution from the nearby shallow shelf, and minor contributions from an arroyo within the Las Salinas drainage basin (sample 535a) and a northeast beach (sample 532b) within the San José drainage area. For the harmonics 16–24 (Fig. 15C), the main source of this beach is the shallow shelf (sample 552s) with 69%, followed by northeast beaches within the San José drainage area (sample 537b and 532b). From these observations it can be assumed that there is contribution of material from the southwest of this area, from the shallow shelf (sample 552s), from the El Alamito arroyo to the southwest, and from the northeast beach and shallow shelf of the San José drainage basin.

For the beach located in the San José drainage basin (sample 537b), the harmonic 2 analysis (Fig. 15A) suggests that this beach is nearly an end member (65%), with some contributions of arroyo material from the El Alamito drainage basin located to the southwest (sample 357a), Las Salinas drainage basin (sample 535a) to the north, and a minor contribution from the nearby shallow shelf (sample 550s). From the harmonics 2–24 (Fig. 15B), the main source of sediment was defined by an arroyo within the Las Salinas drainage basin located to the northeast (sample 535a, 49%), followed by contributions from a southwest beach (sample 528b, 34%), and minor contributions of material from a northeast beach (sample 532b) within the San José drainage basin and shallow shelf from the southwest (sample 549s). For harmonics 16–24 (Fig. 15C), this beach is also an end member (59%), with contribution from the shallow shelf from the south (El Tule area, sample 552s) and El Alamito arroyo (sample 357a). For this beach, it was expected that samples from the San José drainage basin would be the main source. However, the data suggest that the primary sources to this beach are sediments transported from the south and northeast of the beach (sample 537b).

For the beach located at the northeastern portion of the San José drainage basin (sample 532b), harmonic 2 (Fig. 15A) analysis suggests that the main source of sediment is the El Alamito arroyo (sample 357a) with 60%, with contributions from an arroyo to the northeast within Las Salinas drainage basin (sample 535a), 23%, and minor contributions from the local shallow shelf (sample 550s) and a nearby beach to the southwest (sample 537b). From the analysis of harmonics 2–24 (Fig. 15B), this beach is an end member (60%), with some contribution from a beach to the southwest (sample 528b), an arroyo to the northeast within the Las Salinas drainage basin (sample 535a), and the shallow shelf from the southwest (sample 549s). Analysis of harmonics 16–24 (Fig. 15C) also suggests that this beach is an end member (55%), with contributions from the shallow shelf (sample 552s) from El Tule area to the south, a beach from the southwest (sample 537b) within the San José drainage basin, and the El Alamito arroyo from the southwest (sample 357a). From the information of the harmonic combinations, it is inferred that the source of this beach is mostly local material with some contribution of material from beaches, shallow shelf, and arroyos from the southwest (samples 357a, 552s, and 549s). This beach (sample 532b) is the source of the nearby southwest beach (sample 537b).

CONCLUSIONS

Results from this study show that there are two sedimentary provinces, one province, which is on the Pacific side, from Migriño arroyo to San Lucas Cape, is termed the Los Médanos province (Fig. 3), and the second province, which is partially protected from the high energy of the Pacific Ocean and is from San Lucas Cape (the western portion of the Cabo San Lucas Bay) to the northern portion of San José Bay, is termed the Los Cabos province. The sedimentary framework of the Los Médanos province responds mostly to the availability of sand distributed by littoral transport and by the high-energy waves induced by the dominant northwesterly winds, which distribute the sediments along the open coast and from offshore to onshore. Littoral transport is more important for beach nourishment in the Los Médanos province, where drainage basins are fewer and smaller than in Los Cabos province. The sedimentary framework within the Los Cabos province responds mostly to sand supply from local arroyos, where large drainage basins supply material to the littoral zone during periods of high rainfall; it also responds to high-energy waves (induced by southerly winds) that are refracted from the Pacific, which distribute sediments mostly from the southwest to the northeast. Waves arriving in this province are induced by northwesterly and southwesterly winds that are refracted and modified by topography and local bathymetry.

The sediment transport direction in both sedimentary provinces is related to the direction of wave arrival. Along the Los Médanos province, which is more exposed to the Pacific Ocean influence, sediment transport occurs from northwest to southeast, except the southern portion of this province. The Los Arcos area a transitional zone between both sedimentary provinces, where evidence shows sediment transport from west to east and from east to west, influenced also by the dominant refracted waves induced by southerly winds that mostly affect the Los Cabos province. Within Los Cabos province, sediment transport occurs mostly from southwest to northeast along the coast; however, there is evidence that, occasionally, sediment transport occurs from a northeast to southwest direction, related to meteoric events such as hurricanes, which arrive or pass close by the tip of the peninsula and also related to wave refraction. Hurricanes occur almost every year and bring strong wave regimes, strong winds, and large amounts of rain that can flush many smaller arroyo drainage basins of sediment. Our results demonstrate that littoral transport occurs in different directions, likely related to dominant and storm or hurricane wave directions during the year, induced by the dominant northwesterly winds and the scarce but strong southerly winds along the whole south of the peninsula. Several beaches from each of the studied areas become sources for other beaches, by littoral transport, which demonstrates that within both sedimentary provinces, littoral transport is an important factor that nourishes the beaches along the coast.

Within the Los Cabos province, the beaches that were defined as end members, or potential sources (samples 547b, 521b, 523b, 525b, 537b, 532b, 528b, 553b, and 556b), likely have contributions from local arroyos and some contribution of marine material through littoral transport. A more detailed study of any one of these areas would determine the sediment dynamics for each of these end members. The implication is that the resultant compositions of each of the defined end members are themselves a mixture of local sources. The same observation holds for the end-member beaches within the Los Médanos province (samples 315b, 317b, 518b, 495b, 504b, and 519b), which are likely combinations of old aeolian deposits that have been dissected by arroyos.

Within the Los Médanos province, the arroyo and beach sediments display different elongation patterns. The source or sources for beach and shelf sediment in Los Médanos province must be located to the north of this province, which accounts for the contribution of the high percentages of low elongations similar to the shape composition of the El Tule arroyo within the Los Cabos sedimentary province. Results show that the shape composition from Los Médanos provinces are mostly marine influenced, where the shallow and deep inner continental shelf are the major sources, followed by old dissected dunes, and local arroyo materials. Samples along the beach of this province that deviate from this pattern could be related to the mixing of local sources of fluvial sediment and an offshore contribution. Offshore and nearshore sediment movement studies, similar to Evangelista et al. (1996, 2003), would be useful to unravel the complexity of the sediment dynamics observed in this region.

Within Los Médanos sedimentary province, most of the beaches contain between 14% and 27% of low elongated grains (round grains, interval 1) and decreasing percentages to the high elongated grains. This is opposite for the beaches within Los Cabos sedimentary province, which contain fewer grains with low elongation percentage (between 3% and 17%), with exception of beaches on the El Tule area, which contain larger percentages of low elongated grains (between 5% and 32%).

The arroyo material within both provinces contains mostly higher percentages of high elongated grains (interval 10, 10%–20%), and much lower percentages of low elongated grains compared with the beach material (between 0.6% and 9%), with exception of El Tule arroyo, located at Los Cabos province (between 0.9% and 27%), which contains higher percentages of low elongated grains (Fig. 8, sample 524a). From this, it can be concluded that most of the local rocks, which are mostly plutonic rocks (Fig. 2), contain quartz grains with dominant elongated shapes (high elongation).

The overall shape composition of the beach sediments along the coast of Los Cabos province is more variable than in Los Médanos province, which must be related to differences in fluvial-marine sediment contribution. Sources for Los Cabos beaches are local arroyos, surrounding beaches, shallow shelf, and marine sediments from Los Médanos province. Sources for long and wide beaches within Los Médanos province are mostly marine sediments and old dune material dissected by ephemeral runoff and some contribution of material from local arroyos. The Los Médanos sedimentary province does not have mainly local source materials, such as the Los Cabos sedimentary province, because the drainage basins are smaller and because abundant sediment is distributed along the coast by north to south longshore transport, where the probable source is from the north. More studies are needed to identify the main source of Los Médanos province beaches. The shape composition of sediments from the different sedimentary environments, and geographic areas, increases our understanding of the sediment dynamics in the southern coastal zone of the Baja California peninsula.

The results of the present study are similar to those reported in Tortora et al. (1986) and Evangelista et al. (1994, 1996); these results suggest that sediment from smaller drainage basins contributes little to the nearshore and offshore sediment budget, and local sources of sediment dominate the beach material in limited amounts near the local source. Furthermore, the present study in addition to the results of the previously mentioned studies strongly suggest that offshore processes associated with larger storms dominate the sediment record in areas without strong, steady currents.

ACKNOWLEDGMENTS

We acknowledge Consejo Nacional de Ciencia y Tecnología (CONACyT) (J27724T, CGPI208012) for the economic support to this work, Centro Interdisciplinario de Ciencias Marinas, Instituto Politécnico Nacional (CICIMAR-IPN) (CGPI 20010314) for the logistic support, Douglas Nelson from the University of Carolina, who supported the digitizing and statistical programs, Cuauhtemoc Turrent and Lucio Godínez for their participation in collecting some of the samples, Martha Palma who digitized the samples, the Mexican Navy Secretary and crew of the *Juan de La Barrera* and *Suchiate* B1-05 vessels for their support in collecting the shelf material, and Bob Osborne (now deceased) who introduced grain-shape analysis to the first author.

REFERENCES CITED

Camacho-Valdéz, V., 2003, Características Morfodinámicas y Texturales de los Depósitos de Cabo Falso, B.C.S., México [M.S. thesis]: La Paz, Baja California Sur, México, Departamento de Oceanología, Centro Interdisciplinario de Ciencias Marinas, Instituto Politécnico Nacional, 139 p.

Civitelli, G., Corda, L., Evangelista, S., and Full, W.E., 1992, Fourier quartz shape analysis: Application to terrigenous sediments of Laga and Cellino Formations (central Italy): Bollettino Società Geologica d'Italia, v. 111, p. 355–366.

Defense mapping Agency, 1984, Cabo San Lazaro to Cabo San Lucas and Southern Part of Gulf of California (Mexico—West Coast), Chart No. 21014: Surveys from 1874 to 1942: Washington, D.C., Ed. Omega, Hydrographic/topographic Center, 20315, scale 1:667,680.

Ehrlich, R., and Full, W.E., 1987, Sorting out geology—Unmixing mixtures, *in* Size, W., ed., Journal of Mathematic Geology Special Publication 5: New York, Oxford University Press, p. 33–46.

Evangelista, S., Full, W.E., and Tortora, P., 1988, Analisi Della Forma Particelle Sedimentane: Una Applicazione di Clasti Sabbiosi Di Ambrente Costiero: Miniera Società Geologica, v. 34, p. 823–826.

Evangelista, S., Full, W.E., and Tortora, P., 1994, Fourier grain shape analysis as tool to quantify the contribution of the fluvial input to the coastal sedimentary budget: An example from the Port Stephen's area, New South Wales, Australia: Bollettino Società Geologica d'Italia, v. 113, p. 729–747.

Evangelista, S., Full, W.E., and Tortora, P., 1996, Contribution and dispersion of fluvial, beach, and shelf sands in the Bassa Maremma coastal system (central Italy, Tyrrhenian margin): Bollettino Società Geologica d'Italia, v. 116, p. 195–217.

Evangelista, S., Full, W.E., La Monica, G.B., and Nelson, D.D., 2003, Computer modeling of the littoral dynamics of the Circeo-Terracina coastal system (Lazio-Italia): Geologica Romana, v. 37, p. 127–130.

Full, W.E., 1982, Analysis of quartz detritus of complex provenance via analysis of shape [Ph.D. thesis]: Columbia, University of South Carolina, 206 p.

Full, W.E., and Ehrlich, R., 1982, Some approaches for location of centroids of quartz grain outlines to increase homology between Fourier amplitude spectra: Journal of Mathematic Geology, v. 14, no. 1, p. 43–55, doi: 10.1007/BF01037446.

Full, W.E., Ehrlich, R., and Klovan, J.E., 1981, EXTENDED QMODEL—Objective definition of external end-members in the analysis of mixtures: Journal of Mathematic Geology, v. 13, no. 4, p. 331–334, doi: 10.1007/BF01031518.

Full, W.E., Ehrlich, R., and Bezdek, J.C., 1982, FUZZY QMODEL: A new approach for linear unmixing: Journal of Mathematic Geology, v. 14, no. 3, p. 257–268.

Full, W.E., Ehrlich, R., and Kennedy, S.K., 1984, Optimal configuration and information content of sets of frequency distributions: Journal of Sedimentary Petrology, v. 54, no. 1, p. 117–126.

Full, W.E., Ehrlich, R., and Kennedy, S.K., 1985, Optimal configuration and information content of sets of frequency distributions—Reply: Journal of Sedimentary Petrology, v. 55, no. 6, p. 933–934.

Healy-Williams, N., Ehrlich, R., and Full, W.E., 1997, Close-form Fourier analysis: A procedure for extracting ecological information from foraminiferal test morphology: Fourier descriptors and their applications, *in* Lestrel, P.E., ed., Biology: Cambridge, Cambridge University Press, 466 p.

INEGI (Instituto Nacional de Estadística Geografía e Informática), 1985, Carta Topográfica, San José del Cabo, México: Aguascalientes, Instituto Nacional de Estadística Geografía e Informática, F12-2-3-5-6, escala 1:250,000.

INEGI (Instituto Nacional de Estadística Geografía e Informática), 1993, Foto aérea, Línea 67, No. 16, México: Aguascalientes, Instituto Nacional de Estadística Geografía e Informática, F12-2-3-5-6, escala 1:75,000.

INEGI (Instituto Nacional de Estadística Geografía e Informática), 1999, Carta Geológica, La Paz, México: Aguascalientes, Instituto Nacional de Estadística Geografía e Informática, G12-10-11, escala 1:250,000.

Murillo, J.M., Osborne, R.H., and Gorsline, D.S., 1994, Sources of beach sand at Creciente Island, Baja California Sur, México: Fourier grain shape análisis: Ciencias Marinas, v. 20, p. 243–266.

Murillo-Jiménez, J.M., Nava-Sánchez, E.H., Godínez-Orta, L., and León-Manilla, A., 2001, Identificación de fuentes de abastecimiento de las playas de la zona costera sur del estado de Baja California Sur, México: La Paz, Baja California Sur, México, Centro Interdisciplinario de Ciencias Marinas, Instituto Politécnico Nacional (CICIMAR-IPN), Reporte Técnico CONACYT-J27724T y CGPI-208012, 28 p.

Nava-Sánchez, E., Godínez-Orta, L., Murillo-Jiménez, J.M., and Leon-Manilla, A., 2000, Evaluación de Ambientes Sedimentarios Recientes en la Parte Sur del Golfo de California: La Paz, Baja California Sur, México, Centro Interdisciplinario de Ciencias Marinas, Instituto Politécnico Nacional (CICIMAR-IPN), Reporte Técnico DEPI No. 980053, 30 p.

Nelson, D.D., and Full, W.E., 1998, An improved program for the calculation of high resolution Fourier coefficients used for shape analysis: Computers and Geosciences, v. 24, no. 3, p. 237–242, doi: 10.1016/S0098-3004(97)00123-4.

Nelson, D.D., Full, W.E., and Evangelista, S., 1996, FORMA: A program in C to trace object peripheries for two-dimensional shape analysis: Computers & Geosciences, v. 22, no. 6, p. 683–695, doi: 10.1016/0098-3004(96)00011-8.

Osborne, R.H., Ahlschwede, K.A., Broadhead, S.D., Cho, K., Feffer, J.R., Lee, A.C., Liu, J., Magnusen, C., Murillo de Nava, J.M., Robinson, R.A., Yeh, C.-C., and Lu, Y., 1994, The continental shelf: A source for naturally-delivered beach sand, *in* Arcilla, A.S., Stive, M.J.F., and Kraus, N.C., eds., Coastal Dynamics, Proceedings of an International Conference on Large-Scale Experiments in Coastal Research: Reston, Virginia, American Society of Civil Engineers, p. 335–349.

Shackleton, N.J., 1987, Oxygen isotopes, ice volume and sea level: Quaternary Science Reviews, v. 6, p. 183–190, doi: 10.1016/0277-3791(87)90003-5.

Sundborg, A., 1956, The River Klarälven: A study of fluvial processes: Geografiska Annaler, v. 38, p. 125–237, doi: 10.2307/520140.

Tortora, P., Evangelista, S., and Full, W.E., 1986, Fourier grain shape analysis and its application to the southeastern coastline of Australia: Rome, Italy, National Geological Congress of Central Italy Proceedings, 6 p.

Troyo-Dieguez, S., 2003, Oleaje de Viento y Ondas de Infragravedad en la zona costera de Baja California Sur [Ph.D. thesis]: La Paz, Baja California Sur, México, Departamento de Oceanología, Centro Interdisciplinario de Ciencias Marinas, Instituto Politécnico Nacional (CICIMAR-IPN), 211 p.

Manuscript Accepted by the Society 9 August 2006

Detrital apatite geochemistry and its application in provenance studies

Andrew Morton
HM Research Associates Ltd., 2 Clive Road, Balsall Common, West Midlands, CV7 7DW, UK, and *CASP, Department of Earth Sciences, University of Cambridge, West Building, 181a Huntingdon Road, Cambridge CB3 0DH, UK*

Greg Yaxley
Research School of Earth Sciences, The Australian National University, Canberra, ACT 0200, Australia

ABSTRACT

Single-grain, laser-ablation inductively coupled plasma–mass spectrometry analyses of detrital apatites from Pliocene sandstones in the South Caspian Basin (Azerbaijan) and Devonian-Carboniferous sandstones from Clair oil field, west of Shetland (UK), demonstrate that apatite geochemistry has significant potential in provenance analysis. Apatites in Pliocene sandstones deposited by the paleo-Kura River system, which drained the Lesser Caucasus region, were derived largely from mafic to intermediate and alkaline rocks. Apatite populations in Pliocene sediments transported by the paleo-Volga River system, which drained the Russian Platform, show greater compositional diversity and indicate supply from granitoids or other acidic rocks together with subordinate mafic to intermediate and alkaline rocks. Apatites in the Devonian-Carboniferous succession west of Britain were derived predominantly from acidic rocks, either directly from Archean gneisses or indirectly from metasedimentary rocks. In the two case studies, the most useful discriminators of apatite provenance proved to be La/Nd and La + Ce/ΣREE.

Since apatite is stable during burial in sedimentary basins, apatite geochemistry can be used to determine provenance of sandstones from the full range of diagenetic environments, although the instability of apatite during weathering means that the method will be difficult to apply to sandstones with prolonged weathering history. At present, identification of provenance using apatite geochemistry is limited by the lack of a comprehensive database on apatite compositions in some of the potential source rocks, particularly those of metamorphic origin. The role played by sediment recycling is another factor that requires consideration when reconstructing source areas on the basis of apatite compositions.

Keywords: apatite, provenance, rare earth elements, South Caspian Basin, Clair oil field.

INTRODUCTION

Heavy mineral analysis has been widely used to identify, discriminate, and reconstruct sandstone provenance for over a century, as described in reviews by, for example, Hubert (1971), Morton (1985a), and Mange and Maurer (1992). One of the limiting factors for constraining provenance on the basis of heavy mineral data is that most of the minerals found in heavy mineral assemblages occur in a wide variety of different potential source rocks. For example, calcic amphibole occurs in both igneous and metamorphic rocks: igneous rocks containing calcic amphibole range from basic to acidic in composition, and metamorphic sources include those with both igneous and sedimentary protoliths (Deer et al., 1997a). Epidote also has igneous and metamorphic parageneses; it is common in greenschist-facies metabasic rocks, metacarbonates and metapelites, but it also occurs as a hydrothermal or metasomatic alteration product in igneous rocks (Spiegel et al., 2002). Garnet occurs in a wide range of metamorphic rocks at various grades of metamorphism, but it is also present in a variety of igneous lithologies (Deer et al., 1997b). Thus, although some heavy minerals have restricted parageneses, virtually none of the common components of heavy mineral assemblages uniquely identifies specific source lithologies (see Mange and Maurer, 1992).

In many cases, mineral chemistry provides much better constraints on source lithology, since particular mineral compositions can be specific to certain parageneses. Single-grain heavy mineral chemical analysis has, therefore, enabled major advances in provenance reconstruction. Until now, single-grain mineral chemistry has been undertaken using the electron microprobe by either energy-dispersive or wavelength-dispersive X-ray analysis. This method has been applied to virtually all the main components of heavy mineral assemblages, including calcic amphibole (von Eynatten and Gaupp, 1999), chrome spinel (Pober and Faupl, 1988), clinopyroxene (Cawood, 1983; Styles et al., 1989), garnet (Morton, 1985b; Haughton and Farrow, 1988; Hutchison and Oliver, 1998; Hallsworth and Chisholm, 2000; Morton et al., 2004), opaque minerals (Basu and Molinaroli, 1991; Grigsby, 1990), sodic amphibole (Mange-Rajetzky and Oberhänsli, 1982), rutile (Zack et al., 2004), tourmaline (Henry and Guidotti, 1985; Jeans et al., 1993; Morton et al., 2005), and zircon (Owen, 1987).

One mineral that has been largely overlooked in this regard is apatite. Despite evidence that apatite geochemistry is controlled, at least in part, by the composition of the host rock (Fleischer and Altschuler, 1986; Belousova et al., 2002), there have been few attempts to use apatite geochemistry for reconstruction of sediment provenance. The main problem has been that most of the key elements in apatite are present in only trace amounts, and they are therefore below detection limits for conventional electron microprobe analysis. Dill (1994) attempted to circumvent this problem by analyzing bulk apatite separates using inductively coupled plasma–mass spectrometry (ICP-MS). However, this approach has the serious drawback of averaging the detrital apatite compositions, which at best can give only an imprecise view of provenance, and at worst could be seriously misleading. Another problem with this approach is that the bulk apatite separate may contain both detrital and authigenic material (Dill, 1994), and thus the trace-element abundances may not entirely reflect the nature of the detrital source.

The recent advent of laser-ablation inductively coupled plasma–mass spectrometry (LA-ICP-MS) has enabled accurate determination of trace-element abundances on single apatite grains. Belousova et al. (2002) used this method to compare apatite compositions in sediments and some potential host rocks in order to assess the value of apatite geochemistry for geochemical exploration. Bouch et al. (2002) also used LA-ICP-MS in their study of apatite compositions in sandstones, although they focused on the origin of authigenic apatite overgrowths rather than on their detrital cores.

In this paper, we consider the potential role of detrital apatite geochemistry as an indicator of sandstone provenance in two case studies, the Pliocene of the South Caspian Basin (Azerbaijan), and the Devonian-Carboniferous of the Clair oil field (west of Shetland, UK). In both cases, heavy mineral data show the existence of dramatic changes in provenance (Allen and Mange-Rajetzky, 1992; Morton et al., 2003a, 2003b). We investigate whether these differences are reflected in the composition of the detrital apatite populations, and whether apatite compositions can be used to constrain the nature of the hinterland.

APATITE GEOCHEMISTRY

Apatite, $Ca_5(PO_4)_3(OH,F,Cl)$, is a common accessory mineral in virtually all igneous rocks, and it is also a common component of many metamorphic rocks (Chang et al., 1998; McConnell, 1973; Nash, 1984). There is solid solution between fluorapatite, chlorapatite, and hydroxyapatite, and the most common type is, by far, fluorapatite (Chang et al., 1998). In igneous rocks, F contents increase during fractionation (Nash, 1984). F is inversely correlated to Cl, which is consequently more abundant in less-fractionated rock types. Some apatites show appreciable substitution of PO_4 by CO_3: these are termed carbonate-apatites. Apatite is present in sedimentary rocks both as a detrital mineral (predominantly fluorapatite) and as a primary deposit, the latter of which mostly consists of francolite (carbonate-apatite with >1% F).

A variety of elements, including Sr, Y, Mn, U, Th, and the rare earth elements (REE), substitute for Ca, usually in trace amounts (Ayers and Watson, 1993; Belousova et al., 2002; Chang et al., 1998; Nash, 1984). Variations in trace elements appear to be related to whole-rock SiO_2 contents, indicating that degree of host rock fractionation is a major control on apatite compositions in igneous rocks. For example, Y, Mn, and the heavy rare earth elements (HREE) become relatively enriched during fractionation, whereas Sr becomes relatively depleted. Belousova et al. (2002) therefore proposed that apatite host rocks can be discriminated using binary plots of abundances of elements such as Sr-Y and Sr-Mn.

Th contents are also lower in highly fractionated rocks, possibly due to crystallization of monazite, which removes Th and the light rare earth elements (LREE) from the melt (Belousova et al., 2002). The same does not appear to be true for U, which is present in higher abundances in granites and granite pegmatites than in dolerites (median values of 20–25 ppm compared with 3.5 ppm; Belousova et al., 2002). Dill (1994) suggested that the Th-U binary plot was a useful diagram for discrimination of provenance, and this may be partly due to the different behavior of the two elements during fractionation.

Total REE abundances in apatite depend on the REE content of the host rock, and the greatest REE contents apparently occur in apatites from alkaline rocks (Belousova et al., 2002; Chang et al., 1998). The shape of the chondrite-normalized REE pattern is also controlled by the host rock composition (Nash, 1984), in particular the degree of fractionation (Belousova et al., 2002). Apatites from less-fractionated mafic rocks have strong relative LREE enrichment, whereas those from highly fractionated rocks, such as granite pegmatites, show relative LREE depletion. Thus, Belousova et al. (2002) proposed that $(Ce/Yb)_{cn}$ (where cn denotes chondrite-normalized) is a useful index of host rock composition. Fleischer and Altschuler (1986) came to a similar conclusion, although they proposed using a different parameter (La + Ce + Pr/ΣREE). Fleischer and Altschuler (1986) also showed that the gradient shown by the LREE is a useful discriminator of apatites from acidic and alkaline magmas. For example, apatites from granites and granite pegmatites have low La/Nd values, whereas those from alkaline rocks (including pegmatites) have high La/Nd values.

Many apatites have negative Eu anomalies on chondrite-normalized REE plots, which are especially pronounced in more fractionated rocks. According to Budzinski and Tischendorf (1989), this is probably caused by feldspar crystallization, which concentrates Eu^{2+} from the melt. Thus, Belousova et al. (2002) proposed that the magnitude of the Eu anomaly (Eu/Eu*) is another useful measure of host rock composition.

ANALYTICAL METHODS

Apatite grains were extracted from heavy mineral residues acquired by standard gravity settling using bromoform of s.g. 2.80. A narrow grain-size fraction (63–125 μm) was used because this reduces the effects of hydrodynamic fractionation on heavy mineral assemblages (Morton and Hallsworth, 1994). The apatites were mounted in standard size 1 inch epoxy blocks, polished using 0.25 μm diamond paste, and carbon-coated. Major elements were determined on a 4 spectrometer Cameca SX-100 electron microprobe at The Australian National University. An accelerating voltage of 25 kV and a beam current of 20 nA were used. The instrument was calibrated using recognized standard materials.

Trace elements were determined at The Australian National University by laser-ablation ICP-MS (LA-ICP-MS) using an Excimer UV laser (193 nm) and a custom-built sample introduction system (Eggins et al., 1998a) attached to a Hewlett-Packard Agilent 7500 quadrupole mass spectrometer. The laser pulsed at 5 Hz, delivering 70 mJ per pulse onto a 40 μm spot. Identical analytical conditions were used for the calibration and secondary standards.

Ablation was conducted in a $He + H_2$ atmosphere. The ablated material was carried to the plasma in an Ar/He gas stream. The mass spectrometer tuning optimized sensitivity and minimized production of interfering oxide molecules, with $^{232}Th^{16}O/^{232}Th$ always ≤0.5%. Analyses were performed in peak hopping mode with a dwell time of 0.025 s/mass. For each analysis, the gas blank was acquired for 30 s, the laser triggered, and the signal acquired for a further 55 s.

Our analytical protocol was similar to that of Eggins et al. (1998b). We used NIST-612 glass as the primary calibration standard, and secondary standards (BCR2 g and NIST-610 glass) were routinely analyzed as unknowns to check data quality. Batches of analyses of eight unknowns (apatite grains and secondary standards) were bracketed by analyses of the calibration standard allowing monitoring of, and correction for, instrumental drift.

Data reduction used background-corrected count rates and the method established by Longerich et al. (1996). Values of ^{44}Ca were measured to enable use of electron microprobe–determined CaO abundances as the internal reference element. Calibration values for NIST 612 used in the data reduction were those of Eggins (2003). A drift correction based on the analysis sequence and linear interpolation between bracketing NIST 612 analyses was applied to the analyze count rate for each sample. Multiple analyses of the NIST-610 and BCR2 g secondary standards indicated that analytical reproducibility and accuracy were better than 10% for nearly all reported elements, and better than 5% for most, when compared with generally accepted values. The full data set is available from the GSA Data Repository.[1]

CASE STUDY 1: PLIOCENE SANDSTONES OF THE SOUTH CASPIAN BASIN

Geological Background

In the South Caspian Basin (Fig. 1), a thick succession (up to 5–7 km) of clastic sediments known as the "Productive Series" was deposited in a brief interval during the Pliocene, probably between 5.5 and 3.4 Ma (Jones and Simmons, 1996). The Productive Series is composed of a series of "suites" (Kirmaky, Balakhany, etc., see Fig. 2), which are approximately equivalent to formations in international usage. At outcrops in the Apsheron Peninsula area of Azerbaijan, the Productive Series consists of interbedded sandstones and mudstones of broadly fluvio-lacustrine origin (Reynolds et al.,

---

[1]GSA Data Repository item 2007169, Detrital apatite compositions as determined by electron microprobe and LA-ICP-MS, is available on the Web at http://www.geosociety.org/pubs/ft2007.htm. Requests may also be sent to editing@geosociety.org.

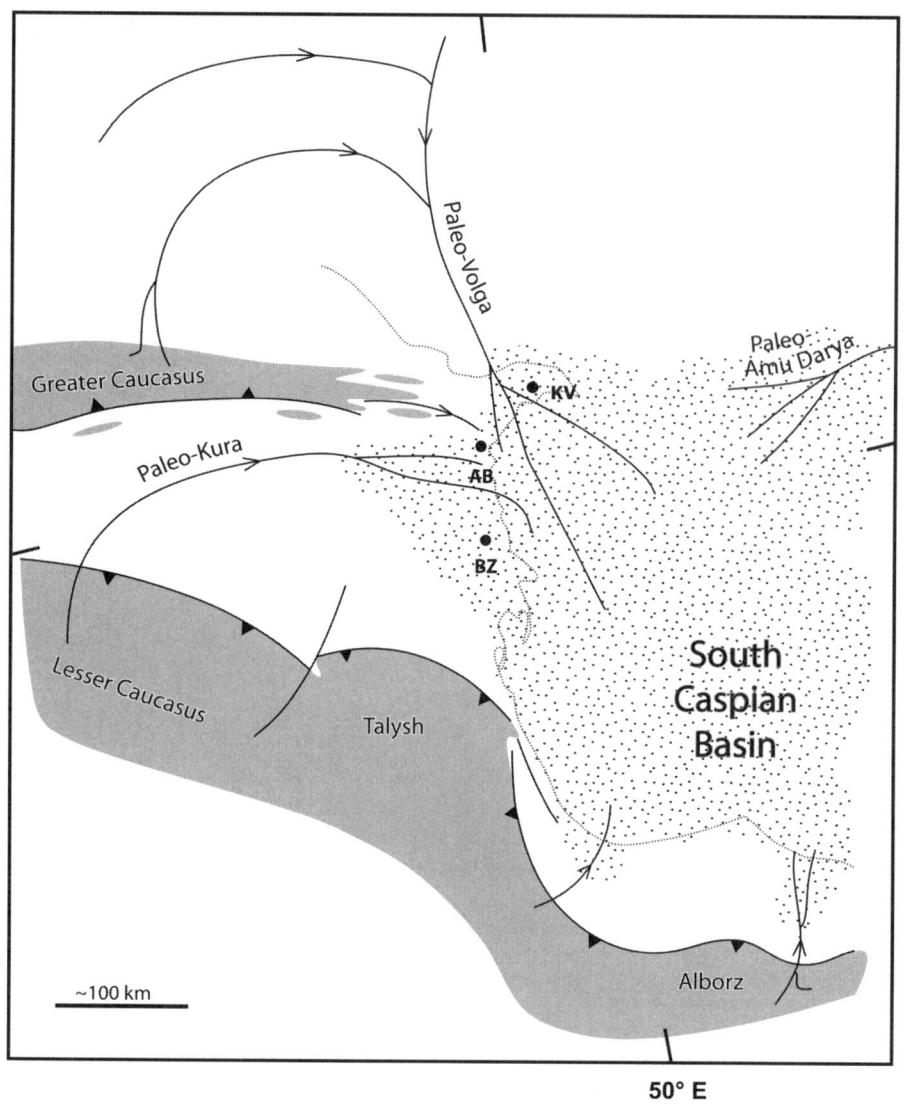

Figure 1. Sketch map of the South Caspian Basin, showing the drainage systems operative during deposition of the Productive Series (adapted from Morton et al., 2003a). The paleo-Kura drained from the west, mainly tapping the Lesser Caucasus, whereas the paleo-Volga drained the Russian Platform to the north, with subsidiary supply from the Greater Caucasus. KV—Kirmaky Valley, AB—Aktapa Bridge, BZ—Babazanan.

1998; Hinds et al., 2004). The Productive Series represents the depositional products of a number of fluvial systems (Fig. 1), including the paleo-Volga, paleo-Amu Darya, and paleo-Kura (Baturin, 1947; Reynolds et al., 1998). In the northern part of the basin, most of the Productive Series sediment was derived from the Russian Platform and Ural Mountains to the north and transported by the paleo-Volga River system, augmented by rivers draining the Greater Caucasus mountain range. The paleo-Amu Darya River, which drained the Pamir and Tian Shan mountain ranges, was the major contributor of sediment to the eastern part of the basin, in Turkmenistan. The paleo-Kura River, which drained the Lesser Caucasus mountain range, supplied sediment to the western part of the basin. The sphere of influence of the paleo-Kura River system expanded with time, reaching as far east as the Apsheron Peninsula during deposition of the Surakhany Suite, which forms the uppermost part of the Productive Series (Morton et al., 2003a).

High-resolution heavy mineral data readily distinguish sandstones of paleo-Volga and paleo-Kura origin (Baturin, 1947; Morton et al., 2003a). Paleo-Kura sandstones are rich in clinopyroxene, calcic amphibole, and epidote, whereas paleo-Volga sandstones are characterized by metasedimentary heavy minerals such as staurolite, kyanite, sillimanite, and garnet. The two sandstone types form distinct fields on plots of provenance-sensitive ratio parameters (Fig. 3), such as apatite:tourmaline (ATi), garnet:zircon (GZi), rutile:zircon (RuZi), and chrome spinel:zircon (CZi). Their garnet populations are also markedly different: paleo-Volga garnet populations are rich in high-Mg, low-Ca types, high-Mg, high-Ca types, and low-Mg types, whereas paleo-Kura garnets are predominantly composed of Fe^{3+}-Ca types (Fig. 3). These markedly different heavy mineral populations reflect fundamental differences in the lithological constitution of the paleo-Volga and paleo-Kura drainage basins.

The paleo-Kura River drained the Lesser Caucasus mountain range, which was the site of a Mesozoic-Cenozoic subduction zone composed of basic igneous (largely volcanic) rocks, together with ophiolites and metamorphic rocks (Kazmin et al., 1986).

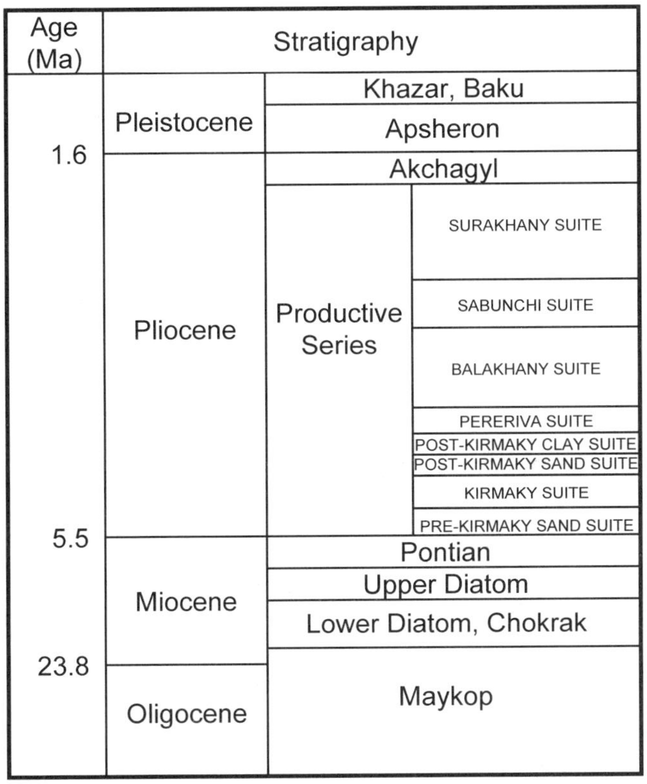

Age (Ma)	Stratigraphy		
1.6	Pleistocene	Khazar, Baku	
		Apsheron	
	Pliocene	Akchagyl	
		Productive Series	SURAKHANY SUITE
			SABUNCHI SUITE
			BALAKHANY SUITE
			PERERIVA SUITE
			POST-KIRMAKY CLAY SUITE
			POST-KIRMAKY SAND SUITE
			KIRMAKY SUITE
			PRE-KIRMAKY SAND SUITE
5.5	Miocene	Pontian	
		Upper Diatom	
		Lower Diatom, Chokrak	
23.8	Oligocene	Maykop	

Figure 2. Summary of the Neogene stratigraphy of the South Caspian Basin, adapted from Morton et al. (2003a).

Both low-Ti, calc-alkaline and high-Ti, alkaline basalts are present. The abundant clinopyroxene in the paleo-Kura heavy mineral assemblage reflects the predominance of basic igneous rocks in the Lesser Caucasus source.

The paleo-Volga drained a large area of the Russian Platform to the north of the Caspian Sea. The modern Volga River, which delivers sediment with heavy mineral characteristics similar to the paleo-Volga (Morton et al., 2003a), has a drainage basin ~1,380,000 km² in area (Kroonenberg et al., 1997). Most of the bedrock in the drainage basin makes up the Phanerozoic sediments that form the cover to the East European craton. The basement underlying this sedimentary pile includes three main blocks (Fennoscandia, Sarmatia, and Volgo-Uralia) that accreted to each other in the early Proterozoic (Claesson et al., 2001; Gorbatschev and Bogdanova, 1993). Precambrian Fennoscandian and Sarmatian basement rocks are exposed in the Baltic region and the Ukraine, respectively, although the latter is outside the drainage basin of the modern Volga. The Ural mountain range, which forms the eastern limit of the Volga drainage basin, consists of Precambrian basement rocks together with Paleozoic volcanics, granitoids, ophiolites, and metasediments that record the accretion of arcs to the craton in the late Paleozoic (Puchkov, 1997). The Phanerozoic sediments on the Russian Platform were presumably derived from a combination of the Precambrian basement blocks and the Uralian orogenic belt.

The Greater Caucasus mountain range was a subsidiary source of sediment to the South Caspian Basin during Productive Series deposition (Morton et al., 2003a). This region is largely composed of a thick Mesozoic clastic succession with minor volcanic rocks (Nalivkin, 1983). However, Paleozoic granitoids and gneisses are exposed in the western part of the mountain belt (Hanel et al., 1992).

Apatite Compositions

Apatite compositional data have been acquired from four samples, two with paleo-Kura heavy mineral characteristics and two with paleo-Volga characteristics. The paleo-Kura samples (KB1-75 and KB2-6) are from the Surakhany Suite (upper part of the Productive Series) at Babazanan and Aktapa Bridge, respectively (Fig. 1). The paleo-Volga samples (S3K5-12 and S3K5-32) are from the Pereriva and Balakhany Suites, respectively, both from outcrops in the Kirmaky Valley (Fig. 1). We compared the apatite compositions using chondrite-normalized REE plots (Fig. 4), plots of La/Nd versus La + Ce/ΣREE (Fig. 5), as proposed by Fleischer and Altschuler (1986), four plots (Sr vs. Y, $[Ce/Yb]_{cn}$ vs. ΣREE, Sr vs. Mn, and Eu/Eu* vs. Y; Figs. 6–9) proposed by Belousova et al. (2002), and the U vs. Th diagram (Fig. 10) proposed by Dill (1994).

Chondrite-normalized REE plots (Fig. 4) show that most of the apatites in the Paleo-Kura samples have moderate LREE enrichment and relatively muted negative Eu anomalies. Apatites that deviate from this pattern are scarce, although one grain has a distinctly different pattern showing LREE depletion and a small positive Eu anomaly. The paleo-Volga apatites show a greater diversity of REE patterns compared with the paleo-Kura apatites. Grains showing LREE enrichment are common, but there is a much greater number showing LREE depletion, and many grains display more pronounced negative Eu anomalies than those seen in the paleo-Kura samples.

Differences in apatite compositions are also apparent on La/Nd versus La + Ce/ΣREE plots (Fig. 5). Most of the paleo-Kura apatites have moderate to high La/Nd and high La + Ce/ΣREE values, and most therefore fall in the fields defined by mafic to intermediate and alkaline host rocks. In contrast, the paleo-Volga apatite populations include a large number of grains derived from acidic sources, characterized by low La/Nd and low La + Ce/ΣREE values. Apatites of alkaline origin (high La/Nd, high La + Ce/ΣREE) are especially scarce in the paleo-Volga samples. Table 1, which shows the relative abundances of apatites that fall into the four fields defined on this diagram, demonstrates the difference in contribution from acidic rocks between the paleo-Kura and paleo-Volga, which have 10%–20% in the former and 55%–76% in the latter. By contrast, alkaline rocks supplied 16%–39% of the paleo-Kura apatites but only 2%–9% in the paleo-Volga.

The paleo-Kura apatites have a more restricted range of Sr and Y contents than those from the paleo-Volga, since the paleo-Volga population includes many apatites with higher Y

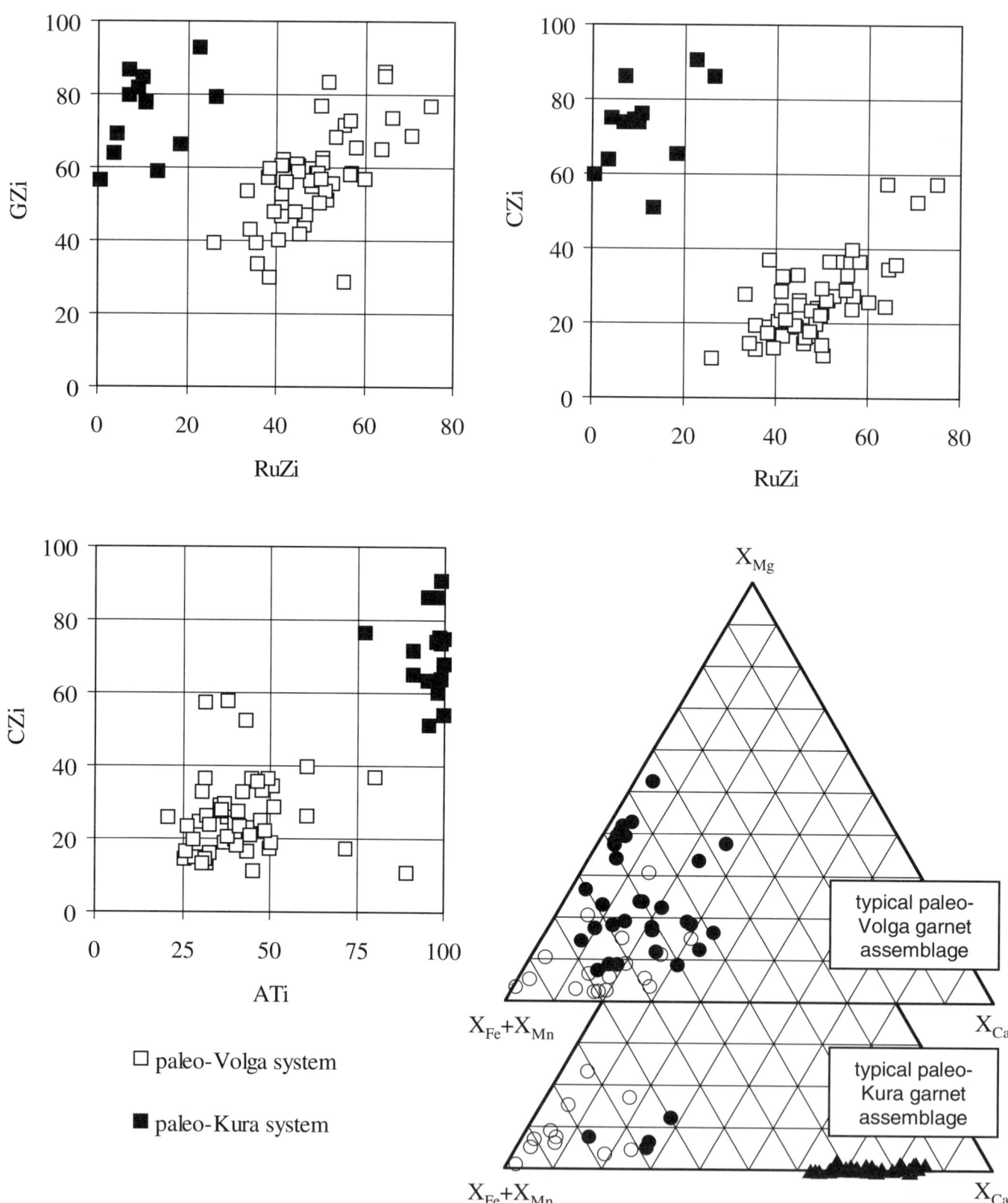

Figure 3. Mineralogical discrimination of paleo-Volga and paleo-Kura sandstones using binary plots of provenance-sensitive heavy mineral ratios and composition of detrital garnet populations (adapted from Morton et al., 2003a). GZi—garnet:zircon index (% garnet in total garnet + zircon), CZi—chrome spinel:zircon index (% chrome spinel in total chrome spinel + zircon), RuZi—rutile:zircon index (% rutile in total rutile + zircon), ATi—apatite:tourmaline index (% apatite in total apatite + tourmaline), X_{Fe}, X_{Mn}, X_{Ca}, X_{Mn}— ionic proportions of Fe, Mg, Ca, and Mn. All Fe was calculated as Fe^{2+}. ●—X_{Mn} < 5%, ○—X_{Mn} > 5%, ▲—Fe^{3+}/Al > 0.1.

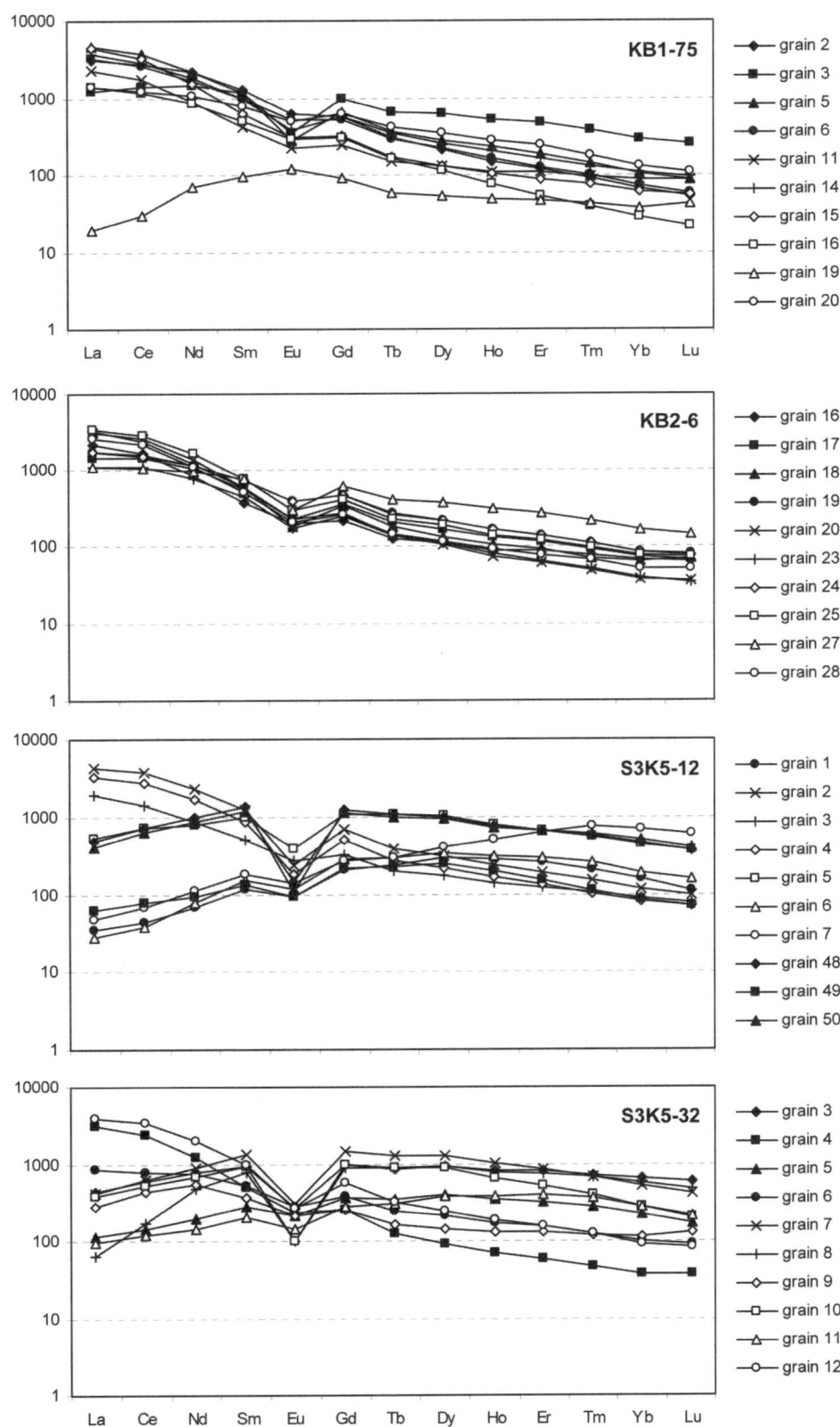

Figure 4. Representative chondrite-normalized rare earth element (REE) profiles of apatites in the South Caspian Basin Productive Series samples. Chondrite-normalizing values are from Taylor and McLennan (1985). Samples KB1-75 (Surakhany Suite, Babazanan) and KB2-6 (Surakhany Suite, Aktapa Bridge) are of paleo-Kura origin, whereas S3K5-12 (Pereriva Suite, Kirmaky Valley) and S3K5-32 (Balakhany Suite, Kirmaky Valley) are from the paleo-Volga (see text for discussion).

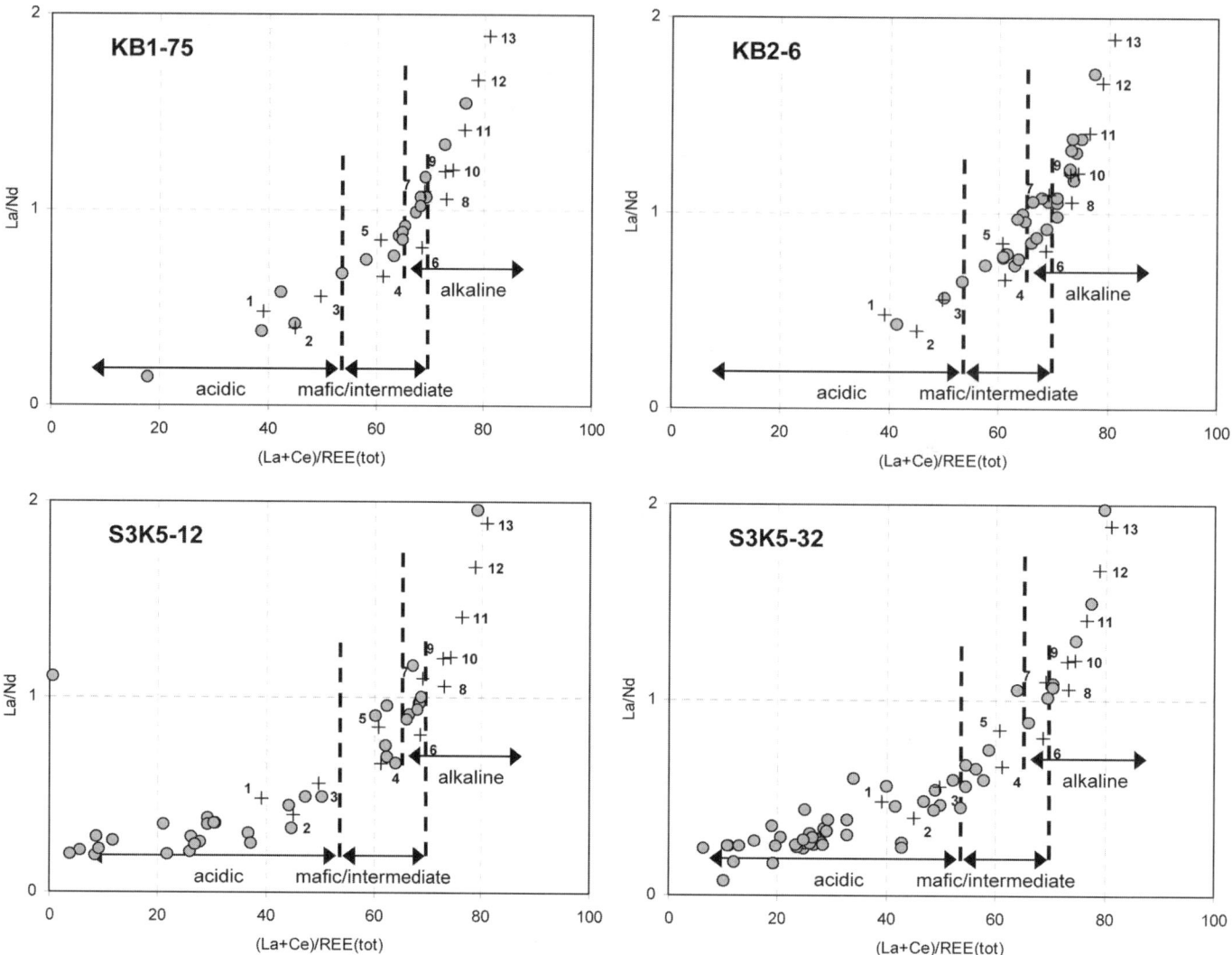

Figure 5. Apatite compositions in South Caspian Basin Productive Series samples plotted on the La/Nd versus La + Ce/ΣREE classification diagram of Fleischer and Altschuler (1986). 1—granite pegmatite, 2—gneiss/migmatite, 3—granite, 4—gabbro, 5—granodiorite, 6—kimberlite, 7—syenite, 8—alkali ultramafic, 9—carbonatite, 10—iron ores, 11—ultramafic, 12—alkaline, 13—alkaline pegmatite.

and lower Sr values (Fig. 6). This implies that a large proportion of the paleo-Volga apatite population is from an evolved source, since Sr contents decrease and Y contents increase during fractionation (Belousova et al., 2002). By analogy, the majority of the paleo-Kura apatite population is likely to be from a relatively unfractionated, more mafic source.

The degree of LREE enrichment is demonstrated by the $(Ce/Yb)_{cn}$ versus ΣREE plot (Fig. 7). Most of the paleo-Kura apatites have strong LREE enrichment, with $(Ce/Yb)_{cn}$ between 10 and 100. Although some paleo-Volga apatites also have strong LREE enrichment, the majority has flat to slightly LREE enriched profiles, with $(Ce/Yb)_{cn}$ = 1–10, and a significant number have LREE depletion, with $(Ce/Yb)_{cn}$ < 1. This suggests that the paleo-Volga source region included highly fractionated rocks, such as granite pegmatites, since these generally show relative LREE depletion (Belousova et al., 2002). There is no apparent difference in total REE contents between the paleo-Kura and paleo-Volga apatites.

The Sr-Mn plot (Fig. 8) mirrors the Sr-Y plot, and the paleo-Volga apatite population shows a greater range in content of both trace elements compared with the paleo-Kura. The maximum Mn content in the paleo-Volga apatites is greater than in the paleo-Kura samples, whereas the minimum Sr values are less in the paleo-Volga apatites than in the paleo-Kura apatites. Since Sr decreases and Mn increases during fractionation, the Sr-Mn plot indicates that the source of the paleo-Volga included a higher proportion of evolved rocks than the paleo-Kura source.

The Eu/Eu* versus Y plot (Fig. 9) shows the presence of a large group of apatites with high Y and large negative Eu anomalies in the paleo-Volga population that is hardly represented in the paleo-Kura population. Since Y contents increase and Eu anomalies

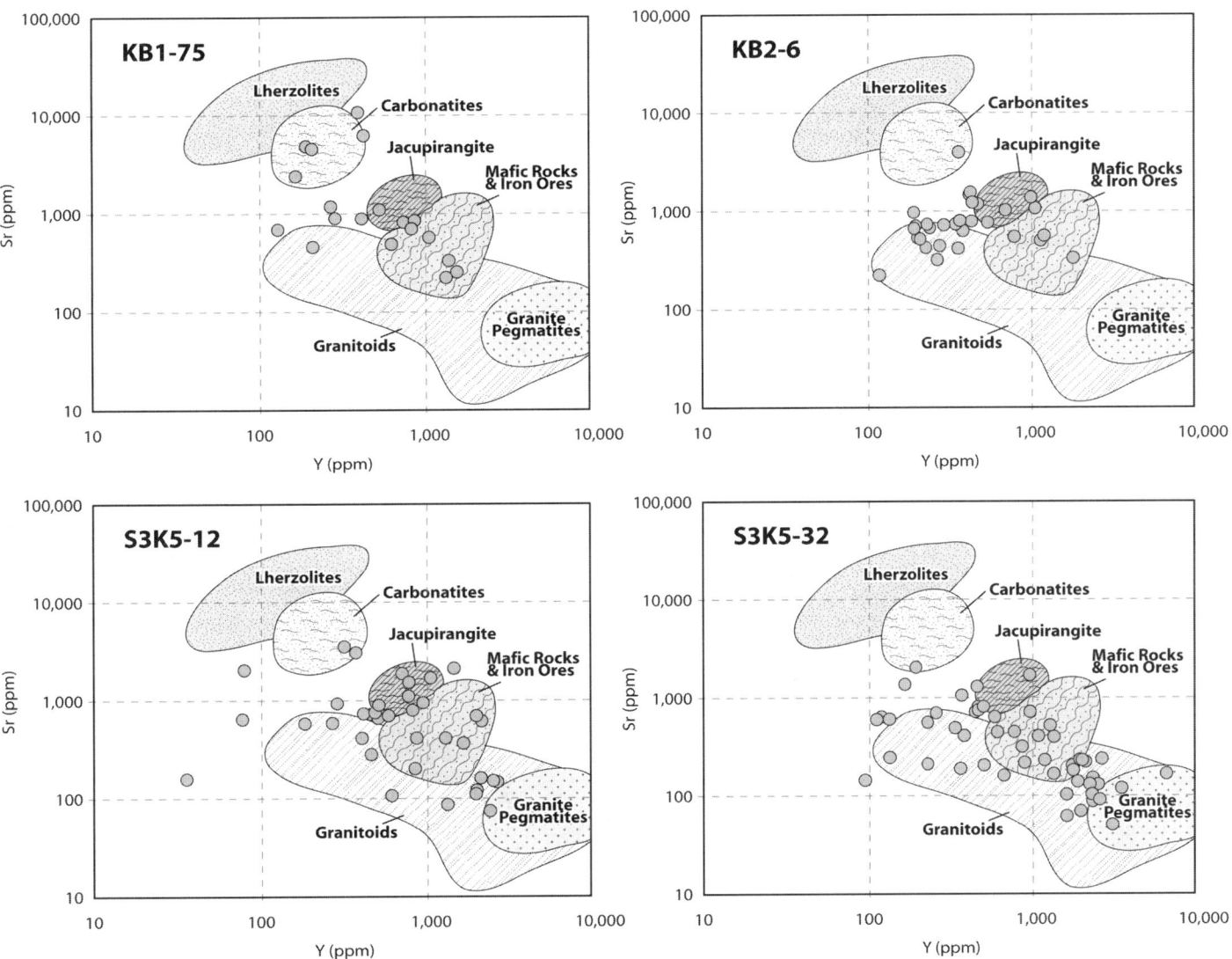

Figure 6. Sr versus Y plot showing apatite compositions in South Caspian Basin Productive Series samples, together with the range of apatite compositions in a variety of rock types determined by Belousova et al. (2002).

become more marked with increasing fractionation, this group of apatites is believed to represent supply from highly evolved rocks, such as granites and granite pegmatites.

The Th-U plot (Fig. 10) also demonstrates the difference in provenance between the paleo-Volga and the paleo-Kura apatites. Virtually all the apatites in the paleo-Kura have Th > U, whereas ~40%–60% of the paleo-Volga apatites have Th < U. Th concentrations decrease during fractionation, probably due to crystallization of monazite (Belousova et al., 2002). Consequently, the apatites with U > Th are likely to represent supply from highly evolved rocks.

Discussion

Compositional data on detrital apatite populations clearly show a difference in provenance between the paleo-Volga and paleo-Kura sediment. This is demonstrated by a variety of plots, all of which show that, compared with the paleo-Kura, paleo-Volga detritus includes a much greater proportion of apatites derived from highly evolved rocks, such as granites, granite pegmatites, or gneisses. Apatites supplied by such highly evolved rocks are recognized by their LREE depletion, as shown by REE profiles and measured by La + Ce/ΣREE and $(Ce/Yb)_{cn}$, as well as their low Sr, high Y, and high Mn contents, their marked negative Eu anomalies, and their low Th/U ratios. Although paleo-Volga sediment also includes apatites derived from less-evolved rocks, these generally have weaker LREE enrichment (as measured by $[Ce/Yb]_{cn}$ and La/Nd) compared with those in the paleo-Kura. The high La/Nd value in the paleo-Kura samples suggests that there was a significant contribution from alkaline rocks, but such rocks appear to have made a minimal contribution to the paleo-Volga system.

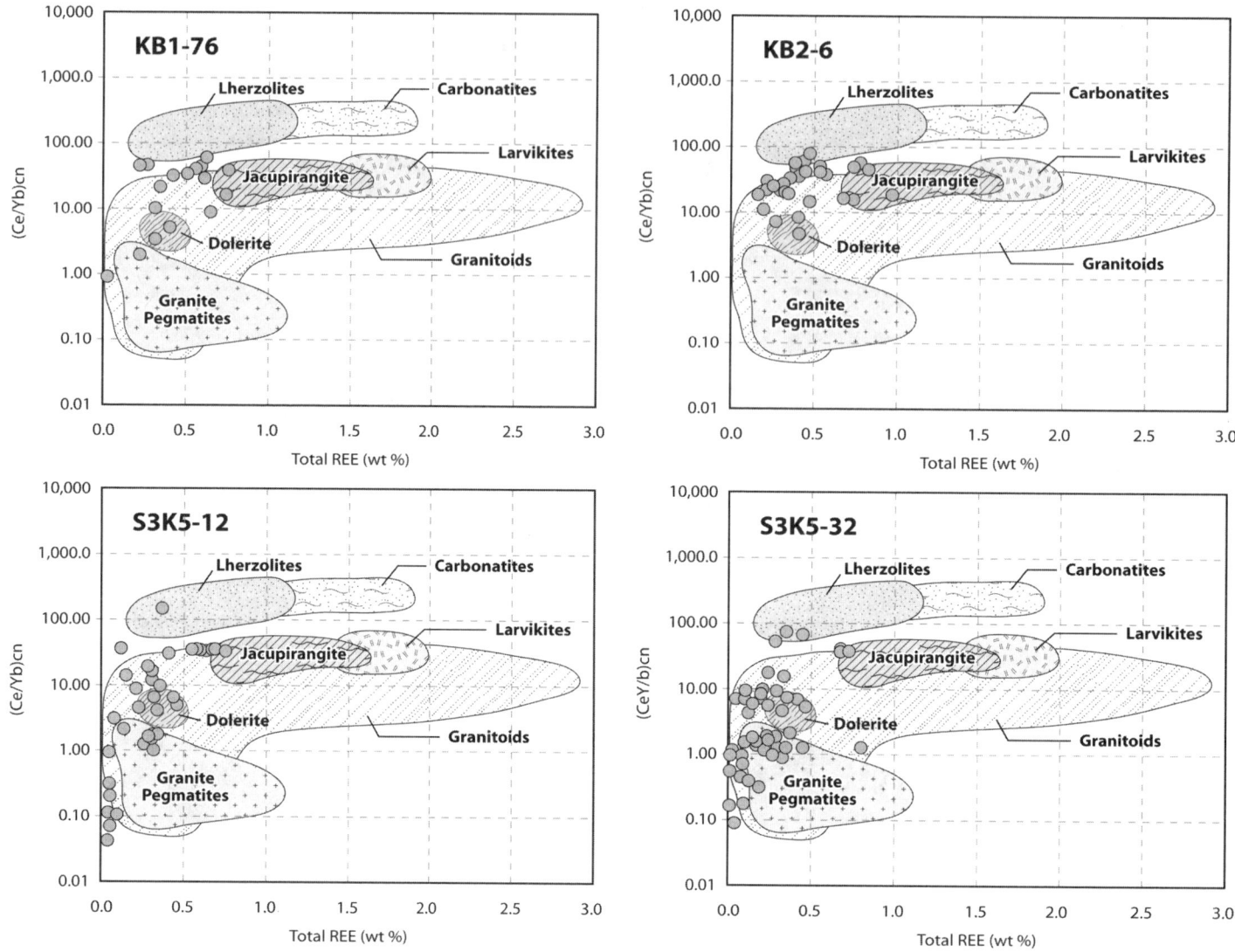

Figure 7. $(Ce/Yb)_{cn}$ versus total rare earth element (REE) plot showing apatite compositions in South Caspian Basin Productive Series samples, together with the range of apatite compositions in a variety of rock types determined by Belousova et al. (2002).

The abundance of apatites derived from mafic to intermediate and alkaline rocks and the scarcity of apatites derived from highly evolved sources in the paleo-Kura sediment agree with the geological evidence from the Lesser Caucasus catchment. The Lesser Caucasus mountain range is dominated by basic igneous rocks of both low-Ti (calc-alkaline) and high-Ti (alkaline) content, together with ophiolites and metamorphic rocks (Kazmin et al., 1986). The ultimate source of the paleo-Volga detritus is less well constrained, since the majority is polycyclic material ultimately derived from the Fennoscandian, Sarmatian, and Volgo-Uralian basement blocks (Morton et al., 2003a). The greater diversity shown by the paleo-Volga apatite populations reflects the larger size of the catchment and the polycyclic nature of the sediment, which inevitably will have introduced sediment from a wide range of sources. The presence of apatites derived from highly evolved rocks indicates that the ultimate source regions contained widespread granites and acidic gneisses, a typical feature of ancient basement terrains.

The relative contribution from different host rocks can be judged using the La/Nd versus La + Ce/ΣREE plot. As discussed already, on the basis of this figure, acidic rocks supplied 10%–20% of the paleo-Kura apatites and 55%–76% of the paleo-Volga apatites, whereas alkaline rocks supplied 16%–39% of the paleo-Kura apatites and 2%–9% of the paleo-Volga apatites (Table 1).

There are some differences between the apatite populations in two assemblages of paleo-Volga origin. The Pereriva Suite sample (S3K5-12) has a smaller number of apatites derived from highly evolved rocks (as indicated both by the

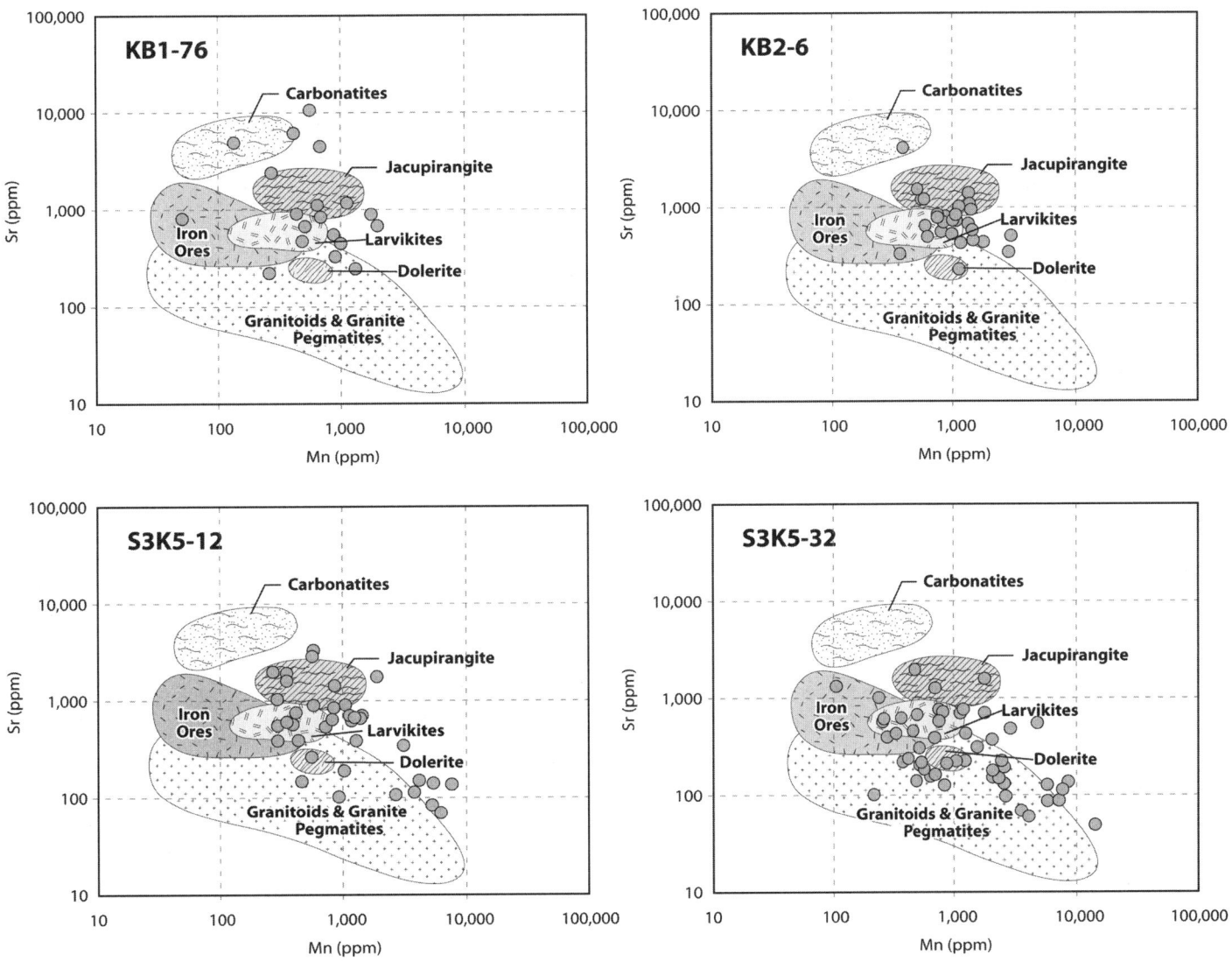

Figure 8. Sr versus Mn plot showing apatite compositions in South Caspian Basin Productive Series samples, together with the range of apatite compositions in a variety of rock types determined by Belousova et al. (2002).

La/Nd versus La + Ce/ΣREE and Th versus U plots) compared with those from the Balakhany Suite (S3K5-32). This may simply reflect heterogeneity in the source terrain, but could also reflect differences in relative supply from the Russian Platform and Greater Caucasus. Heavy mineral data suggest that Greater Caucasus influences were particularly important during deposition of the Pereriva Suite (Morton et al., 2003a).

Belousova et al. (2002) also proposed a method of quantifying the contributions made by a variety of apatite source rocks, using a correlation and regression tree (CART). This CART uses Sr/Y, Th/U, and (Ce/Yb)$_{cn}$ ratios together with abundances of Nd, U, Mg, Eu, and Sr to assign apatites to dolerites, granite pegmatites, granitoids, carbonatites, lherzolites, jacupirangites, larvikites, and iron ores.

As shown in Table 2, the CART provides further evidence for differences in provenance between the paleo-Kura and paleo-Volga apatites. The most obvious difference is shown by the abundance of apatites derived from granites and granite pegmatites, which form between 42% and 63% of the paleo-Volga apatite populations, but only 10%–32% in the paleo-Kura populations. These figures are reasonably similar to those derived from the La/Nd versus La + Ce/ΣREE plot (Table 1). However, the remaining parts of the apatite populations are categorized rather differently compared with the La/Nd versus La + Ce/ΣREE figure. According to the CART, dolerites appear to have made similar contributions to both paleo-Kura (11%–19%) and paleo-Volga (19%–20%) apatite populations, whereas the La/Nd versus La + Ce/ΣREE plot suggests that mafic to intermediate rocks formed a significantly greater relative

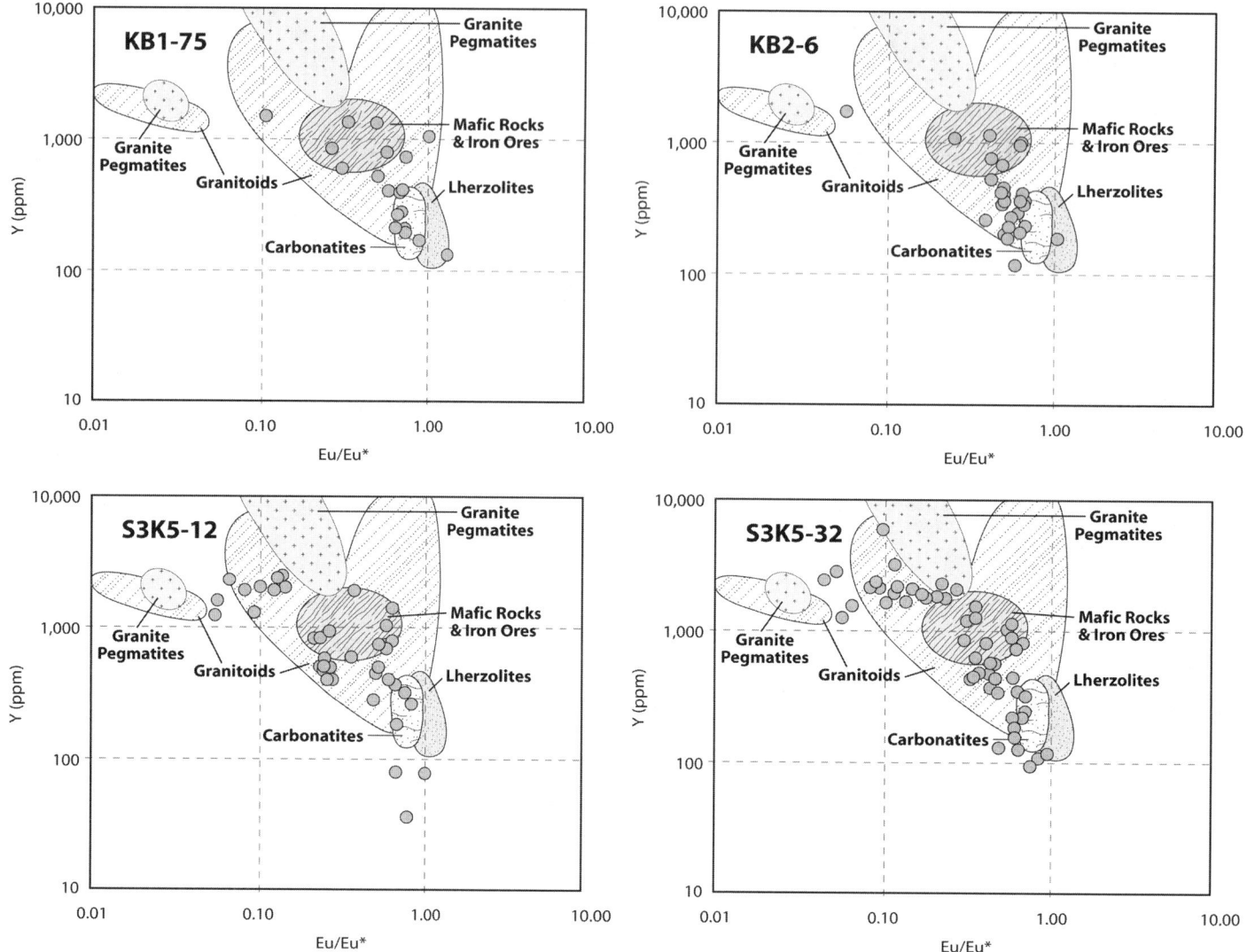

Figure 9. Eu/Eu* versus Y plot showing apatite compositions in South Caspian Basin Productive Series samples, together with the range of apatite compositions in a variety of rock types determined by Belousova et al. (2002).

TABLE 1. CLASSIFICATION OF APATITES ON THE BASIS OF THE La/Nd VERSUS La + Ce/∑REE PLOT (FLEISCHER AND ALTSCHULER, 1986) AND Th/U RATIOS (DILL, 1994)

Sample	Acidic	Mafic/ intermediate	Alkaline–mafic/ intermediate overlap zone	Alkaline	Apatites with U >Th (%)	Number of analyses
Case study 1 (Pliocene, South Caspian Basin)†						
KB1.75	21.1	31.6	31.6	15.8	5.2	19
KB2.6	9.7	32.3	19.4	38.7	6.3	31
S3K5.12	54.8	11.9	31.0	2.4	42.9	42
S3K5.32	75.9	11.1	3.7	9.3	61.1	54
Case study 2 (Devonian/Carboniferous, Clair oil field)‡						
UCG	84.6	7.7	0.0	7.7	86.5	52
Unit VI	86.5	2.7	5.4	5.4	89.1	37
Unit III	82.7	9.7	3.8	3.8	67.3	52
Unit I	87.5	8.3	2.1	2.1	75.0	48

†Sample KB1.75 = Surakhany Suite, Babazanan. Sample KB2.6 = Surakhany Suite, Aktapa Bridge. Sample S3K5.12 = Pererriva Suite, Kirmaky Valley. Sample S3K5.32 = Balakhany Suite, Kirmaky Valley. For locations of sample sites, see Figure 1.
‡All samples are from Well 206/8–8: see Figure 11 for location of this well and the Clair oil field. Figure 13 shows the precise stratigraphic position of the four core samples.

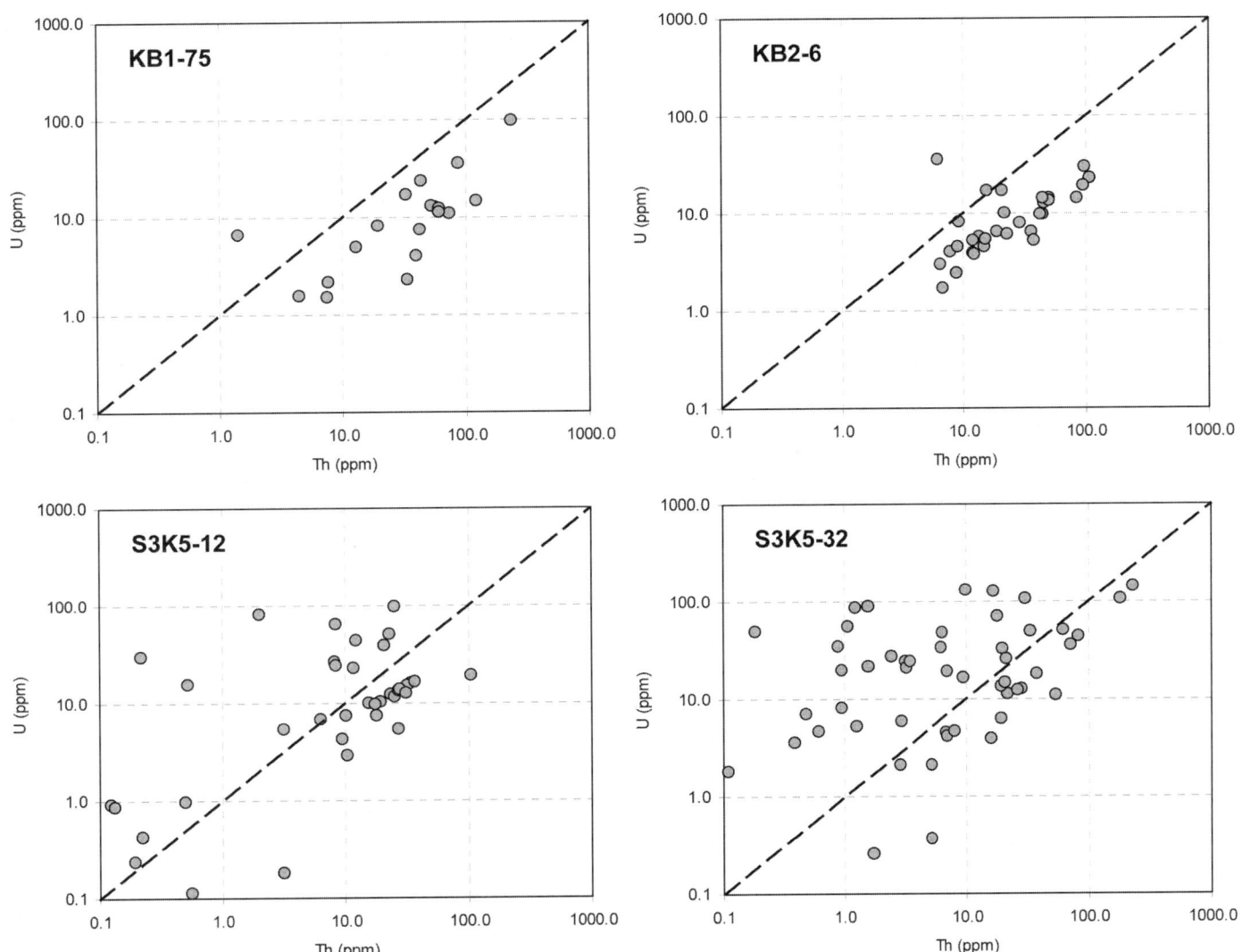

Figure 10. Th versus U plot of apatites in South Caspian Basin Productive Series samples.

TABLE 2. CLASSIFICATION OF APATITES DISCUSSED IN THIS PAPER USING THE CORRELATION AND REGRESSION TREE (CART) OF BELOUSOVA ET AL. (2002)

Sample	Lherzolite	Dolerite	Granite	Granite pegmatite	Iron ore	Carbonatite	Larvikite	Jacupirangite	Number of analyses
Case study 1 (Pliocene, South Caspian Basin)[†]									
KB1.75	26.3	10.5	31.6	0.0	31.6	0.0	0.0	0.0	19
KB2.6	3.2	19.4	9.7	0.0	67.7	0.0	0.0	0.0	31
S3K5.12	2.4	19.5	24.4	17.1	29.3	7.3	0.0	0.0	42
S3K5.32	1.9	18.5	20.4	42.6	14.8	1.9	0.0	0.0	54
Case study 2 (Devonian/Carboniferous, Clair Oilfield)[‡]									
UCG	0.0	32.0	12.0	28.0	6.0	22.0	0.0	0.0	52
Unit VI	0.0	20.0	11.4	37.2	20.0	11.4	0.0	0.0	37
Unit III	0.0	32.7	14.3	24.5	6.1	4.1	16.3	2.0	52
Unit I	0.0	28.3	28.3	26.0	8.7	8.7	0.0	0.0	48

[†]Sample KB1.75 = Surakhany Suite, Babazanan. Sample KB2.6 = Surakhany Suite, Aktapa Bridge. Sample S3K5.12 = Pereriva Suite, Kirmaky Valley. Sample S3K5.32 = Balakhany Suite, Kirmaky Valley. For locations of sample sites, see Figure 1.
[‡]All samples are from Well 206/8–8: see Figure 11 for location of this well and the Clair Oilfield. Figure 13 shows the precise stratigraphic position of the four core samples.

proportion of the paleo-Kura catchment compared with the paleo-Volga. The most notable, and surprising, result of the CART is that it classifies a large proportion of the paleo-Kura apatites (32%–68%) as being derived from iron ores, with lherzolites also making a major contribution (26% in KB1-75). This outcome does not fit with the known geology of the paleo-Kura catchment, unlike the analysis made using the La/Nd versus La + Ce/ΣREE plot. Furthermore, it seems unlikely that such a rare rock type as lherzolite would supply over 25% of any apatite population in a large-scale river system. Although tracing apatites back to particular host rocks using the approach undertaken by Belousova et al. (2002) has merit, at present it generates an unlikely lithological characterization of the paleo-Kura and paleo-Volga source regions.

CASE STUDY 2: DEVONIAN-CARBONIFEROUS SANDSTONES OF THE CLAIR OIL FIELD

Geological Background

The Clair oil field is located west of the Shetland Islands on the UK continental shelf (Fig. 11). The reservoir succession is composed of over 1000 m of Devonian-Carboniferous clastic sediment (predominantly sandstone, with minor siltstone, mudstone, and conglomerate) deposited in a range of fluvial, lacustrine, and aeolian environments (Allen and Mange-Rajetzky, 1992; McKie and Garden, 1996; Nichols, 2005). Allen and Mange-Rajetzky (1992) subdivided the succession into ten units, labeled I–X from base to top. Units I–VI comprise the Lower Clair Group, and VII–X comprise the Upper Clair Group (Fig. 12). Unit X is known to be Carboniferous (Dinantian) in age on the basis of miospores (Blackbourn, 1988), and the top of the Lower Clair Group is believed to be Late Devonian on the basis of holoptychian fish scales (Ridd, 1981; Trewin and Thirlwall, 2002). The rest of the succession is devoid of biostratigraphic markers, although the Devonian-Carboniferous boundary is commonly placed at the boundary between the Lower and Upper Clair Group (e.g., Coney et al., 1993).

Heavy mineral data have proved crucial in the stratigraphic subdivision of the Clair succession (Allen and Mange-Rajetzky, 1992; Morton et al., 2003b). The key parameters used to subdivide the succession are the apatite:tourmaline ratio (ATi), garnet:zircon ratio (GZi), rutile:zircon ratio (RuZi), staurolite:zircon ratio (SZi), and apatite roundness index (ARi). Variations in these parameters in the fully cored Well 206/8-8 are shown in Figure 13, demonstrating that the unit boundaries represent significant changes either in provenance or in sedimentary transport history and depositional environment. The event at the top of unit VI (Lower Clair–Upper Clair boundary) is a first-order heavy mineral event related to a major change in provenance, marked by dramatic changes in ATi, GZi, RuZi, and SZi (Fig. 13). This change in provenance is also seen in clay mineral data (Pay et al., 2000). Variations in heavy mineralogy within the Lower Clair Group relate partly to minor changes in provenance and partly to changes in the extent of aeolian influences. Units III and V have stronger aeolian influences than the rest of the Lower Clair Group (Allen and Mange-Rajetzky, 1992; McKie and Garden, 1996), and these are faithfully reflected in the roundness of apatite (Fig. 13), a mineral that is particularly sensitive to aeolian influences on account of its lack of mechanical durability (Allen and Mange-Rajetzky, 1992). Nichols (2005) considered that Lewisian (Archean) gneisses were the main source of the Lower Clair Group succession, with transport being directed from northwest to southeast. Allen and Mange-Rajetzky (1992), however, suggested that transport was from the northeast, and that the source area supplying the Lower Clair Group sediments included low-grade metasediments as well as high-grade Lewisian gneisses. The major change in provenance at the Lower Clair–Upper Clair boundary is marked by a dramatic increase in abundance of garnet, staurolite, and rutile, consistent with an increase in supply from metasedimentary rocks, such as those comprising the Neoproterozoic to early Paleozoic Moine and Dalradian successions of nearby Scotland and Shetland (Allen and Mange-Rajetzky, 1992). At times, the Upper Clair Group source area may have also extended into East Greenland (Allen and Mange-Rajetzky, 1992).

Apatite Compositions

Apatite compositional data have been acquired from four samples, one from the Upper Clair Group and three from the Lower Clair Group (units I, III, and VI), all from Well 206/8-8 (Fig. 11). The positions of the four samples are shown on the stratigraphic plot of heavy mineral variations (Fig. 13).

Chondrite-normalized REE plots (Fig. 14) show that very few of the apatites in the Clair Group succession have LREE enrichment. Most have relatively flat patterns, with slight depletion in both the LREE and HREE. Some, however, have more extreme LREE depletion, and these tend to show slight HREE enrichment. There are marked variations in the size of the Eu anomalies; some have virtually no anomaly (Eu*/Eu = 1), and others have strong negative anomalies. In general, those with higher REE contents display the largest Eu anomalies (Fig. 15). There is no obvious difference between the samples—all four show a similar range of REE patterns.

Most of the Clair oil field apatites have low La/Nd and low La + Ce/ΣREE values and thereby fall in the acidic field on the La/Nd versus La + Ce/ΣREE plot (Fig. 16). Most of the remaining apatites fall in the intermediate to mafic field, with very few in the alkaline field. There are no significant differences between the four samples.

The four samples also appear to be similar on the Sr-Y plots (Fig. 17); most have relatively high Y and low Sr values, implying the majority are from an evolved source (Belousova et al., 2002). The Upper Clair Group sample is slightly different; it has a higher proportion of apatites with low Y values that fall outside the compositional fields determined by Belousova et al. (2002).

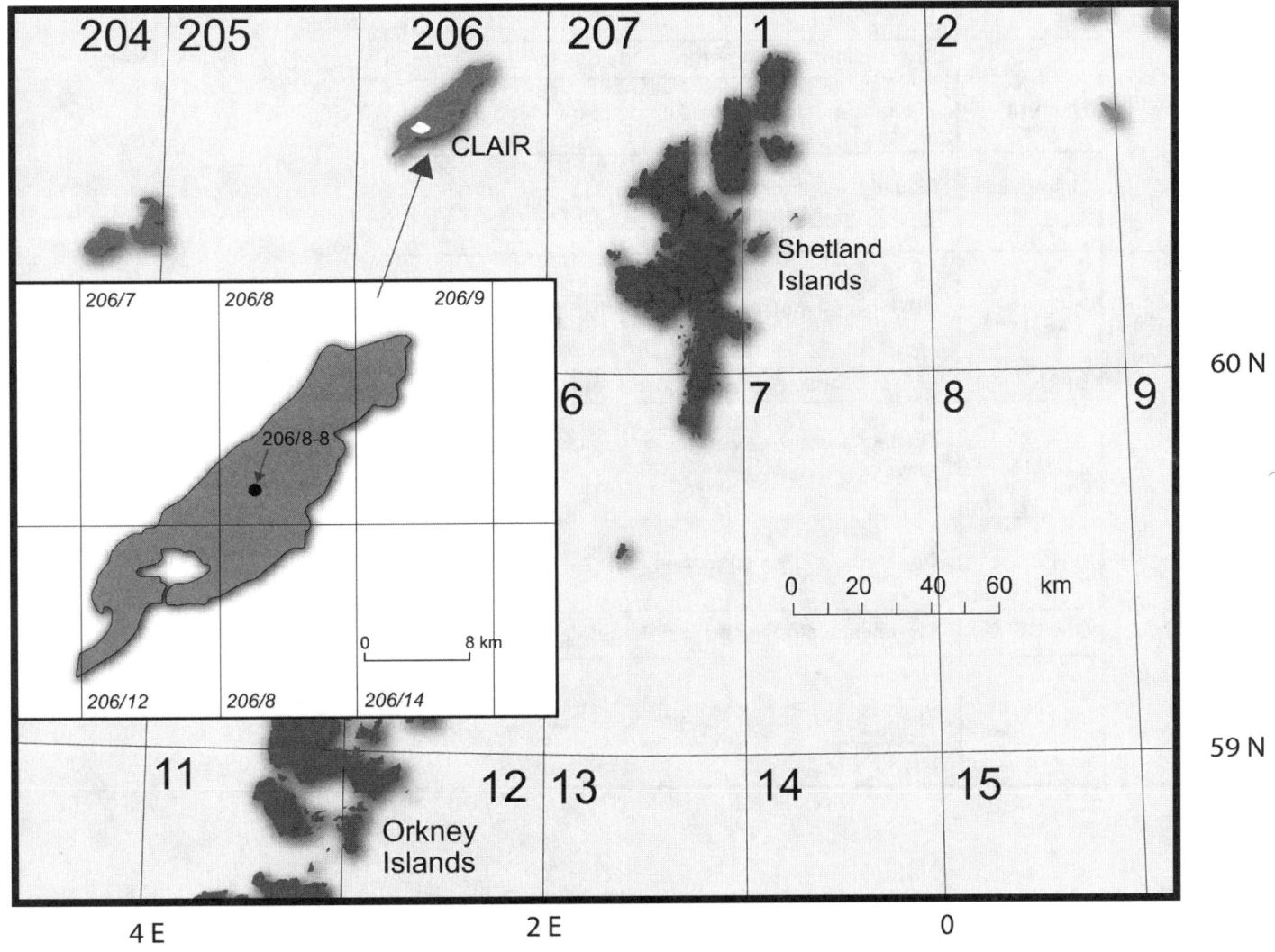

Figure 11. Location map of the Clair oil field, west of Shetland, showing location of Well 206/8-8. Numbers on large-scale map (204, 205, etc.) refer to UK continental shelf quadrants. Numbers on the inset map (206/7, 206/8, etc.) refer to UK continental shelf blocks.

The $(Ce/Yb)_{cn}$ versus ΣREE plot (Fig. 18) shows that the Clair oil field apatites have either flat to slightly LREE enriched profiles, with $(Ce/Yb)_{cn} = 1–10$, or LREE depletion, with $(Ce/Yb)_{cn} < 1$. There is no significant difference in the range of $(Ce/Yb)_{cn}$ values between the four samples, although the unit III population has a greater range of total REE contents compared with the other samples. This diagram confirms that the majority of the Clair oil field apatites are from evolved rocks such as granitoids and granite pegmatites, since these generally show relative LREE depletion (Belousova et al., 2002).

The three Lower Clair Group samples are closely comparable to one another on the Sr-Mn plot (Fig. 19) and form a well-defined cluster corresponding to apatite compositions predicted for granitoids and granite pegmatites. The Upper Clair Group sample is distinctly different and has lower Mn contents that cause a large number of the analyses to plot outside the range of compositions determined by Belousova et al. (2002). This mirrors the situation shown by the Sr-Y plot (Fig. 17).

The samples all have a similar range of Eu/Eu* values and Y contents (Fig. 20); apatites with higher Y contents have larger negative Eu anomalies. These are believed to represent apatites derived from the most highly fractionated rocks such as granite pegmatites. The Upper Clair Group sample differs slightly from the others by having a larger number of grains with low Y that fall outside the compositional fields defined by Belousova et al. (2002).

Most of the apatites in the Clair oil field sandstones have U > Th (Fig. 21), further demonstrating the widespread occurrence of highly evolved rocks in the source area. The two lower samples (units I and III) have somewhat smaller abundances of apatites with U > Th compared with the two upper samples (unit VI and Upper Clair Group), as shown in Table 1.

		Unit	
? Carboniferous	Upper Clair Group	Unit X	Marginal marine and distributary channel sandstones
		Unit IX	Proximal braidplain or fan sandstones
		Unit VIII	Fluvial sandstones including point bar deposits
		Unit VIIB	Fluvial sandstones, overbank fines
		Unit VIIA	Fluvial sandstones
Devonian	Lower Clair Group	Unit VI	Fluvial sandstones, lacustrine sediments
		Unit V	Fluvial sandstones with aeolian reworking
		Unit IV	Fluvial sandstones with minor aeolian reworking
		Unit III	Dry/damp aeolian sandsheet
		Unit II	Fluvial sandstones and conglomerates
		Unit I	Fan conglomerates, fine-grained lacustrine sediments

Figure 12. Stratigraphic subdivision of the Clair Group succession in the Clair oil field showing the predominant lithofacies and depositional environments of each of the units, adapted from Allen and Mange-Rajetzky (1992), McKie and Garden (1996), and Nichols (2005).

Discussion

Despite the evidence for a major change in provenance at the Lower Clair–Upper Clair boundary and more subtle changes through Lower Clair Group deposition, apatite compositions display comparatively little change though the analyzed interval. There are some relatively subtle differences: for example, the unit III sample contains some apatites with relatively high total REE contents, and the unit I and unit III samples have fewer apatites with U > Th compared with unit VI and the Upper Clair Group. The Upper Clair Group apatites have lower overall Mn and Y contents compared with those in the Lower Clair Group. However, compared with the major differences shown by the paleo-Kura and paleo-Volga apatites, such differences appear comparatively trivial.

The geochemical data indicate that the vast majority of the apatites in the Clair Group succession appear to be derived from highly evolved rocks such as granites or granite pegmatites. This is best demonstrated by the La/Nd versus La + Ce/ΣREE plots (Fig. 16), where most apatites fall in the acidic field, but it is also supported by the other plots, which indicate that compositions generally overlap the fields of granites and granite pegmatites as defined by Belousova et al. (2002). The large number of apatites with U > Th also indicates a source dominated by evolved rocks.

The relative contributions from different host rocks can be judged using the La/Nd versus La + Ce/ΣREE plot (Fleischer and Altschuler, 1986) and the CART method proposed by Belousova et al. (2002). On the basis of the La/Nd versus La + Ce/ΣREE plot, 83%–88% of the apatites were derived from acidic rocks, and mafic to intermediate and alkaline rocks are minor contributors (Table 1). The CART (Table 2) provides a different breakdown of source characteristics, suggesting that acidic rocks (granites and granite pegmatites) supplied only 40%–54% of the apatites. The shortfall is made up of dolerite, interpreted as supplying 20%–32% of the apatites, together with iron ores (6%–20%) and carbonatites (4%–22%). From unit III, 16% of the apatites are categorized as larvikitic, presumably reflecting the higher total REE contents in some of the apatites in this sample (Fig. 18).

The Lewisian (Archean) Gneiss Complex is considered to be a major source for the Clair succession (Allen and Mange-Rajetzky, 1992: Nichols, 2005). Lewisian gneisses are predominantly tonalitic, trondhjemitic, or granodioritic in composition, with subordinate granite gneiss sheets and lenses (Park et al., 2002). The breakdown given by the La/Nd versus La + Ce/ΣREE plot (Fleischer and Altschuler, 1986) is therefore much more in

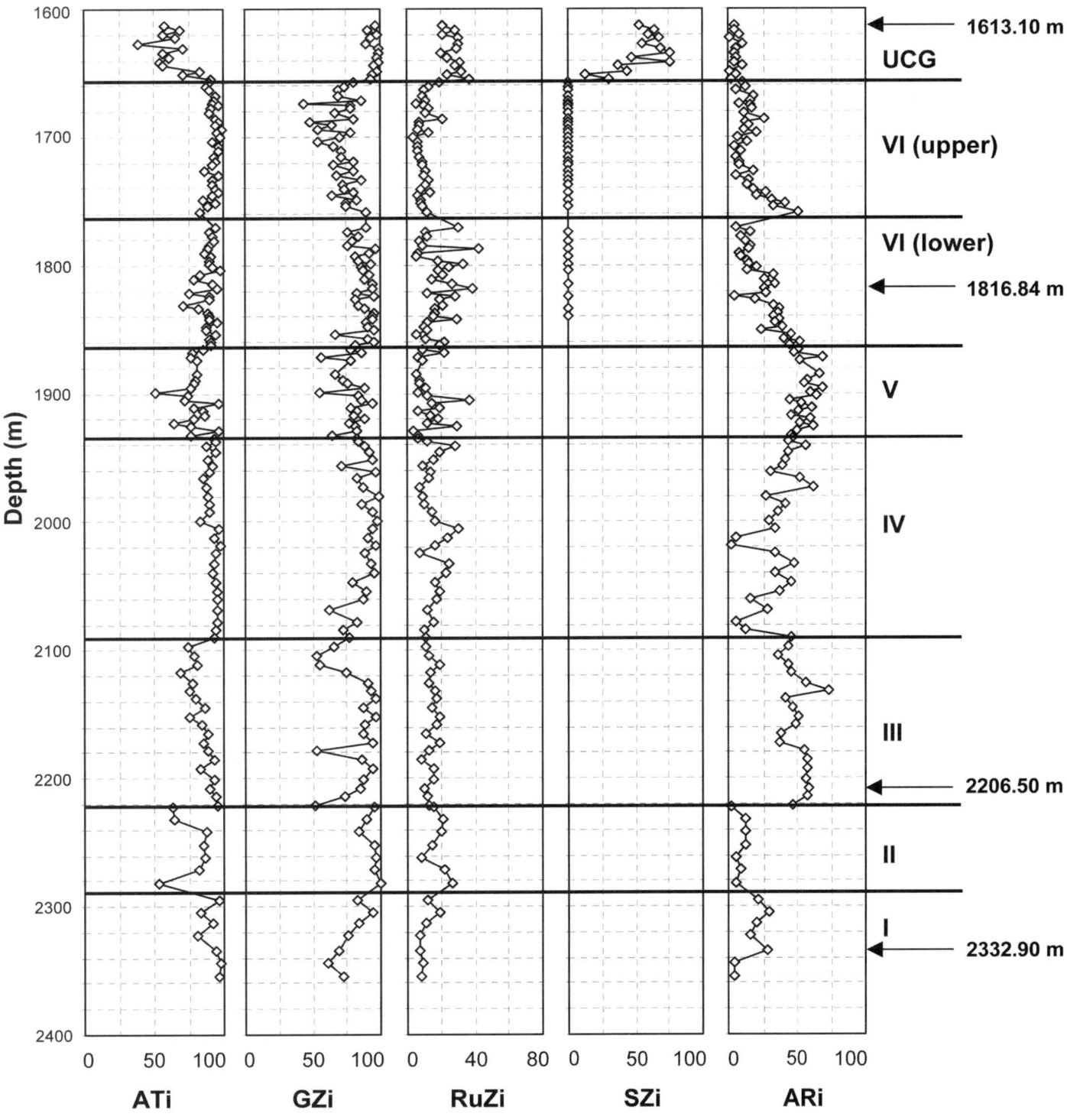

Figure 13. Heavy mineral stratigraphy of the Clair reservoir succession in Well 206/8–8, showing the variations in key provenance and sediment transport history indices. The arrows show the depth location of the four samples used in the apatite geochemical study. UCG—Upper Clair Group (units VII–X) undivided. ATi—apatite:tourmaline index (% apatite in total apatite + tourmaline), GZi—garnet:zircon index (% garnet in total garnet + zircon), RuZi—rutile:zircon index (% rutile in total rutile + zircon), SZi—staurolite:zircon index (% staurolite in total staurolite + zircon), ARi—apatite roundness index (% rounded apatite in apatite population).

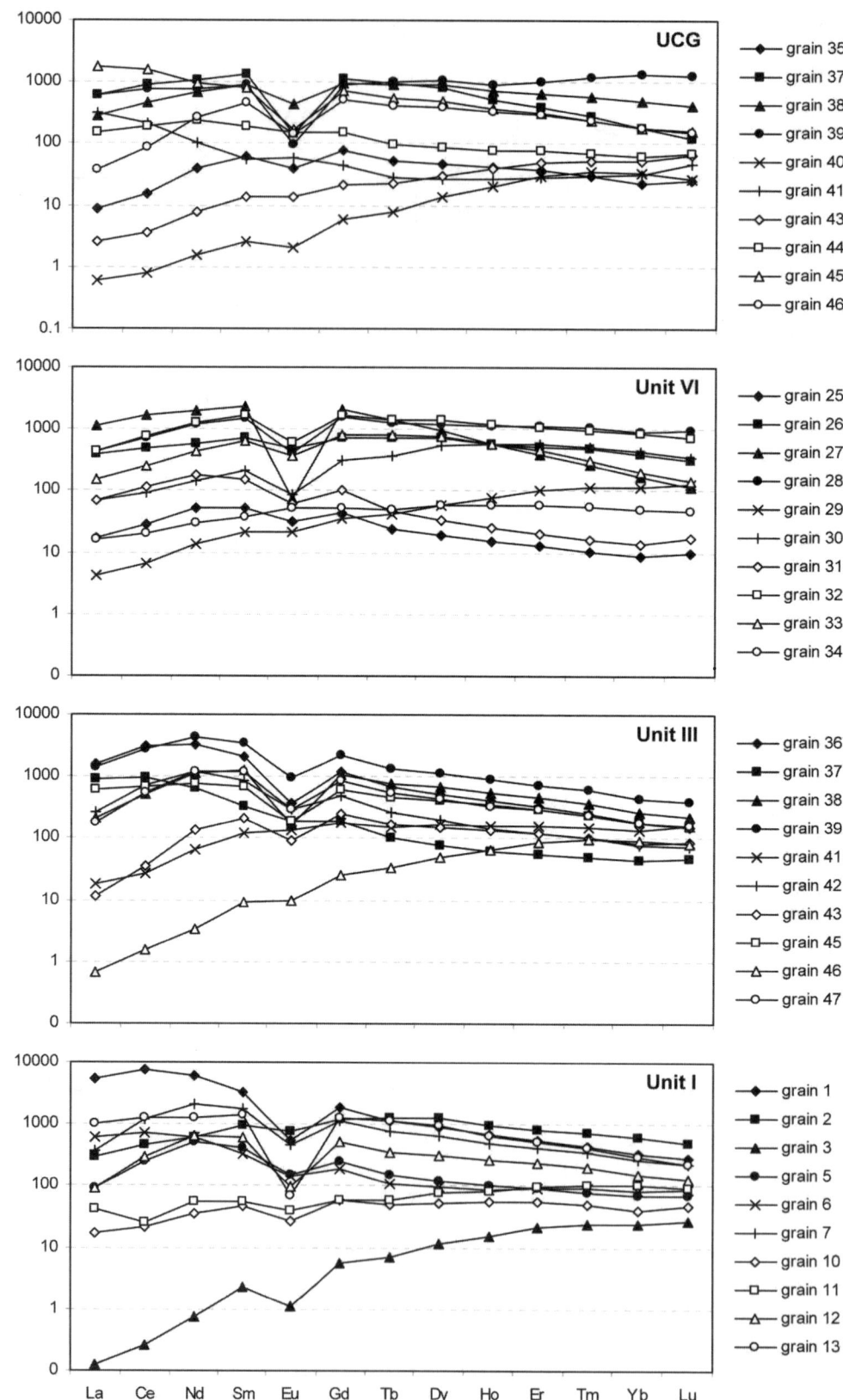

Figure 14. Representative chondrite-normalized rare earth element (REE) profiles of apatites in the Clair oil field samples. Chondrite-normalizing values are from Taylor and McLennan (1985). All samples are from Well 206/8-8 (see Fig. 13): the Upper Clair Group (UCG) sample is from 1613.10 m depth, the unit VI sample is from 1816.84 m, the unit IV sample is from 2206.50 m, and the unit I sample is from 2332.90 m.

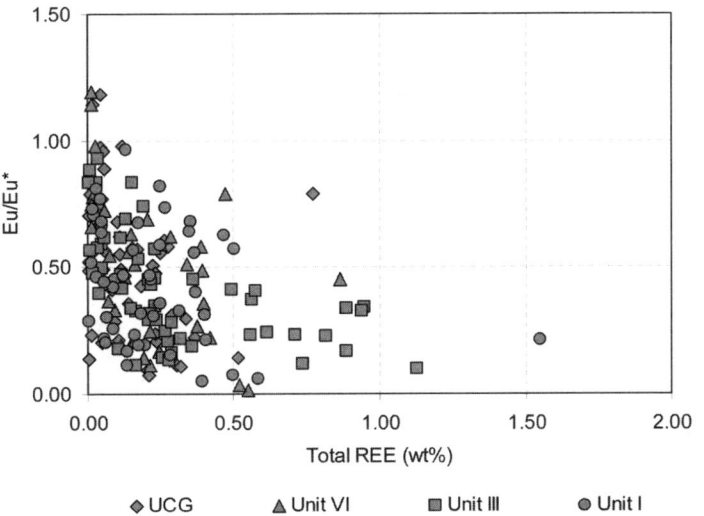

Figure 15. Eu/Eu* versus total rare earth element (REE) plot showing apatite compositions in the Clair oil field samples. Note that the largest Eu anomalies tend to occur in apatites with the highest total REE contents. UCG—Upper Clair Group.

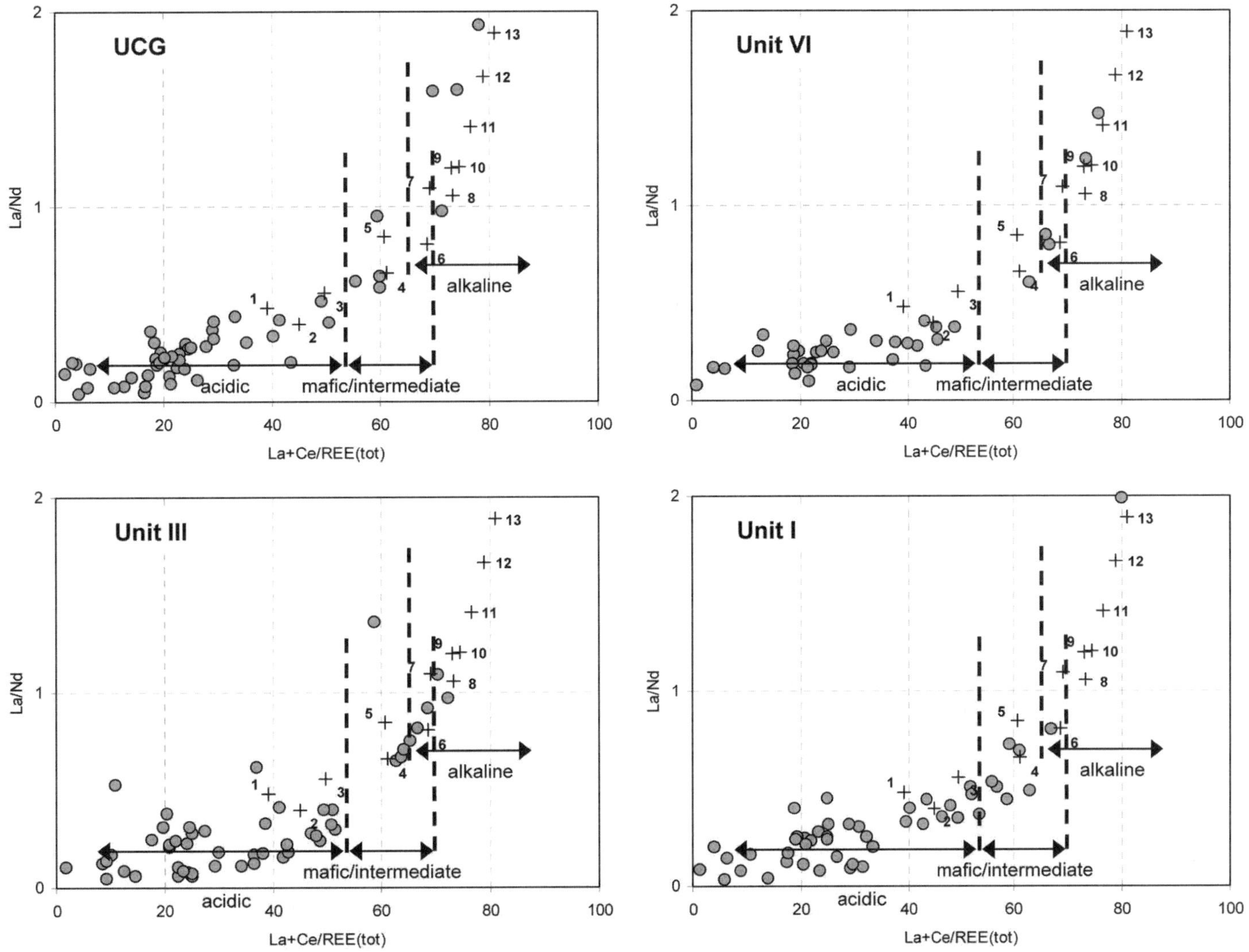

Figure 16. Apatite compositions in the Clair oil field samples plotted on the La/Nd versus La + Ce/ΣREE classification diagram of Fleischer and Altschuler (1986). 1—granite pegmatite, 2—gneiss/migmatite, 3—granite, 4—gabbro, 5—granodiorite, 6—kimberlite, 7—syenite, 8—alkali ultramafic, 9—carbonatite, 10—iron ores, 11—ultramafic, 12—alkaline, 13—alkaline pegmatite. UCG—Upper Clair Group.

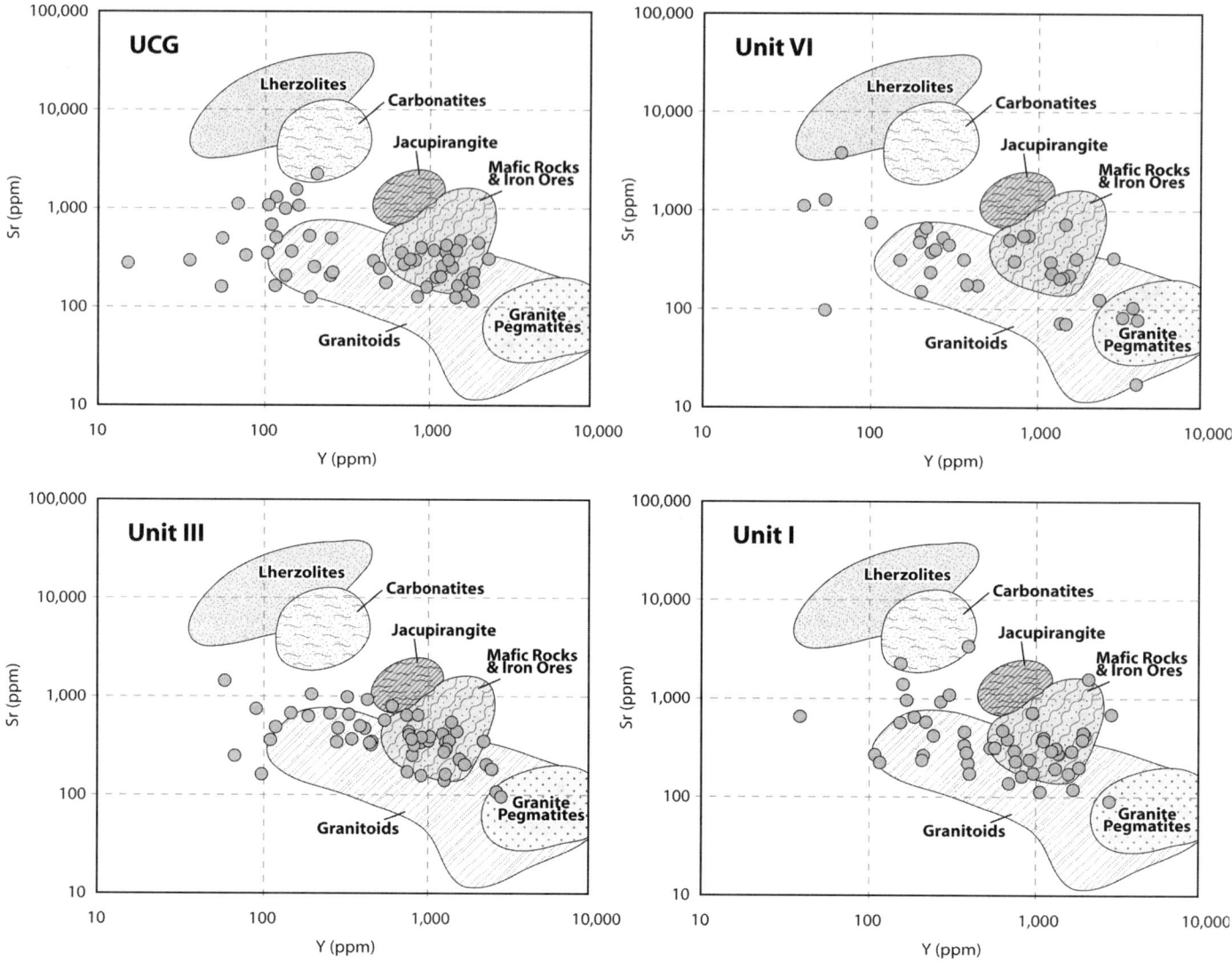

Figure 17. Sr versus Y plot showing apatite compositions in the Clair oil field samples, together with the range of apatite compositions in a variety of rock types determined by Belousova et al. (2002). UCG—Upper Clair Group.

accord with derivation from the Lewisian gneiss than the breakdown using the CART. However, the other potential source of Clair Group sediments (Moine and Dalradian metasediments) could also supply apatite, in this case as a recycled component ultimately derived from the source of the Moine and Dalradian. Archean rocks played a very minor role in the source of the Moine and lower part of the Dalradian (Grampian Group), which were derived predominantly from early to middle Proterozoic rocks (Cawood et al., 2004; Friend et al., 2003). However, Archean sources become more significant in the Appin, Argyll, and Southern Highland Groups, which form the later part of the Dalradian (Cawood et al., 2003). Thus, although the composition of apatites in Moine and Dalradian rocks is presently unknown, at least in the case of the Dalradian, it seems likely that they would have similar compositions to those supplied directly from the Lewisian gneiss. We conclude that, as with the South Caspian Basin example, the La/Nd versus La + Ce/ΣREE plot provides a more realistic assessment of the contributions made by different sources to the detrital apatite budget.

Although there was a major change in source between the Upper and Lower Clair Group successions, from a Lewisian gneiss source (possibly together with low-grade metasediments) in the Lower Clair Group, to a medium- or high-grade metasedimentary (Moine or Dalradian type) in the Upper Clair Group, the apatite compositions show comparatively little change across the boundary. This raises the problems of recycling and the associated difficulty in identifying sources when preexisting sediment or metasediment is the dominant source. In the case of the Upper Clair Group, it is clear that the apatites provided by the metasediments were ultimately derived from basement lithologies similar to those that directly supplied the Lower Clair Group.

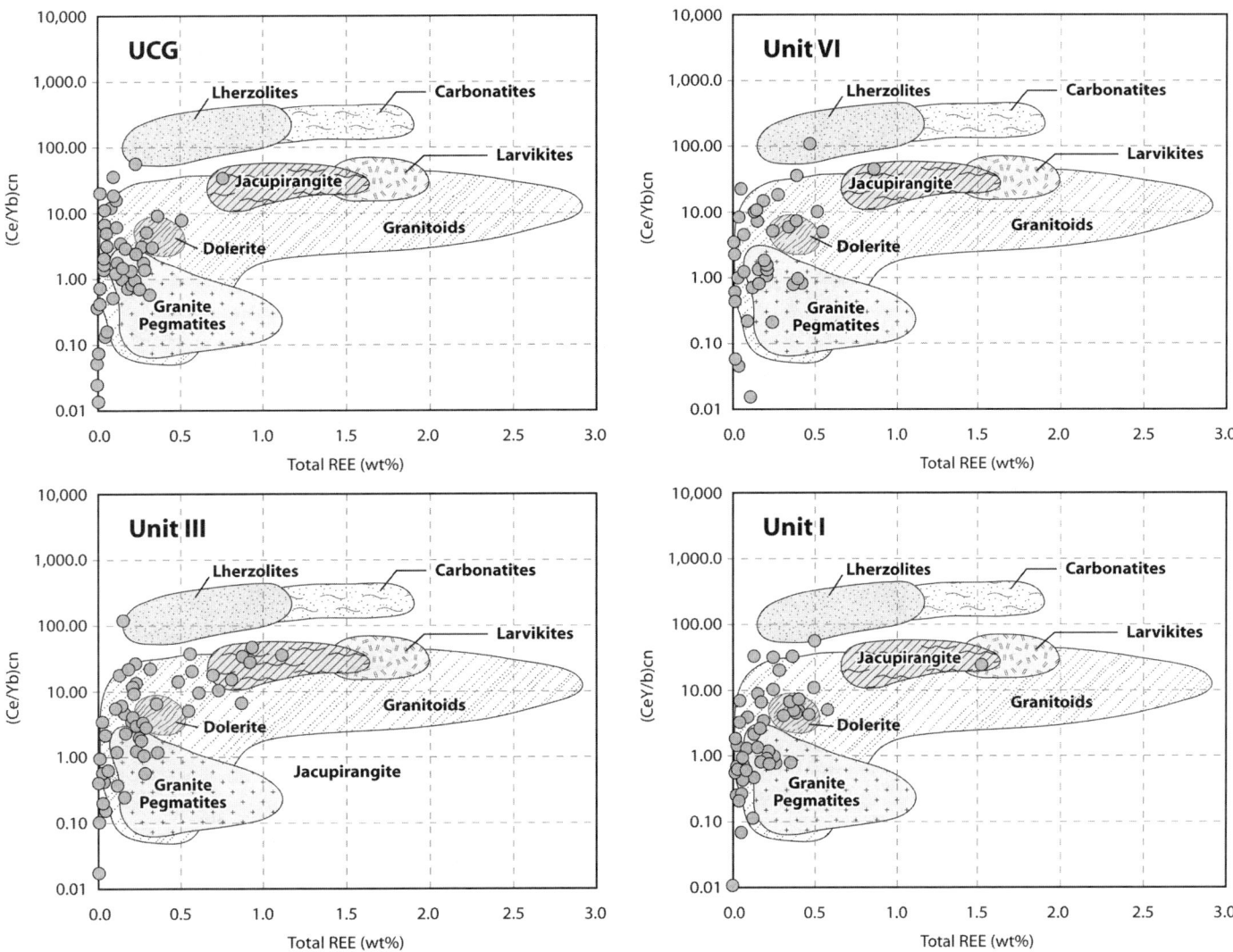

Figure 18. (Ce/Yb)$_{cn}$ versus total rare earth element (REE) plot showing apatite compositions in the Clair oil field samples, together with the range of apatite compositions in a variety of rock types determined by Belousova et al. (2002). UCG—Upper Clair Group.

CONCLUSIONS

The two case studies presented herein show that detrital apatite compositions vary according to the nature of the sediment source. Apatites in sandstones deposited by the paleo-Kura river system have compositions that indicate sediment was supplied largely from mafic to intermediate and alkaline rocks, in accord with the geological evidence from the Lesser Caucasus catchment. Paleo-Volga apatite populations show greater compositional diversity, with supply from granitoids or other acidic rocks together with subordinate mafic to intermediate and alkaline rocks. The wide diversity of apatite compositions in paleo-Volga sediment reflects the large scale of the drainage basin and the polycyclic nature of the source material, which inevitably will have introduced sediment from a wide range of sources, including granites and acidic gneisses. Apatites in the Devonian-Carboniferous succession of the Clair oil field were supplied predominantly by acidic rocks, either directly from Archean gneisses or indirectly from metasedimentary rocks.

A variety of plots may be suitable for assessing differences in provenance, including La/Nd versus La + Ce/ΣREE (Fleischer and Altschuler, 1986); chondrite-normalized REE patterns: Sr versus Y, (Ce/Yb)$_{cn}$ versus ΣREE, Sr versus Mn, and Y versus Eu/Eu* (Belousova et al., 2002); and Th versus U (Dill, 1994). Quantitative estimations of the contributions from a variety of source rocks are possible using the La/Nd versus La + Ce/ΣREE plot (Fleischer and Altschuler, 1986), which seems to give realistic interpretations of source area lithology. The CART approach pioneered by Belousova et al. (2002) has validity; however, at present, their data appear to categorize some apatites incorrectly, thus generating an unlikely lithological breakdown of source regions. One possible explanation for the anomalous outcome is

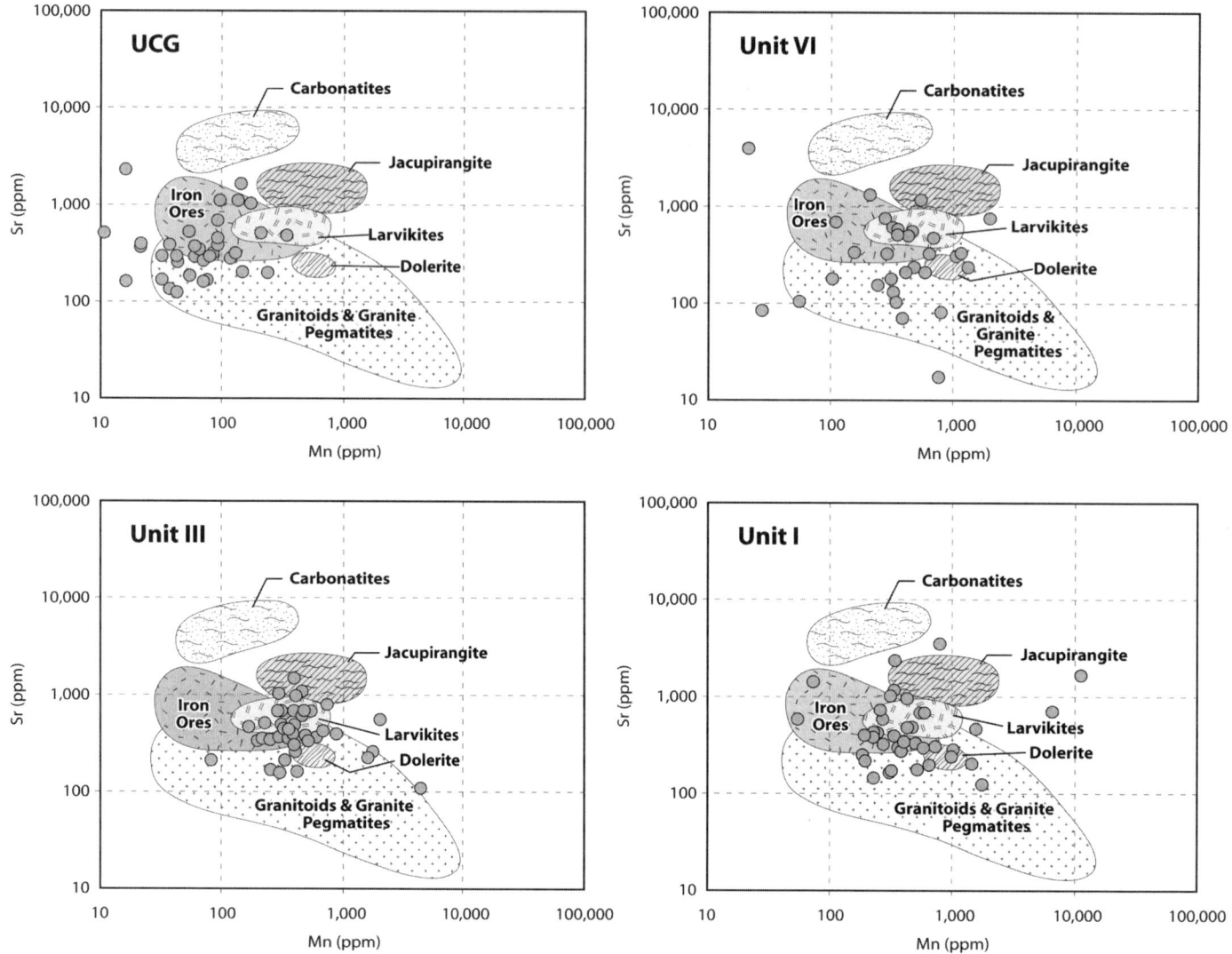

Figure 19. Sr versus Mn plot showing apatite compositions in the Clair oil field samples, together with the range of apatite compositions in a variety of rock types determined by Belousova et al. (2002). UCG—Upper Clair Group

that Belousova et al. (2002) did not acquire a sufficiently comprehensive data set, since many apatite-bearing rock types are not represented, and some of those that have been included are comparatively rare (such as jacupirangite and larvikite). Acquisition of data on a more comprehensive set of apatite-bearing lithologies (including other igneous lithologies and, perhaps more importantly, metamorphics) would help to refine the approach they pioneered. Another unknown in reconstruction of source areas using apatite compositions is the role played by sediment recycling. This requires assessment in individual scenarios by analysis of prospective precursor sediments.

The two case studies described in this paper demonstrate that apatite geochemistry has the potential to become another important tool in the provenance reconstruction toolkit, supplementing the variety of minerals already used widely for this purpose. The stability of apatite during deep burial (Morton and Hallsworth, 2007) makes the method applicable to diagenetically modified sediment, which is a limiting factor for a large number of minerals that are unstable in the subsurface (notably pyroxene, amphibole, and epidote). Apatite is unstable under weathering environments, and thus the method will be difficult to apply in sandstones that have undergone prolonged weathering during transport. The possible modification of the compositional range of apatite populations during weathering, by selective removal of the more unstable apatite varieties, may also be a factor. At present, unequivocal identification of provenance using apatite geochemistry is limited by the lack of a comprehensive database on apatite compositions in some of the potential source rocks, particularly those of metamorphic origin. Therefore, further work is recommended in order to improve the identification of prospective apatite source lithologies.

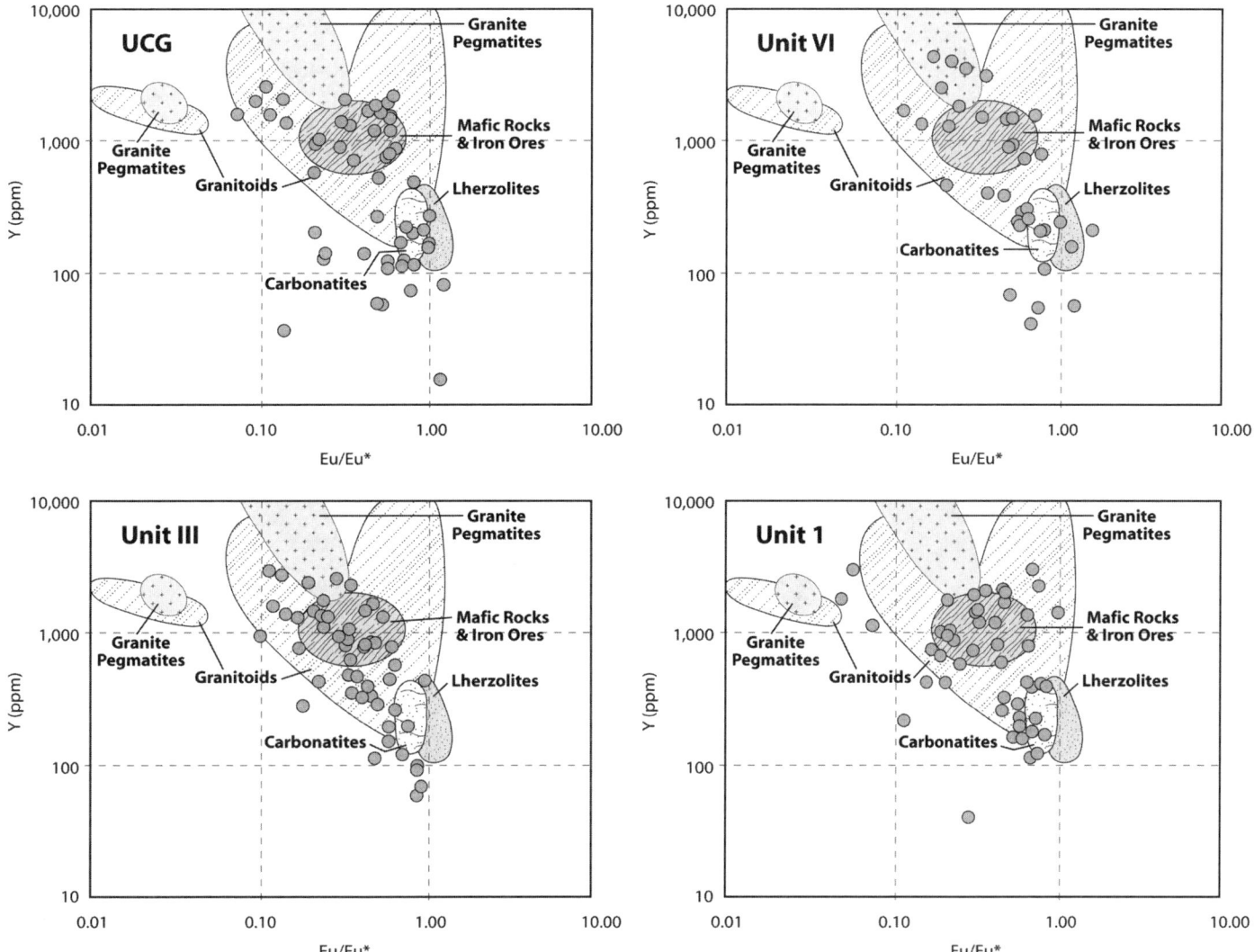

Figure 20. Eu/Eu* versus Y plot showing apatite compositions in the Clair oil field samples, together with the range of apatite compositions in a variety of rock types determined by Belousova et al. (2002). UCG—Upper Clair Group.

ACKNOWLEDGMENTS

We are grateful to CASP for permission to include samples KB1.75 and KB2.6 in the analytical program, and to the Clair oil field partners for permission to show the heavy mineral stratigraphic framework in this paper. We also thank Alessandro Amorosi and Kohki Yoshida for their constructive comments on an earlier version of this paper, Ashley Norris for assistance with the electron microscopy, and Charlotte Allen and Mike Shelley for assistance with the laser-ablation inductively coupled plasma–mass spectrometry (LA-ICP-MS) analyses.

REFERENCES CITED

Allen, P.A., and Mange-Rajetzky, M.A., 1992, Sedimentary evolution of the Devonian-Carboniferous Clair field, offshore northwestern UK: Impact of changing provenance: Marine and Petroleum Geology, v. 9, p. 29–52, doi: 10.1016/0264-8172(92)90003-W.

Ayers, J.C., and Watson, E.B., 1993, Solubility of apatite, monazite, zircon, and rutile in supercritical aqueous fluids with implications for subduction zone geochemistry: Philosophical Transactions of the Royal Society of London, ser. A, v. 335, p. 365–375.

Basu, A., and Molinaroli, E., 1991, Reliability and application of detrital opaque Fe-Ti oxide minerals in provenance determination, in Morton, A.C., Todd, S.P., and Haughton, P.D.W., eds., Developments in Sedimentary Provenance Studies: Geological Society [London] Special Publication 57, p. 55–65.

Baturin, V.P., 1947, Petrographical Analysis of the Geological Past using Terrigene Components: Academy of Science, USSR (in Russian).

Belousova, E.A., Griffin, W.L., O'Reilly, S.Y., and Fisher, N.I., 2002, Apatite as an indicator mineral for mineral exploration: Trace-element compositions and their relationship to host rock type: Journal of Geochemical Exploration, v. 76, p. 45–69, doi: 10.1016/S0375-6742(02)00204-2.

Blackbourn, G.A., 1988, Sedimentary environments and stratigraphy of the Late Devonian–Early Carboniferous Clair Basin, west of Shetlands, in Miller, J., Adams, A.E., and Wright, V.P., eds., European Dinantian Environments: London, Wiley, p. 76–91.

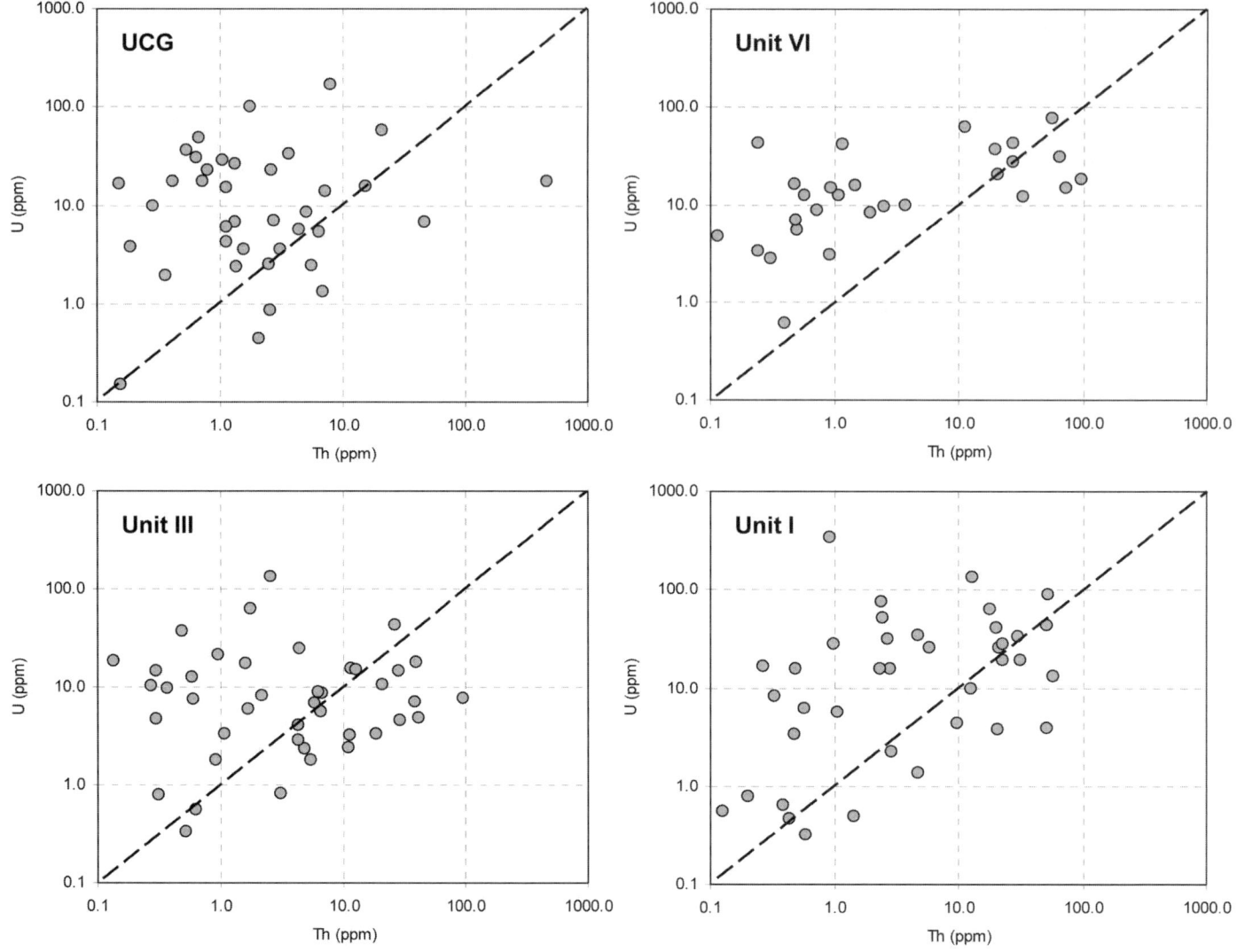

Figure 21. Th versus U plot of apatites in the Clair oil field samples. UCG—Upper Clair Group.

Bouch, J.E., Hole, M.J., Trewin, N.H., Chenery, S., and Morton, A.C., 2002, Authigenic apatite in a fluvial sandstone sequence: Evidence for rare-earth element mobility during diagenesis and a tool for diagenetic correlation: Journal of Sedimentary Research, v. 72, p. 59–67.

Budzinski, H., and Tischendorf, G., 1989, Distribution of REE among minerals in the Hercynian postkinematic granites of Westerzgebirge-Vogtland, GDR: Zeitschrift für Geologische Wissenschaften, v. 17, p. 1019–1031.

Cawood, P.A., 1983, Modal composition and detrital clinopyroxene geochemistry of lithic sandstones from the New England fold belt (east Australia): A Paleozoic forearc terrane: Geological Society of America Bulletin, v. 94, p. 1199–1214, doi: 10.1130/0016-7606(1983)94<1199: MCADCG>2.0.CO;2.

Cawood, P.A., Nemchin, A.A., Smith, M., and Loewy, S., 2003, Source of the Dalradian Supergroup constrained by U-Pb dating of detrital zircon and implications for the East Laurentian margin: Journal of the Geological Society of London, v. 160, p. 231–246.

Cawood, P.A., Nemchin, A.A., Strachan, M., Kinny, P.D., and Loewy, S., 2004, Laurentian provenance and intracratonic setting for the Moine Supergroup, Scotland, constrained by detrital zircons from the Loch Eil and Glen Urquhart successions: Journal of the Geological Society of London, v. 161, p. 861–874.

Chang, L.L.Y., Howie, R.A., and Zussman, J., 1998, Rock-Forming Minerals, Volume 5B: Non-Silicates: Sulphates, Carbonates, Phosphates and Halides (2nd edition): London, The Geological Society, 383 p.

Claesson, S., Bogdanova, S.V., Bibikova, E.V., and Gorbatschev, R., 2001, Isotopic evidence for Palaeoproterozoic accretion in the basement of the East European craton: Tectonophysics, v. 339, p. 1–18, doi: 10.1016/S0040-1951(01)00031-2.

Coney, D., Fyfe, T.B., Retail, P., and Smith, P.J., 1993, Clair appraisal: The benefits of a co-operative approach, in Parker, J.R., ed., Petroleum Geology of Northwest Europe: Proceedings of the 4th Conference: London, The Geological Society, p. 1409–1420.

Deer, W.A., Howie, R.A., and Zussman, J., 1997a, Rock-Forming Minerals, Volume 1A: Orthosilicates (2nd edition): London, The Geological Society, 629 p.

Deer, W.A., Howie, R.A., and Zussman, J., 1997b, Rock-Forming Minerals, Volume 2B: Double-Chain Silicates (2nd edition): London, The Geological Society, 764 p.

Dill, H.G., 1994, Can REE patterns and U-Th variations be used as a tool to determine the origin of apatite in crustal rocks?: Sedimentary Geology, v. 92, p. 175–196, doi: 10.1016/0037-0738(94)90105-8.

Eggins, S.M., 2003, Laser ablation ICP-MS analysis of geological materials prepared as lithium borate glasses: Journal of Geostandards and Geoanalysis, v. 27, p. 147–162.

Eggins, S.M., Kinsley, L.K., and Shelley, J.M.G., 1998a, Deposition and element fractionation processes occurring during atmospheric pressure laser sampling for analysis by ICPMS: Applied Surface Science, v. 127–129, p. 278–286, doi: 10.1016/S0169-4332(97)00643-0.

Eggins, S.M., Rudnick, R.L., and McDonough, W.F., 1998b, The compositions of peridotites and their minerals: A laser-ablation ICP-MS study: Earth and Planetary Science Letters, v. 154, p. 53–71, doi: 10.1016/S0012-821X(97)00195-7.

Fleischer, M., and Altschuler, Z.S., 1986, The lanthanides and yttrium in minerals of the apatite group—An analysis of the available data: Neues Jahrbuch für Mineralogie Monatschefte, v. 10, p. 467–480.

Friend, C.R.L., Strachan, R.A., Kinny, P.D., and Watt, G.R., 2003, Provenance of the Moine Supergroup of NW Scotland: Evidence from geochronology of detrital and inherited zircons from (meta)sedimentary rocks, granites and migmatites: Journal of the Geological Society of London, v. 160, p. 247–257.

Gorbatschev, R., and Bogdanova, S., 1993, Frontiers in the Baltic Shield: Precambrian Research, v. 64, p. 3–21, doi: 10.1016/0301-9268(93)90066-B.

Grigsby, J.D., 1990, Detrital magnetite as a provenance indicator: Journal of Sedimentary Petrology, v. 60, p. 940–951.

Hallsworth, C.R., and Chisholm, J.I., 2000, Stratigraphic evolution of provenance characteristics in Westphalian sandstones of the Yorkshire coalfield: Proceedings of the Yorkshire Geological Society, v. 53, p. 43–72.

Hanel, M., Gurbanov, A.G., and Lippolt, H.J., 1992, Age and genesis of granitoids from the Main-Range and Bechasyn zones of the western Great Caucasus: Neues Jahrbuch für Mineralogie Monatschefte, v. 12, p. 529–544.

Haughton, P.D.W., and Farrow, C.M., 1988, Compositional variation in Lower Old Red Sandstone detrital garnets from the Midland Valley of Scotland and the Anglo-Welsh Basin: Geological Magazine, v. 126, p. 373–396.

Henry, D.J., and Guidotti, C.V., 1985, Tourmaline as a petrogenetic indicator mineral: An example from the staurolite-grade metapelites of NW Maine: The American Mineralogist, v. 70, p. 1–15.

Hinds, D.J., Aliyeva, E., Allen, M.B., Davies, C.E., Kroonenberg, S., Simmons, M.D., and Vincent, S.J., 2004, Sediment in a discharge dominant fluviolacustrine system: The Neogene Productive Series of the South Caspian Basin, Azerbaijan: Marine and Petroleum Geology, v. 21, p. 613–638, doi: 10.1016/j.marpetgeo.2004.01.009.

Hubert, J.F., 1971, Analysis of heavy mineral assemblages, in Carver, R.E., ed., Procedures in Sedimentary Petrology: New York, Wiley, p. 453–478.

Hutchison, A.R., and Oliver, G.J.H., 1998, Garnet provenance studies, juxtaposition of Laurentian marginal terranes and timing of the Grampian orogeny in Scotland: Journal of the Geological Society of London, v. 155, p. 541–550.

Jeans, C.V., Reed, S.J.B., and Xing, M., 1993, Heavy mineral stratigraphy in the UK Trias of the North Sea Basin, in Parker, J.R., ed., Petroleum Geology of NW Europe: Proceedings of the 4th Conference: London, The Geological Society, p. 609–624.

Jones, R.W., and Simmons, M.D., 1996, A review of the stratigraphy of eastern Paratethys (Oligocene-Holocene): Bulletin of the Natural History Museum, v. 52, p. 25–49.

Kazmin, V.G., Sborshchikov, I.M., Ricou, L.-E., Zonenshain, L.P., Boulin, J., and Knipper, A.L., 1986, Volcanic belts as markers of the Mesozoic-Cenozoic active margin of Eurasia: Tectonophysics, v. 123, p. 123–152, doi: 10.1016/0040-1951(86)90195-2.

Kroonenberg, S.B., Rusakov, G.V., and Svitoch, A.A., 1997, The wandering of the Volga delta: A response to rapid Caspian sea-level change: Sedimentary Geology, v. 107, p. 189–209, doi: 10.1016/S0037-0738(96)00028-0.

Longerich, H.P., Jackson, S.E., and Günther, D., 1996, Laser ablation inductively coupled plasma mass spectrometric transient signal data acquisition and analyte concentration calculation: Journal of Analytical Atomic Spectrometry, v. 11, p. 899–904, doi: 10.1039/ja9961100899.

Mange, M.A., and Maurer, H.F.W., 1992, Heavy minerals in colour: London, Chapman and Hall, 147 p.

Mange-Rajetzky, M.A., and Oberhänsli, R., 1982, Detrital lawsonite and blue sodic amphibole in the Molasse of Savoy, France, and their significance in assessing Alpine evolution: Schweizerische Mineralogische und Petrographische Mitteilungen, v. 62, p. 415–436.

McConnell, D., 1973, Apatite—Its crystal chemistry, mineralogy, utilization and geologic and biologic occurrences: New York, Springer-Verlag, 111 p.

McKie, T., and Garden, I.R., 1996, Hierarchical cycles in the non-marine Clair Group (Devonian), UKCS, in Howell, J.A., and Aitken, J.F., eds., High Resolution Sequence Stratigraphy: Innovations and Applications: Geological Society [London] Special Publication 104, p. 139–157.

Morton, A.C., 1985a, Heavy minerals in provenance studies, in Zuffa, G.G., ed., Provenance of Arenites: Dordrecht, Reidel, p. 249–277.

Morton, A.C., 1985b, A new approach to provenance studies: Electron microprobe analysis of detrital garnets from Middle Jurassic sandstones of the northern North Sea: Sedimentology, v. 32, p. 553–566, doi: 10.1111/j.1365-3091.1985.tb00470.x.

Morton, A.C., and Hallsworth, C.R., 1994, Identifying provenance-specific features of detrital heavy mineral assemblages in sandstones: Sedimentary Geology, v. 90, p. 241–256.

Morton, A.C., and Hallsworth, C.R., 2007, Stability of detrital heavy minerals during burial diagenesis, in Mange, M., and Wright, D.K., eds., Heavy Minerals in Use: Developments in Sedimentology (in press).

Morton, A.C., Allen, M.B., Simmons, M.D., Spathopoulos, F., Still, J., Ismail-Zadeh, A., and Kroonenberg, S., 2003a, Provenance patterns in a neotectonic basin: Pliocene and Quaternary sediment supply to the South Caspian: Basin Research, v. 15, p. 321–337, doi: 10.1046/j.1365-2117.2003.00208.x.

Morton, A.C., Spicer, P.J., and Ewen, D.F., 2003b, Geosteering of high-angle wells using heavy-mineral analysis: The Clair field, west of Shetland, UK, in Carr, T.R., Mason, E.P., and Feazel, C.T., eds., Horizontal Wells: Focus on the Reservoir: American Association of Petroleum Geologists, Methods in Exploration, v. 14, p. 249–260.

Morton, A.C., Hallsworth, C.R., and Chalton, B., 2004, Garnet compositions in Scottish and Norwegian basement terrains: A framework for interpretation of North Sea sandstone provenance: Marine and Petroleum Geology, v. 21, p. 393–410, doi: 10.1016/j.marpetgeo.2004.01.001.

Morton, A.C., Whitham, A.G., and Fanning, C.M., 2005, Provenance of Late Cretaceous–Paleocene submarine fan sandstones in the Norwegian Sea: Integration of heavy mineral, mineral chemical and zircon age data: Sedimentary Geology, v. 182, p. 3–28.

Nalivkin, D.V., 1983, Geological map of the USSR and adjacent water-covered areas: Moscow, Ministry of Geology of the USSR, scale 1:2,500,000.

Nash, W.P., 1984, Phosphate minerals in terrestrial igneous and metamorphic rocks, in Nriagu, J.O., and Moore, P.B., eds., Phosphate Minerals: New York, Springer-Verlag, p. 215–241.

Nichols, G.J., 2005, Sedimentary evolution of the Lower Clair Group, Devonian, west of Shetland: Climate and sediment supply controls on fluvial, aeolian and lacustrine deposition, in Doré, A.G., and Vining, B., eds., Petroleum Geology: North-West Europe and Global Perspectives: Proceedings of the 6th Petroleum Geology Conference: London, The Geological Society, p. 957–967.

Owen, M.R., 1987, Hafnium content of detrital zircons, a new tool for provenance study: Journal of Sedimentary Petrology, v. 57, p. 824–830.

Park, R.G., Stewart, A.D., and Wright, D.T., 2002, The Hebridean terrane, in Trewin, N.H., ed., The Geology of Scotland (4th edition): London, Geological Society, p. 45–80.

Pay, M.D., Astin, T.R., and Parker, A., 2000, Clay mineral distribution in the Devonian-Carboniferous sandstones of the Clair field, west of Shetland, and its significance for reservoir quality: Clay Minerals, v. 35, p. 151–162, doi: 10.1180/000985500546549.

Pober, E., and Faupl, P., 1988, The chemistry of detrital chrome spinels and its implications for the geodynamic evolution of the eastern Alps: Geologische Rundschau, v. 77, p. 641–670, doi: 10.1007/BF01830175.

Puchkov, V.N., 1997, Structure and geodynamics of the Uralian orogen, in Burg, J.-P., and Ford M., eds., Orogeny through Time: Geological Society [London] Special Publication 121, p. 201–236.

Reynolds, A.D., Simmons, M.D., Bowman, M.B.J., Henton, J., Brayshaw, A.C., Ali-Zade, A.A., Guliyev, I.S., Suleymanova, S.F., Ateava, E.Z., Mamedova, D.N., and Koshkarly, R.O., 1998, Implications of outcrop geology for reservoirs in the Neogene Productive Series: Apsheron Peninsula, Azerbaijan: Bulletin of the American Association of Petroleum Geologists, v. 82, p. 25–49.

Ridd, M.F., 1981, Petroleum geology west of the Shetlands, in Illing, L.V., and Hobson, G.D., eds., Petroleum Geology of the Continental Shelf of North-West Europe: London, Graham and Trotman, p. 414–425.

Spiegel, C., Siebel, W., Frisch, W., and Berner, Z., 2002, Nd and Sr isotopic ratios and trace element geochemistry of epidote from the Swiss Molasse Basin as provenance indicators: Implications for the reconstruction of the exhumation history of the central Alps: Chemical Geology, v. 189,

p. 231–250, doi: 10.1016/S0009-2541(02)00132-8.

Styles, M.T., Stone, P., and Floyd, J.D., 1989, Arc detritus in the Southern Uplands: Mineralogical charcterisation of a 'missing' terrain: Journal of the Geological Society of London, v. 146, p. 397–400.

Taylor, S.R., and McLennan, S.M., 1985, The Continental Crust: Its Composition and Evolution: Oxford, Blackwell, 312 p.

Trewin, N.H., and Thirlwall, M.F., 2002, Old Red Sandstone, *in* Trewin, N.H., ed., The Geology of Scotland (4th edition): London, The Geological Society, p. 213–249.

von Eynatten, H., and Gaupp, R., 1999, Provenance of Cretaceous synorogenic sandstones in the eastern Alps: Constraints from framework petrography, heavy mineral analysis and mineral chemistry: Sedimentary Geology, v. 124, p. 81–111, doi: 10.1016/S0037-0738(98)00122-5.

Zack, T., von Eynatten, H., and Kronz, A., 2004, Rutile geochemistry and its potential use in quantitative provenance studies: Sedimentary Geology, v. 171, p. 37–58, doi: 10.1016/j.sedgeo.2004.05.009.

MANUSCRIPT ACCEPTED BY THE SOCIETY 9 AUGUST 2006

Geological Society of America
Special Paper 420
2007

Predicting sand character with integrated genetic analysis

William A. Heins[†]
Suzanne Kairo[‡]
ExxonMobil Upstream Research Company, Houston, Texas 77252-2189, USA

ABSTRACT

Many important geotechnical issues (e.g., groundwater supply and contamination, subsurface waste disposal, hydrocarbon exploration and production) require a detailed understanding of porosity and permeability in subsurface clastic formations (= reservoir quality). Reservoir quality depends on the size, shape, and packing of sand grains as they are originally deposited, as well as diagenetic changes during burial. Obtaining enough samples to fully characterize the target formation is prohibitively expensive or physically impossible. Therefore, reservoir quality estimates must be extrapolated from analogues (± sparse samples) or derived from models. Forward-modeling approaches to predicting diagenetic effects on reservoir quality are well established, but they require information about the character of deposited sand, including mean grain size, sorting, matrix content, and composition of diagenetically relevant particles (i.e., all rock fragments, not just lithic fragments). In cases where deposited sand characteristics are not known, they must be estimated. To this end, we advocate an integrated genetic analysis, which simultaneously predicts multiple sand characteristics as a function of many environmental controls, including tectonic setting, provenance lithotype abundance, climate, regional topographic gradient, hinterland transport distance, basin transport distance, basin subsidence rate, and depositional environment. We have implemented this analytic procedure as a Bayesian belief network–based forward model that successfully predicts sand composition and texture in diverse settings, including provenance areas dominated by either volcanic, high-grade metamorphic, or sedimentary lithologic assemblages; climates ranging from tropical to desert; and a range of alluvial/fluvial drainage types represented by small steep drainages as well as continental-scale big rivers.

Keywords: quantitative-provenance-analysis, sediment-evolution-forward-model, reservoir-quality, Bayesian-belief-network.

[†]E-mail: bill.heins@exxonmobil.com.
[‡]Present address: ExxonMobil Production Company, Houston, Texas 77002, USA.

Heins, W.A., and Kairo, S., 2007, Predicting sand character with integrated genetic analysis, *in* Arribas, J., Critelli, S., and Johnsson, M.J., eds., Sedimentary Provenance and Petrogenesis: Perspectives from Petrography and Geochemistry: Geological Society of America Special Paper 420, p. 345–379, doi: 10.1130/2006.2420(20). For permission to copy, contact editing@geosociety.org. ©2007 Geological Society of America. All rights reserved.

INTRODUCTION

Why Predict Sand Character?

Many important geotechnical issues require a detailed understanding of the porosity and permeability of subsurface clastic formations. We will refer to these characteristics as reservoir quality. Some pertinent reservoir quality questions include: (1) What are the draw-down and recharge rates of aquifers that supply domestic and agricultural water? (2) How quickly, and in what direction, will groundwater contaminants flow? (3) What is the reserve size and potential recovery efficiency of an oil field?

In all these cases, obtaining enough samples to fully characterize the porosity and permeability of the target formation is prohibitively expensive or physically impossible. Therefore, estimates of reservoir quality must either be extrapolated from analogues ± sparse samples or derived from models. The environmental and economic costs of faulty estimates can be substantial.

Therefore, it is desirable to have a consistent, systematic, and broadly applicable method to predict reservoir quality in places where direct observations are not possible. Many tools exist to predict reservoir quality *if* sand character (i.e., composition, texture, and matrix content) and burial and thermal history are known (e.g., Lander and Walderhaug, 1999; Perez et al., 1999; Bray et al., 2000; Bonnell and Lander, 2003). These tools are only as good as their input data; if there is no way to directly observe sand character, or if the range of sand character is not fully captured by available samples and analogues, then sand character must be estimated. The accuracy of porosity and permeability predictions is limited by the accuracy of the predicted sand character.

Because reservoir quality depends on the pore network, many features of the solid particles that define the pore network must be known. For example, the grain-size distribution and matrix content determine depositional porosity. Loss of depositional porosity is governed by the ductility of the grains and by the potential for the grains to grow diagenetic cement. An estimate of the bulk mineralogy (e.g., a Gazzi-Dickinson point count) of one grain size (e.g., medium sand) does not provide enough information to effectively model reservoir quality; the sizes and physical characteristics of all grains (especially rock fragments) are critical controls on compaction and cementation.

We advocate integrated genetic analysis as the best way to make sand character predictions because it accounts for the most important factors responsible for sand generation and evolution from the sediment source to the site of ultimate deposition.

Terminology

It is important to note our use of particular geologic terms to eliminate confusion or misconception. We consider provenance to include *all* aspects of the system responsible for sand generation and evolution (Suttner, 1974). We refer to the rocks that are decomposed to produce sand as provenance lithotypes. We use the term hinterland to indicate all portions of Earth's surface that contribute sediment to a particular basin, and basin to indicate the areas that accumulate large volumes of sediment due to subsidence. We use the terms hinterland and basin rather than the currently popular terms source and sink because "source" has a very particular meaning in our industry (i.e., organic-rich sediments that can generate hydrocarbons), whereas hinterland and basin are generally understood in the way we have defined them here. Our usage of hinterland should not be confused with tectonic or structural connotations, but rather taken in its literal sense of "territory behind the coast."

Integrated Genetic Analysis

Integrated genetic analysis is a holistic analytical approach that recognizes the relationships among environmental conditions and the spectrum of sand types that may result as a function of variations in these conditions. There is no unique solution that corresponds with a given sandstone composition and texture, but by analyzing all the controlling factors and processes in concert, "a problem that appears to be hopelessly complex...does not need to be so" (Krynine, 1943, p. 3). It is possible not only to decipher an observed sand's history, but it is also possible to predict the composition and texture that should result from a particular history.

At least since James Hutton, sedimentary geologists have recognized that sand character depends on the entire history of sand generation and evolution. The basic conceptual framework of the sand-producing system was first laid out in Henry Clifton Sorby's Anniversary Address of the President to the Geological Society of London (Sorby, 1880). Periodically since then, other workers have described the systematic genetic relationships in flow charts or schematic representations (e.g., Krynine, 1943; Potter and Siever, 1956; Folk, 1974; Potter, 1978; Johnsson, 1993). The community generally accepts the following basic factors that control sand generation and evolution, as summarized in Figure 1: tectonics, which governs the assemblage of rock types and geomorphic character of the hinterland; weathering, both intensity and duration; transport, including weathering during storage and abrasion; and deposition, including hydrodynamic sorting effects.

Despite general agreement as to the basic factors, there is not yet a holistic and quantitative process to use general environmental knowledge about the sand genetic system to make specific predictions of sand character.

CONCEPTUAL FRAMEWORK AND ANALYTIC PHILOSOPHY

Genetic Context

In order to turn the broadly accepted, but ill-specified, concept model of sand generation and evolution into a tool for quantitative prediction, we define and quantify key aspects of

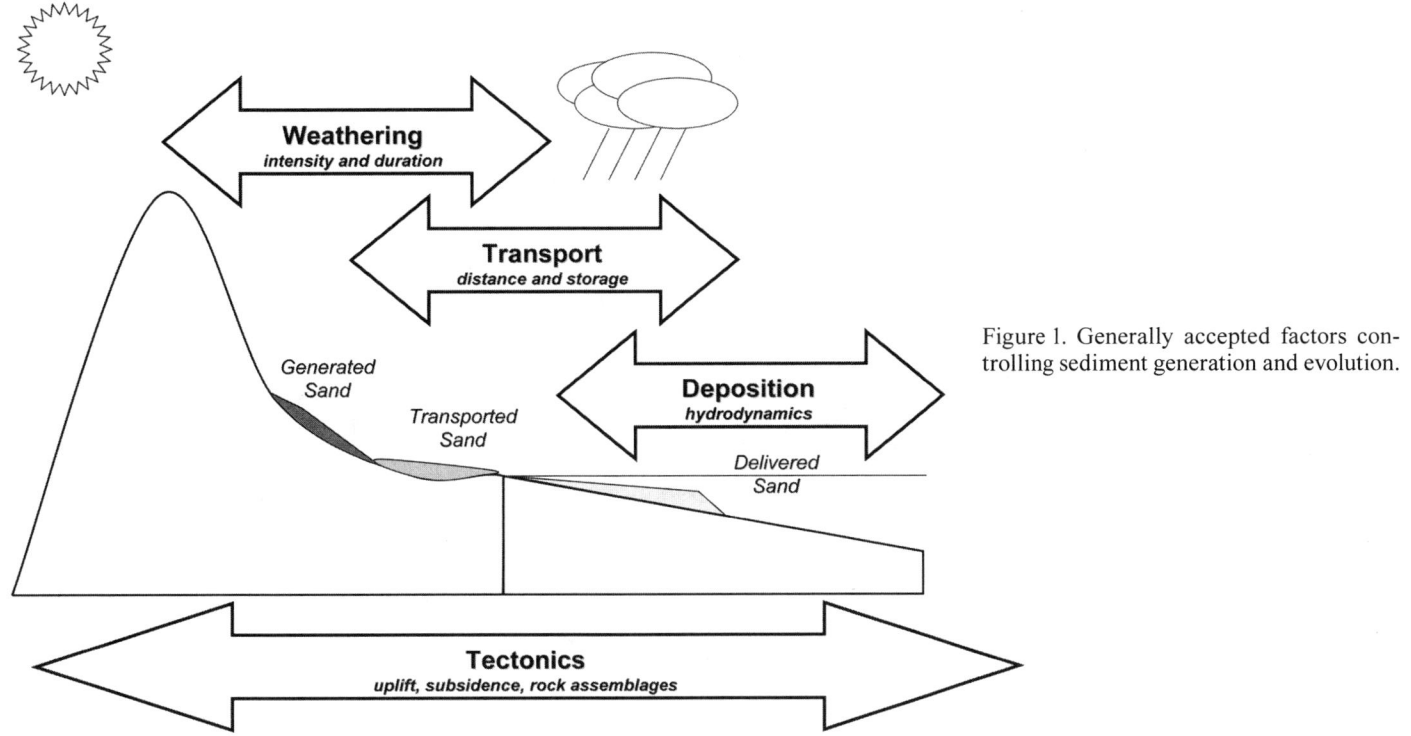

Figure 1. Generally accepted factors controlling sediment generation and evolution.

provenance that control sand character (Fig. 2), including: (1) provenance lithotypes, (2) regional topographic gradient, (3) climate (weathering potential and transport potential), (4) transport distance in the hinterland, (5) transport distance in the basin, (6) basin subsidence, and (7) depositional facies.

Tectonic Setting

The first and most important step in understanding sand generation and evolution is a fundamental analysis of tectonic setting (Dickinson and Yarborough, 1978; Ingersoll and Busby, 1995). Tectonic setting provides a framework in which to interpret or make assumptions about: the rocks from which sediments are derived; the processes by which sediments are generated and evolved; and the depositional environments in which sediments have been deposited. All of the factors except for climate and depositional environment are directly determined by the plate-tectonic setting and geodynamic context. Even climate and depositional environment are moderated by tectonics, through the influence of landmass distribution and topography on global atmospheric and oceanic circulation (e.g., Poulsen et al., 1998; Motoi et al., 2005; Clark et al. 2005) and on local drainage patterns (e.g., Rossetti, 2004; Jones et al., 2004).

The integrated genetic approach presented here diverges from earlier studies correlating tectonics with sand composition (Dickinson and Suczek, 1979; Dickinson et al., 1983; Ingersoll and Busby, 1995) by emphasizing the tectonics of the hinterland as much as, if not more than, the tectonics of the sedimentary basin. Often, the tectonic settings of the basin and hinterland are tightly linked, for example, in foreland basins associated with fold-and-thrust belts (Critelli, 1999; Critelli et al., 2003), but just as often, particularly in the case of very large drainage basins, the tectonic setting of the depositional basin may be substantially disconnected from the hinterland. For example, the Barbados forearc basin clearly receives input from the adjacent South American continent (Faugeres et al., 1997; Mahabir et al., 2004), as well as from the arc system itself; the "anomalously quartzose" (Marsaglia and Ingersoll, 1992) sediments of this basin reflect the assemblage of provenance lithotypes in the hinterland, part of which is a function of a different tectonic setting than the one that formed the basin. We contend that once sand has reached the edge of a sedimentary basin, its composition has already been largely determined by its hinterland and transport heritage.

Schematic Representation

Our Sand Generation and Evolution Model (SandGEM) is shown in Figure 3. Each pair of boxes connected by an arrow in this diagram represents a small portion of the sand generation and evolution system that we can describe (however crudely) with a quantitative relationship that can be calibrated to observations of the real world. The network provides a formal, structured system to account for the complex real-life web of processes and products. The rest of this paper is devoted to explaining why we have chosen these particular relationships to describe the system, what observations support our description of the relationships, and how we can use this model to make predictions about sand character in reservoirs.

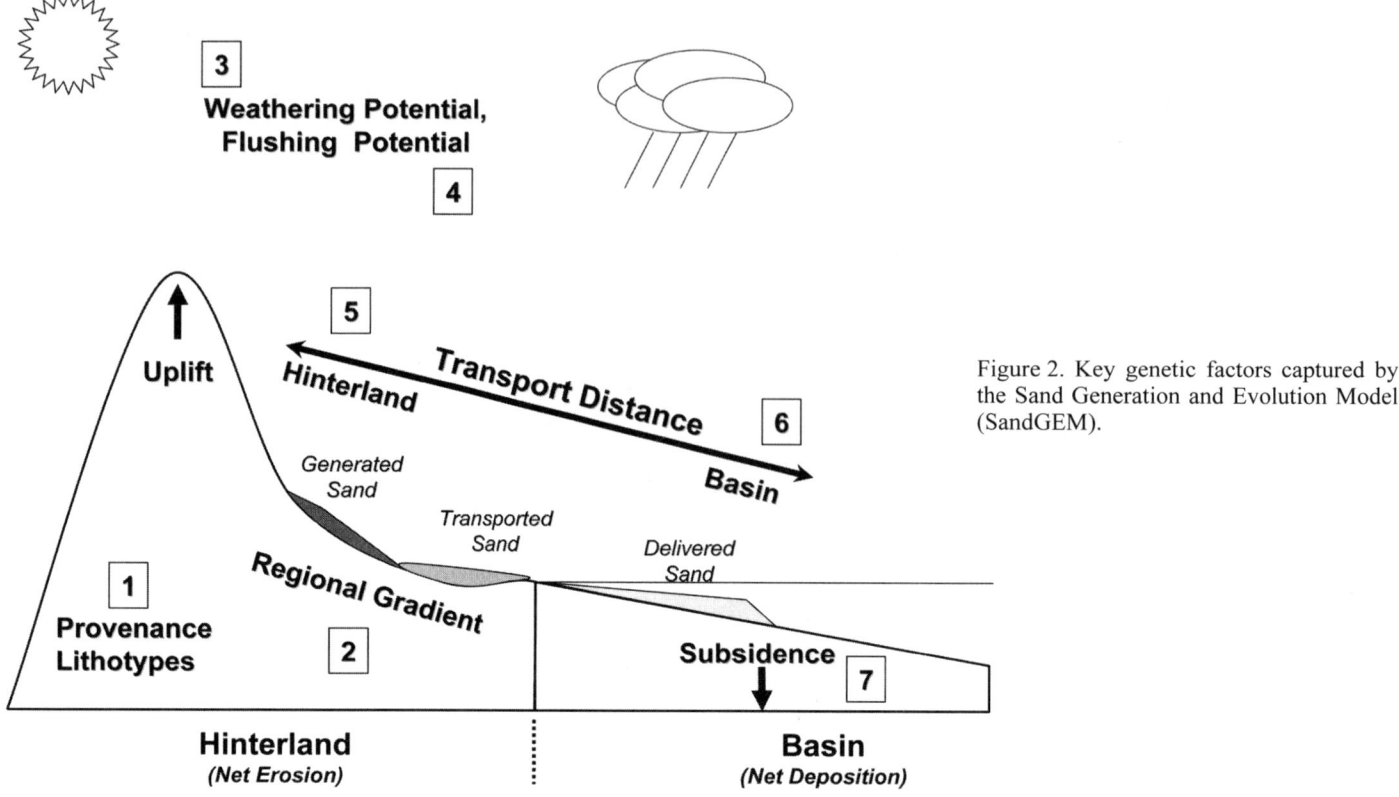

Figure 2. Key genetic factors captured by the Sand Generation and Evolution Model (SandGEM).

Key Variables

Provenance Lithotype

Provenance lithotype (Figs. 2 and 3) is the single most important genetic factor (Heins, 1993). Accumulations of sand that are large enough to be economically significant hydrocarbon reservoirs rarely are derived from only one rock type; we assume that they come from drainage systems that were (or are) large enough and complex enough to encompass numerous lithotypes (typically large second-order to third-order settings, in the sense of Ingersoll et al. [1993]). There are natural associations of provenance lithotypes that are determined by tectonic setting (Kearey and Vine, 1996; Cox and Hart, 1986). In the absence of other information, the Provenance Lithotype assemblage can be inferred from tectonic setting alone (Kairo and Heins, 2004). The relevant tectonic setting is the one that assembled and exposed the provenance lithotypes from which sediments were derived and that determined drainage and basin geometries at the time the reservoir sediments in questions were generated and deposited.

We discriminate among the Lithotypes based on: (1) propensity to create sand-sized detritus (akin to the sand generation index of Palomares and Arribas [1993]); and (2) the relative abundances of quartz, feldspars, micas, rock fragments, and clay among the detritus.

Both of these characteristics depend on the mineralogy and texture of the Provenance Lithotypes.

Regional Topographic Gradient

Regional Topographic Gradient (Figs. 2 and 3) is defined as the average gradient of major rivers that supply sediment to the basin. The regional topographic gradient influences how quickly water flows over and through the landscape and how quickly sediment can be exported from the weathering zone on top of bedrock (Ritter et al., 2002). The velocity of water and sediment influences: the duration, and therefore the cumulative intensity, of initial weathering in the regolith; and the power of the transport system to move sediment (Walling and Webb, 1996).

Climate

Climate (Figs. 2 and 3) is defined by the level and variability of wetness and temperature when and where the sand formed in the hinterland. The interaction of temperature and precipitation over seasons and longer-term climate cycles determines the potential of the environment to do chemical work in transforming provenance lithotypes into sediment and influences the potential of the environment to do physical work in moving the sediments toward the basin.

Temperature has two different effects. Increasing temperatures, if all other factors are equal, will increase mineral-dissolution reaction rates exponentially (Lasaga, 1984). However, increasing temperature also increases evaporation, all other things being equal, so the amount of water available to facilitate dissolution reactions will decrease (Willmott, 1977). Precipitation that is

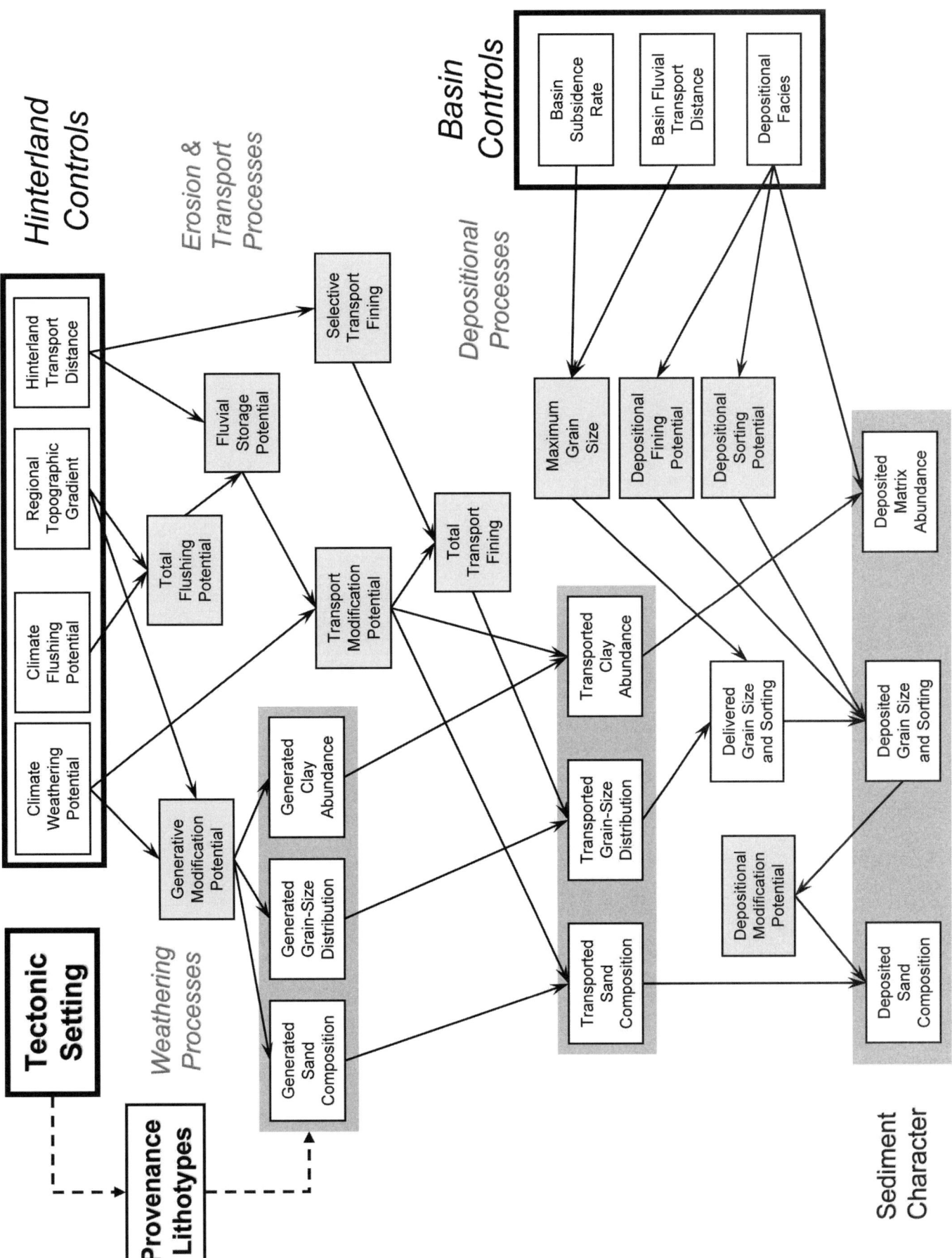

Figure 3. Functional relationships among key genetic factors (dark gray nodes) and sand character as described by the Sand Generation and Evolution Model (SandGEM).

evaporated is also less available to do physical work moving sediment than precipitation that is not evaporated (Wischmeier and Smith, 1978). The seasonal variability of effective precipitation is also important. Cecil and Edgar (2003) observed that sediment transport reaches a maximum when effective precipitation is concentrated into a small fraction of the year; these observations are consistent with the classic observation of Langbein and Schumm (1958) that solid-sediment yields are highest in areas with moderate precipitation, because those climates have the most highly seasonal precipitation patterns.

We capture the time-dependent interaction of temperature and precipitation as Climate Weathering Potential and Climate Flushing Potential (Fig. 3). A method for calculating dimensionless weathering potential and flushing potential indices is described under Network Structure, and in Appendix A. Weathering and flushing potential indices have no intrinsic physical meaning, but they can be compared to levels of dissolved, suspended, and bed load in modern rivers (Curtis et al., 1973; Walling, 1987; Summerfield and Hulton, 1994; Ludwig and Probst, 1996; Milliman, 1997; Hovius, 1998; Schaller et al., 2001; Syvitski et al., 2003; Walling and Fang, 2003) to quantify relationships between climatic styles and sediment yields.

Hinterland Transport Distance

Hinterland Transport Distance (Figs. 2 and 3) is defined as the average distance sediment travels from its site of initial generation to the depositional base level. Hinterland Transport Distance reflects the size of the drainage capture area and the characteristics of the drainage network (Strahler, 1964; Roth et al., 1996). Hinterland Transport Distance is an important control on the potential for the system to store sediment, during which time it can be modified (Johnsson and Meade, 1990; Johnsson, 1993), and on the potential for the system to segregate different grain sizes (Vogel et al., 1992; Robinson and Slingerland, 1998a, 1998b; Gasparini et al., 2004).

Basin Fluvial Transport Distance

Basin Fluvial Transport Distance (Figs. 2 and 3) is defined as the average distance sediment travels across the subsiding depositional basin to the end of the fluvial system. In the case of a fluvial/alluvial basin, the end of the fluvial system is the final point of deposition. The maximum grain size that can be delivered to the basin by a river is, in part, a function of the distance the sediment travels by river within the subsiding basin (e.g., Robinson and Slingerland, 1998a, 1998b).

Rate of Basin Subsidence

Rate of Basin Subsidence (Figs. 2 and 3) is defined as the rate of subsidence of the bottom of the basin (not the sediment-water interface). This rate is approximated by the thickness of sedimentary strata that is preserved in a given time period (essentially the rate of change in structural accommodation). Basin Subsidence controls the regional slope of the depositional basin and is a primary factor in creating accommodation space.

Depositional Facies

Depositional Facies (Figs. 2 and 3) is defined as the deposits of distinct hydrodynamic regimes within generalized environments of deposition. The hydrodynamic regime influences mean grain size, sorting, and the matrix content of deposits (Hsü, 2004). Depositional Facies are the buildings blocks of a depositional environment, but are not necessarily linked to specific environments (e.g., channel deposits can occur in many depositional settings, but they all represent a unidirectional-flow-traction transport plus suspended-load transport.) Absolute grain size is determined by the grain-size mix delivered to the basin; hydrodynamic processes subsequently segregate the delivered mix by facies.

Physical Boundaries

Hinterland

If the boundaries of the hinterland are known, then a precise estimate of provenance lithotype abundance can be made, rather than generically inferred from tectonic setting. In this case, the default approximation is that the modern geologic map (minus units younger than the formation of interest) represents the lithotypes from which the sand was generated. The best possible estimate of hinterland boundary and relative abundance of provenance lithotypes (essentially a paleogeologic map) takes all sources of paleogeographic information into account, including geomorphology (wind gaps, water gaps, etc.), patterns of regional unconformity and facies distribution, paleocurrents, structural styles and patterns, and thermochronology; an example of such a reconstruction is provided by Tokarev (2005) and Tokarev and Gostin (2003).

Basin

The shape and position of the basin in which deposits of interest are found changes through time as the rates and limits of structural subsidence change. The genetic elements must relate in time and space to a distinct body of sediment in the basin, which is defined by formal upper and lower stratigraphic boundaries and distinct lateral boundaries.

Evolutionary Stages

To simplify the naturally complex system of sand derivation and modification while maintaining a genetic approach, we distinguish among three stages in the evolution of sediments: generation, transport, and deposition. In our model, we attempt to characterize texture and composition at each of these stages (Fig. 3).

Generated Sediment

Generated Sediment (Figs. 2 and 3) is an abstraction for the bulk character of all sediment in the regolith and/or soils upon disintegration of provenance lithotypes. The extent to which Generated Sediment differs from the Provenance Lithotypes depends on the intensity and duration of weathering processes, which in turn depend primarily on climate (for intensity) and topography (for duration) (Heins, 1993, 1995).

Transported Sediment

Transported Sediment (Figs. 2 and 3) is an abstraction for the bulk character of all sediments in the hinterland alluvial-fluvial transport network, just prior to delivery to the basin margin (i.e., the boundary between areas of net erosion and net deposition). Transported Sediments are generated sediments that have been reduced in size, had some components removed by dissolution, and had others transformed to clay. The extent to which Transported Sediment differs from Generated Sediment depends on the intensity and duration of weathering processes during storage (Johnsson and Meade, 1990; Johnsson, 1993) and on hydrodynamic segregation during transport (Vogel et al., 1992; Robinson and Slingerland, 1998a, 1998b; Gasparini et al., 2004).

Deposited Sediment

Deposited Sediment (Figs. 2 and 3) forms the prospective reservoir, the character of which we wish to predict. Whereas the transition from Generated to Transported sediment represents the removal and transformation of grains; the transition from Transported to Deposited sediment represents the fractionation of the bulk population of Transported Sediments into different Depositional Facies according to hydrodynamic processes. Deposited Sediment, in each Depositional Facies, differs from Transported Sediment because some grain types are preferentially associated with some grain sizes (Krynine, 1948; Bokman, 1955; Crook, 1960; Kairo et al., 1993). Deposited texture depends on the relative ability of each depositional environment to segregate grains of a particular size.

Sand-Generating Processes

As outlined in Figure 3, we simplify all of the processes that are responsible for converting Provenance Lithotypes into Generated Sediment as the Generative Modification Potential, which is a function of Climate Weathering Potential and Regional Topographic Gradient. Generative Modification Potential is conceptually equivalent to the cumulative chemical weathering index (CCWI) of Grantham and Velbel (1988). Even though CCWI was not an appropriate predictor of sand composition in the particular genetic context examined by Grantham and Velbel (Heins, 1992, 1993), the fundamental reasoning behind the CCWI is sound: generated sediments differ from their parent rocks most significantly when the chemical power of the environment is high (due to higher temperature and/or precipitation) and when the residence time in the environment is high (due to lower topographic gradient).

Transport Processes

Total Flushing Potential

Total Flushing Potential (Fig. 3) quantifies the ability of the hinterland alluvial-fluvial network to move sediment to the basin. Total Flushing Potential is a function of the amount of water available to move sediment (as described by climate flushing potential) convolved with gravitational potential energy (as described by regional topographic gradient). Total Flushing Potential will be higher when Climate Flushing Potential and Regional Topographic Gradient variables are both higher, and vice versa. Between Climate Flushing Potential and Regional Topographic Gradient, Regional Topographic Gradient is considered to be a more important control on Total Flushing Potential, to honor the observations of Walling (1987), which showed that sediment delivery rates are high in mountainous areas across many climatic zones.

Fluvial Storage Potential

Fluvial Storage Potential (Fig. 3) is a measure of the propensity for sediment to be stored in fans, floodplains, terraces, bars, etc., along the alluvial-fluvial transport system of the hinterland. The array of factors that govern the likelihood of a sedimentary particle to be stored is broad, and these factors interact in complex ways (see Ritter et al., 2002, chapter 5, for an overview of relevant literature). An actual calculation of storage probability is only possible at the most conceptual level (Malmon et al., 2003). Nevertheless, fluvial storage is such an important process to determine sediment character (Johnsson and Meade, 1990) that it must be captured in the network.

We abstract all of the controlling factors into the effects of Total Flushing Potential and Hinterland Transport Distance. Fluvial Storage Potential is higher when Total Flushing Potential is lower and Hinterland Transport Distance is longer, and vice versa. Between Total Flushing Potential and Hinterland Transport Distance, Hinterland Transport Distance is considered to be a more important control on Fluvial Storage Potential, because any single flushing episode is highly unlikely to move a particle completely through any but the shortest system; rather, sediment tends to be exchanged between channels and storage sites (Dunne et al., 1998); thus the total number of flushing events, and thus the total amount of time, required to clear a particle through the system increases with Hinterland Transport Distance, regardless of Total Flushing Potential.

Selective Transport Fining

Selective Transport Fining (Fig. 3) is the portion of downstream reduction of grain size that can be attributed to differential transport of grains with different sizes and densities, which we simplify to be a function of Hinterland Transport Distance only. The longer the transport distance, the more opportunities are available for sediment to be temporarily stored; at each step of deposition and remobilization, sediments have an opportunity to be hydrodynamically segregated and different size classes preferentially retained or removed (e.g., Shih and Komar, 1990; Vogel et al., 1992; Robinson and Slingerland, 1998a, 1998b; Gasparini et al., 1999).

Transport Modification Potential

Transport Modification Potential (Fig. 3) is a measure of the ability of the hinterland environment to change Generated Sand

Composition and to produce clay. Transport Modification Potential is a function of Fluvial Storage Potential (the duration of modification) and Climate Weathering Potential (the intensity of modification). Transport Modification Potential is higher when both duration and intensity are higher, and vice versa. Between Climate Weathering Potential and Fluvial Storage Potential, Fluvial Storage Potential is considered to be a more important control on Transport Modification Potential, because of observations that sediment traveling a short time and/or distance is only slightly modified, even in hot, wet climate conditions (Krynine, 1935; Ruxton, 1970), whereas sediment that is stored can be highly modified under a range of climate conditions (Suttner et al., 1981; Johnsson et al., 1988; Robinson and Johnsson, 1997).

Total Transport Fining

Total Transport Fining (Fig. 3) quantifies the ability of the hinterland alluvial-fluvial transport system to reduce the grain size of Transported Sediments by both physical (Selective Transport Fining) and chemical (Transport Modification Potential) processes. Total Transport Fining is higher when either physical or chemical modification is higher. Selective Transport Fining and Transport Modification Potential are considered to be equally responsible for Total Transport Fining.

Depositional Processes

Maximum Grain Size

Maximum Grain Size (Fig. 3) represents an upper limit on the grain size of sediment that can pass from the hinterland to the basin, which we consider to be a function of Rate of Basin Subsidence and Basin Fluvial Transport Distance; together, these factors determine the gradient in the basin, and thus the gravitational potential available to move sediment. Under some circumstances, coarser grain sizes may be sequestered near the basin margin, while the finer fractions of the delivered grain size distribution can be transported to more basin-ward environments of deposition (Robinson and Slingerland, 1998a, 1998b). Longer transport distances and slower subsidence rates (lower gradient) favor more effective trapping of a larger proportion of the coarse tail of the grain-size distribution.

Depositional Fining and Sorting Potentials

Depositional Fining Potential (Fig. 3) quantifies the propensity for a particular Depositional Facies to segregate grains that are finer, on average, than the mean grain size of the Delivered Grain Size and Sorting, whereas Depositional Sorting Potential (Fig. 3) is a measure of the propensity for a particular Depositional Facies to segregate a grain population that is better sorted, on average, than the sorting of the Delivered Grain Size and Sorting.

Deposition Modification Potential

Depositional Modification Potential (Fig. 3) is named in parallel with Generative Modification Potential and Transport Modification Potential to reflect the power of the environment to modify sand composition. Depositional Modification Potential is a function of the ability of Deposited Grain Size and Sorting to reflect the control of grain size on composition (Krynine, 1948; Bokman, 1955; Crook, 1960; Kairo et al., 1993).

MODEL IMPLEMENTATION

Bayesian Inference

We have codified our understanding of the generally accepted relationships between genetic controls and the character of reservoir sands in a Bayesian Belief Network (BBN). Bayes Rule (Bayes, 1763) states that the probability of two things (A,B) occurring together is equal to the probability of A given B, times the probability of B (and vice versa). Alternative symbolic formulations of the rule are given in Equations 1 and 2:

$$P(A,B) = P(A|B)P(B) = P(B|A)P(A) \quad (1)$$

and

$$P(A|B) = P(B|A)P(A)/P(B). \quad (2)$$

Bayes Rule provides a convenient bookkeeping device to keep track of a web of conditional probabilities; if you know the probability of B, given that A is true, and you know the probability of A, then you can calculate a probability of B. There is no limit to how many conditional states you may concatenate.

As a practical matter, formal models using Bayes Rule describe the world in nodes, which exist in discrete states that are comprehensive and mutually exclusive (Norsys, 2006). For example: the weather can be sunny or rainy, the speed of a car can be fast or slow; a road could be straight or curved; the pavement could be wet or dry. The states of the nodes have probabalistic relationships with each other, which can be exploited for predictive purposes. In the current example, the nodes may be arranged as in Figure 4. Nodes can be parent, child, or both. Nodes with arrows coming out are parent nodes. Nodes with arrows coming in are child nodes. Nodes that are parent only (weather condition, road type) are called root nodes or inputs. Nodes that are child only (crash risk) are leaf nodes or outputs. The rest are intermediate nodes, which are both child and parent. The probability of states in each child node depends on the probability of states in the parent node, i.e., the $P(A)$, *and* on a conditional probability table (i.e, the $P[B|A]$). The conditional probability table for the crash risk node of this example is provided in Table 1. The structure of the network describes the fundamental relationships; the values in the conditional probability tables quantify the details of those relationships. The conditional-probability tables are specified by the builder of the network; they can be based on expert opinion or physical models, crudely estimated from sparse data, inferred by sophisticated statistical means from abundant data, or anything in between.

A BBN can accommodate the fact that a single cause may have several effects, each of which may have contrasting, or

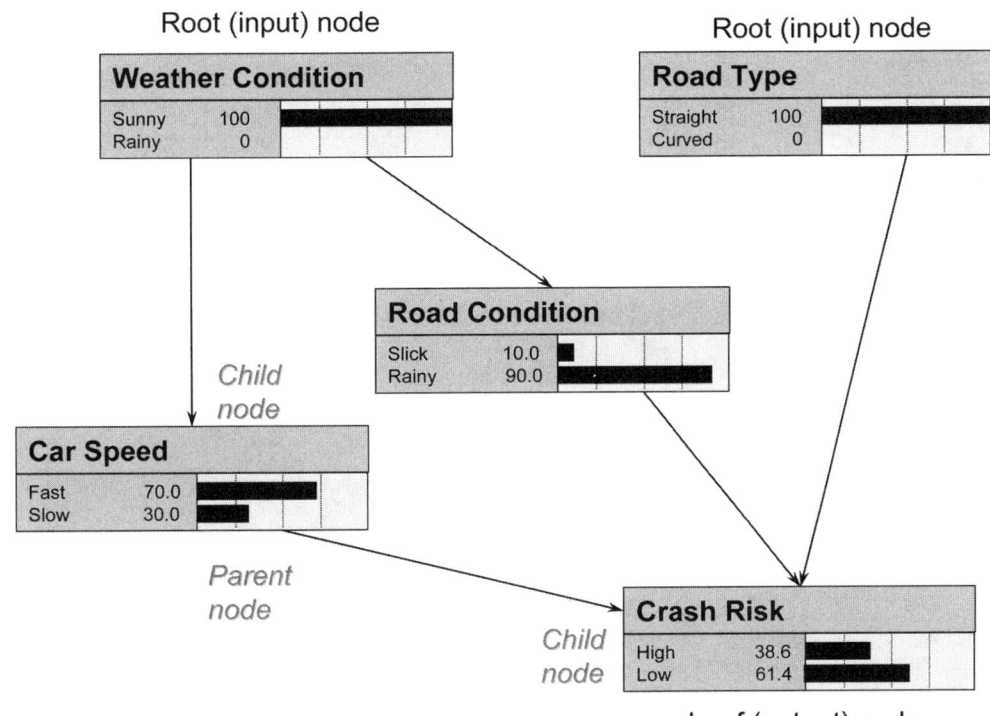

Figure 4. Example Bayesian network describing the relationships among weather condition, road type, and crash risk. Nodes with arrows coming out are "parent" nodes, which provide information about the probability of causes, whereas nodes with arrows coming in are "child" nodes, which provide information about effects. Nodes without parents are known as "root" or "input" nodes, whereas nodes without children are known as "leaf" or "output" nodes.

TABLE 1. CONDITIONAL PROBABILITY TABLE FOR CRASH RISK

Road condition	Road type	Car speed	P (high crash risk)	P (low crash risk)
Slick	Straight	Fast	0.80	0.20
Slick	Straight	Slow	0.40	0.60
Slick	Curved	Fast	0.99	0.01
Slick	Curved	Slow	0.60	0.40
Dry	Straight	Fast	0.50	0.50
Dry	Straight	Slow	0.01	0.99
Dry	Curved	Fast	0.70	0.30
Dry	Curved	Slow	0.40	0.60

TABLE 2. NODES IN THE SAND GENERATION AND EVOLUTION BAYESIAN NETWORK

Node name	Abbreviation for Table 3
Input nodes	
Climate weathering potential	CWP
Climate flushing potential	CFP
Regional topographic gradient	RTG
Hinterland transport distance	HTD
Basin subsidence rate	BSR
Basin fluvial transport distance	BFTD
Depositional facies	DF
Moderately derived nodes (probability depends on few prior nodes)	
Selective transport fining	STF
Depositional fining potential	DFP
Depositional sorting potential	DSP
Maximum grain size	MGS
Generative modification potential	GMP
Total flushing potential	TFP
Fluvial storage potential	FSP
Generated clay abundance	GCA
Generated sand composition	GSC
Generated grain-size distribution	GGS
Transport modification potential	TMP
Total transport fining	TTF
Highly derived nodes (probability depends on many prior nodes)	
Transported clay abundance	TCA
Transported sand composition	TSC
Transported grain-size distribution	TGS
Delivered grain size and sorting	DlGS
Deposited grain size and sorting	DpGS
Deposition modification potential	DMP
Output nodes	
Deposited matrix	DMA
Deposited sand composition	DSC

even contradictory, contributions to the leaf nodes. For example, a weather condition of sunny correlates with better road conditions (which lower crash risk), and with higher car speeds (which increase crash risk). The network provides a convenient means for keeping track of the ultimate relationship of weather condition on crash risk, even in the face of complex interaction of effects.

Network Structure

The schematic diagram of our model presented in Figure 3 and Tables 2 and 3 is also the structure of our BBN. This BBN is constructed with seven input nodes, which represent the key genetic elements governing sand character, and three output nodes which represent sand composition; sand texture; and detrital matrix content.

In between the input and output nodes are a series of intermediate nodes that describe the interactions of the genetic elements and their ultimate influence on the output nodes. Some of the intermediate nodes of the network depend quite directly on the input nodes; they either are children of input nodes, or have only one intermediate node between them and an input node. We informally refer to these as moderately derived nodes. The rest of the intermediate nodes are far removed from the input nodes by several intervening nodes. We informally refer to these as highly

TABLE 3. PARENT-CHILD RELATIONSHIPS AMONG NODES IN THE SAND GENERATION AND EVOLUTION BAYESIAN NETWORK

	STF	DFP	DSP	MGS	GMP	TFP	FSP	GCA	GSC	GGS	TMP	TTF	TCA	TSC	TGS	DIGS	DpGS	DMP	DMA	DSC
CWP					x															
CFP						x														
RTG					x	x														
HTD	x						x													
BSR					x															
BFTD					x															
DF		x	x																x	
STF												x								
DFP																x				
DSP																x				
MGS																x				
GMP								x	x	x										
TFP							x													
FSP								x												
GCA											x									
GSC											x									
GGS											x									
TMP												x	x	x						
TTF														x						
TCA																	x			
TSC																	x			
TGS																x				
DIGS																	x			
DpGS																		x		
DMP																				x

Note: Parent nodes are in the first column; look along each row to see which nodes are children of each parent. Child nodes are in the first row; look down each column to see which nodes are parents of each child.

derived nodes. The intermediate nodes listed in Table 2 are listed in order of increasing distance from the input nodes (equivalent to higher stream numbers of Shreve, 1966). The node "Deposited Grain Size and Sorting" is an output, in the informal sense that it is a feature of sand character that we wish to predict, although it is not a leaf node.

In this model, input and output nodes have precisely defined states with quantitative limits, whereas intermediate nodes related to processes have qualitative states that capture end members and intermediate states along a less-well-defined continuum.

"Tectonic setting" and "provenance lithotype assemblage" are shown in Figure 3 with dotted line connections, because they are not formal nodes. In real life, tectonic setting is the fundamental control on most aspects of the sand generation and evolution system, particularly on the assemblage of rocks exposed at the surface, from which the sediment will be derived. We constructed a library of provenance lithotype assemblages associated with different modern and ancient tectonic settings to use as analogs in cases where time or data restrictions prohibit reconstructing a paleogeologic map. For our purposes, we classified every rock type that can produce significant quantities of sand into 21 exhaustive and mutually exclusive categories (Table 4). We use information about the provenance lithotype assemblage, including relative abundance of each rock type, mineralogic composition (Table 5), and chemical composition (Table 6) to calculate the conditional probabilities that govern the nodes Generated Sand Composition, Generated Grain-Size Distribution, and Generated Clay Abundance.

Generated, Transported, and Delivered Sediments

In the model, we distinguish among generated, transported, and deposited sediments. Each kind of sediment is described by three nodes in the BBN: one for sand composition, one for sand texture, and one for mud. Each of the sand composition nodes has the same 66 states, corresponding to 10% increments in quartz–feldspar–rock fragment (QFR) ternary space (Fig. 5). Although each node state refers to a fixed proportion of Q, F, and R, the actual composition associated with each node state consists of 25 different grain types (Table 7). We calculated the specific 25-component mixture associated with each of the 66 ternary compositions with a separate spreadsheet, outside of the network.

Generated and transported texture-node states correspond to specific grain-size distributions within standard categories from granules to coarse silt (Table 8), whereas deposited texture-node states refer to more generalized verbal descriptions of grain size and sorting (Tables 8 and 9). The number of grain-size distributions or size-sorting categories in each texture node is fixed, but the relative abundance of each grain size within each distribution

TABLE 4. PROVENANCE LITHOTYPE NAMES

Code	Description
P1	Plutonic, ultrabasic
P2	Plutonic, basic, gabbro (sodic plagioclase dominant)
P3	Plutonic, basic, diorite (calcic plagioclase dominant)
P4	Plutonic, intermediate
P5	Plutonic, silicic
P6	Plutonic, sodic ("anorthosite")
P7	Plutonic, potassic ("syenite")
V1	Volcanic, basic
V2	Volcanic, intermediate
V3	Volcanic, silicic
S1	Sedimentary, sandstone, Quartz dominant
S2	Sedimentary, sandstone, Feldspar dominant
S3	Sedimentary, sandstone, Lithic fragments dominant
S4	Sedimentary, shale
S5	Sedimentary, carbonate
M1	Metamorphic, metasandstone
M2	Metamorphic, slate
M3	Metamorphic, metacarbonate
M4	Metamorphic, schist/phyllite
M5	Metamorphic, gneiss, plagioclase-rich
M6	Metamorphic, gneiss, K-feldspar-rich

TABLE 5. PROVENANCE LITHOTYPE MINERALOGICAL CHARACTERISTICS

PL	Bulk fraction macro-quartz			Bulk fraction macro-feldspar			Alkalai feldspar/ total feldspar			Sodic plagioclase/ total plagioclase			Bulk fraction macro-mica			"Black" mica/ total mica			Bulk fraction macro-dense minerals			Bulk fraction micro-crystalline matrix		
	Max	Min	Base	Max	Min	Base	Max	Min	Base	Max	Min	Base	Max	Min	Base	Max	Min	Base	Max	Min	Base	Max	Min	Base
P1	0.000	0.000	0.000	0.100	0.000	0.050	0.000	0.000	0.000	1.000	0.500	0.750	0.000	0.000	0.000	0.000	0.000	0.000	1.000	0.900	0.950	0.000	0.000	0.000
P2	0.000	0.000	0.000	0.900	0.100	0.500	0.000	0.000	0.000	1.000	0.500	0.250	0.000	0.000	0.000	0.000	0.000	0.000	0.900	0.100	0.500	0.000	0.000	0.000
P3	0.000	0.000	0.000	0.900	0.100	0.500	0.000	0.000	0.000	1.000	0.500	0.750	0.000	0.000	0.000	0.000	0.000	0.000	0.900	0.100	0.500	0.000	0.000	0.000
P4	0.330	0.020	0.175	0.930	0.220	0.575	0.350	0.000	0.175	1.000	0.900	0.950	0.100	0.000	0.050	1.000	0.500	0.750	0.300	0.000	0.150	0.000	0.000	0.000
P5	0.330	0.020	0.175	0.930	0.220	0.575	1.000	0.650	0.825	1.000	0.900	0.950	0.100	0.000	0.050	1.000	0.100	0.550	0.300	0.000	0.150	0.000	0.000	0.000
P6	0.100	0.000	0.050	0.960	0.640	0.800	0.350	0.000	0.175	1.000	0.900	0.950	0.100	0.000	0.050	0.100	0.010	0.055	0.200	0.000	0.100	0.000	0.000	0.000
P7	0.100	0.000	0.050	0.960	0.640	0.800	1.000	0.650	0.825	1.000	1.000	1.000	0.100	0.000	0.050	1.000	0.800	0.900	0.200	0.000	0.100	1.000	0.600	0.800
V1	0.000	0.000	0.000	0.300	0.000	0.150	0.400	0.000	0.200	0.500	0.000	0.250	0.050	0.000	0.025	0.000	0.000	0.000	0.100	0.000	0.050	1.000	0.400	0.700
V2	0.150	0.000	0.075	0.300	0.000	0.150	1.000	0.500	0.750	1.000	1.000	1.000	0.100	0.010	0.055	1.000	0.500	0.750	0.100	0.000	0.050	1.000	0.250	0.625
V3	0.300	0.000	0.150	0.180	0.000	0.090	1.000	0.000	0.500	1.000	0.600	0.800	0.050	0.000	0.025	0.100	0.010	0.055	0.100	0.000	0.050	0.020	0.000	0.010
S1	1.000	0.750	0.875	0.600	0.150	0.375	1.000	0.000	0.500	1.000	0.000	0.500	0.100	0.000	0.050	0.200	0.000	0.100	0.100	0.000	0.050	0.100	0.000	0.050
S2	0.750	0.200	0.475	0.500	0.000	0.250	1.000	0.000	0.500	1.000	0.000	0.500	0.200	0.000	0.100	0.500	0.200	0.350	0.200	0.000	0.100	0.200	0.000	0.100
S3	0.900	0.000	0.450	0.900	0.000	0.450?	1.000	0.000	0.500	1.000	0.000	0.500	0.050	0.000	0.025	0.800	0.400	0.600	0.050	0.000	0.025	1.000	0.900	0.950
S4	0.000	0.000	0.000	0.000	0.000	0.000	1.000	0.000	0.500	0.000	0.000	0.000	0.010	0.000	0.005	1.000	0.000	0.500	0.050	0.000	0.025	1.000	0.900	0.950
S5	0.000	0.000	0.000	0.600	0.000	0.300	1.000	0.000	0.500	0.000	0.000	0.000	0.100	0.000	0.050	0.500	0.000	0.250	0.100	0.000	0.050	1.000	0.940	0.970
M1	1.000	0.100	0.550	0.100	0.000	0.050	1.000	0.000	0.500	1.000	0.000	0.500	0.100	0.000	0.050	0.600	0.200	0.400	0.100	0.000	0.050	0.100	0.000	0.050
M2	0.100	0.000	0.050	0.000	0.000	0.000	1.000	0.000	0.500	1.000	1.000	1.000	0.040	0.000	0.020	1.000	0.000	0.500	0.100	0.000	0.050	1.000	0.600	0.800
M3	0.000	0.000	0.000	0.000	0.000	0.000	0.000	0.000	0.000	0.000	0.000	0.000	1.000	0.300	0.650	1.000	0.000	0.500	0.300	0.000	0.150	1.000	0.860	0.930
M4	0.100	0.000	0.050	0.200	0.000	0.100	1.000	0.000	0.500	1.000	1.000	1.000	0.200	0.000	0.100	0.100	0.010	0.055	0.200	0.000	0.150	0.100	0.000	0.050
M5	0.500	0.000	0.250	0.900	0.100	0.500	0.350	0.000	0.175	1.000	0.000	0.500	0.200	0.000	0.100	0.100	0.010	0.055	0.200	0.000	0.100	0.100	0.000	0.050
M6	0.500	0.000	0.250	0.900	0.100	0.500	1.000	0.650	0.825	1.000	0.900	0.950	0.200	0.000	0.100	0.800	0.400	0.600	0.200	0.000	0.100	0.100	0.000	0.050

(for generated and transported), or the numerical value of mean grain size and sorting (for deposited), is calculated with a separate spreadsheet, outside of the network.

Generated Sediment

The three nodes that describe generated sediments are generated sand composition, generated grain-size distribution, and generated clay abundance (Fig. 3; Tables 2 and 3). The node Generated Sand Composition has 66 states. The precise 25-component composition, and the conditional probability, for each state is calculated as a function of Provenance Lithotype relative abundance and Generative Modification Potential.

The node Generated Grain-Size Distribution has nine states (Fig. 6), which represent the coarsest, most likely, and finest possible outcomes produced by weathering the estimated provenance lithotype assemblage of the hinterland under high, moderate, and low states of Generative Modification Potential (Table 10), respectively. The calculations are based on observed grain-size distributions for soils derived from specific Provenance Lithotypes under various climatic and topographic conditions, weighted by the estimated abundance of each Provenance Lithotype. Soil grain-size observations were drawn from a variety of published databases, including Soil Survey Staff (1997), Batjes (2002), Cooper et al. (2005), and the International Soil Research and Information Center's Soil Information System (ISIS, http://lime.isric.nl/index.cfm?contentid = 218, verified 28 December 2005).

The node Generated Clay Abundance has three states: high, moderate, and low (Fig. 6). The conditional probability table is calculated according to: (1) the relative ability of each Provenance Lithotype to generate clay (based on the major-element geochemistry; Table 6), weighted by the relative abundance of each provenance lithotype; and (2) the level of Generative Modification Potential. Higher levels of Generative Modification Potential favor greater clay generation, within the constraints imposed by the aluminum, alkali, and alkali-earth content of the Provenance Lithotype assemblage.

Transported Sediment

The three nodes that describe transported sediments are Transported Sand Composition, Transported Grain-Size Distribution, and Transported Clay Abundance (Fig. 3; Tables 2 and 3).

The node Transported Sand Composition has 66 states. The precise 25-component composition, and the conditional probability, for each state is calculated as a function of the Generated Sand Composition and Transport Modification Potential.

The node Transported Grain-Size Distribution has 27 states (Fig. 7), which represent the coarsest, most likely, and finest possible outcomes produced by modifying the nine generated grain-size distributions under high, moderate, and low states of total transport fining, respectively. The calculated relative abundance of each grain size in each Transported Grain-Size Distribution depends on the relative abundance of that grain size in the correlative Generated Grain-Size Distribution and three Empirical

TABLE 6. PROVENANCE LITHOTYPE MAJOR-ELEMENT OXIDE CHARACTERISTICS (WT%)

PL	SiO_2	TiO_2	Al_2O_3	Fe_2O_3	FeO	MnO	MgO	CaO	Na_2O	K_2O	H_2O	P_2O_5	Reference: Clark (1982)
P1	43.54	0.81	3.99	2.51	9.84	0.21	34.02	3.46	0.56	0.25	0.76	0.05	Table 48, "Peridotite"
P2	48.36	1.32	16.81	2.55	7.92	0.18	8.06	11.07	2.26	0.56	0.64	0.24	Table 47, "Gabbros"
P3	51.86	1.50	16.40	2.73	6.97	0.18	6.12	8.40	3.36	1.33	0.80	0.35	Table 46, "Diorites"
P4	66.80	0.57	15.66	1.33	2.59	0.07	1.57	3.56	3.84	3.07	0.65	0.21	Table 45, "Granodiorites"
P5	72.08	0.37	13.86	0.86	1.67	0.06	0.52	1.33	3.08	5.46	0.53	0.18	Table 45, "Granites"
P6	54.54	0.52	25.72	0.83	1.46	0.02	0.83	9.62	4.66	1.06	0.63	0.11	Table 48, "Anorthosites"
P7	59.41	0.83	17.12	2.19	2.83	0.08	2.02	4.06	3.92	6.53	0.63	0.38	Table 46, "Syenites"
V1	50.83	2.03	14.07	2.88	9.00	0.18	6.34	10.42	2.23	0.82	0.91	0.23	Table 47, "Tholeiitic basalts"
V2	54.20	1.31	17.17	3.48	5.49	0.15	4.36	7.92	3.67	1.11	0.86	0.28	Table 46, "Andesites"
V3	73.66	0.22	13.45	1.25	0.75	0.03	0.32	1.13	2.99	5.35	0.78	0.07	Table 45, "Rhyolites"
S1	78.70	0.25	4.80	1.10	0.30	0.03	1.20	5.50	0.45	0.30	1.30	0.08	Table 87, "Sandstones"
S2	70.00	0.58	8.20	2.50	1.50	0.06	1.90	4.30	0.58	2.10	3.00	0.10	Table 87, "Sandstones from platforms"
S3	66.70	0.60	13.50	1.60	3.50	0.10	2.10	2.50	2.90	2.00	2.40	0.20	Table 87, "Graywackes"
S4	58.90	0.78	16.70	2.80	3.70	0.09	2.60	2.20	1.60	3.60	5.00	0.16	Table 87, "Shales (mainly from geosynclines)"
S5	8.20	0.00	2.20	1.00	0.68	0.07	7.70	40.50	0.00	0.00	0.00	0.07	Table 87, "Carbonate rocks"
M1	78.70	0.25	4.80	1.10	0.30	0.03	1.20	5.50	0.45	0.30	1.30	0.08	Table 87, "Sandstones"
M2	58.90	0.78	16.70	2.80	3.70	0.09	2.60	2.20	1.60	3.60	5.00	0.16	Table 87, "Shales (mainly from geosynclines)"
M3	8.20	0.00	2.20	1.00	0.68	0.07	7.70	40.50	0.00	0.00	0.00	0.07	Table 87, "Carbonate rocks"
M4	62.00	1.00	19.00	2.60	4.70	0.10	2.80	1.50	2.00	3.90	0.00	0.20	Table 95, average of "Phyllite" and "Mica schists"
M5	50.30	1.60	15.70	3.60	7.80	0.20	7.00	9.50	2.90	1.10	0.00	0.30	Table 95, "Amphibolites"
M6	70.70	0.50	14.50	1.60	2.00	0.10	1.20	2.20	3.20	3.80	0.00	0.20	Table 95, "Quartzofeldspathic gneisses"

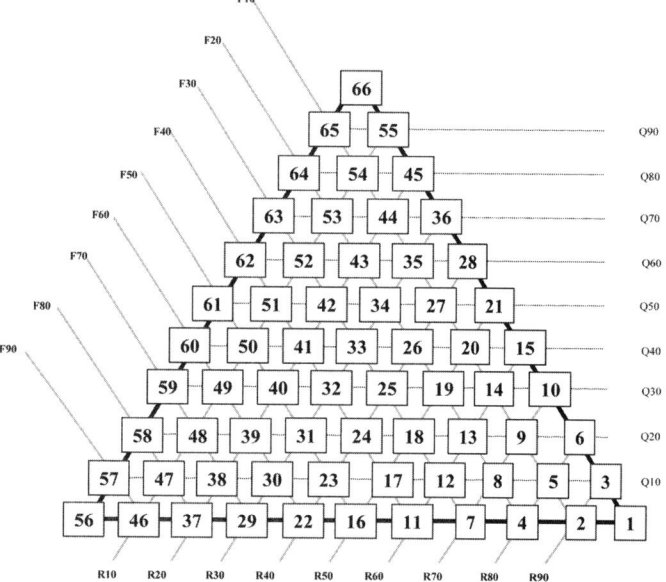

Figure 5. Operational definition of quartz–feldspar–rock fragment (QFR) sand composition states in the Sand Generation and Evolution Model (SandGEM).

TABLE 7. SANDGEM GRAIN TYPES

Code	Description
Q	Quartz (monomineralic, but mono- or polycrystalline)
Fk	Feldspar, potassic; Or >10
Fpna	Feldspar, plagioclase (sodic); Or < 10, An < 50
Fpca	Feldspar, plagioclase (calcic); Or < 10, An > 50
RPp	Rock fragment, plutonic, plagioclase-rich
RPk	Rock fragment, plutonic, K-feldspar-rich
RPh	Rock fragment, plutonic, heavy mineral-rich
RVb	Rock fragment, volcanic, basic
RVi	Rock fragment, volcanic, intermediate
RVs	Rock fragment, volcanic, silicic
RSq	Rock fragment, sedimentary, quartzose sandstone
RSf	Rock fragment, sedimentary, feldspathic sandstone
RSl	Rock fragment, sedimentary, lithic sandstone
RSsh	Rock fragment, sedimentary, shale
RScb	Rock fragment, sedimentary, carbonate
RSct	Rock fragment, sedimentary, chert
RMqf	Rock fragment, metasediment, quartzofeldspathic
RMsh	Rock fragment, metasediment, shale
RMc	Rock fragment, metasediment, carbonate
RMpsc	Rock fragment, metamorphic, phyllite/schist
RMgp	Rock fragment, metamorphic, gneiss, plagioclase-rich
RMgk	Rock fragment, metamorphic, gneiss, K-feldspar-rich
OH	Other grain, "heavy" (dense) minerals
OMb	Other grain, mica, ferromagnesian ("black" or "biotite")
OMw	Other grain, mica, "white"

Note: SandGEM—sand generation and evolution model.

TABLE 8. GRAIN-SIZE CATEGORIES

State	Name	Definition
g	Granules	2000–4000 μm
vc	Very coarse sand	1000–2000 μm
c	Coarse sand	500–1000 μm
m	Medium sand	250–500 μm
f	Fine sand	125–250 μm
vf	Very fine sand	63–125 μm
cs	Coarse silt	45–63 μm

TABLE 9. SORTING CATEGORIES

State	Name	Inclusive graphic standard deviation σ_i (Folk, 1974)
vw	Very well sorted	<0.35
w	Well sorted	0.35–0.50
mw	Moderately well sorted	0.50–0.71
m	Moderately sorted	0.71–1.00
p	Poorly sorted	1.00–2.00
vp	Very poorly sorted	2.00–4.00
xp	Extremely poorly sorted	>4.00

Transform Functions that describe possible evolutionary pathways for each grain size under each level of Total Transport Fining. An example of Empirical Transformation Functions for low Total Transport Fining is provided in Table 11. Each of the Empirical Transformation Functions is calibrated to real world observations of modern stream sediments derived from known sources (D. Novák, 2005, personal commun., ExxonMobil Upstream Research Co.). The conditional probability table for Transported Grain Size Distribution is constructed so that low Total Transport Fining favors the coarsest calculated grain-size distributions, whereas high favors the finest.

The node Transported Clay Abundance has three states (Fig. 7), representing the cumulative amount of clay produced

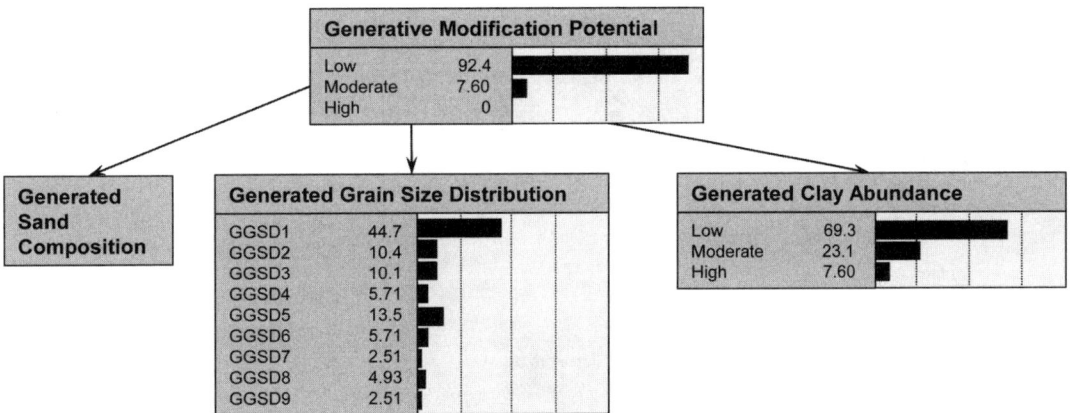

Figure 6. Detailed view of node states for generative modification potential, generated grain-size distribution, and generated clay abundance. Node states of Generated Grain Size Distribution GGSD1-9 represent nine possible grain-size distributions that could be generated from the Provenance Lithotype assemblage of the hinterland. The precise definition of each grain-size distribution is calculated separately, outside the network.

TABLE 10. DEFINITION OF GENERATED GRAIN-SIZE DISTRIBUTIONS

	Generative modification potential		
	Low	Moderate	High
Coarsest plausible	GGSD1	GGSD2	GGSD3
Most likely	GGSD4	GGSD5	GGSD6
Finest plausible	GGSD7	GGSD8	GGSD9

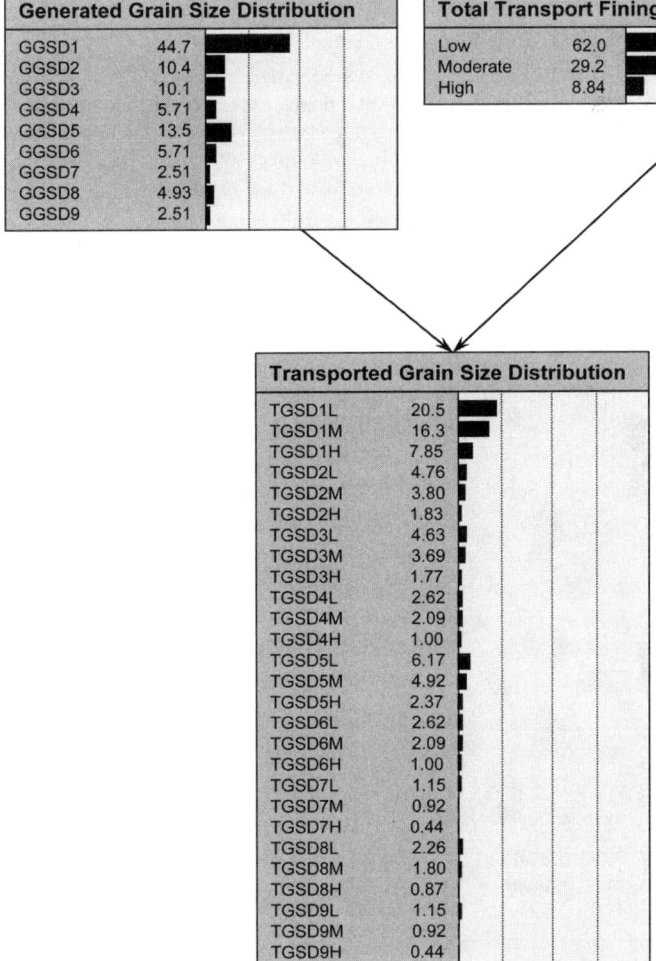

Figure 7. Detailed view of node states for Generated and Transported grain-size distribution, and total transport fining. Node states of Transported Grain Size Distribution TGSD1L-9H represent twenty-seven possible grain-size distributions that could be generated from the nine Generated Grain Size Distributions, if each is subjected to Low, Moderate, or High Total Transport Fining. The precise definition of each grain-size distribution is calculated separately, outside the network.

TABLE 11. EXAMPLE EMPIRICAL TRANSFORM FUNCTION

	g	vc	c	m	f	vf	cs
Fraction that break	0.875	0.750	0.500	0.125	0.000	0.000	0.000
Fraction that go away	0.000	0.000	0.000	0.000	0.125	0.600	0.800
Total fraction transformed	0.875	0.750	0.500	0.125	0.125	0.600	0.800
If they break, where they go							
Granules →		0.5	0.5				
Very coarse →			0.5	0.5			
Coarse →			0.25	0.5	0.25		
Medium →				0.125	0.75	0.125	
Fine →					0.0625	0.875	0.0625
Very fine →							

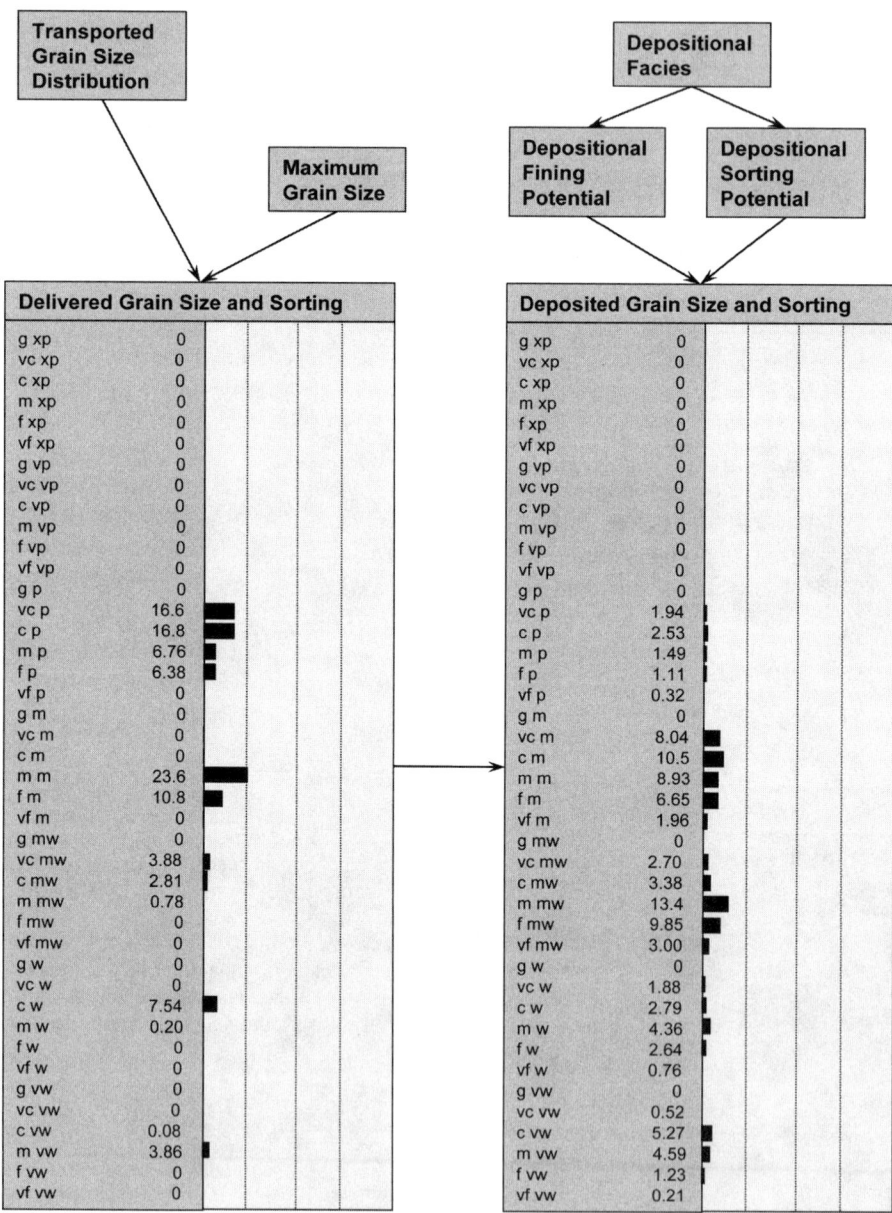

Figure 8. Detailed view of node states for delivered and deposited grain size and sorting. Codes for grain size (g, vc, c, m, f, vf) are defined in Table 8 and those for sorting (xp, vp, p, m, mw, w, vw) are defined in Table 9.

TABLE 12. EXAMPLE OF TRANSPORTED TO DELIVERED GRAIN-SIZE DISTRIBUTION TRANSFORMATION

Transported grain-size distribution	Max grain size	Grain-size distribution after transformation						Mean grain size	Sorting
		g	vc	c	m	f	vf	cs		
TGSD1L	g	8.0	18.0	24.8	22.7	21.9	4.6	8.0	0.96	1.42
TGSD1L	vc	0.0	19.6	27.0	24.6	23.8	5.0	0.0	1.16	1.25
TGSD1L	c	0.0	0.0	33.6	30.6	29.6	6.2	0.0	1.56	1.01
TGSD1L	m	0.0	0.0	0.0	46.1	44.5	9.4	0.0	2.10	0.73

TABLE 13. REGIONAL TOPOGRAPHIC GRADIENT NODE STATES

State	Definition	Examples
High	m/km	Himalayas, Alps, Andes, Sierra Nevada, Papua New Guinea
Moderate	cm/km	Urals, Jura, Caatingas, Appalachians, Ethiopian Highlands
Low	mm/km	Flat continental interiors

during generation and transport. The conditional probability table is calculated according to: (1) the amount of initial clay generation; (2) the clay-generating potential of the Generated Sand Compositions; and (3) the Transport Modification Potential. The highest level of Transported Clay Abundance is favored by high Generated Clay Abundance, high probability of quartz-deficient Generated Sand Composition, and high Transport Modification Potential.

Deposited Sediment

The four nodes that describe deposited sediments are Deposited Sand Composition, Deposited Grain Size and Sorting, and Deposited Matrix Abundance (Fig. 3; Tables 2 and 3).

The node Deposited Sand Composition has 66 states. The precise 25-component composition, and the conditional probability, for each state is calculated as a function of the Transported Sand Composition and Deposition Modification Potential.

The nodes Delivered Grain Size and Sorting and Deposited Grain Size and Sorting each have 42 states (Fig. 8), which represent all combinations of the grain sizes granules through very fine sand (Table 8) and the sorting levels very well to extremely poorly sorted (Table 9). Delivered Grain Size and Sorting is a function of Transported Grain-Size Distribution and Maximum Grain Size. The calculated grain-size distribution associated with each state of Transported Grain-Size Distribution is truncated at the grain size specified by Maximum Grain Size, and the mean grain size and sorting for the new grain-size distribution is calculated (Table 12). The conditional probability table is constructed based on the relative frequency of each grain-size/sorting combination among the calculated values. Deposited Grain Size and Sorting is a function of Delivered Grain Size and Sorting, Depositional Fining Potential, and Depositional Sorting Potential (Fig. 8). Because some grain-size/sorting combinations are not observed in nature (Griffiths, 1951), the conditional probability table is constructed so that unobserved combinations have zero, or extremely low, probabilities. Higher values of Depositional Fining Potential favor finer grain-size/sorting combinations, whereas higher values of Depositional Sorting Potential favor better-sorted grain-size/sorting combinations.

Matrix is considered to be detrital clastic material that is enough smaller than framework grains to occlude intergranular pore throats. We only consider three levels to be important: insufficient to affect permeability, intermediate, sufficient to substantially eliminate permeability. The exact numerical values associated with each level depend on the relative grain size of the system and can be estimated with the methods of Panda and Lake (1994, 1995). The node deposited matrix abundance has three states: low (typically 0%–2%), moderate (typically 2%–10%), and high (typically 10%–25%). Deposited Matrix Abundance is a function of Transported Clay Abundance and Depositional Facies. The conditional probability table is constructed so that higher values of Transported Clay Abundance and Depositional Facies associated with lower values of net-to-gross (e.g., Walker and James, 1992) favor higher values of deposited matrix abundance.

Input Nodes

Regional Topographic Gradient

The states of Regional Topographic Gradient (Table 13; Fig. 9) are high (m/km), moderate (cm/km), and low (mm/km). Generative Modification Potential and Flushing Potential are the children of Regional Topographic Gradient.

Climate: Weathering Potential and Flushing Potential

Weathering Potential and Flushing Potential are characterized by the discrete, qualitative, states high, moderate, and low (Fig. 10, 11). We used information about the level and variability of wetness and temperature to make quantitative estimates of weathering potential index and transport potential index (Appendix A) to guide the selection of the qualitative states.

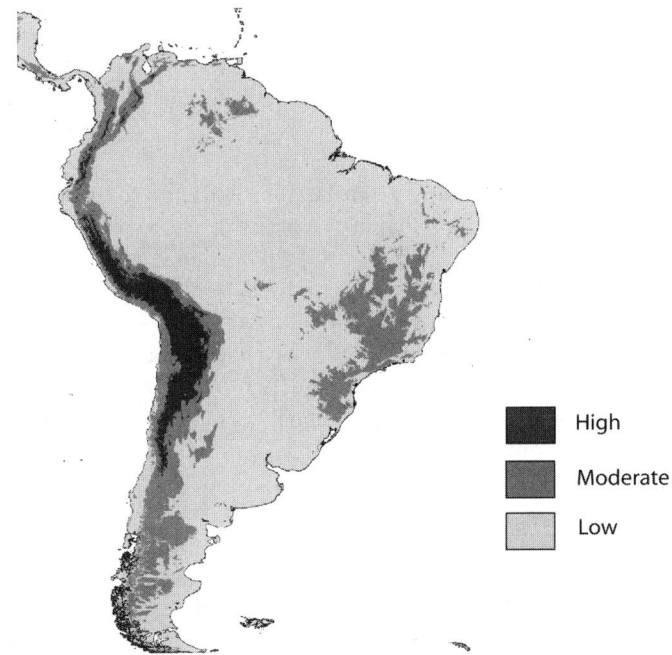

Figure 9. Operational definition of regional topographic gradient in the Sand Generation and Evolution Model (SandGEM), illustrated with a digital elevation model of South America (USGS Hydro1k DEM for South America, visualized at http://edcdaac.usgs.gov/gtopo30/hydro/sa_dem_img.asp [accessed 14 October 2005]).

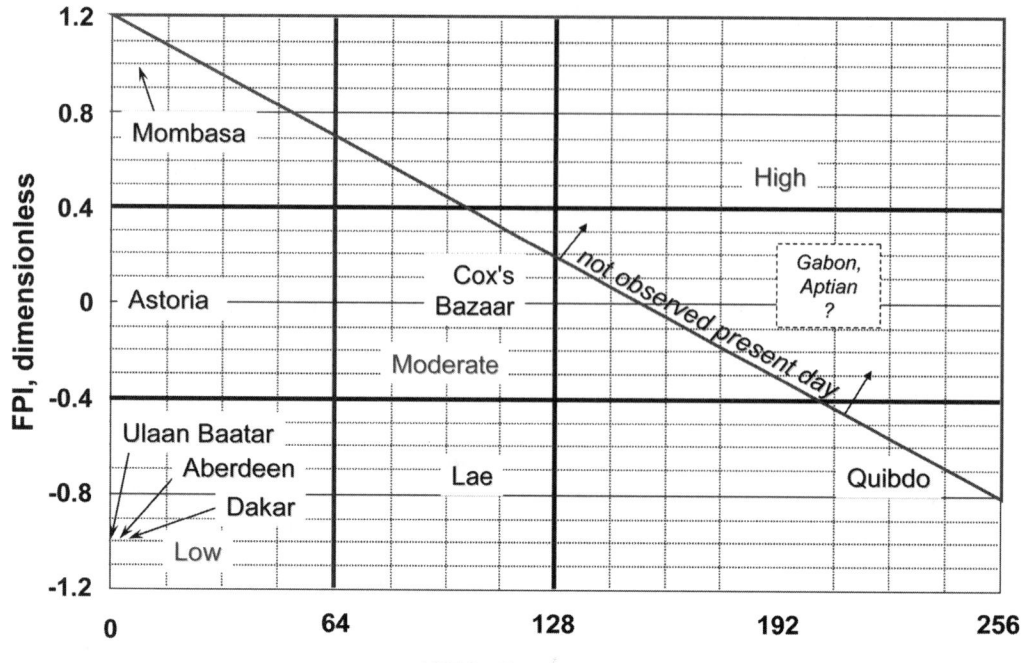

Figure 10. Examples of weathering potential index (WPI) and flushing potential index (FPI) calculated for modern and ancient climate. Locations include: Mombasa, Coast, Kenya; Astoria, Oregon, USA; Ulaan Baatar, Töv, Mongolia; Aberdeen, Scotland, UK; Dakar, Dakar, Senegal, Cox's Bazaar, Chittagong, Bangladesh; Lae, Morobe, Papua New Guinea; Quibdo, Chocó, Colombia. Climate conditions for the mid-Cretaceous were estimated from West et al. (2005).

The Weathering Potential Index (WPI, Appendix A) accounts for two different effects of temperature. Increasing temperatures, if all other factors are equal, will increase mineral-dissolution reaction rates exponentially (Lasaga, 1984). However, increasing temperature also increases evaporation, so the amount of water available to facilitate dissolution reactions will decrease, all other things being equal (Willmott, 1977). WPI provides a bookkeeping device to account for these opposite tendencies. Values of WPI for modern climate (and for plausible ancient climates) range between 0 (least potential to chemically modify silicate minerals) and ~256 (greatest potential).

The Flushing Potential Index (FPI, Appendix A) acknowledges that precipitation that is evaporated is less available to do physical work moving sediment than precipitation that is not evaporated (Wischmeier and Smith, 1978). It also accommodates the observations of Cecil and others (Cecil and Edgar, 2003) that sediment transport reaches a maximum when effective precipitation is concentrated into a small fraction of the year. Values of FPI range from −1.2 (lowest potential to transport sediment) to +1.2 (highest potential).

Weathering and Flushing Potential Indices are dimensionless and have no intrinsic physical meaning, but they can be compared to levels of dissolved, suspended, and bed load in modern rivers (Curtis et al., 1973; Walling, 1987; Summerfield and Hulton, 1994; Ludwig and Probst, 1996; Milliman, 1997; Hovius, 1998; Schaller et al., 2001; Syvitski et al., 2003; Walling and Fang, 2003) in order to guide selection of the high, moderate, and low states in the Weathering and Transport Potential nodes of the network. Values of Weathering and Transport Potential indices (WPI and FPI) that characterize high, moderate, and low levels of Weathering and Transport Potential are summarized in Tables 14 and 15. Figure 10 shows these values graphically, along with modern and ancient examples of each state. Generative Modification Potential and Transport Modification Potential are the children of Climate Weathering Potential. Total Flushing Potential is the child of Climate Flushing Potential.

TABLE 14. WEATHERING POTENTIAL NODE STATES

State	Definition
High	Weathering potential index (WPI) >128
Moderate	64 < WPI < 128
Low	WPI < 64

Note: See Figure 10 for examples.

TABLE 15. FLUSHING POTENTIAL NODE STATES

State	Definition
High	Transport potential index (TPI) > 0.4
Moderate	−0.4 < TPI < 0.4
Low	TPI < −0.4

Note: See Figure 10 for examples.

Hinterland Transport Distance

The states of Hinterland Transport Distance (Table 16; Fig. 11) are short, medium, and long, which equate to first-, second-, and third-order drainages in the sense of Ingersoll et al. (1993). The numerical cutoffs in Table 16 were established after Dutta and Suttner (1986a, 1986b). Fluvial Storage Potential and Selective Transport Fining are the children of Hinterland Transport Distance.

Basin Fluvial Transport Distance

The states of Basin Fluvial Transport Distance are short, intermediate, long, and very long (Table 17; Fig. 11). Maximum grain size is the child of basin fluvial transport distance.

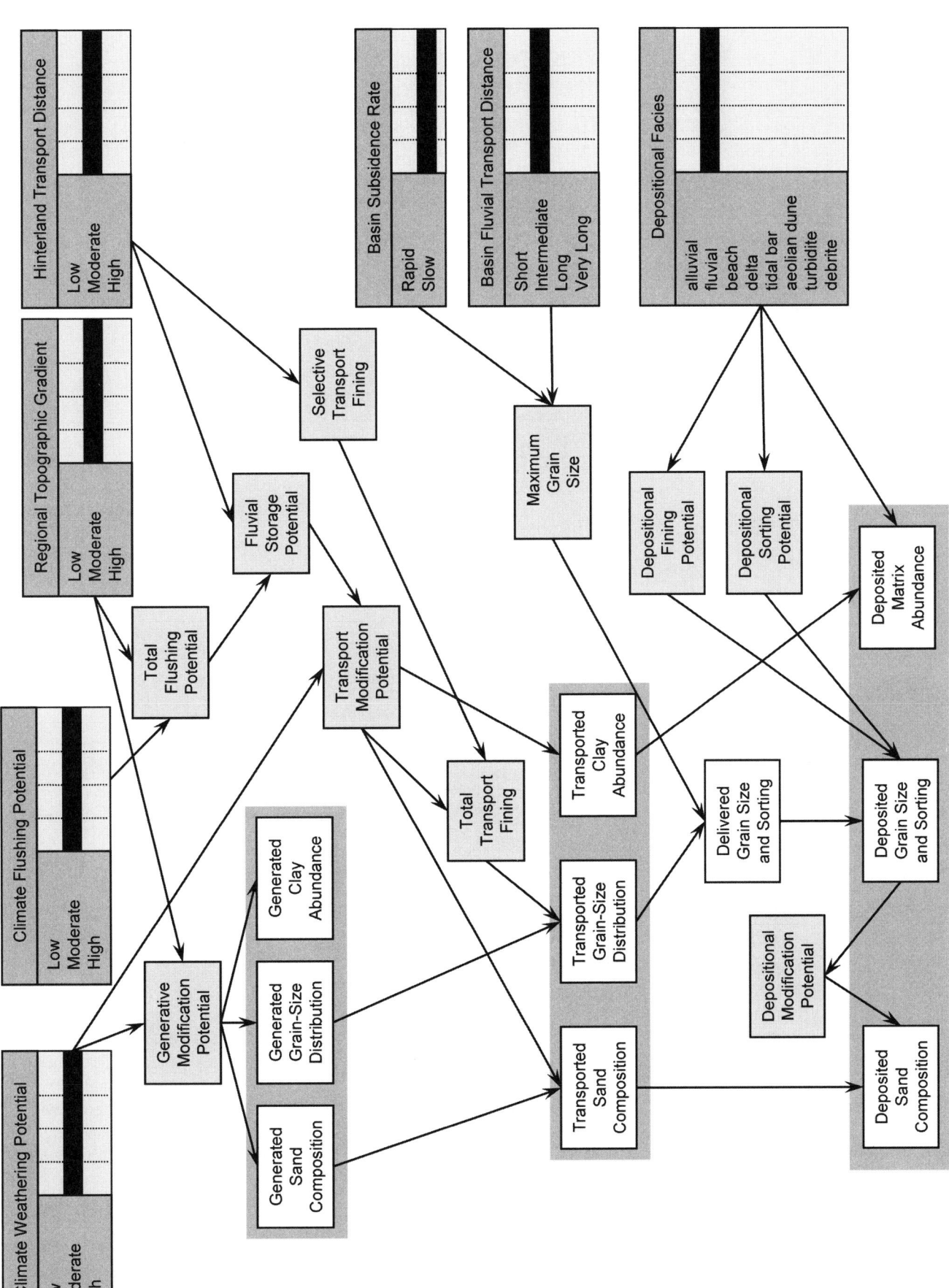

Figure 11. Detailed view of node states for input nodes and their immediate child nodes.

TABLE 16. HINTERLAND TRANSPORT DISTANCE NODE STATES

State	Definition	Example
Long	1000's of km	Mississippi, Orinoco, Nile, Niger, Ganges, Amazon
Intermediate	Several 100's of km	Susquehana, Missouri, Colorado (TX), Ohio
Short	10's to a few 100 km	Juniata (PA), Madison (MT), Oyster Creek (TX)

TABLE 17. BASIN FLUVIAL TRANSPORT DISTANCE NODE STATES

State	Definition (km)	Marine or lacustrine examples[†]
Short	<0	Maximum highstand systems tract
Intermediate	10–50	Intermediate highstand and transgressive systems tracts
Long	50–100	Intermediate lowstand systems tract
Very long	>100	Maximum lowstand systems tract

[†]Sequence stratigraphic nomenclature is appropriate in any setting where base level is determined by a standing body of water with a level that may rise and fall.

TABLE 18. BASIN SUBSIDENCE NODE STATES

State	Definition	Example
Rapid	Many 100s to 1000s m Ma	Malay Basin (Tertiary), Gulf of Mexico (Tertiary), Ridge Basin (Tertiary), Pannonian (Tertiary), South Caspian (Tertiary)
Slow	10s to a few 100s Ma	Sea of Okhotsk (Tertiary), Baltic Sea (modern), eastern margin of U.S. Western Interior Seaway (Cretaceous)

Rate of Basin Subsidence

The states of Rate of Basin Subsidence (Table 18) are rapid and slow. Maximum Grain Size is the child of Rate of Basin Subsidence.

Depositional Facies

Depositional Facies states are summarized in Table 19 and Figure 11. The Depositional Facies included in this model are limited to those likely to be significant hydrocarbon reservoir facies; the list is not intended to be exhaustive. Depositional Fining Potential and Depositional Sorting Potential are the children of Depositional Facies.

Sand-Generating Process Nodes

The states of Generative Modification Potential are high, moderate, and low (Fig. 6). The conditional probability table is constructed so that high Generative Modification Potential is virtually certain when Climate Weathering Potential is high (WPI > 128; see Appendix A) and Regional Topographic Gradient is low (mm/km; Table 12), whereas low Generative Modification Potential is virtually certain when Climate Weathering Potential is low and Regional Topographic Gradient is high. The logic behind conditional-probability assignment is described in Grantham and Velbel (1988).

Transport-Process Nodes

Total Flushing Potential

Total Flushing Potential states are high, moderate, and low. The conditional probability table for this node is constructed so that Total Flushing Potential is virtually certain to be high when Climate Flushing Potential and Regional Topographic Gradient are both high, whereas the Total Flushing Potential is virtually certain to be low when the parent nodes are both low. Between

TABLE 19. DEPOSITIONAL FACIES CHARACTERISTICS

Depositional facies	Most likely fining potential	Most likely sorting potential
Alluvial, confined	Low	Low
Alluvial, unconfined	Low	Low
Fluvial, major channel	Low	Moderate
Fluvial, minor channel	Low	High
Fluvial, point bar	Moderate	Moderate
Fluvial, levee/splay	Moderate	Moderate
Foreshore	Moderate	Moderate
Shoreface	Moderate	High
Delta, distributary channel	Moderate	High
Delta, stream-mouth bar	Moderate	High
Delta, distal front	Moderate	High
Tidal channel bar	High	Moderate
Aeolian dune	High	Moderate
Deep water, turbidite, channel	High	Moderate
Deep water, turbidite, levee	High	High
Deep water, debrite, channel	High	High
Deep water, debrite, levee	High	High

Climate Flushing Potential and Regional Topographic Gradient, Regional Topographic Gradient is considered to be a more important control on Total Flushing Potential.

Fluvial Storage Potential

Fluvial Storage Potential states are high, moderate, and low (Fig. 10). The conditional probability table for this node is constructed so that high Fluvial Storage Potential is virtually certain when Total Flushing Potential is low and Hinterland Transport Distance is long, whereas Total Flushing Potential is virtually certain to be high when Total Flushing Power is high and Hinterland Transport Distance is short. Between Total Flushing Power and Hinterland Transport Distance, Hinterland Transport Distance is considered to be a more important control on Fluvial Storage Potential.

Selective Transport Fining

Selective Transport Fining states are much, some, and none. Hinterland Transport Distance is the only parent of Selective Transport Fining. The longer the transport distance, the more opportunities available for sediment to be temporarily stored; at each step of deposition and remobilization, sediments have an opportunity to be hydrodynamically segregated and different size classes preferentially retained or removed. Total Transport Fining is the child of Selective Transport Fining.

Transport Modification Potential

Transport Modification Potential states are high, moderate, and low. The conditional probability table is constructed so that high Transport Modification Potential is virtually certain when both of the parent nodes are high, whereas low is virtually certain when the parent nodes are both low. Between Climate Weathering Potential and Fluvial Storage Potential, Fluvial Storage Potential is considered to be a more important control on Transport Modification Potential.

Total Transport Fining

Total Transport Fining states are high, moderate, and low (Fig. 7). The conditional probability table is constructed so that high Total Transport Fining is virtually certain when both of the parent nodes are high, whereas low Total Transport Fining is virtually certain when the parent nodes are both low. Selective Transport Fining and Transport Modification Potential are considered to be equally responsible for Total Transport Fining.

Depositional-Process Nodes

Maximum Grain Size

Maximum Grain Size states are granules, very coarse sand, coarse sand, and medium sand (size definitions in Table 8). The conditional probability table is constructed so that longer transport distances and slower subsidence rates (lower gradient) favor more effective trapping of a larger proportion of the coarse tail of the grain-size distribution.

Depositional Fining and Sorting Potential

Depositional Fining Potential and Depositional Sorting Potential states are low, moderate, and high (Fig. 10). The conditional probability table is calibrated to match observations (Table 19). In general, any depositional facies will have some probability for every level of fining and sorting potential, with the most likely value in Table 19 receiving the largest single probability.

Depositional Modification Potential

Depositional Modification Potential states are very coarse–coarse, medium, and fine–very fine; these are considered to be the scale at which compositional variation as a function of grain size will operate. The conditional probability table is set according to grain-size/grain-type relationships derived from a proprietary database of more than 260,000 point counts in which both grain size and grain type have been recorded on a point-by-point basis for samples from a wide range of hydrocarbon reservoirs, adjusted for the Provenance Lithotype assemblage in the hinterland. The simultaneous collection of grain size and grain type on a point-by-point basis is trivial when using an automated data collection system like the one described by Cipriani et al. (this volume).

EXAMPLE PREDICTIONS

The effectiveness of this forward modeling approach can be tested against observations of modern sand, where the genetic factors of tectonic setting, provenance lithotype abundance, climate, uplift, transport distance, etc., can be well documented. Paul Potter and co-workers have accumulated a large set of South American fluvial and beach sand data (Potter, 1993, 1994) that is appropriate for this purpose. ExxonMobil has also funded studies to obtain appropriate data to test the model, including Menacherry (2006) and Menacherry et al. (2006). Next we compare observations made in studies of Brazilian beach sand, Chilean fluvial sand, and Australian alluvial sand with predictions from SandGEM.

Graphical Conventions

The predictions consist of a probability distribution for all of the potential outcomes. The figures that follow graphically depict the probability distribution as a contoured probability density surface, or grid equivalent (Fig. 12). The contours or shading levels represent three levels of probability: Black depicts the single most likely outcome; gray depicts the next most likely outcomes, up to a cumulative probability of 68%, which is equivalent to the area under a standard normal curve within one standard deviation from the mean; and (3) white depicts the next most likely outcomes, up to a cumulative probability of 95%, which is equivalent to the area under a standard normal curve within two standard deviations from the mean.

Brazilian Beach Sand Composition

Observations

The Brazilian beach sands were selected from Potter's Atlantic coastal samples between 0° and 30°S. The data are presented in Table 20, with Potter's original point-count data recast into categories that can be compared to SandGEM predictions. All of the data reported in Potter (1993) and summarized in Potter (1994) were point-counted by Paul Potter from samples he and his co-workers collected, mostly in the early 1980's. Each sample was counted twice: the first count consisted of 200 framework-grain points, counted in the categories of Table 21; the second count consisted of 100 rock-fragment points, counted in the categories of Table 22 (Franzinelli and Potter, 1983). The data presented in Table 20 translate Potter (1993) data into SandGEM categories according to the scheme in Table 23. The beach sands were not characterized for texture, but all samples were collected from medium to fine sand on the berm (Potter, 1986).

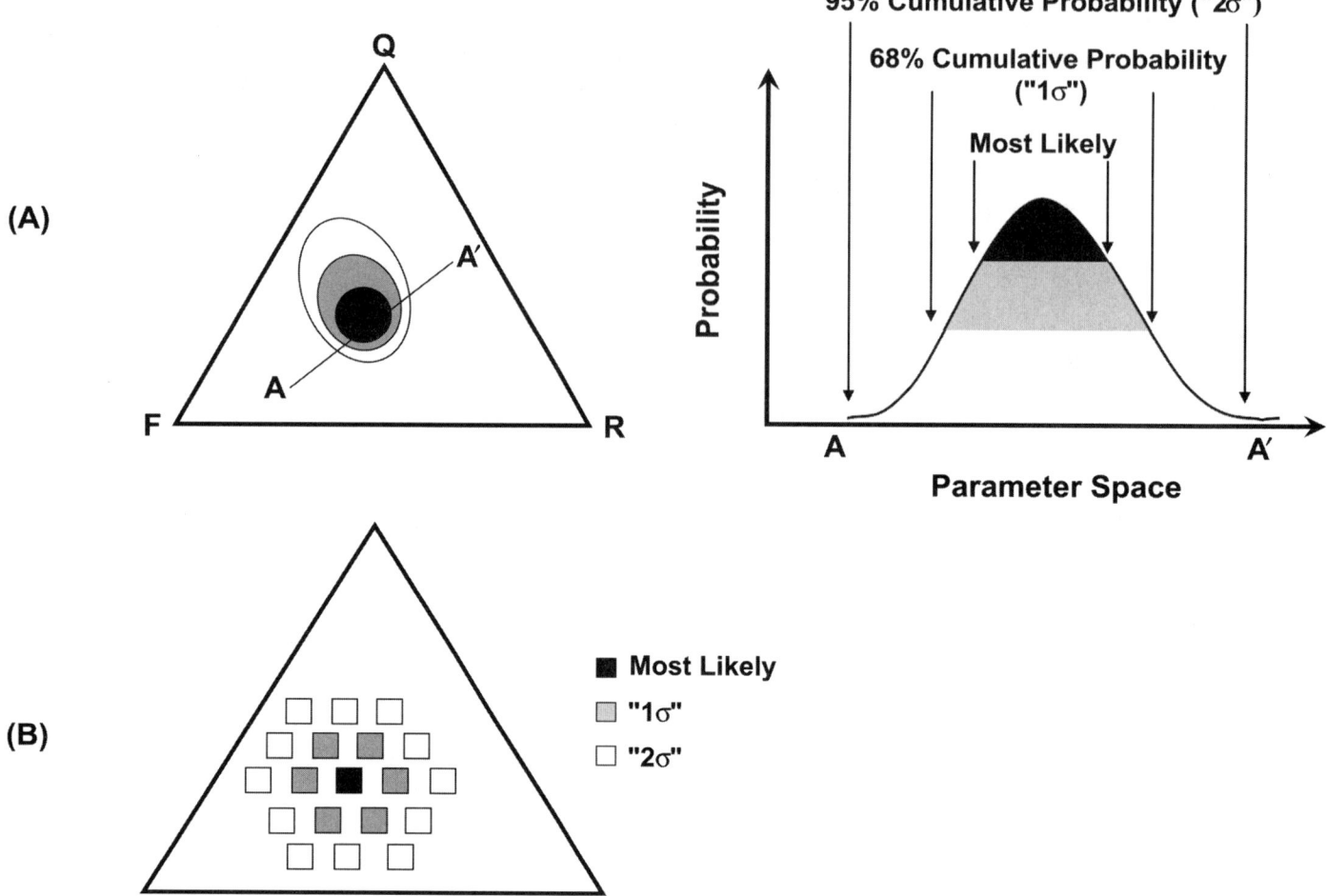

Figure 12. Graphical convention for displaying probability density surfaces in quartz–feldspar–rock fragment (QFR) ternary space (std—standard deviation), as contours (A) or grids (B).

Environmental Conditions

The provenance lithotype assemblage from which these sands were derived is composed (in order of importance) of: low- to medium-grade metasediments and schists; plutons and high-grade gneisses; volcanics; and sedimentary rocks. The precise details are quantified in Table 24. The values in Table 24 were measured by planimetry in Choubert et al. (1976); the precise abundance of lithotypes implied by each (litho)chronostratigraphic unit of the map was calibrated in selected areas with information from 1:1,000,000 geologic maps in the series *Carta Geológica do Brasil ao Milionésimo* (Brazil, Divisão de Geologia e Mineralogia, Ministério das Minas e Energia, Departamento Nacional da Produção Mineral, 1974).

The environmental conditions that governed the generation and evolution of these sediments, as quantified by SandGEM input parameters, are summarized in Table 25. The region is characterized by moderate (compared to all of Earth history) weathering and flushing power, long transport distance over a low topographic gradient, and a small, slowly subsiding basin with a shoreface depositional facies. All of these factors tend to favor a highly mature sediment composition.

Prediction versus Observation

There is good agreement between prediction and observation for sand composition (Figs. 13 and 14; Tables 22, 26, and 27). The model predicts that the single most likely composition is pure quartz and that the probability-weighted average quartz content should be 87.0%. In fact, quartz arenite (Q90–100F0-5R0-5; sensu Dott 1964) is by far the most likely composition (63 of 81 samples), and the average quartz content of all samples is 87.5%. The model predicts that plagioclase should be less abundant than alkali feldspar by a factor of ~4. In fact, in almost all the samples (73 of 81), plagioclase is less abundant than alkali feldspar, with an average Fp:Fk of 0.240. Rock fragments are predicted to be virtually absent. In fact, rock fragments are virtually absent—52 of 81 samples have no rock fragments, and the average rock-fragment content of all the samples is 4.3%. No quantitative data are available about

TABLE 20. OBSERVED BRAZILIAN BEACH SAND COMPOSITION

ID	Lat (°S)	Long (°W)	Q	Fk	Fp	RP	RV	RSs	RSsh	RScb	RSct	RM	OM	OH
660	29.84	49.98	0.85	0.05	0.04	0.00	0.00	0.00	0.01	0.00	0.00	0.04	0.00	0.01
652	29.23	49.74	0.97	0.01	0.00	0.00	0.00	0.00	0.00	0.00	0.00	0.01	0.00	0.01
664	28.97	49.35	0.95	0.02	0.01	0.00	0.00	0.00	0.00	0.00	0.01	0.00	0.00	0.00
750	27.68	48.47	0.94	0.04	0.00	0.00	0.00	0.00	0.00	0.00	0.00	0.00	0.00	0.00
1255	25.58	48.32	0.99	0.01	0.00	0.00	0.00	0.00	0.00	0.00	0.00	0.00	0.00	0.00
1256	25.58	48.37	0.97	0.01	0.01	0.00	0.00	0.00	0.00	0.00	0.00	0.00	0.00	0.01
662	24.34	47.30	0.93	0.02	0.01	0.00	0.00	0.00	0.00	0.00	0.01	0.01	0.00	0.01
661	24.33	47.29	0.94	0.00	0.01	0.00	0.00	0.00	0.00	0.00	0.00	0.00	0.00	0.04
668	24.05	46.37	0.84	0.07	0.01	0.00	0.00	0.00	0.00	0.00	0.00	0.00	0.00	0.07
1284	23.84	46.12	0.83	0.05	0.01	0.00	0.00	0.00	0.00	0.00	0.00	0.00	0.00	0.01
699	23.09	43.76	0.99	0.01	0.00	0.00	0.00	0.00	0.00	0.00	0.00	0.00	0.00	0.00
694	23.03	43.41	0.96	0.01	0.01	0.00	0.00	0.00	0.00	0.00	0.00	0.00	0.00	0.01
952	23.01	43.44	0.85	0.01	0.00	0.00	0.00	0.00	0.00	0.00	0.00	0.00	0.00	0.01
953	23.01	44.51	0.87	0.02	0.00	0.00	0.00	0.00	0.00	0.08	0.00	0.00	0.00	0.03
951	23.00	43.33	0.95	0.03	0.00	0.00	0.00	0.00	0.00	0.00	0.00	0.00	0.00	0.00
949	23.00	43.34	0.94	0.03	0.00	0.00	0.00	0.00	0.00	0.00	0.00	0.00	0.00	0.01
950	23.00	43.39	0.85	0.01	0.00	0.00	0.00	0.00	0.03	0.09	0.00	0.00	0.00	0.02
741	22.96	43.07	0.99	0.01	0.00	0.00	0.00	0.00	0.00	0.00	0.00	0.00	0.00	0.00
702	22.95	42.03	0.09	0.05	0.02	0.00	0.00	0.00	0.00	0.12	0.00	0.01	0.01	0.69
885	22.95	42.95	1.00	0.00	0.00	0.00	0.00	0.00	0.00	0.00	0.00	0.00	0.00	0.00
701	22.93	42.03	0.15	0.03	0.01	0.00	0.00	0.00	0.17	0.00	0.00	0.02	0.00	0.62
715	22.93	42.31	0.97	0.02	0.00	0.00	0.00	0.00	0.00	0.00	0.00	0.00	0.00	0.00
366	22.06	41.06	0.99	0.00	0.00	0.00	0.00	0.00	0.01	0.00	0.00	0.00	0.00	0.00
361	21.69	41.02	0.95	0.02	0.00	0.00	0.00	0.00	0.00	0.02	0.00	0.00	0.00	0.00
360	21.63	41.01	0.83	0.04	0.04	0.00	0.00	0.00	0.00	0.00	0.00	0.00	0.00	0.07
359	21.58	41.06	0.80	0.08	0.02	0.02	0.00	0.00	0.00	0.00	0.00	0.02	0.00	0.05
632	20.65	40.47	0.92	0.00	0.00	0.00	0.00	0.00	0.00	0.08	0.00	0.00	0.00	0.00
633	20.65	40.47	0.30	0.00	0.00	0.00	0.00	0.00	0.00	0.00	0.00	0.00	0.00	0.70
653	20.25	40.19	0.15	0.00	0.00	0.00	0.00	0.00	0.00	0.00	0.00	0.00	0.00	0.85
711	19.60	39.80	0.90	0.02	0.00	0.00	0.00	0.00	0.06	0.00	0.00	0.00	0.00	0.01
710	19.59	39.80	0.91	0.04	0.01	0.01	0.00	0.00	0.00	0.00	0.00	0.00	0.01	0.01
752	17.58	39.20	0.95	0.02	0.01	0.00	0.00	0.00	0.00	0.00	0.00	0.00	0.00	0.01
751	17.57	39.21	0.99	0.00	0.00	0.00	0.00	0.00	0.00	0.00	0.00	0.00	0.00	0.00
1160	17.54	39.22	0.94	0.04	0.01	0.00	0.00	0.00	0.00	0.00	0.00	0.00	0.00	0.00
686	14.81	39.05	0.89	0.02	0.01	0.00	0.00	0.00	0.00	0.00	0.00	0.00	0.00	0.08
687	12.99	38.25	0.81	0.00	0.00	0.00	0.00	0.00	0.00	0.19	0.00	0.00	0.00	0.00
688	12.98	38.25	0.80	0.00	0.00	0.00	0.00	0.00	0.00	0.20	0.00	0.00	0.00	0.00
654	12.02	38.36	0.56	0.01	0.02	0.00	0.00	0.00	0.00	0.35	0.01	0.00	0.01	0.03
339	10.58	36.78	0.97	0.00	0.00	0.00	0.00	0.01	0.01	0.00	0.00	0.00	0.01	0.01
340	10.48	36.53	0.95	0.01	0.00	0.00	0.00	0.00	0.00	0.00	0.00	0.00	0.00	0.03
341	10.47	36.41	0.96	0.01	0.00	0.00	0.00	0.00	0.00	0.00	0.00	0.00	0.00	0.03
342	10.40	36.28	0.98	0.01	0.00	0.00	0.00	0.00	0.00	0.00	0.00	0.00	0.00	0.01
343	10.38	36.26	0.95	0.01	0.02	0.00	0.00	0.00	0.00	0.00	0.00	0.00	0.00	0.01
344	10.32	36.25	0.97	0.01	0.00	0.00	0.00	0.00	0.00	0.00	0.00	0.00	0.00	0.02
666	8.08	34.86	0.19	0.03	0.01	0.00	0.00	0.00	0.00	0.75	0.00	0.00	0.00	0.02
709	8.08	34.86	0.84	0.01	0.00	0.00	0.00	0.00	0.00	0.13	0.01	0.00	0.00	0.01
713	7.10	34.78	0.86	0.00	0.00	0.00	0.00	0.00	0.00	0.14	0.00	0.00	0.00	0.00
712	7.00	34.80	0.81	0.00	0.01	0.00	0.00	0.00	0.00	0.17	0.00	0.00	0.01	0.00
669	6.08	35.10	0.97	0.01	0.01	0.00	0.00	0.00	0.00	0.00	0.00	0.00	0.00	0.00
670	6.06	35.10	0.96	0.02	0.01	0.00	0.00	0.00	0.00	0.00	0.00	0.00	0.00	0.01
671	6.05	35.11	0.84	0.01	0.01	0.00	0.00	0.00	0.00	0.00	0.00	0.00	0.00	0.13
673	6.02	35.13	0.95	0.03	0.00	0.00	0.00	0.00	0.00	0.00	0.00	0.00	0.00	0.01
672	5.99	35.11	0.99	0.00	0.01	0.00	0.00	0.00	0.00	0.00	0.00	0.00	0.00	0.00
674	5.98	35.12	0.96	0.02	0.01	0.00	0.00	0.00	0.00	0.00	0.00	0.00	0.01	0.00
675	5.96	35.16	0.94	0.03	0.01	0.00	0.00	0.00	0.00	0.00	0.00	0.00	0.00	0.01
676	5.94	35.15	0.88	0.00	0.00	0.00	0.00	0.00	0.00	0.00	0.00	0.00	0.00	0.11
677	5.87	35.18	1.00	0.00	0.00	0.00	0.00	0.00	0.00	0.00	0.00	0.00	0.00	0.00
678	5.80	35.18	0.99	0.00	0.00	0.00	0.00	0.00	0.00	0.00	0.00	0.00	0.00	0.01
679	5.78	35.19	0.98	0.01	0.00	0.00	0.00	0.00	0.00	0.00	0.00	0.00	0.00	0.00
680	5.77	35.19	0.98	0.00	0.00	0.00	0.00	0.00	0.00	0.00	0.00	0.00	0.00	0.02
681	5.76	35.19	0.98	0.00	0.00	0.00	0.00	0.00	0.00	0.00	0.00	0.00	0.00	0.01
682	5.75	35.20	0.96	0.01	0.01	0.00	0.00	0.00	0.00	0.00	0.00	0.00	0.00	0.01
683	5.74	35.20	0.96	0.03	0.01	0.00	0.00	0.00	0.00	0.00	0.00	0.00	0.00	0.00
684	5.69	35.21	0.98	0.01	0.00	0.00	0.00	0.00	0.00	0.00	0.00	0.00	0.00	0.01
1148	4.96	37.07	0.71	0.15	0.04	0.00	0.00	0.00	0.00	0.05	0.00	0.00	0.03	0.01
708	4.88	37.27	0.84	0.05	0.04	0.00	0.00	0.00	0.00	0.07	0.00	0.00	0.00	0.00
886	4.26	38.50	0.91	0.03	0.01	0.00	0.00	0.00	0.00	0.00	0.00	0.00	0.00	0.04
1146	4.25	38.02	0.84	0.02	0.00	0.00	0.00	0.00	0.00	0.02	0.00	0.00	0.01	0.11
1150	4.13	38.13	0.85	0.10	0.01	0.00	0.00	0.00	0.00	0.02	0.00	0.01	0.00	0.01
690	3.73	38.52	0.89	0.07	0.02	0.00	0.00	0.00	0.00	0.00	0.00	0.00	0.00	0.01
720	2.90	41.23	0.99	0.01	0.00	0.00	0.00	0.00	0.00	0.00	0.00	0.00	0.00	0.00
685	2.90	41.64	0.83	0.02	0.00	0.00	0.00	0.00	0.00	0.09	0.00	0.00	0.00	0.06
692	2.53	44.27	0.98	0.00	0.01	0.00	0.00	0.00	0.00	0.00	0.00	0.00	0.00	0.00
691	2.15	44.28	0.96	0.01	0.00	0.00	0.00	0.00	0.00	0.00	0.00	0.00	0.00	0.01
754	1.73	46.60	0.96	0.02	0.01	0.00	0.00	0.00	0.00	0.00	0.00	0.00	0.00	0.00
755	1.66	46.18	0.95	0.01	0.00	0.00	0.00	0.00	0.00	0.00	0.00	0.00	0.00	0.02
753	1.55	47.01	0.94	0.02	0.01	0.00	0.00	0.00	0.00	0.00	0.00	0.02	0.00	0.01
853	0.82	46.77	0.99	0.00	0.00	0.00	0.00	0.00	0.00	0.00	0.00	0.00	0.00	0.00
638	0.81	45.77	0.99	0.00	0.00	0.00	0.00	0.00	0.00	0.00	0.00	0.00	0.00	0.01
373	0.63	46.28	0.99	0.00	0.00	0.00	0.00	0.00	0.00	0.00	0.00	0.00	0.00	0.01
513	0.33	48.28	0.96	0.01	0.00	0.00	0.00	0.00	0.00	0.00	0.00	0.00	0.00	0.00

Note: Potter (1993). See Table 7 for grain-type codes in the column headings.

TABLE 21. PETROGRAPHIC CATEGORIES FOR 200-POINT COUNTS OF FRAMEWORK GRAINS IN POTTER (1993)

Petrographic variable	State
Qu	Unit quartz
Qc	Composite quartz
Qt = Qu+Qc	Total quartz
Fk	Potash feldspar
Fp	Plagioclase feldspar
Ft = Fk + Fp	Total feldspar
RFt	Total rock fragments
C	Chert
M2	Micas
HM	Heavy minerals

TABLE 22. PETROGRAPHIC CATEGORIES FOR 100-POINT COUNTS OF ROCK FRAGMENTS IN POTTER (1993)

Petrographic variable	State
P	Plutonic rock fragment
M	Metamorphic rock fragment
V	Volcanic rock fragment
Ar	Argillaceous rock fragment
Ca	Carbonate rock fragment
Sa	Sandstone rock fragment
Si	Siltstone rock fragment
A	Alterites

TABLE 23. TRANSLATION SCHEME FOR POTTER TO SANDGEM PETROGRAPHIC CATEGORIES

Table 18 categories	Potter categories (Tables 19 and 20; Potter, 1993)	SandGEM categories (Table 7)	Common description
Q	Qt	Q	Total quartz (not including chert)
Fk	Fk	Fk	Alkali feldspar
Fp	Fp	Fpna + Fpca	Plagioclase feldspar
RP	RFt * P/sum(P...A)	RPp + RPk + RPh	Plutonic rock fragments
RV	RFt * V/sum(P...A)	RVb + RVi + RVs	Volcanic rock fragments
RSs	RFt * Sa/sum(P...A)	RSq + RSf + RSl	Sandstone rock fragments
RSsh	RFt * (Ar + Si)/sum(P...A)	RSsh	Mudstone rock fragments
RScb	RFt * Ca/sum(P...A)	RScb	Carbonate rock fragments (incl. bioclasts)
RSct	C	RSct	Chert
RM	RFt * M/sum(P...A)	RMqf + RMsh + RMc + RMpsc + RMgp + RMgk	Metamorphic rock fragments
OM	M2	OMb + OMw	Micas
OH	H	OH	"Heavy" (dense) minerals

Note: SandGEM—sand generation and evolution model.

TABLE 24. PROVENANCE LITHOTYPE ABUNDANCE IN BRAZILIAN SHIELD

Provenance lithotype	Relative areal abundance
Plutonic, ultrabasic	0.002
Plutonic, basic, gabbro	0.002
Plutonic, basic, diorite	0.000
Plutonic, intermediate	0.039
Plutonic, silicic	0.039
Plutonic, sodic	0.001
Plutonic, potassic	0.001
Volcanic, basic	0.003
Volcanic, intermediate	0.005
Volcanic, silicic	0.004
Sedimentary, sandstone, quartz-dominant	0.086
Sedimentary, sandstone, feldspar-dominant	0.028
Sedimentary, sandstone, lithic-dominant	0.019
Sedimentary, shale	0.053
Sedimentary, carbonate	0.004
Metamorphic, metasandstone	0.128
Metamorphic, slate	0.062
Metamorphic, metacarbonate	0.034
Metamorphic, schist/phyllite	0.246
Metamorphic, gneiss, plagioclase-rich	0.107
Metamorphic, gneiss, K-feldspar-rich	0.139

Note: Measured from Choubert et al. (1976, sheet 4).

TABLE 25. ENVIRONMENTAL PARAMETERS FOR BRAZILIAN BEACH SAND PREDICTION

Input node	State
Climate weathering potential	Moderate
Climate flushing potential	Moderate
Regional topographic gradient	Low
Hinterland transport distance	Long
Basin fluvial transport	Intermediate
Basin subsidence	Slow
Depositional facies	Shoreface

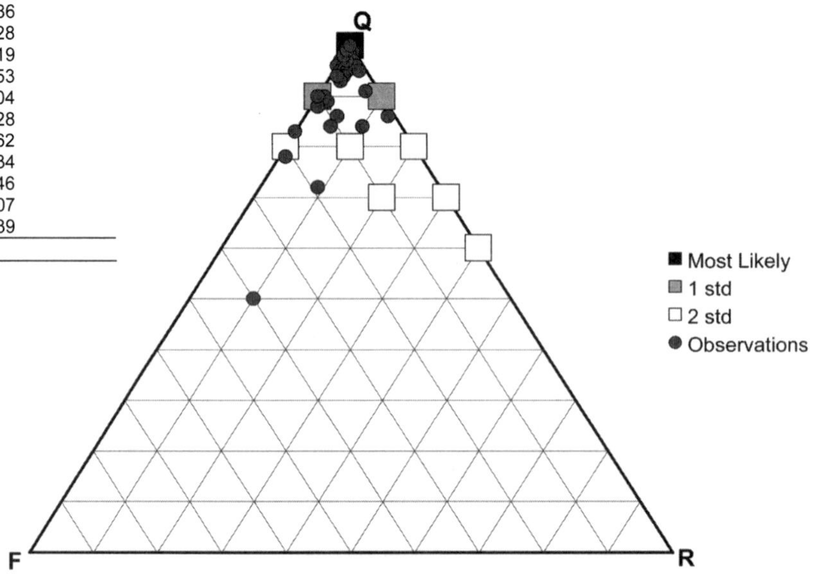

Figure 13. Predicted and observed values for Brazilian beach sand composition in quartz–feldspar–rock fragment (QFR) ternary space (std—standard deviation).

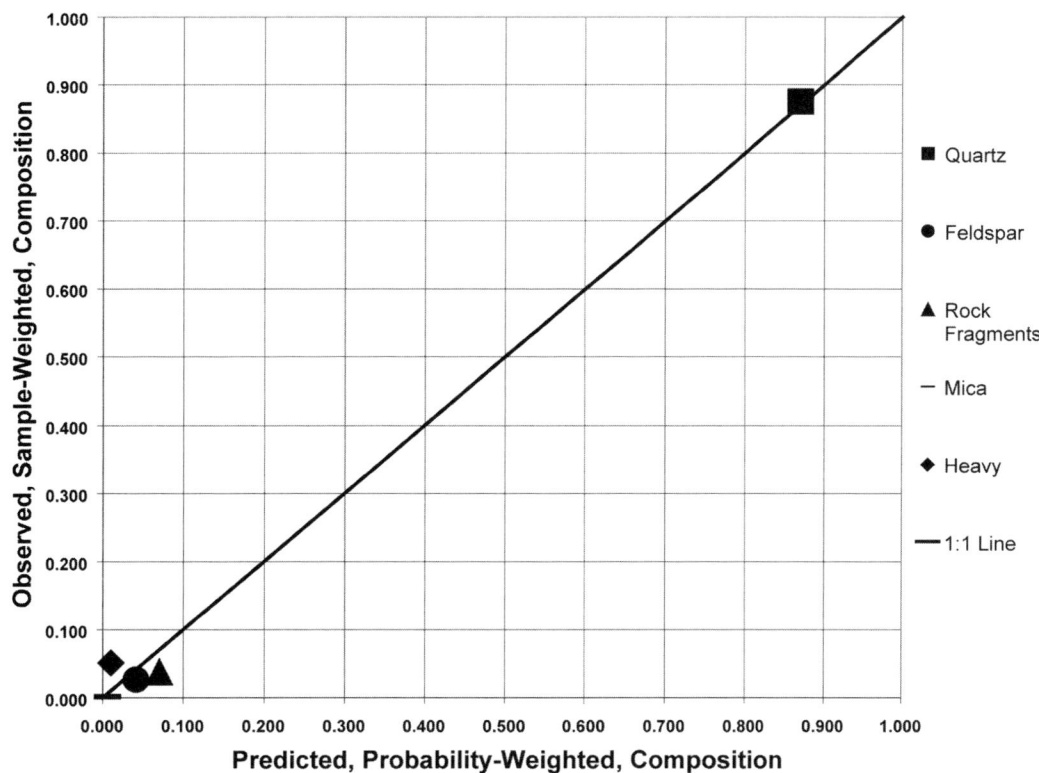

Figure 14. Plot of predicted versus observed relative abundance of grain types for Brazilian beach sand. The *x*-axis values have been weighted by the probability of each predicted composition. The *y*-axis values represent the arithmetic average of values for all samples.

TABLE 26. SANDGEM PREDICTION FOR BRAZILIAN BEACH SAND

ID	$P()^\dagger$	$\Sigma P()^\ddagger$	Q	Fk	Fp	RP	RV	RSs	RSsh	RScb	RSct	RM	OM	OH
66	0.4330	0.4330	0.98	0.01	0.00	0.00	0.00	0.00	0.00	0.00	0.00	0.00	0.00	0.00
65	0.1486	0.5816	0.90	0.08	0.01	0.00	0.00	0.00	0.00	0.00	0.00	0.00	0.00	0.00
55	0.1410	0.7227	0.87	0.01	0.00	0.01	0.00	0.01	0.00	0.00	0.00	0.08	0.01	0.01
45	0.0679	0.7905	0.75	0.01	0.00	0.01	0.01	0.01	0.01	0.00	0.00	0.15	0.02	0.03
64	0.0555	0.8460	0.81	0.14	0.04	0.00	0.00	0.00	0.00	0.00	0.00	0.01	0.00	0.00
36	0.0369	0.8830	0.65	0.01	0.01	0.02	0.01	0.02	0.01	0.00	0.00	0.21	0.03	0.04
54	0.0191	0.9020	0.78	0.08	0.01	0.01	0.00	0.01	0.00	0.00	0.00	0.08	0.01	0.01
28	0.0155	0.9175	0.56	0.01	0.01	0.03	0.01	0.02	0.01	0.00	0.00	0.28	0.03	0.04
44	0.0151	0.9326	0.98	0.01	0.00	0.00	0.00	0.00	0.00	0.00	0.00	0.00	0.00	0.00

Note: SandGEM—sand generation and evolution model. See Table 7 for grain-type codes in the column headings.
$^\dagger P()$ is the individual probability of a predicted composition.
$^\ddagger \Sigma P()$ is the cumulative probability of this composition and all compositions that are more likely.

TABLE 27. COMPARISON OF PREDICTION AND OBSERVATION FOR BRAZILIAN BEACH SAND

	Predicted average (weighted by probability)	Observed average (each sample equally weighted)
Quartz content	0.870	0.875
Feldspar content	0.042	0.026
Rock-fragment content	0.070	0.043
Mica content	0.006	0.001
Heavy mineral content	0.010	0.051
Average plagioclase/alkali feldspar	0.223	0.240

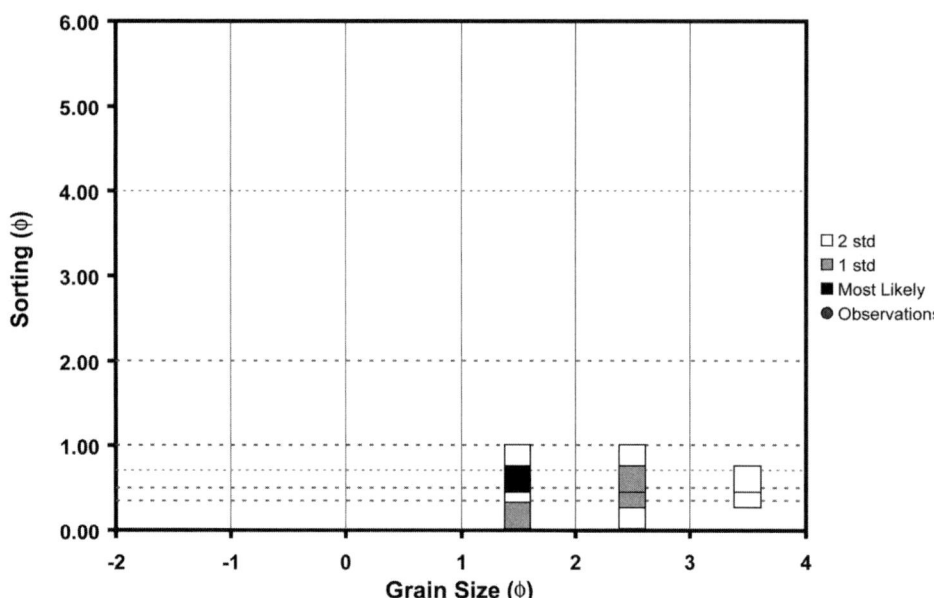

Figure 15. Predicted sand texture for Brazilian beach sand. There are no published measurements to compare with the prediction, but Potter (1986) reported that the sands generally were "medium to fine grained."

TABLE 28. OBSERVED CHILEAN FLUVIAL SAND COMPOSITION

ID	Lat (°S)	Long (°W)	Q	Fk	Fp	RP	RV	RSs	RSsh	RScb	RSct	RM	OM	OH
791	72.70	37.24	0.00	0.00	0.02	0.14	0.73	0.00	0.00	0.04	0.00	0.05	0.00	0.02
837	72.09	36.54	0.09	0.02	0.11	0.00	0.66	0.00	0.00	0.00	0.00	0.00	0.01	0.11
842	71.79	36.24	0.11	0.05	0.06	0.05	0.60	0.00	0.00	0.00	0.00	0.05	0.01	0.07
840	71.72	36.00	0.02	0.00	0.12	0.00	0.72	0.00	0.00	0.00	0.00	0.00	0.00	0.14
839	71.29	35.04	0.00	0.00	0.12	0.01	0.70	0.00	0.00	0.00	0.00	0.00	0.02	0.15
844	71.15	32.82	0.02	0.01	0.09	0.00	0.80	0.00	0.00	0.00	0.00	0.01	0.01	0.06
792	70.73	33.68	0.01	0.00	0.04	0.06	0.61	0.00	0.03	0.07	0.00	0.11	0.00	0.05

Note: From Potter (1993). See Table 7 for grain-type codes in the column headings.

TABLE 29. OBSERVED CHILEAN FLUVIAL SAND TEXTURE

ID	Mean grain size (ϕ)[†]	Sorting (ϕ)[†]
791	−0.471	0.358
837	0.776	0.425
842	0.345	0.357
840	0.835	0.540
839	0.156	0.509
844	1.514	0.485
792	0.929	0.451

Note: Measurements are from point-count data by M.K. DeSantis, University of Cincinnati, 2005.
[†]Mean grain size is quantified by graphic mean. Sorting is quantified by inclusive graphic standard deviation (Folk, 1974, p. 45–46).

the sand texture, but the predicted texture (Fig. 15) corresponds to the qualitative description of grain size provided by Potter (1986).

Chilean Fluvial Sand

Observations

The Chilean fluvial sands were selected from Potter's samples on rivers that have their headwaters in the Andes, between 32 and 38°S; all the samples were taken in the Pampas Central, so the sands do not reflect any contribution from the Cordillera Costera. The compositional data from Potter (1993) are presented in Table 28. Grain-size data (collected for ExxonMobil by M.K. DeSantis at the University of Cincinnati) are presented in Table 29.

Environmental Conditions

The provenance lithotype assemblage from which these sands were derived is composed (in order of importance) of Cenozoic volcanics (~80%), Mesozoic plutons (~10%), and Mesozoic volcanic sediments (~10%). The precise details are quantified in Table 30. The values in Table 30 were measured by planimetry in Choubert et al. (1976); the precise abundance of lithotypes implied by each (litho)chronostratigraphic unit of the map was calibrated in selected areas with information from Zeil (1964).

The environmental conditions that governed the generation and evolution of these sediments, as quantified by SandGEM input parameters, are summarized in Table 31. The region is characterized by low weathering power, moderate flushing power, short transport distance over a high topographic gradient, and a small, slowly subsiding basin with fluvial depositional facies. All of these factors tend to favor highly immature sediment.

Prediction versus Observation

There is good agreement between prediction and observation for sand composition (Figs. 16 and 17; Tables 28, 32, and 33). Quartz and feldspar contents are predicted to be subequal and each below 10%. Rock fragments are predicted to be ~78%, with

TABLE 30. PROVENANCE LITHOTYPE ABUNDANCE IN THE CHILEAN ANDES

Provenance lithotype	Relative areal abundance
Plutonic, ultrabasic	0.000
Plutonic, basic, gabbro	0.000
Plutonic, basic, diorite	0.000
Plutonic, intermediate	0.064
Plutonic, silicic	0.027
Plutonic, sodic	0.000
Plutonic, potassic	0.000
Volcanic, basic	0.450
Volcanic, intermediate	0.377
Volcanic, silicic	0.080
Sedimentary, sandstone, quartz-dominant	0.000
Sedimentary, sandstone, feldspar-dominant	0.000
Sedimentary, sandstone, lithics-dominant	0.000
Sedimentary, shale	0.000
Sedimentary, carbonate	0.000
Metamorphic, metasandstone	0.000
Metamorphic, slate	0.000
Metamorphic, metacarbonate	0.000
Metamorphic, schist/phyllite	0.000
Metamorphic, gneiss, plagioclase-rich	0.000
Metamorphic, gneiss, K-feldspar-rich	0.000

Note: Measured from Choubert et al. (1976, sheet 4).

TABLE 31. ENVIRONMENTAL PARAMETERS FOR CHILEAN RIVER SAND PREDICTION

Input node	State
Climate weathering potential	Low
Climate flushing potential	Moderate
Regional topographic gradient	High
Hinterland transport distance	Short
Basin fluvial transport	Short
Basin subsidence	Slow
Depositional facies	Fluvial major channel

virtually all rock fragments being volcanic. Plagioclase is predicted to be about twice as abundant as alkali feldspar. In fact, the observed sand composition conforms very closely to these predictions. The single most likely predicted sand texture (Fig. 18, Table 34) is medium-grained, moderately well sorted. In fact, every observation (Fig. 18; Table 29) fits this description.

Australian Alluvial Sand

Observations

The Australian dryland sands were collected from the drainage of Umbum Creek, which heads in the Davenport Ranges and deposits sand into Lake Eyre, South Australia (Reilly et al., 2003). Petrographic data from Menacherry (2006) are presented in Table 35.

Environmental Conditions

The provenance lithotype assemblage from which these sands were derived is composed (in order of importance) of Cenozoic sands and silicretes, Mesozoic sandstones, and Proterozoic metasediments, volcanics, and plutons (Menacherry, 2006). The precise details are quantified in Table 36. The environmental conditions that governed the generation and evolution of these sediments, as quantified by SandGEM input parameters, are summarized in Table 37. The region is characterized by low weathering and flushing power, short transport distance over a low topographic gradient, and a small, slowly subsiding basin with confined alluvial depositional facies.

Prediction versus Observation: Sand Composition

There is good agreement between predictions and observations (Figs. 19 and 20; Tables 35, 38, and 39). The sands are predicted to be composed of roughly two-thirds quartz and one-quarter rock fragments, dominated by sedimentary rock fragments. Among the feldspars, plagioclase is predicted to be 2.5

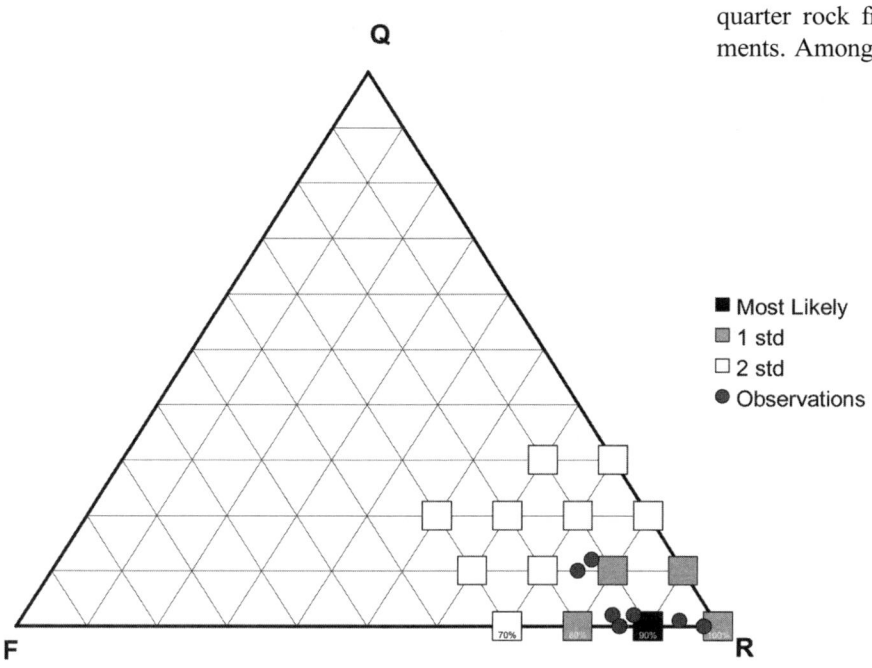

Figure 16. Predicted and observed values for Chilean fluvial sand composition in quartz–feldspar–rock fragment (QFR) ternary space (std—standard deviation).

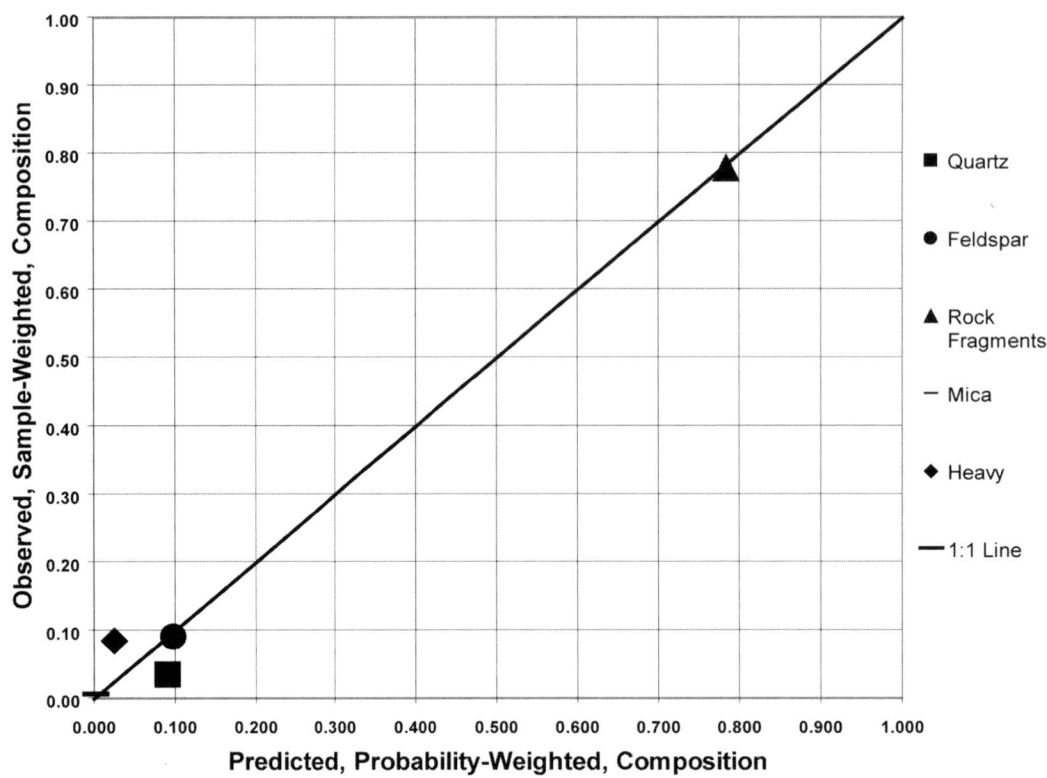

Figure 17. Plot of predicted versus observed relative abundance of grain types for Chilean fluvial sand. The *x*-axis values have been weighted by the probability of each predicted composition. The *y*-axis values represent the arithmetic average of values for all samples.

TABLE 32. SANDGEM PREDICTION FOR CHILEAN SAND COMPOSITION

ID	P()†	ΣP()‡	Q	Fk	Fp	RP	RV	RSs	RSsh	RScb	RSct	RM	OM	OH
2	0.3621	0.3621	0.04	0.03	0.04	0.03	0.86	0.00	0.00	0.00	0.00	0.00	0.00	0.01
5	0.1163	0.4783	0.10	0.04	0.04	0.04	0.71	0.00	0.00	0.00	0.00	0.00	0.00	0.06
3	0.1101	0.5884	0.09	0.00	0.00	0.02	0.87	0.00	0.00	0.00	0.00	0.00	0.00	0.02
4	0.0957	0.6841	0.03	0.06	0.13	0.02	0.76	0.00	0.00	0.00	0.00	0.00	0.00	0.00
1	0.0831	0.7671	0.03	0.00	0.00	0.03	0.94	0.00	0.00	0.00	0.00	0.00	0.00	0.00
6	0.0416	0.8088	0.19	0.00	0.00	0.03	0.73	0.00	0.00	0.00	0.00	0.00	0.00	0.04
9	0.0339	0.8426	0.18	0.04	0.04	0.03	0.58	0.00	0.00	0.00	0.00	0.00	0.01	0.11
8	0.0248	0.8675	0.09	0.07	0.11	0.04	0.62	0.00	0.00	0.00	0.00	0.00	0.00	0.06
7	0.0175	0.8850	0.02	0.10	0.17	0.02	0.69	0.00	0.00	0.00	0.00	0.00	0.00	0.00
12	0.0127	0.8977	0.06	0.09	0.23	0.02	0.58	0.00	0.00	0.00	0.00	0.00	0.00	0.03
10	0.0096	0.9073	0.26	0.00	0.01	0.03	0.58	0.00	0.00	0.00	0.00	0.00	0.01	0.11
14	0.0074	0.9147	0.26	0.05	0.04	0.03	0.51	0.00	0.00	0.00	0.00	0.00	0.01	0.10
13	0.0064	0.9211	0.21	0.09	0.11	0.02	0.57	0.00	0.00	0.00	0.00	0.00	0.00	0.01
18	0.0051	0.9262	0.19	0.11	0.17	0.01	0.45	0.00	0.00	0.00	0.00	0.00	0.00	0.06

Note: SandGEM—sand generation and evolution model. See Table 7 for grain-type codes in the column headings.
†*P*() is the individual probability of a predicted composition.
‡Σ*P*() is the cumulative probability of this composition and all compositions that are more likely.

TABLE 33. COMPARISON OF PREDICTION AND OBSERVATION FOR CHILEAN RIVER SAND COMPOSITION

	Predicted average (weighted by probability)	Observed average (each sample equally weighted)
Quartz content	0.090	0.036
Feldspar content	0.098	0.091
Rock-fragment content	0.784	0.777
Mica content	0.002	0.007
Heavy mineral content	0.025	0.086
Average plagioclase/alkali feldspar	2.491	2.243
Average volcanic/total rock fragments	0.946	0.899

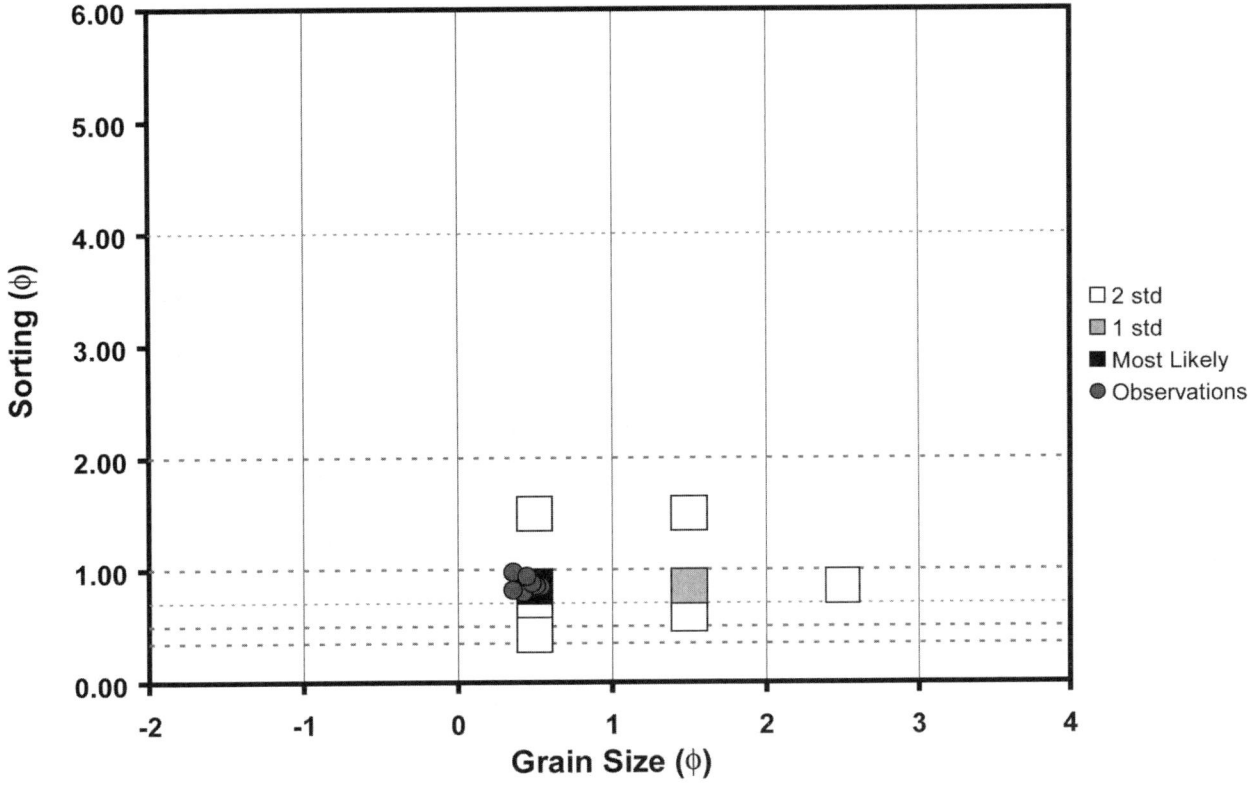

Figure 18. Predicted and observed sand texture for Chilean fluvial sand (std—standard deviation).

TABLE 34. SANDGEM PREDICTION FOR CHILEAN SAND TEXTURE

Mean grain size	Sorting	$P()$[†]	$\Sigma P()$[‡]
Coarse	Moderate	0.234	0.234
Medium	Moderate	0.218	0.452
Coarse	Moderately well	0.165	0.618
Medium	Moderately well	0.117	0.734
Coarse	Poorly	0.050	0.785
Medium	Poorly	0.049	0.834
Fine	Moderate	0.042	0.876
Coarse	Well	0.031	0.907
Fine	Moderately well	0.028	0.935

Note: SandGEM—sand generation and evolution model.
[†]$P()$ is the individual probability of a predicted composition.
[‡]$\Sigma P()$ is the cumulative probability of this composition and all compositions that are more likely.

times more abundant than alkali feldspar. The observations are very close to the predicted values, although plagioclase is actually only ~1.5 times more abundant than alkali feldspar.

Prediction versus Observation: Sand Texture

The single most likely sand texture is medium grained and moderately sorted, with an expected range of variation up to one sorting category better or worse, and one grain size finer (Table 40). All of the observations fall very close to the most likely verbal description, and well within the predicted range of variation (Fig. 21).

Summary of Predictions

The Sand Generation and Evolution Model (SandGEM) successfully predicts sand composition and texture in diverse environments that include: provenance lithotype assemblages dominated by either volcanic, high-grade metamorphic, or sedimentary rocks; tropical to desert climates; and drainage basins that range from local mountain catchments to continental scale. A probabilistic, forward modeling approach that synthesizes quantitative and qualitative understanding of sand generation and evolution processes gleaned from a wide range of disciplines has the ability to make accurate, quantitative predictions of sand character based on observations or estimates of a few key environmental features.

DISCUSSION

Benefits of Integrated Genetic Analysis for Sedimentary Petrology

Traditionally, sedimentary petrologists have taken an inductive approach in which they attempt to infer cause from effect (Pirsig, 1974). Typically, this means examining a restricted subset of a physical sample (e.g., medium sand), with one tool (e.g.,

TABLE 35. OBSERVED AUSTRALIAN ALLUVIAL SAND COMPOSITION AND TEXTURE

ID	Grain size (ϕ)	Sorting (ϕ)	Q	Fk	Fp	RP	RV	RS	RM	OH	OMb	OMw
1(1)	0.472	1.243	0.618	0.021	0.000	0.000	0.000	0.270	0.091	0.000	0.000	0.000
1(2)	1.094	0.986	0.768	0.008	0.000	0.000	0.000	0.179	0.045	0.000	0.000	0.000
2(1)	0.394	0.751	0.713	0.004	0.000	0.000	0.000	0.208	0.075	0.000	0.000	0.000
2(2)	0.609	0.897	0.720	0.008	0.008	0.000	0.000	0.214	0.049	0.000	0.000	0.000
3(1)	1.065	0.794	0.743	0.013	0.000	0.000	0.000	0.198	0.046	0.000	0.000	0.000
3(2)	1.052	1.157	0.671	0.004	0.084	0.000	0.000	0.203	0.038	0.000	0.000	0.000
4(1)	0.559	0.793	0.846	0.000	0.017	0.000	0.000	0.096	0.042	0.000	0.000	0.000
4(2)	0.443	0.924	0.808	0.008	0.000	0.000	0.000	0.142	0.042	0.000	0.000	0.000
6(1)	0.956	0.748	0.860	0.000	0.000	0.000	0.009	0.077	0.055	0.000	0.000	0.000
6(2)	0.874	0.780	0.913	0.026	0.004	0.000	0.000	0.017	0.039	0.000	0.000	0.000
7(1)	0.623	0.785	0.740	0.008	0.000	0.000	0.000	0.132	0.116	0.000	0.000	0.004
7(2)	0.705	0.858	0.757	0.016	0.004	0.000	0.000	0.152	0.070	0.000	0.000	0.000
8(1)	0.860	0.663	0.799	0.004	0.012	0.000	0.000	0.078	0.107	0.000	0.000	0.000
8(2)	0.830	0.812	0.805	0.000	0.004	0.000	0.000	0.087	0.104	0.000	0.000	0.000
9(1)	0.585	0.831	0.821	0.000	0.013	0.000	0.004	0.129	0.033	0.000	0.000	0.000
9(2)	0.560	0.341	0.602	0.008	0.012	0.000	0.016	0.164	0.197	0.000	0.000	0.000
10(1)	0.468	0.758	0.823	0.000	0.000	0.000	0.012	0.095	0.070	0.000	0.000	0.000
10(2)	0.773	0.738	0.781	0.008	0.000	0.000	0.004	0.166	0.040	0.000	0.000	0.000
11(1)	1.152	1.035	0.779	0.004	0.000	0.000	0.000	0.169	0.048	0.000	0.000	0.000
11(2)	0.619	0.746	0.727	0.004	0.004	0.000	0.000	0.200	0.065	0.000	0.000	0.000
12(1)	0.820	0.870	0.579	0.000	0.004	0.000	0.004	0.153	0.260	0.000	0.000	0.000
12(2)	1.474	0.641	0.634	0.009	0.000	0.000	0.017	0.153	0.187	0.000	0.000	0.000
13(1)	0.434	0.947	0.824	0.000	0.000	0.000	0.000	0.084	0.092	0.000	0.000	0.000
13(2)	1.102	0.812	0.547	0.008	0.004	0.000	0.025	0.214	0.202	0.000	0.000	0.000
14(1)	1.203	1.068	0.543	0.004	0.024	0.000	0.012	0.150	0.263	0.000	0.000	0.004
14(2)	0.754	1.069	0.439	0.000	0.016	0.000	0.004	0.102	0.439	0.000	0.000	0.000
15(1)	0.974	0.876	0.715	0.017	0.017	0.000	0.000	0.174	0.077	0.000	0.000	0.000
15(2)	0.880	0.900	0.794	0.000	0.004	0.000	0.000	0.086	0.115	0.000	0.000	0.000

Note: See Table 7 for grain-type codes in the column headings.

TABLE 36. PROVENANCE LITHOTYPE ABUNDANCE IN THE UMBUM CREEK DRAINAGE, AUSTRALIA

Provenance lithotype	Relative areal abundance
Plutonic, ultrabasic	0.000
Plutonic, basic, gabbro	0.000
Plutonic, basic, diorite	0.000
Plutonic, intermediate	0.002
Plutonic, silicic	0.001
Plutonic, sodic	0.000
Plutonic, potassic	0.001
Volcanic, basic	0.006
Volcanic, intermediate	0.003
Volcanic, silicic	0.002
Sedimentary, sandstone, quartz-dominant	0.591
Sedimentary, sandstone, feldspar-dominant	0.000
Sedimentary, sandstone, lithics-dominant	0.003
Sedimentary, shale	0.333
Sedimentary, carbonate	0.026
Metamorphic, metasandstone	0.029
Metamorphic, slate	0.001
Metamorphic, metacarbonate	0.000
Metamorphic, schist/phyllite	0.001
Metamorphic, gneiss, plagioclase-rich	0.001
Metamorphic, gneiss, K-feldspar-rich	0.001

TABLE 37. ENVIRONMENTAL PARAMETERS FOR AUSTRALIAN ALLUVIAL SAND PREDICTION

Input node	State
Climate weathering potential	Low
Climate flushing potential	Low
Regional topographic gradient	High
Hinterland transport distance	Short
Basin fluvial transport	Short
Basin subsidence	Slow
Depositional facies	Alluvial, confined

Gazzi-Dickinson point count, zircon age-spectrum, chemical analysis, etc.) to infer one aspect of provenance sensu latu (e.g., tectonic setting). This is a sensible and economical approach for a broad range of problems, but there is an even broader range of problems for which this approach is inadequate; a short list of such problems in the field of reservoir quality was enumerated at the beginning of this article. For these problems, a deductive approach (inference of effect from cause) is required, because the physical sample is not available or not fully representative.

Integrated genetic analysis provides a framework for conducting more comprehensive, inductive studies, and it also provides a framework within which to integrate observations from restricted deductive studies. Bayesian networks, or other probabilistic quantitative tools, provide a productive avenue to investigate complex systems that previously have been considered intractable.

Productive Lines of Future Research

Climatic Effects on Weathering Intensity

Although intensive research has been focused on mineral dissolution in natural environments (e.g., White et al., 2001) and the laboratory (e.g., Arvidson et al., 2004), there is still a big gap in understanding how climate influences chemical work done on the landscape at large space and time scales, either directly through kinetic controls, or indirectly through influence on biologic controls (e.g., Anderson et al., 2004). For example, it is not intuitively obvious under what climatic condition the most intensive chemical weathering will occur; high temperatures

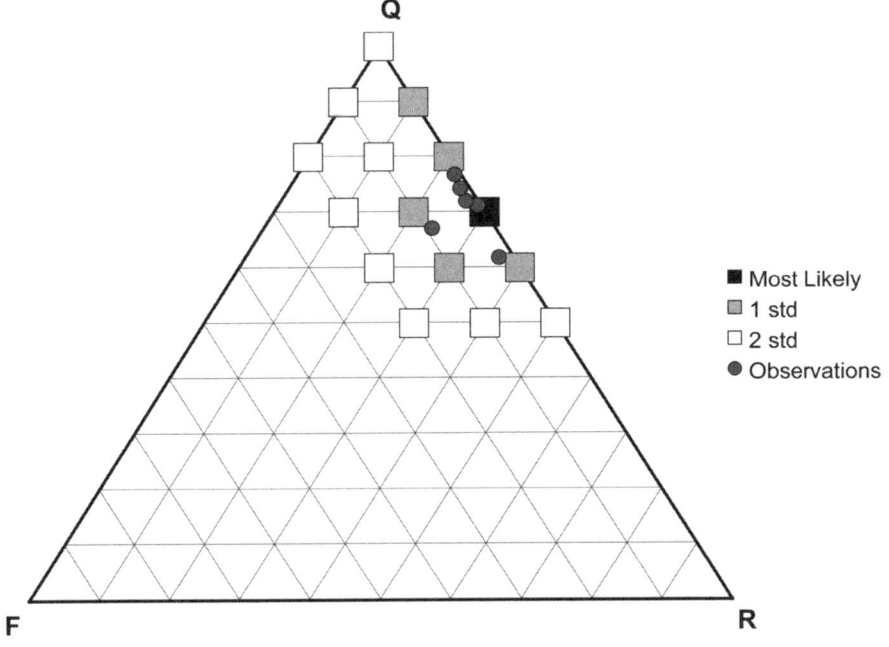

Figure 19. Predicted and observed values for Australian alluvial sand composition in quartz–feldspar–rock fragment (QFR) ternary space (std—standard deviation).

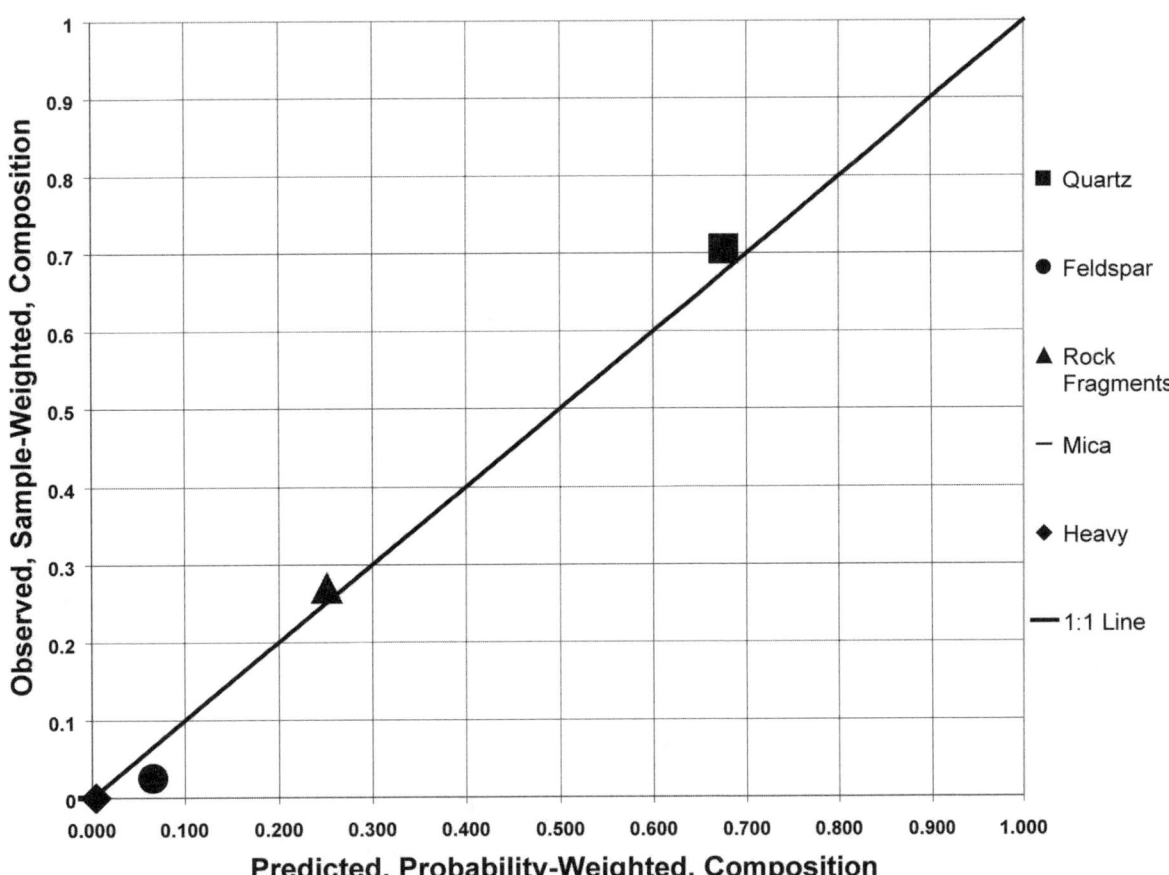

Figure 20. Plot of predicted versus observed relative abundance of grain types for Australian alluvial sand. The *x*-axis values have been weighted by the probability of each predicted composition. The *y*-axis values represent the arithmetic average of values for all samples.

TABLE 38. SANDGEM PREDICTION FOR AUSTRALIAN ALLUVIAL SAND COMPOSITION

ID	P()[†]	ΣP()[‡]	Q	Fk	Fp	RP	RV	RSs	RSsh	RScb	RSct	RM	OM	OH
36	0.3002	0.3002	0.68	0.01	0.00	0.00	0.01	0.17	0.11	0.00	0.00	0.00	0.00	0.01
45	0.1331	0.4333	0.78	0.01	0.00	0.00	0.01	0.12	0.08	0.00	0.00	0.00	0.00	0.00
44	0.0946	0.5279	0.70	0.08	0.00	0.00	0.01	0.12	0.08	0.00	0.00	0.00	0.00	0.00
35	0.0809	0.6087	0.61	0.08	0.00	0.00	0.01	0.17	0.11	0.00	0.00	0.00	0.00	0.01
28	0.0772	0.6859	0.60	0.00	0.00	0.00	0.02	0.20	0.16	0.00	0.00	0.00	0.00	0.00
55	0.0594	0.7453	0.88	0.01	0.00	0.00	0.00	0.06	0.04	0.00	0.00	0.00	0.00	0.00
54	0.0437	0.7889	0.80	0.09	0.00	0.00	0.00	0.06	0.04	0.00	0.00	0.00	0.00	0.00
53	0.0260	0.8150	0.72	0.17	0.01	0.00	0.01	0.05	0.03	0.00	0.00	0.00	0.00	0.00
43	0.0259	0.8408	0.60	0.18	0.02	0.00	0.01	0.11	0.07	0.00	0.00	0.00	0.00	0.00
21	0.0191	0.8599	0.52	0.00	0.00	0.00	0.02	0.20	0.24	0.00	0.00	0.01	0.00	0.00
34	0.0179	0.8778	0.51	0.17	0.02	0.00	0.01	0.16	0.12	0.00	0.00	0.00	0.00	0.00
27	0.0146	0.8924	0.51	0.09	0.01	0.00	0.02	0.20	0.17	0.00	0.00	0.01	0.00	0.00
65	0.0134	0.9059	0.91	0.09	0.00	0.00	0.00	0.00	0.00	0.00	0.00	0.00	0.00	0.00
66	0.0121	0.9179	0.99	0.01	0.00	0.00	0.00	0.00	0.00	0.00	0.00	0.00	0.00	0.00
64	0.0107	0.9286	0.80	0.19	0.01	0.00	0.00	0.00	0.00	0.00	0.00	0.00	0.00	0.00

Note: SandGEM—sand generation and evolution model. See Table 7 for grain-type codes in the column headings.
[†]$P()$ is the individual probability of a predicted composition.
[‡]$\Sigma P()$ is the cumulative probability of this composition and all compositions that are more likely.

TABLE 39. COMPARISON OF PREDICTION AND OBSERVATION FOR AUSTRALIAN ALLUVIAL SAND

	Predicted average (weighted by probability)	Observed average (each sample equally weighted)
Quartz content	0.676	0.706
Feldspar content	0.066	0.025
Rock-fragment content	0.252	0.269
Mica content	0.001	0.000
Heavy mineral content	0.004	0.000
Average plagioclase/alkali feldspar	2.491	1.571
Average sedimentary/total rock fragments	0.912	0.787

TABLE 40. SANDGEM PREDICTION FOR AUSTRALIAN ALLUVIAL SAND TEXTURE

Mean grain size	Sorting	P()[†]	ΣP()[‡]
Coarse	Moderately	0.3499968	0.3500
Coarse	Moderately well	0.2012485	0.5512
Coarse	Poorly	0.1480473	0.6993
Medium	Moderately	0.100074	0.7994
Medium	Moderately well	0.0567723	0.8561
Medium	Poorly	0.0480111	0.9041
Coarse	Well	0.0444685	0.9486
Fine	Moderately	0.0162666	0.9649
Medium	Well	0.0127532	0.9776

Note: SandGEM—sand generation and evolution model.
[†]$P()$ is the individual probability of a predicted composition.
[‡]$\Sigma P()$ is the cumulative probability of this composition and all compositions that are more likely.

and high precipitation both favor increased weathering, but high temperatures also promote faster evaporation and transpiration, so that the amount of water actually available to do chemical work decreases with increasing temperature (Strakhov, 1967). The net effect of the balance between the physical and chemical work of water is also not intuitively obvious; more water should promote more intense chemical weathering, but it also provides more transport power to decrease the duration of weathering. On the other hand, more water may also promote greater vegetation to reduce erosion and increase the duration of weathering (Schumm, 1981). Greater understanding of the fundamental controls on chemical weathering intensity, which determines the initial trajectory of sand generation, will help clarify sand evolutionary processes.

Landscape Evolution

The environmental factors and processes that feed into the Fluvial Storage Potential and Selective Transport Fining nodes of SandGEM currently are addressed by the landscape evolution community (e.g., Gasparini et al., 2006; Hasbargen and Paola, 2003; Pazzaglia, 2004; Willett et al., 2003). Sedimentary petrologists can make great strides in predicting composition and texture (or inferring provenance features from composition and texture) by working on the influence of landscape evolution on the time it takes for sediment to progress from generation to ultimate deposition, and on the character of deposited sediment.

Phytogeographic Evolution

All modern observations of climatic effects on, and landscape interaction with, sediment generation and evolution are made in the context of a fully vegetated world. These observations are probably only relevant back to the early Oligocene and the widespread distribution of grasses. Other significant step-changes in sedimentary response to environmental forcing probably occurred in the Late Cretaceous (extensive colonization of uplands and riparian areas made possible by angiosperms) and Middle Devonian (extensive colonization of coastal lowlands by terrestrial plants). Significantly more work must be done on the paleo-effect of plants on sediment generation and transport (e.g., Fraticelli et al., 2004) before modern sand generation and evolution principles can be applied confidently to the Mesozoic or Paleozoic.

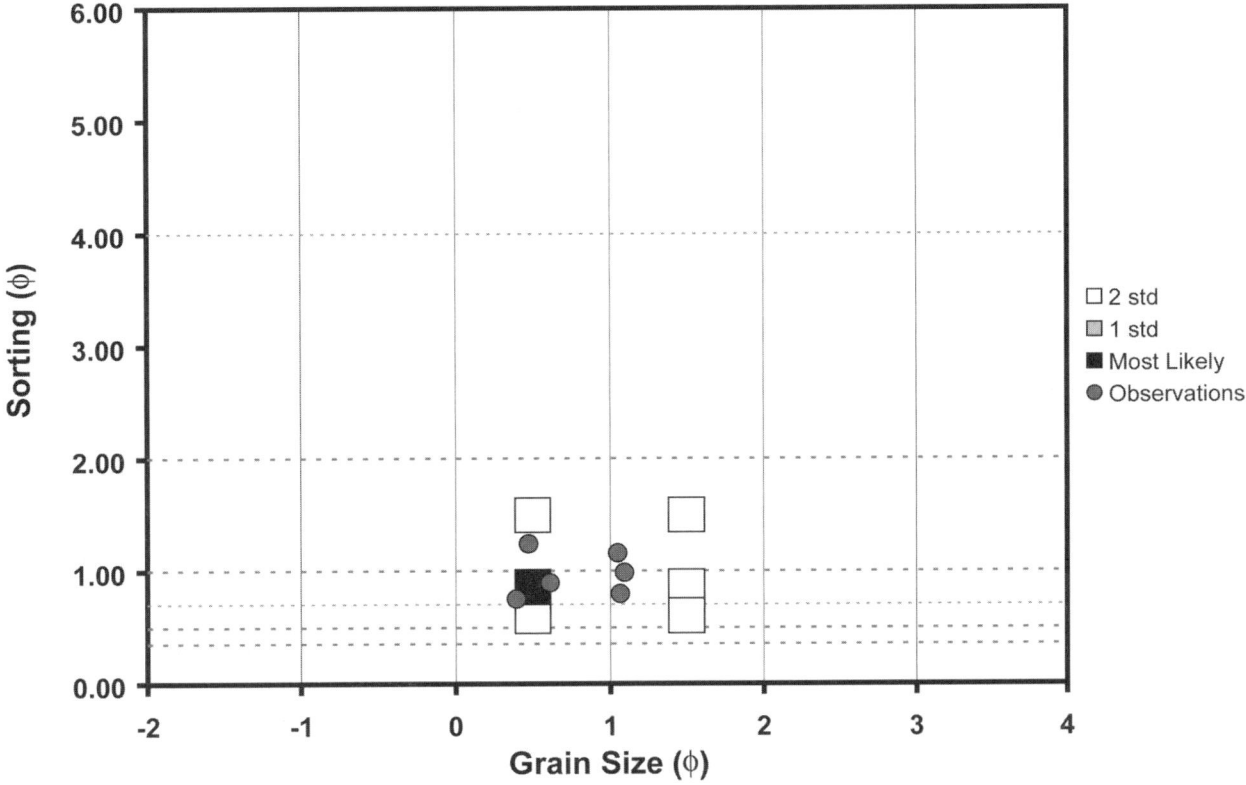

Figure 21. Predicted and observed sand texture for Australian alluvial sand.

APPENDIX A: CALCULATION OF WEATHERING POTENTIAL AND TRANSPORT POTENTIAL INDICES

Weathering Potential

The Weathering Potential Index (WPI) for a modern climate is calculated using mean monthly precipitation and temperature data. It is defined as:

$$WPI = \sum_{i=1}^{12}(P_i - E_i) * e^{-\frac{E_\theta}{R\bar{T}_i}} * 10^9, \quad (A\text{-}1)$$

where P_i = mean monthly precipitation for month i (mm); E_i = potential evaporation in month i (mm) calculated with Hamon's equation (Hamon, 1961; Haith and Shoemaker, 1987); E_θ = average activation energy of silicate mineral dissolution (J·mol^{-1}), with a default value of 48.6 (after Lasaga, 1984; approximately the value for feldspar), which replicates van't Hoff's rule (doubling of reaction rates for every 10 °C increase in temperature) at Earth-surface conditions; R = universal gas constant (J·mol^{-1}·°K^{-1}); and $\bar{T}_i$ = mean temperature for month i (°K)

$$E_i = \frac{2.1 H_i^2 e_s}{\bar{T}_i}, \quad (A\text{-}2)$$

where H_i = hours of sunlight on the Julian day in the middle of month i, and e_s = saturation vapor pressure at $\bar{T}_i$

$$H_i = \frac{24\omega_s}{\pi}, \quad (A\text{-}3)$$

where ω_s = sunset angle on the Julian day in the middle of month i.

$$e_s = k e^{\frac{a(\bar{T}_i - 273.16)}{\bar{T}_i - b}}, \quad (A\text{-}4)$$

where k, a, and b are constants, equal to 0.6112, 17.67, and 29.66, respectively (http://atmos.nmsu.edu/education_and_outreach/encyclopedia/sat_vapor_pressure.htm, verified 14 October 2005).

$$\omega_s = \arccos(-\tan\phi\tan\delta), \quad (A\text{-}5)$$

where ϕ = latitude of the site (+ in Northern Hemisphere, – in Southern Hemisphere) and δ = solar declination (in radians) on the Julian day J_i, at the middle of month i = 23.45 × sin [360/365 × (284 + J_i)].

Transport Potential

The Flushing Potential Index (FPI) for a modern climate is calculated using mean monthly precipitation and temperature data. It is defined as:

$$FPI = \log\left[\sum_{i=1}^{12} n_{i,flushing} \bigg/ \sum_{i=1}^{12} n_{i,non-flushing}\right], \quad (A\text{-}6)$$

where $n_{i,flushing}$ and $n_{i,nonflushing}$ are 1 or 0, depending if the effective precipitation in month i ($P_i - E_i$) is greater or less than a user-specified critical value. When $n_{i,flushing} = 0$, then $n_{i,nonflushing} = 1$, and vice versa. We typically set the critical value at 200 mm.

Example

The layout for a spreadsheet to calculate WPI and FPI, with mean monthly temperature and precipitation observations from Pointe Noire, Congo, is presented in Table A-1.

ACKNOWLEDGMENTS

Our concepts have evolved from the ideas of H.C. Sorby, P.D. Krynine, J.C. Griffiths, R.L. Folk, P.E. Potter, W.R. Dickinson, L.J. Suttner, R.V. Ingersoll, A. Basu, and M.J. Johnsson, among many others.

It would not be possible to employ these concepts without the pioneering work of applying Bayesian networks to geoscience inference done by A. Woronow, K.M. Love, J.F. Scheutte, and C.S. Kim at ExxonMobil Upstream Research Co. The concepts, methods, and tools described herein are the subject of United States and foreign patents by ExxonMobil Upstream Research.

Earlier versions of this manuscript benefited from the comments of R.G. Charles, M.W. French, P.E. Rumelhart, A. Seyedollali, and B.P. West. The final manuscript was greatly improved by the reviews of R.V. Ingersoll and G. Girty.

REFERENCES CITED

Anderson, S.P., Blum, J., Brantley, S.L., Chadwick, O., Chorover, J., Derry, L.A., Drever, J.I., Hering, J.G., Kirchner, J.W., Kump, L.R., Richter, D., and White, A.F., 2004, Proposed initiative would study Earth's weathering engine: Eos (Transactions, American Geophysical Union), v. 85, p. 265, 269.

Arvidson, R.S., Beig, M.S., and Luttge, A., 2004, Single-crystal plagioclase feldspar dissolution rates measured by vertical scanning interferometry: The American Mineralogist, v. 89, p. 51–56.

Batjes, N.H., 2002, A homogenized soil profile data set for global and regional environmental research (WISE, version 1.1): Wageningen, International Soil Research and Information Center, Report 2002/01, 42 p. (see also http://www.isric.org).

Bayes, T., 1763, An essay towards solving a problem in the doctrine of chances: Philosophical Transactions of the Royal Society of London, v. 53, p. 370–418 (see also "Studies in the History of Probability and Statistics: IX. Thomas Bayes's Essay Towards Solving a Problem in the Doctrine of Chances": Biometrika, v. 45, p. 293–315).

Bokman, J., 1955, Sandstone classification: Relation to composition and texture: Journal of Sedimentary Petrology, v. 25, p. 201–206.

Bonnell, L.M., and Lander, R.H., 2003, Reservoir quality prediction in deep water to tight gas sandstones using a process/stochastic modeling approach: American Association of Petroleum Geologists (AAPG) Bulletin, v. 8, p. 1694.

Bray, A.A., Lander, R.H., Watkins, C.A., Lowrey, C.J., and Owen, M., 2000, Characterisation and prediction of clastic reservoir quality; an integrated model for use in exploration, appraisal and production projects: American Association of Petroleum Geologists (AAPG) Bulletin, v. 84, p. 1408.

Cecil, C.B., and Edgar, N.T., 2003, Climate Controls on Stratigraphy: Society for Sedimentary Geology Special Publication 77, 275 p.

Choubert, G., Faure-Muret, A., and Chanteux, P., 1976, Atlas Geologique du Monde: Paris, United Nations Education, Scientific and Cultural Organization, scale 1:10,000,000, 20 sheets.

TABLE A1. SAMPLE WORKSHEET LAYOUT TO CALCULATE WEATHERING POTENTIAL INDEX (WPI) AND FLUSHING POTENTIAL INDEX (FPI) WITH DATA FROM POINTE NOIRE, CONGO

Month	Jan	Feb	Mar	Apr	May	Jun	Jul	Aug	Sep	Oct	Nov	Dec	Year total
Avg daily temp (°C)	26.3	26.7	27	26.9	25.8	23	21.4	21.7	23.5	25.1	25.8	26	24.93333
Avg month precip. (mm)	169.4	205.1	223.3	177.6	88.6	0.8	0.4	1.7	15.6	71.3	179.9	140.6	1274.3
Days/month	31	28	31	30	31	30	31	31	30	31	30	31	
Julian day	16	45	75	105	136	166	197	228	258	289	319	350	
Solar declination (rad)	−0.37	−0.24	−0.04	0.16	0.33	0.41	0.37	0.23	0.04	−0.17	−0.33	−0.41	
Sunset hour angle (rad)	1.60	1.59	1.57	1.56	1.54	1.53	1.54	1.55	1.57	1.59	1.60	1.61	
Avg light/day (h)	12.2	12.2	12.0	11.9	11.8	11.7	11.7	11.8	12.0	12.1	12.2	12.3	
Saturation vapor pressure (kbar)	3.421687	3.503429	3.565851	3.544937	3.321859	2.808525	2.547579	2.594821	2.894735	3.186383	3.321859	3.36148	
Hamon ET	111.6	101.5	111.9	105.3	100.3	82.1	77.7	80.4	88.2	102.0	104.6	110.3	
Effective precip.	57.8	103.6	111.4	72.3	0.0	0.0	0.0	0.0	0.0	0.0	75.3	30.3	450.8
Dissolution-rate multiplier	3.3E-09	3.4E-09	3.5E-09	3.5E-09	3.2E-09	2.7E-09	2.4E-09	2.5E-09	2.8E-09	3.1E-09	3.2E-09	3.3E-09	3.0E-09
Flushing month?	0	1	1	1	0	0	0	0	0	0	0	0	2
Nonflushing month?	1	0	0	0	1	1	1	1	1	1	1	1	10
WPI	192.48	354.03	388.34	250.42	0.00	0.00	0.00	0.00	0.00	0.00	242.71	99.01	1527.00
Log WPI													10.58
FPI													0.69897

Note: ET—Evapotranspiration.

Clark, K.F., 1982, Mineral composition of rocks, in Carmichael, R.S., ed., Handbook of Physical Properties of Rocks, Volume 1: Boca Raton, CRC Press, p. 1–213.

Clark, M.K., House, M.A., Royden, L.H., Whipple, K.X., Burchfiel, B.C., Zhang, X., and Tang, W., 2005, Late Cenozoic uplift of southeastern Tibet: Geology, v. 33, p. 525–528, doi: 10.1130/G21265.1.

Cooper, M., Silveira Mendes, L.M., Luiz Costa Silva, W., and Sparovek, G., 2005, A national soil profile database for Brazil available to international scientists: Soil Science Society of America Journal, v. 69, p. 649–652, doi: 10.2136/sssaj2004.0140.

Cox, A., and Hart, R.B., 1986, Plate Tectonics: How It Works: Palo Alto, Blackwell Scientific Publications, 392 p.

Critelli, S., 1999, The interplay of lithospheric flexure and thrust accommodation in forming stratigraphic sequences in the southern Apennines foreland basin system, Italy: Atti della Accademia Nazionale dei Lincei: Rendiconti Lincei Scienze Fisiche e Naturali, v. 9, no. 10, Part 4, p. 257–326.

Critelli, S., Arribas, J., Le Pera, E., Tortosa, A., Marsaglia, K.M., and Latter, K.K., 2003, The recycled orogenic sand provenance from an uplifted thrust belt, Betic Cordillera, southern Spain: Journal of Sedimentary Research, v. 73, p. 72–81.

Crook, K.A.W., 1960, Classification of arenites: American Journal of Science, v. 258, p. 419–428.

Curtis, W.F., Culbertson, J.K., and Chase, E.B., 1973, Fluvial-sediment discharge to the oceans from the conterminous United States: U.S. Geological Survey Circular 670, 17 p.

Dickinson, W.R., and Suczek, C.A., 1979, Plate tectonics and sandstone compositions: American Association of Petroleum Geologists Bulletin, v. 63, p. 2164–2182.

Dickinson, W.R., and Yarborough, H., 1978, Plate Tectonics and Hydrocarbon Accumulation: Tulsa, Oklahoma, American Association of Petroleum Geologists Course Notes, Volume 1, 139 p.

Dickinson, W.R., Beard, L.S., Brakenridge, G.R., Erjavec, J.L., Fergusun, R.C., Inman, K.F., Knepp, R.A., Lindberg, F.A., and Ryberg, P.T., 1983, Provenance of North American Phanerozoic sandstones in relation to tectonic setting: Geological Society of America Bulletin, v. 94, p. 222–235, doi: 10.1130/0016-7606(1983)94<222:PONAPS>2.0.CO;2.

Dott, R.H., 1964, Wacke, graywacke and matrix—What approach to immature sandstone classification?: Journal of Sedimentary Petrology, v. 34, p. 625–632.

Dunne, T., Mertes, L.A.K., Meade, R.H., Richey, J.E., and Forsberg, B.R., 1998, Exchanges of sediment between flood plain and channel of the Amazon River in Brazil: Geological Society of America Bulletin, v. 110, p. 450–467, doi: 10.1130/0016-7606(1998)110<0450:EOSBTF>2.3.CO;2.

Dutta, P.K., and Suttner, L.J., 1986a, Alluvial sandstone composition and paleoclimate, I. Framework mineralogy: Journal of Sedimentary Petrology, v. 56, p. 329–345.

Dutta, P.K., and Suttner, L.J., 1986b, Alluvial sandstone composition and paleoclimate, II. Authigenic mineralogy: Journal of Sedimentary Petrology, v. 56, p. 346–358.

Faugeres, J.C., Gonthier, E., Bobier, C., and Griboulard, R., 1997, Tectonic control on sedimentary processes in the southern termination of the Barbados Prism: Marine Geology, v. 140, p. 117–140, doi: 10.1016/S0025-3227(96)00102-8.

Folk, R.L., 1974, Petrology of Sedimentary Rocks: Austin, Texas, Hemphill Publishing Co., 182 p. (See also http://www.lib.utexas.edu/geo/folkready/folkprefrev.html.)

Franzinelli, E., and Potter, P.E., 1983, Petrology, chemistry, and texture of modern river sands, Amazon River system: The Journal of Geology, v. 91, p. 23–39.

Fraticelli, C.M., West, B.P., Bohacs, K.M., Patterson, P.E., and Heins, W.A., 2004, Vegetation-precipitation interactions drive paleoenvironmental evolution: Eos (Transactions, American Geophysical Union), v. 85, no. 47, p. F930.

Gasparini, N.M., Tucker, G.E., and Bras, R.L., 1999, Downstream fining through selective particle sorting in an equilibrium drainage network: Geology, v. 27, p. 1079–1082.

Gasparini, N.M., Tucker, G.E., and Bras, R.L., 2004, Network-scale dynamics of grain-size sorting; implications for downstream fining, stream-profile concavity, and drainage basin morphology: Earth Surface Processes and Landforms, v. 29, p. 401–421, doi: 10.1002/esp.1031.

Gasparini, N.M., Bras, R.L., Whipple, K.X, 2006, Numerical modeling of non-steady-state river profile evolution using a sediment-flux-dependent incision model, in Tectonics, Climate, and Landscape Evolution: Geological Society of America Special Paper 398, p. 127–141.

Grantham, J.H., and Velbel, M.A., 1988, The influence of climate and topography on rock-fragment abundance in modern fluvial sands of the southern Blue Ridge Mountains, North Carolina: Journal of Sedimentary Petrology, v. 58, p. 219–227.

Griffiths, J.C., 1951, Size versus sorting in some Caribbean sediments: The Journal of Geology, v. 59, p. 211–243.

Haith, D.A., and Shoemaker, L.L., 1987, Generalized watershed loading functions for stream flow nutrients: Water Resources Bulletin, v. 23, p. 471–478.

Hamon, W.R., 1961, Estimating potential evapotranspiration: Proceedings of the American Society of Civil Engineers: Journal of the Hydraulic Division, v. 87, no. HY3, p. 107–120.

Hasbargen, L.E., and Paola, C., 2003, How predictable is local erosion rate in eroding landscapes?: Geophysical Monograph, v. 135, p. 231–240.

Heins, W.A., 1992, The effect of climate and topography on the composition of modern, pluton-derived sand [Ph.D. thesis]: Los Angeles, University of California, 465 p.

Heins, W.A., 1993, Source-rock texture vs. climate and topography as controls on the composition of modern, plutoniclastic sand, in Johnsson, M.J., and Basu, A., eds., Processes Controlling the Composition of Clastic Sediments: Geological Society of America Special Paper 284, p. 135–146.

Heins, W.A., 1995, The use of mineral interfaces in sand-sized rock fragments to infer ancient climate: Geological Society of America Bulletin, v. 107, p. 113–125, doi: 10.1130/0016-7606(1995)107<0113:TUOMII>2.3.CO;2.

Hovius, N., 1998, Controls on Sediment Supply by Large Rivers, in Shanley, K.W., and McCabe P.J., eds., Relative role of eustasy, climate, and tectonism in continental rocks: Society for Sedimentary Geology Special Publication 59, p. 3–16.

Hsü, K.J., 2004, Physics of Sedimentation (2nd edition): Berlin, Springer-Verlag, 240 p.

Ingersoll, R.V., and Busby, C.J., 1995, Tectonics of sedimentary basins, in Ingersoll, R.V., and Busby, C.J., eds., Tectonics of Sedimentary Basins: Cambridge, Blackwell Science, p. 1–51.

Ingersoll, R.V., Kretchmer, A.G., and Valles, P.K., 1993, The effect of sampling scale on actualistic sandstone petrofacies: Sedimentology, v. 40, p. 937–953, doi: 10.1111/j.1365-3091.1993.tb01370.x.

Johnsson, M.J., 1993, The system controlling the composition of clastic sediments, in Johnsson, M.J., and Basu, A., eds., Processes Controlling the Composition of Clastic Sediments: Geological Society of America Special Paper 284, p. 1–19.

Johnsson, M.J., and Meade, R.H., 1990, Chemical weathering of fluvial sediments during alluvial storage: The Macuapanim Island Point Bar, Solimões River, Brazil: Journal of Sedimentary Petrology, v. 60, p. 827–842.

Johnsson, M.J., Stallard, R.F., and Meade, R.H., 1988, First-cycle quartz arenites in the Orinoco River basin, Venezuela and Colombia: The Journal of Geology, v. 96, p. 263–267.

Jones, M.A., Heller, P.L., Roca, E., Garces, M., and Cabrera, L., 2004, Time lag of syntectonic sedimentation across an alluvial basin; theory and example from the Ebro Basin, Spain: Basin Research, v. 16, p. 467–488, doi: 10.1111/j.1365-2117.2004.00244.x.

Kairo, S., and Heins, W.A., 2004, Predicting sand character with integrated genetic analysis, in Ryan, W.B.F., and Malinverno, A., eds., The Mediterranean Tethys and the development of new concepts of top-down tectonics: 32nd International Geologic Congress, 2004, Abstract Volume pt. 1, abstract number 242-6: Florence, Italy, 32nd International Geological Congress, General Session 21.14.

Kairo, S., Suttner, L.J., and Dutta, P.K., 1993, Variability in sandstone composition as a function of depositional environment in coarse-grained delta systems, in Johnsson, M.J., and Basu, A., eds., Processes Controlling the Composition of Clastic Sediments: Geological Society of America Special Paper 284, p. 263–282.

Kearey, P., and Vine, F.J., 1996, Global Tectonics (2nd edition): Boston, Blackwell Science, 333 p.

Krynine, P.D., 1935, Arkose deposits in the humid tropics; a study of sedimentation in southern Mexico: American Journal of Science, v. 29, p. 353–363.

Krynine, P.D., 1943, Diastrophism and the evolution of sedimentary rocks: Pennsylvania State College Mineral Industries Technical Paper no. 84-a, 21 p.

Krynine, P.D., 1948, The megascopic study and field classification of sedimentary rocks: The Journal of Geology, v. 56, p. 130–165.

Lander, R.H., and Walderhaug, O., 1999, Predicting porosity through simulating sandstone compaction and quartz cementation: American Association of Petroleum Geologists (AAPG) Bulletin, v. 83, p. 433–449.

Langbein, W.B., and Schumm, S.A., 1958, Yield of sediment in relation to mean annual precipitation: Eos (Transactions, American Geophysical Union), v. 39, p. 1076–1084.

Lasaga, A.C., 1984, Chemical kinetics of water-rock interactions: Journal of Geophysical Research, v. 89, p. 4009–4025.

Ludwig, W., and Probst, J.L., 1996, A global modelling of the climatic, morphological, and lithological control of river sediment discharges to the oceans: International Association of Hydrological Sciences-Association internationale des sciences hydrologiques Publication, v. 236, p. 21–28.

Mahabir, K., Khandaker, N.I., Schleifer, S., Flores, D., Maksimonicz, H., and Bang, D., 2004, Geochemical analysis of sandstones, upper Scotland Formation (Eocene), NE Barbados; clues to provenance: Geological Society of America Abstracts with Programs, v. 36, no. 5, p. 234–235.

Malmon, D.V., Dunne, T., and Reneau, S.L., 2003, Stochastic theory of particle trajectories through alluvial valley floors: The Journal of Geology, v. 111, p. 525–542, doi: 10.1086/376764.

Marsaglia, K.M., and Ingersoll, R.V., 1992, Compositional trends in arc-related, deep-marine sand and sandstone; a reassessment of magmatic-arc provenance; with supplemental data 92-36: Geological Society of America Bulletin, v. 104, p. 1637–1649, doi: 10.1130/0016-7606(1992)104<1637:CTIARD>2.3.CO;2.

Menacherry, S., 2006, Source to sink sedimentation and petrology of a modern dryland fluvial reservoir analogue, western Lake Eyre, central Australia [Ph.D. dissertation]: Adelaide, Australian School of Petroleum, University of Adelaide (in press).

Menacherry, S., Lang, S.C., Payenberg, T.H.D., and Heins, W.A., 2006, Source to sink sedimentation in a dryland fluvial system, western Lake Eyre Basin, central Australia: Houston, American Association of Petroleum Geologists Bulletin, v. 90.

Milliman, J.D., 1997, Fluvial sediment discharge to the sea and the importance of regional tectonics, in Ruddiman, W.F., ed., Tectonic Uplift and Climate Change: New York, Plenum Press, p. 239–257.

Motoi, T., Chan, W.L., Minobe, S., and Sumata, H., 2005, North Pacific halocline and cold climate induced by Panamanian Gateway closure in a coupled ocean-atmosphere GCM: Geophysical Research Letters, v. 32, p. 10, doi: 10.1029/2005GL022844.

Norsys Software Corp, 2006, Netica Tutorial: http://www.norsys.com/tutorials/netica/nt_toc_A.htm (accessed 10 February 2006).

Palomares, M., and Arribas, J., 1993, Modern stream sands from compound crystalline sources; composition and sand generation index, in Johnsson, M.J., and Basu, A., eds., Processes Controlling the Composition of Clastic Sediments: Geological Society of America Special Paper 284, p. 313–332.

Panda, M.N., and Lake, L.W., 1994, Estimation of single-phase permeability from parameters of particle-size distribution: American Association of Petroleum Geologists Bulletin, v. 78, no. 7, p. 1028–1039.

Panda, M.N., and Lake, L.W., 1995, A physical model of cementation and its effects on single-phase permeability: American Association of Petroleum Geologists Bulletin, v. 79, no. 3, p. 431–443.

Pazzaglia, F.J., 2004, Landscape evolution models: Developments in Quaternary Science, v. 1, p. 247–274.

Perez, R.J., Chatellier, J.I., and Lander, R., 1999, Use of quartz cementation kinetic modeling to constrain burial histories; examples from the Maracaibo Basin, Venezuela: Revista Latino-Americana de Geoquimica Organica, v. 5, p. 39–46.

Pirsig, R.M., 1974, Zen and the Art of Motorcycle Maintenance: An Inquiry into Values: New York, Morrow, 417 p.

Potter, P.E., 1978, Petrology and chemistry of modern big river sands: The Journal of Geology, v. 86, p. 423–449.

Potter, P.E., 1986, South America and a few grains of sand: Part 1. Beach sands: The Journal of Geology, v. 94, p. 301–319.

Potter, P.E., 1993, Sample location list and data set for "Modern Beach and River Sands of South American" [unpublished manuscript]: Cincinnati, University of Cincinnati, 99 p.

Potter, P.E., 1994, Modern sands of South America; composition, provenance and global significance: Geologische Rundschau, v. 83, p. 212–232, doi: 10.1007/BF00211904.

Potter, P.E., and Siever, R., 1956, Sources of basal Pennsylvanian sediments in the Eastern Interior Basin: Part 3. Some methodological implications: The Journal of Geology, v. 64, p. 447–455.

Poulsen, C.J., Seidov, D., Barron, E.J., and Peterson, W.H., 1998, The impact of paleogeographic evolution on the surface oceanic circulation and the marine environment within the Mid-Cretaceous Tethys: Paleoceanography, v. 13, p. 546–559, doi: 10.1029/98PA01789.

Reilly, M.R.W., Lang, S.C., and Hill, S.M., 2003, Landforms and sediments of Umbum Creek and adjacent plains, Lake Eyre Basin, South Australia, in Roach, I.C., ed., Advances in Regolith, Proceedings of the CRC LEME Regional Regolith Symposia 2003: Adelaide, Cooperative Research Centre for Landscape Evolution and Mineral Exploration, p. 332–335.

Ritter, D.F., Kochel, R.C., and Miller, J.R., 2002, Process Geomorphology (4th edition): Boston, McGraw-Hill, 560 p.

Robinson, R.A.J., and Slingerland, R.L., 1998a, Grain-size trends, basin subsidence and sediment supply in the Campanian Castlegate Sandstone and equivalent conglomerates of central Utah: Basin Research, v. 10, p. 109–127, doi: 10.1046/j.1365-2117.1998.00062.x.

Robinson, R.A.J., and Slingerland, R.L., 1998b, Origin of fluvial grain-size trends in a foreland basin; the Pocono Formation on the central Appalachian Basin: Journal of Sedimentary Research, v. 68, p. 473–486.

Robinson, R.S., and Johnsson, M.J., 1997, Chemical and physical weathering of fluvial sands in an Arctic environment; sands of the Sagavanirktok River, North Slope, Alaska: Journal of Sedimentary Research, v. 67, p. 560–570.

Rossetti, D.F., 2004, Paleosurfaces from northeastern Amazonia as a key for reconstructing paleolandscapes and understanding weathering products: Sedimentary Geology, v. 169, p. 151–174, doi: 10.1016/j.sedgeo.2004.05.003.

Roth, G., La Barbera, P., and Greco, M., 1996, On the description of the basin effective drainage structure: Journal of Hydrology, v. 187, p. 119–135, doi: 10.1016/S0022-1694(96)03090-9.

Ruxton, B.P., 1970, Labile quartz-poor sediments from young mountain ranges in northeast Papua: Journal of Sedimentary Petrology, v. 40, p. 1262–1270.

Schaller, M., von Blanckenburg, F., Hovius, N., and Kubik, P.W., 2001, Large-scale erosion rates from in situ–produced cosmogenic nuclides in European river sediments: Earth and Planetary Science Letters, v. 188, p. 441–458, doi: 10.1016/S0012-821X(01)00320-X.

Schumm, S.A., 1981, Evolution and response of the fluvial system, sedimentologic implications, in Ethridge, F.G., and Flores, R.M., eds., Recent and ancient nonmarine depositional environments; models for exploration: Society of Economic Paleontologists and Mineralogists Special Publication 31, p. 19–29.

Shih, S.M., and Komar, P.D., 1990, Differential bedload transport rates in a gravel-bed stream; a grain-size distribution approach: Earth Surface Processes and Landforms, v. 15, p. 539–552.

Shreve, R.L., 1966, Statistical laws of stream numbers: The Journal of Geology, v. 74, p. 17–37.

Soil Survey Staff, 1997, National Soil Survey Characterization Data: Lincoln, Nebraska, U.S. Department of Agriculture–Natural Resource Conservation Service, National Soil Survey Center, Soil Survey Laboratory.

Sorby, H.C., 1880, On the structure and origin of non-calcareous stratified rocks: Geological Society of London Quarterly Journal, v. 36, p. 46–92.

Strahler, A.N., 1964, Quantitative geomorphology of drainage basins and channel networks, in Chow, V.T., ed., Handbook of Applied Hydrology: New York, McGraw-Hill, p. 439–476.

Strakhov, N., 1967, Principles of Lithogenesis: London, Oliver and Boyd, v. 1, 245 p.

Summerfield, M.A., and Hulton, N.J., 1994, Natural controls of fluvial denudation rates in major world drainage basins: Journal of Geophysical Research, ser. B, Solid Earth and Planets, v. 99, p. 13,871–13,883, doi: 10.1029/94JB00715.

Suttner, L.J., 1974, Sedimentary petrographic provinces: An evaluation, in Paleogeographic Provinces and Provinciality: Society of Economic Paleontologists and Mineralogists Special Publication 21, p. 75–84.

Suttner, L.J., Basu, A., and Mack, G.H., 1981, Climate and the origin of quartz arenites: Journal of Sedimentary Petrology, v. 51, p. 1235–1246.

Syvitski, J.P.M., Peckham, S.D., Hilberman, R., and Mulder, T., 2003, Predicting the terrestrial flux of sediment to the global ocean; a planetary perspective: Sedimentary Geology, v. 162, p. 5–24, doi: 10.1016/S0037-0738(03)00232-X.

Tokarev, V., 2005, Neotectonics of the Mount Lofty Ranges, South Australia [Ph.D. thesis]: Adelaide, University of Adelaide, 272 p.

Tokarev, V., and Gostin, V., 2003, Mt Lofty Ranges, South Australia, in Anand, R.R., and de Broekert, P., eds., Regolith-Landscape Evolution Across Australia: Canberra, Cooperative Research Center for Landscape Evolution and Mineral Exploration, http://crcleme.org.au/Pubs/Monographs/RegLandEvol.html.

Vogel, K.R., van Niekerk, A., Slingerland, R.L., and Bridge, J.S., 1992, Routing of heterogeneous sediments over movable bed: Model verification and testing: Journal of Hydraulic Engineering, v. 118, no. 2, p. 263–279.

Walker, R.G., and James, N.P., 1992, Facies Models, Response to Sea Level Change: Geological Association of Canada GeoText 1, 454 p.

Walling, D.E., 1987, Rainfall, runoff and erosion of the land; a global view, *in* Gregory, K.J., ed., Energetics of Physical Environment: Energetic Approaches to Physical Geography: Chichester, John Wiley and Sons, p. 89–117.

Walling, D.E., and Fang, D., 2003, Recent trends in the suspended sediment loads of the world's rivers: Global and Planetary Change, v. 39, p. 111–126, doi: 10.1016/S0921-8181(03)00020-1.

Walling, D.E., and Webb, B.W., 1996, Erosion and sediment yield; a global overview, *in* Walling, D.E., and Webb, B.W., eds., Proceedings of an International symposium on Erosion and sediment yield; global and regional perspectives: International Association of Hydrological Sciences Publication, v. 236, p. 3–19.

West, B.P., Fraticelli, C.M., Heins, W.A., and Bohacs, K.M., 2005, Paleoclimate Atlas: A Compilation of Silurian through Paleogene Global-Scale Paleogeographic Reconstructions with Parametric Paleoclimate Simulations and Guidelines for Use in Frontier Play Element Prediction: Exxon-Mobil Upstream Research Company Report URC.2005EX.001, 347 p.

White, A.F., Bullen, T.D., Schulz, M.S., Blum, A.E., Huntington, T.G., and Peters, N.E., 2001, Differential rates of feldspar weathering in granitic regoliths: Geochimica et Cosmochimica Acta, v. 65, p. 847–869, doi: 10.1016/S0016-7037(00)00577-9.

Willett, S., Brandon, M.T., Dorsey, R.J., Vendeville, B., and Whipple, K., 2003, Dynamic interactions between tectonics, climate and Earth surface processes, *in* Pollard, D.D., ed., New departures in structural geology and tectonics: Washington, D.C., National Science Foundation, p. 33–42.

Willmott, C.J., 1977, WATBUG: A FORTRAN IV algorithm for calculating the climatic water budget: Publications in Climatology, v. 50, no. 1, 58 p.

Wischmeier, W.H., and Smith, D.D., 1978, Predicting Rainfall Erosion Losses: A Guide to Conservation Planning: U.S. Department of Agriculture Agricultural Handbook 537, 58 p.

Zeil, W., 1964, Geologie von Chile: Berlin, Gebrüder Borntraeger, 233 p.

MANUSCRIPT ACCEPTED BY THE SOCIETY 9 AUGUST 2006

Index

187-S1 core
  geochemistry of, 16, 18, 20–23
  grain-size studies of, 28–35
  lithology of, 16
  maps of, 14, 26
  micropaleontological studies of, 17–18
  provenance of, 16–18, 23
  sedimentology of, 15–17, 27
  stratigraphy of, 16, 28
  ternary plot of, 27
204-S1 core, 26–35
204-S2 core, 26–35
204-S8 core, 26–35
204-S15 core, 14, 18–21, 23
205-S3 core, 26–35
205-S5 core, 26–35
205-S10 core, 26–35
221-S19 core, 26–35
239-S2 core, 26–35
240-S9 core, 26–35

A

Abetone area, 96–104
accretionary prisms, 58, 110, 281
accretionary wedge, 75
Adelosina spp., 18, 20
Adelosina cliarensis, 18, 20
Adelosina mediterranensis, 18
Adige River, 14, 21
Adria Plate
  Apennine orogenesis and, 58, 74–75, 77, 88–90, 109–10
  geological setting of, 38
  Ligure-Piemontese Basin and, 74–75, 77, 88–90
  map of, 223
  roll-back of, 110
  Romagna unit and, 58
  Rossano Basin and, 136
  Tuscan domain and, 58
  Umbrian domain and, 58
African Plate, 74–75, 110
AGIP Frosinone 1 well, 111, 112
Akchagyl Formation, 323
Alameda section, 200, 204–5, 207
Alban Hills, 111
Alborz area, 322
Alcapa Block, 222, 223
Aldealpozo Formation, 201
Alethopteris, 224
algae, 171–75
Aliaga subbasin, 183, 184, 186–94
alkaline volcanism
  in Caspian Basin, 323, 326–28, 330
  in Clair oil field, 330, 337
  in Iberian Range, 215–17
  potassium and, 33
  rubidium and, 33
  yttrium and, 33
  zirconium and, 33
Allgäu Flysch, 53
Alpago area, 38, 41, 42, 50
Alpine fault (New Zealand), 254, 278–81
Alps (European)
  Dolomite Mountains, 2
  glaciers in, 6, 9
  maps of, 2, 14, 223
  Marmolada Mountains, 2
  regional topographic gradient state, 359
  river drainage from
    geochemistry of, 20–21, 23
    Macigno unit source in, 104
    maps of, 2, 14
  Tethys Ocean and, 74–75
Alps (New Zealand), 278, 281, 293
Alto Adige–Trentino region, 2–11
aluminum, 212, 213
aluminum oxide
  in Balagne Nappe, 85
  in Cameros Basin, 210
  chlorite and, 213
  grain size and, 30, 31–33
  in Ligurian units, 85
  in Mecsek-Villány area, 231
  in Moncayo area, 210
  in Po Plain, 31
aluminum oxide/silicon oxide values, 31–33
aluminum/barium values, 33
aluminum/chromium values, 32–34
aluminum/magnesium values, 20–21, 23, 33, 34
aluminum/nickel values, 16, 18–23, 32–34
aluminum/sodium values, 33
aluminum/thallium values, 33
aluminum/vanadium values, 32
aluminum/yttrium values, 33
aluminum/zirconium values, 33
Amazon River, 362
Amendolara Ridge, 139
Ammonia beccarii, 17, 18, 20
Ammonia inflata, 17, 18
Ammonia papillosa, 20
Ammonia parkinsoniana, 17, 18, 20
Ammonia tepida, 17, 18, 20
amphiboles
  in Belluno Flysch, 51–54
  in Claut Flysch, 51–54
  in Clauzetto Flysch, 51–54
  in El Arco area, 302
  grain size and, 30, 31, 33
  in Modena Plain, 58
  in Po Plain, 31
  in provenance studies, 320, 340
Amphistegina, 125
Ampollino Lake, 151
Amu Darya River, 322
anatase, 51
Andes Mountains
  Chilean fluvial sand study, 368–71
  regional topographic gradient state, 359
andesites, 122, 237
angiosperms, 374
anhydrite, 146
Annularia, 224
Antarctic Circumpolar Current, 278, 279
apatite
  calcium and, 237
  CART for, 329–32, 334–40
  in Caspian Basin, 323–32, 339–40
  chemical formula for, 320
  in Clair oil field, 330, 332–40
  europium and, 321
  fractionation and, 320–21
  LA-ICP-MS studies of, 320, 321
  lanthanum/neodymium values for, 321
  in Maestrat Basin, 192
  in pelites, 82
  phosphorus and, 195, 237
  phosphorus pentoxide and, 30
  REEs and, 82, 237, 321
  textural characteristics of, 192
  titanium and, 237
  weathering and, 340
  yttrium and, 30, 195, 237
  zirconium and, 30
Apennines
  accretionary prism in, 58, 110
  Adria Plate and, 58, 74–75, 77, 88–90, 109–10
  African Plate and, 110
  Apulia Plate and, 110
  Argilloso-Arenacea Formation, 110–31
  Brecce della Renga Formation, 110–14
  Emilia region, 14, 26, 71
  Frosinone Formation. *See* Frosinone Formation
  geochemistry of, 14
  geological setting of, 38, 58, 109–10
  glaciers in, 71–72
  Latium-Abruzzi platform, 110, 128, 129
  Ligure-Piemontese Basin, 74–75, 77, 88–90
  Ligurian units. *See* Ligurian units
  maps of
    geologic, 59, 137
    river drainages, 14, 59
    structural, 109, 111
    tectonic, 75
  mineralogical studies of, 14
  Modena Plain bedrock orientation to, 71
  Orbulina Marl. *See* Orbulina Marl
  paleogeographic domains in, 109–10
  river drainage from
    to Modena Plain, 58, 71
    to Po Plain, 20–21, 23, 27, 30–32, 58
  Romagna region, 14, 26
  Rossano Basin, 136, 138–47
  Sila Massif, 150–63
  structure of, 110
  Tethys Ocean and, 58, 74–75
  Tuscan domain. *See* Tuscan domain
  Umbria-Marche-Sabina Basin, 110, 128–30

Umbrian domain, 58, 75, 96
aplite, 230, 234, 237
Appalachian Mountains, 359
Appin Group, 338
Apsheron Formation, 323
Apsheron Peninsula, Productive Series, 321–32, 339
Aptian-Albian Flysch, 54
Apulia Plate, 110
Aranda del Moncayo section, 200, 204–5, 207
arcs
    in Caucasus Mountains formation, 323
    geochemical signals of, 235
    tectonic setting discrimination diagrams for, 215
Ardo River, 40
Arenaceo Conglomeratica Formation, 138, 147
arenites
    in Argilloso-Arenacea Formation, 116–30
    in Balagne Nappe, 78, 79
    in Brazilian beach sand, 364
    in Brecce della Renga Formation, 116–30
    calclithite, 127–30
    composition of, 108
    depositional environment, 249
    in Frosinone Formation, 116–30
    in Gessi Formation, 140–41
    in Iberian Range, 170
    in Ligurian units, 78, 79
    in Modena Plain, 58
    in Orbulina Marl, 125–30
    in Rossano Basin, 139
Argile Scagliose unit, 138
Argille Marnose Salifere Formation, 138, 139
argillites, 256
Argilloso-Arenacea Formation, 110–31
Argyll Group, 338
arkoses, 169, 249
Arnedillo section, 200
Arroyofrio Formation, 201
Artoles Formation, 184, 185
Arvo Lake, 151
ash deposits, 272, 274, 293
Atlantic Ocean, Iberian Plate and, 183, 202
aulacogens, 214
Aurunci Mountains, 111
Ausoni Mountains, 111
Australia
    discrimination diagrams for, 235
    Umbum Creek sand study, 369–75
Australian Plate, 254, 256, 278, 279
Austria, 223
Austroalpine microplate, 53–54, 223
Autapie unit, 75
Azerbaijan
    paleomap of, 322
    South Caspian Basin, 321–32, 339, 362

B
Babazanan, Productive Series, 322–32, 339
Babócsa subunit, 223

bacteria, 171. *See also* cyanobacteria
Baja California peninsula, 298–318
Baksa subunit, 223, 225, 234, 237
Baku Formation, 323
Balagne Nappe, 74–76, 78–88
Balakhany Suite, 323–32, 339
Baltic Sea region, 323, 362
Banks and Otago Peninsula, 293. *See also* Otago area
Barahona Formation, 201
Barakar Formation, 245, 247
Barakar River, 243
Barbados, 347
barium
    in Cameros Basin, 212
    grain size and, 32
    in Ligurian units, 86
    in Mecsek-Villány area, 233
    in Moncayo area, 212
    potassium and, 32, 213
    sample provenance from, 86
(barium+zirconium)/rubidium values, 31
barium/aluminum values, 33
barium/chromium values, 33
barium/rubidium values, 31–34
basalt
    in Balagne Nappe, 79
    in Caucasus Mountains, 322–23
    iron and, 33
    in Ligurian units, 76
    source-rock discrimination diagrams for, 216
    titanium and, 33
    vanadium and, 33
basin
    boundaries of, 350
    definition of, 168, 346
    fluvial transport distance, 350, 352, 353–54, 360–62
    subsidence rate, 350, 352, 353–54, 362
Bay of Biscay Basin, 183
Bayes Rule, 352
beaches, abrasion on, 9
Békés unit, 223
Belluno Basin, 38–39, 53
Belluno Flysch, 38–43, 45–47, 49–54
Beratón section, 200, 204–5, 207
Betic Cordillera, 183
Bhagirathi River, 243
Bhimaram Sandstone Formation, 245, 247
Bihar, 223
biotite
    on Corsica, 79
    kaolinite and, 162
    in Ligurian units, 82
    in Macigno unit, 100–101
    in Maestrat Basin, 186
    in Mecsek-Villány area, 230
    in Neto River Basin, 163
    in ODP 1122 core, 284–85
    in Sila Massif area, 150, 152, 155
    thermal response of, 160

    weathering and, 160
Bisciaro-Schlier, 96
Bocchigliero Complex, 136, 147
Bocco Shale, 76
Boda Siltstone Formation, 223, 225
Bogádmindszent-1 core, 224–30
bogs, coloration from, 249
Bohemian Massif, 222, 223, 237
Boite River, 2–7, 9–11
boron, 182, 195
Bounty Fan system, 278–93, 295–96
Bracco unit, 76, 83
Brazil
    Amazon River, 362
    beach sand study, 363–68
Brazilian Shield, lithotype abundance, 366
Brecce della Renga Formation, 110–14, 118, 121, 123–30
Brenta River, 14, 21
Bressanone, 2, 3
Briançonnais unit, 75
Briozoi and Litotamni Limestones Formation, 110, 112, 128
Brook Street Terrane, 281
brookite, 51
Buenos Aires Basin, 301
Bükk, 223
Bulimina elongata, 17
Buntsandstein Formation, 201–10
Burgo de Osma Formation, 201

C
Caatingas, 359
Cabo Falso, 300
Cabo San Lucas Bay, 299, 300, 305–8, 312–15
Cabo San Lucas Canyon, 299
Calabria
    gneiss in, 10
    maps of, 109, 137, 151
    Neto River Basin, 150–51, 156–59, 163–64
    Rossano Basin, 136, 138–47
    Sila Massif, 150–63
Calabrian Arc
    Gessi Formation and, 146
    maps of, 109
    Rossano Basin and, 136, 146
    Sila Massif and, 150
    Tyrrhenian Basin and, 136
Calamites, 224
calc-alkaline volcanism, 237, 323, 328
Calcare di Base Formation, 136, 138, 139
calcarenites, 99, 110, 112
calcite. *See also* limestone
    abrasion of, 10
    in Belluno Flysch, 45–47
    biofilms and, 171
    in Boite River, 4, 5, 9–10
    in Brecce della Renga Formation, 118
    in Claut Flysch, 44
    in Clauzetto Flysch, 47–48
    cyanobacteria and, 171–75

versus dolomite, 10
  in Gadera River, 4, 5, 9–10
  in Gessi Formation, 140–41
  grain size and, 29–30
  in Iberian Range, 170–77
  in Lamone River, 5
  in Ligurian region, 10
  in Macigno unit, 103
  in Modena Plain, 61–64, 66–70
  in Piave River, 4, 5, 9–10
  in Po Plain, 31
  in Rienza River, 4, 5, 9–10
  shape of, 4, 5, 9, 10
  weathering of, 10
calcium
  apatite and, 237
  manganese and, 320
  in Mecsek-Villány area, 233, 234
  REE and, 237, 320
  strontium and, 320
  thorium and, 320
  uranium and, 320
  weathering and, 212, 236
  yttrium and, 237, 320
calcium oxide
  in Cameros Basin, 210–13
  in Mecsek-Villány area, 231–33, 236
  in Moncayo area, 210–12
  in Po Plain, 31
calclithite, 127–30
caliche, 61–64, 68
Callistocythere spp., 20
Calpionella, 82
Calpionella Limestone, 75, 76, 83
Camarillas Formation, 184–94
Cameros Basin, 183, 184, 195, 200–17
Campbell "drift," 278, 292
Campbell Plateau, 278–79
Candona spp., 20
Canetolo unit, 75
Cansiglio High, 38, 41
Caorame Stream area, 38, 41–42
Caples Terrane, 281
carbonate
  in Boite River, 3–5
  in Brazilian Shield, 366
  in Cameros Basin, 203
  carbonate intrabasinal grains. See CI
  extrabasinal carbonate grains. See CE
  in Gadera River, 3–5
  in Iberian Range, 169–77
  in Moncayo area, 203
  in Piave River, 3–5
  in Rienza River, 3–5, 8
  in Rossano Basin, 139
  in Umbum Creek, 372
carbonate intrabasinal grains. See CI
carbonatite
  in Caspian Basin, 326–31
  in Clair oil field, 331, 334, 337–39, 341
Carpathian-Pannonian area, 222–23, 362

Case Baruzzo Sandstone, 76–78, 80, 82, 84
Caspian Basin, South, 321–32, 339, 362
Cassio unit, 76–78, 80, 82, 84
Castellar Formation, The, 184–94
Catalan Coastal Chain, 183
Caucasus Mountains, 322–23, 328–29, 339
CE
  in Argilloso-Arenacea Formation, 119–20, 122, 130
  in Balagne Nappe, 78
  in Belluno Flysch, 45–47
  in Brecce della Renga Formation, 118, 127, 129–30
  in Claut Flysch, 44
  in Clauzetto Flysch, 47–48
  in Frosinone Formation, 119, 122, 130
  in Gessi Formation, 140–42
  in Ligurian units, 78
  in Modena Plain, 61–63
  in Orbulina Marl, 119
Cecita Lake area, 151, 156
Cellina River Basin, 39
Cerca Stream, 60
cerium, 27, 30, 31
cerium/rubidium values, 33
Cerro Blanco area, 299, 315
Cerro Blanco Basin, 301, 302
Cerro del Dez Formation, 201
Cerro Prieto area, 305, 312
Cerro Prieto Basin, 301
Cervarola-Falterona unit, 96, 97
cesium, 212, 213
(cesium + lanthanum)/ΣREE values. See (lanthanum + cesium)/ΣREE values
cesium/ytterbium values
  in Caspian Basin, 326, 328, 329
  in Clair oil field, 333, 339
  in provenance studies, 321
charophytes, 175
Chatham Rise, 278–79, 293
Chelva Formation, 201
chemical index of alteration (CIA)
  for Camarillas Formation, 188, 194
  for Cameros Basin, 213, 217
  for El Castellar Formation, 188, 194
  factors affecting, 213, 217
  for Maestrat Basin, 188, 194
  for Moncayo area, 213, 217
  for Mora Formation, 188, 194
  for sandstone, 188–89
chemical index of weathering (CIW), 188–89, 194
chert
  in Alps (European), 43
  in Argilloso-Arenacea Formation, 119–20, 122
  in Belluno Flysch, 43, 45–47
  in Boite River, 4, 7
  in Brecce della Renga Formation, 118
  in Claut Flysch, 43–44
  in Clauzetto Flysch, 43, 47–48

  on Corsica, 75
  in Dinarides, 43
  in Frosinone Formation, 119, 122
  in Gadera River, 4
  in Gessi Formation, 141
  in Ligurian region, 10
  in Ligurian units, 82
  in Macigno unit, 100–102
  in Maestrat Basin, 186
  in Modena Plain, 58, 61–63
  in Piave River, 4, 7, 9
  in Rienza River, 4
  in sand maturity index, 9
  shape of, 4, 5
  structure of, 10
  in Waipaoa River Basin, 264
  weathering of, 10
Chianti Mountains, 96
child nodes, 352–53
Chile, fluvial sand study, 368–71
chlorine, 145, 320
chlorite. See also magnesium chlorite
  aluminum oxide and, 213
  in Argilloso-Arenacea Formation, 119–20
  in Belluno Flysch, 45–47
  in Brecce della Renga Formation, 118
  in Claut Flysch, 44
  in Clauzetto Flysch, 47–48
  coloration and, 249
  formation of, 162
  in Frosinone Formation, 119
  in Gessi Formation, 141
  illite and, 213
  kaolinite and, 236
  in Macigno unit, 100–103
  in Mecsek-Villány area, 236, 237
  in ODP 1122 core, 280, 284–85, 290, 292
  in Orbulina Marl, 119
  in Po Plain, 31
  potassium feldspar and, 213
chloritoid, 51
Choč, 223
Chokrak Formation, 323
Chondrites, 42
chromite, 50–54
chromium
  in Balagne Nappe, 85, 87
  in Ligurian units, 83, 85, 87
  in Mecsek-Villány area, 233, 235, 237
  ophiolites and, 14, 31
  in Po Plain, 31
  in Po River Basin, 14
  in provenance studies, 235
  redox state and, 33
  serpentine and, 31
  sorting effects on, 27
  turbidite systems and, 14
  ultramafic rocks and, 31, 33
chromium/aluminum values, 32–34
chromium/barium values, 33
chromium/clay values, 32

chromium/nickel values, 235
chromium/thorium values, 85, 88, 235
chromium/vanadium values, 33
chromium/zirconium values, 235
CI
    in Argilloso-Arenacea Formation, 119–20, 125
    in Balagne Nappe, 78
    in Belluno Flysch, 45–47
    in Brecce della Renga Formation, 118, 123–25, 127, 129–30
    in Claut Flysch, 44
    in Clauzetto Flysch, 47–48
    in Frosinone Formation, 119, 125
    in Gessi Formation, 140–41
    in Ligurian units, 78
    in Modena Plain, 61–63
    in Orbulina Marl, 119, 125
CIA. *See* chemical index of alteration (CIA)
Cidones-Abejar section, 200, 204, 206, 208
CIW, 188–89, 194
Clair oil field, 330–39
Claut area, 38–40, 42
Claut Flysch, 38–40, 42–44, 49–54
Clauzetto Flysch, 38–43, 47–54
clay
    in Maestrat Basin, 188
    minerals, weathering and, 161–62
    in Modena Plain, 66, 68
    in Po Plain, 27–33
    proxies for grain size, 31–35
    in Ravenna, 14
claystone, 259
climate, in SandGEM, 348–50
clinopyroxene, 322–23
Clutha River, 279, 281–84, 290–93, 295–96
coal
    coloration from, 249
    depositional environment, 233
    in Gondwana succession, 242, 248–49
    in Mecsek-Villány area, 235–36
cobalt
    in Cameros Basin, 214
    grain size and, 30, 31
    in Mecsek-Villány area, 233
    in Moncayo area, 214
    in Po Plain, 31
    in provenance studies, 214, 235
    sorting effects on, 27
cobalt/lanthanum values, 214, 235
cobalt/thorium values, 214, 235
"coeval," definition of, 168
Colina River, 38
Colorado Basin, 301, 302
Colorado River, 362
Columbrets Basin, 183
Como region, 54
Conglomerati Irregolari Formation, 138
Congo, Pointe Noire SandGEM worksheet, 376
continental crust, 216, 235

continental margin, 214, 215, 254
contourite, 278, 280
copper, 27, 30, 31
cordierite, 82
Coromandel volcanic zone, 255, 293
correlation and regression tree (CART), 329–32, 334–40
Corsica
    Balagne Nappe, 74–76, 78–88
    maps of, 75
    provenance of, 74–75
    Schistés Lustrées, 75
Cortes de Tajuña Formation, 201
Cortina d'Ampezzo, 2, 3
Costa Azul Basin, 301, 302, 316
Crati River, 151
Cribroelphidium spp., 20
Cribroelphidium pauciloculum, 20
Croatia, 223
Cserdi Formation, 223–35, 237–38
Cuevas Labradas Formation, 201
cumulative chemical weathering index (CCWI), 351. *See also* chemical index of weathering (CIW)
cyanobacteria, 141, 145, 171–77
cyanoliths, 168, 170–72, 175
Cycloforina costata, 18
Cylindrites, 110
Cyprides torosa, 18, 20
Cytheretta adriatica, 20
Cytheretta subradiosa, 20
Cytheridea neapolitana, 20

D
Dacia block, 223
dacites, 79–82, 87, 237
Dalradian succession, 332, 338
Damodar Basin, 243, 244
Damodar River, 243
Damuda Group, 242, 245, 248–49
Danube River, 223
Davenport Ranges, 369–74
Deep Western Boundary Current (DWBC), 278, 279, 292
Delfino Helvetic units, 75
Delivered Grain Size and Sorting, 353–54, 359
Deposited Grain Size and Sorting, 352, 353–54, 359
Deposited Matrix Abundance, 353–54, 359
Deposited Sand Composition, 353–54, 359
Deposited Sediment, 351, 359
Depositional Facies, 350–54, 358, 359, 362
Depositional Fining Potential
    Delivered Grain Size and Sorting and, 352
    Deposited Grain Size and Sorting and, 358, 359
    description of, 352
    node type, 353
    states of, 362, 363
Depositional Modification Potential, 352, 353–54, 359, 363

Depositional Sorting Potential
    Delivered Grain Size and Sorting and, 352
    Deposited Grain Size and Sorting and, 358, 359
    description of, 352
    node type, 353
    states of, 362, 363
Dharmaram Formation, 245, 247
diabase, 58
Diatom Formations, 323
diatoms, 171–73
Dickinson provenance fields, 274
Dinarides, 38–39, 53, 223
Dinnyeberki 9015 core, 224–29
Diósviszló-1 core, 224–26, 228–30, 232
dolerite
    in Caspian Basin, 328, 329, 331
    in Clair oil field, 331, 334, 339
    in provenance studies, 321
dolomite. *See also* limestone
    abrasion of, 10
    in Belluno Flysch, 43, 45–47, 50
    in Boite River, 3–6, 8–11
    versus calcite, 10
    in Claut Flysch, 43–44
    in Clauzetto Flysch, 43, 47–48, 50
    in Gadera River, 3–5, 7, 8–11
    Hetvehely Dolomite Formation, 225
    kaolinite and, 236
    in Lamone River, 5
    Lienz Dolomites, 54
    in Ligurian region, 10
    magnesium/aluminum values and, 33
    in Piave River, 3–6, 8–11
    in Po Plain, 31
    in Rienza River, 3–5, 7, 8–11
    shape of, 4, 5, 9, 10
    weathering of, 10
Dolomite Mountains, 2
dolostone
    in Argilloso-Arenacea Formation, 119–20, 122
    in Belluno Flysch, 45–47
    in Brecce della Renga Formation, 118, 122, 124, 125
    in Claut Flysch, 43–44
    in Clauzetto Flysch, 47–48
    in Frosinone Formation, 122
    in Iberian Range, 169
    in Ligurian units, 82, 87
Drauzig, 223
Drava River Basin, 223, 225
DSDP 594 site, 278, 293
Duero Basin, 183
DWBC, 278, 279, 292

E
East Coast allochthon (New Zealand), 254, 257, 266, 272–74
East European craton, 323
Ebro Basin, 183, 184

echinoids, 125
Egypt, Nile River, 362
El Alamito Basin, 301, 302, 316
El Arco area
    100 m isobath offshore of, 299
    geochronology of, 313
    geological setting of, 298–99
    grain-shape studies in, 304–5, 307, 308–10, 312–13
    map of, 301
    mineralogical studies of, 302
    Palma Seda Basin, 299, 301
El Arco Basin, 301
El Bledito Basin, 301
El Cabo geological subprovince, 298
El Cardonal area
    100 m isobath offshore of, 298
    geochronology of, 298
    geological setting of, 298
    grain-shape studies in, 304, 306, 309, 310, 312, 316
    map of, 301
    Migriño Basin, 298, 301, 302
El Cardonal Basin, 301, 302
El Cardoncito Basin, 301
El Castellar Formation, 184–94
El Diablo Basin, 301
El Guayparin Basin, 301, 302, 315
El Guayparin beach, 315
El Mangle Basin, 301, 302
El Salto Basin
    area of, 298
    geological setting of, 298
    grain-shape studies in, 308, 313
    lithology of, 302
    map of, 301
El Tiburón area
    100 m isobath offshore of, 299
    geological setting of, 299
    grain-shape studies in, 307, 308–9, 311, 312, 314–15
    map of, 301
El Tiburón Basin
    area of, 299
    grain-shape studies in, 308, 314
    lithology of, 302
    map of, 301
El Tule area
    Cerro Blanco, 299–302, 315
    geological setting of, 299
    grain-shape studies in, 308, 309, 311, 312, 315–17
    map of, 301
    Punta Palmillas, 299–302, 315
El Tule Basin, 298–99
Elba Island, 9, 10
Elphidium spp., 17, 18
Elphidium advenum, 20
Elphidium crispum, 20
Elphidium granosum, 18, 20
Elphidium lidoense, 17, 20

Elphidium macellum, 20
Emilia region, 14, 26, 71
Enciso Formation, 201
Enza River, 59
epidote, 51–53, 320, 340
Epi-Ligurian sequence, 58, 59
Ernici Mountains, 111
Erto area, 38–40, 42
Escucha Formation, 201
etch pits, factors affecting, 160–61
Ethiopian Highlands, 359
European Plate
    Apennine orogenesis and, 58, 74–77, 88–90, 109–10
    Korpád Formation and, 236
    Ligure-Piemontese Basin and, 74–77, 88–90
    map of, 223
    Tisza-Dacia unit and, 222
europium
    apatites and, 321
    in Balagne Nappe, 82
    in Cameros Basin, 212, 214
    in CART, 329
    in Caspian Basin, 323, 326, 327, 329, 330
    in Clair oil field, 332–33, 337, 341
    feldspar crystallization and, 321
    fractionation and, 326–27
    in Ligurian units, 82, 84
    in Mecsek-Villány area, 233–34
    in Moncayo area, 212, 214
    in provenance studies, 82, 235, 327
extrabasinal carbonate grains. *See* CE
Eyre, Lake, 369

F
Fanna area, 38, 40–42
feldspar. *See also* potassium feldspar
    in Apennines, 66
    in Argilloso-Arenacea Formation, 122
    in Belluno Flysch, 43, 45–47
    in Boite River, 4, 6
    in Brazilian beach sand, 364, 367
    in Claut Flysch, 43–44
    in Clauzetto Flysch, 43, 47–48
    climate and, 66
    in Clutha River, 290
    in El Arco area, 302
    europium and, 321
    in Frosinone Formation, 122
    in Gadera River, 4
    in Gessi Formation, 140
    glaciers and, 66
    in Gondwana succession, 248
    grain size and, 30–33
    hydrothermal effects in, 209
    in Macigno unit, 100–102
    in Maestrat Basin, 187–88
    in Mecsek-Villány area, 230, 234, 236
    in Modena Plain, 64, 66–71
    in ODP 1122 core, 284–86
    in Pampas Central, 368–70

    in Piave River, 4, 6
    planimeter studies of, 5
    in Rangitata River, 290
    in Rienza River, 4
    in sand maturity index, 9
    in Sila Massif area, 154–61
    thermal response of, 160
    in Umbum Creek, 369–74
    in Waipaoa River Basin, 264
    in Waitaki River, 290
    weathering of, 32, 160–62
feldspar/illite-smectite values, 33
feldspar/kaolinite values, 33
Fennoscandia Block, 323, 328
ferric oxide. *See* iron oxide
Fiorenzuola Mélange, 59
Fiumalbo Shale, 97–99
Flandrian transgression, 313
fluorine, 320
Flushing Potential
    in Bayesian network, 354
    calculation of, 375–76
    Fluvial Storage Potential and, 351, 362
    Generative Modification Potential and, 359
    Hinterland Transport Distance and, 351
    node type, 353
    precipitation and, 360
    Regional Topographic Gradient and, 351, 359
    in river assessment studies, 350
    states of, 359, 360, 362
    in transport process, 351
Fluvial Storage Potential, 351–54, 360, 362
flysch deposits
    Allgäu Flysch, 53
    Aptian-Albian Flysch, 54
    Belluno Flysch, 38–43, 45–47, 49–54
    Claut Flysch, 38–40, 42–44, 49–54
    Clauzetto Flysch, 38–43, 47–54
    Helminthoid Flysch, 75–77
    Lombardian Flysch, 54
    Lydienne Flysch, 76, 78–80, 83
    Monte Cassio Flysch, 77
foraminifers
    in 187-S1 core, 17–18
    in 204-S15 core, 20
    in Balagne Nappe, 79
    Belluno Flysch and, 39, 42, 43
    in Brecce della Renga Formation, 125
    Claut Flysch and, 39, 42, 43
    Clauzetto Flysch and, 39, 43
    in Orbulina Marl, 110, 125
    planktonic zones, 42
Fossa Stream
    diversion of, 58
    map of, 60
    sand composition along, 61, 68, 70
France, Simme nappe, 54
francolite, 320
Friuli Basin, 53
Friuli Platform, 38, 43, 51, 53
Frosinone Formation

Argilloso-Arenacea Formation and, 116
geological setting of, 110
in Latina Valley, 112
lithostratigraphic studies of, 125–27
map of, 111
petrographic studies of, 119, 121–22, 125–30
photographs of, 114
photomicrographs from, 123
provenance of, 113, 122, 127–30
reconstruction of, 128
ternary plots of, 126
frost weathering, 159–60

G

gabbro
    in Caspian Basin, 326
    in Clair oil field, 337
    on Corsica, 75, 79
    in Modena Plain, 58
    in Sila Massif area, 150
Gadera River, 2–5, 7–11
Galve subbasin, 183–84, 186–94
Gan section, 204, 206, 208
Ganga River, 243
Ganges River, 362
Gare de Novella Sandstone. *See* Novella Sandstone
Garicchi Formation, 139
Gariglione Complex, 136
garnet
    in Argilloso-Arenacea Formation, 123
    in Belluno Flysch, 50, 51, 52
    in Clair oil field, 332
    in Claut Flysch, 50, 51, 52
    in Clauzetto Flysch, 50, 51, 52
    in Frosinone Formation, 123
    from Kura River, 322, 324
    in Ligurian units, 82
    in Mecsek-Villány area, 230
    phosphorus pentoxide and, 30
    in provenance studies, 320
    in Sila Massif area, 150
    from Volga River, 322, 324
    yttrium and, 30, 33
    zirconium and, 30, 33
gastropods, 64, 175
Generated Clay Abundance, 353–55, 357, 359
Generated Grain-Size Distribution, 353–55, 357
Generated Sand Composition, 353–55, 359
Generated Sediment, 350–51, 355
Generative Modification Potential, 351–55, 357, 359, 362
Genzana, Mount, 111
Germany
    Allgäu Flysch, 53
    Lienz Dolomites, 54
Gessi Formation, 138–47
Gessoso-Solfifera Formation, 110
glaciers
    in Apennines, 71
    Boite River and, 6

cryoclastism and, 160
feldspar and, 66
Gadera River and, 9
Gondwana succession and, 242
Modena Plain sediment and, 71–72
in New Zealand, 281
Piave River and, 6
Rienza River and, 9
sand composition and, 66
glaucophane, 50
gleying, 159
Globanomalina pseudomenardii, 40
Globobulimina affomos, 17
Globotruncana appenninica, 125
Glossopteris, 245, 250
gneiss
    in Argilloso-Arenacea Formation, 123
    in Balagne Nappe, 79, 87
    in Brazilian Shield, 366
    in Calabria, 10
    in Clair oil field, 337–38
    fluvial transport of, 10
    in Frosinone Formation, 123
    in Gessi Formation, 140
    from Lewisian, 334
    in Ligurian units, 82, 87
    in Mecsek-Villány area, 230, 234
    in provenance studies, 328
    in Sila Massif area, 150
    in Umbum Creek, 372
Gneiss Complex, 334
Godavari River, 243
Gondwana succession, 242–49
Görcsöny-Máriakéménd Ridge, 224, 225
Görgeteg subunit, 223, 234
Gottero Sandstone, 76, 78, 79, 83
grabens, continental margin and, 214
Grampian Group, 338
Gran Sasso, 110, 111, 128
granites
    in Baja California, 300
    in Balagne Nappe, 79, 87
    in Caspian Basin, 326–31
    in Clair oil field, 331, 334, 337–39, 341
    in Gessi Formation, 140
    lanthanum/neodymium values for, 321
    in Ligurian units, 82, 87
    in Mecsek-Villány area, 230, 234
    REE levels in, 326
    in Sila Massif area, 152–53
    source-rock discrimination diagrams for, 216
    tectonic setting of, 234
    uranium in, 321
    weathering of, 160–61
granodiorite, 300, 326, 337
grass, 374
gravel, 66
Greco, Mount, 111
Greenland, Clair oil field and, 332
Grizzaga Stream, 60–61, 63, 65, 67
Guerro Stream

map of, 60
sand composition along, 60, 62, 63, 69, 71
ternary plots of, 65, 67
Gulf of Mexico, 362
gypsarenites, 139, 145
gypsum, 135, 140–46
Gyűrűfű Rhyolite Formation, 223–26, 237

H

Haast Schist, 281, 292
hafnium
    in Cameros Basin, 214, 216
    in Mecsek-Villány area, 233
    in Moncayo area, 214, 216
    in provenance studies, 182, 214, 235
    zircon and, 195, 237
halite, 139
halloysite, 162
Hantkenina, 41
Haynesina spp., 17
Haynesina depressula, 20
Helminthoid Flysch, 75–77
Hesperian Massif, 169, 200, 209, 215. *See also* Iberian Massif
Hetvehely Dolomite Formation, 225
high field strength elements (HFSE)
    in Balagne Nappe, 82, 84
    hafnium. *See* hafnium
    in Ligurian units, 82, 84
    in Mecsek-Villány area, 233, 235–37
    niobium. *See* niobium
    in provenance studies, 84
    tantalum, 86, 182, 195
    thorium. *See* thorium
    titanium. *See* titanium
    uranium, 214, 233, 237
    yttrium. *See* yttrium
    zirconium. *See* zirconium
Hikurangi subduction margin, 254–56, 273
Himalayas, 359
hinterland
    boundaries of, 350
    definition of, 346
Hinterland Transport Distance
    in Bayesian network, 354
    description of, 350
    Fluvial Storage Potential and, 351, 360, 362
    node type, 353
    Selective Transport Fining and, 351, 360, 362–63
    states of, 360
holoptychian fish, 332
Hontoria del Pinar Formation, 201
hornblende, 30
Hortezuelos Formation, 201
humic acids, 161
Hungary
    maps of, 223, 224
    Mecsek-Villány area, 222–38
hurricanes, 317

I

Iberian Massif, 183, 193, 195. *See also* Hesperian Massif
Iberian Plate, 109–10, 183, 201–2
Iberian Range
   climate of, 169
   development of, 201
   geochemistry of, 193
   geological setting of, 169, 182–83, 201
   lithology of, 168
   lithostratigraphic studies of, 184
   maps of, 169, 183, 184, 200
   photomicrographs of sand from, 172, 174, 176
   sand composition in, 170–77
   SEM images of sand from, 173
   volcanism in, 215
Idice River, 14, 20–21, 26
illite
   aluminum and, 213
   in Cameros Basin, 209
   chlorite and, 213
   formation of, 162
   grain size and, 30–33
   kaolinite and, 236
   in Macigno unit, 103
   in Mecsek-Villány area, 236
   in Moncayo area, 209
   in Po Plain, 31, 33
   potassium and, 213
Imon Formation, 201
India
   Ganges River, 362
   Gondwana succession in, 242–50
   map of, 243
Indian Plate, 254, 279
input nodes, 352–53
insolation heating, 160
intermediate nodes, 352–54
Ionian Plate, 109–10
iron
   basaltic rocks and, 33
   in Caspian Basin, 326, 327, 329–32
   in Clair oil field, 331, 334, 337, 338, 341
iron oxide
   clay fractionation and, 30
   grain texture and, 43
   in Mecsek-Villány area, 231–33
   in Modena Plain, 61–63
   in Orbulina Marl, 125
   in Po Plain, 31
   in Sila Massif area, 152, 154
   weathering and, 160–61
iron oxide/silicon oxide values, 31–34
Isarco River, 2
Italy
   Abetone area, 96–104
   Alpago area, 38, 41, 42, 50
   Apennines. *See* Apennines
   Belluno Flysch, 38–43, 45–47, 49–54
   Calabria. *See* Calabria
   Claut Flysch, 38–40, 42–44, 49–54
   Clauzetto Flysch, 38–43, 47–54
   Como region, 54
   Emilia region, 14, 26, 71
   maps of
      bedrock, 3
      cores, 14, 26
      flysch deposits, 38
      geologic, 59, 97, 137, 151
      paleogeographic, 53
      Periadriatic lineament, 38
      river drainages, 2, 14, 26, 38, 59
      stream drainages, 38, 60
      structural, 109, 111
      Tauern Window, 38
      tectonic, 75
      turbidites, 38, 96
      Valsugana thrust system, 38
   Milano region, 54
   Modena Plain, 58–72
   Po Plain. *See* Po Plain
   Romagna region, 14, 26
   Trentino–Alto Adige region, 2–11
   Veneto region, 2–11, 14, 21, 38
Ivrea Verbano unit, 75

J

jacupirangite, 327–29, 331, 338–39
Jakabhegy Sandstone Formation, 225
Jubera section, 200
Juniata River, 362

K

Kamthi Formation, 245, 247
kaolinite
   biotite and, 162
   in Cameros Basin, 209
   chlorite and, 236
   depositional environment, 162
   dolomite and, 236
   factors affecting, 236
   grain size and, 30, 31
   illite and, 236
   in Macigno unit, 103
   magnesium and, 236
   in Mecsek-Villány area, 236
   in Moncayo area, 209
   in Po Plain, 31
   SEM images of, 156
   vermiculite and, 162
kaolinite/feldspar values, 33
Keuper Formation, 201
Khazar Formation, 323
kimberlite, 326, 337
Kirmaky Suite, 323
Kirmaky Valley, Productive Series, 322–32, 339
Kodru, 223
Koel Basin, 243
Korpád Formation, 223–37
Kota Formation, 242–49
Kővágószőlős Formation, 223
Krishna River, 243
Kunság unit, 223, 237
Kura River, 322–32, 339
kyanite, 50, 51

L

La Laguna Basin, 301
Laga Mountains, 111
Laguna Grande Basin, 301
LA-ICP-MS studies, 320, 321
Lamone River
   geochemistry of sediment from, 20, 21
   maps of, 14, 26
   sand composition in, 5
landslides, 160
lanthanum, 30, 31, 214
(lanthanum + cesium)/ΣREE values
   in Caspian Basin, 323, 326–32
   in Clair oil field, 330, 332, 334, 337
   in provenance studies, 321, 338, 339
lanthanum/cobalt values, 214, 235
lanthanum/neodymium values
   in Caspian Basin, 323, 326–32
   in Clair oil field, 330, 332, 334, 337
   in provenance studies, 321, 338, 339
lanthanum/samarium values, 84, 87, 212, 233–35
lanthanum/scandium values, 235
lanthanum/terbium values, 215
lanthanum/thorium values, 214, 216
lanthanum/ytterbium values, 84–85, 212, 233–34
large ion lithophile elements (LILE)
   in Balagne Nappe, 82, 84
   barium. *See* barium
   in Cameros Basin, 212
   cesium, 212, 213
   in Ligurian units, 82, 84
   in Mecsek-Villány area, 233
   in Moncayo area, 212
   potassium. *See* potassium
   rubidium. *See* rubidium
   sample provenance from, 84
   strontium. *See* strontium
larvikites
   in Caspian Basin, 328, 329, 331
   in Clair oil field, 331, 334, 339
Las Margaritas Basin, 301, 302
Las Salinas Basin, 301, 302, 316
laser-ablation inductively coupled plasma-mass spectrometry, 320, 321
Last Glacial Maximum, 71
Latina Valley, 111, 112
Latium-Abruzzi platform, 110, 128, 129
Lazio-Abruzzi platform, 113, 129
Le Mainarde area, 111
lead, 31
leaf nodes, 352–53
Lepini Mountains, 111, 128
Leptocythere spp., 20
Leptocythere bacescoi, 20
Leptocythere levis, 20
Lessini Shelf, 38, 41–42, 53

lherzolites
    in Caspian Basin, 327, 328, 330–32
    in Clair oil field, 331, 338–39, 341
    on Corsica, 75
lichens, 160
Lienz Dolomites, 54
Ligure-Piemontese Basin, 74–75, 77, 88–90
Ligurian region, 10
Ligurian units
    composition of, 58
    geochemistry of, 82–88
    geological setting of, 58, 76
    lithostratigraphic studies of, 76
    maps of, 59, 75
    petrographic studies of, 78, 79–82, 87
    photomicrographs from, 80
    provenance of, 74–77, 82–90
    spider diagrams for, 86
    triangular compositional plots of, 81
Limana Stream area, 38, 41–43, 50
limestone
    in Argilloso-Arenacea Formation, 119–20, 122
    in Balagne Nappe, 79
    in Belluno Flysch, 43, 45–47
    in Boite River, 3, 4, 6, 10–11
    in Brecce della Renga Formation, 118, 122, 124, 125
    Briozoi and Litotamni Limestones Formation, 110, 112, 128
    Calpionella Limestone, 75, 76, 83
    in Claut Flysch, 43–44
    in Clauzetto Flysch, 43, 47–48
    in Frosinone Formation, 119, 122
    in Gadera River, 3, 4, 7, 8, 10–11
    in Gessi Formation, 140–41
    in Iberian Range, 169
    in Lamone River, 5
    in Ligurian units, 82
    in Maestrat basin, 184
    in Maestrat Basin, 186
    in Modena Plain, 58, 61–64, 68–70
    Nagyharsány Limestone Formation, 225
    in Orbulina Marl, 110, 119
    in Piave River, 3, 4, 6, 10–11
    planimeter studies of, 5
    in Rienza River, 3, 4, 7, 8, 10–11
    in Rossano Basin, 136, 139
    San Colombano Limestone, 75–76, 79
    shape of, 4
    Szársomlyó Limestone Formation, 225
lithium, 182, 195
Lombardian Flysch, 54
Longobucco Group, 136–38, 150
Los Arcos Basin, 301
Los Cabos province, 301, 316–17
Los Medanos province, 301, 316–17
Loxoconcha spp., 20
Loxoconcha elliptica, 20
Loxoconcha stellifera, 20
Loxoconcha tumida, 20
lutite, 300
Lydienne Flysch, 76, 78–80, 83
Lyngbya, 171

M
Macigno unit, 96–104
Madison River, 362
Maestrat Basin, 183–94
magnesium, 33, 212, 236, 329
magnesium chlorite, 85, 87
magnesium oxide, 31, 231–33
magnesium/aluminum values, 20–21, 23, 33, 34
Mahadeva Formation, 242–49
Mahanadi Basin, 243
Mahanadi River, 243
Maiella area, 110, 111
Maitai Terrane, 281
Malay Basin, 362
Maleri Formation, 242, 244–49
Mandatoriccio Complex, 136, 147
Mangaheia Group, 257
manganese
    calcium and, 320
    in Caspian Basin, 326, 327, 329
    in Clair oil field, 333–34, 340
    in provenance studies, 320
manganese oxide, 31, 210, 231–33
manganese/strontium values, 320, 326
Manganesiferous Shale, 79
Mangatu Group, 257
Mangatu River, 255, 262–63, 266, 271–72
Mangli Formation, 247
Máriagyűd-1 core, 224–29, 231, 232, 234
Marihiku Terrane, 281
marls
    Marmoreto Marl, 97–100
    Monte Verzi Marls, 79
    Orbulina Marl. *See* Orbulina Marl
    Pievepelago Marl, 96, 97
    in Po Plain, 58
    in Rossano Basin, 139
marlstone, 5
Marmolada Mountains, 2
Marmoreto Marl, 97–100
Marne di Vicchio, 96
Marnoso-Arenacea unit, 96
Marsica area, 111, 116, 129
Matawai Group, 254, 256, 257
matrix, in SandGEM, 359
maximum entropy method, 304
Maximum Grain Size
    Basin Fluvial Transport Distance and, 352, 360
    in Bayesian network, 354
    Delivered Grain Size and Sorting and, 358, 359
    in depositional process, 352
    grain-size distribution and, 358
    node type, 353
    Rate of Basin Subsidence and, 352, 362
    states of, 363
    Transported Grain-Size Distribution and, 358, 359
Maykop Formation, 323
Mecsek Mountains. *See also* Mecsek-Villány area
    cross section of, 225
    geological setting of, 223
    lithology of, 237
    map of, 224
    stratigraphy of, 226
Mecsek-Villány area, 222–38
Médanos Partidos Basin, 301
Media Val Taro unit, 76–78, 82, 84
Medone Stream area, 38, 40–42
mélange
    Fiorenzuola Mélange, 59
    in Ligurian units, 77
    in Modena Plain, 58
    in New Zealand, 257
Meliata Ocean, 53
Messinian salinity crisis, 136, 139, 146
Mexico, Baja California peninsula, 298–318
Mexiglossa, 250
mica
    in Argilloso-Arenacea Formation, 119–20
    in Balagne Nappe, 79
    in Belluno Flysch, 45–47
    in Boite River, 4
    in Brazilian beach sand, 367
    in Brecce della Renga Formation, 118
    in Claut Flysch, 44
    in Clauzetto Flysch, 47–48
    in El Arco area, 302
    in Frosinone Formation, 119
    in Gadera River, 4
    in Gessi Formation, 141
    grain size and, 30, 31
    in Ligurian units, 79–82
    in Macigno unit, 102
    in Mecsek-Villány area, 230, 234, 236
    in Neto River Basin, 163
    in ODP 1122 core, 284, 286–90
    in Orbulina Marl, 119
    in Pampas Central, 370
    in Piave River, 4, 7
    in Po Plain, 31
    provenance from, 291
    in Rienza River, 4
    in sand maturity index, 9
    in Sila Massif area, 154, 161
    in Umbum Creek, 373, 374
    weathering of, 10, 160, 162
Microcodium, 175
Mid-Hungarian line, 222, 223
mid-ocean-ridge basalts (MORBs), 76, 79
migmatite, 326, 337
Migriño Basin, 298, 300, 301, 302
Milano region, 54
Miliolinella elongata, 20

Miliolinella oblonga, 20
Miliolinella subrotunda, 20
mineral-dissolution reaction rates, 348
mineralogical maturity index, 9
miospores, 332
Mississippi River, 362
Missouri River, 362
Modena Plain, 58, 58–72
modified maturity index, 4, 9
Moine succession, 332, 338
Molassa di Castiglione, 136–39
Moldanubian zone, 222, 237
Molyneux Canyon, 279
monazite
    in Belluno Flysch, 50–52
    in Claut Flysch, 50–52
    in Clauzetto Flysch, 50, 52
    crystallization of, 321
    in pelites, 82
    phosphorus and, 237
    phosphorus pentoxide and, 30
    in provenance studies, 50
    thorium and, 237, 321, 327
    yttrium and, 30, 237
    zirconium and, 30
Moncayo area, 200, 202–17
Montana, Madison River, 362
Monte Cassio Flysch, 77
Monte Cervarola Sandstone, 97
Monte Falterona Sandstone, 97
Monte Modino Olistostrome, 97–100, 103
Monte Saccarello unit, 75
Monte Verzi Marls, 79
Montone River, 14, 20, 21, 26
Mora Formation, 184–94
Mórágy Complex, 223, 225, 234–35, 237
Mórágy Hills, 224, 237
Morella subbasin, 183
Morozovella aragonensis, 41, 43
Morozovella formosa formosa, 40, 41, 43
Morozovella velascoensis, 40
Morrone Mountains, 111
Mount Antola unit, 75
Mount Falterona Sandstone, 96
Mount Rizzone Palombini Shale, 76, 77
Mucone River, 150, 151
mudstone
    in Balagne Nappe, 79, 87
    in Belluno Flysch, 45–47
    in Brecce della Renga Formation, 125
    in Claut Flysch, 44
    in Clauzetto Flysch, 47–48
    in Ligurian units, 79–82, 87
    in Modena Plain, 61–64
    in Po Plain, 58
    provenance signature, 257
    in Waimata River Basin, 272
    in Waipaoa River Basin, 254, 259, 272–73
Munecas Formation, 201
Muriel section, 200
Muschelkalk Formation, 201, 202, 209

muscovite
    in Balagne Nappe, 79
    in Clutha River, 290
    in Ligurian units, 82
    in Macigno unit, 100–101
    in Maestrat Basin, 186
    in Mecsek-Villány area, 227, 230, 231, 236
    in Modena Plain, 61–63
    in ODP 1122 core, 284–85
    in Rangitata River, 290
    in Waitaki River, 290

N

Nagyharsány Limestone Formation, 225
Narmada River, 243
Navaccia unit, 75–76, 78, 80, 83
Naviglio Stream, 60
NCE
    in Argilloso-Arenacea Formation, 119–22
    in Balagne Nappe, 78
    in Belluno Flysch, 45–47
    in Brecce della Renga Formation, 118, 121, 127, 129
    in Clauzetto Flysch, 47–48
    in Frosinone Formation, 119, 121, 122
    in Gessi Formation, 140–42
    in Ligurian units, 78
    in Modena Plain, 61–63
    in Orbulina Marl, 119, 121
NCI
    in Argilloso-Arenacea Formation, 119–20, 125
    in Balagne Nappe, 78
    in Belluno Flysch, 45–47
    in Brecce della Renga Formation, 118, 125, 127
    in Claut Flysch, 44
    in Clauzetto Flysch, 47–48
    in Frosinone Formation, 119, 125
    in Gessi Formation, 140–42
    in Ligurian units, 78
    in Modena Plain, 61–63
    in Orbulina Marl, 119, 125
neodymium, 329
neodymium/lanthanum values. See lanthanum/neodymium values
Neto River Basin, 150–51, 156–59, 163–64
Neuropteris, 224
New Zealand
    Bounty Fan system and, 278–93, 295–96
    geological setting of, 254–55, 279–81
    maps of
        Alpine fault, 278
        Bounty Fan system, 278
        East Cape river systems, 255
        geologic, 257, 281
        tectonic setting of, 256, 278
        Waipaoa River system, 254–74
Nicà River, 150, 151
nickel
    in Balagne Nappe, 88

    grain size and, 32
    in Ligurian units, 88
    in Mecsek-Villány area, 233, 235
    ophiolites and, 14, 31
    in Po Plain, 31
    in Po River Basin, 14
    in provenance studies, 235
    redox state and, 33
    serpentine and, 31
    sorting effects on, 27
    turbidite systems and, 14
    ultramafic rocks and, 31, 33
nickel/aluminum values, 16, 18–23, 32–34
nickel/chromium values, 235
nickel/clay values, 32
nickel/vanadium values, 215
Nile River, 362
niobium
    in Balagne Nappe, 85, 86
    grain size and, 30
    in Ligurian units, 85
    in Mecsek-Villány area, 233
    in Po Plain, 31
    in provenance studies, 74, 86, 182
    titanite and, 195
Nizzola Stream, 60, 62, 69, 71
nodes in models, 352–53
noncarbonate extrabasinal grains. See NCE
noncarbonate intrabasinal grains. See NCI
Nonionella turgoda, 17
North American shale composition (NASC), 191
Novella Sandstone, 76, 78, 79, 83
nummulitids, 125

O

ODP 1119 site, 278
ODP 1120 site, 278
ODP 1121 site, 278, 292
ODP 1122 site, 278–80, 283–93, 295–96
Ohio River, 362
Oliete subbasin, 183
olistostromes, 58, 97–100, 103
Oliván Formation, 201
olivine, 50, 51
Oncala Formation, 201
ophiolites
    in Apennines, 74, 76
    Belluno Flysch and, 50
    in Caucasus Mountains, 322
    chromium and, 14, 31
    Claut Flysch and, 50
    Clauzetto Flysch and, 50
    on Corsica, 74–76, 79
    maps of, 38, 75
    in Modena Plain, 58, 68–70
    in New Zealand, 255
    nickel and, 14, 31
    of Vardar Ocean, 50
Ophiomorpha, 42
Orbitolina, 125
Orbulina Marl

age of, 110
Argilloso-Arenacea Formation and, 115
geological setting of, 110, 113
in Latina Valley, 112
lithostratigraphic studies of, 110, 125–27
petrographic studies of, 119, 121, 125, 130
photographs of, 114
photomicrographs from, 124
provenance of, 127–30
in Salto Valley, 113
ternary plots of, 126
in Varri Valley, 113
Orinoco River, 362
Ortona-Roccamonfina line, 109
Ostia Sandstone, 76–78, 82, 84
ostracods, 18, 20, 175
Otago area, 281, 293
output nodes, 352–53
Ovalveolina maccagnoi, 125
Oyster Creek, 362

P
Pacific Plate, 254, 256, 278, 279
Paleotethys Ocean, 222, 223. *See also* Tethys Ocean
Palma Seda area, 302, 313
Palma Seda Basin, 299, 301
Palombini Shale, 75–77, 83–84
Palopoli Molassa unit, 138, 139
Paludi Formation, 138, 147
palynomorphs, 224
Pamir Mountains, 322
Pampas Central, 368–71
Panaro River, 58–71
Panchet Formation, 242, 244–49
Pangea, map of, 223
Pannonian Basin, 362. *See also* Carpathian-Pannonian area
Panthallassa, map of, 223
Papua New Guinea, 10, 359
Papuk, 223
parent nodes, 352–53
Parma River, 59
Parsora Formation, 242–46, 248–50
Patacs Siltstone Formation, 225
Pecopteris, 224
Pécs area, 224
pectinids, 125
pegmatite
in Caspian Basin, 326–31
in Clair oil field, 331, 333, 334, 337–39, 341
in provenance studies, 321
in Sila Massif area, 150
pelite, 82
"penecontemporaneous," definition of, 168
Pennsylvania, Juniata River, 362
Penyagolosa subbasin, 183–94
Perelló subbasin, 183
Pereriva Suite, 323–32, 339
Periadriatic Lineament, 38, 53

peridotites, 50
Pescorocchiano area, 114
Phormidium, 171
Phormidium foveolarum, 175
Phormidium incrustatum, 171
phosphorus
apatite and, 195, 237
monazite and, 237
in provenance studies, 182
in rutile, 237
sorting effects on, 27
zircon and, 237
phosphorus oxide, 30, 233, 237, 320
phyllite
in Argilloso-Arenacea Formation, 119–20
in Boite River, 5–7
in Brazilian Shield, 366
in Brecce della Renga Formation, 118
in Frosinone Formation, 119
in Gadera River, 5, 8
in Gessi Formation, 140–41, 147
in Ligurian units, 82
in Piave River, 5–7
in Rienza River, 5, 8
in Sila Massif area, 150
in Umbum Creek, 372
Piave River, 2–11, 14, 21, 38
Picofrentes Formation, 201
picotite, 82
Piemonte Basin, 75
Pievepelago Marl, 96, 97
plagioclase
in Argilloso-Arenacea Formation, 119–20, 122
in Balagne Nappe, 79
in Belluno Flysch, 43, 45–47
in Boite River, 4, 6
in Brazilian beach sand, 364
in Brecce della Renga Formation, 118
in Claut Flysch, 43–44
in Clauzetto Flysch, 43, 47–48
in Clutha River, 290
in Frosinone Formation, 119, 122
in Gadera River, 4, 8
in Gessi Formation, 140–41
in Ligurian units, 79–82
in Macigno unit, 100–101
in Maestrat Basin, 186, 187–89
in Mecsek-Villány area, 227, 230, 231, 234
in Modena Plain, 61–64
in Neto River Basin, 163
in ODP 1122 core, 284
in Orbulina Marl, 119
in Pampas Central, 369
in Piave River, 4, 6
in Rangitata River, 290
in Rienza River, 4, 8
in Sila Massif area, 155–56, 160, 161
in Umbum Creek, 369–71
in Waitaki River, 290

weathering of, 160
Planolites, 42
plants, 374
plutons
in Brazilian Shield, 366
in Pampas Central, 368–69
Sila Batholith. *See* Sila Batholith
in Umbum Creek, 372
Po Plain
187-S1 core from. *See* 187-S1 core
204-S1 core, 26–35
204-S2 core, 26–35
204-S8 core, 26–35
204-S15 core, 14, 18–21, 23
205-S3 core, 26–35
205-S5 core, 26–35
205-S10 core, 26–35
221-S19 core, 26–35
239-S2 core, 26–35
240-S9 core, 26–35
area of, 27
basin thickness, 58
geochemistry of, 14, 20–21, 23, 30–33
geological setting of, 27, 58
maps of, 14, 26, 75
micropaleontological studies of, 14
mineralogical studies of, 14
provenance of, 71
sequence-stratigraphic analysis of, 6, 14
Po River
187-S1 core and, 15
204-S15 core and, 18
direction of flow, 27
maps of, 14, 26, 59
Poland, Sudetic Basin, 237
Ponthocytere turbida, 20
Pontian Formation, 323
Portugal, maps of, 169, 183, 200
potassium
alkaline volcanism and, 33
barium and, 32, 213
in Cameros Basin, 212–13
cesium and, 213
grain size and, 32
illite and, 213
in Moncayo area, 212–13
rubidium and, 32, 189, 213
weathering and, 212, 236
potassium feldspar
in Argilloso-Arenacea Formation, 119–20, 122
in Balagne Nappe, 79
in Boite River, 6
in Brecce della Renga Formation, 118
in Cameros Basin, 209, 213
chlorite and, 213
in Clutha River, 290
in Frosinone Formation, 119, 122
in Gadera River, 8
in Gessi Formation, 140–41

grain size and, 30
in Ligurian units, 79–82
in Macigno unit, 100–102, 104
in Maestrat Basin, 186, 187–89
in Mecsek-Villány area, 227, 230, 231
in Modena Plain, 61–64
in Moncayo area, 209, 213
in Neto River Basin, 163
in ODP 1122 core, 284–86
in Piave River, 6
in Po Plain, 31
in Rangitata River, 290
in Rienza River, 8
rubidium and, 182, 189
in Sila Massif area, 160
in Waitaki River, 290
weathering of, 160
potassium oxide
in Balagne Nappe, 85
in Cameros Basin, 210
in Ligurian units, 85
in Mecsek-Villány area, 231–34, 236
in Moncayo area, 210
Potonieisporites, 224
Pozalmuro Formation, 201
Pozo de Cota Basin, 301, 302
Pranhita River, 243
Pranhita-Godavari Basin, 243–46, 249, 250
Precambrian shield, 242
Prepiemontese unit, 75
Productive Series, 321–32, 339
Provenance Lithotypes
in Bayesian network, 354
characteristics of, 348
description of, 354
Generated Sediment and, 350, 355
major-element oxide characteristics, 356
mineralogical characteristics of, 355
tectonic setting and, 348
Pseudotriloculina spp., 20
Pseudotriloculina rotunda, 18
Punta Lost Arcos Basin, 301
Punta Palmillas area, 299–302, 315
Punta Palmillas Basin, 301, 315
Pyrenees, 175, 183
pyrite, 31, 230
pyroxene, 340

Q
quartz
abrasion of, 10, 302
in Argilloso-Arenacea Formation, 119–20, 122
in Balagne Nappe, 79
in Belluno Flysch, 43, 45–47
in Boite River, 4, 6
in Brazilian beach sand, 364, 367
in Brecce della Renga Formation, 118
in Cameros Basin, 202
in Cecita Lake area, 156
in Claut Flysch, 43–44
in Clauzetto Flysch, 43, 47–48
in Clutha River, 290
in El Arco area, 302
in Frosinone Formation, 119, 122
in Gadera River, 4, 8
in Gessi Formation, 140–41
glaciers and, 66
in Gondwana succession, 248
grain-size studies of, 31, 33
in Ligurian units, 79–82
in Macigno unit, 100–102
in Maestrat Basin, 186, 187, 189
in Mecsek-Villány area, 227, 230, 231, 234, 237
in Modena Plain, 61–64, 66, 68–71
in Moncayo area, 202
in Neto River Basin, 163
in ODP 1122 core, 284–86
in Orbulina Marl, 119
in Pampas Central, 368–70
in Piave River, 4, 6, 9–11
planimeter studies of, 5
in Po Plain, 31
in Rangitata River, 290
in Rienza River, 4, 8
in sand maturity index, 9
shape of, 4, 5, 302
in Sila Massif area, 155–61
in Umbum Creek, 369, 372–74
in Waipaoa River Basin, 264
in Waitaki River, 290
weathering of, 10, 161, 302
quartz diorite, 230, 231, 234, 237
quartz/illite-smectite values, 33
quartzites, 58, 169
Quinqueloculina spp., 20
Quinqueloculina lata, 18
Quinqueloculina padana, 18
Quinqueloculina seminula, 17, 18, 20
Quinqueloculina stellifera, 20

R
radiolaria
in Argilloso-Arenacea Formation, 122
in Belluno Flysch, 45–47
in Claut Flysch, 44
in Clauzetto Flysch, 47–48
on Corsica, 75, 79
in Frosinone Formation, 122
from Ligure-Piemontese Basin, 74
in Ligurian units, 82, 87
Rajmahal Basin, 243
Rangitata River, 281, 284, 290–92, 295–96
Raniganj Basin, 243, 246–50
Ranzano Basin, 75
rare earth elements (REE)
apatite and, 82, 237, 321
in Balagne Nappe, 82, 86
calcium and, 237, 320
in Cameros Basin, 212, 214
in Caspian Basin, 323, 325–32
chromium. *See* chromium
in Clair oil field, 330, 332–34, 336–39
fractionation and, 320–21
in Ligurian units, 82, 86
in Mecsek-Villány area, 233–34, 236
monazite. *See* monazite
in Moncayo area, 212, 214
in pelites, 82
in provenance studies, 74, 82, 182, 235, 321, 338, 339
scandium. *See* scandium
in shale, 82
thorium. *See* thorium
zircon and, 82, 237
zirconium. *See* zirconium
Raukumara Peninsula, 255, 257
Raukumara Ranges, 273
Ravenna, 14
Reatini Mountains, 111
red beds, 202, 249
redox state, indicators of, 33
Regional Topographic Gradient
in Bayesian network, 354
description of, 348
Flushing Potential and, 351, 359, 362
Generative Modification Potential and, 351, 359, 362
node type, 353
for South America, 359
states of, 359
Reno River
204-S15 core and, 18
geochemistry of sediment from, 20–21
maps of, 14, 26, 59
reservoir quality, 346
rhyolites
in Argilloso-Arenacea Formation, 122
in Balagne Nappe, 79, 87
in Frosinone Formation, 122
in Ligurian units, 79–82, 87
in Mecsek-Villány area, 237
Ridge Basdin, 362
Rienza River, 2–11
rifts, geochemistry of, 182
Rivularia, 171
Rivularia haematites, 173, 175
Robulus clays, 112
Romagna region, 14, 26
Romagna units, 58
Romania, 223
root nodes, 352–53
Rossano Basin, 136, 138–47
Rossano–San Nicola shear zone, 136
Roveto Valley, 111
rubidium
alkaline volcanism and, 33
in Cameros Basin, 212
grain size and, 30–33
in Maestrat Basin, 191, 192

in Mecsek-Villány area, 233, 234–36
in Moncayo area, 212
potassium and, 32, 189, 213
potassium feldspar and, 182, 189
tectonic setting and, 234
weathering and, 236
rubidium/barium values, 31–34
rubidium/cerium values, 33
rubidium/yttrium values, 33–34
rubidium/zirconium values, 33
rubidium/(zirconium+barium) values, 31
rudites, 78, 79, 125
Russian Platform, 322–23, 329
rutile, 51, 237, 332. *See also* ZTR index

S

Sabbie Marmose di Valle unit, 138
Sabbie Marnose di Garicchi unit, 138
Sabini Mountains, 111, 116
Sabunchi Suite, 323
Saint Remo unit, 75
Salti del Diavolo Conglomerate
　age of, 77
　geochemistry of, 84
　lithostratigraphic studies of, 76
　petrographic studies of, 78, 82
　photomicrographs from, 80
Salto Valley, 110, 111, 113, 114
Salzedella subbasin, 183
samarium/lanthanum values, 84, 87, 212, 233–35
samarium/ytterbium values, 84, 87
San Andrés section, 204, 206, 208
San Carlos Basin, 301, 315
San Colombano Limestone, 75–76, 79
San Cristóbal Basin, 301, 302
San José area
　geological setting of, 299–302
　grain-shape studies in, 308, 309, 311, 312, 315–16
　maps of, 300, 301
San José Basin, 298–302, 316
San José del Cabo Bay, 299
San Lorenzo Dam, 298
San Lucas area
　100 m isobath offshore of, 299
　El Arco area and, 298
　El Salto Basin. *See* El Salto Basin
　geochronology of, 299
　geological setting of, 299
　grain-shape studies in, 305–8, 310, 312–14
　maps of, 301, 314
San Lucas Basin, 301
San Martino Formation, 75–76, 83
San Nicola–Rossano shear zone, 136
San Pedro Manrique section, 200, 204, 206, 208
San Siro Shale, 76, 77, 84
sand
　climate and, 348–50, 373–74
　factors affecting, 346–50
　in fluvial systems, 243, 246
　grain-size categories, 356

in Iberian Range, 170–77
mobility of, 302–303
modeling of. *See* Sand Generation and Evolution Model (SandGEM)
in Modena Plain, 60–72
in Neto River Basin, 163
plants and, 374
in Po Plain, 27–33
provenance lithotypes, 348, 354
proxies for grain size, 31–35
rapid production of, 163
in reservoir quality models, 346
sorting categories, 356
study terminology, 168–69, 346
tectonic setting of, 347–48, 354
topography and, 348, 351, 374
Sand Generation and Evolution Model (SandGEM)
　Brazilian beach sand study, 363–68
　Chilean fluvial sand study, 368–71
　Congo, Pointe Noire WPI/FPI worksheet, 376
　description of, 347–63
　probability density charts, 363, 364
　Umbum Creek sand study, 369–75
sand generation index, 5, 348
sand/pelite values, 39
sandstone
　in Argilloso-Arenacea Formation, 125–31
　Bhimaram Sandstone Formation, 245, 247
　in Boite River, 4, 7
　in Brazilian Shield, 366
　in Brecce della Renga Formation, 127
　in Baja California, 300
　in Cameros Basin, 202–17
　Case Baruzzo Sandstone, 76–78, 80, 82, 84
　in Caspian Basin, 323–32, 339–40
　chemical index values for, 188–89
　in Clair oil field, 332–40
　elements associated with, 27
　in Frosinone Formation, 113, 125–31
　in Gadera River, 4
　in Gessi Formation, 140, 146–47
　Gottero Sandstone, 76, 78, 79, 83
　integrated genetic analysis of, 346
　Jakabhegy Sandstone Formation, 225
　from Kura River, 322–32, 339
　in Macigno unit, 99, 103–4
　in Maestrat Basin, 184, 184–94
　maturity index for, 9
　in Mecsek-Villány area, 227, 228–37
　in Modena Plain, 58
　in Moncayo area, 202–17
　Monte Cervarola Sandstone, 97
　Monte Falterona Sandstone, 97
　Mount Falterona Sandstone, 96
　Novella Sandstone, 76, 78, 79, 83
　in Orbulina Marl, 110
　Ostia Sandstone, 76–78, 82, 84
　petrographic studies of, 177
　in Piave River, 4, 7
　planimeter studies of, 4–5

porosity of, 346
provenance of, 129–30, 200–201, 212–14, 339–40
in Rienza River, 4
in Rossano Basin, 136, 139
in Simbruini Mountains, 112
as stratigraphic tool, 246
in Umbum Creek, 372
from Volga River, 322–32, 339
in Waipaoa River Basin, 254, 256, 259
Santerno River, 14, 20–21
Santibañez del Val Formation, 201
saprolite, 152–54, 160
sapropels, 33
Sardinia, maps of, 109, 137
Sarmatia Block, 323, 328
Saterno River, 26
Satpura Basin, 243
Savio River, 14, 20, 21, 26
Savuto River, 150, 151
Saxonian Formation, 201–10
Scaglia Rossa unit, 39, 51
scandium
　in Balagne Nappe, 85
　in Cameros Basin, 214
　in Ligurian units, 85
　in Mecsek-Villány area, 233
　in Moncayo area, 214
　in provenance studies, 74, 214, 235
scandium/lanthanum values, 235
scandium/thorium values, 85, 88, 235
Schistés Lustrées, 75
schists
　in Argilloso-Arenacea Formation, 119–20, 122, 123
　in Balagne Nappe, 79, 87
　in Belluno Flysch, 43
　in Brazilian Shield, 366
　in Brecce della Renga Formation, 118
　in Claut Flysch, 43
　in Clauzetto Flysch, 43
　in Frosinone Formation, 119, 122, 123
　in Gessi Formation, 140–41
　Haast Schist, 281, 292
　in Ligurian units, 79–82, 87
　Schistés Lustrées, 75
　in Sila Massif area, 150
　in Umbum Creek, 372
Schizothrix spp., 171, 175
Scisti Policromi, 96
Scisti Varicolori Line, 96
Scolicia, 42
scree taluses, formation of, 160
Sea of Okhotsk, 362
Secchia River, 58–61, 63–65, 67–71
Sedico-1 well, 41
sediment
　climate and, 348–50, 373–74
　composition of, 26–27
　delivery rates, 351
　plants and, 374

storage of, 350–52
topography and, 348, 351, 374
transportation of, 350–52
Selective Transport Fining, 351–54, 360, 362–63
Semicythurura spp., 20
Semicythurura costata, 20
Semicythurura incongruens, 20
Senio River, 26, 273
Serbia, map of, 223
sericite, 230, 236
serpentine
    in Argilloso-Arenacea Formation, 119–20, 123
    in Balagne Nappe, 79
    in Brecce della Renga Formation, 118
    chromium and, 31
    in Frosinone Formation, 119, 122, 123
    in Gessi Formation, 141
    in Macigno unit, 100–101
    in Modena Plain, 58, 61–63
    nickel and, 31
    in Po Plain, 14, 31
    turbidite systems and, 14
Sesia Lanzo zone, 75
Sestola-Vidiciatico unit, 59
Sestri Voltaggio unit, 75
shale
    in Argilloso-Arenacea Formation, 119–20
    Bocco Shale, 76
    in Brazilian Shield, 366
    in Brecce della Renga Formation, 118
    calcium in, 212
    in Cameros Basin, 203–17
    coloration and, 249
    elements associated with, 30
    Fiumalbo Shale, 97–99
    in Frosinone Formation, 119
    in Gessi Formation, 140–41
    in Gondwana succession, 242
    grain-size studies of, 35
    in Iberian Range, 169
    iron in, 33
    in Macigno unit, 99
    in Maestrat Basin, 184
    Manganesiferous Shale, 79
    in Modena Plain, 58, 61–64, 68–70
    in Moncayo area, 203–17
    Mount Rizzone Palombini Shale, 76, 77
    NASC, spider diagram for, 191
    in ODP 1122 core, 284
    Palombini Shale, 75–77, 83–84
    planimeter studies of, 4–5
    in Po Plain, 58
    provenance of, 212–14
    REE patterns for, 82
    in Rossano Basin, 136, 139
    San Martino Formation, 75–76
    San Siro Shale, 76, 77, 84
    in Umbum Creek, 372
    Val Lavagna Shale, 76, 78–80, 83
    vanadium in, 33

Varicolored Shale, 76–78, 82, 84
Zonati Shale, 79
Shetland Islands, map of, 333
Sicily, maps of, 109, 137
Sicily Channel, map of, 109
siderite, 230
Sierra Nevada, 359
Siklós Formation, 225
Siklósbodony-1 core, 224–26, 228–29, 231–32
Sila Batholith
    emplacement and uplift of, 159
    geological setting of, 136, 150
    Gessi Formation and, 147
    Neto River Basin and, 163
    Rossano Basin and, 138, 147
Sila Massif area, 150–63
Sila unit, composition of, 136
silicate, average activation energy, 375
silicon, 237
silicon oxide, 31, 231–34, 236, 320
silicon oxide/aluminum oxide values, 31–33
silicon oxide/iron oxide values, 31–34
Sillaro River, 14, 20–21, 26
sillimanite, 123, 150
silt
    elements associated with, 27
    in Modena Plain, 66, 68
    in Po Plain, 27–33
    proxies for grain size, 31–35
siltstone
    in Argilloso-Arenacea Formation, 119–20
    in Balagne Nappe, 79
    in Belluno Flysch, 45–47
    Boda Siltstone Formation, 223, 225
    in Boite River, 4, 7
    in Brecce della Renga Formation, 118
    in Claut Flysch, 44
    in Clauzetto Flysch, 47–48
    coloration and, 249
    in Frosinone Formation, 119
    in Gadera River, 4
    in Gessi Formation, 141
    in Gondwana succession, 242
    grain-size studies of, 35
    in Ligurian units, 82
    in Macigno unit, 99, 104
    in Modena Plain, 58, 61–64, 68–70
    in ODP 1122 core, 284
    Patacs Siltstone Formation, 225
    in Piave River, 4, 7
    planimeter studies of, 5
    in Ravenna, 14
    in Rienza River, 4
    in Waipaoa River Basin, 259
Simbruini Mountains, 110–16, 128–29
Simme nappe, 54
Siphonaperta aspera, 18
slate, 366
Slavonia-Drava unit, 223, 226, 237
Slovakia, map of, 223

smectite, 30–33, 103, 236, 273
sodium, 145, 233, 234, 236
sodium oxide, 31–33, 213, 231–34, 236
sodium/aluminum values, 33
Solander Channel/Trough, 278
solution lines, factors affecting, 160
Son Basin, 243–46, 249–50
Son River, 243
Source-to-Sink Initiative, 254
South America
    Amazon River, 362
    Brazil, 363–68
    Brazilian Shield, 366
    Chile, 368–71
    Orinoco River, 362
    SandGEM map of, 359
South Iberian Basin, 183
Southalpine thrust system, 39
Southern Highland Group, 338
Spain
    Cameros Basin, 183, 184, 195, 200–217
    Maestrat Basin, 183–94
    maps of, 169, 183, 184, 200
    Moncayo area, 200, 202–17
    Pyrenees, 175, 183
speleothems, 169
Sphenophyllum, 224
Sphenopteris, 224
Spilamberto quarry, 59, 63, 66, 67
spilite, 61–63
sponges, 44–48
staurolite, 50–53, 332
Sto. Domingo Silo Formation, 201
stromatolites, 168, 175
strontium
    calcium and, 320
    in Cameros Basin, 212
    in CART, 329
    in Caspian Basin, 323–27, 329
    in Clair oil field, 332–33, 338, 340
    fractionation and, 320, 326
    in Mecsek-Villány area, 233, 236
    in Moncayo area, 212
    in Po Plain, 31
    in provenance studies, 182, 327
    weathering and, 236
strontium/manganese values, 320, 326
strontium/yttrium values, 320, 327, 329
subarkose depositional environment, 249
Sub-Ligurian units, 58, 59
Sudetic Basin, 237
sulfur, 31
Supra-Panchet Formation, 242–46, 248–49
Surakhany Suite, 322–32, 339
Susquehana River, 362
swamps, coloration from, 249
syenite, 326, 337
Szársomlyó Limestone Formation, 225
Szilice, 223
Szőlősardó Bódva, 223

T
Tabuenca section, 200, 204–5, 207
Tagliamento River, 38
Taglio River, 60, 61
Tajo Basin, 183, 184
talc, 31
Talchir Formation, 242, 247–49
Talysh area, 322
tantalum, 86, 182, 195
tantalum/ytterium values, 215
Taranto Gulf, 136
Tarndale area, 266, 273
Taro River, 59
Tauern Window, 38, 53
Taupo volcanic zone, 254–56, 272, 278, 293
Te Arai River
  detrital modes for, 266–68, 270–72
  maps of, 255, 257
  point-count data for, 262–63
  sedimentology of, 272, 274
Te Weraroa stream, 257
tectonic setting discrimination diagrams, 214–16
temperature
  evaporation and, 348–50
  mineral-dissolution reaction rates, 348
  thermal stress fatigue, 160
  van 't Hoff's rule for, 375
  in Weathering Potential Index, 360, 375
Tenda Massif, 75
Tera Formation, 201
terbium/lanthanum values, 215
Téseny Formation, 222–37
Tethys Ocean. *See also* Paleotethys Ocean
  Apennine orogenesis and, 58
  Iberian Plate and, 183
  Ligure-Piemontese Basin, 74–75, 77, 88–90
  reconstruction of margins, 89
Texas, Oyster Creek, 362
Textularia spp., 17, 20
Textularia agglutinans, 18
Textularia bocky, 18
thallium/aluminum values, 33
thermal stress fatigue, 160
thorium
  in Balagne Nappe, 88
  calcium and, 320
  in Cameros Basin, 214
  in Caspian Basin, 327, 329–31
  in Clair oil field, 330, 333
  fractionation and, 327
  grain size and, 30
  in Ligurian units, 88
  in Mecsek-Villány area, 233, 237
  monazite and, 237, 321, 327
  in Moncayo area, 214
  in Po Plain, 31
  in provenance studies, 74, 214, 235
  zircon and, 237
thorium/chromium values, 85, 88, 235
thorium/cobalt values, 214, 235
thorium/lanthanum values, 214, 216

thorium/scandium values, 85, 88, 235
thorium/uranium values
  in Cameros Basin, 213–14
  in Caspian Basin, 327, 329–31
  in Clair oil field, 330, 333–34
  in Moncayo area, 213–14
  in provenance studies, 321
  weathering and, 213–14
thorium/yttrium values, 215
Tian Shan Mountains, 322
Tiepido Stream
  diversion of, 59
  map of, 60
  sand composition along, 60, 62, 63, 69, 71
  ternary plots of, 65, 67
Tierga section, 200, 204–5, 207
Tiki Formation, 242, 244–46, 248–49
Tisza River, 223
Tisza-Dacia unit, 222, 223
titanite, 51, 192, 195
titanium
  apatite and, 237
  basaltic rocks and, 33
  in Caucasus Mountains, 328
  grain size and, 33
  in monazite, 237
  in provenance studies, 182
  in rutile, 237
  sorting effects on, 27
  titanite and, 195
  in turbidite systems, 33
  weathering and, 212
titanium oxide
  in Balagne Nappe, 85
  in Cameros Basin, 210
  grain size and, 30
  in Ligurian units, 85
  in Mecsek-Villány area, 231–33, 237
  in Moncayo area, 210
  in Po Plain, 31
titanium/vanadium values, 33–34
Toccone Breccia
  geochemistry of, 83, 86
  lithostratigraphic studies of, 76
  Lydienne Flysch and, 76
  petrographic studies of, 78, 79
  provenance of, 75, 86
Tolaga Group, 257
Tolmino Basin, 53
tonalite, 300
Torlesse Terrane
  composition of, 254, 281
  drainage basins in, 292
  geological setting of, 254
  map of, 281
  provenance of, 274, 281
  in Raukumara Ranges, 273
  triangular compositional plots of, 274
Torrecilla Formation, 201
Torrice Formation, 112, 113, 128
Total Transport Fining, 352, 353–54, 356, 363

tourmaline. *See also* ZTR index
  in Belluno Flysch, 51
  boron and, 195
  in Claut Flysch, 51
  in Clauzetto Flysch, 51
  lithium and, 195
  in Maestrat Basin, 186, 189, 192–93
  textural characteristics of, 192
trachyandesite, 237
Transport Modification Potential
  in Bayesian network, 354
  Depositional Modification Potential and, 352
  description of, 351–52
  Fluvial Storage Potential and, 352, 363
  Generated Sand Composition and, 351–52
  node type, 353
  states of, 357, 363
  Total Transport Fining and, 352, 363
  Transported Clay Abundance and, 359
  Transported Grain-Size Distribution and, 357
  Transported Sand Composition and, 355
  Weathering Potential and, 352, 360, 363
Transport Potential. *See* Flushing Potential
Transported Clay Abundance, 353–54, 356–57, 359
Transported Grain-Size Distribution, 353–59
Transported Sand Composition, 353–55, 359
Transported Sediment, 351, 352, 355
travertine, 168
Travesio area, 38, 41, 42
trees, weathering and, 160
Trentino-Alto Adige region, 2–11
Trento Plateau, 53
Tresinaro Stream, 60, 61, 68
Trevijano section, 200, 204, 206, 208
Triloculina affinis, 20
Triloculina gibba, 18
Triloculina schreiberiana, 18
Triloculina trigonula, 18
Trinchera del Ferrocarril section, 200
Trionto River, 150, 151
Tripoli Formation, 136, 138
tufas, 168–77
turbidite systems
  in Argilloso-Arenacea Formation, 113–31
  in Brecce della Renga Formation, 112, 125
  chromium and, 14
  in Frosinone Formation, 113–31
  on Gessoso-Solfifera Formation, 110
  gypsum and, 135
  in Ligurian units, 76–77
  in Lydienne Flysch, 76
  in Macigno unit, 98–100, 103–4
  maps of, 38, 96
  in Modena Plain, 58
  nickel and, 14
  in ODP 1122 core, 278, 280, 293
  on Orbulina Marl, 110, 125
  in Po Plain, 58
  in Rossano Basin, 136, 139
  serpentine and, 14

titanium in, 33
in Torlesse Terrane, 281
in Torrice Formation, 113
types of beds, 39
zircon in, 27
Turborotalia cerroazulensis frontosa, 41
Turkmenistan, 322
Turony Formation, 222, 225, 226
Turony-1 core, 224–29
Tuscan domain
    Adria Plate and, 58
    Cervarola-Falterona unit, 96, 97
    maps of, 59, 75, 96
    provenance of, 58
    Tuscan Nappe, 96, 97–100, 103
Tuscan Nappe, 96, 97–100, 103
Tuscan Scaglia, 97
Tyrrhenian Sea, 109, 110, 128, 136

U
Ukraine, 223, 323
ultramafic rocks, 31, 33, 326
Umbria-Marche-Sabina Basin, 110, 128–30
Umbrian domain, 58, 75, 96
Umbum Creek sand study, 369–75
Ural Mountains, 322–23, 359
uranium
    calcium and, 320
    in Cameros Basin, 214
    in CART, 329
    in Caspian Basin, 327, 329–31
    in Clair oil field, 333
    fractionation and, 321
    in granites versus dolerites, 321
    in Mecsek-Villány area, 233, 237
    in Moncayo area, 214
    zircon and, 237
uranium/thorium values. *See* thorium/uranium values
Urbion Formation, 201
Utrillas Formation, 201

V
Val Graveglia unit, 76, 83
Val Lavagna Shale, 76, 78–80, 83
Valdemadera section, 200, 204, 206, 208
Valsugana thrust system, 38
Valvulineria perlucida, 17, 18
vanadium
    in Balagne Nappe, 88
    basaltic rocks and, 33
    grain size and, 30–33
    in Ligurian units, 88
    in Mecsek-Villány area, 233
    in Po Plain, 31
    in provenance studies, 182, 235
    redox state and, 33
    in sapropels, 33
    in shale, 33
    sorting effects on, 27
vanadium/aluminum values, 32

vanadium/chromium values, 33
vanadium/clay values, 32
vanadium/nickel values, 215
vanadium/titanium values, 33–34
vanadium/yttrium values, 33
vanadium/zirconium values, 33–34
van't Hoff's rule, 375
Vardar Ocean, 50, 53, 54
Vardar Suture Zone, 53
Varicolored Shale, 76–78, 82, 84
Variscan
    Allgäu Flysch and, 53
    Mecsek-Villány area and, 235–37
    Mórágy Complex and, 237
    Saxonian Formation and, 202
    Sila Batholith and, 136
Variscides, 237
Varri Valley, 111, 113
VECTOR programs, 304
Veneto region, 2–11, 14, 21, 38
Venezuela, Orinoco River, 362
vermiculite, 103, 162
Vignola unit, 66
Villa del Mar Basin, 301, 315
Villány Mountains, 223–26. *See also* Mecsek-Villány area
Vittatina, 224
Volga River, 322–32, 339
Volgo-Uralia Block, 323, 328
Voltri Group, 75
Voltzites, 224

W
Waihora River, 255, 257, 269
Waihuka River, 255, 257, 269
Waimata River
    detrital modes for, 266–71
    geological setting of, 255
    maps of, 255, 257
    point-count data for, 262–63
    sedimentology of, 272–74
    triangular compositional plots of, 266
Waingaromia River
    detrital modes for, 266, 271, 272
    maps of, 255, 257
    point-count data for, 262–63
    triangular compositional plots of, 266
Waipaoa River, 255–56, 256
Waipaoa River Basin, 254–74
Waitaki Canyon, 279
Waitaki River
    drainage geology, 284, 292
    map of, 281
    ODP 1122 site and, 291–93
    petrographic studies of, 284, 290
    point-count data for, 295–96
    submarine canyons from, 279
    ternary plots of, 290–92
water velocity, 348
"Weald facies," 183–84
Weathering Potential

    in Bayesian network, 354
    calculation of, 375
    Generative Modification Potential and, 351, 360, 362
    node type, 353
    precipitation and, 360
    in river assessment studies, 350
    states of, 359, 360
    temperature effects in, 360
    Transport Modification Potential and, 352, 360
West Asturian–Leonese Zone, 195
Wharekopae River, 255

X
xenotime, 50, 51, 52
Xestoleberis communis, 20

Y
Yanguas section, 200, 204, 206, 208
Yerrapalli Formation, 245, 247
ytterbium/cesium values. *See* cesium/ytterbium values
ytterbium/lanthanum values, 84–85, 212, 233–34
ytterbium/samarium values, 84, 87
yttrium
    alkaline volcanism and, 33
    apatite and, 30, 195, 237
    calcium and, 237, 320
    in Caspian Basin, 323–27, 330
    in Clair oil field, 332–34, 338, 341
    fractionation and, 320, 326–27
    garnet and, 30, 33
    grain size and, 33
    hornblende and, 30
    in Mecsek-Villány area, 233, 237
    monazite and, 30, 237
    phosphorus pentoxide and, 30
    in Po Plain, 31
    in provenance studies, 182, 327
    sorting effects on, 27
    zircon and, 30, 33, 237
    zirconium and, 30
yttrium/aluminum values, 33
yttrium/rubidium values, 33–34
yttrium/strontium values, 320, 327, 329
yttrium/tantalum values, 215
yttrium/thorium values, 215
yttrium/vanadium values, 33

Z
zinc, 27, 30, 31
zircon. *See also* ZTR index
    in Belluno Flysch, 51
    in Claut Flysch, 51
    in Clauzetto Flysch, 51
    hafnium and, 195, 237
    in Maestrat Basin, 192
    in pelites, 82
    phosphorus and, 237
    phosphorus pentoxide and, 30
    REEs and, 237

silicon and, 237
sorting effects on, 27
tantalum and, 195
textural characteristics of, 192
thorium and, 237
in turbidite systems, 27
uranium and, 237
yttrium and, 30, 33, 237
zirconium and, 30, 33, 195
zirconium
    alkaline volcanism and, 33
    apatite and, 30
    in Cameros Basin, 214
    grain size and, 30–33
    hornblende and, 30
    in Mecsek-Villány area, 233
    monazite and, 30
    in Moncayo area, 214
    in Po Plain, 31
    in provenance studies, 182, 214, 235
    yttrium and, 30
    zircon and, 30, 33, 195
(zirconium+barium)/rubidium values, 31
zirconium/aluminum values, 33
zirconium/chromium values, 235
zirconium/rubidium values, 33
zirconium/vanadium values, 33–34
zoisite, 50
Zonati Shale, 79
Zoophycos, 110
ZTR index, 52
Zuffa classification of grains, 43, 145